Major Geological, Climatic, and Biological Events

Culmination of mountain-building followed by erosion and moderate, short-lived invasions of continental margins by the sea. Early warming trends were reversed by the middle of the period to cooler and finally to glacial conditions. Subtropical forests gave way to temperate forests and finally to extensive grasslands. Transition from primitive mammals to modern orders and eventually families. Evolution of humans during the last 5–8 million years. (See Figures 20–1 and 20–3, pp. 714 and 716.)

Last great spread of epicontinental seas and shoreline swamps. At the end of the period extensive mountain building cooled the climate worldwide. Angiosperm dominance began. Extinction of archaic birds and many reptiles by the end of the period. (See Figure 14–4, p. 514.)

Climate was warm and stable with little latitudinal or seasonal variation. Modern genera of many gymnosperms and advanced angiosperms appeared. Reptilian diversity was high in all habitats. First birds appeared. (See Figure 14–3, p. 513.)

Continents were relatively high with few shallow seas. The climate was warm; deserts were extensive. Gymnosperms dominated; angiosperms first appeared. Mammal-like reptiles were replaced by precursors of dinosaurs and the earliest true mammals appeared. (See Figure 14–2, p. 509.)

Land was generally higher than at any previous time. The climate was cold at the beginning of the period but warmed progressively. Glossopterid forests developed with the decline of the coal swamps. Mammal-like reptiles were diverse; widespread extinction of amphibians at the end of the period.

Generally warm and humid, but some glaciation in the Southern Hemisphere. Extensive coal-producing swamps with large arthropod faunas. Many specialized amphibians and the first appearance of reptiles. (See Figure 14–1, p. 508.)

Mountain-building produced locally arid conditions, but extensive lowland forests and swamps were the beginning of the great coal deposits. Extensive radiation of amphibians; extinction of some fish lineages and expansion of others.

The land was higher and climates cooler. Freshwater basins developed in addition to shallow seas. The first forests appeared and the first winged insects. There was an explosive radiation of fishes, followed by the disappearance of many jawless forms. The earliest tetrapods appeared. (See Figure 9–1, p. 334.)

The land was slowly being uplifted, but shallow seas were extensive. The climate was warm and terrestrial plants radiated. Eurypterid arthropods were at their maximum abundance in aquatic habitats and the first terrestrial arthropods appeared. The first gnathostomes appeared among a diverse group of marine and freshwater jawless fishes. (See Figure 5–2, p. 198.)

The maximum recorded extent of shallow seas was reached and the warming of the climate continued. Algae became more complex, vascular plants may have been present, and there was a variety of large invertebrates. Jawless fish fossils from this period are fragmentary but more widespread. (See Figure 5–2, p. 198.)

There were extensive shallow seas in equatorial regions. The climate was warm. Algae were abundant and there are records of trilobites and brachiopods. The first remains of vertebrates are found at the end of this period.

Changes in the lithosphere produced major land masses and areas of shallow seas. Multicellular organisms appeared and flourished—algae, fungi, and many invertebrates.

Formation of the earth and slow development of the lithosphere, hydrosphere, and atmosphere. Development of life in the hydrosphere.

Vertebrate Life

THIRD EDITION

Vertebrate Life

F. Harvey Pough
John B. Heiser
William N. McFarland
Cornell University

Macmillan Publishing Company
New York

Collier Macmillan Publishers
London

Macmillan Publishing Company
866 Third Avenue, New York, New York 10022

Collier Macmillan Canada, Inc.

Library of Congress Cataloging-in-Publishing Data

Pough, F. Harvey.
    Vertebrate life / F. Harvey Pough, John B. Heiser, William N. McFarland.—3rd ed.
        p.   cm.
    Rev. ed. of: Vertebrafe life / William N. McFarland . . . [et al.]. 2nd ed. c1985.
    Includes bibliographies and index.
    ISBN 0-02-396360-3
    1. Vertebrates.   2. Vertebrates, Fossil.   I. Heiser, John B. II. McFarland, William N. (William Norman), 1925-   . III. Vertebrate life.   IV. Title.
QL605.P68 1989
596—dc19   88-22092
                                                    CIP

Printing: 1 2 3 4 5 6 7 8   Year: 9 0 1 2 3 4 5 6 7 8

Preface

The third edition of *Vertebrate Life* incorporates major changes that reflect the extraordinary activity in vertebrate biology during the past decade. The most pervasive changes have resulted from the widespread adoption of phylogenetic systematics (cladistics) as the basis for determining the phylogenetic relationships of organisms. The stress that this system of classification places on the importance of monophyletic groupings has ramifications in broad areas of biology. As an objective (although frequently controversial) classification that, in particular, reflects information about the sequence of changes during evolution, cladistics provides an evolutionary framework in which ideas from other biological specialties can be accommodated. As a result, studies of behavior, physiology, and ecology are increasingly being placed in an explicitly evolutionary context, and this common ground has fostered increased interaction among those specialties.

We have adopted a cladistic classification as the basis for organizing the third edition of *Vertebrate Life,* and have included cladograms illustrating the postulated relationships of the different lineages of vertebrates. In doing so, we have tried to reconcile the views of various authorities, and pointed out major areas of disagreement. The cladograms include synopses of the character states on which they are based and citations of the primary sources used. This information will facilitate exploration of different views, and will help faculty and students to modify the phylogenies presented here as new interpretations are published.

As a result of the cladistic perspective of this edition, we have reorganized the large quantity of information about morphology and physiology that was distributed among chapters in earlier editions. Chapter 3 presents a comprehensive review of vertebrate morphology and its evolutionary changes, including new material about embryonic development. Chapter 4 presents a parallel treatment of the general aspects of vertebrate physiology and homeostasis. Topics unique to particular groups are highlighted in the chapters treating those groups.

Another important conceptual change in this edition is the inclusion of a large quantity of material about ecology, behavior, and ecological physiology in response to requests from colleagues who teach courses that cover only living vertebrates. Chapters 16 and 23 present descriptions of the biology of ectotherms and endotherms, respectively, that focus on organismal function in the broadest sense and integrate information about ecology, behavior, and physiology to present a view of the way vertebrates interact with their environments. Chapter 22 summarizes information about the social behavior of mammals in a similarly broad context. We hope that students will find that these treatments illustrate the fascination the authors find in the study of vertebrate biology.

Literature citations have been brought up-to-date, with many references from 1987 and 1988. As before, we have chosen citations on the basis of their helpfulness to students attempting to enter the literature of the subject; review articles are cited where possible, and recent references are used because students can trace earlier work through them.

The task of reviewing all of vertebrate biology is nearly overwhelming, and would have been impossible without the hours of time that colleagues spent helping us. The list is now so long that we have added an acknowledgments section. We are exceedingly grateful to all of them.

Acknowledgments

Writing a book with a scope as broad as this one requires the assistance of many people. We are grateful to the following colleagues for their generous responses to our requests for information and their comments and suggestions: Michael Benton, Bruce Brewer, Sara Cairns, Cynthia Carey, Robert Carroll, Charles Cole, Martin Feder, Robert Full, Malcolm Gordon, James Hopson, Raymond Huey, Kenneth Kennedy, Tom Kemp, George Lauder, Rachel Levin, Deedra McClearn, Amy McCune, Peter Nathanielsz, A. L. Panchen, Karen Reiss, John Reiss, Carol Saunders, Alan Savitzky, Ellen Smith, Margaret Stewart, Stanley Temple, Keith Thomson, Laurie Vitt, David Winkler, and Richard Zweifel.

Dean G. Dillery (Albion College) and James C. Ha (Colorado State University) reviewed the second edition of *Vertebrate Life* and made many suggestions for improvements. William E. Bemis (University of Massachusetts at Amherst), Virginia Hayssen (Smith College), David G. Huckaby (California State University, Long Beach), and Leslie K. Johnson (The University of Iowa) reviewed the entire manuscript of the third edition. Their suggestions have shaped nearly every aspect of this book, and we cannot sufficiently express our gratitude for their efforts.

The illustrations in the third edition have benefitted greatly from the kindness of many colleagues who provided photographs and line drawings: George A. Bartholomew, Albert F. Bennett, Joy Belsky, Philip Bogdonoff, Edward B. Brothers, R. Bruce Bury, Sara J. Cairns, Robert L. Carroll,

Mark A. Chappell, Charles J. Cole, Jack A. Cranford, David M. Dennis, Alan A. Detrich, Lang Elliott, Sharon B. Emerson, John W. Goerg, Peter E. Hillman, Daniel H. Janzen, Farish A. Jenkins, Jr., Cynthia L. Jensen, Frank J. Joyce, Jeffrey W. Lang, Rachel N. Levin, Amy R. McCune, Gail R. Michener, Kenneth A. Nagy, Frederick J. O. Nuss, Charles R. Peterson, Michael A. Recht, Karen Z. Reiss, David Robertshaw, Gerhard Roth, Carol D. Saunders, Robert D. Stevenson, Margaret M. Stewart, Janet M. Storey, Kenneth B. Storey, Theodore L. Taigen, David B. Wake, Marvalee H. Wake, Richard J. Wassersug, and David Winkler.

Lucile Macera painstakingly typed the entire manuscript onto computer disks, Linda Berman meticulously prepared the indices, and Margaret Pough handled ably the final checking of page proofs. Nearly all of the illustrations retained from earlier editions were redrawn and many new illustrations added. We are grateful to the artists: Frances Zweifel, Matthew Zweifel, Mary Dersch, and Network Graphics for their patience and painstaking care. Finally, the production supervisor at Macmillan, Dora Rizzuto, has earned our gratitude for the aplomb with which she has coordinated the activities of three authors and four artists in an enterprise that increased in complexity by the number of people involved.

F. Harvey Pough
John B. Heiser
William N. McFarland

Contents

PART THREE
Terrestrial Ectotherms: Amphibians, Turtles, Crocodilians, and Squamates 338

PART ONE

Vertebrate Diversity, Function, and Evolution

The 50,000 living species of vertebrates inhabit nearly every part of the Earth, and other kinds of vertebrates that are now extinct lived in habitats that no longer exist. Increasing knowledge of the diversity of vertebrates was one of the products of European exploration and expansion that began in the fifteenth and sixteenth centuries. In the middle of the eighteenth century the Swedish naturalist Carolus Linnaeus developed a binomial classification to catalog the varieties of animals and plants. The Linnean system remains the basis for naming living organisms today.

A century later Charles Darwin explained the diversity of plants and animals as the product of natural selection and evolution, and in the early twentieth century Darwin's work was coupled with the burgeoning information about mechanisms of genetic inheritance. This combination of genetics and evolutionary biology is known as the New Synthesis or Neo-Darwinism, and continues to be the basis for understanding the mechanics of evolution. Recent work has broadened our view of evolutionary mechanisms by suggesting, on one hand, that some major events in evolution may be the result of chance rather than selection, and, on the other hand, that natural selection can sometimes extend beyond individuals to related individuals, populations, or even to entire species. Methods of classifying animals also have changed their emphasis during the twentieth century, and classification, which began as a way of trying to organize the diversity of organisms, has become a way of generating testable hypotheses about evolution.

Vertebrate biology and the fossil record of vertebrates have been at the center of these changes in our view of life. Comparative studies of the anatomy, embryology, and physiology of living vertebrates have often supplemented the fossil record. These studies reveal that evolution acts by changing existing structures. All vertebrates have basic characteristics in common that are the products of their common ancestry, and progressive modifications of these characters can trace the progress of evolution. Thus, an understanding of vertebrate form and function is basic to understanding the evolution of vertebrates and the ecology and behavior of living species.

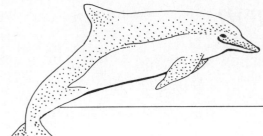

Evolution is central to vertebrate biology because it provides a principle that organizes the diversity that we see among living vertebrates and helps to fit extinct forms into the context of living species. Classification, initially a process of attaching names to organisms, has become a method of understanding evolution. Current views of evolution stress natural selection operating at the level of individuals as a predominant mechanism that produces change over time. Additional mechanisms may involve selection that operates at higher and lower levels of biological organization and chance events. The processes and events of evolution are intimately linked to the changes that have occurred on Earth during the history of vertebrates. These changes have resulted from the movements of continents and the effects of those movements on climates and geography. In this chapter we present an overview of the scene, the participants, and the rules governing the events that have shaped the biology of vertebrates.

The Diversity, Evolution, and Classification of Vertebrates

1

The Vertebrate Story

Mention "animal" and most people will think of a vertebrate. Vertebrates are often abundant and conspicuous parts of people's experience of the natural world. Vertebrates are also very diverse: The 50,000 living species of vertebrates range in size from fishes that weigh as little as 0.1 gram when they are fully mature to whales that weigh nearly 100,000 kilograms. Vertebrates live in virtually all the habitats on earth: Bizarre fishes, some with mouths so large they can swallow prey longer than their own bodies, cruise through the depths of the sea, sometimes luring prey to them with glowing lights. Some 15 kilometers above the fishes, migrating birds fly over the crest of the Himalayas, the highest mountains on Earth. Birds can live at these high altitudes where mammals are incapacitated by lack of oxygen because the lungs of birds have a pattern of airflow that is different from that of mammals and bird lungs are more effective than mammalian lungs at extracting oxygen from the air.

The behaviors of vertebrates are as diverse and complex as their body forms. Vertebrate life is energetically expensive, and vertebrates get the energy they need from food they eat. Carnivores eat the flesh of other animals and show a wide range of methods of capturing prey: Some predators search the environment to find prey, whereas others wait in one place for prey to come to them.

Some carnivores pursue their prey at high speeds, others pull prey into their mouths by suction. In some cases the foraging behaviors that vertebrates use appear to be exactly the ones that maximize the amount of energy they obtain for the time they spend hunting; in other cases vertebrates can appear to be remarkably inept predators. Many vertebrates swallow their prey intact, sometimes while it is alive and struggling, but other vertebrates have very specific methods of dispatching prey: Venomous snakes inject complex mixtures of toxins, and cats (of all sizes from house cats to tigers) kill their prey with a distinctive bite on the neck. Herbivores eat plants: Plants do not run away when an animal approaches, but they are hard to digest and they frequently contain toxic compounds. Herbivorous vertebrates show an array of specializations to deal with the difficulties of eating plants: these specializations include elaborately sculptured teeth and digestive tracts that provide sites in which symbiotic microorganisms digest compounds that are impervious to the digestive systems of vertebrates.

Reproduction is a critical factor in the evolutionary success of an organism and vertebrates show an astonishing range of behaviors associated with mating and reproduction. In general males court females and females care for the young, but these roles are reversed in many species of vertebrates. The forms of reproduction employed by vertebrates range from laying eggs to producing living young. These variations range across almost all kinds of vertebrates—many fishes and amphibians produce live young and a few mammals lay eggs. In fact, only birds show no variation in their reproductive mode; all birds lay eggs. At the time of birth or hatching some vertebrates are entirely self-sufficient and never see their parents, whereas other vertebrates (including humans) have extended periods of obligatory parental care. Extensive parental care is found in seemingly unlikely groups of vertebrates—fishes that incubate eggs in their mouths, frogs that incubate eggs in their stomachs, and birds that feed their nestlings a fluid called crop milk that is very similar in composition to mammalian milk.

The diversity of living vertebrates is fascinating, but the species now living are only a small proportion of the species of vertebrates that have existed. For each living species there may be as many as ten extinct species, and some of these have no counterparts among living forms. The dinosaurs, for example, that dominated the Earth for 180 million years are so entirely different from any living animals that it is hard to reconstruct the lives they led. Even mammals were once more diverse than they are now: the Pleistocene saw giants of many kinds—ground sloths as big as modern rhinoceroses and raccoons and rodents as large as bears. Humans are closely related to the great apes (especially to chimpanzees and gorillas) and much of the biology of humans is best understood in the context of our vertebrate heritage. In the modern world the fate of other species of vertebrates is very much affected, for good or ill, by human decisions, and our responsibilities to other vertebrates cannot be ignored.

The story of vertebrates is fascinating: Where they originated, how they evolved, what they do, and how they work provides endless intriguing details. In preparation to tell this story we must introduce some basic information: What the different kinds of vertebrates are called and how they are classified, how evolution works, and what the world within which the story of vertebrates unfolded was like. In this chapter we provide an overview of the vertebrates and the processes of evolution and environmental change that have shaped them.

The Different Kinds of Vertebrates

Describing and classifying the variety of vertebrates, living and extinct, has become more complicated recently than it used to be as a result of a change in the criteria used for recognizing natural groups of organisms. Most of us have grown used to classifying vertebrates as jawless fishes, cartilaginous fishes, bony fishes, amphibians, reptiles, birds, or mammals. Those names are familiar; each conjures up an image of a particular kind of animal. When someone said "reptile" we thought of turtles, alligators, crocodiles, lizards, and snakes. The term "reptile" communicated information about particular animals, and the same was true of the terms "jawless fishes" or "birds."

The animals that we recognized by those names still exist, of course, but some go by different names and are grouped differently now. The reason for the change is an increased emphasis on the proposition that *groups* of animals can be identified only if they share a common evolutionary lineage. Basically the old method of classification (which can be called evolutionary systematics) lumped together animals that had very different evolutionary histories and produced groups that contained unrelated evolutionary lineages. The class Reptilia is the major example of the consequences of this sort of lumping, because the evolutionary lineages that gave rise to the living turtles and crocodilians separated about the same time as the divergence of mammals from both of those groups. That is, crocodilians are no more closely related to turtles than they are to mammals, and it makes no sense to place turtles and crocodilians in one class (Reptilia) and mammals in a class of their own (Mammalia). Similarly, crocodilians and birds are quite closely related, whereas snakes and lizards are only distantly related to crocodilians and birds. Consequently, classifying crocodilians as the order Crocodilia of the class Reptilia, birds as the class Aves, and snakes and lizards in the order Squamata of the class Reptilia does not reflect what we actually know to be true of the evolution of those groups: It separates the two closely related groups (birds and crocodilians) and lumps the more distantly related ones (crocodilians and lizards plus snakes).

The principles that underlie the new approach to classification, which is popularly known as cladistics, are explained in more detail in a subsequent section. They are important concepts and the application of cladistics is making the study of evolution more rigorous than it has been heretofore. The natural groups formed by cladistics are also easier to understand than the artificial groups we are used to, except that we are familiar with the names of the artificial groups and the names of the new groups are strange. At this point we need to establish a basis for talking about particular animals by naming them, and to relate the old, familiar names to the new, less familiar ones.

Figure 1–1 shows the major kinds of vertebrates and the relative numbers of living species and Table 1–1 compares the names of old and new groupings of vertebrates and shows the chapters in which the major discussions of the different groups will be found. In the following sections we describe briefly the different kinds of living vertebrates. The names used in classifications of animals follow certain rules that are explained in Box 1–1.

Fishes Without Jaws: Hagfish and Lampreys

Unique among living vertebrates because they lack jaws, the few species of hagfishes and lampreys occupy an important position in the study of vertebrate evolution. Hagfishes and lampreys have traditionally been grouped as agnathans (*a* = without, *gnath* = jaw) or cyclostomes (*cyclo* = round, *stoma* = mouth), but they actually represent two independent evolutionary lineages. Hag-

fishes appear to be genuinely primitive vertebrates, whereas lampreys are degenerate in many respects. That is, the evolutionary lineage of lampreys once had derived characters that are not present in lampreys. The agnathous condition of lampreys, however, is primitive.

Lampreys and hagfishes are elongate, scaleless, and slimy and have no internal hard tissues. They are scavengers and parasites and are specialized for those roles. Hagfishes are marine and occur in deep water, whereas lampreys are migratory forms that live in oceans and rivers.

Cartilaginous Fishes: Sharks, Rays, and Ratfish

Sharks have a reputation for ferocity that most of the 700 species would have difficulty living up to. Many species are small (15 centimeters or less), and the whale shark, which grows to 10 meters in length, is a filter-feeder that subsists on plankton it strains from the water. Rays are dorsoventrally flattened, frequently bottom dwellers that swim with undulations of their extremely broad pectoral fins. Ratfish are bizarre marine fishes with long, slender tails and buck-toothed faces that look rather like rabbits. The name "Chondrichthyes" (*chondro* = cartilage, *ichthyes* = fish) refers to the cartilaginous skeletons of these fishes.

Primitive Bony Fishes: Bowfin, Gars, and Others (Actinopterygians)

The fishes are so diverse that any attempt to characterize them briefly is doomed to failure. Two broad categories can be recognized, the ray-finned fishes (actinopterygians; *actino* = ray, *ptero* = wing or fin) and the lobe-finned fishes (sarcopterygians; *sarco* = lobe).

The primitive actinopterygian fishes we have included in this category are called chondrosteans, holosteans, and primitive neopterygians in various

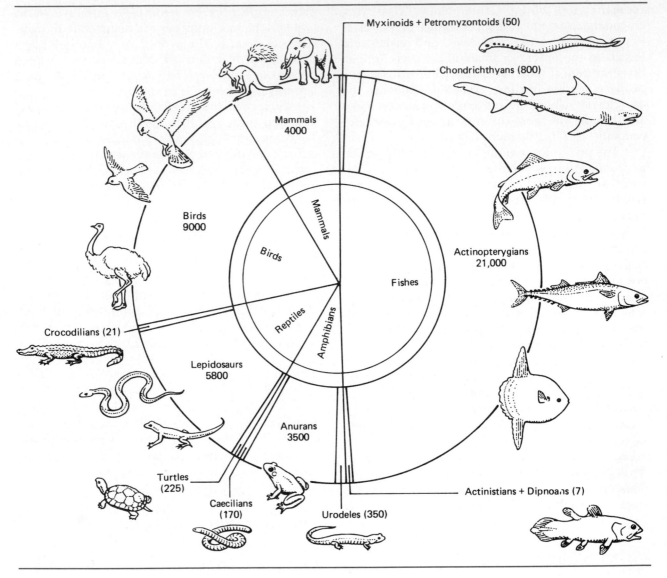

Figure 1—1. Diversity of vertebrates. The areas within the diagram correspond to the approximate numbers of living species in each group.

systems of classification. Most have cylindrical bodies, thick scales, and jaws armed with sharp teeth. These fishes seize prey in their mouths with a sudden rush or gulp; they lack the specializations of the jaw apparatus that allow advanced bony fishes to use more complex feeding modes.

Advanced Bony Fishes: Teleost Fishes

More than 20,000 species of fishes fall into this category and they cover every imaginable range of body sizes, habitats, and habits from seahorses to the giant ocean sunfish. Most of the familiar fishes are in this category—the bass and panfish you have fished for in fresh water, and the sole (a kind of flounder) and redfish you have eaten in restaurants. Modifications of the jaw apparatus have allowed many teleosts to be highly specialized in their feeding habits; the cichlids of the African rift lakes discussed in Box 1–3 are an example of the versatility of teleosts.

Lobe-Finned Fishes: Lungfishes and the Coelacanth

These are the sarcopterygian fishes. They are the living fishes most closely related to terrestrial vertebrates (the tetrapods, *tetra* = four, *pod* = foot). (The actual closest relatives of tetrapods are the extinct osteolepiform fishes, but lungfishes are very similar in many respects.) The lobe-finned fishes are heavy-bodied and slow-moving. The four genera of lungfishes live in fresh water and the coelacanth is marine.

Salamanders, Caecilians, and Frogs

These three groups of tetrapods are popularly known as amphibians (*amphi* = double, *bios* = life) in recognition of their complex life histories, which often include an aquatic larval form (the *larva* of a salamander or caecilian and the *tadpole* of a frog) and a terrestrial adult. The living amphibians may be derived from two different lineages, but the similarities they share make the collective term "amphibians" a convenient way to refer to the three kinds of animals. In particular, all amphibians have bare skins (that is, lacking scales, hair, or feathers) that are important in the exchange of water, ions, and gases with their environment. Salamanders are elongate animals, mostly terrestrial and usually with four legs; caecilians are legless aquatic or burrowing animals; and anurans (frogs, toads, treefrogs) are short-bodied tetrapods with large heads and large hind legs used for walking, jumping, and climbing.

Turtles

Turtles are probably the most immediately recognizable of all tetrapods. The shell that encloses a turtle has no exact duplicate among other vertebrates, and the morphological modifications associated with the shell make turtles extremely peculiar tetrapods. They are, for example, the only tetrapods with the shoulders (pectoral girdle) and hips (pelvic girdle) inside the ribs.

The Tuatara, Lizards, and Snakes

These three kinds of tetrapods can be recognized by their scale-covered skin as well as by characteristics of the skull. The tuatara, a stocky-bodied animal found only on some islands near New Zealand, is the sole living remnant of a lineage of animals called sphenodontians that were more diverse in the Mesozoic. In contrast, lizards and especially snakes are now at the peak of their diversity. Snakes are nearly as distinctive and widely recognized as turtles, but some lizards are legless like snakes.

Alligators and Crocodiles

These impressive tetrapods (one living species has the potential to grow to a length of 7 meters) are descendents of the same lineage (the Archosauromorpha) that produced the dinosaurs and the birds. Crocodilians, as they are known collectively, are semiaquatic predators with long snouts armed with numerous teeth. Their skin contains many bones (osteoderms, *osteo* = bone, *derm* = skin) that lie beneath the scales and provide a kind of armor plating. Crocodilians are noted for the parental care they provide for their young.

Birds

The birds are a lineage of archosauromorphs that evolved flight in the Mesozoic. The ability of birds to fly is based on feathers that provide the surfaces that create lift and propulsion, and feathers are the

Table1–1. Comparison of the traditional classification of living vertebrates with phylogenetic systematics (cladistics). Extinct forms are omitted. This table does not reflect all of the complexities of the two classifications; see the chapters indicated for details.

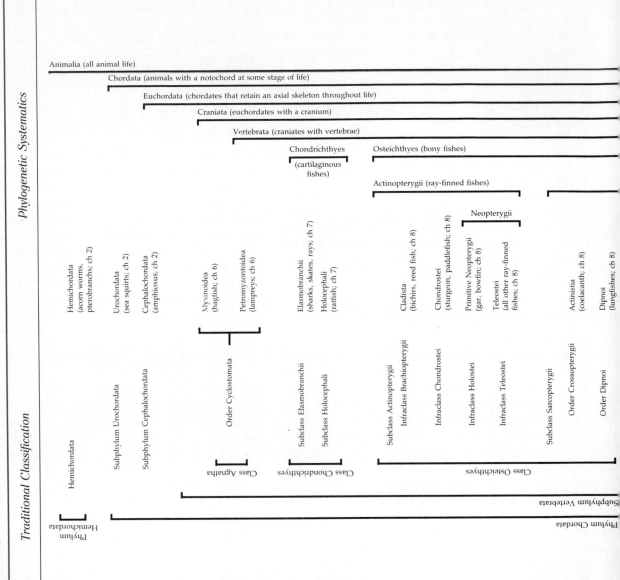

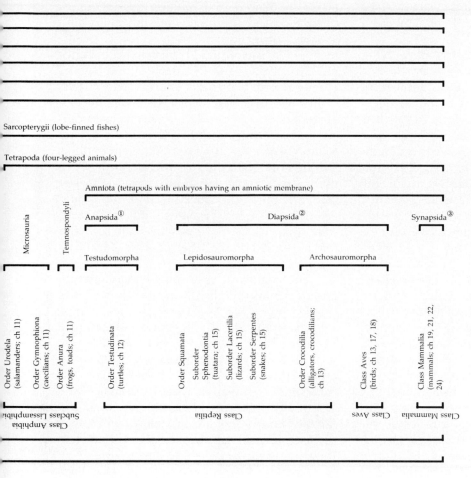

Sarcopterygii (lobe-finned fishes)

Tetrapoda (four-legged animals)

Amniota (tetrapods with embryos having an amniotic membrane)

Microsauria

Temnospondyli

Anapsida①

Diapsida②

Synapsida③

Testudomorpha

Lepidosauromorpha

Archosauromorpha

Order Urodela
(salamanders; ch 11)

Order Gymnophiona
(caecilians; ch 11)

Order Anura
(frogs, toads; ch 11)

Order Testudinata
(turtles; ch 12)

Order Squamata

Suborder
Sphenodontia
(tuatara; ch 15)

Suborder Lacertilia
(lizards; ch 15)

Suborder Serpentes
(snakes; ch 15)

Order Crocodilia
(alligators, crocodilians;
ch 13)

Class Aves
(birds; ch 13, 17, 18)

Class Mammalia
(mammals; ch 19, 21, 22, 24)

Class Amphibia
Subclass Lissamphibia

Class Reptilia

Class Aves

Class Mammalia

① Anapsida—amniotes without a temporal fenestra in the skull.

② Diapsida—amniotes with two temporal fenestrae in the skull.

③ Synapsida—amniotes with one temporal fenestra in the skull.

9

Box 1–1. Names and Endings

The diversity of vertebrates is enormous and the terminology that is used to classify this diversity is correspondingly extensive and often confusing. However, there is some underlying method to the system and knowing a few rules helps in understanding and remembering names.

Nomenclature is the process of applying names to groups of organisms. A **taxon** is a group of organisms that has been defined on the basis of characteristics that all the members have in common. The plural form of taxon is **taxa**. Taxa are arranged in a hierarchical order extending from the most inclusive through increasingly smaller divisions to the lowest unit of classification, the species. Many of the names are based on Greek and Latin words that describe some prominent feature of the group being studied. The Osteichthyes, for example, are the bony fishes; *oste(o)* is derived from the Greek word for bone and *ichthyes* from the Greek word for fish. The same Greek and Latin words recur frequently, and the same terms are often used to describe morphology and even physiology. Consequently, learning several dozen Greek and Latin roots greatly simplifies the process of remembering names of animals and structures. To help you learn these roots, we have translated many of the names as they appear in the text, and a list of Greek and Latin terms appears inside the back cover. A reference such as *Dictionary of Word Roots and Combining Forms* by D. J. Borror (1960) will provide additional details.

Frequently, very similar names are used to designate different taxonomic levels in the hierarchy, and the form of the final syllable of the name indicates its taxonomic level. Groups of vertebrates are often referred to by the name of a higher order taxon (usually a family or order), and the ending of the name is an important clue to how inclusive the reference is intended to be. The names of families of animals always end in -idae. For example, Hominidae is the family of primates in which humans are placed. Subfamilies are sometimes described, and the names of subfamilies end in -inae. Orders often end in -oidea, as in Anthropoidea, the order of primates that includes humans. This is not invariable, however; the names of orders of birds end in -formes (Passeriformes, the perching birds), and the superorders of fishes end in -omorpha as in Clupeomorpha (the herrings). Some taxa do not follow these rules; they have different endings that do not fit into the system described. Mammals, for example, have orders ending in -ata (Monotremata, the duck-billed platypus and echindas), -ates (Primates, the primates), -omorpha (Lagomorpha, the rabbits and hares), -ia (Rodentia, the rodents), and other forms.

The names of taxa are frequently anglicized (put into English form) for convenience in speech and writing. Family names become -id (-ids in the plural); "hominids" refers to all members of the family Hominidae, a "hominid" is one species of the family or one individual of a species. Orders become -oids (anthropoids), -thyans (osteichthyans), and -forms (passeriforms). Note that the names of families and orders are capitalized when they are used formally and are not capitalized when they are anglicized.

Species are grouped into genera. Each species has its own name. (To be technical, the name of a species is the "specific epithet." That does not mean a curse, although it sounds as if it does.) A genus is a group of species believed to be closely related by descent from a common ancestral species. The domestic dog, for example, is *Canis familiaris* and the wolf is *Canis lupus*. *Canis* is the generic name (name of the genus) and *familiaris* and *lupus* are species names (specific epithets). The genus *Canis* includes other species. such as the

Table 1–2. Species of living canids. This list is arranged alphabetically by genus and alphabetically by species within genera. The species of a genus are considered to be more closely related to each other than they are to species in other genera. Note that the common names do not reflect either generic groupings or closeness of relationship.

Genus and Species	Common Name
Alopex lagopus	Arctic fox
Atelocynus microtis	Small-eared dog
Canis adustus	Jackal
C. aureaus	Golden jackal
C. dingo	Dingo
C. familiaris	Domestic dog
C. latrans	Coyote
C. lupus	Wolf
C. mesomalus	Black-backed jackal
C. rufus	Red wolf
C. simensis	Abyssinian wolf
Cerdocyon thous	Crab-eating fox
Chrysocyon brachyurus	Maned wolf
Cuon alpinus	Dhole
Dusicyon culpaeus	South American fox
Fennecus zerda	Fennec
Lycaon pictus	African hunting dog
Nyctereutes procyonoides	Raccoon dog
Otocyon megalotis	Big-eared fox
Speothos venaticus	Bush dog
Urocyon cinereourgenteus	Gray fox
Vulpes fulva	Red fox
V. macrotis	Swift fox

coyote (*C. latrans*), the golden jackal (*C. aureus*), and the black-backed jackal (*C. mesomelas*). Note that the names of genera and species are written in italics (or underlined in handwritten text). The generic name is always capitalized and the species name is never capitalized, even when it is derived from the name of a person. The generic name can be abbreviated to its first letter after it has been written out on its initial appearance.

All of the species of *Canis* are members of the family Canidae, and other canids include the arctic fox (*Alopex lagopus*), the red fox (*Vulpes fulva*), the swift fox (*V. macrotis*), the gray fox (*Urocyon cinereoargentatus*), the raccoon dog (*Nyctereutes procyonides*), the African hunting dog (*Lycaon pictus*), and the maned wolf (*Chrysocyon brachyurus*) (Table 1–2). You will note that common names do not tell you anything about relationships: Animals called foxes, wolves, and dogs occur in several genera and are no more closely related to each other than they are to other members of the family Canidae.

In formal references, the name of the author who described a species may be appended to the name of the species. For example, *Crassigyrinus scoticus* Watson means the species that D. M. S. Watson named *Crassigyrinus scoticus*. It sometimes happens that the same name is applied to different species by different authors who are unaware of each other's work, or different names may be applied to the same species. Including the name of the author makes it clear that you are referring to the species that Watson called *Crassigyrinus scoticus*, not to some other species that might inadvertently have been given the same name.

distinguishing characteristic of birds. Indeed, some fossils of *Archaeopteryx*, the earliest bird known, were originally classified as dinosaurs because no marks of the feathers were visible. Birds are conspicuous, often vocal, and active during the day (diurnal). As a result they have been studied extensively and much of our information about the behavior and ecology of terrestrial vertebrates is based on studies of birds.

Mammals

The living mammals can be traced back through the synapsid lineage to an origin in the late Paleozoic from some of the earliest fully terrestrial tetrapods. Modern mammals include about 8000 species, most of which are placental (eutherian) mammals. Their name comes from the placenta, a structure that transfers nutrients from the mother to the embryo and removes the waste products of the embryo's metabolism. Most of the familiar animals of the world are placentals. Marsupials dominate the mammalian fauna only in Australia. Kangaroos, koalas, and wombats are familiar Australian marsupials. The strange monotremes, the duck-billed platypus and the echidnas, are mammals—they produce milk to feed their young—but the young are hatched from eggs, not born alive like marsupials and placentals.

Evolution

Evolution is the process that has shaped the vertebrate story, and it is the underlying principle of biology. An understanding of the principles and processes of evolution is essential to appreciating the diversity of vertebrates because that diversity is the direct result of evolution.

Scientific ideas are shaped by the society and philosophical system in which they form, and in turn they may reshape society and philosophy (Box 1–2). That process has been a conspicuous part of the development of evolutionary theory in western societies, most recently in the conflicts over teaching of evolution in schools. Many parts of western thought had their origin in Greek philosophy, but evolution was not among them. The Greek conception of nature was a static one, and the idea of evolutionary change in species of plants and animals was foreign to Greek thought. Platonic philosophy was based on the concept of an abstract ideal; earthly manifestations of that archetype (individual animals or plants, for example) were imperfect copies of its form. This attitude was incorporated into Christian theology as the view that an eternal, inviolable essence of every worldly object existed in the mind of God. A strict interpretation of the Bible insists that everything on Earth was created by God in its present form, and therefore nothing can change or has changed.

This view of life as unchanging was incorporated into biological thought and is still manifested in some biological practices. For example, the Rules of Zoological Nomenclature require that when a new species is named, one specimen of the new species must be designated the type specimen, or **holotype** of the species. The holotype is deposited in a museum, where it serves as a permanent record of the new species. The original role of the holotype was to show what the species was like. If you had another individual that you thought might be a member of the same species, you could compare it to the holotype and decide. Thus the holotype was a single example that was considered to define the essence of the new species. In effect it was the designated representative of the Platonic ideal of the species.

Naming a new species still requires a holotype, but the role of the holotype has changed significantly, and in addition to the holotype it is usual to deposit several additional specimens called **paratypes** or the **paratypic series**. The paratypes are intended to show the range of variation of the species in such characteristics as body size, color, and pattern. The holotype now defines the species in those rare cases in which two unnamed species have been confused and both are represented in the paratypic series. The name goes to the species represented by the holotype and the other species receives a new name.

The recognition that variation within species

Box 1–2. Evolving Views of Evolution

Darwin's book, *The Origin of Species by Means of Natural Selection*, sold out in a single day when it was issued in November 1859. The fervor created by Darwin's theory of evolution continues to the present, and it has affected not just biology, but also the philosophical foundations of Western society. However, Darwin's realization that organisms are not immutable but instead change through time had been preceded by other concepts that the world was not static. The idea, which had prevailed through the dark ages—that Earth was an unchanging entity—began to fall with the suggestions of Copernicus and Kepler that the Earth was not the center of the universe. The demonstrations by Newton and others that many physical events followed precise mathematical relationships and, particularly, the undeniable demonstration by paleogeologists that extinct organisms were recorded as fossils in the rocks placed in question one of the fundamental tenets of Christianity—that all life originated from a divine creation and was consequently immutable.

Thus the idea that organisms were related and, perhaps, even descendents of earlier ancestors was not novel to many biologists in the middle of the nineteenth century. Earlier scientists had voiced the possibility that animals might change or evolve over time. Darwin's theory of evolution was accepted rapidly in the scientific community and within 20 years had become the major framework of biological thought. What features of Darwin's theory of evolution made it scientifically acceptable when earlier theories, such as that of Lamarck, had failed?

First, the evidence presented by Darwin in support of evolution was overwhelming. It covered fields as diverse as comparative anatomy, embryology, paleontology, geology, and animal and plant breeding. This time evolution could not be dismissed as a speculative concept. Darwin documented its reality with explicit logically argued examples—descent from a common ancestor more readily explained the traits of organisms than a multitude of special creations. Second, to explain the origin of a new species Darwin provided a mechanism, **natural selection**, that was based on the inherent variation that he found so characteristic within each species. In a natural population, selective forces, which might be environmental factors such as food availability, temperature, or moisture conditions or biotic factors such as predation or competition between individuals, would favor those individuals that were most fit over those individuals that were less fit *in the conditions prevailing at the time*. Although often perceived as the theory of survival of the fittest, as if evolution were a constant battle to the death between beasts, Darwin's concept of favored individuals was not so much survival of individuals as it was survival of the individual's progeny. He clearly saw that variation conferring a selective advantage would be passed on to future generations. Unfavorable variations would be diminished in future generations and ultimately eliminated.

Central to Darwin's theory was the belief that evolution proceeded by the accumulation of small heritable changes, not large sudden changes, and that selective forces acted on the individual. Furthermore, it was Darwin's contention that evolution acted without design—heritable traits accumulated randomly and natural selection depended on prevailing conditions. Of course, Darwin was unaware of the mechanisms of genetic inheritance because Gregor Mendel's work was not reported until 1866 and was not widely disseminated until

Box 1—2. (Continued)

the early twentieth century. Nevertheless, Darwin realized that in some manner discrete changes or mutations that affected morphology and other aspects of an animal's biology, such as its behavior, occurred in individuals and were inherited. Darwin realized that variation within a species provided the framework on which natural selection could operate to produce new species. Evolution was seen to proceed not only by the loss or culling of unnecessary traits, but also by the selection of randomly accumulated variation in traits (the appearance of new traits by mutation and recombination). New attributes did not arise out of need for them, as postulated by Lamarckian evolution, but through the relentless operation of natural selection on the accumulation of variation among individuals of a species.

Although Darwin's voluminous works led to the rapid scientific acceptance of evolution, his theory of natural selection met with resistance. It was not until the 1920s that accumulating evidence, especially from the newly developing field of genetics, led the scientific community to become supporters of natural selection. The merger of Darwinian selection and genetic theory is referred to as **neo-Darwinism** or the **modern synthesis** after a book by Julian Huxley (1942) entitled *Evolution: The Modern Synthesis*. Many books presented data showing that point mutations and genetic recombination were the source of variation and that evolution (changes in gene frequency) proceeded generally in small steps and resulted from natural selection acting on genetic variation (Fisher 1930; Haldane 1932; Dobzhansky 1937, 1970; Mayr 1942, 1963, 1976; Simpson 1944, 1953; Stebbins 1950). Such processes were considered sufficient to explain the origin of higher taxa if they acted over prolonged intervals. This view of the evolutionary process is now generally identified as **microevolution** or **phylogenetic gradualism**.

Recently, the hypothesis that evolution proceeds through the slow accumulation of small genetic mutations and/or gene recombinations has been challenged by several biologists who contend that speciation seen in the fossil record may not appear to be gradual but that new species, or even novel higher taxa, may appear suddenly (Eldredge and Gould 1972, Stanley 1975, Gould and Eldredge 1977, Vrba 1980). Underlying their forcefully presented viewpoint is the fact that a gradual change or transition from one species to another is often missing in the fossil record. A gap frequently exists between recognizably related but distinct forms. Indeed, in the rare case when a species is represented by a long sequence of fossils, its characteristics usually show variation but no directional change of the sort expected if natural selection were operating. Rather than proceeding through a constant accumulation of small changes in structure, physiology, and behavior, evolution appears to proceed in a series of jumps followed by static pauses (Figure 1–2).

The competing theories of micro- and macroevolutionary processes of speciation have become popularly referred to as **gradualism** and **punctuated equilibrium**. Gradualists would expect a species to accumulate structural changes even in a more or less stable environment, whereas punctuationalists would expect a species to remain in structural equilibrium unless the environment changed significantly. Central in both views is the fact that environment, although broadly stable over reasonable periods of time, does oscillate continually and therefore incessantly stresses each individual.

Punctuated equilibrium provides an ex-

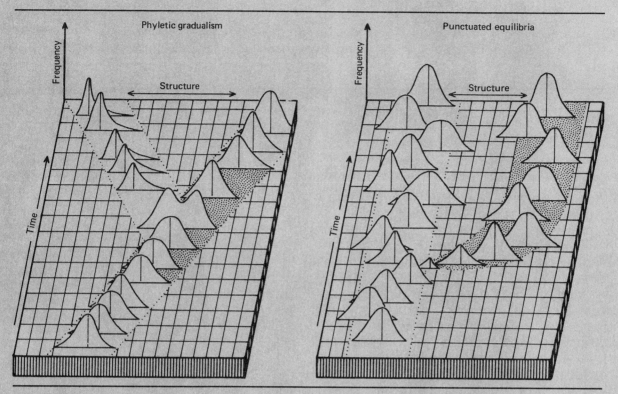

Figure 1–2. Examples of microevolution or gradual changes and of macroevolution or punctuated changes in species through time. The axes of the graphs plot character state against time. The various curves show the frequency distribution of the character state within the population at specific times. The tendency to speciate, or at least divergence in character state between populations, is shown by the separation into two lineages. A speciation event occurs when the two populations become reproductively isolated as indicated by shading. (Modified after E. S. Vrba, 1980, *South African Journal of Science* 76:61–84.)

planation for the existence of a recognizable species through time. If species arise suddenly through rapid genetic structural adjustments and then remain in stable equilibrium until the next punctuation, they represent distinct entities with a proscribed structure and period of existence.

It is difficult to know whether either of these views is exclusively correct. The rapid accumulation of data to bolster one view or the other has revealed that both patterns of evolutionary change have occurred. The mechanism or mechanisms by which species form remains an actively debated area of evolutionary biology and one of the most fundamental problems in biology as a whole.

is biologically important is a major change from the idea that one specimen by itself could define a species, and it reflects the most significant change in biological thought that occurred in the nearly two and a half millennia between the writings of Plato (about 380 B.C.) and the work of Darwin and Wallace in the nineteenth century. In a Platonic view the variations within a species are the imperfections of different copies of the archetype and they have no significance. To an evolutionary biologist those variations are the raw material of evolution.

Biological Variation

The idea that no two people (except for identical twins) are exactly the same is a familiar one. Humans use individual variation every day in contexts that range from recognizing friends to fingerprinting criminal suspects. Individual variation is not confined to morphological variation like facial contours or fingerprints; it extends to any characteristic that can be measured.

These sorts of variation are examples of **phenotypic** variation. The phenotype of an organism means its form in a very broad sense. For example, phenotype can refer to color (redheads versus brunettes), size or shape (tall versus short), or performance capacity (fast versus slow runners). Phenotypes can also be defined on the basis of molecular characteristics: hemoglobin B is the normal form of the beta chain of the adult human hemoglobin molecule and hemoglobin S is the form of the beta chain that leads to sickle-cell anemia. The two forms differ in only one amino acid reside—a valine appears in hemoglobin S in place of a glutamine.

Behavioral phenotypes also can be defined: Guppies are most familiar as aquarium fish, but they occur wild in streams in the New World tropics. In Trinidad guppies occupy stream habitats with relatively few predators and other streams where predators are abundant, and populations in these habitats differ in male color and in the response of females to courtship by bright-colored

and plain males (Breden and Stoner 1987). The bright colors of male guppies are part of the courtship display, but they also make males more visible to predators. In low-predation areas most male guppies are bright colored, but in areas of high predation they are plain. Experiments have shown a genetically determined difference in the responses of female guppies from streams with high and low predation to courtship by bright and plain males: female guppies from low-predation habitats choose to mate with bright-colored males, whereas females from high-predation areas choose plain males. This response is apparently mediated by the chances that the offspring will survive. Bright-colored males sire bright-colored male offspring. In low-predation habitats, bright-colored male offspring are advantageous because they are attractive to females and run little risk of being eaten by a predator. However, in high-predation habitats bright-colored male offspring are vulnerable to predation and females with the genetically determined behavioral trait of mating with plain males are more successful.

These types of phenotypic variation are manifestations of genetic variation. Evolution is defined as change through time in the frequencies of different forms of a gene in the gene pool of a population. Different forms of a gene are called **alleles** and different alleles produce different phenotypes. The **genotype** of an individual is its genetic composition. Remember that in sexually reproducing organisms an individual receives one allele from its mother and one from its father. To use the example of sickle-cell anemia, there are three possible genotypes: If an individual receives the normal form of the beta chain allele from both parents it will be Hb_BHb_B, that is, a **homozygous genotype**, and it produces the normal beta chain phenotype. An individual that receives one allele from its mother and a different allele from its father would be Hb_BHb_S, that is, a **heterozygote**. Heterozygotes for the sickle-cell allele have an advantage in areas where malaria is endemic, apparently because red blood cells that contain hemoglobin S collapse (sickle) when a malaria parasite enters them, and this sickling causes a decrease in the concentration of potassium inside the red cell and the eventual death of the parasite (Friedman 1978). An individual that receives the sickle-cell allele from both parents will have the genotype Hb_SHb_S: These homozygous individuals are at a disadvantage because the hemoglobin molecules sickle whenever the oxygen concentration in the blood drops, as it may, for example, during physical exercise.

Evolution depends on the existence of variation in genotypes among individuals that is (1) manifested in the phenotype and (2) inherited. Natural selection changes the relative frequency of different alleles in a population, and these changes in alleles are reflected in changes in the frequencies of different phenotypes. It is easier to think of phenotypes in discussions of evolution because phenotypes are visible, whereas alleles are not. However, it is alleles that are inherited.

Variation in genotypes results from the combined actions of several genetic processes in sexually reproducing organisms. During meiosis and the formation of gametes (sperm and ova) the pairs of chromosomes that make up the genome of an individual are first duplicated and then separated into gametes that contain only one of each pair of chromosomes. (This is the **haploid** chromosomal complement.) When two gametes unite to form a zygote (fertilized egg) the **diploid** chromosome number is regained: Two chromosomes of each pair are present, one from the father and one from the mother. Alleles are not entirely independent entities; their function is affected by other alleles present in the genotype. Consequently, the genetic **rearrangement** that accompanies sexual reproduction is an important source of phenotypic variation. **Mutation** is another source of variation, of smaller magnitude than rearrangement but nonetheless important. Mutation can result from errors in copying the genetic code during the initial duplication of chromosomes during meiosis, or it can be caused by events that precede meiosis, such as exposure of the sex cells to ionizing radiation. Other cellular events can cause changes in the linear sequence of genetic loci on chromosomes. Because the action of alleles is influenced by neighboring alleles, these changes in sequence can affect the phenotype.

Natural Selection

You will have noticed that of the mechanisms discussed so far, only mutation changes the frequencies of different alleles. Mutations are such rare events that by themselves they would not produce measurable changes in allelic frequencies. Thus none of these genetic processes is responsible for evolution. Evolution results from the action of natural selection on different phenotypes, and natural selection works through **differential reproduction**, that is, the contribution of different phenotypes to the gene pool of the next generation.

The term **fitness** is used to describe the relative contribution of different individuals to future generations. If individual A produces 100 offspring that survive to reproduce and individual B produces only 90 offspring, individual B is only 90 percent as successful as individual A. The fitness of the most successful individual (A in this example) is defined as 1.00, and the fitness of other individuals are defined in relation to A. Thus the fitness of B would be 0.90. That means that the alleles represented in the genotype of individual B would be underrepresented in the next generation by 10 percent compared to alleles in the genotype of A. In this example, selection would be said to *select against* B and the selection coefficient would be 0.10.

Modes of Selection

Natural selection can affect the distribution of phenotypes in a population in three different ways (Figure 1–3). **Directional selection** discriminates against individuals at one extreme of the variation in a phenotypic character. For example, small individuals might be at a disadvantage compared to individuals of normal or larger-than-normal size

Figure 1–3. Three kinds of selection: directional, stabilizing, and disruptive. In each case the vertical axis is the proportion of individuals and the horizontal axis shows the range of variation. The outlines of the upper diagrams show the initial variation within the population. The individuals in the shaded areas are at a disadvantage relative to the rest of the population. The lower diagram shows the variation in the populations after selection ($\bar{x}$, average value).

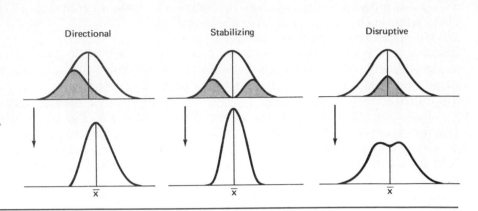

(Figure 1–3a). The effect of directional selection is to move the mean value of the character in the direction of the most fit phenotypes from generation to generation.

Stabilizing selection discriminates against individuals that have extreme variation in the phenotypic character in either direction (Figure 1–3b). It favors individuals that have values that lie close to the mean, and it can result in a reduction of the amount of variation within a population, but it does not change the mean value.

Disruptive selection is the reverse of stabilizing selection. That is, individuals at both extremes of the range of variation are favored and individuals near the mean are at a disadvantage (Figure 1–3c). Disruptive selection increases the variation but does not change the mean value.

Note that these descriptions have assumed that only one mode of selection affects the trait being considered. In fact, more than one mode of selection can occur simultaneously, and both the mean value of the trait and the amount of variation can change simultaneously. Natural selection is difficult to demonstrate under natural conditions because many different and sometimes conflicting selective forces operate simultaneously. Nonetheless, some examples are known, and John Endler (1986) has reviewed the evidence for natural selection in the wild.

Levels of Selection

One of the active areas of discussion and research in evolution is the question of the levels at which selection acts. Evolution is defined as a change in the frequencies of alleles, but does selection act directly on alleles, or on phenotypes (that is, individuals) that have those alleles as part of their genotype, or at still higher levels of biological organization?

The best-known examples of selection act through differential reproduction of individuals, but other kinds of selection are possible. **Genic selection** describes selection for individual alleles. Normally, a heterozygous individual produces equal numbers of gametes with each of its two alleles. However, situations can occur in which one allele is transmitted to more than half of the viable gametes of a heterozygous individual. This phenomenon is called segregation distortion or meiotic drive. In that situation the allele that is overrepresented in the viable zygotes is more fit (leaves more offspring) than the underrepresented allele.

Interdemic selection is another possibility. A **deme** is a local population of a wide-ranging species. Usually, the frequencies of alleles differ slight from deme to deme. If populations that differ in allelic frequency become extinct at different rates or give rise to new populations at different rates, selection might act on all the alleles in a deme.

That idea can be extended to consider that different species might have characteristics that made them more or less likely to become extinct. For example, species with small geographic ranges may be wiped out by local changes in the environment, such as floods or droughts, whereas species with wide geographic ranges are less vulnerable to the effects of local changes. The same line of reasoning can be applied to other characteristics of a species—narrow versus broad choice of habitats, for example, or specialized versus generalized feeding habits. The rate of speciation might also be subject to selection. The assumption here is that an evolutionary lineage that forms new species (speciates) rapidly is less likely to become extinct than is a lineage that speciates slowly. If those characteristics of species are genetically determined, they might be subject to **species level selection**. This is still a controversial hypothesis; further discussion can be found in Jablonski (1986).

Inclusive Fitness

We have defined fitness as the relative genetic contribution of different individuals to future generations, and described it in terms of the number of offspring produced by an individual. That is an oversimplification in the sense that individuals of a species have alleles in common, and the more closely related two individuals are, the higher will be the proportion of their shared alleles. Siblings, for example, have an average of 50 percent of their alleles in common and half-siblings (one parent the same and one different) share 25 percent of their alleles. Remember that it is alleles that are transmitted from generation to generation, and you can see that it is possible for an individual to increase its own fitness (that is, to transmit alleles identical

to its own) by assisting a related individual to reproduce successfully. For example, when your sibling reproduces, he or she transmits half the alleles you share to that offspring (which is your niece or nephew), and 25 percent of the alleles in the genotype of the offspring are the same as alleles in your genotype (50 percent of alleles shared between siblings x 50 percent of the genotype contributed by each parent = 25 percent of the alleles in the genotype of the offspring). Your own offspring has 50 percent of the alleles in your genotype, twice as many as your niece or nephew. All else being equal, then, two nieces or nephews are the equivalent of one offspring of your own in terms of transmitting your alleles to future generations. The same calculations can be applied to increasingly distant genetic relationships. The important point is that reproduction of relatives contributes to the fitness of an individual. In some situations helping relatives to reproduce successfully may be the best way for an individual to increase its own fitness. This principle of **inclusive fitness** is believed to underlie many of the altruistic behaviors of birds and mammals (Chapters 18 and 22).

Variation and Evolution

Phenotypic and genotypic variation is the material on which natural selection operates, but not all kinds of variation are subject to selection and characteristics of the biology of certain species can increase or decrease the importance of selection. In the first place, variation must be **heritable** if selection is to act on it. That is, variation in phenotypic characters must have an underlying genetic basis and parents that manifest a particular character must produce offspring that also show that character. Not all variation does have a genetic basis, and many characters that do have a genetic basis are also affected by the environment. The body size of turtles at hatching is an example of this interaction between genetic and environmentally induced variation. Newly hatched turtles receive no parental care: They must dig their way out of nests in the soil, find their way to water, and begin to catch their own food entirely by their own efforts. Probably it is desirable for a female turtle to produce the largest hatchlings she can because larger hatchlings can dig more strongly, move to water faster, and capture a wider variety of prey than can small individuals. The size of a hatchling turtle is correlated with the size of the egg it came from, and some female turtles lay larger eggs than others. All else being equal, females that lay large eggs should produce large hatchlings and be more fit than females that lay smaller eggs. However, the availability of water in the nest also affects the body size of hatchlings—nests in moist soil give rise to larger hatchlings than do nests in dry soil, and the hatchlings from the moist nests can also crawl and swim faster than hatchlings from a dry nest. This component of variation in body size of hatchlings has no genetic basis and cannot be acted on directly by natural selection. If selection does act on this type of variation (and this sort of selection has not been demonstrated) it would have to be indirect, via a reduction in the fitness of female turtles that chose dry nest sites.

Variation That Is Subject to Natural Selection

Some female turtles have a genotype that results in their putting a large quantity of yolk into each egg so that they produce large eggs, that is, **individual variation**. Individual variation is the type of variation one usually thinks of in terms of natural selection—some individuals are larger than others, or faster, or more colorful, or more aggressive. The list of characters can be extended indefinitely, and those kinds of differences result from the luck of the draw when gametes combine to form a zygote. The effects of genetic recombination and mutation lead to different genotypes and different phenotypes, and selection can act on those differences.

Individual variation is usually continuous. That is, some animals are small, others are large, and most are somewhere between those extremes.

Continuous variation is the situation illustrated in Figure 1–3. A second type of variation is discontinuous: Instead of having a bell-shaped curve of frequencies of different forms, discontinuous variation can be sorted out into discrete categories. **Polymorphism** (*poly* = many, *morph* = form) is a common type of discontinuous variation. Figure 1–4 illustrates pattern polymorphism in a Puerto Rican frog, the coquí. Some individuals are uniformly colored, some have stripes along the sides of the body, and some have a mottled pattern of dark blotches. These different patterns are genetically determined and offspring inherit the patterns of their parents. One can find individuals with all those patterns in any population of the frogs, but some patterns are more common in one kind of habitat than another, probably because certain patterns are especially cryptic (hard for predators to detect) in particular habitats. In grasslands, for example, the striped patterns occur in high frequency, whereas in forests the unicolored and mottled patterns are more common. In grassy areas predators see the frogs against a background of grass stems, and the striped pattern may blend well with the straight, light-colored stems. In forests predators often see the frogs on a background of fallen leaves; there are very few sharp, straight lines in a forest and the mottled and solid patterns of the frogs are probably more cryptic than stripes. The differences in frequencies of the patterns in different habitats indicate that natural selection has acted on the gene pool to change the frequencies of alleles producing the patterns.

Sexual dimorphism (*di* = two) is a special case of polymorphism in which males and females differ. Sexual dimorphism results from the different roles that males and females play in reproduction and the differences in the selective pressures that affect the fitness of individuals of the two sexes. The difference in sexual roles begins with the investment of males and females in gametes: males produce sperm, females produce ova. Sperm are small, energetically cheap, and quick to produce, whereas ova may contain large quantities of yolk and require a substantial investment of time and energy by the female.

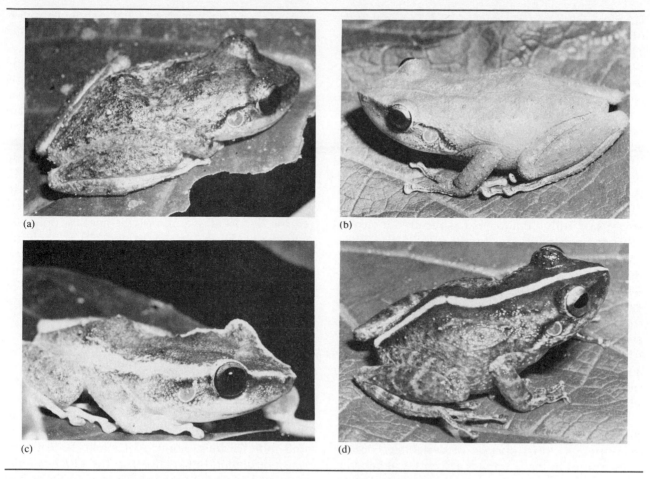

Figure 1–4. Polymorphic patterns of the Puerto Rican coquí. These are four of the more than 20 different patterns that have been described: (a) mottled; (b) unicolor; (c) broad lateral stripes; (d) broad dorsal stripe. Striped patterns (such as c and d) are more common in grassland habitats than they are in the forest. [Photographs by Margaret M. Stewart (a,b,d) and F. Harvey Pough (c).]

As a result of this asymmetry of investment, the reproductive strategies that increase the fitness of males and females can be quite different. Because a male can readily produce more sperm he can usually increase his fitness by mating with as many females as possible—each mating has only a small cost and he can mate repeatedly with little delay between matings. The situation for the female is quite different: by the time she is ready to mate she has invested substantial energy and time in producing mature ova. Furthermore, in many species the female provides most of the parental care. As a result, it is costly in time and energy for a female to breed repeatedly. Instead, her fitness is increased by mating with the best possible male.

Sexual dimorphism often reflects these differences in the selective forces acting on males and females: In *The Descent of Man and Selection in Re-*

lation to Sex Darwin pointed out that males are often larger than females and more aggressive. They may engage in ritualized or actual combat with other males, and may form dominance hierarchies in which high-ranking males have opportunities to court females and low-ranking males do not. He noted that females are often coy and must be courted by a male before they will mate. Many secondary sexual characteristics of males such as bright colors, the presence of horns or antlers, and vocalization, are examples of sexual dimorphism that are employed in combat or courtship. The role of female choice in determining the reproductive success of individual males has received increased attention recently. The bright-colored and plain-colored Trinidad guppies discussed in the preceding section provide an example of the role of female choice in determining the reproductive success of males, and additional examples will be found in subsequent chapters. Eberhard (1985) has suggested that the enormous elaboration of male genitalia of vertebrates and invertebrates reflects sexual selection operating via female choice for males that are especially good at stimulating females during courtship.

Geographic variation describes phenotypic differences that occur across the geographic range of a species. Most species of vertebrates occupy large ranges that encompass a diversity of habitats, and populations of these species consist of thousands to hundreds of millions of individuals. A few highly successful vertebrate species, such as humans, rats, and herring, attain populations numbering into the billions or even trillions of individuals, and some are so widely distributed that their ranges encompass many millions of square kilometers. Within a single genus of birds, the falcons, one finds extremes ranging from the Mauritius falcon (*Falco punctatus*) with a total population of fewer than 20 individuals and a geographic range restricted to the island of Mauritius in the Indian Ocean, to the peregrine falcon (*F. peregrinus*), which, although numbering only in the few thousands of individuals, occurs as a breeding bird on every continent except Antarctica and on many of the oceanic islands of the world.

Wide-ranging species usually show considerable geographic variation in their biological traits and are often referred to as **polytypic species**. A polytypic species is one that taxonomists divide into two or more subspecies (geographic variants sufficiently distinctive to be recognized), whereas a **monotypic species** has biological traits that are so uniform over its entire range that recognizable subspecies cannot be delimited. The peregrine varies considerably in size and color over its wide range and has been divided into as many as 22 subspecies. The osprey, or fish hawk (*Pandion haliaetus*) which is nearly as wide-ranging as the peregrine, is virtually a monotypic species, only two subspecies having been named. It is doubtful, however, that there is such a thing as a monotypic species in the sense that all populations have the same frequencies of alleles, except, perhaps, in species with a very limited geographic range and small population.

Factors Promoting Speciation

Evolution consists of changes in frequencies of alleles within populations. If two populations of a species accumulate more and more differences in allelic frequencies over time, they become more and more distinct. Some alleles may be lost from one population but retained in the other population, and mutation may introduce new alleles into each population that are not represented in the other. How long can this process continue before the populations are no longer one species but two?

To answer that question we need a definition of a species. For sexually reproducing organisms, which includes the overwhelming majority of vertebrates, the definition of a **biological species** that has been proposed by Ernst Mayr remains most general: "Species are groups of actually (or potentially) interbreeding natural populations which are reproductively (genetically) isolated from other such groups." Reproductive isolation means that each species consists of a group of individuals that contribute to and share in a common gene pool. The genetic implications of reproductive isolation mean that species are the fundamental units of evolutionary change, because the genes of no other population can influence the evolutionary potential of this isolated gene pool. Species, or isolated populations of a species, are the units of biological organization that evolve. It is for these reasons that the species has remained a central unit of organization in studies of evolution, ecology, and population biology.

Each species, then, is the expression of a genetic system that evolves independently from other species. **Speciation** is the process by which populations of organisms become reproductively isolated from other species. This view of species stresses the fact that members of a species not only evolve independently of other species but also must be able to survive in the face of competition from members of other species.

The Role of Isolation in Speciation As long as populations are in contact, alleles can move among them. Even polytypic species have **gene flow** that maintains genetic continuity between populations at the extremes of the range of the species via intervening populations. A new allele created by a mutation in one part of the species range can eventually spread through the entire species, and an allele that is lost in one deme will eventually be replaced by gene flow from the gene pool of the species. Gene flow is both a constraining and a creative force in evolution (Slatkin 1987). On one hand, gene flow prevents local populations from accumulating enough genetic differences to evolve into different species, but at the same time gene flow allows superior alleles and combinations of alleles to spread through an entire species.

The Importance of Population Size in Evolution
How speciation might proceed in large interbreeding populations has perennially puzzled evolutionists. A large population that occupies extensive areas can usually be divided into subpopulations, or demes. Large continental areas are seldom homogeneous, but rather are a series of patchy habitats because of variation in geology and vegetation. Members of a deme are more likely to breed with members of their own deme than with members of adjacent demes, if for no other reason than proximity. The result is restricted gene flow within the species and, presumably, a reduction in heterozygosity or genetic variation in the deme. If the deme is large the effect of interbreeding will be slight, but if it is small the reduction in genetic variation may be significant. Occasional outbreeding with individuals from other demes will tend to maintain heterozygosity within the deme: an average of one individual exchanged between populations per generation is sufficient to prevent divergence.

Superimposed on this concept of semi-isolated demes with limited outbreeding is the amount or the intensity of natural selection. If, for example, environmental change, predation, or competition for resources is intense within a deme, individuals with the most appropriate phenotypes for the local conditions will be favored and their genotypes will increase within the population. If gene flow between demes remains low, genetic divergence of demes can occur.

Perhaps more critical is the mechanism of **genetic drift** and its relationship to evolutionary processes. Genetic drift is a random process that can lead to the loss or the fixation of alleles in any population. The process is more rapid in small, populations than in large ones. Genetic drift can lead to the loss of genetic variation in populations and divergence between populations by chance alone. Evolution, therefore, represents a delicate balance between mechanisms—if a deme is small genetic drift can reduce variation, as can intense selection. Genetic drift, however, is random with respect to any allele and natural selection is not.

Gene flow prevents a species composed of freely interbreeding populations from accumulating enough genetic differences among populations to speciate. Some kind of block to the free exchange of genes must be interposed between populations of a species before it can split into daughter species. In other words, some kind of spatial or **geographic isolation** is generally required before **reproductive isolation** can evolve between segments of a population. This process is called

allopatric speciation (*allo* = different, *patris* = native country).

The rate of evolution is most likely to be rapid in small populations, and small breeding populations can result from various circumstances including dispersion of a few adults to a new location, the sudden local isolation of a new deme by some environmental event, or a sudden reduction in the size of a population (a population crash). These periods of diminished population size are referred to as **bottlenecks,** and they potentially stimulate rapid changes in the frequencies of alleles in the population. Furthermore, the small population resulting from the bottleneck will contain only those alleles present in the individuals that pass through the bottleneck, and these are the only alleles that will be represented in the descendant population. A similar phenomenon, called the **founder effect,** describes a situation in which a new isolated population is founded by a few individuals. For example, a flock of birds might be blown out to sea by a storm and ultimately land on an island. If conditions on the island are suitable, the waifs could breed and establish a new population that will contain only the alleles represented in the genotypes of the founding individuals. This is the sequence of events that is thought to have produced the bird faunas of many oceanic islands including Hawaii and the Galapagos Islands.

A Classic Example of Geographic Isolation: The Galapagos Finches How geographic isolation promotes speciation is emphasized on oceanic islands, as Darwin himself observed in his examination of the endemic forms of animal life in the Galapagos Islands, which lie about 1000 kilometers off the coast of Ecuador. Darwin's finches (subfamily Geospizinae) have become the classic example of evolution at the species level (Figure 1–5). These 14 species have evolved rapidly from a single, ancestral population of seed-eating birds that arrived from South America in relatively recent geological time—perhaps less than 500,000 years ago.

From time to time individuals spread to other islands in the Galapagos from the original, founding population. In these new environments, with ecological opportunities for which no competition from other species existed, these new founders were able to repeat the process of establishment and eventual adaptation to local conditions. In time many of these isolated populations accumulated sufficient genetic differences so that, when secondary contact was established on islands that had already been colonized, they proved to be reproductively isolated from one another and could coexist as sympatric species, each with an ecological niche of its own (Grant 1986). Today, each island has as many as three to ten species, depending on the diversity of the vegetation and ecological opportunities.

Studies of diversity among the finches reveal that morphological variation has emphasized change in body size and in the shape of the bill. Three primary subgroups are recognized: the warbler-like finches, the tree finches, and the ground finches (Figure 1–5). Differences in beak size and shape represent specialization for different diets and ways of obtaining food.

Investigations of genetic differences among the finches produce species relationships that are basically similar to the arrangement proposed by David Lack in 1947. Although evolution has been rapid, allopatric speciation seems the simplest explanation for the diversity of Darwin's finches, especially when one considers the relatively high chances of dispersal of a founder population to a nearby island. A recent study by Gibbs and Grant (1987) reveals how rapidly the morphology of a species of finch can change in response to changes in the environment. The Galapagos Islands are characterized each year by wet and dry seasons. In exceptional years, however, the wet season can be prolonged and bring heavy rain. When that happens the plants flourish and food becomes abundant. These exceptional years are the result of climatic effects that reach far beyond the Galapagos—they are El Niño years, the name given to the periodic invasion of the eastern South Pacific by warm water masses. In 1982–83, a decade-long drought in the Galapagos was broken by a record El Niño year, the most intense in recent history. Gibbs and Grant had been monitoring a popula-

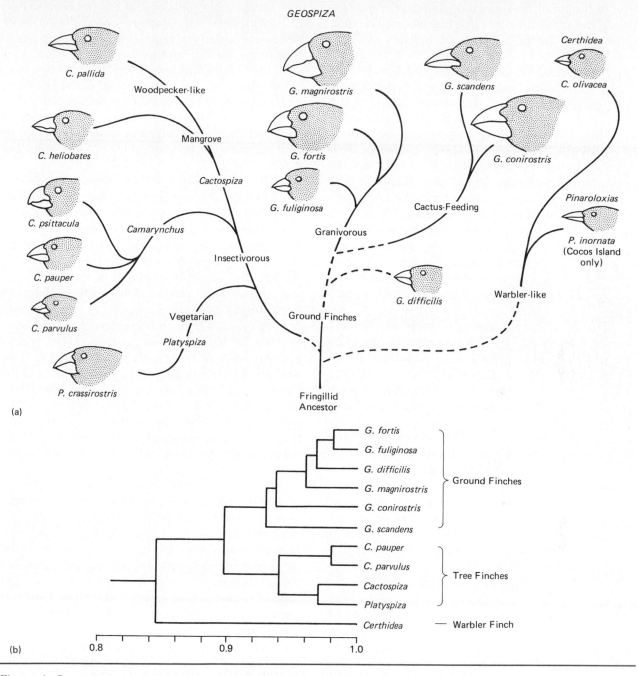

Figure 1–5. Differences among Darwin's finches and the likely rate and sequence of evolution. (a) The evolutionary sequence postulated by David Lack (1947, *Darwin's Finches*, Cambridge University Press, Cambridge) was based mostly on morphological criteria. It is very similar to the phylogeny based on a recent genetic analysis of similarity proposed by S. Y. Yang and J. L. Patton (1981, *Auk* 98:230–242) as shown in (b). Differences occur only within the ground finches.

tion of the medium-size ground finch, *Geospiza fortis*, on the small island of Daphne Major. The morphology of the birds changed between 1976–77 and 1984–85. Before the El Niño event, the population consisted of large-bodied and heavy-billed birds, but during and following the wet season both body size and bill size diminished. The differences in body and bill size have a genetic basis, and they also relate directly to success in feeding. In a drought, when food is scarce, the larger birds apparently are favored because they can crack the large, hard seeds that are the most abundant food for the finches. During wetter times the small birds are apparently successful because small, soft seeds are abundant. Here is a clear example of changing environmental conditions leading to the selection of genetically controlled morphologies that favor survival. The genotype of these finches must carry enough heterozygosity to express either phenotype.

For land-based animals such as finches, oceanic islands are obvious geographic features for the isolation and speciation of populations. The ocean is a barrier to dispersal from a continental source, but it is not an absolute barrier. Habitats on mountaintops are isolated from each other in much the same way as islands. It is easy to see how a surrounding lowland forest or desert may be an effective barrier to frequent dispersal of animals that are adapted to live only in the highland situation.

The rapid evolution of numerous related species, such as Darwin's finches, each of which represents a different adaptive type, is a recurrent process in the evolution of living organisms (Box 1–3). It has been documented in all classes of vertebrates from fishes to mammals. Usually, the rapid evolution of groups of species results from new environmental opportunities, each of which requires a different set of adaptations for successful exploitation. This phenomenon is called **adaptive radiation**. However, evidence from studies of protein evolution, which reflects genetic change, lends support to the view that molecular evolution proceeds at a fairly constant rate, especially when considered over long periods of time (Box 1–4).

Isolation on Continents It has not been so easy to explain how speciation has occurred on continents, particularly in regions where environmental conditions remain uniform over vast areas, such as in the tundra and boreal forests of the Northern Hemisphere or the Amazonian forest of South America. It now appears that a mechanism similar to isolation on islands has been at work repeatedly and that much speciation of land vertebrates on continents can be accounted for by geographic isolation.

The Pleistocene period—approximately the last 2 million years of geological history—has been characterized by repeated climatic fluctuations, which have been particularly expressed as changes in mean annual temperature and rainfall. Geological and biogeographic evidence supports the conclusion that island-like refugia of the tundra and taiga ecosystems persisted in northern latitudes even during maximum glaciations, whereas during interglacials these refugia have merged as continuous habitats of great expanse. Thus, species populations that were widely distributed in these uniform habitats during interglacials became fragmented and isolated in refugia during glaciations. This is precisely the sequence of events needed for speciation, and the present distributions of a number of bird and mammal species in boreal North America indicate that they have evolved in this way.

Similar events have also been postulated for environments outside the regions of active glaciation, for example, in the forest biomes south of the glaciated region of North America. Evidently, biomes south of the ice have not remained stable in position or species composition, but have fluctuated with the climatic cycle and fractured and reformed many times during the Pleistocene. As many as 10 to 17 of the 46 species of wood warblers (Parulidae) occurring in North America can be explained as the products of isolation in forest refugia created during the Pleistocene (Figure 1–9).

The changes in temperature and precipitation during the Pleistocene have been great enough to have effected changes in most biomes of the world, including those in the tropics. The

Box 1–3. A Special Case of Adaptive Radiation: Species Flocks

The East African rift lakes provide us with an interesting unsolved problem in evolutionary biology—the **explosive radiation** of a multitude of species within a single family of fishes, the Cichlidae. This widely distributed family of tropical and subtropical freshwater fishes contains more than 1300 species, of which nearly 1000 are endemic to the African rift lakes, especially the large lakes—Victoria, Malawi, and Tanganyika (Table 1–3). Something special has occurred in these lakes—the formation of what are termed **species flocks** (Greenwood 1984, Ribbink 1984). The species of cichlids in each lake are very numerous (speciose), closely related, often highly endemic (each species restricted to a single lake), and dominate the fish communities within each lake (over 50 percent of the total species). This dominance in number of species stands in contrast to the fish communities of the surrounding African rivers, where the cichlid components of the fish fauna seldom exceed 15 to 20 percent.

How do so many species evolve from a single ancestor and manage to coexist within a single lake system? Can 200 species of African cichlids possibly originate from a limited number of ancestors in sympatry? Or was allopatry somehow achieved between populations and reproductive isolation followed? There is no clear answer to this evolutionary puzzle, and it is further complicated by the relative youth of the rift lakes, for the extreme diversity must have evolved in less than 2 million years, the probable age of the oldest lake.

The African lake cichlids have been the subject of studies of taxonomy, functional morphology, behavioral ecology, and genetic affinities (for an overview, see Echelle and Kornfield 1984). What can be learned from these studies?

Consider the cichlids of Lake Victoria. The 250 endemic species are morphologically conservative; they differ only slightly in general body shape (Figure 1–6). In fact, many species are so similar morphologically that they cannot

Table 1–3. Comparison of the numbers of cichlids and other species of fishes in the rift lakes and nearby rivers of East Africa.

Habitat	Age of Habitat (Millions of Years)	Number of Species		Percent Cichlids
		Cichlids	*Others*	
Lakes				
Tanganyika	2	136	111	55
Malawi	1.5	500	42	92
Victoria	0.75	200	38	84
Rivers				
Nile	*	10	105	9
Zambezi	*	20	90	18

*The rivers are older than the rift lakes.

Box 1–3. (Continued)

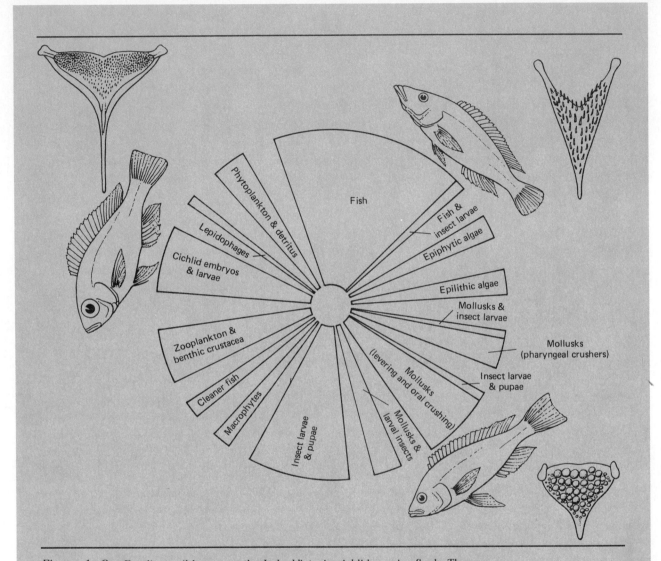

Figure 1–6. Feeding guilds among the Lake Victoria cichlid species flock. The size of each pie section approximates the number of species in each feeding guild. The species pictured provide examples of how similar body shape is between guilds. The differences in pharyngeal teeth show that slight changes in morphology can allow closely related species to exploit different feeding zones. (Modified after G. W. Coulter et al., 1986, *Environmental Biology of Fishes* 17:161 -183 and G. Fryer and T. D. Iles, 1972, *The Cichlid Fishes of the Great Lakes of Africa: Their Biology and Evolution*, Oliver & Boyd, Edinburgh.)

be easily distinguished from each other. (Nearly identical species of this sort are called **sibling species**.) The sibling species of cichlids have been shown to be reproductively isolated and to be discontinuously distributed in patches throughout Lake Victoria. Of the 250 cichlid species in the lake, only a few are widely distributed, and in extreme cases an entire species may be restricted to a single rock outcrop that occupies less than 2500 square meters of lake bottom. Most of the species are distinctively marked and highly colored. In some instances polymorphic colors and patterns exist within single species, and each morph may have slightly different habitat requirements. How do 250 rather similar species coexist without excluding each other through competition for food or other resources?

In spite of their general morphological similarity, the Lake Victoria cichlids have distinctive feeding specializations. At least ten different feeding **guilds** (a group of species that all use similar foods, such as the plankton-eating guild, the mollusk-eating guild, or the algae-eating guild) are recognized. Primary specializations for these different modes of feeding are seen in the modifications of jaw structure, jaw teeth, and the pharyngeal teeth. Important here is the demonstration that rather slight changes in morphology can support major shifts in feeding habits and, perhaps, lead to the rapid invasion of new adaptive zones (Greenwood 1984). Such rapid invasions of new adaptive zones that initially result from very slight changes in morphological structures are the essential elements for punctuated equilibrium. Members of a feeding guild generally show slight differences in niche requirements that may reduce interspecific competition. For example, the nine species in Lake Victoria that belong to the fish-eating guild all occur at different depths.

The cichlids of the African rift lakes are primarily mouth brooders, the eggs being protected and aerated within the oral cavity of the female (Fryer and Iles 1972). After hatching, the young form schooling clouds and feed on plankton in the immediate vicinity of the parent, retreating into the refuge of their mother's mouth when signaled to do so by her or when threatened by predators. This unusual mode of parental care is probably central to the formation of extensive species flocks, for it tends to localize offspring and impede dispersal.

The close morphological similarities and cohesive behaviors of the African cichlids confirm that they have evolved from only a few ancestral species. Their evolution within Lake Victoria must have occurred in less than 750,000 years. Although the rift lakes are fairly old compared to most lakes, it is likely that the lakes experienced much lower water levels during the Pleistocene glacial episodes, and Lake Victoria may have dried to a few isolated ponds. Lakes Malawi and Tanganyika are deeper and probably persisted through the glacial episodes, perhaps as a series of partially isolated basins (McCune et al. 1984). The possibility that geographic isolation occurred between populations on several occasions followed by reformation of the lake cannot be ruled out. Geographic isolation is the essential mechanism of most gradual or microevolutionary processes.

If the species flocks of African lake cichlids evolved recently, as their phenotypic similarities suggest, they should be genetically cohesive. Examination of the blood serum proteins of ten haplochromines (a closely defined subgroup of cichlids within Lake Victoria)

Box 1–3. (Continued)

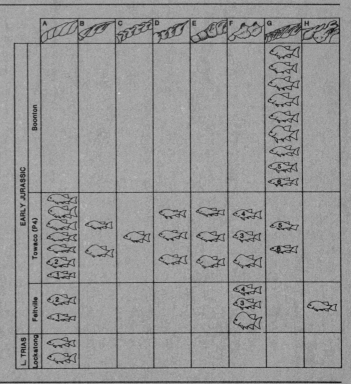

Figure 1–7. Pattern of distribution of *Semionotus* species through the history of the Newark Basin. The diagram shows the diversity of *Semionotus* in four geological formations: the Lockatong formation of the late Triassic and the Feltville, Towaco, and Boonton formations of the early Jurassic. Each of the four samples is from a single sedimentary cycle and represents a radiation of *Semionotus* in the lake that occupied the basin at that time. The species in each cycle are grouped by the morphology of their dorsal ridge scales as shown at the top of the diagram. The complexity of the dorsal ridge scales increases from left to right. Note that similar body forms evolved repeatedly. (Source: A. R. McCune, 1987, *Palaios*, 2, in press.)

show virtually no interspecific differences (Sage et al. 1984). In contrast, the ecologically similar American sunfishes, whose precise origins are unclear because of the absence of a fossil record, have speciated in allopatric isolation and show high serum protein divergences between species (Avise and Smith 1977) and only slightly greater morphological diversity than the Lake Victoria cichlids. The lack of divergence in serum protein (akin to overall genetic diversity) suggests that speciation was rapid in the cichlids and that evolutionary processes such as genetic drift have not had time to accumulate allelic differences. In fact, genetic diversity tends to be very low in all the fish species flocks that have been examined, not just the African rift lake cichlids (Humphries 1984). This lack of genetic variation compared to the morphological diversity suggests that minor genetic changes have operated to produce the diversity of mouth parts, digestive systems, and body size that seem to characterize most fish species flocks.

The causes of the explosive formation of species flocks remains obscure. Many hypotheses have been proposed, but they remain to be thoroughly tested (Mayr 1984, Sage et al. 1984). The cichlids are living proof that vertebrates are capable of rapidly exploiting new adaptive zones when conditions are right, but the formation of species flocks is not limited to the African rift lakes. The cyprinid fishes of Lake Lanao in the Philippines, the cottid fishes of Lake Baikal in the USSR, the pupfishes of Laguna Chichancanab in Mexico, the killifishes in the genus *Orestias* in Lake Titicaca in South America, and Darwin's finches in the Galapagos Islands represent other examples of vertebrate species flock formation. Similar rapid or explosive radiation events have taken place in the distant past, as documented by the sequential radiations into apparent species flocks by primitive neopterygian fishes, the semionotids, in the Jurassic rift lakes that existed in the general area of what is now New Jersey (McCune et al. 1984). In this example the lakes in which the radiations occurred persisted for about 21,000 years. The Newark basin contains

an excellent stratigraphic record of these lakes that covers some 20 million years. Fossilized in the lake sediments are the beautifully preserved remains of these ancient semionotid fishes. Comparisons of the fish community assemblages within each depositional period clearly indicates that they were species flocks—closely related, localized to the area, and speciose. As evaporation of the lakes took place, habitat restriction forced the members of the species flocks to retreat to the rivers that drained the system. Presumably, only those forms that were adapted to riverine conditions survived to reinvade the lakes as they reformed, and the cycle was repeated. Whatever the primary causes of speciation, it is clear from the morphological data that each successive lake had its own flock of semionotids (Figure 1–7).

Box 1–4. Tempo in Evolution: The Molecular Clock Hypothesis

The fossil record of vertebrates is incomplete, especially in older deposits, because time and geologic events have folded, metamorphosed, and eroded away many of the fossil-bearing rocks. Of necessity, therefore, estimates of the tempo of vertebrate evolution have been based largely on groups for which the geological record is most complete. In spite of discontinuities, the fossil evidence documents that some kinds of vertebrates have evolved in spurts. Some paleontologists describe vertebrate evo-

lution as a dynamic process in which speciation proceeds rapidly at some times and slowly at others. This oscillation in the rate of evolution is the theory of punctuated equilibrium (Box 1–2).

It was surprising, therefore, when Zuckerkandl and Pauling (1962) postulated that substitutions in the amino acid sequences of certain proteins of vertebrates occurred at a constant rate (Table 1–4). If this hypothesis is correct, it could be an important tool for evo-

Box 1–4. (Continued)

Table 1–4. Estimated time required for a 1 percent change in the amino acid sequence of different proteins.

Type of Protein	Millions of Years
Histones	60–400
Collagen	36
Dehydrogenases	13–55
Myoglobin	6
Hemoglobin	3.3–3.7
Immunoglobins	0.7–1.7

Source: A. C. Wilson et al., 1977, *Annual Review of Biochemistry* 46:573–639.

lutionary research because it would be possible to estimate the time that the phylogenetic lineages of two groups of animals separated by measuring the number of amino acid substitutions in their proteins. For example, if the amino acid composition of hemoglobin is compared among different taxa, the number of amino acid differences between any pair of them should be proportional to the time of evolutionary divergence if the differences have accumulated at a constant rate. The greater the amino acid differences, the farther back in time the taxa separated. Furthermore, because changes in amino acid sequences in proteins reflect changes in the nucleotide sequences of DNA, they indirectly measure occurrences of point mutations in the gene complex. This hypothesis, called the **molecular clock**, has sub-

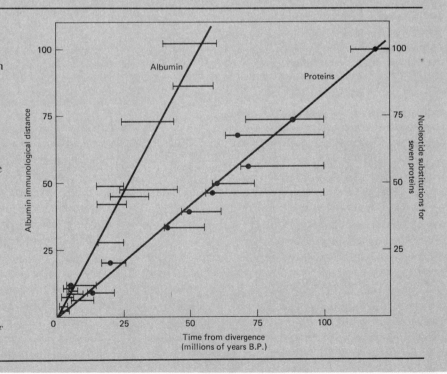

Figure 1–8. Constancy in the rate of molecular evolution of mammalian proteins. Left: Rates of albumin evolution in carnivorous and hoofed mammals. Right: Nucleotide substitutions since the time of divergence in living pairs of mammals. Closed symbols are the mean values for seven different proteins. Horizontal lines through the closed symbols show the estimated ranges of the times of divergence for each species pair. (Modified from C. H. Langley and W. M. Fitch, 1974, *Molecular Evolution* 3:161–177 and A. C. Wilson et al., 1977, *Annual Review of Biochemistry* 46:573–639.)

sequently been extended to include the rates of change in the base pairs of DNA from the nucleus and from mitochondria.

If the molecular clock hypothesis is correct, a close correlation should result when the number of amino acid substitutions for proteins of different vertebrates are regressed against their time of divergence. This relationship is evident for some species: The blood serum albumins of numerous pairs of mammals have been tested for immune cross reactions to estimate the number of amino acid substitutions among the albumin molecule's approximately 580 amino acids. When the proteins from different pairs of mammalian carnivores and ungulates are examined for immune response and the immunological differences are plotted against divergence time of each pair of taxa, as estimated from the fossil record, a good correlation is obtained (Figure 1–8).

However, further studies have shown that the molecular clock does not necessarily keep good time: Different proteins evolve at different rates (Table 1–4), and nuclear and mitochondrial DNA have different rates of evolution (Vawter and Brown 1986). Histones and structural proteins like collagen are conservative and show only low rates of substitution, whereas proteins involved in maintaining homeostasis, like hemoglobin, and in providing protection, such as the immunoglobulins which serve as antibodies against foreign antigens, tend to evolve very rapidly. A heavy chain immunoglobulin with 500 amino acids residues could substitute as many as six to seven residues per million years. Furthermore, the rate of change in DNA varies by a factor of 5 among different phylogenetic groups, and within a single group, the rate of DNA evolution changes with time (Britten 1986). For example, birds have lower rates of DNA evolution than do rodents, and the rate of change in DNA among primates has decreased during their evolution.

Thus, the molecular clock hypothesis has not entirely fulfilled the expectations of its early proponents. There is no general rate of change in DNA that can be applied to all phylogenetic lineages over the entire course of evolution. However, measurements of biochemical differences within particular lineages over shorter periods have been immensely valuable in providing information about specific evolutionary relationships. Some of the most intriguing and controversial examples come from the application of molecular techniques to studies of the origin of humans (Chapter 24).

Amazonian forest has apparently been broken up repeatedly into patches that persisted for a time as isolated forests and then coalesced to form a continuous biome again (Figure 1–10). Periods of drought have long been considered the cause of this disruption of continuous forest, but a recent interpretation suggests that floods were a major factor in the evolution of the Amazonian forest (Räsänen et al. 1987). Similar shifts in the distribution of savanna and forest biomes in Australia, also associated with long-term climatic changes, may account for much recent speciation among birds on that continent. Most biogeographers and paleoclimatologists now agree that climate and the major biotic associations of plants and animals have fluctuated repeatedly on a worldwide scale

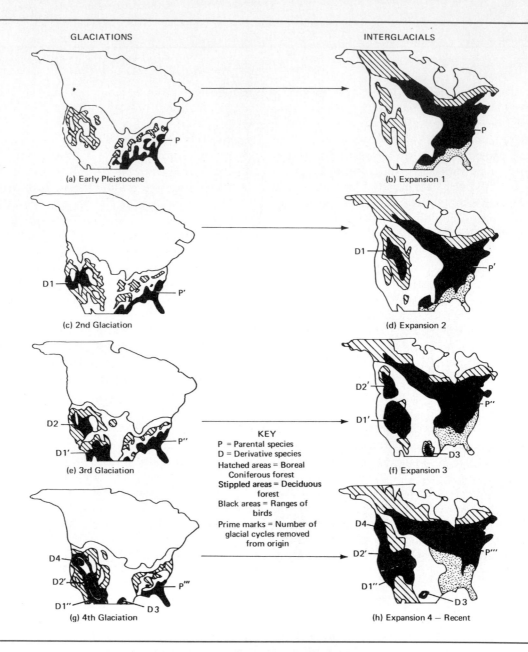

GLACIATIONS

INTERGLACIALS

(a) Early Pleistocene

(b) Expansion 1

(c) 2nd Glaciation

(d) Expansion 2

(e) 3rd Glaciation

(f) Expansion 3

KEY
P = Parental species
D = Derivative species
Hatched areas = Boreal
    Coniferous forest
Stippled areas = Deciduous
    forest
Black areas = Ranges of
    birds
Prime marks = Number of
    glacial cycles removed
    from origin

(g) 4th Glaciation

(h) Expansion 4 — Recent

Figure 1–9. Model of speciation in wood warblers (Parulidae) based on the black-throated green warbler group in the genus *Dendroica.* Each glaciation (left) isolated another population in the southwestern refugia of North America long enough for it to become unable to breed with descendents of the parent population during subsequent periods of expanded range (right). Thus glaciation produced five species in North America where a single species occurred in the early Pleistocene. (Modified from R. Mengel, 1964, *The Living Bird* 3:9–43.)

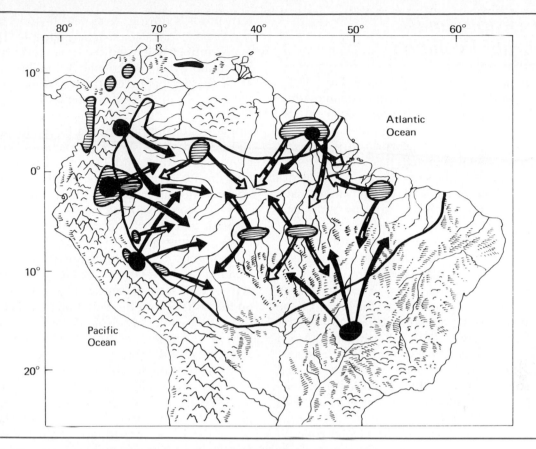

Figure 1–10. Proposed effects of changes during the Pleistocene on lowland forest vertebrates of the Amazon Basin. The black line shows the area within which forests are thought to have been fragmented. Within this region birds (black areas) and lizards (shaded areas) are thought to have been isolated in forest remnants. The arrows show postulated paths of reinvasion as the forests returned. (Based on P. E. Vanzolini, 1972, in *Tropical Forest Ecosystems in Africa and South America: A Comparative Review*, edited by B. J. Meggers et al., Smithsonian Institution Press, Washington, D.C.)

during the Pleistocene. The result has been the repeated fragmentation of biomes, alternating with re-formation of more continuous, continental distributions, plus some change in species composition of the biomes. Whereas geographic isolation on continents, as a requisite for speciation, was once difficult to explain, the Pleistocene and earlier cyclic events may now provide explanations. A few years ago ornithologists thought that

most bird species dated from the Pliocene and were several million years old. Now most of them feel that speciation has been very rapid during the Pleistocene, especially among small birds with limited dispersal abilities—characteristics equally applicable to many other vertebrates. However, some aspects of refuge theory are currently being questioned, especially in the case of tropical species (Connor 1986).

Earth History and Vertebrate Evolution

The phenomena of isolation and secondary contact that are important in the evolution of individual species like Galapagos finches and eastern warblers have also been important in the larger story of vertebrate evolution. The world in which vertebrates have been evolving has changed enormously and repeatedly since the origin of vertebrates in the Cambrian, and these changes have affected vertebrate evolution directly and indirectly. An understanding of the sequence of changes in the positions of the continents and the significance of those positions in terms of climates and interchange of faunas is a central part of understanding the vertebrate story. A summary table of these events is presented in the front of the book, and details are given in Chapters 5, 9, 14, and 20. The history of life on Earth has occupied four **geological eras**, the Precambrian, Paleozoic, Mesozoic, and Cenozoic. These eras are divided into **periods**, and within the Cenozoic the periods are divided into **epochs**. The history of vertebrates extends from the Cambrian to the present, a time span that is sometimes called the **Phanerozoic**.

Movements of land masses have been a feature of Earth's history, at least since the Precambrian (Box 5–1). The direction of vertebrate evolution has been in large part molded by continental drift. By the early Cambrian a recognizable scene had appeared—the seas had formed, continents floated atop the Earth's mantle, life had become complex, and an atmosphere of oxygen had formed, signifying that the photosynthetic production of food resources had become a central phenomenon of life.

The continents still drift today—North America is moving westward and Australia northward at approximately 4 centimeters per year. Because the movements are so complex, their sequence, their varied directions, and the precise timing of the changes are difficult to summarize. By viewing the movements broadly, however, a simple pattern unfolds during vertebrate history: *fragmentation–coalescence–fragmentation* (Figure 1–11).

Continents existed as separate entities over 500 million years ago. Some 300 million years ago all of these separate continents combined to form Pangaea, birthplace of the terrestrial vertebrates. Persisting and drifting northward as an entity, this single huge continent began to break apart about 100 million years ago. Its separation occurred in two stages, first into Laurasia and Gondwana, then into a series of units that have drifted and become the continents we know today.

The complex movements of the continents through time have had major effects on many phenomena significant to the evolution of vertebrates. The most obvious is the relationship between the location of land masses and their **climates**. At the start of the Paleozoic much of Pangaea was located on the equator and this situation persisted through the middle of the Mesozoic. Solar radiation is most intense at the equator, and climates at the equator are correspondingly warm. During the Paleozoic and much of the Mesozoic large areas of land enjoyed tropical conditions, and terrestrial vertebrates evolved and spread in these tropical regions. By the late Mesozoic much of the land mass of the Earth had moved out of equatorial regions, and most climates in the Northern and Southern Hemispheres are now temperate instead of tropical.

A less obvious effect of the position of continents on terrestrial climates comes about through changes in patterns of **ocean circulation.** For example, the Arctic Ocean is now largely isolated from the other oceans and it does not receive warm water via currents flowing from more equatorial regions. High latitudes are cold because they receive less solar radiation than do areas closer to the equator, and the Arctic Basin does not receive enough warm water to offset the lack of solar radiation. As a result, the Arctic Ocean is permanently frozen and cold climates extend well southward across the continents. The cooling of climates in the Northern Hemisphere at the end of the Mesozoic that had such a drastic effect on the dino-

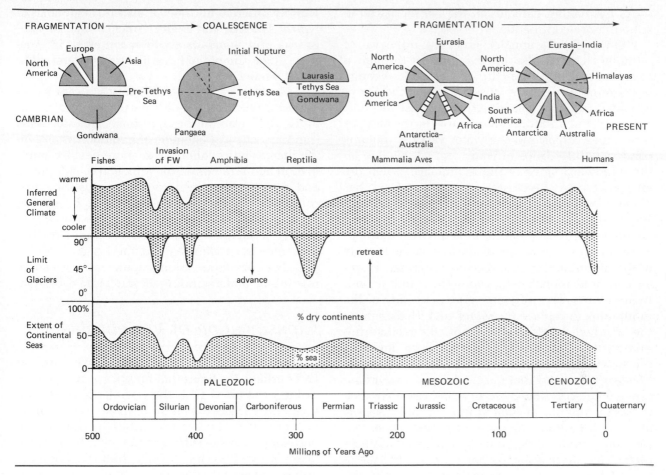

Figure 1–11. A summary of continental drift, climate, and the extent of epicontinental seas from the Ordovician to the present.

saurs is partly the result of the changes in oceanic circulation at that time.

Another factor that influences climates is the relative level of the continents and the seas. At some periods of the history of the Earth, most recently in the late Mesozoic and again in the first part of the Cenozoic, large parts of the continents have been flooded by shallow seas. These **epicontinental seas** extended across the middle of North America and the middle of Eurasia in the Cretaceous and early Cenozoic. Water has a great capacity to absorb heat as environmental tempera-

tures rise and to release that heat as temperatures fall. The heat capacity of water buffers temperature change in areas of land near bodies of water. Areas with **maritime climates** do not get very hot in summer or very cold in winter, and they are usually moist because water that evaporates from the sea falls as rain on the land. **Continental climates**, which characterize areas of land far from the sea, are usually dry with cold winters and hot summers. The draining of the epicontinental seas at the end of the Cretaceous was a second factor that probably contributed to the demise of the dinosaurs by making climates in the Northern Hemisphere more continental.

On a slightly smaller scale, the upheaval of mountain chains also affects climates through a phenomenon known as the **rain shadow**. A range of mountains near the coast, such as the Sierra Nevadas in California, intercepts moist air as it moves inland and forces it upward. As the air rises it becomes cooler and the water vapor it contains condenses into drops of water and falls as rain on the windward sides of the mountains. When the air passes over the crests of the mountains it has already lost much of the moisture it contained, and the leeward side of the mountains is in a rain shadow and receives little moisture. In addition, as the air mass descends in altitude on the lee side of the mountain range, it becomes warmer. Warm air can hold more water vapor than cold air, but there is no source of water on the lee side of the mountains to replace the water that fell as rain on the windward side. As a result, the relative humidity of the air becomes progressively lower as the air streams down the lee side of the mountains. The hot, dry wind that descends from a mountain range when conditions are right is called a chinook in North America, a *foehn* in Europe, and a *simoom* in Asia. Mountain ranges are uplifted when continental plates collide, and the effects on local climates can affect animals. For example, the radiation of large grazing mammals in South America in the Cenozoic followed the formation of the Andes Mountains and the replacement of forest by grassland in their rain shadow.

In addition to changing climates, continental drift has formed and broken land connections between the continents. Isolation of different lineages on different land masses has produced dramatic examples of parallel and convergent evolution of similar types of organisms. These are best shown by the convergences of mammals in the Cenozoic, a period when the Earth's continents were more widely separated than they ever have been before or since. Sabre-tooth carnivores evolved among placental mammals in the Northern Hemisphere and among marsupials in South America, and grazing animals were represented by entirely different lineages of placental mammals in North America and South America and by kangaroos and wombats (both marsupials) in Australia. (See Chapter 21 for illustrations and additional details.)

Much of evolutionary history appears to depend on whether or not a particular lineage was in the right place at the right time. This **stochastic** (random, chance) element of evolution is assuming increasing prominence as more detailed information about the times of extinction of old groups and radiation of new groups is suggesting that competitive replacement of one group by another is not the usual mechanism of large-scale evolutionary change. The movements of continents and their effects on climates and the isolation or dispersal of animals are taking an increasingly central role in our understanding of vertebrate evolution.

Classification of Vertebrates

The diversity of vertebrates (50,000 living species and perhaps ten times that number of species now extinct) combines with the prevalence of convergent and parallel evolution to make the classification of vertebrates an extraordinarily difficult undertaking. Yet classification has long been at the heart of evolutionary biology. Initially, classification of species was seen as a way of managing the diversity of organisms, much as an office filing system manages the paperwork of the office. Each species could be placed in a pigeonhole marked with its name, and when all species were in their pigeonholes, one could comprehend the diversity of vertebrates. This approach to classification was satisfactory as long as species were regarded as static and immutable: Once you had placed a species in the filing system it was there to stay.

Acceptance of the fact of evolution in the nineteenth century made a classification system that provided an individual pigeonhole for each species inadequate—now it was necessary to express evolutionary relationships among species by incorporating **phylogenetic** (*phylo* = a tribe, *genesis* = origin) information in the system of classification.

Ideally, a classification system should not only attach a label to each species, it should also encode the phylogenetic relationship between that species and other species. Modern techniques of **systematics** (the phylogenetic classification of organisms) have moved beyond the role of filing systems and have become methods for generating testable hypotheses about evolution.

Classification and Evolution

Certainly, all of us would agree that birds and fishes are different, but how can you define how different they are? It is not just that a fish swims and a bird can fly that makes them different, for some birds do swim and some fish can fly. More telling is the presence of scales on fish and feathers on birds, for in this morphological difference we can begin at least to separate fishes from birds. However, not all fishes have scales, and birds have scales on their legs. Birds respire through their lungs, fish through their gills, each a morphologically distinct structure that serves the same function. If you include a lizard, you can say that it has scales like a fish and lungs like a bird, and although it does not have wings, it does have legs. You might suspect that a lizard's four legs are the equivalent of a bird's hind legs and wings, but are they the equivalent of a fish's paired fins? All three animals possess a dorsal hollow nerve chord, a notochord, and at some time in their lives they also have pharyngeal pouches: All of these structures are characteristic of a large group of animals called chordates, so the bird, fish, and lizard are all chordates.

How would you begin if you were asked to distinguish the three animals on an objective basis? No doubt you would start by listing characters that they share and do not share. The first two steps are straightforward: All three share the chordate characters of notochord, nerve cord, and pharyngeal pouches, so they are chordates. Also, they all have bone, so they must be vertebrates, a subdivision of chordates.

The remaining characters are more difficult because they are not all shared in common. Only the fish possesses gills and fins. Also because only the bird has wings, you could terminate your distinctions here and say that by using only gills and wings you can categorize the three species into three separable groups: fish (gills, no wings), bird (wings, no gills), and lizard (no gills or wings). However, the fact that each animal has characters in common with the others means that they are related through descent. With the information available can you determine how closely they are related? Which of the three possible combinations pictured in Figure 1–12 best describes their relatedness and the probable polarity (direction) of the evolution of those characters and the pattern of their descent? The fish and lizard share only scales; the fish and bird share only the basic vertebrate character states (notochord, nerve cord, pharyngeal pouches, and bone). The lizard and bird, however, share scales and hind legs, and if wings and forelimbs are very similar then they also share what zoologists define as tetrapod appendages. You could conclude that the lizard is more closely related to the bird than to the fish because it shares more characters in common with the bird than

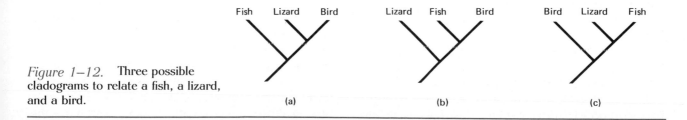

Figure 1–12. Three possible cladograms to relate a fish, a lizard, and a bird.

with the fish. The conclusion is tenuous nevertheless because wings, legs, and fins might be equated, even though they appear to be very different, if it can be shown that they are derived from the same ancestral structure. If that were true the lizard would share one unique character (scales) with the fish and the bird, all the other characters would be basic chordate or vertebrate characters and it would be impossible to decide on the basis of these characters alone whether the lizard is more closely related to the fish or to the bird. A dilemma exists, even though you know intuitively that a lizard is more like a bird than a fish and, therefore, Figure 1–12a best describes the true relationship.

More information is required to be objective about the classification. Examination of the nature of bird feathers and of lizard scales would reveal that they are embryologically similar and derived from the same tissues, but fish scales are of quite a different origin. Furthermore dissection of the lizard's and bird's appendages would expose very similar bones that support the limbs, bones that the fish's fins lack. What you grasp intuitively— that lizards and birds are closely related—would be substantiated. Systematists would say that lizards are a sister group to the birds, or vice versa. But which of these vertebrates is the most primitive (retains the most ancestral characters), and which is the most advanced (has the most derived characters)? Stated in other terms, what is the **polarity** (direction) of transformation of these characters? The answer to that question requires additional information. Certainly increasing the total number of character states used in the analysis would help, but comparing the characters of the three with an **outgroup**, an organism that is known to be primitive and distantly related, like a lamprey (one of the jawless fishes described earlier) would best establish the polarity of change. Fossil evidence would be useful in guiding your analysis but not in totally resolving the question of descent because of the incompleteness of the fossil record and its lack of information about soft tissues such as lungs and gills. An examination of the embryonic development of the organisms might be helpful because embryonic stages often reveal similarities not seen in adults. The appearance of pharyngeal gill pouches at some stage in the development of all chordates is a good example of this phenomenon.

This exercise emphasizes that an extensive knowledge of the characters is necessary to arrange different organisms into a meaningful classification, and the more characters that can be used, the better. Second, ancestral characters like the notochord cannot be used to separate organisms because they are shared by nearly all descendants. Third, if a major purpose of classifying organisms is to determine their relatedness, it follows that shared derived (advanced) characters reveal more about the recency of descent between organisms than do their primitive characters.

This example began with taxonomy, the separation or classification of organisms. It then graded into systematics, the establishment of evolutionary patterns. Taxonomy includes three processes: the arrangement of organisms into useful groupings, usually on the basis of similar character states (classification), the assignment of names to the organisms that make up a classification (nomenclature), and the arrangement of a classification into a form, often a dichotomous key, that can be used effectively to identify specimens (identification). Systematic biology grew out of the widening interest in evolution, especially over the last 50 years. It is directed at revealing both patterns and processes in evolution. [See Duellman (1985) for a further discussion.] Over the past three decades the application of modern techniques to systematic problems has produced an explosion of new information, new interpretations of evolution, and new controversies. Because the heart of systematics is evolutionary pattern, a basic goal of systematics is to establish the evolutionary histories of characters and organisms, and that goal demands that organisms be classified. The opening example showed that classifying even simple groupings of animals is not an easy undertaking. The classification systems that biologists have used have changed dramatically with the growth of systematic biology and have become more objective and less dependent on intuition. In the next section we examine systems of classification.

Linnaean System of Classification

Our system of classifying organisms traces back to methods established by the naturalists of the seventeenth and eighteenth centuries, especially those of Carl von Linné, a Swedish naturalist, better known by his Latin pen name, Carolus Linnaeus. The Linnaean system employs **binomial nomenclature** to designate species and arranges species into hierarchical categories (**taxa**) for classification.

Binomial Nomenclature   The scientific naming of species became standardized when Linnaeus's monumental work, *Systema Naturae*, was published in sections between 1735 and 1758. He attempted to give an identifying name to every known species of plant and animal. His method involves assigning a generic name, which is a Latin noun, Latinized Greek, or a Latinized vernacular word, and a second species name, usually a Latin adjective, or similar derivative. Some familiar examples include *Homo sapiens* for human beings (*homo* = human, *sapiens* = wise), *Passer domesticus* for the house sparrow (*passer* = sparrow, *domesticus* = belonging to the house), and *Canis familiaris* for the domestic dog (*canis* = dog, *familiaris* = of the family).

Why use Latin words? Aside from the historical fact that Latin was the early universal language of European scholars and scientists, it has provided a uniform usage that has continued to be recognized worldwide by scientists regardless of their vernacular language. The same species may have different colloquial names in the same language. For example, *Felis concolor* ("the uniformly colored cat") is known in various parts of North America as cougar, puma, mountain lion, American panther, painter, and catamount. It has other common names in other languages, but mammalogists of all nationalities recognize the name *Felis concolor* as referring to a specific kind of cat.

Hierarchical Groups: The Higher Taxa   Linnaeus and other naturalists of his time developed what they called a natural system of classification in which all similar species are grouped together in one **genus** (plural **genera**), based on shared characters that define the genus. The concept of characters being either primitive or derived was not part of the Linnean system. Thus a genus could be based on shared primitive characters. The most commonly used characters were anatomical, because they are the ones most easily preserved in specimen form. Thus all dog-like species—various wolves, coyotes, and jackals—were grouped together in the genus *Canis* because they all share certain anatomical features, such as an erectile mane on the neck and a skull with a long, prominent sagittal crest on which massive temporal muscles originate for closure of the jaws. Linnaeus's method of grouping species in classification proved to be functional because it was based on anatomical (and to some extent on physiological and behavioral) similarities and differences. Unknowingly, he used taxonomic characters that we understand today are genetically determined biological traits that, within limits, express the degree of genetic similarity or difference among groups of organisms. However, Linnaeus lived before there was any knowledge of genetics and the mechanisms of inheritance, and his interpretation of why his system works was completely wrong.

Subsequent development of biological classification has employed seven basic taxonomic categories, listed below in decreasing order of inclusiveness:

Kingdom
   Phylum (= Division in plants)
      Class
         Order
            Family
               Genus
                  Species

These are the taxonomic levels (categories) that are employed in most of the primary and secondary literature of vertebrate biology.

Phylogenetic and Evolutionary Interpretation of Classification

A fundamental change in the principles of classifying organisms came with the acceptance of evolutionary theory in the late nineteenth century. Biologists came to realize that species are not separate, unchanging creations, but rather that they have evolved through natural selection acting on the inherited variability of individuals within a species. The relationships among species were no longer considered to be mere abstractions designated as genera and other higher taxa. Instead, scientists began to see that species are related in the sense that they have descended, generation by generation, from common ancestors.

Thus, similar species in a genus share common, genetically determined patterns of form and function because they are all derived from a single, common ancestral species. In the same way, all the genera in a family share certain common, familial characters because they are derived from the same ancestral stock. The same principle applies to increasingly remote ancestry right through the higher taxa of classification. Although the characters shared by members of a taxon are both primitive *and* derived, only the latter are useful in establishing relationships. Shared primitive characters may be found in other taxa simply because they have been retained from ancient ancestors and thus provide no information about relatedness. Today, vertebrates are grouped into higher taxa according to their phylogeny—how closely they are related in their ancestry.

Determining Phylogenetic Relationships

In order to apply the phylogenetic principle to classification, the derived characters that systematists use to group species into higher taxa must be inherited from a common ancestry, that is, they must be **homologous** similarities. Unfortunately, the word "homologous" has at least three distinct meanings in evolutionary biology, only one of which is the same as its use in everyday speech.

(1) In the narrowest sense of the word, homologous means descended from the same derived structure. (2) In a slightly broader sense, homologous can mean descended from the same primitive structure. (3) In the usage employed by molecular geneticists (most of whom are not familiar with evolutionary biology) homologous means similar without any implications of evolutionary descent. The third definition is not widely used in evolutionary biology, although it does appear in cases where DNA sequences are used to infer evolutionary relationships. A reader must be careful to determine which of the possible meanings of homology the author intended (Gould 1988).

An example can illustrate the difference between the first and second definitions: The tetrapod forelimb consists of one bone in the upper limb (the humerus) and two bones (the radius and ulna) in the lower arm. This is the primitive condition for tetrapods and all tetrapod forelimbs have this structure. By definition 2, all tetrapod forelimbs are homologous structures. That includes forelimbs as different as those of humans, horses, and birds. Birds have extensively modified the primitive tetrapod forelimb and hand to form wings. Wings are a derived character of birds, and no other vertebrates have exactly the same modifications of the forelimb bones. Different kinds of birds have wings of different shapes—short and broad, long and narrow, and so on. By the most restrictive definition of homology (definition 1), the wings of one species of bird are homologous to the wings of another species of bird because both are descended from the wing (derived forelimb) of an ancestral bird. By this restricted definition, however, the wings of birds are *not* homologous to the forelimbs of other tetrapods because the next step back takes you to the primitive tetrapod limb.

Structures that perform the same function without being homologous are called **homoplastic** (*plast* = form, shape). The application of this term, too, depends on the fineness of detail that is being considered. The wings of birds and bats are both forelimbs specialized for flight (Figure 1–13). By definition 1 they are not homologous because they

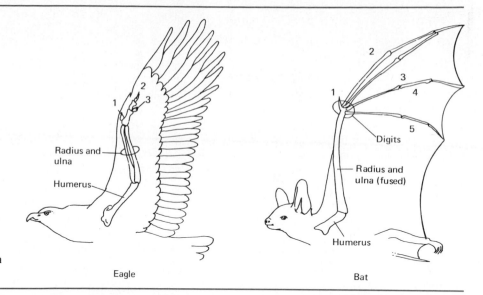

Figure 1–13. Wing bones of a bird and a bat.

Eagle

Bat

Radius and ulna

Humerus

Digits

Radius and ulna (fused)

Humerus

are not derived from a common ancestor of birds and bats that had a forelimb specialized for flight; therefore, the wings of birds and bats would be called homoplastic. By definition 2, the wings of birds and bats would be considered homologous because they are derived from the primitive tetrapod limb.

Some structures that are derived from the same primitive character state have become so modified during the course of evolution that they are no longer very similar in form or function in the various descendants. This evolutionary process is called **transformation**, and transformed structures can be extremely difficult to identify unless there is a good fossil record of their ancestries or the details of their embryonic development can be worked out. The sound-transmitting apparatus in the ears of vertebrates serves to illustrate the kind of problems involved in detecting these transformations (Figure 1–14).

Hearing is an important sense for vertebrates, but different problems in sound reception exist on land and in water. Sound waves in air are relatively weak and ordinarily have little direct effect in producing vibrations in the fluid (endolymph) of the inner ear. Some sort of sound-amplifying system is necessary for vertebrates living on land. Some tetrapods solved this problem by incorporating the hyomandibular bone into a new bony structure called the **columella** or **stapes** that transmitted sound waves from the external tympanic membrane covering the first pharyngeal cleft to the oval window of the inner ear, where the amplified vibrations were transmitted to the endolymph. Earlier in evolution, the hyomandibular element had served mainly as a support for the jaws of fishes, but it happened to be positioned between the first gill cleft and the otic capsule in an appropriate place to transmit sound waves. This transformation is clearly evident from comparative embryological studies. Although many modern amphibians depart from this generalized method of sound amplification and transmission from the external to the inner ear regions, birds, crocodilians, lizards, and turtles retain its essential feature: a single bony element, the columella.

Mammals are different. They have three middle ear ossicles involved in sound transmission. The stapes is the innermost bone, with its expanded foot against the oval window, as in other tetrapods. In addition, there is the **incus**, a middle element, and the **malleus**, with its expanded sur-

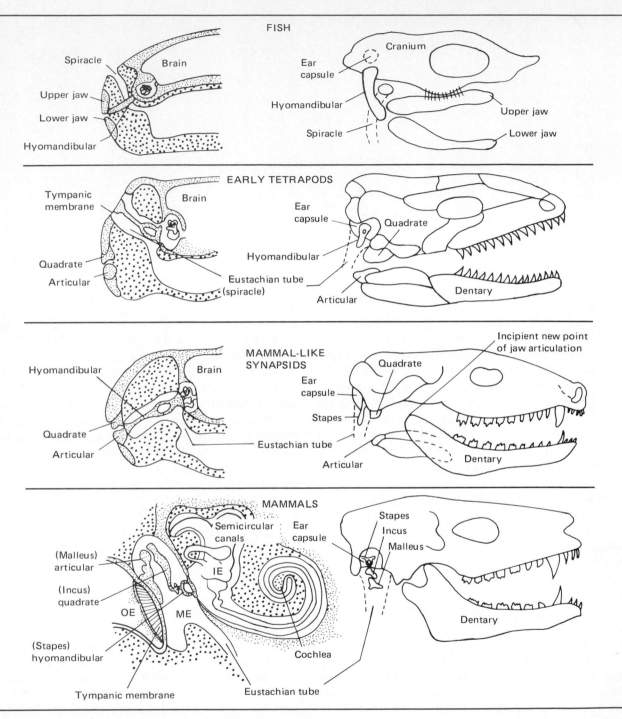

Figure 1–14. Evolution of the sound-transmitting structures in the ears of vertebrates. OE, outer ear; ME, middle ear; IE, inner ear.

face against the eardrum. What are the evolutionary origins of the malleus and incus? They are derived from ancestral tetrapod jaw bones. During the evolution of mammals a new method of articulating the jaws gradually developed between the dentary bone of the lower jaw and the squamosal bone of the upper jaw. The articular bone of the lower jaw became transformed into the mammalian malleus and the quadrate bone of the skull was transformed into the incus. Recognition that the malleus is homologous (definition 2) with the articular and the incus with the quadrate came from embryological studies that showed that the early developmental stages are the same. This is a good example of the use of ontogenetic comparisons to decipher the direction of change in character states during evolution. Later paleontological discoveries of fossil mammal-like synapsids revealed some of the intermediate evolutionary stages of these transformations (Chapter 19).

Homoplastic similarities include **convergence** and **parallelism**. It is well known that distantly related species with similar modes of life may have superficially very similar morphological structures and body shapes as the result of the same selective forces acting on different genotypes (convergence). Examples of convergence among vertebrates include the similarity in body shape of a pelagic fish (shark), an aquatic lepidosauromorph (ichthyosaur), and an aquatic mammal (porpoise). The fusiform body shapes of these animals have been derived independently.

Traditional and Cladistic Classifications

All methods of classifying organisms are based on similarities among the included species, whether or not the similarities reflect common ancestry. **Phenetic** methods, such as classical numerical taxonomy, do not explicitly address evolutionary relationships. Pheneticists group organisms strictly by the number of characters they have in common. Characters are not weighted to reflect their relative importance, and as a result the presence of two

arches in the skulls of lizards and birds would be treated equally to the presence of the oxygen-transport pigment hemoglobin in both groups. Phenetic classification is ideal for some types of data, such as biochemical information about the similarities and differences in the proteins or DNA sequences of species. With these sorts of data there is no basis for assuming that one difference is more important than another and assigning greater weight to it.

Morphological, behavioral, and physiological characteristics of organisms provide a different sort of information: it is possible to make a judgment that some differences are more significant than others. For example, only a few kinds of vertebrates have skulls with a pattern of two arches, but nearly all vertebrates have hemoglobin. Consequently, knowing that the species in question share the two-arched skull pattern tells you more about the closeness of their relationship than knowing that they have hemoglobin, and you would give more weight to the skull characters than to the hemoglobin. Today, two major evolutionary phylogenetic systems are in use for these sorts of data—the older systematics which we will refer to as **traditional** or **evolutionary systematics** following Charig's (1982) terminology, and the newer **phylogenetic systematics**, also known as **cladistics**. The goal of each of these approaches to systematics is to establish the pattern of evolutionary descent of organisms, but the methods differ.

We have adopted a cladistic classification of vertebrates for this edition of *Vertebrate Life*. This change has resulted in a substantial regrouping of taxa recognized in earlier editions of this book, and the same changes are occurring in other books and in the journals in which the primary literature of vertebrate biology is published. Familiar taxonomic groups such as cyclostomes and reptiles are not valid by the methods of cladistics, and the names no longer have any technical meaning, although they continue to be used colloquially. Reptiles, for example, are what evolutionary systematists call a **structural grade**—different organisms with structural and functional characters that bind them together despite the fact that they represent different evolutionary lineages. Cladists define these familiar groupings of vertebrates as **paraphyletic groups** that are invalid in cladistic classifications. To help clarify what is meant by cladistics and to emphasize how it differs from evolutionary systematics, we provide a simple example in this section. Readers who desire greater detail can refer to *Phylogenetic Patterns and the Evolutionary Process* by N. Eldredge and J. Cracraft (1980), *Phylogenetics: the Theory and Practice of Phylogenetic Systematics* by E. O. Wiley (1981), and *Systematics and Biogeography: Cladistics and Vicariance* by G. Nelson and N. Platnick (1981).

Classification and the Phylogeny of Vertebrates

A classification of the vertebrates by evolutionary systematics (Figure 1–15) and a classification based on cladistic systematics (Figure 1–16) are only vaguely similar, although each is based on similar analyses. If both evolutionary systematists and cladists establish their hypothesized phylogenies using similar methods, how and why do they differ?

The two systematic schemes differ because of the criteria used to define the taxa that make up the cladogram, and that difference is fundamental. Cladists establish and name a taxonomic group within a cladogram solely on the basis of monophyly (single origin). For example, every vertebrate depicted in Figure 1–16 that is ascendent from a chosen branch point (i.e., a clade or lineage of ascent) along the cladogram is related through the shared derived characters (**synapomorphies**) that define that branch point. The name assigned to this branch point and the particular set of shared derived characters defines all members in the clade. Evolutionary systematists are happy to accept monophyletic taxa, but they also are willing to include groups that are not monophyletic within a named taxon, or exclude a group from a monophyletic lineage and assign it to another taxon.

An example emphasizes the differences between the two schemes. The cladogram depicted in Figure 1–16 is our cladistic construction (that is, our hypothesis) of the evolutionary relationships of the major living groupings of vertebrates. There are 13 dichotomous branches leading from the origin of the vertebrates from other chordates to the mammals. A cladist ranks the organisms by sequencing the cladogram from primitive to advanced dichotomies, and assigning names for the shared derived characters that define each branch point. Thus, the name Gnathostomata represents all vertebrate animals that have jaws; that is, every taxon to the right of the number 3 in Figure 1–16 is included in the Gnathostomata, every taxon to the right of number 4 is included in the Teleostomii, and so on.

This sequence of taxa within taxa is different from the traditional classification by evolutionary systematics explained earlier and in Box 1–1. Only eight taxonomic names are usually assigned to the living major groups of vertebrates. Furthermore, the rankings assigned by evolutionary systematists do not represent an ascendent sequence of the taxa, but rather cut across the monophyletic categories (Table 1–1). Thus the Diapsida, which includes the Archosauromorpha (birds, crocodilians) and the Lepidosauromorpha (the tuatara, snakes, and lizards), is a monophyletic and inclusive group and all members of it share the derived condition of having a specific arrangement of fenestrae (openings) in the skull. When the birds, crocodiles, and squamates are further separated (through additional comparisons of derived characters), cladistic convention requires that the new groupings (Archosauromorpha and Lepidosauromorpha) must be assigned lower taxonomic ranks. They cannot be elevated to an equal or higher taxonomic rank than Diapsida because they share derived character states with all other diapsids.

Evolutionary systematists, in contrast, often elevate taxonomic rankings above their position in the phylogeny. Ranking birds as the class Aves, and grouping crocodilians, squamates, and turtles in the class Reptilia denies the exclusive significance of shared derived character states. Evolutionary systematists elevate the birds to a taxonomic ranking equal to the other groups of

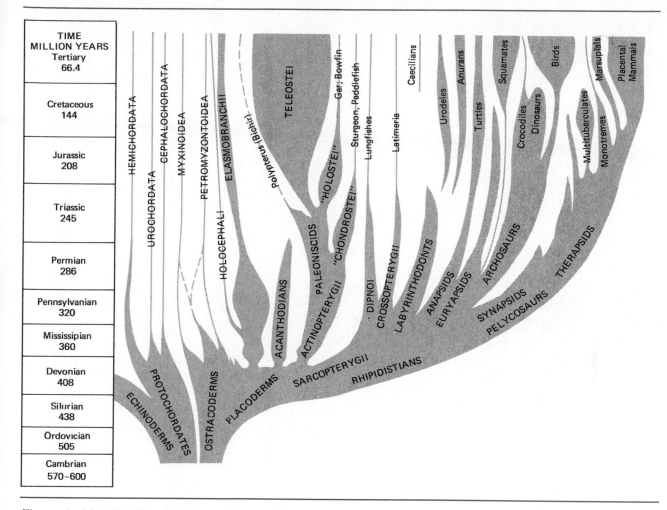

TIME MILLION YEARS	
Tertiary	66.4
Cretaceous	144
Jurassic	208
Triassic	245
Permian	286
Pennsylvanian	320
Mississipian	360
Devonian	408
Silurian	438
Ordovician	505
Cambrian	570–600

Figure 1–15. Traditional phylogenetic tree of the vertebrates. Width of branches indicates the relative number of recognized genera for a given time level on the vertical axis (time in millions of years indicates beginning of geological periods).

vertebrates because they attribute great significance to the presence in birds of feathers and endothermy, and they group the diapsid crocodilians and squamates with the turtles, which are anapsids and morphologically distinct, because they share many ancestral characters. This method of classification, which assigns equal taxonomic status to paraphyletic groups, is useful if one desires to emphasize structural grades, or groupings of organisms that share broad structural and functional patterns of organization, but it may confuse

the actual branching patterns with the extent of divergence. The same comparison can be made for the old class Agnatha, in which evolutionary systematists include the living hagfishes and lampreys. In reality, however, the major feature that hagfishes and lampreys have in common is the lack of jaws. Lampreys otherwise are more like advanced fishes, whereas hagfishes retain a suite of ancestral characters, conditions that are immediately acknowledged in the cladistic classification. There are proponents of both views (see Charig

Phylogenetic Relationships of Living Vertebrates

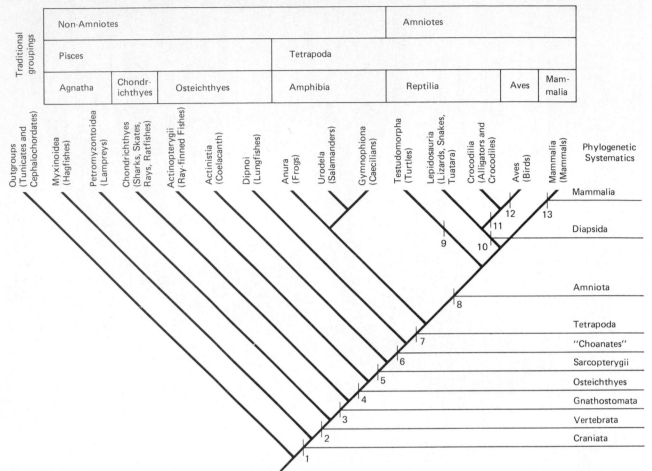

1. Distinct head region skeleton incorporating anterior end of notochord, one or more semicircular canals, brain consisting of three regions, paired kidneys, gill bars, neural crest tissue.
2. Arcualia or their derivatives form vertebrae, two or three semicircular canals.   3. Jaws formed from mandibular arch, teeth containing dentine, three semicircular canals, branchial skeleton internal to gill membranes, branchial arches contain four elements on each side plus one unpaired ventral median element, paired fins with internal skeleton and muscles supported by girdles in the body wall.   4. Presence of lung or swim bladder derived from the gut, unique pattern of dermal bones of shoulder, unique characters of jaw and branchial muscles.   5. Unique supporting skeleton in fins.   6. Choanae* present, advanced paired appendage structure, conus arteriosus of heart partly divided, unique dermal bone pattern of braincase, loss of interhyal bone.   7. Distinctive paired pectoral and pelvic limbs characterized by a single bone in the proximal portion of the limb and two bones in the distal portion.   8. A distinctive arrangement of extraembryonic membranes (the amnion, chorion, and allantois).   9. Dermal bones form a shell enclosing the trunk and fused with part of the axial skeleton.   10. Skull with a dorsal temporal fenestra, upper temporal arch formed by triradiate postorbital and triradiate squamosal bones.   11. Presence of a fenestra anterior to the orbit of the eye, orbit shaped like an inverted triangle.   12. Feathers, loss of teeth, metabolic heat production used to regulate body temperature (endothermy).   13. Hair, lower jaw formed only by dentary bone, mammary glands, independent development of endothermy.

*The homology of the choanae of dipnoans with those of tetrapods is questioned by many authorities, thus the name "Choanates" may not be appropriate but the grouping appears to be valid.

Figure 1-16. This diagram shows the probable relationships among the major groups of vertebrates. The boxes across the top of the diagram show the traditional groupings of the taxa, and the names along the right side show how the lineages are grouped by phylogenetic systematics. Note that these groupings are nested progressively; that is, all sarcopterygians are osteichthyians, all osteichthyians are gnathostomes, all gnathostomes are vertebrates, all vertebrates are craniates, and so on. The numbers indicate derived characters that distinguish the lineages.

1982), but the older view is giving way to a purely cladistic classification.

An Example of Cladistic Classification

The shared derived characters of vertebrates (see Chapter 2) indicate that the Vertebrata is a monophyletic taxon, and all lineages of vertebrates should be traceable to a common ancestor, probably in the Cambrian. There are about 50,000 living vertebrates, and the process of extinction and the vagaries of fossilization have left huge gaps in the records of vertebrate lineages. Given this diversity and the spotty nature of the fossil record, it is understandable that systematists often disagree about the relationships among vertebrates. Against such odds it is unlikely that the true phylogeny of the vertebrates will ever be known. In our view, however, the closest approach to a true phylogeny that can be hoped for is based on the methods of cladistics.

Cladistics simply means branching patterns. A cladistic classification to represent the phylogeny of vertebrates, therefore, arranges animals on the basis of their historical divergences from a common ancestral species. Because the transitional fossils that would normally identify a common ancestral species are usually missing, cladistic classification is based on comparisons of character states of the animals that are available for study. Simply stated, animals with similar derived characters are considered more closely related than animals that do not share the characters. The results of such an analysis should produce a cladogram that approximates the phylogeny of the animals considered, but formidable problems arise in practice. Because of processes such as convergent evolution, loss or reversal of characters, and parallelism, similarities and differences can easily be misinterpreted. The greatest problem in creating groupings is the difficulty of determining which character states are primitive and which are derived.

In 1966, Willi Hennig summarized his philosophy of classification by forcefully emphasizing that a phylogeny can be reconstructed only on the basis of shared derived characters (**synapomorphies**, *syn* or *sym* = together, *apo* = away from, i.e., derived from, *morph* = form). The presence of shared ancestral characters, what Hennig called **symplesiomorphies** (*plesi* = near), is useful only in that it helps to highlight what characters are synapomorphies. Symplesiomorphies tell us nothing about degrees of relatedness. It is the rigorous application of this criterion, what might be called Hennig's rule, that particularly characterizes cladistics.

The jargon associated with cladistics tends to cloud its importance. To clarify the meaning of the major terms, let us consider the examples presented in Figure 1–17. Any of the three dendrograms represents an equally possible phylogeny for the three taxa identified as 1, 2, and 3. The open bars connecting As or Bs represent shared ancestral characters, or symplesiomorphies. Notice that the symplesiomorphies do not distinguish which two of the three taxa are most closely related. The three taxa can be separated, however, on the basis of the two derived states a and b. Neither of the derived characters a or b is shared

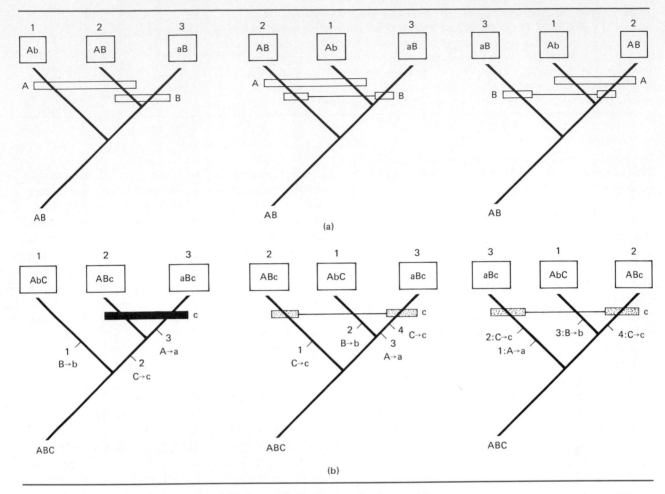

Figure 1—17. Cladograms involving the distribution of two (above) and three (below) character states in three taxa (1, 2, and 3). The three ancestral states are represented by the capital letters A, B, and C; the derived conditions by a, b, and c. Open bars connect shared ancestral characters (symplesiomorphies) darkened bars connect shared derived characters (synapomorphies) and stippled bars represent independent origins of the same derived character state. The numbers in (b) identify changes from the ancestral character state to the derived condition.

by a subset of the three taxa (e.g., by 1 and 2, or by 1 and 3, or by 2 and 3); they cannot be used to determine common ancestry. Both a and b, however, uniquely define taxon 3 and taxon 1, respectively. The representation of a singular derived character in a taxon is termed **autapomorphy** (*aut* = self, unique).

In Figure 1–17b a third character state has been added to the analysis; c indicates a derived character state. Notice that the ancestral character state C occurs only in taxon 1 and the derived condition c in both taxa 2 and 3. The most *parsimonious* phylogeny (the phylogeny that requires the fewest number of evolutionary changes) is represented by the left dendrogram in Figure 1–17b—it requires only three changes to produce the derived character states a, b, and c from the ancestral character states of A, B, and C. The other two phylogenies are possible, but they would require that the derived condition c originated independently in taxon 2 and taxon 3. In the left dendrogram, taxon 3 represents the most derived condition because it is autapomorphous in character states a and c and, as a result, shares only the ancestral character state B with taxon 2. Taxon 1 is called the primitive **sister group** (closest relative) of taxa 2+3 and taxon 2 is the primitive sister group of taxon 3.

Cladistic methodology rigidly excludes the possibility that taxon 1 and taxon 3 or taxon 1 and taxon 2 are more closely related than 2 and 3, because they lack synapomorphies. However, an evolutionary systematist might lump taxa 1 and 2 together to emphasize their primitiveness compared to taxon 3. This grouping would require elevating taxon 3 to a rank higher than the cladogram predicts. A case in point is the lumping of the living crocodiles, lizards, and snakes, which are all diapsids, into the **paraphyletic** (*para* = beside or beyond, *phyl* = tribe) class Reptilia, and the birds, which are also diapsids, into a separate but equally ranked class Aves. To compound the confusion, the turtles, which are anapsids and not diapsids, are included within the class Reptilia. Such artificial taxonomic assemblages are defended on the grounds that the shared ancestral characters define the group best (as a grade) and that the highly derived archosaurian diapsids, the birds, should

be recognized with an equal rank. Although there are good reasons to compare broad similarities in distantly related vertebrates, comparisons should not sacrifice the phylogenetic integrity of a classification. A phylogeny can be accurately described only by the inclusion of all members of a **monophyletic** (*mono* = one, single) group. The exclusion of any member of the group or the inclusion of any nonmember diminishes the information contained in the phylogeny.

Although there are many reasons for accepting a cladistic classification, it is most appropriate because it can remove subjective guesses about relationships from classification and replaces them with an objective and rigidly defined set of rules. To be sure, even cladistic classifications change with time as shared derived characters are better defined, and as the significance of different characters is reevaluated.

Conclusion

In this chapter we have introduced the major groups of vertebrates that we will be considering in detail through the rest of the book. The process of evolution is responsible for the diversity of vertebrates, and phylogenetic systematics provides a method of interpreting evolutionary history. The evolution of vertebrates has taken place in a changing world, and we will see that drifting continents have influenced many parts of the vertebrate story.

Summary

The 50,000 species of living vertebrates span a size range from less than a gram to more than 100,000 kilograms and live in habitats from the bottom of the sea to the tops of mountains. This extraordinary diversity is the product of 500 million years of evolution. Evolution means a change in the relative frequencies of alleles in the gene pool of a species. Heritable variation among the individuals of a species is the raw material of evolution, and natural selection is the mechanism that produces evolutionary change. Natural selection works by

differential reproduction, and fitness describes the relative contribution of different individuals to future generations. Most selection probably operates at the level of individuals, but it is possible that selection also operates at the level of alleles, populations, and even species.

In addition to individual variation, species often exhibit sexual dimorphism and geographic variation. Sexual dimorphism reflects the different selective forces that act on males and females of a species as a result of the asymmetry of reproductive investment by the two species. In most cases males maximize their fitness by mating with as many females as possible, whereas females should attempt to choose the best possible mate. Geographic variation results from varying environmental conditions in different parts of the geographic range of a species. When a local population of a species is isolated from the rest of the species, genetic differences accumulate. If genetic differences become extensive, individuals of the isolated population may be unable to breed with individuals of the main population when they reestablish contact. This process is called allopatric speciation.

The Earth has changed dramatically during the half billion years of vertebrate history. Continents were fragmented when vertebrates first appeared; coalesced into one enormous continent, Pangaea, about 300 million years ago; and began to fragment again about 100 million years ago. This pattern of fragmentation–coalescence–fragmentation has resulted in isolation and re-contact of major groups of vertebrates on a worldwide basis. On a continental scale the advance and retreat of glaciers in the Pleistocene caused homogeneous habitats to split and merge repeatedly, isolating populations of widespread species and leading to the evolution of new species.

Phylogenetic systematics, usually called cladistics, classifies animals on the basis of shared derived character states. Natural evolutionary groups can be defined only on the basis of these derived characters; retention of ancestral characters does not provide information about evolutionary lineages. Application of this principle produces groupings of animals that reflect evolutionary history as accurately as we can discern it and forms a basis for making hypotheses about evolution.

References

Avise, J. C. and M. H. Smith. 1977. Gene frequency comparisons between sunfish (Centrarchidae) populations at various stages of evolutionary divergence. *Systematic Zoology* 26:319–335.

Borror, D. J. 1960. *Dictionary of Word Roots and Combining Forms.* Mayfield Publishing Co., Mountain View, Calif.

Breden, F. and G. Stoner. 1987. Male predation risk determining female preference in the Trinidad guppy. *Nature* 329:831–833.

Britten, R. J. 1986. Rates of DNA sequence evolution differ between taxonomic groups. *Science* 231:1393–1398.

Charig, A. 1982. Systematics in biology: A fundamental comparison of some major schools of thought. Pages 363–440, in *Problems of Phylogenetic Reconstruction*, edited by K. A. Joysey and A. E. Friday. Academic Press, New York.

Connor, E. F. 1986. The role of Pleistocene forest refugia in the evolution and biogeography of tropical biotas. *Trends in Ecology and Evolution* 1:164–168.

Dobzhansky, T. 1937. *Genetics and the Origin of Species.* Columbia University Press, New York.

Dobzhansky, T. 1970. *Genetics of the Evolutionary Process.* Columbia University Press, New York.

Duellman, W. E. 1985. Systematic zoology: slicing the Gordian knot with Ockham's razor. *American Zoologist* 25:751–762.

Eberhard, W. G. 1985. *Sexual Selection and Animal Genitalia.* Harvard University Press, Cambridge.

Echelle, A. C. and I. Kornfield (eds). 1984. *Evolution of Fish Species Flocks.* University of Maine Press, Orono.

Eldredge, N. and J. Cracraft. 1980. *Phylogenetic Patterns and the Evolutionary Process.* Columbia University Press, New York.

Eldredge, N. and S. J. Gould. 1972. Punctuated equilibria: an alternative to phyletic gradualism. Pages 82–115 in *Models in Paleontology*, edited by T. J. M. Schopf. Freeman, Cooper, San Francisco.

Endler, J. A. 1986. *Natural Selection in the Wild.* Monographs in Population Biology, No. 21. Princeton University Press, Princeton, N. J.

Fisher, R. A. 1930. *The Genetical Theory of Natural Selection.* Clarendon Press, Oxford.

Friedman, M. J. 1978. Erythrocyte mechanism of sickle cell resistance to malaria. *Proceedings of the National Academy of Sciences (USA)* 75:1994–1997.

Fryer, G. and T. D. Iles. 1972. *The Cichlid Fishes of the Great Lakes of Africa: Their Biology and Evolution.* Oliver & Boyd, Edinburgh.

Gibbs, H. L. and P. R. Grant. 1987. Oscillating selection on Darwin's finches. *Nature* 327:511–513.

Gould, S. J. 1988. The heart of terminology. *Natural History* 97(2):24–31.

Gould, S. J. and N. Eldredge. 1977. Punctuated equilibrium: the tempo and mode of evolution reconsidered. *Paleobiology* 3:115–151.

Grant, P. R. 1986. *Ecology and Evolution of Darwin's Finches*. Princeton University Press, Princeton, N.J.

Greenwood, P. H. 1984. What is a species flock? Pages 13–25 in *Evolution of Fish Species Flocks*, edited by A. A. Echelle and I. Kornfield. University of Maine Press, Orono, Maine.

Haldane, J. B. S. 1932. *The Causes of Evolution*. Longmans, Green, New York.

Hennig, W. 1966. *Phylogenetic Systematics*. University of Illinois Press, Urbana, Ill.

Humphries, J. M. 1984. Genetics of speciation in pupfishes from Laguna Chichancanab, Mexico. Pages 129–139 in *Evolution of Fish Species Flocks*, edited by A. A. Echelle and I. Kornfield. University of Maine Press, Orono, Maine.

Huxley, J. S. 1942. *Evolution, the Modern Synthesis*. Allen & Unwin, London.

Jablonski, D. 1986. Background and mass extinctions: the alternation of evolutionary regimes. *Science* 231:129–133.

Lack, D. 1947. *Darwin's Finches*. Cambridge University Press, Cambridge.

Mayr, E. 1942. *Systematics and the Origin of Species*. Columbia University Press, New York.

Mayr, E. 1963. *Animal Species and Evolution*. Belknap Press, Cambridge, Mass.

Mayr, E. 1976. *Evolution and the Diversity of Life*. Harvard University Press, Cambridge, Mass.

Mayr, E. 1984. Evolution of fish species flocks: a commentary. Pages 3–11 in *Evolution of Fish Species Flocks*, edited by A. A. Echelle and I. Kornfield. University of Maine Press, Orono, Maine.

McCune, A. R., K. S. Thomson and P.E. Olsen. 1984. Semionotid fishes from the Mesozoic Great Lakes of North America. Pages 27–44 in *Evolution of Fish Species Flocks*, edited by A. A. Echelle and I. Kornfield. University of Maine Press, Orono, Maine.

Nelson, G. and N. Platnick. 1981. *Systematics and Biogeography: Cladistics and Vicariance*. Columbia University Press, New York.

Räsänen, M. E., J. S. Salo, and R. J. Kalliola. 1987. Fluvial perturbance in the western Amazon Basin: regulation by long-term sub-Andean tectonics. *Science* 238:1398–1401.

Ribbink, A. J. 1984. Is the species flock concept tenable? Pages 21–25 in *Evolution of Fish Species Flocks*, edited by A. A. Echelle and I. Kornfield. University of Maine Press, Orono, Maine.

Sage, R. D., P. V. Loiselle, P. Basasibwaki, and A. C. Wilson. 1984. Molecular versus morphological change among cichlid fishes of Lake Victoria. Pages 185–197 in *Evolution of Fish Species Flocks*, edited by A. A. Echelle and I. Kornfield. University of Maine Press, Orono, Maine.

Simpson, G. G. 1944. *Tempo and Mode in Evolution*. Columbia University Press, New York.

Simpson, G. G. 1953. *The Major Features of Evolution*. Columbia University Press, New York.

Slatkin, M. 1987. Gene flow and the geographic structure of natural populations. *Science* 236:787–792.

Stanley, S. M. 1975. A theory of evolution above the species level. *Proceedings of the National Academy of Sciences (USA)* 72:646–650.

Stebbins, G. L. 1950. *Variation and Evolution in Plants*. Columbia University Press, New York.

Vawter, L. and W. M. Brown. 1986. Nuclear and mitochondrial DNA comparisons reveal extreme rate variation in the molecular clock. *Science* 234:194–196.

Vrba, E. S. 1980. Evolution, species and fossils: how does life evolve? *South African Journal of Science* 76:61–84.

Wiley, E. O. 1981. *Phylogenetics: the Theory and Practice of Phylogenetic Systematics*. Wiley, New York.

Zuckerkandl, E. and L. Pauling. 1962. Molecular disease, evolution and genetic heterogeneity. Pages 189–225 in *Horizons in Biochemistry*, edited by M. Kasha and B. Pullman. Academic Press, New York.

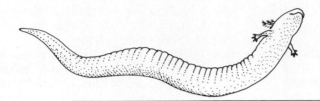

The principles of evolution and natural selection that we described in the preceding chapter apply to all forms of life, not just to vertebrates. What are the derived characters that make vertebrates unique? In this chapter we explain what characteristics are necessary to define a vertebrate, and describe the systems that make a vertebrate self-sustaining. We also address the question of the origin of vertebrates: What does a consideration of the derived characters of vertebrates tell us about the phylogenetic relationships of vertebrates? If we could identify the sister group of vertebrates, would that information help us to explain the process by which vertebrates arose from invertebrate ancestors? We compare the most widely accepted view of the origin of vertebrates with several alternatives that may provide different insights about the events and mechanisms that were responsible for their evolution.

The Origin of Vertebrates

2

Some Familiar Facts About Vertebrates

Vertebrates derive their name from the serially arranged vertebrae, the axial endoskeleton that vertebrates share as a common diagnostic character (Figure 2–1). Anteriorly skeletal elements have been elaborated into a cranium or skull, which houses various sense organs and a complex brain. An older name for the vertebrates is Craniata. In fact, the distinctive vertebrate cranium and tripartite brain may have evolved before the vertebral column and they are, perhaps, more characteristic of vertebrates than is the backbone.

Vertebrates also share some fundamental morphological features with certain marine invertebrates and with them are classified in the phylum Chordata. The common chordate structures are the **notochord, dorsal hollow nerve cord, pharyngeal slits,** and a **muscular postanal tail**. Only the nerve cord remains as a functional entity in the adult stage of many vertebrates, but all four chordate features are evident at some stage in the development of all vertebrates. This is generally true also of the invertebrate-like urochordates (tunicates or ascidians) and cephalochordates (acraniates or lancelets).

We now have the minimum information needed to define a **vertebrate**. A vertebrate is a chordate animal that has a cartilaginous or bony endoskeleton. The axial components of this endoskeleton consist of a cranium housing a brain divided into three basic parts and a vertebral column through which the nerve cord passes. No other animals possess this constellation of fundamental characters.

The Significance of Similarity and Differences

Each group of vertebrates differs from all others in some fundamental way, but all share the common chordate–vertebrate characters. What is the meaning of the underlying similarity? Since Darwin's *Origin of Species* we have understood that the sharing of fundamental similarities, or **homologies**, among widely different groups of species indicates that they evolved from a common ancestor that possessed the same features. In general, the more homologies two species share, the more closely related they are.

In modern cladistic analysis a relationship between different taxonomic groups can be established only by using shared, derived characters (homologous character states) as the basis of a classification. Primitive character states, in contrast, are features that are shared by all members of a taxonomic grouping. The notochord, for example, is a primitive character state for vertebrates because all chordates share the trait. The notochord is a shared derived character that separates chordates from other deuterostomes, such as the echi-

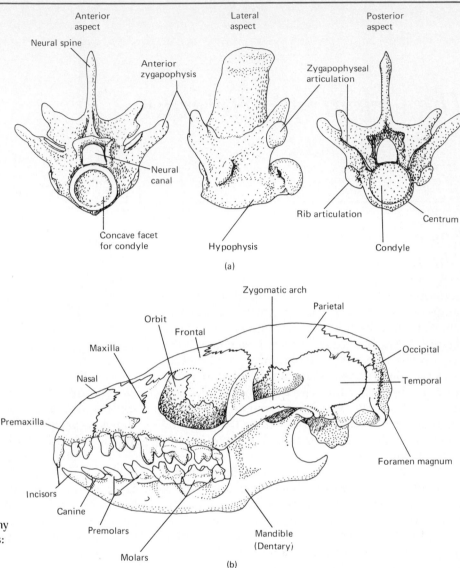

Anterior aspect

Neural spine

Anterior zygapophysis

Lateral aspect

Posterior aspect

Zygapophyseal articulation

Neural canal

Concave facet for condyle

Hypophysis

Rib articulation

Centrum

Condyle

(a)

Zygomatic arch

Parietal

Orbit

Frontal

Maxilla

Nasal

Occipital

Temporal

Premaxilla

Foramen magnum

Incisors

Canine

Premolars

Mandible (Dentary)

Molars

(b)

Figure 2–1. Examples of bony structures typical of vertebrates: (a) vertebra; (b) skull.

noderms, which do not possess a notochord. In considering evolutionary relationships between organisms, the word *primitive* implies an *ancestral* condition, not a condition of inferior quality.

What is the meaning of the diversity within a lineage of related groups and species? The differences relate to adaptation to different environmen-tal conditions or opportunities. Each species has an ecological **niche** that is different from all others and that is often expressed by specialized body form and function. The diversity of species is in-dicative of the genetic responsiveness of each group to environmental differences.

Evolution and adaptation are major themes of

this book. We shall therefore direct attention to the following sorts of questions. What were the historical, ancestral precursors of any structure or behavior under consideration? How does the structure or behavior promote survival and reproduction of the organism in its natural environment? Was the original function of the structure or behavior the same as its current function?

Adaptations can be tricky subjects to write about, because an adaptation, presumably, has some useful function in the life of the organism possessing it. Because humans are purposive animals, perceive the means to ends, and anticipate results prior to their achievement, some philosophers and scientists of an earlier era ascribed a divine purpose as the cause for useful adaptations and for evolution (Singer 1959). This philosophy is called **teleology**.

When biologists discuss adaptations they refer to alterations in structure or function that result from natural selection operating on the genetic variability of organisms. These alterations confer improved fitness for survival and reproduction on the altered individuals. Adaptation emerges, therefore, without the existence of a prior purpose for it. This scientific explanation of adaptations has been called **teleonomy** (Williams 1966).

In more than a metaphorical sense the history of vertebrate evolution is reflected in the structure and organization of the human body. Most of us are fairly conversant with the human body and can therefore most easily compare our structural and functional details to other vertebrates. We can consider this assertion by examining some of our unique and not-so-unique features (Box 2–1).

The Basic Vertebrate Body Plan

What is the minimum functional organization that we can call vertebrate? Do any vertebrates actually conform to this generalized plan? Or is it an abstraction based on an a priori assumption that evolution has proceeded from simple to complex organization?

The symmetry of a hypothetical ancestral vertebrate would certainly be bilateral with definite head and tail ends. The basic internal organization would be a tube-within-a-tube arrangement with the major internal organs lying within a body cavity or coelom (Figure 2–3). In these respects the ancestral vertebrate is not unlike many higher invertebrate animals, which also possess these general body features—mollusks, annelids, and arthropods.

Required Functional Systems

To qualify as a vertebrate, the hypothetical animal would also have to possess the diagnostic chordate–vertebrate characters that we considered briefly in the preceding section. Bone is also a unique vertebrate tissue, which evolved as early as the late Cambrian; however, because not all vertebrates have bony skeletons, we must omit it from the list of required features.

So far we have identified only the fundamental features of vertebrate morphology. Other organs and structures are required to make even the simplest vertebrate into a self-maintaining reproducing organism. There are ten organ systems involved in the vital functions of all vertebrates. These systems are listed in Table 2–2 and described more completely in Chapter 3.

A Search for the Relatives of the Vertebrates

Which animals are most closely related to vertebrates? The fact that no fossils are intermediate between the earliest vertebrates known, the **ostracoderms**, and any other group of animals presents difficulties for a biologist interested in the origin of the vertebrates. Indirect evidence must be used to infer evolutionary relationships between the vertebrates and other animals.

Fortunately, evidence from comparative anatomy and embryology of living forms, evidence based on the principle of homology, is helpful. Following the principle of homology, it is reasonable

Box 2–1. The Human as a Typical Vertebrate

Table 2–1 lists the anatomical and behavioral traits of humans. The most crucial structural development for the evolution of the human condition was acquisition of a fully upright posture and strict bipedal locomotion. Other peculiarities of human anatomy, such as the manipulative hand, the S-shaped vertebral column, and the oversized brain, follow from that posture or are coadaptations with it. The major behavioral traits in human adaptive achievement have been tool making and language, which in turn have been dependent upon evolution of the human hand and brain (Simpson 1966, 1969).

The scientific name that humans have assumed—*Homo sapiens*—suggests the trait that we consider unique, our wisdom. We owe our vaunted intelligence to a large, complex brain, the intricate workings of which are just beginning to be understood. Especially we owe it to our forebrain. The cerebral hemispheres, which have become enlarged with respect to the rest of the brain, cover most of it and have become highly folded, thereby greatly increasing the surface area available for associational neurons (Figure 2–2a). The most important component of the cerebrum—the roof or **neopallium** (*neo* = new, *pallium* = cloak)—is a structure that humans share with all other mammals.

Probably the neopallium began to enlarge in the precursors of mammals in the late Paleozoic, around 250 million years ago. The neopallium itself, however, derives from a still

Table 2–1. Distinctive traits of *Homo sapiens*.

Anatomical	Behavioral and Psychological*
1. Normal posture upright	1. Curiosity, imitation, attention, memory, imagination
2. Legs longer than arms	
3. Toes short, the first usually longest and not divergent	2. Ability to improve adaptive nature of behavior by rational thought
4. Vertebral column with S curve	3. Uses and makes tools in great variety
5. Hands prehensile and thumb strongly opposable	4. Consciousness of self
	5. Makes mental abstractions and develops related symbolism—especially language
6. Body mostly bare with only short, sparse, inconspicuous hair	6. Culture and social organization unique in complexity
7. Joint for neck in middle of base of skull	
8. Brain uniquely large for body size with very large, complex cerebrum	
9. Face short, almost vertical under front of brain	
10. Jaws short, with rounded dental arch	
11. Canines usually no larger than premolars	

Source: G. G. Simpson, 1966, *Science* 152:474–478

*Based on Charles Darwin, 1871, *The Descent of Man.*

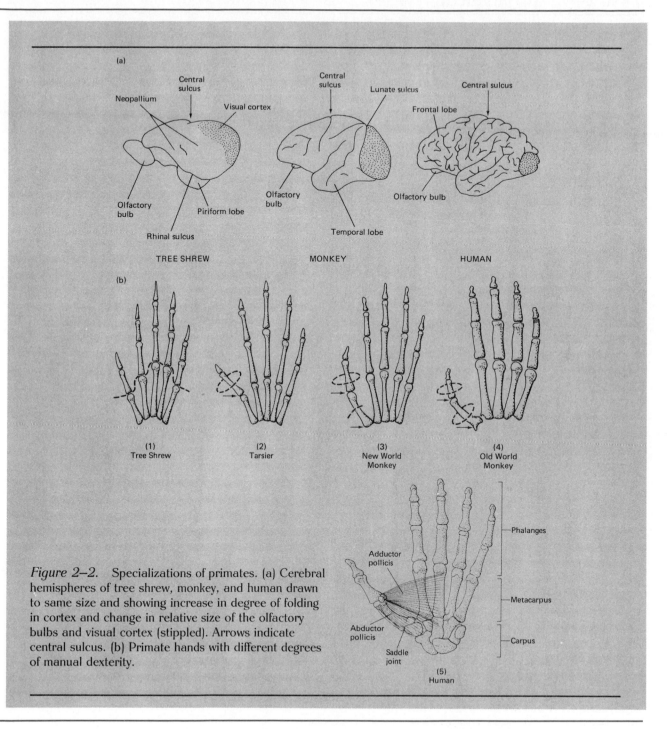

Figure 2–2. Specializations of primates. (a) Cerebral hemispheres of tree shrew, monkey, and human drawn to same size and showing increase in degree of folding in cortex and change in relative size of the olfactory bulbs and visual cortex (stippled). Arrows indicate central sulcus. (b) Primate hands with different degrees of manual dexterity.

Box 2–1. (Continued)

older structure, the olfactory forebrain or **prosencephalon** (*pro* = before, *encephalon* = brain). Like the midbrain and the hindbrain, the prosencephalon existed in the earliest known vertebrates, about 500 million years ago.

The human hand is a tool-using and tool-making organ, and the development of human culture and civilization has been based on our ability to make things and manipulate things with our hands. Our technology is quite firmly based on the evolution of our hands. The human hand is also used for social communication. Emotive communication especially is likely to be manual. We salute the flag, we point accusing fingers, we shake our fists in anger, and we swear in a variety of finger signs. Good speakers continually add emotional content to their speeches with hand gestures.

The human hand performs four basic movements—divergence, convergence, prehensility, and opposability. The first two are general mammalian characteristics. **Divergence** is the ability to spread the toes. **Convergence**, or cupping of the hand, is achieved by flexion at the joints of the metacarpals and phalanges. Two convergent paws can be used like one prehensile hand, and many mammals—squirrels and sea otters are examples—manipulate their food in two convergent paws while eating. **Prehensility**, which refers to the ability to wrap the fingers around an object, has been perfected by primates in association with **brachiation**, swinging by hands and arms through branches of trees. It has been independently evolved in other arboreal mammals such as the opossums. Opposability is the ability to pass the thumb across the palm while rotating it around its long axis to allow the ventral ball to touch the

tips of other fingers. Although many primates perform this movement, the bone and muscular structures associated with it are best developed in humans (Figure 2–2b).

During the course of evolution humans capitalized on prehensility and opposability, inherited from their tree-dwelling ancestors, to develop a power grip and a precision grip that allowed the development of tool using and toolmaking. These capabilities have merely been added to the older movements of the primitive mammalian paw by subtle modifications of bony structure and muscles in association with appropriate changes in the nervous system.

The evolutionary sequence in development of the human hand can be seen by comparing the skeletons of various primates (Figure 2–2b). The trend toward the human hand began 60 to 65 million years ago among the early arboreal ancestors of the primates. All these specialized primate and human ways of working the hand are based on a skeletal pattern of the manus—the phalanges, metacarpals, and carpals—that trace back in time to the earliest progenitors of terrestrial vertebrates.

We could consider other human features in the same fashion and would discover that our uniqueness is relative and is based on some prior condition in our ancestors. In the same manner, also, we could examine any species of vertebrate and show how it consists of specialized traits added to or modified from an older form. All living organisms are mixtures—**mosaics**—of special characters peculiar to a few forms and of ancestral traits that are widely distributed among groups of distantly related species.

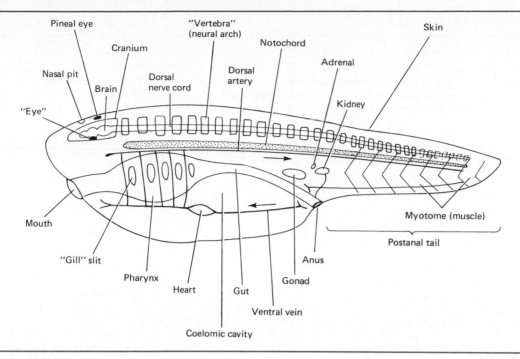

Figure 2–3. Hypothetical body plan of an ancestral vertebrate.

to expect to find the closest relatives of vertebrates among the invertebrate chordates, the **urochordates** or the **cephalochordates**, which possess a notochord, dorsal hollow nerve cord, pharyngeal slits, and postanal tail. No other group of animals shows closer structural and developmental affinities with vertebrates, although many earlier evolutionists believed that the vertebrates arose from mollusks or arthropods, and some still do.

Amphioxus as a Model Prevertebrate

The marine cephalochordates, particularly the well-studied amphioxus or lancelet (*Branchiostoma lanceolatum*), possess the basic chordate characters in diagrammatic form. Significantly, their mode of filter feeding is similar to that of the larval stage of lampreys (Chapter 6). These facts indicate a relationship to vertebrates and suggest that the mor-

phology of cephalochordates might provide hints about the chordate relatives of the vertebrates.

The Cephalochordata consists of some 14 species, all of which are small, fusiform, superficially fish-like, aquatic animals usually under 5 centimeters long. Lancelets are widely distributed in oceanic waters of the continental shelves, and, although they can swim freely, they are burrowing, sedentary animals as adults. The main features of the lancelet's morphology relate to its methods of locomotion and feeding (Figure 2–4).

A significant characteristic of amphioxus is its fish-like locomotion, resulting from the contraction of serially arranged **myotomes**, blocks of striated muscle fibers positioned dorsolaterally along both sides of the body and separated by sheets of connective tissue, the **myocommas**. Contractions of the myotomes bend the body in a way that results in forward propulsion. (For a discussion of swimming, see Chapter 6.) An incompressible, elastic rod, the **notochord**, extends the full length of the

Table 2–2. Basic vertebrate systems.

System	Basic Functions	Major Components
Integumentary	1. Protect underlying tissues from injury. 2. Prevent excessive loss or absorption of water and the consequent effect on tissues. 3. Aid excretion and absorption of specific metabolites and ions. 4. Almost all sense organs are derived in part from the integument.	Skin: Composed of epidermis above and the dermis below (Greek, *epi* = upon; *derma* = skin) and the derivatives of these two layers, such as scales, feathers, and hair.
Skeletal	1. Provide a framework for all body systems. 2. Provide attachments for muscles, tendons, and fascia. 3. Enclose and protect vital organs. 4. Serve as a reserve storehouse for minerals.	Bones, cartilage, and ligaments. These tissues are divided into: 1. Axial skeleton: skull, vertebral column, and (when present) ribs. 2. Appendicular skeleton: pectoral and pelvic girdles and limbs in some vertebrates.
Muscular	1. Movement of body and body parts. 2. Maintenance of posture. 3. Internal transport and expulsion (movement of food through digestive tract, blood through vessels, germ cells through reproductive tract, bile from gallbladder, urine from kidneys, feces from alimentary canal). 4. Homeostatic adjustments such as size of opening of the pupil of the eye, the pylorus, the anus, blood vessels; heat production in some vertebrates.	Smooth (nonstriated) muscles of involuntary control found primarily in wall of digestive tract, genital ducts, and blood vessels. Cardiac muscle of involuntary control restricted to the heart. Striated muscles generally under voluntary control found attached to the skeleton so intimately that the name "musculoskeletal system" is often applied; tendons (the connective tissue bands that bind striated muscle to bone).
Digestive	1. Capture and physical/chemical disintegration of food. 2. Absorption, detoxification, alteration, storage, and controlled release of the products of digestion and metabolism.	Alimentary canal: mouth and oral cavity with associated teeth, tongue, and jaws (when present). Pharynx (associated intimately with the respiratory system). Esophagus. Stomach (when present). Intestine divided and specialized in various ways. Accessory glands: salivary (when present). Liver [responsible for most of the functions listed in (2)]. Pancreas.
Circulatory	1. Transport of materials to and from cells. 2. Transport, formation, and storage of blood cells for oxygen transport, defensive, and immunogenic functions. 3. Drain fluids from between cells and return it to the regular circulatory system from which it leaked.	Heart. Arteries (from the heart to the tissues). Arterioles (small arteries). Capillaries (extremely small vessels connecting arterioles and venules). Venules (small veins). Veins (from tissues to the heart). Spleen (and other sites in various vertebrates, but always intimately associated with the digestive tract and/or skeletal system). Lymphatic system.

Table 2–2.   (*continued*)

System	Basic Functions	Major Components
Respiratory	1. Exchange of gases (primarily intake of oxygen and discharge of carbon dioxide) between the organism and its environment (water or air). 2. Various accessory functions from production of sound to nest building.	Lungs, gills, and/or skin, depending on which groups of vertebrates are under discussion; lungs and gills are derived from and intimately connected with the pharyngeal region of the digestive system.
Excretory	Chemical (and to a lesser extent physical) homeostasis or maintenance of a constant internal environment by (1) excreting toxic and metabolic waste products, especially those containing nitrogen; (2) maintaining proper water balance; (3) maintaining proper concentration of salts and other substances in the blood; (4) maintaining proper acid–base equilibrium in body fluids.	Kidneys and excretory ducts (tube-like passages) variously aided by the gills, lungs, skin, and/or intestines. The mode of development and use of common ducts makes this and the reproductive system inseparable morphologically so that the two are often referred to as the urogenital system.
Reproductive	Formation of zygotes by the union of two gametes to produce new individuals of the same biological variety.	Primary sex organs in the form of male (testes) or female (ovaries) gonads. Secondary sex organs concerned with transport of gametes from their site of formation to their site of union. Accessory sex organs assuring union of gametes, such as glands and external genitalia.
Endocrine	Regulation and correlation/integration of body activities through chemical substances (*hormones*) carried by the blood. As opposed to the method of action of the nervous system, the endocrine system is slower acting—being limited by the rate of blood flow—but it is capable of long, continuous action.	A large number of cell types discharge secretions that have regulatory effects on other cells. In more primitive vertebrates, these cells tend to be widely scattered in other tissues. More advanced vertebrates have discrete aggregations of these cells to form endocrine glands.
Nervous	Regulation and correlation/integration of body activities through conduction within and between individual nerve cells or neurons, which eventually cause a response in some other system (especially muscular contractions). The nervous system is fast acting, and conduction may be faster than 90 meters per second.	Central nervous system (CNS): brain, spinal cord. Peripheral nervous system (PNS): craniospinal nerves, which exit from the protective skeletal sheath of the cranium and vertebrae and may be either of a voluntary nature (to striated muscles) or involuntary (to smooth muscles); nerves of the latter type are often referred to collectively as the autonomic nervous system. Sensory nerves from either complex sense organs (e.g., eye, ear) or simple receptors (e.g., cutaneous sensory nerves) enter the CNS via the craniospinal nerves.

cephalochordate body and prevents the body from shortening when the myotomes contract. As a result, the myotomal contractions bend the body. The notochord of amphioxus extends from the tip of the snout to the end of the tail, projecting well beyond the region of the myotomes at both ends. This condition apparently aids in burrowing.

Amphioxus filters small food particles from a stream of water drawn through its mouth and pharynx by the movements of cilia. As in all animals that use cilia for filtering, a large surface area is required to obtain sufficient food. The pharyngeal apparatus occupies more than half of the body length. Its walls are perforated by up to 200

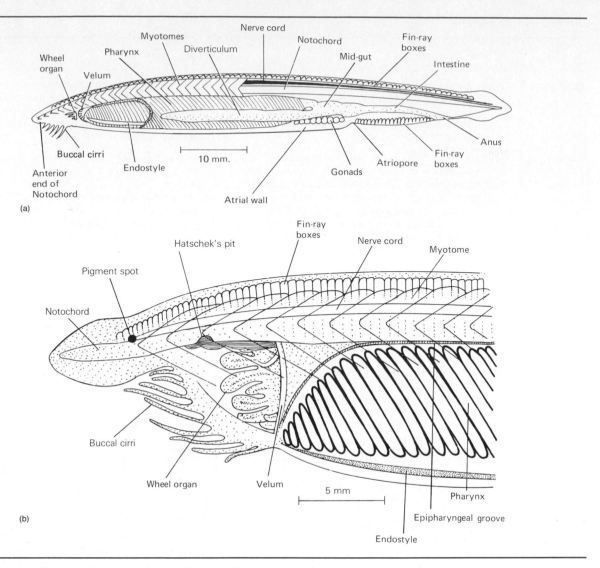

Figure 2–4. Cephalochordates: (a) lancelet, amphioxus—a longitudinal parasagittal section with the posterior myotomes removed; (b) detail of the anterior end of amphioxus showing the structures involved in filter-feeding.

oblique, vertical slits on which cilia are located. The slits are separated by bars with internal skeletal rods. The sidewall of the perforated pharynx is covered by the protecting walls of the **atrium** which exits to the outside of the animal through a posterior **atriopore**. Solid particles entrained in the water current encounter a variety of organs that separate edible from inedible matter and conduct the entrapped food to the digestive tract. **Buccal cirri**, attached to the margin of the **oral hood** in front of the mouth, possess sensory cells and form a funnel-like sieve that prevents the entry of large particles. The **velum** presumably functions as an additional screening mechanism against unwanted items. Food particles are caught partly on a complex set of ciliated tracts known as the **wheel organ**. Mucus from **Hatschek's pit** is discharged onto the wheel organ, and food particles caught in this material are swept by cilia toward the mouth and enter with the main stream of water.

The **endostyle**, another mucus-secreting organ located on the floor of the pharynx, consists of ciliated cells alternating with mucus-secreting cells and produces sticky threads of mucus in which food particles become entangled. Ciliary currents then draw the food-carrying mucus strands up the sides of the pharynx and into a median, dorsal **epipharyngeal groove** in which cilia drive the material into the midgut.

Feeding mechanisms that consist of trapping particles and entraining them in mucus are widespread among invertebrates and the structures involved may be homologous. The tunicates have an incurrent mouth, excurrent pharyngeal slits, cilia, and a mucus-secreting endostyle. The use of cilia and mucus cells on *external* structures occurs on the proboscis of the Hemichordata or acorn worms, as well as on the oral tentacles of other groups of invertebrates, the pterobranchs, brachiopods, bryozoans, and phoronid worms, and on the ambulacral system of some echinoderms. Among vertebrates, the food-filtering mechanism of the larva of lampreys is the same, but a muscular pump is used to move the water instead of cilia.

Amphioxus shows other fundamental similarities to the plan of vertebrate organization. It has a closed circulatory system in which slow waves of contraction drive the blood forward in the ventral blood vessels and backward in the dorsal ones. Below the hind end of the pharynx there is a large sac, the **sinus venosus**, which collects blood from all parts of the body, like the vertebrate heart, although it is not the exclusive pumping organ. Amphioxus reveals no indication of a complex brain, possessing only an anterior expansion of the dorsal nerve cord into a **cerebral vesicle**.

Despite many homologous similarities, basic differences indicate that amphioxus and its relatives are not in the direct evolutionary lineage of the vertebrates. The nerve cord has a series of paired spinal nerves that contain both sensory fibers (carrying information from sense organs) and motor fibers (sending impulses to stimulate muscles). Structures formerly called ventral spinal nerves are really specialized muscle fibers. In this respect Amphioxus is unlike vertebrates. The excretory system of amphioxus consists of short excretory tubules that terminate in flask-shaped cells that drain body wastes from the body fluids bathing them. The open end of each tubule drains into the atrium. These structures have been called **protonephridia**, but they are probably not homologous to the kidneys of vertebrates. The lack of a distinct head and the unique extension of the notochord to the anterior end of the rostrum also make cephalochordates unlikely candidates as animals that could evolve the large, vertebrate-type brain. Also, in amphioxus there are no homologs of the eyes, ears, nose, or other cephalic sense organs of vertebrates. Some biologists believe that the cephalochordates are the closest living relatives of the vertebrates, but others do not. Cephalochordates provide insights about the likely ancestral vertebrate body plan, but cephalochordates and vertebrates represent divergent paths of evolution from a common ancestry far back in time.

Other Relatives

Tunicate larvae are tadpole-like animals with pharyngeal slits, notochord, and dorsal hollow nerve tube fully developed. In addition, they possess a muscular, postanal tail, which moves in a fish-like swimming pattern (Figure 2–5). The larvae are sufficiently generalized in their chordate and other structural features to suggest that they represent only slightly modified living forms of chordates related to the vertebrate line of evolution. However, tunicate larvae, after a brief free-swimming existence, metamorphose into adult animals that bear no resemblance to vertebrates. Adult tunicates are anchored or free-living animals with body plans that are so divergent from those of vertebrates that it is hard to imagine an evolutionary transformation from one to the other. The same

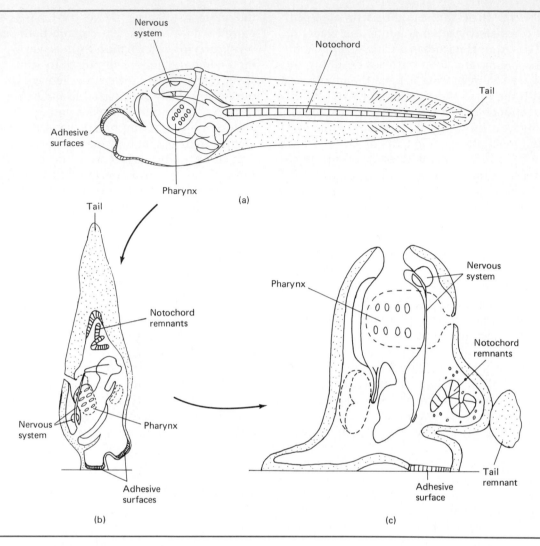

Nervous system

Notochord

Tail

Adhesive surfaces

Pharynx

(a)

Tail

Notochord remnants

Nervous system

Pharynx

Adhesive surfaces

(b)

Pharynx

Nervous system

Notochord remnants

Tail remnant

Adhesive surface

(c)

Figure 2–5. Tunicates, showing the free-swimming larva, an attached larva undergoing metamorphosis, and the sessile adult.

applies to the acorn worms and pterobranchs. Pharyngeal slits and a probably nonhomologous section of hollow nerve tube are the only chordate-like features they possess.

By what evolutionary process could the larval traits of the tunicate tadpole be passed on to succeeding generations of adult animals? A resolution of this difficulty was proposed by W. Garstang in 1928. He suggested that vertebrates originated from tunicate-like larvae that failed to metamorphose but nevertheless developed functional gonads and reproduced: the genotype could be passed on without the necessity of a separate adult morph in the life cycle (Figure 2–6).

Garstang termed this evolutionary process **paedomorphosis** (*paedo* = child, *morph* = form).

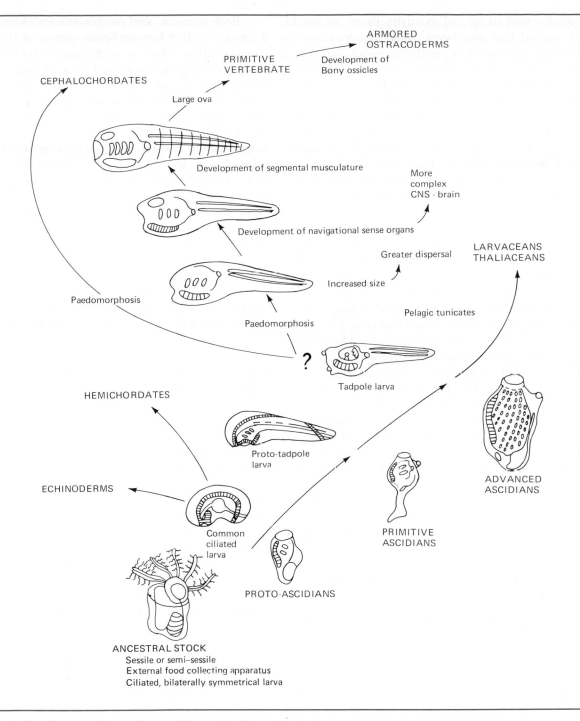

ARMORED
OSTRACODERMS

PRIMITIVE
VERTEBRATE   Development of
Bony ossicles

CEPHALOCHORDATES

Large ova

Development of segmental musculature

More
complex
CNS - brain

Development of navigational sense organs

LARVACEANS
THALIACEANS

Greater dispersal

Paedomorphosis

Increased size

Pelagic tunicates

Paedomorphosis

?

Tadpole larva

HEMICHORDATES

Proto-tadpole
larva

ADVANCED
ASCIDIANS

ECHINODERMS

Common
ciliated
larva

PRIMITIVE
ASCIDIANS

PROTO-ASCIDIANS

ANCESTRAL STOCK
Sessile or semi–sessile
External food collecting apparatus
Ciliated, bilaterally symmetrical larva

Figure 2–6.   Garstang's hypothesis of the origin of vertebrates by
paedomorphosis from an ancestral lophophorate.

sense organs. Dentine and enamel-like tissue may have originated as a protective and electrically insulative coating around the electroreceptors in the head of prevertebrates that enhanced electrosensitive detection of prey (Northcutt and Gans 1983).

Environment in Relation to the Origin of Vertebrates

By the late Silurian, armored ostracoderms and primitive jawed fishes were abundant in both freshwater and marine environments. For years students of evolution have argued whether the vertebrates had a freshwater origin or marine origin. Under what conditions did the first vertebrates evolve?

It seems reasonable that animals essentially vertebrate in character existed in the early to middle Cambrian seas. They were small (probably under 10 centimeters in length), had all the primary chordate characters fully developed, and in addition, had a tripartite brain, possibly a cartilaginous cranium and neural arches surrounding the nerve cord, a primitive glomerular system (perhaps like the embryonic nephrons of some present day vertebrates), and body fluids similar to seawater. They may also have had small, dermal ossicles of calcium phosphate that served as mineral stores. By the late Cambrian dermal bony armor had evolved in the ostracoderms.

Later vertebrates became adapted to estuarine conditions at the mouths of rivers, where the waters were less saline than in the sea. The osmotic stress imposed by the variable salt concentration in estuaries may have been a powerful selective force operating on these vertebrates that led to a reduction in their body fluid concentrations and the evolution of a renal system effective in water removal. This new environment may also have affected calcium and phosphate regulation, leading to a further function of the dermal bony tissues. We know, for example, that calcium decreases cell membrane permeability, and its presence in dilute seawater can greatly increase the osmoregulatory ability of marine organisms. In fact, some marine fishes penetrate freshwater streams that have a high calcium content. Similarly, a marine fish that could carry a readily mobilized store of calcium in its tissues might have been able to live in fresh water. But were the first vertebrates marine or freshwater organisms?

Suggestions of a Freshwater Origin

The Cambrian and Ordovician seas were rich environments and fossils of most major groups of animals have been found only in marine deposits of these periods. The first land plants appear in the Silurian, and the first land animals in the Devonian. Geologists have found few rocks earlier than those of the Silurian that were formed in freshwater deposits. Consequently, we know little about the freshwater biota of the Ordovician period or of earlier times. Nonetheless, Alfred Sherwood Romer, a paleontologist, and Homer Smith, a renal physiologist, were staunch advocates for the freshwater origin of the first vertebrates (Romer 1967, Smith 1953).

Romer's proposal of a freshwater origin of vertebrates was based on the fossil record. Ostracoderm fragments found in the middle Ordovician Harding Sandstone of Colorado occur in association with fossils of marine origin. Their location in the sediments and their worn condition were interpreted by Romer to indicate that these fragments had washed downstream from rivers or lakes into estuaries and were fossilized in marine sediments. Romer was impressed by another aspect of the fossil record: at the time he formulated his ideas, vertebrate remains were unknown from the rich fossil-bearing rocks of the Cambrian and were rare in Ordovician rocks. If vertebrates arose in the sea, Romer wondered why they left no significant fossil record before the Silurian. If, however, vertebrates evolved in fresh water, Romer reasoned, their absence from the exclusively marine Cambrian and Ordovician formations is explained. Their sudden appearance in a variety of forms in the late Silurian and Devonian corresponds with the earliest occurrence of abundant freshwater sediments in the geological record. Romer thus proposed that active swimming by means of a muscular postanal tail evolved in the early vertebrates in response to the downriver sweep of stream currents.

Homer Smith's studies of vertebrate kidney structure and function led him to support Romer's idea of a freshwater origin. The blood and body fluids of most marine organisms contain salts at concentrations similar to those of the seawater that bathes them. Any alteration of this concentration equilibrium will produce a flow of water from the more dilute to the more concentrated fluid—a process called **osmosis** (see Chapter 4). In bony fishes, body fluid concentrations are very dilute compared to seawater . In freshwater habitats this lowering of body fluid concentration, although reducing the inward osmotic flow of water, nevertheless cannot stop hydration of the body completely. A freshwater fish therefore takes up water and must have a mechanism for excreting the excess water and for retaining needed ions. The vertebrate glomerular kidney is well suited for these functions, and Smith proposed that dilute body fluids and the glomerular kidney evolved as adaptations to freshwater conditions.

Evidence for a Marine Origin

In spite of Romer's and Smith's arguments, evidence for a marine origin of vertebrates is now overwhelming. All protochordate and deuterostome invertebrate phyla are exclusively or primitively marine forms. Because of chordate affinities to these phyla, there must have been a marine transitional form in the evolutionary line leading to the vertebrates. All known Cambrian and Ordovician vertebrates occur as marine fossils. The Harding Sandstone Formation, which Romer and others interpreted to be estuarine, is now known to extend over many thousands of square kilometers and is more reasonably interpreted as an offshore shelf deposit. Denticles from early Ordovician rocks came from a type of sandstone that

is formed only under true marine conditions, and the same is true for all of the more recently discovered fragments of early ostracoderms from North American locations (Repetski 1978).

The exclusively marine hagfishes (*Myxiniformes*) have body fluids that are similar in salt concentration to seawater , as do the tunicates and other deuterostomes. Presumably the first vertebrates were also in osmotic equilibrium with the sea. Nevertheless, the hagfishes have a well-developed glomerular kidney and so do the marine cartilaginous fishes, which are also osmotically similar to seawater . Therefore, a functional glomerulus is not necessarily associated with the need for powerful osmotic regulation. The glomerulus produces a fluid from which red blood cells and blood proteins have been excluded. The tubule cells can return water, salts, glucose, and other useful materials to the circulatory system, while allowing toxic substances, nitrogenous wastes, and excess salts to remain in the urine. The kidney system therefore separates certain ions and molecules, including water, from others. Its primary function is excretory, not osmoregulatory, and the kidney would be valuable to either a freshwater or a marine organism. In this view, whether a vertebrate is adapted to the sea, to fresh water, or to life on land, the glomerular kidney is valuable and could have evolved in a marine vertebrate or in a freshwater one. Once a powerful and efficient glomerular filtration system had been achieved, however, only slight modifications of the kidney would be needed to produce an efficient water-excreting device for osmotic regulation in rivers and lakes.

The Significance of the Armor of Ostracoderms

Homer Smith postulated that the dermal bony plates of ostracoderms were a barrier against the intake of water, acting as an auxiliary mechanism in osmoregulation. The bony plates, however, are porous and probably were overlain by epidermis. Romer doubted the armor could have been an effective barrier to permeation by water.

The plates probably were protective armor to thwart the attacks of powerful predators. Several observations suggest this function. The body design of early ostracoderms indicates that they were sluggish creatures and probably swam only for short distances before coming to rest on the bottom. They had no true paired appendages and therefore little maneuverability to avoid predators. Lacking jaws, they had little in the way of defense mechanisms. The only defense against predation would be protective armor to discourage the onslaughts of grasping or biting feeders.

The ostracoderm fossils are frequently found in association with fossil eurypterids, giant scorpion-like arthropods. Eurypterids were dominant predators of the Cambrian, Ordovician, and Silurian. They first were marine, but became abundant in fresh waters during the Silurian. They were found in the same habitats as the ostracoderms, and it is likely that they were prominent among the predators that exerted the selective forces responsible for the evolution of dermal bony armor in these small, vulnerable fishes.

Conclusion

The living vertebrates share a suite of ancestral characters with several groups of invertebrates and chordates. Many of these characters are manifested only in embryonic or larval stages and are not present in adults. That pattern suggests that the origin of vertebrates might have occurred through the process of paedomorphosis whereby an organism becomes reproductively mature while retaining juvenile, larval, or embryonic characters. Alternative hypotheses stress the derived characters of vertebrates and suggest different possible origins. Whatever the process by which vertebrates evolved, fossil evidence points to marine habitats as being the most likely site of vertebrate evolution.

Summary

The history of vertebrates covers a span of more than 500 million years. We think of the human as the most highly evolved vertebrate, specialized in many structures—hands, feet, vertebral column, cerebrum—but the structure and organization of the human body have been determined by a long and complex course of evolution. When we strip away the special features of humans and compare the result with other vertebrates, we can identify a basic body plan. Presumably ancestral, the plan consists of a bilateral, tubular organization, with such characteristic features as the notochord, pharyngeal slits, dorsal hollow nerve cord, vertebrae, and cranium. One of the protochordates, amphioxus, and the ammocoete larva of lampreys provide glimpses of what the earliest vertebrates may have been like.

The earliest known vertebrates are the ostracoderms, primitive jawless aquatic animals that are related to lampreys and hagfishes and first appear in the late Cambrian and Ordovician. They were completely encased in heavy dermal bony armor. No fossils have been found that are intermediate between ostracoderms and any of the presumed invertebrate progenitors of the first vertebrates, but some idea of the possible origin of vertebrates from invertebrates can be inferred by comparison of living forms. The notochord, pharyngeal slits, and dorsal hollow nerve cord are shared with certain protovertebrate animals, and it is most probable that vertebrates arose from chordates with the general features of amphioxus. Garstang's theory of vertebrate evolution from a tunicate larval stage is commonly accepted, but remains unproven. Further back in evolutionary history the chordates are more closely allied to the Echinoderms and certain lophophorate groups than to any other invertebrate phyla. The first animals that can be called vertebrates probably evolved in Cambrian seas. Although this sequence of events is not an absolutely proven history of vertebrate origins, it is a good example of how biologists build evolutionary hypotheses from comparative studies of living and fossil animals.

References

Barrington, E. J. W. 1965. *The Biology of Hemichordata and Protochordata*. W. H. Freeman, San Francisco.

Benton, M. J. 1987. Conodonts classified at last. *Nature* 325:482–483.

Berril, N. J. 1955. *The Origin of Vertebrates*. Oxford University Press, Oxford.

Garstang, W. 1928. The morphology of the Tunicata and its bearing on the phylogeny of the Chordata. *Journal of the Microscopical Society* 72:51–87.

Gould, S. J. 1977. *Ontogeny and Phylogeny*. Belknap Press, Cambridge, Mass. A most readable historic overview and biological analysis. But see B. A. Pierce and H. M. Smith, 1979, Neoteny or paedogenesis? *Journal of Herpetology* 13(l):119–121 for continuing problems with terminology.

Halstead, L. B. 1969. *The Pattern of Vertebrate Evolution*. Oliver & Boyd, Edinburgh. A stimulating opinionated short book by a student of the earliest vertebrates.

Higgins, A. 1983. The conodont animal. *Nature* 302 (5904):107.

Jeffries, R. P. S. 1986. *The Ancestry of the Vertebrates*. British Museum of Natural History, Dorset Press, Dorchester, England.

Lindstorm, M. 1964. *Conodonts*. Elsevier, Amsterdam.

Lovtrup, S. 1977. *The Phylogeny of the Vertebrates*. Wiley, London.

Mikulic, D. G. , D. E. G. Briggs, and J. Kluessendorf. 1985. A Silurian soft-bodied biota. *Science* 228.715–717.

Nielsen, C. 1985. Animal phylogeny in the light of the trochea theory. *Biological Journal of the Linnean Society* 25:243–229.

Northcutt, R. G. and C. Gans. 1983. The genesis of neural crest and epidermal placodes: a reinterpretation of vertebrate origins. *Quarterly Review of Biology* 58:1–28.

Repetski, J. E. 1978. A fish from the Upper Cambrian of North America. *Science* 200:259–531.

Romer, A. S. 1967. Major steps in vertebrate evolution. *Science* 158:1629–1637.

Schaeffer, B. 1987. Deuterstome monophyly and phylogeny. *Evolutionary Biology* 21:179–235.

Simpson, G. G. 1966. The biological nature of man. *Science* 152:474–478. A revealing essay by an outstanding American paleontologist on human traits and their origins.

Simpson, G. G. 1969. *Biology and Man*. Harcourt Brace Jovanovich, New York.

Singer, C. J. 1959. *A History of Biology*, 3rd revised edition. Abelard-Schuman, London. A widely read general history that describes developments in human knowledge of vertebrates from Aristotle to Darwin.

Smith, H. 1953. *From Fish to Philosopher*. Little, Brown, Boston. A delightful, easy-to-read, speculative book by one of the leading proponents of a freshwater origin of vertebrates.

Tarlo, L. B. H. 1960. The invertebrate origins of the vertebrates. *21st International Geological Congress* 22: 113–123.

Williams, G. C. 1966. *Adaptation and Natural Selection, a Critique of Some Current Evolutionary Thought*. Princeton University Press, Princeton, N.J.

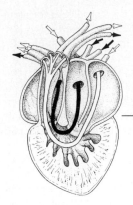

We saw in the first chapter that evolution is a process of change in the frequencies of alleles in the gene pool of a species. The frequencies of different alleles determine the relative numbers of various genotypes, and the genotypes are expressed as phenotypes. This relationship of gene to structure is expressed during development: derived structures rarely appear *de novo*; rather, they are produced by modifications of patterns of embryonic development. This mechanism imparts a high degree of continuity to the body forms of vertebrates. During the life of an individual a fertilized egg (a single cell) grows into the billions of cells that form an adult vertebrate. In the process the cell differentiates into a wide variety of specialized cells. An understanding of embryonic development and the development and function of specialized organs is needed to appreciate the changes that have occurred during the evolution of vertebrates. When all the necessary information is at hand, derived characters can be traced to their ancestral origins in the fossil record and in embryonic development as we did for the example of the mammalian middle ear discussed in Chapter 1.

Vertebrate Organ Systems and Their Evolution

3

The Unity of Vertebrate Structure

The well-known renal physiologist Homer Smith summed up a century of accumulation of detailed morphological and physiological research by scores of biologists with the title of his 1953 book *From Fish to Philosopher*. The continuity of vertebrate organ systems and their functional attributes had indeed been fully substantiated from their origins in some creature we would all recognize as fish-like to the condition in the human organism. It is important to understand the profundity of these similarities in order to appreciate the importance of the differences between one vertebrate clade and its sister taxon. Without appreciation for the basic structure and function of the organ systems, the variations often seem trivial when in fact they separate the lineages in question in ecological space and evolutionary history. An appreciation of the vertebrate organ systems requires basic knowledge of their development in the individual from the **zygote** (fertilized egg) to the adult form. This requires examination of the sequential processes of **embryogenesis** and **organogenesis**. Next must come an understanding of the basic functions of organ systems and how the associated organs carry out those functions. Finally, we would benefit from understanding the evolutionary changes in each system. Unfortunately, only a few of the organ systems have left any fossil record. Our only

recourse is to examine the character states of the various organs in living vertebrates as they correspond to our overall evaluation of vertebrate relationships. Here we tread on thin ice, for any evolutionary hypotheses generated by this process can be only as good as our hypothesis of vertebrate relationships. In taxa with few or highly specialized living members we face the additional challenge of distinguishing whether we are dealing with primitive states of organ systems or with derived specializations. In the case of many soft tissue, physiological, and behavioral characteristics of vertebrates, our skills at phylogenetic outgroup analysis are in their infancy. Mindful of these problems, let us examine the basic patterns of development of vertebrates, the differentiation of organ systems and their function.

Embryogenesis

The vertebrate zygote is a complete eukaryotic cell, in many ways rather typical of such cells. One difference from most other eukaryotic cells sets the stage for subsequent development—a zygote is endowed with a specialized, inert nutritive material called **yolk**. The large eggs of birds have a large quantity of concentrated yolk, whereas the tiny (150 micrometers in diameter) human ovum has a sparse, dilute yolk. The 2 or 3 millimeter diameter eggs of many amphibians, lungfish, and primitive bony fish (e.g., sturgeon) have a dilute yolk, more

79

concentrated at one end of the zygote than the other. This contrasts with the generally much smaller eggs of many teleosts, which are less than 1 millimeter in diameter and are endowed with a concentrated yolk.

Eggs and zygotes are classified as **oligolecithal** (*oligo* = little; *lekithos* = egg yolk), as in cephalochordates and marsupial and placental mammals; **mesolecithal** (*meso* = intermediate), as in lampreys, amphibians, lungfish and sturgeon; or **macrolecithal** (*macros* = large) as in hagfish, sharks, rays, teleosts, turtles, lizards, snakes, crocodiles, birds, and monotremes. Oligolecithal eggs are also **isolecithal** (*isos* = equal), the thin yolk being evenly distributed throughout the embryo. Eggs with more yolk, whether meso- or macrolecithal, have a higher concentration of yolk at one end of the zygote and are therefore **telolecithal** (*teleos* = end). The end with the higher yolk content is known as the **vegetal pole**; the opposite end is called the **animal pole** for reasons that do not become obvious until the zygote begins division.

The early division of a vertebrate zygote occurs without growth of the embryo and serves to change the zygote, which is a single but abnormally large cell, to a multicellular embryo with cells of a size more typical for the species (Figure 3–1). This cell division without embryo growth is called **cleavage** and it results in an embryonic stage called the **blastula** (*blastos* = a germ or bud). Cleavage is more than just cell division. Because there is rapid and unabated synthesis of nuclear material but no periods of cell growth between successive cell divisions, the nuclear-to-cytoplasmic ratio typical of adult cells is eventually reached. The higher nuclear-to-cytoplasmic ratio of late cleavage cells may be important for effective genetic control of cellular differentiation; embryonic gene expression first occurs at this time. The sequence and spatial arrangement of cell divisions appear regular, especially early in development, suggesting that individual cells may have distinct fates at this stage. Such is not the case, however; development proceeds normally even if many cells are removed from the early cleavage stages. Hans

Spemann found in the early decades of this century that a single cell removed from the early blastula could produce an entire embryo capable of successful development. Cleavage in vertebrates is thus said to be **indeterminate** and results in **totipotent** cells for a significant number of cell divisions.

The blastulae in various lineages of vertebrates are quite different due to the effects of yolk on the process of cleavage. Apparently, the division of yolk-free cytoplasm is readily accomplished, but the presence of yolk slows cleavage in proportion to its concentration. Thus the oligolecithal zygotes of cephalochordates and therian mammals cleave evenly and completely to yield a blastula of nearly even-sized cells. Complete cleavage of the zygote is termed **holoblastic** cleavage (*holos* = entire). Mesolecithal zygotes cleave more quickly at the animal pole than at the vegetal pole, resulting in a blastula composed of many small cells at one end and fewer and larger yolk-filled cells at the opposite (vegetal) pole. This pattern is a direct reflection of the distribution of yolk in telolecithal eggs, but because the entire zygote cleaves, it is still holoblastic. More dramatic still is cleavage of macrolecithal zygotes. Here cell division within the highly concentrated yolk is inhibited completely and cleavage leads to a thin disk of cells (**blastodisc**) atop an enormous uncleaved yolk, **meroblastic** cleavage (*meros* = a part). The embryo is represented initially by the blastodisc. By processes that differ in various macrolecithal groups of vertebrates (indicating convergent evolution), the blastodisc proliferates an **extra embryonic membrane**. This membrane forms the **yolk sac**, which surrounds the inert yolk with blood vessels that transport nutrients to the developing embryo atop the yolk mass. It is now apparent why the cleaving animal hemisphere acquired that name—the definitive animal appears here. The yolk of the vegetal hemisphere is slowly absorbed to provide the energy and structural materials for the development of the embryo.

The blastula of a vertebrate may be a hollow sphere of cells, the central cavity of which is known as the **blastocoel** (*koilos* = hollow), or an

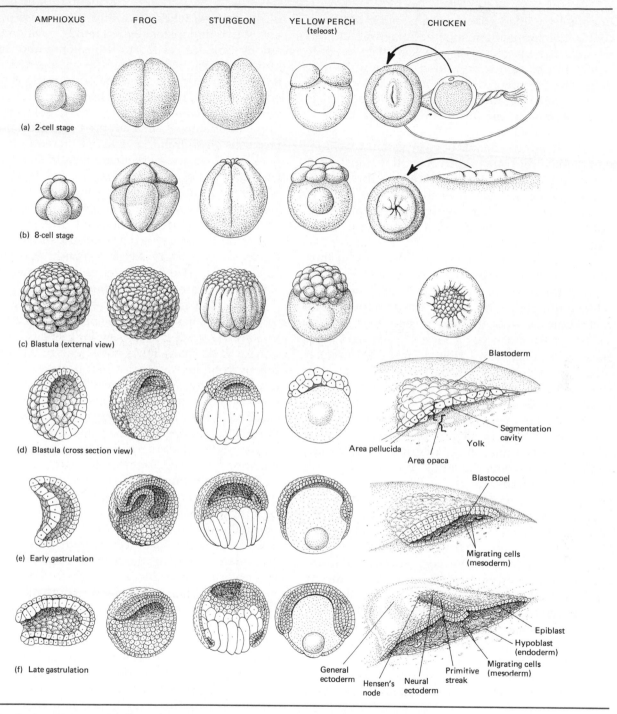

AMPHIOXUS   FROG   STURGEON   YELLOW PERCH (teleost)   CHICKEN

(a) 2-cell stage

(b) 8-cell stage

(c) Blastula (external view)

(d) Blastula (cross section view)

Blastoderm

Segmentation cavity

Area pellucida

Area opaca

Yolk

(e) Early gastrulation

Blastocoel

Migrating cells (mesoderm)

(f) Late gastrulation

Epiblast

Hypoblast (endoderm)

Migrating cells (mesoderm)

General ectoderm

Hensen's node

Neural ectoderm

Primitive streak

Figure 3–1. Cleavage (a–d) and gastrulation (e and f) of a cephalochordate and four vertebrate zygotes formed from eggs with scant, evenly distributed yolk (left) through those from eggs with a large amount of concentrated yolk at one end, the vegetal pole (right).

unequally divided sphere with a restricted blasto-coel in the animal hemisphere. In meroblastic vertebrates the blastocoel is represented by a space where the blastodisc has lifted off the undivided yolk. The importance of this hollow becomes evident as the vertebrate embryo enters its next stage of development, the formation of the **gastrula** (= belly or stomach).

Although the details of gastrulation vary, fundamentally the process involves the transformation of the blastula to a structure with three layers known as **germ layers**. Much has been made of the distinctions among the three cell layers differentiated during gastrulation. They are given names and assigned fates in the developing embryo that produce an impression of fixity in their roles in embryogenesis. However, this apparent fixity is the result of later interactions. Experimental manipulations show that there is little or no restriction on what tissue any cell from any layer in the gastrula can participate in forming. For us what is significant are their normal fates in a developing vertebrate embryo (Figure 3–2) as they interact, influence, and are influenced by the cell layers surrounding them, and the fact that these fates have been very conservative throughout vertebrate evolution. In all vertebrates the outermost layer of the gastrula, the **ectoderm** (*ecto* = outside; *derm* = skin) forms the adult superficial layers of skin, the most anterior and most posterior parts of the digestive tract and the nervous system including most of the sense organs such as the eye and the ear. The rest of the lining of the digestive tract, as well as the lining of the glands associated with the gut, and most respiratory surfaces of vertebrates are formed by the innermost layer, the **endoderm** (*endo* = within). The muscles, skeleton, connective tissues, circulatory and urogenital systems are formed by the middle layer, the **mesoderm** (*mesos* = middle), which is usually the last of the three layers to form.

Formation of the germ layers is initiated by radical changes in the spatial relationships of the embryonic cells as layers of cells buckle and fold and groups of cells migrate relative to other groups. The complete rearrangement of the cells of the embryo takes place through processes that are three-dimensionally complex. Differences among taxa are related to the amount and distribution of yolk as well as to the phylogenetic histories of the groups. The majority of the activity occurs at a single spot, the **blastopore**, where layers of proliferating cells buckle and fold inward into the blastocoel. In meroblastic gastrulae, surface cells stream as less coherent masses into the pocket-like blastocoel that results from meroblastic cleavage. The blastopore forms, or is in the vicinity of, the adult anus. Vertebrates and other chordates share the blastopore-to-anus developmental pathway with echinoderms and a few minor invertebrate phyla, all of which are called deuterostomes (*deutero* = secondary; *stoma* = mouth). It is on the basis of this shared character of development that echinoderms are linked to chordates.

Although gastrulation has set the stage for further development, it remains for the next stage to start the processes of growth, cellular differentiation, and organogenesis that lead to the formation of complex structures that begin to look like adult anatomy. **Neurulation** takes its name from the formation of the nervous system, which is associated with major changes in superficial and internal cell arrangements. A population of mesodermal cells known as **chordamesoderm** aggregates early in development to form an elongate rod, the **notochord**, that defines the axis of the embryonic body. The ectoderm overlying the forming notochord begins to form two longitudinal folds with a middorsal furrow between them. The crests of the ectodermal folds grow toward one another, forcing the furrow, its floor, and its sidewalls into the dorsal mesoderm flanking the notochord. When these **neural folds** meet, they fuse, sealing a hollow tube of now-isolated ectoderm below the surface of the embryo. This **neural tube** will become the central nervous system of the adult. It has been shown experimentally that in the absence of chordamesoderm the neural tube will not form. Embryologists describe such a relationship by saying that chordamesoderm induces formation of the neural tube.

The mesoderm remaining on either side after

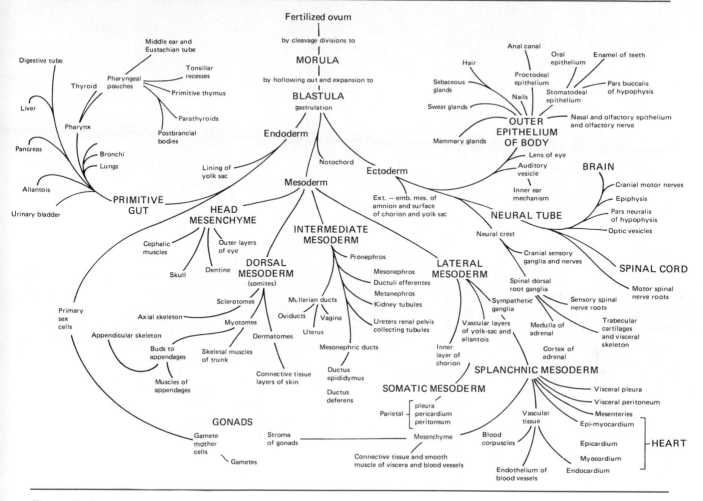

Figure 3–2. Schematic diagram showing basic vertebrate organ structures as derived from embryonic endoderm, mesoderm, or ectoderm. The diagram indicates origin of epithelial part of organ only. Most organs have supporting structures of mesodermal origin. Neural crest contributes to many more structures than can be illustrated here; see Table 3-1. (Modified after B. M. Patten, 1964, *Foundations of Embryology*, McGraw-Hill, New York.)

the differentiation of the notochord becomes divided into three morphologically distinct cell populations. These are best viewed by stripping away the ectoderm along the dorsolateral aspect of the embryo (Figure 3–3). Immediately flanking the notochord and neural tube is the **paraxial mesoderm**.

In the trunk this becomes segmented into bilaterally paired blocks of mesoderm called somites. Lateral to the somites is a long, thin strip of **intermediate mesoderm**, which may also be segmented. Lateral to the intermediate mesoderm is the sheet-like expanse of **lateral plate mesoderm**.

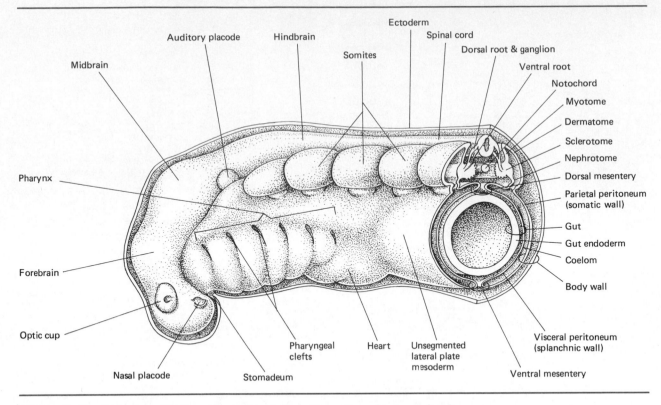

Figure 3–3. Diagrammatic three-dimensional view of a portion of a generalized vertebrate embryo showing segmentation of the mesoderm in the trunk region and pharyngeal development. (After E. S. Goodrich, 1930, *Studies on the Structure and Development of Vertebrates*, Macmillan, London.)

If we were to look at a representative cross-section through the neurula, we could see some of the internal morphogenesis of these three populations of cells. We find that the entire mesoderm is hollow: Each somite has a cavity and the more extensive and continuous space in the lateral plate defines a superficial or **somatic** layer immediately beneath the ectoderm and a deep or **splanchnic** layer (*splanchnos* = viscera) surrounding the gut. This space in the lateral plate is the body cavity or **coelom**. Soon after the somites form from the paraxial mesoderm, three distinct regions can be identified in each somite. The dorsal portion, just beneath the ectoderm, forms the **dermatome** (*tomos* = a slice or section), which will produce the deep portions of the skin. Deep to the dermatome and

across the slit-like cavity of the somite is the forming **myotome** (*myo* = muscle), which will form adult voluntary muscles. Cells located next to the notochord on the inner face of the somite form the **sclerotome** (*sclero* = hard), which will become vertebrae. The intermediate mesoderm eventually becomes clearly separated from the paraxial mesoderm and thickens into a **nephric ridge** (*nephros* = kidney), which develops into the kidney. Much later a second ridge forms from the medial border of the kidneys. This is the **genital ridge** and will form the gonads. The ducts serving these organs developed from the intermediate mesoderm are often shared between the excretory and reproductive systems harkening back to their common embryonic origin.

Organogenesis

During neurulation cells begin to differentiate, losing a part of their totipotency as they assume more specific roles. In general, the cells of the endoderm form a close-knit lining for the hollow gut and the tubular structures that develop from the gut and penetrate mesodermal spaces. Most endodermal cells are specialized for secretion or absorption. Cells of the ectoderm cover the outside of the body, forming a protective layer. Both the endoderm and the ectoderm are characterized as epithelia because of their tight intercellular junctions and sheet-like structure. The rest of the cells, the mesodermal derivatives, form the inner matrix of the organism—a highly varied task accomplished by an enormous diversity of cells. Mesodermal cells that form the linings of the walls of the coelom, the blood vessels and heart, or the kidney and reproductive ducts resemble those of skin surface or gut linings. However, mesoderm cells that form blood, bone or muscle or the general structural scaffolding of organs or the deep layers of the skin have a very different morphology.

The specialized roles of cells are reflected in their morphology—in the organelles that are present and the extent of their elaboration, and in the relationship of the cell to its neighbors. Coherent groupings of cells with specialized functions form **tissues**. There are five basic tissues in vertebrates: *epithelial, connective, blood, muscular,* and *nervous*. These tissues are combined in a wide assortment of permutations to form larger functional units with more complex functions called **organs**. In vertebrates, organs often have all or most of the five basic tissues and contributions from two or occasionally all three of the germ layers. The basic functions of vertebrate life are supported by groups of organs united into one of the ten vertebrate organ systems (Table 2–2).

Wandering Cell Populations

Not all cells from the three germ layers coalesce and separate from adjacent cells as a unit and subsequently differentiate into organs and organ systems as so far described. In fact, some of the most important and characteristic tissues of vertebrates break free of their parental germ layer and migrate as individual cells through the embryo to remote sites. These loose, wandering cells, as well as any other cells that are not arranged in an epithelium, are called **mesenchyme**. This term refers to the morphology of those cells, and it must not be confused with *mesoderm*, which is a germ layer that is only one possible source of mesenchymal cells. Because of the difficulty of following individual cells, we are still learning about the behavior and significance of these wandering cells.

The endoderm may contain cells that attain a distinctively larger size, break loose from the other cells of the yolk sac, and migrate dorsally around the gut wall and up the newly formed dorsal mesenteries. From here they spread laterally to invade the genital ridges on either side of the body. They become the eggs or sperm of the adult. The location and behavior of these **primitive sex cells** are of unknown evolutionary significance.

Cells from many regions of the mesoderm become mesenchymal, migrate, and contribute to components of locally forming tissues including cartilage, bone, smooth and some voluntary muscle, blood cells and vessels, and lymph vessels and glands. Many of these cells persist long into posthatching or postnatal life as undifferentiated progenitors of connective tissue cells. An important early site of migration and differentiation of mesenchymal cells is in the **skeletogenous septa** (Figure 3–4). Many segmental elements of the vertebrate skeleton form at the intersections of these thin sheets of connective tissue. The myosepta are oriented at right angles to the notochord and separate adjacent myotomes. The dorsal skeletogenous septum extends from the notochord to the middorsal line, splitting around the neural tube, and the ventral skeletogenous septum extends from the notochord ventrally around the gut or caudal blood vessels to the midventral line. In gnathostomes a horizontal skeletogenous septum runs laterally from the notochord to intersect the dermis of the skin along a midlateral line on each side of the body. The intersections of these septa form skeletogenous areas, and the vertebrate endoskel-

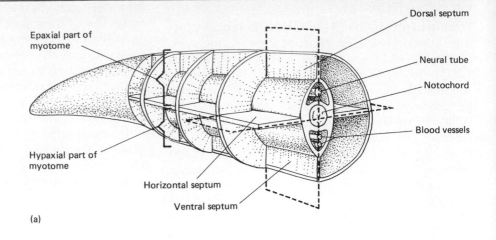

Epaxial part of myotome

Dorsal septum

Neural tube

Notochord

Blood vessels

Hypaxial part of myotome

Horizontal septum

Ventral septum

(a)

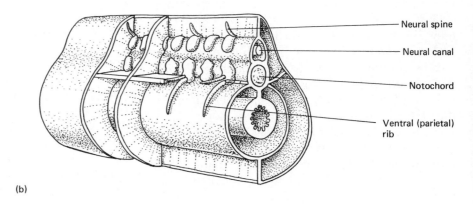

Neural spine

Neural canal

Notochord

Ventral (parietal) rib

(b)

Figure 3–4. The skeletogenous septa of gnathostomes: (a) diagrammatic representation of the orientation of the various mesenchymal septa; (b) differentiation of the septa and the endoskeletal elements that grow at the various intersections in a generalized fish-like vertebrate; (c) human thoracic vertebra and rib, which develop in the same skeletogenous septa intersections.

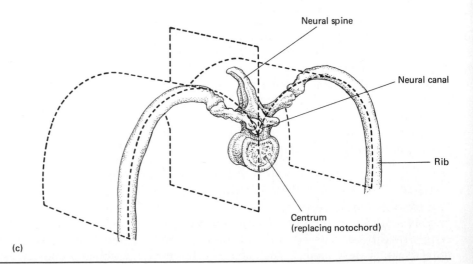

Neural spine

Neural canal

Rib

Centrum (replacing notochord)

(c)

eton of cartilage and bone may be local specialization, of what may well have been the primitive vertebrate internal skeleton of unmineralized connective tissue. When it is present, the sensory lateral line and its specialized scales form where the horizontal septum and the dermis meet; the neural spines form where the myosepta meet the dorsal septum; hemal spines arise where myosepta meet the ventral septum; dorsal or intermuscular ribs form where myosepta cross the horizontal septum; and ventral or subperitoneal ribs are produced where the lateral wall of the coelom and myosepta intersect. Around the notochord the sclerotomic mesoderm condenses and forms a vertebra at the intersection of every myoseptum with the horizontal, dorsal and ventral septa. The myosepta are intersegmental, and the sclerotomal condensation and the resulting vertebra are also intersegmental. Thus, vertebrae alternate with developing muscle masses. The muscles eventually insert on the vertebrae and form the flexible spinomuscular basis of lateral undulatory locomotion.

Perhaps the most significant of the wandering cell lineages in vertebrate development are the **neural crest cells** derived from the ectoderm. These cells are unique to the craniates and are key players in the formation of nearly every unique, derived character that sets vertebrates apart from all other organisms (Table 3–1). Le Douarin (1982) lists 42 different structures to which neural crest cells contribute. Neural crest cells become distinct at the time of neural tube formation, arising from cells at the angle between the developing neural tube and the margin of the remaining ectoderm which is closing over the neural tube. They are first clearly visible lying above and to either side of the neural tube. From here they rapidly disperse laterally and ventrally, ultimately settling and differentiating in diverse locations throughout the embryo (Figure 3–5). Derivatives of the neural crest form almost all of the peripheral nervous system (including the cranial nerve ganglia, the autonomic system, and the Schwann cells, which make the myelin sheath of peripheral nerves), several endocrine glands, pigment cells, and perhaps most remarkably, the skeleton and connective tissues of

the head. The complexities of their movement and their interactions with other cells are still not thoroughly understood, and remain a focus of intense interest although their contribution to skeletogenesis was first described in 1898 by Julia Platt. It seems that the origin of neural crest cells was a key factor in the evolution of vertebrates themselves.

A series of neurogenic epidermal thickenings called placodes are found only in the head region. Some are far dorsal, adjacent to the neural crest, whereas others are more ventral, just above the pharyngeal pouches. The cells of the placodes contribute to the sense organs (nose, eye, ear, lateral-line mechano- and electroreceptors, taste buds) and portions of the cranial sensory ganglia. In addition, portions of the placodes may migrate; the entire lateral line system of the trunk is formed by migration of cells from the head. Although the placodes are not part of the neural crest, it has been suggested that they and the neural crest cells evolved from a common ancestral cell line present in a prevertebrate ancestor.

The Pharyngula

We have emphasized the similarity of developmental phenomena in vertebrates. Discovery of these common features occupied the entire careers of many of the best biologists of the nineteenth century. Developmental differences between the major clades of vertebrates are enormous both because of uniquely derived patterns within each clade and because of the effects on cell division and migration of the quantity and location of yolk. However, the pharyngula stage of embryogenesis has a universality that transcends differences in cleavage, gastrulation, extra-embryonic membrane formation, and neurulation. The stage is named for the characteristic development of **pharyngeal pouches**. The primitive vertebrate feature of pharyngeal (or gill) slits makes at least a fleeting appearance in the embryos of all vertebrates. At this stage vertebrate embryos have their greatest similarity. After the pharyngula stage the distinctive characters of the adult organism begin to establish

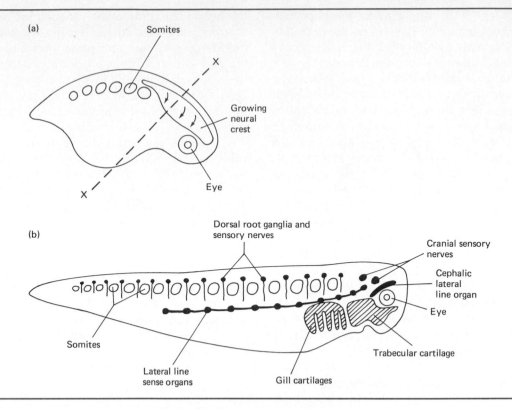

Figure 3–5. Migration of neural crest from the neural tube to contribute to several unique vertebrate structures: (a) lateral view of early embryo; (b) lateral view of more advanced embryo showing some neural crest derivatives (all black and hatched structures); (c) quail embryo sectioned through the head at the level of the mesencephalon as indicated in (a). Note the epithelial nature of the ectoderm and pharyngeal lining and the mesenchymal nature of the neural crest and paraxial mesoderm; (d) higher magnification of (c). [(c) and (d) courtesy of K. Z. Reiss.]

themselves and the embryos once again diverge in morphology.

The most conspicuous new feature of the pharyngula stage is the sculpturing of the sides of the posterior portion of the head into six to nine columns of solid tissue separated by deep furrows. The columns are the **pharyngeal** or **visceral arches**, the furrows, **pharyngeal grooves**. They form when endoderm of the pharyngeal lining evaginates in a series of pouches toward the exterior and visceral grooves of the ectoderm form indentations directly opposite the pouches on the external body wall.

When the evaginations of endoderm meet the invaginations of the ectoderm, they unite to form thin partitions called **closing plates**. The closing plates in nonamniotes perforate to become the gill slits. Most are transient, and they are obliterated in adult tetrapods, but the anteriormost maintains its thin membranous nature as the eardrum or **tympanum**. The linings of the pharyngeal pouches give rise to half a dozen or more glandular structures often associated with the lymphatic system, including the thymus gland, parathyroid glands, carotid bodies, and tonsils.

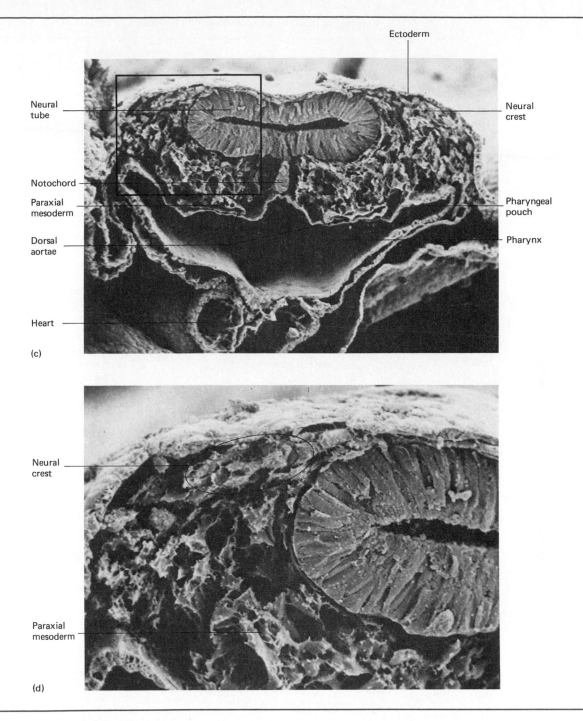

Ectoderm

Neural
tube

Neural
crest

Notochord

Paraxial
mesoderm

Pharyngeal
pouch

Dorsal
aortae

Pharynx

Heart

(c)

Neural
crest

Paraxial
mesoderm

(d)

Figure 3–5 (Continued)

Table 3–1. Embryonic origins of shared derived characters of vertebrates.

Character	Embryonic Origin
Integument and skeleton	
Skin of face and ventral neck (dermis, smooth muscle, and adipose tissue)	Neural crest
Pigment cells	Neural crest
Cephalic armor	Neural crest
Tooth papillae	Neural crest
Anterior neurocranium, sensory capsules (sclera of the eye), and fragments of the cranial vault	Neural crest
Dorsal fin mesenchyme	Neural crest
Nervous system	
Cranial nerves with sensory ganglia, including satellite cells	Neural crest, epidermal placodes
Trunk nerves with sensory ganglia, including satellite cells	Neural crest
Peripheral motor ganglia, including sympathetic and parasympathetic ganglia and plexuses	Neural crest
Forebrain	? Neural crest
Schwann cells (myelin sheath cells) of peripheral nerves	Neural crest
Meninges of prosencephalon and part of mesencephalon	Neural crest
Paired sense organs	
Nose	Epidermal placodes
Eyes (sclera?, lens, ciliary muscles, cornea)	Neural crest, epidermal placodes
Ears	Epidermal placodes
Lateral-line mechanoreceptors and electroreceptors	Epidermal placodes
Gustatory organs (taste buds, olfactory receptors)	Neural crest, epidermal placodes, endoderm
Pharynx and digestive tract	
Derivatives of visceral arch skeleton	Neural crest
Pharyngeal muscle	? Paraxial mesoderm
Connective tissues of pharyngeal muscles	Neural crest
Smooth muscle of gut	Lateral plate mesoderm
Calcitonin cells of ultimobranchial bodies and thyroid gland	Neural crest
Chromaffin cells of interrenal bodies and adrenal medulla	Neural crest
Connective tissue component of the pituitary, lacrymal, salivary, thyroid, parathyroid, and thymus glands	Neural crest
Circulatory system	
Gill capillaries	Lateral plate mesoderm
Muscularized aortic arches and their connective tissue components	Neural crest
Carotid sensory bodies in aortic arch	Neural crest
Muscular heart	Lateral plate mesoderm

Source: C. Gans and R. G. Northcutt, 1983, *Science* 220:268–274, N. Le Douarin, 1982, *The Neural Crest*, Cambridge University Press, Cambridge.

Each pharyngeal arch contains an arterial blood vessel carrying blood from the ventral aorta and developing heart. The vessel passes dorsally through the arch to unite with the other arch arteries to form the main conduit of blood to the body, the dorsal aorta. In fishes the arch vessels supply the gills, in tetrapods one arch is modified to supply the lungs. Each pharyngeal arch also contains the gill (or visceral) skeleton, typically consisting of an articulated crescent of pharyngo-, epi-, cerato-, and hypobranchial elements on each side and a midventral basibranchial. These elements support the gills of fishes and larval amphibians. In gnathostomes the skeletal elements of two anterior arches are modified to form jaws, the defining character of the group and a key feature in vertebrate evolution. Finally, each arch contains pharyngeal (or branchiomeric) muscle. In nonamniotes this muscle is used for respiration as well as for feeding, whereas in amniotes it is concerned primarily with food processing, head movement and, in mammals, facial expression.

Recent experimental studies of chickens suggest that all vertebrate voluntary musculature, including that associated with the pharynx, is derived from paraxial mesoderm (Noden and deLahunta 1985). This result conflicts with the traditional view that pharyngeal muscle, like all other musculature associated with the gut, is derived from splanchnic mesoderm of the lateral plate. If this result is confirmed in studies of other vertebrates, as seems likely to be the case, the traditional distinction between the visceral (gut-associated) and somatic (notochord-associated) organ systems of vertebrates will require reevaluation. In either case, it is clear that the strong muscularization of the vertebrate pharynx is associated with an evolutionary switch from ciliary movement of feeding currents through the gill clefts to muscular pumping of feeding currents. The elastic gill skeleton and the increased vascularization are also pharyngeal arch correlates of this functional shift.

The principal shared derived characters of adult vertebrates all have their origins in shared derived characters of vertebrate embryos—the neural crest, epidermal placodes, and muscular pharynx. The vertebrates entered a new adaptive realm with the novel feeding shifts which occurred at their origin. These shifts had profound effects on food detection and ingestion and put new selective pressures on metabolic functions, especially food processing and gas exchange. The relationship of these morphological and physiological changes of the unique characters of vertebrates and their embryonic origins are outlined in Table 3–1.

Protection, Support, and Movement

Motility characterizes vertebrates, and the ability to move requires muscles and an endoskeleton. Motility brings vertebrates into contact with a wide range of environments and objects in those environments, and the external protective covering must be tough but flexible. Bone, the mineralized tissue that we consider characteristic of the skeletons of living vertebrates, had its origin in this protective integument. The history of the shift of bone from superficial to internal skeleton parallels the history of increasing complexity of the vertebrate body plan, especially that of the organ systems involved in protection, support, and movement.

The Integument

The integument is a single organ, one of the largest of the body, making up 15 to 20 percent of the body weight in many vertebrates and much more in armored forms. It includes a series of derivatives of the skin, such as glands, scales, feathers, hair, spines, horns, and hoofs. The skin protects the body and receives information from the outside world (Bereiter-Hahn et al. 1986). The major divisions of the vertebrate skin (Figure 3–6) are the **epidermis** (the superficial cell layer derived from embryonic ectoderm) and the unique vertebrate **dermis** (the deeper cell layer of mesodermal and

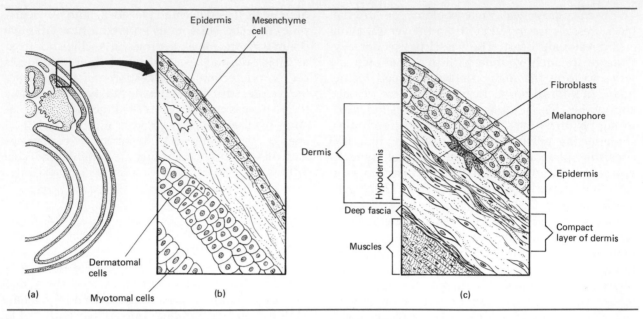

Figure 3–6. Development and differentiation of the vertebrate integument.
(a) The somite and ectoderm contribute to the adult skin. (b) In addition, wandering cells from the neural crest and later the mesenchyme cells contribute to the differentiation of the integument. (c) The differentiated skin as described in the text. [After M. H. Wake (editor), 1979, *Hyman's Comparative Vertebrate Anatomy*, 3rd edition, University of Chicago Press, Chicago.]

neural crest origin). Both rest on the mesodermal subcutaneous tissue (**hypodermis)** that overlies the muscles and bones.

Epidermis The boundary between a vertebrate and the environment is the epidermis. As such, it is of paramount importance in protection, exchange, and sensation. Nevertheless, it may be only a few cells thick in fishes. In nonamniotes (except for adult amphibians) the entire epidermis, even the surface cells, is alive and metabolically active. It often has secretory unicellular or multicellular glands.

In adult amphibians and other tetrapods the epidermis is an avascular cell layer with active and proliferating cells only in its deepest part and dead cells near the surface that are regularly shed. The layer of deep, active cells (**stratum germinativum)**

obtains its metabolic needs from the underlying vascular dermis. Epithelial cell divisions at this interface force daughter cells toward the surface. These daughter cells produce and accumulate protein granules (keratohyalin) and related substances, which slowly change as the cells die and are forced nearer the skin's surface. By the time cells reach the surface they are adherent sacs of **soft keratin**, a tough, pliable hydrophobic protein.

The cohesiveness between cells of the tetrapod epidermis during all stages of development and degeneration imparts marvelous characteristics to the tissue: it is living, growing, regenerative—and at the same time dry, abrasive, resistant, and expendable. Nonamniotes, especially fishes and aquatic amphibians, are subject to osmotic movement of water across the epidermis, and the living cells of the epidermis may play an active role in minimizing cutaneous water flux in conjunction with specialized cells of the gills (discussed in detail in Chapter 4). Terrestrial vertebrates are faced with evaporative water loss. Deposition of extracellular and intracellular lipids in the keratinized (dead) layers of the epidermis appears to establish a water barrier that greatly reduces cutaneous water loss. The details of the formation of an epidermal lipid water barrier differ in lizards, birds, and mammals, indicating multiple separate evolutions of cutaneous barriers. Epidermal derivatives are enormously varied among vertebrates—poison glands, feathers, hair, hoofs, and horns. The interaction of epidermis and dermis appears to be essential for formation of these complex appendages of the integument. An example is the production by the epidermis of **enamel**, the hardest tissue evolved by vertebrates. Mature enamel consists almost entirely of calcium phosphate-fluoride crystals in closely packed rods. The rods are secreted by basal epidermal cells and may attain considerable lengths in thick enamels. They are produced only by epidermal cells sharing a basement membrane complex with dermis cells that are producing **dentine**, another mineralized tissue. No other region of epidermis produces anything similar to enamel in spite of the basic similarity of most epidermal cells.

Dermis The dermis, unlike the epidermis, is unique to vertebrates and is composed primarily of extracellular products. An extracellular basement membrane welds epidermis to dermis and is important in the relationship between the two layers. The dermis is elastic, loose, and thin over joints and in areas of mobility. In areas of continual friction with the outside world (the soles of feet, prehensile tails, flippers), the dermis is thick, firm, and immobile. Its deepest layer of interlaced bundles and sheets of collagen fibers is produced by sparsely scattered cells (fibroblasts). The interweaving and the amount of elastic fibers determine the tensile strength of skin. In sharks and many mammals this strength is extraordinary, and it is primarily this layer that is used to make leather. Occasionally, smooth muscle fibers occur in the dermis. Their contraction in mammals produces skin wrinkling such as that of the nipple of the mammary glands or the male scrotum.

Blood vessels and nerves course through the lower dermis to their destination in the superficial dermis, a thin layer composed of more delicate and loosely arranged collagen and elastic fibers. Here arterioles break up to form a complex web of capillaries closely applied to the dermis–epidermis junction. These fuse again as venules, which in birds and mammals exit from the superficial dermis, often in parallel with an incoming arteriole; there may be direct connections between arterioles and venules (arteriovenous anastomoses). In birds and mammals two or more flat networks of vessels may occur at different depths within the dermis and hypodermis parallel to the body surface. These vascular networks are best developed in mammals lacking thick fur (for example, swine and humans), in naked regions of hairier mammals (the perineum, that is, around the urogenital/anal openings; the scrotum; the nostrils) or in the egg-incubating brood patches of many birds. Lizards also have well-developed vascular networks in the dermis. The amount of blood and rate of flow within these vessels can be varied by central neural and hormonal control or directly by local temperature. Under conditions of excess internal heat, the vessel plexus dilates and warm blood is cooled by its close contact with the skin surface. When the skin is chilled, most of the capillaries constrict, minimizing blood flow to the skin. The close juxtaposition of arterioles and venules, which continue to carry small amounts of life-supporting blood to the epidermis, acts to conserve heat by countercurrent exchange.

In tetrapods the dermis houses the majority of sensory structures and nerves associated with the sensations of temperature, pressure, and pain. Some free nerve endings penetrate the epithelium, but the majority terminate in the dermis as specialized end organs made up of connective tissue capsules of varying thickness.

Some neural crest cells come to rest in the dermis, in the deepest layers of the epidermis, and in the interface between the dermis and epidermis.

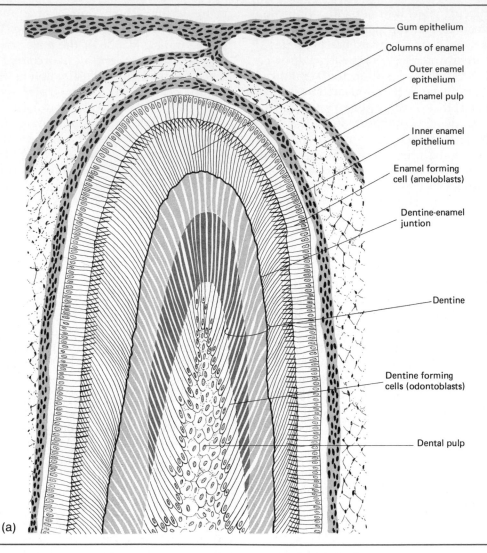

Gum epithelium

Columns of enamel

Outer enamel epithelium

Enamel pulp

Inner enamel epithelium

Enamel forming cell (ameloblasts)

Dentine-enamel juntion

Dentine

Dentine forming cells (odontoblasts)

Dental pulp

(a)

Figure 3–7. Organization of vertebrate mineralized tissues: (a) enamel and dentine as seen developing in a tooth; (b) bone from a section of the shaft of a long bone of a mammal.

These cells differentiate into **melanocytes** and produce granules of melanin that may be yellow, rusty red, brown, or black in color. Through long, narrow, cytoplasmic extensions of the melanocytes, melanin is injected into adjacent cells that lack melanin-producing enzymes. Additional hues may be produced by structural colors overlying melanin, by **chromatophores** other than the dark, melanin-containing ones, and by vascularization of the skin that gives shades from pink to scarlet. Skin coloration that incorporates the vascular system has the advantage of being rapidly variable so that red display areas may become brilliant in excited individuals. Skin coloration that involves the

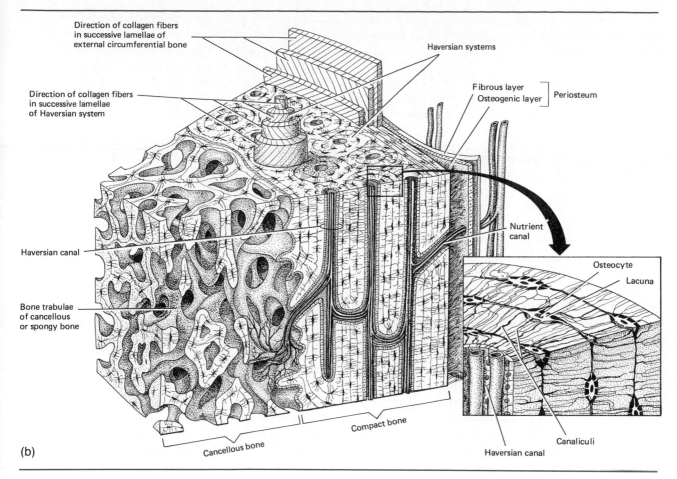

Labels in figure:

Direction of collagen fibers in successive lamellae of external circumferential bone

Direction of collagen fibers in successive lamellae of Haversian system

Haversian canal

Bone trabulae of cancellous or spongy bone

(b)

Cancellous bone

Compact bone

Haversian systems

Fibrous layer
Osteogenic layer
} Periosteum

Nutrient canal

Osteocyte

Lacuna

Canaliculi

Haversian canal

Figure 3–7 (Continued)

nervous system and cells where pigments rapidly aggregate or spread out under neural influence are the ultimate in color changing mechanisms. Many fish and some amphibians and lizards are capable of such exceptional change.

Hypodermis   The subcutaneous tissue or hypodermis is not functionally a part of the skin but lies between the skin and the muscles and bones. This region contains collagenous and elastic fibers, and in several lineages of vertebrates it contains a major energy store of the body: subcutaneous fat or oils. Fat storage reaches its maximum develop-

ment in mammals and birds from cold environments. Penguins, seals, sea lions, porpoises, and whales have hypodermal fat in the form of insulative blubber. A unique hypodermis characteristic of mammals is the extensive system of subcutaneous muscles that move the skin relative to underlying tissues. The fly-disturbing jiggle of a horse's skin is an example of its function. In many carnivores, ungulates, and especially in humans and the other derived primates, subcutaneous muscle from the pharyngeal muscles reaches exceptional development in the facial region. It produces facial expressions, movement of the external ears, opening and closing the eyelids, and the ability to purse the lips and suck—a vital attribute to early mammalian life.

Mineralization   The vascular and collagenous networks of the dermis play an important role in **mineralization**. Many fiber rich connective tissues such as the dermis have the capacity to induce and promote the crystallization of calcium salts (carbonates or phosphates) within the fibers of their extracellular matrix, the process of **calcification**. Such mineralization of connective tissue provides rigid mechanical support and a rich store of essential minerals but appears to inhibit diffusion of substances through the mineralized extracellular matrix. This leaves the connective tissue cells isolated from blood vessels. They begin to starve and suffocate from lack of nutrients and oxygen brought to the tissue by the circulation and are slowly poisoned by their own metabolic wastes, which can no longer diffuse into the circulatory system to be carried away. Vertebrates have two methods of achieving the advantages of mineralization while maintaining the vitality of the cells responsible for calcification, **dentine formation** and bone formation or **ossification** (Figure 3–7). In both processes a specific crystalline form of calcium phosphate, hydroxyapatite, is involved.

Dentine formation is restricted to the dermis and usually to the region of the basement membrane complex at the interface with the epidermis. In this process the cells promote mineralization along narrow cytoplasmic extensions on the side of the cell away from the nearest blood vessel. As crystals of hydroxyapatite form along these cytoplasmic extensions, the dentine-forming cells retreat toward blood vessels. They must, of course, elaborate longer cytoplasmic extensions to maintain dentine contact, and these become embedded in the highly mineralized dentine. The resulting mineralized tissue is dense, hard, and contains nonmineralized regions only in the thin tubules where the cytoplasmic extensions continue to deposit hydroxyapatite. These living projections maintain metabolic processes through their continuity with the rest of the cell. The cell bodies eventually retreat to clump around the blood vessel in a cavity within the mineralized tissue, the **pulp cavity**. Although dentine formation is usually thought of only in conjunction with gnathostome teeth, the placoid and cosmoid scales of gnathostome fishes, fossil thelodont scales, and the heterostracan and osteostracan head shields and scales all contain dentine tissue. Thus dentine formation appears to have been a general dermis-wide characteristic of the earliest vertebrates.

Ossification is a second mineralization process peculiar to vertebrates that allows continued vitality of the mineralizing cells. Ossification begins with cells forming a specialized network throughout a nonmineralized matrix. Cytoplasmic extensions make contact with those of other cells thus indirectly joining to cells in perivascular (close to a blood vessel) positions. Directly or indirectly every cell has cytoplasmic channels for metabolic exchange no matter how distant it may be from regional vascularization. Mineralization proceeds as bone matrix is deposited by the cells approximately equidistantly from neighboring blood vessels in the dermal plexus. These eventually form a latticework of branching and anastomosing needles of hydroxyapatite interwoven with the blood vessels (Fawcett 1986). The interconnected cells stimulate continued mineral deposition on the initial latticework until they become incarcerated in the matrix. These bone cells are able to remain metabolically active via their cytoplasmic supply lines to adjacent populations of cells. The unmineralized pockets in which the cells reside are the *lacunae* and the channels for the cell extensions the *canaliculi* . The metabolic activity of a bone cell continues through the period of **intramembranous ossification** (the building up of flat plate-like bones in the dermis as just described) and extends to reabsorption and remodeling of the internal microstructure of the bone throughout life. This ability to absorb and redeposit bone allows vertebrates to grow, to mobilize stores of minerals, and to heal damage to the mineralized tissue. When such **membrane bones** are restructured the mineralized matrix and bone cell networks are organized into layers and the blood vessels straightened out. The resulting **lamellar bone** is arranged in concentric layers around vascular channels to form cylindrical units called **Haversian systems**. Intramembranous bone was characteristic of the superficial armor of the Paleozoic jawless vertebrates. Unlike dentine formation, however, ossification invaded deeply-lying connective tissues to form the endoskeleton of later vertebrates.

The Endoskeleton

An endoskeleton is one of the basic characteristics of chordates. The **notochord** continues as the basic structural element of internal support for vertebrates. However, it is often obliterated during development by more familiar elements of the internal skeleton, the **vertebrae**. Phylogenetically older than vertebrae is the **cranium**, or skull, and appearing with it at the dawn of vertebrate life was the **visceral (pharyngeal) skeleton** of the gill arches and their derivatives.

Later contributions to the endoskeleton came with the evolution of ribs and the skeleton of the appendages. The cranium, visceral skeleton, notochord, vertebrae, and ribs are together called the **axial skeleton**. The paired fin or limb skeletons and girdles make up the **appendicular skeleton**. The capacity to form mineralized tissues, the essence of the vertebrate endoskeleton, shows a clear historic pattern of step-by-step penetration of deep-lying connective tissues, beginning with phylogenetically primitive dermal armor, progressing to cranial ossification, then to vertebral, and finally to rib and appendicular ossification. Skeletal ossification may primitively have been limited to superficial neural-crest-containing tissues, only later extending to deeper tissues of presumed neural crest origin and finally to tissues of presumed mesodermal origin. Perichondrally (on the surface of cartilage) mineralized neural (dorsal) and hemal (ventral) elements along the notochord were the first indications of vertebrae. Lateral plate skeletal derivatives (ribs, paired appendages) also acquired the capacity to support ossification of connective tissues. Each region of the axial and appendicular skeleton progressively became more and more completely ossified during the evolution of the vertebrates, producing the bony skeletons of amniotes as we see them today.

One of the processes that becomes increasingly important phylogenetically is that of **endochondral ossification**. In this ontogenetic process a cartilaginous model of the adult structure develops and is subsequently invaded by blood vessels and bone-forming cells. The cartilage is eroded and replaced by bone. Ultimately, the entire structure, which retains the general shape of the original cartilage, is ossified. This process is quite obviously different from the formation of dermal bone or intramembranous ossification. Although these processes are distinct and appear to have long separate phylogenetic histories, the final result, the bony tissue of the dermal or endochondral skeleton, is identical. Only developmental studies can determine how a particular bone was formed.

Skull The **skull**, which houses the brain and sense organs, is the most complicated portion of the skeleton. The cranium of vertebrates has evolved through additions from multiple ontogenetic sources, often repeatedly modified, to the point that recognition of homology is difficult at best. The skull is formed by three basic components: the **dermatocranium**, the **chondrocranium**, and the visceral arches or **splanchnocranium**.

The **dermatocranium** is in many ways the simplest of the skull's components. It is made up of dermal or intramembranous bones of the head integument (including the oral and pharyngeal lining). It covers the other portions of the skull. Ossification first appears in the fossil record in these superficial tissues, and their record is better than that for the other primitively unmineralized skull elements. The dermatocranium is made up of the roof of the skull, the superficial area around the orbits, the gill covers, the roof of the mouth (the palate and portions of the floor of the skull), and contributions to the jaws, which in their derived character state are formed primarily (e.g., birds) or entirely (mammals) by dermatocranium. The dermatocranium has been secondarily abandoned in

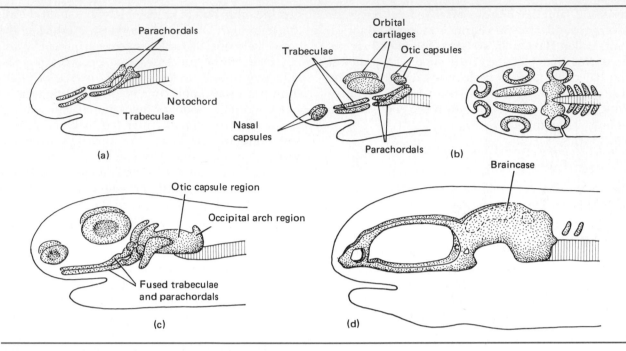

Figure 3–8. Early development of the chondrocranium. (a) through (d) are progressively more advanced stages of development. In (b) lateral view left, dorsal view right. The trabecular and parachordal cartilages are usually formed first, followed by the sensory capsules and the occipital arch. These centers of chondrification fuse to form the braincase. (Modified after W. W. Ballard, 1964, *Comparative Anatomy and Embryology*, Ronald Press, New York.)

living jawless fish and chondrichthyans, but in other vertebrates it surrounds, unites with, or replaces portions of the other components of the skull so closely as to have usurped the functions of chondrocranial and visceral elements.

The **chondrocranium** is the deepest lying and probably phylogenetically the oldest portion of the skull. As its name implies, the entire structure begins phylogenetically and embryologically as cartilage that subsequently is replaced, partly or entirely, by endochondral ossification. A few living taxa retain a cartilaginous chondrocranium in adult life. Because it is often not formed by cartilage in adult vertebrates, some authors prefer to call it the **neurocranium.** The chondrocranium (Figure 3–8) is made up of portions of the floor of the cranial

cavity that begin as a pair of cartilages flanking the anterior tip of the notochord (the **parachordals**). Other contributions come from the posterior wall of the brain cavity surrounding the spinal cord and from a pair of longitudinal cartilaginous bars (the **trabeculae**) located anteriorly and derived from the neural crest. The chondrocranium is completed by the cartilaginous capsules that form around two of the sense organs, the olfactory and otic capsules. The third pair of sense organs derived from placodes, the eyes, remain mobile and their connective tissue capsule becomes the sclera of the eyeball. These contributions to the chondrocranium by so many of the unique derived characters of vertebrates indicates the antiquity of the structure and its essential place in the origin of the vertebrates.

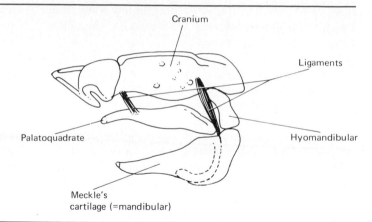

Figure 3–9. Relationships of the hyoid arch in supporting the jaw hinge as seen in the shark *Scyllium.* The hyomandibular, which acts as a strut to the jaw hinge, allows the jaw to be thrust forward for grasping and to enlarge the gape.

The final component, the **splanchnocranium**, comes from an unusual set of skeletal elements, the only hard ones to form in the wall of the gut (embryonic splanchnopleure), and are again derived from neural crest. Differentiating late during the pharyngula stage, the cartilaginous elements are all clearly related to the basic seven fleshy septa between pharyngeal gill slits or pouches. Each of these skeletal visceral arches is made up of several elements—usually five per side. In gnathostomes the anterior two pairs of arches become the jaws and their supporting structures (Figures 3–9 and 7–1), often undergoing partial or complete endochondral ossification. Where functional gills exist in adult vertebrates, they are supported by the more posterior arches. When gills are lost and respiration is via lungs, the posterior arches are much less significant but contribute to the larynx and trachea. The contribution of the visceral skeleton to the evolutionary success of the vertebrates cannot be overstated: The primary derived characters of gnathostomes originate from the visceral arch skeleton.

Notochord The notochord, long recognized as one of the most constant and characteristic features of vertebrates, is an outwardly simple, stiff but flexible rod. It resists shortening of the body during segmental muscle contraction and trans-

lates unilateral contractions into the graceful bending curves of sinusoidal lateral undulations—the primitive pattern of vertebrate locomotion. The notochord is a unique structure among the tissues of vertebrates. In some vertebrates it begins as a row of flat circular cells stacked like coins for the length of the body. At maturity it is made up of large, closely packed cells distended with fluid-filled vacuoles. This core of notochordal tissue is wrapped in a complex fibrous sheath that accounts for the rigidity of the rod. The physical characteristics of the notochord result from the incompressible nature of the fluids bound within the fibrous sheath. The sheath also represents the attachment surface between the notochord and the myomere connective tissue.

The notochord of all craniates ends just posterior to the bud of the pituitary gland. Posteriorly the notochord continues to the tip of the fleshy portion of the tail. In many fossil and living fishes the notochord as described (called an unrestricted notochord) appears as the endoskeletal body axis in the adult. In other fishes and in tetrapods the sheath of the notochord and the adjacent skeletogenous septa become involved in the development of structures that pinch and otherwise distort the notochord. In several clades of vertebrates the notochord becomes segmentally constricted into an axial series of nubbins (portions of the intervertebral disks in mammals) or obliterated altogether by derived axial endoskeletal elements, the vertebrae.

Vertebrae and Ribs   In the majority of gnathostomes the axial skeleton is made up of a series of intersegmentally arranged **vertebrae**. These dense, usually mineralized units offer strength and a greater number and variety of attachment points for muscles than does the simple sheath of the notochord. The series of vertebrae from the skull to the tip of the tail constitutes the **vertebral column** or **spine**, an organ of complex function. The spine provides rigidity to the body, especially in large terrestrial vertebrates where it becomes disproportionately massive. In swimming vertebrates the vertebral column retains the primitive function of resisting shortening during lateral undulation. To do this and yet permit flexibility, the spine is not made up entirely of rigid mineralized vertebrae. The vertebrae alternate with flexible cartilaginous **intervertebral disks**. These disks, which are partly remnants of the notochord, are bound between vertebrae by muscle and ligament attachments running from vertebra to vertebra. In terrestrial vertebrates the compressional forces of bilateral muscle contractions and the complex interlocking of their vertebrae often restrict the number of planes of movement possible in the spine. These interlocking surfaces provide a rigid support for the suspended weight of the body musculature and viscera.

Vertebrae are composite structures made up of mineralized cartilages or of both endochondral and intramembranous bony elements formed in the various skeletogenous septa intersecting at the central body axis. Six major components are usually recognizable: The **centrum** is a solid cylindrical spool that surrounds and often completely replaces or incorporates the notochord and makes up the body of the vertebra. A **neural arch** grows dorsally to cover the neural (spinal) cord like a rigid tent. A **hemal arch** grows ventrally (on postanal centra only) and similarly encloses the principal caudal blood vessels. **Neural** and **hemal spines** are bony blades that project into the dorsal

and ventral skeletogenous septa, respectively, and provide sites for attachment of muscles and ligaments. Finally, a variable number of bilaterally symmetrical **apophyses** project from the vertebrae and attach via muscles and ligaments to other skeletal elements.

Ribs are functionally and anatomically related to the vertebral apophyses. They provide sites for muscle attachment and strengthen the body wall. Ribs form in intersecting skeletogenous septa, generally articulating with vertebral apophyses forming in the same septa. Myosepta are one of the skeletogenous septa participating in rib formation and thus ribs are intersegmental structures. **Intermuscular** or **dorsal ribs** form at the myoseptal and horizontal skeletogenous septum intersection. Where the myosepta intersect the connective tissue surrounding the coelom, **subperitoneal** or **ventral ribs** form between the coelomic wall and the hypaxial muscles. The ventral ribs are thought to be phylogenetically older than the dorsal ribs but both types occur together in many bony fishes along with additional rib-like intermuscular bones peculiar to these fishes.

Vertebrae and ribs are only moderately specialized in most fishes, usually in conjunction with unique derived characters of a particular taxon such as hearing in minnows and their relatives. The trend in the vertebrae and ribs of tetrapods has been toward differentiation into five regions from anterior to posterior: **cervical, thoracic, lumbar, sacral,** and **caudal**. Enhancing mobility of the head while continuing protection of the spinal cord is the function of the cervical vertebrae. The two anteriormost cervical vertebrae are the **atlas** and the **axis**. The sacral vertebrae fuse with the pelvic girdle and transfer force to the appendicular skeleton. The trunk vertebrae are differentiated into two regions. The anterior region is the thoracic group with articulated ribs that function in respiratory movements in most amniotes as well as in support and protection for the lungs and heart. The **sternum** is a midventral, usually segmental structure, common to all tetrapods, that links the distal ends of right and left thoracic ribs. It ossifies well only in birds and mammals, and its evolutionary origins are obscure. The next region posteriorly is the lumbar group, with abbreviated ribs fused to the centra and supporting an extensive muscular sling for the pliant, mobile viscera of the posterior coelom. Posterior to the sacral region, the caudal vertebrae are generally much like those of fishes but with short spinous processes. The phylogenetic trend toward increased specialization of the vertebrae has proceeded independently in different lines as shown by the variable number of vertebra of each region in closely related taxa. The fact that a taxon as diverse as the mammals almost invariably has seven cervical vertebrae is surprising, and is the only notable exception to a general pattern of variation seen in tetrapods. Regional specialization of the axial endoskeleton reflects the appearance of specialized subdivisions of the dorsal musculature, for the skeleton is closely linked to the muscles as the skeletomuscular complex.

Appendicular Skeleton   Many of the evolutionary changes seen in elements of the axial skeleton of vertebrates are reflections of the vast reorganizations that have occurred in several stages of the evolution of the appendicular skeleton. Although median fins, especially the caudal fin, are essential to fishes, only the paired pectoral and pelvic appendicular structures have maintained a continuity of function across the fish to tetrapod boundary.

Paired appendages appear in all of the early clades of jawed fishes. Their anatomical origins remain obscure, with no clear embryological clues such as those to the origin of jaws. In jawed fishes, especially advanced ray-finned forms, the pectoral girdle is robust and articulates directly with the axial skeleton via the posterior part of the skull. The pelvic appendages retain the more primitive role of stabilizers but gain mobility often in conjunction with migration and attachment to the pectoral girdle.

In tetrapods the pelvic girdle becomes fused directly to modified sacral vertebrae and the hind limbs function as the primary propulsive mechanism. The pectoral girdle is most often free of di-rect articulation with the axial skeleton and is often less robust than the pelvic girdle.

Homologies of elements of the appendicular skeletons of living vertebrates are obscure, although there is a high degree of early embryonic similarity. Note that the similarity between the elements of the anterior and posterior appendages is not due to homology in the usual sense of that term. They may, however, be considered as **serially homologous** structures, derived from similar epigenetic forces acting in different body segments.

The fins and girdles of fishes are more variable than can be explained by the number of clades of fishes. The primitive form of the endoskeletal elements of the paired fins of most fishes is a semicircular girdle embedded in the ventral musculature providing articulation for a small number of fan-like **basal elements** (Figure 3–10a). The basals support one or more ranks of cylindrical **radials**, which usually articulate with one or more ray-like structures that support most of the surface of the fin web. The thickness of this web and the development of intrinsic fin musculature determines the flexibility and mobility of the fin surface. These characteristics are generally poorly developed in primitive vertebrates such as sharks and well developed in rays and some advanced bony fishes.

The basic tetrapod limb (Figure 3–10b) is made up of the limb girdle and five segments articulating end to end: the propodium (humerus in the forelimb and femur in the hind limb), the epipodium (radius plus ulna and tibia plus fibula), the mesopodium (carpus and tarsus), the metapodium (metacarpus and metatarsus), and the phalanges (fingers and toes).

The Muscular System

The skeleton's functions of support and locomotion would be impossible without the muscular system with which the endoskeleton has so intimately coevolved. The musculature of vertebrates is generally divided into **involuntary** (**smooth**) muscles derived from the splanchnic mesoderm and **voluntary** (**striated**) or **skeletal** muscles, of

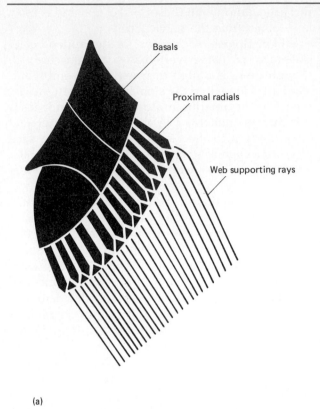

APPENDICULAR SEGMENTS		PECTORAL	PELVIC
GIRDLE		Scapula (shoulder blade)	Pelvis (hip bone)
PROPODIUM		Humerus (upper arm)	Femur (thigh)
EPIPODIUM		Ulna/Radius (forearm)	Tibia/Fibula (lower leg)
MESOPODIUM		Carpus (wrist)	Tarsus (ankle)
METAPODIUM		Metacarpus (hand)	Metatarsus (foot)
PHALANGES		Phalanges (fingers)	Phalanges (toes)

(a)   (b)

Figure 3–10.   Basic divisions of the endoskeletal supports of vertebrate paired appendages: (a) most fishes; (b) tetrapods as exemplified by the primitive mammalian condition with common and anatomical terminology compared.

which most, if not all, are derived from the myotomes of paraxial mesoderm.

The gross morphology of most of the muscle of fishes is similar, but it is radically different from that of tetrapods (Figure 3–11). Fishes retain obvious axial segmentation of the musculature, reflecting that of the embryonic somites. The generally short muscle fibers of fishes run from the posterior face of one myoseptum to the anterior face of the next posterior myoseptum. Thus the muscles seem to be in simple blocks. However, in adult fishes each myomere and its adjacent and well-developed myosepta are complexly folded fore and aft so that a given myomere extends anteriorly and posteriorly over several body seg-ments (and thus vertebrae). As a result, each intervertebral joint is acted upon by several sequentially arranged and innervated myomeres. The functional outcome is a smoothly graded bending of the spine as each motor nerve from head to tail fires down the sides of the swimming fish. The superficial midlateral muscle fibers in such an arrangement are nearly parallel to and at the level of the vertebral column. The deeper fibers, although running from myoseptum to myoseptum, may be strongly and regularly angled with respect to the longitudinal body axis due to the complex folding of the myosepta. The complex three-dimensional arrangement of fibers matches the amount of contraction of fibers close to the

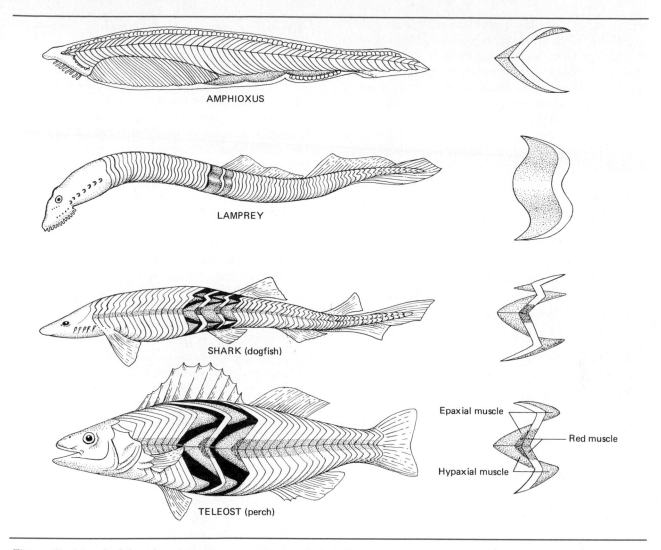

AMPHIOXUS

LAMPREY

SHARK (dogfish)

TELEOST (perch)

Epaxial muscle

Red muscle

Hypaxial muscle

Figure 3–11. In fishes the myotomes are more complexly folded in advanced chondrichthyans and osteichthyans than in jawless fishes. Acraniates like amphioxus have still less complex myotomal morphology. Unshaded, surface appearance below skin; light shading, deep portions of myotome associated with axial skeleton but not visible from the surface; dark shading, areas of concentration of red muscle fibers. (Modified from A. G. Kluge et al., 1977, *Chordate Structure and Function*, 2nd edition, Macmillan, New York; and Q. Bone and N. B. Marshall, 1982, *Biology of Fishes*, Blackie, Glasgow.)

vertebral column (and thus near the center of curvature) with the amount of contraction of those farthest from it, with a resultant efficiency not possible in an all-fibers-parallel arrangement. Differences in the three-dimensional arrangement of muscle fibers also occur in tetrapods. However, in terrestrial vertebrates individual muscles are likely to differ in fiber orientation from neighboring muscles rather than having substantial differences of fiber orientation within the same muscle. One reason for differences in tetrapod muscle fiber orientation is the variable way the muscles relate to the skeleton. Muscles may attach directly to the connective tissue sheath of a bone, but it is more common for muscles to attach to the skeleton through **tendons**, tightly packed bundles of the long fibrous protein collagen. Tendons, whether in fishes or tetrapods, allow the force of muscle contraction to be applied at restricted insertions on a bone so that many muscles, through their tendons, can act on a small skeletal element. **Ligaments** are often highly elastic so they store energy when they are stretched by forces and can return a joint to its original position without additional energy when the muscular contraction ceases.

Skeletomuscular Function

The skeleton, activated by muscles that may work at a distance through tendons, is essentially a system of levers. The skeletomuscular apparatus is subject to the same physical constraints as simple levers: Levers that maximize power act over short distances and are slow, whereas levers that maximize speed of action act over long distances and are weak. Sometimes the skeletomuscular systems are arranged into complex, interconnecting lever units known as **kinematic chains**. Prime examples of kinematic chains are the jaw opening and protrusion mechanisms of fishes and the kinetic head morphology of snakes. If one link in such a chain is fixed (by muscle contractions that brace it immovably against the axial skeleton), it limits the range of movements possible by other elements in the kinematic chain. When the fixed element is freed, however, the motions of the remaining parts of the kinematic chain of levers may be very different. By fixing in position different elements of such a chain at different times an animal can obtain a very wide range of motions. The force outputs from a limited number of skeletal and muscular elements is thus greatly diversified.

Energy Acquisition and Support of Metabolism

Vertebrates are among the most intensive consumers of energy on Earth. The energy vertebrates use is gleaned from the environment as food that must be processed to release energy and nutrients. This processing is the primary function of the digestive system.

After food has been digested and assimilated into the body, it must be transported to the tissues where the energy it contains is released and some of its chemical constituents are incorporated into the body tissues of the animal. Oxygen is required for this process, and the functions of gas exchange surfaces and the circulatory system are closely intertwined with those of the digestive system.

Feeding and Digestion

The trophic process of vertebrates is divisible into feeding and digestion. Feeding includes getting food into the oral chamber, some oral or pharyngeal processing, and swallowing. Digestion includes the breaking down of complex compounds to small molecules that are absorbed across the wall of the gut. Both feeding and digestion are two-part processes; each has a physical component and a chemical component, although the physical components dominate in vertebrate feeding, the chemical ones in digestion.

Most vertebrates feed on large particles relative to the size of their heads. In aquatic environments suction feeders take advantage of the viscous properties of water by creating a flow of water that carries food into the mouth. No similar mechanism exists among terrestrial vertebrates. The anterior teeth or the beaks of most vertebrates are used to capture prey and to sever pieces of food items. Most vertebrates do little or no mechanical processing of food in the mouth before swallowing. However, some chemical processing may take place. Terrestrial vertebrates have salivary glands that secrete a watery lubricant that probably aids swallowing. The salivary secretions of many vertebrates contain enzymes that begin the process of digestion. These secretions have been elaborated into venoms that kill prey in insectivorous mammals, certain lizards, and, of course, several lineages of snakes. A variety of venoms are known in fishes from many evolutionary lineages, but they all appear to be used defensively. None is known to be used in subduing potential prey and venoms are rarely associated with the teeth of fishes, perhaps because fishes do not have salivary glands.

Swallowing is accomplished by contraction of the splanchnic musculature of the gut wall. These muscles, the mechanism of swallowing, and peristalsis (traveling waves of circular muscle contraction in the gut) are derived characters of the vertebrates. Other chordates use ciliary action throughout the gut to move food, digestate, and solid wastes.

The **stomach**, probably a derived character of gnathostomes, is in general a rather simple organ in gross anatomy if it is present at all. Its lining is often abruptly different from that of the **esophagus**, from which it is separated by a muscular sphincter. The stomach of most vertebrates is capable of considerable distention. This is carried to an extreme in some deep-sea fishes: the swallowers (family Saccopharyngidae, order Elopomorpha) have a pharynx, esophagus, and stomach of such distensibility that they can swallow fishes twice their own length.

The size and structure of the stomach is related to dietary habits. Vertebrates that feed on fluids have reduced or nonexistent stomachs; these include jawless fishes such as lampreys and nectar-feeding birds and mammals. Herbivores, which feed primarily on the digestion-resistant leaves and stems of plants, often have greatly enlarged and frequently multichambered stomachs. As a rule, herbivores also have elongate intestines that permit the slow release of nutrients from plants.

A glandular epithelium lines the stomach of vertebrates. The glands in the stomach lining secrete hydrochloric acid and the proteolytic enzyme pepsin, which functions only in an acid medium.

The muscles of the stomach wall churn and mix the food forming a particle-laden solution known as chyme. Chyme passes to the small intestine through the pyloric sphincter or valve for further digestion and absorption.

The intestine is straight only in living jawless fishes. The intestines of most vertebrates are many times longer than the body and are coiled or folded. Long intestines are associated with herbivorous food habits. Sharks, skates, primitive bony fishes including lungfishes, and at least one fossil (a placoderm) illustrate an alternative method of increasing the area of intestinal surface in contact with the chyme—the **spiral valvular intestine**. From its phylogenetic distribution this distinctive structure is apparently primitive, at least within gnathostomes. The spiral valve consists of a broad sheet of the secretory and absorptive lining of the intestine that protrudes into the lumen (cavity) of the gut; the base of the fold extends from near the pylorus to near the end of the gut in a tight spiral path along the wall of the intestine. Both long intestines and spiral valves increase the surface area of the intestine, augmenting its function of absorption. Additional specializations, especially well developed in mammals, further increase the surface area of the intestine by large- and small-scale foldings of the lining. In birds and mammals finger- or leaf-like projections composed of epithelial cells around a core of mesoderm form **villi**. The gut epithelial cells themselves have closely packed **microvilli** on the lumenal surface of each cell. These structures increase the surface area of the intestine greatly: the total surface area of the human intestine is about 300 square meters, about the size of a regulation basketball court.

The small intestine secretes mucus and probably enzymes that break down the cell walls of microorganisms. Accessory digestive organs adjacent to the gut produce enzymes and related substances that break chyme down to the simple sugars and amino acids that can be absorbed by the gut.

The **liver** is the largest gland in the body of any vertebrate, reaching a maximum of 25 percent of the body mass in some sharks. In these fishes it is a storage reservoir for oils in addition to its normal functions. In all vertebrates the liver produces digestive secretions, processes absorbed nutrients and metabolites, and detoxifies harmful substances. These functions are facilitated by the position of the liver in the circulatory system: It receives blood coming from intestinal walls carrying the molecules absorbed from the digestate. The secretory function of the liver is twofold; it produces **endocrine** secretions for release into the blood and **exocrine** secretions (defined as those carried to the outside of the body—in this case to the lumen of the gut). The exocrine secretions are combined in the **bile**, which is part waste (neutralized toxins) and part digestive juice. The digestive components are not enzymatic, but rather fat-emulsifying organic molecules that act like detergents upon entering the gut. They form submicroscopic droplets of fats a few molecules in size, allowing them to be absorbed by the intestinal epithelium. Without these bile secretions, fats and oils in foods cannot be absorbed and would simply pass through the tract. Bile enters the intestine via the bile duct just posterior to the pyloric valve. A gallbladder may or may not occur as a blind sac along the bile duct. The distribution of the gallbladder among vertebrates, its size, and the overall rate of bile secretion, correlate positively with the amount of fats consumed.

The **pancreas** is the second of the accessory digestive glands along the intestine. Although not involved in as many complex biochemical processes as the liver, the pancreas (when present as a definitive gland) has both endocrine and exocrine functions segregated quite sharply by cell type. A discrete pancreas is absent in lampreys and most bony fishes. However, cells with pancreatic function are scattered in the mesenteries or embedded in the wall of the intestine or the liver of these vertebrates. Pancreatic enzyme secretions are essential for the final breakdown of carbohydrates to simple sugars, and fats to glycerol and fatty acids. Most important are as many as five potent pancreatic protease and nuclease enzymes that complete the digestion begun in the stomach.

These enzymes would, of course, digest the pancreas if they were not secreted as inactive precursor molecules that are activated upon mixing with the lumenal contents. The bile and pancreatic juice are also rich in bicarbonate that neutralizes the stomach acid; thus, digestion in the gut occurs at neutral pH. A final suite of digestive enzymes is incorporated in the membranes of the microvilli of the intestinal epithelial cells. Digestion occurs on the surfaces of these microvilli, which also possess the carrier molecules necessary for active transport of glucose and amino acids into the cells and eventually to the circulatory system of the gut. The products of fat digestion are biochemically reconfigured and shunted to the lacteals, specialized branches of the lymphatic system of the gut wall. From there fats are introduced to the general circulation, bypassing the liver.

One (in mammals) or a pair (in birds) of **ceca** (singular, cecum) are found at the junction of the large and small intestine. Some bony fishes have similar ceca in the region of the pyloric valve. Ceca appear to be fermentation chambers for gut microorganisms that digest cellulose. These symbiotic microorganisms break down plant cell components, especially carbohydrates, which may then be available for absorption. The **large intestine** or **colon**, a unique character of tetrapods, is primarily a region of absorption, mucus being the only important secretion. Its surface is less complicated than that of the small intestine. The colon ends in a short **rectum** that generally has an involuntary sphincter formed by smooth muscle and a voluntary sphincter formed by striated muscle. The rectum may open to the outside directly through the anus (advanced bony fishes and most mammals) but usually opens into a small chamber, the **cloaca**, shared with the opening of the urogenital ducts. A cloaca is found in the embryos of all vertebrates, and it is considered a distinctive primitive character of vertebrates. **Feces** are the result of removal of water, ions, and some digested organic molecules from the chyme. The remaining undigested food residue, bile secretions, microorganisms, and actively secreted wastes such as heavy metals are formed and passed by the rectum. In mammals, anal glands frequently add lubricants as well as pheromones to the numerous volatile substances already present in the feces. Thus, feces are used in communication, especially to mark territorial boundaries.

Respiration and Ventilation

Energy is obtained from the metabolic oxidation of nutrients absorbed across the gut wall. This process requires the delivery of oxygen to every living cell. Respiration is a multistage process that requires ventilation of respiratory surfaces where oxygen is absorbed from the environment, transportation of oxygen in the blood, and sometimes storage of oxygen in the tissues.

Ancestral chordates probably relied on **cutaneous respiration**—oxygen absorption across a thin aquatic skin—a mechanism that continues to be important for many vertebrates. The pharyngeal gill slits, initially part of the feeding system, had a large surface area and were constantly exposed to flowing water. Early in vertebrate evolution the endodermal lining of the pharyngeal arches developed a complex highly folded surface, the **gill lamellae**, richly invested with blood vessels. The gills of living fishes are variations on this theme. Within the bony fishes the relative breadth of the lamellar attachment to the fleshy gill septum distinctly diminishes from the condition seen in the sharks and skates (the elasmobranchs, "plate gills"). An intermediate degree of attachment is seen in primitive ray-finned bony fishes; complete loss of the interlamellar septum and development of long, free gill filaments is characteristic of most teleost bony fishes.

In addition to functional gills, many bony fishes have accessory respiratory organs that are derived from the gut. The most common and phylogenetically oldest of these structures appears to be the primitive swim bladder, the homolog of the tetrapod lung. These blind-ended, thin walled vascular sacs arise from outpocketings of the gut just posterior to the pharyngeal pouches. In fishes and amphibians the hyoid apparatus is used to ventilate the lungs. The oral cavity is expanded, sucking air into the mouth, and then the floor of the mouth is raised, squeezing the air into the lungs. This method of lung ventilation is called a force pump or positive pressure mechanism.

Amniotes use a suction pump or negative pressure mechanism of lung ventilation. Movement of the ribs (in birds and squamates), internal organs (turtles and crocodilians), or a muscular diaphragm (mammals) increases the volume of the **pleural cavity**, which houses the lungs. This expansion creates a negative pressure (that is, below atmospheric pressure) in the pleural cavity and sucks air into the lungs. Air is expelled (exhaled) by compressing the pleural cavity, primarily through the elastic return of the rib cage and contraction of the elastic lungs. The lungs of most amniotes are divided into numerous blind chambers (**alveoli**) where gas exchange occurs. Alveoli are best developed in mammalian lungs, but some squamates also have well-developed alveoli. The posterior lung of some squamates is undivided and nonvascular. Similar structures are highly developed in birds and are known as air sacs, some of which penetrate hollow bones or adhere closely to flight muscles. The respiratory portion of the lung of birds is also distinct. A complex system of channels allows air to flow in one direction past exchange surfaces.

Vocalization The regular large tidal flow of air through the pharynx and trachea to the lungs has set the stage for tetrapod vocalizations. Tetrapods restrict and sometimes redirect the flow of respiratory air so that it is forced through openings occluded by thin but often muscular membranes that vibrate and produce sounds. Usually, a **larynx** (a derived character of tetrapods that incorporates elements of the hyoid and other pharyngeal arch elements) at the junction of the pharynx and trachea is the site of sound production. In birds, however, a second site, the **syrinx**, has taken over sound production. The syrinx is formed by tracheal cartilages, muscles, diverticula of air sacs,

Figure
single-c
exempli

The walls of the capillaries are generally one cell thick. The cells making up the capillary walls and lining the interior of the rest of the circulatory system are called **endothelial** cells. Some endothelial cells in the capillaries of the intestine, pancreas, and endocrine glands are perforated by small holes, each closed by a thin diaphragm through which dissolved substances are believed to pass. The majority of endothelial cells have vesicles that transport dissolved substances across the capillary wall.

In addition to capillaries, the smallest veins (venules) may have walls thin enough to be involved in exchange with surrounding tissues. All the other vessels function only in the distribution and collection of blood. Arteries have walls made up of three layers: an internal endothelium, a middle layer of involuntary (smooth) muscle cells, and an outer layer of fibrous connective tissue. The muscle cells in the middle layer are usually arranged circumferentially and the connective tissue fibers in the outer walls are oriented longitudinally. The muscle layer is thick in arteries, whereas the connective tissue layer is better developed in veins.

Arteries and veins usually run side by side in a supply trunk related to a particular organ or set of tissues. Veins are usually more numerous than arteries, and the volume of the venous system is larger than the volume of the arterial system. Blood pressure in the venous system is lower than arterial pressure, partly because of this greater volume and partly because of the pressure drop that occurs as the blood passes through the resistance created by the narrow capillaries. Blood flows more slowly in veins than in arteries, and some veins have valves that prevent backflow.

Arteries are more elastic than veins because the rubbery protein elastin is present in the middle layer of the artery wall. Indeed, the main arteries of large mammals contain so much elastin that they appear yellow. Near the heart these vessels distend during ventricular contraction (systole) and elastically recoil during ventricular refilling (diastole). This expansion and contraction of the arteries smooths out the pulsatile flow of blood produced by the pumping of the heart. Elastin is unknown among invertebrates and has not been found in hagfish or lampreys, but it occurs in all gnathostomes (Sage and Gray 1979, 1981a,b). Sharks have elastin with a presumably primitive amino acid composition, and ray-finned fishes have an elastin that is different from the elastins of sharks and also from that of lungfishes, coelacanths, and tetrapods. These changes mirror the basic pattern of vertebrate phylogeny and chronicle the increase in systemic blood pressure in gnathostome evolution.

Some vessels, known as **portal vessels**, are interposed between two capillary beds. In those instances arterial blood passes through the first capillary bed and is collected in a vein that subsequently breaks down into a second capillary bed in a different tissue or organ. Finally, the blood is collected in a systemic vein and carried to the heart. Portal vessels promote sequential processing of blood. The hepatic portal vein, for example, is interposed between the capillary bed of the gut and the sinusoids of the liver. Substances absorbed from the gut are transported directly to the liver where toxins are rendered harmless and some nutrients are further processed or removed for storage. The hepatic portal system appears in all living vertebrates and is considered a primitive character for vertebrates. A renal portal system occurs only in some nonamniote gnathostomes and is apparently a derived character of gnathostomes.

Lymphatic vessels resemble veins but have thinner walls and are not part of the closed blood circulatory system. The lymphatic system is a one-way system of vessels that largely parallels the blood vessels. Lymphatic vessels originate as blind-ended tubes in the tissues and connect to the major veins. Plasma that is forced out of capillaries into the tissues by hydrostatic pressure is collected by the lymphatic system and channeled back into the circulatory system. Specialized lymph drainage in the intestine via the lacteals has already been discussed in the context of absorption of fat. The lymphatic systems of most nonamniotes have muscular areas that contract and keep the lymph flowing. In amniotes lymph is moved by the compression of lymphatic vessels as muscles and tissues contract and move, and a series of valves keeps the flow moving toward the circulatory system. **Lymph nodes** lie along the course of the lymphatic vessels of mammals and a few birds. Lymph in the nodes is exposed to phagocytic cells that cleanse it of foreign material, and lymphocytes, part of the immune system, are added to the lymph and thus distributed via the circulatory system to all parts of the body.

Basic Vascular Circuits The primitive vertebrate circulatory plan consists of midline vessels with the major arteries in a dorsal position and the veins ventral to them (Figure 3–13a). The heart pumps blood anteriorly in the **ventral aorta** (the only major ventral artery) to gill capillaries, where it is oxygenated. The blood then enters the **dorsal aorta** and flows posteriorly into arteries, arterioles, and finally capillaries. Blood leaving the capillaries collects in venules and then in veins of increasing size, returning to the heart in the **posterior cardinal veins**. A fish-like or **single circulation** of this sort is a low-pressure system because of the drop in blood pressure across the resistance provided by the gill capillaries. The entire output of the heart passes through the gills, and by the time blood returns to the heart it may have passed through as many as three capillary beds (i.e., gills, general body tissues, and either the liver or kidneys after passing through a portal system).

With the advent of lungs vertebrates evolved a **double circulation** in which the **pulmonary** circuit supplies the lungs with deoxygenated blood and the **systemic** circuit supplies oxygenated blood to the body. The heart is the pump for both circuits, and is divided into right (pulmonary) and left (systemic) halves. The division of the heart into right and left sides is an actual morphological separation in archosaurs (crocodilians and birds) and mammals. In lungfishes, amphibians, squamates, and turtles there is no permanent morphological separation, but complex interactions between heart morphology and flow patterns maintain the separation of oxygenated and deoxygenated blood.

The double circulation of tetrapods can be pictured as a figure eight with the heart at the intersection of the loops. One loop is the pulmonary circuit, and the other is the systemic circuit. This morphology permits the blood pressure to be low in the pulmonary circuit (where the blood flows through delicate capillaries during gas exchange) and high in the systemic circuit (where blood must be pumped long distances and through muscles that are contracting and squeezing the blood vessels).

The Heart As one would expect, the transition from a single circulation to a double circulation required major changes in the morphology of the heart and major vessels. The heart of fishes is a muscular tube folded upon itself and constricted into four sequential chambers: the **sinus venosus**, the **atrium**, the **ventricle**, and the **conus arteriosus** (or bulbus cordis). The sinus venosus and conus arteriosus are reduced or absent in the hearts of tetrapods and their prominence in fishes probably relates to the low peripheral pressure of the single circulation. The atrial and ventricular chambers are extensively modified in tetrapods.

The anatomical limits of the heart can be defined by the presence of **cardiac muscle**. Although the heart is controlled by the autonomic nervous system (as are most organs that contain smooth muscles), the fibers of cardiac muscle do not resemble smooth muscle but are striated like skeletal muscle. Unlike voluntary muscle, however, the cells of cardiac muscle appear to have one nucleus and distinct terminal junctions, the **intercalated disks** that are visible in histological preparations.

Blood enters the tubular fish heart via multiple right and left systemic veins and usually by a hepatic vein from the liver. The **sinus venosus** is a thin-walled sac with a few cardiac muscle fibers. It is filled by pressure in the veins and pulsatile drops in pressure in the pericardial cavity as the heart beats. Suction produced by muscular contraction draws blood anteriorly into the **atrium**, which has valves at each end that prevent backflow. It thus functions as a one-way pump that fills the main pump, the **ventricle**. A cross section of

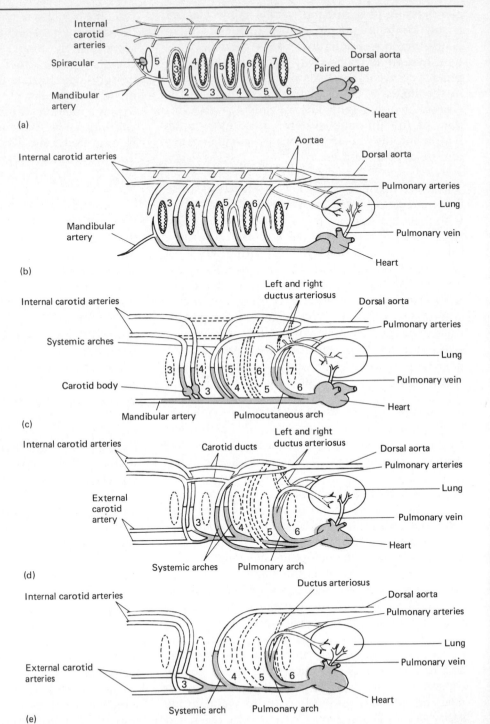

Figure 3–13. Evolution of the aortic arches. (a) Generalized scheme for a gill breathing primitive vertebrate. Chondrichthyans and actinopterygians have numerous independently derived specializations that do not depart drastically from this basic plan. (b) The lungfish and (c) the adult anuran demonstrate variations on the double circulation and dual respiratory adaptations of these vertebrates. (d) Primitive amniote condition as found in a lizard. (e) Advanced amniote condition as found in a mammal. Upper course of numbers, gill pouches; lower course, aortic arches.

the ventricle of a fish shows thick, muscular walls and a spongy interior with a small lumen. Contraction of the ventricle forces the blood into the **conus arteriosus**. Teleosts lack a conus arteriosus formed by cardiac muscle, but have an elastic bulbus cordis that is formed by smooth muscle and is perhaps better regarded as a modified part of the ventral aorta. The conus and bulbus serve as elastic reservoirs that damp the pulses of blood pressure that are produced by contractions of the ventricle, and they have a valvular construction that prevents blood from flowing back into the ventricle.

Evolution of Vertebrate Cardiovascular Systems
The circulatory systems in lungfish and amphibians are quite similar. The resemblances probably reflect the presence in both taxa of dual respiratory mechanisms that vary in their dominance as environmental conditions change. The lungfish has both gills and lungs; adult amphibians have lungs and important cutaneous and mouth (buccopharyngeal) respiratory surfaces. Details for individual genera vary, but the general trends in structural evolution of lungfish and amphibian cardiovascular systems include (1) evolution of a double circulation with systemic and pulmonary circuits, (2) alteration of the primitive gill arch complement involving reduction in number of arches and specialization of individual aortic arch blood distributions, and (3) morphological and functional division of the heart to serve the two circuits and the increasingly specialized aortic arches.

The atrium of the heart is divided into right and left chambers—either anatomically by an interatrial septum or functionally by flow patterns. Blood from the systemic veins flows into the right side of the heart and that from the lungs to the left. The ventricle shows a variable subdivision which correlates with the physiological importance of lung respiration to the species (Figure 3–14). In frogs the ventricle is undivided, but the position within the ventricle of a particular parcel of blood before ventricular contraction appears to determine its fate upon leaving the contracting ventricle. The short conus arteriosus contains a spiral

valve of tissue which differentially guides the blood from the left and right sides of the ventricle to the aortic arches. The anatomical relationships are such that blood from the pulmonary veins enters the left atrium, which injects it on the left side of the common ventricle. The spongy muscular lumen minimizes the mixing of right- and left-side blood. Contraction of the ventricle tends to eject blood in laminar streams that spiral out of the pumping chamber carrying the left-side blood into the ventral portion of the spirally divided conus. This half of the conus is the one from which the carotid (head supplying) and systemic aortic arches arise.

Thus, when the lungs are actively ventilated, oxygen-rich blood returning from them to the heart is selectively distributed to the tissues of the head and body. Systemic blood entering the right atrium is similarly directed, but into the dorsal half of the spiral-valved conus. It goes to the pulmonary arch (pulmocutaneous arch in amphibians), destined for oxygenation in the lungs. However, when the skin is the primary site of gaseous exchange as it is when a frog is underwater, the highest oxygen content is in the systemic veins that drain the skin. The lungs may actually be net users of oxygen, and due to vascular constriction little blood passes through the pulmonary circuit. Because the ventricle is undivided and the majority of the blood is arriving from the systemic circuit, the ventral section of the conus receives blood from an overflow of the right side of the ventricle. The scant unoxygenated left atrial supply to the ventricle also flows through the ventral conus. Thus, the most oxygenated blood coming from the heart flows to the tissues of the head and body during this shift in primary respiratory surface, a phenomenon possible only because of the undivided ventricle. The variability of the cardiovascular output in dipnoans and amphibians is an essential part of their ability to use alternative respiratory surfaces effectively.

Although they generally use only lung respiration, lizards, snakes, turtles, and crocodilians also have a wide variety of cardiopulmonary morphologies. A small sinus venosus supplies the

pass through the capillary endothelium. **Lymph** is derived from interstitial fluid and is augmented by the addition of some of the larger proteins characteristic of circulating plasma and by certain white blood cells from the lymph nodes (lymphocytes). Cerebrospinal fluid and the similar aqueous humor of the eye have a low content of protein and different proportions of small organic compounds and inorganic salts compared to blood plasma. The cerebrospinal fluid is the result of active and highly selective transport of portions of the blood plasma by the vascular membranes that cover the brain.

Nearly half of the volume of the blood (depending upon the species) is made up of erythrocytes. These red blood cells owe their red color to high concentrations of **hemoglobin**, the iron-containing globin protein that functions as a carrier for oxygen. The structure of erythrocytes varies among vertebrates. Most mammals have eliminated the nucleus and many other cell organelles from mature erythrocytes, so that technically they are not living cells. Other vertebrates retain nucleated erythrocytes but the nuclei seem functionally inert. At least one species of fish (the Antarctic ice fish, *Channichthys*) has eliminated erythrocytes altogether, transporting oxygen exclusively in its plasma in sufficient quantities to support its lethargic life. The sites of erythrocyte formation differ interspecifically and also change during the ontogeny of an individual. Adult sites of **hematopoiesis** (blood cell formation) include the blood vessels of fishes, the gut wall, kidneys and liver of most vertebrates, and a few specialized organs especially characteristic of tetrapods: the **spleen**, the **thymus**, the **lymph nodes,** and the **red bone marrow**. In adult mammals erythrocytes are almost exclusively produced in the bone marrow. Mature vertebrate erythrocytes appear incapable of replication or even self-repair, and there is a considerable turnover of these cells. New ones are produced and released into the circulation and old or damaged ones are scavenged from the blood, generally by the same hemopoietic tissues.

Erythrocytes contribute to homeostasis in a myriad of ways. The affinity of hemoglobin for oxygen (that is, how tightly oxygen molecules are held by hemoglobin) is changed by genetic, physical, and chemical effects. The most familiar illustration of the contribution of the oxygen affinity of hemoglobin to homeostasis is the way in which oxygen release is adjusted to match the metabolic requirements of different tissues.

The oxygen affinity of hemoglobin is reduced by acidity and increased by alkaline conditions. In other words, hemoglobin releases oxygen more readily at low pH than at higher pH. Carbon dioxide, which is produced by cellular metabolism, diffuses into the blood plasma and from there into the erythrocytes. Erythrocytes are rich in the enzyme **carbonic anhydrase**, which catalyzes the reaction between water and carbon dioxide that yields carbonic acid (H_2CO_3). Carbonic acid then dissociates to form a proton (H^+) and a bicarbonate ion (HCO_3^-). The protons liberated by the dissociation increase the acidity of the blood, and the lower pH reduces the affinity of hemoglobin for oxygen, resulting in the release of oxygen from the hemoglobin. This mechanism adjusts the delivery of oxygen to the metabolic needs of different tissues because carbon dioxide production is proportional to oxygen consumption.

In the gills or lungs the process is reversed: hemoglobin that has released its oxygen in the tissues is exposed to high oxygen pressure and binds oxygen, releasing protons in the process. The increased concentration of protons drives the reaction in the reverse direction, recombining the protons with bicarbonate ions to form carbonic acid. Carbonic anhydrase in the erythrocytes catalyzes the formation of carbon dioxide and water from the carbonic acid, and the carbon dioxide diffuses across the wall of the capillary into the water or air.

Cells specialized to promote clotting of blood (**thrombocytes**) are present in all vertebrates except mammals, where they are replaced by noncellular **platelets**. These blood elements react to surfaces they do not normally encounter by adhering to the unfamiliar surface and to each other. Specific blood proteins react to damaged tissues as well, polymerizing as a network of fibers that enmesh cellular components of blood to form a wound-plugging clot.

White blood cells (**leucocytes**) are less abundant in the blood and more diverse than erythrocytes. There are at least five types of white blood cells and they are living, complete cells. Leucocytes form the first line of defense against invasion by pathogenic bacteria and other foreign particles, they are involved in inflammation responses, and they support antigen–antibody activities of the immune system. Unlike erythrocytes, which function entirely within the vascular channels, leucocytes squeeze between the endothelial cells of capillaries and venules and spend most of their lives moving freely through the loose connective tissues.

Immune System A significant aspect of the survival of an organism is the ability to distinguish between *self* and *nonself* so that invasion by foreign substances can be prevented. Although all animals show some capacity to recognize self at a molecular level, only the vertebrates possess an integrated cell and plasma-interstitial fluid **antibody immunity** that resists almost all organisms and toxins that damage tissues and does so with great *specificity* and *memory* (Cooper 1985). When an **antigen** (*anti* = against; *genos* = birth) or foreign substance invades the vertebrate body for the first time it initiates a multiphase immune response. Several types of white blood cells that come in contact with the antigen attempt to engulf it (*phagocytosis*), and various freely circulating enzymes specific for digesting bacterial cell walls may become involved. Relatively nonspecific responses such as these are known throughout the animal kingdom. However, in the vertebrates other processes are under way at the same time. Globulin protein molecules attach to the surface of antigens but to no other surface. As these **immunoglobulins** adhere to antigens, they produce aggregates too large to remain in solution. Once bound to its specific antigen, an immunoglobin activates enzymes that circulate in the plasma and interstitial fluid. The results of this activation attract and stimulate phagocytic white blood cells and induce lysis of antigen membranes. The activation process also stimulates local blood flow and increases fluid and plasma protein leakage into the interstitial space in the vicinity of the antigen. Finally, coagulation of some dissolved interstitial fluid proteins blocks movement of the invading antigen through the tissues.

As remarkable as this highly specific reaction is, it does not persist for long periods or spread widely upon first exposure to an antigen. However, a second invasion of the same antigen stimulates a greatly enhanced immune reaction. The reaction to a second exposure is so rapid that it implies memory of the antigen in the form of preexisting specific antibodies matching those which appeared during the first invasion. What is the nature of vertebrate immunoglobulin antibodies, and how are they remembered?

By chemically tagging antigens with fluorescent dyes it has been possible to discover the cells in birds and mammals responsible for producing antibodies. They are white blood cells known as **B lymphocytes** because they were first discovered in the bursa of Fabricius, a lymphoid outgrowth of the cloaca of birds. B cells are widely dispersed in the lymphoid tissue of vertebrates. They receive fragments of antigen passed to them by phagocytic cells of the lymph nodes and produce immunoglobulins that are secreted into the body fluids. Not all lymphocytes that receive fragments of antigens produce antibodies, however. Some take in the antigen and then divide repeatedly to form a clone of cells. Apparently, these particular B lymphocytes transmit to their daughter cells precise information about the antigen, for the clone derived from them becomes the basis for immune system memory. The cells of the clone disperse widely in the lymphatic tissue of the body and when they are again stimulated by the same antigen, rapidly begin to produce their special antibody.

Another class of lymphocytes, the T lymphocytes, may receive antigen from phagocytes but neither produce antibodies directly nor serve as antibody memory. These T cells, so called because in mammals they are resident for a portion of their life in the **thymus gland**, look exactly like B cells when the latter are not actively producing antibodies for release. They can be distinguished only by their behavior. T cells respond to antigen stimulation in three distinct ways that apparently relate to three different kinds of T lymphocytes that are morphologically identical. One group of T cells responds in a way similar to that of B cells, but instead of elaborating antibodies, they reproduce entire clones of cytotoxic cells. These killer T cells travel via the circulatory system until they contact their specific antigen. They bind tightly to it and inject digestive enzymes into the antigen. A second group of T cells, the helper T cells, stimulate the activity of B cells, other T cells, and phagocytic white blood cells. The third group of T cells are called suppressor T cells because they are believed to function as a negative feedback on other aspects of the immune reaction, keeping it from becoming excessively severe. These cells are also thought to be significant in helping the immune system to recognize self, and their malfunction is implicated in many autoimmune diseases.

Although both immunoglobulin antibodies and probable cytotoxic cells are found in jawless fishes, there is no evidence of the high level of differentiation of cell function characteristic of the system in birds and mammals. The degree of differentiation of these lymphocytes in ectothermic gnathostomes is not yet known, and the evolution of the unique vertebrate immune system remains uncertain.

Kidneys

The cells of an organism can exist over only a narrow range of solute concentrations of the body fluids. Another area of narrow tolerance is in the accumulation of wastes. The small nitrogen-containing compounds that result from protein catabolism are toxic and are an especially important category of wastes. The vertebrate kidney, with a tubular/vascular structure, unlike excretory organs in other organisms, has evolved unique capacities for homeostatic control of water balance and waste excretion. Both functions go hand in hand, but for many years evolutionary physiologists have argued about which was the ancestral function of the kidney. Because it now seems certain that early vertebrates lived in a marine environment, kidneys are unlikely to have originated as water-excreting organs. Removal of nitrogenous and other wastes is likely to have been a primordial function of the kidney.

Two sets of anatomical terms, which unfortunately are not mutually exclusive, have been applied to the kidneys of vertebrates (Table 3–2). One set describes an ontogenetic (developmental) series of kidney types—**pronephros**, **mesonephros,** and **metanephros**—in order of their appearance in the embryonic development of amniotes. Another set of terms for kidneys is descriptive of their arrangement, drainage, and internal structure as seen in the adults of different vertebrate groups: **holonephros** (hypothetical, not actually known in any group), **opisthonephros** (nonamniotes), and **metanephros** (amniotes). Ontogenetically, and presumably phylogenetically, the earliest kidney tubules were segmentally arranged and opened through funnel-like ciliated mouths to the coelom, from which they drained fluid that was derived from the interstitial fluid (Figure 3–15). This fluid was conveyed along an **archinephric duct** which connected each segment's tubule to the region of the anus. Here the ducts from each side opened to release the urine to the exterior. A later stage involved association of the individual tubules with circulatory capillary beds where wastes passed directly to the tubules. As this system became more and more efficient, the open connections to the coelom were lost but more numerous tubules crowded into each section. This stage is essentially that of the mesonephros in ontogeny and the opisthonephros in adult nonamniotes. Finally, the tubules became highly compacted and extremely numerous, acquired a new draining mechanism that formed as an outgrowth of the cloaca and

Table 3–2. Types of kidneys of vertebrates.

Ontogenetic Classification	Phylogenetic Classification
Pronephros	**Holonephros**
Develops in anteriormost portion of the nephrogenic tissue in all vertebrates, segmental organization, drained by the archinephric duct, functional in embryos of fishes and amphibians, present but not functional in early embryos of amniotes.	A hypothetical structure, not found in any extant adult vertebrate. A single tubule per body segment for the entire length of the nephrogenic tissue, drained by the archinephric duct.
Mesonephros	**Opisthonephros**
Develops in the middle portion of the nephrogenic tissue of all vertebrates, drained by the archinephric duct, has no obvious segmentation, tubules are more numerous than segments, functional in embryos of all gnathostomes, but only briefly in embryos of amniotes.	Develops in middle and subsequently in the posterior portion of the nephrogeneic tissue, drained by the archinephric duct, and additional posterior accessory ducts of nephrogenic origin, no segmentation, numerous tubules, functional in adults of all nonamniotes.

Metanephros

Develops in the posterior portion of the nephrogeneic tissue of amniotes, drained by a cloacal outgrowth (the ureter), very large number of tubules with no evidence of segmentation, functional in late embryos and adults of all amniotes.

---

achieved the capacity to concentrate the urine. These features are characteristic of amniote metanephric kidneys. The mammalian kidney is further specialized in having acquired an organized structure permitting an exceptionally dynamic range of urine concentrations. In spite of these evolutionary changes, the basic unit of all adult vertebrate kidneys is essentially the same.

The adult kidney consists of hundreds to millions of tubular **nephrons**, each of which produces urine. The primary functions of the nephron are cleansing the blood of excess water, waste metabolites, and foreign substances. To accomplish this the blood is first filtered through the **glomerulus**, a structure unique to vertebrates (Figure 3–16). Each glomerulus is composed of a leaky arterial capillary tuft encapsulated within a sieve-like filter. Arterial blood pressure forces fluid into the nephron to form an **ultrafiltrate** composed of blood minus blood cells and larger molecules. The ultrafiltrate is then processed to return essential metabolites (glucose, amino acids, and so on) and water to the general circulation. When necessary, the walls of the nephron actively excrete toxins. In mammals, where the most highly differentiated nephrons have been described, these actions take place in the **proximal convoluted tubules** (PCT) and **distal convoluted tubules** (DCT) of the nephron (Figure 3–17). Finally, wastes are excreted as urine to the exterior.

The cells of the PCT have an enormous luminal surface area produced by long, closely spaced microvilli, and the cells contain many energy-producing mitochondria. These structural features reflect the function of the PCT in rapid, massive transport of sodium from the lumen of the tubule to the peritubular space and capillaries; sodium transport is followed by passive movement

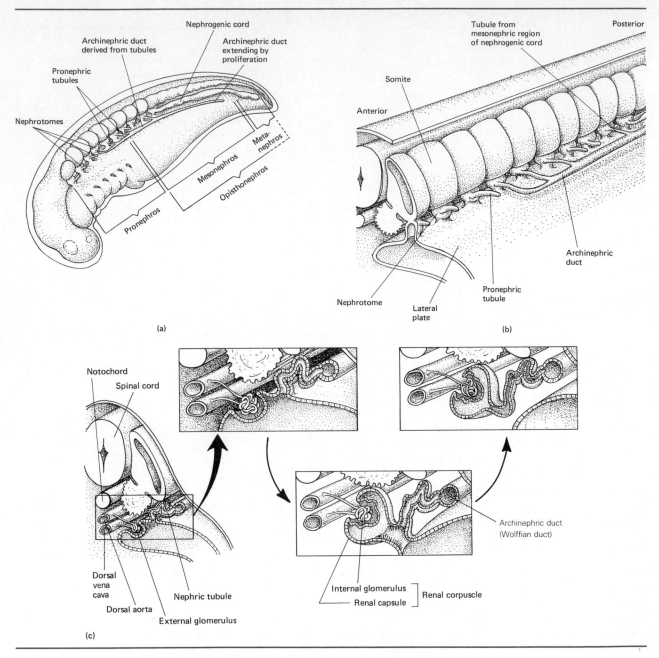

Figure 3–15. Kidney and nephron development and evolution. (a) Nephrotome regions in a generalized amniote embryo with region of adult nonamniote opisthonephros origin superimposed (see Table 3–2). (b) The relationship of the nephrotome and its developing tubules and archinephric duct to the somites, lateral plate, and coelom. (c) The presumed condition in primitive vertebrates (left) was replaced by an intermediate condition seen in the pronephri of the larvae of some living fish (center), and finally by the condition seen in all living adult vertebrates (right).

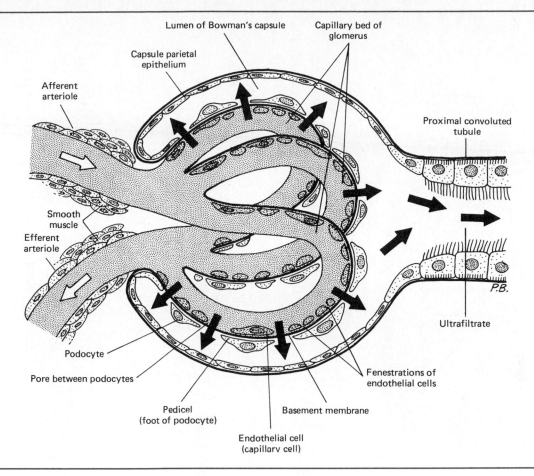

Figure 3–16. Detail of a typical mammalian glomerulus illustrating the morphological basis for its function.

of chloride to the peritubular space to neutralize electric charge (Figure 3–18). Water then flows osmotically in the same direction. Farther down the nephron, the cells of the thin segment of the loop of Henle are wafer-like and contain fewer mitochondria. The descending limb permits passive flow of sodium and water, and the ascending limb actively removes sodium from the ultrafiltrate. Finally, cells of the collecting tubule appear to be of two kinds. Most seem to be suited to the relatively impermeable state characteristic of periods of sufficient body water. Other cells are mitochondria-rich and have a greater surface area. They are probably the cells that respond to the presence of antidiuretic hormone (ADH) from the pituitary gland, triggered by insufficient body fluid. Under ADH influence the collecting tubule actively exchanges ions, pumps urea, and freely permits the flow of water from the lumen of the tubule to the concentrated peritubular fluids (Chapter 4).

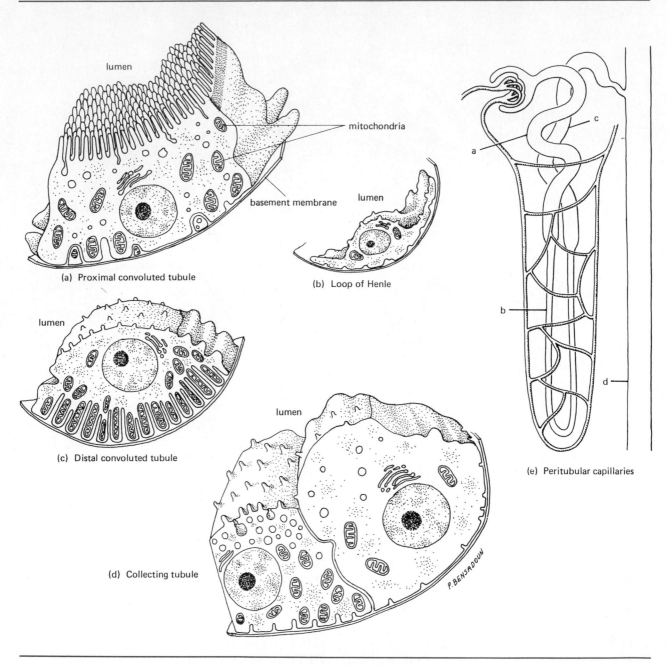

Figure 3–17. The structure of the cells lining the walls of the mammalian nephron: (a) proximal convoluted tubule; (b) loop of Henle; (c) distal convoluted tubule; (d) collecting tubule. In part (e) the letters correspond to the detailed drawings.

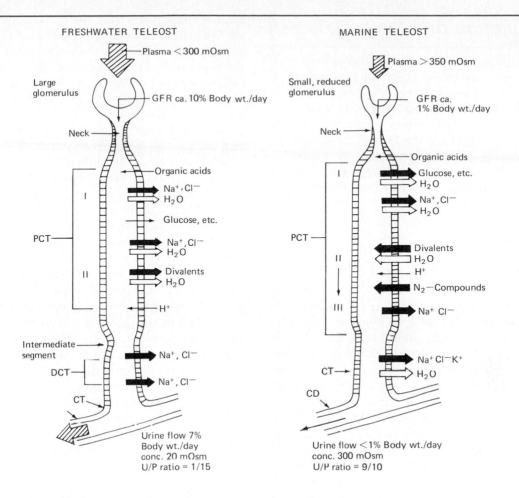

FRESHWATER TELEOST

Plasma < 300 mOsm

Large glomerulus

GFR ca. 10% Body wt./day

Neck

Organic acids

Na^+, Cl^-
H_2O

Glucose, etc.

PCT

I

Na^+, Cl^-
H_2O

II

Divalents
H_2O

H^+

Intermediate segment

Na^+, Cl^-

DCT

Na^+, Cl^-

CT

Urine flow 7% Body wt./day
conc. 20 mOsm
U/P ratio = 1/15

MARINE TELEOST

Plasma > 350 mOsm

Small, reduced glomerulus

GFR ca. 1% Body wt./day

Neck

Organic acids

Glucose, etc.
H_2O

I

Na^+, Cl^-
H_2O

PCT

Divalents
H_2O

II

H^+

N_2—Compounds

III

$Na^+ Cl^-$

CT

$Na^+ Cl^- K^+$
H_2O

CD

Urine flow < 1% Body wt./day
conc. 300 mOsm
U/P ratio = 9/10

Figure 3–18. Schematic comparison of nephron structure and functions in a freshwater and marine teleost fish. GFR is the glomerular filtration rate at which the ultrafiltrate is formed, expressed in percentage of body weight per day. PCT is the proximal convoluted tubule (in fishes sometimes referred to as the proximal tubule). Two segments (I and II) of the PCT are recognized in both freshwater and marine teleosts. Segment III of the PCT of marine teleosts is sometimes equated with the DCT (distal convoluted tubule) of freshwater teleosts. Darkened arrows represent active movements of substances, open arrows passive movement, and hatched arrows indicate by their size the relative magnitude of fluid flow. Note that Na^+ and Cl^- are reabsorbed in the PCT segment I and in the CT (collecting tubule) in both freshwater and marine teleosts; water flows osmotically across the PCT in both freshwater and marine teleosts but only across the CT of marine teleosts. Water permeability of the CT (and also the DCT) is therefore low in freshwater teleosts. U/P: ultrafiltrate to blood plasma concentration ratio— a measure of the concentrating power of a nephron.

Coordination and Integration

Physiological homeostasis at the level of individual organs is only part of the organismal homeostasis. Organ functions must be coordinated to work in concert if the action of one is not to cancel the action of another or extend an adjustment beyond the narrow tolerances of the tissues. Endocrine secretions distributed to target organs by the cardiovascular system and signals transmitted by the nervous system are significant coordinators in this regard.

The Nervous System: Brain Anatomy, Evolution, and Size

The basic unit of neuroanatomy is the **neuron**. Neurons are made up of nerve cell bodies with thin, sometimes very long, extensions, the **dendrites** and **axons**, which extend from the cell body and transmit impulses to and from it. Axons, generally the longest and least branched of these extensions, are often encased in a fatty insulating coat, the **myelin sheath**, which increases the conduction velocity of the nerve impulse. These sheaths are derived from Schwann cells that originate from the neural crest. An axon carries impulses over long distances to another nerve cell or population of nerve cells. Generally, axons in transit from one population of cells to another are collected together like wires in a cable. Such collections of axons in the **peripheral nervous system** (PNS) are called nerves; within the **central nervous system** (CNS) they are called tracts and compose most of the white matter (so called because of the color of the myelin). Nerve cell bodies are often clustered together, usually in groups with similar connections or functions, and are called ganglia (in the PNS) and nuclei (not to be confused with the cellular nucleus) that make up most of the gray matter (in the CNS).

The nerves of the PNS are segmentally arranged, exiting from the spinal cord between the vertebrae. Each spinal nerve complex is made up of somatic sensory fibers from the body wall, visceral sensory fibers from the gut wall, visceral motor fibers to the muscles and glands of the gut, and somatic motor fibers to the muscles and glands of the body wall. The nerve cell bodies of both types of sensory fibers are collected in a segmental series of spinal ganglia adjacent to the spinal cord. They are derived from neural crest cells. These various types of fibers sort themselves out into the peripheral nerves in distinctive combinations that are consistent with our hypotheses of phylogenetic sequence. Lampreys have separate motor and sensory nerves that do not unite to form a complex spinal nerve. The most derived condition, special branches of combined fibers destined for a topographic area, is characteristic of amniotes. The cranial nerves are a special case of spinal nerve modification, including the isolation of fibers relating to different anatomical regions into separate nerves. The spinal nerve fibers that relate exclusively to the involuntary functions of the vertebrate body form two motor divisions that, together with their corresponding sensory nerves, are known as the **autonomic nervous system**. However, the autonomic system is not clearly separated from the CNS or the PNS. The **sympathetic motor division** of the autonomic system generally stimulates an organ to react in a manner appropriate to a stressful set of circumstances (e.g., prepare for rapid energy expenditures). The **parasympathetic division** is antagonistic in its effect, causing the organs innervated to function appropriately for peaceful conditions (e.g., rest or digestion).

The **spinal cord** is the simplest part of the CNS, yet its cellular interconnections and functional relationships are complicated. The gross structure is that of a hollow tube (the lumen, or neurocoel, may be very restricted) with an inner core of gray matter and an outer layer of white matter. The spinal cord receives sensory inputs, transmits and integrates them with other portions of the CNS, and sends motor output as appropriate.

Primitively, the spinal cord had considerable autonomy, even in such complex movements as swimming. Fish can continue to produce coordinated swimming movements when the brain is

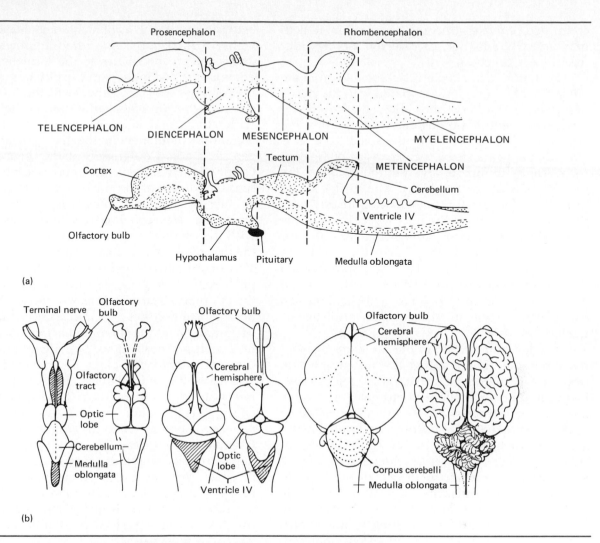

Figure 3–19. Vertebrate brain. (a) Diagrammatic developmental stage in generalized vertebrate brain showing principal divisions and structures described in the text. Lateral view above; sectioned view below. (b) Representative vertebrate brains seen in dorsal view. All brains are drawn to approximately the same total length, which emphasizes relative differences in regional development. From left to right: *Scymnus*, a shark; *Gadus*, a teleost; *Rana*, a frog; *Alligator*, a crocodilian; *Anser*, a goose; and *Equus*, a horse, as an example of an advanced modern mammal.

severed from the spinal cord. The trend in vertebrate evolution has been toward more complex circuits within the spinal cord and between the spinal cord and the brain. With these connections has come an ever-increasing dependence of spinal cord functions on control by higher CNS centers.

The neuroanatomy of the brain is exceedingly complex (Figure 3–19). The vertebrate brain is composed of five distinct regions, containing both gray and white matter, each region serving a different basic function.

Posteriorly, two regions differentiate from the embryonic brain region associated with the developing ear. The most posterior, the **myelencephalon** or **medulla oblongata**, is primarily an enlarged anterior extension of the spinal cord. It is altered in appearance by the expansion of the neurocoel cavity to form a large fluid-filled space, the fourth ventricle of the brain. The nuclei of the medulla oblongata form synapses primarily with the sensory organs of the muscles and skin of the head. However, the myelencephalon has one distinctive set of nuclei, those associated with the sense organs of the inner ear. All impulses from the receptor cells of the balance (vestibular) and the hearing (cochlear) regions of the mammalian ear synapse first in these medulla oblongata nuclei.

The anterior portion of the posterior region of the embryonic brain, called the **metencephalon**, develops an important dorsal outgrowth, the **cerebellum**. The cerebellum coordinates and regulates motor activities whether they are reflexive, such as maintenance of posture, or directed motor activities, such as escape movements. The nuclear or gray matter of the cerebellum receives nerve impulses from the acoustic area of the myelencephalon (in particular impulses relayed from the vestibular nuclei). It also receives impulses from the complex system of muscle and tendon stretch receptors and indirectly from the skin, optic centers, and other coordinating brain centers.

The central embryonic brain region, the **mesencephalon**, develops in conjunction with the eye and plays a part in the functioning of the visual system. The roof of the mesencephalon is known as the **tectum** and receives input from the optic nerve. The floor of this region contains fiber tracts that pass anteriorly and posteriorly to other regions of the brain as well as to nuclei concerned with eye movements.

A rather small region, the **diencephalon**, develops from the most anterior of the embryonic brain regions. In amniotes it is a major relay station between sensory areas of many modalities and the higher brain centers. A ventral outgrowth of the diencephalon contributes to the formation of the dominant endocrine organ, the pituitary gland, or hypophysis, which together with the floor of the diencephalon (hypothalamus) forms the primary center for neural–hormonal coordination and integration. Another endocrine gland, the pineal organ, is a dorsal outgrowth of the diencephalon. Originally it was a median photoreceptor (light-sensitive organ) of primitive vertebrates.

Finally, the most anterior region of the adult brain, the **telencephalon**, develops in association with the olfactory capsules. While functioning as the first nuclei of olfactory synapse, the telencephalon of primitive vertebrates also coordinates inputs from other sensory modalities. In mammals the primary seat of sensory integration and nervous control occurs in the telencephalon. The neuronal cells in the anterior olfactory nuclei of the telencephalon are variously enlarged into masses of gray matter that are often differentiated into distinctive layers and subdivisions with complex interconnections (Figure 3–20). Phylogenetic differences include the increase in size of the telencephalon relative to the rest of the brain, differentiation of nuclear (gray matter) areas, and the appearance of new nuclear types; corticalization, in which the gray matter emerges on the surface of the telencephalon, forcing the white matter to a central position; and finally, the convolution of the most recently evolved portion of the brain in advanced mammals.

In primitive tetrapods (Figure 3–20a) the telencephalon is a small portion of the total brain. Nonetheless, the telencephalon is differentiated into three deep-lying regions: the basal nuclei (= corpus striatum), the paleopallium, and the archipallium. In primitive amniotes (Figure 3–20b) expansions and migrations of the paleopallium and archipallium have forced the basal nuclei to an internal position. In a more advanced stage (Figure 3–20c) the telencephalon constitutes the single major portion of the brain. A new nuclear region, the **neopallium**, has made its appearance and corticalization is well advanced. In birds (Fig-

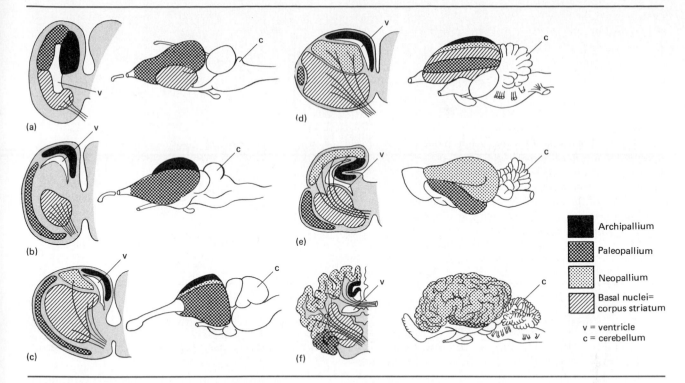

Figure 3–20. Various stages in the differentiation of the telencephalon of tetrapods in cross-sectional half-brain and lateral full-brain diagrams: (a) primitive tetrapod with relatively large contribution of nontelencephalon brain regions to total brain weight; (b) later stage, probably represented by primitive amniotes; (c) advanced stage, probably arrived at by several lineages; (d) in birds the telencephalon and cerebellum make up the major portions of the brain; (e) primitive mammals; (f) advanced mammals. Lines representing cut nerve tracts indicate an association center's connections with the brain stem.

Legend:
- Archipallium
- Paleopallium
- Neopallium
- Basal nuclei= corpus striatum
- v = ventricle
- c = cerebellum

ure 3–20d) the paleopallium, always associated with olfaction, is very small but the basal nuclei associated with innate, reflexive behaviors are enormous. These basal nuclei are covered with a relatively thin coat of archipallium and what is thought to be neopallium.

Two phenomena in the evolution of the mammalian brain make it unique among vertebrate brains. First is the development of an external mat-like covering of gray matter from the more primitive central nuclear condition. This change places the nerve cell bodies on the outside of the brain, where additions to their number will not cause dis-ruption of the tracts leading to them. Second is the domination of a region of gray matter distinctive in its cell types and cellular organization, the neo-pallium. Birds, although nearly as brainy as many mammals on a ratio of brain weight to body weight, have comparatively little of this type of neural tissue. The mammalian neopallium is a site of sensory projection, of integration, of memory, and of the construction and reconstruction of past or even hypothetical patterns of sensory input. It is ultimately the seat of intelligence. The neopallium and paleopallium of primitive mammals (Figure 3–20e) are the two major divisions of the te-

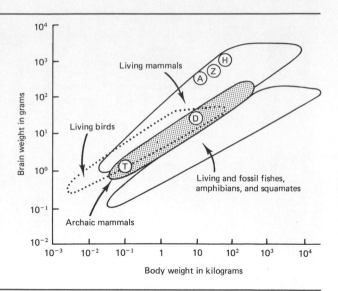

Figure 3–21. Relative brain weight of vertebrates. Logarithm of brain weight plotted against logarithm of body weight based on measurements of 198 living vertebrate species and endocranial cast volumes of some fossils. T, *Triconodon* of the Jurassic; D, *Didelphis* the living Virginia opossum; A, *Australopithecus africanus*; Z, *Australopithecus boisei*; H, *Homo sapiens*. (Modified after H. J. Jerison, 1973, *Evolution of the Brain and Intelligence*, Academic Press, New York.)

lencephalon, and the archipallium is forced into a curled internal position. The majority of connections to the brain stem run directly to the neopallium, unlike the condition seen in advanced non-amniotes, primitive tetrapods, and birds, where the basal nuclei serve as major association areas. In advanced mammals (Figure 3–20f) the neopallium dominates the entire telencephalon and becomes highly folded. It communicates not only with the brain stem but also directly with its contralateral cerebral hemisphere via the corpus callosum.

Most vertebrate clades have evolved species of increasing body size during their histories. Numerous locomotory, homeostatic, and defensive benefits derive from large size. However, large size increases the demands on integrative systems. In large animals the tension and position of the additional muscle fibers must be monitored and their firing controlled, and the increased skin surface brings with it more sensory organs. These demand more nerve cells in the central nervous system, especially in the brain. Nevertheless, in vertebrates brain size does not increase in exact proportion to body mass. Rather, as body mass doubles, brain mass increases by a factor of about 1.6.

To compare the brain sizes of vertebrates, which vary from the teleost fish *Schindleria praematurus*, weighing only 2 milligrams, to the blue whale (*Balaenoptera musculus*), weighing over 160,000 kilograms, it is appropriate to compare brain mass per unit of body mass. When examined in this way, the brains of living birds and mammals are approximately 15 times larger than those of other vertebrates of the same body size. To visualize the relationship between various vertebrate taxa, including extinct groups, the data for each taxon can be enclosed in a polygon (Figure 3–21). Even the early mammals had four to five times as much brain tissue as other vertebrate taxa of similar body size. Obviously, an increase in relative brain-to-body mass ratios occurred in the evolution of modern mammals.

Spinal and Cranial Nerves

From the spinal cord in each body segment issue two nerves that supply motor (efferent) and sensory (afferent) fibers to the **somatic** and **visceral**

Table 3–3. Cranial nerves and their primary function.

Number	Name	Function		Mixed Nerves	
		Somatic Motor	Somatic Sensory	Visceral Motor	Visceral Sensory
0	Terminalis		+?		
I	Olfactory		+		
II	Optic		+		
III	Oculomotor	+			
IV	Trochlear	+			
V	Trigeminal		+	+	
VI	Abducens	+			
VII	Facial		+	+	+
VIII	Acoustic		+		
IX	Glossopharyngeal		+	+	+
X	Vagus		+	+	+
XI	Accessory			+	
XII	Hypoglossal	+			

components of that segment. Somatic components include the skeletal muscles; visceral components, the gut and circulatory musculature. Although these nerves vary among vertebrates, a ventral and a dorsal root tract are always recognizable. In the primitive condition the ventral root supplied only motor nerves to the myotomes or trunk muscles of each body segment (somatic components). The dorsal root was a mixed nerve that contained both visceral motor and sensory nerves.

Ventral root nerves are conservative and have retained throughout vertebrate evolution the metameric relationship between the muscle-forming myotome (somite) and the nerve supply to that somite. If a particular somite is lost (e.g., some head somites), so is the ventral nerve; if a myotome is modified and moved to serve other functions (e.g., the appendicular or limb muscles of tetrapods), the nerve preserves the metamerism and exits from the segment of the cord embryonically associated with that myotome. The dorsal root nerves appear to be more plastic; interseg-

mental capture is common. The dorsal root nerve contains the visceral motor fibers of the gut and the sensory nerves of the skin, two parts of the body which lack segmentation.

Armed with this concept of a primitive spinal nerve, we can ask if the cranial nerves represent modified spinal nerves. If so, we have additional evidence that the vertebrate head and jaws were derived from a fusion and modification of anterior segments. Ten to twelve pairs of cranial nerves are recognized in vertebrates (Table 3–3). Their numbers refer to the anterior-to-posterior sequence of exit of the nerves from the mammalian brain, and their names broadly describe what the nerves do (e.g., oculomotor = eye mover) or their appearance (e.g., trigeminal = a nerve with three major branches).

Three of the cranial nerves, the olfactory (I), the optic (II), and the auditory (VIII), are special sensory nerves associated with separate portions of the tripartite brain. The cranial nerves considered most clearly derived from ancient, anterior spinal ventral roots are nerves III, IV, VI, and XII (Table 3–3). Like the ventral root spinal nerves, these cranial nerves send motor fibers to the somatic muscles of the head (i.e., eye and tongue). What was the fate of the dorsal roots of the ancient anterior spinal nerves? The glossopharyngeal nerve (IX) is analogous to that of a primitive dorsal root spinal nerve. It supplies sensory fibers, as well as visceral motor fibers (and is therefore a mixed nerve), to the first gill arch of fish-like, jawed vertebrates. Because the glossopharyngeal nerve is located just posterior to the hyoid arch of fishes, we assume that it was associated with the third visceral arch of gnathostome ancestors. The facial nerve (VII), which is a mixed nerve, innervates the hyoid apparatus or suspensorium derived from a second visceral arch. The trigeminal (V), again a mixed nerve, innervates the jaws, which are believed to have been derived from the most anterior visceral arch. Thus cranial nerves V, VII, and IX, on the basis of their mixed sensory and motor components, most likely represent dorsal roots of ancient anterior spinal nerves from body segments that have fused and can be aligned with specific visceral arches 1, 2, and 3.

Attempts to align the ventral root components of these presumed old spinal nerves (cranial nerves III, IV, and VI) with specific visceral arches are highly speculative. Similar arguments can be used to explain the fate of the primitive spinal nerves associated with more posterior gill arches. For example, the vagus nerve (X) probably represents a coalescence of several dorsal roots, and the hypoglossal (XII), the remnant of one or more ventral roots.

It is possible that one or more anterior body segments, which included the original mouth, have been lost. A small cranial nerve that parallels the olfactory nerve was described in 1894 by F. Pinkus in lungfish and it occurs in many, but not all, vertebrates. Called the nervus terminalis (Table 3–3), it is assigned the numerical position 0. Perhaps the terminalis represents the dorsal root vestige of a terminal body segment long lost or so modified that it is unrecognizable.

The Sense Organs

Sense organs are classified on the basis of the type of stimulus to which they react. **Chemoreceptors** are stimulated by chemical substances. **Mechanoreceptors** are sensitive to mechanical deformation even at the level of distortion of a single cell's stereocilia. **Radioreceptors** respond variously to electromagnetic disturbances of a broad spectral range, including thermal radiation. In mammals, the major senses include the classic five senses of taste, touch, sight, smell, and hearing. Most vertebrates share these and a few have additional senses or have greatly refined one or more of the familiar five.

Taste Taste buds made up of specialized sensory epithelial cells and associated nerves are restricted to the moist membranes of the walls of the mouth, the throat, and especially the tongue (Figure 3–22) of tetrapods, but many osteichthyans have taste buds on their heads, fins, or specialized oral structures called barbels (e.g., catfishes). Taste buds are stimulated when dissolved molecules contact their specialized surfaces. Although there are many similarities between the senses of taste and smell, they have different embryonic origins, innervations, sensitivities, and types of molecules to which they respond.

Touch At least four differently structured nerve ending–epithelium–connective tissue combinations are also known to be sensitive to stimuli ranging from light touch to heavy pressure and are found in tetrapods. In mammals touch receptors may also be combined with specialized hairs, the vibrissae, which grow on the muzzle, around the eyes, or on the lower legs. A specialized set of mechanoreceptors, the lateral-line system, occurs in aquatic nonamniotes (discussed in Chapter 8).

Vision Vision is considered the distance sense *par excellence* of tetrapods and is equally important to many fishes. Visual systems are sensitive to those wavelengths of electromagnetic radiation that reach the surface of the earth with the least

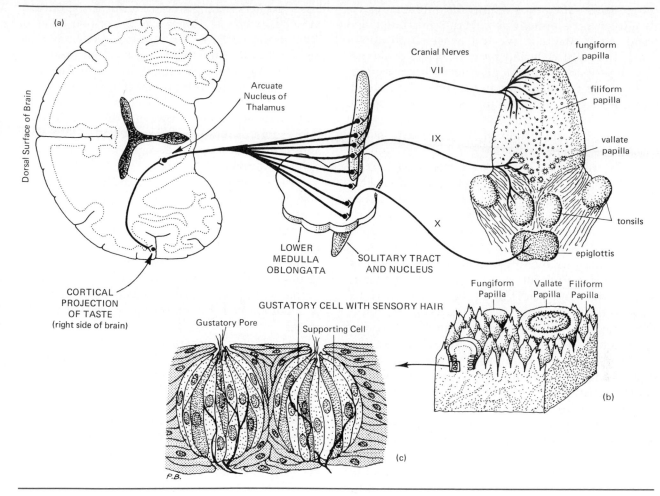

Figure 3–22. Sense of taste as illustrated by mammals: (a) tongue of a human showing the topography and types of lingual papillae. The distribution of innervation and route of projection onto the border of the lateral fissure of the cerebrum are shown for one side of the tongue; (b) enlargement of region of tongue illustrating details of papillae structure and position of taste buds; (c) further enlargement of taste buds on a fungiform or vallate papilla showing cellular organization and sensory hairs, a common morphology of vertebrate taste buds. (Modified after F. H. Netter, 1962, *The CIBA Collection of Medical Illustrations,* volume I, *Nervous System,* CIBA Publications, Summit, N.J.)

interference by the atmosphere and hydrosphere. Radiation of these wavelengths, which we call light, has several properties that make it a superior transmitter of information over distance. Light travels great distances in air with little attenuation (absorption). It travels in a straight line in a homogeneous medium. Light of different wavelengths is differentially absorbed, transmitted, and reflected so that the spectral (wavelength) distribution of light reflected from an object is altered. These properties of light give nearly unambiguous cues about the structure and texture of an object. Another significant attribute of light is its speed, which is instantaneous compared to the speed of transmission and action of the vertebrate neuromuscular system.

The receptor field of the vertebrate eye is arrayed in a hemispherical sheet—the **retina**. Each point on the retina corresponds to specific neural connections and a different visual axis in space. Thus, a vertebrate can determine where a target is in at least two-dimensional space and whether it is stationary or moving. Because of neuronal interactions, vertebrate eyes can also detect sharp beginnings and endings of visual stimuli (ons, offs, and edges) with great precision.

The eyes of vertebrates (Figure 3–23) consist of a light-shielded container, the sclera and choroid coats. These prevent stimulation of the eye by light from multiple directions. The eye has a variable entrance aperture, the pupil surrounded by the iris, to control the amount of light that enters. A focusing system, the cornea, lens, and ciliary body, converges the rays of light on the photosensitive cells of the retina.

The retina (Figure 3–24) originates as an outgrowth of the brain. The exceptional nature of the eye as an extension of the brain is not as generally appreciated as it deserves to be. For example, amphibians process distant stimuli in the elaborate neural network of the retina. The optic nerve sends signals to the brain with already encoded distinctions between predators, prey, and stationary objects in the environment. Wiring patterns from the mammalian optic nerve indicate considerable retinal processing; there are approximately 100 million photoreceptor cells in the retina but only 1 million axons in the optic nerve.

Lizards and birds have better visual acuity than most fishes and mammals. In part, their ability to distinguish between two close points in space, which would appear to vertebrates with lesser acuity as a single point, is due to their nearly pure cone retinas. **Cones** are one of two types of light-sensing cells in the vertebrate retina and are distinguished from the other type, the **rods**, by morphology, photochemistry, and neural connections. In addition to high visual acuity, cones are the basis for color vision. Cones have relatively low photosensitivity, however; they are at least two orders of magnitude less sensitive than rods and thus fail to function at night and in other dim-light conditions.

Rods are morphologically distinct from cones and differ physiologically—they are exceptionally sensitive to dim light. Several rods synapse with a single neural element, increasing the chances that dim light will stimulate the neuron, but they achieve this increased sensitivity at the cost of visual acuity. An analogy can be made with photographic films—increased sensitivity to dim light is accompanied by increased graininess of the image.

The loss of acuity in a nocturnal eye is considerable. Whereas the diurnal pigeon can distinguish between two lines producing retinal images less than l micrometer apart, a rat requires a separation of more than 20 micrometers. Try to concentrate on the form or details of objects seen at the edge of the visual field of your eye—the peripheral rod–rich areas of the human retina—and the poor acuity of rod vision will be apparent.

The eyes and retina produce a two- or three-dimensional representation of the world (Figure 3–25). Each cone or closely adjacent group of rods on the retina corresponds precisely to a point or series of adjacent points in space. The muscles controlling the orientation of the eye and the accommodation of the lens give sufficient information for very accurate neural reconstruction of the visually perceived world.

Olfaction The olfactory receptor apparatus is basically the same in all macrosmatic (strongly olfactory) vertebrates. In fishes the ectodermally derived olfactory organs are usually independent of the pharynx; separate incurrent and excurrent canals open on the snout. The nasal epithelium is

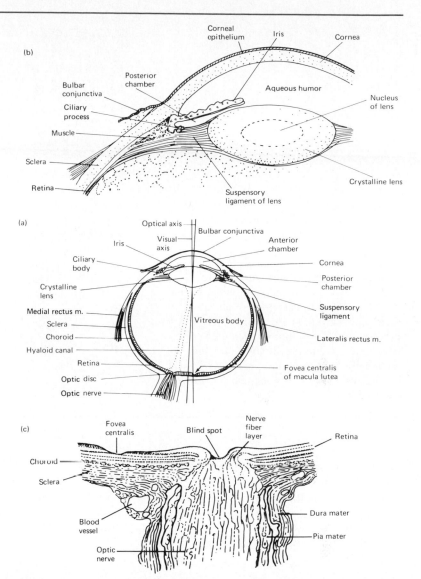

Figure 3–23. Vertebrate eye as illustrated by mammals. (a) The eye sectioned horizontally in the plane of the optic axis and optic nerve. In addition to the structures discussed in the text, (b) details structures of the anterior portion of the eye primarily responsible for image focus (cornea, lens, ciliary body) and quantitative control of light (iris). (c) Enlargement of the fundus of the eye showing the specialized sensory and neural regions found there. Note especially the continuity of the brain covering meninges (pia mater, arachnoid, and dura mater) to the protective and nutritive coats of the eye (sclera and choroid). (Modified after J. E. Crouch, 1969, *Text-Atlas of Cat Anatomy,* Lea & Febiger, Philadelphia.)

folded into a complex rosette of great surface area but does not have bony support. In mammals the sensory cells reside on the turbinals and nasal septum high in the nasal cavity. The lower turbinals and nasal epithelium are concerned with temperature and water regulation (Figure 3–26).

The total area of the olfactory surfaces may exceed that of the entire body surface in highly olfactory-oriented vertebrates. A single molecule can excite a receptor cell, and only a few molecules are needed to elicit a behavioral response. A molecule of odorant probably fits more or less tightly into one or more of the different sensory pits along the sensory hairs, which occur on the surface of olfactory epithelia. Once seated, a molecule lingers, producing prolonged stimulation that allows

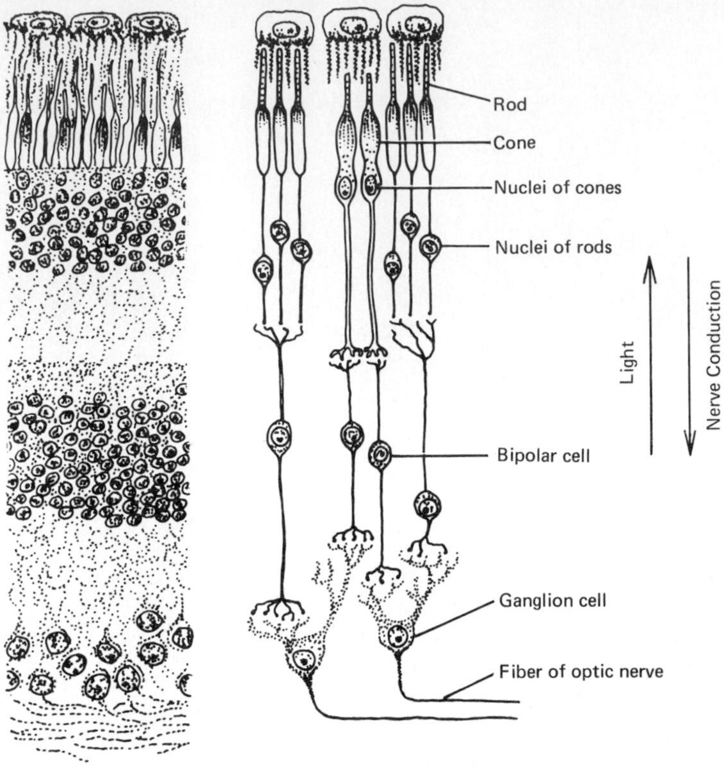

Figure 3–24. Histological (left) and diagrammatic (right) structure of the mammalian retina as seen in humans. The direction of light passage and neural conduction are indicated. The different types, great number, and interconnections of nerve cells in the retina are the basis for the considerable information processing that occurs within the eye. (See text for further discussion.)

perception of the relative concentrations of stimulating substances. However, the lingering nature of the stimulus limits rapid temporal or spatial perception of the olfactory surroundings. Many mammals snort or sneeze during olfactory sensing, perhaps to flush air as completely as possible from the upper regions of their nasal cavities. The subtle differences in odor that mammals can detect are probably due to an ability to distinguish the hundreds of combinations possible from no more than a dozen basic odors. Such combinations would produce the individual differences in body odor that seem important in maintenance of many mammalian social groups. The physiological state of an individual can often be determined from the odors it produces, especially in the feces and urine. Excretions are used by a wide variety of vertebrates to mark their territories or home ranges.

One dominant group of freshwater fishes, the Ostariophysi (minnows and their 6000 relatives; Chapter 8) share a unique type of epidermal cell which exudes an alarm substance when the skin is abraded. Adjacent fish smell the substance and scatter, diving for the bottom in apparent antipredator behavior in direct response to the chemical signal.

Pheromones are chemical signals produced by an individual that affect the behavior and/or physiology of conspecifics. Minnow alarm substances are an example. The odors of feces and urine undoubtedly have such effects. Many male ungulates (hoofed mammals) sniff or taste the urine of females. This behavior is usually followed by *flehmen*, a behavior in which the male curls the upper lip and often holds his head high, probably inhaling. A specialized structure of the anterior palate,

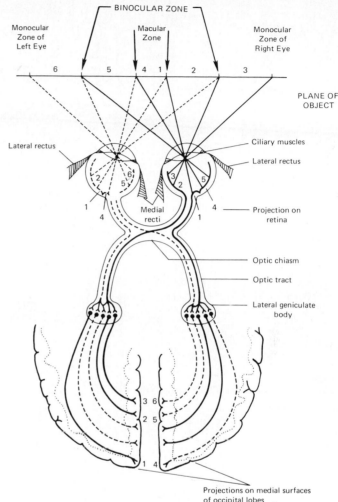

BINOCULAR ZONE

Monocular
Zone of
Left Eye

Macular
Zone

Monocular
Zone of
Right Eye

6   5   4 1   2   3

PLANE OF
OBJECT

Lateral rectus

Ciliary muscles

Lateral rectus

Medial
recti

Projection on
retina

Optic chiasm

Optic tract

Lateral geniculate
body

3 6

2 5

1 4

Projections on medial surfaces
of occipital lobes

Figure 3–25. Neural connections and point for point correspondence between visual field and cerebral projections in the human brain. The one-dimensional representation of the line shown, as received by the brain, may be translated into a three-dimensional reconstruction of the visual world (depth perception). This is accomplished by information from the extrinsic ocular muscles on the degree of convergence or divergence of the eyes (such as from the lateral and medial rectus muscles shown) or by similar information from the ciliary muscles on the degree of lens distortion (accommodation) necessary for focusing. Many vertebrates lack significant binocular vision and have fewer visual system cues for depth perception.

called the **vomeronasal** or Jacobson's organ, contains branches of the olfactory nerve and is used during *flehmen*. Probably this behavior allows males to determine the stage of the reproductive cycle of a female.

How does olfaction rate as a precise, directional distance sense? The great sensitivity of the olfactory epithelium and the relative stability of many volatile substances suggest that stimuli can be perceived from considerable distances, but lo-cating the source may be difficult. An animal using olfaction faces a challenging task in orienting and approaching the source. Nonetheless, transmission of olfactory stimuli is not impeded by most obstacles, and olfactory signals can be used day and night. Olfaction, despite its relative slowness and poor resolution for spatial interpretation, offers durability and an enormous variety of signals to vertebrates, and its use is especially well elaborated in the fishes and mammals.

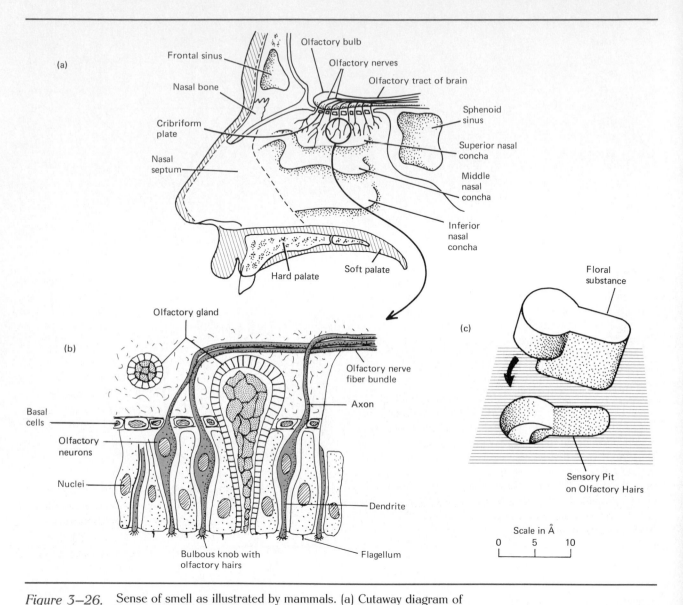

Figure 3–26. Sense of smell as illustrated by mammals. (a) Cutaway diagram of sagittal (medial) section of the human nasal cavity showing the distribution of the olfactory nerves and the non-olfactory nasal conchae (=turbinals). (b) The cellular organization of the olfactory epithelium is enlarged. Note that the sensory cells are the actual nerve cells that penetrate the epithelium with a modified dendrite. (c) Diagram of a hypothetical basis for sensitivity in olfaction. The three-dimensional shape of a substance determines whether or not it will lock into a sensory pit on the olfactory hairs. A sufficient number of fits results in sensory nerve firing. Each basic odor is associated with a different three-dimensional shape and set of sensory pits. [(c) Modified after J. Amoore, 1962, *Proceedings of the Toilet Goods Association, Scientific Section Supplement* 37:1–20.]

Audition   In contrast to the structural similarity in most of the sensory organs of touch, taste, sight, and smell of most fishes and the tetrapods, their organs of hearing are in many ways distinctive, especially those of mammals (Figure 3–27). The auditory apparatus of a typical mammal has three major functional units. One unit, the external ear, enhances the energy content and directional characteristics of sound stimuli. The second unit, the middle ear, amplifies and transduces sound stimuli into fluidborne analogs of the original sound. The third unit (the inner ear, which is common to vertebrates) discriminates the frequency and intensity of sound and transmits this information to the central nervous system in the form of neurally encoded firing patterns.

The **pinna** (external ear) and narrowing external auditory meatus of mammals concentrate sound from the relatively large area encompassed by the external opening of the pinna to the small, thin, tympanic membrane. The pinna is unique to mammals, although it has a feathery analog in certain owls. The auditory sensitivity of a terrestrial mammal is reduced if the pinnae are removed.

The middle ear, found in all tetrapods, greatly increases auditory sensitivity. Under most conditions the sounds tetrapods attend to are airborne, but the inner ear, where sound stimuli become encoded into neural impulses, is a series of fluid-filled membrane-lined chambers. It requires considerably more energy to set the fluids of the inner ear in motion than most airborne sounds could impart directly. The middle ear receives the relatively low energy of airborne sound waves on its outer membranous end, the **tympanic membrane** or **eardrum**, and produces analogous vibrations in the fluids of the inner ear.

Two devices that enhance sound detection are employed in the mammalian middle ear, which is the most highly derived middle ear known. First, the area of the eardrum is about 20 times greater than that of the oval window, which separates the air-filled middle ear from the fluid-filled inner ear. In all tetrapods the gap between the tympanum and the oval window is bridged by bony elements, the ear ossicles. If mammals had a single bridging element, such as the columella of birds, we would expect something on the order of a 20-fold increase in the force per unit area between sound reception at the tympanum and vibrational input at the oval window. However, instead of a single bone, mammals have a chain of three ossicles linking the tympanum with the oval window. The **malleus**, **incus**, and **stapes** are interconnected and controlled by a series of fine muscles. Although there has been much debate about the exact function of the ossicles, they seem to have a mechanical advantage that boosts the force of vibration on the oval window and provides a much broader range of frequency sensitivity than a single transducing element.

If the middle ear were an airtight cavity and the pressure in the environment was different from that of the middle ear, the tympanum of tetrapods would be distended and put under increased tension, losing some of its responsiveness to sound pressure waves. The auditory tube, derived from the first embryonic gill pouch, connecting the mouth with the middle ear alleviates this problem. The tympanum is forced into the middle ear cavity when a sound wave impinges on it, momentarily compressing the air in the cavity. As the tympanum moves outward it temporarily creates reduced pressure in the middle ear. The pressure differences hinder free vibration of the tympanum and its attached ossicles by damping tympanic oscillations. Mammals have minimized these damping pressure differences by an enlargement of the middle ear cavity into a **tympanic bulla** (plural: *bullae*). This hollow bubble of bone creates a middle ear volume so large that the volume changes produced by oscillations of the tympanum result in insignificant pressure variations.

The auditory bullae are especially well developed in some desert rodents, including the kangaroo rats of North America and the jerboas of Africa and Asia, where the bullae may constitute one-third of the length of the skull and have a volume greater than that of the braincase. These nocturnal rodents are particularly sensitive to low-frequency sounds, such as those made by owls in flight or by snakes moving across sand. Douglas

sion of electrical versus chemical information and in terms of time, nervous activity being rapid (seconds or less) and endocrine activity slow (minutes to hours). The release of epinephrine (adrenalin) from the adrenal gland is followed by a multitude of responses in different target areas—increase in the strength of heartbeat, increased blood pressure, erection of hair (piloerection) in mammals, and inhibition of smooth muscle activity in the gut. Critical to the length of delay in response is the distance of the target organs from the site of hormone release. In the early days of endocrinology hormones were identified as chemical messengers that were released from ductless glands and transported by the circulatory system until they reached a specific target organ. As a result the response was slow. There are many examples of exactly these kinds of hormones (Table 3–4).

Neurobiology and endocrinology have merged considerably in recent decades, however, with the discovery of **neurohumors**, hormones that act as chemical intermediates in the transfer of electrical activity from cell to cell. The classic example is acetylcholine, which is released from many nerve cell axonal terminals into a synapse (junction between nerve cells or nerve cells and muscles), and rapidly diffuses across the synapse to produce a response, usually a depolarization or hyperpolarization in the next nerve cell (the postsynaptic cell). Thus, in most cases neural activity in vertebrates and other metazoans involves not only electrical transmission, but also the release of a neurohumor, or neurotransmitter substance, to pass information from cell to cell. Neurohumors cross synaptic junctions by chemical diffusion, which is slow relative to nerve conduction. The width of the synaptic junction is on the order of micrometers, and the synaptic time delay of neurohumoral diffusion from the presynaptic release point to the postsynaptic target sites can be measured in milliseconds. Hormones can therefore no longer be considered solely chemical messengers to distant target organs.

Vertebrates have been central to investigations of the endocrine system (Table 3–4). The fact that some type of chemical might have an effect on morphology or behavior was first demonstrated by the classic experiments of A. A. Berthold around 1849, when he demonstrated that castration of young male cockerels inhibited development of the characteristic comb and wattle of the rooster and that reimplanting the testes allowed normal development of maleness. At the turn of the century a series of investigations demonstrated that extracted chemicals from various organs caused specific effects when injected into the bloodstream of mammals. Perhaps the most significant was the discovery by Banting and Best (1922) that extracts from the distinctive cells of the Islets of Langerhans in the pancreas were critical in the control of sugar metabolism. When these extracts were injected, they successfully lowered high blood glucose levels in diabetic dogs. The importance of this demonstration and discovery of the hormone insulin has had untold clinical value in the treatment of human diabetes.

Modern endocrinology, although considered a subdiscipline of physiology, has developed into a major field of inquiry. Table 3–4 outlines some of the major hormones known to function in vertebrates. Almost every month a new vertebrate brain neurohormone is identified—they number in the hundreds.

The **pituitary gland** and **hypothalamus**, or the **pituitary axis** as it is often called, are present in all living vertebrates (Figure 3–28). Often called the master gland, the pituitary is essentially where nervous activity from the peripheral and central nervous systems converge to provide major regulation of hormone release. Nerve activity directed into the hypothalamus causes the release of peptides (releasing factors) that are transported by the blood to the anterior, intermediate, or posterior pituitary. Arriving in one of the lobes of the pituitary, they cause the secretion of **tropins** (hormones that induce the release of other hormones in other endocrine glands). The release of thyroid hormone serves as a model (see Figure 3–28). Once thyroid hormone concentration rises in the blood it begins a feedback loop, thus inhibiting the secretion of

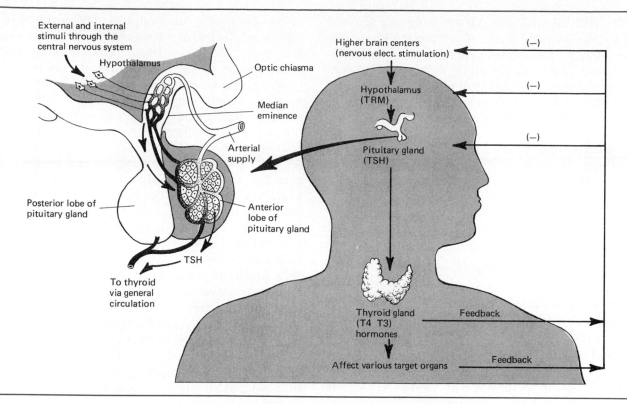

Figure 3–28. Pituitary axis and the release of thyroid hormones as an example of negative feedback to control hormonal release. High circulating levels of thyroid hormones in blood inhibit the release of thyroid releasing hormone from the hypothalamus and inhibit the release of thyroid stimulating hormone from the anterior lobe of the pituitary gland (indicated by minus signs), thereby lowering the rate of secretion of thyroid hormones from the thyroid gland.

more thyrotropins and regulating the level of thyroid hormone in circulation. Most hormones involve negative feedback loops of this sort.

Endocrine secretions are predominantly involved in the control and regulation of energy use, storage and release, as well as energy allocation to special functions at critical times (Table 3–4). As a reflection of the role that hormones play in homeostasis, the majority of hormones that function in general metabolism, nutritional needs, or osmoregulation are counteracted by another hormone that acts in exactly the opposite direction. Often

both hormones are the secretions of the same endocrine gland, usually from distinctive cell types. The regulation of these antagonistic hormones is usually through a negative feedback system whereby blood levels of the substance in question directly or indirectly shift secretion rates of the antagonistic hormones.

The existence of endocrine hormones predates the origin of vertebrates. In fact, hormones from invertebrates as well as vertebrates often can be classified as belonging to chemical families, such as the catecholamines (epinephrine, norepinephrine, and dopamine) and the neurohypophysial physins (oxytocin, vasopressin, and their various homologs). Hormone families share a similar chemical structure, but subtle chemical differences cause them to have different effects on their target organs. Within the vertebrates the structures of hormones have changed through time. The trend in the evolution of vertebrate endocrine glands has been consolidation from the widely scattered clusters of cells common in fishes to more definitive vascularized endocrine organs characteristic of amniotes.

Continuity of Life: The Reproductive System

Reproduction is vital to life, but it is subject to powerful environmental influences. Reproduction shows great variation among vertebrates. Here we concentrate on a few aspects of the reproductive biology of vertebrates that hold true throughout the group, have special generality, or illustrate trends consistent with phylogenetic hypotheses of relationships. Ecological and detailed phylogenetic discussions are postponed until later chapters.

Gametogenesis and Gamete Conduits

During embryogenesis, paired genital ridges differentiate along the length of several somites from the medial border of the intermediate mesoderm next to the nephrotome; primitive sex cells migrate from the gut wall into the ridges and the stage is set for differentiation of the gonads (Figure 3–29). Of the three cell types in the primordial gonad, only the **primordial sex cells** will produce gametes, but whether the gametes will be ova or sperm depends on the differentiation of the other two cell types. One of these is the epithelial lining of the coelomic cavity, and the second is composed of cells at least some of which had already differentiated and functioned in the Bowman's capsule

of the embryonic mesonephros. These nephron cells de-differentiate, becoming less specialized and more like cells in early stages of embryonic development. De-differentiation is an uncommon phenomenon in vertebrate development. The primordial sex cells migrate to the region of the presumptive genital ridge. The coelomic epithelium organizes and thickens upon the arrival of these cells and forms the gonadal cortex. The primordial sex cells multiply as well and form a primordial gonad medulla. The relative number of each type of cell is considered to be under the control of sex-determining genes because it is the relative activities of these two cell populations that determine gonadal sex. Cortical cells attract migrating sex cells but inhibit cell division, whereas medullary cells attach cell processes to sex cells and stimulate both their migratory activity and cell division. The influence of cortex results in the formation of an ovary, that of the medulla forms a testis. The definitive gonad, whether male or female, is both an endocrine gland producing circulating hormones and a specialized exocrine gland that produces an unusual product: highly specialized cells—sperm or ova.

Sex Determination

The gender of most mammals is expressed by genitalia and other secondary sex characteristics and is generally obvious from birth. In other classes of vertebrates, however, genitalia and sexually dimorphic structures are absent or not expressed until maturity is achieved (van Tienhoven 1983). To discuss sex determination in all vertebrates, we must bear in mind the basic definition of sex: the female sex produces eggs; the male sex produces sperm (Naftolin 1981). Egg production requires a functional ovary and sperm production a functional testis; each structure has a distinct cytological appearance. In many kinds of vertebrates both an ovary and a testis are present in the same individual. By definition these individuals are both male and female, a seemingly bizarre condition, for in mammals individual sex is segregated into

Spinal cord
Somite
Notochord
Aorta
Coelom
Gut cavity
(a)
Embryo cross section
Glomerular capsule
Germ cell
Mesentery
coelom
Endoderm
Gut cavity
Mesenephros
Gonad
Archinephric duct
Oviduct
(b)
Indifferent state
Glomerular capsule
Rete cords
Primary sex cords
Germ cells
Germinal epithelium
Ovary
Fallopian tube
Uterus
Vagina
(c)
Glomerular capsule
Primary sex cords
Rudimentary tunica albuginea
Secondary sex cords (ovarian epithelium)
Germ cell
Germinal epithelium
Ovary
Epididymis
Testis
Vas deferens
Bladder
Seminal vesicle
Prostate
(d)
Glomerular capsule
Rete cords
Testis cords (seminiferous tubules)
Tunica albuginea
Germinal epithelium
Germ cells
Testis

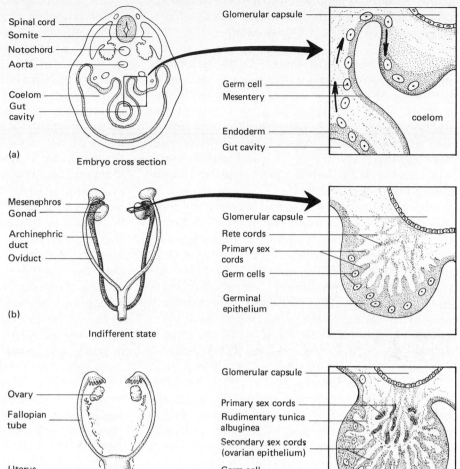

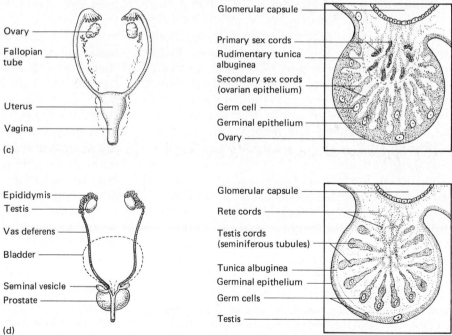

Figure 3–29. Generalized scheme of modification of the indifferent vertebrate gonad into an ovary or a testis. The gonadal structure whether primary cortex or medulla is derived from embryonic mesoderm. (a) The primordial germ cells, which give rise to eggs or sperm, are initially located in embryonic endoderm and migrate through the mesenteries during development into the indifferent gonad. (b) The germ cells arrange themselves between the medullary and cortical tissues of the indifferent gonad. (c) In formation of the ovary the cortical tissue predominates. (d) In the testis the medullary tissue predominates.

strong female and male roles. If some vertebrate individuals can function as mixed sexes, what is the purpose of rigid sex segregation?

Sex and Sex Chromosomes   In all vertebrates the gonad is initially indifferent and capable of producing an ovary or a testis (Figure 3–29). Individuals with both types of gonads present and functional are called **hermaphrodites**; individuals with either ovaries or testes are **gonochorists**. Hermaphroditism is common in nonamniotic vertebrates, especially fishes, but virtually absent among amniotes.

In amphibians and fishes distinctive sex chromosomes have seldom been demonstrated. Breeding experiments clearly indicate, however, that male- and female-determining genes are distributed over several chromosomes (Bull 1983). Because of the morphological compatibility these chromosomes of many fishes and amphibians can cross over and exchange genes. As a result, intersexuality and several types of hermaphroditism, including functional sex reversal, are widespread in nonamniotes (van Tienhoven 1983; Box 8–5).

In amniotes, however, sex reversals are atypical. The sex of a mammal or bird is genetically prescribed by the sex chromosome pairs. In humans, for example, zygotes of females contain 23 similar (homomorphic) chromosome pairs, but in males one chromosome (Y) is much smaller than its mate (X). Males therefore are chromosomally XY, a condition called heteromorphic. In males the Y chromosome carries a gene dominant for maleness (Gordon and Ruddle 1981, Haseltine and Ohno 1981). Sex chromosomal anomalies are well studied in humans. Individuals with only one X chromosome (XO) are female, although infertile. Individuals with sex chromosomal constitutions of XXY, XXXY, XXXXY, and XXXYY are phenotypic males, but sterile. Males of XYY are fertile. In mammals, the presence of the Y chromosome directs the medullary layer of the indifferent gonad to differentiate as a testis. In contrast, its absence and the presence of an X chromosome causes the cortical layer to develop as an ovary.

Sex chromosomes influence only the sexual development of the gonads, that is, the primary sex characters. Precisely how the sex genes act is unclear. Some evidence suggests that an antigen is secreted by the XY gonad to initiate gonadal development in mammals (Ohno 1979). More recent results suggest that this male antigen is probably involved in spermatogenesis and not in the initiation of gonadal development. Differentiation of testes in mammals, nevertheless, does seem to be controlled by a gene located on the male Y chromosome, and its absence leads to the development of female gonads. In humans, for example, females with an XY chromosomal complement exist because they lack the male-determining portion of the Y chromosome. And males with an XX chromosomal complement are also known because they have the appropriate male-determining portion of the Y chromosome translocated onto one of the two X chromosomes (Kolata 1986). What seems to be consistent is that a gene located on the Y chromosome initiates male gonadal development, and female gonadal development results from its absence. Once a gonad has had its primary sex declared as female or male, the sex hormone estrogen or testosterone is produced. These hormones affect the development of the secondary sex characters—all the structural and behavioral differences that exist between male and female (Ehrhardt and Meyer-Bahlburg 1981, Wilson et al. 1981). In humans the genitalia, breasts, hair patterns, and differential growth patterns are secondary sex characters. Horns, antlers, differences in plumage length, and dimorphic color patterns are familiar differences that we associate with sex in other vertebrates.

Sex hormones can influence an individual's secondary sex characteristics. The development of so-called freemartins in cattle provides a natural example of this phenomenon. Twinning sometimes results in the mixing of fetal blood supplies. If one twin is male (XY) and the other female (XX), the female embryo is influenced by the testosterone secretions of the developing male. At birth the female calf (a freemartin) shows mixed male–

female traits and is infertile. The genetic male calf is normal. The evidence suggests that in mammals testosterone leads to maleness, even in potentially genetic females (Bardin and Catterall 1981). Experimental evidence supports this conclusion. Castrating an early embryo of a placental mammal results in the development of female secondary sex characteristics, whether the embryo is XX or XY. Administration of testosterone to the castrated embryo, however, induces the development of male secondary sex characteristics (Bull 1983). Gonadal differentiation is not directly influenced by the hormone treatment, however, for XX embryos develop ovaries and XY embryos testes regardless of the hormonal effects on secondary sex characteristics.

Heteromorphic sex chromosomes occur in birds, where the female is the heteromorphic sex and the male the homomorphic sex. To distinguish this difference, geneticists use alternative symbols: ZW for female and ZZ for male. The dissimilar chromosome W is dominant and directs the indifferent gonad to develop as an ovary. In the shelled (cleidoic) egg equivalent of freemartinism—double-yolked eggs where the embryonic blood supplies mix—ZW females are normal and ZZ males are sterile intersexes. The conclusion is that estrogen effects are dominant over the effects of testosterone in these vertebrates.

Several questions about vertebrate sex-determining mechanisms are pertinent to a consideration of vertebrate evolution and characters. Why has sex determination been fixed rigidly in birds and mammals by heteromorphic sex chromosomes? Why is the heterogametic sex female in birds and male in mammals?

In birds and mammals many species form large groups in which individual behaviors often involve distinct male and female roles. In mammals, fixed sex determination may have been essential for the exploitation of the behavioral complexity that was made possible by the increased integrative capacities associated with larger brain size. Certainly, it would be disruptive to reproduction if the activities and morphologies associated with maleness and femaleness in birds and mammals were labile. If, for example, male baboons wavered in fending off predators while females hesitated to herd the young to safety, the result could reduce reproductive success. Thus, although it is unlikely that any single factor explains fixation of sex determination, we suspect that the behavioral complexity of birds and mammals is a significant component.

The opposite sexual expression of heterogametic individuals in mammals on one hand and birds on the other is intriguing. In gonochoristic vertebrates, the heterogametic sex is physiologically dominant. Carrying young *in utero* as placental mammals do, however, places strictures on the determination of sex. Ursula Mittwoch (1973) pointed out that during pregnancy maternal estrogens can cross the maternal–fetal barrier. For a female fetus (XX) this presents no problem, but a male (XY) would be swamped by estrogenic effects (femaleness) unless a strong masculinizing agent existed. Mittwoch has suggested that this crossing of the placental barrier by maternal hormones explains the dominance of XY masculinizing effect in directing the primary sex differentiation of mammals. Monotremes, mammals that lay eggs, are believed to lack a Y chromosome: females are XX and males are X. In birds the egg isolates the embryo, whether male or female, from direct maternal influence. Genetic sex, expressed in the relative safety of the egg, could equally well be determined by female heteromorphy (WZ) or by male heteromorphy (XY). Why female heteromorphy occurs in birds is not clear from this reasoning—quite possibly it was a matter of evolutionary chance.

Female and Male Reproductive Organs Ovaries differentiate as relatively simple organs with an undifferentiated connective tissue **stroma** in which are embedded many **follicles**. Follicles begin as large primary sex cells completely surrounded by a covering of epithelial cells from the germinal cortex. As they mature the follicular cell layer becomes much larger, often forming a spherical, sometimes fluid-filled organ within the ovary nurturing the developing egg and producing the hormone estrogen. The follicle stimulates the development of yolk in the egg. Egg yolk constitutes a major energy investment in reproduction for many vertebrates, either because of the number of eggs produced or because of the amount of yolk in each egg. When the eggs mature the follicle ruptures, releasing the completed egg (**ovulation**). In most vertebrates the eggs are released into the coelomic cavity. Advanced ray-finned fishes have very large, hollow sac-like ovaries and follicular rupture is toward the interior, so that the ovary becomes distended with ovulated eggs. The epithelial cells of the collapsed follicle remain in the stroma of the ovary after ovulation. In many vertebrates these cells form a second hormone-producing gland, the **corpus luteum** (plural copora lutea). The corpus luteum produces the hormone progesterone, which stimulates changes in the relevant portions of the female for retention of the developing young. In birds where no retention of ova occurs, corpora lutea do not develop. The production of eggs by the ovary may be almost continuous (humans), seasonal (the vast majority of vertebrates), or occur only once in a lifetime (semelparous fishes such as the eel, and certain salmon native to the North Pacific).

The testes differentiate as compact organs made up of interconnecting **seminiferous tubules** where sperm develop. Sperm differentiate from prolifically dividing sex cells and are supported, nourished, and conditioned by cells that remain permanently attached to the tubule walls, the **supporting** or **Sertoli** cells. Unlike the follicular cells of the ovary, the testicular supporting cells are not endocrine but strictly sperm-nurturing cells. The hormone testosterone is produced by clusters of **interstitial cells** between the seminiferous tubules in tetrapods.

Sperm are among the most highly specialized of animal cells, and their development takes place in the sex cells embedded in the walls of the supporting cells. Mature vertebrate sperm consist of a head and a long flagellum-like tail. The head is made up of a terminal **acrosomal cap** of egg-penetrating enzymes over the highly condensed haploid nucleus. A connecting piece joins the head to the tail. The latter is made up of a middle piece with a sheath of mitochondria wrapped around a core of flagellar motor tubules and stiffening fibers that continue as the principal portion of the tail. No cytoplasm in the conventional sense remains in a sperm cell; it is specialized to deliver nuclear information to the ovum.

The reproductive tracts of vertebrates represent the gamete conduits to the external environment. The primitive gnathostome condition involves two sets of ducts that carry the gametes to the exterior. One set of conduits are the **archinephric (Wolffian) ducts** that form on each side of the coelom by the fusion and posterior growth of the embryonic pronephric tubules (Figure 3–15). No duct transports sperm in jawless vertebrates; sperm erupt from the testis and move through the coelom to pores that open to the outside near the cloaca. In male gnathostomes the archinephric duct transports sperm. In a few species (primitive ray-finned fishes, gymnophionans, and some other amphibians) sperm and urine are carried to the exterior through the archinephric duct. However, in the majority of gnathostomes the kidney tubules involved are modified exclusively for sperm transport or eliminated, the seminiferous tubules connecting directly to the archinephric duct. The archinephric duct also repeatedly evolves toward exclusive sperm transport functions as the opisthonephric kidney develops accessory urinary ducts (cartilaginous fishes, most nonamniote tetrapods) or the metanephric kidney evolves with its unique ureter (amniotes). The lungfishes and advanced ray-finned fishes have very posteriorly positioned testes, which thus have connections with the urine-carrying archinephric duct only very posteriorly (lungfish) or not at all (teleosts). Short new ducts, unrelated to those of any other vertebrate gonad, transport sperm in these fishes. Like other vertebrates, bony fishes thus establish separation of sperm and urine. The terminus of the exclusively sperm-carrying archinephric ducts (or in some fishes of their distinct sperm ducts) is often incorporated into specialized **intromittent organs** that permit sperm to be directly introduced into the female reproductive tract. These organs may be derived from pelvic fins (claspers of cartilaginous fishes), anal fins (the gonopodia of fishes like guppies and swordtails), gill covers (the gonopodia of some ray-finned fishes), or from the walls of the cloaca (the penises and hemipenes of tetrapods).

The female reproductive tract develops in parallel to that of the archinephric duct, except in jawless fishes in which both sexes lack conduits. The distinctive paired ducts, called the **oviducts** or **Mullerian ducts**, of females form by a longitudinal splitting of the archinephric duct (in chondrichthyans and salamanders) or by an invagination of the peritoneum over the kidney (in primitive bony fishes and the remaining tetrapods). Teleost fishes form analogous but not homologous ducts by posterior extension of the saccular ovaries, thus producing a conduit system that has no coelomic segment. The Mullerian ducts form early in embryonic development and are generally located beside the archinephric ducts before the gonads differentiate. Thus one or the other set of ducts degenerates during sexual differentiation.

The anterior end of the oviduct opens into the coelom, except in teleosts. Ova enter the **ostium** of the oviduct after they have ruptured from the ovary into the coelom. The oviducts are variously specialized in different groups to accommodate the great range of reproductive patterns seen in vertebrates. In forms with simple external fertilization of extruded eggs (**oviparous** species) the oviducts are simple tubes. The oviducts of most amniotes are specialized to produce albumen, and shells (cartilaginous fishes, most amniotes) or jelly coatings (amphibians, some bony fishes) for eggs that are to be extruded. (The term "oviparity" is also correctly applied to these species.) A wide variety of vascular and secretory specialization is found in the **uteri** (singular uterus) of forms in which both fertilization and development are internal (**viviparity**). The degree of development and type of uterine specializations depend on whether additional energy, over and above that contained in the yolk, is provided to the embryo.

Summary

The complex activities of vertebrates are supported by an equally complex morphology. Interactions between different tissues and structures are central to the embryonic development of a vertebrate and to its function as an organism. Patterns of embryonic development are generally phylogenetically conservative, and many of the shared derived characters of vertebrates can be traced to their origins in the early embryo. In particular, the neural crest cells, which are unique to vertebrates, participate in the embryonic origin of a number of derived characters of vertebrates.

An adult vertebrate can be viewed as a number of systems that interact continuously. The integument separates a vertebrate from its environment and participates in regulating the exchange of matter and energy between the organism and the environment. Support and movement are the province of the skeletomuscular system, and both are necessary for the effective function of the food-gathering and food-processing systems. Respiration and circulation carry metabolic substrates and oxygen to the tissues and remove wastes. The nitrogenous waste products of protein metabolism are eliminated by the integument of primitive forms and by the renal system of derived vertebrates. The renal system also participates in regulating salt and water balance and maintaining the pH of the blood. Coordination of these activities is accomplished by the nervous and endocrine systems, and the reproductive system transmits genetic information from generation to generation. In this chapter we have described the phylogenetic and embryonic origins of the major structures and systems of vertebrates, and discussed their modifications during vertebrate evolution.

References

Banting, F. G. and C. H. Best. 1922. The internal secretion of the pancreas. *Journal of Laboratory and Clinical Medicine* 7:251–266.

Bardin, C. W. and J. F. Catterall. 1981. Testosterone: a major determinant of extragenital sexual dimorphism. *Science* 211:1285–1294.

Bereiter-Hahn, J., A. G. Matoltsy, and K. S. Richards (editors). 1986. *Biology of the Integument 2: Vertebrates*. Springer-Verlag, New York.

Bull, J. J. 1983. *Evolution of Sex Determining Mechanisms*. Benjamin-Cummings, Menlo Park, Calif.

Burggren, W. W. 1987. Form and function in reptilian circulations. *American Zoologist* 27:5–19.

Cooper, E. L. 1985. Comparative immunology. *American Zoologist* 25:649–664.

Ehrhardt, A. A. and H. F. L. Meyer-Bahlburg. 1981. Effects of prenatal sex hormones on gender related behavior. *Science* 211:1312–1218.

Fawcett, D. W. 1986. *A Textbook of Histology*. 11th edition. W. B. Saunders, Philadelphia.

Gans, C. and R. G. Northcutt. 1983. Neural crest and the origin of vertebrates: A new head. *Science* 220:268–274.

Gordon, J. W. and F. H. Ruddle. 1981. Mammalian gonadal determination and gametogenesis. *Science* 211:1265–1271.

Haseltine, F. P. and S. Ohno. 1981. Mechanics of gonadal differentiation. *Science* 211:1272–1278.

Kolata, G. 1986. Maleness pinpointed on Y chromosome. *Science* 234:1076–1077.

Le Douarin, N. 1982. *The Neural Crest*. Cambridge University Press, Cambridge.

Mittwoch, U. 1973. *Genetics of Sex Determination*. Academic Press, New York.

Naftolin, F. 1981. Understanding the basis of sex differences. *Science* 211:1263–1264.

Noden, D. M. and A. deLahunta. 1985. *The Embryology of Domestic Animals*. Williams & Wilkins, Baltimore.

Ohno, S. 1979. *Major Sex-Determining Genes*. Springer-Verlag, Berlin.

Sage, H. and W. R. Gray. 1979. Studies on the evolution of elastin. I. Phylogenetic distribution. *Comparative Biochemistry and Physiology* 64B:313–327.

Sage, H. and W. R. Gray. 1981a. Studies on the evolution of elastin. II. Histology. *Comparative Biochemistry and Physiology* 66B:13–22.

Sage, H. and W. R. Gray. 1981b. Studies on the evolution of elastin. III. The ancestral protein. *Comparative Biochemistry and Physiology* 68B:473–480.

Smith, H. W. 1953. *From Fish to Philosopher*. Little, Brown, Boston.

van Tienhoven, A. 1983. *Reproductive Physiology of Vertebrates*. 2nd edition. Cornell University Press, Ithaca, N. Y.

Webster, D. 1966. Ear structure and function in modern mammals. *American Zoologist* 6:451–466.

Wilson, J. D., F. W. George and J. E. Griffin. 1981. The hormonal control of sexual development. *Science* 211:1278–1284.

In the preceding chapter we described some of the structural complexities of vertebrates, and here we consider the way those structures work. Vertebrates are complicated organisms: in particular they maintain very different conditions inside their bodies from the conditions in the environment immediately external to them. The concentrations of water, ions, and molecules inside the body of a vertebrate have profound effects on biochemical and physiological mechanisms, and they must be regulated within specific limits. Temperature affects both the biochemical processes of life and the cellular environment within which those processes take place. Regulation of their internal conditions (homeostasis) is a central part of the biology of vertebrates. Structure and function are usually tightly coupled— changing the form of a structure is likely to change its function as well. Thus the evolution of vertebrate morphology has been accompanied by changes in the ways that vertebrates work. Some kinds of specializations of vertebrates are mutually exclusive, whereas others are mutually compatible and can be combined in the same organism. The relationships between structure and function are often reflected in broad aspects of the biology of vertebrates and directly affect their ecology and behavior. These relationships are also intimately related to the characteristics of the environments in which vertebrates live: for example, many of the problems faced by terrestrial and aquatic vertebrates are quite different. In this chapter we describe the basic aspects of homeostasis of vertebrates and how they have changed in major evolutionary steps such as the transition from aquatic to terrestrial life or from ectothermy to endothermy.

Homeostasis and Energetics: Water Balance, Temperature Regulation, and Energy Use

The Internal Environment of Vertebrates

Seventy to eighty percent of the body mass of most vertebrates is water, and many of the chemical reactions that release energy or synthesize new compounds take place in an aqueous environment. Some ions are cofactors that control the rates of metabolic processes; others are involved in the regulation of pH, the stability of cell membranes, or the electrical activity of nerves. Metabolic substrates and products must diffuse from sites of synthesis to the sites of utilization. Almost everything that happens in the body tissues of vertebrates involves water, and maintaining the concentrations of water and solutes within narrow limits is a vital activity.

Temperature, too, is critical in the function of vertebrates. Biochemical reactions are temperature-sensitive. In general, the rates of reactions increase as temperature increases, but not all reactions have the same sensitivity to temperature. Furthermore, the permeability of cell membranes and other features of the cellular environment are sensitive to temperature. A metabolic pathway is a series of chemical reactions in which the product of one reaction is the substrate for the next, yet each of these reactions may have a different sensitivity to temperature. In that situation a change in temperature can mean that too much or too little substrate is produced to sustain the next reaction in the series. To complicate the process of regulation of substrates and products even more, the chemical reactions take place in a cellular milieu that is itself changed by temperature. Clearly, the integration of metabolic pathways is greatly simplified if an organism can limit the range of temperatures its tissues experience, and temperature regulation is another critical element in energy flow.

From the perspective of environmental physiology, an organism can be described as a self-sustaining interaction with the environment. Energy is the basis of this interaction. Vertebrates gain energy from the environment as food or as heat; they use energy for activity, growth, and reproduction; and they release energy as waste products or as heat. Regulation of water, ions, and temperature is a key element in the self-sustaining nature of the interactions. Energy is the basis of the activity of vertebrates, and calculations of the energy expended and the energy gained in specific activities can illuminate features of the behavior of organisms. In this chapter we describe some of the mechanisms that vertebrates use to regulate their exchanges of water and heat with the environment, and the significance of those mechanisms in the use of energy by vertebrates.

153

Exchange of Water and Ions

In a sense, an organism can be thought of as an aqueous solution of organic and inorganic substances contained within a leaky membrane, the body surface. Exchange of material and energy with the environment is essential to the survival of the organism, and much of that exchange is regulated by the body surface. The significance of differential permeability of the skin to various compounds is particularly conspicuous in the case of aquatic vertebrates, but it applies to terrestrial vertebrates as well. Both active and passive processes of exchange are used by vertebrates to regulate their internal concentrations in the face of varying external conditions.

Regulation of Ions and Body Fluids

The first vertebrates, the ostracoderms, probably had ion levels like their marine invertebrate ancestors, which, presumably, were like those of most living marine invertebrates. The salt concentrations in the body fluids of many marine invertebrates are similar to seawater, as are those of hagfishes (Tables 4–1 and 4–2). In contrast, blood salt levels are greatly reduced in all other vertebrates, a characteristic shared only with invertebrates that have penetrated estuaries, fresh waters, or the terrestrial environment.

The presence of solutes in seawater or blood plasma lowers the kinetic activity of water. Therefore, water flows from a dilute solution (high kinetic activity of water) to a more concentrated solution (low kinetic activity), a phenomenon termed **osmosis**. The osmotic concentrations of various animals and of seawater are shown in Table 4–1. In most marine invertebrates and the hagfish, the body fluids are in osmotic equilibrium with seawater; that is, they are **isosmotic** to seawater. Body fluid concentrations in marine teleosts and lampreys are between 350 and 450 milliOsmoles per liter (mOsm/liter). Therefore, water flows outward from their blood to the sea. In chondrichthyans the osmolality of the blood is higher than that of seawater, and water flows from sea to blood. These osmotic differences are specified by the terms **hyposmotic** (marine teleosts and lampreys) and **hyperosmotic** (coelacanth and chondrichthyans). Freshwater fishes are hyperosmotic to the medium, but through reduction in NaCl their blood osmolality is lower than that of their marine counterparts (Table 4–1). The physiological integrity of teleosts is constantly threatened by either (1) osmotic loss of water and salt gain when in seawater, or (2) osmotic gain of water and loss of salts when in fresh water. [For additional details see Evans (1980) and Nishimura and Imai (1982).]

Most fishes are **stenohaline** (steno = narrow, haline = salt): They inhabit either fresh water or seawater and survive only modest changes in salinity. Because they remain in one environment, the magnitude and direction of the osmotic gradient to which they are exposed is stable. Some fishes, however, are **euryhaline** (eury = wide): They inhabit both fresh water and seawater and tolerate large changes in salinity. In euryhaline species the osmotic and salt gradients are reversed as they move from one medium to the other.

Freshwater Organisms: Teleosts and Amphibians Several mechanisms are involved in the osmotic regulation of vertebrates that live in fresh water (Figure 4–1). The body surface of fishes has low permeability to water and to ions. However, fishes cannot entirely prevent osmotic exchange. Gills, which are permeable to gases, are also permeable to water. As a result, most water and ion movements take place across the gill surfaces. Water is gained by osmosis and ions are lost by diffusion. To compensate for this influx of water, the kidney produces a copious volume of urine. To reduce salt loss, the urine is diluted by actively reabsorbing salts. Indeed, urine processing in a freshwater teleost provides a simple model of vertebrate kidney function.

The large glomeruli of freshwater teleosts allow production of a copious flow of urine, but the glomerular ultrafiltrate is isosmotic to the blood and contains essential blood salts. To conserve

Table 4–1. Representative concentrations of the sodium and chloride and osmolality of the blood in vertebrates and marine invertebrates. Ion concentrations are expressed in millimoles per liter of water; all values are reported to the nearest 5 units. Osmolality reported in milliOsmoles (mOsm; 1 Osm = 1 mole dissolved particles per kilogram water).

Type of Animal	mOsm	Na^+	Cl^-	Other Major Osmotic Factor	Source
Seawater	~1000	475	550		
Fresh water	< 10	~ 5	~ 5		
Marine invertebrates					
Coelenterates, mollusks, etc.	~1000	470	545		1
Crustacea	~1000	460	500		1
Marine vertebrates					
Hagfishes	~1000	535	540		2
Lamprey	~ 300	120	95		2
Teleosts	< 350	180	150		3
Coelacanth	<1000 to 1180	180	200	Urea 375	4,10
Elasmobranch (bull shark)	1050	290	290	Urea 360	5
Holocephalian	~1000	340	345	Urea 280	9
Freshwater vertebrates					
Polypterids	200	100	90		3
Acipenserids	250	130	105		3
Primitive neopterygians	280	150	130		3
Dipnoans	240	110	90		3
Teleosts	< 300	140	120		3
Elasmobranch (bull shark)	680	245	220	Urea 170	5
Elasmobranch (freshwater rays)	310	150	150		6
Amphibians*	~ 250	~100	~80		7
Terrestrial vertebrates					
Reptiles	350	160	130		8
Birds	320	150	120		8
Mammals	300	145	105		8

*Ion levels and osmolality highly variable, but tend toward 200 mOsm in fresh water.

Sources:
1. W. T. W. Potts and G. J. Parry, 1964, *Osmotic and Ionic Regulation in Animals,* Macmillan, New York.
2. J. D. Robertson, 1954, *Journal of Experimental Biology* 31:424–442.
3. M. R. Urist et al., 1972, *Comparative Biochemistry and Physiology* 42:393–408.
4. G. E. Pickford and F. G. Grant, 1964, *Science* 155:568–570; R. W. Griffith et al., 1975, *Journal of Experimental Zoology* 192:165–171.
5. T. B. Thorson et al., 1973, *Physiological Zoology* 46:29–42.
6. T. B. Thorson et al., 1967, *Science* 158:375–377.
7. P. J. Bentley, 1971, *Endocrines and Osmoregulation,* Zoophysiology and Ecology Series, volume 1, Springer-Verlag, New York.
8. C. L. Prosser, 1973, *Comparative Animal Physiology,* 3rd edition, W. B. Saunders, Philadelphia.
9. L. J. Read, 1971, *Comparative Biochemistry and Physiology* 39A:185–192.
10. D. H. Evans, 1979, *Comparative Physiology of Osmoregulation in Animals,* edited by G. M. O. Maloiy, Academic Press, New York.

Table 4–2. Intracellular concentration of major inorganic ions in marine invertebrates and vertebrates. (Concentration values are in millimoles per liter; compare with Table 4–1.)*

	Na^+	Cl^-	K^+	Ca^{2+}	Mg^{2+}
Seawater	475	550	10	10	53
Marine invertebrates	54–325	54–380	48–175	3–89	8–96
Vertebrates					
Hagfish	122	107	117	2	13
All others	8–45	11–30	83–185	2–9	7–11

*Note that the monovalent ions Na^+ and Cl^- are reduced relative to seawater, and K^+ increased in all animals. For divalent cations, especially Mg^{2+}, a reduction is found in all vertebrates but not in all marine invertebrates.

salt, ions are reabsorbed across the proximal and distal convoluted tubules. Because the distal convoluted tubule is impermeable to water, the urine becomes less concentrated as ions are removed from it. Ultimately, the urine becomes hyposmotic to the blood. In this way the water that was absorbed across the gills is removed and ions are conserved. Nonetheless, some ions are lost in the urine in addition to those lost by diffusion across the gills. Salts from food compensate for some of this loss. In addition, freshwater teleosts have special cells located in the gills that absorb sodium and chloride ions from fresh water. This absorption of ions is an up-hill process. That is, ions are moved by active transport against a concentration gradient, and this requires energy.

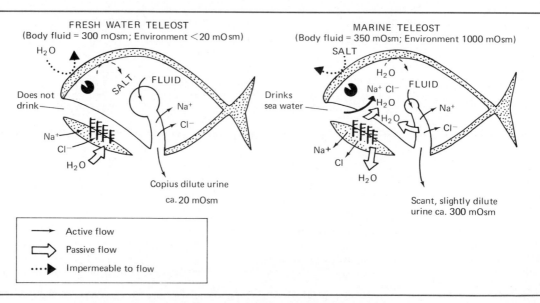

Figure 4–1. General scheme of the osmotic and ionic gradients encountered by freshwater and marine teleosts.

Freshwater amphibians face the same osmotic problems as do freshwater fishes. The entire body surface of amphibians is involved in the active uptake of ions from the water. Acidity inhibits this active transport of ions in both amphibians and fishes, and inability to maintain internal ion concentrations is one of the causes of death of these animals in habitats acidified by acid precipitation.

Marine Organisms: Teleosts and Other Fishes The osmotic and ionic gradients of vertebrates in seawater are the reverse of those experienced by freshwater vertebrates. Seawater is more concentrated than the body fluids of vertebrates, so water flows out of the body by osmosis and ions diffuse inward.

Teleosts The integument of marine fishes, like that of freshwater teleosts, is highly impermeable, so that most osmotic and ion movements occur across the gills (Figure 4–1). The kidney glomeruli are small and the glomerular filtration rate is low. Less urine is formed, and the water lost in urine is reduced. Marine teleosts lack a water-impermeable distal convoluted tubule. As a result, urine leaving the nephron is less copious but more concentrated than in freshwater teleosts, although it is always hyposmotic to blood. To compensate for osmotic dehydration, marine teleosts do something unusual—they drink seawater. From this briny beverage sodium and chloride ions are actively absorbed across the lining of the gut, and water flows osmotically into the blood. Estimates of seawater consumption vary, but many species drink in excess of 25 percent of their body weight per day and absorb 80 percent of this ingested water into their extracellular spaces. Drinking seawater to compensate for osmotic water loss increases the influx of sodium and chloride ions. To compensate for this salt loading, discrete cells, called **chloride cells**, located in the gills, actively pump sodium and chloride ions outward against a large concentration gradient.

Other Fishes Hagfishes minimize their problems with ion balance by regulating only divalent ions and reduce osmotic water movement by being nearly isosmotic to seawater. Chondrichthyans and coelacanths also minimize osmotic flow by maintaining the osmotic concentration of the body fluid close to that of seawater. These animals retain urea, a waste product from metabolism of protein, to produce osmolalities that are usually slightly hyperosmotic to seawater (Table 4–1). As a result, chondrichthyans gain water by osmotic diffusion across the gills and need not drink seawater. This net influx of water permits larger kidney glomeruli, as in freshwater teleosts, to produce high filtration rates and therefore rapid cleansing of the blood. Urea, however, is very soluble and diffuses through most biological membranes. The gills of chondrichthyans are highly impermeable to urea, and the kidney tubules actively reabsorb it. With low internal ion concentrations chondrichthyans experience ion influxes across the gills as do marine teleosts. Unlike marine teleosts, the gills of chondrichthyans have low ion permeabilities (less than 1 percent that of teleosts). Chondrichthyans therefore generally do not require highly developed salt-excreting cells in the gills. Rather, they achieve ion balance by secreting from the rectal gland a fluid that is isosmotic to body fluids but contains higher concentrations of sodium and chloride ions.

Freshwater Elasmobranchs and Marine Amphibians Some elasmobranchs are euryhaline; sawfish, some stingrays, and bull sharks are examples (Thorson 1970, Thorson et al. 1973). In seawater bull sharks retain high levels of urea, but in fresh water their blood urea levels decline. Stingrays in the family Potamotrygonidae spend their entire lives in fresh water. Urea is present only at very low blood concentration. Blood sodium and chloride ion concentrations are 35 to 40 percent below those in sharks that enter fresh water and only slightly above levels typical of freshwater teleosts (Table 4–1). The potamotrygonids have existed in the Amazon basin perhaps since the Tertiary, and their reduced ion-osmotic gradient may reflect their long adaptation to fresh water. When exposed to increased salinity, potamotrygonids do not increase urea as do euryhaline elasmobranchs, even though the enzymes required to produce urea are present. Apparently, their long evolution in fresh water has led to an increase in the permeability of their gills to urea and reduced the ability of their kidney tubules to reabsorb it. The presence of high concentrations of urea in the blood of fishes appears to relate to marine, not fresh water habits.

Most amphibians are found in freshwater or terrestrial habitats. One of the few species that occurs in salt water is the crab-eating frog, *Rana cancrivora*. This frog inhabits intertidal mudflats in southeast Asia and is exposed to 80 percent seawater at each high tide. During seawater exposure, the frog allows its blood ion concentrations to rise and thus reduces the ionic gradient. In addition, proteins are deaminated and the ammonia rapidly converted to urea and released into the blood. Blood urea rises from 20 millimoles (mmol)/liter to 30 mmol/liter and the frogs become hyperosmotic to the surrounding water. In this sense *Rana cancrivora* functions like an elasmobranch and absorbs

water osmotically. Frog skin, unlike that of elasmobranchs, is permeable to urea and as a result urea is rapidly lost. To compensate for this loss, the urea-synthesizing enzymes are very active and concentrated. Like most tadpoles, those of *Rana cancrivora* lack urea-synthesizing enzymes. They possess extrarenal salt-excreting cells in the gills and maintain their blood hyposmotic to seawater in the same manner as marine teleosts.

Osmoregulation and the Evolutionary History of Fishes The sequence of appearance of different groups of freshwater vertebrates in the fossil record correlates with blood osmolality and major salts (Table 4–3). Each group has retained freshwater representatives throughout all subsequent periods of geologic time. Most marine fishes derive from freshwater ancestors, the myxinoids probably being the sole exception. Evidence concerning the presumed origin of chondrichthyans in fresh water is equivocal. Assuming that their ancestors showed some degree of adaptation to fresh water by internal reduction of ions, each reinvasion of

Table 4–3. Correlation between the first appearance of different fish groups and the blood osmolality and salt (Na$^+$ and Cl$^-$) concentrations of living representatives. (Osmolalities and salt concentrations are referenced in Table 4–1.)

Habitat Type and Vertebrate	First Appearance in Habitat Type	Blood Concentration in Living Species (mOsm)	Na + Cl (mM/liter)
Fresh water			
Paleoniscoids	Early Devonian	200	190
Dipnoans	Early Devonian	240	200
"Chondrosteans"	Pennsylvanian	250	235
Primitive Neopterygians	Late Permian–Early Triassic	280	280
Teleosts	Late Cretaceous	<300	260
Potamotrygonids	Pliocene (?)	310	300
Bull sharks, stingrays, etc.	Recent	680	465
Marine		157	
Myxinoids	Ordovician–Silurian	Isosmotic with seawater	1000
Elasmobranchs	Late Devonian	Hyperosmotic to seawater	580
Coelacanths	Early Triassic	Hyperosmotic to seawater?	380
Teleosts	Late Cretaceous	350	330

the sea must have required reversal in the osmotic adjustments of body fluids and cell ion sensitivities to seawater (Lutz 1975).

The order of appearance of groups of vertebrates in the marine fossil record correlates with an increase in the major blood ions (Table 4–3). Two of the most ancient marine groups retained urea as an osmotically active substance. But the oldest group, the hagfishes, have blood salt concentrations that we believe represent the condition typical of the first vertebrates. Their high internal ion levels support the idea that hagfishes have always been marine. Teleosts, which may derive from marine holosteans and are the most recent invaders of the sea, have ion levels above their freshwater predecessors, but have not utilized urea to eliminate osmotic water loss. Rather, they expend additional energy in ion and osmoregulation. Although the reasons are unknown, maintaining low ion concentrations in body fluids and cells must be beneficial. Even the coelacanths and chondrichthyans have allowed their body fluid ion concentrations to rise only slightly toward seawater concentrations, in spite of their long residence in the sea. This also suggests that benefits are derived from reduced ion concentrations. Why?

Peter Lutz has proposed that the answer lies in the interplay between cell activities and ion concentrations. Cellular enzymes require ions as cofactors. Ion concentrations can alter the rates of biochemical reactions and, therefore, cell activity. Indeed, intracellular ion concentrations are rigidly maintained by energy-demanding ion pumps located in the cell membrane. These ion pumps also operate to maintain cell volume and transmembrane potentials, properties also vital to cell function. In colonizing a hyposmotic environment, an organism confronts osmotic hydration as well as a loss of body ions. Simple reduction in the concentration of body fluids lessens the threat of hydration and salt loss. However, for a metazoan to reduce its concentration, the cells must suffer the consequences, for they remain isosmotic to the body fluids. Lowering ion concentrations therefore dictates accompanying changes in those cell processes that depend on specific ion levels to maintain function. Apparently, these intracellular adjustments to different ion concentrations require considerable evolutionary time. Cell chemistry appears to be a conservative trait and perhaps has set limits to some aspects of vertebrate evolution. Analysis of the evolution of fishes, as Lutz points out, suggests this possibility.

Nitrogen Excretion by Vertebrates

In considering the water relations of vertebrates, we should consider how different vertebrates process nitrogenous wastes. This subject is a fascinating example of how the biochemical activities of vertebrates and their ecology interact.

Vertebrates require food in proportion to their activity. In utilizing carbohydrates and fats (composed of carbon, hydrogen, and oxygen) vertebrates produce carbon dioxide and water, which are easily voided. Proteins and nucleic acids are another matter, for they contain nitrogen (N). When protein is metabolized, the nitrogen is enzymatically reduced to ammonia (NH_3) through a process called deamination. Ammonia is very diffusable and soluble in water but also extremely toxic to cell activities. Rapid excretion of NH_3 is therefore crucial. Differences in how NH_3 is excreted are partly a matter of the availability of water and partly the result of vertebrate phylogeny. Nitrogen is eliminated by most vertebrates as NH_3, as urea, or as uric acid (Figure 4–2).

Fishes and Amphibians   Many aquatic invertebrates excrete NH_3 directly, as do vertebrates with gills, permeable skins, or other permeable membranes that contact water. Excretion of nitrogenous wastes as NH_3 is termed **ammonotelism**, excretion as urea is **ureotelism**, and excretion as uric acid is **uricotelism**. Sharks, for example, synthesize urea from NH_3 in a cellular enzymatic process called the **urea cycle**. Urea synthesis requires greater expenditure of energy than does ammonia production. The value of ureotelism, therefore, lies in the benefits derived from urea itself, not in energy economy.

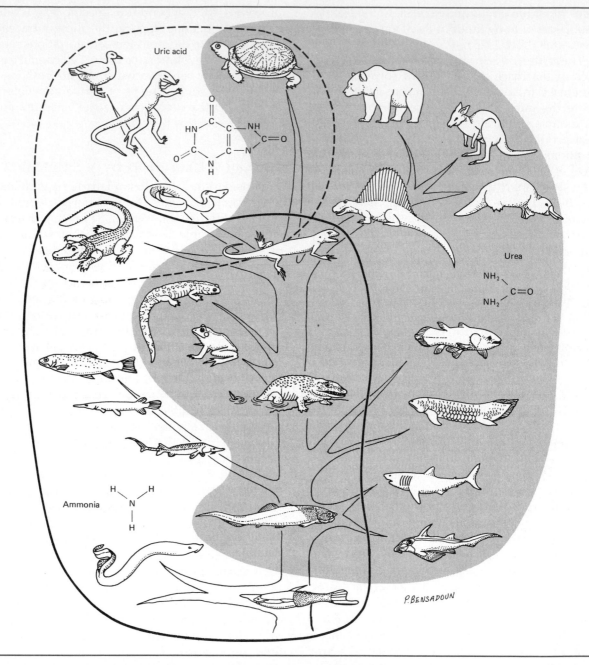

Figure 4–2. Phylogenetic distribution of the three major nitrogenous wastes in vertebrates. The types of wastes excreted by extinct vertebrates are unknown; examples merely provide visual continuity to the phylogeny. (Modified from B. Schmidt-Nielsen, 1972, pages 79–103 in *Nitrogen Metabolism and the Environment*, edited by J. W. Campbell and L. Goldstein, Academic Press, London.)

Urea has two advantages. First, its retention by some marine vertebrates counters osmotic dehydration. A second function of urea synthesis is the detoxification of NH_3 under environmental circumstances that prevent its rapid elimination. The less toxic urea can be concentrated in urine, thus conserving water.

Ureotelism probably evolved independently several times. Perhaps ureotelism developed in early freshwater fishes to avoid osmotic dehydration as they reinvaded the sea, as exemplified by chondrichthyans and coelacanths. In other instances ureotelism may have evolved in response to desiccation. Ureotelism was preadaptive for invasion of the land, because it allowed nitrogen to be retained in a detoxified state until sufficient water was available for excretion.

A fascinating example of this preadaptation was discovered in the ammonotelic African lungfish, *Protopterus*, by the late Homer Smith. During drought lungfishes aestivate and slowly oxidize their carbohydrate, fat, and protein stores to sustain life. The NH_3 is detoxified and urea accumulates, increasing body fluid osmolality. This reduces the vapor pressure across the lung surface and water loss through evaporation. When rains return, the lungfishes rapidly take up water and excrete the accumulated urea. An analogous process takes place in some spadefoot toads that inhabit arid environments (see Chapter 16). Ureotelism may have been similarly advantageous to the first Devonian tetrapods.

Mammals The capacity of the mammalian kidney to conserve water, rid the body of nitrogenous and other wastes, and maintain a narrow ion and acid–base variation is essential to mammalian life. Only with the concentrating powers of the mammalian kidney could mammals have invaded so many diverse and severe environments. Understanding the mammalian kidney is a key factor in understanding the success of mammals.

The mammalian kidney is composed of millions of nephrons, the basic microanatomical units of kidney structure recognizable in all vertebrates. Each nephron is composed of a **glomerulus** that filters the blood and a long tubular conduit in which the chemical composition of the filtrate is altered. The mammalian kidney is capable of producing a urine more concentrated than that of any nonamniote and, in most cases, more concentrated than that of diapsids as well (Table 4–4). This ability enormously reduces water loss and the need for water, and is important for several lineages of mammals that live in arid habitats.

The basic problem in urine concentration involves removal of water from an ultrafiltrate, leaving behind the concentrated excretory residue. Because cells are unable to transport water directly, they use osmotic gradients to manipulate movements of water molecules. In addition, the cells lining the nephron must actively reabsorb substances important to the body's economy from the ultrafiltrate and also secrete toxic substances into it. The single layer of cells lining the nephron differs along the length of the nephron in permeability, molecular and ion transport activity, and reaction to the hormonal and osmotic environments in the surrounding body fluids. The nephron's activity may be considered a six-step sequential process, each step localized in regions having special cell characteristics and distinctive variations in the osmotic environment.

The first step is production of an ultrafiltrate at the glomerulus (Figure 4–3). The ultrafiltrate is isosmotic with blood plasma and resembles whole blood after the removal of (1) cellular elements, (2) substances with a molecular weight of 70,000 or greater (primarily proteins), and (3) substances with molecular weights between 15,000 and 70,000, depending on the shapes of the molecules. An average filtration rate for resting humans approximates 120 milliliters of ultrafiltrate per minute. Obviously, a primary function of the nephron is reduction of the ultrafiltrate volume—to excrete 170 liters (45 gallons) of glomerular filtrate per day is impossible.

The second step in the production of the urine is the action of the **proximal convoluted tubule** (PCT, Figure 4–3) in decreasing the volume of the ultrafiltrate. The PCT cells have greatly enlarged lumenal surfaces that actively transport sodium

Table 4—4. Maximum urine concentrations of tetrapods.

Species	Maximum Observed Urine Concentration (mOsm/liter)	Approximate Urine: Plasma Concentration Ratio
Shingle-back lizard (*Trachydosaurus rugosus*)	300	0.95
Pelican (*Pelecanus erythrorhynchos*)	700	approx. 2
Savanna sparrow (*Passerculus sandwichensis*)	2000	4.4
Human (*Homo sapiens*)	1430	4
Bottlenose Porpoise (*Tursiops truncatus*)	1600–1800	approx. 5
Hill kangaroo (*Macropus robustus*)	2730	7–8
Camel (*Camelus dromedarius*)	2800	8
White rat (*Rattus norvegicus*)	2900	8.9
Cat (*Felis domesticus*)	3250	9.9
Pack rat (*Neotoma albigula*)	4250	11 (est.)
Marsupial mouse (*Dasycercus cristicauda*)	approx. 4000	12 (est.)
Kangaroo rat (*Dipodomys merriami*)	approx. 4650	12 (est.)
Vampire bat (*Desmodus rotundus*)	4650	14
Australian hopping mouse (*Notomys alexis*)	9370	22

Sources: P.J. Bentley, 1959, *Journal of Physiology* 145:37–47; M.S. Gordon et al., 1982, *Animal Physiology*, 4th edition, Macmillan, New York; R.L. Malvin and M. Rayner, 1958, *American Journal of Physiology*, 214:187–191; R.E. MacMillen, 1972, *Symposium of the Zoological Society of London* 31:147–174; W.N. McFarland and W.A. Wimsatt, 1976, *Comparative Biochemistry and Physiology* 28:985–1007; K. Schmidt-Nielsen, 1964, *Desert Animals*, Oxford University Press, Oxford.

and perhaps chloride ions from the lumen to the exterior of the nephron. Water flows osmotically through the PCT cells in response to the removal of sodium chloride. By this process about two-thirds of the salt is reabsorbed in the PCT, and the volume of the ultrafiltrate is reduced by the same amount. Although it is still very nearly isosmotic with blood, the substances contributing to the osmolality of the urine after it has passed through the PCT are at different concentrations than in the blood.

The next alteration occurs in the descending limb of the **loop of Henle**, where the thin, smooth-surfaced cells of this segment freely permit diffusion of sodium and water. Because the descending limb passes through tissues of increasing osmolality, water is lost from the urine and it becomes more concentrated. In humans the osmolality of the fluid in the descending limb may reach 1200 mOsm/liter. Mammals producing a more concentrated urine achieve correspondingly higher concentrations. By this mechanism the volume of the forming urine is reduced to 25 percent of the initial filtrate volume, but it is still large. In an adult human, for example, 25 to 40 liters of fluid per day reach this stage, yet only a few liters will be urinated.

The fourth step takes place in the ascending limb of the loop of Henle, which possesses cells with large, numerous densely packed mitochondria. The ATP produced by these organelles is utilized in actively removing sodium from the forming urine. Because these cells are impermeable to water, the volume of urine does not decrease and it enters the next segment of the nephron hyposmotic to the body fluids. Although this sodium-pumping, water-impermeable, ascending limb does not concentrate or reduce the volume of the forming urine, it sets the stage for these important processes.

The very last portion of the nephron changes in physiological character, but the cells closely resemble those of the ascending loop of Henle. This region, the **distal convoluted tubule** (DCT), is permeable to water. The osmolality surrounding the DCT is that of the body fluids, and water in the entering hyposmotic fluid flows outward and equilibrates osmotically. This process reduces the fluid volume to 5 to 20 percent of the original ultrafiltrate.

The final touch in the formation of a scant, highly concentrated mammalian urine occurs in the **collecting tubules**. Like the descending limb of the loop of Henle, the collecting ducts course through tissues of increasing osmolality, which withdraw water from the urine. The significant phenomenon associated with the collecting duct, and to a lesser extent with the DCT, is its conditional permeability to water. During excess fluid intake, the collecting duct demonstrates low water permeability; only half of the water entering it may be reabsorbed and the remainder excreted. In this way a copious, dilute urine can be produced. When a mammal is dehydrated the collecting ducts and the DCT become very permeable to water and the final urine volume may be less than 1 percent of the original ultrafiltrate volume. In certain desert rodents so little water is contained in the urine that it crystallizes almost immediately upon micturition.

A polypeptide called **antidiuretic hormone**, ADH (also known as **vasopressin**), is produced by specialized neurons in the hypothalamus, stored in the posterior pituitary, and released into the circulation whenever blood osmolality is elevated or blood volume drops. When present in the kidney, ADH increases the permeability of the collecting duct to water and facilitates water reabsorption to produce a scant concentrated urine. The absence of ADH has the opposite effects. Alcohol inhibits the release of human ADH, induces a copious urine flow, and this frequently results in dehydrated misery the following morning.

The key to concentrated urine production clearly depends on the passage of the loops of Henle and collecting ducts through tissues with increasing osmolality. These longitudinal osmotic gradients are formed and maintained within the mammalian kidney as a result of its structure (Figure 4–4), which sets it apart from the kidneys of other vertebrates. Particularly important are the structural arrangements within the kidney medulla of the descending and ascending segments of the loop of Henle and its blood supply, the **vasa recta**. These elements create a series of parallel tubes with flow passing in opposite directions in adjacent vessels (countercurrent flow). As a result, sodium secreted from the ascending limb of the loop of Henle diffuses into the medullary tissues to increase their osmolality, and this excess salt is distributed by the countercurrent flow to create a steep osmotic gradient within the medulla (see Figure 4–3b). The final concentration of a mammal's urine is determined by the amount of sodium accumulated in the fluids of the medulla. Physiological alterations in the concentration in the medulla result primarily from the effect of ADH on the rate of blood flushing the medulla. When ADH is present, blood flow into the medulla is retarded and salt accumulates to create a steep osmotic gradient. Another hormone, aldosterone, from the adrenal gland increases the rate of sodium secretion into the medulla to promote an increase in medullary salt concentration.

In addition to these physiological means of concentrating urine, a variety of mammals have morphological alterations of the medulla. Most mammals have two types of nephrons: those with a cortical glomerulus and abbreviated loops of Henle that do not penetrate far into the medulla and those with juxtamedullary glomeruli, deep within the cortex, with loops that penetrate as far as the papilla of the renal pyramid (Figure 4–4c). Obviously, the longer, deeper loops of Henle experience large osmotic gradients along their lengths. The flow of blood to these two populations of nephrons seems to be independently controlled. Juxtamedullary glomeruli are more active in regulating water excretion; cortical glomeruli function in ion regulation. Finally, some highly adapted desert rodents have exceptionally long renal pyramids. Thus, the loops of Henle and the

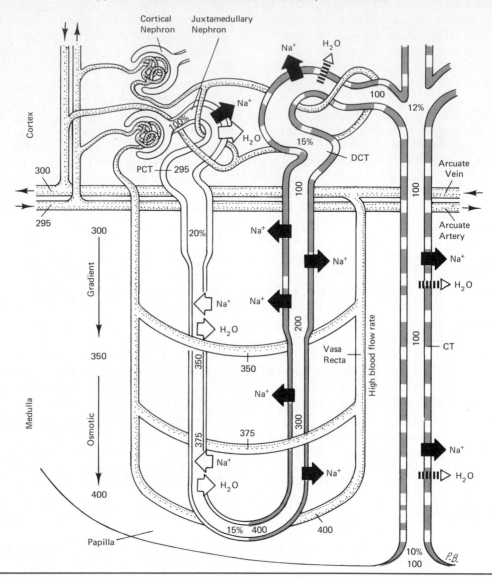

(a) BODY HYDRATED — ADH ABSENT — COPIOUS, DILUTE URINE

Figure 4–3. Diagram showing how the mammalian kidney produces dilute urine when the body is hydrated and concentrated urine when the body is dehydrated. Black arrows indicate active transport and white arrows passive flow. The numbers represent the approximate milliosmolality of the fluids in the indicated regions. Percentages are the volumes of the forming urine relative to the volume of the initial ultrafiltrate. (a) When blood osmolality drops below normal concentration (about 300 milliOsmoles per liter), excess body water is excreted. (b) When osmolality rises above normal, water is conserved. (Based on F. H. Netter, 1973, *The CIBA Collection of Medical Illustrations,* volume 6, CIBA Publications, Summit, N. J.)

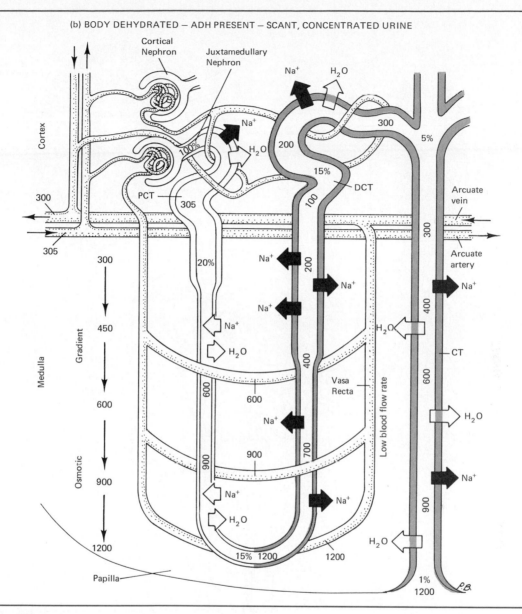

(b) BODY DEHYDRATED — ADH PRESENT — SCANT, CONCENTRATED URINE

Figure 4–3 (Continued)

vasa recta are extended and can produce large differences in osmolality from the cortical to the papillary ends. The maximum concentrations of urine measured from a given species of mammal correlate well with the length of its renal pyramids.

Fossils of early mammals suggest that they were primarily insectivores and carnivores with a high energy demand. This diet would be rich in protein which, when metabolized, would produce large amounts of urea. Considerable water would

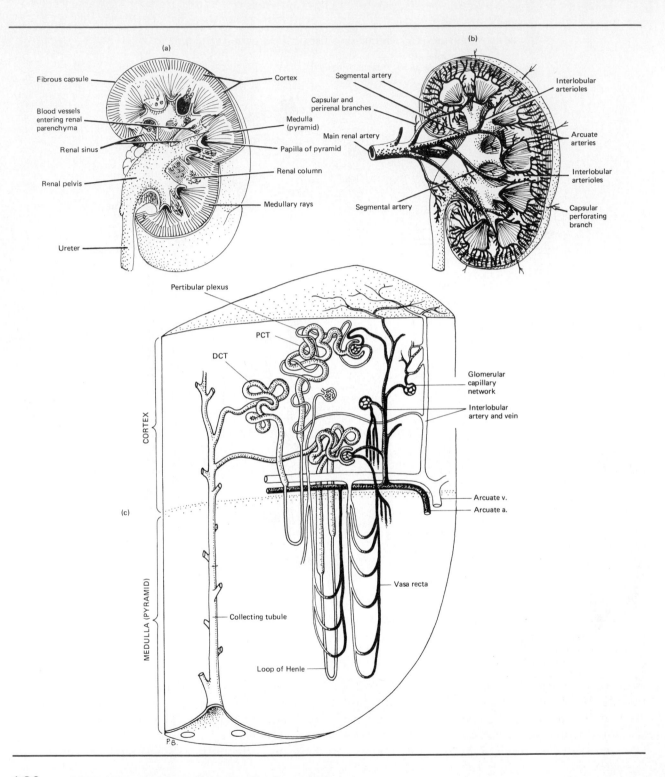

Table 4–5. Distribution of nitrogenous end products among diapsids and turtles.

Group	Percent of Total Urinary Nitrogen		
	Ammonia	Urea	Salts of Uric Acid
Squamates			
Tuatara	3–4	10–28	65–80
Lizards and snakes	Small	0–8	90–98
Archosaurs			
Crocodilians	25	0–5	70
Birds	6–17	5–10	60–82
Turtles			
Aquatic	4–44	45–95	1–24
Desert	3–8	15–50	20–50

be required to void this nitrogenous waste unless a means of concentrating urea was available. The unique concentrating power of the mammalian kidney may have been an early response to the accumulation of metabolic wastes from high levels of activity and the water demands of endothermy.

Diapsids and Turtles All of the living representatives of the diapsid lineage are uricotelic, and uric acid and its salts account for 80 to 90 percent of urinary nitrogen in most species. Turtles, also, excrete a variable proportion of their nitrogenous wastes as salts of uric acid (Table 4–5).

The kidneys of diapsids and turtles lack the long loops of Henle that allow mammals to reduce the volume of urine and raise its osmotic concentration to several times the osmotic concentration of the blood plasma. Urine from the kidneys of diapsids and turtles consists of a moderately dilute solution of uric acid and ions. It is isosmotic with the blood plasma, or even somewhat hyposmotic to the blood. However, uric acid differs from urea in being only slightly soluble in water. It will precipitate from a dilute solution, and that is what happens when urine from the ureters enters the cloaca or bladder. (Many diapsids lack a urinary bladder entirely; others have an ephemeral bladder that is lost shortly after they hatch; and some diapsids and probably all turtles have a functional bladder throughout life.) The uric acid combines with ions in the urine and precipitates as a whitish mass that includes sodium, potassium, and ammo-

Figure 4–4. Gross morphology of the mammalian kidney exemplified by that of a human. (a) Structural divisions of the kidney and proximal end of the ureter; (b) renal artery and its subdivisions in relation to the structural components of the kidney. The renal vein (not shown) and its branches parallel those of the artery. (c) Enlarged diagram of a section extending from the outer cortical surface of the apex of a renal pyramid, the renal papilla. The general relationship of the nephrons and blood vessels to the gross structure of the kidney can be visualized by comparing these diagrams. (Based on F. H. Netter, 1973, *The CIBA Collection of Medical Illustrations*, volume 6, CIBA Publications, Summit, N.J., and H. W. Smith, 1956, *Principles of Renal Physiology*, Oxford University Press, New York.)

nium salts of uric acid and also contains ions held by complex physical forces. When the uric acid and ions precipitate from solution, the urine becomes less concentrated. In other words, water is released, and this water is reabsorbed into the blood. In this respect, excretion of nitrogenous wastes as uric acid is even more economical of water than is excretion of urea because the water used to produce urine is reabsorbed and reused.

Water is not the only substance that is reabsorbed from the cloaca, however. Many diapsids and turtles also reabsorb sodium ions and return them to the bloodstream. At first glance, that seems a remarkably inefficient thing to do. After all, energy was used to create the blood pressure that forced the ions through the walls of the glomerulus into the urine in the first place, and now more energy is being used in the cloaca to drive the active transport that returns the ions to the blood. The animal has used two energy-consuming processes and it is back where it started, with an excess of sodium ions in the blood—why do that?

The answer to the paradox lies in a third water-conserving mechanism that is present in many diapsids and turtles, salt-secreting glands that provide an extrarenal pathway that disposes of salt with less water than the urine. In at least four groups of diapsids (lizards, snakes, crocodilians, and birds) some species possess glands specialized for the elective transport of ions out of the body (Peaker and Linzell 1975, Minnich 1982). Salt glands are widespread in lizards, where they have been found in eight families. In all these families it is the lateral nasal glands that excrete salt. The secretions of the glands empty into the nasal passages, and a lizard expels them by sneezing or by shaking its head. In birds, also, it is the lateral nasal gland that has become specialized for salt excretion. The glands are situated in or around the orbit, usually above the eye. Salt glands have been found in 14 orders of modern birds. Marine birds (pelicans, albatrosses, penguins) have well-developed salt glands, as do many freshwater birds (ducks, loons, grebes), shorebirds (plovers, sand-

pipers), storks, flamingos, carnivorous birds (hawks, eagles, vultures), upland game birds, the ostrich, and the roadrunner. Depressions in the supraorbital region of the skull of the extinct aquatic birds *Hesperornis* and *Ichthyornis* suggest that salt glands were present in these forms as well.

In sea snakes (Hydrophiidae) and elephant-trunk snakes (Acrochordidae), the posterior sublingual gland secretes a salty fluid into the tongue sheath, from which it is expelled when the tongue is extended. The Homalopsinae is a subfamily of the Colubridae that contains rear-fanged aquatic snakes from the Indoaustralian region. In some species of homalopsines the premaxillary gland secretes salt. Salt-secreting glands on the dorsal surface of the tongue have been identified in several species of crocodiles, in a caiman, and in the American alligator.

The diversity of glands involved in salt excretion among diapsids indicates that this specialization has evolved independently in various groups. At least four different glands are used for salt secretion by diapsids, indicating that a salt gland is not an ancestral character for the group, and the differences between crocodilians and birds and between snakes and lizards suggest that salt glands are not ancestral either for archosaurs or for squamates.

Finally, in sea turtles and in the diamondback terrapin, an emydid turtle that inhabits estuaries, the lachrymal gland secretes a salty fluid around the orbits of the eyes. Photographs of nesting sea turtles frequently show clear paths streaked by tears through the sand that adheres to the turtle's head. Those tears are the secretions of the salt glands.

Despite their different origins and locations, the functional properties of salt glands are quite similar. They secrete fluid containing primarily sodium or potassium cations and chloride or bicarbonate anions in high concentrations (Table 4–6). Sodium is the predominant cation in the salt gland secretions of marine vertebrates, and potassium is present in the secretions of terrestrial lizards, espe-

Table 4–6. Salt gland secretions from diapsids and turtles.

Species and Condition	Ion Concentration (mmol/liter)		
	Na^+	K^+	Cl^-
Lizards			
Desert iguana (*Dipsosaurus dorsalis*), estimated field conditions	180	1700	1000
Fringe-toed lizard (*Uma scoparia*), estimated field conditions	639	734	465
Snakes			
Sea snake (*Pelamis platurus*), salt-loaded	620	28	635
Homalopsine snake (*Cerberus rhynchops*), salt-loaded	414	56	—
Crocodilian			
Saltwater crocodile (*Crocodylus porosus*), natural diet	663	21	632
Birds			
Blackfooted albatross (*Diomeda nigripes*), salt-loaded	800–900	—	—
Herring gull (*Larus argentatus*), salt-loaded	718	24	—
Turtles			
Loggerhead sea turtle (*Caretta caretta*), seawater	732–878	18–31	810–992
Diamondback terrapin (*Malaclemys terrapin*), seawater	322–908	26–40	—

cially herbivorous species such as the desert iguana. Chloride is the major anion, and herbivorous lizards may also excrete bicarbonate ions.

The total osmotic concentration of the salt gland secretion may reach 2000 mOsm/liter—more than six times the osmotic concentration of urine that can be produced by the kidney. This efficiency of excretion is the explanation of the paradox of active uptake of salt from the urine. As ions are actively reabsorbed, water follows passively, so an animal recovers both water and ions from the urine. The ions can then be excreted via the salt gland at much higher concentrations, with a proportional reduction in the amount of water needed to dispose of the salt. Thus, by investing energy in recovering ions from urine, diapsids and turtles with salt glands can conserve water by excreting ions through the more efficient extrarenal route.

Uricotelic Frogs A final note emphasizes how flexible pathways of nitrogen excretion are and how intertwined with water economy they remain. Terrestrial amphibians have long been considered ureotelic, aquatic forms ammonotelic. However, J. P. Loveridge (1970) discovered that, during the dry season, a period when most frogs retire to a burrow and aestivate, the South African frog, *Chiromantis xerampelina*, remains above ground. Even more surprising, its excretions contain uric acid—biochemically, *Chiromantis* is like a lizard. It has subsequently been shown that the South American frog, *Phyllomedusa sauvagei*, responds in the same way to aridity (Shoemaker et al. 1972). There is a lesson in these unusual findings: evolutionary convergence works on all levels of biological organization—anatomical, behavioral, physiological, biochemical—and its direction is determined by interactions with the environment.

Responses to Temperature

Vertebrates occupy habitats from cold polar latitudes to hot deserts. To appreciate this adaptability, we must consider how temperature affects a vertebrate such as a fish that has little capacity to maintain a difference between its body temperature and the temperature of the water around it (a poikilotherm). Organisms have been described as "bags of chemicals catalyzed by enzymes." This view, although narrow, emphasizes that organisms are subject to the laws of physics and chemistry. Because temperature influences the rates at which chemical reactions proceed, temperature vitally affects the life processes of organisms. Most chemical reactions double or triple in rate for every rise of 10°C. We describe this change in rate by saying that the reaction has a Q_{10} of 2 or 3, respectively. A reaction that does not change rate with temperature has a Q_{10} equal to 1 (Figure 4–5).

The **standard metabolic rate** (SMR) of an organism is the minimum rate of oxygen consumption needed to sustain life. That is, the SMR includes the costs of ventilating the lungs or gills, of pumping blood through the circulatory system, of transporting ions across membranes, and all the other activities that are necessary to maintain the integrity of an organism. The SMR does not include the costs of activities like locomotion. The SMR is temperature-sensitive, and that means that the cost of living is affected by changes in body temperature. If the SMR of a fish is 2 milliliters of oxygen per minute at 10°C and the Q_{10} response 2, then the fish will consume 4 milliliters of oxygen per minute at 20°C and 8 ml/min at 30°C. This change represents a substantial increase in the cost

of maintenance. Most fishes cannot withstand sudden temperature changes exceeding 10°C, but adjust to temperature changes of 20°C or more if the changes occur over several hours. The resting metabolism of a goldfish at different temperatures is shown in Figure 4–5. In this figure curves W and C are the acute (rapid) response curves for two groups of fish: one maintained for 1 week at 30°C (warm acclimated = W), and a second maintained for 1 week at l0°C (cold acclimated = C). Examination of these curves reveals several features of importance.

1. The rate of oxygen consumption is strongly affected by temperature. At low temperatures the rate of oxygen consumption is reduced. This response is typical of most poikilotherms and of a wide variety of enzymatic reactions. Because oxygen consumption of a resting fish represents the energy costs of all its maintenance processes, the data indicate that virtually all life processes must slow as T_a falls. As T_a rises, the metabolic rate increases. Metabolism may show a temperature optimum where its rate peaks, a response that is analogous to the temperature optima of most enzymes (Figure 4–5).
2. Oxygen consumption rates, when measured at the same temperature, are greater for the cold-acclimated fish than for the warm-acclimated fish. When warm-acclimated fish are cooled from a temperature of 30° to 10°C, the oxygen consumption rate declines nearly fivefold. But in fish acclimated to cold their metabolism is depressed only twofold. These adjustments to temperature are referred to as temperature compensation. If a series of goldfish are acclimated to different temperatures and then each fish's oxygen consumption is measured at its temperature of acclimation, an acclimation response curve can be established. This chronic curve has a flatter overall shape than the acute curves.

Fishes that encounter large seasonal changes in ambient temperature usually compensate, and

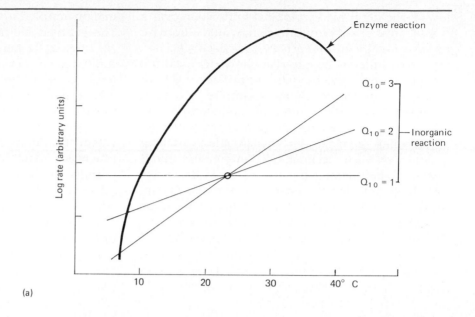

(a)

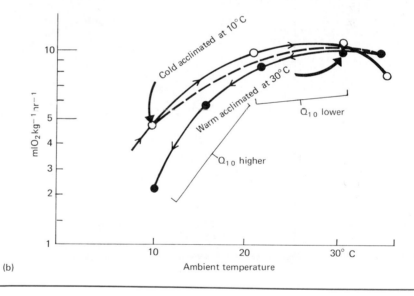

(b)

Figure 4–5. Rate–temperature responses in nonliving and living systems: (a) typical Q_{10} responses of inorganic reactions and an enzyme-catalyzed organic reaction; (b) acute (solid lines) and chronic (dashed lines) responses of cold- and warm-acclimated goldfish. (Modified from M. Kanungo and C. L. Prosser, 1959, *Journal of Cellular and Comparative Physiology* 54:259–264.)

thus partly escape what has been called the tyranny of temperature. Compensation occurs in many rate processes: heart rate, respiratory rate, nerve conduction velocities and discharge frequen-

cies, and so on. When a fish is first exposed to cold, the heart rate slows, but with chronic exposure the heart slowly adjusts its rate upward. In ecological terms, temperature compensation allows poikilothermic vertebrates to function over a wider range of temperatures than would otherwise be possible. [See Prosser (1986) for details.]

Active metabolic rate also shows temperature compensation, as demonstrated in fishes by F. E. J. Fry and his students. For example, J. R. Brett acclimated salmon to different temperatures and then measured their resting and their active metabolic rates (at maximum sustainable swimming speeds) at each acclimation temperature (Brett 1971). By subtracting the resting from the active rate he established the energy required for swimming. This difference, defined by Fry as the **scope for activity**, varies at each acclimation temperature (Figure 4–6). Salmon have highest scope when acclimated at 15°C. Coincident with this peak in scope is a peak in swimming speed. Thus, salmon display thermal optima; that is, a specific acclimation temperature at which they swim fastest. These physiological examples of thermal optima are often reflected in behavior. Given the opportunity to select from different temperatures, fishes tend to seek out those temperature ranges at which they perform best—the range of their thermal optima (Reynolds and Casterlin 1980).

The effect of temperature compensation is to extend the temperature range over which poikilotherms remain active, an ecological benefit that permits them to exploit their environment more effectively. Because temperature compensations take time, they are effective in countering slow seasonal changes in temperature and are most prevalent in aquatic vertebrates, where water buffers sudden changes in temperature. In contrast, terrestrial vertebrates encounter sudden fluctuations as well as seasonal changes in temperature

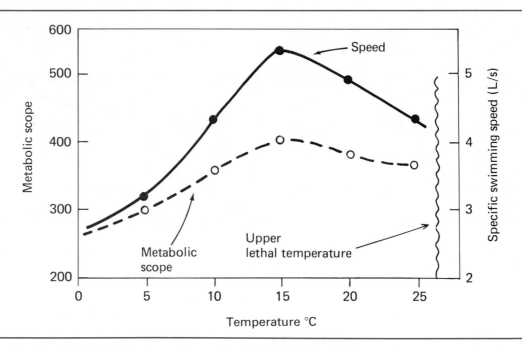

Figure 4–6. Relationship of swimming speed and scope for activity to various acclimation temperatures in salmon. Active metabolic rate minus standard metabolic rate is defined as the metabolic scope for activity. Number of body lengths a fish swims per second is defined as size-specific swimming speed. (Data from J. R. Brett, 1971, *American Zoologist* 11:99–113.)

and utilize divergent mechanisms to cope with the fluctuating temperatures.

Control of Body Temperature: Ectothermy and Endothermy

Because the rates of many biological processes are affected by temperature, it would probably be advantageous for any animal to be able to control its body temperature. However, the high heat capacity and heat conductivity of water make it difficult for most fishes or aquatic amphibians to maintain a temperature differential between their bodies and their surroundings. Air has both a lower heat capacity and a lower conductivity than water, and the body temperatures of most terrestrial vertebrates are at least partly independent of the temperature of their environment. Some aquatic vertebrates also have body temperatures substantially above the temperature of the water around them. Maintaining these temperature differentials requires thermoregulatory mechanisms, and these are well developed among vertebrates.

The classification of vertebrates as poikilotherms (*poikil* = variable, *therm* = heat) and homeotherms (*homeo* = the same) was widely used through the middle of the twentieth century, but this terminology has become less useful as our knowledge of the temperature-regulating capacities of a wide variety of animals has become more sophisticated. Poikilothermy and homeothermy describe the variability of body temperature, and they cannot readily be applied to groups of animals. For example, mammals have been called homeotherms and fishes poikilotherms, but some mammals become torpid at night or in the winter and allow their body temperatures to drop 20°C or more from their normal levels, whereas many fishes live in water that changes temperature less than 2°C in an entire year. That example presents the contradictory situation of a homeotherm that experiences ten times as much variation in body temperature as a poikilotherm.

Because of complications of that sort, it is very hard to use the words "homeotherm" and "poik-

ilotherm" in a rigorous way. Some mammalogists and ornithologists still retain those terms, but biologists concerned with the temperature regulation of other animals prefer the terms **ectotherm** and **endotherm**. They are *not* synonymous with poikilotherm and homeotherm because, instead of referring to the variability of body temperature, they refer to the source of energy used in thermoregulation. Ectotherms (*ecto* = outside) gain their heat from external sources—by basking in the sun, for example, or by resting on a warm rock. Endotherms (*endo* = inside) depend upon metabolic production of heat to raise their body temperatures. The source of the heat used to maintain body temperatures is the major difference between ectotherms and endotherms. Terrestrial ectotherms like squamates and turtles and endotherms like birds and mammals all have activity temperatures in the range 30 to 40°C (Table 4–7).

Endothermy and ectothermy are not mutually exclusive mechanisms of temperature regulation, and many animals use them in combination. In general, birds and mammals are endothermal, but some species make extensive use of external sources of heat. For example, roadrunners are predatory birds that live in the deserts of the southwestern United States and adjacent Mexico. On cold nights roadrunners apparently become hypothermic, allowing their body temperatures to fall from the normal level of 38 to 39°C down to 33 to 35°C. In the morning they bask in the sun, raising the feathers on the back to expose an area of black skin in the interscapular region. Calculations indicate that a roadrunner can save 132 joules per hour by using solar energy instead of metabolism to raise its body temperature. Similarly, most fishes are ectotherms, but several fast-swimming fishes such as tuna and sharks use countercurrent blood flow to retain the heat produced by their muscles as they swim. A bluefin tuna swimming in water at 19.3°C had a maximum muscle temperature of 31.4°C, and a mako shark had a muscle temperature of 27.2°C in water that was 21.2°C. Thus, generalizations about the body temperatures and thermoregulatory capacities of vertebrates must be made cautiously, and the actual

Table 4–7. Representative body temperatures of vertebrates. Body temperatures are those that the animals maintain when they are able to thermoregulate normally.

Group	Body Temperature
	Primarily Ectothermal Groups
Fishes	
Most fishes	Little different from water temperature
Warm-bodied fishes (tunas, some sharks)	About 30°C in water of 20°C
Amphibians	
Aquatic	Little different from water temperature
Terrestrial	Usually slightly below air temperature because of evaporative cooling; some amphibians raise their body temperatures 5–10°C above air temperature by basking
Amniotic ectotherms	
Turtles and crocodilians	From close to water temperature to about 35°C while thermoregulating
Squamates	From 20–25°C for forest-dwelling tropical species to 35–42°C for thermoregulating desert lizards
	Primarily Endothermal Groups
Birds	40–41°C
Mammals	
Monotremes*	28–30°C
Marsupials	33–26°C
Placentals	36–38°C

*Sloths, which are placentals, have body temperatures in this range.

mechanisms used to regulate body temperature must be studied carefully.

Ectothermal Thermoregulation

From the time of Aristotle onward, lizards, snakes, and amphibians have paradoxically been called cold-blooded while they were thought to be able to tolerate extremely high temperatures. Salamanders frequently seek shelter in logs, and when a log is put on a fire, the salamanders it contains may come rushing out. Observations of this phenomenon gave rise to the belief that salamanders live in fire. In the first part of the twentieth century biologists were using similar lines of reasoning. In the desert lizards often sit on rocks. If you approach a lizard it runs away, but touching the rock shows that it is painfully hot. Clearly, the reason-

ing went, the lizard must have been equally hot. Biologists marveled at the heat tolerance of lizards, and statements to this effect are found in the authoritative textbooks of the period.

A study of thermoregulation of lizards by Raymond Cowles and Charles Bogert (1944) demonstrated the falsity of earlier observations and conclusions. They showed that reptiles regulate their body temperatures with considerable precision, and the level at which the temperature is regulated is characteristic of a species. The implications of this discovery in terms of the biology of amphibians and reptiles are still being explored.

Energy Exchange Between an Organism and Its Environment A brief discussion of the pathways by which thermal energy is exchanged between a living organism and its environment is necessary to understand the thermoregulatory mechanisms employed by terrestrial animals. There are several pathways by which thermal energy can be gained or lost by an organism, and by adjusting the relative flow through various pathways an animal can warm, cool, or maintain a stable body temperature (Tracy 1982).

Figure 4–7 illustrates pathways of thermal energy exchange. In the case of an ectotherm, the sun is the source of virtually all the energy used to regulate body temperature; however, solar energy can reach the animal in several ways. Direct solar **radiation** impinges on an animal when it is standing in a sunny spot. In addition, solar energy is reflected from clouds and dust particles in the atmosphere and from other objects in the environment and reaches the animal by these circuitous routes. The wavelength distribution of the energy in all these routes is the same—the portion of the solar spectrum that penetrates the Earth's atmosphere. About half this energy is contained in the visible wavelengths of the solar spectrum (400 to

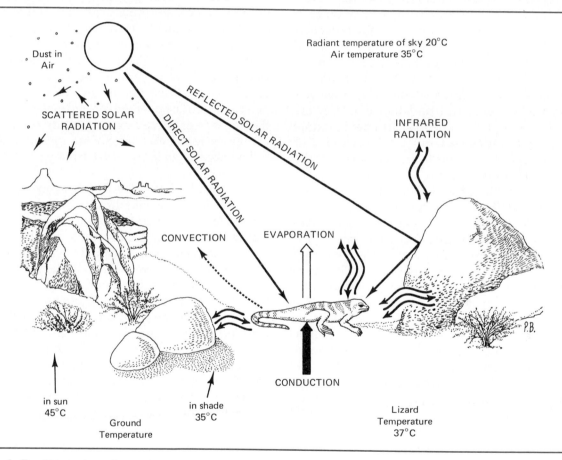

Figure 4–7. Energy is exchanged between a terrestrial organism and its environment by many pathways. These are illustrated in simplified form by a lizard resting on the floor of a desert arroyo. Small adjustments of posture or position can change the magnitude of the various routes of energy exchange and give a lizard considerable control over its body temperature.

700 nanometers) and the other half is in the ultraviolet and infrared regions of the spectrum.

Energy exchange in the **infrared** is an important part of the radiative heat balance. All objects, animate or inanimate, radiate energy at wavelengths determined by their absolute temperatures. Objects in the temperature range of animals and the Earth's surface (roughly $-20°C$ to $+50°C$) radiate in the infrared portion of the spectrum. Animals continuously radiate heat to the environment and receive infrared radiation from the environment. Thus, infrared radiation can lead to either heat gain or loss, depending on the relative temperature of the animal's body surface and the environmental surfaces and on the radiation characteristics of the surfaces themselves. In the example in Figure 4–7 the lizard is cooler than the sunlit rock in front of it and receives more energy from the rock than it loses to the rock. However, the lizard is warmer than the shaded side of the rock behind it and has a net loss of energy in that exchange. The radiative temperature of the clear sky is about 20°C, so the lizard loses energy by radiation to the sky.

Heat is exchanged between objects in the environment and the air via **convection**. If an animal's surface temperature is higher than air temperature, convection leads to heat loss; if the air is warmer than the animal, convection is a route of heat gain. In still air convective heat exchange is accomplished by convective currents formed by local heating, but in moving air forced convection replaces natural convection and the rate of heat exchange is greatly increased. In the example shown the lizard is warmer than the air and loses heat by convection.

Conductive heat exchange resembles convection in that its direction depends on the relative temperatures of the animal and environment. Conductive loss occurs between the body and the substrate where they are in contact. It can be modified by changing the surface area of the animal in contact with the substrate and by changing the rate of heat conduction in the parts of the animal's body that are in contact with the substrate. In this example the lizard gains heat by conduction from the warm ground.

Evaporation of water occurs from the body surface and from the pulmonary system. Each gram of water evaporated represents a loss of about 2450 joules (the exact value changes with temperature). Evaporation almost always occurs from the animal to the environment, and thus represents a loss of heat. The inverse situation, condensation of water vapor on an animal, would produce heat gain, but it rarely occurs under natural conditions.

Metabolic heat production is the final pathway by which an animal can gain heat. Among ectotherms metabolic heat gain is usually trivial in relation to the heat derived directly or indirectly from solar energy. There are a few exceptions to that generalization, and some of them are discussed later. Endotherms, by definition, derive most of their heat energy from metabolism, but their routes of energy exchange with the environment are the same as those of ectotherms and must be balanced to maintain a stable body temperature.

Behavioral Control of Body Temperatures by Ectotherms The behavioral mechanisms involved in ectothermal temperature regulation are quite straightforward and are employed by insects, birds, and mammals (including humans) as well as by ectothermal vertebrates (Avery 1979). Lizards, especially desert species, are particularly good at behavioral thermoregulation. Movement back and forth between sun and shade is the most obvious thermoregulatory mechanism they use. Early in the morning or on a cool day, lizards bask in the sun, whereas in the middle of a hot day they have retreated to shade and make only brief excursions into the sun. Sheltered or exposed microhabitats may be sought out—in the morning when a lizard is attempting to raise its body temperature, it is likely to be in a spot protected from the wind. Later in the day when it is getting too hot it may climb into a bush or onto a rock outcrop where it is exposed to the breeze, and its convective heat loss is increased.

The amount of solar radiation absorbed by an animal can be altered by changing the orientation of its body with respect to the sun, the body contour, and the skin color. All of these mechanisms are used by lizards. An animal oriented perpendicular to the sun's rays intercepts the maximum amount of solar radiation, and one oriented parallel to the sun's rays intercepts minimum radiation. Lizards adjust their orientation to control heat gained by direct solar radiation. Many lizards are capable of spreading or folding the ribs to change the shape of the trunk. When the body is oriented perpendicular to the sun's rays and the ribs are spread, the surface area exposed to the sun is maximized and heat gain is increased. Compressing the ribs decreases the surface exposed to the sun and can be combined with orientation parallel to the rays to minimize heat gain. Horned lizards provide a good example of this type of control (Heath 1965). If the surface area that a horned lizard exposes to the sun directly overhead when it sits flat on the ground with its ribs held in a resting position is considered to be 100 percent, the maximum surface area the lizard can expose by orientation and change in body contour is 173 percent and the minimum is 28 percent. That is, the lizard can change its radiant heat gain more than sixfold solely by changing its position and body shape.

Color change can further increase a lizard's control of radiative exchange. Lizards darken by dispersing melanin in melanophore cells in the skin and lighten by drawing the melanin into the base of the melanophores. The lightness of a lizard affects the amount of solar radiation it absorbs in the visible part of the spectrum, and changes in heating rate (in the darkest color phase compared to the lightest) are from 10 to 75 percent.

Lizards can achieve a remarkable independence of air temperature as a result of their thermoregulatory capacities (Avery 1982). Lizards occur above the timberline in many mountain ranges, and during their periods of activity are capable of maintaining body temperatures 30°C or more above air temperature. While air temperatures are near freezing, these lizards scamper about with body temperatures as high as those species that inhabit lowland deserts.

The repertoire of thermoregulatory behavior seen in lizards is greater than that of many other ectothermal vertebrates. Turtles, for example, cannot change their body contour or color, and their behavioral thermoregulation is limited to movements between sun and shade and in and out of water. Crocodilians are very like turtles, although young individuals may be able to make minor changes in body contour and color. Most snakes cannot change color, but some rattlesnakes lighten and darken as they warm and cool.

During the parts of a day when they are active, desert lizards maintain their body temperatures in a zone called the **activity temperature range**. This is the region of temperature within which a lizard carries out its full repertoire of activities—feeding, courtship, territorial defense, and so on. For many species of desert lizards the activity temperature range is as narrow as 4°C, but for other ectotherms it may be as broad as 10°C. Different species of lizards have different activity temperature ranges. The thermoregulatory activities of a lizard are directed toward keeping it within its activity temperature range, but the precise temperature it maintains within this range depends on a variety of internal and external conditions. For example, many ectotherms maintain higher body temperatures when they are digesting food than when they are fasting. Female lizards may maintain different body temperatures when they are carrying young than they do at other times, and ectotherms with experimentally induced bacterial infections show a behavioral fever achieved by maintaining a higher-than-normal body temperature by behavioral means (Kluger 1979).

Not all lizards regulate their body temperatures closely. Some lizards that live in the understory vegetation of tropical forests, for example, do not raise their body temperatures above air temperatures. Differences in the intensity and availability of solar radiation in different seasons, different habitats, or even at different times of day can alter the balance of costs and benefits of thermoregulatory behavior (Huey 1982). These ecological aspects of thermoregulation are discussed in Chapter 15.

Physiological Control of the Rate of Change of Body Temperature by Ectotherms   A new dimension was added to studies of ectothermal thermoregulation in the 1960s by the discovery that ectotherms can use physiological mechanisms to adjust their rate of temperature change (Bartholomew 1982). The original observations, made by George Bartholomew and his associates, showed that several different kinds of large lizards were able to heat faster than they cooled when exposed to the same differential between body and ambient temperatures. Subsequent studies by other investigators extended these observations to turtles and snakes. From the animal's viewpoint, heating rapidly and cooling slowly prolongs the time it can spend in the normal activity range.

Fred White and his colleagues have demonstrated that the basis of this control of heating and cooling rates lies in changes in peripheral circulation. Heating the skin of a lizard causes a localized vasodilation of dermal blood vessels in the warm area. Dilation of the blood vessels, in turn, increases the blood flow through them, and the blood is warmed in the skin and carries the heat into the core of the body. Thus, in the morning, when a cold lizard orients its body perpendicular to the sun's rays and the sun warms its back, local vasodilation in that region speeds transfer of the heat to the rest of the body.

The same mechanism can be used to avoid overheating. The Galapagos marine iguana is a good example (White 1973). Marine iguanas live on the bare lava flows on the coasts of the islands. In midday, beneath the equatorial sun, the black lava becomes extremely hot—uncomfortably, if not lethally hot for a lizard. Retreat to shade of the scanty vegetation or into rock cracks would be one way the iguanas could avoid overheating, but the males are territorial and those behaviors would mean abandoning their territories and probably having to fight for them again later in the day. Instead, the marine iguana stays where it is and uses physiological control of circulation and the cool breeze blowing off the ocean to form a heat shunt that absorbs solar energy on the dorsal surface and carries it through the body and dumps it out the ventral surface.

The process is as follows: In the morning the lizard is chilled from the preceding night and basks to bring its body temperature to the normal activity range. When its temperature reaches this level the lizard uses postural adjustments to slow the increase in body temperature, finally facing directly into the sun to minimize its heat load. In this posture the forepart of the body is held off the ground (Figure 4–8). The ventral surface is exposed to the cool wind blowing off the ocean, and a patch of lava under the animal is shaded by its body. This lava is soon cooled by the wind. Local vasodilation is produced by warming the blood vessels: It does not matter whether the heat comes from the outside (from the sun) or from inside (from warm blood). Warm blood circulating from the core of the body to the ventral skin warms it and produces vasodilation, increasing the flow to the ventral surface. The lizard's ventral skin is cooler than the rest of its body—it is shaded and cooled by the wind, and in addition it loses heat by radiation to the cool lava in the shade created by the lizard's body. In this way the same mechanism that earlier in the day allowed the lizard to warm rapidly is converted to a regulated heat shunt that rapidly transports solar energy from the dorsal to the ventral surface and keeps the lizard from overheating. In combination with postural adjustments and other behavioral mechanisms, such as the choice of a site where the breeze is strong, these physiological adjustments allow the lizard to remain on station in its territory all day.

The behavioral and physiological thermoregulatory mechanisms of ectotherms are intimately intertwined. Although we have tried to simplify our presentation by discussing them separately, it is essential to realize that neither behavioral, nor physiological, nor morphological thermoregulatory mechanisms function by them-

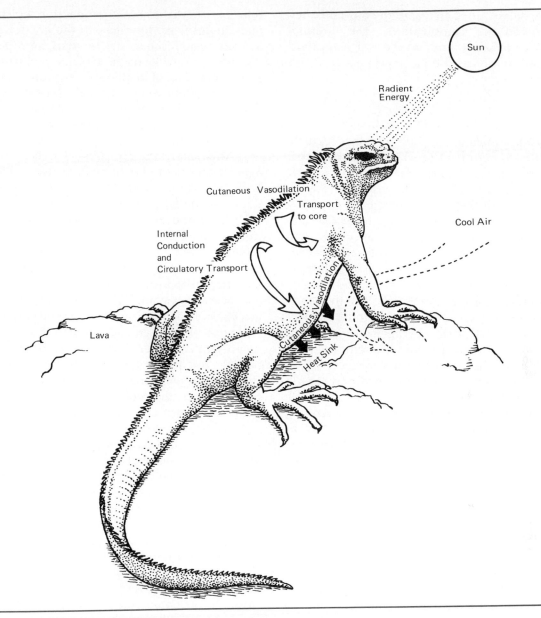

Figure 4–8. The Galapagos marine iguana uses a combination of behavioral and physiological thermoregulatory mechanisms to shunt heat absorbed by its dorsal surface out its ventral surface. (Modified from F. N. White, 1973, *Comparative Biochemistry and Physiology* 45A:503–513.)

selves. They are used in combination, and they have evolved in combination. The thermoregulation of a lizard (and, as we will see, a bird or mammal) involves all these mechanisms simultaneously.

Endothermal Thermoregulation

Birds and mammals are endotherms that regulate their high body temperatures by mechanisms that precisely balance heat loss and heat gain. Heat production by metabolism is one aspect of this regulation. An endotherm can change the intensity of its heat production by varying metabolic rate over a wide range. In this way, an endotherm maintains a constant high body temperature by adjusting heat production to equal heat loss from its body under different environmental conditions.

Endotherms produce metabolic heat in several ways. Besides the obligatory heat production derived from the basal or resting metabolic rate, there is the heat increment of feeding, often called the **specific dynamic action** or **effect** of the food ration. This added heat production after ingestion of food apparently results from the energy requirements for assimilation and varies in amount depending on the type of foodstuff being processed. It is highest for a meat diet and lowest for a carbohydrate diet.

The voluntary actions of skeletal muscle produce large amounts of heat, especially during locomotion, which can result in a heat production exceeding the basal metabolic rate by 10- to 15-fold. This muscular heat can be advantageous for balancing heat loss in a cold environment, or it can be a problem requiring special mechanisms of dissipation in hot environments that approach or exceed the body temperature of the animal. **Shivering**, the generation of heat by muscle fiber contractions in an asynchronous pattern that does not result in gross movement of the whole muscles, is an important mechanism of heat production. Birds and mammals also possess mechanisms for nonshivering thermogenesis.

Because endotherms usually live under conditions in which ambient temperatures are lower than the regulated body temperatures of the animals themselves, heat loss to the environment is a more usual circumstance than heat gain, although heat gain can be a major problem in deserts. Balancing of heat loss is therefore one of the most important regulatory functions of an endotherm, and birds and mammals employ their plumage or hair in a very effective way as insulation against heat loss.

Any material that traps dead air is an insulator against conductive heat transfer. Hair and feathers both provide insulation by trapping air, and the depth of the layer of trapped air can be adjusted by raising and lowering the hair or feathers. We humans have goose bumps on our arms and legs when we are cold because our few remaining hairs rise to a vertical position in an ancestral mammalian attempt to increase our insulation.

These physiological responses to a heat load are controlled from neurons located in the hypothalamus of the brain. The neural center functions as a thermostat, sensing when body temperature is either too high or low, to control the appropriate response. In some mammals, as in hibernators, the hypothalamic thermostat can be reset for a lower control temperature, and if ambient temperature is low, body temperature declines to a new response level. In ectotherms the hypothalamic thermostat functions to control behaviors that place the animal in favorable circumstances (e.g., moving to a preferred ambient temperature away from excess heat, or orienting to maximize heat loss or gain).

Mechanisms of Endothermal Thermoregulation
Body temperature and metabolic rate must be considered simultaneously to understand how endotherms maintain their body temperatures at a stable level in the face of environmental temperatures that may range from -70 to $+40°C$. Most birds and mammals conform to the generalized diagram in Figure 4–9.

Each species has a definable range of ambient temperatures (T_1 to T_4) over which the body temperature can be kept at a stable level by using physiological and postural adjustments of heat

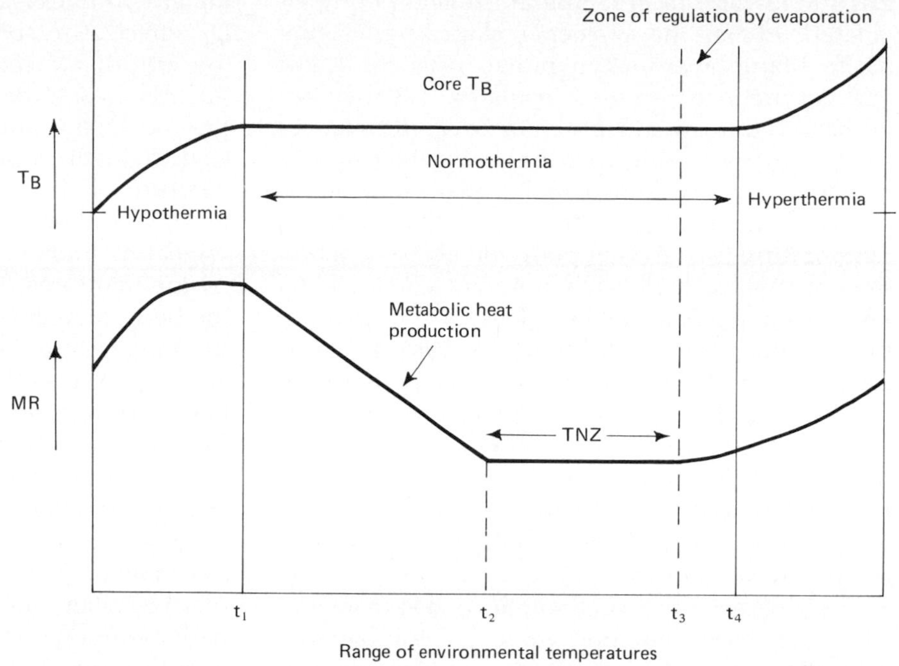

Figure 4–9. Generalized pattern of changes in body temperature and metabolic heat production of an endothermic homeotherm in relation to environmental temperature. Core T_b is the normal body temperature and varies somewhat for different mammals and birds. T_1, incipient lower lethal temperature; T_2, lower critical temperature; T_3, upper critical temperature; T_4, incipient upper lethal temperature; TNZ, thermoneutral zone.

loss and heat production. This ambient temperature range is called the **zone of tolerance**. Above this range the animal's ability to dissipate heat is inadequate, and both the body temperature and metabolic rate increase as ambient temperature increases until the animal dies. At ambient temperatures below the zone of tolerance the animal's ability to generate heat to balance heat loss is exceeded, body temperature falls, metabolic rate declines, and cold death results. Large animals usually have lower values for T_1 and T_2 than small animals because heat is lost from the body surface, and large animals have smaller surface-to-mass ratios than those of small animals. Similarly, well-insulated species have lower values for T_1 and T_2 than those of poorly insulated ones, but thinly insulated species usually have higher values for T_3 than those of heavily insulated ones.

The **thermoneutral zone** (T_2 to T_3) is the range of ambient temperatures within which the metabolic rate of an endotherm is at its basal level and thermoregulation is accomplished by changing the rate of heat loss. The thermoneutral zone is also called the zone of physical thermoregulation because an animal uses processes such as fluffing or sleeking its hair or feathers, postural changes such as huddling or stretching out, and changes in blood flow (vasoconstriction or vasodilation) to exposed parts of the body (feet, legs, face) to adjust its loss of heat.

The larger an animal is and the thicker its insulation, the lower will be the temperature it can withstand before physical processes become inadequate to balance its heat loss. The **lower critical temperature** (T_2) is the point at which an animal must increase metabolic heat production to maintain a stable body temperature. In the **zone of chemical thermogenesis**, the metabolic rate increases as the ambient temperature falls. The quality of the insulation determines how much additional metabolic heat production is required to offset a change in ambient temperature. That is, well-insulated animals have relatively shallow slopes for the graph of increasing metabolism below the lower critical temperature, and poorly insulated animals have steeper slopes (see Figure 23–3). Many birds and mammals from the Arctic and Antarctic are so well insulated that they can withstand the lowest temperatures on earth (about $-70°C$) by increasing their basal metabolism only threefold. Less-well-insulated animals may be exposed to temperatures below their **lower lethal temperature** (T_1). At that point metabolic heat production has reached its maximum rate and it is still insufficient to balance heat loss to the environment. Under those conditions the body temperature falls, and the Q_{10} effect of temperature on chemical reactions causes the metabolic rate to fall as well. A positive-feedback condition is initiated in which falling body temperature reduces heat production, causing a further reduction in body temperature. Death from hypothermia (low body temperature) follows.

Endotherms are remarkably good at maintaining stable body temperatures in cool environments, but they have difficulty at high ambient temperatures. The **upper critical temperature** (T_3) represents the point at which heat loss has been maximized by using all of the physical processes that are available to an animal—exposing the poorly insulated areas of the body and maximizing cutaneous blood flow. If these mechanisms are insufficient to balance heat gain, the only option vertebrates have is to use evaporation of water by panting, sweating, or gular flutter. The temperature range from T_3 to T_4 is the **zone of evaporative cooling**. Many mammals sweat, a process in which water is released from sweat glands on the surface of the body. Evaporation of the sweat cools the body surface. Other animals pant, breathing rapidly and shallowly so that evaporation of water from the respiratory system provides a cooling effect. Many birds use a rapid fluttering movement of the gular region to evaporate water for thermoregulation. Panting and gular flutter require muscular activity, and some of the evaporative cooling they achieve is used to offset the increased metabolic heat production they require.

At the **upper lethal temperature** (T_4) evaporative cooling cannot balance the heat flow from a hot environment, and body temperature rises. The Q_{10} effect of temperature produces an increase in the rate of metabolism, and metabolic heat production raises the body temperature, increasing the metabolic rate still further. This process can lead to death from hyperthermia (high body temperature).

The difficulty that endotherms experience in regulating body temperature in high environmental temperatures may be one of the reasons that the body temperatures of most endotherms are in the range 35 to 40°C. Most habitats seldom have air temperatures that exceed 35°C. Even the tropics have average yearly temperatures below 30°C. Only in open situations where solar radiation impinges directly on a mammal or a bird can the environment reverse the normal heat gradient. Thus the high body temperatures maintained by mammals ensure that in most situations the heat gradient is from animal to environment. Still higher body temperatures, around 50°C, for example, could ensure that mammals were always warmer than their environment. There are upper limits to the body temperatures that are feasible, however. Many proteins denature near 50°C. During heat stress, some birds and mammals may tolerate body temperatures of 45 to 46°C for a few hours, but only some bacteria, algae, and a few invertebrates exist at higher temperatures. This is another case in which the direction of vertebrate evolution has been established by a balance between biotic needs and physicochemical realities.

Advantages of a High Body Temperature

Mammals and birds have high metabolic rates, at least six times the SMR of ectotherms. Although it is energetically expensive, there are benefits to regulating body temperature at a high rather than a low level that are independent of the mechanisms of thermoregulation discussed in the preceding section.

We have already noted that the biochemistry of vertebrates involves thousands of interacting enzyme-catalyzed reactions, most of which are temperature sensitive. A constant internal temperature is required to obtain maximum chemical coordination among these reactions. In addition, the higher the body temperature, the more rapid the response of cells to organismal needs. Although it is capable of acting at very cold temperatures (as in arctic fishes), the CNS functions more rapidly at high temperatures. As one example, neurotransmitters such as acetylcholine and norepinephrine act by diffusing from their site of release across the synaptic junction to the postsynaptic receptor surface. Because diffusion is a physical process, its rate increases as temperature increases. A high body temperature enhances the rate of information processing, a competitive advantage often neglected when considering the success of mammals and birds. Very rapid responses can be vital in catching prey and avoiding predators. Some ectotherms enjoy the same neurological benefits when they are warm, but very rapid responses on cool nights can occur only in endothermal homeotherms. In addition, muscle viscosity declines at high temperatures. This reduction in internal friction may result in more rapid, forceful contraction and faster response times.

Thus, endothermal homeothermy has some obvious advantages over ectothermy. (Ectothermy has its own advantages, which are discussed in Chapter 16.) Ectothermy is the ancestral condition for vertebrates. How did endothermy evolve?

The Evolution of Endothermy

Endothermy has evolved from an ancestral ectothermal condition at least twice in the history of vertebrates—in birds and in mammals. Some evidence suggests that pterosaurs (flying archosaurs of the Mesozoic) might also have been endothermal, and if this is true, it would represent a third independent origin of endothermy. How would that transition occur?

The difference in the sources of heat used by ectotherms and endotherms creates a paradox when one tries to understand how an evolutionary lineage shifts from ectothermy to endothermy. Ec-

totherms rely on heat from outside their bodies, and the major specializations of ectothermal thermoregulation facilitate exchange of heat with the environment. The body surfaces of ectotherms have little insulation, probably because insulation would interfere with the gain and loss of heat. Metabolic rates of ectotherms are low, and ectotherms normally do not obtain sufficient heat from metabolism to warm the body significantly. (The warm-bodied tunas and sharks are exceptions to this generalization because of their large body sizes, high levels of activity, and specializations of the circulatory system.) Thus, the thermoregulatory mechanisms of ectotherms are based on low metabolic rates, little insulation, and rapid exchange of heat with the environment.

Endothermal thermoregulation has exactly the opposite characteristics. The high metabolic rates of endotherms produce large quantities of heat, and that heat is retained in their bodies by the insulation provided by hair or feathers. Endothermal thermoregulation consists largely of adjusting the layer of insulation so that heat loss balances the heat produced by high rates of metabolism.

An evolutionary shift from ectothermy to endothermy appears to encounter a catch-22 situation: A high metabolic rate is no use unless an animal has insulation to retain metabolically produced heat, because without insulation the heat is rapidly lost to the environment. However, insulation serves no purpose for an animal without a high rate of metabolism because there is little internally produced heat for the insulation to conserve. Indeed, insulation can be a handicap for an ectotherm, because it prevents it from warming up. Raymond Cowles demonstrated that fact in the 1930s when he made small fur coats for lizards and measured their rates of warming and cooling. The potential benefit of a fur coat for a lizard is, of course, its effect of keeping the lizard warm as the environment cools off. However, the lizards in Cowles' experiments never achieved that benefit of insulation, because when they were wearing fur coats they were unable to get warm in the first place.

Those well-dressed lizards illustrate the paradox of the evolution of endothermy: Insulation is ineffective without a high metabolic rate, and the heat produced by a high metabolic rate is wasted without insulation. By this line of reasoning, neither one of the two essential features of endothermy would be selectively advantageous for an ectotherm without the previous development of the other. So how did endothermy evolve?

Probably endothermy evolved as a by-product of selection for one or more other activities that involved some of the same characteristics that are needed by endotherms. Ideas about the evolution of endothermy are unavoidably speculative, but they can be tested to some extent by asking if the intermediate stages that are postulated could have been functional in thermoregulation. For example, feathers, which provide the insulative layer that permits birds to be endotherms, are derived from the scales that covered the bodies of Mesozoic archosaurs. One hypothesis for the origin of feathers suggests that an intermediate stage in their evolution consisted of elongate scales. An animal with scales of this type could orient its body while it was warming so that sunlight penetrated between the scales. When it was warm enough, a change of orientation would convert the scales to a series of parasols, blocking heat uptake (Regal 1975). This sort of scale probably would be suitable for the sort of ectothermal thermoregulation exemplified by the Galapagos marine iguana, and it might also have the basic features needed to provide insulation if the scales trapped a layer of air when they were lowered to press against each other. An animal with a body covering that had achieved some insulative capacity would be able to retain metabolically produced heat and might derive some benefit from a high metabolic rate (Regal 1985).

A different sequence of events has been suggested to account for the origin of mammalian endothermy. The synapsid lineage from which mammals are derived shows skeletal changes that seem to indicate a progressive increase in locomotor capacity (Chapter 19). If the animals were indeed becoming more active, it is plausible that the morphological changes were accompanied by an increasing metabolic rate needed to sustain that activity, and that heat production by muscles during activity raised the body temperature (Bennett and Ruben 1979). In that situation, hair that provided insulation could help to maintain the elevated body temperature for some period after activity ceased.

Thus, in the evolution of mammalian endothermy metabolic heat production (as a by-product of locomotor activity) might have preceded insulation, whereas in birds insulation (a by-product of ectothermal thermoregulation) might have preceded high rates of metabolism. Other plausible evolutionary scenarios can be proposed to explain how endothermy could evolve in birds or mammals, and the ones suggested here may be incorrect. However, they do have the merit of illustrating an important point—the selective pressures that were responsible for the origin of a trait are not necessarily the ones that are responsible for its present utility. In the case of the evolution of endotherms from ectotherms, the conflicting requirements of the two modes of thermoregulation are so different that some factors other than thermoregulation were almost certainly involved in the initial stages.

Energy Utilization: Patterns Among Vertebrates

The complexity of many features of vertebrates relates to their characteristic mobility, which, in turn, has allowed them to occupy a commanding position in most ecosystems they inhabit. This prominence has a cost, however, for activity requires energy, and energy often is limited. Vertebrates usually require more energy than most metazoan animals, and those vertebrates that physiologically regulate their body temperature at high levels require much more energy. As a result, vertebrates expend much of their active time in the search for food. As a consequence, the impact of vertebrates on the environment (that is, their demand on resources) is great.

Aerobic and Anaerobic Metabolism

Vertebrates are obligatory aerobes, and oxidize their basic foods (carbohydrates, fats, and proteins) to carbon dioxide and water. To accomplish this task, oxygen must be supplied to the mitochondria of each cell, where the final stages of oxidation and ATP formation take place.

The advantages of aerobic metabolism lie in the high yield of ATP produced per gram of food oxidized. Nevertheless, activity performed at too high a level cannot be sustained for long before fatigue sets in. Many invertebrates supply their energy needs via anaerobic metabolism. In this process foods are incompletely oxidized, the amount of ATP generated per gram of food is lower, and the food residues are excreted. As a result, anaerobic animals can be considered wasteful, but the process does allow animals to live in oxygen-depleted environments, a feat that aerobes cannot accomplish.

All vertebrates accomplish the initial steps of the cellular breakdown of foods anaerobically (Figure 4–10). Thus, glucose is ultimately converted to pyruvate, and the energy released by this glycolytic process is converted in part into ATP. Generally, the pyruvate is decarboxylated and oxidized to carbon dioxide and water to yield additional ATP. In active skeletal (striated) muscle, the pyruvate may accumulate faster than oxygen can be supplied for its complete oxidation. As a result, the tissue becomes anoxic (devoid of oxygen) and the accumulation of pyruvate inhibits further breakdown of glucose and thus slows ATP generation. By converting pyruvate to lactic acid, however, glycolysis can proceed until depletion of cellular energy stores causes fatigue. This process is called anaerobic metabolism because oxygen is not used as an electron acceptor.

A vertebrate can swim, fly, or run at maximum rate only for a limited time until it drops from fatigue. This is an important ecological point: There is a time limit as well as a mechanical limit placed on vertebrate motility. Most behaviors take place at levels of energy expenditure that exceed resting metabolic levels only slightly. For example, the lengthy migrations that many vertebrates undertake require much energy, but they are performed at speeds that do not exceed the limits of aerobic metabolism. In contrast, the brief pursuits of prey by predator, territorial defense, and nuptial gyrations between mates may require ATP to be synthesized faster than can be accomplished by aerobic metabolic pathways alone.

All vertebrates have the capacity for both aerobic and anaerobic metabolism, and the balance between the pathways differs in various species. In general, endotherms have greater capacity for aerobic metabolism than do ectotherms, but even endotherms use anaerobic pathways to supplement aerobic production of ATP during intense activity. High levels of aerobic metabolism require high rates of transport of oxygen to active tissues, and animals with high aerobic metabolic capacities have large hearts, high rates of blood flow, high hematocrits, and high concentrations of hemoglobin in the blood (Table 4–8).

Physiological characteristics of muscles specialized for aerobic and anaerobic metabolism also differ. Red muscle takes its color from the myoglobin it contains. Myoglobin is a protein that binds oxygen and speeds its diffusion from blood to mitochondria, and muscles with high concentrations of myoglobin are well vascularized and have high activities of the enzymes associated with aerobic metabolic pathways. White muscle lacks myoglobin, is poorly vascularized, and has high activities of enzymes associated with anaerobic metabolic pathways. The distribution of red and white muscle in many vertebrates tells much about the sorts of activities those muscles support. For example, domestic chickens in a barnyard walk around all day, but they fly only if they are startled, and then for only a few seconds. The leg muscles of chickens are composed of red muscle (the dark meat) with high aerobic capacity. The flight (breast) muscles are white and are largely anaerobic. In tunas the swimming muscles are red (dark meat tuna), and other trunk muscles are white (light meat tuna).

The metabolic capacity of an animal is the sum

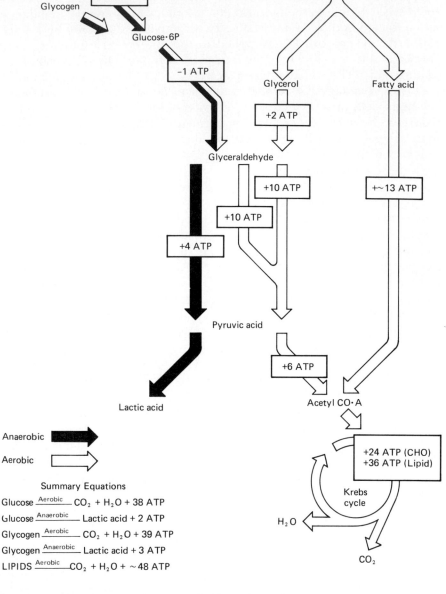

Glucose

−1 ATP

Glycogen

Glucose·6P

−1 ATP

Lipid

Glyceraldehyde

Glycerol

Fatty acid

+2 ATP

+10 ATP

+~13 ATP

+10 ATP

+4 ATP

Pyruvic acid

+6 ATP

Acetyl CO·A

Lactic acid

+24 ATP (CHO)
+36 ATP (Lipid)

Krebs cycle

H_2O

CO_2

Anaerobic

Aerobic

Summary Equations

Glucose $\xrightarrow{\text{Aerobic}}$ CO_2 + H_2O + 38 ATP

Glucose $\xrightarrow{\text{Anaerobic}}$ Lactic acid + 2 ATP

Glycogen $\xrightarrow{\text{Aerobic}}$ CO_2 + H_2O + 39 ATP

Glycogen $\xrightarrow{\text{Anaerobic}}$ Lactic acid + 3 ATP

LIPIDS $\xrightarrow{\text{Aerobic}}$ CO_2 + H_2O + ~48 ATP

Boxes and equations show approximate yield of ATP in moles per six carbon segment of substrate

Figure 4–10. Schematic diagram of the main elements in aerobic and anaerobic energy metabolism. (From M. S. Gordon, G. A. Bartholomew, A. D. Grinnell, C. B. Jorgensen, and F. N. White, 1982, *Animal Physiology: Principles and Adaptations,* Macmillan, New York.)

Table 4–8. Representative values of the heart and blood for vertebrate classes.

Species	Body Mass (kg)	Heart Mass (percent of body mass)	Cardiac Output (ml/kg · min)	Blood Pressure (mm Hg)	Hematocrit (percent)	Hemoglobin (g/100 ml blood)
Fishes: carp	1	0.15	9	43	31	10.5
Lizards: iguana	1	0.19	58	75	31	8.4
Mammals: dog	14	0.65	150	134	46	14.8

of its aerobic and anaerobic energy production. A lizard (the desert iguana) and a mammal (the kangaroo rat) are both able to produce about 0.015 millimole of ATP per gram during 30 seconds of activity, but about 70 percent of that ATP comes from aerobic pathways in the mammal compared to 76 percent from anaerobic pathways for the lizard (Table 4–9). The ecological significance of that difference lies in the ability of the two animals to sustain activity. The oxygen and metabolic substrates used for anaerobic metabolism are carried to the muscles by the circulatory system. For anaerobic metabolism the substrate is glycogen, which is stored in the cell, and when that glycogen has been used up, anaerobic metabolism must stop. Thus, aerobic metabolism can continue nearly indefinitely, whereas anaerobic metabolism can produce large quantities of ATP, but for only

Table 4–9. Estimated aerobic and anaerobic contributions to 30 seconds of activity for a mammal (the kangaroo rat, *Dipodomys merriami*) and a lizard (the desert iguana, *Dipsosaurus dorsalis*).

Species	Millimoles ATP per Gram Body Mass		
	Aerobic	Anaerobic	Total
Dipodomys	0.0098 (70%)	0.0043 (30%)	0.0141
Dipsosaurus	0.0044 (24%)	0.0142 (76%)	0.0186

Source: J. A. Ruben and D. E. Battalia, 1979, *Journal of Experimental Zoology* 208:73–76.

a brief time before the substrate is depleted and the animal is exhausted. Animals with high aerobic capacities can be seen as specialized for sustained activity, whereas animals with low aerobic capacities are specialized for burst activity.

Metabolic Levels in Vertebrates

The amount of energy that vertebrates use is reflected by the amount of oxygen they consume. In general, the minimum rate at which animals consume oxygen is set by the level necessary to support life. The maximum rate is determined by the respiratory and circulatory systems, that is, the rate at which oxygen and nutrients can be supplied to cells (Weibel 1984). The lower aerobic limit, often referred to as the **standard metabolic rate** (SMR), is set by the minimum energy required to maintain life in an organized state. As a rule, vertebrates seldom operate at their SMR. Therefore, it is necessary to define the conditions under which the SMR is measured. When the resting oxygen consumption is measured following a meal, the metabolism will be 5 to 30 percent higher than the SMR because of the costs of digestion. Other factors, such as visual or mechanical disturbance, cause significant increases in oxygen uptake by inducing stress, and low oxygen or high carbon dioxide levels have pronounced effects on energy metabolism. Because increasing temperature usually increases the rates of chemical reactions, including aerobic metabolism, it can have an overriding influence on organisms.

The Effect of Body Size SMR is reported in terms of the volume of oxygen consumed per unit of time at standard pressure and temperature (STP). Obviously, large animals consume more oxygen than small animals (Figure 4–11). To allow comparisons of animals of different sizes, the SMR is adjusted for body size to yield a mass-specific SMR (rate of oxygen consumption/body mass). It is immediately apparent that within vertebrates the mass-specific SMR decreases as body size increases. The slope of the regression of SMR on body mass is roughly similar for different kinds of vertebrates, and also for a wide variety of invertebrates. This relationship has intrigued biologists for over 100 years, but a clear, unambiguous explanation has proved elusive.

Several biologists have pointed out that the increase in mass-specific SMR with decreased size might result from the relatively larger body surface area of small animals. Surface area is proportional to the 2/3 power (0.67) of the body mass. About 100 years ago Max Rubner showed that the heat

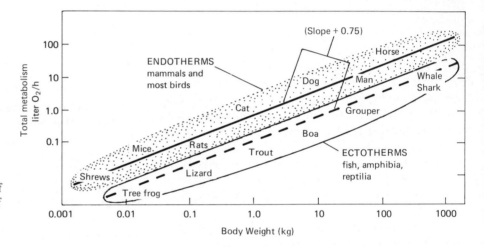

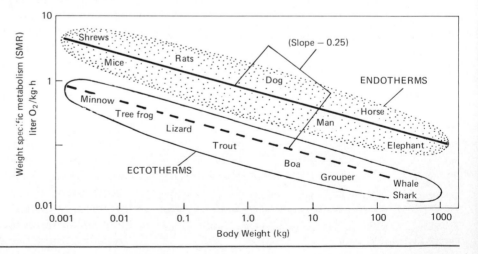

Figure 4–11. Comparison of body size and metabolic rate of ectotherms and endotherms. Upper graph: the log of the total metabolism per hour plotted against the log of body mass yields a straight regression line with a slope that usually lies between 0.65 and 0.85 and is conventionally considered to be equal to 0.75. Lower graph: total metabolism converted to oxygen consumption per unit of body mass per hour to give a mass-specific metabolic rate (SMR). The plot of the log of SMR against the log of body mass yields a slope that is conventionally considered to be −0.25.

loss across each unit of body surface of small and large dogs was the same (about 100 kilocalories per square meter), even though the mass-specific SMR was higher in smaller dogs. To explain this paradox he reasoned that small dogs must lose heat more rapidly than larger dogs because of their relatively high surface area. To maintain body temperatures small dogs compensate for their increased heat loss by an increase in mass-specific SMR. Rubner felt the regression of SMR on mass was sufficiently close to a slope of 0.67 that the phenomenon was explained by body geometry.

For many vertebrates this slope has value near 0.75, that is, the SMR varies as the 3/4 power and not the 2/3 power of body mass (Kleiber 1961). Subsequently theoretical arguments have been advanced to support Kleiber's rule. A mass exponent of 0.75 can be derived from the mechanics of locomotion (McMahon 1973) and from the geometry of four dimensions (Blum 1977). Recently the mass exponent of 0.67 has reemerged as the predicted value for comparisons of different-sized individuals of a single species (Heusner 1982). Feldman and McMahon (1983) have suggested that both 0.75 and 0.67 are valid exponents, the first applying to comparisons of different species and the second being appropriate for comparisons of individuals within species. Although the mechanistic basis of the phenomenon remains unexplained, two points are clear: Metabolic rate is related to body size in all vertebrates, and that relationship has profound ecological and evolutionary consequences.

Although the exact values of the slopes relating metabolism to body mass are still subject to debate, it is clear that the slopes are less than 1. The metabolic rate can be thought of as the energy requirement of an animal. Because the slope of metabolism versus mass is less than 1, doubling the size of an animal does not double its energy requirement. To understand that relationship, assume that the mass exponent for metabolism is 0.75 and consider an animal weighing 2.5 kilograms and another animal of the same kind that weighs 5 kilograms. The metabolic rates of the two animals will be proportional to their body masses

raised to the 0.75 power. Thus,

$$\text{MR of animal 1} = (2.5 \text{ kg})^{0.75} = 1.99$$

and

$$\text{MR of animal 2} = (5 \text{ kg})^{0.75} = 3.34$$

The energy requirements of the larger animal are only 1.68 times greater than those of the small animal. In ecological terms, that means that a small animal has a greater energy requirement *for its size* than does a large animal, although a large animal needs more energy *in total* than does a small animal.

At least two regression lines are required to fit the SMR data for all vertebrates (Figure 4–11). The lower SMRs of ectotherms result partly from their lack of internal heat generation to maintain a high body temperature. The SMR of an ectotherm averages one-sixth that of an endotherm of the same size. The cost of maintaining a high body temperature demands that birds and mammals consume more food than ectotherms of similar size. But vertebrates are motile animals and, whether ectothermic or endothermic, all require more energy when active.

The Energy Cost of Activity

Most vertebrates move by swimming, flying, running, or walking. Predictably, the rate of metabolism during locomotion increases with increased velocity. The energy cost of moving a unit mass a unit distance differs for each mode of locomotion. Specifically, for travel over the same distance—independent of the speed—swimming costs less than flying, and flying costs less than running. Why?

Because SMR is a function of body size it can be expected that sustained active metabolism will also relate to body size. The mass-specific **active metabolic rates** (AMR) of smaller individuals are greater than those of larger individuals (Figure 4–12). Furthermore, for each size the specific AMR is linearly related to velocity. It is possible, therefore, to calculate the cost of movement independent of speed. Only body size must be taken

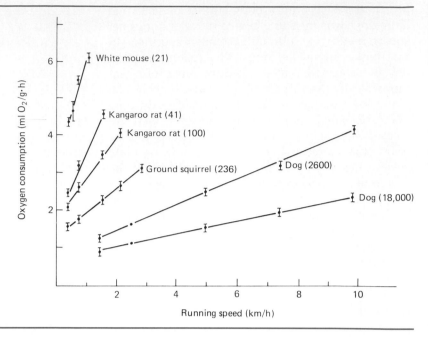

Figure 4–12. Costs of running at different speeds for dogs and rodents. Body masses are in grams. Note that the regressions are linear for any individual over a range of speeds, but the slopes decrease with increasing body size. (Modified after C. R. Taylor et al., 1970, *American Journal of Physiology* 219:1104–1107.)

into account. For example the units in Figure 4–12 are in milliliters of oxygen per gram per hour (ml O_2 /g·h) and in kilometers per hour (km/h). The slope of each line is ml O_2/g·h/km/h, which reduces to ml O_2/g·km, which is the metabolic cost of moving 1 gram over 1 kilometer. Plotting this cost against body mass reveals that each mode of locomotion is represented by a separate regression (Figure 4–13). These results show that swimming costs less than flying, and flying less than walking or running. Each vertebrate, of course, has a body form and physiology suited to its particular mode of locomotion. The importance of such variation is illustrated by the excessively high cost of swimming for humans, nonswimmers by nature.

The cost of locomotion is about the same whether a vertebrate is an ectotherm or endotherm. Although this may appear paradoxical, the energy required to move a unit mass a unit distance is invariable. The energy a fish or whale needs to move 1 kilogram a given distance requires an equal minimum amount of force. To be sure, precise studies reveal slight differences in the costs of movement. Some fishes are more efficient than others and some birds are especially effective flyers, but the point of these studies is the similarity within swimmers, flyers, and runners, regardless of their phylogenetic position.

The question is why swimming is a less expensive mode of locomotion than flying or running and why flying is less expensive than running. As yet the solutions are not clear. The relatively low costs of swimming contradict intuitions about locomotion. The explanations probably lie in the biomechanics of locomotion. For example, a fish moves by sinuous undulations. Every time a trunk muscle contracts, thrust is produced and simultaneously the body is positioned for the next propulsive contraction. However, in a tetrapod forward motion is created only when the limb is thrust backward. To return the limb to a forward position for the next propulsive motion requires that different muscles expend additional energy on the limb without producing thrust (Alexander 1981). These considerations may in part explain the energetic differences among the three major modes of locomotion. Recent measurements of garter snakes—land undulators—show that their

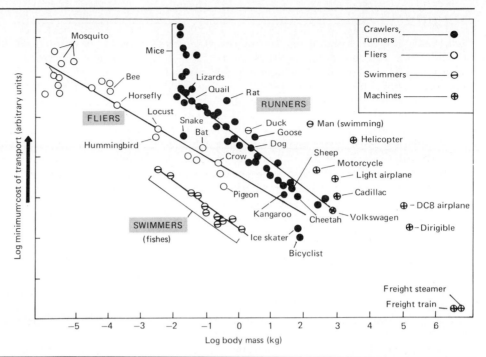

Figure 4–13. Costs of transport for swimmers, fliers, and runners. Each locomotory mode tends to fall along a separate regression line. Note that a duck swims with leg movements similar to those used in walking and a snake crawls with undulations similar to those used by swimmers. (Modified from V. A. Tucker, 1975, *American Scientist* 63:413–419.)

cost falls between that of swimmers and runners. This result fits predictions because snakes undulate like swimmers but, in addition, must overcome greater friction than if suspended in water.

Conclusion

Organisms can be regarded as self-sustaining interactions with the environment. To be self-sustaining, an organism must regulate its internal conditions (homeostasis) and obtain energy from the environment. Temperature, water, and ions are the most conspicuous elements of homeostatis by vertebrates and the mechanisms involved in their regulation are intimately intertwined. For example, vertebrates that must conserve water and excrete salt nonetheless reabsorb ions from the urine across the walls of kidney tubules or from the bladder. This process actually conserves water because the ions are subsequently excreted at higher

concentrations (that is, with a smaller expenditure of water).

Paradoxes of that sort appear when one organ system is considered in isolation; they are resolved when individual organ systems are viewed in the context of the entire organism and its interaction with its environment. General links between the morphology and physiology of vertebrates and the ways in which they interact with their environments are often readily apparent: The body fluid concentrations of different groups of aquatic vertebrates in relation to the osmotic concentrations of their environments provide one example, and the patterns of blood flow through the hearts of ectotherms and endotherms, another. This mechanistic approach to understanding how organisms work can be carried to the level of species and individuals; examples of homeostasis by particular kinds of vertebrates in challenging environments are discussed in more detail in Chapters 16 and 23.

Summary

Vertebrates, like other organisms, are composed mostly of water. Inorganic and organic solutes are dissolved in the water, and the complex biochemical processes that make organisms self-sustaining require regulation of the water content and solute concentrations of their tissues and cells. Most vertebrates have osmotic concentrations between 250 and 350 milliOsmoles per kilogram of water, whereas fresh water is usually below 10 milliOsmolal and seawater is about 1000 milliOsmolal. A rough correspondence exists among groups of fishes between the time they have been in freshwater or marine habitats and how closely their internal osmotic concentrations match their environment. Sodium and chloride are the major osmotically active compounds in seawater and in most vertebrates. The body surfaces of fishes and amphibians are permeable to water.

Freshwater teleosts and amphibians have osmotic and ion concentrations higher than their surroundings. Consequently, they must cope with an inward osmotic flow of water and an outward diffusion of ions. They produce copious, dilute urine to excrete water and expend energy to take up ions from the external medium. Marine teleosts are less concentrated than seawater; they lose water by osmosis and gain salt by diffusion. These fishes drink seawater and use active transport to excrete ions. Hagfishes, elasmobranchs, and coelacanths have osmotic concentrations close to that of seawater, but ionic concentrations that are different from those of their environment. As a result, osmotic water movement is low, but energy is used to regulate solute concentrations.

Deamination of proteins during metabolism produces ammonia, which is toxic. Ammonia is very soluble in water and aquatic vertebrates excrete ammonia as their main nitrogen-containing waste product (ammonotelism). Terrestrial animals do not have enough water available to be ammonotelic. Mammals convert ammonia to urea, which is nontoxic and very soluble. The capacity of the mammalian kidney to produce concentrated urine allows mammals to excrete urea (ureotelism) without an excessive loss of water. The kidneys of diapsids and turtles do not have the urine-concentrating capacity of mammalian kidneys, and these animals convert much of the ammonia to uric acid (uricotelism). Uric acid is not very soluble, and it combines with ions to form urate salts that precipitate in the cloaca. As the salt precipitates, water is released, and uricotely is very economical of water. Some uricoteles save even more water by using extra-renal routes of salt secretion (salt glands) to eliminate sodium and chloride in solutions that may exceed 2000 milliOsmoles.

Temperature profoundly affects the biochemical processes that sustain vertebrates, and thermoregulatory mechanisms are widespread. Few fishes and amphibians can maintain a temperature difference between their bodies and the water around them, but some fast-swimming tunas and sharks have muscle temperatures that are 10°C above water temperature. Many terrestrial vertebrates have the capacity to regulate their body temperature. Ectotherms rely on sources of heat from outside the body for thermoregulation, balancing heat gained and lost by radiation, conduction, convection, and evaporation. This is a complex and effective process; many ectotherms maintain stable body temperatures substantially above ambient temperatures while they are thermoregulating. Endotherms use metabolically produced heat and manipulate insulation to balance the rates of heat production and loss. Endothermal thermoregulation confers considerable independence of environmental conditions but is energetically expensive. The mechanisms of ectothermal and endothermal thermoregulation are quite different, and an evolutionary transition from ectothermy to endothermy would be complex. Nonetheless, that transition occurred at least twice, once in the evolution of birds and once in the evolution of mammals. The evolutionary origins of the two essential components of endothermy—insulation and a high metabolic rate—were probably different from their current significance.

Vertebrates are active animals, and that activity requires a high level of energy expenditure. The energy requirements of vertebrates depend on body size and mode of thermoregulation, and on what the animals do. Locomotion is an energetically expensive activity: Flying is more expensive than swimming, and running is more expensive than either. Vertebrates use two pathways of metabolic energy production—aerobic metabolism and anaerobic metabolism (glycolysis). Aerobic metabolism requires a circulatory system that can transport oxygen and metabolic substrates to active tissues, whereas anaerobic metabolism relies on the glycogen stores present in the cell. Both can produce ATP at high rates, but only aerobic metabolism can be sustained for long periods.

References

Alexander, R. McN. 1981. The gaits of tetrapods: adaptations for stability and economy. *Symposia of the Zoological Society of London* 48:269–287. Academic Press, London.

Avery, R. A. 1979. *Lizards—A Study in Thermoregulation.* University Park Press, Baltimore.

Avery, R. A. 1982. Field studies of reptilian thermoregulation. Pages 93–166 in *Biology of the Reptilia*, volume 12, edited by C. Gans and F. H. Pough. Academic Press, London.

Bartholomew, G. A. 1982. Physiological control of body temperature. Pages 167–211 in *Biology of the Reptilia*, volume 12, edited by C. Gans and F. H. Pough. Academic Press, London.

Bennett, A. F. and J. A. Ruben. 1979. Endothermy and activity in vertebrates. *Science* 206:649–655.

Blum, J. J. 1977. On the geometry of four dimensions and the relationship between metabolism and body mass. *Journal of Theoretical Biology* 64:599–601.

Brett, J. R. 1971. Energetic responses of salmon to temperature. *American Zoologist* 11:99–113.

Cowles, R. B. and C. M. Bogert. 1944. A preliminary study of the thermal requirements of desert reptiles. *Bulletin of the American Museum of Natural History* 83:261–296.

Evans, D. H. 1980. Osmotic and ionic regulation by freshwater and marine fishes. Pages 93–122 in *Environmental Physiology of Fishes*, edited by M. A. Ali. Plenum Press, New York.

Feldman, H. A. and T. A. McMahon. 1983. The 3/4 mass exponent for energy metabolism is not a statistical artifact. *Respiratory Physiology* 52:149–163.

Heath, J. E. 1965. Temperature regulation and diurnal activity in horned lizards. *University of California Publications in Zoology* 64:97–136.

Heusner, A. A. 1982. Energy metabolism and body size. I. Is the 0.75 mass exponent of Kleiber's equation a statistical artifact? *Respiration Physiology* 48:1–12.

Huey, R. B. 1982. Temperature, physiology, and the ecology of reptiles. Pages 25–91 in *Biology of the Reptilia*, volume 12, edited by C. Gans and F. H. Pough. Academic Press, London.

Kleiber, M. 1961. *The Fire of Life: An Introduction to Animal Energetics.* Wiley, New York.

Kluger, M. J. 1979. Fever in ectotherms: evolutionary implications. *American Zoologist* 19:295–304.

Loveridge, J. P. 1970. Observations on nitrogenous excretion and water relations of *Chiromantis xerampelina* (Amphibia, Anura). *Arnoldia* 5:1–6.

Lutz, P. L. 1975. Adaptive and evolutionary aspects of the ionic content of fishes. *Copeia* 1975:369–373.

McMahon, T. 1973. Size and shape in biology. *Science* 179:1201–1204.

Minnich, J. E. 1982. The use of water. Pages 325–395 in *Biology of the Reptilia*, volume 12, edited by C. Gans and F. H. Pough. Academic Press, London.

Nishimura, H. and M. Imai. 1982. Control of renal function in freshwater and marine teleosts. *Federation Proceedings* 41:2355–2360.

Peaker, M. and J. L. Linzell. 1975. *Salt Glands in Birds and Reptiles.* Cambridge University Press, Cambridge.

Prosser, C. L. 1986. *Adaptational Biology: Molecules to Organisms.* Wiley, New York.

Regal, P. J. 1975. The evolutionary origin of feathers. *Quarterly Review of Biology* 50:35–66.

Regal, P. J. 1985. Commonsense and reconstructions of the biology of fossils: *Archaeopteryx* and feathers. Pages 67–74 in *The Beginnings of Birds, Proceedings of the International* Archaeopteryx *Conference, Eichstätt 1984*, edited by M. K. Hecht, J. H. Ostrom, G. Viohl, and P. Wellnhofer. Jura Museum, Eichstätt, West Germany.

Shoemaker, V. H., D. Balding, and R. Ruibal. 1972. Uricotelism and low evaporative water loss in a South American frog. *Science* 175:1018–1020.

Thorson, T. B. 1970. Freshwater stingrays, *Potomotrygon* spp.: failure to concentrate urea when exposed to saline medium. *Life Sciences* 9:893–900.

Thorson, T. B., C. M. Cowan, and D. E. Watson. 1973. Body fluid solutes of juveniles and adults of the euryhaline bull shark, *Carcharinus leucas*, from freshwater and saline environments. *Physiological Zoology* 46:29–42.

Tracy, C. R. 1982. Biophysical modeling in reptilian physiology and ecology. Pages 275–321 in *Biology of the Reptilia*, volume 12, edited by C. Gans and F. H. Pough. Academic Press, London.

Weibel, E. R. 1984. *The Pathway for Oxygen.* Harvard University Press, Cambridge, Mass.

White, F. N. 1973. Temperature and the Galapagos marine iguana—insights into reptilian thermoregulation. *Comparative Biochemistry and Physiology* 45A:503–513.

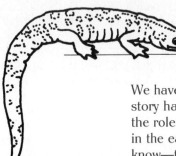

We have emphasized the importance of the stage on which the vertebrate story has been played in shaping its plot, and in this chapter we illustrate the role of the environment by considering the conditions that prevailed in the early Paleozoic. The world then was very different from the one we know—the continents were in different places, climates were different, and there was little structurally complex life on land. All of these elements played a role in setting the stage for the origin of vertebrates.

Geology and Ecology During the Origin of Vertebrates

Earth History, Changing Habitats, and Vertebrate Evolution

Careful examination of vertebrate evolution requires considering when and where specific events occurred. But records of evolutionary events are seldom complete; therefore, we must work with probabilities that one interpretation is more likely than another. For the major lineages of vertebrates, the fossil record is sufficiently complete to provide a firm and consistent chronology of origins.

Determining *where* and under *what* ecological conditions various vertebrates evolved is, however, a much more complex problem. How can one explain the enigmatic fact that marsupial mammals are mainly characteristic of Australia and South America? Why were horses and camels absent from North America when the Spaniards first arrived, yet the fossil record indicates that they probably originated there? There remains much to learn about where different vertebrates arose, but the initial answers are startling.

To evaluate where and in what kinds of habitats specific vertebrate groups have evolved, we must consider concepts of modern geology (Box 5–1). How continental movements relate to the origin of each class of vertebrates is considered in this and several subsequent chapters that introduce each major group of vertebrates.

Continental Positions in the Early Paleozoic

In late Paleozoic times vertebrates were well established in the seas, fresh waters, and land areas of Pangaea (Hallam 1973). But where were the continents located in the Cambrian (Bambach et al. 1980, Goodwin 1981), especially from the Ordovician to the Silurian, when vertebrates first become common in the fossil record? The geological data suggest that Pangaea did not coalesce until the Carboniferous. The earlier positions of the continents are approximated for Cambrian, Ordovician, and Silurian times in Figure 5–2. This figure emphasizes several features of importance about vertebrate origins:

1. Gondwana, which straddled the paleoequator and extended into both north and south temperate regions (beyond 30 degrees north and south latitudes), began a relentless drift southward. That cold conditions increased in much of Gondwana is demonstrated by the presence of glacially scoured rocks, of late Ordovician age, in what is now central Africa and southern South America (Boucot and Gray 1982). By the mid-Silurian, Gondwana overlay the South Pole.
2. In contrast, throughout the Cambrian, Ordovician, and Silurian, Laurentia and Kazakhstania were located in warm-equatorial lati-

195

Box 5–1. Why the Continents Move

The continents literally float on other elements of Earth's crust (Figure 5–1). The average density of the granitic and sedimentary rocks that form the continents is 2.6 to 2.7 grams per cubic centimeter. This is less than the density of the basaltic rocks of the underlying lithosphere or seafloor (2.8 to 3.0 grams per cubic centimeter). A continent sinks into the mantle until it displaces its own weight, much like an ice cube in a glass of water.

Because the boundary between mantle and crust can be deformed, the crust and overlying continents can move. In the Atlantic, Pacific, and Indian Oceans mountain ridges protrude from the seafloor. Basaltic materials flow upward from the mantle to create these ridges. At varied distances from the ridge, **subduction** (withdrawal of the seafloor into the Earth's mantle) takes place along oceanic trenches, producing a horizontal movement of the seafloor from ridge to trench. Because the continents float on top of basaltic plates that are continuous with the seafloor, the continents are dragged along as the seafloor spreads.

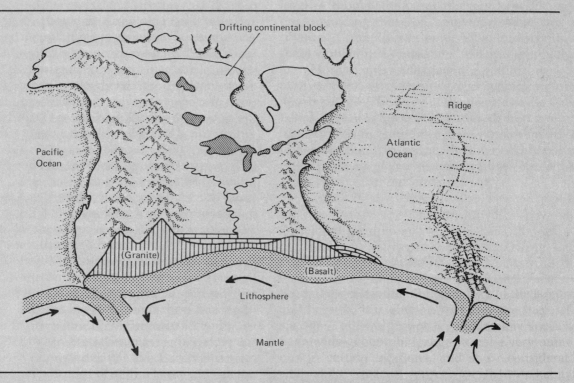

Figure 5–1. Generalized geological structure of a continent. The continents float on a basaltic crust, and the entire lithosphere floats atop the underlying, noncrystalline mantle. Arrows show the general movements of the crustal elements and their interaction with the mantle that causes continental drift.

The actual mechanism driving seafloor spreading, that is, the addition of material at the ridges and its subduction at the trenches, is a matter of considerable speculation (Takeuchi et al. 1970). Whatever the mechanism, it is subject to change, and a change in the rate of seafloor spreading is reflected in continental geology. For example, when rates of seafloor spreading increase, ridge material accumulates and raises the sea level relative to the continents to produce shallow, epicontinental seas. It also increases the rate at which the continents drift. The rate of seafloor spread, therefore, has profoundly influenced the course of vertebrate evolution.

Most paleogeologists accept the idea that the six major landmasses recognized today were once grouped together as a single supercontinent. [See Ben-Avraham (1981) and Wilson (1972) for details.] This early continent was christened **Pangaea** by Alfred Wegener, who as early as 1915 suggested that this landmass broke in late Paleozoic or early Mesozoic time into discrete sections which slowly drifted to their present positions. Two lines of evidence attest to the existence of Pangaea during the Paleozoic—sedimentary geology and paleomagnetism.

Areas of extensive sedimentation, termed **geosynclines**, tend to border or form low-lying margins of the continents. Appropriate rearrangement of our modern continents into a single land mass should incorporate these ancient Paleozoic geosynclinal sediments in such a way that they lie on the margins of Pangaea. The fitting together of this geological jigsaw puzzle suggests that Pangaea united six sepa-rate ancient continents that probably existed from Cambrian through Silurian time (Figure 5–2). Five of these primitive and separate continental blocks included most of North America, Greenland, Scotland, and part of the northwestern USSR (named **Laurentia**); of the USSR west of the Ural mountains, Scandinavia, and much of central Europe (**Baltica**); of central southern Asia (**Kazakhstania**); of **Siberia**; and of **China** south of Mongolia and including all of Indochina. The sixth ancient continent consisted of a single massive landmass—**Gondwana**—that included South America, the southeastern United States, Africa, Antarctica, Australia, India, Tibet, Iran, Saudi Arabia, Turkey, and southern Europe.

Paleomagnetic inferences about past continental positions rely on the fact that the spatial orientation of magnetic minerals in rocks is fixed in reference to the Earth's magnetic field at the time when a particular rock was formed from molten material. The declination and inclination (position in three-dimensional space) of magnetic minerals *in situ* can reveal the past orientation of a given formation relative to the north and south magnetic poles (paleolatitude). Using these magnetic orientations as a compass, it is possible to approximate the position of ancient continents with respect to the magnetic poles. For example, a part of Africa during the Ordovician–Silurian was located near the South Pole (Figure 5–2). Paleomagnetic evidence alone is compelling, and fossil distributions also support the historical reality of Pangaea and the movement of the continents.

Box 5–1 (Continued)

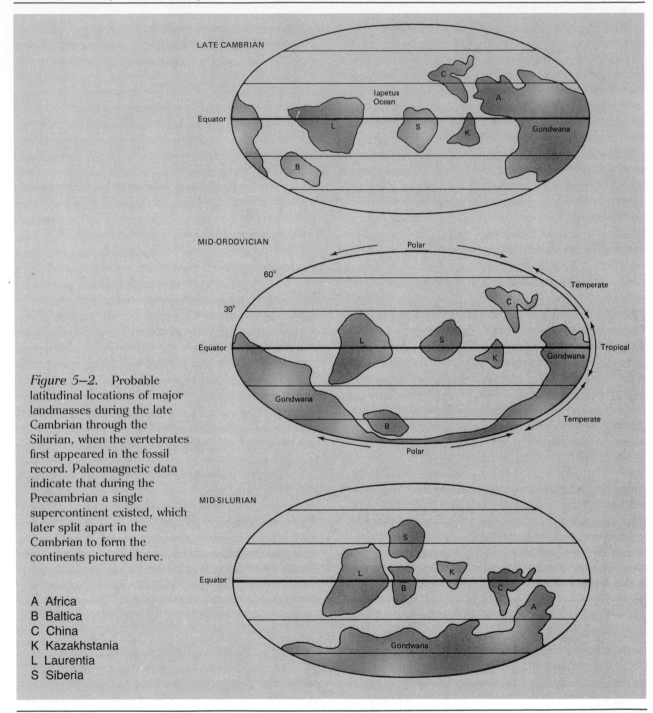

Figure 5–2. Probable latitudinal locations of major landmasses during the late Cambrian through the Silurian, when the vertebrates first appeared in the fossil record. Paleomagnetic data indicate that during the Precambrian a single supercontinent existed, which later split apart in the Cambrian to form the continents pictured here.

A Africa
B Baltica
C China
K Kazakhstania
L Laurentia
S Siberia

tudes. For a shorter part of this long time period, China, Baltica, Siberia, and the Australia–Antarctica portions of Gondwana remained in warm equatorial locations. The occurrence of biotically produced limestones, dolomites, and coral formations in North America, Europe, and North Australia are indicative of past tropical conditions. These types of deposits are produced today mostly in shallow tropical warm seas within 30 degrees of the equator.

The earliest vertebrate fossils occur in late Cambrian to Silurian marine sediments in North American, Eurasia, and in Australia. Because of their paleoequatorial location, these first vertebrates were characteristic of *warm* shallow continental seas (Bray 1985, Halstead 1985, Janvier 1985). Few early Paleozoic marine sediments are associated with Gondwana except in eastern Australia and the western parts of South America. Presumably, the generally high altitude of Gondwana prevented the sea from covering the continent.

Interestingly, only one ostracoderm is known from Ordovician–Silurian marine deposits of eastern Australia (Hallam 1973). Why this is so remains obscure, but possibly it reflects the great open-water distance and isolation of Australia from North America and Eurasia. Probably the numerous benthic and shallow-dwelling ostracoderms of Laurentia and Kazakhstania were not well adapted to dispersion across the deep seas that separated these equatorial landmasses. We conclude that vertebrates must have evolved first somewhere in North America or western Europe under warm (or tropical) marine conditions that uniquely favored their existence and subsequent radiation.

The Early Habitat of Vertebrates

How well can we describe the habitat of these earliest vertebrates? The association of late Cambrian, Ordovician, and many of the Silurian ostracoderms with brachiopods, crinoids, and corals—all marine invertebrates—attests to an origin in shallow warm seas (Morris and Whittington 1979). During the late Cambrian and Ordovician, North America was largely covered by a shallow continental sea, apparently providing the necessary conditions for evolution of a chordate into a vertebrate. We can say little about the microhabitats concerned, but the general habitat, at least, of the first fossilized forms was probably the seafloor. Further, geochemists and geophysicists suggest that the Paleozoic sea was ionically much as it is today (Nicolls 1965). The early ostracoderms therefore faced physiological problems, with respect to the physical and chemical properties of seawater, similar to those faced by modern tunicates, echinoderms, and pterobranchs. They must have lived in highly saline, shallow, warm, and most likely highly transparent seas. These conditions are matched today in many shallow tropical seas.

Heterotrophic animals like vertebrates ultimately depend on plants as a primary source of energy. In addition, many of the important characteristics of the habitats in which vertebrates live are determined by plants. What kinds of plants existed in the early Paleozoic? Simple, single-celled organisms originated in Precambrian seas—the cyanobacteria are examples. By the Ordovician, however, more complex multicellular green and red algae had evolved (Banks 1970). Phytoplankton such as diatoms and dinoflagellates, a major source of food in aquatic habitats today, was abundant in Ordovician seas. Most fossilized green and red algae were lime secretors, but fossils of non-lime-secreting algae suggest that noncalcareous multicellular algae were also important in these ancient shallow seas.

Similarly, complex freshwater plants of Ordovician and Silurian age are poorly known. The invasion of fresh waters by early marine animals was probably initiated in Silurian times (Valentine 1973). Quite likely it accompanied the adaptation of marine plants to fresh water or, at least, followed their invasion closely. Not until the early Devonian did multicellular algae become abundant in freshwater deposits (Banks 1970). Many of these are much like the modern genus *Chara* and, indeed, are considered by many botanists to be in the same genus.

In general, then, an abundance of simple plants, including diatoms, cyanobacteria, green algae, and other algae, was present from the Ordovician through the Devonian when vertebrates evolved in the sea and radiated in marine and fresh waters. The point of major importance is that these early habitats contained an abundant renewable source of energy. The challenge for vertebrates—as it has always been—was to exploit that resource.

Early Paleozoic Climates

The early Paleozoic is characterized by thick limestone deposits along the edges and in the interior of the continents. This distribution of marine sediments indicates that overall land profiles were low and seas were invasive (Bray 1985). Paleoclimatologists conclude that the entire earth may have been cool in the Cambrian, and extensive glaciation occurred at the close of the Proterozoic era. Continental glaciation, often considered a sign of extensive cooling, is not seen in the geological record again until the late Ordovician when the African sector of Gondwana moved across the South Pole (Figure 5–2; Pearson 1978). But throughout the last 500 million years or more the earth has probably displayed generally the same type of climate as it does today, wherein the polar regions are cooler than equatorial locations (Boucot and Gray 1982) and tropical, temperate, and boreal climates can develop (Figure 5–3). No doubt paleoclimates showed seasonal changes resulting from the yearly procession of the earth about the sun. Latitudinal drift of the continents (north or south of the equator) must have dramatically affected long-term continental climates. From the Ordovician and on into the Triassic the movements of the continents positioned much of North America, and to a lesser extent western Europe, close to the paleoequator (Figures 5–2 and 9–1). Over this long

span of time these areas were exposed to a warm and stable year-round climate, as tropical latitudes are today. Only areas outside the tropics would have been subjected to strong seasonal changes in climate, as they also are today.

Continental movements have also modified climates by creating and obliterating entire oceans and changing patterns of oceanic circulation. A dramatic example is the mild climate of Great Britain that results largely from the northward transport of warm equatorial waters by the Gulf Stream. Mountains formed by the collision of continental plates also changed climates. One example is the effect of the Andes of South America in producing the tropical forests and dry deserts of that continent. During the Cambrian and especially the early Ordovician, mountain building occurred, perhaps as a result of initial contacts between Laurentia and Baltica.

These early mountain-building episodes were less extensive than those that followed at the end of the Ordovician, Silurian, and Devonian epochs when the ancient continental blocks began to coalesce to form Pangaea. The coalescence of Laurentia and Baltica to form Laurussia (the Old Red Continent, named after the oxidized reddish sandstones that characterize its sediments) was accompanied by a major period of mountain building that is identified as the **Caledonian orogeny.** The creation of Laurussia raised its altitude well above sea level, and as a result stream deposits became common. The ostracoderms fossilized in the Ordovician and early Silurian strata of what was to become Laurussia are associated exclusively with sediments deposited in shallow marine environments. The fossils and sediments recorded from this early period into the late Silurian and early Devonian document that the ostracoderms underwent a major radiation (Halstead 1985, Janvier 1985) and that many of them managed to penetrate the physiological barrier posed by freshwater conditions (Bray 1985). Very similar but independent radiations of ostracoderms from marine into freshwater habitats occurred in the early Devonian in Siberia and in south China (Halstead 1985). At present these separate radiations cannot be attrib-

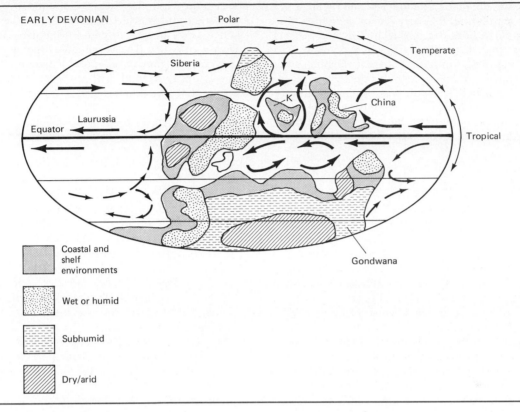

EARLY DEVONIAN

Coastal and shelf environments

Wet or humid

Subhumid

Dry/arid

Figure 5–3. Probable climates and ocean circulation in the early Devonian. Note that general climatic conditions were not unlike today—both tropical to temperate and humid to dry conditions existed.

uted to a single cause such as the uplifting of the land, but the periods were characterized by orogenies and each of these ancient lands were subjected to subtropical and tropical conditions. Of course, these early Silurian vertebrates could not long have endured freshwater conditions unless their invasion of the rivers was preceded by plants and an associated fauna of invertebrates upon which they could feed.

Summary

Here we have considered evidence for continental drift—the theory that the continental landmasses have drifted across the face of the earth. Continental movement is particularly important to an appreciation of the complexity of vertebrate evolution. Understanding this complexity requires a knowledge of the phylogeny of vertebrates (which we consider in detail in succeeding chapters), and also when, where, and under what conditions each group originated. During the early Paleozoic, when the first vertebrates appear in the fossil record, the ancient landmasses were coalescing into a giant single continent called Pangaea. The ancient marine sediments in which the first vertebrates are found were located near the paleoequator and deposited in what must have been tropical seas.

Parts of Pangaea, especially the warmer parts, were

repeatedly inundated by shallow epicontinental seas. The climate throughout the early Paleozoic, which from the late Cambrian until the end of the Devonian covers slightly more than 150 million years, was equable. This extremely long period of fairly constant climatic conditions must have promoted high productivity and favored most forms of life. Simple freshwater plants occur in Ordovician deposits, but complex freshwater forms did not become common until the Devonian. It is likely that ostracoderms and several types of invertebrates did not penetrate into fresh waters until plants were firmly established and productivity was high. Therefore, the extensive radiation of early freshwater fishes was largely a Devonian event. The vertebrates, then as now, were highly motile animals adept at exploiting a rich and diverse supply of food. The tenets of modern geology teach us that this climatic equability, which was so important to the evolution and radiation of the early vertebrates, was largely the result of chance.

References

Bambach, R. K., C. R. Scotese, and A. M. Ziegler. 1980. Before Pangaea: the geographies of the Paleozoic world. *American Scientist* 68(1):26–38. A thesis that suggests that the various continents during the early Paleozoic were separated for long periods of time by oceans, and therefore differs considerably from the maps shown in our chapters.

Banks, H. P. 1970. *Evolution and Plants of the Past*. Wadsworth, Belmont, Calif.

Ben-Avraham, Z. 1981. The movement of the continents. *American Scientist* 69(3):291–299.

Boucot, A. J. and J. Gray. 1982. Paleozoic data of climatological significance and their use for interpreting Silurian-Devonian climate. Pages 189–198 in *Climate in Earth History*, Geophysics Study Committee, National Research Council, National Academy Press, Washington, D.C. A description of how Earth's climate became more uniform from the Ordovician through the Carboniferous.

Bray, A. A. 1985. The evolution of the terrestrial vertebrates: environmental and physiological considerations. *Philosophical Transactions of the Royal Society of London* B309:289–322.

Goodwin, A. M. 1981. Precambrian perspectives. *Science* 213:55–61.

Hallam, A. (editor). 1973. *Atlas of Palaeobiogeography*. Elsevier, New York. A collection of papers that cover a variety of problems relating to the distribution of fossil organisms on the earth. Especially good for Paleozoic and Mesozoic fossils.

Halstead, L. B. 1985. The vertebrate invasion of fresh water. *Philosophical Transactions of the Royal Society of London* B309:243–258.

Janvier, P. 1985. Environmental framework of the diversification of the Osteostraci during the Silurian and Devonian. *Philosophical Transactions of the Royal Society of London* B309:259–272.

Morris, S. C. and H. B. Whittington. 1979. The animals of the Burgess shale. *Scientific American* 241(1): 122–133. A beautifully illustrated article showing many of the invertebrate fossils associated with Cambrian sediments.

Nicolls, G. D. 1965. The geochemical history of the oceans. Chapter 20 in *Chemical Oceanography*, volume 2, edited by J. P. Riley and G. Skirrow. Academic Press, New York.

Pearson, R. 1978. *Climate and Evolution*. Academic Press, New York.

Takeuchi, H., S. Uyeda, and H. Kanamori. 1970. *Debate About the Earth*. Freeman, Cooper, San Francisco. A short, simple treatise on the complex data and arguments that underpin the theory of continental drift.

Valentine, J. W. 1973. *Evolutionary Paleoecology of the Marine Biosphere*. Prentice-Hall, Englewood Cliffs, N.J. A general treatise that considers problems associated with reconstructing ancient habitats. Especially significant are discussions of how skeletal materials can be used to assess the ecology of extinct organisms.

Wilson, J. T. (editor). 1972. *Continents Adrift*. W. H. Freeman, San Francisco. A series of articles from *Scientific American* that reviews the developments of the theory of continental movements and plate tectonics.

Aquatic Vertebrates: Cartilaginous and Bony Fishes

Vertebrates originated in the sea, and more than half the living vertebrates are the products of evolutionary lineages that have never left an aquatic environment. Water now covers 73 percent of the Earth's surface (the percentage has been higher in the past) and provides habitats extending from deep oceans and lakes to fast-flowing streams and tiny pools in deserts. Fishes have adapted to all of these habitats, and the nearly 22,000 species of fishes are the subject of this portion of the book.

Life in water poses challenges for vertebrates and offers many opportunities. Aquatic habitats are some of the most productive on Earth, and energy is plentifully available in many of them. Other aquatic habitats, such as the deep sea, have no *in situ* production of food and the animals that live in them are dependent on energy that comes from elsewhere. The physical structure of aquatic habitats has a similar range: some aquatic habitats (coral reefs are an example) have enormous structural complexity, whereas others (like the open ocean) have virtually none. The diversity of fishes reflects specializations for this variety of habitats.

The diversity of fishes and the habitats they live in has offered unparalleled scope for variations in life history. Some fishes produce millions of eggs that are released into the water to drift and develop on their own, other species of fish produce a few eggs and guard both the eggs and the young, and many fishes give birth to living young. In some species of fish males are larger than females, in others the reverse is true, some species have no males at all, and a few species of fish change sex partway through life. Feeding mechanisms have been a central element in the evolution of fishes, and the specializations of modern fishes extend from species that swallow prey longer than their own bodies to species that extend their jaws like a tube to suck up minute invertebrates from tiny crevices. In this part of the book we consider the evolution of this extraordinary array of vertebrates and the ecological conditions in the Devonian that contributed to the next major step of evolution, the origin of terrestrial vertebrates.

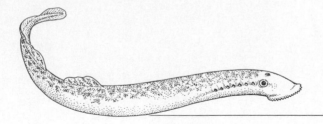

The earliest vertebrates known were aquatic filter feeders, but they
represented an important advance over the protochordate filter feeders
described in Chapter 2 because they used muscular contractions instead of
cilia to move water. A muscular pump can move more water than cilia,
and the early vertebrates were able to grow larger than the protochordates.
Bone, a distinctive form of mineralized tissue, was a second innovation of
the earliest vertebrates. The bony armor that encased these animals probably
gave some protection from predators and also may have been a store of
minerals that contributed to homeostasis. With these advances, combined
with mobility, the earliest vertebrates were able to radiate into adaptive
zones that had previously been unoccupied. We know a remarkable amount
about the anatomy of some of these early vertebrates because the internal
structure of their bony armor reveals the positions and shapes of many
parts of their soft anatomy. The brains and cranial nerves of these most
primitive vertebrates were remarkably like the brain and cranial nerves of a
living vertebrate, the lamprey. In this chapter we trace the earliest steps in
the radiation of vertebrates some 500 million years ago.

Earliest Vertebrates

6

The First Evidence of Vertebrates

A major advance was made by the vertebrates with the evolution of the physiological ability to lay down a skeleton composed of calcium phosphate (bone). At first this ability seems to have been limited to superficial, dermal tissues but it soon extended to endoskeletal tissues of presumed mesodermal origin. Bony materials are more likely to be fossilized than soft tissues, and the fossil record of vertebrates subsequent to the evolution of bone is extensive.

The oldest fossil vertebrates occur in the late Cambrian and middle Ordovician, between 40 and 125 million years prior to the time when vertebrate fossils became abundant (Repetski 1978). The best known of these early forms occur in the late Silurian to middle Devonian Old Red Sandstone in southwestern England and Wales and in similar rock found in Scotland, Norway, and Spitzbergen, dating from 400 million years ago. The microscopic structure of the early specimens is similar to that of the dermal bony plates of certain later forms (Figure 6–1). The similarities indicate that these animals, characterized by laterally placed eyes, double nostrils, a complex of bone and bone-like tissues, and all the other features recognized as basic to the ancestral stock of vertebrates, existed in the Cambrian and Ordovician prior to the Silurian radiation of jawless and jawed

fishes. These bony fragments also tell us that bone had evolved early in vertebrate history, casting doubt on an old idea that a cartilaginous skeleton is primitive.

The fossil record of the first vertebrates mainly whets our appetites for more information. It reveals little about the course of evolution from the earliest Cambrian vertebrates to the sudden appearance of a variety of agnathous and jawed forms in the late Silurian. Also, it provides no clues about the evolution of vertebrate organization from an invertebrate progenitor. Nor does it shed light on the evolution of jawed vertebrates from their agnathous predecessors.

Many biologists conclude that vertebrate animals were the last major group of animals to evolve. The absence of vertebrate fossils in rocks older than the late Cambrian, however, allows only speculation about the earlier history of vertebrates. Most of the other animal phyla appeared 50 million years earlier, in the oldest Cambrian rock sufficiently undistorted by geophysical forces to yield good fossil materials. The best guess is that the earliest vertebrates were rather uncommon, small, soft-bodied marine forms whose existence was unlikely to be recorded by the rare circumstances that lead to fossilization of soft parts.

The first fossil vertebrates are encountered in the late Cambrian and Ordovician and are unimpressive fragments. But when sectioned and examined microscopically (Figure 6–1), these fragments show an internal three-layered bony

205

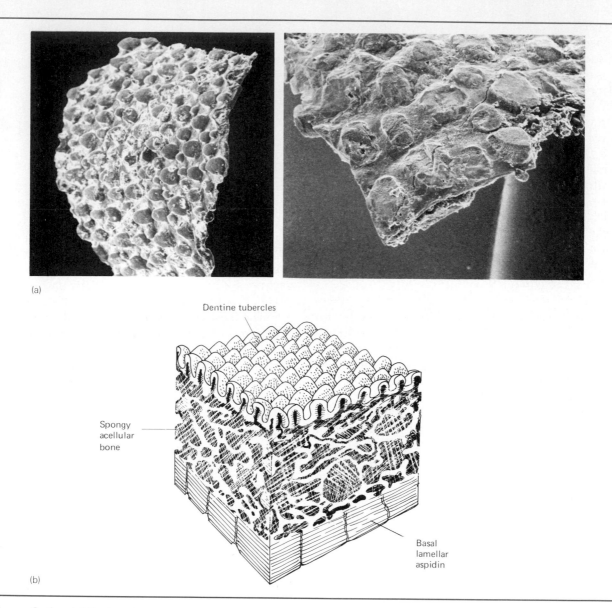

Dentine tubercles

Spongy
acellular
bone

Basal
lamellar
aspidin

(a)

(b)

Figure 6–1. (a) Earliest known remains of vertebrate origin. Scales of the
heterostracan *Anatolepis* from the late Cambrian have definitely been identified from
Oklahoma, Washington, and Wyoming. By the early Ordovician, *Anatolepis* was
widely distributed in North America and to Spitzbergen, and related genera
occurred in Australia. Two views of the same specimen show (left) tongue-like
surface ornaments, which probably pointed posteriorly and (right) an edge showing
a break through the plate revealing the solid surface layer, the cavity-rich middle
layer, and the thick internal lamellar layer. (Scanning electron micrographs at
approximately 220x courtesy of J. E. Repetski, U.S. National Museum.) (b) three-
dimensional block diagram of heterostracan dermal bone. (Based on B. Stahl,
1974, *Vertebrate History*, McGraw-Hill, New York; and L. Halstead, 1969, *The
Pattern of Vertebrate Evolution*, Oliver & Boyd, Edinburgh.)

structure of considerable complexity already as advanced as that of vertebrate remains of a later date. If such histological complexity evolved gradually, this complexity suggests that these vertebrates had undergone considerable evolution *before* the earliest fossil evidence known was laid down. Bone formation requires the coordination of at least three types of tissues: fibroblast cells that lay down the highly organized collagen framework into which calcium phosphate is deposited as hydroxyapatite, cells that produce hormones that regulate the production of hydroxyapatite, and unique scleroblast cells that deposit the minerals in the collagen framework. It seems doubtful that these various tissues evolved in a sufficiently short time to produce the apparently sudden appearance of vertebrate fossils. If that is so, why are there no earlier fossils of vertebrates?

Perhaps conditions for fossilization simply did not exist during the earliest period of vertebrate evolution. A moment's reflection allows us to eliminate this hypothesis: Delicate organisms such as jellyfish medusae were fossilizing during the Precambrian and leaving such excellent impressions that internal anatomy can be deciphered. Contemporaneous with fragments of the first vertebrates are brachiopods of enormous variety. Some of these invertebrates form their shell of hydroxyapatite, the same calcium phosphate salt found in bone. They have fossilized perfectly.

Perhaps the lack of early vertebrate fossils is due to the particular habitat in which they lived, which was inimical to fossilization. Many students of vertebrate evolution, especially A. S. Romer (1966), have proposed that vertebrates evolved in freshwater streams and rivers (as we discussed in Chapter 2). These are areas of erosion, not deposition, and hence would not have provided a good fossil record. However, x-ray diffraction of the crystalline structure of the fragments and their surrounding matrices indicates that these specimens were fossilized where they are found today. Spectrographic analysis allows precise determination of the chemical elements incorporated into the fossilized material. Boron is an important element in determining the salinity of the environment during sedimentation. Additional ecological reconstruction comes from the older technique of stratigraphic analysis, which compares existing fossils in one formation with other and perhaps better understood assemblages of fossils elsewhere.

X-ray diffraction, boron analysis, and stratigraphic techniques have led to the conclusion that the earliest fishes lived, died, and were deposited in a marine environment where some critical factor prevented the fauna from becoming abundant or diverse. Boron concentrations suggest that this critical factor was variable salinity. The most logical environment to fit all the evidence is a euryhaline estuarine zone. If so, some early vertebrates were capable of tolerating the stress of variable salinity, as are many modern fishes.

We have yet to answer the question: Why do vertebrates appear so late in the fossil record? The most likely hypothesis (but the least susceptible to verification) is that the earliest vertebrates and their evolution occurred in an environment that prevented the fossilization of soft tissues.

What Were the Earliest Vertebrates Like?

Two early sites have yielded most of the oldest vertebrate fossils (Repetski 1978, Romer 1968), but without microscopic examination the fossils would hardly have been recognized. One site is a small nearshore island in the Baltic Sea and the adjacent coast as far as Leningrad. The second site includes the Harding Sandstone Formation, extending from Arkansas to Montana. The more widespread Silurian fossils are similar to those of the Ordovician but are occasionally articulated and recognizable as fish. The most diverse assemblages of fossils are in the North American Silurian (Box 6–1).

In addition to bone, these Silurian organisms possessed another innovation, the switch from a ciliary mode to a muscular pumping mode of filter feeding. Ciliary filter feeding is common among invertebrates (Chapter 2). In these invertebrates (a

Box 6–1. Reconstructing the First Vertebrates

Although upper Cambrian hydroxyapatite fragments are accepted as vertebrate remains, articulated pieces of early vertebrates which give us a basis for reconstructing the organisms are very rare. No complete individuals are known until the Silurian/Devonian boundary. Only three geological formations, one each in Australia, North America, and South America, have yielded partial vertebrates of Ordovician age. The reasons seem clear. These heterostracan vertebrates were externally armored with a large number of small closely fitting, polygonal bony plates. Very special and rapid conditions of burial in beds destined to suffer minimal distortion in the subsequent 470 million years are required to keep such small plates articulated. These required conditions are understandably rare. It appears that a few shallow coastal marine, perhaps even tidal flat, environments provided the required conditions. Recently two attempts at complete reconstructions of middle and late Ordovician vertebrates have been made (Ritchie in Rich and van Tets 1985, Elliott 1987). Both *Arandaspis* from central Australia and *Astraspis* from the eastern slopes of the Rocky Mountains were 13 to 14 centimeters long and had symmetrical tails (Figure 6–2). They were completely encased in small ornamented plates, each plate 3 to 5 millimeters in maximum dimension. Although the scales abutted one another in the head and gill region, from about midbody posteriorly they overlap as do scales in more recent fishes. These bony plates show specializations for sensory canals, special protection around the eye, and in the reconstruction of the North American specimen, as many as eight gill openings on each side of the head.

If we take these characters to be primitive for craniates, then the large head shield plates, single gill opening, and hypocercal tail of later heterostracans must be derived characteristics of this jawless fish radiation. Both reconstructions have well-developed eyes. If the eyes of hagfish are not secondarily degenerate, those paradoxical members of the modern fauna may have ancestors that predate the reconstructed Ordovician fishes. Full description of the gill region in the Australian and South American fossils and the rostral and mouth region of any of these Ordovician vertebrates will prove to be most interesting.

Figure 6–2. Reconstructions of two Ordovician vertebrates: (a) *Arandaspis* from Australia; (b) *Astraspis* from North America. [(a) Modified from R. V. Rich and G. F. van Tets, 1985, *Kadimkara*, Pioneer Design Studio, Lilydale, Victoria, Australia; (b) from D. K. Elliott, *Science* 273:190–192.]

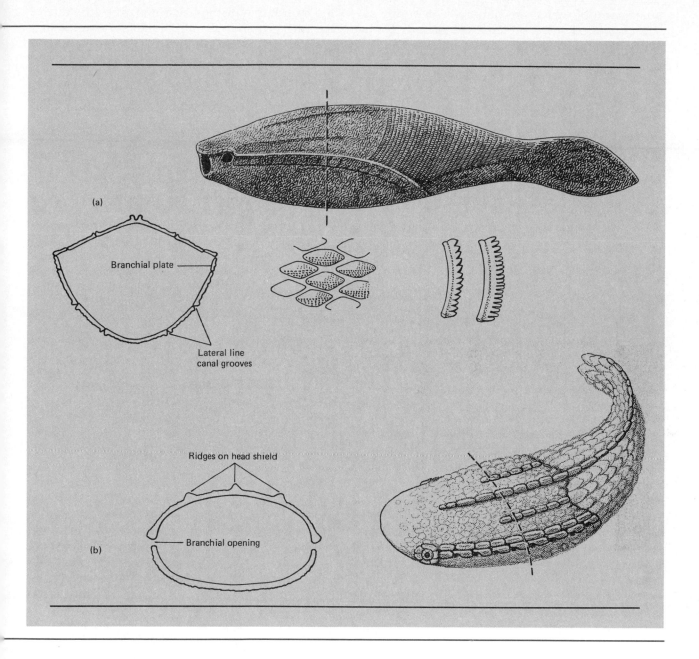

(a)

Branchial plate

Lateral line
canal grooves

Ridges on head shield

Branchial opening

(b)

great many of which are nonmotile or nearly so) water is wafted past food-snaring structures by the activity of great numbers of ciliated cells. Cilia, although they can move considerable volumes of water rapidly, are not especially effective at moving volumes a distance away from the incurrent opening. Food-rich water must be very close to a ciliated filter feeder. Living jawless vertebrates, however, suck water into the pharynx using muscles to contract the oral chamber or to move the tongue and fleshy tentacles or hood surrounding the mouth in a piston-like action. The structure of the pharynx and gills of Silurian vertebrates suggests that they did the same. This novel method of moving water for feeding as well as for respiration is not subject to the same restrictions as ciliary pumping. Water is moved in rapid, powerful pulses and considerable suction can be exerted to engulf large particles even if they are at some distance from the oral opening. This feeding mode may have permitted larger, more active organisms to evolve. Most early vertebrates were less than 20

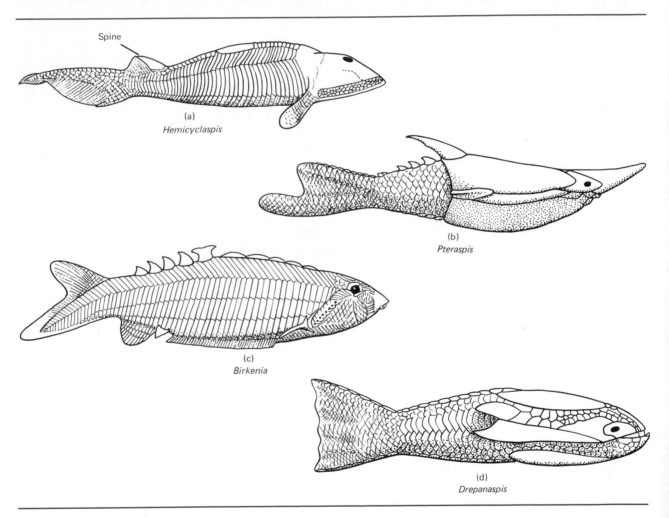

Figure 6–3. Four representative "ostracoderms." Cephalaspids [= Monorhina, (a) and (c)]; Pteraspids [= Diplorhina, (b) and (d)].

centimeters long, but this represents a considerable size increase compared to the average invertebrate. Living invertebrate suspension feeders have very much higher flow rates than those of most suspension-feeding vertebrates, but the motile vertebrates may have moved to areas, often ephemeral, where food is highly concentrated (Mallatt 1984).

Earliest Known Vertebrates

The earliest vertebrates were aquatic animals collectively called **ostracoderms** (*ostrac* = shell, *derm* = skin). At least two distinct groups, each with recognizable subdivisions, are among these remarkable life forms of the Paleozoic (Figure 6–3). The major ostracoderm groups are the **Pteraspida** (*ptera* = wing, *aspid* = shield) and the **Cephalaspida** (*cephal* = head).

Ranging from about 10 centimeters in most species to more than 50 centimeters in length for a few forms (Figure 6–3), the ostracoderms lacked jaws, although some of them may have had peculiar movable mouth parts not found in any other vertebrates. Typically, their mouths were fixed circular or slit-like openings that appear to have functioned as intakes that filtered small food particles from the water or from bottom detritus. They also lacked paired appendages with any structural similarity to those of other vertebrates. Their respiratory apparatus consisted of a variable number of separate pharyngeal gill pouches that opened along the side of the head independently or through a common passage. The notochord was the main axial support throughout adult life. No vertebrae have ever been found in ostracoderms. Because ostracoderms and the living cyclostomes (hagfishes and lampreys) share these characteristics, they have long been placed together in the class Agnatha as representatives of a primitive level of vertebrate organization. However, because no useful information about interrelationships is contained in shared primitive characters, this taxon is in considerable dispute. Much remains to be learned about the pattern of early vertebrate evolution. As we shall see, the ostracoderms were

specialized in the development of their bony armor, and the cyclostomes were specialized for burrowing or parasitic/predatory modes of existence.

Figure 6–4 shows some possible modern interpretations of the relationships of early vertebrates. If lampreys are the sister group to the Galeaspids + Osteostracans + Anaspids and Gnathostomes (solid lines), the similarities often drawn between lampreys and certain anaspids (i.e., *Jaymoytius*) are superficial. If they are closely related (dashed lines), the characters of true paired fins and an endoskeleton that extends posteriorly to the tail must have been lost by lampreys. Similarly, if anaspids and lampreys are the sister group of osteostracans, the evolution of these anaspid features must have paralleled their evolution in gnathostomes. No resolution of these differing concepts of early vertebrate evolution is currently possible.

The Pteraspids (= Heterostracans, = Diplorhina) and the First Vertebrate Life

The earliest jawless vertebrates are found in the fossil record until the very end of the Devonian. Because of their curious shelled appearance, they are called Pteraspids in some classifications and the Heterostraci (*hetero* = different, *strac* = shell) in other classifications (Figure 6–5). They varied in size from 10 centimeters to 2 meters and were encased anteriorly by bony articulating pieces that extended to the anus in some groups (Moy-Thomas and Miles 1971). Posterior to the anus was a short, probably mobile tail covered by smaller, protruding barb-like plates. The head shell had an ornamented dorsal plate, one or more lateral plates, and several large ventral elements. The shell was composed of three distinct layers, and in all but two genera the shell appears to have continued to grow as the animal increased in size. Impressions on the inside of the dorsal plate suggest that the brain had two separate olfactory bulbs. Because it is assumed that these were connected with two separate nasal openings, these

Phylogenetic Relationships of the Craniata

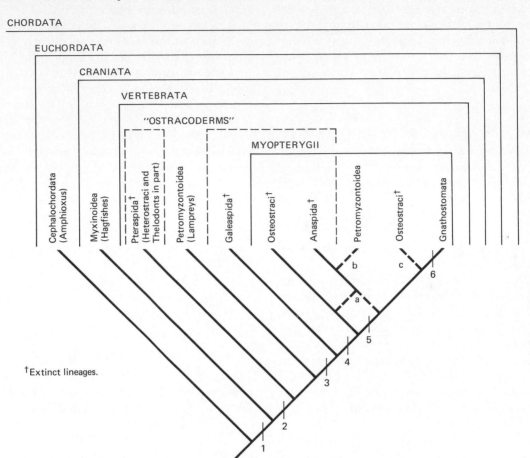

earliest vertebrates are grouped as the Diplorhina (*diplo* = two, *rhin* = nostril) in some classifications. Thus "Pteraspida," "Heterostraci," and "Diplorhina" are synonyms for these earliest known vertebrates.

Openings for two eyes are lateral, one on each side of the head shield. In the middle of the dorsal plate is a small opening for a third, median eye or **pineal organ**. The mouth is near the end of the body (terminal) but opens ventrally in many forms. The lower lobe of the tail is disproportionately large. This lower lobe contains the axial sup-

port element (the notochord), a tail fin construction called **hypocercal**. The body is generally round in cross section, like that of a tadpole, and early pteraspids show little sign of stabilizing projections. Possibly these fishes were erratic swimmers and resembled tadpoles by swimming with something less than precisely controlled locomotion. While feeding they may have oriented head down and plowed their jawless mouth through the bottom sediments.

As might be guessed, evolutionary trends in the pteraspids led to the improvement of loco-

1. Craniata: Neural crest cells, coelom formed by split in unsegmented lateral plate, highly differentiated somites, gills supported by a distinctive skeleton, a distinct head region with the following characters: tripartite division of brain with cranial nerves differentiated from neural tube, segmental nerves, paired optic, auditory, and probably olfactory organs, one or more semicircular canals, cranium incorporating the anterior end of the notochord and enclosing brain and paired sensory organs. In other regions of the body are a system of distinctive endocrine glands, lateral line system, probable electrosensitivity, well-developed heart, paired kidneys, and at least 15 additional derived characters.   2. Vertebrata: presence of arcualia that fuse with additional elements in the adults of most forms to produce vertebrae, physiological capacity to form bone in the dermis, two or three semicircular canals, eyes developed to essentially the modern level of sophistication, and 20 additional derived characters. 3. Perichondral bone (at least in head), cellular dermal bone.   4. Myopterygii: Heterocercal tail, paired fins with internal musculature developed from a lateral plate that extends from behind the gills to the region of the cloaca. 5. (Characters tentative pending resolution of the position of lampreys.) Lateral skeletogenous septum and the resultant unique trunk segmentation, endoskeletal fin radials and dermal fin rays at least in the tail, posterior paired fins in position of pelvic appendages.   6. Gnathostomata: Jaws formed of bilateral palatoquadrate (upper) and mandibular (lower) cartilages of the mandibular visceral arch at some stage of development, teeth containing dentine, modified hyoid gill arch, branchial elements internal to gill membranes, branchial arches contain four elements on each side plus one unpaired ventral median element, paired trabeculae contribute to cranium, three semicircular canals, internal supporting girdles associated with pectoral and pelvic fins, myelinated nerve fibers, calcium carbonate statoconia or otoliths, and 34 additional derived characters. (Based on J. G. Maisey, 1986, *Cladistics* 2:201–256.)

Figure 6–4.   Phylogenetic relationships of the Craniata. This diagram shows the probable relationships among the major groups of craniates. Extinct lineages are marked by a dagger (†). The numbers indicate derived characters that distinguish the lineages. The best corroborated relationships are shown by solid lines, the dashed lines labeled (a) and (b) show two possible alternative relationships for lampreys, and the dashed line labeled (c) shows an alternative relationship for the osteostracans. The quotation marks indicate that "ostracoderms" is a paraphyletic group.

motor capabilities. During their later history, the cross section of benthic species flattened ventrally but remained arched or rounded dorsally (Figure 6–5). The head shield developed solid lateral wing-like stabilizing projections called **cornua** (horns). The head shield shortened, and except for a dorsal ridge of plates the bony covering became restricted to the anterior end. Although specialized edges around the mouth for biting and grasping did not evolve, some of the oral plates developed enlarged tooth-like projections that may have been used for scraping.

The pteraspid lineage also gave rise to species of bizarre appearance. Some developed enormous cornua that may have acted as water-planing surfaces (hydrofoils) to produce lift for the heavy head when swimming. In addition, some had two narrow sled-like runners on the ventral surface, which presumably held the head above the substrate. Several forms developed dorsally directed mouths and a much reduced skeleton. Perhaps these species fed at the surface by filtering large quantities of plankton-rich water. Forms are also known with a long tooth-bearing projection extending from the

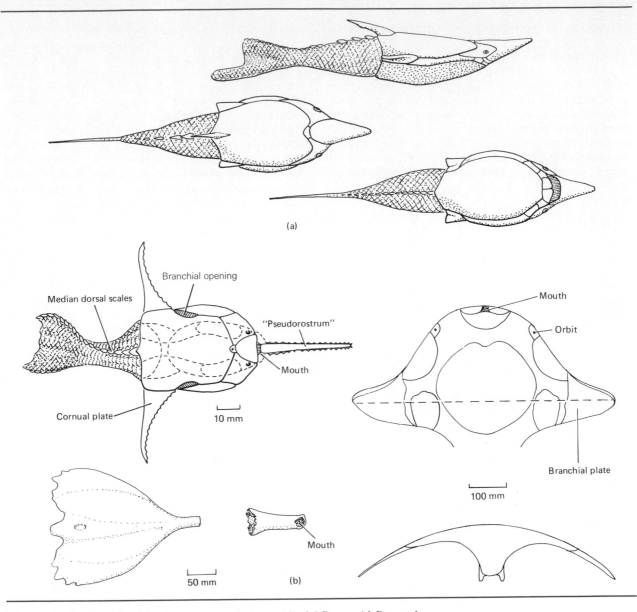

Figure 6–5. Details of the bony armor of pteraspids. (a) Pteraspid *Pteraspis
rostrata,* a typical, relatively unspecialized species. Reconstruction of an
18-centimeter specimen from the lower Devonian. Lateral view above, dorsal
surface left, ventral surface right. (b) Specialized pteraspids (in dorsal view).
Clockwise from the upper left: a sawfish-like form *Doryaspis,* lower Devonian; form
with enlarged cornua (horns) and ventral runners (cross section of head region
below), *Pycnosteus,* middle Devonian; and eyeless, tube-snouted form *Eglonaspis,*
lower and middle Devonian. (Modified after J. A. Moy-Thomas and
R. S. Miles, 1971, *Paleozoic Fishes,* W. B. Saunders, Philadelphia.)

edge of the mouth, rather like that of the living sawfish. The function of this rostrum is difficult to understand because the mouth was dorsal to the saw, just opposite to morphologically similar forms among today's vertebrates. Perhaps the saw was used to stir up organisms from the bottom.

About the middle of the Silurian, when the diverse array of pteraspids entered the fossil record, another distinct but poorly known lineage of jawless vertebrates appeared. Rather than large plates, these small fish (Figure 6–6) were covered by numerous tiny **denticles**, small tooth-like structures not unlike those of living sharks. Isolated denticles from the Ordovician may belong to these fishes. No articulated specimens are known before the mid-Silurian, and by the mid-Devonian they were extinct. These vertebrates were fusiform (torpedo shaped), with ridges that correspond to the dorsal and anal fins in more advanced fishes and a hypocercal tail. In addition, broad-based flanges projected from their sides where anterior paired appendages occur in later vertebrates. Like the pteraspids, they had the primitive vertebrate characters of lateral eyes, a pineal opening, and a jawless mouth. Several names have been given to this group of fishes, all referring to characteristics of the scales: **Thelodonti** (*thelo* = nipple, *dont* = tooth) and **Coelolepida** (*coel* = hollow, *lepida* = scale) are the names most commonly used.

Coelolepids were between 10 and 20 centimeters long. They were dorsoventrally flattened anteriorly and laterally compressed posteriorly. Living fishes with this shape feed by swimming along the bottom of the sea. Coelolepids, therefore, may have fed like the pteraspids by skimming organic deposits off the bottom into their small mouths with the aid of muscular suction. Apparently, coelolepids were most numerous in coastal estuaries, but eventually also radiated into fresh water. Where they occurred together, the small, lightly armored coelolepids were probably behaviorally very different from the larger, ar-

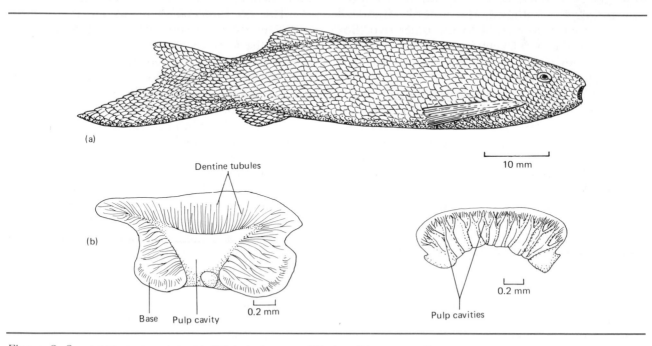

Figure 6–6. (a) Typical coelolepid, *Phlebolepis*, upper Silurian; (b) cross sections of two types of coelolepid scales. (Modified after J. A. Moy-Thomas and R. S. Miles, 1971, *Paleozoic Fishes*, W. B. Saunders, Philadelphia.)

Box 6–2 (Continued)

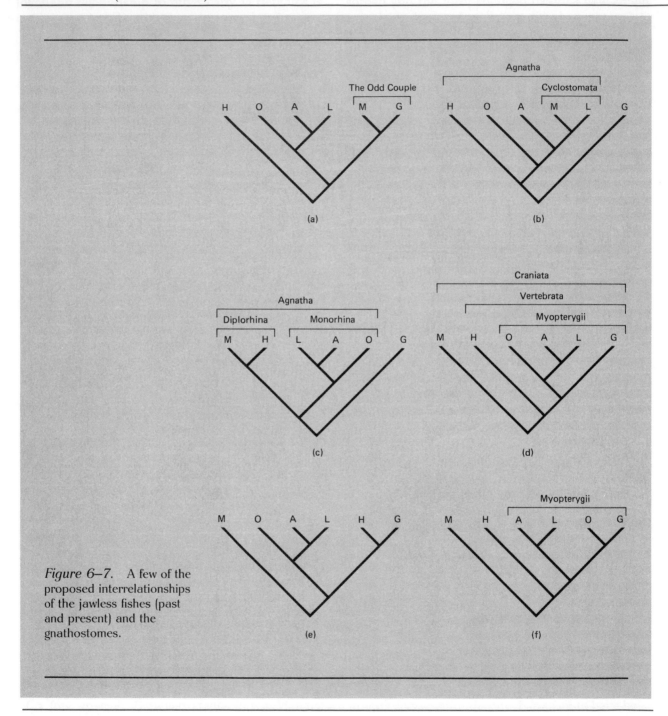

Figure 6–7. A few of the proposed interrelationships of the jawless fishes (past and present) and the gnathostomes.

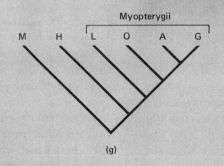

Myopterygii

M   H   L   O   A   G

A — Anaspids
G — Gnathostomes
H — Heterostracans
L — Lampreys
M — Myxinids (hagfish)
O — Osteostracans

Figure 6–7 (Continued)   (g)

there is little evidence that hagfish characters are degenerate rather than primitive. The latter authors derive hagfish from the base of the evolution of craniates, sometimes expressly identifying them as the sister group of vertebrates but not themselves vertebrates (Figure 6–7d).

Focusing on the lamprey side of living jawless-fish relationships, most authors accept an anaspid/lamprey sister group arrangement but differ on whether these two are closely related to osteostracans (Figure 6–7d and e) or whether the osteostracans are closest to the gnathostomes (Figure 6–7f). This judgment depends entirely upon which characters one values and which direction of the character state is considered to be derived in making interpretations. As an example, if acellular bone represented by gnathostome dentine is considered a derived condition from primitive cellular bone, then, contrary to almost every other proposed arrangement, heterostracans become the sister group of gnathostomes since *all* of their mineralized tissue is acellular. A recent entry into the your-guess-is-as-good-as-mine field of attempting to determine interrelationships of jawless fishes and gnathostomes (Figure 6–7g; Maisey 1986) sees anaspid/lamprey relationships as equivocal but offers the possibility of an anaspid/gnathostome sister relationship. We cautiously accept this view in our phylogeny (Figure 6–3).

In 1889, American paleontologist Edward Drinker Cope wrote: "We are embarrassed in the endeavor to present the relations of the earliest and lowest Vertebrata by want of knowledge of their structure . . ." (see Forey 1984). The embarrassment continues.

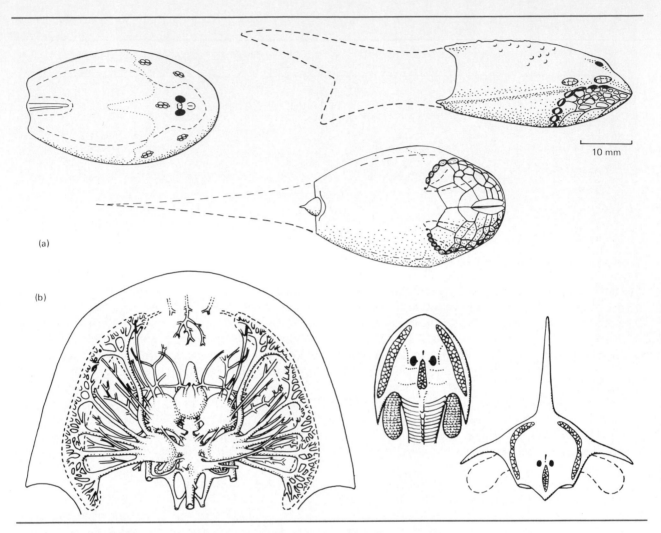

Figure 6–8. Details of osteostracans. (a) Primitive osteostracan, *Tremataspis*, upper Silurian, in lateral (right), dorsal (left), and ventral (below) reconstructions. (b) More specialized osteostracans (l-r): reconstruction of the brain and cranial nerves of *Kiaeraspis*, lower Devonian, showing the detailed information obtainable from impressions left on the inner surface of the head shield; *Tyriaspis* of the upper Silurian and the long rostrum *Boreaspis* of the lower Devonian. (Modified primarily after J. A. Moy-Thomas and R. S. Miles, 1971, *Paleozoic Fishes*, W. B. Saunders, Philadelphia.)

positioned *anterior* to the otic capsule, and the rest were concentrated compactly in the posterior part of the head shield. Astonishingly, the internal anatomy of the brain and nervous system of osteo-stracans 425 million years old is very similar to that found in the modern lamprey.

Along the dorsolateral edges of the head shield, and sometimes in the center behind the pineal opening, are peculiar fields of thin, irregular, and separate small plates (Figure 6–8). These fields form depressions connected to the cranial cavity by huge canals that run through the shield and into the inner ear. Whether these were forerunners of the lateral line, electroreceptors, or even electrogenic organs, as the polygonal nature of the small plates has suggested to some, is unknown. The osteostracans became abundant and diverse during the Devonian, even though competing with older groups and surviving in the presence of jawed vertebrates. In part, their success may relate to these mysterious, unique adjuncts to their nervous system.

An apparently geographically isolated group of early Devonian jawless fishes has come to light during the past 15 years in southern China. A large diversity of these osteostracan look-alikes have been described as the **Galeaspida** (*gale* = a helmet). They differ from the osteostracans in the large slit-, bean-, or even heart-shaped opening on the dorsal surface of their head shield. This opening connected with the pharynx and may have functioned as an inhalant canal. The exact interrelationships of the galeaspids is not yet clear (Janvier 1984).

In sediments from the late Silurian through the Devonian a third group of cephalaspids, the **anaspids** (*an* = without), are found. All about 15 centimeters long, these freshwater fishes had minnow-like body proportions (Figure 6–9) resembling the probably unrelated coelolepids. Like their osteostracan relatives, the anaspids had a single median nasal opening anterior to the pineal foramen. Narrow scale rows (when they were present) covered the body in a manner similar to those along the posterior part of the osteostracans, but the flat scales were constructed of acellular layers. The head, however, was covered in most species by a complex of small plates or was naked. Anaspids also differed from the osteostracans in having a hypocercal tail. They are considered to have been bottom detrital feeders that fed in a head-down position reminiscent of that proposed for the heterostracans. Their stabilizing dorsal, anal, and lateral projections or folds, the spines

and scutes associated with these projections, and the compressed shape of their fusiform bodies probably allowed an agility and locomotor capacity not known in the heterostracans or osteostracans.

During the late Silurian and Devonian all known extinct jawless vertebrates coexisted (Figure 6–10). Whatever the nature of the pre-heterostracan vertebrate, we can point to the evolution of muscular filter-feeding coupled with the rewards of increased mobility and the protection that dermal bone afforded as important characters. Together, these features triggered a proliferation of variations on the vertebrate theme that spread into the waters of the world. Wherever photosynthesis gave rise to small particulate matter capable of being sucked up and digested, vertebrates competed successfully with the established invertebrate lineages. This basic agnathan body plan also gave rise to eyeless, tube-snouted heterostracans, and nearly naked, sucker-mouthed anaspids that left their sucking marks on other organisms.

Here we observe for the first time a phenomenon repeated over and over again in the history of vertebrate life: A basic modification of the vertebrate framework appears, and a flood of forms using this new modification in conjunction with specializations explodes onto the scene. From a generalized form, vertebrate life radiates in scores of directions to exploit the resources that this innovation has made available.

Living Agnathans

Two distinct groups of recent fishes are jawless and have scarcely any fossil record. Two lampreys, *Hardistiella* (from Montana) and *Mayomyzon* (from Illinois), are known from the Carboniferous, and an undescribed hagfish (Carroll 1987) and a second possible relative (Janvier 1981) have been found in the same deposits as *Mayomyzon*.

Living agnathans have primitive characters that harken back to the origin of vertebrates: they lack jaws and have no paired appendages to aid them in their locomotion. Some, however, are unlike any other fish, living or fossil, in being spe-

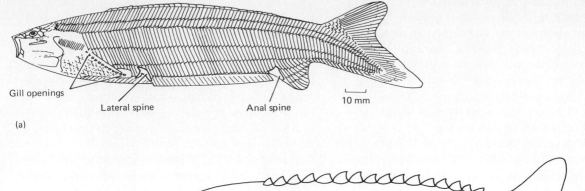

Gill openings

Lateral spine

Anal spine

10 mm

(a)

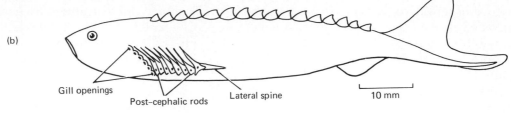

(b)

Gill openings

Post-cephalic rods

Lateral spine

10 mm

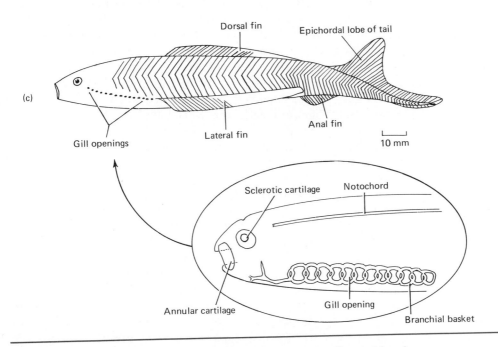

Dorsal fin

Epichordal lobe of tail

(c)

Gill openings

Lateral fin

Anal fin

10 mm

Sclerotic cartilage

Notochord

Annular cartilage

Gill opening

Branchial basket

Figure 6–9. Reconstruction of upper Silurian fishes generally considered anaspids. (a) *Pharyngolepis*; (b) *Lasanius*; (c) *Jamoytius* showing (inset) internal structures known from the head region. (Modified after J. A. Moy-Thomas and R. S. Miles, 1971, *Paleozoic Fishes*, W. B. Saunders, Philadelphia.)

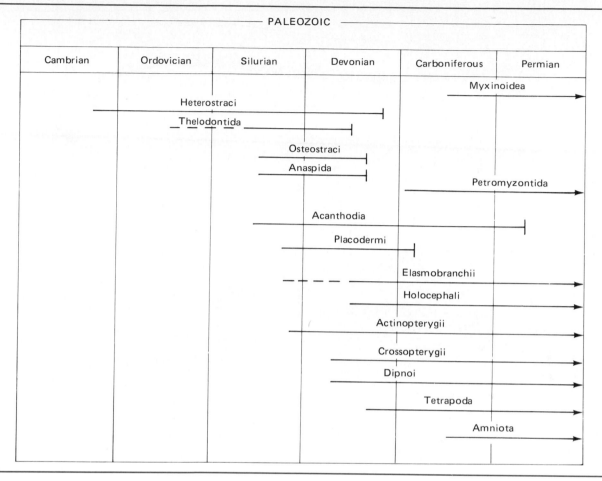

Figure 6–10. Occurrence in the fossil record of the Paleozoic vertebrates.

cialized as obligate ectoparasites of other verte-brates. Because they possess round jawless mouths these fishes have often been combined in the **Cyclostomata** (*cyclo* = a circle, *stoma* = mouth), lampreys in the Petromyzontidae and hagfish or slime-hags in the Myxinidae. The groups show such great differences in morphology as a result of their long phylogenetic separation and their different habits and habitats that we place them in separate taxa of questionable rela-tionship (Figure 6–3).

Hagfishes

The hagfishes (Figure 6–11) are entirely marine. The 15 to 20 recognized species in five or six genera have nearly worldwide distribution, primarily on continental shelves (Brodal and Fange 1963). Hag-fishes are never caught much above the bottom, often in deep water. Some live in colonies, each individual in a mud burrow with a volcano-like mound at the entrance. Burrowing polychaete worms form the most common item in the diet of

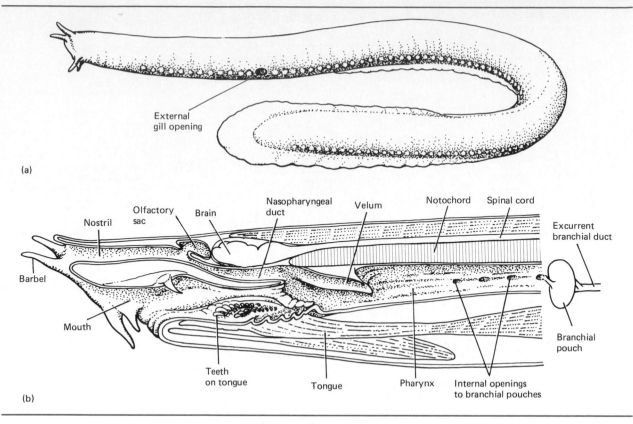

Figure 6–11. Hagfish: (a) lateral view; (b) sagittal section of the head region. (Modified from D. Jensen, 1966, *Scientific American* 214(2):82–90).

many species, and they probably live a mole-like existence finding their prey beneath the ooze. They must be active when out of their burrows, for they are quickly attracted to bait and moribund fishes caught in gill nets. Small morphological differences between populations indicate that hagfish are not wide ranging, but rather tend to live and breed locally.

Elongated, scaleless, and pinkish to purple in color, hagfishes have a single terminal nasal opening that connects via a broad tube directly with the pharynx. The eyes are degenerate and covered with a thick skin. The mouth is surrounded by six tentacles supported by cartilage that can be radially erected and swept to and fro by movements of the head when the hagfish is searching for food

in the open. Within the mouth two multicusped, horny plates border the sides of a protrusible tongue-like structure. These plates spread apart when protruded and fold together, the cusps interdigitating in a pincer-like action when retracted. The feeding apparatus of hagfish has been described as "extremely efficient at reeling in long worms," as the keratin plates alternately flick in and out of the oral cavity (Mallatt 1985). When feeding on something the size and shape of another fish, hagfishes concentrate their pinching efforts on surface irregularities, such as the gills or the anus, where they can more easily grasp the flesh. Once attached, they tie a knot in their tail and pass it forward along their body until they can brace themselves against their prey and tear off the flesh in their pinching grasp. Hagfishes take only dead or dying vertebrate prey and they often begin by eating only enough flesh to enter the coelomic cavity, where dining on soft parts is possible. Once a food parcel reaches the hagfish's gut it is enfolded in a mucoid bag secreted by the gut wall. This membrane is permeable to digestive enzymes and digestate and is excreted as a neat wrapper around the feces. No functional significance is known for this curious feature.

Different genera and species of hagfishes have variable numbers of external gill openings. From 1 to 15 openings occur on each side (Figure 6–11), but they do not correspond to the number of internal gills. The external openings occur as far posterior as the midbody, although the pouch-like gill chambers are just posterior to the head. The long tubes leading from the gills fuse to reduce the number of external openings to which they lead. The posterior position of the gill openings may be related to burrowing.

The internal anatomy of hagfishes also is peculiar. They have no vertebral anlage, very primitive kidneys, and only one semicircular canal on each side of the head. This last character has been the subject of much technical debate. Northcutt (1985) has pointed out that in spite of what the membranous structures look like, neurologically they are double sensory patches just as in lampreys, which have two semicircular canals. Hagfishes have long been thought to lack a lateral-line system. Recent work suggests that *Eptatretus* has traces of the system, but whether this is a primitive condition or a secondary reduction of ancestrally well-developed structures is not known.

Opening through the body wall to the outside are large mucous glands that secrete enormous quantities of mucus and tightly coiled proteinaceous threads. The latter straighten on contact with seawater to entrap the slimy mucus close to the hagfish's body. This obnoxious defense mechanism is apparently a deterrent to predators, but also seems to be less than appreciated by the hagfish. When danger is past a hagfish draws a knot and squeezes out of the mass of mucus, then sneezes sharply to blow its nasal passage clear.

In contrast to all other vertebrates, hagfishes have accessory hearts in the caudal region in addition to the heart near the gills. They have capacious blood sinuses and very low blood pressure. In addition, the several hearts of the hagfishes are aneural, meaning that their pumping rhythm is intrinsic rather than coordinated via the central nervous system. The blood vascular system demonstrates few of the immune reactions characteristic of other vertebrates, and its osmotic concentration is approximately the same as that of seawater (see Chapter 4). Examination of the gonads suggests that at least some species are hermaphroditic, but nothing is known of mating. The eggs are oval and over a centimeter in length. Encased in a tough clear covering, the yolky eggs are secured to the sea bottom by hooks and hatch into small, completely formed hagfish, thus bypassing any larval stage. Adult hagfish are generally under a meter in length.

Lampreys

Although similar in size and shape to hagfishes, the 30 species in nine genera of lampreys (Figure 6–12) are in other respects radically different from hagfish (Hardisty and Potter 1971). All but the most specialized lampreys are **anadromous**; that is, they ascend rivers and streams to breed. Some of the most specialized species are known only from fresh water and have abbreviated adult lives that function solely as reproductive stages inasmuch as the adults neither feed nor migrate. Lampreys have a worldwide distribution except for the tropics and high polar regions. Anadromous species that spend some of their life in the sea attain the greatest size, although 1 meter is about the upper limit. The smallest species are less than one-fourth of this size.

Little is known of the habits of adult lampreys because they are generally observed only during reproductive activities or when captured with their host. They attach to the body of another vertebrate by suction and rasp a shallow, seeping wound through the integument of the host with horny spines and the action of a protrusible tongue-like

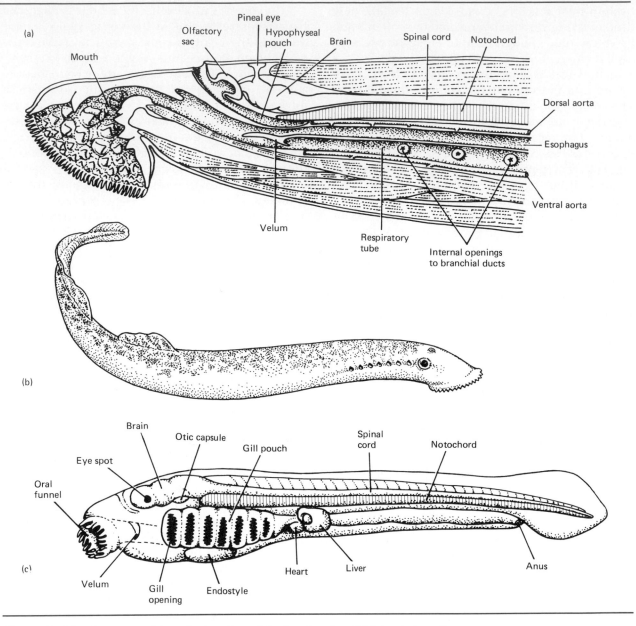

Figure 6–12. Lampreys: (a) sagittal section of the head region; (b) lateral view of an adult; (c) larval lamprey (ammocoete).

structure. In captivity, lampreys swim sporadically with exaggerated, rather awkward lateral undulations. The single nasal opening is high on the head and continues as a blind-end tube beneath the brain in close proximity to the pituitary gland. The eyes are large and well developed as is the pineal body, which lies under a pale spot just posterior to the nasal opening. In contrast to the hagfishes, lampreys have two semicircular canals on each side of the head, a condition shared with the extinct osteostracan agnathans. The nerves, which exit segmentally along the length of the spinal cord, are completely separated into dorsal and ventral nerve roots, a lamprey peculiarity shared by no other living vertebrate. In addition, the heart is not aneural as in hagfishes, but is innervated by the parasympathetic nervous system. In lampreys these nerves cause cardiac acceleration and not deceleration as do the neural regulators of heart rate in all other vertebrates.

Despite their anatomically well-developed senses, no clear picture has emerged of how a lamprey locates or initially attaches to its prey. The round mouth and tiny esophagus are located at the bottom of a large fleshy funnel the inner surface of which is studded in a species-specific pattern with horny conical spines. The protrusible tongue-like structure is covered with similar keratin spines, and together these structures allow tight attachment and rapid abrading of the host's integument. An oral gland secretes an anticoagulant. Feeding is probably continuous when a lamprey is attached to its host.

Lampreys generally do not kill their hosts but detach, leaving a weakened animal with an open wound. At sea, lampreys have been found feeding on several species of whales and porpoises in addition to fishes. Swimmers in the Great Lakes, after having been in the water long enough for their skin temperature to drop, have reported initial attempts by lampreys to attach to their bodies. The bulk of an adult lamprey's diet consists of body fluids of fishes. The digestive tract is reduced, as is appropriate for an animal that feeds on such a rich and easily digested diet as blood and tissue fluids.

Seven pairs of gills open to the outside just behind the head. Chloride cells in the gills and well-developed kidneys regulate ions, water, and nitrogenous wastes and maintain the osmolality of the body fluids, allowing the lamprey to exist in a variety of salinities. Female lampreys produce hundreds to thousands of eggs about a millimeter in diameter and devoid of any specialized covering such as that found in the hagfish. Like the hagfish, however, lampreys have no duct to transport the specialized products of the gonads to the outside of the body. Instead, the eggs and sperm fill the coelom, and contractions of the body wall expel them from pores located near the openings of the urinary ducts.

The spawning of lampreys is complicated. After temperature-triggered migration to the upper reaches of streams where the current flow is moderate and the streambed composed of cobbles and gravel, the lampreys construct nests to receive the spawn. Males, later joined by females, select a site, attach themselves by their mouths to the largest rocks in the area, and thrash about violently. Smaller rocks are thus dislodged and carried a short distance by the current. The nest is complete when a pit is rimmed upstream by large stones and downstream by a mound of smaller stones and sand that acts as an eddy producer. Water in the nest is well oxygenated by turbulence and does not flow strongly in a single direction. The weary nest builders spend the last of their energy depositing eggs—a process that may take 2 days. The female attaches to one of the upstream rocks laying eggs and the male wraps around her, fertilizing them. At last the female exhausts herself and dies, and the male dies a day or two later.

The larvae hatch in about 2 weeks. Radically different from their parents, they were originally described as a distinct genus, *Ammocoetes* (Figure 6–12). This name has been retained as a vernacular name for the larval form. A week to 10 days after hatching, the tiny 6- to 10-millimeter-long ammocoetes leave the nest. They are pink, worm-like organisms with a large fleshy oral hood and nonfunctional eyes hidden deep beneath the skin. Currents carry the ammocoetes downstream to backwaters and quiet banks, where they burrow into the soft mud and take up a sedentary filter-feeding life for 3 to 7 years. The protruding oral hood funnels the muscularly pumped water through the pharynx, where food particles are entrapped in mucus and subsequently swallowed. If conditions permit, the ammocoete may spend its entire larval life in the same burrow without any major morphological or behavioral change until it is 10 centimeters or more in length and several years old. Metamorphosis into the eyed, silver-gray, and usually parasitic stage begins in mid-summer, but downstream migration may not occur until the following spring. Adult life is usually no more than 2 years, and many lampreys return to spawn after one year in a lake or at sea. Some nonparasitic lamprey species transform and leave their burrows to spawn immediately and die.

During this century man and the lamprey have increasingly become at odds. Although the sea lamprey, *Petromyzon marinus*, seems to have been indigenous to Lake Ontario, it was unknown from the other Great Lakes of North America before 1921. The St. Lawrence River flowing from Lake Ontario proved no barrier to colonization by lampreys, and the rivers and streams that fed into Lake Ontario provided acceptable conditions for landlocked populations to develop. For some reason until the 1920s lampreys presented little problem, and their populations remained at a low level. It is known that by slowly creeping upward using their sucking mouth, lampreys can negotiate a waterfall at least 2 meters high during spawning migrations as long as the flow of water is not strong. The 50-meter height of Niagara Falls was, however, too much for the most amorous lamprey. The Welland Canal connecting Lakes Erie and Ontario and bypassing the falls was opened in 1829, but it was not until 100 years later that lampreys were known to be spawning successfully above the falls in Lake Erie's drainage basin.

Since the 1920s lampreys have expanded rapidly across the entire Great Lakes basin. The surprising fact is not that they were able to invade the upper Great Lakes but that it took them so long to initiate the invasion. Environmental conditions that vary between the lakes may provide the answer to this curious delay. Lake Erie is the most eutrophic and warmest of all the lakes and has the least appropriate feeder streams. Many of these streams run through flat agricultural lands that have been under intensive cultivation since early in the nineteenth century. The streams are highly silted and frequently have had their courses changed by human activities. Because of the terrain, flow is slow and few rocky or gravel bottoms occur. Perhaps lampreys simply could not find appropriate spawning sites in Lake Erie to develop a strong population.

Once they reached the upper end of Lake Erie, however, they quickly gained access to the other lakes. By 1946 they were known from all the Great Lakes. There they found suitable conditions and were able to expand unchecked until sporting and commercial interests became alarmed at the reduction of economically important fish species, such as lake trout, burbot, and lake whitefish. Chemical lampricides, electrical barriers at the mouths of spawning streams, and mechanical weirs at similar sites have all been employed to bring the Great Lakes lamprey populations down to their present level. Although the populations of large fish species, including those of commercial value, are recovering, it may never be possible to discontinue these antilamprey measures, costly though they are. Human mismanagement (or initial lack of management) of the lamprey has been to our own disadvantage. The story of the demise of the Great Lakes fishery is but one of hundreds in the recent history of vertebrate life where human failure to understand and appreciate the interlocking nature of the biology of our nearest relatives has led to gross changes in our environment.

What Does "Primitive" Mean?

From the description of the biology of the lamprey and hagfish, it is obvious that they are successful species. The two groups are moderately speciose, have wide geographic ranges, can be very numerous in certain locations, and show considerable adaptability in food sources (hagfish) or habitats (lampreys). The lampreys have invaded habitats previously populated by much more derived forms of fishes. The lampreys have changed little since the Carboniferous: The hagfish and lamprey have the largest accumulation of verifiably antique features of any living vertebrates. By any criterion they are the most primitive living fishes, yet they are not pathetic remnants, they are demonstrably competitive and successful. What, then, does "primitive" mean?

A primitive character may be identified in any aspect of the biology of an organism—morphology, physiology, or behavior. The primitive character state is recognized as one that is similar to (or believed to be similar to) that in the ancestors of the taxon under consideration. In fact, it is imprecise to call any species or taxon primitive or derived because all species are mosaics of primitive and derived characters. In general, the usage "primitive species" refers to those which retain numerous primitive features. Likewise, "advanced species" would be those with significant derived features. The primitive species we have been considering are, of course, modern fishes that have retained an array of primitive characters. The very phenomenon of their retention through millions of years points at the least to the neutral selective value of the characters and more probably to their positive advantage in the way of life these fishes have pursued. The antiquity of a character does not diminish its selective value. Presumably a derived character, if carrying out the same function in a very similar but superiorly efficient way, will replace a primitive character through competition between populations. When a derived character opens up new or different functional possibilities, the directness of competition is blunted and may not even exist. Hence multiple populations may coexist using the resources of the environment in different, noncompetitive ways—one population dependent on the function of a derived character state, another utilizing the primitive character perhaps in much the same way as the common ancestor of both. The primitive character state is not inferior or necessarily less adaptive, it is simply older.

The terms "primitive" and "derived" are relative and refer to the time of appearance of the character state in the history of the lineage. Thus a character that we would consider derived for the Paleozoic fishes we have been considering, such

as a tooth-bearing endoskeletal jaw, would be the primitive character state for mammals, which did not appear until the mid-Mesozoic. It is also important not to confuse the terms "primitive" and "derived," which refer to temporal appearance in a lineage, with the terms "generalized" and "specialized," which refer to the *functional* aspects of the characters of species. Generalized features of an organism are those with the ability to perform a variety of functions. Specialized features are those with a restricted range of functions. Thus, a primitive character could be either generalized (as is the segmentally arranged axial musculature of the lamprey) or specialized (for example, the jawless mouth and fleshy hood of these parasitic vertebrates). Likewise, a derived character can be generalized (the omnivorous dentition of humans is an example) or specialized (the sectorial or meat-slicing teeth of the cats). Frequently, "derived" and "specialized" are both used to describe a single character, but it should be understood that distinct criteria qualify a character for each of these designations. No selective value judgment is implied by the terms generalized or specialized. The environmental context determines whether a generalized or a specialized feature will be most effectively passed to subsequent generations.

Bone and the Early Vertebrates

Presumably, the radiation of the vertebrates began in the late Precambrian or early Cambrian, for ostracoderms possessing the suite of characters identifying vertebrates appear in the fossil record in the late Cambrian. These fishes lacked paired appendages and jaws found in later fossil remains, but they had a unique vertebrate character—bone. A bony skeleton greatly enhanced chances for fossilization and our record of vertebrate life begins with the remains of these tissues. The initial position of bone was in a dermal exoskeletal armor. This encasing shell appears to have caused the early vertebrates much the same problem as the marine crustaceans experience: how to grow within a mineralized skin. The radiation of the ostracoderms is the history of evolutionary experiments to solve the growth-in-a-suit-of-armor dilemma. From the distinctive nature of the various armors, some paleontologists feel that bone probably evolved separately in each major group of ostracoderms (Carroll 1987).

The earliest recorded solution was that of the pteraspids, where either large articulating plates did not form until maximum size was attained, or numerous centers of bone formation enlarged circumferentially to abut and fuse into plates or a solid shield only as the animal reached maximum size. All these forms achieve an essentially solid carapace over the anterior one-third of their body, pierced by the mouth and a single pair each of eyes and external gill openings. On the interior of this armor the paired impressions of bilateral nasal sacs can frequently be discerned. No bone is known to have formed in endoskeletal structures and no bone cells are found within the osseous tissue, nor were true paired appendages or well-developed dorsal or anal fins evolved.

Inadequately known because they are more lightly armored are the coelolepids of uncertain affinity. Although they had distinctive small separate scales, each with a pulp cavity like a tooth, the other characters established by the fossils are those considered primitive for vertebrates and thus of no value in determining coelolepid relationships. The mosaic of small, closely spaced scales that grow throughout life represents a second solution to the growth-in-armor problem.

The remaining bone-bearing ostracoderms are clearly interrelated on the basis of shared derived characters. The cephalaspids—both the benthic, heavily armored osteostracans with their nongrowing shield and the more pelagic, lightly armored anaspids with overlapping, marginally growing scales—have a single, median nasal opening associated with the pituitary. This is presumed to be a derived condition, a conclusion supported by the osteostracans' delayed appearance in the fossil record, as much as 100 million years after the appearance of the first vertebrate fossils. Some cephalaspids also independently evolved a partially ossified endoskeleton, bone-cell spaces within the ossified tissue matrix, and paired pectoral fins. These characters also evolved in jawed vertebrates, where they were greatly elaborated (Forey 1984).

Bone was not and is not a universal characteristic of vertebrates, as the living jawless vertebrates demonstrate. This avoidance of the growth-in-armor problem leaves little to fossilize. An enigma to the phylogeneticist, the lamprey and hagfish, whether derived from cephalaspids or not, are fascinating creatures whose many primitive features offer much fertile material for future research.

Summary

Fossil evidence indicates that vertebrates evolved in a marine environment during the Cambrian. We know little about the group until some forms evolved bony dermal armor. The evolution of bone, muscular pump filter feeding, and increased motility led to at least two distinct groups of early vertebrates. Earliest were the Pteraspida or Diplorhinae, also known as heterostracans; later the Cephalaspida or Monorhinae, typified by the osteostracans, appeared. The extensive radiation of these forms demonstrates the numerous successful solutions to growth inside an armored skin. Only two types of survivors from these early agnathan radiations exist today: the hagfishes and the lampreys. Nevertheless, living agnathans illustrate the extreme specialization of which the primitive vertebrate body plan is capable.

References

Brodal, A. and R. Fange (editors). 1963. *The Biology of Myxine*. Universitetsforlaget, Oslo, Norway. A compilation of data and references on what many zoologists consider the most primitive living vertebrates.

Carroll, R. L. 1987. *Vertebrate Paleontology and Evolution*. W. H. Freeman, New York. The most up-to-date and comprehensive review work on the subject. A modern treatment to follow Romer (1966).

Elliott, D. K. 1987. A reassessment of *Astraspis desiderata*, the oldest North American vertebrate. *Science* 237:190–192.

Forey, P. L. 1984. Yet more reflections on agnathan-gnathostome relationships. *Journal of Vertebrate Paleontology* 4(3):330–343.

Gorbman, A. and A. Tamarin. 1985. Early development of olfactory and adenohypophyseal structures of agnathans and its evolutionary implications. Pages 165–185 in *Evolutionary Biology of Primitive Fishes*, edited by R. E. Foreman et al. Plenum Press, New York.

Halstead, L. B. 1982. Evolutionary trends and the phylogeny of the agnatha. Pages 159–196 in *Problems in Phylogenetic Reconstruction*, Systematics Association Special Volume 21, edited by K. A. Joysey and A. E. Friddy.

Hardisty, M. W. and I. C. Potter (editors). 1971 to 1982. *The Biology of Lampreys*. Academic Press, New York. Four volumes survey much of what is known about these fascinating vertebrates.

Janvier, P. 1981. The phylogeny of the Craniata, with particular reference to the significance of fossil "agnathans." *Journal of Vertebrate Paleontology* 1:121–159.

Janvier, P. 1984. The relationships of the Osteostraci and Galeaspida. *Journal of Vertebrate Paleontology* 4:344–358.

Maisey, J. G. 1986. Heads and tails: A chordate phylogeny. *Cladistics* 2:201–256. A detailed review of the pertinent literature on nontetrapod chordates making well-balanced conclusions.

Mallatt, J. 1984. Feeding ecology of the earliest vertebrates. *Zoological Journal of the Linnean Society* 82:261–272.

Mallatt, J. 1985. Reconstructing the life cycle and the feeding of ancestral vertebrates. Pages 59–68 in *Evolutionary Biology of Primitive Fishes*, edited by R. E. Foreman et al. Plenum Press, New York.

Moy-Thomas, J. A. and R. S. Miles. 1971. *Paleozoic Fishes*. W. B. Saunders, Philadelphia. The classic, concise, and well-illustrated review of knowledge of early vertebrate evolution.

Northcutt, R. G. 1985. The brain and sense organs of the earliest vertebrates: reconstruction of a morphotype. Pages 81–112 in *Evolutionary Biology of Primitive Fishes*, edited by R. E. Foreman et al. Plenum Press, New York.

Olson, E. C. 1971. *Vertebrate Paleozoology*. Wiley-Interscience, New York.

Repetski, J. E. 1978. A fish from the upper Cambrian of North America. *Science* 200:529–531. Contains references to several reports on the earliest vertebrate remains.

Rich, P. V. and G. F. van Tets. 1985. *Kadimakara, Extinct Vertebrates of Australia*. Pioneer Design Studio, Lilydale, Victoria, Australia.

Romer, A. S. 1966. *Vertebrate Paleontology*, 3rd edition. University of Chicago Press, Chicago.

Romer, A. S. 1968. *Notes and Comments on Vertebrate Paleontology*. University of Chicago Press, Chicago.

Schaeffer, B. and K. S. Thomson. 1980. Reflections on agnathan–gnathostome relationships. Pages 19–22 in *Aspects of Vertebrate History*, edited by L. L. Jacobs, Museum of Northen Arizona Press, Flagstaff, Ariz.

Yalden, D. W. 1985. Feeding mechanisms as evidence for cyclostome monophyly. *Zoological Journal of the Linnean Society* 84:291–300.

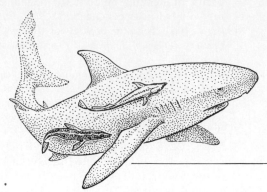

As we have seen, the earliest vertebrates lacked jaws and true paired appendages. The next major step in vertebrate evolution was the appearance of jaws and paired fins, and it occurred relatively soon after the first evidence of vertebrates in the fossil record. Jaws play a number of roles in the biology of living vertebrates, but their major use is in feeding. Paired fins were a second important innovation because they gave a swimming vertebrate precise control of steering. The diversity of predatory specializations available to a vertebrate with jaws and precise steering is great, and the appearance of these characters signaled a new radiation of vertebrates. The cartilaginous fishes (sharks, rays, and ratfish) are the descendents of this radiation, and they combine derived characters such as a cartilaginous skeleton with a generally primitive anatomy. Sharks have undergone three major radiations, which can be broadly associated with increasingly specialized feeding mechanisms, and living sharks are a diverse and successful group of fishes. In this chapter we consider the origin of jaws and paired appendages and the roles that these two innovations have played in the success of cartilaginous fishes.

The Rise of Jawed Vertebrates and the Radiation of the Chondrichthyes

7

The First Appearance of Jaws and Unique Gnathostome Characters

Fossils of vertebrates with jaws are known from the mid-Silurian. It may seem strange that a major new morphological feature like jaws should arise before the extensive radiation of agnathans occurred instead of arising from some later product of that radiation. This pattern of evolution, however, is seen over and over again in an examination of vertebrate life: major new innovations arise from less specialized members of a lineage.

Some ecological niches were so suited to the agnathan body plan that no evolutionary advance yet realized has been able to replace it. Lampreys and hagfishes provide evidence of this fact. The great majority of agnathans, however, succumbed to what is generally thought to have been competition from jawed vertebrates. Actually, no one knows exactly what characteristics of the jawed vertebrates may eventually have led to the near exclusion of the agnathans.

Albert Sherwood Romer stated in *The Vertebrate Body* that "perhaps the greatest of all advances in vertebrate history was the development of jaws and the consequent revolution in the mode of life of early fishes." Why did Romer elevate jaws to such an eminent position? Jaws allow behaviors that otherwise would be difficult, if not impossi-

ble, to perform. The presence of jaws around the mouth, manipulated by muscles, allows an organism to grasp objects firmly. When the jaws are armed with teeth their grip becomes surer. Teeth with sharp cutting edges allow food particles to be reduced to edible size, and large, flattened, and opposed teeth allow grinding of harder foods. When vertebrates evolved jaws, therefore, new food sources became available. Jaws apparently placed the early gnathostomes in a commanding position, for many increased in size and, during the Devonian, gnathostomes appear to have replaced many more primitive vertebrates.

The functions of jaws are not limited to capturing and chewing prey. A grasping, movable jaw permits a new behavior—manipulation of objects—that enters many aspects of the life of vertebrates. Jaws can be used to dig holes or to carry pebbles or vegetation to build nests or to grasp mates during courtship and juveniles during parental care. No wonder Romer placed so much importance on jaws.

Jawed fishes appear in the fossil record fully developed, without intermediates (Moy-Thomas and Miles 1971, Romer 1966). The Silurian fossils consist of detached spines, scales, teeth, and jaws. Jaws gave strength and form to the rim of the mouth that was not possible in agnathan mouths. In addition, from their first appearance jaws were solidly braced against the cranium (and thus, via the rest of the skeleton, connected to the entire

233

body), yet they are mobile and are fitted with a specialized musculature.

The first jaws in the fossil record give little insight into their evolutionary history. Our belief that jaws originated through modification of the gill arch skeleton comes from detailed comparative anatomical and embryological studies. The neural crest cells seem to have been the key tissue in the evolution of jaws and of many of the other derived characters of gnathostomes. Some of the neural crest cells migrate to the visceral arches to form the bilaterally paired upper palatoquadrate and lower mandibular cartilages that form the jaws (Figure 7–1). Between the jaws and the rest of the branchial skeleton lies the second visceral arch, known as the hyoid arch, which is primitively associated with supporting the jaws and bracing them against the cranium. The neural crest cells also contribute to the branchial skeleton, primitively consisting of five bilaterally paired series of hinged skeletal elements, four in each arch on each side extending from the vertebral column to the midventral line, where a single unpaired element unites the bilateral arches. This gill skeleton is entirely internal to the ectodermal gills, in contrast to the skeletal supports of the gills in jawless fish that lie external to the endodermally derived gills.

A series of distinctive additional derived characters of the cranium, trunk, and mechanosensory apparatus is evident in fossil gnathostomes and retained in living forms (Table 7–1).

Numerous other traits distinguish living gnathostomes from hagfish and lampreys, although these are generally not detectable in fossils. They include the greater rotational action of the external ocular muscles; the myelination of nerve fibers; and two types of the contractile protein actin, one specific to smooth muscle and the other to striated muscle (lamprey actins seem to be of one type only). The soft anatomy of gnathostomes includes several common elements unknown in jawless fishes: a distinct stomach, a spiral valvular intestine, a renal portal venous flow, distinct oviducts and mesonephric ducts, a pancreas with both en-

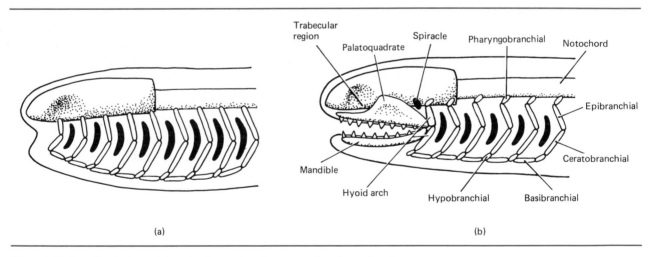

(a)   (b)

Figure 7–1. Evolution of the vertebrate jaw from anterior visceral arches: (a) agnathous condition; (b) gnathostomous condition. The fate of the first visceral arch (or arches) is unclear. An anterior arch probably formed the trabecular region while a more posterior arch gave rise to the jaws. Note that an additional arch (the hyoid) acts as support for the jaw hinge. The gill slit between the jaws and the hyoid arch was reduced to a spiracle, as seen in some modern chondrichthyans and bony fishes.

Table 7–1. Derived characters of gnathostomes found in fossil and living forms.

Cranial Characters

1. Paired trabeculae derived from the neural crest contribute to the floor of the cranium beneath the enlarged forebrain
2. Cranium enlarged anteriorly to end in a precerebral fontanelle
3. Incorporation of one or more occipital neural arches into the rear of the cranium
   3a. Cranium enlarged posteriorly, extending the position of the foramen magnum posteriorly
4. Addition of a third (horizontal) semicircular canal
5. Presence of calcium carbonate stataconia or otoliths
6. Development of a postorbital process on the chondrocranium separating the functions of supporting the jaws and enclosing the eye
7. Basicranial muscles that originate on the cranium and insert on the branchial arches

Trunk Characters

8. A horizontal septum of connective tissue dividing the trunk musculature into dorsal (epaxial) and ventral (hypaxial) units
9. Neural and haemal arches (usually mineralized) regularly appear along notochord

Sensory Characters

10. A unique and evolutionarily conservative pattern of cephalic lateral-line canals
11. Lateral line on the trunk region flanked or enclosed by specialized scales

---

Source: J. G. Maisey, 1986, *Cladistics* 2:201–256.

docrine and exocrine functions, and a spleen. A number of endocrine hormones, discrete endocrine glands, and storage/mobilization mechanisms for metabolites are derived characters of gnathostomes.

Locomotor function was improved by a well-developed heterocercal tail and fin webs supported by collagenous finrays. Undoubtedly overshadowing these is the most outstanding shared derived character of the gnathostomes besides the jaw: paired pectoral and pelvic appendages with internal supporting girdles.

The Origin of Fins

Jaws are an advantage only when applied to an object. Suction can draw objects into the mouth over modest distances, but mostly the body must be guided to the graspable object. This sounds simple, but guidance of a body in three-dimensional space is complicated. Yaw (swinging to the right or left) combines with pitch (tilting up or down) to make accurate contact with a target difficult. Roll (rotation around the body axis) must be countered for effective jawed grasping. Especially if the target moves, perhaps evasively, quick adjustments of roll, pitch, and yaw are necessary. It is little wonder that the development of strong mobile fins was coincident with the evolution of jaws.

Fins act as hydrofoils applying pressure to the surrounding water. Because water is practically incompressible, force applied by a fin in one direction against the water is opposed by an equal force in the opposite direction (Figure 7–2). Thus fins can resist roll if pressed on the water in the direction of the roll. Fins projecting horizontally near the anterior end of the body similarly counteract pitch. Yaw is controlled by vertical fins along the middorsal and midventral lines. Fins serve other functions as well. They increase the area of the tail for greater thrust during propulsion. Presented at angles to a flow of water, they produce lift. Spiny fins are used in defense, and become systems to inject poison when combined with glandular secretions. Colorfully marked fins are used to send visual signals. Collapsed and unfolded abruptly and often startlingly by muscular action, fins are used to increase the apparent size of their possessor. Fins obviously provide many benefits.

Fin structure of early fishes was variable. Agnathans had spines or enlarged scales derived from dermal armor that functioned like fins. Osteostracans had paddle-like pectorally located structures without internal supports. Some anaspids had long fin-like sheets of tissue running along the flanks. The earliest paired fins, although a universal feature of gnathostomes, were dissimilar in details of structure. Two early groups of gnathostomes, the **acanthodians** and the **placod-**

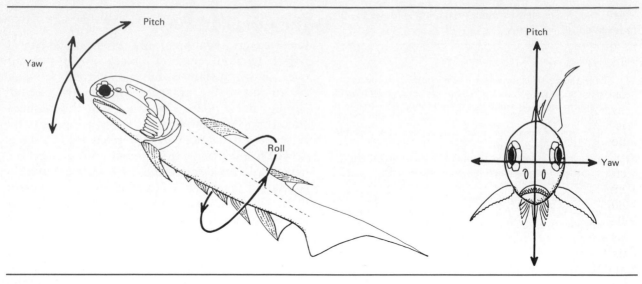

Figure 7–2. An acanthodian, *Climatius*, as seen in lateral and frontal views to show the orientation of pitch, yaw, and roll and the fins that counteract these movements.

erms, illustrate this. Acanthodians had a variable number of spines, sometimes with attached membranes, extending along the ventrolateral aspects of the trunk. The internal skeleton of acanthodian fins (Figure 3–10) was composed of at least two rows of tightly packed rod-shaped cartilages associated with horny, thread-like rays that did not protrude far into the fin. Placoderm pectoral appendages, in contrast, had numerous rod-shaped elements and horny fin rays extending into the fin. Clearly fins, although always in approximately the same positions, were radically different in internal and external construction and number.

There is no fossil evidence of the origin of fins, especially the paired pectoral and pelvic fins that were significant in later stages of vertebrate evolution. Although some early workers fancied a resemblance between the pectoral fins and the gill arches, a branchial origin is extremely doubtful. Fins are derived from mesoderm and have somatic, not visceral, musculature and innervation.

Long held in high regard was the fin-fold theory of the origin of paired appendages. The fin-fold theory used as prototypes certain anaspids such as *Jaymoytius* (Figure 6–9) that had a pair of long-based lateral flaps extending from gills to anus. Early acanthodians had pairs of spines in the same region. In fact, this was merely an artificial assemblage of organisms with no phylogenetic relationship. By comparing this hypothetical series to the fin folds of amphioxus, discrete fins were pictured as originating from continuous fin folds. These fin folds were thought to be paired laterally but single dorsally and posteriorly. Broken into short segments and reduced in number, fin folds were said to have been the origin of the fins seen in today's fishes.

Because no fossil evidence to document these events has been uncovered, the fin-fold theory has become less appealing in recent years. In line with our current understanding of evolutionary principles, it seems best to consider fins so beneficial that multiple evolutions have occurred. A multiple evolution with similar results is to be expected when only a limited variety of fin positions and shapes can provide the hydrodynamic advantages we attribute to fins. Thus the fins of ostracoderms may be convergent with those of gnathostomes.

The girdles and basal elements of the paired appendages of living gnathostomes seem, on the basis of considerable anatomical and embryological evidence, to be homologous.

A well-developed heterocercal caudal fin is almost as universal a feature of all early jawed fishes as were paired fins. Comparably developed caudal fins were found in some jawless fish, although other jawless fish had hypocercal tails. An abruptly up- or downturned notochord produces an increased depth of the caudal fin, a structure that is important in rapid acceleration (Webb and Smith 1980). In addition, the gnathostomes with their collagenous fin rays had a stiffened caudal web of considerable area further enhancing acceleration. This kind of unsteady burst swimming is important in predator avoidance and provides significant economies in terms of locomotor energy when bursts of acceleration are alternated with glides. All fishes with a fin-strengthening upturned or downturned axial skeleton tip have a noncollapsible caudal fin effective in burst swimming.

The gnathostomes appear as four distinctive clades all flowering in the Devonian—the age of fishes. One of these clades, the **placoderms**, was isolated from the other three (Chondrichthyans, Acanthodians, and Osteichthyans) despite sharing many derived gnathostome characters. The jaw muscles of placoderms are different from the jaw muscles of the other three groups, placoderms have nothing comparable to the teeth of the other gnathostomes, and the skeletal anatomy of the paired fins of placoderms lacks homologies with those of other gnathostomes. The placoderms seem to have left no descendants in the modern fauna. A second clade, the **Chondrichthyes**, clearly related to all other gnathostomes, evolved distinctive reductions and specializations of dermal armor, internal calcification, jaw and fin mobility, and reproduction. These chondrichthyans have successfully survived to the present.

The final two clades of fishes, the Acanthodians and Osteichthyes (together = Teleostomes), may be closely related and form the root of all subsequent vertebrate evolution (Chapter 8). Before turning to this majority of fish species past and present, let us examine the placoderms and chondrichthyans to see the variety that early gnathostomous vertebrate life produced.

Placoderms: The Armored Fishes

Among the earliest gnathostomes in the fossil record is a confusingly diverse assemblage of generally heavily armored fishes, the **placoderms** (Figure 7–3). R. L. Carroll (1987) has pointed out that the placoderms are without modern analogs, and their massive external armor makes interpretation of their ways of life particularly difficult. Although placoderms clearly share an impressive list of derived characters with other gnathostomes, several elements of their morphology appear to isolate placoderms from all other jaw-bearing vertebrates. The most profound of these characters is the position of the jaw musculature. In all other gnathostomes the jaw muscles lie external to the jaw's skeletal elements. In those placoderms where it can be determined, it appears that the main mass of the jaw musculature is *medial* to the palatoquadrates. If this is true for placoderms in general, the common ancestor of placoderms and all other gnathostomes may not have had a functional jaw, and jaws may have evolved more than once among primitive fishes. Placoderms also lack teeth that correspond to those of any other gnathostome. They have a hyoid arch that is not clearly involved in the same suspensory function as in other gnathostomes and is distinctively different in arrangement and number of elements from that of other vertebrates.

Superficially the earliest placoderms, the arthrodires, resembled heterostracans and osteostracans in appearance and habitat (Figure 7–3a). As the name placoderm (*placo* = plate, *derm* = skin) implies, they were covered with a thick, often ornamented bony shield over the anterior one-half to one-third of their bodies. By their extinction in the early Carboniferous, some placoderms had muscular, mobile pectoral fin structures that must

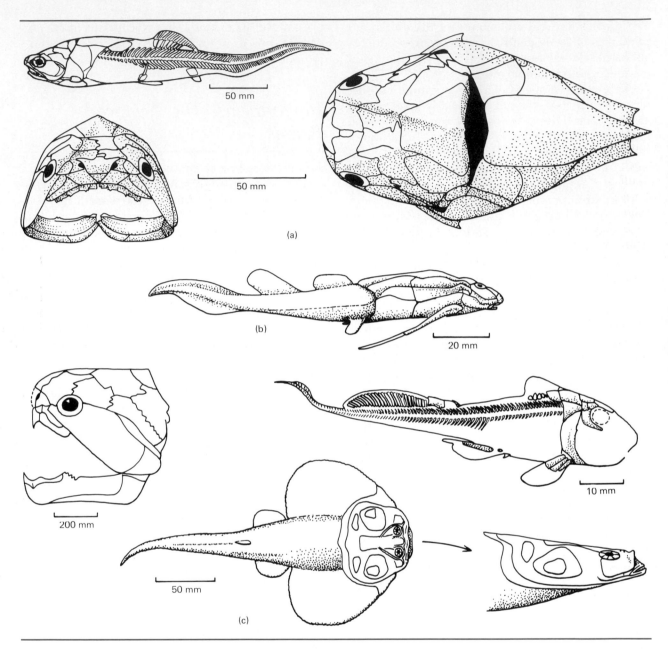

Figure 7–3. Placoderms: (a) lateral, frontal, and dorsal views of an arthrodire, *Coccosteus*, middle Devonian; (b) the peculiar placoderm *Bothriolepis*, with a jointed exoskeleton that supported pectoral appendages; (c) three widely varying types of placoderms: left, the giant predator *Dunkleosteus*, upper Devonian; right, the chimaera-like *Rhamphodopsis* (see also Figure 7–11b); and below, the ray-like *Gemuendina*. (Modified after J. A. Moy-Thomas and R. S. Miles, 1971, *Paleozoic Fishes*, W. B. Saunders, Philadelphia.)

have contributed to their active predatory existence. Other placoderms, the antiarchs, developed pectoral appendages into stiff props by encasing all soft tissue in joined, bony tubes reminiscent of arthropod appendages (Figure 7–3b).

In early placoderms the head shield, which was formed by numerous large plates, was separated from the mosaic of small dermal bones that shielded the trunk by a narrow gap. A mobile connection between the anterior vertebrae and the skull allowed the head to be lifted. This craniovertebral joint permitted the mouth to be opened wider than would lowering only the mandible, or to be opened when the lower jaw was pressed against the seafloor as a placoderm quietly awaited the approach of its prey. During their evolution, predatory arthrodire placoderms developed a curious specialization further increasing the gape. The space between the head and the thoracic shields widened and a pair of joints evolved, one above each pectoral fin on a line that passed through the older craniovertebral joint of the axial skeleton. This arrangement provided great flexibility between the shields and allowed an enormous head-up gape. In addition, it probably increased respiratory efficiency and improved steering control. **Arthrodires**, the name given to this order of placoderms, pays tribute to this strange specialization: *arthros* (a joint) plus *dira* (the neck).

Placoderms were mostly creatures of the Devonian. During that period placoderms radiated into a large number of lineages and types. *Dunkleosteus* was a voracious, 10-meter-long, predatory arthrodire. *Bothriolepis*, an antiarch, supported itself on stilt-like pectoral fins. Another group of placoderms had the palatoquadrate firmly attached to the cranium and complex, solid tooth plates for crushing shellfish. In some placoderms, sexually dimorphic pelvic appendages suggest that internal fertilization occurred, probably coupled with complex courtship. Other groups show a tendency for dorsoventral flattening, eyes on top of the head, and subterminal mouths, all indicating benthic specialization. *Gemuendina* bears a striking resemblance to modern skates, although

it was completely armored with a mosaic of small plates and could not have used its broad pectoral fins in the skate's undulating manner of locomotion.

Characteristic of all placoderms is the absence of teeth of a modern type. Slightly modified dermal bones lined the jaw cartilages of placoderms and, though they had long knife-like cutting edges and strong pick-like points for slicing and piercing prey, they were subject to wear and breakage without replacement. The jaws in placoderms, although an outstanding milestone in vertebrate evolution, were often immovably bound to the cranium or tightly articulated to the rest of the head shield. This prevented their participation in any sucking action, a very successful feeding process as evidenced by its success in the agnathans and again in the vast majority of living jawed fishes. According to a growing consensus among paleoichthyologists, placoderms became extinct in the lower Carboniferous without giving rise to any surviving forms.

Chondrichthyes: The Cartilaginous Fishes

The sharks and their relatives first appear in the fossil record in the early Devonian. Since then, morphological refinements in some of their systems have evolved to levels surpassed by few other living vertebrates; yet they retain many primitive elements of their basic anatomy, and sharks have long been used to exemplify a primitive vertebrate body form. Identified by a cartilaginous skeleton, living forms can be divided into two groups: those with a single gill opening on each side of the head and those with multiple gill openings on each side. The first group is called the **Holocephali** because of the undivided appearance of the head that results from having a single gill opening. Their common names of ratfish and chimaera come from their bizarre form: a long flexible tail, a fish-like body, and a head with big eyes and buckteeth that resembles a caricature of a rabbit.

Phylogenetic Relationships of Placoderms, Acanthodians, Chondrichthyans, and Teleostomes

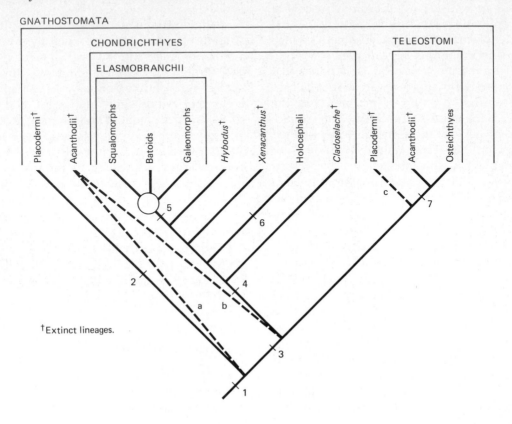

†Extinct lineages.

The second group is the **Elasmobranchii**, meaning plate-gilled. Elasmobranchs include the sharks, cylindrical forms with five to seven gill openings on each side of the head. (They are known in older literature as Pleurotremata or Selachii). Sharks are acknowledged as two distinct clades, the squalomorphs and the galeomorphs. A second group of elasmobranchs are dorsoventrally flattened with five gill slits on the ventral surface (the skates and rays, Hypotremata or Batoidea) (Greenwood et al. 1973).

Evolutionary Specializations of Elasmobranchs

In spite of the rather good fossil record, the phylogeny of cartilaginous fishes remains unclear. Early elasmobranchs, like modern species, were diverse in habits and habitats. Their initial radiation from a common ancestor emphasized changes in teeth, jaws, and fins. Apparently, the feeding and locomotor apparatus evolved at different rates within different lineages. In some lineages, ad-

1. Jaws formed of bilateral palatoquadrate (upper) and mandibular (lower) cartilages of the mandibular visceral arch at some stage of development, teeth containing dentine, modified hyoid gill arch, branchial elements internal to gill membranes, branchial arches contain four elements on each side plus one unpaired ventral median element, paired trabeculae contribute to cranium, three semicircular canals, internal supporting girdles associated with pectoral and pelvic fins, myelinated nerve fibers, calcium carbonate statoconia or otoliths, and 34 additional derived characters. 2. Placodermi: tentatively placed as the sister group of all other gnathostomes, but see Gardiner (1984) for a different view. Provisionally united by the following derived characters: A specialized joint in the neck vertebrae, a unique arrangement of dermal skeletal plates of the head and shoulder girdle, a distinctive articulation of the upper jaw, a unique pattern of lateral line canals on the head. 3. Chondrichthyes plus Teleostomi (Eugnathostoma): Epihyal element of second visceral arch modified as the hyomandibula, which is a supporting element for the jaw. 4. Chondrichthyes: Unique peri-chondral and endochondral mineralization as prismatic hydroxyapatite tesserae, placoid scales, unique teeth and tooth replacement mechanisms, distinctive characters of the basal and radial elements of the fins, inner ear labyrinth opens externally via the endolymphatic duct, distinctive features of the endocrine system. 5. Elasmobranchii: Pectoral fin with three basal elements, the anteriormost of which is supported by the shoulder girdle, characteristics of the nervous system, cranium, and gill arches, 25 additional derived characters. 6. Holocephali: Hyostylic jaw suspension, gill arches beneath the braincase, dibasal pectoral fin, dorsal fin articulates with anterior elements of the axial skeleton, 24 additional derived characters of living chimaeras. 7. Teleostomi: Hemibranchial elements of gills not attached to interbranchial septum, bony opercular covers, branchiostegal rays, five additional characters. (Based on G. V. Lauder and K. F. Liem, 1983, *Bulletin of the Museum of Comparative Zoology* 150:95–197; J. G. Maisey, 1986, *Cladistics* 2:201–256; and B. G. Gardiner, 1984, *Bulletin of the British Museum (Natural History) Geology* 37:173–427.)

Figure 7–4. Phylogenetic relationships of placoderms, acanthodians, chondrichthyans, and teleostomes. This diagram shows the probable relationships among the major groups of primitive gnathostomes. Extinct lineages are marked by a dagger (†). The numbers indicate derived characters that distinguish the lineages. The best corroborated relationships are shown by solid lines; the dashed lines labeled (a) and (b) show other relationships that have been proposed for acanthodians, and the dashed line labeled (c) shows a possible alternative relationship for placoderms. The circle indicates that the relationships of the lineages at that node cannot yet be defined.

vanced dentition was coincident with primitive fin structures, whereas the opposite combination is seen in others. As a result, fossil elasmobranchs display confusing mosaics of primitive and advanced characters.

Through time different lineages of elasmobranchs tended to accumulate similar but not identical modifications in their feeding and locomotor structures, presumably because of similar selective pressures. This pattern of similar adaptations in related lineages is an example of parallel evolution: When similar selective forces act on similar body forms and developmental mechanisms, certain modifications appear independently and often repeatedly in the course of time.

Adaptations found in several lineages are known as **general** or **broad adaptations**. Examples are paired appendages, jaws, and use of muscular pump filter feeding. Broad adaptations have penetrating effects on the organisms' integration of behavior, physiology, and morphology. Broad adaptations define an organizational level in a

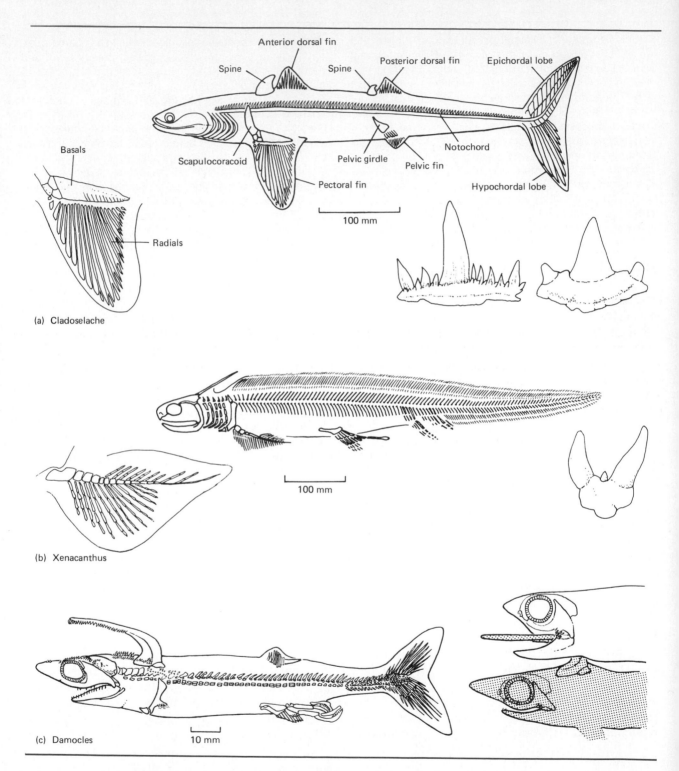

(a) Cladoselache

Basals

Radials

Anterior dorsal fin

Spine

Posterior dorsal fin

Spine

Epichordal lobe

Scapulocoracoid

Pelvic girdle

Pelvic fin

Notochord

Pectoral fin

Hypochordal lobe

100 mm

(b) Xenacanthus

100 mm

(c) Damocles

10 mm

PART TWO   Aquatic Vertebrates: Cartilaginous and Bony Fishes

horizontal (not phylogenetic) classification. Species showing characters of a particular organizational level belong to the same **grade**. A grade may contain different phylogenetic lineages. Each phylogenetic lineage is called a **clade**. For elasmobranchs we can define three grades in their evolution, but at present it seems that only one of the clades was involved in the more recent evolutionary history of elasmobranchs (Figure 7–4) (Greenwood et al. 1973, Carroll 1987).

The Earliest Elasmobranch Radiation

The earliest elasmobranchs are identified by the form of the teeth common to the majority of the species. The teeth (Figure 7–5) are basically three-cusped with little root development. Although there is evidence of bone around their bases, the teeth are primarily dentine structures capped with enamel. The central cusp is the largest in *Cladoselache*, the best known genus, and smallest in *Xenacanthus*, a more specialized form.

Cladoselache was shark-like in appearance, about 2 meters long when fully grown, with large fins and mouth and five separate external gill openings. The mouth opened terminally and the chondrocranium had several large areas for the tight ligamentous attachment of the upper jaw cartilage, the palatoquadrate. One area of attachment was at the symphysis between the right and left halves of the palatoquadrate, a second extended up from the palatoquadrate behind the eye, and a third via a process to the ear region. The jaw also

obtained some support from the second visceral arch, the hyoid arch. The name **amphistylic** (*amphi* = both, *styl* = pillar or support) is applied to this mode of multiple sites of upper jaw suspension. The gape was large, the jaws extending well behind the rest of the skull. The three-pronged teeth were probably especially efficient for feeding on fishes that could be swallowed whole or severed by the knife-edge cusps.

Wear of teeth, which renders them less functional, is a problem faced by all vertebrates. The earliest sharks, and also some acanthodians and early bony fishes, solved this problem in a unique way. Each tooth on the functional edge of the jaw was but one member of a tooth whorl, attached to a ligamentous band that coursed down the inside of the jaw cartilage deep below the fleshy lining of the mouth. Aligned in each whorl in a file directly behind the functional tooth were a series of developing teeth. In modern sharks essentially the same dental apparatus is present. Tooth replacement is rapid: young sharks under ideal conditions replace each lower jaw tooth every 8.2 days and each upper jaw tooth every 7.8 days. If *Cladoselache* replaced its teeth, as seems likely, a significant advantage in feeding mechanics is indicated for elasmobranchs compared to their placoderm contemporaries.

The body of *Cladoselache* was supported only by a notochord, but cartilaginous neural arches gave added protection to the spinal cord. The fins of *Cladoselache* consisted of two dorsal fins, paired pectoral and pelvic fins, and a well-developed forked tail. The first and sometimes the second dorsal fins were preceded by stout spines, trian-

Figure 7–5. Early elasmobranchs: (a) *Cladoselache* with details of the pectoral structure and teeth of the "*Cladodus*" type; (b) *Xenacanthus*, a freshwater elasmobranch with details of its archipterygial pectoral fin structure and peculiar teeth; (c) left, male *Damocles serratus*, a 15-centimeter shark from the late Carboniferous showing sexually dimorphic nuchal spine and pelvic claspers; right, male (below) and female (above) as fossilized, possibly in courtship position. (Modified after J. A. Moy-Thomas and R. S. Miles, 1971, *Paleozoic Fishes*, W. B. Saunders, Philadelphia; and R. Lund, 1985, *Journal of Vertebrate Paleontology* 5:1–19, and 1986, *Journal of Vertebrate Paleontology* 6:12–19.)

gular in cross section and thought by some to have been covered by soft tissue during the life of the shark. The dorsal fins were broad triangles with an internal structure consisting of a triangular basal cartilage and a parallel series of long radial cartilages that extended to the margin of the fin. The pectoral fins were larger but similar in construction.

Among the early radiations of sharks, almost every species seems to have had a different sort of internal pectoral fin arrangement (Figure 7–5), but all possessed basal elements that anchored the pectoral fins in place. From their structure the pectoral fins appear to have been hydrofoils with little capacity for altering their angle of contact with the water. The pelvic fins were smaller, but otherwise shaped like the pectorals. Some species show evidence of **claspers**: male copulatory organs. No anal fin is known; it is also lacking in many modern sharks.

The caudal fin of *Cladoselache* is distinctive (Figure 7–5). Externally symmetrical, its internal structure was asymmetrical and contained a band of subchordal elements resembling the hemal arches that protect the caudal blood vessels in modern sharks. Long unsegmented radial cartilages extended into the hypochordal (lower) lobe of the fin. At the base of the caudal fin were paired lateral keels that are identifying characteristics of modern rapid pelagic (open water) swimmers.

The integument included only a few scales, but these resembled the teeth. Cusps of dentine were covered with enamel and contained a cellular core or pulp cavity. Unlike a tooth, each scale had several pulp cavities to match its several cusps. These scales were limited to the fins, the circumference of the eye, and within the mouth behind the teeth. Their structural similarities leave little doubt that the teeth of early sharks and other vertebrates in general are derived from specialized elements of the integument.

We can piece together the lives of many of the early elasmobranchs from their morphology and fossil localities. Probably pelagic predators, most early sharks, and *Cladoselache* in particular, would have swum after their prey in a sinuous manner, engulfing them whole or slashing them with their dagger-like teeth. The lateral keels on the tail base indicate that *Cladoselache* was a fast swimmer. The lack of body denticles and calcification in many structures suggest a tendency to reduce weight and thereby increase buoyancy.

Reproduction of some forms involved internal fertilization as evidenced by pelvic claspers, implying that a behavioral system existed to ensure successful mating. If the possession of claspers is a primitive character shared with the placoderms, as some authors contend, the lack of claspers in *Cladoselache* may mean that the male of this comparatively well-known fossil form remains unidentified. Alternatively, the early sharks may not have had claspers. The claspers of placoderms and of later chondrichthyans are structurally different, and this may indicate separate, parallel evolution of pelvic copulatory organs.

Recent descriptions of two species of 15-centimeter-long sharks from the lower Carboniferous of Montana may indicate just how complex reproductive behavior was in early elasmobranchs. Males can be identified by pelvic claspers, sharp rostra, and by an enormous forwardly curved mid-dorsal spine firmly embedded just behind the head (Figure 7–5c). Richard Lund considered that the great degree of sexual dimorphism and the finding of many more males than females of at least one of these species was compatible with a hypothesis of male display during courtship and female choice, perhaps in a regular male display site. He also suggested that one of the specimens discovered might be a pair in a precopulatory courtship position (Figure 7–5c) with the blunt-snouted female grasping the male's nuchal spine in her jaws.

One of the score or so of genera produced in this early radiation of elasmobranch evolution is *Xenacanthus* , which had a braincase, jaws, and jaw suspension very similar to those of *Cladoselache*. But there the resemblance ends. The xenacanths were freshwater bottom dwellers of rather slow habit. The idea that they were bottom dwellers derives from the similarity of their fins to those of living Australian lungfish (which are bottom dwellers) and their heavily calcified cartilaginous

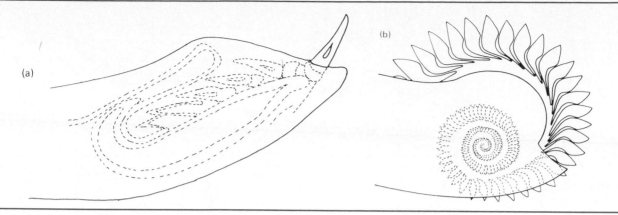

Figure 7–6. (a) Diagramatic cross section of the jaw of a living shark showing the single functional tooth backed by a band of replacement teeth in various stages of development; (b) lateral view of the symphysial (middle of the lower jaw) tooth whorl of the edestoid cladodont *Helicoprion*, showing the chamber into which the life-long production of teeth spiraled.

skeleton that would have decreased their buoyancy. The xenacanths appeared in the Devonian and survived until the Triassic, when they died out without leaving descendents.

Another group of chondrichthyan fishes had a suite of morphological characteristics seen also among other fusiform, fast-swimming marine sharks. These characters included pectoral fins and stiff, symmetrical, deeply forked tails. These sharks, the **edestoids**, were morphologically distinct from the main lines (clades) of elasmobranch evolution, especially because of their peculiar dentition. Most of the tooth whorls of edestoids were greatly reduced, but the symphysial (central) tooth row of the mandible was tremendously enlarged and each tooth interlocked with adjacent teeth at its base. Apparently, several members of this tooth whorl were functional at the same time. Blunt for crushing in some forms and compressed to create a series of knife-edge blades in others, the mandibular tooth row bit against small flat teeth associated with a poorly developed palatoquadrate. Most edestoids replaced their teeth rapidly, the oldest worn teeth being shed from the tip of the mandible. In contrast, *Helicoprion* retained all of its teeth in a specialized chamber into which the life-

long production of dentition spiraled (Figure 7–6b). Perhaps teeth no longer efficient in size or shape provided a solid foundation for the functional teeth.

The Second Elasmobranch Radiation

Further elasmobranch evolution involved reorganization in feeding and locomotor systems. These modifications began in the Carboniferous and lasted until the late Cretaceous. To identify these new adaptations, we shall describe a late and well-known genus of the Triassic and Cretaceous, *Hybodus*, which has left complete skeletons 2 meters in length (Figure 7–7a).

The diverse dentition of hybodont sharks seems pivotal to their success. The anterior teeth were sharp-cusped and appear to have been used for piercing, holding, and slashing softer foods. The posterior teeth were stout, blunt versions of the anterior teeth. Instead of becoming functional one at a time, they appeared above the fleshy lining of the mouth in batteries consisting of several teeth from each individual tooth whorl. A modern

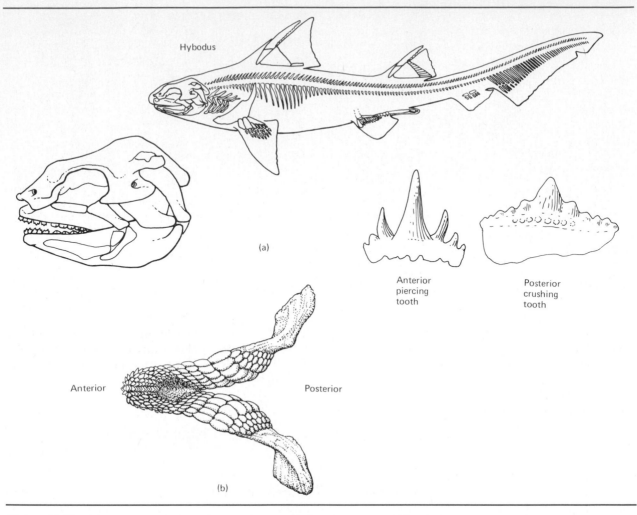

Hybodus

(a)

Anterior
piercing
tooth

Posterior
crushing
tooth

Anterior

Posterior

(b)

Figure 7–7. (a) Second radiation of elasmobranch evolution represented by *Hybodus*. Detail of head skeleton and teeth. (Modified after J. A. Moy-Thomas and R. S. Miles, 1971, *Paleozoic Fishes*, W. B. Saunders, Philadelphia.) (b) Upper jaw of the living hornshark, *Heterodontus*, with a dentition similar to that of many elasmobranchs in the second radiation during the late Paleozoic and early Mesozoic.

form with a similar dentition (Figure 7–7b) indicates how the mouth of a *Hybodus*-like shark must have looked: the living horn sharks of the genus *Heterodontus*, which have *Hybodus*-like dentition, feed on small fish, crabs, shrimp, sea urchins, clams, mussels, and oysters. The sharp teeth near the symphysis seize and dispatch soft-bodied food, but shelled foods are thoroughly crushed by pavement-like posterior teeth before being swallowed. Because they lived in rich shallow marine environments, the variety of foods available to hybodont sharks must have been abundant and varied.

The coming together of landmasses to form

Pangaea brought together faunas that had been living and evolving in isolation on the shelves of distant continents for millions of years (see Chapter 9). If this meant competition and extinction for many forms, as the fossil record suggests, a euryphagous (*eury* = broad, *phagos* = to eat) vertebrate that could feed on a wide variety of organisms would have an advantage over a stenophagous (*stenos* = narrow) form. Perhaps *Hybodus* and its relatives were preadapted in their diets to survive the radical changes in invertebrate faunas that characterized the Permo-Triassic transition.

Also characteristic of the hybodonts were their fins. The pectoral girdle remained divided into separate right and left halves, but the articulation between girdle and the fin consisted of three narrowed plate-like basals instead of the long series seen in earlier sharks. This tribasal arrangement was also found in the pelvic fins, and both pairs of fins were more detached from the body than in earlier sharks, supported on a mobile stalk composed of the three basals. Mobility of the distal portion of the paired fins was also increased. The cartilaginous radials did not extend to the fin margin and were segmented along their shortened length.

Proteinaceous, flexible structures called **ceratotrichia** extended from the outer radials to the margin of the fin. Intrinsic fin muscles produced flexion (arching the fin from anterior to posterior and along its long axis). This greater mobility allowed the paired fins to be used hydrodynamically in ways that seem impossible with the fin construction characteristic of *Cladoselache*. By assuming different shapes, the pectorals could produce lift anteriorly, aid in turning, or function as simple hydrofoils. Along with changes in the paired fins, the caudal fin assumed new functions and an anal fin appeared. Caudal fin shape was altered by reduction of the hypochordal lobe, division of its radials, and addition of flexible ceratotrichia. This construction is known as **heterocercal** (*hetero* = different, *kerkos* = tail). Undulated from side to side, the fin twisted due to water pressure so that the flexible lower lobe trailed behind the stiff upper one. This distribution of force produced forward thrust that, combined with the variable planing surfaces produced by flexible pectorals, could counter the shark's tendency to sink or could lift it from a benthic resting position. All gnathostomes from the early fossil record had heterocercal caudal fins. The evolutionary success of the heterocercal fin compared to a straight or ventrally bent fin may relate to this generation of vertical forces, however slight they might have been (Webb and Smith 1980). Whatever mechanism the earlier sharks used to remain afloat, it is likely that the dynamic design of the fins in this second radiation of sharks allowed them more behavioral flexibility.

Other morphological changes in the sharks of the second major radiation include the appearance of a complete set of hemal arches that protected the arterial and venous trunks running below the notochord, well-developed ribs, and narrow, more pointed dorsal fin spines closely associated with the leading edge of the dorsal fins. These spines were deeply inserted in the muscle mass, exposed above the skin, ornamented with ridges and grooves, and studded with numerous barbs on the posterior surface indicating defensive functions. Claspers are common to all species, leaving little doubt about the development of courtship and internal fertilization. In addition, male *Hybodus* had one or two pairs of hooked spines above the eye that may have been used as claspers during copulation.

Hybodus and its relatives resembled their presumed *Cladoselache*-like ancestors in having terminal mouths, amphistylic jaw suspension, unconstricted notochords, and multicusped teeth, but a direct line cannot be drawn between the two in time or in morphology. Some forms considered related to *Hybodus* had *Cladoselache*-like dentition combined with tribasal pectoral fins; others developed a very tetrapod-like support for highly mobile pectoral fins. Because the caudal fin was reduced, they probably moved around on the seafloor using their limb-like pectoral fins. Another form, known only from a 5-centimeter juvenile, had a paddle-shaped rostrum one-third its body length. Other types were 2.5-meter giants with blunt snouts and enormous jaws. Despite their variety and the fact that they flourished during the Mesozoic, this second radiation of elasmobranchs became increasingly rare and disappeared from earth at the close of that era.

The Modern Radiation: Sharks and Rays

As early as the Triassic, and perhaps even in the late Carboniferous, fossilized indications of the modern radiation of elasmobranchs appear. By the Jurassic, sharks of modern appearance had evolved, and the Cretaceous contains genera that are still living. Paleontologists are not in agreement about the origin of modern sharks. They may have evolved from *Hybodus*-like sharks, but a few details of the morphology of *Hybodus*-like sharks seem to preclude them from the lineage of modern sharks. Perhaps more likely is the evolution of the modern sharks, skates, and rays (the **Neoselachii**) from a *Cladoselache*-like lineage that acquired characteristics in parallel with *Hybodus* and its relatives. The most obvious difference between most members of the earlier radiations and modern sharks is the almost ubiquitous rostrum or snout that overhangs the ventrally positioned mouth in living forms. The technical characters distinguishing modern sharks are more subtle.

Modern elasmobranchs have an enlarged hyomandibular cartilage, which braces the posterior portion of the palatoquadrate and attaches firmly but movably to the otic region of the cranium (Figure 7–8). A second connection to the chondrocranium is via paired palatoquadrate projections to either side of the brain case just behind the eyes and attached to it by elastic ligaments. Jaw suspension of this type is known as **hyostylic** (see the discussion of amphistylic suspension on page 243). Hyostyly permits multiple jaw positions, each appropriate to different feeding opportunities (Moss 1984).

The right and left halves of the pectoral girdle are fused together ventrally into a single U-shaped scapulocoracoid cartilage. Muscles run from the ventral coracoid portion to the symphysis of the lower jaw and function in opening the mouth. The advantages of the jaws of modern elasmobranchs are displayed when the upper jaw is protruded. Muscles swing the hyomandibular laterally and anteriorly to increase the distance between the right and left jaw articulations and thereby increase the volume of the orobranchial chamber. When it is performed rapidly, this movement sucks water and food forcefully into the mouth. This expansion of the mouth was not possible with an amphistylic jaw suspension because the palatoquadrate was attached to the chondrocranium.

With hyomandibular extension, the palatoquadrate is protruded to the limits of the elastic ligaments on its orbital processes. This protrusion allows delicate plucking of benthic foodstuffs. Protrusion also drops the mouth away from the head to allow a modern shark to bite an organism much larger than itself despite its large, sensitive rostrum. The dentition of the palatoquadrate is specialized; the teeth are stouter than those in the mandible and often recurved and strongly serrated. When feeding on large prey a shark opens its mouth, sinks its lower and upper teeth deeply into the prey, and protrudes its upper jaw ever more deeply into the slash initiated by the teeth. As the jaws reach their maximum initial penetration, the shark throws its body into exaggerated lateral undulations, resulting in a violent side-to-side shaking of the head. The head movements bring the serrated upper teeth into action as saws to sever a large piece of flesh from the victim. Mobility in the head skeleton, known as **cranial kinesis**, allows consumption of large food items. Cranial kinesis permits inclusion of large items in the diet of vertebrates, such as elasmobranchs, without excluding smaller, more diverse foodstuffs. Throughout their evolutionary history the Chondrichthyes have been consummate carnivores. In the third adaptive radiation of the elasmobranchs, locomotor, trophic, sensory, and behavioral characteristics evolved together in the mid-Mesozoic to produce forms still dominating the top levels of modern marine food webs. This

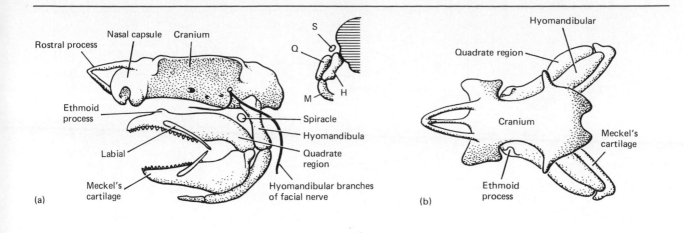

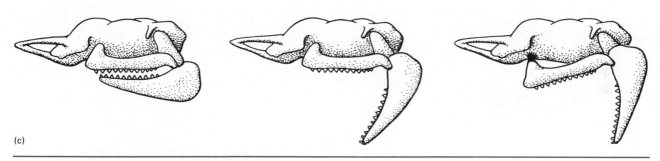

Figure 7–8. Anatomical relationships of the jaws and chondrocranium of living hyostylic sharks based on *Scyllium* and *Carcharhinus*. (a) Lateral and cross-sectional views of the head skeleton of *Scyllium* with the jaws closed. (b) Dorsal view of *Carcharhinus*. (c) During jaw opening and upper jaw protrusion the hyomandibula rotates from a position parallel to the long axis of the cranium to a position nearly perpendicular to that axis. S, spiracle; Q, quadrate region of the palatoquadrate; H, hyomandibula; M, mandible. (a) Modified after E. S. Goodrich, 1930, *Studies on the Structure and Development of Vertebrates*, Macmillan, London; (b) and (c) modified after S. A. Moss, 1984, *Sharks: An Introduction for the Amateur Naturalist*, Prentice-Hall, Englewood Cliffs, N. J.

position has gone hand in hand with the evolution of gigantism, one advantage of which is avoiding predation. The 339 species of sharks (Figure 7–9) and 424 species of skates and rays known today are large organisms, even for vertebrates (Gilbert 1982). An average-sized shark is about 2 meters long, an average ray half that length. Nevertheless a few interesting miniature forms have evolved

and inhabit mostly deeper seas off the continental shelves (Box 7–1).

In spite of their enormous range in size, all modern elasmobranchs have common skeletal characteristics that earlier shark radiations lacked. The continuous notochord was replaced in the modern radiation by cartilaginous centra that calcify in several distinctive ways. Between centra,

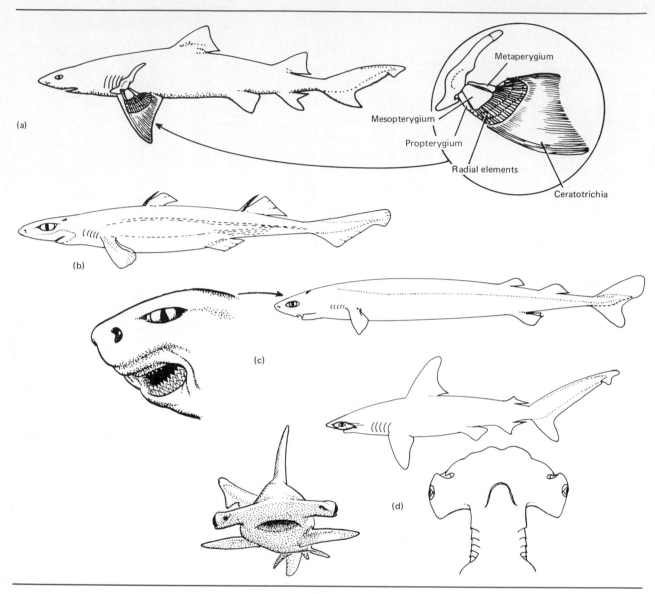

Figure 7–9. Some examples of modern sharks. (a) *Negaprion brevirostris*, the lemon shark, is widely used in elasmobranch research. It inhabits warm waters of the Atlantic frequented by bathers and divers and can attain a size sufficient to threaten humans. The internal anatomy of the pectoral girdle and fin are shown superimposed in their correct relative positions. (b) *Etmopterus vierens*, the green dogfish, is a miniature shark only 25 centimeters in length, yet it feeds on much larger prey items. (c) *Isistius brasiliensis*, the cookie cutter shark, is another miniature species whose curious mouth (left) is able to take chunks from fish and cetaceans much larger than itself. (d) Hammerhead shark (*Sphyrna*) in lateral, ventral, and frontal views.

Box 7–1. Food for Sharks

Living elasmobranchs offer us an opportunity to examine the diversity of food habits and ecological specializations within a group of closely related vertebrates. An interesting perspective on ecological principles that apply to many vertebrates can be obtained by examining the feeding habits of the smallest and largest elasmobranchs.

The green dogfish, *Etmopterus virens*, attains a length of only 25 centimeters (Figure 7–9b). The black and gray bodies of these schooling sharks are punctuated with an elaborate pattern of green-glowing photophores. More than half of their food is squid and octopus, the eyes and beaks of which are sometimes so large that one wonders how they could have passed through the jaws and throat. Stewart Springer concludes that "green dogfish hunt in packs and may literally swarm over a squid or octopus much larger than they are, biting off chunks . . . and perhaps maintaining the integrity of their school visually through their distinctive lighting system."

Another miniature shark, *Isistius brasiliensis*, is even more bold in its feeding on large prey (Figure 7–9c). This brilliantly luminiscent shark lives in tropical deep waters in what is known as the **deep scattering layer** (DSL), which is a concentrated band composed by many species of vertically migrating animals. Many surface predators descend to the DSL in feeding forays; porpoises and tuna, which may weigh 400 kilograms, are among the voracious vertebrates that visit the DSL. These giant, fast-swimming predators often suffer wounds of curious origin, the shape of a silver dollar and about 1 centimeter in depth reminiscent of the hole left in dough by a cookie cutter. Only recently have these wounds been matched in size and shape to the jaws of *Isis-*

tius. Whether the photophores of a 40-centimeter *Isistius* play a part in some deception of the large predators on which it feeds or are used as social cues awaits observation from a deep submersible vehicle.

On the other end of the modern elasmobranch size spectrum are the largest living fishes: the basking shark and the whale shark. The basking shark, *Cetorhinus maximus*, 10 meters long or more, lives in subpolar and temperate seas feeding exclusively on zooplankton, such as millimeter-long copepods. A feeding basking shark swims with its mouth wide open. Over a thousand erectile, whiplike denticles on the inner surface of the gill arches strain from the surface waters the hundreds of kilograms of food consumed each day. A typical adult basking shark filters about 1500 cubic meters of seawater per hour as it swims along with its mouth open (Priede 1984). Studies of the bioenergetics of basking sharks indicate that there is not enough food energy in the water column in winter to sustain these behemoths. During the winter months when plankton stocks are at a low, basking sharks disappear and are thought to rest on the bottom after shedding their gill rakers for the season. In contrast, whale sharks, *Rhiniodon typus*, which grow to 20 meters, are more tropical and able to feed on plankton year around. Unique branchial specializations have converted their gills to function as enormous sieves as well as respiratory surfaces. These solitary, white-spotted and striped Goliaths have been observed to feed head up and tail down at the surface. They lift above the water in the middle of a shoal of small fish until all the water in their orobranchial chamber has drained out the gill slits. With gills closed and mouth open the whale shark sinks tail down until surface water

Box 7—1 (Continued)

floods over the rim of its terminal mouth, bringing with it sustenance for its 10,000 kilogram body.

In 1976 just outside the Hawaiian Islands and again in 1984 off southern California, two specimens of another very curious giant shark were inadvertently tangled in gear and hauled aboard ships. These 4.5-meter-long, rather soft-bodied and apparently slow-moving sharks were completely unknown to science and have been named *Megachasma pelagios*, popularly known as megamouth. Megamouth has an enormous head and astonishingly wide terminal mouth filled with tiny teeth similar to the teeth of the basking and whale sharks. The cavernous mouth is lined with reflective crystals and may be bioluminescent. The stomach of the first specimen was filled with deep water shrimp and that of the second with assorted plankton, including numerous remains of deep-water jellyfish. As yet it is not known how megamouth feeds, but one can imagine a glowing blue-green mouth cruising slowly through the gloom of the deep attracting the small life forms found there like moths to a flame, and then engulfing them.

Although it seems paradoxical that the largest vertebrates feed on tiny motes floating in the sea and the smallest species tackle organisms many times their own size, there are good ecological reasons behind these phenomena. Sunlight is the ultimate source of energy for life on earth, but no vertebrate has the ability to capture this energy directly. The enumeration of the indirect sources through which solar energy passes is called a **food chain** or, because few vertebrates feed on a single food source, more properly, a **food web**. At each point of transfer in a web a great deal of energy is dissipated. On the average only about 10 percent of the energy in each step is passed to the next. Clearly, a vertebrate feeding on predatory tuna, as tiny *Isistius* does, is far removed from the primary source of energy. Because their individual and population requirements are comparatively low, the feeding methods employed by *Isistius* are effective. If a whale shark fed primarily on tuna, it would simply not be able to find enough food to maintain its enormous bulk. The solution for whale sharks and all extremely large vertebrates—most dinosaurs, moas, elephants, blue whales—has been to feed near the sunny side of the food web by eating plants or organisms that eat plants.

---

spherical remnants of the notochord fit into depressions on the opposing faces of adjacent vertebrae. Thus, the axial skeleton is a laterally flexible structure with rigid central elements swiveling on ball-bearing joints of calcified cartilage and notochordal remnants. In addition to the neural and hemal arches, extra elements not found in the axial skeleton of other vertebrates (the intercalary plates) protect the spinal cord above and the major arteries and veins below the centra. Although centra occur in many other gnathostomes, those of living sharks are so distinctive in morphology and fossil history that they are probably independently evolved.

Shark scales also changed. The scales of modern sharks are single-cusped and have a single pulp cavity. Scales in more primitive sharks may have begun development similarly but often fused to form larger scales later in life. Modern sharks add more individual scales to their skin as they grow and new scales may be larger in proportion to the increase in size of the shark. The size, shape, and arrangement of these **placoid scales** reduces turbulence in the flow of water adjacent to the body surface and therefore increases the efficiency of swimming. It seems that this simplification of denticles may relate to increased locomotor efficiency.

A characteristic that differentiates chondrichthyans from other living jawed fishes is the absence of a gas-filled bladder. Elasmobranchs (and holocephalians) use their liver to counteract the weight of their dermal denticles, teeth, and calcified cartilages. H. D. Baldridge has found the average tissue densities of sharks with their livers removed to be 1.062 to 1.089 grams per milliliter (g/ml). Because seawater has a density of about 1.030 g/ml, one might conclude that a shark not swimming to stay afloat would sink. The liver of sharks, however, is well known for its high oil content, which gives shark liver tissue a density of about 0.95 g/ml. The liver may contribute as much as 25 percent of the body weight. By adjusting the oil content and size of their livers through growth and reabsorption, sharks can adjust their buoyancy as well as store energy. A 4-meter tiger shark (*Galeocerdo cuvieri*) weighing 460 kilograms on land may weigh as little as 3.5 kilograms in the sea. Not surprisingly, benthic sharks have livers with fewer and smaller oil vacuoles in their cells.

Long before it was realized that sharks may weigh very little in their natural habitat due to liver buoyancy, a very different mechanism was credited with preventing sharks from sinking. The peculiar hydrodynamics of a laterally oscillated heterocercal tail were thought to produce an upward lift as well as a forward thrust. Anterior surfaces such as the pectoral fins and perhaps the snout, if properly oriented, will produce lift as the shark moves through the water. Thus, although gravity pulls a shark downward, two areas of lift—one anterior and one posterior—were considered to act on a swimming shark, permitting it to float. Two objections may be raised to this long-used explanation. First, through the work of Baldridge and others we now know that a motionless pelagic shark may be very nearly neutrally buoyant. Any lift, either anterior or posterior, during forward

locomotion would therefore cause a shark to rise in the water column—something we know does not necessarily occur. Second, the amount of lift resulting from the heterocercal tail and the pectoral fins is proportional to the forward velocity; the gravitational force is constant. Hence a shark would be neutrally buoyant and able to remain at the same height in the water column only over a narrow range of swimming speeds. If the shark swam too slowly, it would sink; if it swam too rapidly, it would rise.

K. S. Thomson has analyzed the swimming action of several living shark species (Northcutt 1977, Moss 1984). Because dynamic adjustments of the muscles attached to the ceratotrichia of the hypochordal lobe can independently change the lobe's angle of attack, the heterocercal tail can deliver thrust over a wide range of angles, not simply forward and up as supposed previously. The more ventral portions of the heterocercal caudal fin attach farther anteriorly on the axial skeleton than their dorsal counterparts and therefore react to the rearward passage of a lateral undulation at an earlier time. In addition, portions of the heterocercal caudal have strong muscular and connective tissue elements that actively alter the shape of the various lobes to produce differing hydrodynamic effects. Thomson believes that in general the thrust produced by the heterocercal tail projects forward and through the plane of the insertions of the pectoral fins on the trunk. Sharks apparently have great control of their position in the water column. Indeed, Thomson thinks that the highly controllable heterocercal tail of living sharks allows them to develop extremely powerful dives and climbs in the water over a wide range of speeds, permitting sharks to make oblique attacks and shear off flesh from large prey. Students of fossil fishes have been slow to apply this functional analysis to the many forms with heterocercal tails.

Although not unique, the sensory systems of modern sharks, skates, and rays are certainly refined and diverse (Hodgson and Mathewson 1978, Sweet et al. 1983). Sharks may detect prey via **mechanoreceptors** of their **lateralis system**, an interconnected series of superficial tubes, pores, and

Box 7–2. Electroreception by Elasmobranchs

The ability to detect electric fields is found in many fishes, especially elasmobranchs. On the heads of sharks, and on the heads and pectoral fins of rays, are structures known as the **ampullae of Lorenzini**. The ampullae are sensitive electroreceptors (Figure 7–10). The canal that connects the receptor to the surface pore is filled with an electrically conductive gel, and the wall of the canal is nonconductive. Because the canal runs for some distance beneath the epidermis, the sensory cell can detect a difference in electrical potential between the tissue in which it lies (which reflects the adjacent epidermis and environment) and the distant pore opening. Thus it can detect electrical fields, which are changes in electrical potential in space. The primitive electroreceptor cell is a modification of the hair cells of the lateral line (Figure 8–12). Electroreceptors of elasmobranchs respond to minute changes in the

electrical field surrounding an animal. They act like voltmeters, measuring a difference in electric potentials at discrete locations across the body surface. Voltage sensitivities are remarkable: ampullary organs have thresholds lower than 0.01 microvolt per centimeter, a level of detection achieved only by the best voltmeters.

Elasmobranchs use their electric sensitivity to detect prey and, possibly, for navigation. All muscle activity generates electric potential: motor nerve cells produce extremely brief changes in electrical potential, and muscular contraction generates changes of longer duration. In addition, a steady potential issues from an aquatic organism as a result of the chemical imbalance between the organism and its surroundings.

A. J. Kalmijn (1974) demonstrated that sharks (*Scyliorhinus caniculus*) can detect the electrical potentials emitted by flounders. The

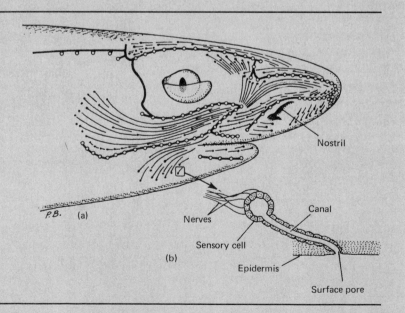

Figure 7–10. Ampullae of Lorenzini: (a) distribution of the ampullae on the head of a spiny dogfish, *Squalus acanthius*. Open circles represent the surface pores and the black dots are the positions of the sensory cells; (b) a single ampullary organ.

sharks were able to find a live flounder, even when it was buried under the sand (Figure 7–11a), and enclosing the flounder in an agar chamber (which prevents the escape of olfactory cues but is transparent to electrical potentials) did not alter the shark's reaction (Figure 7–11b). However, pieces of dead fish, which had little electrical activity but were odorifer-

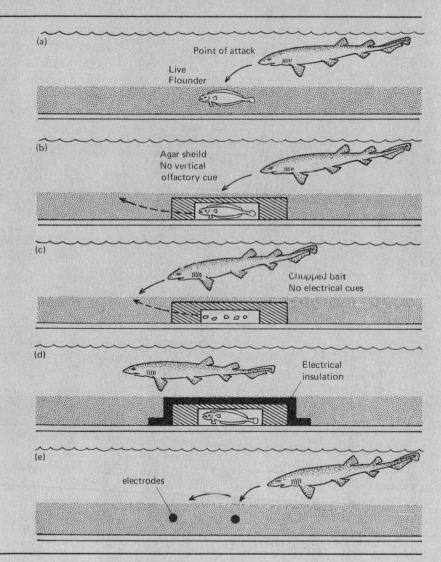

Figure 7–11. Kalmijn's experiments illustrating the electrolocation capacity of elasmobranchs. (Modified from A. J. Kalmijn, 1974, in *Handbook of Sensory Physiology,* volume 3, part 3, edited by A. Fessard, Springer-Verlag, New York.)

Box 7–2 (Continued)

ous, were not detected (Figure 7–11c), and enclosing a live flounder in an electrically insulting film also prevented detection (Figure 7–11d). Electrodes that simulated the bioelectric field of a flounder induced the shark to attack, even in the absence of olfactory cues (Figure 7–11e). From these elegant experiments it was clear that neither visual nor olfactory cues were necessary, and electrical activity in the absence of any other stimuli was sufficient to elicit a response. Subsequently, Kalmijn (1982) has demonstrated electroreception by free-ranging sharks in the wild.

Electroreception might be used for navigation as well. The electromagnetic field at the Earth's surface produces tiny voltage gradients, and a shark swimming could encounter gradients as large as 0.4 microvolt per centimeter. In addition, ocean currents generate electric gradients as large as 0.5 microvolt per centimeter as they carry ions through the Earth's magnetic field. The use of these potentials for navigation has not been demonstrated, but their magnitude places them well within the range of sensitivity of elasmobranchs.

---

patches of sensory cells distributed over the head and along the sides that respond to vibrations transmitted through the water. The basic units of mechanoreceptors are the **neuromast organs**, a cluster of sensory and supporting cells that are found on the surface and within the lateral-line canals. The **ampullae of Lorenzini**, mucus-filled tubes with sensory cells and afferent neurons at their base, are exquisitely sensitive to electrical potentials and can even detect prey from their weak electrical fields (Box 7–2).

Chemoreception is another important sense. In fact, sharks have been described as swimming noses, so acute is their sense of smell. Hammerhead sharks of the genus *Sphyrna* (Figure 7–9) may have enhanced the directionality of their olfactory apparatus by placing the nostrils far apart on the odd lateral expansions of their heads.

Finally, vision is important to the feeding behavior of sharks. Especially well developed are mechanisms for vision at low light intensities at which humans would find vision impossible. This sensitivity is due to a rod-rich retina and cells with numerous plate-like crystals of guanine that are located just behind the retina in the choroid layer.

Called the **tapetum lucidum**, the crystals function like microscopic mirrors to reflect light back through the retina and increase the chance that light will be absorbed. This mechanism, although of great benefit at night or in the depths, has obvious disadvantages in the bright sea surface of midday. To regulate the amount of bright light, cells containing the dark pigment melanin expand over the reflective surface to occlude the tapetum lucidum and absorb all light not stimulating the retina on first penetration. With so many sophisticated sensory systems, it is not surprising that the brains of many species of sharks are proportionately heavier than the brains of other fishes and approach the brain-to-body weight ratios of some tetrapods.

Anecdotal and circumstantial evidence suggests that sharks regularly use their various sensory modalities in an ordered sequence in locating, identifying, and attacking prey. Olfaction is often the first of the senses to alert a shark of potential prey, especially when the prey is wounded or otherwise releasing body fluids. Such signals are relatively nondirectional in a still, fluid environment, but currents give directionality to odors and a shark employs its sensitive sense of smell to guide its swimming up current through an increasing odor gradient. Because of its exquisite sensitivity, a shark can effectively use smell as a long-distance sense. Not as useful over such potentially great distances but much more directional over a wide range of environmental conditions is another distance sense—vibration sensitivity. The lateralis system and the sensory areas of the inner ear are related forms of mechanoreceptors highly efficient in detecting vibrations such as those produced by a struggling fish. The vibration sense is effective in drawing sharks from considerable distances to a sound source. This has been demonstrated in macabre sea rescue operations in which sharks were apparently attracted by the helicopter rotor vibrations that fall into the same frequency range as those of a struggling fish. Under more controlled conditions, hydrophones can be used to attract sharks by broadcasting vibrations with frequencies like those produced by a struggling fish.

Whether by olfaction or vibration detection, once a shark is close to the stimulus source, vision takes over as the primary prey detection modality. If the prey is easily recognized visually, a shark may proceed directly to an attack. Unfamiliar prey is treated differently, as studies aimed at developing shark deterrents discovered. A circling shark may suddenly turn and rush toward unknown prey. Instead of opening its jaws to attack, however, the shark bumps or slashes the surface of the object with its rostrum. Opinions differ as to whether this is an attempt to determine texture through mechanoreception, to make a quick electrosensory appraisal, or to use the rough placoid scales to abrade the surface releasing fresh olfactory cues. Following further circling and apparent evaluation of all sensory cues from the potential prey, the shark may either wander off or attack. In the latter case the rostrum is raised, the jaws protruded, and in the last moments before contact many sharks draw an opaque eyelid, the **nictitating membrane** across the eyes to protect them. At this point it appears that sharks shift entirely to electroreception to track prey during the final stage of attack. This hypothesis was developed while studying the attacks by large sharks on bait suspended from boats and observed from submerged protective cages. After the occlusion of its eyes by the nictitating membrane, an attacking shark was frequently observed to veer from the bait and bite some inanimate, generally metallic object in the close vicinity (including the observer's cage, much to the dismay of the divers). Apparently, olfaction and vibration senses are of little use at very close range, leaving only electroreception to guide the attacking shark. The unnatural environment of the bait stations included stronger influences on local electric fields than the bait, and the sharks attacked these artificial sources of electrical activity.

Even when no such confusions are in the surroundings, some species of sharks, the great white shark (*Carcharodon carcharias*) foremost among them, often do not devour prey immediately. Adult great white sharks specialize in large, primarily mammalian, vertebrate prey that are themselves powerful and often well armed with sharp teeth and claws. Sharks that feed on this sort of prey usually take a single bite and then resume circling at a safe distance until the wounded prey bleeds to death before continuing to feed.

Much of the success of the modern grade of elasmobranchs may be attributed to their sophisticated breeding mechanisms. Internal fertilization is universal. The pelvic claspers have a solid skeletal structure that may increase their copulatory effectiveness. During copulation (Figure 7–12a) a single clasper is bent at 90 degrees to the long axis of the body, and a dorsal groove present on each clasper comes to lie directly under the cloacal papilla from which sperm exit. The single flexed clasper is inserted into the female's cloaca and locked there by an assortment of barbs, hooks, and spines near the clasper's tip. Male sharks of small species secure themselves *in copulo* by wrapping around the female's body. Large sharks swim side by side, their bodies touching or enter copulation in a sedentary position with their heads on the substrate and their bodies angled upward (Tricas and Le-Feuvre 1985). Some male sharks and skates bite the female's flanks or hold on to one of her pectoral fins with their jaws. In these species females

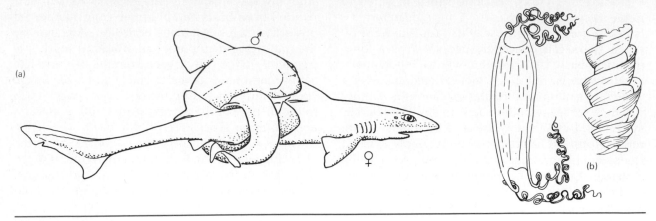

Figure 7–12. (a) Copulation in the smooth dogfish *Scyliorhinus*. Only a few other species of sharks and rays have been observed *in copulo* but all assume postures so that one of the male's claspers can be inserted into the female's cloaca. (b) The egg cases of two oviparous sharks, *Scyliorhinus* (left) and *Heterodontus* (right). (Not to same scale.)

may have skin on the back and flanks twice as thick as the skin of a male the same size. Whatever the position taken by the pair, when the single clasper is crossed to the contralateral side and securely inserted into the female cloaca, sperm from the genital tract are ejaculated into the clasper groove. Simultaneously, a muscular subcutaneous sac extending anteriorly beneath the skin of the male's pelvic fins contracts. This **siphon sac** has a secretory lining and is filled with seawater by the pumping activity of the male's pelvic fins before copulation. Seminal fluid from the siphon sac washes sperm down the groove into the female's cloaca, from which point the sperm swim up the female's reproductive tract.

The female has specialized structures at the anterior end of the oviducts, the **nidimental glands**, which secrete a proteinaceous shell around the fertilized egg. Most elasmobranch eggs are large (the size of a chicken yolk or larger) and contain a very substantial store of nutritious yolk. **Oviparous** (*ovum* = egg, *pario* = to bring forth) elasmobranchs have very large egg cases with openings for seawater exchange and protuberances to tangle or wedge themselves into protected portions of the substrate (Figure 7–12b). During the 6- to 10-month developmental period the zygote obtains nutrition exclusively from the yolk. Inorganic molecules including dissolved oxygen are taken from a flow of water induced through swimming-like movements made by the developing fish. Upon hatching, the young are generally miniature replicas of the adults and seem to lead a life much as they do when mature.

A most significant step in the evolution of elasmobranch reproduction was prolonged retention of the fertilized eggs in the reproductive tract. Today, many **ovoviviparous** (*vivus* = alive) species retain the developing young within the oviducts until they are able to lead an independent life. Reduction in the nidimental gland's shell production, and increased vascularization of the oviducts and yolk sacs, are the only notable differences between oviparous and ovoviviparous forms. All nutrition comes from the yolk, and only inorganic ions and dissolved gases are exchanged between the maternal circulation and that of the developing young. The eggs often hatch within the oviducts, and the young may spend as long in their mother after hatching as they did within the shell. As many as 100, but more often about a dozen, young are born at a time.

A natural step from the ovoviviparous condition is full **viviparity** or **matrotrophy** (*matro* = mother, *troph* = nourishment), the situation in which the nutritional supply is not limited to the yolk. Elasmobranchs have independently evolved matrotrophy several times. Some elasmobranchs develop long spaghetti-like extensions of the oviduct walls that penetrate the mouth and gill openings of the young and secrete a milky nutritive substance. Other species simply continue to ovulate, and the young that hatch in the oviducts feed on these eggs. The most common and most complex form of viviparity among sharks is the yolk sac placenta, whereby each embryo obtains nourishment from the maternal uterine bloodstream via the highly vascular yolk sac of the embryo. Our knowledge of vertebrate reproductive modes has recently undergone an enormous growth. The terminology traditionally applied and introduced above is not fully accurate when applied to other classes of vertebrates. **Lecithotrophic viviparity** (*lecith* = egg, *troph* = nourishment) more precisely defines the phenomenon of ovoviviparity in some instances, and **placentotrophic viviparity** (*placenta* − a placenta) is precise usage when a functional placenta occurs.

No matter which of the various routes of nourishment have brought young elasmobranchs to their free-living size, once the eggs are laid or the young born there is no evidence of further parental investment in them through care, protection, or feeding. In fact, elasmobranchs have long been considered solitary and asocial, but this view is changing. Accumulating field observations, often from aerial surveys (Kenney et al. 1985) or by scuba divers in remote areas, indicate that elasmobranchs of many species may aggregate in great numbers periodically, perhaps annually. More than 60 giant basking sharks were observed milling together and occasionally circling in head-to-tail formations in an area off Cape Cod in summer, and an additional 40 individuals were nearby. Over 200 hammerhead sharks have been seen near the surface off the eastern shore of Virginia in suc-

cessive summers. Divers on seamounts that reach to within 30 meters of the surface in the Gulf of California have observed enormous aggregations of hammerheads schooling in an organized manner around the seamount tip. Some observations include behavior thought to be related to courtship (Moss 1984). More than 1000 individuals of the blue shark (*Prionace glauca*) have been observed near the surface over canyons on the edge of the continental shelf off Ocean City, Maryland. Fishermen are all too familiar with the large schools of spiny dogfish (genus *Squalus*) that seasonally move through shelf regions, ruining fishing by destroying gear, consuming bottom fish and invertebrates and displacing commercially valuable species. These dogfish schools are usually made up of individuals that are all the same size and the same sex. The distribution of schools is also peculiar: Female schools may be inshore and males offshore or male schools may all be north of some point and the females south.

A similar sort of geographic sex segregation by hammerhead shark schools has recently been suggested to be the result of differential feeding preferences in habitats with different biological productivities (Klimley 1987). Females feeding on richer grounds grow faster, thus being larger at sexual maturity than males and better able to support embryonic young successfully. Such arguments do not explain why males should live in less productive habitats and grow more slowly. Our understanding of these phenomena is slim, but it is clear that not all elasmobranchs are solitary all of the time.

For the most part we have been examining the characteristics of **pleurotremate** elasmobranchs—the sharks with gill openings (*trem*) are on the sides (*pleur*) of the head. These forms number about 339 living species. Although similar in overall appearance, the living sharks actually come from two long isolated lineages. The more primitive in their general anatomy (especially the smaller size of their brain) are the 80 species of **squaloid** sharks. Squaloids include the spiny and green dogfish, the cookie-cutter shark, and the basking and megamouth sharks. These species usually live in cold, deep water. The other 250 or so species of sharks are members of the **galeoid** lineage that includes the hornshark, the nurse and carpet sharks, the whale shark, the mackerel sharks (including the great white shark), the carcharhinid sharks, and the hammerhead sharks. Galeoid sharks are the dominant carnivores of shallow, warm, species-rich regions of the oceans.

Surprising to many is the fact that the **hypotremate** elasmobranchs—the skates and rays—are more diverse than are the sharks. Approximately 424 living species of skates and rays are currently recognized. These fishes are closely interrelated and have a long history of phylogenetic isolation from both clades of living sharks. The suite of specializations characteristic of skates and rays relates to their early assumption of a benthic, **durophagous** (*duro* = hard, *phagus* = to eat) habit. The teeth are almost universally hard, flat crowned plates that form a pavement-like dentition. The mouth may be highly and rapidly protrusible to provide a powerful suction used to dislodge shelled invertebrates from the substrate.

Skates and rays are derivatives of the modern shark radiation adapted for benthic habitats: they have radial cartilages that extend to the tips of their pectoral fins. However, these fins are greatly enlarged and the anterior–most basal elements fuse with the chondrocranium in front of the eye or with each other in front of the rest of the head. These specializations are related to the very different mode of locomotion employed by skates and rays.

Locomotion in these dorsoventrally compressed fishes is accomplished by undulation of the massively enlarged pectoral fins. This great flexibility of the fins results from a reduction in the number of placoid scales. The placoid scales so characteristic of the integument of a shark are absent from large areas of the bodies and pectoral fins of skates and rays. The few remaining denticles are often greatly enlarged to form sharp, stout bucklers along the dorsal midline. Perhaps to compensate for this relative nudity, or perhaps as a predatory ruse, many of these skates and rays cover themselves with a thin layer of sand. They spend hours partially buried and nearly invisible except for the dorsally prominent eyes and spiracles through which they survey their surroundings and take in oxygenated respiratory water. More advanced forms, such as stingrays (family Dasyatidae), have a very few greatly elongate and venomous modified placoid scales at the base of the tail—an apparently sufficient defense for their otherwise entirely naked bodies. The most highly specialized rays (of the family Mobulidae) are derived from these entirely naked types but spend little of their time resting on the bottom. Using powerfully extended pectorals, these devilfish or manta rays (up to 6 meters in width) swim through the open sea with flapping motions of their pectoral fins. Skates and rays are primarily benthic invertebrate feeders (occasionally managing to capture small fishes), but the largest rays, like the largest sharks, are plankton strainers (Box 7–1). Of unknown significance is the phenomenon of sexually dimorphic dentition in many benthic invertebrate feeding rays. Different dentitions coupled with the generally larger size of females may effectively reduce competition for food resources between the sexes, but no difference in stomach contents has been found. Alternatively, the teeth of the males may be important in properly holding onto and stimulating the females during copulation.

Skates (family Rajidae) also display another unique feature for Chondrichthyes: They have specialized tissues in their long tails that are capable of emitting a weak electric discharge. Each species appears to have a unique pattern of discharge, and the discharges may identify conspecifics in the gloom of the seafloor. The electric rays and torpedo rays (family Torpedinidae) have modified gill muscles that produce electrical discharges up to 200 volts.

A Second Radiation of Chondrichthyans: Holocephali

Most living chondrichthyans are contained in the Elasmobranchii, but a small portion are grouped as ratfish or chimaeras (subclass Holocephali, Figure 7–13a). There is little agreement about holocephalan phylogeny. Two distinct fossil groups have been proposed as ancestors. A group of peculiar placoderms, the ptyctodontids, is known from the mid-Devonian. Rarely exceeding 20 centimeters in length, they showed a reduction in the extent and number of head and thoracic shield plates. A short palatoquadrate bound to the cran-

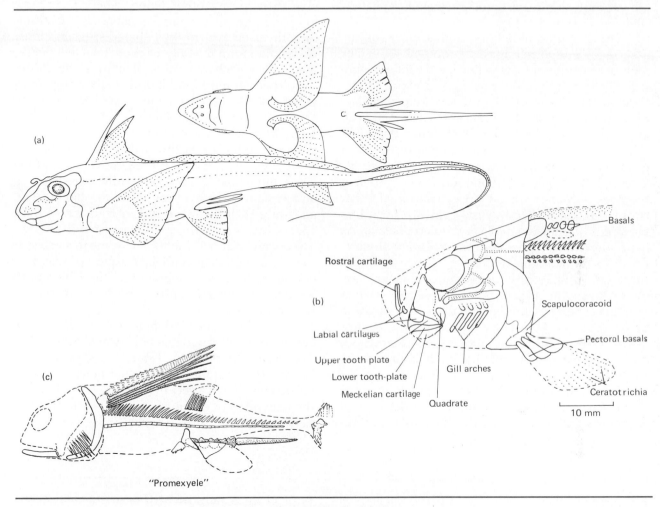

Figure 7–13. (a) Common chimaera *Hydrolagus colliei*, a living holocephalan, in lateral and ventral views. (Modified after H. B. Bigelow and W. C. Schroeder, 1953, *Fishes of the Western North Atlantic*, part 2, Sears Foundation for Marine Research, Yale University, New Haven, Conn.) (b) *Ctenurella*, an upper Devonian ptyctodont placoderm claimed by some paleontologists to be related to chimaeras; see also Figure 7–3c. (Modified after J. A. Moy-Thomas and R. S. Miles, 1971, *Paleozoic Fishes*, W. B. Saunders, Philadelphia.) (c) *Phomeryele*, a Carboniferous iniopterygian shark proposed by other paleontologists as a relative to the modern ratfishes. (Modified after P. H. Greenwood et al., 1973, *Interrelationships of Fishes*, Academic Press, New York.)

ium carried a single pair of large upper tooth plates that were opposed by a smaller mandibular pair. The gills were covered by a single operculum. Postcranial characters, such as fin spines, paired appendages, claspers and caudal development, are strongly reminiscent of living holocephalans (Figures 7–3c and 7–13a).

The first undoubted holocephalans are of Jurassic age, and other fossils proposed as earlier members of the chimaera lineage do little to link the ptyctodont placoderms and modern forms. Ptyctodonts are more like modern holocephalans than the most primitive holocephalans, which are in some cases rather shark-like. Since the 1950s Rainer Zangerl and coworkers have described a group of bizarre and obviously specialized forms (Figure 7–13c). These Iniopterygia (*inion* = back of the neck, *pteron* = wings) have characteristics that convinced Zangerl they were evidence for a link between the earliest sharks and holocephalans. As in modern holocephalians, the palatoquadrate is fused to the cranium (autostylic suspension), but the teeth, unlike those of modern chimaeras, are in replacement families like those of elasmobranchs. The earliest holocephalans are thought to be of upper Carboniferous age, too old to be descendent from the contemporaneous iniopterygians. We currently think that both groups arose from primitive sharks. If a connection is made between holocephalans and primitive-grade elasmobranchs, the ptyctodont placoderms and chimaeras will represent one of the most outstanding examples of convergent evolution in groups from different eras.

Whatever the origin of the chimaeras, the 25 living forms (none much over a meter in length) have a soft anatomy more similar to sharks and rays than to any other living fishes. They have long been grouped with elasmobranchs as Chondrichthyes because of their shared specializations. Generally found in water of over 80 meters depth, the Holocephali move into shallow water to deposit their 10-centimeter horny shelled eggs from which hatch miniature chimaeras. They feed on shrimp, gastropod mollusks, and sea urchins. Their locomotion is produced by lateral undulations of the body that throw the long tail into sinusoidal waves and by fluttering movements of the large, mobile pectorals. The solidly fused nipping and crushing tooth plates grow throughout life, adjusting their height to the wear they suffer. Of special interest are the armaments: a poison gland associated with the stout dorsal spine in some species, and mace-like cephalic claspers of males. A curious detour in vertebrate life, the Holocephali are, like the other Chondrichthyes, worthy of further study.

From the array of adaptations present in modern chondrichthyans, it is not surprising that they have survived unchanged since the Mesozoic. The surprising fact is that they are not more diverse or found in a wider variety of habitats. To understand their limitations we must examine their competition—the bony fishes (Chapter 8).

Summary

A major evolutionary innovation of the vertebrates was the appearance in the Silurian of fishes with jaws and paired appendages that had internal skeletal supports. The first of these jawed fishes, the acanthodians (Chapter 8), were soon joined by the diverse but phylogenetically isolated placoderms. Remains of these first gnathostomes yield few clues to the origins of jaws and paired appendages. However, patterns of embryonic development indicate that jaws and their supports evolved from modified anterior skeletal elements related to the gill supports. The modern fauna includes descendants of these early jawed fish: the Chondrichthyes, which include sharks, skates, rays, and chimaeras. The evolution of chondrichthyans shows periods of relatively rapid changes in feeding and locomotor mechanisms followed by radiations, subsequent further morphological changes and reradiations, culminating in modern elasmobranchs and holocephalians. The specializations of these fishes offer us an opportunity to compare and contrast the evolutionary results of an alternative to the dominant lineage of vertebrate evolution.

References

Carroll, R. L. 1987. *Vertebrate Paleontology and Evolution.* W. H. Freeman, New York.

Gilbert, P. W. (editor). 1982. *Oceanus* 24(4). Well-illustrated, up-to-date reviews for the nonspecialist on shark biology.

Greenwood, P. H., R. S. Miles, and C. Patterson (editors). 1973. *Interrelationships of Fishes.* Academic Press, New York.

Hodgson, E. S. and R. F. Mathewson (editors). 1978. *Sensory Biology of Sharks, Skates, and Rays.* Technical Information Division, Naval Research Laboratory, Washington, D.C.

Kalmijn, A. J. 1974. The detection of electric fields from inanimate and animate sources other than electric organs. Pages 147–200 in *Handbook of Sensory Physiology,* volume 3, part 3, edited by A. Fessard. Springer-Verlag, New York.

Kalmijn, A. J. 1982. Electric and magnetic field detection in elasmobranch fishes. *Science* 218:916–918.

Kenney, R. D., R. E. Owen and H. E. Winn. 1985. Shark distributions off the Northeast United States from marine mammal surveys. *Copeia* 1985:220–223.

Klimley, A. P. 1987. The determinants of sexual segregation in the scalloped hammerhead shark, *Sphyrna lewini. Environmental Biology of Fishes* 18:27–40.

Moss, S. A. 1984. *Sharks: An Introduction for the Amateur Naturalist.* Prentice-Hall, Englewood Cliffs, N. J. One of the best popular accounts of these fascinating vertebrates.

Moy-Thomas, J. A. and R. S. Miles. 1971. *Paleozoic Fishes.* W. B. Saunders, Philadelphia.

Northcutt, R. G. (editor). 1977. Recent advances in the biology of sharks. *American Zoologist* 17(2). Symposium papers review a wide variety of topics with extensive references to the important literature.

Priede, I. G. 1984. A basking shark (*Cetorhinus maximus*) tracked by satellite together with simultaneous remote sensing. *Fisheries Research* 2:201–216.

Romer, A. 1966. *Vertebrate Paleontology,* 3rd edition. The University of Chicago Press, Chicago.

Sweet, W. J. A. J., R. Nieuwenhuys, and B. L. Roberts. 1983. *The Central Nervous System of Cartilaginous Fishes.* Springer-Verlag, New York.

Tricas, T. C. and E. M. LeFeuvre. 1985. Mating in the reef white-tip shark *Triaenodon obesus. Marine Biology* 84:233–237.

Webb, P. W. and R. Smith. 1980. Function of the caudal fin in early fishes. *Copeia* 1980(3):559–562.

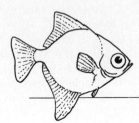

By the end of the Silurian the agnathous fishes described in Chapter 6 had diversified and the cartilaginous gnathostomes treated in Chapter 7 were in the midst of their first radiation. The stage was set for the appearance of the final and most successful group of fishes, the bony fishes. The first fossils of bony fishes come from the late Silurian and their radiation was in full bloom by the middle of the Devonian. Once again progressive specialization of feeding mechanisms is one of the key features of the evolution of a major group of vertebrates. An increasing flexibility among the bones of the skull and jaws allowed the ray-finned fishes to exploit a wide range of prey types and predatory modes. Specializations of locomotion, habitat, behavior, and life histories have accompanied the specializations of feeding mechanisms, and bony fishes are the largest and most diverse group of living vertebrates. Their body forms, behaviors, and mechanical functions are intimately related to the characteristics of the aquatic habitat and to the properties of water as an environment for life.

Dominating Life in Water: Teleostomes and the Major Radiation of Fishes

8

Living in Water

Seventy-three percent of the surface of the Earth is covered by water. Most of this water is held in the great ocean basins, which are populated by vertebrates, especially the bony fishes. The fresh-water lakes and rivers of the planet hold a negligible percentage of all the water on earth—about 0.01 percent. This is much less than that tied up in the atmosphere, ice, and groundwater, but it is exceedingly rich biologically. Rivers and lakes are complex habitats with short histories on a geological time scale. Nevertheless, Earth's lakes and rivers have individual life spans long enough for evolutionary processes to occur within their isolated shores. Nearly one-third of all bony fishes, the largest taxon of vertebrates, have evolved and live exclusively in fresh waters. Thus, watery environments gave rise to the first vertebrates, supported the first great steps in vertebrate evolution, and today they provide habitats for more kinds of vertebrates than any other environment.

In spite of this, water is a demanding medium for vertebrate life. Water can hold only about one twentieth as much oxygen as an equal volume of air. Biological and chemical processes can lower the oxygen content of waters inhabited by fishes. Water is also about 830 times as dense as air and 80 times more viscous than air, making movement in water energetically expensive. Water's mass means that pressures increase at depth. Other physical characters limit the amount of light and change the speed of sound transmission posing serious limitations on sensory modalities. It must be kept in mind that water is the ancestral medium of vertebrate life and adaptation to it dominated the first 150 to 200 million years of vertebrate evolution.

A characteristic feature of vertebrates is their exceptional motility which is reflected behaviorally in their diverse, usually agile, sometimes rapid, and often graceful locomotor abilities. Because of this ability to move about, vertebrates occupy a dominant ecological position in the various ecosystems in which they live. Fishes have responded evolutionarily to the demanding characteristics of their watery environment. Although each major clade seems to have solved environmental challenges in somewhat different ways, an examination of the specializations of the bony fishes to life in water illustrates one of their keys to evolutionary success—versatility. High rates of energy utilization, which are required to sustain the motion of fishes and their generally high activity levels, depend on an efficient respiratory system, the gills, which facilitate high rates of oxygen uptake for the oxidative combustion of food. For most fishes, locomotion is accomplished by undulatory contractions of the body muscles, a motility mechanism that has carried over into terrestrial vertebrates. Guidance is usually visual, but special senses include the lateral line (for detection of low-

265

frequency pressure waves) and electroreception, a sense highly developed in some fishes.

In this chapter we describe several basic specializations that have allowed fishes to become active and successful aquatic vertebrates. Many of the structures and functions of fishes set the stage for life on land.

Obtaining Oxygen in Water: Gills

Most aquatic vertebrates possess gills, evaginations from the body surface over which respiratory gases are exchanged. Gills are subject to abrasive damage when they are exposed externally, as they are in some aquatic salamanders. In fishes gills are protected within covered pharyngeal pockets (Figure 8–1). In osteichthyans water is generally pumped across the gills in one direction during two stages by the action of muscles that expand the buccal cavity and the opercular plates that cover the gills. The buccal flaps just inside the mouth and the flaps at the posterior and ventral margin of the opercula act as one-way valves to direct the movement of water from the pharynx across the gills to the exterior. The respiratory surfaces of the gill are delicate, vascularized projections from the lateral side of each gill arch (Figure 8–1b). Two columns, made up of numerous gill filaments, extend from each gill arch. The tips of these filaments from adjacent arches can be extended to meet, and thus water as it exits the buccal cavity must intimately contact the vascularized surfaces. Actual gas exchange takes place across numerous microscopic projections, the secondary lamellae (Laurent and Dunel 1980).

The pumping action of the mouth and opercular cavities sustains a positive pressure across the gills so that the respiratory current is only slightly interrupted during each pumping cycle (Figure 8–1c). In pelagic fishes such as mackerel, sharks, tuna, and swordfish, the ability to pump water across the gills has been lost or greatly reduced. A respiratory current is created by swimming with the mouth slightly opened, a method known as **ram ventilation**, but the fishes must perpetually swim. A great many other fishes capable of rapid sustained locomotion facultatively switch to ram ventilation during prolonged swimming bouts.

The vascular arrangement in the teleost gill is an engineering marvel, for it maximizes oxygen exchange. Each gill filament is supplied with two arteries, an afferent vessel running from the gill arch to filament tip and an efferent vessel returning blood to the arch (Farrell 1980). Each secondary lamella is a blood space covered only by a thin epithelium (through which gas exchange occurs) connecting the afferent to the efferent vessel (Figure 8–2a). Blood flow through the lamellae is counter to the flow of water across the gill. This structural arrangement, known as a **countercurrent exchanger**, assures that as much oxygen as possible diffuses into the gill (Figure 8–2b). Active pelagic fishes such as tunas, which sustain activity over extended periods of time, have strengthened gill filament structure, relatively larger gill exchange areas, and a higher oxygen-carrying capacity per milliliter of blood than sluggish benthic fishes, such as toadfishes and flatfishes (Table 8–1).

Terrestrial vertebrates seldom encounter low oxygen concentrations, for air contains 20 to 21 percent oxygen by volume. In aquatic habitats the amount of oxygen dissolved in water is much less than that present in a similar volume of air (a liter of air contains 210 milliliters of oxygen, whereas a liter of fresh water contains only 10.29 milliliters of oxygen). Increased temperature reduces the solubility of oxygen in water. This is particularly serious in tropical fresh waters, where high temperatures and the metabolism of microorganisms and phytoplankton often produce anoxic ("without oxygen") conditions.

Fishes that live in these conditions cannot obtain enough oxygen via gills and have accessory respiratory structures that enable them to breath air. The electric eel, in addition to gills, has an extensive series of highly vascularized papillae in the pharyngeal region. The eels rise to the surface to gulp oxygen-rich air, which diffuses across the papillae into the blood. Other fishes swallow air and extract oxygen through vascularized regions

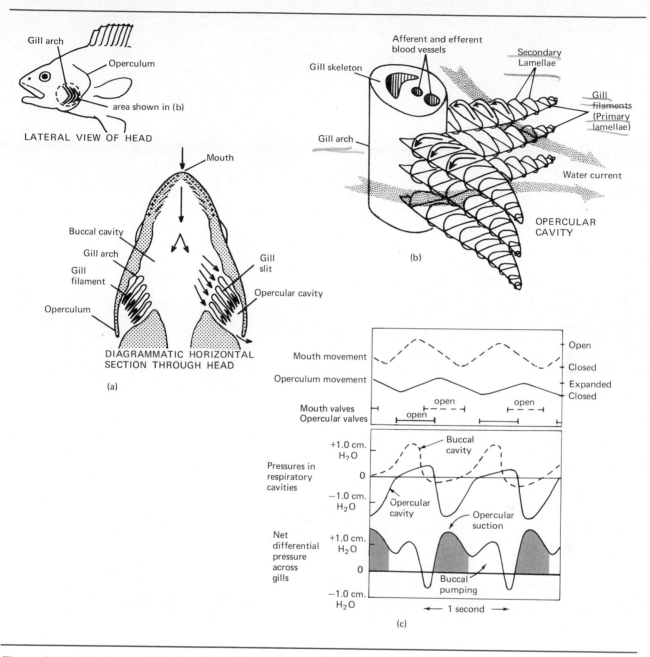

Figure 8–1. Anatomy and functional morphology of teleost gills: (a) position of gills in head and general flow of water; (b) water flow (shaded arrows) and blood flow (solid arrows) patterns through the gills; (c) water pressure changes across the gills during various phases of ventilation. (Modified after G. M. Hughes, 1963, *Comparative Physiology of Vertebrate Respiration*, Harvard University Press, Cambridge, Mass.)

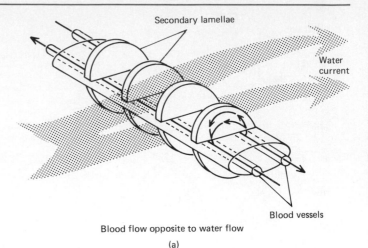

Secondary lamellae

Water current

Blood vessels

Blood flow opposite to water flow

(a)

Figure 8–2. Countercurrent exchange in the gills of actinopterygian fishes. (a) The direction of water flow across the gill opposes the flow of blood through the secondary lamellae. (b) This results in a higher oxygen loading tension in the blood and lower oxygen tension in the water leaving the gills. (c) If water and blood flowed in the same direction over and within the secondary lamellae, a lower oxygen tension would occur in the blood leaving the gills. (Modified after G. M. Hughes, 1963, *Comparative Physiology of Vertebrate Respiration*, Harvard University Press, Cambridge, Mass.)

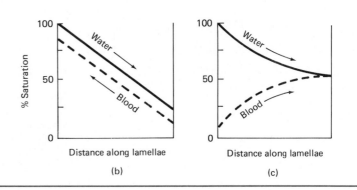

of the gut. The anabantid fishes of tropical Asia (bettas, gouramies) have vascularized chambers in the rear of the head, called labyrinths, which function in a similar manner. Many fishes are facultative air breathers, that is, oxygen uptake switches from gills to accessory organs when oxygen in the surrounding water becomes low. Others, like the electric eel and the anabantids, are obligatory air breathers, for the gills alone are inadequate to obtain sufficient oxygen even if the surrounding water is oxygen saturated. When denied access to air these fishes drown.

South American and African lungfish are obligate air breathers that drown if they are held under water. Lungfish breathe air with the aid of their lungs, which are derived evaginations of the gut. Similar structures are typical of all osteichthyans, perhaps extending back to the early Devonian acanthodians. The early Devonian freshwater fishes probably encountered stresses like those faced by tropical freshwater fishes today. Early lungs may also have served as floats to produce a more nearly neutral density, as does the homologous swim bladder of teleosts.

Locomotion in Water

Swimming results from sequential contractions of the muscles along one side of the body and simultaneous relaxation of those of the opposite

Table 8–1. The relation between general level of activity, rate of oxygen consumption, respiratory structures, and blood characteristics in fishes of three activity levels.

Activity Level	Species of Fish	Oxygen Consumption (ml O_2 g^{-1} h^{-1})	No. Secondary Gill Lamellae mm^{-1} of Primary Gill Lamella	Gill Area (mm^2 g^{-1} of body mass)	Oxygen Capacity (ml O_2 100 ml^{-1} of blood)	Hemoglobin (g 100 ml^{-1} of blood)
High	Mackerel* (*Scomber*)	0.73	31	1160	14.8	37.1
Intermediate	Porgy (*Stenotomus*)	0.17	26	506	7.3	32.6
Sluggish	Toadfish† (*Opsanus*)	0.11	11	197	6.2	19.5

*Modified carangiform swimmers; swim continuously.
†Benthic fish.

side. Thus, a portion of the body momentarily bends and the bend is propagated down the body. When moving, therefore, a fish oscillates from side to side. These lateral undulations are most visible in elongate fishes, such as lampreys and eels (Figure 8–3). Identical undulations are observed in salamanders, snakes, and many lizards. The basic movement in aquatic locomotion, therefore, has been carried on in much of vertebrate life.

If a variety of fishes are observed, different propulsive movements can be described. In 1926, Charles Breder classified the undulatory motions of fishes into three types: **anguilliform**—typical of highly flexible fishes capable of bending into more than half a sinusoidal wavelength; **carangiform**—undulations limited mostly to the caudal region, the body bending into less than half a wavelength; and **ostraciiform**—body inflexible, undulation of the caudal fin (Figures 8–3 and 8–8). These basic categories have been extensively subdivided and redefined since 1926 (Lindsey 1978, Webb and Blake 1985) but are useful beginning points for understanding locomotion in water.

By recording body motions on film scientists have described the details of undulating locomo-

tion in different vertebrates to determine how they swim. Many of the specializations of body form, surface structure, fins, and muscle arrangement increase the efficiency of the different modes of swimming. Although the ultimate analysis of these efficiencies involves the complex mathematics of fluid mechanics and hydrodynamics, they are described here in qualitative terms.

A swimming fish must overcome the effect of gravity by producing lift and the drag of water by producing thrust (Figure 8–4). Most of us are familiar with how gravity, lift, drag, and thrust interact. We have all felt the wind tug on our hand when we held it out the window of a moving car. Held horizontal our hand (analogous to a foil represented by a fin on a fish or a wing of a bird or airplane) moves through the air without experiencing upward or downward forces on it. Angled up into the wind our hand rises effortlessly, angled down it declines. If the hand's angle to the wind is too great, the hand is forced backward. To develop forward motion of the hand against the wind an additional element is required: an expenditure of muscular energy to produce forward thrust. Fishes also expend muscular energy to pro-

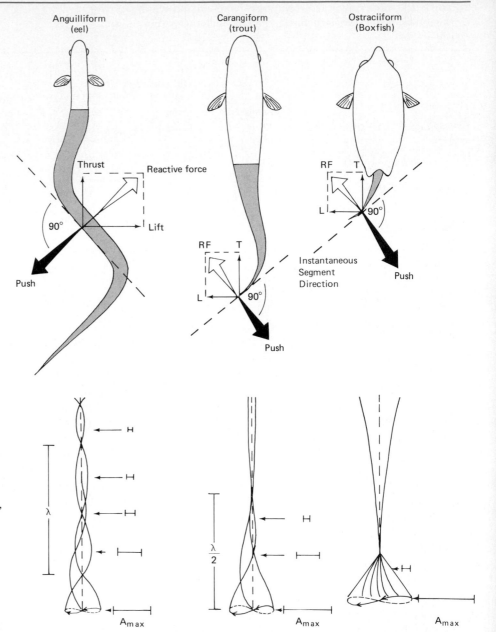

Figure 8–3. Basic movements of swimming fishes. Upper, outlines of some major swimming types showing regions of body that undulate (shaded); lower, diagrammatic waveforms created by undulations of points along the body and tail. A_{max} represents the maximum lateral displacement of any point. Note that A_{max} increases posteriorly; λ is the wavelength of the undulatory wave. Ostraciiform swimming, as initially defined by C. M. Breder, refers to a limited number of very specialized fishes, like box fishes and trunk fishes. A variety of fishes swim by propulsion from various fins rather than the body (see Figure 8–8).

duce forward movement by pushing on the water around them.

Overcoming Gravity: The Generation of Vertical Lift Because the average tissue density of typical fishes exceeds that of water, they sink unless their total density is reduced with a float or vertical lift is created by propulsive movements. Many pelagic fishes such as some tunas, mackerels, swordfishes (all known collectively as scombroid fishes), and

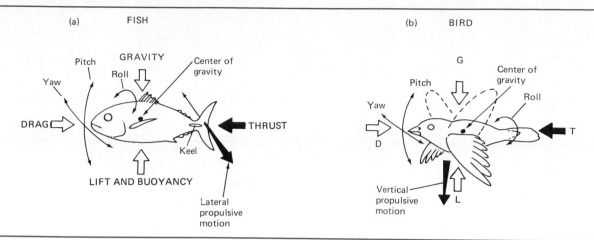

Figure 8–4. Comparison of forces associated with locomotion for a swimming and a flying vertebrate. (a) Motion of the caudal fin produces a lateral movement far from the center of gravity. The reactive force to this movement causes the head to yaw in the same direction as the tail. (b) In a bird the major propulsive stroke of the wings is downward. Since both the propulsive stroke and the reactive force (lift) act near the center of gravity the bird does not pitch. An up-and-down motion of the entire body occurs, however, because vertical lift is produced during the downward stroke.

sharks are negatively buoyant. Tunas and many sharks must swim constantly to produce sufficient vertical lift to overcome gravity. In some cases lift is accomplished by extending large wing-like pectoral fins at a positive angle of attack to the water that flows over them. Many bony fishes, however, are neutrally buoyant and maintain position without body undulations. Only the pectoral fins backpaddle to counter a forward thrust produced in reaction to the water ejected from the gills during each respiratory cycle. Here neutral buoyancy results from an internal float—a gas-filled swim bladder lying below the spinal column (Figure 8–5a). Fishes that are capable of hovering in the water tend to have well-developed swim bladders.

The swim bladder is located extraperitoneally and ventral to the vertebral column. It is a gas-filled sac that arises as an evagination from the embryonic gut. In marine teleosts, the bladder occupies about 5 percent of the body volume, in freshwater teleosts, 7 percent—volume differences just sufficient to produce neutral buoyancy in each

medium. Thus a teleost neither rises nor sinks and needs to expend little energy to maintain position. To maintain neutral buoyancy, however, a constant bladder volume is required.

As a fish swims vertically in the water column its body is subjected to varying pressures from the overlying water column. Increased pressure at depth tends to collapse the swim bladder; decreased pressure toward the surface allows the bladder to expand. A mechanism is needed to maintain constant volume if the bladder is to act as a float and to allow unimpeded vertical movements. This is accomplished by moving gas into it upon descent and removing gas upon ascent. The swim bladder wall, composed of interwoven collagen fibers, is virtually impermeable to gas diffusion, so gas leaks out slowly. More primitive teleosts, such as bony tongues, eels, herrings, anchovys, and the minnows, salmons, and their kin, retain a connection, the **pneumatic duct**, between gut and swim bladder (Figure 8–5). These fishes, referred to as **physostomes** (*phys* = a blad-

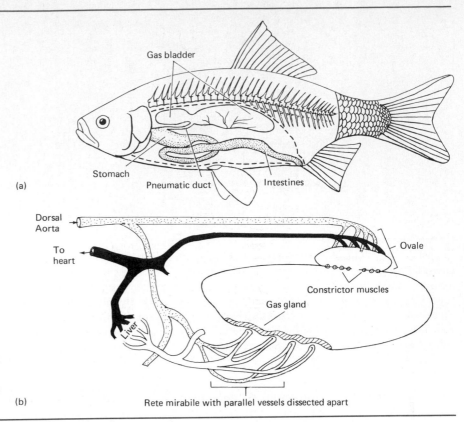

Gas bladder

Stomach

Pneumatic duct

Intestines

(a)

Dorsal
Aorta

To
heart

Liver

Ovale

Constrictor muscles

Gas gland

Rete mirabile with parallel vessels dissected apart

(b)

Figure 8–5. Swim bladder of actinopterygians: (a) the swim bladder is dorsal in the coelomic cavity just beneath the vertebral column; (b) vascular connections of a physoclistous swim bladder.

der, *stom* = mouth), can gulp air at the surface to fill the bladder and can burp gas upon ascent. Although the bladder functions as a float, many physostomes also use the bladder as an accessory respiratory structure, that is, as a lung. The pneumatic duct serves the same purpose as the windpipe or trachea of terrestrial vertebrates. In adult advanced teleosts, the pneumatic duct is absent, a condition termed **physoclistic** (*clist* = closed); gas cannot be released through the gut as it can in physostomes.

Regulation of constant swim bladder volume upon descent requires secretion of gas into the bladder in both physostomes and physoclists. Many teleosts possess a structure, the **gas gland**, located in the anterior ventral floor of the swim bladder (Figure 8–5b). Underlying the gas gland is the highly vascular **rete mirabile** ("marvelous net"), a complex, organized, overlapping array of blood vessels. This structure moves gas from blood to bladder, especially the oxygen carried by the hemoglobin of the red blood cells. Gas secretion occurs in many deep-sea fishes at considerable depths and against the enormous gas pressure within the bladder that is produced by the surrounding water pressure. Consider a mesopelagic lanternfish at 1000 meters depth. Each 10 meters of water increases the pressure on the fish by 1 atmosphere. The gas pressure within the swim bladder is 101 atmospheres, yet the oxygen pressure in the arterial blood and the surrounding water does not exceed 0.21 atmosphere (essentially the same as in surface waters). To secrete oxygen into the swim bladder requires that the oxygen pressure in the blood be multiplied 500 times.

sulting acidity of the blood in the rete mirabile causes hemoglobin to release oxygen into solution. Because of the anatomical relations of the rete mirabile, which folds back upon itself in a **counter-current multiplier** arrangement (Figure 8–5b), oxygen released from the hemoglobin accumulates and is retained within the rete mirabile until its pressure exceeds the oxygen pressure within the swim bladder. When this occurs, oxygen diffuses into the bladder, adjusting its volume. The longer the capillaries of the rete mirabile, the greater the multiplication of gas pressure achievable.

To compensate for gas expansion during ascent, physoclists simply open a muscular valve, called the **ovale**, located dorsally in the posterior region of the bladder adjacent to a capillary bed. The higher internal pressure of gases in the bladder (especially oxygen) causes diffusion into the blood.

Many deep-sea fishes have oily deposits of low density in the gas bladder or have reduced or lost the gas bladder and instead have lipids distributed in rich deposits throughout the body. The lipids provide static lift, as do the oils in shark livers. Because a smaller volume of the bladder contains gas, the amount of secretion required for a given vertical descent is less. Nevertheless, a long rete mirabile is needed to secrete oxygen at the high pressures required and the gas gland in deep-sea fishes is very large. Mesopelagic fish that migrate large vertical distances depend more on lipids such as wax esters than on gas for buoyancy. Their close relatives that do not undertake such extensive vertical movements depend more on swim bladders for buoyancy. [For additional details, see Bone and Marshall (1982).]

Overcoming Drag: The Generation of Thrust Exactly how do undulations produce thrust? The factors involved and how they interact to produce motion are extremely complex, and complete hydrodynamic explanations have yet to be achieved. Fishes swim forward by pushing backward on the water. For every *active* force there is an opposite *reactive* force (Newton's third law of motion). The overall reactive force is directed forward and at an angle to the side. Undulations produce not only an active force pushing backward, but also a lateral force. How can we analyze the effect of these multiple forces?

The push against the water that a fish makes with its body or fins has both direction and a magnitude of force and thus can be analyzed as a vector. The force of the water pushing back on the fish is also a vector in the opposite direction and (in an ideal system) of equal strength and is called the *reactive force*. Analysis of the reactive force into its geometric components reveals a forwardly directed part of the reactive force on the fish's body, the *thrust* required to overcome drag. The second component of the reactive force by vector analysis definition acts at right angles to thrust, that is, laterally to the side of the fish, and may be called the *side force* (Webb 1984). Side forces impart a lateral displacement motion to the fish's head through the axial skeleton tending to produce *yaw*. In many fishes *yaw* is countered by the inertia of a large rigid head that offers resistance to lateral displacement.

In slender flexible fishes such as eels, wave velocity along the body is uniform and always exceeds the swimming velocity, but the amplitude of each undulation increases from head to tail. Anguilliforms and carangiforms increase swimming speed by increasing the frequency of their body undulations—the tail beat frequency. Increasing the frequency of body undulations increases speed because it applies more power (force per unit time) to the water. Yet different fishes achieve very different speeds through body undulations; some are slow (eels) and others are very fast (tunas).

The eel's long body limits speed because it induces too much drag due to the friction of water on the surface of the fish. Fishes that swim rapidly are less flexible, and force from the contraction of anterior muscle segments is transferred through ligaments to the caudal peduncle and the tail. Morphological specializations of this swimming mode reach their zenith in fishes like tunas, where the caudal peduncle is slender and the tail greatly expanded vertically (Magnuson 1978).

The angle of attack of the undulating body or

tail fin is critical to efficient generation of thrust. Hydrodynamic studies reveal that the force produced by a foil moving through water at constant velocity is proportional to the angle of attack. The greater the angle of attack, the more effective the generation of thrust. But there are limits to the magnitude of the angle, because drag increases exponentially with the angle of attack. At some critical angle (Figure 8–6) the thrust generated suddenly decreases and drag greatly increases (Webb 1975, 1978, 1984). To be effective a fin must be set at a low angle of attack to the water, usually somewhere between 10 and 20 degrees to its path of motion.

The evolution of a hinged coupling between the caudal fin and the caudal peduncle distinguishes carangiform swimming. In eels the angle of attack at a given point on the body changes continuously throughout each power stroke (Figure 8–7). Thrust generation is maximal when body segments pass across the midline of the path of motion, and fall to zero at maximum displacement (Webb 1975) (Figure 8–7a). In carangiform swimmers the angle of attack changes in a similar manner, but the angle is constant over a larger part of each power stroke (Figure 8–7b), and this results

in more thrust. The critical interrelationships involved are emphasized by the ability of most carangiforms to adjust the caudal fin angle by tightening and loosening the pull on the tendons connecting the tail to the trunk.

Other fishes seldom flex the body to swim, but undulate the median fins (referred to as **balistiform** swimming). Usually, several complete waves are observed along the fin (Figure 8–8), and very fine adjustment in the direction of motion can be produced.

Many fishes (for example, ratfish, surf perches, surgeonfishes, wrasses, and parrot fishes) generally do not oscillate the body or median fins, but row the pectoral fins to produce movement (**labriform** swimming). Most fishes show combinations of the basic swimming patterns. Intermediates between anguilliform, carangiform, ostraciiform, balistiform, and labriform movements exist, and most fishes can independently move paired fins to adjust body position. The muscles and nerves that move the fins are derivatives of the segmental muscles and spinal motor nerves that produce body undulations (Roberts 1981). In fact, the nerve–muscle complexes that move the limbs of tetrapods have

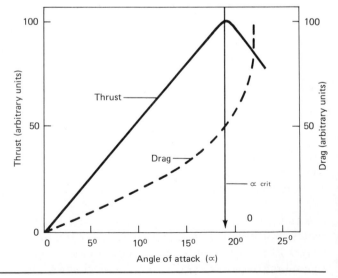

Figure 8–6. Effect of angle of attack on thrust and drag; α_{crit} is the angle at which a fin or wing (foil) stalls. The critical angle will vary with the shape of the foil and its speed through the fluid medium (water or air). Note that drag increases exponentially.

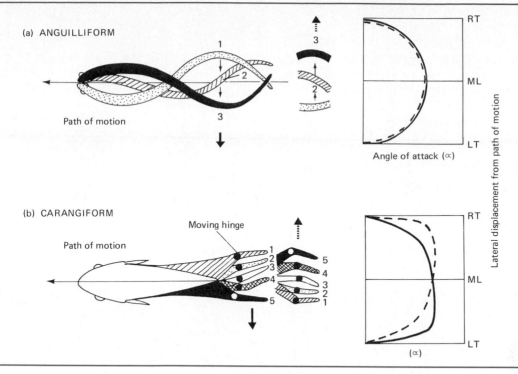

(a) ANGUILLIFORM

1
2
3

Path of motion

3
2

Angle of attack (α)

RT

ML

LT

(b) CARANGIFORM

Moving hinge

Path of motion

1
2
3
4
5

5
4
3
2
1

(α)

RT

ML

LT

Lateral displacement from path of motion

Figure 8–7. Attack angle of the propulsive surface throughout a complete undulatory cycle in two distinctive swimming modes: (a) anguilliform swimmers in which the angle of attack varies continuously throughout each undulatory cycle; (b) carangiform swimmers with a nearly constant angle of attack through an undulatory cycle.

a similar origin. As different as swimming, crawling, running, and flying appear in vertebrates, the machinery that drives their locomotion has a common origin—the segmental neuromuscular complex.

Improving Thrust: Minimizing Drag Drag is of two forms: **viscous drag** from friction between a fish's body and the water, and **inertial drag** from pressure differences created by the fish's displacement of water. Inertial drag is low at slow speeds, but increases rapidly with speed. Viscous drag is relatively constant over a range of speeds. Viscous drag is affected by surface smoothness, inertial drag by body shape. Streamlined (teardrop)

shapes produce minimum inertial drag when their maximum width is about one-fourth of their length and occurs about one-third of the length from the leading tip (Figure 8–9). It is not surprising that the shape of many rapidly swimming vertebrates closely approximates these dimensions. Too thin a body has high viscous drag because it has a large surface area relative to its muscle mass, and too thick a body induces high inertial drag because thick bodies displace a large volume of water as they move forward. Usually, fast-swimming fishes have small scales or are scaleless, resulting in a smooth body contour and reduced drag. (For a variety of reasons many slow-swimming fishes also are scaleless, emphasizing that universal general-

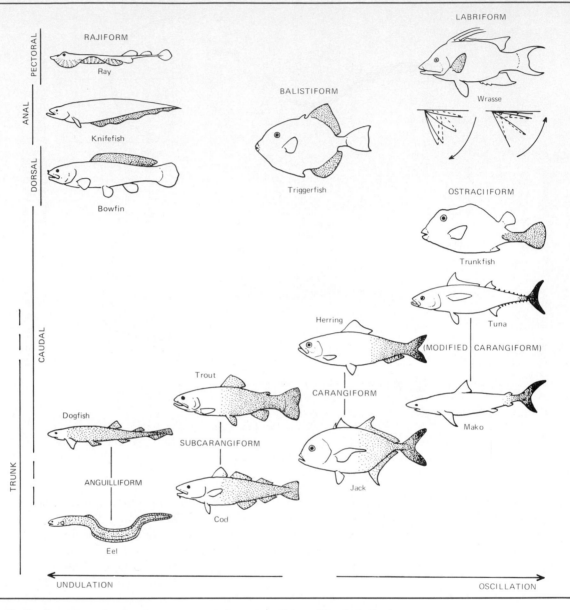

Figure 8–8. Location of swimming movements in various fishes. Stipple indicates areas of body undulated or moved in swimming. The names, such as carangiform, are used to describe the major types of locomotion found in fishes and are not a phylogenetic identification of all fishes using a given mode. (Modified after C. C. Lindsey, 1978, *Fish Physiology*, volume 7, Academic Press, New York.)

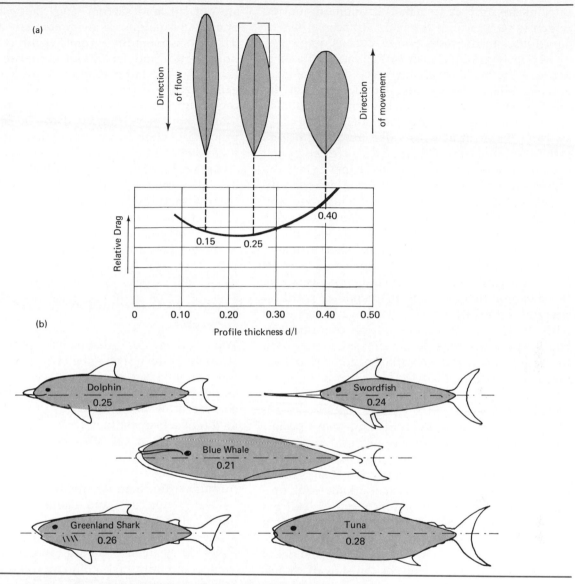

Figure 8–9. Effect of body shape on drag. (a) Streamlined profiles with width (*d*) equal to approximately one-fourth of length (*l*) minimize drag. The examples are for solid smooth test objects with thickest section about two-fifths of the distance from the tip. (b) Width to length ratios (*d/l*) for several swimming vertebrates. Like the test objects these vertebrates tend to be circular in cross section. Note that the ratio is near 0.25 and that the general body shape approximates a fusiform shape. (Modified after H. Hertel, 1966, *Structure, Form, Movement,* Krausskopf, Mainz, West Germany.)

izations are difficult to make.) In addition, mucus, which is so typical of the body surfaces of fishes, contributes to the reduction of viscous drag.

Drag is minimized in carangiform swimmers because the body is relatively stiff and retains an efficient shape during lateral undulations of the tail. The tail, however, in its thrusting movements against the water creates turbulent vortices of swirling water in a fish's wake, resulting in one form of inertial drag. The total drag created by the caudal fin also depends on its shape. When the **aspect ratio** of the fin (dorsal-to-ventral length divided by the anterior-to-posterior width) is large, the amount of thrust produced relative to drag is high. The stiff sickle-shaped fin of scombroids (mackerels, tunas) and of certain sharks (mako, great white) results in a high aspect ratio and efficient forward motion. Even the cross section of these caudal fins assumes a streamlined teardrop shape, which further reduces drag. Many forms with these specializations swim continuously. High-aspect-ratio fins do not present enough area to the water to produce sufficient force to initiate rapid accelerations.

Swimming in which only the caudal peduncle and fin undulate is usually called modified carangiform motion. In scombroids and many pelagic sharks the caudal peduncle is narrow dorsoventral but is relatively wide. Often the peduncle in carangids is studded laterally with bony plates called scutes. These structures present a knife-edge profile to the water as the peduncle undulates from side to side and contribute to the reduction of drag on the laterally sweeping peduncle. The importance of these seemingly minor morphological changes is underscored by the tail stalk of whales and porpoises, which is also narrow and has a double knife-edge profile. But the tail stalk is narrow laterally, unlike the caudal peduncle of scombroid fishes. In these cetaceans the tail and stalk undulate dorsoventrally rather than laterally. These strikingly similar specializations of modified carangiform swimmers—whether shark, scombroid, or cetacean—produce an efficient conversion of muscle contractions into forward motion.

The caudal fins of trout, minnows, and perch are not stiff and seldom have high aspect ratios. These subcarangiform swimmers can change caudal fin area to modify propulsive thrust and to produce vertical movements of the posterior part of the body. The latter action is achieved by propagating an undulatory wave up or down the flexible caudal fin, as occurs in the median fins of balistiform swimmers (Figure 8–8). In carangiforms propulsion often proceeds in bursts, usually from a standstill with rapid acceleration initiated by special neural systems, the Mauthner cells (see Box 8–1). The caudal peduncle of these fishes, unlike that of modified carangiform swimmers, is laterally compressed and deep. Under these circumstances the peduncle contributes a substantial part of the total force of propulsion.

Water and the Sensory World of Fishes

Water possesses distinctive properties that strongly influence the behaviors of fishes and other aquatic vertebrates. Water is some 830 times as dense as air at sea level. It is enlightening to consider how water compares to air in the propagation of sensory stimuli that vertebrates normally use—light, chemical substances, sound, and displacement.

Light is attenuated more when passing through water than air. Because light is absorbed by water molecules and scattered by suspended particles, objects become invisible at a distance of a few hundred meters even in the very clearest water. In clean air, distance vision is virtually unlimited and only dust, pollutants, and water vapor reduce distance vision in the atmosphere. In a similar way particles (silt, plankton) and many substances dissolve in water and reduce visual range.

Nevertheless, fishes generally possess well-developed eyes. A major difference between aerial and aquatic vision relates to focusing light rays on the retina to produce a sharp image. Air, by definition, has an **index of refraction** of 1.00. The cornea in the eye of both terrestrial and aquatic vertebrates has an index of refraction of about 1.37,

The complex behavioral adaptations shown by actinopterygians are associated with morphologically complex nervous systems. Studies of the nervous system of fishes are still in their infancy, but already clear is the distinctiveness of the central nervous system of ray finned fishes.

Some especially well-developed actinopterygian neural adaptations are widespread in aquatic vertebrates. A neuro-locomotor spe-

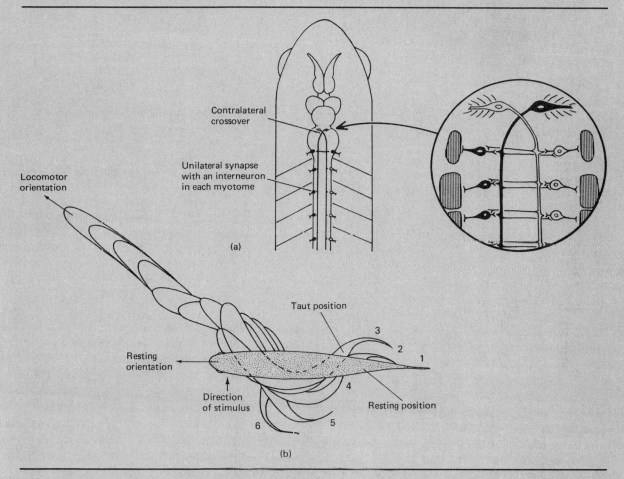

Figure 8–10. (a) Anatomical relationships for the Mauthner neurons and associated structures in a teleost; (b) body snap or startle response of a teleost (trout) as seen from above in tracings from a motion picture.

Box 8–1 (Continued)

cialization, the **Mauthnerian system**, present in many fishes and in tailed phases of amphibians, is well developed in the majority of teleosts, but absent in adult sharks and rays (Faber and Korn 1978, Eaton 1984). Located in the medulla oblongata of the brain are two giant nerve cells, one on either side of the midline. Each cell body is accompanied by two en-

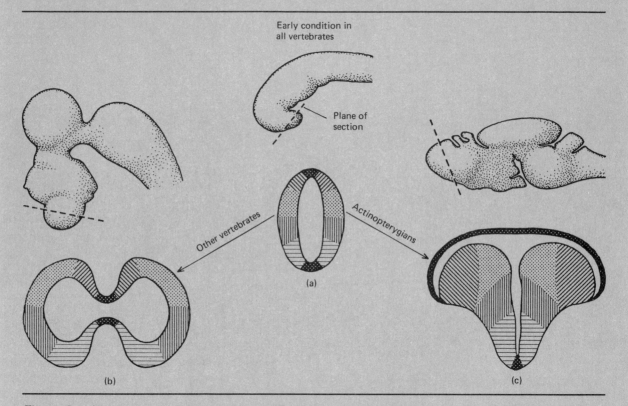

Figure 8–11. Unique development of the forebrain in actinopterygians. (a) The anterior-most portion of the neural tube (the future telencephalon of the brain) of actinopterygians early in its differentiation. (b) All other vertebrates, including sarcopterygians and their tetrapod descendents, undergo a rapid thickening of the sidewalls which results in a ballooning laterally of the telencephalic walls and the formation of paired, hollow cerebral hemispheres. (c) In actinopterygians the walls also greatly thicken, but this results in eversion of the dorsal portions as the upper walls curl outward, stretching the primordial dorsal roof of the central cavity (the ventricle) to a thin covering over a ventricle of large surface area but narrow width.

larged dendrites that synapse with cranial nerve VIII, the acoustic nerve. A single, heavily myelinated giant axon issues from each cell, crosses to the opposite side of the medulla, and then descends the full length of the spinal cord. Each giant axon synapses with motor neurons in every segment on one side of the body (Figure 8–10).

Stimulation of one of the Mauthner cells results in a rapid, unilateral, forceful contraction of the trunk and tail myomeres. This reaction is similar and perhaps identical to the body snap or startle response that many teleosts exhibit when frightened by a sudden noise, mechanical shock, or change in illumination. The simultaneous, strong contraction of the muscles on one side of the body rapidly propels the fish forward. The value of such a startle response is obvious: a predator incautious enough to produce stimulae activating the Mauthnerian system may lose potential prey to this startle reaction.

Distinctive development of the forebrain (the telencephalon or anterior most of the five basic craniate brain regions) is a shared derived character that may define the ray-finned fishes (Nieuwenhuys 1982). In all other vertebrates the walls of the anterior end of the neural tube balloon outward on each side during development (Figure 8–11). The final result is the formation of hollow cerebral hemispheres. In actinopterygians the walls thicken, stretching the roof thinner as they evert. The result is solid cerebral hemispheres and an expansive hollow cloak of ventricle. Functional correlates of this structural difference are unknown. However, in examination of a series of ray-finned fishes' brains it is clear that there is a marked increasing differentiation of the forebrain from *Polypterus* to sturgeons to *Amia* and the teleosts. Neural connections representing olfactory input become less important, structural complexity of the nerve cell architecture and interconnections increases, and connections to other distant regions of the brain increase. This pattern in actinopterygians is analogous to the evolution of the mammalian forebrain.

which is close to the index of refraction of water (1.33). Light rays are bent as they pass through a boundary between media with different refractive indices. The amount of bending is proportional to the difference in indices of refraction. Because the index of refraction of the cornea is substantially different from that of air, light rays are bent as they pass from air into the cornea. As a result the cornea is an important part of the focusing system of the eye of terrestrial vertebrates. This relationship does not hold true in water. The refractive index of the cornea is too close to that of water for the cornea to have much effect in focusing light entering the eye. Terrestrial vertebrates rely on the lens, which often is flattened and pliable, for detailed focus. Fishes possess a less pliable spherical lens with high refractive power to focus images on the retina, as do aquatic mammals, such as cetaceans. Otherwise, a fish's eye is similar to the eyes of terrestrial vertebrates—water primarily limits the distance over which objects can be seen.

Fishes have taste bud organs in the mouth and around the head and anterior fins, as well as receptors of general chemical sense that detect substances that are only slightly soluble in water, and olfactory organs on the snout that detect soluble substances. Sharks and salmon are capable of detecting odoriferous compounds at less than 1 part per billion. Of course, chemicals must reach the receptors by diffusion and by transport in currents of water. Feeding and searching responses in many fishes are based on the detection of dissolved substances that indicate nearby or upstream prey. Adult migrating salmon are directed to their stream of origin from astonishing distances by a chemical signature from the home stream permanently imprinted when they were juveniles. Plugging the nasal olfactory organs destroys the ability to home (Cromie 1982).

Mechanical receptors provide the basis for detection of displacement—touch, sound, pressure, and motion. Like all vertebrates, fishes possess an internal ear (labyrinth organ), including the semicircular canals, which inform the animal of changes in speed and direction of motion. They also possess, at the base of the canals, gravity detectors that inform them of positional orientation—allowing them to distinguish up from down. Most vertebrates also possess in this labyrinth complex an auditory region sensitive to sound pressure waves (Tavolga et al. 1981). These diverse functions of the labyrinth are dependent on basically similar types of sense cells, the **hair cells** (Figure 8–12). In fishes and aquatic amphibians clusters of hair cells and associated support cells form **neuromast organs** that are dispersed over the surface of the head and body. In jawed fishes neuromast organs often form a series of canals on the head, and one or more canals pass along the sides of the body onto the tail. This unique surface receptor system of fishes and aquatic amphibians is referred to as the **lateral-line system**.

Detection of Water Displacement: The Lateral Line Neuromasts are distributed in two configurations: within tubular canals or exposed in epidermal depressions. Many kinds of fishes have both. Hair cells have a kinocilium placed asymmetrically in a cluster of microvilli and are arranged in pairs with the kinocilia positioned on opposite sides of adjacent cells. A neuromast contains many of these hair cell pairs; each neuromast

unit is serviced by two afferent lateral-line nerves; one transmits impulses from hair cells with kinocilia in one orientation, the other carries impulses from cells with kinocilia positions reversed by 180 degrees.

All kinocilia and microvilli are embedded in an oval gelatinous structure, the **cupula** . Displacement of the cupula causes the kinocilia to bend. The resultant deformation either excites or inhibits its afferent nerve discharge. Each hair cell pair, therefore, encodes unambiguously the direction of cupula displacement. The excitatory output of each pair has a maximum sensitivity to displacement along the line joining the kinocilia and falling off in other directions. The net effect of cupula displacement is to increase the firing rate in one afferent nerve and to decrease it in the other nerve. These changes in lateral-line nerve firing rates inform a fish of the direction of displacement forces on different surfaces of its body.

What forces produce cupula displacement? Because water currents displace the cupulae, currents directed on different portions of the body can be detected. Cutting the afferent nerves abolishes any response. Water currents of only 0.025 millimeter per second are detected by the exposed neuromasts in the aquatic clawed frog, *Xenopus laevis,* with maximum response to currents of 2 or 3 millimeters per second. Similar responses occur in fishes. The lateral-line organs also respond to low frequency sound, but controversy exists as to whether sound is a natural lateral-line stimulus. Many fishes also hear low frequencies with the ear, even after the lateral-line organs have been destroyed. Sound induces traveling pressure waves in the water and also causes local water displacement as the pressure wave passes. It has been difficult to be sure whether neuromast output results from the accompanying water motions on the body surface or the sound's compressional wave.

Several surface-feeding fishes and *Xenopus* provide a vivid example of how the lateral-line organs can function under natural conditions. These species find insects on the water surface from the surface waves created by the prey's motions. In a series of clever experiments, E. Schwartz (see Fes-

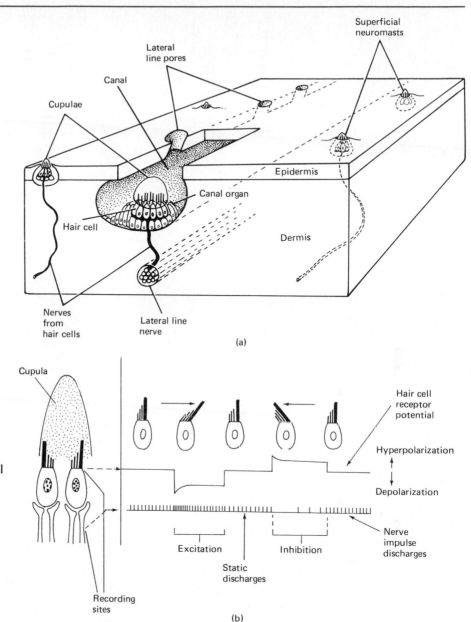

Figure 8–12.
(a) Semidiagrammatic representations of the two configurations of lateral-line organs in fishes. Diagram is of superficial neuromasts and canal organs of a scaleless fish.
(b) Hair cell deformations and their effect on hair cell transmembrane potential (receptor potential) and afferent nerve cell discharge rates. (Modified after A. Flock, 1967, in *Lateral Line Detectors*, edited by P. Cahn, Indiana University Press, Bloomington, Ind.)

Labels for (a): Superficial neuromasts; Lateral line pores; Canal; Cupulae; Canal organ; Epidermis; Hair cell; Dermis; Nerves from hair cells; Lateral line nerve; (a)

Labels for (b): Cupula; Hair cell receptor potential; Hyperpolarization; Depolarization; Nerve impulse discharges; Excitation; Inhibition; Static discharges; Recording sites; (b)

sard 1974) has demonstrated that each neuromast group on the head of the killifish, *Aplocheilus lineatus*, provides information about surface waves coming from a different direction (Figure 8–13). Thus, the nasally located units are sensitive to waves arriving toward the front of the head, the supraorbital groups detect laterally arriving waves, and the postorbital groups respond to waves arriving from a caudal direction. All groups, however, show stimulus field overlaps. Bilateral inter-

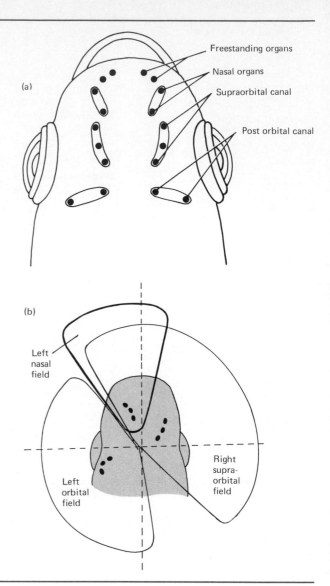

Freestanding organs

Nasal organs

Supraorbital canal

Post orbital canal

(a)

(b)

Left
nasal
field

Left
orbital
field

Right
supra-
orbital
field

Figure 8–13. Distribution of the lateral-line canal organs on (a) the dorsal surface of the head of the killifish *Fundulus notatus* and (b) the perceptual fields of the head canal organs in another killifish *Aplocheilus lineatus.* The wedge-shaped areas indicate the relative directional sensitivity for each group of canal organs. Note that fields overlap on each side as well as on the same side of the body. (Modified after E. Schwartz, 1974, in *Handbook of Sensory Physiology*, volume 3, part 3, edited by A. Fessard, Springer-Verlag, New York.)

actions between neuromast groups are indicated by the fact that extirpation of an organ from one side of the head disturbs the directional response to stimuli arriving from several directions.

Presumably, the large numbers of neuromasts on the heads of some fishes assist in orienting their postures and motions with respect to water disturbances. Several researchers have suggested that this might be important for schooling fishes in sen-

sing vortex trails in the wake of adjacent schoolmates. However, many of the fishes that form extremely dense schools lack lateral-line organs along the flanks (herrings, atherinids, mullets, and so on) and retain only the cephalic canal organs. This reduction in sensory elements would reduce the constant noise from turbulence that must be present within fish schools. The well-developed cephalic canal organs concentrate sensitivity to water motion in the head region, where it is needed to sense the degree of turbulence into which the fish is swimming. Over periods of time the reduction in drag through avoidance of turbulence could achieve a considerable metabolic swimming economy, as well as an even spacing between school members (Lindsey 1978).

Electric Discharge and Electroreception The beginning of human experience with electricity, other than lightning, may have resulted from contact with the torpedo ray in the Mediterranean region and electric catfish in the Nile. These fishes and the electric eel of South America can discharge sufficient electricity to stun other animals. The

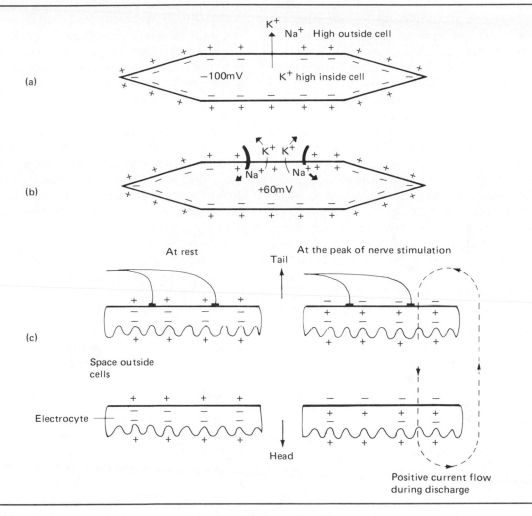

Figure 8–14. Use of transmembrane potentials of modified muscle cells by electric fishes to produce a discharge. (a) At rest; (b) stimulated, sodium diffuses into the cell and potassium diffuses out; (c) differential movement of ions across the rough and smooth surfaces of electrocyte cells produces a directional current flow. By arranging electrocytes in series, electric fishes can generate very high voltages. Electric eels, for example, have 6000 electrocytes in series and produce potentials in excess of 600 volts.

source of the electric current is modified muscle tissue. The cells of this modified muscle, called **electrocytes**, have lost the capacity to contract but are specialized for generating an ion current flow (Figure 8–14). When at rest the membranes of muscle cells or neurons are electrically charged, with the intracellular fluids about -100 millivolts relative to the extracellular fluids (Figure 8–14a). K^+ is maintained at a high and Na^+ at a low internal concentration by the action of a Na^+/K^+ cell membrane pump. At rest, permeability of the membrane to K^+ exceeds the Na^+ membrane permeability. As a result, K^+ diffuses outward faster than Na^+ diffuses inward (arrow) and sets up the 100-millivolt resting potential. When appropriately stimulated the permeability of the membrane to Na^+ increases (Figure 8–14b). As a result a large, rapid, but fleeting local influx of Na^+ is generated. This large influx inverts the membrane potential and excites adjacent membranes to depolarize. As a result, the disturbance propagates over the cell surface. In an electrocyte, which is a modified noncontracting muscle cell, one cell surface is rough and the opposite surface smooth (Figure 8–14c). Innervation is associated with the smooth surface, and only the smooth surface depolarizes. The resulting Na^+ flux across the smooth surface into the cell and the K^+ leakage across the rough surface and out of the cell yield a net positive current in one direction.

Because electrocytes are arranged in stacks or in series, like the batteries in a flashlight, the discharge potentials across each stack can add to produce higher voltages. Synchrony of discharge also is required, however, and this is produced by simultaneity of nerve impulses arriving at each electrocyte. In fishes such as the African electric catfish and the South American electric eel, considerable potentials can be generated (in excess of 300 and 600 V, respectively). It was realized during Darwin's time that the unusual anatomical structures of electrocytes were present in several species of fishes that did not produce electric shocks. Because of the anatomical similarities, these organs were considered to generate electric currents but not until the 1950s was their weak electric nature demonstrated (Bennett 1971a, Bass 1986). Electric fishes are now classified into two groups, the strongly electric and the weakly electric fishes (Figure 8–15).

Most electric fish species are limited to tropical fresh waters of Africa and South America. Electric marine forms are very limited: the torpedo ray (*Torpedo*), the ray genus *Narcine*, and some skates

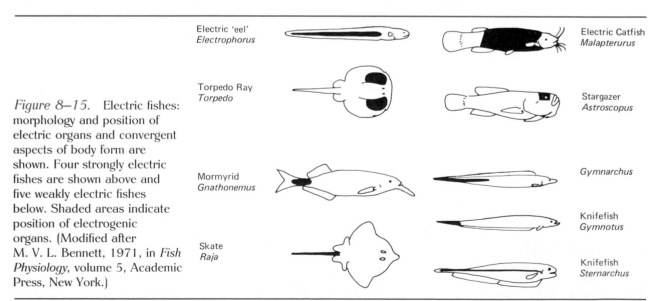

Figure 8–15. Electric fishes: morphology and position of electric organs and convergent aspects of body form are shown. Four strongly electric fishes are shown above and five weakly electric fishes below. Shaded areas indicate position of electrogenic organs. (Modified after M. V. L. Bennett, 1971, in *Fish Physiology*, volume 5, Academic Press, New York.)

Electric 'eel'
Electrophorus

Electric Catfish
Malapterurus

Torpedo Ray
Torpedo

Stargazer
Astroscopus

Mormyrid
Gnathonemus

Gymnarchus

Knifefish
Gymnotus

Skate
Raja

Knifefish
Sternarchus

among elasmobranchs, and only the stargazers (family Uranoscopidae) among the teleosts are known to be electric. In the strongly electric fishes, the discharge is of sufficient intensity to be of use in defensive and predatory behavior. In all other electric fishes, the discharge voltages are too weak to be of direct defensive or offensive value. It is now realized that weakly electric fishes use their discharges for electrolocation and social communications.

When a fish discharges its electric organ an electric field is established in its immediate vicinity (Figure 8–16). Because of the high energy costs of maintaining a continuous dc (direct current) discharge, electric fishes produce a discontinuous pulsating discharge and field. Some species produce a lifelong constant-frequency discharge, others emit a variable-frequency discharge. Most weakly electric fishes pulse at rates between 50 and 300 hertz (cycles per second), but in the sternarchid knifefishes of South America the frequency reaches 1700 hertz, which is the most rapid continuous firing rate known for any vertebrate muscle or nerve.

In the marine electric fishes known as stargazers, which produce modest voltages, a specific behavioral function has yet to be demonstrated. It is reported that their discharge is proportional to the size of the prey taken but is insufficient in voltage to stun the prey. Perhaps the discharge functions in the social facilitation of feeding by signaling nearby stargazers. The conductivity of seawater is so high that the electric field produced even by a strong discharge is essentially short-circuited and, therefore, limited in range. In fresh water, where electric conductivity is much lower, the electric field from even weak discharges may extend outward for several meters and will be distorted by the presence of either conductive or highly resistant objects. Distortions will result in an increase or decrease of the electric potential distributed across specific regions of the fish's body surface. Rocks, for example, are highly resistive but other fishes, invertebrates, and plants are conductive. An electric fish might thus detect the presence and position of such objects by sensing where on its body maximum distortion of its electric field occurs. If the fish could sense the discharges of

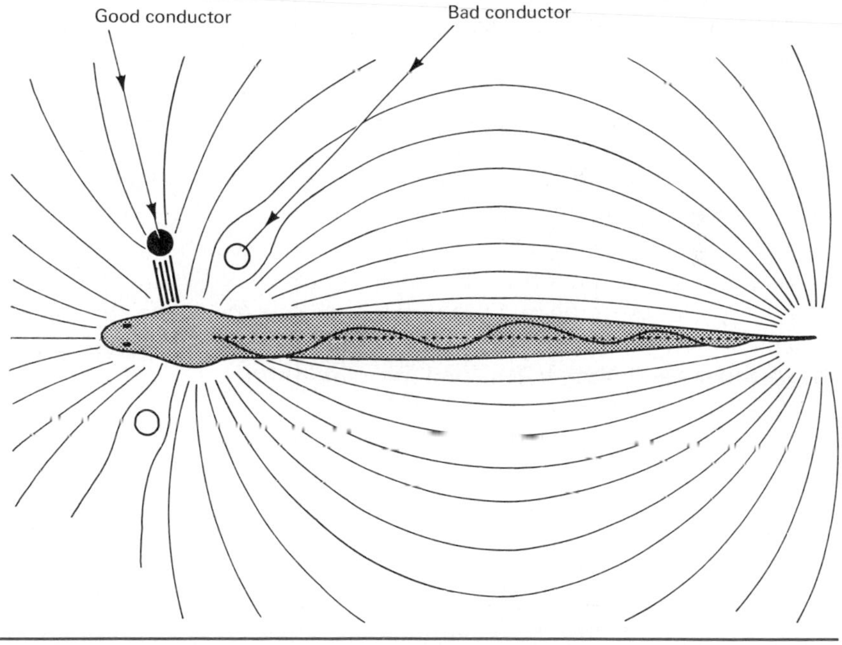

Good conductor   Bad conductor

Figure 8–16. Distortion of the electric field (lines) surrounding an electric fish *(Gymnarchus)* by conductive and nonconductive objects. Conductive objects concentrate the field on the skin of the fish where the increase in electrical potential is detected by the electroreceptors. Nonconductive objects spread the field and diffuse potential differences along the body surface.

another electric fish, the possibility for electrocommunication exists. To accomplish either electrolocation or electrocommunication, it is necessary to sense the electric field.

The skin of weakly electric teleosts contains special sensory receptors: **ampullary** organs and **tuberous** organs (Fessard 1974). These organs detect tonic (steady) and phasic (rapidly changing) discharges, respectively. Each type represents modification of lateral-line neuromast receptors. Electroreceptors, like lateral-line receptors, have double innervation, an afferent channel sending impulses to the brain and an efferent channel that causes inhibition of the receptors (Bennett 1971b). Theodore Bullock and his collaborators have shown that during each electric organ discharge (EOD) an inhibitory command is sent to the electroreceptors and the fish is rendered insensitive to its own EOD. Between pulses, electroreceptors sense distortion in the electric field or the presence of a foreign electric field. The electric organs and receptors of weakly electric fish provide a sixth sense. The usefulness of this electric sensory system is apparent, because the African and South American electric fishes usually inhabit turbid waters where vision is limited to short distances. Perhaps more important, most electric fishes tend to be nocturnal.

Although weakly electric fishes are restricted almost entirely to tropical fresh waters, electrogenesis and reception are not restricted to a single group of aquatic vertebrates (Figure 8–17) nor is the function of these systems limited to navigation and predation. The capacity of aquatic vertebrates to detect electric fields is an ancient characteristic of the sensory systems of vertebrates. Precisely localized activity occurs in the brain of the lamprey in response to electric fields (Bodznick and Northcutt 1981), and it seems likely that the earliest vertebrates also possessed electroreceptive capacity. All fish-like vertebrates of lineages evolved before the earliest neopterygians (represented by living gars and *Amia*) have electroreceptor cells characterized by a prominent kinocilium. The receptor cell excites an afferent nerve whenever the environment around the kinocilium is negative in charge relative to the cell. This afferent neuron excitation eventually reaches the medial region of the posterior third of the brain of these primitive clades. The neopterygians lost electrosensitivity, and teleosts demonstrate at least two separate evolutions of electroreceptors, which are in all ways distinct from those of other vertebrates: Teleost electroreceptors lack a kinocilium, the afferent nerve fires when the outside is positive relative to the inside, and the electroreceptor projection is on the lateral aspect of the rhombencephalon. At least one mammal, the duck-billed platypus, uses electroreception to detect prey (Scheich et al. 1986).

Although electroreception is widespread, electrogenic vertebrates are less common but electrogenesis occurs in phyletically diverse groups. In species where electroreception and electrogenesis occur together, the functions of the systems surpass electrolocation and extend into the exotic realm of bioelectric communication. Studying weakly electric fishes in their natural habitats, Carl Hopkins (1980) recorded characteristics of EODs that must be interpreted as specializations to facilitate communication between individuals both intra- and interspecifically. EODs vary with habit and habitat. The EODs of species that aggregate or live in shallow, narrow streams are generally of high frequency and short duration, physical characteristics rendering them less susceptible to interference from neighbors. In territorial species and in males of species showing sexual differences in electric discharges, the EODs are clear announcements of longer duration.

Seasonal, sex hormone–mediated EODs distinguish immature individuals, ripe females, and sexually active males in some species. The EODs can be compared to frog and bird song, visual courtship and agonistic displays by lizards, or mammalian scent marking of territories, depending on the circumstances and electric fish species considered. Some electric fishes, considered a single homogeneous species on conventional systematic grounds, can be clearly separated into sibling species on the basis of their EODs. Study of the evolution and adaptive radiation of electrocommunication has provided important comparative

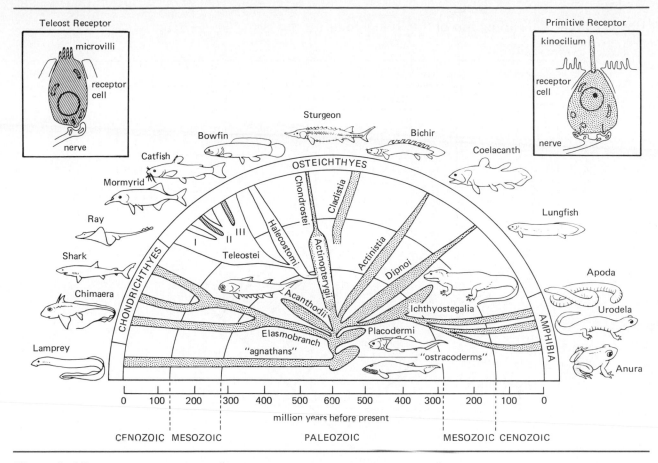

Figure 8—17. Phylogenetic distribution of electrosensitivity. Stippling and crosshatching identify two classes of electroreceptors as indicated in the upper right and left panels. (Produced with the assistance of Dr. Carl Hopkins.)

data for the understanding of animal communication.

The Appearance of Teleostomes

The Devonian is known as the Age of Fishes because all major lineages of fishes, extant and extinct, coexisted in the fresh and marine waters of the planet during its 48-million-year duration. Most groups of gnathostomous fishes made their first appearance during the period including the most species-rich and morphologically diverse lineage of vertebrates, the Teleostomi (Figure 8–18) encompassing acanthodians and bony fish.

Acanthodians

The earliest jawed fishes in the fossil record are called acanthodians because of the stout spines (*acanthi*) anterior to their well-developed dorsal, anal, and often numerous paired fins (Denison 1979). The first fossil remains of acanthodians are

Phylogenetic Relationships of the Teleostomi

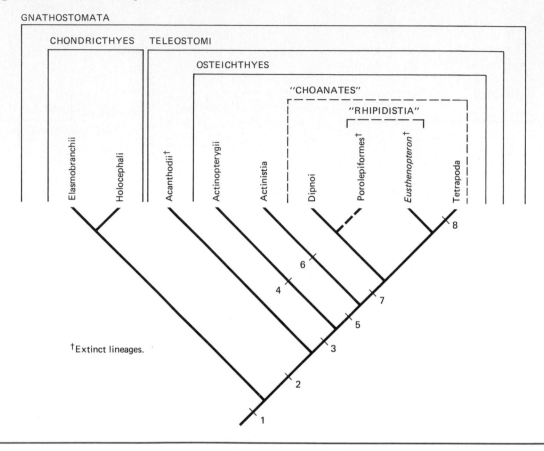

†Extinct lineages.

isolated spines from the early Silurian. Later in the Silurian near-shore marine deposits contain more fin spines, scales, and teeth. Each of these elements is distinct from any comparable structures in other vertebrates. If they alone were available for study, the acanthodians would be considered as isolated from other vertebrate clades as are the placoderms. Such is not the case, however. When more complete specimens were discovered (early Devonian through the early Permian) they had the unmistakable spines, scales, and teeth of the Silurian fossils coupled with a mosaic of dermal, axial, and appendicular characters very like those of the majority of other gnathostomes and greatly different from those of any known jawless fish. Many of these characters seem to be primitive for gnathostomes, as we might expect from such early forms. Nevertheless, these characters have been used to suggest relationships between acanthodians and (1) placoderms (on the basis of shared characters in the number of ossification centers in the jaws, gill arches, and shoulder girdles); (2) chondrichthyans (special border scales along the lateral line, details of the cranial and gill skeleton); and (3) osteichthyans (operculum, gill soft tissue and skeleton, branchiostegal rays, otoliths, and nu-

1. Chondrichthyes plus Teleostomi (Eugnathostoma): Epihyal element of second visceral arch modified as the hyomandibula, which is a supporting element for the jaw. 2. Teleostomi: Hemibranchial elements of gills not attached to interbranchial septum, bony opercular covers, branchiostegal rays, five additional characters. 3. Osteichthyes: Presence of lepidotrichia, differentiation of the muscles of the branchial region, presence of a lung or swimbladder derived from the gut, a unique pattern of ossification of the dermal bones of the shoulder girdle, medial insertion of the mandibular muscle on the lower jaw. 4. Actinopterygii: Basal elements of pectoral fin enlarged, median fin rays attached to skeletal elements that do not extend into fin, single dorsal fin, scales with unique arrangement, shape, interlocking mechanism, and histology, and at least six additional derived characters. 5. Sarcopterygii: Fleshy pectoral and pelvic fins have a single basal skeletal element, muscular lobes at the bases of those fins, enamel on surfaces of teeth, unique characters of jaws, articulation of jaw supports, gill arches, and shoulder girdles. 6. Actinistia: Double jaw articulation, rostral organ, ossified swimbladder, loss of maxilla and branchiostegal. 7. "Choanates": Internal oral openings from the olfactory passages. The homology of the choanae of dipnoans with those of tetrapods has not been established; thus the name "choanates" may be inappropriate but the following shared derived characters indicate that the grouping of these lineages is valid: Pelvic girdle forms pubic and ischial processes, pectoral and pelvic appendages each have two primary joints (shoulder/hip and elbow/knee) with unique articulations, conus arteriosus of heart partly divided, unique dermal bone pattern of braincase, loss of interhyal bone, and at least 12 other characters (the exact number depends on which fossil forms are included). 8. Tetrapoda: Four limbs characterized by a single bone in the proximal portion and two bones in the distal portion. (Based on G. V. Lauder and K. F. Liem, 1983, *Bulletin of the Museum of Comparative Zoology* 150:95–197; and J. G. Maisey, 1986, *Cladistics* 2:201–256.)

Figure 8–18. Phylogenetic relationships of the Teleostomi. This diagram shows the probable relationships among the major groups of teleostomes. Extinct lineages are marked by a dagger (†). The numbers indicate derived characters that distinguish the lineages. Some authorities place the origin of *Eusthenopteron* between the Actinopterygii and the Actinistia. The dashed line for Porolepiformes indicates uncertainty about the affinities of this group, and the quotation marks around "Rhipidistia" indicate that this is a paraphyletic grouping.

merous other functional and morphological characters, especially of the cranium and jaw). We follow Lauder and Liem (1983) in associating acanthodians as the sister group of the Osteichthyes until more convincing interpretations are produced. The **Teleostomi** (acanthodians + osteichthyans) are defined by a unique mechanism of opening the mouth by lowering the mandible by movements of a hyoid apparatus transmitted to the lower jaw by ligaments, the presence of an ossified dermal operculum, a closely associated new element in the hyoid arch, the interhyal bone, and branchiostegal rays. All these elements become important in the later evolution of osteichthyan feeding mechanisms.

Acanthodians were usually not more than 20 centimeters in length, although some 2-meter species are known. These marine and freshwater fishes were often clad in small, square-crowned scales each of which grew in size as the animal grew. The head was large and blunt and housed large eyes. The mouth was also large, and in many species the jaws were studded with teeth.

Unlike the teeth of all other gnathostomes, however, acanthodian teeth are not known to have been regularly replaced, and they lack an enamel-like surface. In some acanthodians a few scales were enlarged to form a cover for the gills. In others, the head scales were lost completely except along the courses of cephalic sensory canals. The brain was encased in a well-developed cartilaginous box, which also housed three semicircular canals. There was a vertebral column, evidenced by remains of neural and hemal arches, but no centra. The well-developed heterocercal tail, the fusiform body, and the arrangement of the fins indicate that acanthodians were good swimmers and probably were not bottom dwellers.

This basic body form lasted throughout acanthodian history, but several trends in their evolution are evident. Some early forms had robust spines in the position of the pectoral and pelvic fins. Between these were other pairs of spines of variable size. These spines were not embedded deeply in the body. Often the pectoral spines were associated with ventral dermal plates in a supportive girdle (Figure 8–19). In later species, this dermal skeleton, along with some intercollated spines and areas of scales were lost, and the remaining spines became set more deeply in the body. In

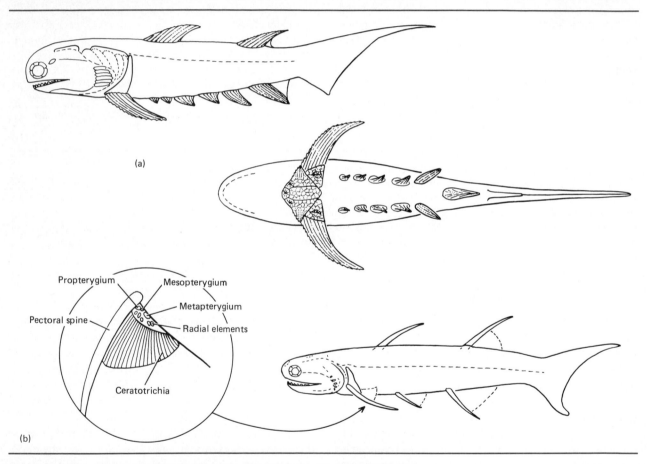

Figure 8–19. Reconstructions of acanthodians: (a) primitive type: *Climatius*, lower Devonian, with multiple superficially attached spines (lateral view above, ventral view below); (b) more advanced type: *Ischnacanthus*, lower Devonian, with fewer, more deeply embedded spines. (Inset) Detail of pectoral fin and spine of *Acanthodes*, lower Carboniferous.

addition, there was refinement of the respiratory apparatus and, in more than one lineage, a loss of teeth. Jaw articulation and gill changes accompanying the loss of teeth indicate that edentulous types were able to open their mouths and orobranchial chambers very widely. These same species have long gill rakers and probably were surface-to-midwater plankton feeders that swam in eel-like fashion with their enormous mouths open to strain small organisms from the water.

Such feeding specializations would have placed acanthodians at a key position in the aquatic and marine food webs of the latter part of the Paleozoic: transferring energy from zooplanktonic herbivores of tiny size into the relatively large pieces of biomass of their own bodies and providing prey for the many piscivorous forms of the Age of Fishes. Acanthodians may have been very abundant in Devonian seas but during the early Permian the last forms disappeared, leaving the waters of the world to the Chondrichthyes and the most species-rich clade of vertebrates yet evolved—the bony fishes.

The Earliest Osteichthyes and the Major Groups of Bony Fishes

Fragmentary remains of bony fishes are known from the late Silurian. It is not until the Devonian that more complete remains are found. These animals resemble acanthodians in details of the head structure. The similarities may point to a common teleostome ancestor for acanthodians and osteichthyans in the early Silurian.

Remains of the bony fishes representing a radiation of forms already in full bloom appear in the lower to middle Devonian. Two major and distinctive osteichthyan (*osteo* = bone, *ichthys* = fish) types possessed locomotor and trophic characters that made them dominant fishes during the Devonian. Some evolved specializations that led to land vertebrates. Most, however, retained fish-like characteristics and from among them rose the modern bony fishes—the largest group of living vertebrates (Table 8–2).

Articulated, abundant remains of the two basic types of osteichthyans are known from the middle Devonian: the Sarcopterygii (*sarcos* = fleshy, *pterygium* = fin; Figure 8–20a to d) and the Actinopterygii (*actinos* = stout ray; Figure 8–20e and f).

Characteristic of both groups of Osteichthyes were similar derived patterns of lateral-line canals, similar opercular and pectoral girdle dermal bone elements, fin webs supported by bony dermal rays, and a gas-filled diverticulum of the esophagus functioning as an accessory respiratory organ and buoyancy device. Many of the forms had two dermal bones tightly associated with the palatoquadrate forming the upper biting edge of the mouth (a premaxilla and a maxilla) typically with enamel-coated teeth fused to them. A neurocranium with anterior and posterior ossified sections separated by a fissure allowed movement between the two halves of the skull in many of the forms. The presence of bone is not a unifying osteichthyan characteristic because agnathans, placoderms, and acanthodians also possess true bone in diverse regions of their bodies and the chondrichthyans show considerable evidence of derived bone loss or supression. The name Osteichthyes was coined before the occurrence of bones in other primitive vertebrates was realized. Likewise, the various names long in use for the living actinopterygian subgroups imply an increase in the ossification of the skeleton as an evolutionary trend (for example, chondrosteans, the "cartilaginous bony fishes," gave rise to a radiation often called the holosteans, "entirely bony fishes," which culminated in teleosteans, "final bony fishes"). The fossil record indicates that a regular sequence of increasing ossification did not occur. A tendency to reduce ossification, especially in the skull and scales, is, in fact, apparent when the full array of early Osteichthyes is compared with their modern descendents.

Early osteichthyans are not presently describable by a single widely acceptable phylogeny (Bemis et al. 1987). Although relationships among primitive ray-finned fishes are generally agreed upon because they are supported by several

Table 8–2. Classification and geographic distribution of Osteichthyes, the bony fishes.*

Class Osteichthyes (bony fishes), about 21,000 living species
   Subclass Sarcopterygii (fleshy-finned fishes), 7 living species
      Order Dipnoi (lungfishes), 6 living species; Southern Hemisphere, fresh water
      Order "Rhipidistia"†
      Order Actinistia [Coelacanthiformes] (coelacanths), 1 living species; Indian Ocean
        islands, deep water marine
   Subclass Actinopterygii (ray-finned fishes), 20,850 living species
      Order "Paleonisciformes"†
      Order Polypteriformes [Cladistia] (bichirs), 11 living species; African, fresh water
      Order Acipenseriformes (sturgeons and paddlefishes), 25 living species; Northern
        Hemisphere, coastal and fresh water
    Infraclass Neopterygii,‡ 20,814 living species
      Order Lepisosteiformes [Ginglymodi] (gars), 7 living species; North and Central
        America, fresh and brackish water
      Order Amiiformes (bowfins), 1 living species; North America, fresh water
     Series Teleostei (numbers of living species are conservative minima)
      Superorder Osteoglossomorpha (bony tongues), 206 living species; worldwide tropical,
        fresh water
      Superorder Elopomorpha (tarpons and eels), 633 living species; worldwide, mostly
        marine
      Superorder Clupeomorpha (herrings and anchovies), 331 living species; worldwide,
        especially marine
     Subseries Euteleostei, 19,636 living species
      Superorder Ostariophysi (catfish and minnows), 6,050 living species; worldwide, fresh
        water
      Superorder "Protacanthopterygii" (trouts and relatives), 320 living species; temperate
        Northern and Southern Hemisphere, fresh water
      Superorder Scopelomorpha = "Myctophiformes" (lanternfishes and relatives), 677
        living species; worldwide, mesopelagic (middle depth), marine
      Superorder Paracanthopterygii (cods and anglerfishes), 1,160 living species; Northern
        Hemisphere, primarily marine
      Superorder Acanthopterygii (spiny-rayed fishes), 10,349 living species, including the
        Atherinomorpha (silversides), 1,080 living species; worldwide, surface-dwelling,
        fresh water and marine; and Perciformes (perches), 7,800 living species; worldwide,
        primarily marine

     *Taxa in quotation marks are known to be artificial, but interrelationships are
unresolved.
    †Extinct.
    ‡The subdivision of the Neopterygii varies greatly from author to author.

shared derived characteristics, the phylogenetic relationships of the fleshy finned forms are controversial. This controversy is important because from within the Sarcopterygii arose the tetrapods and with them terrestrial vertebrate life. For many years the sarcopterygians were thought to be composed of two sister groups: the lungfishes or Dipnoi (*di* = double, *pnoe* = breathing), and the Crossopterygii (*cross* = a fringe or tassel, *pterygium* =

fin). Primitive Sarcopterygii have similar body shapes and sizes (20 to 70 centimeters), two dorsal fins, an epichordal lobe on the heterocercal caudal fin, and paired fins with a fleshy, scaled, and bony central axis. The paired fins' rays extend in a feather or compound leaf-like manner in contrast to the fan-like form of actinopterygian paired fins.

    The jaw muscles of sarcopterygians were massive by comparison with those of actinopterygians,

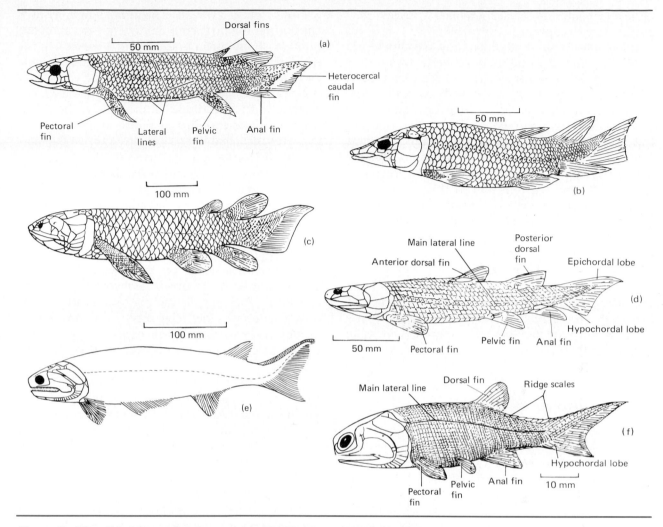

Figure 8–20. Primitive osteichthyans. (a and b) Dipnoans, (c and d) other
Sarcopterygii, (e and f) Actinopterygians. (a) Relatively unspecialized dipnoan
Dipterus, middle Devonian; (b) long-snouted dipnoan *Griphognathus*, late Devonian;
(c) laterally compressed porolepiform *Holoptychius*, late Devonian to early
Carboniferous; (d) cylindrical osteolepiform *Osteolepis*, middle Devonian; (e) fine-
scaled actinopterygian *Cheirolepis*, middle to late Devonian; (f) typical early
actinopterygian *Moythomasia*, late Devonian. (Modified from J. A. Moy-Thomas and
R. S. Miles, 1971, *Paleozoic Fishes*, W. B. Saunders, Philadelphia.)

and the size of these muscles influenced the details
of cranium, hyoid, and dermal skull characters
that set sarcopterygians apart from the
actinopterygians. Finally, the early sarcopterygians were coated with a peculiar layer of dentine-

like material that spread right across the sutures
between dermal bones and shows indications of
being periodically reabsorbed. The very intricate
interconnecting cavities characteristic of this cosmine are thought by some authorities to have con-

tained an elaborate array of sensory organs, perhaps electroreceptors.

Despite these similarities only the Dipnoi are generally agreed to be monophyletic by most recent workers. The other Sarcopterygii have been variously combined as a single sister group, the **Crossopterygii** (*cross* = a fringe or tassel), now considered by most workers to be polyphyletic; or two equally ancient sister groups, **Rhipidistians** (entirely extinct forms) and **Actinistians** (for the sole surviving form *Latimeria* and its undisputed fossil relatives). Currently even this set of subdivisions has been brought into question by the revision of the rhipidistians into two or more separate clades. Although the details of these arguments (mostly dealing with head and paired fin anatomy) need not concern us, it is important to recognize that active discussion characterizes the study of the vertebrates closest to tetrapods. We shall take a further look at fossil sarcopterygians when we examine the origin of their sister taxon—the tetrapods.

The Evolution of the Actinopterygii

Basal actinopterygians include the paleoniscoids (in reference to a variety of primitive forms now extinct), polypterids or cladistians (a small group of living fishes that provide the best model for understanding the extinct paleoniscoids), and the chondrosteans (living forms that are very different from the early fossil forms). Although fragments of late Silurian Actinopterygii exist, complete fossil skeletons are not found earlier than the mid to late Devonian. Compared with sarcopterygians, they were small fishes (5 to 25 centimeters) with a single dorsal fin, a strongly heterocercal forked caudal fin with no fin web dorsal to the axial skeleton, paired fins with long, thin (not fleshy) bases, large eyes, and a reduced snout (Figure 8–20e and f). The interlocking scales, although thick like those of sarcopterygians, were otherwise distinct in gross as well as histological structure and growth pattern. Structures supporting the fins were parallel arrays of closely packed radials. The number of bony rays supporting the fin membrane was greater than the number of supporting radials and were clearly derived from elongated scales aligned end to end. Two morphological aspects of the paleoniscoids deserve special attention: specializations for locomotion and for feeding. Unfortunately, the living cladistians and chondrosteans are so specialized in these respects that they shed little light on the biology of early actinopterygians.

The lower jaw of paleoniscoids was supported by the hyomandibular and was snapped closed in a scissors-like action by the adductor mandibulae muscles to drive small conical teeth into prey (Figure 8–21). This muscle originated in a narrow enclosed cavity between the maxilla and the palatoquadrate and inserted on the lower jaw near its articulation with the quadrate (Schaeffer and Rosen 1961). As a result, the bite was swift but the force created was low. The close-knit dermal bones of the cheeks permitted little expansion of the orobranchial chamber beyond that required for respiration.

Paleoniscoids were successful for 200 million years. No evidence supports a monophyletic origin of the fishes called paleoniscoids. The different types share numerous characters (Figure 8–22), but these all may be primitive. Near the end of the Paleozoic, paleoniscoids showed signs of change. The upper and lower lobes of the caudal fin were often nearly symmetrical and all fin membranes were supported by fewer bony rays—about one for each internal supporting radial in the dorsal and anal fins. Greater versatility in fin movements may have been gained by this morphological reorganization. Certainly the trends in the caudal morphology of advanced paleoniscoids permitted considerable control over the increasingly flexible fins.

The dermal armor of late Paleozoic paleoniscoids was also reduced. The changes in fins and armor may have been complementary—more mobile fins mean more versatile locomotion, and increased ability to avoid predators may have permitted a reduction in heavy armor. This reduction of weight could have further stimulated the evolution of increased locomotor ability. Evolution of

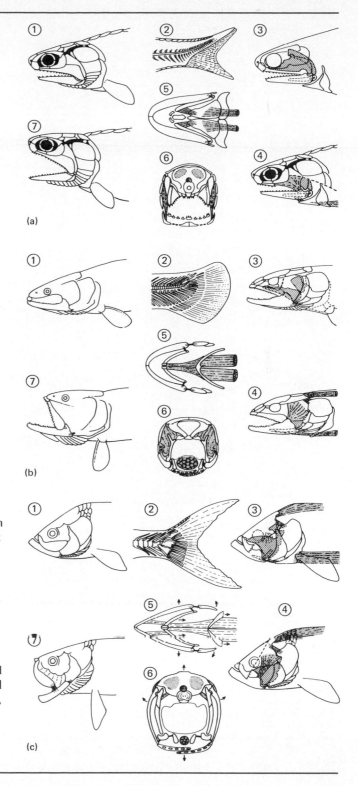

Figure 8–21. Morphological characters of three stages of actinopterygian evolution. (a) Early actinopterygian jaw size as indicated by gape (compare a1 and a7), hyomandibular support (a3), musculature (in ventral view [a5], cross section of the head [a6], and cut away lateral view [a4]), and caudal construction (a2). Morphology is based on that reconstructed for *Moythomasia,* a paleoniscid. (b) Neopterygian jaw size and partially circular gape (b1 and b7), nearly vertical hyomandibular orientation (b3), enlarged jaw musculature (ventral view [b5], cross section of head [b6], and cut away view, [b4]), and caudal construction (b2). Morphology is that of *Amia,* the living bowfin. (c) Teleostean jaw mobility to form a circular opening (c1 and c7), as illustrated in an elopomorph fish, the vertical (sometimes even anteriorly directed) hyomandibular (c3), the complex jaw musculature (ventral view [c5], cross section of head [c6], and cut away view [c4]), and the homocercal caudal construction (c2). Morphology is based on *Megalops,* the living tarpon.

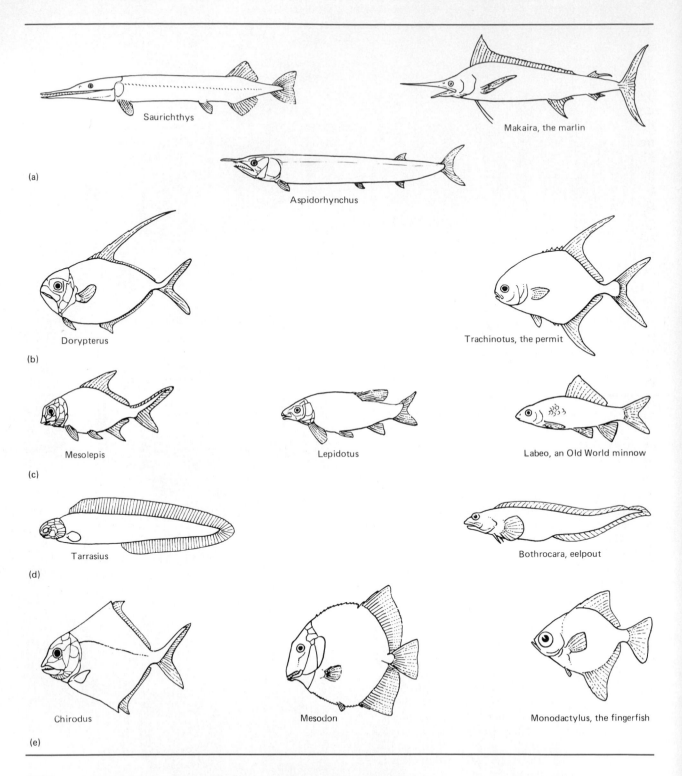

Saurichthys

Makaira, the marlin

Aspidorhynchus

(a)

Dorypterus

Trachinotus, the permit

(b)

Mesolepis

Lepidotus

Labeo, an Old World minnow

(c)

Tarrasius

Bothrocara, eelpout

(d)

Chirodus

Mesodon

Monodactylus, the fingerfish

(e)

this positive feedback type was probably enhanced by perfection of the swim bladder as a delicately controlled hydrostatic device.

Better locomotion enhances predatory capacity. The food-gathering apparatus of several clades underwent radical changes to produce, in the upper Permian, a new clade of actinopterygians—the **neopterygians**. For over a century the most primitive neopterygians have been called holosteans, but they are no longer considered a coherent taxon with unique shared defining characters; those they share are considered primitive for the neopterygians as a whole. Their jaw mechanism was characterized by a short maxilla, and a freeing of the posterior end of the maxilla from the other bones of the cheek (Figure 8–21). Because the cheek no longer was solid, the hyomandibular could swing *laterally*, thus increasing the volume of the orobranchial chamber in a rapid motion to produce a powerful suction useful in capturing prey. The power of the sharply toothed jaw could be increased because the adductor muscle was not limited in size by a solid bony cheek. No longer enclosed, the jaw muscle mass expanded dorsally through the space opened by the freeing of the maxilla. In addition, an extra lever arm—the coronoid process—developed at the site of insertion of the adductor mandibulae, adding torque and thus power to the closure of the mandible of many of the more forceful biters.

The bones of the gill cover (operculum) were connected to the mandible so that expansion of the orobranchial chamber aided in opening the mouth. The anterior, articulated end of the maxilla developed a ball-and-socket joint with the neurocranium. Because of its ligamentous connection to the mandible, the free posterior end of the maxilla was rotated forward as the mouth opened. This directed the maxilla's marginal teeth forward, aiding in grasping prey. The folds of skin covering the maxilla changed the shape of the gape from a semicircle to a circular opening. This enhanced the directionality of suction and also eliminated a possible side door escape route for small prey. The primary result of these changes was probably a reduction in the size of the opening of the expandable orobranchial cavity that increased the suction produced at the mouth.

Thus, the first neopterygians had considerable trophic and locomotor advantages. The neopterygians first appeared in the Permian (Figure 8–23) and became the dominant fishes of the Mesozoic. During the Jurassic, and perhaps in the late Triassic, basal neopterygians gave rise to fishes with feeding and locomotor specializations. These fishes constitute the largest vertebrate radiation, the **Teleostei**. Although teleosts probably evolved in the sea, they soon radiated into fresh water. By the late Cretaceous, teleosts had replaced most of the more primitive neopterygians, and most of the 200 to 300 families of modern teleosts had evolved. Their first specializations involved changes in the fins.

In adult teleosts, the caudal fin is supported by a few enlarged and modified hemal spines attached to the tip of the abruptly upturned vertebral column. Modified posterior neural arches—the uroneurals—add additional support to the dorsal

---

Figure 8–22. Convergence in specializations of body form in actinopterygians from the paleoniscoid, holostean, and teleostean grades of evolution. Paleoniscoids of Carboniferous to early Triassic age are shown in the left column, Holosteans of late Triassic to Cretaceous age in the middle column, and living teleosts are shown on the right. No attempt to show the fishes to scale or to strictly match habitats (when they are known) has been made, but the detail of morphological convergence is readily apparent. (a) Piscivorous fishes with long bill-like rostra and/or jaws; (b) fork-tailed strong swimmers with trailing fins; (c) broad finned bottom-feeding fishes; (d) eel-like fishes with confluent dorsal, caudal, and anal fins; (e) laterally compressed, maneuverable fishes.

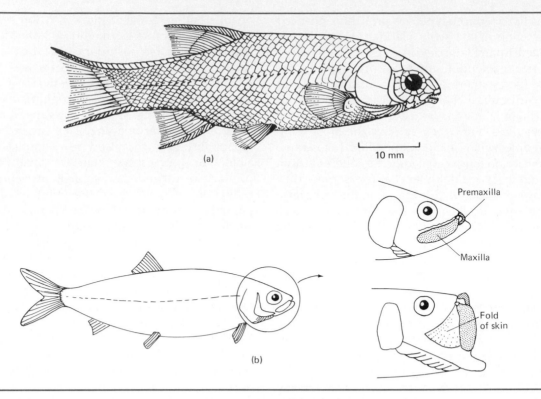

Figure 8–23. (a) Early neopterygian *Acentrophorus* of the Permian illustrating the generalized type from which the holostean radiation began. (b) *Leptolepis* an early Jurassic teleost with enlarged mobile maxillae, which form a nearly circular mouth when the jaws are fully opened. Membranes of skin close the gaps behind the protruded bony elements. Modern herrings have a similar jaw structure.

side of the tail. These uroneurals are a derived character of the teleosts (Lauder and Liem 1983). The caudal fin is symmetrical and of great flexibility. This type of caudal structure is known as **homocercal**. In conjunction with a gas bladder, a homocercal tail allows a teleost to swim horizontally without using its paired fins as control surfaces. Drag is reduced by folding the fins close to the body. Relieved of a lift function, the paired fins of teleosts became more flexible, mobile, and diverse in shape, size, and position. There is hardly a function for which teleost fins have not become specialized—from food getting to courtship, from sound production to walking and even flying. As with their predecessors, improvements in teleostean locomotion often seem to have brought about reduction in armor. Modern teleosts are thin-scaled (by Paleozoic and Mesozoic standards) or naked. The few heavily armored exceptions generally show a secondary reduction in locomotor abilities.

Teleosts, like most fishes that preceded them, also evolved improvements in their feeding apparatus. Trophic specializations in the earliest teleosts involved only a slight loosening of the premaxillae, so that they moved during jaw opening to accentuate the round mouth shape. One early clade of teleosts showed an enlargement of the free-swinging posterior end of the maxilla to form a nearly circular mouth when the jaws were fully opened (Figure 8–23). Later in the radiation of the teleosts distinctive changes in the jaw apparatus permitted a wide variety of feeding modes based on suction produced by their highly integrated jaw, gill arch, cranium, and body structure and function.

The Evolution of Jaw Mechanisms in Actinopterygians

Most of the main themes in actinopterygian evolution involve changes in the opening and closing mechanisms of the jaws as these functions relate to the transition from a simple prey grabbing device to a highly sophisticated suction device. Suction is important in active prey capture underwater. A rapid approach by a predator toward prey pushes a bow wave of water in front of the predator. This diverts water around and away from the mouth. Potential prey thus can be deflected around the predator's body and away from the grasp of its jaws. If anteroposterior streamlines leading directly into the mouth could be created, this would eliminate the bow wave and prey would be directed into instead of around the mouth. This is exactly what happens in neopterygians. By rapidly increasing the volume of their orobranchial chamber, they create suction which draws a stream of water into the mouth.

The primitive **Actinopterygii**, such as the paleoniscid *Moythomasia* (Figure 8–21), were not much different from the acanthodians in the basic organization and function of their jaws during feeding. The tooth-studded dermal premaxillae and maxillae of the upper jaw and dentaries of the lower jaw worked much like a snap trap to grab prey. In water a significant force must be exerted to open the mouth against the resistance of the medium. Primitive actinopterygians achieved this in two simultaneous and complementary ways: (1) they lifted the cranium, to which the upper jaw was firmly attached by solid bony cheeks, by the action of epaxial muscles attached to the posterodorsal margin of the skull; and (2) they lowered the lower jaw by applying a posterodorsal force to

a ligament that inserted on the lower jaw near its articulation with the upper jaw.

The force that rotates the lower jaw around its articulation and opens the mouth is generated in a peculiar, indirect way. Hypaxial muscles pull posteriorly on the pectoral girdle at the same time as similar bands of muscle between the ventral elements of the hyoid arch and the pectoral girdle contract. These actions pull posteroventrally on the anterior projections of the hyoid apparatus in the floor of the mouth. This causes the posterior portions of the hyoid to rotate dorsally. The ligament whose action opens the jaw is attached to these posterior hyoid elements and thus it is the movement of the hyoid that opens the lower jaw. Why such an indirect mechanism? The retraction of the hyoid has another component: It spreads the elements of the hyoid arch laterally, increasing the distance between the lateral walls of the orobranchial cavity and producing suction just as the mouth opens. Primitively, the extent of expansion was limited because the jaws, cheeks, and opercula of early actinopterygians were solidly encased in dermal bone with only a little flexibility between elements. The dorsal elements of the skull roof had little mobility. The long upper jaw was supported against the cranium with the help of the enlarged hyomandibular which angled obliquely posteroventrally from its articulation on the cranium to far behind the eye near the articulation point for the lower jaw. The long oblique upper element of the hyoid arch also restricted the lateral expansion of the ventral elements of the hyoid.

The muscles that close early actinopterygian jaws are more direct and the action simpler than the opening process. The mandibular adductor mass has several portions with fibers running at different angles providing different leverage advantages. The muscle mass originates in the narrow space between the dermal cheek bones and **suspensorium** (the composite of hyomandibular and palatoquadrate bone derivatives which form the hinge joint from which the lower jaw is suspended from the cranium). The muscles thus have little space to expand as their fibers contract, leading to a fairly weak muscle system. The adductor inserts on the lower jaw just anterior to its articulation with the suspensorium.

A major reorganization of the jaw mechanisms occurred with the evolution from the early actinopterygians of the **Neopterygii**. In the transition to these more advanced actinopterygians, the adductor muscles for the mandible became larger and more complex. They inserted on a specialized coronoid process of the dentary, resulting in increased power in closing the jaws. This was possible only because of loosening of the connections between the rigid cheek bones which had encased and limited the size of the adductor muscles in earlier ray-finned fishes. The stem neopterygians, illustrated by *Amia* (Figure 8–21), had an epaxial head lifting and a pectoral girdle/hyoid lower jaw opening mechanism very like that of their ancestors. In addition they had a second, independent jaw opening mechanism. Like the pectoral girdle/hyoid mechanism, it is indirect and acts by way of another ligament attached to the lower jaw below and behind its articulation with the suspensorium. Many of the dermal bones of the cheek and operculum in neopterygians are free to move relative to others, due to gaps between the elements of the cheek. This provides space for the enlargement of the jaw adductor musculature. The opercular series of bones rotates around a newly evolved socket-like articulation with the suspensorium. Rotation is powered by contraction of short opercular levator muscles originating on the back of the cranium and inserting on the dorsal border of the operculum. This rotation causes a posterodorsal movement of the ventralmost part of the opercular series. The force of this movement is transmitted via a ligament to the mandible which swings open in response. Thus in the Neopterygii, two unrelated, biomechanically independent pathways for lowering the jaw permit the evolution of functional variation and specialization.

Two other distinctions separate the jaws of primitive actinopterygians from those of the early neopterygians. One is a further consequence of opening up the cheeks to accommodate the mandibular adductor muscles: the maxilla no longer is tied to the bones surrounding the eye and cheek. It is, in fact, quite free to move and rotates on a medial peg at its articulation with the premaxilla. When the lower jaw opens, the posterior end of the maxilla rotates abruptly downward, making the mouth opening round. This efficiently directs water responding to suction into the mouth from directly in front of the fish where prey is aligned. A second innovation is the enhanced power of suction itself. This is in part due to a near vertical orientation of the hyomandibula, which more freely swings laterally as the ventral hyoid elements splay apart under the pull of the muscles attached to the pectoral girdle. In neopterygians feeding is accomplished by lifting the head, depressing of the floor of the orobranchial cavity, and expanding the head laterally: The mouth opens rapidly with a strong directed suction.

The teleosts (Figure 8–21) were at first essentially identical in feeding mechanism to the earlier neopterygians with one exception: Instead of the maxilla alone rotating as the mouth opened, a new rotation site replaces that seen in *Amia*: the premaxillae rotate on the tip of the skull, their toothed margins joining the maxilla in moving down and forward. This perfected the rounding of the open mouth even further. It also set the stage for at least two separate radiations of long sliding processes of the premaxillae that run up the midline of the snout when the jaws are closed. When the mouth is open the processes slide down, allowing the tooth-bearing portions of the jaw to be greatly extended—a phenomenon known as **protrusibility**. Although protrusibility is generally associated with perciform fishes (Figure 8–24), it occurs in the related atherinid and paracanthopterygian fishes and in the unrelated cypriniform ostariophysans as well. Jaw morphology differs significantly among fishes with protrusible jaws, clearly indicating that some of these protrusibilities have evolved independently.

All protrusion mechanisms involve complex ligamentous attachments that allow the ascending processes of the premaxilla to slide forward on the top of the cranium without dislocation. In addition, since no muscles are in position to pull the premaxillae forward, they must be pushed by leverage from behind. Two sources provide the necessary leverage. The premaxillae may be protruded by the opening of the lower jaw through ligamentous ties to the posterior tip of the premaxillae. Second, leverage can be provided by complex movements of the maxillae, which become isolated from the rim of the mouth by long, often toothed posterior projections of the premaxillae.

With so many groups converging on the same complex function, the adaptive significance of protrusibility would seem to be great. Surprisingly, there is no single hypothesis of advantage that has much experimental support (Lauder and Liem 1981). Protrusion may enhance the hydrodynamic efficiency of the circular mouth opening of primitive teleosts, but this seems insufficient to explain the complex anatomical changes necessary for protrusibility. Protrusion may aid in gripping prey, or the mouth's mobile jaws may be fitted to the substrate during feeding while the body remains in the horizontal position required for rapid escape from one's own predators. In many fishes with protrusible jaws, prey are clearly sucked into the mouth, but what part of the suction is produced by the protrusion itself depends greatly on the relative timing of the action of all parts of the system. No consistent pattern of steps in the feeding sequence indicates that protrusion is a significant contributor to suction efficiency. Although again no universal conformity to the ascribed functions can be seen in teleosts with protrusible premaxillae, another set of advantages may lie in the functional independence of the upper jaws relative to other parts of the feeding apparatus. Thus some fishes, such as the silversides and killifishes, can either greatly protrude, moderately protrude, or not protrude the upper jaw while opening the mouth and creating suction. The modulations direct the mouth opening and major axis of suction either ventrally, straight ahead, or dorsally, allowing the fish to feed from substrate, water column, or surface with equal ease. The independent movement of the protrusible upper jaw also permits closure of the mouth through maximum extension of the premaxillae while the orobranchial cavity is still expanded. Thus engulfed prey are entrapped before the orobranchial cavity volume has been reduced through evacuation of water from the gill openings. Perhaps the most broadly applicable hypothesis for the strong selection for jaw protrusion is that shooting out the jaws in front of the head allows a predator to approach the prey with a portion of its feeding apparatus more rapidly than can fishes that lack protrusion. Protrusion can add significant velocity to the final crucial moment of a predator's approach. The shooting out of protrusible jaws has been measured to increase the approach velocity of the predator by 39 to 89 percent in the crucial last instant.

Another innovation in the feeding apparatus of teleosts with multiple evolutions (as evidenced by diversity of anatomical arrangements) is the evolution of powerful mobile **pharyngeal jaws**. Primitively, ray-finned fishes have numerous dermal tooth plates in the pharynx. These plates are aligned with but not fused to both dorsal and ventral skeletal elements of the gill arches. A general trend in fusion of these tooth plates and their association with and frequent fusion to a few specific gill arch elements above and below the esophagus can be traced in the Neopterygii. Primitively, these consolidated pharyngeal jaws are not very mobile and are used primarily to hold and manipulate prey in preparation for swallowing it whole. In the ostariophysan minnows and their relatives, the suckers, the primary jaws are toothless but protrusible and the pharyngeal jaws are greatly enlarged and can chew against a horny pad on the base of the skull. These feeding and digestive specializations allow extraction of nutrients from thickwalled plant cells and represent one of the largest radiations of herbivores among ray-finned fishes.

In the Neoteleostei the muscles associated with the branchial skeletal elements supporting the pharyngeal jaws have undergone radical evolution, resulting in a variety of powerful movements of the pharyngeal jaw toothplates. Not only are the movements of these second jaws completely unrelated to the movements and functions

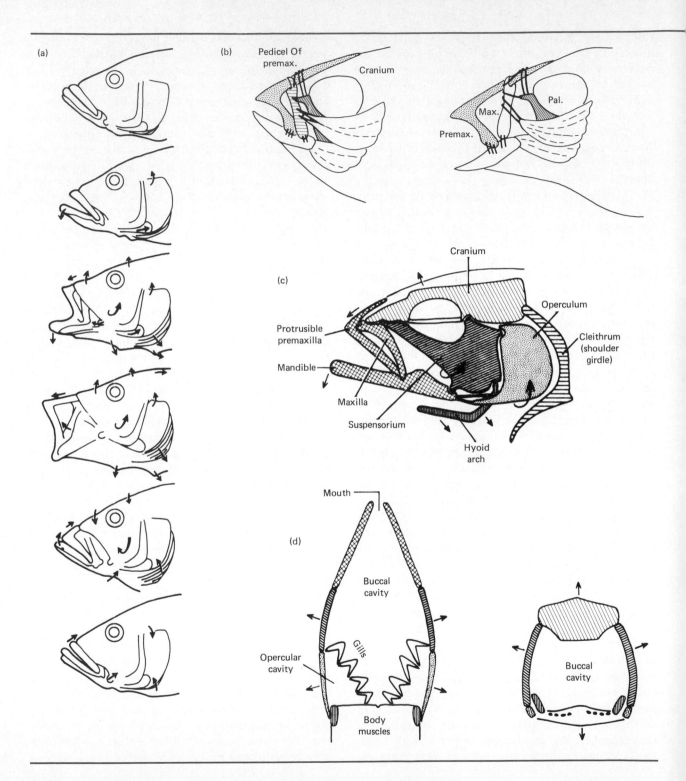

(a)

(b) Pedicel Of premax.
Cranium

Pal.
Max.
Premax.

(c)
Cranium

Protrusible premaxilla

Operculum

Cleithrum (shoulder girdle)

Mandible

Maxilla

Suspensorium

Hyoid arch

Mouth

(d)

Buccal cavity

Gills

Opercular cavity

Buccal cavity

Body muscles

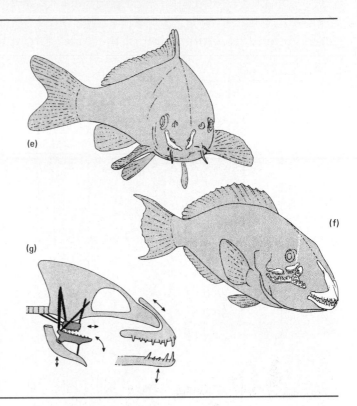

Figure 8–24. Jaw protrusion in suction feeding. (a) Top to bottom: The sequence of jaw movements in an African cichlid fish, *Serranochromis*. (Modified from K. F. Liem in A. G. Kluge et al., 1977, *Chordate Structure and Function*, 2nd edition, Macmillan, New York.) (b) Muscle, ligament, and bone movements during premaxillary protrusion. (c) Skeletal movements and ligament actions during jaw protrusion. (After G. V. Lauder, 1980, *Biofluid Mechanics* 2:161–181.) (d) Frontal section (left) and cross section (right) of buccal expansion during suction feeding. (e and f) Position of pharyngeal jaws in a carp (e) and a wrasse (f). (g) Movements of the primary and pharyngeal jaws of a wrasse during feeding.

of the primary jaws, but the upper and lower tooth plates of the pharyngeal jaws move quite independently of each other. With so many separate systems to work with, it is little wonder that some of the most extensive adaptive radiations among teleosts have been in fishes endowed with protrusible primary jaws and specialized mobile pharyngeal jaws.

Living Actinopterygii: Ray-Finned Fishes

With an estimated 20,850 living species (Nelson 1984), the living actinopterygians present a fascinating, even bewildering, diversity of forms of vertebrate life. Because of their numbers we are forced to examine them in a disproportionately brief survey compared to other vertebrate taxa. We focus on the primary characteristics of the group and its evolution.

The study of the phylogenetic relationships of actinopterygians entered its current active state in 1966 with a major revised scheme of teleostean phylogeny proposing several new relationships. The following two decades have seen a phenomenal growth in our understanding of the interrelationships of ray finned fishes (Rosen 1982). Nevertheless, certain of the relationships indicated in Figure 8–25 are uncertain and should be considered as working hypotheses (Lauder and Liem 1983).

Polypterids and Chondrosteans

Paleoniscoids were replaced during the early Mesozoic by neopterygians, but a few skeletally degenerate or specialized forms have survived. The most primitive surviving lineage of actinop-

Phylogenetic Relationships of the Actinopterygii

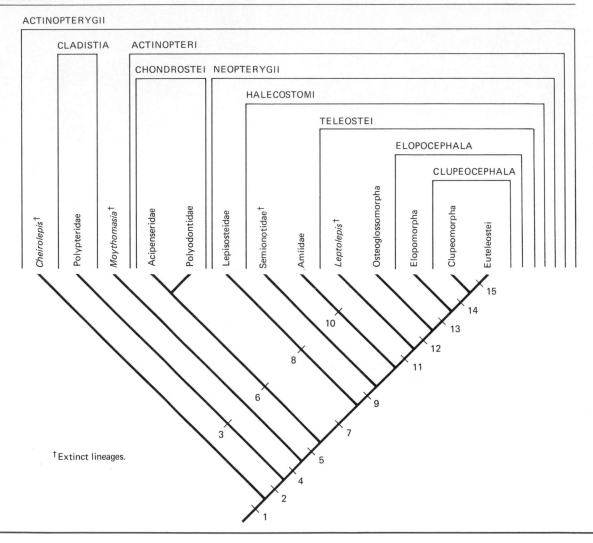

† Extinct lineages.

terygian fishes is the **Polypteriformes**. They are similar to paleoniscoids in many ways. However, polypteriforms have enough specializations to obscure their relationships to other fishes. Sometimes called the Cladistia, the Polypteriformes include 11 species of African bichirs and reedfish (Figure 8–26). Like the primitive sturgeons and paddlefish, these modest-size (less than a meter), slow-moving fishes have modified heterocercal tails. However, polypteriforms differ from these other basal actinopterygians in having well-ossified skeletons. In addition to a full complement of dermal and endochondral bones, polypteriforms are covered by thick, interlocking, multilayered

1. Actinopterygii: Basal elements of pectoral fin enlarged, median fin rays attached to skeletal elements that do not extend into fin, single dorsal fin, scales with unique arrangement, shape, interlocking mechanism, and histology, and at least six additional derived characters. 2. Cladistia plus Actinopteri: A specialized dentine (acrodin) forms a cap on the teeth, details of posterior braincase structure, specific basal elements of the pelvic fin are fused, and numerous features of the soft anatomy of living forms that cannot be verified for fossils. 3. Polypteridae (Cladistia): Unique dorsal fin spines, facial bone fusion and pectoral fin skeleton and musculature. 4. Actinopteri plus fossils such as *Moythomasia* and *Mimia:* Derived characters of the dermal elements of the skull and pectoral girdle and fins. 5. Actinopteri: A spiracular canal formed by a diverticulum of the spiracle penetrating the postorbital process of the skull, other details of skull structure, three cartilages or ossifications in the hyoid below the interhyal, swimbladder connects dorsally to the foregut, fins edged by specialized scales (fulcra). 6. Chondrostei: Fusion of premaxillae, maxillae, and dermopalatines, unique anterior palatoquadrate symphysis. 7. Neopterygii: Dorsal and anal fins' rays reduced to equal the number of endoskeletal supports, upper lobe of caudal fin containing axial skeleton reduced in size to produce a nearly symmetrical caudal fin, upper pharyngeal teeth consolidated into tooth-bearing plates, characters of pectoral girdle and skull bones. 8. Lepisosteidae (Ginglymodi): Vertebrae with convex anterior faces and concave posterior faces (opisthocoelus), toothed infraorbital bones contribute to elongate jaws. 9. Halecostomi: Modifications of the cheek, jaw articulation, and opercular bones including a mobile maxilla. 10. Amiidae (Recent Halecomorphi): Jaw articulation formed by both the quadrate and the symplectic bones. 11. Teleostei: Elongate posterior neural arches (uroneurals), unpaired ventral pharyngeal toothplates on basibranchial elements, premaxillae mobile, details of skull foramina, jaw muscles, and axial and pectoral skeleton. 12. Recent Teleosts: Presence of an endoskeletal basihyal, four pharyngobranchials and three hypobranchials, median toothplates overlying basibranchials and basihyals. 13. Elopocephala: Two uroneural bones extend anteriorly to the second ural (tail) vertebral centrum, abdominal and anterior caudal epipleural intermuscular bones present. 14. Clupeocephala: Pharyngeal toothplates fused with endoskeletal gill arch elements, neural arch of first caudal centrum reduced or absent, distinctive patterns of ossification and articulation of the jaw joint. 15. Euteleostei: Presence of an adipose fin posteriorly on the mid-dorsal line, presence of nuptial tubercles on the head and body, paired anterior membranous outgrowths of the first uroneural bones of the caudal fin. (These characters are usually lost in the most derived euteleosts.) (Based on G. V. Lauder and K. F. Liem, 1983, *Bulletin of the Museum of Comparative Zoology* 150:95–197; B. G. Gardiner, 1984, *Bulletin of the British Museum (Natural History) Geology* 37:173–427; and J. G. Maisey, 1986, *Cladistics* 2:201–256.)

Figure 8–25. Phylogenetic relationships of the Actinopterygii. This diagram shows the probable phylogenetic relationships among the major groups of actinopterygians. A dagger (†) indicates extinct lineages. The numbers indicate derived characters that distinguish the lineages.

scales. These **ganoid** scales are covered with a coat of ganoin, an enamel-like tissue characteristic of primitive actinopterygians. Larval bichirs (*Polypterus*) have external gills, possibly a primitive condition for Osteichthyes. *Erpetoichthys,* the reed fish, is eel-like. All polypteriforms are predatory, and their jaw mechanics provide our best model of the original actinopterygian condition. Their peculiar flag-like dorsal finlets and the fleshy bases of their pectoral fins must be considered specializations, but so little is known about their natural history that the significance of these features cannot be appreciated.

The **Chondrostei** or **Acipenseriformes** include

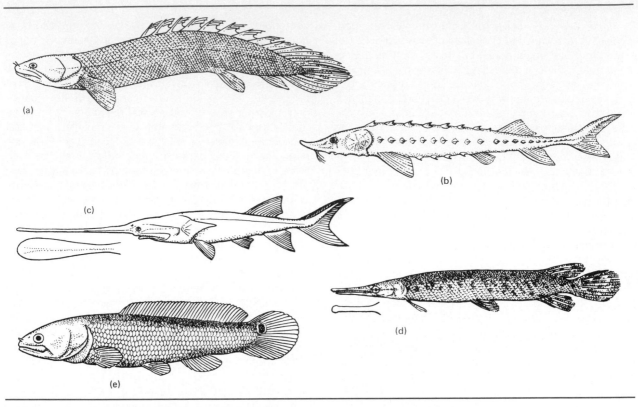

Figure 8–26. Living fishes of the primitive paleoniscoid and chondrostean clades of actinopterygian evolution (a–c) and surviving primitive neopterygians (d and e). (a) *Polypterus*, the genus of bichirs; (b) *Acipenser*, one of the genera of sturgeons; (c) *Polyodon spathula*, one of two living species of paddlefish; (d) *Lepisosteus*, the genus of gars; (e) *Amia calva*, the bowfin.

two families of specialized primitive actinopterygians. The 23 species of sturgeon, family Acipenseridae (Figure 8–26), are large (1- to 6-meter), active, benthic fishes. They lack endochondral bone and have lost much of the dermal skeleton of more primitive actinopterygians. Sturgeons have a strongly heterocercal tail armored with a specialized series of scales extending from the dorsal margin of the caudal peduncle along the upper edge of the caudal fin. Most sturgeons have five rows of enlarged armor-like scales along the body. The protrusible jaws of sturgeons make them effective suction feeders. The mode of jaw protraction is unique and independently derived. Sturgeons are

found only in the Northern Hemisphere and are either **anadromous** (ascending into fresh water to breed) or entirely freshwater in habit. Commercially important both for their rich flesh and as a source of the best caviar, they have been severely depleted by intensive fisheries in much of their range.

The two surviving species of paddlefish, Polyodontidae, are closely related to the sturgeons but have a still greater reduction of dermal ossification. Their most outstanding feature is a greatly elongate and flattened rostrum, which extends nearly one-third of their 2-meter length. The rostrum is richly innervated with ampullary organs that are believed to detect minute electric fields. Despite the common notion that this paddle is used to stir food from muddy river bottoms, the American paddlefish is a planktivore. Paddlefish feed by actively swimming with their prodigious mouths agape and straining crustaceans and small fishes from the water column, using modified gill rakers as strainers. The two species of paddlefish have a disjunct zoogeographic distribution: one is found in the Yangtze River valley of China, the other in the Mississippi River valley of the United States. Fossil paddlefish are known from western North America.

Primitive Neopterygians

The two living genera of primitive neopterygians are currently limited to North America and represent widely divergent types. The seven species of gars, *Lepisosteus* (Figure 8–26), are medium- to large-sized (1- to 4-meter) predators of warm-temperate fresh and brackish (estuarine) waters. The elongate body, jaws, and teeth are specialized features, but their interlocking multilayered scales are similar to those of many Paleozoic and Mesozoic actinopterygians. Gars feed on other fishes taken unaware when the seemingly lethargic and excellently camouflaged gar dashes alongside them and, with a sideways flip of the body, grasps them with needle-like teeth. Sympatric with gars is the single species of bowfin, *Amia calva*. The head skeleton is not specialized by long prey-holding jaws like that of the gar but points toward the condition in more advanced fishes in its modifications as an effective suction device (Lauder 1980). *Amia* prey on almost any organisms smaller than their own 0.5- to 1-meter length. Scales of the bowfin are comparatively thin and made up of a single layer of bone as in teleost fishes; however, the asymmetric caudal fin is very similar to the heterocercal caudal fin of more primitive fishes.

Teleosteans

Most living fishes are teleosts. They share many characters of caudal and cranial structure and are grouped into four clades of varying size and diversity.

The **Osteoglossomorpha**, which appeared in Cretaceous seas, are now restricted to tropical fresh waters. *Osteoglossum* (Figure 8–27) is a 1-meter-long predator from the Amazon. *Arapaima* is an even larger Amazonian predator (the largest freshwater fish, reaching a length of at least 3 meters and perhaps 4.5 meters), and *Mormyrus* is representative of the small African bottom feeders that use weak electric discharges to communicate with conspecifics. As dissimilar as they may seem, the osteoglossomorph fishes are united by unique osteological characters of the mouth and feeding mechanics.

The **Elopomorpha** (Figure 8–28), distinguishable on the basis of detailed analysis of the skeleton, appeared by early Cretaceous times. A unique character of elopomorphs is the specialized leptocephalus larva: These larvae, adapted to long life near the open ocean surface, permit wide dispersal even though the adults may be restricted to shallow inshore habitats. Elopomorphs include tarpons (Megalopidae), ladyfish (Elopidae), and bonefish (Albulidae), all of which are tropical and sub-tropical sport fishes, and several families of eels, Anguilliformes. This relationship is based primarily on similarities of the larvae. Fossils useful for defining relationships within the Elopomorpha have not yet been discovered.

Most elopomorphs are marine and eel-like, but some species are tolerant of fresh waters. The common American eel, *Anguilla rostrata*, has one of the most spectacular life histories of any fish (Norman and Greenwood 1975). After growing to sexual maturity (perhaps 10 to 12 years) in rivers, lakes, and even ponds, the **catadromous** (downstream migrating) eels enter the sea. The North Atlantic eels apparently continue their migration until reaching the Sargasso Sea. Here they are thought to spawn and die, presumably at great depth. The eggs and newly hatched leptocephalus larvae float to the surface and drift in the currents. Larval life extends until the larvae reach continental margins, where they transform into miniature eels and ascend rivers to feed and mature.

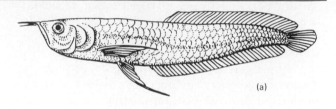

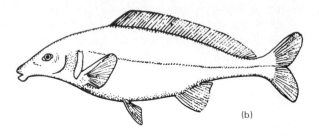

Figure 8–27. Living osteoglossomorphs (a) *Osteoglossum*, the arawana from South America; (b) *Mormyrus*, an elephant-nose from Africa.

Most of the **Clupeomorpha** are specialized for feeding on minute plankton sucked into a specialized mouth and gill straining apparatus. They are silvery, mostly marine schooling fishes of great commercial importance. Common examples are herrings, shad, sardines, and anchovies. Several clupeomorphs are anadromous; the springtime migrations of American shad (*Alosa sapidissima*) from the North Atlantic into rivers in eastern North America involve millions of individuals.

The vast majority of living teleosts belong to the fourth clade, the **Euteleostei**, which evolved before the upper Cretaceous (Figure 8–29). Their basal stock is represented today by the specialized ostariophysans and the generalized salmoniforms.

The **Ostariophysi**, the predominant fishes of the world's fresh waters, seem to be at the very base of the euteleostean radiation but have a distinctive derived character. Their name refers to small bones (*ostar* = a little bone) that connect the swim bladder (*physa* = a bladder) with the inner ear (Figure 8–30). Using the swim bladder as an amplifier and the chain of bones as conductors, this **Weberian apparatus** greatly enhances hearing sensitivity of these fishes. The ostariophysans are more sensitive to sounds and have a broader frequency range of detection than other fishes (Pop-

per and Coombs 1980). Although all Ostariophysi possess Weberian ossicles, they are a very diverse taxon and include the characins of South America and Africa, the carps and minnows found on all inhabitable continents except South America and Australia, and the catfishes found on all continents with flowing fresh waters and in many shallow marine areas.

About 80 percent of the fish species in fresh water are ostariophysans. As a group they display diverse traits. For example, many ostariophysans have protrusible jaws and are adept at obtaining food in a variety of ways. In addition, pharyngeal teeth act as second jaws. Many forms have fin spines or special armor for protection, and the skin typically contains glands that produce substances used in olfactory communication. Although they have diverse reproductive habits, most lay sticky eggs or otherwise guard the eggs, preventing their loss downstream.

The esocid and salmonid fishes (Figure 8–31) include important commercial and game fishes. These fishes have often been lumped into a taxon, the "Protacanthopterygii" (Figure 8–31a), but the basis for this classification is shared primitive euteleostean characters that are not valid for determining phylogenetic relationships. The most prim-

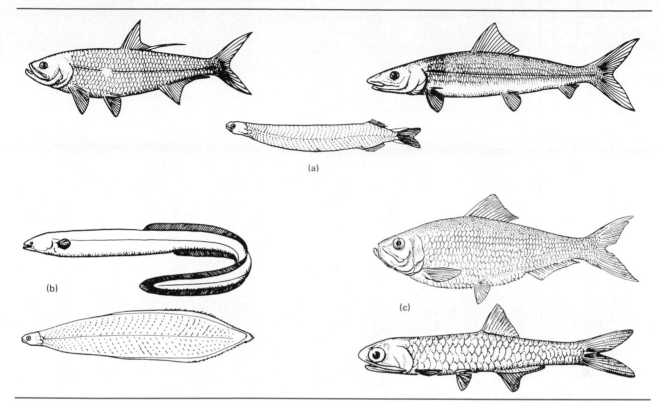

Figure 8–28. Living teleosts of isolated phylogenetic position: (a) Elopomorpha represented by a tarpon (left) and a bonefish (right) and a typical fork-tailed leptocephalus larva (below); (b) anguilliform elopomorphs represented by the common eel, *Anguilla rostrata* (above) and its leptocephalus larva (below); (c) Clupeomorpha represented by a herring (above) and an anchovy (below).

itive living euteleosteans are the esocids. These Northern Hemisphere temperate freshwater game fishes include pickerel, pikes, muskellunges and their relatives. The anadromous salmon usually spend their adult lives at sea, but the closely related trouts live in fresh waters. The Southern Hemisphere galaxiids are also primitive euteleosts and live in similar habitats to salmonids. The **Scopelomorpha** include deepwater lanternfishes, the myctophiforms (Figure 8–31b), which resemble the salmonids but have derived characters of gill arch musculature, jaw structure, and fins. Tiny, light-producing organs called **photophores** are arranged on their bodies in species- and even sex-specific patterns. They probably function as signals to conspecifics in the dim light of the deep sea, where resources, including mates, may be difficult to find.

High mobility of the jaws and the development of protective, lightweight spines in the median fins have evolved in several groups of teleosts. Several fishes, including cods and anglerfishes, are grouped as the **Paracanthopterygii** (Figure 8–32), although their similarities may represent convergence. However, the **Acanthopterygii**, or true spiny-rayed fishes, which dominate the surface and shallow marine waters of the world, appear to be monophyletic. Among the acanthop-

Phylogenetic Relationships of the Euteleostei

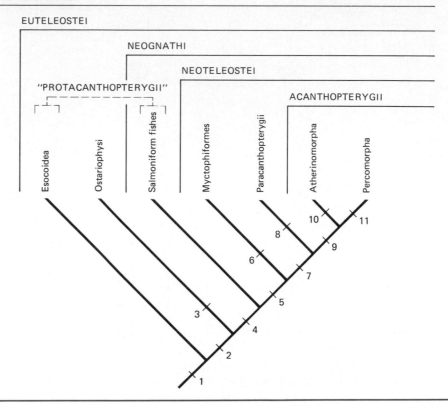

terygians, the atherinomorphs have protrusible jaws and specializations of form and behavior that suit them to shallow marine habitats, although some occur in fresh water. This group includes the silversides, grunions, flyingfishes, halfbeaks, as well as the egg-laying and live-bearing cyprinodonts. Killifish are a familiar example of egg-laying cyprinodonts, and the live bearers include the guppies, mollies, and swordtails commonly maintained in home aquaria.

The majority of species of acanthopterygians and the largest order of fishes is the Perciformes, with 7800 living species (Nelson 1984). A few of the well-known members are snooks, sea basses, sunfishes, perch, darters, dolphins, snappers, grunts, porgies, drums, cichlids, mullets, barra-cudas, tunas, billfish, and almost all the fishes found on coral reefs (see Figure 8–35).

Reproduction of Actinopterygians

Reproductive modes of actinopterygians show a greater diversity than is known in any other vertebrate taxon. Despite this diversity, the vast majority of ray-finned fishes are **oviparous** (producing eggs that develop outside the body of the mother). Within oviparous teleosts, marine and freshwater species show contrasting specializations. Most marine teleosts release large numbers of small, buoyant, transparent eggs into the water. These eggs are fertilized externally and left to develop and hatch while drifting in the open sea. The

1. Euteleostei: Presence of an adipose fin posteriorly on the mid-dorsal line, presence of nuptial tubercles on the head and body, paired anterior membranous outgrowths of the first uroneural bones of the caudal fin. (These characters are usually lost in the most derived euteleosts.) 2. Ostariophysi plus Salmoniform fishes plus Neoteleostei: Loss of the toothplate on the fourth basibranchial. 3. Ostariophysi: Haemal spines of the anterior caudal vertebrae fused to the centra, presence of cells in the epidermis that produce a fright pheromone, reduction or loss of certain skull bones, most forms have a Weberian apparatus developed from modified ribs, vertebrae, and swimbladder. 4. Neognathi (Salmoniform fishes plus Neoteleostei): Increased contact of first vertebra with skull bones. 5. Neoteleostei: A rostral ethmoid cartilage, specialized tooth attachment to jaw bones, new muscles and muscle insertions associated with the branchial arches and upper jaw or operculum, characters of the jaw suspension, fusion of a toothplate to the third epibranchial, reduction of the caudal skeleton. 6. Myctophiformes: Reduction of the fourth pharyngobranchial and its toothplate, third pharyngobranchial is the largest toothed element in the upper pharyngeal jaw and has acquired a portion of the musculature of the fourth arch. 7. Paracanthopterygii plus Acanthopterygii: Expansion of ascending and articular premaxillary processes. 8. Paracanthopterygii: Caudal skeleton with a full neural spine on the second preural centrum followed by two epurals posteriorly. 9. Acanthopterygii: Insertion of the branchial retractor muscle on the third pharyngobranchial element only, characters of the pharyngobranchials and epibranchials, symphysial and toothed portions of the premaxilla capable of significant antero–ventral movement. 10. Atherinomorpha: Unique jaw protrusion mechanism, reduction of infraorbital bones, loss of fourth pharyngobranchial. 11. Percomorpha: Pelvic girdle firmly joined to pectoral girdle, pelvic fins with one spine and five soft rays, a flange on the second circumorbital bone forms a shelf beneath the eye. (Based on G. V. Lauder and K. F. Liem, 1983, *Bulletin of the Museum of Comparative Zoology* 150:95–197.)

Figure 8-29. Phylogenetic relationships of the Euteleostei. This diagram shows the probable relationships among the major groups of modern teleosts. The numbers indicate derived characters that distinguish lineages. The quotation marks around "Protacanthopterygii" show that this is a paraphyletic grouping.

larvae that hatch from these small eggs are also small and usually have little yolk reserve. They begin preying on microplankton soon after hatching. Marine larvae are generally very different in appearance from their parents, and many larvae have been described for which the adult forms are unknown. Such larvae are often specialized for life in the oceanic plankton, feeding and growing while adrift at sea for weeks or months, depending on the species. The larvae eventually settle into the juvenile or adult habitats appropriate for their species. It is not yet generally understood whether arrival at the appropriate adult habitat (deep-sea floor, coral reef, or river mouth) is an active or passive process on the part of larvae. However, the arrival does coincide with metamorphosis from larval to juvenile morphology in a matter of hours to days.

This strategy of producing planktonic eggs and larvae exposed to a prolonged and risky pelagic existence appears to be wasteful of gametes. Nevertheless, complex life cycles of this sort are the principal mode of reproduction of marine fishes. Several hypotheses have been proposed to explain this paradox (Thresher 1984). One view involves the selective advantages gained by reduction of some types of predation on their zygotes that fishes may achieve by spawning pelagically. Predators that would capture the zygotes may be abundant in the parental habitat but absent from the pelagic realm. Pelagic spawning fishes often migrate to areas of strong currents to spawn, or

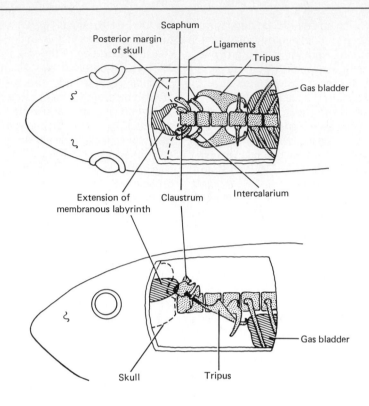

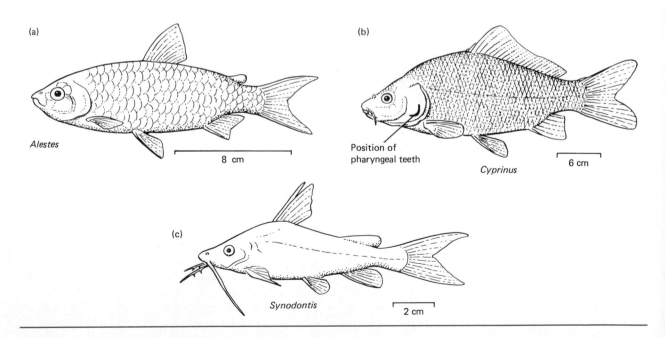

(a) *Alestes*

8 cm

(b) Position of pharyngeal teeth

Cyprinus

6 cm

(c) *Synodontis*

2 cm

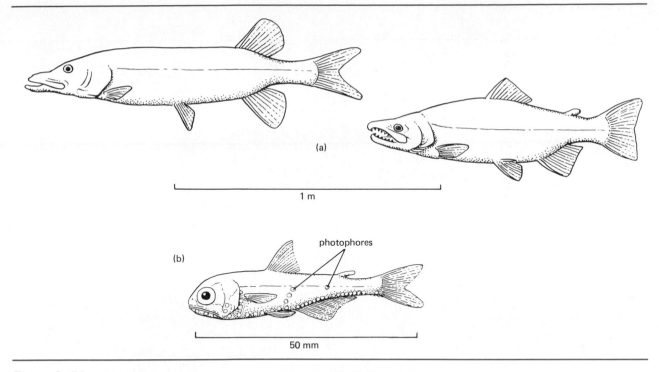

1 m

photophores

(b)

50 mm

Figure 8–31. (a) Primitive euteleosts represented by the pike (left) and the salmon (right); (b) myctophiform fishes, represented by a lanternfish. Species of lanternfishes differ in the number and arrangement of light-producing photophores concentrated on their ventral surface.

spawn in synchrony with maximum monthly or annual tidal currents, thus assuring rapid offshore dispersal of their zygotes. In shallow, clear waters these fishes may also reach a daily peak of spawning at dusk when particulate plankton feeders are rare in the water column. All these timing synchronies can be argued to support the nearshore zygote predator hypothesis. A second hypothesis explaining the advantages of pelagic spawning involves the high biological productivity of the sunlit surface of the pelagic environment. Microplankton of the appropriate size for food of pelagic larvae (bacteria, algae, protozoans, and minute crustaceans) are abundant where sufficient nutrients

Figure 8–30. Ostariophysan fishes have a sound-detection system, the Weberian apparatus, that is a modification of the swim bladder and the first few vertebrae and their processes. Sound (pressure) waves impinging on the fish cause the air bladder to vibrate. The tripus is in contact with the air bladder; as the bladder vibrates the tripus pivots on its articulation with the vertebra. This motion is transmitted by ligaments to the intercalarium and scaphium. Movement of the scaphium compresses an extension of the membranous labyrinth against the claustrum, stimulating the auditory region of the brain. Typical ostariophysans include (a) characins, (b) minnows, and (c) catfish.

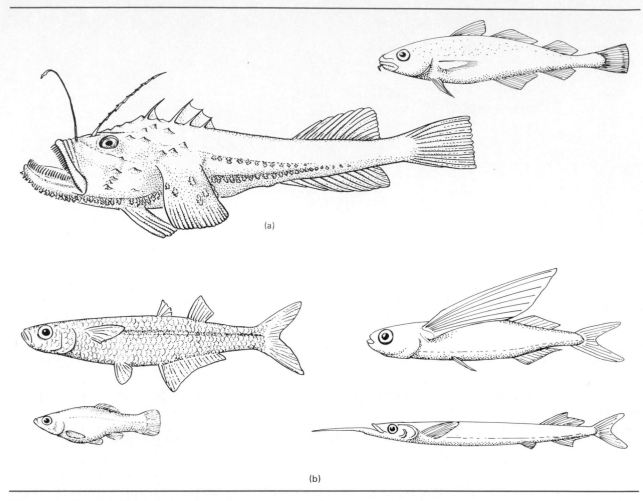

Figure 8–32. (a) So-called Paracanthopterygians represented by the cod (right) and the goosefish angler (left). Not to scale. (b) Atherinomorph fishes represented by (clockwise from upper left) an Atlantic silverside (*Menidia*), a flying fish, a halfbeak, and a live-bearing killifish, the Amazon molly (*Poecilia*). Not to scale. The Atlantic silverside and the Amazon molly have unusual sex determination patterns (see pages 318 and 320).

reach sunlit waters. If energy is limiting to the adult population, a successful strategy might be to produce eggs with a minimum of energy reserve (yolk) that hatch into specialized larvae able to utilize this pelagic productivity. In support of this hypothesis it can be pointed out that several inshore marine fishes migrate to areas of nutrient inflow or upwelling to spawn, indicating the value of pelagic productivity. Basic differences in the productivity of the tropical Atlantic and Pacific exist, and in areas of lower productivity pelagic spawning fishes have larger, yolk-rich eggs.

A final hypothesis springs from the fact that floating, current-borne eggs and larvae are more widely dispersed than would otherwise be the case. This would make more probable the broad distribution of the species and increase the chances of the young arriving at all possible patches of appropriate adult habitat. A widely dispersed species is not vulnerable to local environmental changes that could extinguish a species with a restricted geographic distribution. Perhaps the predominance of pelagic spawning species in the marine environment reflects the results of millions of years of extinctions of species with less dispersal-promoting habits.

Some marine species lay adhesive eggs on rocks, plants, or in gravel or sand, where they may be guarded by parents. Nests vary from depressions in sand or gravel to elaborate constructions of woven plant material held together by parental secretions. Species of marine fishes that construct nests are smaller than species that are pelagic spawners, perhaps because small species are better able to find secure nest sites than are large species. Parental guarding of eggs may be unwittingly assisted by other organisms near, on, or in which the eggs are laid. Among these are stinging anemones, mussels, crabs, sponges, and tunicates. Perhaps the ultimate in protection of eggs is portage by one of the parents. Species are known that carry eggs on fins, under lips, in the mouth or gill cavities, or on specialized protuberances, skin patches, or even in pouches.

In spite of this diversity of reproductive habits among marine teleosts, the fact remains that the vast majority of species produce eggs that are shed into the environment with little further parental investment. These eggs tend to be small relative to the adult fish's size and produced in prodigious numbers. An interesting consequence of this reproductive characteristic of most marine and some freshwater teleosts is that the number of individuals breeding in any given year (breeding stock size) bears no clear relationship to the number of individuals in the next generation (recruited stock year class strength). Thus, a breeding season rich with spawning adults may produce few or no offspring which survive to breed in subsequent seasons if environmental factors prevent the survival of eggs, larval, or juveniles. Conversely, a few breeding adults could, under exceptionally benign environmental conditions, parent a very large number of offspring which survive to maturity. This is possible because of the large number of eggs a single female is capable of producing. The characteristic low predictability of future stock size based on current stock size has been a major stumbling block to effective fisheries management. Since so much of a population's size depends on the environment experienced by eggs and larvae—conditions not usually obvious to fishermen or scientists—it is difficult to demonstrate the effects of overfishing in its early stages or the direct results of conservation efforts. Too much of successful recruitment depends on chance (stochastic) events in the pelagic environment. Increasing effort is being placed on understanding water column conditions conducive to egg, larval, and juvenile success and the monitoring and incorporation of these parameters into fisheries modeling.

Tragically, as we realize the importance of water column environmental conditions to be among the most critical to the youngest stages of marine life, we are altering that environment at an unprecedented rate. Coastal waters are receiving unnaturally large inputs of nutrients, especially nitrogen, from human activities: fertilizer runoff, domestic animal waste, urban sewage, industrial outfalls, and fossil fuel produced atmospheric nitrates. These inputs cause supernormal algae growth and loss of oxygen and light to the water column resulting in unsuccessful reproduction of many marine organisms.

In contrast to their marine counterparts, freshwater teleosts produce and care for fewer, larger, nonplanktonic eggs that produce hatchlings with adult-like body forms and behaviors. This reproductive strategy in fresh water has been related to the flowing and ephemeral characteristics of upland waters, which could easily flush a less well-developed fish from its preferred habitat. However, exclusively environmentally determined hypotheses for the predominance of parental care of small clutches of larger demersal eggs in fresh water have recently been questioned (Gebhardt

1987). Only one-third of 60 families of fishes that are thought to have evolved entirely in fresh water (**primary freshwater fishes**) include species that display parental care of the eggs or young, whereas more than one-half of 70 families of freshwater fishes derived from marine environments (**secondary freshwater fishes**) include species with parental care. Thus, a phylogenetic component to the reproductive patterns of freshwater fishes cannot be ignored.

Large yolk-rich demersal eggs may be primitive for actinopterygians, parental care having evolved independently and frequently throughout the evolution of ray-finned fishes. Producing pelagic eggs and larvae may be a derived characteristic of euteleosts. To understand more precisely the myriad of variations on these two basic themes of actinopterygian reproduction, it will be necessary to know more about the early life history of fishes in the wild (Box 8–2).

Sex Reversal and Life History Strategies of Actinopterygians

The life history of a number of teleosts holds some surprises for those most familiar with mammalian development and the distinct and permanent sexuality of each individual. In at least a dozen separate lineages of teleosts, primarily among the Acanthopterygii, an individual's ultimate functional sex is not determined at the time of fertilization of the egg as it is in birds, mammals, many lizards, and snakes. The functional sex of many teleost fishes changes during their sexual maturity; a number of fishes are functional, simultaneous hermaphrodites (capable of producing eggs and sperm at the same time). The study of the environmental and social contexts of teleosts' bizarre sexual patterns has significantly advanced understanding of how strong a selective force maximizing an individual's total lifetime reproductive output can be.

Although vertebrates typically begin early development of the reproductive organs and ducts of both sexes, one or the other usually soon begins to dominate. Birds and mammals have a special-ized mechanism that determines the direction of sexual differentiation: a pair of sex-determining heteromorphic chromosomes. In contrast, few teleosts have specialized sex chromosomes or even an accumulation of sex-related genes linked to a limited number of chromosomes.

Whenever environmental factors affect the reproductive capacity of the sexes differently, the stage is set for natural selection to favor differentiation of one sex over the other. Because the availability of mates is one important characteristic of the environment, most circumstances would not be expected to lead to the entire population becoming unisexual. Nevertheless, all-female species have been described: they use sperm from other closely related species to trigger zygote development but do not actually incorporate the genes into the developing embryo. The best known of the all-female populations are a half-dozen different forms of mollies, viviparous cyprinodonts from Mexico. All-female populations seem universally derived from hybridization of other local species and have limited distribution. An ecological basis for this strange phenomenon is not yet generally agreed upon (Turner 1983). However, another peculiar phenomenon, the changing of sex ratios of differentiating juveniles with season and latitude, has recently been explained in ecological terms.

There is an enormous size differential between vertebrate male gametes (sperm) and the female gametes (ova). A small female may be able to produce only a few eggs in her small ovaries, whereas even a small male has ample gonadal tissue to produce a prodigious quantity of sperm. Fishes are very small relative to adults at hatching but they continue to grow throughout life, often attaining sexual maturity long before reaching their largest known size. This means a female's egg production increases over successive growth and reproductive seasons (Figure 8–34). For example, female carp *Cyprinus carpio* hatch at a few millimeters, reach sexual maturity in 4 to 5 years at about 430 millimeters in length and are then capable of laying at least 36,000 eggs 1 millimeter in diameter. Carp may live 20 years, reaching 18 to 23 kilograms and a length of 1.2 meters. A 10-

Box 8–2. What a Fish's Ears Tell About Its Life

Following minute fish eggs or translucent larvae in the open sea or turbid rivers seems an insurmountable problem. However, an indirect method of tracing the details of life history of an individual fish of practical use for many species is promising (Campana and Neilson 1985). A characteristic of Osteichthyes is the presence of compact mineralized structures suspended in the interior of the inner ear. These structures are especially well developed in the majority of teleosts, where these **otoliths** are often curiously shaped, fitting into the spaces of the membranous labyrinth very exactly and growing in proportion to the growth of the fish. They are important in orientation and locomotion, and they are formed in most teleosts during late incubation.

Otoliths grow in concentric layers much like the layers of an onion (Figure 8–33). This structure appears to reflect daily growth. The relative width, density, and interruptions of the layers show the environmental conditions the individual encountered from hatching, including variations in temperature and food capture. A day-by-day record of the individual is written in its otoliths. While still fraught with problems of interpretation, the study of otolith microstructure is significantly advancing our understanding of reproduction and early life history of fishes (Thresher 1988).

Figure 8–33. Scanning electron micrograph of an otolith from a juvenile French grunt, *Haemulon flavolineatum.* The central area represents the focus of the otolith from which growth proceeds. The alternating dark and light rings are daily growth increments. Note that the width of the growth increments varies, signifying day-to-day variation in the rate of growth. (Photograph courtesy of E. Brothers.)

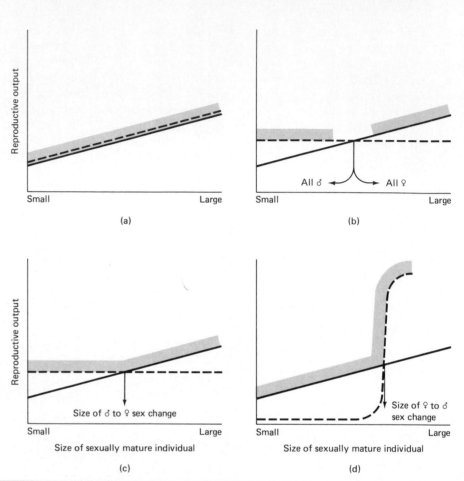

Figure 8–34. Size advantage hypothesis of sex determination and sexual patterns in organisms with labile sexuality such as teleosts. Solid line, female fecundity; dashed line, male fecundity; shaded band, predicted pattern of sexuality. (a) Gonorchorism (separate sexes) in iteroparous (multiple breeding season) species where males compete directly to fertilize eggs; (b) gonorchorism in semelparous (single breeding season) species where sex determination complements size at reproduction; (c) protandrous sequential hermaphroditism in iteroparous species with random mating; (d) protogynous sequential hermaphroditism in iteroparous species with female choice for larger dominant males. (Modified in part from R. R. Warner, 1984 , *American Scientist* 72(2):128–136).

kilogram, 850-millimeter female carp contained 2,208,000 eggs. Thus a doubling in female length produced more than a 60-fold increase in gametes. No one seems to have counted the sperm in the testis of a carp. It is probable that a 400-millimeter male would contain enough sperm to fertilize every egg that an 800-millimeter female could produce. Sex seems to be genetically determined in carp, and the sex ratio is about 1:1.

Environmental forces affect short-lived fish species differently than long-lived ones like carp. The Atlantic silverside, *Menidia menidia*, hatches in spring or summer, grows until conditions deteriorate in its habitat along the North American east coast in late fall, overwinters without growth, then spawns beginning in May through July of its second year. Most do not survive a second winter. A silverside hatched early in the season will have a long growth period and attain a length of 8 centimeters, whereas those hatched in August are much smaller when growth ceases in winter. Females hatched from early spawnings have a reproductive advantage over later hatched females when it comes time to reproduce in the following year: They are larger and produce more eggs. The date of hatching and hence size at reproduction has no measurable effect on male silversides, which produce an excess of sperm no matter whether they are small or large. Silversides have a dual system of sex determination: primarily genetic sex determination in northern populations and primarily environmental sex determination in southern populations (Conover and Heins 1987). In southern populations, those sexually differentiating at cool temperatures (which correlate with spring season and a long growing period ahead) become females. Those differentiating at the higher temperatures of mid to late summer become males. Environmental sex determination permits each individual to contribute a maximum of zygotes in the following year's breeding: Individuals likely to attain large size are female, those likely to remain small are male. Farther north the sex ratio of young silversides is 1:1 no matter at what temperature they grow; genetic sex determination dominates. This seems appropriate, for in the North the breeding season is so compressed that little difference in feeding and growth separates early and late hatches. Additionally, the more variable seasons of higher latitudes could leave an environmentally sex-determined population with all one sex after a particularly hot or cold brief summer.

Mating systems also affect patterns of sex determination. *Gonostoma gracile* is a primitive marine neoteleostean living in bathypelagic depths (below 1000 meters) where food is relatively scarce and populations thinly dispersed. Males identify females by the pattern of bioluminescent photophores on their flanks in a manner similar to myctophid lanternfish. It seems unlikely that multiple males attempt to fertilize a single female at the same time and also unlikely that females choose from a collection of competing males. If these are correct assumptions about the characteristics of the mating system, males gain no advantage by being large and females are at a disadvantage when they are small because they are capable of producing only a few gametes during the breeding season. Through the first year of life and the first breeding season individuals are less than 60 millimeters in length and all are males. Immediately after the first breeding season the individuals between 50 and 70 millimeters in length are intersexes and specimens larger than 70 millimeters are all females. Individuals appear to maximize their lifetime reproductive capacity by functioning as males when they are small. At this time they can produce many sperm—probably more than needed for the limited number of eggs available—but could only produce a few eggs. They change sex when they have attained a size at which the number of eggs they can produce is greater than the number of fertilizations they could find as males.

Similar patterns of functional sex change (**sequential hermaphroditism**) are known from teleosts such as the acanthopterygian porgies (family Sparidae). In some species of pelagic spawning porgies small individuals are males and larger ones are females, with intermediate sizes undergoing a male-to-female sex change; this pattern of sex change is called **protandrous** hermaphroditism (*protos* = first, *andros* = male). Size of the individual social system and environment are all of importance in understanding the adaptive nature of such variable patterns of sexuality. The hypothesis developed to explain these sex changes is based on a **size advantage model**. Obviously, the exact size when sex change should take place is determined by relative ability to contribute to zygotes, but this is very difficult to predict a priori in the cases so far discussed.

A third variation of the standard sexuality pattern of vertebrates that occurs in teleosts is **protogynous** hermaphroditism (*protos* = first, *gyn* = female). Here a rather precise size at which a change from female to male should occur can be predicted. Protogyny occurs in species where females exhibit a preference for spawning with males larger than themselves. This is often correlated with male–male competition, especially male efforts to dominate a territory that includes exclusive access to some limited resource. Every female in the local population would theoretically show a preference to spawn with the male capable of such domination. Thus that male's ability to contribute

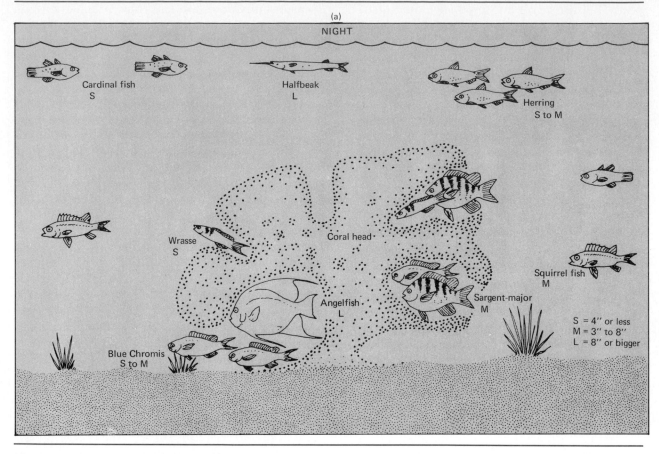

(a)

Figure 8–35. Fish population and activity differences on a Caribbean coral reef (a) at midnight and (b) midday.

to zygotes is greater than lesser males or any individual female.

Protogyny is widespread among coral reef fishes, where only a few spawning sites on the reef are appropriately situated in currents that will rapidly carry the fertilized eggs safely offshore for their pelagic phase. Females should become males at the size at which they can dominate the limiting resource. If they change too soon, they will lose chances to lay eggs but not be able to fertilize eggs. If they delay changing, they miss a portion of the time of greatest reproductive success that sex change might provide them. Examinations of pop-

ulations of protogynous fishes have made it possible to predict when they should change sex based on the size of dominant males.

A curious twist on this pattern is found among protogynous reef species with high population densities. Two ontogenetic types of males are maintained in these populations: large sexually dimorphic males that have changed sex, and smaller males that do not change sex and look similar to the females. These small males may have much larger testes for their size than do the largest males. In a dense population where many females crowd to one resource (e.g., a point of the reef

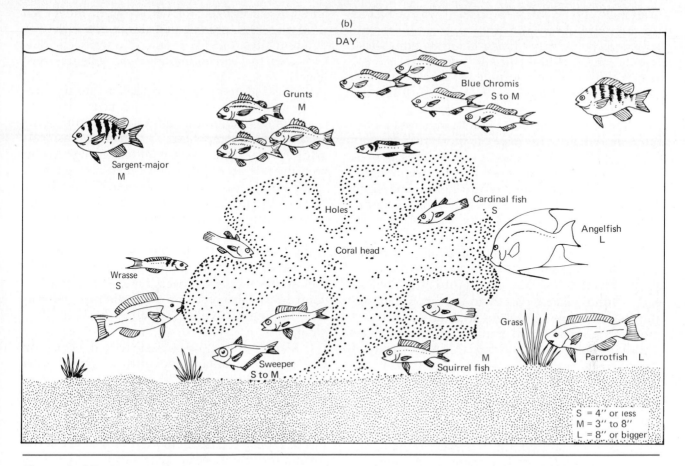

(b)

DAY

Grunts
M

Blue Chromis
S to M

Sargent-major
M

Holes

Cardinal fish
S

Angelfish
L

Coral head

Wrasse
S

Grass

Sweeper
S to M

Squirrel fish
M

Parrotfish   L

S = 4" or less
M = 3" to 8"
L = 8" or bigger

Figure 8–35   (Continued)

exposed to strong offshore currents) these small males sneak in to fertilize eggs during spawnings of females and territorial males.

In male-dominant social systems of fishes with sedentary habits, harems are established consisting of a single large male and several smaller females to which he has exclusive breeding access. Experimental removal of the male initiates behavioral changes in the largest and dominant female: Within a few hours she begins behaving like a male. In 1 to 2 weeks protogynous sex change is complete and the former female is producing sperm.

Specialization and Coexistence Among Fishes in Coral Reef Communities

A coral reef is one of the most spectacular displays of animal life. The diversity of invertebrate and vertebrate animals is unknown in such concentration elsewhere. The vertebrate component is drawn almost exclusively from a single taxon, the acanthopterygian teleosts.

Acanthopterygians exhibit unparalleled diversity in feeding modes, all of which relate to their precise control over jaw actions and body positions

(Marshall 1971). This precision was achieved through strong evolutionary pressures on the interactions between feeding and locomotion. These interactions are illustrated by coral reef fishes, which constitute one of the most species-rich vertebrate communities (Sale 1980). Over 600 species, most of which are acanthopterygians, may be found on a single reef. E. S. Hobson notes that the most primitive spiny-rayed fishes are predators (for example, squirrelfishes, cardinalfishes; refer to Figure 8–35). They disperse over the reef at night to feed, but during the daylight hours congregate in caves and holes in the reef—they are **nocturnal** (Hobson 1975). Other predators, such as the groupers, feed heavily at dawn and dusk (that is, are **crepuscular**). They stalk prey, use the reef to conceal their approach, and rely on a large mouth to seize prey that are fully exposed to attack. As an early evolutionary response to predation, Hobson suggests that many reef invertebrates performed their activities at night and remained concealed during the day. In response to this nocturnality of prey, early acanthopterygians evolved the capacity to feed at night. The large sensitive eyes of these nocturnal predators are highly functional at low light intensities.

A major advance among reef acanthopterygians was the evolution of fishes specialized to take food items hidden in the complex reef surface, generally by suction or a forceps action of their protrusible jaws. This mode of predation demanded sensory specializations, the most important being high visual acuity, which can only be achieved in the bright light of day. In addition, delicate positioning was required to direct the jaws. These interacting selection pressures produced fishes capable of manuevering through the reef in search of food. So accurate is their locomotion, visual surveillance, and memory for hiding places and escape routes that these fishes can expose themselves and feed in broad daylight (that is, are **diurnal**). The refined feeding specializations of diurnal reef fishes allow them to extract small invertebrates from their daytime hiding places or snip coral polyps or nibble sponges. In some species, mouths, dentition, and some digestive systems are specialized for herbivory; in others for removing small organic particles from a variety of different sites, including sand, the water column, and even the bodies, mouths, and gills of larger fishes.

Released from heavy predation during daytime, many of these fishes have evolved gaudy colors that communicate information to conspecifics and other competitors. At dusk as these colorful diurnal fishes seek nighttime refuge in the reef, the nocturnal fishes leave their hiding places to replace them in the water column (Figure 8–35). Most of the fish community can be thought of as divided into species that are either nocturnal or diurnal. The temporal precision with which each species leaves and enters the protective cover of the reef day after day indicates an important ecological function. The space, time, and trophic resources available on a reef are partitioned, permitting the great diversity of vertebrate life present on coral reefs.

Many reef fishes are closely related, belonging to but a few families and often to genera with many coexisting species. The cardinalfishes, damselfishes, angelfishes, butterflyfishes, wrasses, and parrotfishes are examples. The great number of closely related species is thought to be the result of the repeated partial isolation of sections of the great pantropical ocean. (The Atlantic Ocean was isolated from the Pacific a mere 4 million years ago by the uplift of Central America.) Many reef fish genera have fossil records extending back over 50 million years. In this time sea levels have varied, volcanic islands have grown and sunk, and continents and archipelagos have drifted with the spreading of the seafloor. The reef fishes, almost all of which have pelagic eggs and/or larvae, have found themselves alternately isolated in some corner of the vast tropical sea or close enough for their larvae to drift and mingle with those of other species drawn from an enormous geographic range. Periods of isolation probably fostered incipient speciation, which often must have been occurring in multiple isolated reef systems simultaneously. The result has been a fish fauna of great variety, new species having entered the fauna after each

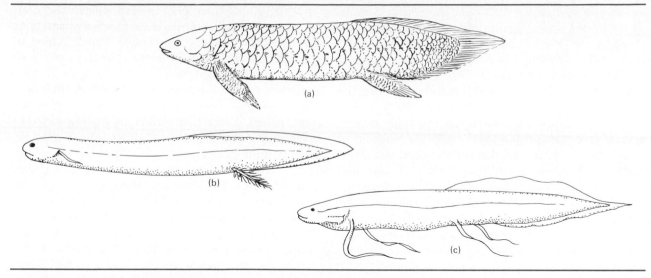

Figure 8–36. Living Dipnoans: (a) Australian lungfish, *Neoceratodus forsteri*; (b) South American lungfish, *Lepidosiren paradoxa*, male; note the specialized pelvic fins of the male during the breeding season; (c) African lungfish, *Protopterus*.

period of isolation. Currently, the fauna is in a period of low isolation—the geographic ranges of many species extend from the east coast of Africa to Hawaii and a few reach Mexico. Many of the species are very similar. Such groups of closely related species share the majority of their adaptations through mutual inheritance of unique derived characters. One would expect these species groups to experience intense competition between their members. Such competition should ultimately lead either to elimination of most of the closely related species or to driving the related species to exploit different resources despite their similarity (Watt 1987). Surprisingly, neither of these results of competition seems to have occurred. Rather, storms and predation appear never to permit the competitive interactions to run full course.

Living Sarcopterygii: Lobe-Finned Fishes

Although they were abundant in the Devonian, the number of species of sarcopterygians dwindled

in the late Paleozoic and Mesozoic. Their early evolution was vigorous and resulted in a significant radiation in fresh and marine waters and (by their descendents) on land. Today only four genera remain: the dipnoans (*Protopterus* in Africa, *Lepidosiren* in South America, *Neoceratodus* in Australia), and the deep-water actinistian *Latimeria* of the Comoro Archipelago off East Africa (Figures 8–36 and 8–37). Nevertheless, these fishes are of importance to us, for terrestrial vertebrates originated within the lineage. We will discuss fossil sarcopterygians in more detail with their sister group, the tetrapods.

Dipnoans

The earliest fossil dipnoans, which were marine, are distinct from the other osteichthyans. The Dipnoi are distinguishable by the lack of articulated tooth-bearing premaxillary and maxillary bones and the fusion of the palatoquadrate to the undivided cranium. The teeth are scattered over the palate and fused into tooth ridges along the lateral palatal margins. Powerful adductor muscles of the lower jaw spread upward over the neurocranium. Throughout their evolution, this **durophagous** (feeding on hard foods) crushing apparatus has persisted. During the Devonian, lungfishes evolved a body form quite distinct from the other Osteichthyes. The anterior dorsal fin was lost, the remaining median fins fused around the posterior third of the body, the caudal fin, originally heterocercal, became symmetrical, and the mosaic of small dermal bones of the earliest dipnoan skulls (often covered by a continuous sheet of cosmine, an enameloid substance) evolved a pattern of fewer large elements without the cosmine cover. Since the Devonian, the dipnoan feeding apparatus has changed very little. Living dipnoans, therefore, probably are not very different from their ancestors (Thomson 1969, Bemis et al. 1987).

The monotypic Australian lungfish, *Neoceratodus forsteri*, is morphologically most similar to Paleozoic and Mesozoic Dipnoi. Like all other living dipnoans, *Neoceratodus* is restricted to fresh waters; naturally occurring populations are limited to southeastern Queensland. The Australian lungfish may attain a length of 1.5 meters and a reported weight of 45 kilograms. Although a powerful fighter when netted, it is a slow-moving plant and invertebrate eater. It swims by body undulations or slowly walks across the bottom of a pond on its pectoral and pelvic appendages. Chemical senses seem important to lungfishes, and their mouths are reported to contain numerous taste buds. The nasal passages are located near the upper lip, with the incurrent openings on the rostrum just outside the mouth and the excurrent openings within the oral cavity. Thus, each gill ventilation draws water across the nasal epithelium. Under normal circumstances, *Neoceratodus* respires almost exclusively via its gills and uses its single lung only when stressed. Little is known of its behavior. Although they go through a complex courtship, which perhaps includes male territoriality and are very selective about the vegetation upon which they lay their adhesive eggs, no parental care has been observed after spawning. The jelly-coated eggs, only 3 millimeters in diameter, hatch in 3 to 4 weeks,

but the young have proven very elusive and nothing is known of their juvenile life. Surprisingly little is known about the single South American lungfish, *Lepidosiren paradoxa*, but the closely related African lungfishes, *Protopterus*, with four recognized species, are better understood, their ecology having been studied more than other lungfishes. These two genera are distinguished by different numbers of weakly developed gills. Because their gills are very small, these lungfishes drown if they are prevented from using their paired lungs. Nevertheless, the gills are important in eliminating carbon dioxide. These thin-scaled, heavy-bodied snake-like fishes, 1 to 2 meters long, have unique filamentous and highly mobile paired appendages. For a time after their discovery 150 years ago lungfish were considered to be specialized urodele amphibians, which they superficially resemble because the scales of the African and South American species are covered by epidermis. Although the skeletons of these lungfishes are mostly cartilaginous, their toothplates are heavily mineralized and fossilize readily.

One habit of some species of African dipnoans, **aestivation**, considerably increases the chance of fossilization. Similar in some ways to hibernation, aestivation is induced by drying of the habitat rather than by cold. African lungfishes frequent areas that flood during the wet season and bake during the dry season—habitats not available to actinopterygians except by immigration during floods. The lungfishes enjoy the flood periods, feeding heavily on a wide variety of animals (especially mollusks) and growing rapidly, but unlike other fishes, which leave the area during periods of drought, the African lungfishes remain and aestivate. When the flood waters recede, the lungfish digs a vertical burrow in the mud that ends in an enlarged chamber and varies in length in proportion to the size of the animal, the deepest being less than 1 meter. As drying proceeds, the lungfish becomes more lethargic and breathes air from the burrow opening. Eventually, even the water of the burrow dries up, and the lungfish enters the final stages of aestivation. In the deep

(a)

20 mm

(b)

Figure 8–37. Representative Actinistia (coelacanths): (a) *Rhabdoderma*, a Carboniferous actinistian; (b) *Latimeria chalumnae*, the living coelacanth.

chamber, the lungfish remains folded into a U-shape with its tail over its eyes, and the heavy mucoid secretions the fish has produced since entering the burrow condense and dry to form a protective envelope around its body. Only an opening at its mouth remains to permit breathing.

Although the rate of energy consumption during aestivation is very low, metabolism continues, using muscle proteins as an energy source (see Chapter 4). Lungfishes normally spend less than

6 months aestivating, but they have been revived after 4 years of enforced aestivation. When the rains return, the withered and shrunken lungfish becomes active and feeds voraciously on mollusks, crustaceans, and fishes. In less than a month it regains its previous size.

Aestivation is not a recent adaptation of dipnoans. Fossil burrows containing lungfish tooth plates have been found in Carboniferous and Permian deposits of North America and Europe.

Without the unwitting assistance of the lungfishes, which initiated fossilization by burying themselves, such fossils might not exist.

Actinistians

Actinistians are unknown before the middle Devonian. Their hallmarks are an unlobed first dorsal fin and the unique, symmetrical, three-lobed tail with a central fleshy lobe that ends in a fringe of rays (Figure 8–37). Actinistians also differ from all other sarcopterygians in the head bones (they lack, among other elements, a maxilla), in details of the fin structure, and in the presence of a curious rostral organ. Following rapid evolution during the Devonian, the actinistians show a history of evolutionary stability. Devonian actinistians differ from the more recent Cretaceous fossils mostly in the degree of skull ossification. Some early actinistians lived in shallow fresh waters, but the fossil remains of these and other osteichthyans during the Mesozoic are largely marine. The osteichthyans radiated into a variety of niches, but the actinistians retained their peculiar form. Fossil actinistians are not known after the Cretaceous, and until 50 years ago they were thought to be extinct.

In 1938, an African fisherman bent over an unfamiliar catch and nearly lost his hand to its ferocious snap. Imagine the astonishment of the scientific community when J. L. B. Smith of Rhodes University announced that the catch was a living actinistian! This large fish was so similar to Mesozoic fossil coelacanths that its systematic position was unquestionable. Smith named this living fossil *Latimeria chalumnae* in honor of Ms. Courtenay-Latimer, who recognized it as unusual and brought the specimen to his attention (Smith 1956).

Despite public appeals, no further specimens of *Latimeria* were captured until 1952. Since then more than 150 specimens ranging in size from 75 centimeters to slightly over 2 meters and weighing from 13 to 80 kilograms have been caught, all in the Comoro Archipelago between Madagascar and Mozambique. Coelacanths are hooked near the bottom, usually in 260 to 300 meters of water about 1.5 kilometers offshore. Strong and aggressive, *Latimeria* is steely blue-gray with irregular white spots and reflective golden eyes. The reflective eyes result from a tapetum lucidum that enhances visual ability in dim light. A large cavity in the midline of the snout communicates with the exterior by three pairs of rostral tubes enclosed by canals in the wall of the chondrocranium. These tubes are filled with gelatinous material and open to the surface through a series of six pores. The rostral organ is almost certainly an electroreceptor (Bemis and Hetherington 1982). Whatever senses are utilized, *Latimeria* is a predator, for stomachs have contained fishes and cephalopods.

A fascinating glimpse of the life of the coelacanth was reported by Hans Fricke and his colleagues who used a small submarine to observe the fish (Fricke et al. 1987; Fricke 1988). They saw six coelacanths at depths between 117 and 198 meters off a short stretch of the shoreline of one of the Comoro Islands. Coelacanths were seen only in the middle of the night and only on or near the bottom. Unlike living lungfish, the coelacanths did not use their paired fins as props or to walk across the bottom. However, when they swam the pectoral and pelvic appendages were moved in the same sequence as tetrapods move their limbs.

The discovery of *Latimeria* has confirmed earlier reconstructions based on coelacanth fossils (Thomson 1986). A case in point is the mode of coelacanth reproduction. In 1927, D. M. S. Watson described two small skeletons from inside the body cavity of *Undina*, a Jurassic coelacanth, and suggested that coelacanths gave birth to their young. Because copulatory structures have never been found on any coelacanth fossil, some dismissed Watson's specimen as a case of cannibalism. Female *Latimeria* containing up to 19 eggs each 9 centimeters in length have now been captured. R. W. Griffith and K. S. Thomson surmised that Watson was correct because of the small number of eggs and their lack of a shell to provide osmotic protection. While C. L. Smith and others at the American Museum of Natural History were dissecting a 1.6-meter specimen they discovered five advanced young, each 30 centimeters long, in the single oviduct. Internal fertilization must occur, but how copulation is achieved is unknown. In spite of some excellent fossils and numerous specimens of the modern coelacanths to study, there has never been stable agreement about the relationships of the Actinistia with other gnathostomes. Workers have disagreed about the position of the coelacanths more than of most other vertebrate taxa. In part this is due to the surprising combinations of morphology and physiology shown by *Latimeria*, which has many derived characters, some of which are most similar to Chondrichthyes, others to the Dipnoi, some to Actinopterygii, and also a curious collection of unique features (McCosker and Lagios 1979).

Summary

Because of its physical properties, water is a demanding medium for vertebrate life. Nevertheless, the greatest number of vertebrate species, the vast majority of them Osteichthyes, are found exclusively in the planet's oceans, lakes, and rivers. Highly efficient respiratory systems, the gills, facilitate rapid oxygen uptake to sustain the activity of fishes. Their locomotion is generally accomplished with undulations produced by contraction of the body muscles. Sensory guidance for activity is often visual, but other sensory systems also are highly refined. Fishes detect low-frequency pressure waves via the lateral-line system or navigate and communicate through electroreception.

At their first appearance in the fossil record, osteichthyans, the largest vertebrate taxon, are separable into distinct lineages. The Sarcopterygii (fleshy-finned fishes: lungfishes, actinistians, and other lobe-finned fishes) and the Actinopterygii (ray-finned fishes) show indications of common ancestry. Living sarcopterygian fishes offer exciting glimpses of adaptations evolved in Paleozoic environments. Actinopterygian fishes were distinct as early as the Devonian. Actinopterygians inhabit the 73 percent of the Earth's surface that is covered by water and are the most numerous and speciose lineages of vertebrates. Several levels of development in food-gathering and locomotory structures characterize actinopterygian evolution. The radiations of these levels are represented today by relict groups: cladistians (bichirs and reedfish), chondrosteans (sturgeons and paddlefish), and the primitive neopterygians (gars and *Amia*). The most derived level—teleosteans—may number close to 21,000 living species with two groups, ostariophysans in fresh water and acanthopterygians, characteristically in seawater, constituting a large proportion of these species. Teleostean specializations of morphology, behavior, and life history are so numerous and diverse that their dominant position in the aquatic and marine ecosystems of the world is understandable.

References

Bass, A. H. 1986. Electric organs revisited: evolution of vertebrate communication and orientation organs. Pages 13–70 in *Electroreception*, edited by T. H. Bullock and W. Heiligenberg. Wiley, New York.

Bemis, W. E., W. W. Burggren, and N. E. Kemp (editors). 1987. *The Biology and Evolution of Lungfishes*. Alan R. Liss, New York. An excellent compendium of our current understanding of these interesting fishes with a thorough guide to the extensive literature on the group.

Bemis, W. E. and T. E. Hetherington. 1982. The rostral organ of *Latimeria chalumnae*. Morphological evidence of an electroreceptive function. *Copeia* 1982:467–471.

Bennett, M. V. L. 1971a. Electric organs. Pages 347–491 in *Fish Physiology*, volume 5, *Sensory Systems and Electric Organs*, edited by W. S. Hoar and D. J. Randall. Academic Press, New York. This review extensively describes the properties of the electrocytes, their distribution in fishes, and their neural control.

Bennett, M. V. L. 1971b. Electroreception. Pages 493–574 in *Fish Physiology*, volume 5. *Sensory Systems and Electric Organs*, edited by W. S. Hoar and D. J. Randall. Academic Press, New York. A detailed description of electroreceptors in elasmobranchs and teleosts, the review covers physiology, morphology, distribution, function, and evolution.

Bodznick, D. and R. G. Northcutt. 1981. Electroreception in lampreys: evidence that the earliest vertebrates were electroreceptive. *Science* 212:465–467.

Bone, Q. and N. B. Marshall. 1982. *Biology of Fishes*. Blackie and Son, Glasgow.

Campana, S. E. and J. D. Neilson. 1985. Microstructure of fish otoliths. *Canadian Journal of Fisheries and Aquatic Science* 42:1014–1032.

Conover, D. O. and S. W. Heins. 1987. Adaptive variation in environmental and genetic sex determination in a fish. *Nature* 326:496–498.

Cromie, W. J. 1982. Born to navigate. *Mosaic* 13(4):17–23. A review of natural navigation research which focuses on fishes.

Denison, R. H. 1979. *Acanthodii*. In *Handbook of Paleoichthyology*, volume 5. Gustav Fischer Verlag, Stüttgart, West Germany.

Eaton, R. C. (editor). 1984. *Neural Mechanisms of Startle Behavior*. Plenum Press, New York. Contains the most recent review of the Mauthner cell system.

Faber, D. S. and H. Korn (editors). 1978. *Neurobiology of the Mauthner Cell*. Raven Press, New York.

Farrell, A. P. 1980. Vascular pathways in the gill of ling cod, *Ophiodon elongatus*. *Canadian Journal of Zoology* 58:796–806.

Fessard, A. (editor) 1974. Electroreceptors and other specialized receptors in lower vertebrates. Pages 59–124 in *Handbook of Sensory Physiology*, volume 3, part 3. Springer-Verlag, New York. Included are chapters on lateral-line receptors and electroreceptors.

Fricke, H. 1988. Coelacanths, the fish that time forgot. National Geographic 173(6):824–838.

Fricke, H., O. Reinicke, H. Hofer, and W. Nachtigall. 1987. Locomotion of the coelacanth *Latimeria chalumnae* in its natural environment. *Nature* 324:331–333.

Gebhardt, M. D. 1987. Parental care: a freshwater phenomenon? *Environmental Biology of Fishes* 19(l):69–72.

Hobson, E. S. 1975. Feeding patterns among tropical reef fishes. *American Scientists* 63(4):382–392.

Hopkins, C. 1980. Evolution of electric communication channels of mormyrids. *Behavioral Ecology and Sociobiology* 7:1–13.

Lauder, G. V. 1980. Evolution of the feeding mechanism in primitive actinopterygian fishes: a functional anatomical analysis of *Polypterus, Lepisosteus* and *Amia. Journal of Morphology* 163:283–317.

Lauder, G. V. and K. F. Liem. 1981. Prey capture by *Luciocephalus pulcher*: Implications for models of jaw protrusion in teleost fishes. *Environmental Biology of Fishes* 6:257–268.

Lauder, G. V. and K. F. Liem. 1983. The evolution and interrelationships of the actinopterygian fishes. *Bulletin of the Museum of Comparative Zoology* 150:95–197. Specifics may be found in G. V. Lauder, 1980. On the evolution of the jaw adductor musculature in primitive gnathostome fishes, *Breviora* 460:1–10; 1980, Hydrodynamics of prey capture by teleost fishes, *Biofluid Mechanics* 2:161–181.

Laurent, P. and S. Dunel. 1980. Morphology of gill epithelia in fish. *American Journal of Physiology* 238:147–149 (R).

Lindsey, C. C. 1978. Form, function, and locomotory habits in fish. Pages 1–100 in *Fish Physiology*, volume 7, *Locomotion*, edited by W. S. Hoar and D. J. Randall. Academic Press, New York.

Magnuson, J. J. 1978. Locomotion by scombroid fishes: hydromechanics, morphology, and behavior. Pages 239–313 in *Fish Physiology*, volume 7, *Locomotion*, edited by W. S. Hoar and D. J. Randall. Academic Press, New York.

Marshall, N. B. 1971. *Explorations in the Life of Fishes*. Harvard University Press, Cambridge, Mass. Selected topics in fish biology analyzed in a comparative manner.

McCosker, J. E. and M. D. Lagios (editors). 1979. The biology and physiology of the living coelacanth. *California Academy of Sciences Occasional Papers* 134:1–175.

Moy-Thomas, J. A. and R. S. Miles. 1971. *Paleozoic Fishes*. 2nd edition. W. B. Saunders, Philadelphia.

Nelson, J. S. 1984. *Fishes of the World*, 2nd edition. Wiley, New York. A technical family-by-family characterization of living and (in a more abbreviated treatment) fossil fishes.

Nieuwenhuys, R. 1982. An overview of the organization of the brain of actinopterygian fishes. *American Zoologist* 22(2):287–310.

Norman, J. R. and P. H. Greenwood. 1975. *A History of Fishes*, 3rd edition. Ernest Benn, London. A very readable general introduction to fish biology.

Popper, A. N. and S. Coombs. 1980. Auditory mechanisms in teleost fishes. *American Scientist* 68(4):429–440.

Roberts, B. L. 1981. The organization of the nervous system of fishes in relation to locomotion. In *Ver-*

tebrate Locomotion, edited by M. H. Day. *Symposia of the Zoological Society of London* 48:11–136. Academic Press, New York.

Rosen, D. E. 1982. Teleostean interrelationships, morphological function and evolutionary inference. *American Zoologist* 22(2):261–273.

Sale, P. F. 1980. The ecology of fishes on coral reefs. *Annual Review of Oceanography and Marine Biology* 18:367–421.

Schaeffer, B. and D. E. Rosen. 1961. Major adaptive levels in the evolution of the actinopterygian feeding mechanism. *American Zoologist* 1(2):187–204. A technical analysis of jaw and jaw muscle function at different stages in ray-finned fish evolution.

Scheich, H. G. Langner, C. Tidemann, R. B. Coles, and A. Guppy. 1986. Electroreception and electrolocation in platypus. *Nature* 319:401–402.

Smith, J. L. B. 1956. *Old Four Legs*. Longman, London. A reminiscence of the discovery of *Latimeria* by its original describer.

Tavolga, W. N., A. N. Popper, and R. R. Fay (editors). 1981. *Hearing and Sound Communication in Fishes*. Springer-Verlag, New York.

Thomson, K. S. 1969. The biology of the lobe-finned fishes. *Biological Review* 44:91–154. Reconstructions of the morphology, physiology, and behavior of these nearly extinct fishes. A more recent, if not widely accepted view is K. S. Thomson, 1980. The ecology of Devonian lobe-finned fishes, pages 187–222 in *The Terrestrial Environment and the Origin of Land Vertebrates*. Edited by A. L. Panchen. Academic Press, New York.

Thomson, K. S. 1986. Marginalia: a fishy story. *American Scientist* 74(2):169–171.

Thresher, R. E. 1984. *Reproduction in Reef Fishes*. T. F. H. Publications, Neptune City, N.J.

Thresher, R. E. 1988. Otolith microstructure and the demography of coral reef fishes. *Trends in Ecology and Evolution* 3:78–80.

Turner, B. J. (editor). 1983. *Evolutionary Genetics of Fishes*. Plenum Press, New York.

Watt, K. E. F. 1987. Deep questions about shallow seas. *Natural History* 96:60–65.

Webb, P. W. 1975. Hydrodynamics and energetics of fish propulsion. *Bulletin of the Fisheries Research Board of Canada* 190:1–159. A comprehensive review of hydrodynamic considerations of fish locomotion. The approach is amply illustrated, but intended for advanced students.

Webb, P. W. 1978. Hydrodynamics: nonscombroid fishes. Pages 189–237 in *Fish Physiology*, volume 7, *Locomotion*, edited by W. S. Hoar and D. J. Randall. Academic Press, New York.

Webb, P. W. 1984. Form and function in fish swimming. *Scientific American* 251(1):72–82.

Webb, P. W. and R. W. Blake. 1985. Swimming. Pages 110–128 in *Functional Vertebrate Morphology*, edited by M. Hildebrand, D. M. Bramble, K. F. Liem, and D. B. Wake. Belknap Press, Cambridge, Mass.

The last time we glanced at the land, back in the Cambrian (Chapter 5), the prospect was pretty bleak. The ground surface was largely barren, and where simple plants like algae had managed to gain a terrestrial foothold, they were probably limited to the vicinity of water. The prospects of finding a meal were as discouraging as the scenery; as far as we are aware, no invertebrates had invaded the terrestrial habitat in the Cambrian. Now more than 100 million years have passed; it is the middle of the Devonian, and life on land is burgeoning. Multicellular plants are beginning to lift their stems off the ground, providing food and hiding places for terrestrial invertebrates, and every prospect pleases.

Geology and Ecology of the Origin of Tetrapods

9

Continental Geography in the Late Paleozoic

Coalescence of the ancient continents was nearly complete in Devonian time (Figure 9–1); only a narrow sea, an extension of the ancestral Tethys, separated Gondwana from ancestral North America, Europe, Siberia, Kazakhstania, and China. The arm of the Tethys Sea between Laurussia and Gondwana did not close completely until the late Carboniferous. Epicontinental seas were more restricted in the Devonian than they had been in the early Paleozoic. One major transgression—which covered much of southern Europe, middle North America, and regions that would ultimately form Asia—had occurred in the middle Devonian. The Acadian and the beginning of the Hercynian orogenies resulted in continental uplift and deep folding of those parts of Laurussia that are now located in eastern North America and Scotland. Several floristic regions characterize the late Devonian (Banks 1979, see Figure 9–1): zone 1—a tropical or subtropical region of enormous extent that contained mosses and simple herbaceous vascular plants in the early Devonian and a flora dominated by arborescent club mosses (lycopods) and heterosporous plants in the late Devonian. The Siberian flora is considered a phytogeographic subunit of the tropical flora. Zone 2—the Australian flora, which displays close affinities to the plants

of zone 1. However, the flora of Gondwana (zone 3) was distinct.

Devonian Climates

As we have seen, in Devonian times bony fishes dominated both marine and freshwater habitats. Intensive predation and competition for food and suitable habitats probably resulted. In the late Devonian the earliest tetrapods originated from one group of fishes—the lobefins (sarcopterygians). Amphibious habits must have provided a selective advantage. Whether the advantage was associated with a reduction in predation, a new food source, or reproduction is a question we shall discuss. But leaving the water, even for short periods, exposed vertebrates to new and formidable difficulties. What climates prevailed during this period of Earth's history? What terrestrial organisms existed when tetrapods ventured onto land?

In Chapter 5 we suggested that the climate of equatorial latitudes in the early Paleozoic was tropical and reasonably uniform. A single major climatic interruption that lowered temperatures worldwide by several degrees is indicated by extensive glaciation in the Ordovician. Its effect on the early marine ostracoderms is hard to assess because the tropical seas in which ostracoderms originated would have buffered the effect of climatic change. In the Silurian and Devonian, the

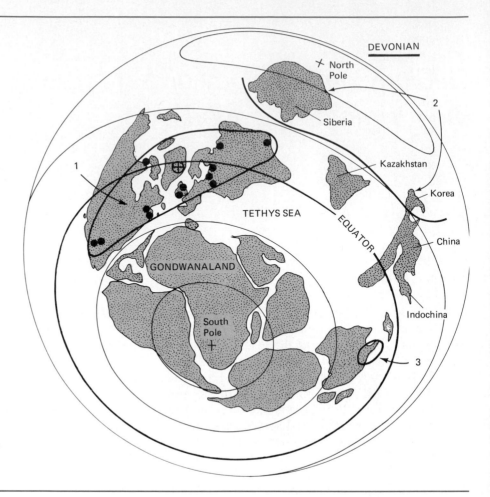

DEVONIAN

North Pole

Siberia

Kazakhstan

Korea

China

Indochina

TETHYS SEA

EQUATOR

GONDWANALAND

South Pole

Figure 9–1. Continental positions during the late Devonian. The distribution of Devonian ostracoderms (•) lies within floral zone 1. Also in this region are the earliest vertebrates of probable freshwater origin. Jawless vertebrates radiated widely in the fresh waters of zone 1 during the Devonian. The first tetrapods, the ichthyostegids, are recorded from Greenland, Nova Scotia, and possibly from Australia.

climate was again warmer in the tropics and undisturbed by glaciation.

North America, Europe, western Asia, and much of Indochina and Australia lay in the tropics and subtropics between 30°S and 30°N latitudes (Figure 9–1). Evaporation of water from the warm equatorial Tethys Sea provided abundant rainfall on the continents. Lakes, rivers, and streams were extensive in North America and Europe in the late Silurian and early to middle Devonian (Beerbower 1985). Conditions were ripe for the radiation of Devonian freshwater fishes and their amphibious descendents.

Terrestrial Habitats of the Devonian

Plant and animal life was abundant and diverse in shallow-water habitats during the early Devonian. Plant colonization and radiation were extensive in fresh water, and animal radiation was well under way. On land, however, the emergence of plants and animals had only begun. The first land plants were primitive nonvascular bryophytes (mosses and liverworts) and simple vascular plants, probably derived from green algae (Pratt et al. 1978,

Chapman 1985). Like mosses today, they needed moist conditions. Macrofossils of the first vascular plants, which were short, leafless, and possessed only simple branching stems, are found in late Silurian strata in Ireland, Wales, and Bohemia. In the Silurian this area was close to the equator and its climate was tropical and moist (Figure 9–1).

The transition to land demanded broadly similar adjustments for plants and for animals. For both the first major barrier was exposure to increased water loss through evaporation (Raven 1985). Although the first macrofossil vascular plant, *Cooksonia*, occurs in late Silurian strata, protovascular land plants may have existed as early as the late Ordovician. Microfossils of plants with cell anatomy and spores typical of vascular plants occur in the fossil record of this period (Gray 1985). Vessel-like banded tubes in these Ordovician fossils have been shown to contain the strengthening and waterproofing compound lignin, suggesting that these plants were able to conduct water from one part of the plant to another (Niklas and Pratt 1980). Water impermeability in land plants is provided mainly by two waxy substances, cuterin on leaves and stems and suberin over roots. Only vascular land plants have these chemicals, and they were present in the earliest vascular land plants (Niklas 1979, 1980; Chapman 1985). It is likely that the earliest land plants slowly encroached on the land by invading intertidal areas, river and lake banks, swamps, and ultimately upland marshes. Even before this invasion it is probable that various prokaryotes and protists preceded the plants onto the land and may well have made the terrestrial habitats more suitable for land plants (Campbell 1979). It is possible even that a simple terrestrial flora existed as early as the middle to late Ordovician (Retallack 1985, Wright 1985). The late Silurian and early Devonian were typified by an increasingly diverse land flora, but one that was small in profile (most plants were less than 50 centimeters in height) and essentially limited to moist habitats.

The effect of this floristic invasion of the land was of great significance, for plants began to stabilize sedimentation processes in the areas where they lived (Beerbower 1985). Associated with this early land flora was a simple arthropod fauna (Almond 1985, Rolfe 1985), consisting of arachnid-like animals and myriapods (millipedes and their kin). Probably, myriapods fed on dead plant material as they do today, and they would have been key elements in decomposition chains contributing to the development of humus soils. A vegetative mat provided sites in which the earliest tetrapods might have deposited eggs, hidden, or preyed on early arthropods. These earliest land ecosystems were relatively simple and often included monospecific stands of plants (Edwards 1980). Nevertheless, community interaction was apparently intense, for damage caused by pathogenic fungi and arthropod bites can be recognized in fossil plants. Predation and competition, it is believed, acted as such a powerful force during the early Devonian that by the middle Devonian an arborescent or tree-like community had evolved. No longer were land communities limited to only a few centimeters above the soil (Swain 1978, Swain and Cooper-Driver 1981).

By the middle Devonian the terrestrial flora consisted of bryophytes, large horsetails (sphenophytes), and a variety of scale trees (lycopods), indicating that terrestrial habitats were moist, for all of these plants required water for reproduction. Upland habitats in the early to mid-Devonian appear to have been largely devoid of plants (Scott 1980). Without the depositional stability provided by plants, erosion was rapid, and uplands may not have persisted as long as they do today. An important innovation in the Devonian flora, the appearance of heterosporous plants (unequal spore sizes and incipient female–male differences), occurred in the early to middle Devonian. This signaled one of the first adaptations to the invasion of more arid lands (see Chapter 14). The first fossil seed is found among the direct ancestors of the gymnospermous plants in late Devonian strata (Taylor 1981). Although the seed's origin is unknown, the accompanying progymnospermous fossils were gigantic in comparison to their earlier

Devonian progenitors. *Archaeopteris* measured 30 meters high, had deeply planted roots, and like its descendents the gymnosperms, had developed strong secondary growth in the form of wood for support. Thus, by the end of the Devonian plant communities had become arborescent, perhaps partly to escape their predators and to effect better dispersal of spores and early seeds (Chaloner and Sheerin 1979).

Insects, represented by springtails (collembolans) and bristletails (thysanurans), occur in a few Devonian sediments, but they are not abundant in fossil deposits until the Carboniferous. It is significant that flying arthropods did not exist. Access to the upper story or canopy of these late Devonian forests required climbing. Many of the plants had spines, scales, and other processes that provided impediments to climbing and gave protection from potentially herbivorous arthropods.

Because the Devonian land profile was mostly low, early terrestrial communities were swamp-like. Reproduction and distribution of plants were dependent on vegetative processes and spore transport which require moist soils. Only in seed-bearing plants, where the embryo is protected from desiccation, could plant dispersal and the colonization of arid areas have effectively taken place. Because the seed-bearing gymnosperms were rare, Devonian terrestrial plants were mainly limited to the margins of water. Early terrestrial organisms must have strayed little from the major watercourses that crisscrossed the land.

A unique feature of the middle to late Devonian flora should be stressed. Most plants were uniformly distributed between 30°S and 30°N latitude (Figure 9–1), where the climate was predominantly tropical (uniform). In Siberia, however, a distinctive flora of late Devonian age shows seasonal growth rings. Drift theory locates Siberia between 30 and 60°N latitude in the Devonian. This Siberian flora probably represents one of the first extensive terrestrial plant communities adapted to cooler conditions (Ziegler et al. 1981). In the Carboniferous, however, the distinction became blurred, possibly because Siberia drifted farther southward (Raymond et al. 1985).

Summary

Tetrapods originated during the Devonian, as Pangaea began to coalesce into a single landmass and equable climates prevailed. Slight continental uplift reduced the transgression of shallow epicontinental seas during this period, and swamp-like freshwater habitats were common. The earliest terrestrial plants required moist soils for growth and reproduction, and therefore were limited to the margins of waters. Early arthropods (crustaceans, insects, and arachnids) became abundant in the late Devonian and provided a new food resource on land. Whether in response to this new terrestrial resource or to the abundant freshwater predators, Devonian fishes, probably the osteolepiforms, gave rise to the first tetrapods. Anatomically and physiologically this feat was complex and required millions of years. The long equability in climate, due in part to the location of much of Laurussia in tropical latitudes, favored this transition from water to land.

References

Almond, J. E. 1985. The Silurian–Devonian fossil record of the Myriapoda. *Philosophical Transactions of the Royal Society of London* B309:227–237.

Banks, H. P. 1979. Floral assemblage zones in the Siluro-Devonian. Pages 1–24 in *Biostratigraphy of Fossil Plants: Successional and Paleoecological Analysis*, edited by D. D. Dilcher and T. N. Taylor. Dowden, Hutchinson and Ross, Stroudsburg, Pa.

Beerbower, R. 1985. Early development of continental ecosystems. Pages 47–91 in *Geological Factors and the Evolution of Plants*, edited by B. H. Tiffney. Yale University Press, New Haven, Conn.

Campbell, S. 1979. Soil stabilization by a prokaryotic desert crust; implications for Pre-Cambrian land biota. *Origin of Life* 9:335–345.

Chaloner, W. G. and A. S. Sheerin. 1979. *Devonian Ma-*

crofloras, Special Papers in Paleontology, No. 23. London Palaeontological Society, London, pages 145–161.

Chapman, D. J. 1985. Geological factors and biochemical aspects of the origin of land plants. Pages 23–45 in *Geological Factors and the Evolution of Plants*, edited by B. H. Tiffney. Yale University Press, New Haven, Conn.

Edwards, D. 1980. Early land floras. Pages 55–85 in *The Terrestrial Environment and the Origin of Land Vertebrates*, edited by A. L. Panchen. Academic Press, London.

Gray, J. 1985. The microfossil record of early land plants: advances in understanding of early terrestrialization, 1970–1984. *Philosophical Transactions of the Royal Society of London* B309:167–195.

Niklas, K. J. 1979. An assessment of chemical features for the classification of plant fossils. *Taxon* 28:505–511.

Niklas, K. J. 1980. Palaeobiochemical techniques and their applications to palaeobotany. *Progress in Phytochemistry* 6:143–182.

Niklas, K. J., and L. Pratt. 1980. Evidence for lignin-like constituents in early Silurian (Llandoverian) plant fossils. *Science* 209:396–397.

Pratt, L. M., T. C. Phillips, and J. M. Dennison. 1978. Evidence of nonvascular land plants from the early Silurian (Llandoverian) of Virginia, U.S.A. *Review of Paleobotany and Palynology* 25:121–149.

Raven, J. A. 1985. Comparative physiology of plant and arthropod land adaptation. *Philosophical Transactions of the Royal Society of London* B309:273–288.

Raymond, A., W. C. Parker, and S. F. Barrett. 1985. Early Devonian phytogeography. Pages 129–167 in *Geological Factors and the Evolution of Plants*, edited by B. H. Tiffney. Yale University Press, New Haven, Conn.

Retallack, G. J. 1985. Fossil soils as grounds for interpreting the advent of large plants and animals on land. *Philosophical Transactions of the Royal Society of London* B309:105–142.

Rolfe, W. D. I. 1985. Early terrestrial arthropods: a fragmentary record. *Philosophical Transactions of the Royal Society of London* B309:207–218.

Scott, A. C. 1980. The ecology of some upper Paleozoic floras. Pages 87–115 in *The Terrestrial Environment and the Origin of Land Vertebrates*, edited by A. L. Panchen. Academic Press, London.

Swain, T. 1978. Plant-animal coevolution: a synoptic view of the Paleozoic and Mesozoic. Pages 1–19 in *Biochemical Aspects of Plant and Animal Coevolution*, edited by J. B. Harboarne. Academic Press, London.

Swain, T. and G. Cooper-Driver. 1981. Biochemical evolution in the early land plants. Pages 103–134 in *Paleobotany, Paleoecology, and Evolution*, volume 1, edited by K. J. Niklas. Praeger Scientific, New York.

Taylor, T. N. 1981. *Paleobotany*. McGraw-Hill, New York.

Wright, V. P. 1985. The precursor environment for vascular plant colonization. *Philosophical Transactions of the Royal Society of London* B309:143–145.

Ziegler, A. M., R. K. Bambach, J. T. Parrish, S. F. Barrett, E. H. Gierlowski, W. C. Parker, A. Raymond, and J. J. Sepkoski. 1981. Paleozoic biogeography and climatology. Pages 213–266 in *Paleobotany, Paleoecology, and Evolution*, volume 2, edited by K. J. Niklas. Praeger Scientific, New York.

PART THREE

Terrestrial Ectotherms: Amphibians, Turtles, Crocodilians, and Squamates

The spread of plants and then invertebrates across the land provided a new habitat for vertebrates. The evolutionary transition from water to land is complex because water and air have such different properties: Aquatic animals are supported by water; terrestrial animals need skeletons and limbs. Aquatic animals extract oxygen from a unidirectional flow of water across the gills; terrestrial animals breathe air that they must pump in and out of sac-like lungs. Aquatic animals face problems of water and ion balance as the result of osmotic flow; terrestrial animals lose water by evaporation. Even sensory systems like eyes and ears work differently in water and air. The transition from aquatic to terrestrial habitats must have been facilitated by characteristics of fishes that were functional both in water and in air, although the functions may not have been exactly the same in the two fluids.

Once in terrestrial habitats, vertebrates radiated into some of the most remarkable animals that have ever lived. The dinosaurs are the best known of these, but many smaller groups contained forms that were just as bizarre, although not as large as many of the dinosaurs. One contribution of phylogenetic systematics has been the emphasis it has placed on the relationship of birds, crocodilians, and dinosaurs. This perspective suggests that the complex behaviors we consider normal for birds might be ancestral characters of their lineage. Sure enough, living crocodilians display parental care that is quite like that of birds (allowing for the differences in the size and morphology of birds and crocodilians), and evidence is accumulating that at least some dinosaurs also showed extensive parental care and probably other behaviors we now associate with birds.

Mass extinctions have punctuated the evolution of vertebrates, and the extinction of the dinosaurs at the end of the Mesozoic is the best known of these. As mass extinctions go, the one that occurred in the Cretaceous was minor—barely detectable in fact—but the inherent appeal of dinosaurs has made it a *cause célèbre* in the debate over the causes of mass extinctions.

The distinction between ectotherms (animals that obtain the heat needed to raise their body temperatures from outside the body) and endotherms (animals that use metabolic heat production for thermoregulation) has important functional considerations that cut across phylogenetic lineages. In some respects crocodilians are more like turtles and squamates (lizards, snakes, and related forms) than like the birds that are their closest living relatives. The relationship between an ectothermal organism and its physical environment (solar radiation, air temperature, wind speed, and humidity) is often an important factor in its ecology and behavior. One consequence of relying on outside sources of energy for thermoregulation is efficient use of metabolic energy, and ectotherms transform a high proportion of the food they eat into their own body tissue. This characteristic gives them a unique position in the flow of energy through terrestrial ecosystems.

In this part of the book we describe the radiation of vertebrates into terrestrial habitats in the Paleozoic and Mesozoic, and various theories to account for the extinction of dinosaurs and for mass extinctions generally, and consider the advantages and disadvantages of being an ectotherm in the modern world.

By the middle Devonian the stage was set for the appearance of terrestrial vertebrates, and the first vertebrate to set fin on land was a sarcopterygian (lobe-finned fish). The sarcopterygians, introduced with the other bony fishes in Chapter 8, did not participate in the evolutionary success of the ray-finned fishes. Indeed, the only surviving sarcopterygians are the lungfishes and the coelacanth. However, all the terrestrial vertebrates are descendents of sarcopterygians.

The far-reaching structural changes required for life on land were barely complete when some lineages of tetrapods became secondarily aquatic, returning to freshwater habitats. However, other lineages became increasingly specialized for terrestrial life. Once again, a radiation of vertebrates was associated with progressive changes in the jaws that allowed new ways of feeding and simultaneous changes in the limbs that appear to have increased the agility of predators. These structural changes were widespread, but only one of the terrestrial lineages of Paleozoic tetrapods made the next major transition in vertebrate history with the appearance of the embyronic membranes that define the amniotic vertebrates.

Origin and Radiation of Tetrapods in the Late Paleozoic

10

An Interrupted Story

The history of nonamniotic tetrapods falls into three parts: The origin of tetrapods from sarcopterygian (lobe-finned) fishes between the middle and late Devonian can be inferred from features of the skull, teeth, and limbs. Cladistic analyses of the relationships of the sarcopterygians to tetrapods have only recently been undertaken, and they also point to the sarcopterygians as the most likely sister group of the tetrapods. This conclusion confirms and strengthens the classical view of tetrapod relationships that was based largely on primitive characters shared by sarcopterygians and tetrapods.

A second stage in the history of nonamniotic tetrapods was their radiation into different lineages and different ecological types, from the late Paleozoic to the middle Mesozoic. By the late Devonian and early Carboniferous tetrapods had split into two lineages that are distinguished in part by the way the roof of the skull is fastened to the posterior portion of the braincase. One of these lineages includes temnospondyls, which were the largest and longest-lasting group of Paleozoic nonamniotic tetrapods; some lineages of temnospondyls extended into the Jurassic. The second lineage contains a diverse array of animals, including some that were small and limbless, and others that were large and had well-developed limbs and girdles. Two groups of large terrestrial animals in this lineage, the anthracosauroids and diadectomorphs, appear to be the most likely candidates for the sister group of the Amniota, but the information currently available does not allow a clear choice between them.

A third step in the history of nonamniotic tetrapods was the origin of the living amphibians, the salamanders, anurans, and caecilians. The living amphibians may be derived from one or perhaps from two branches of the temnospondyl lineage, but no transitional fossils have yet been identified.

The first amniotic vertebrates appeared in the late Carboniferous. They were small, agile animals that show modifications of the skeleton and jaws, suggesting that they fed on terrestrial insects. The diversity of nonamniotic tetrapods waned during the late Permian and Triassic. Simultaneously amniotic tetrapods radiated into many of the terrestrial life zones that had been occupied by nonamniotes, as well as developing some specializations that had not previously been seen among tetrapods.

Osteolepiform Fishes

Coincident with the appearance of the Dipnoi in the early Devonian, the earliest known osteolepiform fossils are found. The Osteolepiformes were slender-bodied, broad-headed forms with thick

Box 10–1. The Bendable Braincase

The braincase of vertebrates has two important functions: Of course, it encloses and protects the brain, but in addition it is a site of attachment for muscles and for some of the dermal bones of the skull. In all osteolepiforms the braincase was separated into anterior and posterior portions by a joint at the level of the mesencephalic and metencephalic parts of the brain (Figure 10–1a). The anterior part of the braincase contained the olfactory apparatus, the olfactory hemispheres, the thalamic structures, the tectum, and the dorsum sellae. The posterior portion enclosed the metencephalon and myelencephalon and the inner ear.

The joint between the two parts of the braincase differed in various genera of osteolepiforms. *Ectosteorhachis* had two sets of articulating surfaces: the dorsal pair resembled

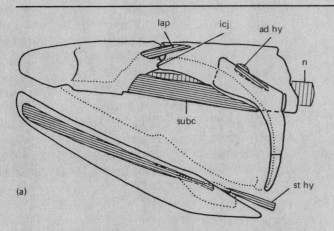

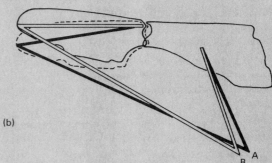

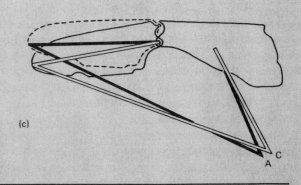

Figure 10–1. Kinetic skull of osteolepiforms. (a) The braincase of *Ectosteorhachis* shows the division into anterior and posterior parts that was characteristic of osteolepiforms. The dotted lines indicate the cheek region. The anterior portion of the braincase swung upward as the jaws opened (b) and downward as they closed (c). The outlined area labeled A shows the position of the snout at rest; B shows the elevation of the snout as the mouth opened, and C shows the depression of the snout as the mouth closed. Code: adhy, adductor hyomandibularis; icj, intracranial joint; lap, levator arcus palatini; n, notochord; subc, subcephalic; st hy, sternohyoideus. (From K. S. Thomson, 1967, *Zoological Journal of the Linnean Society*, 46:223–253.)

ball-and-cup joints, and the ventral pair fit into facets on the inner surface of the posterior portion of the braincase. This joint would allow only dorsoventral rotation; no lateral movement was possible. The anterior unit of the skull (the anterior portion of the braincase and the anterior skull roof and the cheek and palate) would have rotated in a vertical plane relative to the posterior unit (the posterior part of the braincase and the posterior skull roof). The skull roof was divided by a suture that corresponded in position to the underlying division between the anterior and posterior portions of the braincase. The large hyomandibular provided the link between the lower jaw and the anterior unit of the skull.

In Thomson's (1969) reconstruction of the kinesis of the osteolepiform skull, the anterior unit rotated dorsally as the mouth opened, increasing the gape. As the mouth closed the anterior unit rotated ventrally, perhaps helping to embed the palatal fangs in the prey. The hyomandibular, palate, and anterior portion of the skull can be viewed as forming an interlocked triangle that pivoted about its attachment to the rear portion of the skull (Figure 10–1b).

The muscles primarily responsible for raising the anterior portion of the skull were the levator arcus palatini and the coraco-mandibularis (Figure 10–1c). The subcephalic was the major muscle, producing the downward rotation when the jaw closed. As the mouth opened, contraction of the muscles that opened the jaws forced the hyomandibular to rotate forward. This rotation was transmitted via the palate to the anterior unit of the skull and, aided by contraction of the levator arcus palatini, caused the snout to flex dorsally (triangle A in Figure 10–1b). As the mouth closed, the anterior unit of the skull was rotated downward by contraction of the subcephalic muscle (triangle C in Figure 10–1c).

Skull kinesis in which the snout rotates dorsally as the mouth opens and ventrally as the mouth closes occurs widely among lizards, although the anatomical details are different from those of osteolepiforms (Chapter 15). One function of this kinesis is thought to be the advantage of bringing the upper and lower jaws into contact with prey at the same time, thereby reducing the chance that the prey will bounce away from its first contact and escape. Another possibly significant aspect of the kinesis of osteolepiform skulls is the apparent ability to use contraction of the subcephalic muscle to rotate the anterior portion of the snout downward when the mouth was closed to apply a crushing force to prey. This function might have been significant, because the jaw-closing muscles of osteolepiforms produced a quick snap but were not oriented so as to apply force to prey that was held in the mouth (see Box 10–2).

scales, a single external nostril on each side, and variable caudal structure. They shared many features with the Devonian dipnoans (Figure 8–20), but are distinguished by the division of the braincase (**neurocranium**) into anterior and posterior portions movably articulated behind the orbits. The dorsoventral flexion in the middle of the head allowed changes in the orientation of the open mouth (Box 10–1). Importantly, many osteolepiforms had ring-shaped vertebral centra with various accessory ossifications associated with the spinal nerves; similar vertebrae are found in the earliest tetrapods. Although all osteolepiforms were basically free-swimming predators of shallow waters, many were specialized for life at the water's edge (Thomson 1969).

Tetrapod Relationships

The fossil record provides evidence of the origins of terrestrial vertebrates. Important paleontological discoveries of early tetrapods have been made in sediments of the middle and late Devonian age in Greenland and Australia (Warren and Wakefield 1972, Campbell and Bell 1977, Jarvik 1980, Panchen 1980). These areas, now far from the equator, were warm tropical regions during the Devonian.

Two major groups of fishes have been considered the closest relatives of terrestrial vertebrates— the lungfishes and the osteolepiforms. Both have had their supporters, and superficially either seems a possible sister group of tetrapods. Lungs, which we have seen were an early development in vertebrates, were present in both groups. In both clades the fins were modified in the direction of limbs that, to judge from living sarcopterygians, were moved in the alternating pattern common in tetrapod locomotion (Fricke et al. 1987).

Closer examination, especially of the fin skeleton and the skull, points to osteolepiforms as the most likely sister group of tetrapods. The question is not settled, however, and the hypothesis that tetrapods arose from lungfish has recently been revived (Rosen et al. 1981) and vigorously criticized (Jarvik 1981, Holmes 1985, Carroll 1987). Lungfish are highly specialized, and Devonian fossils are scarcely less specialized than the living species. In particular, the dermal bones of the lungfish skull have a different pattern from those of early tetrapods and the braincase of lungfishes is a solid structure, not divided like that of osteolepiforms and *Ichthyostega*. Furthermore, lungfish lack teeth on the margins of the jaws. In all but the earliest lungfish the teeth are in the form of large plates on the palate. In addition, the skeleton of a lungfish's fin consists of a central axis with symmetrical rays forming a leaf-shaped structure or **archipterygium** (Figure 10–2). No obvious homologies are apparent between this limb structure and the limb seen in the earliest known tetrapod.

In contrast, comparison of the morphology of osteolepiforms with early tetrapods reveals extensive similarities, suggesting that tetrapods probably arose from this group sometime in the middle Devonian. Early osteolepiforms had an archipterygium like that of lungfish, but in later forms the central axis was shortened and branches were confined to the anterior margin. Homologies between the limb bones of osteolepiforms and early tetrapods are clearly apparent. The skulls of osteolepiforms and tetrapods also are similar. The major differences are found in the overall proportions of the skull and the bones of the nasal region (Figure 10–3). In osteolepiforms the nasal region is short and covered by a mosaic of small bones, whereas early tetrapods have a much longer snout and a pair of large nasal bones.

A third similarity between osteolepiforms and early tetrapods lies in the structure of the teeth. Both osteolepiforms and tetrapods appear to have been predators; they were armed with sharp, sturdy teeth. In cross section the teeth of both groups show the **labyrinthodont** pattern of complex infoldings of the walls of the pulp cavity (Figure 10–4).

Examination of the ichthyostegids from late Devonian deposits in Greenland has provided additional support for the origin of tetrapods from osteolepiforms. In addition to the homologies already discussed, which are visible in advanced Paleozoic tetrapods, ichthyostegids have fish-like features that were lost in most later forms. Unlike any other known tetrapods, *Ichthyostega* had a caudal fin that was supported by fin rays dorsal to the neural spines of the caudal vertebrae. Finally, the discovery of *Ichthyostega* solved a problem that had been a stumbling block for paleontologists who favored osteolepiforms as the sister group of tetrapods. In osteolepiforms the braincase was formed in two units which could move in relation to each other, and there was a depression in the floor of

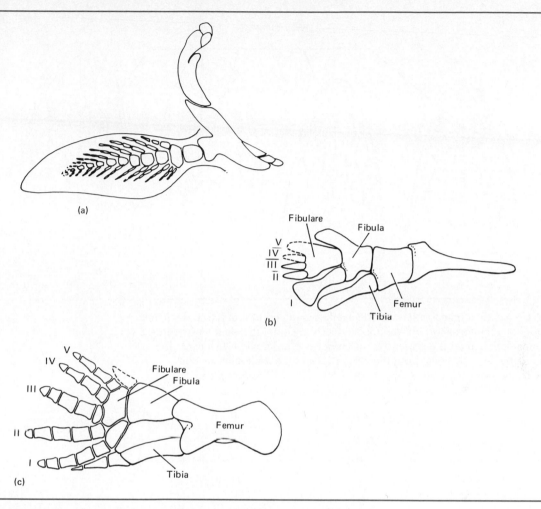

Figure 10–2. Limbs of an early tetrapod and two sarcopterygians. (a) Lungfish; (b) osteolepiform; (c) *Ichthyostega*. [(a) From A. S. Romer and T. S. Parsons, 1977, *The Vertebrate Body*, 5th edition, W. B. Saunders, Philadelphia; (b) and (c) from E. Jarvik, 1980, *Basic Structure and Evolution of Vertebrates*, Academic Press, London.]

the rear part of the braincase that probably accommodated the anterior part of the notochord. Neither feature was known in tetrapods until *Ichthyostega* was discovered. In *Ichthyostega* the braincase was a solid structure, as it is in all other tetrapods, but the suture between the two parts is visible. Furthermore, the notochordal canal persisted in the posterior part of the braincase.

Evolution of Terrestrial Vertebrates

Tantalizingly incomplete as the skeletal evidence is, it is massive compared to the information about the ecology of osteolepiforms and early tetrapods. It is not possible even to be certain if the evolution

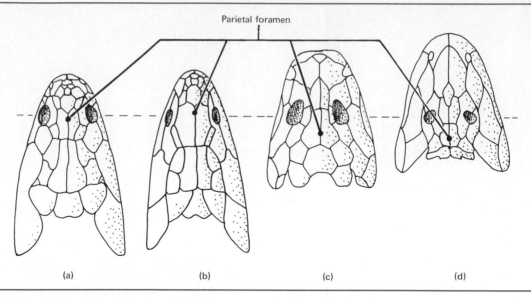

Parietal foramen

(a)   (b)   (c)   (d)

Figure 10–3. Skull proportions of osteolepiforms (a and b) and tetrapods (c and d). In both osteolepiforms and tetrapods the facial region of the skull lengthened and the elements behind the parietal foramen became progressively shorter. The changes can be seen by comparing the proportions of the skull anterior and posterior to the eyes. (a) *Osteolepis,* middle Devonian; (b) *Eusthenopteron,* late Devonian; (c) *Ichthyostega,* late Devonian; (d) *Eryops,* early Permian.

Figure 10–4. One of the morphological features shared by osteolepiforms and Paleozoic tetrapods was the labyrinthodont tooth. A complex folding of the walls of the pulp cavity produced the labyrinthine structure that identifies this sort of tooth. (a) External view showing striations on outside of tooth; (b) cross section of a tooth from the osteolepiform *Polyplocodus;* (c) cross section of a tooth from the temnospondyl *Benthosuchus.* (From A. P. Bystrow, 1935, *Acta Zoologica* 16:65–141.)

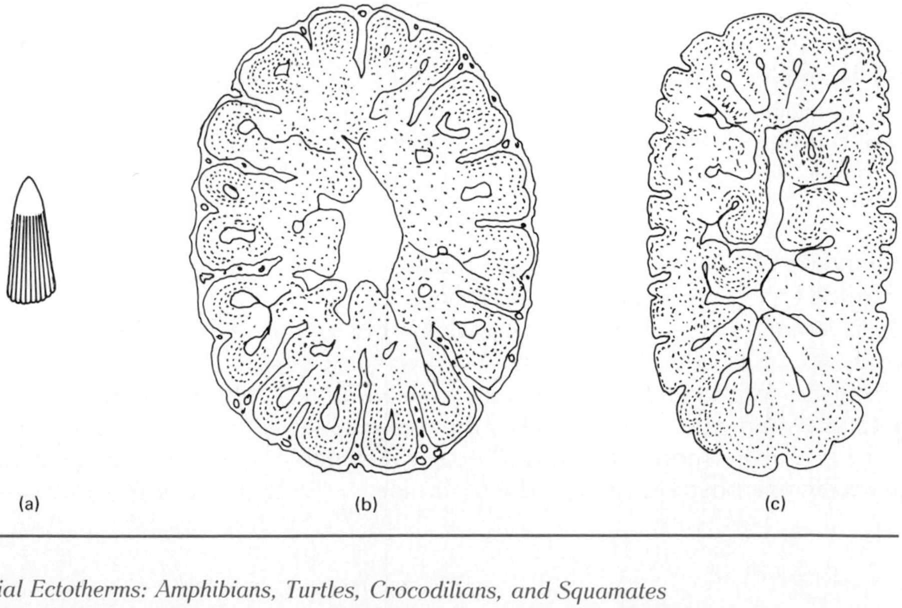

(a)   (b)   (c)

of tetrapods occurred in purely freshwater habitats (Panchen 1977, Thomson 1980). Greenland and Australia, the two land masses from which Devonian tetrapods are known, are believed to have been separated by marine environments in the Devonian. If that is correct, osteolepiforms must have been able to traverse those seas.

What sorts of lives did the osteolepiforms lead? What was their habitat like? What were their major competitors and predators? What were the earliest tetrapods able to do that osteolepiforms could not? This is the sort of information we need to assess the selective forces that shaped the evolution of tetrapods. Again the gap between the middle Devonian osteolepiforms and the late Devonian tetrapods is frustrating. If we knew the sequence in which the differences between osteolepiforms and tetrapods appeared, we could infer much about the lives of the first tetrapods and the selective value of the differences. Unfortunately, the first tetrapods of which we have fossils, *Ichthyostega* and its relatives, are thoroughgoing amphibians when they appear in the fossil record.

How Does a Land Animal Evolve in Water?

Certain inferences can be drawn from the fossil material available. We start from the basis that any animal must function in its habitat. If it does not, it does not leave descendants and its phylogenetic lineage becomes extinct. Evolutionary change occurs because the naturally occurring variation in organisms is subject to selection. Any evolutionary trend is perceived because progressive changes conferred an advantage on the individuals possessing them; it is merely a fortunate coincidence that some features may be preadaptive for a different (in this case a terrestrial) way of life. Our challenge is to interpret the fossil record in the light of these principles and infer what we can about the changing lives of earliest tetrapods.

Osteolepiforms were large fish, some as much as 4 meters long, with heavy cylindrical bodies and large teeth (Figure 8–20d). They probably either

stalked their prey or lay in ambush and made a sudden rush. One can picture an osteolepiform prowling through the dense growth of plants on the bottom of a Devonian pond or estuary. The flexible lobe fins would support it as it waited motionless for prey, and the evolutionary change from the elongate archipterygium of the early osteolepiforms to the stouter limb of later ones might reflect selection for strengthening the limb for this sort of movement and to bear the weight of the fish in very shallow water.

Although there is no direct fossil evidence of lungs in osteolepiforms, we believe they were present because every related group (dipnoans, actinistians, and amphibians) is known to have lungs or the remnants of lungs. In the warm swampy areas osteolepiforms inhabited, oxygen levels in the water were probably low much of the time. Osteolepiforms presumably could breathe atmospheric oxygen by swimming to the surface and gulping air, or in shallow water, by propping themselves on their pectoral fins to lift their heads to the surface.

Evolution of Tetrapod Characters in an Aquatic Habitat

The features we have discussed so far are very general features that apply to most osteolepiforms and in some cases to all sarcopterygians. What were the selective forces that could have started some osteolepiforms along an evolutionary lineage that culminated in the tetrapods?

One of the most striking differences between osteolepiforms and ichthyostegids is the extension of the facial region of the snout in the tetrapods (Figure 10–3). Examination of a variety of osteolepiforms suggests that this trend already existed among the fishes. In living aquatic vertebrates, a long snout is associated with utilization of a sudden sideward movement of the head to capture food. Crocodilians are the best examples of this technique, although it is observed in other vertebrate classes as well. A longer snout could be advantageous for fishes that depend on waiting in ambush for prey, because it would increase the volume in front of the fish in which prey could be captured. A fish's morphology, however, prevents it from taking full advantage of a long snout compared to a tetrapod with a snout of the same length. In fishes the attachment of the pectoral girdle to the rear of the skull stiffens the anterior trunk and reduces the sideward oscillation of the head as a fish swims (Gosline 1977). Because its pectoral girdle is attached to the rear of its skull, a fish does not have a functional neck. A fish cannot turn its head; it bends or rotates its whole body. Thus its capacity for a sudden sideward snap is less than that of a tetrapod. A neck is produced when the pectoral girdle loses its attachment to the rear of the skull, and the greater the separation of the pectoral girdle and skull, the more flexible the neck becomes.

It seems possible that the separation of the pectoral girdle from the skull was an early stage in the evolution of tetrapods. The ability to strike sideward to capture prey through a larger arc than could an osteolepiform with a fish-like pectoral girdle may have been an early advantage leading to exploitation of a new ecological niche by tetrapods.

Freeing the pectoral girdle from its attachment to the skull would have little effect on its locomotor function. When a tetrapod was immersed in water the limbs bore little weight and would function equally well whether the girdles were fused to the skull or not. In shallow water the forelimbs bore some of the weight of the body and, through selection for this ability, the muscles holding the pectoral girdle to the trunk might increase in mass, eventually reaching the ichthyostegid condition.

What Were the Advantages of Terrestrial Activity?

The reasoning outlined above proposes an evolutionary scenario by which selection for features that were advantageous to animals in an aquatic habitat might have produced an organism with enough preadaptations for terrestrial life to have emerged onto land. However, it does not explain why an aquatic organism might have taken that step. All of the morphological features that were functional for life on land were originally functional in the aquatic habitat in which they evolved. Why did some vertebrates leave the aquatic habitat for which they were so well suited and begin to exploit a radically different habitat on land?

This question has fascinated biologists for a century or more, and there is no shortage of theories (Romer 1958, Szarski 1962, Schaeffer 1965, Bray 1985). Their ideas, although frequently presented as "the" explanation for the evolution of tetrapods, are seldom mutually exclusive, and can often be combined.

The classic theory was put forth most forcefully by A. S. Romer. He proposed that the Devonian was a time of seasonal droughts. Shallow ponds that formed during the rainy season often evaporated during the dry season, stranding their inhabitants in rapidly shrinking bodies of stagnant water. Fishes trapped in such situations are doomed unless the next rainy season begins before the pond is completely dry. Romer suggested that certain osteolepiforms had limbs sufficiently well developed to allow them to crawl from these drying ponds and move overland to other, larger bodies of water in which their chances of surviving the dry season were better. Romer believed that millions of years of selection of the osteolepiforms best able to escape death by finding their way to permanent water would produce a lineage showing increasing ability on land.

This theory has been criticized on several grounds. It is questionable whether intermittent catastrophic selection of the type Romer suggested could have the effect he envisioned. Further, a fish that succeeds in moving from a drying pond to one that still holds water has enabled itself to go on leading the life of a fish—that seems a backward way to evolve a terrestrial animal.

Various alternative theories have been proposed that stress positive selective values associated with increasing terrestrial activity to osteolepiforms. One emphasizes the contrast between terrestrial and aquatic habitats in the Devonian. The water was swarming with a variety of fishes that had radiated to fill a multiplicity of ecological niches. Active, powerful predators and competitors abounded. In contrast, the land was free of vertebrates. Any osteolepiform that could occupy terrestrial situations had a predator- and competitor-free environment at its disposal. The exploitation of this habitat can be seen as proceeding by gradual steps.

Juvenile osteolepiforms might have congregated in shallow water, as do juveniles of living fishes, to escape the attention of larger predatory fishes that are restricted to deeper water. At the edge of a lake or estuary many morphological and physiological characteristics of terrestrial vertebrates would have been useful to a still-aquatic osteolepiform. Warm water holds little oxygen, and shallow pond edges are likely to be especially warm during the day. Thus, lungs are important to a vertebrate in that habitat, whether it be fish or tetrapod. Similarly, legs would have borne the weight of the animal in the absence of water deep enough to float it. In shallow water an air-breathing, upstanding fish could have lifted its head above the water, and the change in the shape of the lens of the eye associated with the differences in the refractive indices of water and air could have started to occur. These behavioral and morphological features can be found among a number of living fishes, such as the mudskipper, climbing perch, and walking catfish that make extensive excursions out of the water, even climbing trees and capturing food on land.

Starting from fishes that snapped up terrestrial invertebrates that fell into the water, one can envision a gradual progression of increasingly agile forms capable of exploiting the terrestrial habitat for food as well as for shelter from aquatic predators. Terrestrial agility might have developed to the stage at which small osteolepiforms moved overland from the pond of their origin to other ponds. Many vertebrates include a dispersal stage in their life history, usually in the juvenile period. In this stage individuals spread from the place of their origin to colonize suitable habitats, sometimes long distances from their starting point. This type of behavior is so widespread among living

vertebrates that we may guess that it occurred in some form in the osteolepiforms as well.

Perhaps the main selection for terrestrial life occurred in the juvenile stages of osteolepiforms. The adults were large animals, and it is hard to imagine an adult osteolepiform 1 meter long being sufficiently agile to capture a terrestrial invertebrate such as a scorpion scuttling about in its own terrestrial habitat. It is much easier to visualize a 15-centimeter-long juvenile osteolepiform chasing the scorpion under a log and grabbing it. A small body size would have greatly simplified the difficulties of support, locomotion, and respiration in the transition from an aquatic to a terrestrial habitat.

The Radiation and Diversity of Nonamniotic Paleozoic Tetrapods

For 150 million years, from the late Devonian to the early Jurassic, nonamniotic tetrapods radiated into an unimaginable variety of terrestrial and aquatic forms. It is no wonder that the fossil record is confusing. Parallel and convergent evolution were widespread, and it is hard to separate phylogenetic relationships from convergent evolution to similar ways of life. Large gaps in the fossil record, especially the lack of early representatives of many groups, still obscure relationships.

The application of cladistic methods to interpreting the phylogeny of nonamniotic tetrapods has produced major revisions of Romer's (1966) arrangement. Romer had distinguished three subclasses, the Labyrinthodontia (orders Anthracosauria, Temnospondyli, and Ichthyostegalia), the Lepospondyli (orders Aïstopoda, Microsauria, and Nectridea), and the Lissamphibia (the modern orders Anura, Urodela, and Caecilia). His classification was based largely on primitive characteristics, however, and examination of derived character states has suggested radically different alignments.

Phylogenetic Relationships of the Tetrapoda

SARCOPTERYGII

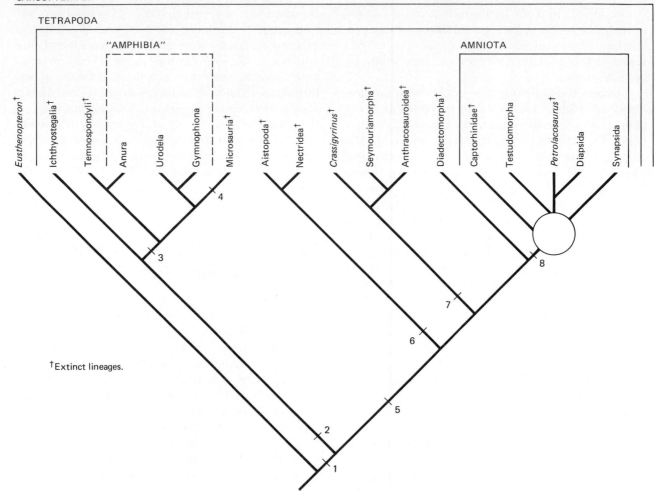

†Extinct lineages.

1. Tetrapoda: Four limbs characterized by a single bone in the proximal portion and two bones in the distal portion. **2.** Ichthyostegalia: Fully ossified, akinetic (immobile) skull roof, ribs with uncinate processes. **3.** Temnospondyli and Microsauria: Braincase attached to skull roof by suture between exoccipitals and postparietals, characteristics of the palate. **4.** Microsauria: Unique articulation between skull and atlas vertebra, absence of otic notch. **5.** Kinesis retained between skull roof and cheeks, ventral otic fissure lost, characteristics of the palate and snout. **6.** Aistopoda and Nectridea: Characters of the vertebrae.

7. *Crassigyrinus,* Seymouriamorpha, and Anthracosauroidea: Characters of the skull and a distinctive pattern of pits and ridges on the dermal bones. **8.** Amniota: A distinctive arrangement of extra-embryonic membranes (the amnion, chorion, and allantois). Analyses of the relationships of some of the other groups shown here are not possible. (Based on T. R. Smithson, 1985, *Zoological Journal of the Linnean Society* 85:317–410; and M. J. Heaton and R. R. Reisz, 1986, *Canadian Journal of Earth Science* 23:402–418.)

No single phylogenetic scheme has replaced the classical view, but most arrangements have features in common. Smithson (1982, 1985) stressed the importance of characteristics of the skull as forming a primary dichotomy within tetrapods. This dichotomy separates a lineage consisting of the ichthyostegalians + temnospondyls + microsaurs + the living amphibians (anurans, caecilians, and urodeles) from a second lineage that includes aïstopods + nectrideans + palaeostegalians + loxommatoids + anthracosaurs + diadectomorphs + amniotes (Figure 10–5).

The kinetic braincase of osteolepiforms was divided into anterior and posterior parts by the ventral otic fissure, which corresponds to the embryonic division between the prechordal and parachordal cartilages. The entire area of attachment between the posterior braincase and the skull roof of osteolepiforms was derived from the auditory capsules, the opisthotics (Figure 10–6).

The anthracosaur lineage retained the attachment of the braincase seen in osteolepiforms, and the contact between the braincase and the skull roof of anthracosaurs was formed by the auditory capsules. In contrast, in the temnospondyl lineage a new medial connection developed between the exoccipitals of the posterior braincase and the postparietal bones of the skull roof. Thus, in temnospondyls the connection between braincase and skull is derived from the embryonic parachordal cartilages.

This dichotomy between temnospondyl and anthracosaur lineages is repeated in other features of the skull as well as in other skeletal features: The skulls of temnospondyls were flat, akinetic, and the ventral otic fissure was retained, whereas the skulls of anthracosaurs were deep, kinetic, and

the ventral otic fissure was eliminated. The ribs of temnospondyls bear uncinate processes (posterior projections from the ribs to which axial muscles attach), and the ribs of anthracosaurs lack uncinate processes.

Ichthyostegalia

Although *Ichthyostega* is the earliest tetrapod of which we have a record of the axial skeleton, it is far beyond the transitional stage (Figure 10–7). By the late Devonian, ichthyostegids were not only fully at the amphibian structural grade, but they had started to diversify—three genera of ichthyostegids have been distinguished in the Greenland deposits.

Most of the structural features distinguishing aquatic from terrestrial animals are a consequence of the differences between water and air. The high frictional resistance of water makes a streamlined body form important for all but the most sedentary aquatic animals, but only the speediest terrestrial animals need to cope with significant air resistance. Water buoys an animal up, but air gives no such support. Resistance to gravity is of minor importance in the skeleton of a fish—the demands of locomotion are paramount. In contrast, resistance to the downward pull of gravity is a major factor shaping the skeleton of a terrestrial vertebrate.

When a terrestrial vertebrate stands, its body hangs from the vertebral column which, like the arch of a suspension bridge, supports the weight of the trunk and transmits it to the ground by two sets of vertical supports, the girdles and legs. The vertebral column must be rigid and the girdles and legs sturdy and firmly connected with the vertebral column. Without all of these features, a ter-

---

Figure 10–5. Phylogenetic relationships of the Tetrapoda. This diagram shows the probable relationships among the major groups of tetrapods. Extinct lineages are marked by a dagger ([†]). The numbers indicate derived characters that distinguish the lineages. The quotation marks around "Amphibia" indicate that it may be a paraphyletic group. The circle indicates that the relationships of the lineages at that node cannot yet be defined. (From T. R. Smithson, 1985, *Zoological Journal of the Linnean Society* 85:317–410.)

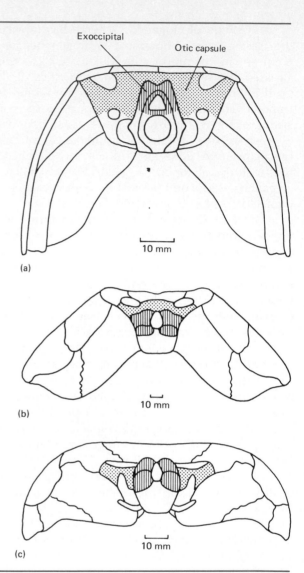

Figure 10–6. Anthracosaurs and temnospondyls are distinguished by the mode of attachment of the braincase and skull roof. In osteolepiforms (a, *Eusthenopteron*) the connection was formed by the auditory capsules, and this connection was retained in the anthracosaur lineage (b, *Megalocephalus*). In the temnospondyl lineage a new connection was formed by the exoccipital bone (c, *Greererpeton*). (From T. R. Smithson, 1982, *Zoological Journal of the Linnean Society* 76:29–90.)

restrial vertebrate cannot stand. In osteolepiforms like *Eusthenopteron* (Figure 10–8a), each segment of the body contained two sets of bones that together formed the vertebral centrum. The anterior elements (intercentra) were wedge-shaped in lateral view and crescentic when viewed from the end. They lay beneath the notochord and extended upward around it. The posterior elements were a pair of small bones (pleurocentra) that lay on the dorsal surface of the notochord. The neural arches of ad-

jacent vertebrae did not articulate. The structure of the vertebrae of *Ichthyostega* (Figure 10–8b) was very similar to that of the osteolepiforms, but articulations between the neural arches transmitted forces from one vertebra to the next, providing the support needed by a terrestrial animal. These articulations were still more extensive in later forms such as the Permian dissorophid *Eryops* (Figure 10–8c).

In the pectoral girdle of *Ichthyostega* the post-

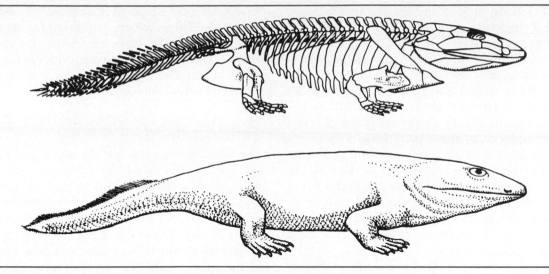

Figure 10–7. Skeleton and restoration of the possible external appearance of *Ichthyostega*. A number of fish-like characters are evident in *Ichthyostega*, most notably scales and a tail fin supported by fin rays. Other features arc fully tetrapod-like, especially the well-developed ribs, girdles, and limbs. (From E. Jarvik, 1955, *The Scientific Monthly*, 80:141–154.)

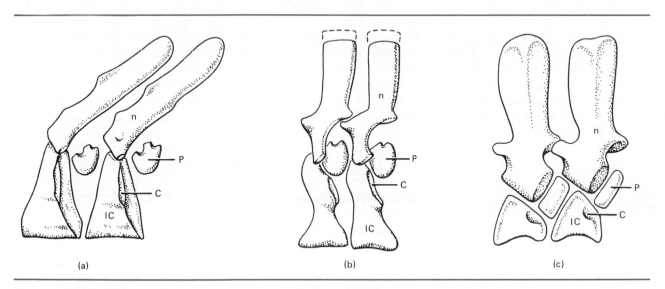

Figure 10–8. Vertebral structure: (a) the osteolepiform *Eusthenopteron*; (b) *Ichthyostega*; (c) *Eryops*. n, neural arch; p, pleurocentrum; ic, intercentrum; c, articulation for head of rib. (From A. S. Romer, 1966, *Vertebrate Paleontology*, 3rd edition, University of Chicago Press, Chicago.)

temporal and supracleithral bones, which in fishes attach the cleithrum and clavicle to the skull, had disappeared completely. In the ventral midline a new bone, the sternum, braced the ventral part of the pectoral girdle, and the scapulocoracoid provided a strong dorsal attachment for the muscles that bound the pectoral girdle to the trunk. In the pelvic region, the two bony ventral plates of fishes had been replaced by three paired bones that form the pelvic girdles of all tetrapods—the pubis, ischium, and ilium. The pubis and ischium of the left and right sides met in the ventral midline in a firm symphysis while the ilium on each side articulated with the sacral rib and vertebra, forming a rigid connection from the spine through the hind legs to the ground.

In a large, heavy-bodied animal like *Ichthyostega*, a rib cage is necessary to support the body when the animal lies on the ground. Without that rigid, bony support the weight of the body would squash the internal organs and collapse the lungs. *Ichthyostega* had a well-developed rib cage that might have developed at the osteolepiform stage, where it would support the body as the animal ventured into shallow water. Indeed, the rib cage of *Ichthyostega* was so much more robust than the rib cages of later tetrapods that it may have had an additional function in *Ichthyostega* that was lacking in other Paleozoic tetrapods. Possibly the broadly overlapping ribs, which would have been linked by sheets of muscle in life, provided stiffness in the trunk region that compensated for the relatively unmodified structure of the vertebrae.

Two external features were retained from the osteolepiform ancestors—a tail fin supported by fin rays, and a scaly body covering. (Although fossilization of soft tissues such as skin is rare, a number of examples have been found among Paleozoic tetrapods, and these indicate that many, possibly most, had a scaly covering, at least on the ventral surface of the body.)

Although ichthyostegids show numerous primitive features that place them closer to the fish–tetrapod transition than any other known fossils, they have several specialized features. For example, the intertemporal bone has been lost in ichthyostegids, although it is present in many later tetrapods. Thus, we are presently forced to base our inferences about the first tetrapods and their origins on animals which were not only long separated in time from the transition, but which had branched off on their own.

Temnospondyli

The temnospondyls were the most speciose non-amniotic tetrapods of the Paleozoic. Many temnospondyls were stocky, short-legged, heavy-bodied, large-headed semiaquatic predators that probably fed on fishes and other tetrapods (Figure 10–9). *Eryops* of the late Paleozoic was about 1.6 meters long, and its large head comprised one-fifth of its total length. Its contemporary *Cacops* was only about 40 centimeters long, but its head was nearly one-third the total length of its body. The skeleton of *Cacops* indicates that it was quite terrestrial; the limb girdles are sturdy and the legs and toes are long in proportion to the body. Dissorophid temnospondyls such as *Cacops* appear to represent the height of temnospondyl terrestrial development. This was a diverse family; many of its members, including *Cacops*, were large and had plates of dermal bone protecting the trunk region.

Other temnospondyls were small, aquatic forms and may be close to the lineage from which anurans are derived (Milner 1982). Some species of *Doleserpeton*, *Tersomius*, and *Amphibamus* had the distinctive pedicellate tooth structure that characterizes the modern amphibians (Box 11–1). The branchiosaurs (Figure 10–10) were paedomorphic temnospondyls less than 10 centimeters in total length. Branchiosaurs were probably related to the dissorophids but may not be a monophyletic lineage. Paedomorphosis occurs among a number of families of living salamanders (see Chapter 11), and the branchiosaurs might represent more than one origin of paedomorphic temnospondyls. However, two recent studies have suggested that branchiosaurs are monophyletic and that they are the sister group of the living amphibians (de Queiroz and Cannatella 1987, Trueb and Cloutier 1987).

Many of the temnospondyls of the late Paleo-

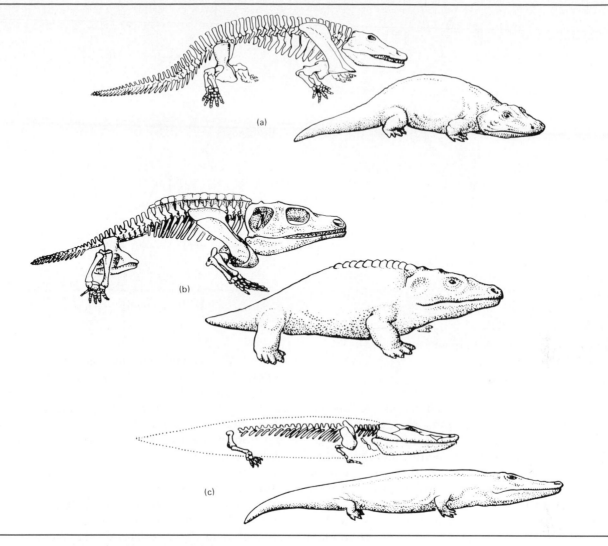

Figure 10–9. Temnospondyls: (a) *Eryops*, a Permian dissorophid with sturdy legs and a rounded body; (b) Permian dissorophid *Cacops*; (c) *Cyclotosaurus* from the late Triassic. [(a) From E. A. Colbert, 1969, *Evolution of the Vertebrates*, Wiley, New York; (b) from B. J. Stahl, 1974, *Vertebrate History: Problems in Evolution*, McGraw-Hill, New York; (c) from C. L. Fenton and M. A. Fenton, 1958, *The Fossil Book*, Doubleday, New York.]

zoic and early Mesozoic were very large aquatic forms with short limbs and reduced ossification. The capitosaurs and metoposaurs of the Triassic had skulls nearly a meter long. *Cyclotosaurus* from the late Triassic was an aquatic form with small legs and a dorsoventrally flattened body (Figure 10–9c). These impressive temnospondyls may have represented the top of the aquatic food chain,

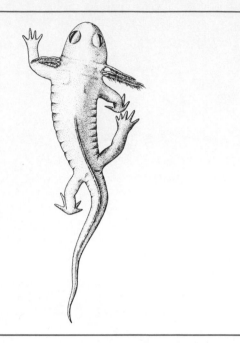

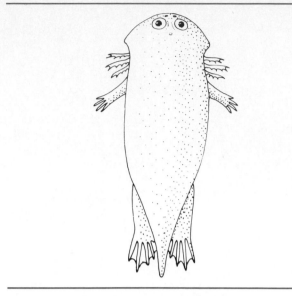

Figure 10—11. *Gerrothorax*, a plagiosaur.

Figure 10—10. Paedomorphic temnospondyl, *Branchiosaurus*.

feeding on their smaller relatives as well as any terrestrial animals incautious enough to try to swim in their vicinity.

Some of the most interesting ecological specializations among temnospondyls are found in aquatic forms. The Plagiosauridae (several genera which are probably not as closely related as their inclusion in one family suggests) and the Brachyopidae are bizarre, flat-headed, short-snouted temnospondyls. *Gerrothorax*, a late Triassic plagiosaurid, was about 1 meter long (Figure 10–11). The body was very flat and armored dorsally and ventrally. The dorsal position of the eyes suggests that *Gerrothorax* lay in ambush on the bottom of the pond until a fish or another temnospondyl swam close enough to be seized with a sudden rush and gulp. The broad head and wide mouth would enable *Gerrothorax* to attack large prey. The broad, short skull and retention of external gills in at least some forms suggest that the brachyopids and pla-

giosaurids evolved by paedamorphosis from the larvae of temnospondyls. Because brachyopids and plagiosaurids were separated in both distance and time, it is likely that they represent at least two separate derivations, and quite possibly genera within the two families were independently derived.

The trematosaurids are among the most remarkable temnospondyls. As early as the Permian some temnospondyls reversed the trend to a broad, flat skull and evolved the elongate snout characteristic of specialized fish eaters (Figure 10–12). Forms such as *Archegosaurus* may later have given rise to even more specialized fish eaters including *Aphaneramma*, or the later forms may represent a convergent evolutionary lineage. The trematosaurids are found in early Triassic marine beds. Salt water is an unusual habitat for temnospondyls. The scaly covering that trematosaurs probably retained from osteolepiform ancestors could have helped to protect the adults from the osmotic stress of either fresh or salt water, but what of the presumed larval stages? Were trema-

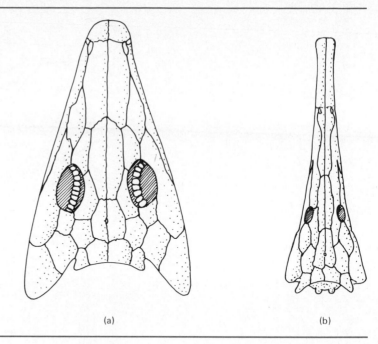

Figure 10–12. Specialized long-snouted fish-eating dissorophids like *Archegosaurus* (a) appeared in the early Permian. The trematosaurs (b, *Aphaneramma*) were still more specialized marine forms of the Triassic.

(a)   (b)

tosaurs viviparous like a few living amphibians, or could their larvae tolerate at least moderate salinities as do tadpoles of a few frogs and toads? Furthermore, if trematosaurs were able to invade the sea, why didn't other temnospondyls?

Microsauria

The microsaurs can be distinguished from other members of the temnospondyl lineage by the modification of the first cervical vertebra into a unipartite atlas-axis complex that is unlike the multipartite atlas-axis complex of other tetrapods. Two groups of microsaurs can be distinguished: The tuditanomorphs were almost exclusively terrestrial and the microbrachomorphs were largely aquatic. The name Microsauria (= small reptile) is inappropriate, but this confusion with early amniotes is understandable. Some microsaurs had moderately elongate bodies with small but functional limbs and deep, sturdy skulls (Figure 10–13). Like the amniotes they were probably small, active terrestrial predators that specialized in feeding on in-

vertebrates. Some microsaurs had body proportions very like those of modern lizards and salamanders.

Aïstopoda

The Aïstopoda is among the smallest groups of nonamniotic tetrapods of the Paleozoic, containing only six genera of legless forms (Figure 10–14). Most information comes from fossils from the late Carboniferous and early Permian. Aïstopods had very long bodies, some had more than 200 vertebrae in the trunk, and they lacked both limbs and limb girdles. The skulls of aïstopods had an open structure, like those of snakes and some lizards. The flexible skulls of snakes and lizards are associated with the ability to swallow large prey, and the same may have been true of aïstopods. Flexible skulls are not very effective for digging burrows through soil, and fossorial snakes and lizards usually have rigid skulls. The open nature of aïstopod skulls suggests that these were not burrowing animals. They might have lived on the surface of the

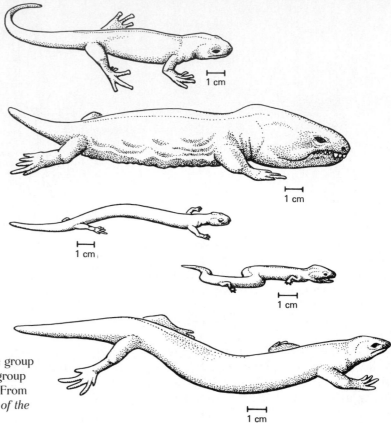

Figure 10–13. Microsaurs were a diverse group of tetrapods. Many were terrestrial, but the group also included aquatic and burrowing forms. (From R. L. Carroll and P. Gaskill, 1978, *Memoirs of the American Philosophical Society* 126:1–211.)

ground, foraging in leaf litter, or they might have been aquatic.

Nectridea

Nectrideans also were elongate aquatic animals, but in nectrideans the elongation occurred mainly by lengthening the tail. Unlike aïstopods, nectrideans had only 15 to 26 trunk vertebrae. In *Urocordylus*, an early nectridean, the tail was twice as long as the body (Figure 10–14). The neural and hemal arches of the tail vertebrae were expanded and probably supported a flat muscular tail fin. In the urocordylids the skull was sharply pointed and the legs were small. In advanced members of the

lineage the ossification of the limbs and girdles was reduced.

A second lineage, represented by the family Keraterpetontidae, contained animals that were dorsoventrally flattened with broad skulls in which the tabular bones were elongated. In adult *Diploceraspis* the tabular bones formed crescentic structures often called horns that at their greatest extent were more than five times the width of the anterior part of the skull. The ontogenetic development of the skull can be traced from the normal-looking oval skull of small individuals to fully developed horns in adults. The mouth was small and the eyes were far forward and directed upward, suggesting that the horned nectrideans may have

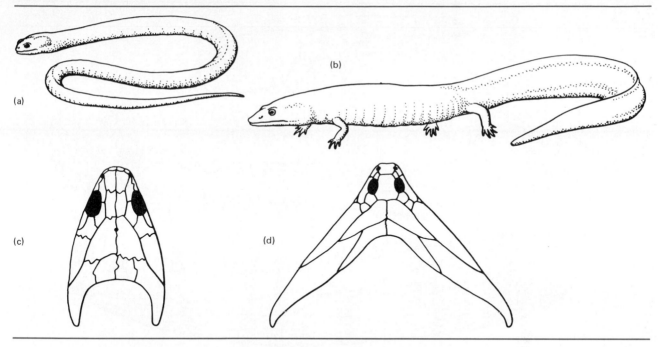

Figure 10–14. (a) The legless aïstopod *Ophiderpeton* was about 75 centimeters long. (b) The aquatic nectrideans were more varied than the aïstopods. One type, illustrated by *Sauropleura*, had an elongated body and small legs, a laterally flattened tail, and a sharply pointed snout. The tabular bones of "horned nectrideans," (c) *Keraterpeton* and (d) *Diploceraspis*, were greatly elongated. [(a) and (b) From A. R. Milner and (c) from A. C. Milner, both in *The Terrestrial Environment and the Origin of Land Vertebrates*, edited by A. L. Panchen, 1980, Academic Press, London; (d) from J. R. Beerbower, 1963, *Bulletin of the Museum of Comparative Zoology* 130:31–108.]

lived on the bottoms of bodies of water. These peculiar animals have aroused considerable speculation but little in the way of satisfactory suggestions about the function of the horns. Were they covered in life with a flap of skin that extended back to the shoulder? The fossil evidence does not show such a skin flap, but one would not necessarily expect it to be preserved. If there was a flap, was it a hydrofoil to help the animal swim, or a highly vascularized tissue that extracted oxygen from water? Possibly a ligament ran from the tabular horn to the pectoral girdle, stabilizing the head and reducing lateral oscillation during swimming in a manner analogous to the connection between the skull and pectoral girdle of fishes.

Paleostegalia and Loxommatoidea

These two taxa together may constitute a sister group to the other anthracosauroids; their shared derived characters are teeth with anterior and posterior keels and a series of enlarged teeth in the dentary bone. Our knowledge of paleostegalians is currently based on the genus *Crassigyrinus scotius*, which is known from early Carboniferous de-

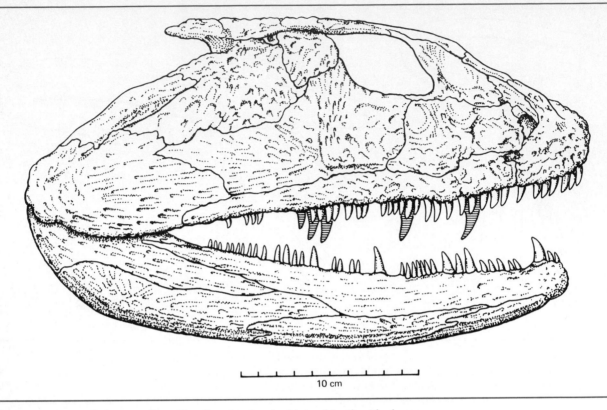

Figure 10–15. Paleostegalians like *Crassigyrinus* had lateral teeth with sharp anterior and posterior keels as well as large teeth in the palate. (From A. L. Panchen, 1985, *Philosophical Transactions of the Royal Society of London* B309:505–568.)

posits in Scotland (Figure 10–15). *Crassigyrinus* has several primitive characters that it shares with osteolepiforms and with *Ichthyostega*, including the absence of an occipital condyle, the presence of an anterior tectal bone, and the retention of a preopercular bone that was actually larger in *Crassigyrinus* than in *Ichthyostega* (Panchen 1985). Retention of these primitive characters in an animal that lived as late as the Carboniferous is unexpected.

Loxommatoids are known from the middle and late Carboniferous. The postcranial skeleton of loxommatoids is nearly unknown, but their skulls were either elongate and crocodile-like or broader and more alligator-like. Both kinds of loxommatoids had long, slender teeth that suggest a diet of fishes. Paleostegalians and loxommatoids were aquatic predators that probably lived in large bodies of water.

Anthracosauria and Diadectomorpha

The phylogenetic relationships of the anthracosaurs and diadectomorphs have not yet been deciphered. The past half century has seen many proposals of new groupings such as the seymouriamorphs, the batrachosaurs, the neostegalians, and the herpetospondyls that appeared to bring some order to these fossils, but none has enjoyed

Phylogenetic Relationships of the Anthracosauroidea Diadectomorpha

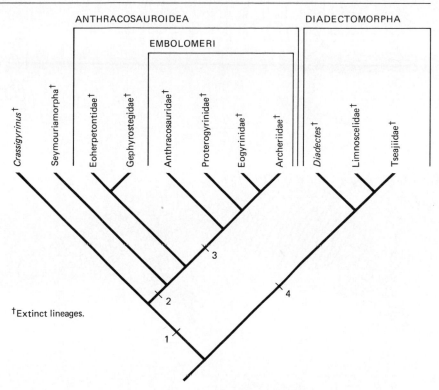

1. *Crassigyrinus,* Seymouriamorpha, and Anthracosauroidea: Characters of the skull and a distinctive pattern of pits and ridges on the dermal bones, lanceolate teeth, and a series of enlarged teeth in the dentary.   2. Seymouriamorpha and Anthracosauroidea: Suture between the tabular and parietal bones.   3. Embolomeri: Fenestrae on inner surface of lower jaw, 32 or more presacral vertebrae.   4. Diadectomorpha: Lateral shelf on ilium. (Based on T. R. Smithson, 1985, *Zoological Journal of the Linnean Society* 85:317–410; and A. L. Panchen, 1985, *Philosophical Transactions of the Royal Society of London,* Series B 309:505–568.)

Figure 10–16.   Phylogenetic relationships of the Anthracosauroidea and Diadectomorpha. This diagram shows the probable relationships among the major groups of anthracosaurs and diadectomorphs. Extinct lineages are marked by a dagger ([†]). The numbers indicate derived characters that distinguish the lineages. (From T. R. Smithson, 1985, *Zoological Journal of the Linnean Society* 85:317–410.)

lasting acceptance. At present the relationship suggested by Smithson (1985) is our best guess about the phylogeny of these animals (Figure 10–16).

A relationship has long been assumed to exist between the anthracosaurs and the amniotes, but no shared derived characters have yet been identified that support this grouping. The posterior

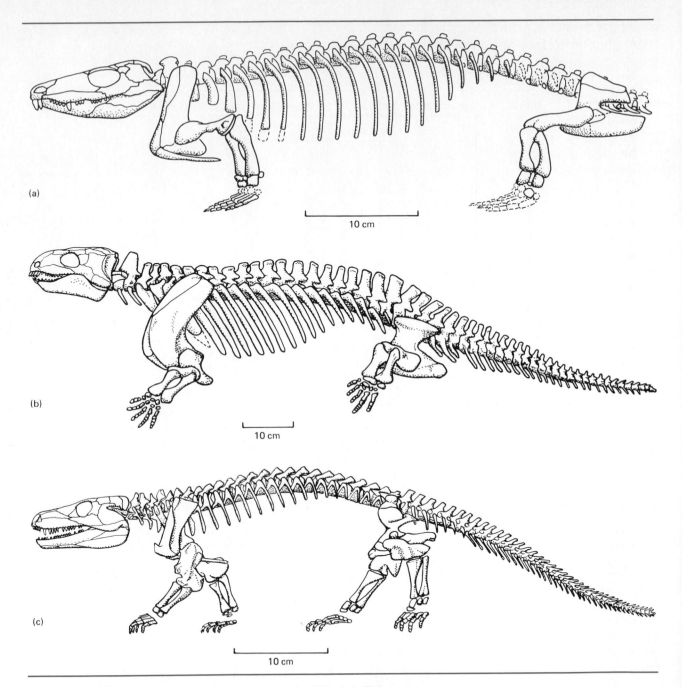

Figure 10–17. Terrestrial nonamniotic tetrapods of the late Paleozoic:
(a) *Tseajaia*; (b) *Diadectes*; (c) *Seymouria*. (From R. L. Carroll, 1969, in *Biology of the Reptilia*, volume 1, edited by C. Gans, A. d'A. Bellairs, and T. S. Parsons, Academic Press, London.)

part of the skull, especially the otic region and the stapes, has been used in arguments for and against a relationship between anthracosaurs and amniotes. The diadectomorphs further complicate the relationships among the late Carboniferous nonamniotes and the amniotes. The diadectomorphs form a monophyletic lineage that can be characterized by the presence of a prominent lateral shelf on the ilium. Three families are included in the Diadectomorpha—the Limnoscelidae, Tseajaiidae, and Diadectidae.

Diadectomorpha *Diadectes* was a stocky, heavy-bodied tetrapod of the early Permian about 2 meters in total length (Figure 10–17). It had well-ossified limbs and girdles, and a short, sturdy vertebral column. *Diadectes* was unusual in having laterally expanded cheek teeth with broad grinding surfaces and chisel-shaped front teeth. This specialized dentition suggests that *Diadectes* was an herbivore. If so, it was possibly the first terrestrial herbivorous tetrapod. Its contemporaries, the limnoscelids and tseajaiids, had sharp-pointed teeth and were apparently carnivorous. They were as large as 1.5 meters in length and appear to have been terrestrial or semiaquatic.

Seymouriamorpha The seymouriamorphs are known from the early Permian of North America. *Seymouria* (Figure 10–17), which takes its name from the town of Seymour, Texas, was about 1 meter long and appears to have been a fully terrestrial animal. The limbs and girdles were well developed and the vertebral column was relatively short. However, lateral-line canals have been described in the skull of an adult *Seymouria*, suggesting the existence of an aquatic larval stage. Other seymouriamorphs were aquatic even as adults. Discosauriscids, which are found in Permian deposits in central and eastern Europe and China, are known only from aquatic larval or paedomorphic forms, and the kotlassids from the late Permian and Triassic of the USSR were also aquatic.

Eoherpetontidae and Gephyrostegidae The gephyrostegids and eoherpetontids were relatively small terrestrial animals with short trunks and long limbs (Figure 10–18). They lacked a lateral-line system and had large piercing teeth, which suggest that they fed on arthropods and small vertebrates. The animals grouped as embolomeres include the families Proterogyrinidae, Anthracosauridae, Eogyrinidae, and Archeriidae. *Proterogyrinus*, from the middle Carboniferous of North America, appears to have been terrestrial; it had well-developed limbs and a relatively short trunk. In contrast, *Eogyrinus* and *Archeria* were elongate aquatic animals that may have been as much as 4 meters in length.

The Amniotic Egg

A major difference between modern amphibians and the remaining tetrapods is the occurrence of an **amniotic egg** in the latter group. The amniotic (or cleidoic) egg is sometimes referred to as the "land egg," but this is a misnomer. Many species of living amphibians and some fishes have anamniotic eggs (*an* = without) that develop quite successfully on land, and terrestrial invertebrates also lay anamniotic eggs. Even the differences in moisture requirements of anamniotic and amniotic eggs are not great. Both must have relatively moist conditions to avoid desiccation. Nonetheless, the amniotic egg is a derived character that distinguishes the two major groups of tetrapods—amniotes and nonamniotes.

The amniotic egg, as we know it, is characteristic of turtles, squamates, crocodilians, birds, monotremes, and in modified form, of therian mammals as well. It is assumed to have been the reproductive mode of Mesozoic diapsids, and fossilized dinosaur eggs are relatively common in some deposits. An amniotic egg is a remarkable example of biological engineering (Figure 10–19). The shell, which may be leathery or calcified, provides mechanical protection while allowing movement of respiratory gases and water vapor. The

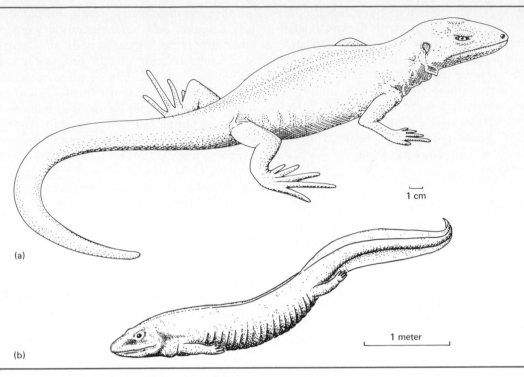

Figure 10–18. (a) *Gephyrostegus*, a late Carboniferous anthracosauroid, was terrestrial. It had sturdy legs supporting a deep body. (b) *Eogyrinus*, an embolomere of the same period, was aquatic. Its legs were small and its body was cylindrical. [(a) From R. L. Carroll, 1972, *Handbuch der Palaeoherpetologie*, part 5B, pages 1–19 Gustav Fischer Verlag, Stuttgart; (b) from A. R. Milner, 1980, *The Terrestrial Environment and the Origin of Land Vertebrates*, edited by A. L. Panchen, Academic Press, London.]

albumin (egg white) gives further protection against mechanical damage and provides a reservoir of water and protein. The large yolk is the energy supply for the developing embryo. At the beginning of embryonic development, the embryo is represented by a few cells resting on top of the yolk. As development proceeds these multiply, and endodermal tissue surrounds the yolk and encloses it in a yolk sac that is part of the developing gut. Blood vessels differentiate rapidly in the tissue of the yolk sac and transport food and gases to the embryo. By the end of development, only a small amount of yolk remains, and this is absorbed before or shortly after hatching.

In these respects the amniotic egg does not differ greatly from the anamniotic eggs of amphibians and fishes. The significant differences lie in three other extraembryonic membranes—the **chorion, amnion,** and **allantois**. The chorion and amnion develop from outgrowths of the body wall at the ends of the embryo. These two pouches spread outward and around the embryo until they meet. At their junction, the membranes merge and leave an outer membrane, the chorion, which surrounds the embryo and yolk sac, and an inner membrane, the amnion, which surrounds the embryo itself. The allantoic membrane develops as an outgrowth of the hind gut posterior to the yolk sac and lies

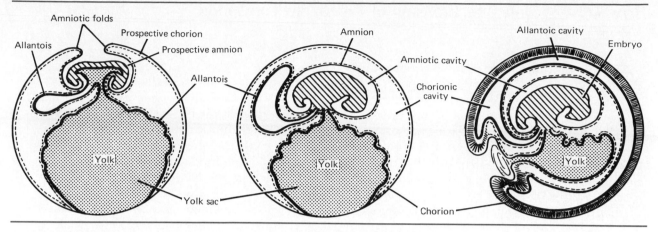

Figure 10–19. Amniotic egg. (From T. W. Torrey, 1962, *Morphogenesis of the Vertebrates*, Wiley, New York.)

within the chorion. It is a respiratory organ and a storage place for nitrogenous wastes produced by the metabolism of the embryo. The allantois is left behind in the egg when the embryo emerges, and the nitrogenous wastes stored in it do not have to be reprocessed.

Early Amniotes

In the Late Carboniferous there was an evolutionary event which, although it did not involve vertebrates, was to have a tremendous influence on the course of their evolution. This event was the radiation of insects in terrestrial habitats. Insects had appeared in the Devonian, and we have speculated about the role they might have played in the evolution of the earliest tetrapods. Through the Mississippian, however, the diversity of terrestrial insects appears to have remained limited, and the insects themselves may have been confined to areas near water. In the early Pennsylvanian the situation changed. The fossil record indicates that there was an abrupt expansion of many orders of insects, including dragonflies, stoneflies, and roaches. It is probable that the diversity of insects in the fossil record at this time reflects their spread into a variety of terrestrial habitats. The ra-

diation of terrestrial insects was probably a response to the increasing quantity and diversity of terrestrial vegetation in the Carboniferous.

It is difficult to escape the conclusion that each of these groups, which today are often so interdependent (for example, bees seeking a food source and angiosperms requiring pollinators) rapidly exploited this new symbiotic relationship to their mutual benefit. How did these changes affect vertebrates? One need only think of how dependent many modern birds, mammals, lizards, and amphibians are on insects to realize that their mutual evolution has profoundly influenced vertebrate evolution.

Most terrestrial vertebrates at that time were probably carnivorous. (No adult amphibian among living forms is herbivorous, and there is little evidence in the fossil record to suggest that Paleozoic tetrapods were herbivores. *Diadectes* is the only likely exception to this generalization.) Carnivorous vertebrates could not respond directly to the energy supply offered by terrestrial plants, but they could and apparently did respond to the opportunities presented by the radiation of insects. Probably for the first time in evolutionary history there was an adequate food supply to support fully terrestrial vertebrate predators.

Two forces that shaped the radiation of am-

Box 10–2. Jaw Mechanisms of Paleozoic Tetrapods

Osteolepiforms were predatory fish, capturing prey with a snap of the jaws. Speed of jaw closure is the critical element in this sort of predation; the jaws must close fast enough to trap prey before it can escape. In osteolepiforms the muscles that closed the jaws originated on the lateral surface of the braincase and inserted on the posterior half of the lower jaw (Figure 10–20a). The major component of the force of the jaw muscles of osteolepiforms would have been directed anteriorly. This arrangement means that the muscles have their maximum effect when the jaw is open. As the jaw swings closed it becomes more nearly parallel to the force vector of the muscles, and the muscles can exert little force in the direction parallel to their fibers. This type of jaw mechanism is called a **kinetic-inertial** system. The muscles exert a powerful force on the open jaw, which produces the quick snap, and prey is impaled on the teeth as a result of the momentum of the jaw as it swings closed. Vertebrates with a kinetic-inertial jaw mechanism have long, sharp teeth (often in the palate as well as in the jaws) that penetrate the prey on impact and hold it. They have little ability to

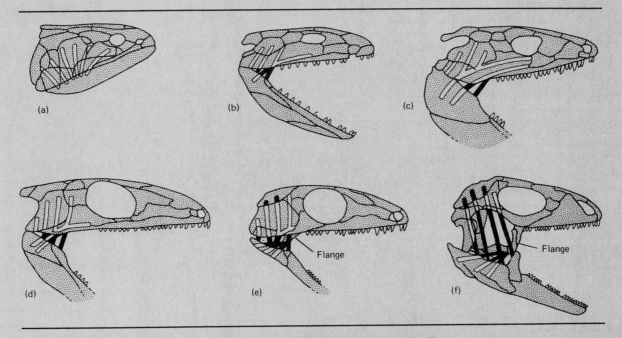

Figure 10–20. Evolution of the jaw muscles in tetrapods; (a) the osteolepiform *Ectosteorachis*; (b) *Ichthyostega*; (c) *Palaeogyrinus*; (d) *Gephyrostegus*; (e) *Paleothyris*; (f) The modern lizard *Iguana*. Light lines indicate muscles running behind a layer of bone, and dark lines show muscles that are exposed on the surface of the head. (From R. L. Carroll, 1969, *Biological Review* 44:393–432.)

crush prey, however, because the jaw muscles can exert relatively little force when the mouth is nearly closed. The skull kinesis of osteolepiforms might have compensated for this limitation of the jaw muscles.

Ichthyostega appears to have had a jaw morphology very like that of osteolepiforms and probably had a kinetic-inertial mechanism (Figure 10–20b). This mechanism is also seen in the large temnospondyls and many of the anthracosauroids that had large palatal fangs. However, some of the anthracosaurs showed a modification of the jaw morphology that was carried further in the earliest amniotes and represents a different mechanism of jaw closure, the **static pressure system**. In these animals (Figure 10–20c) the jaw-closing muscles differentiated into two distinct groups with different sites of origin. The temporal muscles continued to originate on the temporal region of the skull and to insert on the posterior por-

tion of the lower jaw, very much as they had in osteolepiforms. The function of these muscles, also, would have been similar to that of their counterparts in osteolepiforms—rapid closing of the jaws to trap prey. A second group of muscles originated from a newly developed flange on the pterygoid bones in the roof of the mouth and inserted on the lower jaw. When the mouth was closed the fibers in this pterygoideus muscle would have a large force component perpendicular to the long axis of the jaw, allowing a crushing force to be applied to objects held in the mouth. This crushing ability may have been a key factor in the ability of early amniotes to exploit the radiation of insects, which have hard exoskeletons. An insect that might bounce away from the impact of the large fangs of a kinetic-inertial jaw could be crushed and eaten by an animal with the static pressure system seen in the late anthracosaurs and early amniotes.

niotes were directly related to the exploitation of this new energy source. One was the evolution of a more effective jaw mechanism specifically adapted to feeding on insects (Box 10–2). The evolution of more effective jaws was accompanied by changes in body structure that permitted more effective locomotion on land.

Carboniferous and Permian Tetrapods

Several groups of vertebrates, both nonamniotes and amniotes, evolved specializations for terrestrial life in the Carboniferous. We have already mentioned the microsaurs and the large dissorophid temnospondyls like *Cacops*. In the anthracosauroid lineage the gephyrostegids and eoherpe-

tontids were relatively small terrestrial predators, and the limnoscelids and tseajaiids were large predators. *Diadectes* may have been the first terrestrial herbivore.

Amniotes began to radiate into many of those same life zones in the late Carboniferous and early Permian (Figure 10–21). Morphological specializations of early amniotes can be associated with different ways of life (Carroll 1982). The most primitive amniotes, captorhinids such as *Eocaptorhinus* (Figure 10–21d), had head-plus-body lengths of only 15 centimeters, but later captorhinids were larger, and the largest had skulls more than 40 centimeters long. Most captorhinids had more than one row of marginal teeth, and some of the later forms had as many as six rows. They appear to have been slow-moving predators that probably fed on arthropods. Protorothyrids and diapsids

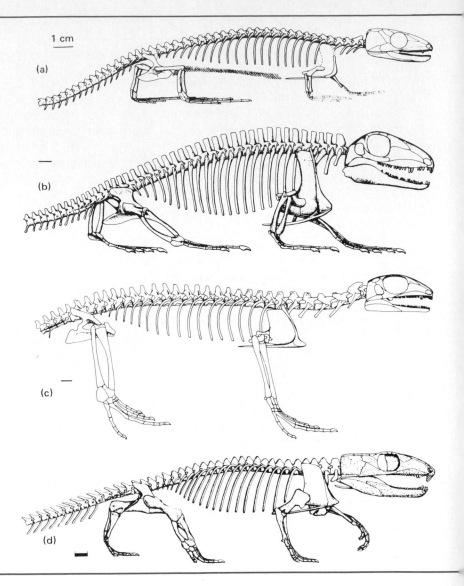

1 cm

(a)

(b)

(c)

(d)

Figure 10–21. Paleozoic amniotic tetrapods: (a) *Hylonomus lyelli*, a protorothyrid; (b) *Haptodus garnettensis*, a pelycosaur; (c) *Petrolacosaurus kansensis*, a diapsid; (d) *Eocaptorhinus laticeps*, a captorhinid; (e) *Mesosaurus brasiliensis*, a mesosaur; (f) *Milleretta rubidgei*, a millerosaur; (g) *Barasaurus besairiei*, a procolophonid; (h) *Pareiasaurus karpinskyi*, a pareiasaur. The scale mark is 1 centimeter. (From R. L. Carroll, 1982, *Annual Review of Ecology and Systematics*, 13:87–109.)

appear to form a natural group characterized by keels on the pleurocentra, slender limbs, and long slender feet (Heaton and Reisz 1986). Protorothyrids like *Hylonomus* (Figure 10–21a) were the same size as early captorhinids, but much more agile. Their heads were small in proportion to their bodies, and they may have fed on small fast-moving prey such as insects. Protorothyrids survived into the middle Permian with only a slight increase in size. The early diapsids (Figure 10–21c) were also about 15 centimeters in head–plus–body length, but they had longer legs, longer necks, and were more lightly built than the former groups. They had a large number of small teeth, like the protorothyrids, and may have fed on insects. Diapsids changed little during the Paleozoic, but in the Me-

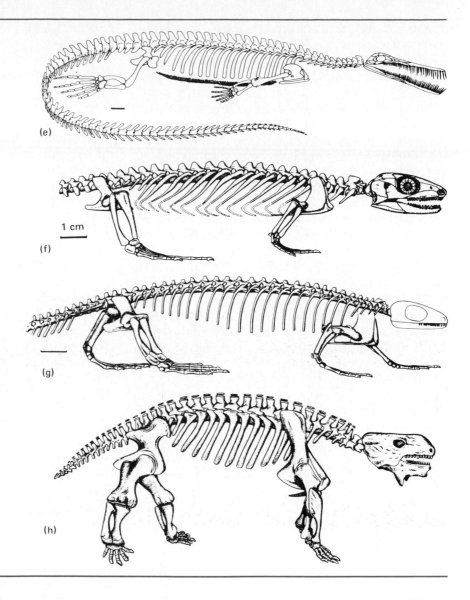

Figure 10–21 (Continued)

sozoic they radiated extensively (see Chapter 13).

The early pelycosaurs (Figure 10–21b) were about the same size as the protorothyrids, but they had substantially larger heads and fewer but larger teeth that may have been specialized for piercing and tearing large prey animals. The change in the angle of the rear of the skull from vertical to sloping may have given the jaw muscles a mechanical advantage in resisting the struggles of prey held in the jaws. Later pelycosaurs were much larger than the first representatives of the lineage (see Chapter 19).

The mesosaurs (Figure 10–21e) are one of the classic pieces of evidence for continental drift. Their fossils are known only from middle Permian deposits in South Africa and from beds of the same age in South America. In the Permian, these areas were adjacent parts of Gondwana. Mesosaurs are the first amniotes known to have developed specializations for aquatic life; the large hind feet were probably webbed and the long, laterally compressed tail probably was used for swimming. The ribs are densely ossified and heavy; they may have increased the specific gravity of the animal, helping it to dive. The long snout was filled with slender teeth, and it is hard to picture that dentition being effective in capturing fish. Current theories speculate that mesosaurs used their teeth to strain from the water the small crustaceans that are abundant in the same deposits in which mesosaurs are found.

Millerosaurs, which lived in the middle and late Permian, were very like the protorothyrids in body size and general form, although millerosaurs had longer legs (Figure 10–21f). Like the protorothyrids, millerosaurs had relatively small skulls and simple conical teeth that suggest a diet of insects. The procolophonids of the late Permian and Triassic (Figure 10–21g) initially had a large number of small, peg-like teeth. In later members of this group the number of teeth was reduced and each tooth was laterally expanded. The dentition appears to have been specialized for crushing or grinding, and procolophonids might have been herbivorous.

The pareiasaurs are known from the middle and late Permian of southern Africa, western Europe, the USSR, and China. In contrast to all the preceding groups, pariesaurs were large animals, approaching 3 meters in length (Figure 10–21h). Their limb girdles and limbs suggest that they moved with the fore-and-aft limb movement that is typical of massive mammals such as elephants. The teeth were laterally compressed and leaf-shaped, like the teeth of modern herbivorous lizards.

The morphological and ecological diversity of the amniotes that had developed by the late Paleozoic testifies to the success of this clade. More-or-less coincident with the diversification of terrestrial amniotes was a contraction in the variety of terrestrial nonamniotes. Terrestrial temnospondyls and anthracosauroids were at their peak in the Permian; the groups that survived through the Triassic were mostly aquatic forms such as the capitosaurs, trematosaurs, and metoposaurs. From the start of the Mesozoic onward, terrestrial habitats were dominated by a series of radiations of amniotic tetrapods.

Summary

The origin of tetrapods from osteolepiform fishes in the Devonian is inferred from similarities in the bones of the skull and braincase, vertebral structure, and limb skeletons. Lungfishes appear to be the sister group of the osteolepiforms + tetrapods. The fossil record of tetrapod origins has a gap of some 30 million years in the Devonian. This gap has obscured the sequence of changes in the evolutionary transition from osteolepiforms to tetrapods.

Paleozoic tetrapods can be divided into two clades on the basis of the structure of the skull. In one lineage, which includes the anthracosauroids and amniotes, the skull roof attaches to the posterior part of the braincase via the auditory capsules. In the temnospondyl lineage the connection is made by the exoccipital bones, which are part of the posterior end of the braincase. This lineage contains the earliest known tetrapods, the ichthyostegalians, as well as the temnospondyls and mi-crosaurs. Modern amphibians—the salamanders, anurans, and caecilians—may be derived from this lineage.

The anthracosauroid lineage included terrestrial and aquatic forms that radiated during the Carboniferous. Many of the anthracosauroids became extinct in the Permian, and only a few persisted into the Triassic. The temnospondyls were predominantly aquatic, and some were as large as modern crocodilians. Temnospondyls radiated extensively in the late Carboniferous and Permian, and several lineages extended through the Triassic into the middle Jurassic.

The amniotic egg, with its distinctive extra-embryonic membranes, is a derived character that distinguishes the nonamniotes (fishes and amphibians) from the amniotes (turtles, squamates, crocodilians, birds, and mammals). Amniotes are probably derived from the anthracosauroid lineage, although the phylogenetic relationships among the anthracosaurs, diadectomorphs, and amniotes are not yet fully understood. The earliest amniotes were small animals, and their appearance coincided with a major radiation of terrestrial insects in the Carboniferous. Progressive modifications of post-cranial skeletons of early amniotes appear to show increased agility, and simultaneous changes in the jaws may be related to predation on insects. By the end of the Carboniferous, amniotes had begun to radiate into most of the terrestrial life zones that had been occupied by nonamniotes, and only the relatively aquatic groups of nonamniotic tetrapods maintained much diversity through the Triassic.

References

Bray, A. A. 1985. The evolution of terrestrial vertebrates: environmental and physiological considerations. *Philosophical Transactions of the Royal Society of London* B309:289–322.

Campbell, K. S. W. and M. W. Bell. 1977. A primitive amphibian from the late Devonian of New South Wales. *Alcheringa* 1:369–381.

Carroll, R. L. 1982. Early evolution of reptiles. *Annual Review of Ecology and Systematics* 13:87–109.

Carroll, R. L. 1987. *Vertebrate Paleontology and Evolution*. W. H. Freeman, New York.

de Queiroz, K. and D. C. Cannatella. 1987. The monophyly and relationships of the Lissamphibia. *American Zoologist* 27:60A.

Fricke, H., O. Reinicke, H. Hofer, and W. Nachtigall. 1987. Locomotion of the coelacanth *Latimeria chalumnae* in its natural environment. *Nature* 329:331–333.

Gosline, W. A. 1977. The structure and function of the dermal pectoral girdle in bony fishes with particular reference to ostariophysines. *Journal of Zoology (London)* 183:329–338.

Heaton, M. J. and R. R. Reisz. 1986. Phylogenetic relationships of captorhinomorph reptiles. *Canadian Journal of Earth Science* 23:402–418.

Holmes, E. B. 1985. Are lungfishes the sister group of tetrapods? *Biological Journal of the Linnean Society* 25:379–397.

Jarvik, E. 1980. *Basic Structure and Evolution of Vertebrates*. Academic Press, London.

Jarvik, E. 1981. (Review of) Lungfishes, tetrapods, paleontology, and plesiomorphy. *Systematic Zoology* 30:378–384.

Milner, A. R. 1982. Small temnospondyl amphibians from the middle Pennsylvanian of Illinois. *Palaeontology* 25:635–664.

Panchen, A. L. 1977. Geographical and ecological distribution of the earliest vertebrates. Pages 723–738 in *Major Patterns in Vertebrate Evolution*, edited by M. K. Hecht, P. C. Goody, and B. M. Hecht. NATO Advanced Study Series. Plenum Press, New York.

Panchen, A. L. 1980. The origin and relationships of the anthracosaur amphibia from the late Paleozoic. Pages 319–350 in *The Terrestrial Environment and the Origin of Land Vertebrates*, edited by A. L. Panchen. Academic Press, London.

Panchen, A. L. 1985. On the amphibian *Crassigyrinus scotius* Watson from the Carboniferous of Scotland. *Philosophical Transactions of the Royal Society of London* B309:505–568.

Romer, A. S. 1958. Tetrapod limbs and early tetrapod life. *Evolution* 12:365–369.

Romer, A. S. 1966. *Vertebrate Paleontology*. 3rd edition. University of Chicago Press, Chicago.

Rosen, D. E., P. L. Forey, B. G. Gardiner, and C. Patterson. 1981. Lungfishes, tetrapods, paleontology, and plesiomorphy. *Bulletin of the American Museum of Natural History* 167:159–276.

Schaeffer, B. 1965. The rhipidistian–amphibian transition. *American Zoologist* 5:267–276.

Smithson, T. R. 1982. The cranial morphology of *Greererpeton burkemorani* Romer (Amphibia: Temnospondyli). *Zoological Journal of the Linnean Society* 76:29–90.

Smithson, T. R. 1985. The morphology and relationships of the Carboniferous amphibian *Eoherpeton watsoni* Panchen. *Zoological Journal of the Linnean Society* 85:317–410.

Szarski, H. 1962. The origin of the amphibia. *Quarterly Review of Biology* 37:189–241.

Thomson, K. S. 1969. The biology of the lobe-finned fishes. *Biological Review* 44:91–154.

Thomson, K. S. 1980. The ecology of Devonian lobe-finned fishes. Pages 187–222 in *The Terrestrial Environment and the Origin of Land Vertebrates*, edited by A. L. Panchen. Academic Press, London.

Trueb, L. and R. Cloutier. 1987. Historical constraints on lissamphibian osteology. *American Zoologist* 27:33A.

Warren, J. W. and N. A. Wakefield. 1972. Trackways of tetrapod vertebrates from the Upper Devonian of Victoria, Australia. *Nature (London)* 238:469–470.

The amphibians may be derived from two different lineages of the nonamniotic Paleozoic vertebrates discussed in Chapter 10, but certain shared characters of amphibians—especially a moist, permeable skin—have channeled their evolution in similar directions. Frogs are the most successful amphibians, and it is tempting to think that the variety of locomotor modes permitted by their specialized morphology may be related to their success: frogs can jump, walk, climb, and swim. In contrast to frogs, salamanders retain the primitive tetrapod locomotor pattern of lateral undulations combined with limb movements. The greatest diversity among salamanders is found in the family Plethodontidae, which is characterized by a specialized feeding mechanism that involves projecting the tongue to capture prey on its sticky tip.

The range of reproductive specializations of amphibians is nearly as great as that of fishes, a remarkable fact when one remembers that there are more than five times as many species of fishes as amphibians. The ancestral reproductive mode of amphibians probably consisted of laying large numbers of eggs that hatched into aquatic larvae, and many amphibians still reproduce this way. An aquatic larva gives a terrestrial species access to resources that would not otherwise be available to it. Modifications of the ancestral reproductive mode include bypassing the larval stage, viviparity, and parental care of eggs and young, including females that feed their tadpoles.

11

Salamanders, Anurans, and Caecilians

Amphibians

Salamanders, anurans, and caecilians are usually grouped under the common name "amphibians," although it is not clear that the three groups share a common origin (Box 11–1). The relationships of modern amphibians have long been an enigma because there are no fossils that demonstrate a transition between some Paleozoic group and any of the three modern orders. The oldest fossils that may represent modern amphibians are isolated vertebrae of Permian age that appear to include both salamander and anuran types. The earliest relatively complete amphibian fossil may be *Triadobatrachus*, a possible anuran from the early Triassic (Figure 11–1). The skull is frog-like—broad, with the orbits enlarged into the cheek and temporal regions. The axial skeleton is less frog-like than the skull, but the vertebral column is short and the ilium extends forward. The posterior vertebrae do not form a urostyle, however, and the anterior end of the ilium is not fused to a sacral rib (compare to Figure 11–5). The absence of a long tail suggests that *Triadobatrachus* relied on limb movements for swimming and was probably able to jump. Fossils later than *Triadobatrachus* come from the Jurassic and Cretaceous and are as specialized as modern forms. In most cases they can be tentatively assigned to modern families.

All living adult amphibians are carnivorous, and relatively little morphological specialization is associated with different dietary habits within each order. Amphibians eat almost anything they are able to catch, kill, and swallow. In aquatic forms the tongue is broad, flat, and relatively immobile, but some terrestrial amphibians can protrude the tongue from the mouth to capture prey. The size of the head is an important determinant of the maximum size of prey that can be taken, and sympatric species of salamanders frequently have markedly different head sizes, suggesting that this is a feature that reduces competition. Frogs in the tropical American genus *Ceratophrys*, which feed largely on other frogs, have such large heads that they are practically walking mouths.

The anuran body form probably evolved from a more salamander-like starting point. Both jumping and swimming have been suggested as the mode of locomotion that made the change advantageous. Salamanders and caecilians swim as fish do by passing a sine wave down the body. Anurans have inflexible bodies and swim with simultaneous thrusts of the hind legs. Some paleontologists have proposed that the anuran body form evolved because of the advantages of that mode of swimming. An alternative hypothesis traces the anuran body form to selection for an amphibian that could rest near the edge of a body of water and escape aquatic or terrestrial predators with a rapid leap followed by locomotion on either land or water.

The earliest fossils of all the orders of modern

Box 11–1. The Subclass Lissamphibia?

The three orders of living amphibians are often grouped in the subclass Lissamphibia (*liss* = smooth). That taxonomic arrangement implies that the salamanders, frogs, and caecilians have a common ancestor and that they are more closely related to each other than to any other group. What is the evidence for that arrangement?

Parsons and Williams (1963) pointed out that most of the differences among the three orders of amphibians are related to the different modes of locomotion of salamanders, anurans, and caecilians and thus represent specializations of the three lineages. Those differences are striking, and they may have obscured less obvious similarities among the groups. The similarities are not related to functional characteristics of the animals. Recent reviews of the question have focused on several derived characters that appear to be common to the three orders of amphibians, although not all the characters are unique to amphibians (Gardiner 1982, 1983; Duellman and Trueb 1986).

1. *Pedicellate teeth.* Nearly all modern amphibians have teeth in which the crown and base (pedicel) are composed of dentine and are separated by a narrow zone of uncalcified dentine or fibrous connective tissue. A few amphibians lack pedicellate teeth, including salamanders of the genus *Siren* and frogs of the genera *Phyllobates* and *Ceratophrys*, and the boundary between the crown and base is obscured in some other genera. Pedicellate teeth also occur in some actinopterygian fishes, which are not thought to be related to amphibians.

2. *Operculum–plectrum complex.* Most amphibians have two bones that are involved in transmitting sounds to the inner ear. The columella (plectrum) is derived from the hyoid arch and is present in salamanders and caecilians and in most frogs. The operculum develops in association with the fenestra ovalis of the inner ear. The columella and operculum are fused in anurans and caecilians and in some salamanders.

3. *Papilla amphibiorum.* All amphibians have a special sensory area, the papilla amphibiorum, in the wall of the sacculus of the inner ear. The papilla amphibiorum is sensitive to frequencies below 1000 hertz

---

amphibians are very like living forms (Carroll 1987). The oldest frog, *Vieraella*, from the early Jurassic of Argentina shows affinities to two modern families, the Ascaphidae and Discoglossidae. Some salamanders are known from the late Jurassic and caecilians from the Paleocene. Clearly, the modern orders of amphibians have had separate evolutionary histories for a long time. The continued presence of such common characteristics as a permeable skin in the three orders after at least 250 million years suggests that the shared characteristics are critical in shaping the evolutionary success of modern amphibians. In other characters, such as reproduction, locomotion, and defense, they show tremendous diversity.

Salamanders

The salamanders (order Urodela or Caudata) have the most generalized body form and locomotion of the living amphibians. Salamanders are elongate, and all but a very few species of completely aquatic salamanders have four functional limbs (Figure 11–2). Their walking gait is probably similar to that

(cycles per second), and a second sensory area, the papilla basilaris, detects sound frequencies above 1000 hertz.

4. *Structure of the skin and the importance of cutaneous gas exchange.* All amphibians have mucous glands that keep the skin moist. A substantial part of an amphibian's exchange of oxygen and carbon dioxide with the environment takes place through the skin. All amphibians also have poison (granular) glands in the skin.

5. *Green rods.* Salamanders and frogs have a distinct type of retinal cell, the green rods. Caecilians apparently lack green rods, but the eyes of caecilians are extremely reduced and the green rods may have been lost.

6. *Structure of the levator bulbi muscle.* This muscle is a thin sheet in the floor of the orbit that is innervated by the fifth cranial nerve. It causes the eyes to bulge outward, thereby enlarging the buccal cavity. This muscle is present in salamanders and anurans and in modified form in caecilians.

7. *Fat bodies.* All three orders of amphibians have fat bodies that develop from the ger-

minal ridge and are located near the gonads.

8. *Skull roof.* The postfrontal bone is absent, and the parietal bones cover the roof of the temporal region of the skull.

Not all workers agree that these similarities necessarily indicate a common ancestry, and other structures that might reasonably be expected to reveal a common origin are not similar in the three orders of amphibians. For example, David Wake studied the vertebral column and concluded that there was no evidence for a common origin of the three orders (Wake 1970). Substantial differences between a salamander *(Triturus)* and a frog *(Bombina)* in the embryonic origin of endoderm and the primordial germ cells have been claimed (Nieuwkoop and Satasurya 1976). Robert Carroll has pointed out a number of ways in which salamanders, anurans, and caecilians differ from each other (Carroll and Currie 1975, Carroll and Holmes 1980, Carroll 1987). Recently two independent studies have suggested that lissamphibians are the monophyletic sister group of the branchiosaur temnospondyls (de Queiroz and Cannatella 1987, Trueb and Cloutier 1987).

employed by the earliest tetrapods. It utilizes the lateral bending characteristic of fish locomotion in concert with leg movement. The approximately 350 species of salamanders are almost entirely limited to the northern hemisphere; their southernmost occurrence is in northern South America. North and Central America have the greatest diversity of salamanders—more species of salamanders are found in Tennessee than in all of Europe and Asia combined. Paedomorphosis is widespread among salamanders and several families of aquatic salamanders are constituted solely of such

paedomorphic forms (Table 11–1). These can be recognized by the retention of larval characteristics, including larval tooth and bone patterns, the absence of eyelids, retention of a functional lateral-line system, and (in some cases) retention of external gills.

The largest living salamanders are the Japanese and Chinese giant salamanders, which reach lengths of 1 meter or more. The related North American hellbenders grow to 60 centimeters. All are members of the Family Cryptobranchidae and are paedomorphic and permanently aquatic. As

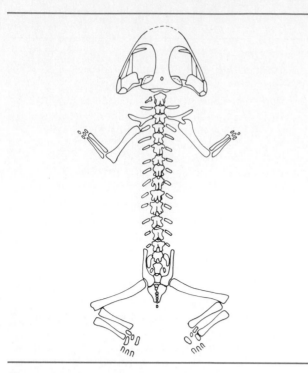

Figure 11–1. *Triadobatrachus*, a fossil from early Triassic sediments in Madagascar, shows some anuran-like characters. (From R. Estes and O. A. Reig, 1973, in *Evolutionary Biology of Anurans,* edited by J. L. Vial, University of Missouri Press, Columbia, Mo.)

their name indicates (*crypto* = hidden, *branchus* = gill), they do not retain external gills, although they do have other larval characteristics. Another large aquatic salamander, the mudpuppy (*Necturus*), is paedomorphic and does retain external gills. Mudpuppies occur in lakes and streams in eastern North America.

Several evolutionary lineages of salamanders have become adapted to life in caves (**troglodyty**). The constant temperature and moisture of caves makes them good salamander habitats, and food is supplied by cave-dwelling invertebrates. The brook salamanders (*Eurycea*, family Plethodontidae) include a number of species that form a continuum from those with fully metamorphosed adults inhabiting the twilight zone near cave

mouths to fully paedomorphic forms in the depths of caves of sinkholes. The Texas blind salamander, *Typhlomolge*, is a highly specialized troglodyte—blind, white, with external gills, extremely long legs, and a flattened snout used to probe underneath pebbles for food. The unrelated European olm (*Proteus*, family Proteidae) is another troglodyte that has converged on the same body form.

Terrestrial salamanders like the North American mole salamanders (*Ambystoma*) and the European salamanders (*Salamandra*) have aquatic larvae that lose their gills at metamorphosis. The most fully terrestrial salamanders, the lungless plethodontids, include species in which the young hatch from eggs as miniatures of the adult and there is no aquatic larval stage.

Feeding Specializations of Plethodontid Salamanders Lungs seem an unlikely organ for a terrestrial vertebrate to abandon, but among salamanders the evolutionary loss of lungs has been a successful tactic. The family Plethodontidae is characterized by the absence of lungs and contains more species with a wider geographic distribution than any other family of salamanders. Furthermore, many plethodontids have evolved specializations of the hyobranchial apparatus that allow them to protrude the tongue a considerable distance from the mouth to capture prey. This ability has not evolved in salamanders with lungs, probably because the hyobranchial apparatus in these forms is an essential part of the system that ventilates the lungs.

The modifications of the respiratory system and hyobranchial apparatus that allow tongue protrusion appear to be linked with a number of other characteristics of the biology of plethodontids (Roth and Wake 1985). These associations can be seen most clearly in the bolitoglossine plethodontids, which have the most specialized tongue-projection mechanisms (Figure 11–3). Bolitoglossine plethodontids (*bolis* = dart, *glossa* = tongue) can project the tongue a distance equivalent to their head plus trunk length and can pick off moving prey. This ability requires fine visual discrimination of distance and direction, and the eyes of bol-

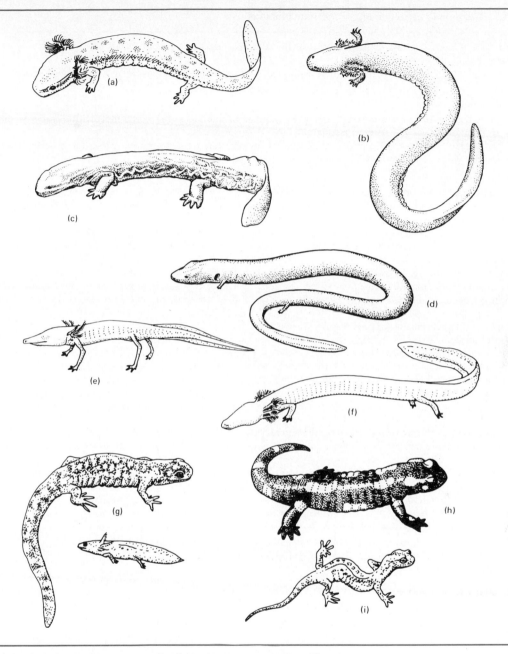

Figure 11–2. Salamander body forms reflect differences in life histories and habitats. Aquatic salamanders: (a) mudpuppy (*Necturus*); (b) siren (*Siren*); (c) hellbender *(Cryptobranchus)*; (d) congo eel (*Amphiuma*). Specialized cave-dwellers: (e) Texas blind salamander (*Typhlomolge*); (f) olm (*Proteus*). Terrestrial salamanders: (g) tiger salamander (*Ambystoma*) and its aquatic larval form; (h) fire salamander (*Salamandra*); (i) slimy salamander (*Plethodon*).

Table 11–1. Taxonomic arrangement of living amphibians.

Order Urodela (or Caudata): the salamanders (approximately 350 species)
  Suborder Sirenoidea
    Family Sirenidae: elongate aquatic salamanders with gills and lacking the pelvic girdle and hindlimbs (3 species in North America)

  Suborder Cryptobranchoidea
    Family Hynobiidae: small to medium-size terrestrial or aquatic salamanders (100 to 200 mm) with external fertilization of the eggs and aquatic larvae (32 species in Asia)
    Cryptobranchidae: very large (750 mm) to enormous (1520 mm) aquatic salamanders with external fertilization of the eggs (1 species in North America and 2 in Asia)

  Suborder Salamandroidea
    Family Proteidae: paedomorphic aquatic salamanders with external gills (6 species in North America and Europe)
    Dicamptodontidae: small to large (100 to 350 mm) semiaquatic salamanders with aquatic larvae (3 species in North America)
    Amphiumidae: very large (1000 mm) elongate aquatic salamanders lacking gills (3 species in North America)
    Salamandridae: small to medium-size terrestrial and aquatic salamanders (49 species in Europe and Asia, 4 in North America)
    Ambystomatidae: small to large terrestrial salamanders with aquatic larvae (30 species in North America)
    Plethodontidae: very small (27 mm) to medium-size aquatic or terrestrial salamanders, some with aquatic larvae, others with direct development (219 species in North, Central, and South America, plus 1 species in Europe)

Order Gymnophiona: the caecilians (approximately 170 species)
    Family Ichthyophiidae: moderately large (to 500 mm) terrestrial caecilians with aquatic larvae (41 species in southeast Asia)
    Rhinatrematidae: small (to 300 mm) terrestrial caecilians, believed to have aquatic larvae (9 species in South America)
    Scolecomorphidae: moderately large terrestrial caecilians, possibly viviparous (7 species in Africa)
    Caeciliidae: very small (100 mm) to very large (1.5 m) terrestrial caecilians with both oviparous and viviparous species, no aquatic larval stage (91 species in Central and South America, Africa, India, and the Seychelles Islands)
    Typhlonectidae: large (to 750 mm) aquatic caecilians with viviparous reproduction (19 species in South America)

Order Anura: the frogs, toads, treefrogs, etc. (approximately 3500 species)
    Family Leiopelmatidae: small (50 mm) semiaquatic frogs with internal fertilization (3 species in New Zealand, 1 in North America)
    Discoglossidae: small to medium-size (50 to 85 mm) frogs with aquatic larvae (14 species in Europe and Asia)
    Rhinophrynidae: a burrowing frog with aquatic larvae (1 species in Central America)
    Pipidae: specialized aquatic frogs; *Xenopus, Hymenochirus,* and some species of *Pipa* have aquatic larvae, other species of *Pipa* have eggs that develop into juvenile frogs (26 species in South America and Africa)

Source: Based on W. E. Duellman and L. Trueb, 1986, *Biology of Amphibia*, McGraw-Hill, New York; M. H. Wake, 1986, *Mémoires de la Societé de Zoologie Français*, number 43, pp. 21–38.

itoglossines are placed more frontally on the head than the eyes of less specialized plethodontids. Furthermore, the eyes of bolitoglossines have a large number of nerves that travel to the ipsilateral (= same side) visual centers of the brain as well as the strong contralateral (= opposite side) visual projection that is typical of salamanders. As a result of this neuroanatomy, bolitoglossines have a complete dual projection of the binocular visual fields to both hemispheres of the brain and can make very exact and rapid estimation of the distance of a prey object from the salamander.

Tongue projection is reflected in diverse aspects of the life history characteristics of plethodontid salamanders. Aquatic larval salamanders employ suction feeding, opening the mouth and expanding the throat to create a current of water that carries the prey item with it. The hyobranchial apparatus is an essential part of this feeding mechanism, and some elements of the hyobranchial ap-

Table 11—1 (Continued)

Pelobatidae: short-legged terrestrial frogs with aquatic larvae (83 species in North America, Asia, Europe, and northern Africa)

Pelodytidae: small (50 mm) terrestrial frogs with aquatic larvae (2 species in Europe and Asia)

Myobatrachidae: very small (20 mm) to large (115 mm) aquatic and terrestrial frogs with diverse reproductive modes (99 species in Australia, Tasmania, and New Guinea)

Heleophrynidae: medium-size frogs that live in mountain streams and have aquatic larvae (3 species in extreme southern Africa)

Sooglossidae: small terrestrial frogs that lay eggs on land; the eggs hatch into juvenile frogs or into nonfeeding tadpoles that are carried on the back of an adult (3 species in the Seychelles Islands)

Leptodactylidae: very small (12 mm) to enormous (250 mm) frogs from all habitats and with diverse modes of reproduction (710 species in southern North America, Central and South America, and the West Indies)

Bufonidae: very small (20 mm) to enormous (250 mm) mostly terrestrial toads; most have aquatic larvae but some species of *Nectophrynoides* are viviparous (335 species in North, Central, and South America, Africa, Europe, and Asia)

Brachycephalidae: very small terrestrial frogs that probably have direct development (2 species in southeastern Brazil)

Rhinodermatidae: small terrestrial frogs that lay eggs on land; the tadpoles of *Rhinoderma rufum* are transported to water, whereas those of *R. darwini* complete development in the vocal sacs of the male (2 species in southern Chile and Argentina)

Pseudidae: aquatic frogs with enormous tadpoles (250 mm for *Pseudis paradoxa*) that metamorphose into medium-size (to 70 mm) adults (4 species in South America)

Hylidae: most hylids are arboreal, but a few species are aquatic or terrestrial; most species have aquatic larvae, but the marsupial frogs show several variations, including direct development of juvenile frogs (630 species in North, Central, and South America, Europe, Asia, and Australia)

Centrolenidae: mostly small arboreal frogs with aquatic larvae that live in streams (65 species in Central and South America)

Dendrobatidae: small terrestrial frogs, many of which are brightly colored and extremely toxic; terrestrial eggs hatch into tadpoles that are transported to water by an adult (117 species in Central and South America)

Ranidae: medium-size (50 mm) to enormous (300 mm) aquatic or terrestrial frogs; most have aquatic tadpoles but several genera show direct development (667 species in North, Central, and South America, Europe, Asia, and Africa)

Hyperoliidae: small to medium-size mostly arboreal frogs with aquatic larvae (206 species in Africa, Madagascar, and the Seychelles Islands)

Rhacophoridae: very small to large mostly arboreal frogs; some species have filter-feeding aquatic larvae, whereas others lay eggs in holes in trees and have larvae that do not feed (186 species in Africa, Madagascar, and Asia)

Microhylidae: small to medium-size terrestrial or arboreal frogs; many have aquatic larvae, but some species have nonfeeding tadpoles and others have direct development (279 species in North, Central, and South America, Asia, Africa, and Madagascar)

---

paratus become well developed during the larval period. These larval specializations of the hyobranchial apparatus are quite different from those associated with the tongue-projection mechanism of adult plethodontids. Furthermore, larval salamanders have laterally placed eyes and the optic nerves project mostly to the contralateral side of the brain. Thus, the morphological specializations of aquatic plethodontid larvae that make them successful in that ecological role are different from the specializations of adults, and this situation may create a conflict of selective forces.

The bolitoglossines do not have aquatic larvae, and the morphological specializations of adult bolitoglossines appear during embryonic development. In contrast, the hemidactyline plethodontids have aquatic larvae and have considerable ability to project the tongues, but even as adults hemidactylines retain the large first ceratobranchial that appears in the larvae. This is a mechan-

Figure 11–3. Prey capture by a bolitoglossine salamander, *Hydromantes.* (Courtesy of G. Roth.)

ically less efficient arrangement than the large second ceratobranchial of bolitoglossines, and the ability of hemidactylines to project their tongues is correspondingly less than that of bolitoglossines. Thus, the development of a specialized feeding mechanism by plethodontid salamanders has had ramifications on such diverse aspects of their biology as respiratory physiology and life history, and demonstrates that organisms evolve as whole functioning units, not as collections of independent characters.

Social Behavior of Plethodontid Salamanders Plethodontid salamanders can be recognized externally by the paired nasolabial grooves that extend ventrally from the external naris to the lip of the upper jaw (Figure 11–4). These grooves are an important part of the chemosensory system of plethodontids. As a plethodontid salamander moves about, it repeatedly presses its snout against the substrate. Fluid is drawn into the grooves and moves upward to the external nares, into the nasal chambers, and over the chemoreceptors of the vomeronasal organ. Many plethodontid salamanders are territorial and use scent to mark their territories. Male salamanders of some species are aggressive toward conspecific males. Robert Jaeger has studied the territorial behavior of the red-backed salamander, *Plethodon cinereus*, a common species in woodlands of eastern North America. These small salamanders can be kept in cages in the laboratory and fed fruit flies.

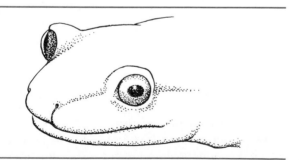

Figure 11–4. Nasolabial grooves of a plethodontid salamander.

Male red-backed salamanders establish territories in the cages. A resident male salamander marks the substrate of its cage with pheromones (chemical substances that are released by an individual and stimulate responses by other individuals of the species). A salamander can distinguish between substrates it has marked and those marked by another male salamander or by a female salamander. Male salamanders can also distinguish between the familiar scent of a neighboring male salamander and the scent of a male they have not previously encountered, and they react differently to those scents.

In laboratory experiments, red-backed salamanders select their prey in a way that maximizes their energy intake: When equal numbers of large and small fruit flies are released in the cages, the salamanders first capture the large flies. This is the most profitable foraging behavior for the salamanders because it provides the maximum energy intake per capture. In a series of experiments, Jaeger and his colleagues showed that territorial behavior and fighting can interfere with the ability of salamanders to select the most profitable prey (Jaeger et al. 1983). These experiments used "surrogate salamanders" that were made of a roll of moist filter paper the same length and diameter as a salamander. The surrogates were placed in the cages of resident salamanders to produce three experimental conditions: a control surrogate, a familiar surrogate, and an unfamiliar surrogate. In the control experiment, male red-backed salamanders were exposed to a surrogate that was only moistened filter paper; it did not carry any salamander pheromone. For both of the other groups the surrogate was rolled across the substrate of the cage of a different male salamander to absorb the scent of that salamander before it was placed in the cage of a resident male.

The experiments lasted 7 days; the first 6 days were conditioning periods and the test itself occurred on the seventh day. For the first 6 days, the resident salamanders in both of the experimental groups were given surrogates bearing the scent of another male salamander. The residents thus had the opportunity to become familiar with the scent of that male. On the seventh day, however, the familiar and unfamiliar surrogate groups were treated differently. The familiar surrogate group once again received a surrogate salamander bearing the scent of the same individual it had been exposed to for the previous 6 days, whereas the resident salamanders in the unfamiliar surrogate group received a surrogate bearing the scent of a different salamander, one to which they had never been exposed before. After a 5-minute pause, a mixture of large and small fruit flies was placed in each cage, and the behavior of the resident salamander was recorded.

The salamanders in the familiar surrogate group showed little response to the now-familiar scent of the other male salamander. They fed as usual, capturing large fruit flies. In contrast, the salamanders that were exposed to the scent of an unfamiliar surrogate began to give threatening and submissive displays, and their rate of prey capture decreased as a result of the time they spent displaying. In addition, salamanders exposed to unfamiliar surrogates did not concentrate on catching large fruit flies, so the average energy intake per capture also declined. The combined effects of the reduced time spent feeding and the failure to concentrate on the most profitable prey items caused an overall 50 percent decrease in the rate of energy intake for the salamanders exposed to the scent of an unfamiliar male.

The ability of male salamanders to recognize the scent of another male after a week of habituation in the laboratory cages suggests that they would show the same behavior in the woods. That is, a male salamander could learn to recognize and ignore the scent of a male in the adjacent territory, while still being able to recognize and attack a strange intruder. Learning not to respond to the presence of a neighbor might allow a salamander to forage more effectively, and it may also help to avoid injuries that can occur during territorial encounters. Resident male red-backed salamanders challenge strange intruders, and the encounters involve aggressive and submissive displays and biting. Most bites are directed at the snout of an opponent, and may damage the nasolabial grooves. Salamanders with damaged nasolabial grooves apparently have difficulty perceiving olfactory stimuli. Twelve salamanders that had been bitten on the snout were able to capture an average of only 5.8 fruit flies in a 2-hour period compared to an average of 18.6 flies for 12 salamanders that had not been bitten. In a sample of 144 red-backed salamanders from the Shenandoah National Forest, 11.8 percent had been bitten on the nasolabial grooves, and these animals weighed less than the unbitten animals, presumably because their foraging success had been reduced (Jaeger 1981).

The possibility of serious damage to an important sensory system during territorial defense provides an additional advantage for a red-backed salamander of being able to distinguish neighbors (which are always there and are not worth attacking) from intruders (which represent a threat and should be attacked). The phenomenon of being able to recognize territorial neighbors has been called "dear enemy" recognition, and may be generally advantageous because it minimizes the time and energy that territorial individuals expend on territorial defense and also minimizes the risk of injury during territorial encounters. Similar "dear enemy" recognition has been described among territorial birds that show more aggressive behavior upon hearing the songs of strangers than they do when hearing the songs of neighbors.

Anurans

In contrast to the limited number of species of salamanders and their restricted geographic distribution, the anurans (*an* = without, *uro* = tail) include nearly 3500 species and occur on all of the continents except Antarctica. Specialization of the body for jumping is the most conspicuous mor-

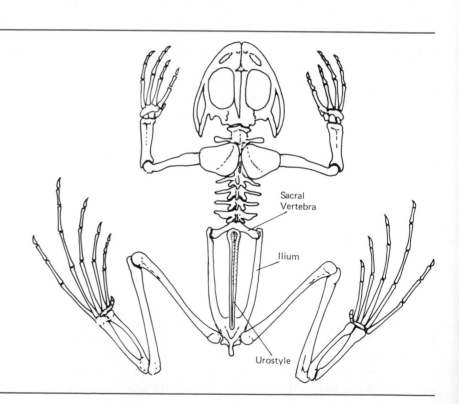

Figure 11–5. Anuran skeleton showing numerous specializations for saltatory locomotion.

phological feature of anurans: The hind legs are elongate and the tibia and fibula are fused. The hind limbs and muscles form a lever system that can catapult an anuran into the air (Figure 11–5). Numerous morphological changes are associated with this type of locomotion: A powerful pelvis strongly fastened to the vertebral column is clearly necessary, as is a stiffening of the vertebral column. The ilium is elongate and reaches far anteriorly, and the posterior vertebrate are fused into a solid rod, the **urostyle**. The pelvis and urostyle render the posterior half of the trunk rigid. The vertebral column is short, with only five to nine presacral vertebrae, and these are strongly braced by zygapophyses that restrict lateral bending. The strong forelimbs and flexible pectoral girdle absorb the impact of landing. The eyes are large and are placed well forward on the head, giving binocular vision.

Specializations of the locomotor system can be used to distinguish many different kinds of anurans, and the difficulty is finding names for them—the diversity of anurans exceeds the number of common names that can be used to distinguish various specialties. Animals called frogs usually have long legs and move by jumping. Many species in the family Ranidae have this body form, and very similar jumping frogs are found in other families as well. Semiaquatic forms are moderately streamlined and have webbed feet (Figure 11–6a). Stout-bodied terrestrial anurans that make short hops instead of long leaps are often called toads. They usually have blunt heads, heavy bodies, relatively short legs, and little webbing between the toes. This body form is represented by members of the family Bufonidae, and very similar body forms are found in other families, including the spadefoot toads of western North America and the horned frogs of South America (Figure 11–6b to d). Spadefoot toads take their name from a keratinized structure on the hind foot that they use for digging. The horned frogs have extremely large heads and mouths. They feed on small vertebrates, including birds and mammals, but particularly on other frogs. The tadpoles of horned frogs

also are carnivorous and feed on other tadpoles. Many burrowing frogs have pointed heads, stout bodies, and short legs (Figure 11–6e).

Arboreal frogs usually have large heads and eyes, and often slim waists and long legs (Figure 11–6f). Arboreal frogs in many different families move by quadrupedal walking and climbing as much as by leaping. Many arboreal species in the families Hylidae and Rhacophoridae have enlarged toe disks and are called treefrogs. The surfaces of the toe pads consist of an epidermal layer with peg-like projections separated by spaces or canals (Figure 11–7). Mucous glands distributed over the disks secrete a viscous solution of polymers in water, and excess fluid drains away through the canals. Arboreal species of frogs appear to have at least two methods of sticking to the surfaces on which they climb: Interlocking of the projections of the toe disks with irregularities of rough surfaces appears to account for the ability of treefrogs to climb tree trunks, whereas capillary attraction allows the toe disks to stick to smooth surfaces like leaves (Emerson and Diehl 1980). The mucus secreted by the glands on the disks wets the disk and the leaf surface and establishes a meniscus at the interface between air and fluid at the edges of the toes. Some arboreal frogs have such effective adhesion that they can walk upside down on the bottom surface of a sheet of glass. Expanded toe disks are not limited exclusively to arboreal frogs; some terrestrial species that move across smooth, slippery surfaces also have toe disks.

Specialized aquatic anurans in the family Pipidae are dorsoventrally flattened and have thick waists, powerful hind legs, and large hind feet with extensive webs (Figure 11–6g). Many aquatic anurans also have well-developed lateral-line systems and sensory structures at the tips of their fingers.

Several aspects of the natural history of anurans appear to be related to their different modes of locomotion. In particular, short-legged species that move by hopping are frequently wide-ranging predators that cover large areas as they search for

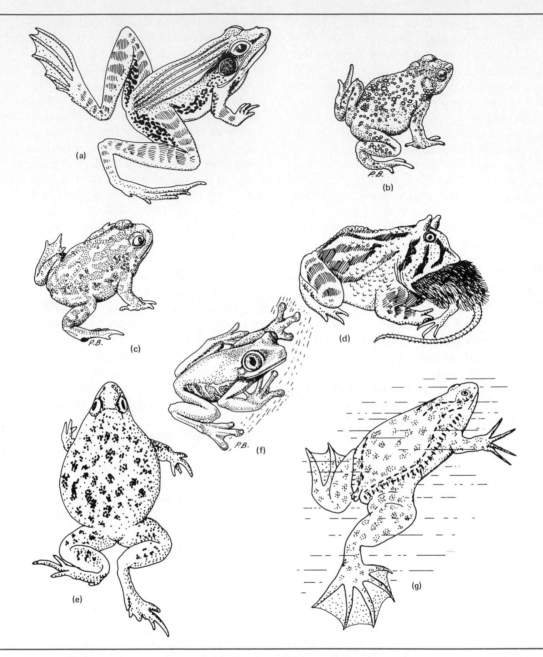

Figure 11–6. Anuran body forms reflect specializations for different habitats and different methods of locomotion. Semiaquatic form (a), an African ranid, *Ptychadena*. Terrestrial anurans: (b) true toad (*Bufo*); (c) spadefoot toad (*Scaphiopus*); (d) horned frog (*Ceratophrys*). A burrowing species (e), an African ranid, *Hemisus*. Arboreal frog (f) the Central American hylid *Agalychnis*. Specialized aquatic frog: (g) African clawed frog (*Xenopus*).

Figure 11–7. Toe disks of a hylid frog. [(a) Appeared in *Biological Journal of the Linnaean Society*, vol. 13 (1980). Photographs courtesy of Sharon B. Emerson.]

food. This behavior exposes them to predators, and their short legs prevent them from fleeing rapidly enough to escape. Many of these anurans have potent defensive chemicals that are released from glands in the skin when they are attacked. Species of frogs that move by jumping, in contrast to those that hop, are usually sedentary predators that wait in ambush for prey that passes their hiding places. These species are usually cryptically colored, and they often lack chemical defenses. If they are discovered by a predator, they rely on a series of rapid leaps to get away. Anurans that forage widely encounter different kinds of prey from those that wait in one spot, and differences in dietary habits may be associated with differences in locomotor mode. Aquatic species of anurans use suction feeding to engulf food in the water, but most semiaquatic and terrestrial species have highly specialized sticky tongues that can be flipped out to trap prey and carry it back to the mouth (Figure 11–8).

Caecilians

The third group of living amphibians is the least known and does not even have an English common name. These are the caecilians (order Gymnophiona), legless burrowing or aquatic amphibians that occur in tropical habitats around the world (Figure 11–9). In addition to their loss of limbs caecilians are characterized by having solid skulls that may resemble those of Carboniferous and Permian microsaurs. The eyes of caecilians are covered by skin or even by bone, but the retinae of many species have the layered organization that is typical of vertebrates and appear to be functional as photoreceptors. Conspicuous dermal folds (annuli) encircle the bodies of caecilians. The primary annuli overlie vertebrae and myotomal septa and reflect body segmentation. Many species of caecilians have dermal scales in pockets in the annuli; scales are not known in the other groups of living amphibians. A second unique feature of caecilians is

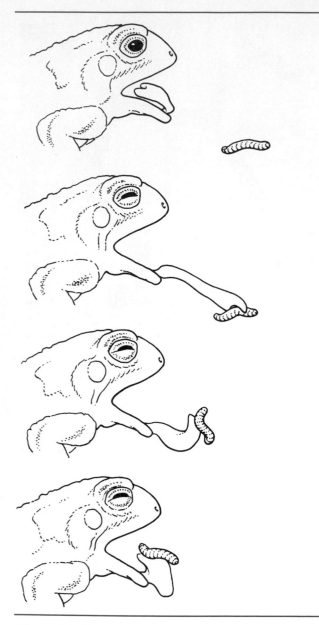

Figure 11–8. Prey capture by a toad.

a pair of protrusible tentacles between the eye and nostril. Some structures that are associated with the eyes of other vertebrates have become associated with the tentacles of caecilians. One of the eye muscles, the retractor bulbi, has become the retractor muscle for the tentacle; the levator bulbi moves the tentacle sheath; and the Harderian gland lubricates the channel of the tentacle. It is likely that the tentacle is a sensory organ that allows chemical substances to be transported from the animal's surroundings to the Jacobson's organ on the roof of the mouth. Caecilians feed on small or elongate prey—termites, earthworms, and larval and adult insects—and the tentacle may allow them to detect the presence of prey when they are underground. Females of some species of caecilians brood their eggs, whereas other species give birth to living young. The embryos of terrestrial species have long, filamentous gills and the embryos of aquatic species have sac-like gills.

Diversity of Life Histories of Amphibians

Of all the characteristics of amphibians, none is more remarkable than the variety they display in modes of reproduction and parental care. It is astonishing that the range of reproductive modes among the 4000 species of amphibians far exceeds that of any other group of vertebrates except for fishes, which outnumber amphibian species by more than 5 to 1. Most species of amphibians lay eggs: The eggs may be deposited in water or on land and they may hatch into aquatic larvae or into miniatures of the terrestrial adults. The adults of some species of frogs carry eggs attached to the surface of their bodies. Others carry their eggs in pockets in the skin of the back or flanks, in the vocal sacs, or even in the stomach. In still other species the females retain the eggs in the oviducts and give birth to living young. Many amphibians have no parental care of their eggs or young, but in many other species a parent remains with the eggs and sometimes with the hatchlings, transports tadpoles from the nest to water, and in a few species an adult even feeds the tadpoles.

Figure 11–9. Caecilians, order Gymnophiona: (a) adult, showing body form; (b) a female coiled around her eggs. Embryos of terrestrial (c) and aquatic (d) species. [(b) Modified from H. Gadow, 1909, *Amphibia and Reptiles*, Macmillan, London; (c) and (d) modified from E. H. Taylor, 1968, *The Caecilians of the World*, University of Kansas Press, Lawrence, Kans.]

Caecilians

The reproductive adaptations of caecilians are as specialized as their body form and ecology (Wake 1977, 1986). Internal fertilization is accomplished by a male intromittent organ that is protruded from the cloaca. Some species of caecilians lay eggs, and the female may coil around the eggs, remaining with them until they hatch (Figure 11–9). Viviparity is widespread in the order, however, and three of the five families include species in which the eggs are retained in the oviducts and the female gives birth to living young. Recent studies by Marvalee Wake have provided fascinating details about this process. At birth young caecilians are 30 to 60 percent of their mother's body length. A female *Typhlonectes* 500 millimeters long may give birth to nine babies, each 200 millimeters long.' Wake found that the initial growth of the fetuses is supported by yolk contained in the egg at the time of fertilization, but this yolk is exhausted long before embryonic development is complete. In *Typhlonectes* the fetuses have absorbed all of the yolk in the eggs by the time they are 30 millimeters long. Thus, the energy they need to grow to 200 millimeters (a 6.6-fold increase in length) must be supplied by the mother. The energetic demands of producing nine babies, each one increasing its length 6.6 times and reaching 40 percent of the mother's length at birth, must be considerable.

It appears that the fetuses obtain this energy by scraping material from the walls of the oviducts with specialized embryonic teeth. Wake observed that the epithelium of the oviduct proliferates and forms thick beds surrounded by ramifications of connective tissue and capillaries. As the fetuses exhaust their yolk supply, these beds begin to secrete a thick, white, creamy substance that has been called uterine milk. When their yolk supply has been exhausted, the fetuses emerge from their egg membranes, uncurl, and align themselves lengthwise in the oviducts. The fetuses apparently bite the walls of the oviduct, stimulating secretion and stripping some epithelial cells and muscle fibers that they swallow with the uterine milk. Wake found that small fetuses are regularly spaced along the oviducts. Large fetuses have their heads spaced at intervals, although the body of one fetus may overlap the head of the next. She suggested that this spacing gives all the fetuses access to the secretory beds.

Gas exchange appears to be achieved by apposition of fetal gills to the walls of the oviducts. All the terrestrial species have fetuses with a pair of triply branched filamentous gills. In preserved specimens the fetuses frequently have one gill extending above the head and the other stretched along the body. In the aquatic genus *Typhlonectes*, the gills are sac-like but are usually positioned in the same way. Both the gills and the oviductal wall are highly vascularized, and it seems likely that exchange of gases, and possibly of small molecules such as metabolic substrates and waste products, takes place across the adjacent gill and oviduct. The gills are absorbed before birth, and cutaneous exchange may be important for fetuses late in development.

Differences in the details of fetal dentition in different species of caecilians suggest to Wake that this specialized form of fetal nourishment may have evolved independently in different phylogenetic lines. Analogous methods of supplying energy to fetuses are known in some elasmobranch fishes.

Salamanders

Most groups of salamanders utilize internal fertilization; two primitive families, the Hynobiidae and Cryptobranchidae, retain external fertilization. Internal fertilization in salamanders is accomplished not by an intromittent organ but by the transfer of a packet of sperm (the **spermatophore**) from the male to the female (Figure 11–10). The form of the spermatophore differs in various species of salamanders, but all consist of a sperm cap on a gelatinous base. The base is a cast of the interior of the male's cloaca, and in some species it reproduces the ridges and furrows in accurate detail.

Courtship Courtship patterns are important species-isolating mechanisms, and they show great interspecific variation. The description that follows of the courtship of the red-spotted newt is as close to a typical pattern as any one species can be. In this species, courtship and egg laying are both aquatic. Courtship starts with the male nudging the cloacal region of the female with his snout (Figure 11–11a). If the female is receptive, the male climbs onto her back, grasping her pectoral region with his forelegs, and then working his way forward until he is clasping her shoulders with his hind legs. By bending his body almost double, he rubs the female's head with his chin (Figure

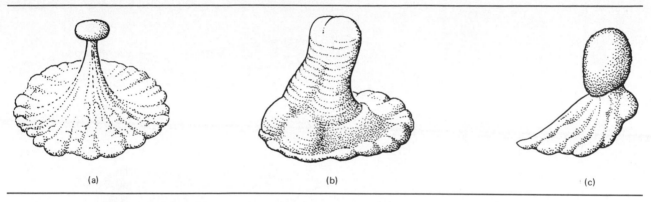

Figure 11–10. Spermatophores from (a) red-spotted newt, *Notophthalmus viridescens;* (b) dusky salamander, *Desmognathus fuscus;* (c) two-lined salamander, *Eurycea bislineata.* (Modified from G. K. Noble, 1931, *The Biology of the Amphibia,* McGraw-Hill, New York.)

11–11b). **Hedonic glands,** located on the surface of the male's body, are believed to secrete a substance that stimulates the female during this stage. The male moves off the female's back and leads her forward (Figure 11–11c). The female's attention seems to be fixed on the cloacal region of the male, and in many species the male's cloacal region takes on a brilliant color in the breeding season. The male may wave his tail at the female as if he were wafting the secretions of hedonic glands toward her. The female nudges the male's cloaca, and this seems to be the stimulus for the male to deposit a spermatophore on the pond bottom. The male continues to walk forward, leading the female directly over the spermatophore. As her cloaca passes over it, she picks the cap off with her cloacal lips. The gelatinous covering dissolves in the cloaca, releasing the sperm.

A complex behavioral pattern of this sort includes visual, tactile, and chemical cues, and all of them can be species specific. The effectiveness of courtship patterns as isolation mechanisms is seen in the widespread occurrence of simultaneous breeding of different species of salamanders in the same pools. In northern New York, for example, two species of mole salamanders breed simultaneously in early spring. The Jefferson salamander,

Ambystoma jeffersonianum, has a courtship pattern much like that described for *Notophthalmus* in which a single male clasps a female. The spotted salamander, *A. maculatum,* has a different pattern entirely. Several males court a female simultaneously. Courtship by a single male may not stimulate a female sufficiently to cause her to breed. The males cluster around the female, nudging her with their snouts, swimming off to deposit a spermatophore a short distance away, then returning again to nudge the female. Eventually, the female moves to the spermatophores that have been deposited by the males and picks up one or more. The courtship behavior of these two species is so different that it is probably sufficient to isolate them even without the visual and chemical differences between them.

Eggs and Larvae In most cases, salamanders that breed in water lay their eggs in water. The eggs may be laid singly, or in a mass of transparent gelatinous material. The eggs hatch into gilled aquatic larvae which, except in paedomorphic forms, transform into terrestrial adults. Some families, including the lungless salamanders (Plethodontidae), have a number of species that have dispensed in part or entirely with an aquatic larval

Figure 11–11. Courtship behavior of salamanders: (a–c) red-spotted newt, *Notophthalmus viridescens;* (d) red salamander, *Pseudotriton ruber.* [(a) and (c) drawn from photographs by Kentwood D. Wells; (b) modified from S. C. Bishop, 1941, *The Salamanders of New York,* New York State Museum Bulletin, 324, Albany; (d) modified from J. A. Organ and D. J. Organ, 1968, *Copeia* 1968:217–223.]

stage. Plethodontid courtship on land is basically the same as in water, although the female may maintain closer contact with the male during the stage just preceding deposition of a spermatophore, straddling his tail as she follows him in a posture called the tailwalk (Figure 11–11d). The terrestrial environment does not offer the male the opportunity of wafting a species-identifying scent toward the female at this stage, and the tailwalk may be a modification of the pattern that has the same function. The dorsal surface of the tail has a concentration of glands, and there may be chemoreceptors on the ventral surface of the female's body.

Some plethodontid salamanders lay eggs in water, but many lay their eggs in moist microhabitats on land. The dusky salamander, *Desmognathus fuscus*, lays its eggs beneath a rock or log near water, and the female remains with them until after they have hatched. The larvae have small gills at hatching, and may either take up an aquatic existence or move directly to terrestrial life. The red-backed salamander, *Plethodon cinereus*, lays its eggs in a hollow space in a rotten log or beneath a rock. The embryos have gills, but these are reabsorbed before hatching and the hatchlings are miniatures of the adults.

A few salamanders give birth to living young. The European salamander *(Salamandra salamandra)* produces 20 or more small larvae, each about one-twentieth the length of an adult. The larvae are released in water and have an aquatic stage that lasts about 3 months. The closely related alpine salamander *(S. atra)* gives birth to one or two fully developed young about one-third the adult body length. A female alpine salamander produces as many eggs as a European salamander, but only one egg in each oviduct develops. The remaining eggs break up into a mass that provides food for the developing embryo.

Paedomorphosis Paedomorphosis is the rule in families like the Cryptobranchidae and Proteidae and characterizes most troglodytes. It also appears as a variant in the life history of species of salamanders that usually metamorphose, and can be a facultative response to conditions in aquatic or terrestrial habitats. The life histories of two species of salamanders from eastern North America provide examples of the flexibility of paedomorphosis (Reilly 1987).

The small-mouthed salamander, *Ambystoma talpoideum*, is the only species of mole salamander in eastern North America that displays paedomorphosis, although a number of species of *Ambystoma* in the western United States and in Mexico are paedomorphic. Small-mouthed salamanders breed in the autumn and winter, and during the following summer some larvae metamorphose to become terrestrial juveniles. These animals become sexually mature by autumn and return to the ponds to breed when they are about a year old. Ponds in South Carolina also contain paedomorphic larvae that remain in the ponds through the summer and mature and breed in the winter. Some of these paedomorphs metamorphose after breeding, whereas others do not metamorphose and remain in the ponds.

The situation in red-spotted newts is more complex, with four different life-history patterns possible. In one pattern aquatic adults lay eggs that hatch into aquatic larvae. After a few months these larvae metamorphose into a bright orange terrestrial salamander called a red eft. The eft stage lasts for 1 to 8 years, after which the efts return to the ponds and transform into aquatic adults. The orange color changes to olive-green dorsally with a yellow ventral surface, the skin becomes smooth, and a fleshy tail fin develops. In a second pattern of life history, larvae metamorphose into aquatic juveniles that transform into aquatic adults after about 2 years. In the third pattern larvae metamorphose into a form called a branchiate. The branchiates initially retain many larval characters, including gills and gill slits, labial folds, a dorsal body fin, and an opercular fold. They become sexually mature in their second year, and over a period of several years they slowly lose their larval characters and transform into aquatic adults. One can imagine a fourth life-history pattern that would consist of a fully paedomorphic larva that retained all its larval characters and never metamorphosed. That pattern does occur, but apparently it is rare: Only one paedomorphic individual was found in a sample of 1600 newts.

Local environmental conditions apparently determine which life-history pattern is most advantageous. Many of the populations of red-spotted newts that omit the eft stage occur in the coastal plain where the terrestrial habitat is hot, dry, and inhospitable to salamanders and the ponds are large and long lasting. In contrast, efts often are abundant in cool, hilly regions where the terrestrial habitat is moist but ponds are small and may be ephemeral. Many ponds in these habitats are formed by beaver dams, and they have a lifetime of only a few decades before they become filled with silt and turn into meadows. A terrestrial eft stage can colonize new ponds, and that ability may be important in such habitats.

The flexibility of the onset of metamorphosis and of sexual maturity seen in these salamanders allows the pattern of life history to be shaped by environmental conditions in different habitats. However, the two species use different developmental mechanisms to produce ecologically similar results. In small-mouthed salamanders the timing of sexual maturity is fixed at about 1 year, and metamorphosis can occur before or after sexual maturity. In red-spotted newts the timing of metamorphosis is fixed and almost always occurs a few months after hatching, but the onset of sexual maturity may occur after only 2 years, or it may be delayed as much as 8 years.

Anurans

Anurans are the most familiar amphibians, largely because of the vocalizations associated with their reproductive behavior. It is not even necessary to get outside a city to hear them. In springtime a weed-choked drainage ditch beside a highway or a trash-filled marsh at the edge of a shopping center parking lot is likely to attract a few toads or treefrogs that have not yet succumbed to human usurpation of their habitat.

Vocalizations Anuran calls are diverse; they vary from species to species, and most species have two or three different sorts of calls used in different situations. The most familiar calls are the ones usually referred to as mating calls, although a less specific term such as **advertisement calls** is preferable. These calls range from the high-pitched *peep* of a spring peeper to the nasal *waaah* of a spadefoot toad or the bass *jug-o-rum* of a bullfrog. The characteristics of a call identify the species and sex of the calling individual. Many species of anurans are territorial, and in these it is possible that territorial defenders identify each other individually by voice.

A species-specific call is a conservative evolutionary character, and among related taxa there is often considerable similarity in the calls. Superimposed on the basic similarity are the effects of morphological factors, such as body size, as well as ecological factors that stem from characteristics of the habitat. Most toads (*Bufo*) have a trilled mating call, but the pitch of the call varies with the body size. In the oak toad (*B. quercicus*), which has a body length of only 2 or 3 centimeters, the dominant frequency of the call is 5200 hertz. In the larger southwestern toad (*B. microscaphus*), which is 8 centimeters long, the dominant frequency is lower, 1500 hertz, and the giant toad (*B. marinus*), with a body length of nearly 20 centimeters, has the lowest pitched call of all, 600 hertz.

In tropical habitats, where ten or more species of anurans may call simultaneously from the same breeding area, division of the auditory spectrum between species is very marked. In general, each species has a dominant frequency that is distinct from the dominant frequency of all other species in the area. Neurophysiological recordings from the brains of anurans have shown that individual cells in the auditory region are closely tuned to the dominant frequency of the call of the species. Those cells respond strongly to sounds of a species' own frequency and give very little response to sounds of slightly different frequencies.

Female frogs are responsive to the advertisement call of their own species for a very brief period when their eggs are ready to be laid. The hormones associated with ovulation are thought to sensitize specific cells in the auditory pathway. In some cases intraspecific differences in neural and behavioral responses to advertisement calls have been demonstrated. The Puerto Rican coquí (*Eleutherodactylus coqui*) gets its name from its two-note advertisement call—*co-qui*. The *co* note is a constant frequency of about 1100 hertz; the *qui* note sweeps upward in frequency from 1800 to 2100 hertz. Male coquíes call at night from leaf surfaces and tree trunks in the forest. When a calling male coquí is approached by a second calling male, the resident drops the *qui* note from its call and emits only the *co* note. If the intruder continues to approach, the resident may attack, butting and biting the intruder. In behavioral experiments, female coquíes approached speakers playing recordings of the full *co-qui* call or of the *qui* note alone, but showed little response to the *co* note when it was played by itself. In contrast, male coquíes switched from the *co-qui* call to *co* notes when they heard a recording of either a *co-qui* or the *co* itself. Thus, the male frogs responded to the *co* note and the females to the *qui* note (Narins and Capranica 1978).

Intracellular recordings from auditory fibers in the eighth nerve showed a neural basis for the different responses of male and female coquíes. The male frogs had many neurons that fired in response to the frequencies in the *co* note, but only a few that responded to the higher frequencies of the *qui* note. In female coquíes the situation was reversed; they had many units sensitive to the *qui* notes and few sensitive to *co* notes (Narins and Capranica 1976).

Mixed choruses of anurans are very common in the mating season; a dozen or more species may breed simultaneously in one pond. A female's response to her own species' mating call is one isolating mechanism, and it is supplemented by a variety of behavioral and probably chemical differences between species. Calling site is one important difference. Among the Australian leptodactylids of the genus *Crinia*, for example, *C. sloanei* calls while floating in the water, *C. signifera* calls from beneath overhanging vegetation near shore, and *C. parinsignifera* calls from grass clumps out of water. Presumably, a female frog seeks a male in the usual calling site for her species, and this behavior would reduce the chances of encountering a male of the wrong species.

(a)

(b)

Figure 11–12. Male túngara frog, *Physalaemus pustulosus*, vocalizing. Air is forced from the lungs (a) into the vocal sacs (b). (Photographs courtesy of Theodore L. Taigen.)

Costs and Benefits of Vocalization The vocalizations of male frogs are costly in two senses. The actual energy that goes into call production can be very large, and the variations in calling pattern that accompany social interactions among male frogs in a breeding chorus can increase the cost per call (see Box 11–2). Another cost of vocalization for a male frog is the risk of predation. A critical function of vocalization is to permit a female frog to locate a male, but female frogs are not the only animals that can use vocalizations as a cue to find male frogs—predators of frogs also find that calling males are easy to locate. The túngara frog *(Physalaemus pustulosus)* is a small terrestrial leptodactylid that occurs in Central America (Figure 11–12). Stanley Rand and Michael Ryan have studied the costs and benefits of vocalization for this species (Ryan 1985).

Túngara frogs breed in small pools, and breeding assemblies range from a single male to choruses of several hundred males. The advertisement call of a male túngara frog is a strange noise, a whine that sounds as if it would be more at home in an arcade of video games than in the tropical night. The whine starts at a frequency of 900 hertz and sweeps downward to 400 hertz in about 400 milliseconds (Figure 11–13). The whine may be produced by itself, or it may be followed by one or several *chucks* with a dominant frequency of 250 hertz. When a male túngara frog is calling alone in a pond it usually gives only the whine portion of the call, but as additional males join a chorus, more and more of the frogs produce calls that include chucks. By playing recordings of the whine calls to male frogs in breeding ponds, Rand was able to make them shift to giving calls that included chucks. That observation suggested that it was the presence of other calling males that stimulated frogs to make their calls more complex by adding chucks to the end of the whine.

What advantage would a male frog in a chorus gain from using a whine-chuck call instead of a whine? Rand suggested that the complex call might be more attractive to female frogs than the simple call. He tested that hypothesis by placing female túngara frogs in a test arena with a speaker

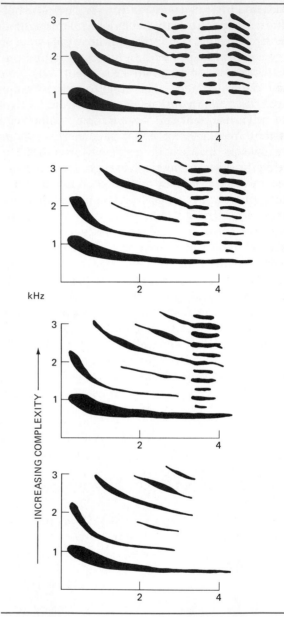

Figure 11–13. Sonograms of the advertisement call of *Physalaemus pustulosus*. The calls increase in complexity from bottom (a whine only) to top (a whine followed by three chucks). A sonogram is a graphic representation of a sound: Time is shown on the horizontal axis and frequency on the vertical axis. (Modified from M. J. Ryan, 1985, *The Túngara Frog*, University of Chicago Press, Chicago.)

at each side. One speaker broadcast a whine call and the second speaker broadcast a whine-chuck. Rand released female frogs individually in the center of the arena and noted which speaker they moved toward. As he had predicted, most of the female frogs (14 of the 15 he tested) chose the speaker broadcasting the whine-chuck call.

If female frogs are attracted to whine-chuck calls in preference to whine calls, why do male frogs give whine-chuck calls only when other males are present? Why not always give the most attractive call possible? One possibility is that whine-chuck calls require more energy than whines, and males save energy by using whine-chucks only when competition with other males makes the energy expenditure necessary. However, Ryan measured the energy expenditure of calling male túngara frogs and found that energy cost was not related to the number of chucks. Another possibility is that male frogs giving whine-chuck calls are more vulnerable to predators than frogs giving only whine calls. Túngara frogs in breeding choruses are preyed upon by frog-eating bats, *Trachops cirrhosus,* and the bats locate the frogs by homing on their vocalizations.

In a series of playback experiments Ryan and Merlin Tuttle placed pairs of speakers in the forest and broadcast vocalizations of túngara frogs. One speaker played a recording of a whine and the other a recording of a whine-chuck. The bats responded as if the speakers were frogs: They flew toward the speakers and even landed on them. In five experiments at different sites, the bats approached speakers broadcasting whine-chuck calls twice as frequently as those playing simple whines (168 approaches versus 81). Thus, female frogs are not alone in finding whine-chuck calls more attractive than simple whines—predators of frogs also respond more strongly to the complex calls. Predation can be a serious risk for male túngara frogs. Ryan and his colleagues measured the rates of predation in choruses of different sizes. The major predators were frog-eating bats, a species of opossum *(Philander opossum),* and a larger species of frog *(Leptodactylus pentadactylus);* the bats were the most important predators of the túngara frogs.

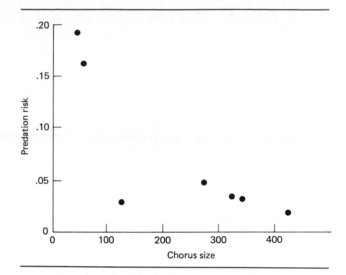

Figure 11–14. The risk for an individual male *Physalaemus pustulosus* of being caught by a predator is high in small choruses, and decreases with increasing chorus size. (From M. J. Ryan, 1985, *The Túngara Frog,* University of Chicago Press, Chicago.)

Large choruses of frogs did not attract more bats than small choruses, and consequently the risk of predation for an individual frog was less in a large chorus than in a small one (Figure 11–14). Predation was an astonishing 19 percent of the frogs per night in the smallest chorus and a substantial 1.5 percent per night even in the largest chorus. When a male frog shifts from a simple whine to a whine-chuck call, it increases its chances of attracting a female, but it simultaneously increases its risk of attracting a predator. In small choruses the competition from other males for females is relatively small and the risk of predation is relatively large. Under those conditions it is apparently advantageous for a male túngara frog to give simple whines. However, as chorus size increases, competition with other males also increases while the risk of predation falls. In that situation the advantage of giving a complex call apparently outweighs the risks.

Box 11–2. The Energy Cost of Vocalization by Frogs

The vocalizations of frogs, like most acoustic signals of tetrapods, are produced when air from the lungs is forced over the vocal cords, causing them to vibrate. Contraction of trunk muscles provides the pressure in the lungs that propels the air across the vocal cords, and these contractions require metabolic energy. Measurement of the actual energy expenditure by frogs during calling is technically difficult because a frog must be placed in an airtight metabolism chamber to measure the amount of oxygen it consumes, and that procedure can frighten the frog and prevent it from calling. Theodore Taigen and Kentwood Wells (1985) at the University of Connecticut overcame that difficulty in studies of the gray treefrog, *Hyla versicolor*, by taking the metabolism chambers to the breeding ponds. Calling male frogs were placed in the chambers early in the evening and then left undisturbed. With the stimulus of the chorus around them, frogs would call in the chambers. Their vocalizations were recorded with microphones attached to each chamber, and the amount of oxygen they used during calling was determined from the decline in the concentration of oxygen in the chamber over time (Figure 11–15).

The rates at which individual frogs consumed oxygen were directly proportional to their rates of vocalization (Figure 11–16). At the lowest calling rate, 150 calls per hour, oxygen consumption was barely above resting levels. However, at the highest calling rates, 1500 calls per hour, the frogs were consuming oxygen even more rapidly than they did during high levels of locomotor activity. Examination

Figure 11–15. Gray treefrog *(Hyla versicolor)* in a metabolism chamber beside a breeding pond. A microphone in the chamber records the frog's calls, and a thermocouple measures the temperature inside the chamber. Gas samples are drawn from the tube for measurements of oxygen consumption. (Photograph courtesy of Theodore L. Taigen.)

of the trunk muscles of the male frogs, which hypertrophy enormously during the breeding season, revealed biochemical specializations that appear to permit this high level of oxygen consumption during vocalization (Taigen et al. 1985).

The advertisement call of the gray treefrog is a trill that lasts from 0.3 to 0.6 second. During their studies, Wells and Taigen found that gray treefrogs gave short calls when they were in small choruses, and lengthened their calls when many other males were calling near them (Wells and Taigen 1986). It has subsequently been shown that long calls are more attractive to female frogs than short calls (Klump and Gerhardt 1987), but the long calls require more energy. A 0.6-second call requires about twice as much energy as a 0.3-second call, and the rate of oxygen consumption during calling increases as the length of the calls increases. That relationship suggests that a male gray treefrog that increases its call duration to be more attractive to female frogs must pay a price for its attractiveness with a higher rate of energy expenditure. Is that increased energy cost important in determining the chances that a male frog will find a mate?

Indirect evidence suggests that the energy cost of calling might limit the time a male gray treefrog could spend in a breeding chorus. The treefrogs call for only 2 to 4 hours each night, and they lose weight during the several nights they spend in a chorus. Wells and Taigen tested some of these hypotheses. They were able to simulate the effects of different chorus sizes by playing tape recordings of vocalizations to frogs. The frogs matched their own calls to the recorded calls they heard—short responses to short calls, medium to medium calls, and long responses to long calls. As the

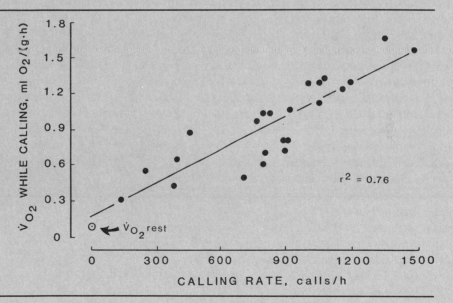

Figure 11–16. Rates of oxygen consumption of frogs calling inside metabolism chambers (on the vertical axis) as a function of the rate of calling (on the horizontal axis). The energy expended by a calling frog increases linearly with the number of times it calls per hour. (From T. L. Taigen and K. D. Wells, *Journal of Comparative Physiology,* B155:163–170.)

Box 11–2 (Continued)

length of their calls increased, the frogs reduced the rate at which they called. The reduction in rate of calling approximately balanced the increased length of each call, and the overall calling effort (the number of seconds of vocalization per hour) changed little. Thus, it appears that male gray treefrogs compensate for the higher cost of long calls by giving fewer of them. However, that compromise may not entirely eliminate the problem of high energy costs for frogs giving long calls. Males giving long calls at slow rates spent fewer hours per night calling than did frogs that produced short calls at higher rates.

The high energy cost of calling offers an explanation for the pattern of short and long calls produced by male gray treefrogs. The length of time that an isolated male can call may be the most important determinant of his success in attracting a female. In that situation giving short calls, which conserve energy stores and allow the frog to call for several hours every night, seems to be the best strategy. For a male in a large chorus, however, competition with other males is intense and giving a more attractive call may be important, even if the male can call for only a short time.

Modes of Reproduction   Fertilization is external in most anurans; the male uses his forelegs to clasp the female in the pectoral region (**axillary amplexus**) or pelvic region (**inguinal amplexus**). Amplexus may be maintained for several hours or even days before the female lays eggs. During this period the male and female are in close contact. It seems likely that tactile and chemical cues play a role in species identification at this stage. Internal fertilization has been demonstrated in the Puerto Rican coqui and may be widespread among frogs that lay eggs on land (Townsend et al. 1981). Fertilization must also be internal for the few species of anurans that give birth to living young.

Anurans show even greater diversity in their modes of reproduction than urodeles. Similar reproductive habits have clearly evolved independently in different groups. The underlying selective mechanism appears to have been the advantage gained by reducing the amount of energy invested per offspring that survives to maturity. This end has been achieved in various ways by different groups of frogs, including parental care of the eggs (and in some cases by care of the tadpoles as well), by increasing the size of the egg

and thus the size of the hatchlings, by preparing a favorable nest site, or by some combination of these mechanisms. The discussion that follows is organized only in terms of increasingly extensive parental care of the young. It is not meant to imply an evolutionary sequence; the more elaborate forms of parental care did not necessarily evolve via the less complex forms.

The primitive anuran reproductive pattern is thought to consist of laying large numbers of small eggs. This pattern is retained in two of the most widespread genera of anurans, *Rana* and *Bufo*. For example, toads may lay as many as 10,000 eggs in one clutch, and the female puts about half the energy in her body into the eggs. She leaves the breeding pond immediately thereafter and accumulation of sufficient fat reserves to produce next year's eggs occupies her for the rest of the season. In temperate regions this reproductive pattern is almost universal among anurans, but in tropical areas as many as 80 percent of the species of anurans have other patterns.

One method of increasing the proportion of eggs that hatch successfully is to give them protection from predators, and a number of methods

Figure 11–17. Reproductive modes of anurans; see the text for discussion. (a) Eggs laid over water, *Centrolenella*; (b) eggs in a nest of foam, *Physalaemus*; (c) eggs carried by the adult, *Rhinoderma*; (d) tadpoles carried by the adult, *Colostethus*. Eggs carried on the back of an adult: (e) *Hemiphractus*; (f) *Pipa*. [(c) From G. K. Noble, 1931, *The Biology of the Amphibia,* McGraw-Hill, New York; (e) from W. E. Duellman, 1970, *The Hylid Frogs of Middle America,* Monograph of the Museum of Natural History, 1, The University of Kansas, Lawrence, Kans.; (f) from M. Lamotte and J. Lescure, 1977, *La Terre et la Vie* 31:225–311.]

of accomplishing this have evolved. Many arboreal frogs (represented in Figure 11–17a by *Centrolenella*) lay their eggs in the branches of trees overhanging water. The eggs undergo their embryonic development out of the reach of aquatic egg pred-ators, and when the tadpoles hatch they drop into the water and take up an aquatic existence. Other frogs, such as *Physalaemus pustulosus* (Figure 11–17b), achieve the same result by constructing foam nests that float on the water surface. The female emits a copious mucous secretion that she beats into a foam with her hind legs, and the eggs are laid in the foam mass. When the tadpoles hatch, they drop through the foam into the water.

Although these methods reduce egg mortality, the tadpoles are subjected to predation and competition. Some anurans avoid both problems by finding or constructing breeding sites free from competition and predation. Many treefrogs, for example, lay their eggs in the water that accumulates in bromeliads—epiphytic tropical plants that grow in trees and are morphologically specialized to collect rainwater. A large epiphyte may hold several liters of water, and the frogs pass through egg and larval stages in that protected microhabitat. Many tropical frogs lay eggs on land near water. The eggs or tadpoles may be released from the nest sites when pond levels rise after a rainstorm. Other frogs construct pools in the mud banks beside streams. These volcano-shaped structures are filled with water by rain or seepage and provide a favorable environment for the eggs and tadpoles. Some frogs have eliminated the tadpole stage entirely. These lay large eggs on land that develop directly to little frogs. This pattern is characteristic of about 20 percent of all anuran species.

Parental Care   Parental care is widespread among anurans (Wells 1981). Even among the forms with a primitive breeding pattern, the males of some species maintain territories and attack almost any small animal that intrudes. For example, male green frogs *(Rana clamitans)* defend areas 2 or 3 meters in diameter, and any eggs laid in that area are almost certain to be fertilized by the territorial male. The presence of a male guarding this territory may offer a degree of protection to his eggs. Adults of many species of frogs guard the eggs specifically. In some cases it is the male, in others the female, and in most cases it is not clearly known which sex is involved because external sex identification is difficult in many anurans. Some of the frogs that lay their eggs over water remain with them. Some species sit beside the eggs, others rest on top of them. Removing the guarding frog frequently results in the eggs desiccating and dying before hatching or being eaten by predators (Taigen et al. 1984, Townsend et al. 1984). Many of the terrestrial frogs that lay direct-developing eggs remain with the eggs and will attack an animal that approaches the nest.

Some of the poison-dart frogs of the American tropics deposit their eggs on the ground, and one of the parents remains with the eggs until they hatch into tadpoles. The tadpoles adhere to the adult and are transported to water (Figure 11–17d). Females of the Panamanian frog *Colostethus inguinalis* carry their tadpoles for more than a week and the tadpoles increase in size during this period (Wells 1980). The largest tadpoles being carried by females had small amounts of plant material in their stomachs, showing that they had begun to feed while they were still being transported by their mother. Females of another Central American poison-dart frog, *Dendrobates pumilio*, release their tadpoles in small pools of water, and then return at intervals to the pools to deposit unfertilized eggs that the tadpoles eat (Weygoldt 1980).

Other anurans, instead of remaining with the eggs, carry the eggs with them. The male of the European midwife toad *(Alytes obstetricans)* gathers the egg strings about his hind legs as the female lays them. He carries them with him until they are ready to hatch, at which time he releases the tadpoles into water. The male of the terrestrial Darwin's frog *(Rhinoderma darwinii)* of Argentina snaps up the eggs the female lays and carries them in his vocal pouches, which extend back to the pelvic region (Figure 11–17c). The embryos pass through metamorphosis in the vocal sacs and emerge as fully developed froglets. Males are not alone in caring for eggs. The females of a group of treefrogs carry the eggs on their back, either in an open oval depression, a closed pouch, or in individual pockets (Figure 11–17e). The eggs develop into miniature frogs before they leave their mother's back. A similar specialization is seen in the completely aquatic Surinam toad, *Pipa*. In the breeding season the skin of the female's back thickens and softens. In egg laying the male and female in amplexus swim in vertical loops in the water. On the upward part of the loop the female is above the male and releases a few eggs which fall onto his ventral surface. He fertilizes them and, on the downward loop, presses them against the female's back. They sink into the soft skin and a cover forms over each egg, enclosing it in a small capsule (Figure 11–17f). The eggs develop through metamorphosis in the capsules.

Tadpoles of the two species of the Australian frog genus *Rheobatrachus* are carried in the stomach of the female frog. The female swallows eggs or newly hatched larvae and retains them in her stomach through metamorphosis. *Rheobatrachus silus* was the first species in which this behavior was described, and it is accompanied by extensive morphological and physiological modifications of the stomach. These changes include distension of the proximal portion of the stomach, separation of individual muscle cells from the surrounding connective tissue, and inhibition of hydrochloric acid secretion, perhaps by prostaglandin released by the tadpoles (Tyler 1983). In January 1984, a second species of gastric brooding frog, *R. vitellinus*, was discovered in Queensland. Strangely, this species lacks the extensive structural changes in the stomach that characterize the gastric brooding of *R. silus* (Leong et al. 1986). The striking differences between the two species suggest the surprising possibility that this bizarre reproductive mode might have evolved independently.

Only a few anurans are viviparous. The females of some African bufonids retain the eggs in the oviducts and give birth to baby toads, and the golden coquí (*Eleutherodactylus jasperi,* a Puerto Rican leptodactylid) has a similar mode of reproduction (Wake 1978, 1980).

The Ecology of Tadpoles   Although many species of frogs have evolved reproductive modes that bypass an aquatic larval stage, a life history that includes a tadpole has certain advantages. A tadpole is a completely different animal from an adult anuran, both morphologically and ecologically.

Tadpoles are as diverse in their morphological and ecological specializations as adult frogs and occupy nearly as great a range of habitats (Figure 11–18). Tadpoles that live in still water usually have ovoid bodies and tails with fins that are as large as the muscular part of the tail, whereas tadpoles that live in fast-flowing water have more streamlined bodies and smaller tail fins. Semi-terrestrial tadpoles wiggle through mud and leaves and climb on damp rock faces; they are often dorsoventrally flattened and have little or no tail fin, and many tadpoles that live in bromeliads have a similar body form. Direct developing tadpoles have large yolk supplies, and reduced mouthparts and tail fins. The mouthparts of tadpoles also show variation that is related to diet (Figure 11–19). Filter-feeding tadpoles that hover in midwater lack keratinized mouthparts, whereas species that graze from surfaces have small beaks that are often surrounded by rows of denticles. Predatory tadpoles have larger beaks that can bite pieces from other tadpoles. Funnel-mouthed surface-feeding tadpoles have greatly expanded mouthparts that skim material from the surface of the water.

Tadpoles of most species of anurans are filter-feeding herbivores, whereas all adult anurans are carnivores that catch prey individually. Because of these differences, tadpoles can exploit resources that are not available to adult anurans. Richard Wassersug (1975) has suggested that it is this advantage that has led many species of frogs to retain the complex pattern of life history in which an aquatic larva matures into a terrestrial adult. Wassersug pointed out that many aquatic habitats experience annual flushes of primary production when nutrients washed into a pool by rain or melting snow stimulate the rapid growth of algae. The energy and nutrients in this algal bloom are transient resources that are available for a brief time to organisms that are able to exploit them.

Tadpoles, Wassersug suggests, are excellent eating machines. All tadpoles are filter feeders, and feeding and ventilation of the gills are related activities. The stream of water that moves through the mouth and nares to ventilate the gills carries

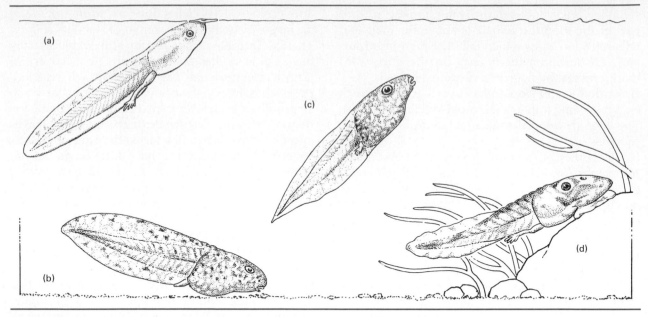

Figure 11–18. Body forms of tadpoles: (a) *Megophrys minor,* a pelobatid. The mouthparts unfold into a platter over which water and particles of food on the surface are drawn into the mouth. (b) *Rana aurora,* a ranid. A generalized feeder that nibbles and scrapes food from surfaces. (c) *Agalychnis callidryas,* a hylid. A midwater suspension feeder shows the large fins and protruding eyes that are typical of midwater tadpoles. It maintains its position in the water column with rapid undulations of the terminal part of its tail. (d) An unidentified species of *Nyctimystes,* a hylid. A stream-dwelling tadpole that adheres to rocks in swiftly moving water with a sucker-like mouth while scraping algae and bacteria from the rocks. The low fins and powerful tail are characteristic of tadpoles living in swift water.

with it particles of food. As the stream of water passes through the branchial basket, small food particles are trapped in mucus secreted by epithelial cells. The mucus, carrying particles of food with it, is moved from the gill filters to the ciliary grooves on the margins of the roof of the pharynx and then transported posteriorly to the esophagus (Figure 11–19e).

Although all tadpoles filter food particles from a stream of water that passes across the gills, the method by which the food particles are put into suspension differs among species. Some tadpoles filter floating plankton from the water. Tadpoles of this type are represented in several families of

anurans, especially the Pipidae and Hylidae, and usually hover in the water column. Midwater-feeding tadpoles are out in the open, where they are vulnerable to predators and they show various characteristics that may reduce the risk of predation. Tadpoles of the African clawed frogs are nearly transparent, and they may be hard for predators to see. Some midwater tadpoles form schools that, like schools of fishes, may confuse a predator by presenting it with so many potential prey that it has difficulty concentrating its attack on one individual.

Many tadpoles are bottom feeders and scrape bacteria and algae off the surfaces of rocks or the

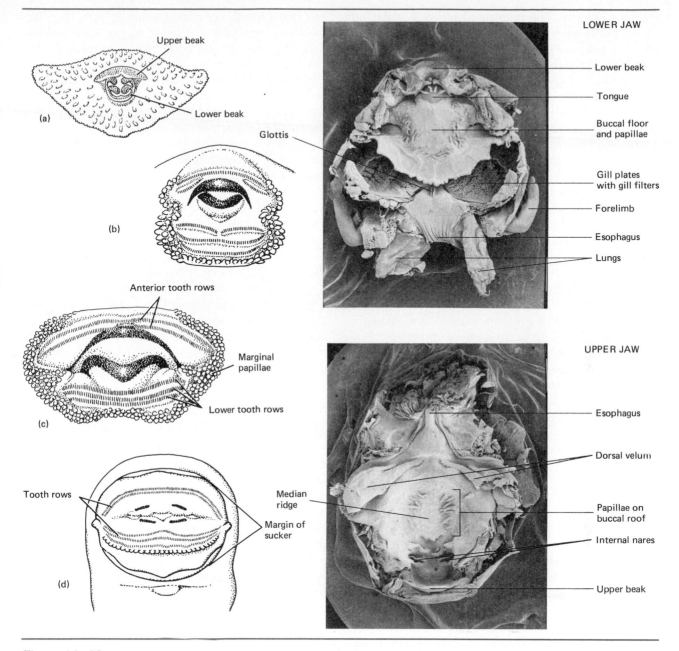

Figure 11–19. Mouths of tadpoles: (a) *Megophrys minor;* (b) *Rana aurora;*
(c) *Agalychnis callidryas;* (d) *Nyctimystes* sp.; (e) Scanning electron micrograph of the
inside of the mouth and buccal region of a tadpole (*Alsodes monticola,* a
leptodactylid). [(e) courtesy of Richard J. Wassersug.]

leaves of plants. The rasping action of keratinized mouthparts frees the material and allows it to be whirled into suspension in the water stream entering the mouth of a tadpole, and then filtered out by the branchial apparatus. Some bottom-feeding tadpoles like toads and spadefoot toads form dense aggregations that create currents that lift particles of food into suspension in the water. These aggregations may be groups of siblings. Toad (*Bufo americanus)* and cascade frog tadpoles (*Rana cascade)* are able to distinguish siblings from nonsiblings and they associate preferentially with siblings (Blaustein and O'Hara 1981, Waldman 1982). Recognition is probably accomplished by olfaction, and toad tadpoles can distinguish full siblings (both parents the same) from maternal half-siblings (only the mother the same), and they can distinguish maternal half-siblings from paternal half-siblings.

Some tadpoles are carnivorous and feed on other tadpoles. The large Central American frog *Leptodactylus pentadactylus* that preys on smaller species of anurans such as the túngara frog has a carnivorous tadpole that preys on tadpoles of other species of anurans. Predatory tadpoles have large mouths with a sharp keratinized beak. Predatory individuals appear among the tadpoles of some species of anurans that are normally herbivorous. Some species of spadefoot toads in western North America are famous for this phenomenon.

Carnivorous tadpoles are also found among some species of frogs that deposit their eggs or larvae in bromeliads. These relatively small reservoirs of water may have little food for tadpoles. It seems possible that the first tadpole to be placed in a bromeliad pool may feed largely on other frog eggs—either unfertilized eggs deliberately deposited by the mother of the tadpole as is the case for the poison-dart frog *Dendrobates pumilio,* or fertilized eggs subsequently deposited by unsuspecting female frogs.

The feeding mechanisms that make tadpoles such effective collectors of food particles suspended in the water allow them to grow rapidly, but that growth contains the seeds of its own termination. As tadpoles grow bigger they become less effective at gathering food because of the changing relationship between the size of food-gathering surfaces and the size of the body of a tadpole. The branchial surfaces that trap food particles are two-dimensional. Consequently, the food-collecting apparatus of a tadpole increases in size approximately as the square of the linear dimensions of the tadpole. However, the food the tadpole collects must nourish its entire body, and the volume of the body increases in proportion to the cube of the linear dimensions of the tadpole. The result of that relationship is a decreasing effectiveness of food collection as a tadpole grows—the body it must nourish increases in size faster than its food-collecting apparatus.

The morphological specializations of tadpoles are entirely different from those of adult frogs, and the transition from tadpole to frog involves a very complete metamorphosis in which tadpole structures are broken down and their chemical constituents are rebuilt into the structures of adult frogs.

Amphibian Metamorphosis

The importance of thyroid hormones for amphibian metamorphosis was discovered quite by accident in the early twentieth century by the German biologist Friedrich Gudersnatch. He was able to induce rapid precocious metamorphosis in tadpoles by feeding them extracts of beef thyroid glands. Some of the details of the interaction of neurosecretions and endocrine gland hormones have been worked out, but no fully integrated explanation of the mechanisms of hormonal control of amphibian metamorphosis is yet possible.

The most dramatic example of metamorphosis is found among anurans, where almost every tadpole structure is altered. Anuran larval development is generally divided into three periods: (1) During premetamorphosis tadpoles increase in size with little change in form; (2) in prometamorphosis the hind legs appear and growth of the body continues at a slower rate; and (3) during metamorphic climax the forelegs emerge and the tail regresses. These changes are stimulated by the actions of thyroxine, and production and release of thyroxine is controlled by a product of the pituitary gland, thyroid stimulating hormone (TSH).

Structural changes in the median eminence may be part of the timing system of anuran metamorphosis. The median eminence is the portion of the brain that connects the hypothalamus to the pituitary. In premetamorphosis it is a small structure with a loose network of capillaries. During prometamorphosis the median eminence enlarges as the pituitary separates from the hypothalamus, and the capillary bed is organized into a portal system. The increased efficiency of the portal system may improve the transport of the neurosecretory products of the hypothalamus to the pituitary. In mammals secretion of TSH by the pituitary is stimulated by a hypothalamic neurosecretion, thyroid releasing factor, which is carried from the hypothalamus to the pituitary by the portal system of the median eminence. This mechanism may apply to anurans as well.

The details of production and release of thyroxine during metamorphosis of tadpoles are not well understood, and binding and metabolism of the thyroxine by tissues introduce further complications. Nonetheless, the action of thyroxine on larval tissues is profound. Its effects are both specific and local. In other words, it has a different effect in different tissues, and that effect is produced by the presence of thyroxine in the tissue; it does not depend on induction by neighboring tissues. The particular effect of thyroxine in a given tissue is genetically determined, and virtually every tissue of the body is involved (Table 11–2). In the liver, for example, thyroxine stimulates the enzymes responsible for the synthesis of urea (the urea cycle enzymes) and starts the synthesis of serum albumin. In the eye it induces the formation of rhodopsin. When thyroxine is administered to the striated muscles of a tadpole's developing leg, it stimulates growth; but when administered to the

Table 11–2. Some of the morphological and physiological changes induced by thyroid hormones during amphibian metamorphosis.

Body form and structure
    Formation of dermal glands
    Restructuring of mouth and head
    Intestinal regression and reorganization
    Calcification of skeleton

Appendages
    Degeneration of skin and muscle of tail
    Growth of skin and muscle of limbs

Nervous system and sense organs
    Increase in rhodopsin in retina
    Growth of extrinsic eye muscles
    Formation of nictitating membrane of the eye
    Growth of cerebellum
    Growth of preoptic nucleus of the hypothalamus

Respiratory system
    Degeneration of the gill arches and gills
    Degeneration of the operculum that covers the gills
    Development of lungs
    Shift from larval to adult hemoglobin

Organs
    Pronephric resorption in the kidney
    Induction of urea-cycle enzymes in the liver
    Reduction and restructuring of the pancreas

Source: Based on B. A. White and C. S. Nicoll, 1981, in *Metamorphosis, a Problem in Developmental Biology*, edited by L. I. Gilbert and E. Freeden, Plenum Press, New York.

striated muscles of the tail, it stimulates the breakdown of tissue. When a larval salamander's tail is treated with thyroxine, only the tail fin disappears; but thyroxine causes the complete absorption of the tail of a frog tadpole.

Metamorphosis in salamanders is relatively undramatic compared to the process in anurans. Extensive changes occur at the molecular and tissue level, but the loss of gills and absorption of the tail fin are the obvious external changes. In contrast, the metamorphosis of a tadpole to a frog involves readily visible changes in almost every part of the body. The tail is absorbed and recycled into the production of adult structures. The small

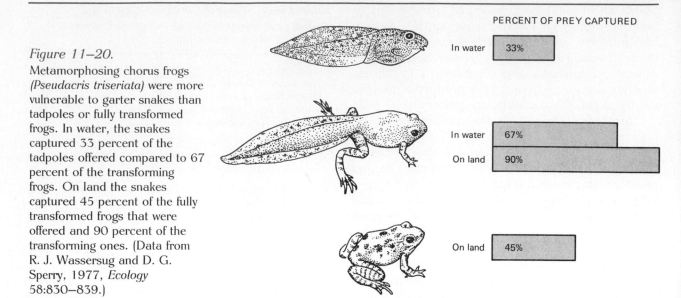

Figure 11–20.
Metamorphosing chorus frogs *(Pseudacris triseriata)* were more vulnerable to garter snakes than tadpoles or fully transformed frogs. In water, the snakes captured 33 percent of the tadpoles offered compared to 67 percent of the transforming frogs. On land the snakes captured 45 percent of the fully transformed frogs that were offered and 90 percent of the transforming ones. (Data from R. J. Wassersug and D. G. Sperry, 1977, *Ecology* 58:830–839.)

tadpole mouth that accommodated algae broadens into the huge mouth of an adult frog. The long tadpole gut, characteristic of herbivorous vertebrates, changes to the short gut of a carnivorous animal. Respiration is shifted from gills to lungs, and partly metamorphosed froglets can be seen swimming to the surface to gulp air.

Metamorphic climax begins with the appearance of the forelimbs and ends with the disappearance of the tail. This is the most rapid part of metamorphosis, occupying only a few days after a larval period that lasts for weeks or months. One reason for the rapidity of metamorphic climax may lie in the vulnerability of larvae to predators during this period. A larva with legs and a tail is neither a good tadpole nor a good frog: The legs inhibit swimming and the tail interferes with jumping. As a result, predators are more successful at catching anurans during metamorphic climax than they are in prometamorphosis or following the completion of metamorphosis. Metamorphosing chorus frogs *(Pseudacris triseriata)* were most vulnerable to garter snakes when they had developed legs and still retained a tail. Both tadpoles (with a tail and no legs)

and metamorphosed frogs (with legs and no tail) were more successful than the metamorphosing individuals at escaping from snakes (Figure 11–20). Life-history theory predicts that selection will act to shorten the periods in the lifetime of a species when it is most vulnerable to predation, and the speed of metamorphic climax may be a manifestation of that sort of selection.

Water Relations of Amphibians

Amphibians have a glandular skin that lacks external scales and is highly permeable to water. Both the permeability and glandularity of the skin have been of major importance in shaping the ecology and evolution of amphibians. Mucous glands are distributed over the entire body surface and secrete mucopolysaccharide compounds. The primary function of the mucus is to keep the skin moist and permeable. For an amphibian, a dry skin means reduction in permeability to water and gases. That, in turn, reduces oxygen uptake and the ability of the animal to use evaporative cooling to maintain its body temperature within equable limits. Experimentally produced interference with mucous gland secretion can lead to lethal overheating in frogs undergoing normal basking activity.

Permeability of Amphibian Skin

Both water and gases pass readily through amphibian skin. In biological systems, permeability to water is inseparable from permeability to gases, and amphibians depend upon cutaneous respiration for a variable but significant part of their gas exchange. Although the skin permits passive movement of water and gases, it controls the movement of other compounds. Sodium is actively transported from the outer surface to the inner, and urea is retained by the skin. These characteristics are important in the regulation of osmotic concentration and in facilitating uptake of water by terrestrial species. The internal osmotic pressure of amphibians is approximately two-thirds that characteristic of most other vertebrates. The primary reason for the dilute body fluids of amphibians is low sodium content—approximately 100 milliequivalents (mEq) compared to 150 mEq in other vertebrates. Amphibians can tolerate a doubling of the normal sodium concentration, whereas an increase from 150 mEq to 170 mEq is the maximum humans can tolerate.

A watery animal with a permeable skin seems an unlikely candidate for success in an arid habitat, and most amphibians are restricted to moderately moist microhabitats. Anurans have been by far the most successful invaders of arid habitats. All but the harshest deserts have substantial anuran populations, and in different parts of the world, different families have converged on similar specializations. Avoiding the harsh conditions of the ground surface is the most common mechanism by which amphibians have managed to invade deserts and other arid habitats. Anurans and salamanders in deserts may spend 9 or 10 months of the year in moist retreat sites, sometimes more than a meter underground, emerging only during the rainy season and compressing feeding, growth, and reproduction into just a few months (Chapter 16).

A different pattern of adaptation to arid conditions is seen in a few treefrogs. The African rhacophorid *Chiromantis xerampelina* and the South American hylid *Phyllomedusa sauvagei* lose water through the skin at a rate only one-tenth that of most frogs. *Phyllomedusa* has been shown to achieve this low rate of evaporative water loss by using its legs to spread the lipid-containing secretions of dermal glands over its body surface in a complex sequence of wiping movements (Shoemaker and McClanahan 1975). These two frogs are unusual also because they excrete nitrogenous wastes as salts of uric acid rather than as urea (see Chapter 4). This uricotelism provides still more water conservation.

Behavioral Control of Evaporative Water Loss

For animals with skins as permeable as those of most amphibians, the main difference between rain forests and deserts may be how frequently they encounter a water shortage. The Puerto Rican coquí lives in wet tropical forests; nonetheless, it has elaborate behaviors that reduce evaporative water loss during its periods of activity (Pough et al. 1983). Male coquíes emerge from their daytime retreat sites at dusk and move 1 or 2 meters to calling sites on leaves in the understory vegetation. They remain at their calling sites until shortly before dawn, when they return to their daytime retreats. The activities of the frogs vary from night to night, depending on whether it rained during the afternoon. On nights after a rainstorm, when the forest is wet, the coquíes begin to vocalize soon after dusk and continue until about midnight, when they fall silent for several hours. They resume calling briefly just before dawn. When they are calling, coquíes extend their legs and raise themselves off the surface of the leaf (Figure 11–21a). In this position they lose water by evaporation from the entire body surface.

(a)

(b)

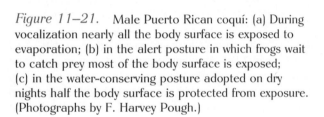

Figure 11–21. Male Puerto Rican coquí: (a) During vocalization nearly all the body surface is exposed to evaporation; (b) in the alert posture in which frogs wait to catch prey most of the body surface is exposed; (c) in the water-conserving posture adopted on dry nights half the body surface is protected from exposure. (Photographs by F. Harvey Pough.)

(c)

On dry nights the behavior of the frogs is quite different. The males move from their retreat sites to their calling stations, but they call only sporadically. Most of the time they rest in a water-conserving posture in which the body and chin are flattened against the leaf surface and the limbs are pressed against the body (Figure 11–21c). A frog in this posture exposes only half its body surface to the air, thereby reducing its rate of evaporative water loss. The effectiveness of the postural adjustments is illustrated by the water losses of frogs in the forest at El Verde, Puerto Rico, on dry nights. Frogs in one test group were placed individually in small wire mesh cages that were placed on leaf surfaces. A second group was composed of unrestrained frogs sitting on leaves. The caged frogs spent most of the night climbing around the cages trying to get out. This activity, like vocalization, exposed the entire body surface to the air, and the caged frogs had an evaporative water loss that averaged 27.5 percent of their initial body mass. In contrast, the unrestrained frogs adopted water-conserving postures and lost an average of only 8 percent of their initial body mass by evaporation.

Experiments showed that the jumping ability of coquíes was not affected by an evaporative loss of as much as 10 percent of the initial body mass, but a loss of 20 percent or more substantially decreased the distance frogs could jump (Beuchat et al. 1984). Thus, coquíes use behavior to limit their evaporative water losses on dry nights to levels that do not affect their ability to escape from predators or to capture prey. Without those behaviors, however, they would probably lose enough water by evaporation to affect their survival.

Uptake and Storage of Water

The mechanisms that amphibians use for obtaining water in terrestrial environments have received less attention than those for retaining it. Amphibians do not drink water. Because of the permeability of their skins, species that live in aquatic habitats face a continuous osmotic influx of water that they must balance by producing urine. Species in arid habitats rarely encounter enough liquid water in one place to drink it, and if they should find a puddle they can quickly absorb the water they need through their skins. The impressive adaptations of terrestrial amphibians are ones that facilitate rehydration from limited sources of water. One such adaptation is the **pelvic patch**. This is an area of highly vascularized skin in the pelvic region that is responsible for a very large portion of an anuran's cutaneous water absorption. Toads that are dehydrated and completely immersed in water rehydrate only slightly faster than those placed in water just deep enough to wet the pelvic area. In arid regions, water is frequently available as a thin layer of moisture on a rock, or as wet soil. The pelvic patch allows an anuran to absorb this water.

The urinary bladder plays an important role in the water relations of terrestrial amphibians, especially anurans. Amphibian kidneys produce urine that is hyposmotic to the blood, so the urine in the bladder is dilute. Amphibians can reabsorb water from urine to replace water they lose by evaporation, and terrestrial amphibians have larger bladders than aquatic species. Storage capacities of 20 to 30 percent of the body mass of the animal are common for terrestrial anurans, and some species have still larger bladders: The Aus-

tralian desert frogs *Notaden nicholsi* and *Neobatrachus wilsmorei* can store urine equivalent to about 50 percent of their body mass, and a bladder volume of 78.9 percent of body mass has been reported for the Australian frog *Helioporus eyrei*.

Behavior is as important in facilitating water uptake as it is in reducing water loss. James Dole (1965, 1967) studied a population of leopard frogs, *Rana pipiens*, that spent their summer activity season in grassy meadows where they had no access to ponds. The frogs spent the day in retreats they created by pushing vegetation aside to expose moist soil. In the retreats, the frogs rested with the pelvic patch in contact with the ground, and tests showed that the frogs were able to absorb water from the soil. On nights when dew formed, many frogs moved from their retreats and spent some hours in the early morning sitting on dew-covered grass before returning to their retreats. Leopard frogs in the meadow showed a daily pattern of water gain and loss during a period of several days when no rain fell: In the morning the frogs were sleek and glistening with moisture, and they had urine in their bladders. That observation indicates that in the morning the frogs had enough water to form urine. By evening the frogs had dry skins, and little urine in the bladder, suggesting that as they lost water by evaporation during the day they had reabsorbed water from the urine to maintain the water content of their tissues.

By the following morning the frogs had absorbed more water from dew and were sleek and well-hydrated again. Net gains and losses of water were shown by daily fluctuations in body masses of the frogs: In the mornings they were as much as 4 or 5 percent heavier than their overall average mass, and in the evenings they were lighter than the average by a similar amount. Thus, these terrestrial frogs were able to balance their water budgets by absorbing water from moist soil and from dew to replace the water they lost by evaporation and in urine. As a result they were independent of sources of water like ponds or streams and were able to colonize meadows and woods far from any permanent sources of water.

Poison Glands and Other Defense Mechanisms

Although there is evidence that the secretions of the mucous glands of some species of amphibians are irritating or toxic to predators, the chemical defense system is located primarily in the poison glands (Figure 11–22). These glands are concentrated on the dorsal surfaces of the animal, and defense postures of both anurans and urodeles present the glandular areas to potential predators. A wide variety of irritating and, in some cases, exceedingly toxic compounds are produced by amphibians. The compounds produced by different groups reflect their phylogenetic relationship, and skin toxins have been helpful in classification in some taxa (Box 11–3).

Many amphibians advertise their distasteful properties with conspicuous **aposematic**, or warning, colors and behaviors. A predator that makes the mistake of seizing one is likely to spit it out because it is distasteful. The toxins in the skin may also induce vomiting that reinforces the unpleasant experience for the predator. Subsequently, the predator will remember its unpleasant experience and avoid the distinctly marked animal that produced it. Some toxic amphibians combine a cryptic dorsal color with an aposematic ventral pattern. Normally, the cryptic color conceals them from predators, but if they are attacked they adopt a posture that displays the brightly colored ventral surface (Figure 11–23).

Red efts are classic examples of aposematic animals. Their color and behavior make them extremely conspicuous and they contain tetrodotoxin, a potent neurotoxin. Touching an eft to your lips produces an immediate unpleasant numbness and tingling sensation, and the behavior of animals that normally prey on salamanders indicates that it affects them the same way. As a result, an eft that is attacked by a predator is likely to be rejected before it is injured. After one or two such experiences, a predator will no longer attack efts. Support for the belief that this protection may operate in nature is provided by the observation that 4 of 11 wild-caught bluejays (*Cyanocitta cristata*) refused to attack the first red eft they were offered in a laboratory feeding trial (Tilley et al. 1982). That behavior suggests that those four birds had learned to avoid red efts before they were captured. The remaining seven birds attacked at least one eft, but dropped it immediately. After one or two experiences of this sort, the birds made retching movements at the sight of an eft and refused to attack.

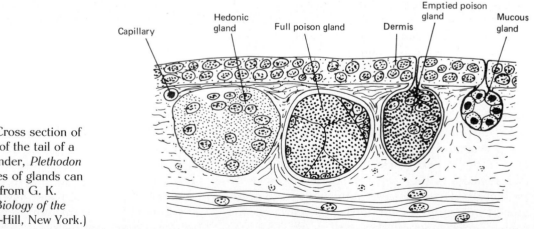

Figure 11–22. Cross section of skin from the base of the tail of a red-backed salamander, *Plethodon cinereus*. Three types of glands can be seen. (Modified from G. K. Noble, 1931, *The Biology of the Amphibia*, McGraw-Hill, New York.)

A great diversity of pharmacologically active substances has been found in the skins of amphibians. Some of them are extremely toxic and others less toxic but capable of producing unpleasant sensations when a predator bites an amphibian. Biogenic amines such as serotonin and histamine, peptides such as bradykinin, and hemolytic proteins have been found in frogs and salamanders belonging to many families. Many of these substances, such as bufotoxin, physalaemin, and leptodactyline, are named for the animals in which they were discovered.

Cutaneous alkaloids are abundant and diverse among the poison-dart frogs, the family Dendrobatidae, of the New World tropics (Myers and Daly 1983). More than 200 new alkaloids have been described from dendrobatids, mostly species in the genera *Dendrobates* (about 50 species) and *Phyllobates* (five species). Most of these frogs are brightly colored and move about on the ground surface in daylight, making no attempt at concealment.

The name poison-dart frogs refers to the use by South American Indians of the toxins of some of these frogs to poison the tips of the blowgun darts used for hunting. The use of frogs in this manner appears to be limited to three species of *Phyllobates* that occur in western Colombia, although plant poisons like curare are used to poison blowgun darts in other parts of South America (Myers et al. 1978). A unique alkaloid, batrachotoxin, occurs in the genus *Phyllobates*. Batrachotoxin is a potent neurotoxin that prevents the closing of sodium channels in nerve and muscle cells, leading to irreversible depolarization and producing cardiac arrhythmias, fibrillation, and cardiac failure.

The bright yellow *Phyllobates terribilis* is the largest and most toxic species in the genus. The Emberá Choco Indians of Colombia use *Phyllobates terribilis* as a source of poison for their blowgun darts. The dart points are rubbed several times across the back of a frog, and set aside to dry. The Indians handle the frogs carefully, holding them with leaves—a wise precaution because batrachotoxin is exceedingly poisonous. A single frog may contain up to 1900 micrograms of batrachotoxin, and less than 200 micrograms is probably a lethal dose for a human if it enters the body through a cut. Batrachotoxin is also toxic when it is eaten. In fact, the investigators inadvertently caused the death of a dog and a chicken in the Indian village in which they were living when the animals got into garbage that included plastic bags in which the frogs had been carried. Cooking destroys the poison and makes prey killed by darts anointed with the skin secretions of *Phyllobates terribilis* safe to eat.

Mimicry

The existence of unpalatable animals that deter predators with aposematic colors and behaviors offers the opportunity for other species that lack noxious qualities to take advantage of predators that have learned by experience to avoid the aposematic species. In this phenomenon, known as **mimicry**, the mimic (a species that lacks noxious

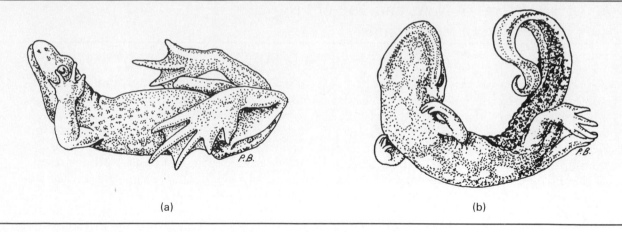

(a) (b)

Figure 11–23. Aposematic displays of amphibians present bright colors that predators can learn to associate with the animals' toxic properties. (a) The European fire-bellied toad has a cryptically colored dorsal surface and a brightly colored underside that is displayed in the *unken* reflex when the animal is attacked. (b) The Hong Kong newt has a brownish dorsal surface and a mottled red and black venter that is revealed by its aposematic display. [(a) Modified from H. Gadow, 1909, *Amphibia and Reptiles*, London; (b) from a photograph by E. D. Brodie, Jr.]

properties) resembles a noxious model and that resemblance causes a third species, the dupe, to mistake the mimic for the model (Pasteur 1982). Some of the best known cases of mimicry among vertebrates involve salamanders (Pough 1988). One that has been investigated involves two color morphs of the common red-backed salamander, *Plethodon cinereus.*

Red-backed salamanders normally have dark pigment on the sides of the body, but in some regions an erythristic (*erythr* = red) color morph is found that lacks the dark pigmentation and has red-orange on the sides as well as on the back (Figure 11–24). These erythristic morphs resemble red efts, and could be mimics of efts. Red-backed salamanders are palatable to many predators, and mimicry of the noxious red efts might confer some degree of protection on individuals of the erythristic morph. That hypothesis was tested in a series of experiments (Brodie and Brodie 1980). Salamanders were put in leaf-filled trays from which they

could not escape, and the trays were placed in a forest where birds were foraging. The birds learned to search through the leaves in the trays to find the salamanders. This is a very life-like situation for a test of mimicry because some species of birds are important predators of salamanders. For example, the salamanders *Plethodon cinereus* and *Desmognathus ochrophaeus* made up 25 percent of the prey items brought to the nest by hermit thrushes (*Catharus guttatus*) in western New York.

Three species of salamanders were used in the experiments, and the number of each species was adjusted to represent a hypothetical community of salamanders containing 40 percent dusky salamanders (*Desmognathus ochrophaeus*), 30 percent red efts, 24 percent striped red-backed salamanders, and 6 percent erythristic red-backed salamanders. The dusky salamanders are palatable to birds and are light brown; they do not resemble either efts or red-backed salamanders and they served as a control in the experiment. The striped red-backed

Figure 11–24. The red eft (a) is aposematically colored and toxic to predators. The red-backed salamander (b) is palatable. The erythristic color morph of the red-backed salamander (c) is a mimic of the red eft. [(a) and (c) Photographs by F. Harvey Pough; (b) photograph by Michael J. Hopiak, courtesy of the Cornell University Herpetology Collection.]

salamanders represent a second control: The hypothesis of mimicry of red efts by erythristic salamanders leads to the prediction that the striped salamanders, which do not look like efts, will be eaten by birds, whereas the erythristic salamanders, which are as palatable as the striped ones but which do look like the noxious efts, will not be eaten.

A predetermined number of each kind of salamander was put in the trays and birds were allowed to forage for 2 hours. At the end of that time the salamanders that remained were counted. As expected, only 1 percent of the efts had been taken by birds, whereas 44 to 60 percent of the palatable salamanders had disappeared (Table 11–3). As predicted, the birds ate fewer of the erythristic form of the red-backed salamanders than they ate of the striped form.

These results indicate that the erythristic morph of the red-backed salamander does obtain some protection from avian predators as a result of its resemblance to the red eft. In this case the resemblance is visual, but mimicry can exist in any sensory modality to which a dupe is sensitive, and olfactory mimicry by amphibians might be effective against predators such as shrews and snakes, which rely on scent to find and identify prey. This possibility has scarcely been considered, but careful investigations may yield fascinating new examples.

Table 11–3. Differential survival of salamanders exposed to foraging birds.

Experimental Design

Hypothesis: The erythristic morph of *Plethodon cinereus* is a mimic of the red eft.

Predictions:
1. Birds will not eat efts because the efts are noxious.
2. Birds will readily eat dusky salamanders, which are palatable and not mimetic.
3. Birds will eat the striped *Plethodon,* which are also palatable and not mimetic.
4. Birds will mistake the erythristic *Plethodon* for efts and will not eat them.

Results

Percent of Salamanders Gone from Trays

| | | *Plethodon* | |
Red Efts	Dusky Salamanders	Striped	Erythristic
1.0	52.6	60.1	43.9

Interpretation

The predictions of the hypothesis were supported:
1. Birds did not eat the noxious red efts (prediction 1).
2. Birds did eat the palatable, nonmimetic dusky salamanders (prediction 2).
3. Birds ate the striped morph of *Plethodon* (prediction 3).
4. Birds ate significantly fewer of the mimetic morph of *Plethodon* than of the striped morph (prediction 4).

Source: Based on E. D. Brodie, Jr., and E. D. Brodie III, 1980, *Science* 208:181–182.

Summary

The phylogenetic relationships of the three orders of amphibians are poorly known. Several shared derived characters suggest the possibility of a common origin for the three groups, but there is little agreement about their closest relative. The three orders of amphibians have evolved very diverse body forms and ways of life.

Locomotor adaptations distinguish the orders. Salamanders (order Urodela or Caudata) usually have short, sturdy legs that are used with lateral undulation of the body in walking. Aquatic salamanders use lateral undulations of the body and tail to swim, and some specialized aquatic species are elongate and have very small legs. Frogs and toads (order Anura) are characterized by specializations of the pelvis and hindlimbs that permit both legs to be used simultaneously to deliver a powerful thrust used both for jumping and for swimming. Many anurans walk quadrupedally when they move slowly and some are agile climbers. The caecilians (order Gymnophiona) are so little known that they lack a common name. They are legless tropical amphibians; some are burrowers and others are aquatic.

The diversity of reproductive modes of amphibians exceeds that of any other group of vertebrates except the fishes. Fertilization is internal in all but the most primitive salamanders, but most frogs rely on external fertilization. All caecilians have internal fertilization. Many species of amphibians have aquatic larvae. Tadpoles, the aquatic larvae of anurans, are specialized for life in still or flowing water, and some species of frogs deposit their tadpoles in very specific sites such as the pools of water that accumulate in the leaf axils of bromeliads or other plants. The specializations of tadpoles are entirely different from the specializations of frogs, and metamorphosis causes changes in all parts of the body. Direct development that omits the larval stage is also widespread among anurans and is often combined with parental care of the eggs. Viviparity occurs in all three orders.

In many respects the biology of amphibians is determined by properties of their skin. Hedonic glands are key elements in reproductive behaviors, poison glands protect the animals against predators, and mucous glands keep the skin moist, facilitating gas exchange. Above all, the permeability of the skin to water limits most amphibians to microhabitats in which they can control water gain and loss. That sounds like a severe restriction, but in the proper microhabitat amphibians can utilize the permeability of their skin to achieve a remarkable degree of independence of standing water. Thus, the picture that is sometimes presented of amphibians as animals barely hanging on as a sort of evolutionary oversight is misleading. Only a detailed examination of all facets of their biology can produce an accurate picture of amphibians as organisms.

An examination of that sort reinforces the view that the skin is a dominant structural characteristic of amphibians. This is true not only in terms of the limitations and opportunities presented by its permeability to water and gases, but also as a result of the intertwined functions of the skin glands in defensive and reproductive behaviors. The structure and function of the skin may be primary characteristics that have shaped the evolution and ecology of amphibians.

References

Beuchat, C. A., F. H. Pough, and M. M. Stewart. 1984. Response to simultaneous dehydration and thermal stress in three species of Puerto Rican frogs. *Journal of Comparative Physiology* B154:579–585.

Blaustein, A. R. and R. K. O'Hara. 1981. Genetic control of sibling recognition? *Nature* 290:246–248.

Brodie, E. D. Jr. and E. D. Brodie III. 1980. Differential avoidance of mimetic salamanders by free-ranging birds. *Science* 208:181–182.

Carroll, R. L. 1987. *Vertebrate Paleontology and Evolution.* W. H. Freeman, New York.

Carroll, R. L. and P. J. Currie. 1975. Microsaurs as possible apodan ancestors. *Zoological Journal of the Linnaean Society* 57:229–247.

Carroll, R. L. and R. Holmes. 1980. The skull and jaw musculature as guides to the ancestry of salamanders. *Zoological Journal of the Linnean Society* 68:1–40.

de Queiroz, K. and D. C. Cannatella. 1987. The monophyly and relationships of the Lissamphibia. *American Zoologist* 27:60A.

Dole, J. W. 1965. Summer movements of adult leopard frogs, *Rana pipiens* Schreber, in northern Michigan. *Ecology* 46:236–255.

Dole, J. W. 1967. The role of substrate moisture and dew in the water economy of leopard frogs, *Rana pipiens. Copeia* 1967:141–149.

Duellman, W. E. and L. Trueb. 1986. *Biology of Amphibians.* McGraw-Hill, New York.

Emerson, S. B. and D. Diehl. 1980. Toe pad morphology and mechanisms of sticking in frogs. *Biological Journal of the Linnean Society* 13:199–216.

Gardiner, B. G. 1982. Tetrapod classification. *Zoological Journal of the Linnaean Society* 74:207–232.

Gardiner, B. G. 1983. Gnathostome vertebrae and the classification of the Amphibia. *Zoological Journal of the Linnaean Society* 79:1–59.

Jaeger, R. G. 1981. Dear enemy recognition and the costs of aggression between salamanders. *American Naturalist* 117:962–974.

Jaeger, R. G., K. C. B. Nishikawa, and D. E. Barnard. 1983. Foraging tactics of a terrestrial salamander: costs of territorial defense. *Animal Behaviour* 31:191–198.

Klump, G. M. and H. C. Gerhardt. 1987. Use of nonarbitrary acoustic criteria in mate choice by female gray tree frogs. *Nature* 326:286–288.

Leong, A. S.-Y., M. J. Tyler, and D. J. C. Shearman. 1986. Gastric brooding: a new form in a recently discovered Australian frog of the genus *Rheobatrachus. Australian Journal of Zoology* 34:205–209.

Myers, C. W. and J. W. Daly. 1983. Dart-poison frogs. *Scientific American* 248(2):120–133.

Myers, C. W., J. W. Daly, and B. Malkin. 1978. A dangerously toxic new frog (Phyllobates) used by Embera Indians of western Colombia, with discussion of blowgun fabrication and dart poisoning. *Bulletin of the American Museum of Natural History* 161: 307–366.

Narins, P. M. and R. R. Capranica. 1976. Sexual differences in the auditory system of the tree frog *Eleutherodactylus coqui*. *Science* 192:378–380.

Narins, P. M. and R. R. Capranica. 1978. Communicative significance of the two-note call of the treefrog *Eleutherodactylus coqui*. *Journal of Comparative Physiology* A127:1–9.

Nieuwkoop, P. D. and L. A. Satasurya. 1976. Embryological evidence for a possible polyphyletic origin of recent amphibians. *Journal of Embryology and Experimental Morphology* 35:159–167.

Parsons, T. and E. Williams. 1963. The relationship of modern Amphibia: a reexamination. *Quarterly Review of Biology* 38:26–53.

Pasteur, G. 1982. A classificatory review of mimicry systems. *Annual Review of Ecology and Systematics* 13:169–199.

Pough, F. H. 1988. Mimicry of vertebrates: are the rules different? *American Naturalist* 131 Supplement:S67–S102.

Pough, F. H., T. L. Taigen, M. M. Stewart, and P. F. Brussard. 1983. Behavioral modification of evaporative water loss by a Puerto Rican frog. *Ecology* 64:244–252.

Reilly, S. M. 1987. Ontogeny of the hyobranchial apparatus in the salamanders *Ambystoma talpoideum* (Ambystomatidae) and *Notophthalmus viridescens* (Salamandridae): the ecological morphology of two neotenic species. *Journal of Morphology* 191:205–214.

Roth, G. and D. B. Wake. 1985. Trends in the functional morphology and sensorimotor control of feeding behavior in salamanders: an example of the role of internal dynamics in evolution. *Acta Biotheoretica* 34:175–192.

Ryan, M. J. 1985. *The Túngara Frog: A Study in Sexual Selection and Communication.* University of Chicago Press, Chicago.

Seale, D. B. 1988. Amphibia. Pages 467–552 in *Animal Energetics*, volume 2, edited by F. J. Vernberg and T. J. Pandian, Academic Press, New York.

Shoemaker, V. and L. McClanahan. 1975. Evaporative water loss, nitrogen excretion and osmoregulation in phyllomedusine frogs. *Journal of Comparative Physiology* 100:331–345.

Taigen, T. L., F. H. Pough, and M. M. Stewart. 1984. Water balance of terrestrial anuran eggs (*Eleutherodactylus coqui*): importance of parental care. *Ecology* 65:248–255.

Taigen, T. L. and K. D. Wells. 1985. Energetics of vocalization by an anuran amphibian (*Hyla versicolor*). *Journal of Comparative Physiology* B155:163–170.

Taigen, T. L., K. D. Wells, and R. L. Marsh. 1985. The enzymatic basis of high metabolic rates in calling frogs. *Physiological Zoology* 58:719–726.

Tilley, S. G., B. L. Lundrigan, and L. P. Brower. 1982. Erythrism and mimicry in the salamander *Plethodon cinereus*. *Herpetologica* 38:409–417.

Townsend, D. S., M. M. Stewart, and F. H. Pough. 1984. Male parental care and its adaptive significance in a neotropical frog. *Animal Behaviour* 32: 421–431.

Townsend, D. S., M. M. Stewart, F. H. Pough, and P. F. Brussard. 1981. Internal fertilization in an oviparous frog. *Science* 212:465–471.

Trueb, L. and R. Cloutier. 1987. Historical constraints on lissamphibian osteology. *American Zoologist* 27:33A.

Tyler, M. J. (editor). 1983. *The Gastric Brooding Frog.* Croom Helm, Beckenham, Kent, England.

Wake, D. B. 1970. Aspects of vertebral evolution in the modern Amphibia. *Forma et Functio* 3:33–60.

Wake, M. H. 1977. The reproductive biology of caecilians: an evolutionary perspective. Pages 73–101 in *The Reproductive Biology of Amphibians*, edited by D. H. Taylor and S. I. Guttman. Plenum Press, New York.

Wake, M. H. 1978. The reproductive biology of *Eleutherodactylus jasperi* (Amphibia, Anura, Leptodactylidae) with comments on the evolution of live-bearing systems. *Journal of Herpetology* 12:121–133.

Wake, M. H. 1980. The reproductive biology of *Nectophrynoides malcolmi* (Amphibia, Bufonidae) with comments on the evolution of reproductive modes in the genus *Nectophrynoides*. *Copeia* 1980:193–209.

Wake, M. H. 1986. A perspective on the systematics and morphology of the Gymnophiona (Amphibia). *Mémoires de la Societé de Zoologie Français.* Number 43:21–38.

Waldman, B. 1982. Sibling association among schooling toad tadpoles, field evidence and implications. *Animal Behaviour* 30:700–713.

Wassersug, R. J. 1975. The adaptive significance of the tadpole stage with comments on the maintenance of complex life cycles in anurans. *American Zoologist* 15:405–417.

Wells, K. D. 1980. Evidence for growth of tadpoles during parental transport in *Colostethus inguinalis*. *Journal of Herpetology* 14:428–430.

Wells, K. D. 1981. Parental behavior of male and female frogs. In *Natural Selection and Social Behavior: Recent Research and New Theory*, edited by R. D. Alexander and D. W. Tinkle. Chiron Press, New York, pp. 184–197.

Wells, K. D. and T. L. Taigen. 1986. The effect of social interactions on calling energetics in the grey treefrog *(Hyla versicolor)*. *Behavioral Ecology and Sociobiology* 19:9–18.

Weygoldt, P. 1980. Complex brood care and reproductive behavior in captive poison-arrow frogs, *Dendrobates pumilio*. *Behavioral Ecology and Sociobiology* 7:329–332.

Turtles are at once the most recognizable and most enigmatic of vertebrates. Some aspects of their morphology place them close to the early amniotic vertebrates of the late Paleozoic that were discussed in Chapter 10, but in other respects turtles are so specialized that no intermediate character states can be identified to link them to other vertebrates. The shell of a turtle is, of course, its most distinctive character but the rest of the skeleton is equally modified and the earliest fossil turtles (from the Triassic) are nearly as specialized as modern forms.

Turtles provide a contrast to amphibians in the relative lack of diversity in their life histories. All turtles lay eggs and none exhibit parental care. Turtles show morphological specializations associated with terrestrial, freshwater, and marine habitats, and marine turtles make long-distance migrations that rival those of birds. Probably, turtles and birds use many of the same navigation mechanisms to find their way. Most turtles are long-lived animals with relatively poor capacities for rapid population growth, and many, especially sea turtles and large tortoises, are endangered by human activities. Some efforts to conserve turtles have apparently been frustrated by a peculiarity of the embryonic development of some species of turtles—the sex of an individual is determined by the temperature it is exposed to in the nest. This experience emphasizes the critical importance of information about the basic biology of animals to successful conservation and management.

Turtles

12

Everyone Recognizes a Turtle

Turtles found a successful approach to life in the Triassic and have scarcely changed since. The shell, which is the key to their success, has also limited the diversity of the group (Table 12–1). For obvious reasons flying or gliding turtles have never existed, and even arboreality is only slightly developed. Shell morphology reflects the ecology of turtle species: The most terrestrial forms, the tortoises of the family Testudinidae, have high domed shells and elephant-like feet (Figure 12–1a). Smaller species of tortoises may show adaptations for burrowing. The gopher tortoises of North America are an example. The forelegs are flattened into scoops and the dome of the shell is reduced. The Bolson tortoise, recently discovered in northern Mexico, constructs burrows a meter or more deep and several meters long in the hard desert soil. These tortoises bask at the mouths of their burrows, and when a predator appears they throw themselves down the steep entrance tunnels of the burrows to escape, just as an aquatic turtle dives off a log. The pancake tortoise of Africa is a radical departure from the usual tortoise morphology (Figure 12–1b). The shell is flat and flexible because its ossification is much reduced. This turtle lives in rocky foothill regions and scrambles over the rocks with nearly as much agility as a lizard. When threatened by a predator, the pancake tortoise crawls into a rock crevice and uses its legs to wedge itself in place. The flexible shell presses against the overhanging rock and creates so much friction that it is almost impossible to pull the tortoise out.

Other terrestrial turtles have moderately domed **carapaces** (upper shells), like the box turtles of the family Emydidae (Figure 12–1c). This is only one of several kinds of turtles that have evolved flexible regions in the **plastron** (lower shell) that allow the front and rear lobes to be pulled upward to close the openings of the shell. Aquatic turtles have low carapaces that offer little resistance to movement through water. The family Emydidae contains a large number of pond turtles (Figure 12–1d), including the painted turtles and the red-eared turtles usually seen in pet stores and anatomy and physiology laboratory courses.

The snapping turtles (family Chelydridae) and the mud and musk turtles (family Kinosternidae) prowl along the bottoms of ponds and slow rivers and are not particularly streamlined (Figure 12–1f and g). The mud turtle has a hinged plastron, but the musk and snapping turtles have very reduced plastrons. They rely on strong jaws for protection. A reduction in the size of the plastron makes these species more agile than most turtles, and musk turtles may climb several feet into trees, probably to bask. If a turtle falls on your head while you are canoeing, it is probably a musk turtle.

The soft-shelled turtles (family Trionychidae) are specialized for fast swimming (Figure 12–1e).

419

Table 12–1. Living families of turtles.

Family Pleurodira (about 50 species)
    Chelidae: about 30 species of very small (15 cm) to medium-size (40 cm) aquatic turtles from Australia, New Guinea, and South America
    Pelomedusidae: about 19 species of small (30 cm) to large (80 cm) aquatic turtles found in South America and Africa

Family Cryptodira (about 175 species)
    Cheloniidae: six species of large to very large (1.25 meters) marine turtles with a worldwide distribution in warm and temperate oceans
    Dermochelyidae: a single species of very large sea turtle (1.5 meters) found in tropical, temperate, and subarctic seas; lacks epidermal scutes on the shell
    Dermatemydidae: a single species of aquatic turtle about 45 cm in length found in southern Mexico and northern Central America
    Chelydridae: two species of North American aquatic turtles; the snapping turtle attains a shell length of nearly 50 cm, and the alligator snapper grows to 80 cm
    Kinosternidae: about 20 species of small to medium-size aquatic turtles found in North, Central, and South America
    Emydidae: the largest family of turtles with about 85 aquatic and terrestrial species ranging in size from small to medium; occur on all continents except Australia, Antarctica, and sub-Saharan Africa
    Platysternidae: a single species of small aquatic turtle from southeast Asia
    Testudinidae: about 35 living species (and 8 recently extinct) of terrestrial tortoises; occur on all continents except Australia and Antarctica
    Carettochelyidae: one species, the New Guinea plateless turtle; lacks epidermal scutes on the shell
    Trionychidae: about 22 species of small to medium-size aquatic turtles with reduced ossification of the shell; occur in North America, Africa, and Asia

---

Source: Based on P. C. H. Pritchard, 1979, *Encyclopedia of Turtles*, T. F. H. Publications, West Neptune, N.J.

The ossification of the shell is greatly reduced, lightening the animal, and the feet are large with extensive webbing. Soft-shelled turtles lie in ambush for prey partly buried in debris on the bottom of the pond. Their long necks allow them to reach considerable distances to seize the invertebrates and small fish on which they feed.

The two suborders of living turtles can be traced through fossils to the Mesozoic (Gaffney et al. 1987). The **cryptodires** (*crypto* = hidden, *dire* = neck) retract the head into the shell by bending the neck in a vertical S-shape. The **pleurodires** (*pleuro* = side) retract the head by bending the neck horizontally. All of the turtles discussed so far have been cryptodires, and these are the dominant suborder of chelonians. Cryptodires are the only turtles now found in most of the Northern Hemisphere, and there are aquatic and terrestrial cryptodires in South America and terrestrial ones in Africa. Only Australia has no cryptodires. Pleurodires are now found only in the Southern Hemi-

sphere, although they had worldwide distribution in the late Mesozoic and early Cenozoic. *Stupendemys*, a pleurodire from the Pliocene of Venezuela, is the largest turtle known; it had a carapace more than 2 meters long. All the living pleurodires are at least semiaquatic, but some fossil pleurodires had high, domed shells that suggest they may have been terrestrial. The most terrestrial of the living pleurodires are probably the African pond turtles (Figure 12–1h), which readily move overland from one pond to another.

The snake-necked pleurodire turtles (family Chelidae) are found in South America, Australia, and New Guinea (Figure 12–1i). As their name implies, they have long, slender necks. In some species the length of the neck is considerably greater than that of the vertebral column. These forms feed on fish that they catch with a sudden dart of the head. Other snake-necked turtles have much shorter necks. Some of these feed on mollusks and have enlarged palatal surfaces used to crush shells.

Figure 12–1. Body forms of turtles: (a) tortoise *(Testudo)*; (b) pancake tortoise *(Malacochersus)*; (c) terrestrial box turtle *(Terrapene)*; (d) pond turtle *(Pseudemys)*; (e) The softshell turtle *(Trionyx)*; (f) mud turtle *(Kinosternon)*; (g) alligator snapping turtle *(Macroclemys)*; (h) African pond turtle *(Pelusios)*; (i) Australian snake-necked turtle *(Chelodina)*; (j) South American matamata *(Chelys)*; (k) loggerhead sea turtle *(Caretta)*; (l) leatherback sea turtle *(Dermochelys).*

The same specialization for feeding on mollusks is seen in certain cryptodire turtles.

An unusual feeding method among turtles is found in a pleurodire, the matamata of South America (Figure 12–1j). Large matamatas reach shell lengths of 40 centimeters. They are bizarre-looking animals. The shell and head are broad and flattened, and numerous flaps of skin project from the sides of the head and the broad neck. To these are added trailing bits of adhering algae. The effect is exceedingly cryptic. It is hard to recognize a matamata as a turtle even when you see one in clear water sitting on the slate bottom of an aquarium; against the mud and debris of a muddy river bottom they are practically invisible. In addition to obscuring the shape of the turtle, the flaps of skin on the head are sensitive to minute vibrations in water caused by the passage of a fish. When it senses the presence of prey, the matamata abruptly opens its mouth and expands its throat. Water rushes in carrying the prey with it, and the matamata closes its mouth, expels the water, and swallows the prey. The matamata lacks the horny beak that other turtles use for seizing prey or biting off pieces of plants.

Living marine turtles are cryptodires. The families Cheloniidae and Dermochelyidae show more extensive specialization for aquatic life than any freshwater turtle. All have the forelimbs modified as flippers. Cheloniids retain epidermal scutes on the shell (Figure 12–1k). The largest of the sea turtles of the family Cheloniidae is the loggerhead, which once reached weights exceeding 400 kilograms. The largest marine turtle, the leatherback, reaches shell lengths of more than 2 meters and weights exceeding 600 kilograms (Figure 12–1l). The dermal ossification has been reduced to bony platelets embedded in the skin. This is a pelagic turtle that ranges far from land, and it has a wider geographic distribution than any other ectothermal amniote: Leatherback turtles penetrate far into cool temperate seas, and have been recorded in the Atlantic from Newfoundland to Argentina and from Norway to South Africa, and in the Pacific from the USSR to Tasmania. In cold water leatherback turtles maintain body temperatures substantially above water temperature, using countercurrent heat exchangers to retain the heat released by muscular activity (Chapter 4). Leatherback turtles have recently been reported to dive to depths of more than 1000 meters (Mrosovsky 1987). One dive that drove the depth recorder off-scale is estimated to have reached 1200 meters. If that is true, it exceeds the deepest dive recorded for a sperm whale (1140 meters) and is probably a record for air-breathing vertebrates. Leatherback turtles feed largely on jellyfish, whereas the smaller hawksbill sea turtles eat sponges that are defended by spicules of silica (glass) as well as a variety of chemicals (including alkaloids and terpenes) that are toxic to most vertebrates (Meylan 1988).

Phylogenetic Relationships of Turtles

Turtles show a combination of primitive features and highly specialized characters that are not shared with any other group of vertebrates, and their phylogenetic affinities are not clearly known. The turtle lineage probably originated among the early amniotes of the late Carboniferous. Like those animals, turtles have **anapsid** ("without an arch") skulls, but the shells and postcranial skeletons of turtles are unique.

The earliest turtles are found in late Triassic deposits in Germany and Thailand. These animals had nearly all of the specialized characteristics of modern turtles, and shed no light on the phylogenetic affinities of the group. *Proganochelys* from the German Triassic beds was larger than most living turtles, with a high, arched shell nearly a meter long. The marginal teeth had been lost and the maxilla, premaxilla, and dentary bones were probably covered with a horny beak as they are in modern turtles. The skull of *Proganochelys* retained the supratemporal and lacrimal bones and the lacrimal duct, and the palate had rows of denticles; all these structures have been lost by modern turtles. The plastron of *Proganochelys* also contained some bones that have been lost by modern turtles, and the vertebrae of the neck lack specializations that would have allowed the head to be retracted into the shell.

Turtles with neck vertebrae specialized for retraction are not known before the Cretaceous, but differences in the skulls and shells allow the pleurodire and cryptodire lineages to be traced back to the late Triassic (shells of pleurodires) and late Jurassic (skulls of cryptodires) (Carroll 1987). The otic capsules of all turtles beyond the proganochelids are enlarged, and the jaw adductor muscles bend posteriorly over the otic capsule (Figure 12–2). The

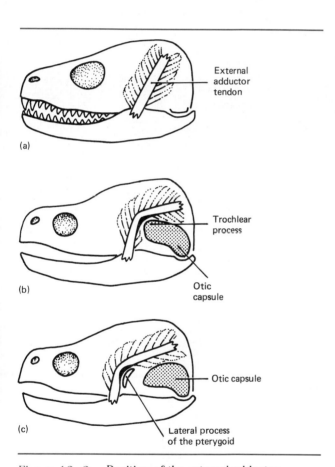

(a)

External adductor tendon

(b)

Trochlear process

Otic capsule

(c)

Otic capsule

Lateral process of the pterygoid

Figure 12–2. Position of the external adductor tendon in (a) primitive anapsids, (b) cryptodiran turtles, and (c) pleurodiran turtles. (From E. S. Gaffney, 1975, *Bulletin of the American Museum of Natural History* 15:387–436.)

muscles pass over a pulley-like structure, the **trochlear process**. In cryptodires the trochlear process is formed by the anterior surface of the otic capsule itself, whereas in pleurodires it is formed by a lateral process of the pterygoid. Fusion of the pelvic girdle to the carapace and plastron distinguishes pleurodires from cryptodires, which have a suture attaching the shell to the girdle.

Turtle Structure and Function

Turtles are among the most bizarre vertebrates: Covered in bone, with the limbs inside the ribs, and with horny beaks instead of teeth—if turtles had become extinct at the end of the Mesozoic they would rival dinosaurs in their novelty. However, because they survived they are regarded as commonplace, and are used (inappropriately, because they are so specialized) in comparative anatomy courses to represent ectothermal amniotes. In fact, the morphology and physiology of turtles are specialized, and they are ecologically quite different from the other ectothermal amniotes, the crocodilians (which are archosaurs) and the lepidosaurs (the tuatara, lizards, snakes, and amphisbaenians).

Shell and Skeleton

The shell is the most distinctive feature of a turtle (Figure 12–3). The carapace is composed of dermal bone that grows from 59 separate centers of ossification. Eight plates along the dorsal midline form the neural series and are fused to the neural arches of the vertebrae. Lateral to the neural bones are eight paired costal bones, which are fused to the broadened ribs. The ribs of turtles are unique among tetrapods in being external to the girdles. Eleven pairs of peripheral bones, plus two unpaired bones in the dorsal midline, form the margin of the carapace. The plastron is formed largely from dermal ossifications, but the entoplastron is derived from the interclavicle, and the paired epiplastra anterior to it are derived from the clavicles. Processes from the hypoplastron fuse with the first

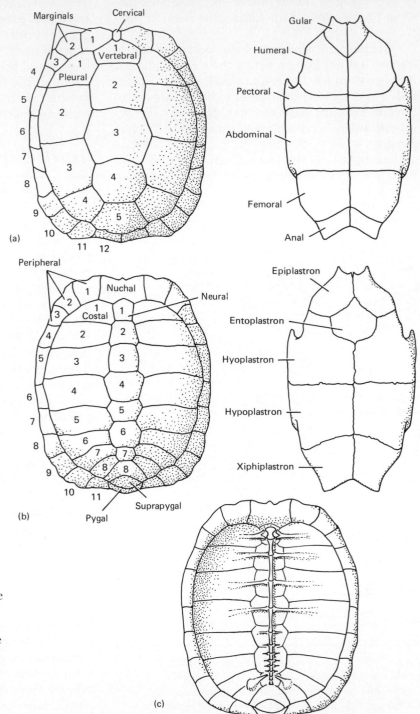

Figure 12–3. Shell and vertebral column of a turtle: (a) epidermal scutes of the carapace (left) and plastron (right); (b) dermal bones of the carapace (left) and plastron (right); (c) vertebral column of a turtle, seen from the inside of the carapace. Note that anteriorly the ribs articulate with two vertebral centra. (From R. Zangerl, 1969, in *Biology of the Reptilia,* volume 1, edited by C. Gans, A. d'A. Bellairs, and T. S. Parsons, Academic Press, London.)

and fifth pleurals, forming a rigid connection between the plastron and carapace.

The bones of the carapace are covered by horny scutes of epidermal origin that do not coincide in number or position with the underlying bones. The caparace has a row of five central scutes, bordered on each side by four lateral scutes. Eleven marginal scutes on each side turn under the edge of the caparace. The plastron is covered by a series of six paired scutes.

Flexible areas, called hinges, are present in the shells of a number of turtles. The most familiar examples are the North American and Asian box turtles (*Terrapene* and *Cuora*) in which a hinge between the hyoplastral and hypoplastral bones allows the anterior and posterior lobes of the plastron to be raised to close off the front and rear openings of the shell. Mud turtles (*Kinosternon*) have two hinges in the plastron: The anterior hinge runs between the epiplastra and the entoplastron (which is triangular in kinosternid turtles rather than diamond shaped) and the posterior hinge is between the hypoplastra and xiphiplastra. In the pleurodire turtle *Pelusios* a hinge runs between the mesoplastra and the hypolastra. Some species of tortoises have plastral hinges; in *Testudo* the hinge lies between the hyoplastra and xiphiplastra as it does in *Kinosternon*, but in another genus of tortoise, *Pyxis*, the hinge is anterior and involves a break across the entoplastron. The African forest tortoises (*Kinixys*) have a hinge on the posterior part of the carapace. The margins of the epidermal shields and the dermal bones of the carapace are aligned, and the hinge runs between the second and third pleural bones and the fourth and fifth costals. The presence of hinges is sexually dimorphic in some species of tortoises. The erratic phylogenetic occurrence of kinetic shells and differences among closely related species indicates that shell kinesis has evolved many times in turtles.

Asymmetry of the paired epidermal scutes is quite common among turtles, and modifications of the bony structure of the shell are seen in some families. Soft-shelled turtles lack peripheral ossifications and epidermal scutes. The distal ends of the broadened ribs are embedded in flexible connective tissue, and the caparace and plastron are covered with skin. The New Guinea river turtle (*Carretochelys*) is also covered by skin instead of scutes, but in this species the peripheral bones are present and the edge of the shell is stiff. The leatherback sea turtle (*Dermochelys*) has a carapace formed of cartilage with thousands of small polygonal bones embedded in it, and the plastral bones are reduced to a thin rim around the edge of the plastron. In the pancake tortoise (*Malachochersus*) the neural and pleural ossifications are greatly reduced, but the epidermal plates are well developed.

Modern turtles have only 18 presacral vertebrae, 10 in the trunk and 8 in the neck. The centra of the trunk vertebrae are elongated, and each centrum lies beneath one of the dermal bones in the dorsal midline of the shell. The centra are constricted in their centers and fused to each other. The neural arches in the anterior two-thirds of the trunk lie between the centra as a result of anterior displacement, and the spinal nerves exit near the middle of the preceding centrum. The ribs are also shifted anteriorly; they articulate with the anterior part of the neurocentral boundary, and in the anterior part of the trunk where the shift is most pronounced the ribs extend onto the preceding vertebra.

Cryptodires have two sacral vertebrae (the 19th and 20th vertebrae) with broadened ribs that meet the ilia of the pelvis. In pleurodires the pelvic girdle is firmly fused to the dermal carapace by the ilia dorsally and by the pubic and ischial bones ventrally, and the sacral region of the vertebral column is less distinct. The ribs on the 17th, 18th, 19th, and sometimes the 20th vertebrae are fused to the centra and end on the ilia or the ilia-carapacial junction.

The cervical vertebrae of cryptodires have articulations that permit the S-shaped bend used to retract the head into the shell. Specialized condyles (ginglymes) function as hinges that permit vertical rotation. This type of rotation, **ginglymoidy**, is peculiar to cryptodires, but the anatomical details vary within the group. In most families the hinge is formed by two successive ginglymoidal joints between the 6th and 7th and the 7th and 8th cervical vertebrae. The lateral bending of the necks of pleurodire turtles is accomplished by ball-and-socket or cylindrical joints between adjacent cervical vertebrae.

The Heart

The circulatory systems of tetrapods can be viewed as consisting of two circuits: The systemic circuit carries oxygenated blood from the heart to the head, trunk, and appendages, whereas the pulmonary circuit carries deoxygenated blood from the heart to the lungs. The blood pressure in the systemic circuit is higher than the pressure in the pulmonary circuit, and the two circuits operate in series. That is, blood flows from the heart through the lungs, back to the heart, and then to the body. The morphology of the hearts of birds and mammals makes this sequential flow obligatory, but the hearts of turtles and of squamates have the ability to shift blood between the pulmonary and systemic circuits.

This flexibility in the route of blood flow can be accomplished because the ventricular chambers in the hearts of turtles and squamates are in anatomical continuity instead of being divided by a septum like the ventricles of birds and mammals. The pattern of blood flow can best be explained by considering the morphology of the heart and how intracardiac pressure changes during a heartbeat. [For more details, see White (1968), Johansen and Burggren (1980), or Burggren (1987).] Figure 12–4 shows a schematic view of the heart of a turtle or squamate. The left and right atria are completely separate, and three subcompartments can be distinguished in the ventricle. A muscular ridge in the core of the heart divides the ventricle into two spaces, the **cavum pulmonale** and the **cavum venosum**. The muscular ridge is not fused to the wall of the ventricle, and thus the cavum pulmonale and the cavum venosum are only partly separated. A third subcompartment of the ventricle, the **cavum arteriosum**, is located dorsal to the cavum pulmonale and cavum venosum. The cavum arteriosum communicates with the cavum venosum through an intraventricular canal. The pulmonary artery opens from the cavum pulmonale, and the left and right aortic arches open from the cavum venosum.

The right atrium receives deoxygenated blood from the body via the sinus venosus and empties into the cavum venosum, and the left atrium receives oxygenated blood from the lungs and empties into the cavum arteriosum. The atria are separated from the ventricle by flap-like atrioventricular valves that open as the atria contract, and then close as the ventricle contracts, preventing blood from being forced back into the atria. The anatomical arrangement of the connections between the atria, their valves, and the three subcompartments of the ventricle are cru-

---

Figure 12–4. *(Facing page)* Blood flow in the heart of a turtle. (a) As the atria contract oxygenated blood (open arrows) from the left atrium (LA) enters the cavum arteriosum (CA) while deoxygenated blood (dark arrows) from the right atrium (RA) first enters the cavum venosum (CV) and then crosses the muscular ridge (MR) and enters the cavum pulmonale (CP). The atrioventricular valve (AVV) blocks the intraventricular canal (IVC) and prevents mixing of oxygenated and deoxygenated blood. (b) As the ventricle contracts the deoxygenated blood in the cavum pulmonale is expelled through the pulmonary arteries; the AVV closes, no longer obstructing the ICV; and the oxygenated blood in the cavum arteriosum is forced into the cavum venosum and expelled through the aortic arches. The adpression of the wall of the ventricle to the muscular ridge prevents mixing of deoxygenated and oxygenated blood. (c) Summary of the pattern of blood flow through the heart of a turtle. (Modified from N. Heisler et al., 1983, *Journal of Experimental Biology* 105:15–32.)

cial, because it is those connections that allow pressure differentials to direct the flow of blood and to prevent mixing of oxygenated and deoxygenated blood.

When the atria contract the atrioventricular valves open, allowing blood to flow into the ventricle. Blood from the right atrium flows into the cavum venosum, and blood from the left atrium

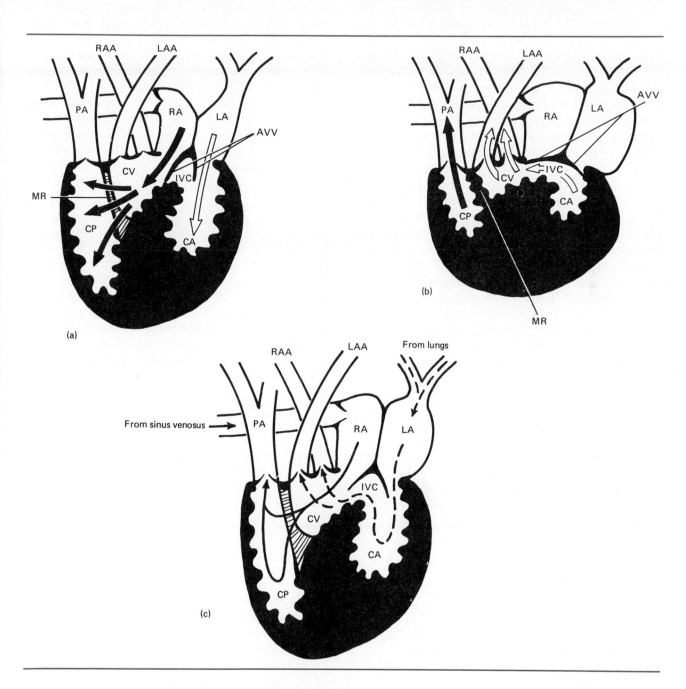

flows into the cavum arteriosum. At this stage in the heart beat the large median flaps of the valve between the right atrium and the cavum venosum are pressed against the opening of the intraventricular canal, sealing it off from the cavum venosum. As a result, the oxygenated blood from the left atrium is confined to the cavum arteriosum. Deoxygenated blood from the right atrium fills the cavum venosum and then continues over the muscular ridge into the cavum pulmonale.

When the ventricle contracts, the blood pressure inside the heart increases. Ejection of blood from the heart into the pulmonary circuit precedes flow into the systemic circuit because resistance is lower in the pulmonary circuit. As deoxygenated blood flows out of the cavum pulmonale into the pulmonary artery, the displacement of blood from the cavum venosum across the muscular ridge into the cavum pulmonale continues. As the ventricle shortens during contraction, the muscular ridge comes into contact with the wall of the ventricle and closes off the passage for blood between the cavum venosum and cavum pulmonale.

Simultaneously, blood pressure inside the heart increases, and the flaps of the right atrioventricular valve are forced into the closed position, preventing backflow of blood from the cavum venosum into the atrium. When the valve closes, it no longer blocks the intraventricular canal. Oxygenated blood from the cavum pulmonale can now flow through the intraventricular canal and into the cavum venosum. At this stage in the heart beat, the wall of the ventricle is pressed firmly against the muscular ridge, separating the oxygenated blood in the cavum venosum from the deoxygenated blood in the cavum pulmonale.

As the pressure in the ventricle continues to rise, the oxygenated blood in the cavum venosum is ejected into the aortic arches. This system effectively prevents mixing of oxygenated and deoxygenated blood in the heart, despite the absence of a permanent morphological separation of the two circuits.

Respiration

Primitive amniotes probably used movements of the rib cage to draw air into the lungs and to force it out, and lizards still employ that mechanism. The fusion of the ribs of turtles with their rigid shells makes that method of breathing impossible. Only the openings at the anterior and posterior ends of the shell contain flexible tissues. The lungs, which are large, are attached to the carapace dorsally and laterally. Ventrally, the lungs are attached to a sheet of nonmuscular connective tissue that is itself attached to the viscera (Figure 12–5). The weight of the viscera keeps this diaphragmatic sheet stretched downward.

Turtles produce changes in pressure in the lungs by contraction of muscles that force the viscera upward, compressing the lungs and expelling air, followed by contraction of other muscles that increase the volume of the visceral cavity, allowing the viscera to settle downward. Because the viscera are attached to the diaphragmatic sheet, which in turn is attached to the lungs, the downward movement of the viscera expands the lungs, drawing in air. In tortoises both inhalation and exhalation require muscular activity. The viscera are forced upward against the lungs by the contraction of the transverse abdominus muscle posteriorly

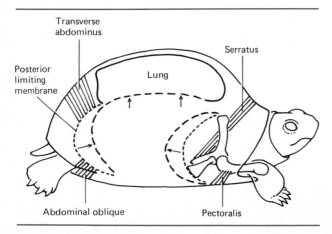

Figure 12–5. Schematic view of the lungs and respiratory movements of a tortoise. (Modified from C. Gans and G. M. Hughes, 1967, *Journal of Experimental Biology* 47:1–20.)

and the pectoralis muscle anteriorly. The transverse abdominus inserts on the cup-shaped connective tissue, the **posterior limiting membrane**, that closes off the posterior opening of the visceral cavity. Contraction of the transverse abdominus flattens the cup inward, thereby reducing the volume of the visceral cavity. The pectoralis draws the shoulder girdle back into the shell, further reducing the volume of the visceral cavity.

The inspiratory muscles are the abdominal oblique, which originates near the posterior margin of the plastron and inserts on the external side of the posterior limiting membrane, and the serratus, which originates near the anterior edge of the carapace and inserts on the pectoral girdle. Contraction of the abdominal oblique pulls the posterior limiting membrane outward, and contraction of the serratus rotates the pectoral girdle outward. Both of these movements increase the volume of the visceral cavity, allowing the viscera to settle back downward and causing the lungs to expand. The in-and-out movements of the forelimbs and the soft tissue at the rear of the shell during breathing are conspicuous.

The basic problem of respiring within a rigid shell are the same for most turtles, but the mechanisms show some variation. For example, aquatic turtles can use the hydrostatic pressure of water to help move air in and out of the lungs. In addition, many aquatic turtles are able to absorb oxygen and release carbon dioxide to the water. The pharynx and cloaca appear to be the major sites of aquatic gas exchange. In 1860, in his "Contributions to the Natural History of the U. S. A.," Louis Agassiz pointed out that the pharynx of softshell turtles contains fringe-like processes and suggested that these structures are used for underwater respiration. Subsequent study has shown that when softshell turtles are confined under water they use movements of the hyoid apparatus to draw water in and out of the pharynx, and that pharyngeal respiration accounts for most of the oxygen absorbed from the water. The Australian turtle *Rheodytes leukops* uses cloacal respiration. Its cloacal orifice is as much as 30 millimeters in diameter, and the turtle holds it open. Large **bursae** (sacs) open from the wall of the cloaca, and the bursae have a well-vascularized lining with numerous projections (**villi**). The turtles pump water in and out of the bursae at rates of 15 to 60 times per minute. Captive turtles rarely surface to breathe, and experiments have shown that the rate of oxygen uptake through the cloacal bursae is very high.

Patterns of Circulation and Respiration

The morphological complexity of the hearts of turtles and of squamates allows them to adjust blood flow through the pulmonary and systemic circuits to meet short-term changes in respiratory requirements. The key to these adjustments is changing pressures in the systemic and pulmonary circuits.

Recall that in the turtle heart deoxygenated blood from the right atrium normally flows from the cavum venosum across the muscular ridge and into the cavum pulmonale. As the ventricle contracts, the blood pressure inside the heart increases and blood is first ejected into the pulmonary artery because the resistance to flow in the pulmonary circuit is normally less than the resistance in the systemic circuit. However, the resistance to blood flow in the pulmonary circuit can be increased by muscles that narrow the diameter of blood vessels. When this happens, the delicate balance of pressure in the heart that maintained the separation of oxygenated and deoxygenated blood is changed. When the resistance of the pulmonary circuit is essentially the same as that of the systemic circuit, blood flows out of the cavum pulmonale and cavum venosum at the same time and some deoxygenated blood bypasses the lungs and flows into the systemic circuit (Figure 12–6). This process is called a "right-to-left intracardiac shunt." "Right-to-left" refers to the shift of deoxygenated blood from the pulmonary into the systemic circuit, and "intracardiac" means that it occurs in the heart rather than by flow between the major blood vessels. [Left-to-right shunts also occur, but they are beyond the scope of this dis-

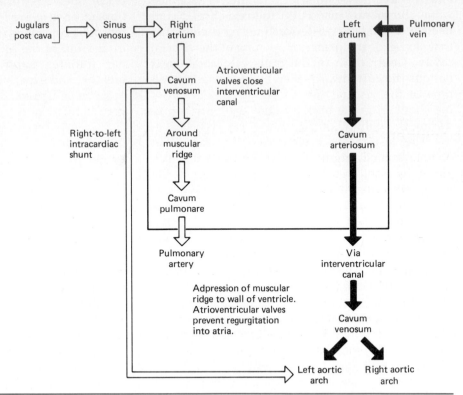

Figure 12–6. Diagram of the right-to-left shunt of blood in the heart of a turtle. Light arrows show deoxygenated blood and dark arrows show oxygenated blood. The box encloses the cycle of events during normal blood flow. (Compare to Figure 12–4.) When pressure in the pulmonary circuit increases, some deoxygenated blood from the cavum venosum flows into the left aortic arch instead of into the pulmonary arteries. (Modified from M. S. Gordon et al., 1982, *Animal Physiology: Principles and Adaptations,* 4th edition, Macmillan, New York.)

cussion. More information can be found in Burggren (1987).]

Why would it be useful to divert deoxygenated blood from the lungs into the systemic circulation? The ability to make this shunt is not unique to turtles—it occurs also among squamate reptiles and in crocodilians. The heart morphology of squamates is very like that of turtles and the same mechanism of changing pressures in the pulmonary and systemic circuits is used to achieve an intracardiac shunt. Crocodilians have hearts in which the ventricle is permanently divided into right and left halves by a septum, and they employ a different mechanism to achieve a right-to-left shunt.

The widespread occurrence of blood shunts among amniotic ectotherms suggests that the ability has important consequences for the animals, and one of these has already been discussed in Chapter 4: Lizards and crocodilians use a right-to-left shunt of blood for thermoregulation. By increasing systemic blood flow as they are warming, they increase the transport of heat from the limbs and body surface into the core of the body, thereby warming more rapidly.

Warren Burggren (1987) reviewed a number of other possible functions for blood shunts, and concluded that the most general advantage may lie in the ability it provides to match patterns of lung ventilation and pulmonary gas flow. Squamates, crocodilians, and turtles normally breathe intermittently, and periods of lung ventilation alternate with periods of **apnea** (''no breathing''). Turtles are particularly prone to periods of apnea because their method of lung ventilation means they cannot breathe when they withdraw their heads and legs into their shells. Aquatic turtles, like other aquatic animals, are apneic when they dive, and right-to-left blood shunts may occur during diving. Limiting blood flow to the lungs during periods of apnea may permit more effective use of the pulmonary store of oxygen.

Temperature Relations of Turtles

Turtles are ectotherms, and like lizards and crocodilians, they can achieve a considerable degree of stability in body temperature by regulating their exchange of heat energy with the environment. Turtles basking on a log in a pond are a familiar sight in many parts of the world, and this basking probably has primarily a thermoregulatory function. The body temperatures of basking pond turtles are higher than water and air temperatures, and these higher temperatures may speed digestion, growth, and the development of eggs. In addition, aerial basking may help aquatic turtles to rid themselves of algae and leeches. A few turtles are quite arboreal; these turtles have small plastrons that provide considerable freedom of movement for the limbs. The big-headed turtle (*Platysternon megacephalum*) from southeast Asia lives in fast-flowing streams at high altitudes and is said to climb on rocks and trees to bask. In North America musk turtles (*Sternotherus*) bask on overhanging branches and drop into the water when they are disturbed.

Terrestrial turtles can thermoregulate by moving between sun and shade. Small tortoises warm and cool quite rapidly, and they appear to behave very much like other small ectotherms in selecting suitable microclimates for thermoregulation. Familiarity with a home range may facilitate this type of thermoregulation. A study conducted in Italy compared the thermoregulation of resident Hermann's tortoises (animals living in their own home ranges) with individuals that were brought to the study site and tested before they had learned their way around (Chelazzi and Calzolai 1986). The resident tortoises warmed faster and maintained more stable shell temperatures than did the introduced

tortoises. More effective temperature regulation might benefit a number of activities of the tortoises, including digestion, growth, finding mates, and maturing eggs.

The large body size of many tortoises provides a considerable thermal inertia, and large species like the Galapagos and Aldabra tortoises heat and cool slowly. The giant tortoises of Aldabra Atoll (*Geochelone gigantea*) allow their body temperatures to rise to 32 to 33°C on sunny days and they cool to 28 to 30°C overnight. Aldabra tortoises weigh 60 kilograms or more, and even for these large animals, overheating can be a problem. The difficulty is particularly acute for some tortoises on Grande Terre (Swingland and Frazier 1979, Swingland and Lessells 1979). During the rainy season each year a portion of this population moves from the center of the island to the coast. The migrant turtles gain access to a seasonal flush of plant growth on the coast, and migrant females are able to lay more eggs than females that have remained inland. However, shade is limited on the coast and the rainy season is the hottest time of the year. Tortoises must restrict their activity to the vicinity of patches of shade, which may be no more than a single tree in the midst of a grassy plain. During the morning tortoises forage on the plain, but as their temperatures rise they move back to the shade of the tree. Competition for shade appears to result in death from overheating of smaller tortoises, especially females.

Marine turtles are large enough to achieve a considerable degree of endothermy (Spotila and Standora 1985). A body temperature of 37°C was recorded by telemetry from a green sea turtle swimming in water that was 29°C. The leatherback turtle is the largest living turtle; adults may weigh more than 600 kilograms. It ranges far from warm equatorial regions, and in the summer can be found off the coasts of New England and Nova Scotia in water as cool as 8 to 15°C. Body temperatures of these turtles appear to be 18°C or more above water temperatures, and a countercurrent arrangement of blood vessels in the flippers may contribute to retaining heat produced by muscular activity.

Ecology and Behavior of Turtles

Turtles are long-lived animals. Even small species like the painted turtle do not mature until they are 7 or 8 years old, and they may live to be 14 or older. Larger species of turtles live longer. Estimates of centuries for the life spans of tortoises are exaggerated, but large tortoises and sea turtles may live as long as humans, and even box turtles may occasionally live for 50 years (Gibbons 1976). These longevities make the life histories of turtles hard to study. Furthermore, a long lifetime is usually associated with a low replacement rate of individuals in the population, and species with those characteristics are at risk of extinction when hunting or habitat destruction reduces their numbers. Conservation efforts for sea turtles and tortoises are especially important areas of concern.

Social Behavior and Courtship

Tactile, visual, and olfactory signals are employed by turtles during social interactions. Many of the pond turtles of North America have distinctive stripes of color on their heads, necks, and on the fore- and hindlimbs, and the tail. These patterns are used by herpetologists to distinguish the species, and they may be species-isolating mechanisms for the turtles as well. During the mating season male pond turtles swim in pursuit of other turtles, and the color and pattern on the posterior limbs may enable males to identify females of their own species. Pheromones may also play a role in species identification. Long claws on the forefeet distinguish males of many species of pond turtles from females, and during courtship the male turtle swims backward in front of the female, vibrating his claws against the sides of her head (Figure 12–7).

Among terrestrial turtles, the behavior of tortoises is best known (Auffenberg 1977). Many tortoises vocalize during courtship; the sounds they produce have been described as grunts, moans, and bellows. The frequencies of the calls that have been measured range from 500 to 2500 hertz. Some tortoises have glands that become enlarged during the breeding season and appear to produce pheromones. The secretion of the subdentary gland found on the underside of the jaw of tortoises in the North American genus *Gopherus* appears to identify both the species and the sex of an individual. During courtship, the males and females of the Florida gopher tortoise rub their subdentary gland across one or both forelimbs, and then extend the limbs toward the other individual, which may sniff at them. Males also sniff the cloacal region of other tortoises, and male tortoises of some species trail females for days during the breeding season. Fecal pellets may be territorial markers; fresh fecal pellets from a dominant male tortoise have been reported to cause dispersal of conspecifics.

Tactile signals used by tortoises include biting, ramming, and hooking. These behaviors are used primarily by males, and they are employed against other males and also against females. Bites are usually directed at the head or limbs, whereas ramming and hooking are directed against the shell. The epiplastral region is used for ramming, and in some species the epiplastral bones of males are elongated and project forward beneath the neck. A tortoise about to ram another individual raises itself on its legs, rocks backward, and then plunges forward, hitting the shell of the other individual with a thump that can be heard for 100 meters in large species. During hooking the epiplastral projections are placed under the shell of an adversary, and the aggressor lifts the front end of its shell and walks forward. The combination of lifting and pushing hustles the adversary along and may even overturn it.

Movements of the head appear to act as social signals for tortoises, and elevating the head is a signal of dominance in some species. Herds of tortoises have social hierarchies that are determined largely by aggressive encounters. Ramming, biting, and hooking are employed in these encounters, and the larger individual is usually the winner, although experience may play some role. These social hierarchies are expressed in behaviors

Figure 12–7. Social behavior of turtles. (a) Male painted turtle *(Chrysemys picta)* courting a female by vibrating the elongated claws of his forefeet against the sides of her head. (b) The head-raising dominance posture of a Galapagos tortoise *(Geochelone).* [(b) Modified from S. F. Schafer and C. O. Krekorian, 1983, *Herpetologica* 39:448–456.]

such as access to food or forage areas, mates, and resting sites. Dominance relationships also appear to be involved in determining the sequence in which individual tortoises move from one place to another. The social structure of a herd of tortoises can be a nuisance for zoo keepers trying to move the animals from an outdoor pen into an enclosure for the night, because the tortoises resist moving out of their proper rank sequence.

Little information is available about the social behavior of freshwater and marine turtles because they are harder to study than terrestrial species.

However, underwater observations by Julie Booth of green sea turtles *(Chelonia mydas)* on the Great Barrier Reef of Australia revealed some of the social behaviors of these animals (Booth and Peters 1972). A small population of green turtles is resident in the lagoon at Fairfax Island, and this population is augmented by migrants that arrive in the breeding season. Male green turtles patrol the lagoon and attempt to court any female they encounter. The behavior of the female determines whether mating occurs. An unreceptive female turtle signals her unwillingness to mate by swimming away from a male. If the male pursues her, the female turns to face the male and assumes a refusal posture, hanging vertically in the water with her plastron facing the male and her limbs widespread. Male turtles usually swim off when a female adopts the refusal posture. Male turtles investigate other large objects in the water, including divers. Julie Booth noted that when male turtles swam toward her, she was able to discourage them by assuming the female turtle refusal position. The most remarkable aspect of the social behavior of these turtles is the presence of a section of the lagoon that appears to function as a female reserve. The bottom of the lagoon in this area is occupied only by female turtles resting on the sand. Male turtles patrol around the area and may swim across it at the surface of the water, but they do not attempt to initiate courtships with females in the reserve. As soon as a female leaves the reserve, however, she is courted by the waiting male turtles.

Navigation and Migration

Pond turtles and terrestrial turtles usually lay their eggs in nests they construct within their home ranges. The mechanisms of orientation they use to find nesting areas are probably the same ones they use to find their way among foraging and resting areas. Familiarity with local landmarks is an effective method of navigation for these turtles, and they may also use the sun for orientation. Sea turtles have a more difficult time, partly because the open ocean lacks conspicuous landmarks, and also because feeding and nesting areas are often widely separated. Most sea turtles are carnivorous. The leatherback turtle feeds on jellyfish, ridley and loggerhead turtles eat crabs and other benthic invertebrates, and the hawksbill turtle uses its beak to scrape encrusting organisms (sponges, tunicates, bryozoans, mollusks, and algae) from reefs. Juvenile green turtles are carnivorous, but the adults feed on vegetation, particularly turtle grass (*Thalassia testudinium*), which grows in shallow water on protected shorelines in the tropics. The areas that provide food for sea turtles often lack the char-

acteristics needed for successful nesting, and many sea turtles move long distances between their feeding grounds and their breeding areas.

The ability of sea turtles to navigate over thousands of kilometers of ocean and find their way to nesting beaches that may be no more than tiny coves on a small island is a remarkable phenomenon. The migrations of sea turtles, especially the green turtle, have been studied for decades by Archie Carr and his associates (Carr 1967, 1987; Carr et al. 1978; Mortimer and Carr 1987). Turtles captured at breeding sites in the Caribbean and Atlantic Oceans have been individually marked with metal tags since 1956, and tag returns from turtle catchers and fishing boats have allowed the major patterns of movements of the populations to be established (Figure 12–8). Similar studies by investigators in other parts of the world are beginning to shed light on movements of other species of turtles (Meylan 1981).

Four major nesting sites of green turtles have been identified in the Caribbean and South Atlantic, one at Tortuguero on the coast of Costa Rica, one on Aves Island in the eastern Caribbean, one on the coast of Surinam, and one on Ascension Island between South America and Africa. Male and female green turtles congregate at these nesting grounds during the nesting season. The male turtles remain offshore, where they court and mate with females, and the female turtles come ashore to lay eggs on the beaches. A typical female green turtle at Tortuguero produces three clutches of eggs about 12 days apart. About a third of the female turtles in the Tortuguero population nest in alternate years, and the remaining two-thirds of the turtles follow a 3-year breeding cycle. The coast at Tortuguero lacks the beds of turtle grass on which green turtles feed, and the turtles come to Tortuguero only for nesting. In the intervals between breeding periods, the turtles disperse around the Caribbean. The largest part of the Tortuguero population spreads northward along the coast of Central America. The Miskito Bank off the northern coast of Nicaragua appears to be the main feeding ground for the Tortuguero colony. A smaller number of turtles from Tortuguero swim

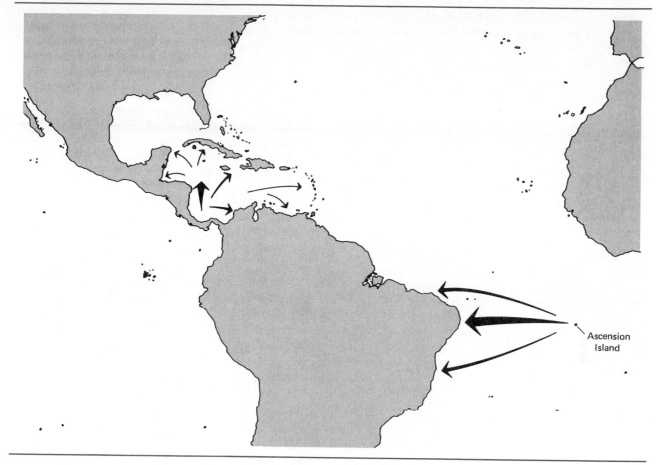

Figure 12–8. Migratory movements of green sea turtles *(Chelonia mydas)* in the Caribbean and South Atlantic. The population that nests on beaches in the Caribbean is drawn from feeding grounds in the Caribbean and Gulf of Mexico. The turtles that nest on Ascension Island feed along the coast of northern South America.

south along the coast of Panama, Colombia, and Venezuela.

Probably the most striking example of the ability of sea turtles to home to their nesting beaches is provided by the green turtle colony that nests on Ascension Island, a small volcanic peak that emerges from the ocean 2200 kilometers east of the coast of Brazil. The island is less than 20 kilometers in diameter—a tiny target in the vastness of the South Atlantic. The Ascension Island population has its feeding grounds on the coast of Brazil. The

navigation mechanisms used by the turtles are largely unknown. Migrating and homing birds use a variety of orientation mechanisms, including the ability to detect the magnetic field of the earth, to perceive polarized light, to use the sun and stars for orientation, and to hear very low frequency sounds. Sea turtles probably have a similarly wide repertoire of mechanisms for navigation, but they are more difficult to study than birds and little information is available.

The choice of Ascension Island as a nesting

site is puzzling—Why should the turtles swim so far to lay their eggs? Archie Carr has suggested that the current Ascension Island green turtle colony has inherited a behavior that evolved some 70 million years ago when the South Atlantic Ocean was first formed by the separation of Africa from South America (Carr and Coleman 1974). The continents have been separating at a rate of some 2 centimeters per year, and the mid-ocean ridge between the continents has spawned a succession of volcanic islands. Ascension, which is about 7 million years old, is only the most recent of these islands; a sea mount about 15 kilometers west of Ascension Island that is now submerged to a depth of 1500 meters is probably the remnant of the previous island. According to Carr's theory, green turtles started nesting on these volcanic islands about 70 million years ago, when the behavior required a seaward journey of only some 300 kilometers. As the spreading has continued, old islands have eroded and sunk beneath the sea and new islands have been spawned from the mid-ocean ridge. As one island became unusable, the turtles are assumed to have kept swimming seaward, transferring their breeding site to the next island in the volcanic chain. Of course, each island has been farther from the Brazilian mainland than the previous island because the South Atlantic has been getting wider, and the turtles now face a journey of more than 2000 kilometers from their feeding grounds to the nesting beaches. This hypothesis emphasizes the importance of remembering that the conditions under which traits evolved are not necessarily the same as those in which they function today. Present utility and evolutionary origin are just as likely to be different for behavioral characters as they are for morphological ones.

Nesting Behavior

All turtles are oviparous. Female turtles use their hind limbs to excavate a nest in sand or soil, and deposit a clutch that ranges from four or five eggs in small species to more than 100 eggs for the largest sea turtles. The period of embryonic development of turtles has been reported to extend from 28 days for an Asian species of softshell turtle to 420 days for the African leopard tortoise; 40 to 60 days is typical for many species (Ewert 1985). Turtles in the families Cheloniidae, Dermochelyidae, and Chelydridae lay eggs with soft, flexible shells, as do most species in the families Emydidae and Pelomedusidae. The eggs of turtles in the families Carettochelyidae, Chelidae, Kinosternidae, Testudinidae, and Trionychidae have rigid shells. In general, soft-shelled eggs develop more rapidly than hard-shelled eggs.

Environmental Effects on Egg Development Temperature, wetness, and the concentrations of oxygen and carbon dioxide can have profound effects on the embryonic development of turtles (Packard and Packard 1988). The temperature of a nest affects the rate of embryonic development, and excessively high or low temperatures can be lethal. The recent discovery that the sex of some turtles, lizards, and crocodilians is affected by the temperature they experienced during embryonic development has important implications for understanding patterns of life history and for conservation of these species. Temperature dependent sex determination has been demonstrated in some (but not all) families of turtles, in two species of lizards, and in crocodilians (Bull 1980, Ferguson and Joanen 1982, Webb and Smith 1984, Hutton 1987). In turtles, low temperatures during incubation usually produce males and high temperatures produce females; the switch from one sex to the other occurs within a span of 3 or 4°C (Figure 12–9). In crocodilians and lizards low temperatures produce females and high temperatures usually yield males. Temperatures of natural nests are not completely stable, of course. There is some daily temperature variation superimposed on a seasonal cycle of changing environmental temperatures. In turtles the middle third of embryonic development is the critical period for sex determination; the sex of the embryos depends on the temperatures they experience during those few weeks. When eggs are exposed to a daily temperature cycle, the high point of the cycle is most critical for sex determination.

As a consequence of the narrowness of the

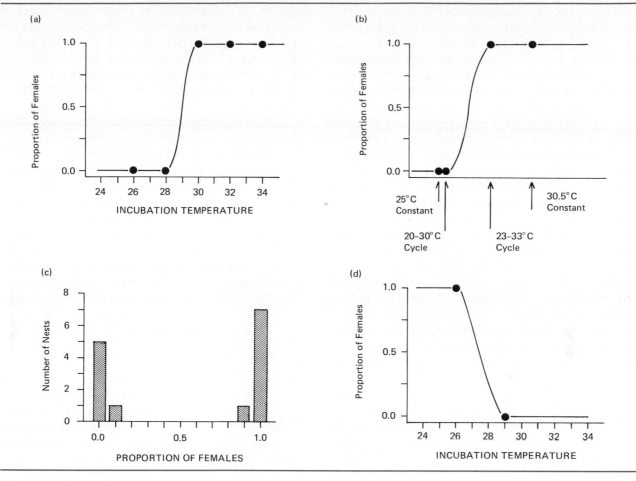

Figure 12–9. Temperature-dependent sex determination. (a) Eggs of the European pond turtle *Emys orbicularis* hatch into males when they are incubated at 26 or 28°C and into females at 30°C or above. (b) The North American map turtle *Graptemys ouachitensis* shows the same pattern. A temperature that cycles between 20 and 30°C produces males, whereas a temperature cycle of 23 to 33°C produces females. (c) Natural nests of map turtles produce predominantly males or females, depending on the nest temperature. (d) Eggs of the lizard *Agama agama* also show temperature-dependent sex determination, but the male and female determining temperatures are the opposite of those in turtles—for the lizard low temperatures produce females and high temperatures produce males. [(a) and (d) From J. J. Bull, 1980, *Quarterly Review of Biology* 55:13–21; (b) and (c) from J. J. Bull and R. C. Vogt, 1979, *Science* 206:1186–1188.]

thermal windows that are involved in sex determination and the variation that exists in environmental temperatures, both sexes are produced under field conditions, but not necessarily in the same nests (Vogt and Bull 1984, Spotila et al. 1987). A nest site may be cooler in late summer when it is shaded by vegetation than it was early in the spring when it was exposed to the sun. Thus, eggs laid early in the season would produce females, whereas eggs deposited in the same place later in the year would produce males. Temperature can also differ between the top and the bottom of a nest. For example, temperatures averaged from 33 to 35°C in the top center of dry nests of American alligators in marshes, and males hatched from eggs in this area. At the bottom and sides of the same nests, average temperatures were from 30 to 32°C and eggs from those regions hatched into females.

Temperature-dependent sex determination has important implications for efforts to conserve endangered species. Nests of the American alligator in marshes are cooler than those built on levees; females hatch from marsh nests and males from nests built on levees. The sex ratio of hatchlings from 8000 eggs collected from natural nests was five females to one male, reflecting the relative abundance of marsh and levee nests. It is not clear whether this large imbalance of the sexes represents a normal condition for alligators, or if it is the result of recent changes in the availability of nest sites.

Conservation efforts have been confounded by temperature-dependent sex determination in sea turtles. A number of programs have depended on collecting eggs from natural nests and incubating them under controlled conditions. The hatchling turtles are released at sea, thereby eliminating the mortality that hatchlings from natural nests experience in crossing the beach. Unfortunately, these unnaturally uniform conditions of incubation can result in producing hatchlings of only one sex (Mrosovsky and Yntema 1981).

The amount of moisture in the soil surrounding a turtle nest is another important variable during embryonic development of the eggs. The wetness of a nest interacts with temperature in sex determination and also influences the rate of embryonic development and the size and vigor of the hatchlings that are produced (Gutzke and Paukstis 1983, Packard and Packard 1986, Miller et al. 1987). Dry substrates induced the development of some female painted turtles at low temperatures (26.5 and 27.0°C) that would normally have produced only males. The wetness of the substrate did not affect the sex of turtles from eggs incubated at 30.5 and 32°C: All of the hatchlings from these eggs were females, as would be expected on the basis of temperature-dependent sex determination alone.

Wet conditions produce larger hatchlings than dry conditions, apparently because water is needed for metabolism of the yolk. When water is limiting, turtles hatch early and at small body sizes, and their guts contain a quantity of yolk that was not used during embryonic development. Hatchlings from nests under wetter conditions are larger and contain less unmetabolized yolk. The large hatchlings that emerge from moist nests are able to run and swim faster than hatchlings from drier nests, and as a result they may be more successful at escaping from predators and at catching food.

Hatching and the Behavior of Baby Turtles

Turtles are self-sufficient at hatching, but in some instances interactions among the young may be essential to allow them to escape the nest. Sea turtle nests are quite deep, the eggs may be buried 50 centimeters beneath the sand, and the hatchling turtles must struggle upward through the sand to the surface. After several weeks of incubation the eggs all hatch within a period of a few hours, and the hundred or so baby turtles find themselves in a small chamber at the bottom of the nest hole. Spontaneous activity by a few individuals sets the whole group into motion, crawling over and under each other. The turtles at the top of the pile loosen sand from the roof of the chamber as they scramble about, and the sand filters down through the mass of baby turtles to the bottom of the chamber.

Periods of a few minutes of frantic activity are interspersed by periods of rest, possibly because the turtles' exertions reduce the concentration of oxygen in the nest and they must wait for more oxygen to diffuse into the nest from the surrounding sand. Gradually, the entire group of turtles moves upward through the sand as a unit until it reaches the surface. As the baby turtles approach the surface, high sand temperatures probably inhibit further activity, and they wait a few centimeters below the surface until night when a decline in temperature triggers emergence. All of the babies emerge from a nest in a very brief period, and all of the babies in different nests that are ready to emerge on a given night leave their nests at almost the same time, probably because their behavior is cued by temperature. The result is the sudden appearance of hundreds or even thousands of baby turtles on the beach, each one crawling toward the ocean as fast as it can.

Simultaneous emergence is an important feature of the reproduction of sea turtles, because the babies suffer severe mortality crossing the few meters of beach and surf. Terrestrial predators await their appearance—crabs, foxes, raccoons, and other predators gather at the turtles' breeding beaches at hatching time. Some of the predators come from distant places to prey on the baby turtles. In the surf, sharks and bony fish patrol the beach. Few, if any, baby turtles would get past that gauntlet if it were not for the simultaneous emergence that brings all the babies out at once and temporarily swamps the predators.

Turtles exhibit no parental care, and the long period of embryonic development renders their nests vulnerable to predators. Females of many species of turtles scrape the ground in a wide area around the nest when they have finished burying their eggs. This behavior may make it harder for predators to identify the exact location of the nest. Major breeding sites of sea turtles are often on islands that lack mammalian predators that could excavate the nests. Another important feature of a nesting beach is provision of suitable conditions for the hatchling turtles. Newly hatched sea turtles are small animals; they weigh 25 to 50 grams, which is less than 0.05 percent of the body mass of an adult sea turtle. The enormous disparity in body size of hatchling and adult turtles is probably accompanied by equally great differences in their ecological requirements and their swimming abilities. Many of the major sea turtle nesting areas are upstream from the feeding grounds, and that location may allow currents to carry the baby turtles from the breeding beaches to the feeding grounds.

We know even less about the biology of baby sea turtles than we do about the adults. The phenomenon of the ''lost year'' following hatching has been a long-standing puzzle in the life cycle of sea turtles (Carr 1987). For example, green turtles hatch in the late summer at Tortuguero. Observations of hatchlings indicate that when they emerge from their nests they crawl rapidly toward the lightest area of sky. On a beach this form of orientation leads them directly to the water. As soon as they are in the surf, the baby turtles swim strongly seaward. Wave action appears to be necessary for this response; if waves are absent, the turtles hesitate and their swimming appears to be nondirected. The swimming frenzy of newly hatched green turtles may last for 24 hours, and probably carries them well away from the beach and into currents that run parallel to the shore.

The turtles disappear from sight as soon as they are at sea, and they are not seen again until they weigh 4 or 5 kilograms. Apparently, they spend the intervening period floating in ocean currents. Material drifting on the surface of the sea accumulates in areas where currents converge, forming drift lines of flotsam that include sargassum (a brown algae) and the vertebrate and invertebrate fauna associated with it. These drift lines are probably important resources for juvenile sea turtles (Carr 1987).

Conservation of Turtles

Many species of turtles have slow rates of growth and require long periods to reach maturity. These are characteristics that predispose a species to the risk of extinction when changing conditions increase the mortality of adults or drastically reduce recruitment of juveniles into the population. The plight of large tortoises and sea turtles is particularly severe, partly because these species are among the largest and slowest growing of turtles, and also because other aspects of their biology expose them to additional risk. The conservation of tortoises and sea turtles is a subject of active international concern (Bjorndal 1981, Bury 1982).

The largest living tortoises are found on the Galapagos and Aldabra Islands. The relative isolation of these small and (for humans) inhospitable landmasses has probably been an important factor in the survival of the tortoises. Human colonization of the islands has brought with it domestic animals such as goats and donkeys that compete with the tortoises for the limited quantities of vegetation to be found in these arid habitats, and dogs, cats, and rats that may prey on tortoise eggs and on baby tortoises.

The limited geographic range of a tortoise that occurs only on a single island makes it vulnerable to local habitat destruction, but also provides a circumscribed area in which the species must be protected. Sea turtles range over thousands of square kilometers of ocean, including international waters and also coastal areas that are under the jurisdictions of many different nations. Protection of sea turtles is a very difficult proposition. The limited number of breeding sites used by many species adds to the difficulty. The most dramatic example is the Kemp's ridley turtle (*Lepidochelys kempi*) in which the entire population appears to nest on a single beach in Tamaulipas, Mexico. The number of turtles nesting has declined precipitously: In 1947 more than 40,000 females nested on the beach, but in recent years the nesting population has consisted of 200 to 500 individuals, and the Kemp's ridley is considered the most endangered species of sea turtle.

Attempts to save sea turtles are in progress all over the world, sponsored by a variety of governmental agencies, private organizations, and even by dedicated individuals. The problems they face are massive, and the most effective methods to employ are still subject to disagreement (Pritchard 1980, Ehrenfeld 1981). For example, is controlled exploitation of sea turtles more feasible than an outright ban on the use of sea turtles and their products? Do turtle farming operations benefit conservation by producing captive-bred individuals, or do they indirectly harm natural populations by sustaining a demand for turtle products that would otherwise vanish?

Beyond those questions, which have their origin in the ways people in rich and poor nations respond to the often-conflicting demands of earning a living versus conserving sea turtles, is another series of questions that arise from our inadequate knowledge of the biology of sea turtles. For example, is it a wise management practice to dig up nests of turtle eggs from the nesting beaches and incubate them in a protected area? Predation on eggs in natural nests can be high, but sea turtles display temperature-dependent sex determination, and the widespread technique of incubating eggs in plastic foam boxes appears to produce predominantly male hatchlings. Is the practice of "headstarting" sea turtles beneficial? In this technique, baby turtles are kept in captivity for some weeks or months and allowed to grow before they are released at sea. This method avoids the very high losses of baby turtles to predators that occurs when the newly hatched babies make their own way down the beach and into the sea. However, imprinting of the characteristics of the beach and the adjacent water might occur as the baby turtle makes its own way to sea. If that is the case, headstarting may prevent the imprinting that is essential for successful navigation of an adult turtle back to the nesting beach. We simply cannot evaluate the effects of these manipulations because we do not know enough about the biology of sea turtles. Even the assumption that sea turtles return to breed on the beaches where they hatched has not been confirmed. No one has yet devised a method of marking hatchling turtles that will allow them to be identified many years later when they have grown 2000-fold or more to adult size.

These questions emphasize the central role of information about all aspects of the biology of organisms in successful conservation plans. This sort of information is not easy to obtain for any species of organism, and sea turtles are more difficult to study than most animals. Nonetheless, pitfalls lie in wait for even the best-intended management techniques, and conservation, like any other area of applied biology, must begin from a thorough understanding of the basic biology of the organisms involved.

Summary

Turtles show a mixture of primitive characters and unique derived characters that has made their phylogenetic relationships difficult to determine. The earliest turtles known, fossils from the Triassic, have nearly all the features of modern turtles. The first Triassic forms were not able to withdraw their heads into the shell, but this ability appeared in the two major lineages of living turtles, which were established by the late Triassic. The cryptodire turtles retract the head with a vertical flexion of the neck vertebrae, whereas the pleurodires use a sideward bend.

Turtles are among the most morphologically specialized vertebrates. The shell is formed of dermal bone that is fused to the vertebral column and ribs. In most turtles the dermal shell is overlain by a horny layer of epidermal scutes. The limb girdles are inside the rib cage. Breathing presents special difficulties for an animal that is encased in a rigid shell: Exhalation is accomplished by muscles that squeeze the viscera against the lungs, and inhalation is accomplished by muscles that increase the volume of the visceral cavity, thereby allowing the lungs to expand. The heart of turtles (and of squamates as well) is able to shift blood between the pulmonary and systemic circuits in response to the changing requirements of gas exchange and thermoregulation.

The social behavior of turtles includes visual, tactile, and olfactory signals used in courtship. Dominance hierarchies shape the feeding, resting, and mating behaviors of some of the large species of tortoises. All species of turtles lay eggs, and none provides parental care. Coordinated activity by hatchling sea turtles may be necessary to enable them to dig themselves out of the nest, and simultaneous emergence of baby sea turtles from their nests helps them to evade predators as they rush down the beach into the ocean. Sea turtles migrate tens, hundreds, and even thousands of kilometers between their feeding areas and their nesting beaches. The navigation mechanisms they use are largely unknown, but their homing ability is astonishingly precise.

The life history of many turtles makes them vulnerable to extinction. Slow rates of growth and long periods required to reach maturity are characteristic of turtles in general and of large species of turtles in particular. Tortoises and sea turtles are especially threatened, and conservation efforts are in progress in many countries. Recently discovered features of the basic biology of turtles have important implications for conservation efforts. For example, many species of turtles show temperature-dependent sex determination. That is, the sex of an individual turtle is determined by the temperature it experiences in the egg during embryonic development. Some conservation efforts undertaken before this phenomenon was appreciated resulted in the production and release of thousands of hatchling baby turtles, nearly all of which were probably males. Additional information about basic aspects of the biology of turtles is critically important to guide efforts to sustain existing populations and to reestablish populations that have been lost.

References

Auffenberg, W. 1977. Display behavior in tortoises. *American Zoologist* 17:241–250.

Bjorndal, K. A. (editor). 1981. *Biology and Conservation of Sea Turtles*. Smithsonian Institution Press, Washington, D.C.

Booth, J. and J. A. Peters. 1972. Behavioural studies on the green turtle (*Chelonia mydas*) in the sea. *Animal Behaviour* 20:808–812.

Bull, J. J. 1980. Sex determination in reptiles. *Quarterly Review of Biology* 55:13–21.

Burggren, W. W. 1987. Form and function in reptilian circulations. *American Zoologist* 27:5–19.

Bury, R. B. (editor). 1982. *North American Tortoises: Conservation and Ecology*. Wildlife Research Report 12. Fish and Wildlife Service, United States Department of the Interior, Washington, D.C.

Carr, A. 1967. *So Excellent a Fishe*. Natural History Press, Garden City, N.Y.

Carr, A. 1987. New perspectives on the pelagic stage of sea turtle development. *Conservation Biology* 1:103–121.

Carr, A., M. H. Carr, and A. B. Meylan. 1978. The ecology and migrations of sea turtles. 7. The west Caribbean green turtle colony. *Bulletin of the American Museum of Natural History* 162:1–46.

Carr, A. and P. J. Coleman, 1974. Seafloor spreading and the odyssey of the green turtle. *Nature* 249:128–130.

Carroll, R. L. 1987. *Vertebrate Paleontology and Evolution*. W. H. Freeman, New York.

Chelazzi, G. and R. Calzolai. 1986. Thermal benefits from familiarity with the environment in a reptile. *Oecologia* 68:557–558.

Ehrenfeld, D. 1981. Options and limitations in the conservation of sea turtles. Pages 457–463 in *Biology and Conservation of Sea Turtles*, edited by K. A. Bjorndal. Smithsonian Institution Press, Washington, D.C.

Ewert, M. A. 1985. Embryology of turtles. Pages 76–267 in *Biology of the Reptila*, volume 14, edited by C. Gans, F. Billet, and P. F. A. Maderson. Wiley, New York.

Ferguson, M. W. J. and T. Joanen. 1982. Temperature of egg incubation determines sex in *Alligator mississippiensis*. *Nature* 296:850–853.

Gaffney, E. S., J. H. Hutchison, F. A. Jenkins, Jr., L. J. Meeker. 1987. Modern turtle origins: the oldest known cryptodire. *Science* 237:289–291.

Gibbons, J. W. 1976. Aging phenomena in reptiles. Pages 454–475 in *Experimental Aging Research*, edited by M. F. Elias, B. E. Eleftheriou, and P. K. Elias. EAR, Bar Harbor, Maine.

Gutzke, W. H. N. and G. L. Paukstis. 1983. Influence of the hydric environment on sexual differentiation of turtles. *Journal of Experimental Zoology* 226:467–469.

Hutton, J. M. 1987. Incubation temperatures, sex ratios and sex determination in a population of Nile crocodiles (*Crocodylus niloticus*). *Journal of Zoology (London)* 211:143–155.

Johansen, K. and W. Burggren. 1980. Cardiovascular function in the lower vertebrates. Pages 61–117 in *Hearts and Heart-Like Organs*, volume 1, edited by G. Bourne. Academic Press, London.

Meylan, A. 1981. Sea turtle migration—evidence from tag returns. Pages 91–100 in *Biology and Conservation of Sea Turtles*, edited by K. A. Bjorndal. Smithsonian Institution Press, Washington, D.C.

Meylan, A. 1988. Spongivory in hawksbill turtles: a diet of glass. Science 239:393–395.

Miller, K., G. C. Packard, and M. J. Packard. 1987. Hydric conditions during incubation influence locomotor performance of hatchling snapping turtles. *Journal of Experimental Biology* 127:401–412.

Mortimer, J. A. and A. Carr. 1987. Reproduction and migrations of the Ascension Island green turtle (*Chelonia mydas*). *Copeia* 1987:103–113.

Mrosovsky, N. 1987. Leatherback turtle off scale. *Nature* 327:286.

Mrosovsky, N. and C. L. Yntema. 1981. Temperature dependence of sexual differentiation in sea turtles: implications for conservation practices. Pages 271–280 in *Biology and Conservation of Sea Turtles*, edited by K. A. Bjorndal. Smithsonian Institution Press, Washington, D.C.

Packard, M. J. and G. C. Packard. 1986. Effect of water balance on growth and calcium mobilization of embryonic painted turtles (*Chrysemys picta*). *Physiological Zoology* 59:398–405.

Packard, G. C. and M. J. Packard. 1988. Physiological ecology of reptile eggs. Pages 523–605 in *Biology of the Reptilia*, volume 16, edited by C. Gans and R. B. Huey. Alan Liss, Philadelphia.

Pritchard, P. C. H. 1980. The conservation of sea turtles: practices and problems. *American Zoologist* 20:609–617.

Spotila, J. R. and E. A. Standora. 1985. Environmental constraints on the thermal energetics of sea turtles. *Copeia* 1985:694–702.

Spotila, J. R., E. A. Standora, S. J. Morreale, and G. J. Ruiz. 1987. Temperature dependent sex determination in the green turtle (*Chelonia mydas*): effects on the sex ratio on a natural nesting beach. *Herpetologica* 43:74–81.

Swingland, I. R. and J. G. Frazier. 1979. The conflict between feeding and overheating in the Aldabran giant tortoise. Pages 611–615 in *A Handbook on Biotelemetry and Radio Tracking*, edited by C. J. Amlaner, Jr. and D. W. MacDonald. Pergamon Press, Oxford.

Swingland, I. R. and C. M. Lessells. 1979. The natural regulation of giant tortoise populations on Aldabra Atoll. Movement polymorphism, reproductive success and mortality. *Journal of Animal Ecology* 48:639–654.

Vogt, R. C. and J. J. Bull. 1984. Ecology of hatchling sex ratio in map turtles. *Ecology* 65:582–587.

Webb, G. J. W. and A. M. A. Smith. 1984. Sex ratio and survivorship in the Australian freshwater crocodile *Crocodylus johnstoni*. Pages 319–355 in *The Structure,* *Development and Evolution of Reptiles*, edited by M. W. J. Ferguson. *Symposia of the Zoological Society of London*, No. 52.

White, F. N. 1968. Functional anatomy of the heart of reptiles. *American Zoologist* 8:211–219.

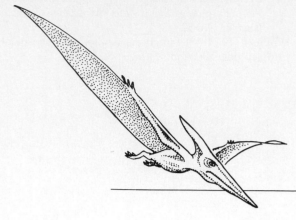

About the same time that turtles were evolving in the Triassic the most successful lineage of amniotic vertebrates, the Diapsida, was starting its radiation. The most spectacular diapsids were the dinosaurs, but the lineage also gave rise to a majority of the species of living terrestrial vertebrates. Birds are diapsids, as are the squamates (lizards, snakes, and amphisbaenians). A variety of other, lesser known forms fills the roster of diapsids, including crocodilians, ichthyosaurs, and pterosaurs, to name only a few.

The dinosaur fauna of the Mesozoic was unlike anything that has existed before or since. Many dinosaurs were enormous—some herbivores were quadrupedal and all of the carnivores were bipedal. The enormous body sizes of dinosaurs makes it difficult to recreate the details of the lives they led, because we have no living models of truly large vertebrates. Even an elephant is only as large as a small to medium-size dinosaur.

A second group of diapsids, the lepidosauromorphs, radiated into a variety of animals in the Mesozoic and then had another radiation that produced somewhat different animals that survive today. The living lepidosauromorphs are such successful animals in their own right that they will be discussed in the next chapter; however, a discussion of the Mesozoic world would be incomplete without considering the secondarily aquatic marine forms (ichthyosaurs, plesiosaurs, and placodonts), and these were probably lepidosauromorphs.

The remarkable success of large diapsids in the Mesozoic ended at the close of that era with the extinction of the dinosaurs. That mass extinction has attracted more than its share of attention because dinosaurs have great popular appeal—as mass extinctions go, it was minor. However, it does provide a good opportunity to consider in detail the merits of two types of explanations of mass extinctions, the gradualist versus the catastrophism schools of thought.

Mesozoic Diapsids: Dinosaurs, Birds, Crocodilians, and Others

13

The Mesozoic Fauna

The Mesozoic Era, frequently referred to as the Age of Reptiles, extended for some 180 million years from the close of the Paleozoic 245 million years ago to the beginning of the Cenozoic only 66 million years ago. Through this vast period evolved a worldwide fauna that diversified and radiated into most of the adaptive zones occupied by all the terrestrial vertebrates living today and some that no longer exist (for example, the enormous herbivorous and carnivorous tetrapods called dinosaurs). Although the dinosaurs are the most familiar representatives of the Age of Reptiles, they are only one of many groups.

Inevitably such a huge number of animals is complicated and confusing, not only on first acquaintance but even after study. Parallel and convergent evolution were widespread in Mesozoic tetrapods. Long-snouted fish eaters evolved repeatedly, as did heavily armored quadrupeds and highly specialized marine forms. A trend to bipedalism was general, and a secondary reversion to quadrupedal locomotion is seen in many forms. Knowledge of phylogenetic relationships is in a state of flux, and the scheme outlined in Figure 13–1 will undoubtedly need revision as additional material is analyzed. Current views of the ecology of dinosaurs are likewise undergoing a radical revision. Classic ideas have been based on a naive impression of the ecology and behavior of croco-

dilians and lizards. Paleontologists have assumed the Mesozoic tetrapods led sedentary lives, and those ideas have colored their interpretations of posture and behavior. Recent reexamination of the morphological evidence suggests that some Mesozoic tetrapods were considerably more active than had been realized, and these reevaluations have changed our views of their ecology and behavior.

This chapter commences with a brief review of the phylogenetic relationships of Mesozoic tetrapods and some aspects of functional morphology and major evolutionary trends. More detailed information on these topics and additional illustrations of members of the groups discussed can be found in the references cited at the end of the chapter. Following a consideration of some aspects of the ecology of dinosaurs, we consider their disappearance at the end of the Cretaceous.

Phylogenetic Relationships Among Diapsids

Our understanding of the phylogenetic relationships of a number of groups of Mesozoic tetrapods has been in a state of flux. Many of these forms (thecodonts, crocodilians, pterosaurs, dinosaurs, squamates, and rhynchosaurs) had skulls with two temporal openings (a **diapsid** skull). Romer (1966)

Phylogenetic Relationships of the Diapsida

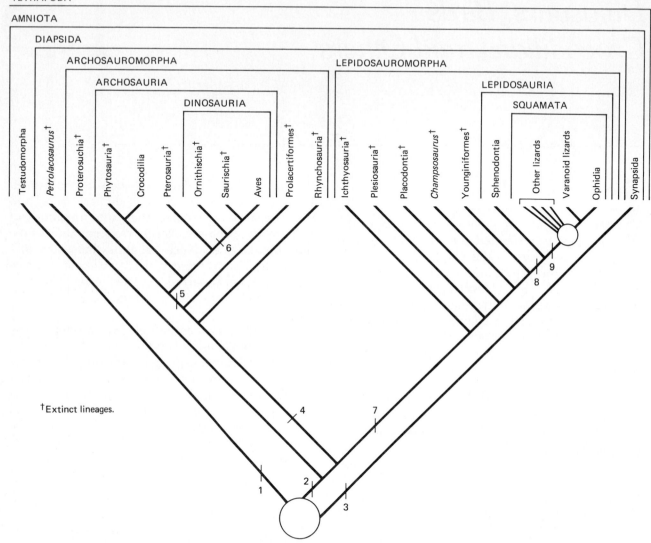

†Extinct lineages.

The phylogenetic relationships among testudomorphs, diapsids, and synapsids are not clearly known and are represented here as undetermined; character states of the three groups are presented first. **1.** Testudomorpha: Plastron and carapace formed by dermal bones fused with parts of the axial skeleton, skull anapsid (a primitive character state for amniotes). **2.** Diapsida: Skull with a dorsal temporal fenestra, upper temporal arch formed by triradiate postorbital and triradiate squamosal, modifications of the snout, characteristics of the nervous system including a true Jacobson's organ present at least in embryos and with olfactory bulbs anterior to the eyes. **3.** Synapsida: Skull with a lateral temporal fenestra (synapsid). **4.** Archosauromorpha: Characteristics of the snout, stapes lacking a foramen, characteristics of the vertebrae, humerus, and feet. **5.** Archosauria: Presence of an antorbital fenestra, orbit shaped like an inverted triangle, teeth laterally compressed, fourth trochanter on femur. **6.** Dinosauria: Characteristics of the palate, pectoral and pelvic girdles, hand, hindlimb, and foot. **7.** Lepidosauromorpha: Postfrontal enters border of upper temporal fenestra, characteristics of the vertebrae, ribs, and sternal plates. **8.** Lepidosauria: Determinant growth with epiphyses on the articulating surfaces of the long bones, characteristics of the skull, pelvis, and feet. **9.**

Squamata: Fusion of bones in snout region, characteristics of the palate and skull roof, vertebrae, ribs, pectoral girdle, and humerus. (Based on M. J. Benton, 1985, *Zoological Journal of the Linnaean Society* 84:97–164; J. Gauthier, 1986, pages 1–55 in *The Origin of Birds and the Evolution of Flight,* edited by K. Padian, *Memoirs of the California Academy of Sciences,* Number 8; and H.-D. Sues, 1987, *Zoological Journal of the Linnaean Society* 90:109–131.)

Figure 13–1. Phylogenetic relationships of the Diapsida. This diagram shows the probable relationships among the major groups of diapsids. Extinct lineages are marked by a dagger (†). The numbers indicate derived characters that distinguish the lineages. The circles indicate that the relationships of the lineages at those nodes cannot yet be defined.

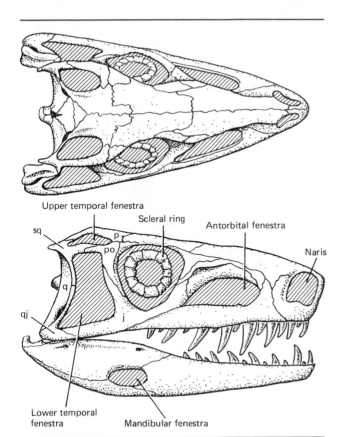

Upper temporal fenestra

Scleral ring

Antorbital fenestra

sq

p

po

Naris

q

qj

j

Lower temporal fenestra

Mandibular fenestra

believed that two lineages of tetrapods independently evolved the diapsid condition, whereas other authors considered the diapsids a single group. The latter view now prevails (Benton 1985a), and the Diapsida is considered a monophyletic lineage that includes most of the major groups of Mesozoic tetrapods as well as the living crocodilians, birds, the tuatara (*Sphenodon*), and squamates (lizards, snakes, and amphisbaenians) (Figure 13–1).

The Diapsida, as it is currently constituted, is distinguished by a number of derived characters, several of which pertain to fenestrae in the skull. The name diapsid means "two arches" and refers to the presence of an upper and a lower fenestra in the temporal region of the skull (Figure 13–2). More distinctive than these openings is the morphology of the bones that form the arch separating them. This upper temporal arch is composed of a three-pronged postorbital bone and a three-pronged squamosal. The lower temporal arch is

Figure 13–2. The skull of *Euparkeria,* a primitive archosaur, shows the two arches in the temporal region that characterize the diapsid condition, and the antorbital fenestra. (From A. S. Romer, 1966, *Vertebrate Paleontology,* 3rd edition, University of Chicago Press, Chicago.)

formed by the jugal and quadratojugal bones. This lower arch has been lost repeatedly in the radiation of diapsids, and the upper arch also is missing in some forms. Living lizards and snakes clearly show the importance of those modifications of the skull in permitting increased skull flexion (**kinesis**) during feeding, and the same significance may attach to loss of the arches in some extinct forms. In addition to the two temporal fenestrae, advanced diapsids have a suborbital fenestra on each side of the head anterior to the eye, and the presence of this fenestra modifies the relationships among the bones of the palate and the side of the skull.

Two characteristics of the olfactory system distinguish diapsids from turtles and amphibians: A Jacobson's organ appears during the embryonic development of diapsids as a ventromedial outpocketing of the nasal cavity. Squamates and the

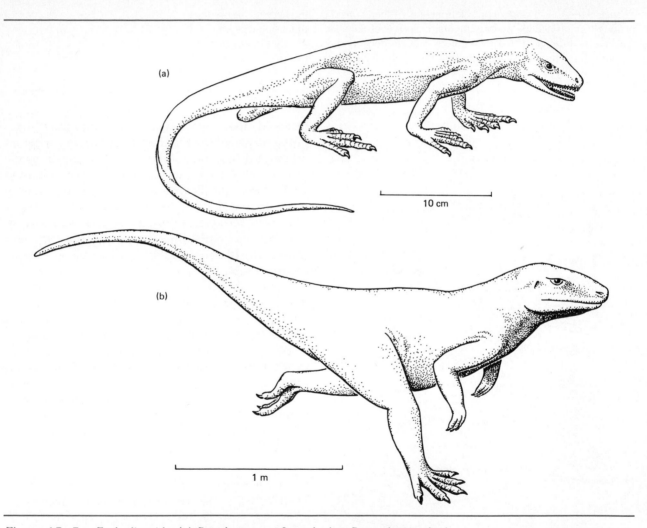

(a)

10 cm

(b)

1 m

Figure 13–3.   Early diapsids: (a) *Petrolacosaurus* from the late Pennsylvanian had forelimbs and hindlimbs that were about the same length; (b) *Euparkeria* had hindlimbs that were much longer than its forelimbs and was probably bipedal. [(a) From R. R. Reisz, 1977, *Science* 196:1091–1093.]

tuatara retain a functional Jacobson's organ as adults. The roof of the Jacobson's organ of these animals is lined with vomeronasal epithelium, and innervated by the main olfactory bulb. The Jacobson's organ is part of the chemosensory system, and may be involved particularly with social interactions and mating. In birds and crocodilians the Jacobson's organ develops in the embryo but disappears in the adults. (The vomeronasal organs of turtles and amphibians are morphologically different, but they are probably homologous with the Jacobson's organ of diapsids.) In addition, diapsids have well-developed olfactory bulbs that are situated anterior to the eyes and are linked to the forebrain (telencephalon) by a stalk-like olfactory tract. In turtles and amphibians the olfactory bulbs are in direct contact with the telencephalon and lack olfactory tracts.

The earliest diapsid known is *Petrolacosaurus* from late Carboniferous deposits in Kansas. It is a moderately small animal, 60 to 70 centimeters from snout to tail-tip, with a long neck, large eyes, and long limbs (Figure 13–3). It gives the impression of having been an agile terrestrial animal that may have fed on large insects and other arthropods. The advanced diapsids can be split into two groups, the **Archosauromorpha** and the **Lepidosauromorpha** (Figure 13–1). The archosauromorphs include the living crocodilians and birds, and the extinct rhynchosaurs, phytosaurs, dinosaurs, pterosaurs, and a number of late Permian and Triassic forms. The lepidosauromorphs include the extinct Younginiformes and the living tuatara and squamates plus their extinct relatives. In addition, three groups of specialized marine tetrapods (the placodonts, plesiosaurs, and ichthyosaurs) that were formerly placed in the subclass Euryapsida are tentatively considered to be lepidosauromorphs. The skulls of these animals have a dorsal temporal opening, but lack a lower temporal fenestra, and the postorbital and squamosal bones do not have the three-pronged shape characteristic of diapsids. However, these patterns are within the range of modifications of the basic diapsid skull that is seen among other members of the clade.

The Archosauromorpha: Dinosaurs and Their Relatives

The Archosauromorpha includes the most familiar of the Mesozoic diapsids, the dinosaurs, as well as their close relatives, the crocodilians, phytosaurs, birds, and pterosaurs (Figure 13–1). Less well-known groups of archosauromorphs are the rhynchosaurs and Prolacertiformes. The archosauromorphs are distinguished by a number of characteristics of the skull and axial skeleton (Benton 1985a).

The Rhynchosauria

The rhynchosaurs were squat, heavily built tetrapods as much as 2 meters long (Figure 13–4). The distinctive specializations of rhynchosaurs lie in the structure of their jaws and teeth. In late Triassic forms the teeth in the upper jaw were borne on two maxillary tooth plates, each of which had several rows of teeth and was divided into two parts by a deep longitudinal groove. The lower jaw also bore two rows of teeth, one on the crest of the jaw and one lower down on the inner (lingual) side of the jaw. When the mouth was closed, the lower jaw fit snugly into the groove like the blade of a penknife folding into its handle. The premaxilla formed a heavy beak, and the jaw-closing muscles were powerful, giving the head a triangular shape.

Rhynchosaurs probably diverged from the diapsid stock in the Permian. The earliest rhynchosaurs known are small, lizard-like animals from the Triassic of South Africa. By the late Triassic, large rhynchosaurs were dominant members of many fossil faunas accounting for approximately half of all the terrestrial animals found (Benton 1983a). The abundance of rhynchosaurs in the late Triassic has been compared to that of antelopes in modern African savanna faunas.

Rhynchosaurs became extinct about 17 million years before the end of the Triassic, and a few million years after that dinosaurs were the dominant terrestrial animals. What could account for such a rapid (in geological time) disappearance of

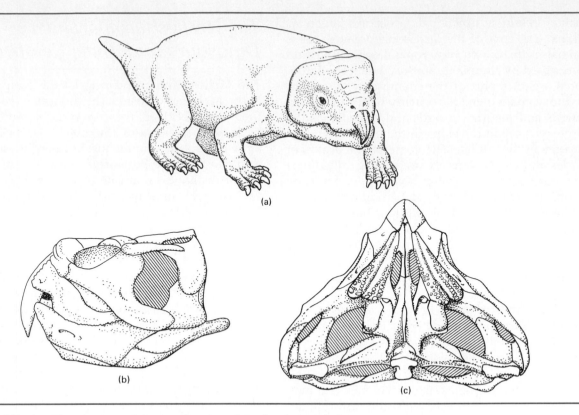

Figure 13–4. Rhynchosaurs: (a) reconstruction of the South American rhynchosaur *Scaphonyx*; (b) side view and (c) palatal view of the skull of the Scottish rhynchosaur *Hyperodapedon*. (From M. J. Benton, 1983, *Quarterly Review of Biology*, 58:29–55.)

a successful group of animals with a worldwide distribution? Theories of competitive replacement of one group of animals by another better adapted group have long dominated the interpretation of paleontological information. However, as more detailed information becomes available this view is becoming less tenable as a general proposition, although it may apply to particular cases. Most of the time the temporal sequence of events appears to show that a once-dominant group waned, and shortly thereafter (in geological time) a new group burgeoned (Benton 1983b,c). That temporal mismatch appears to apply to the disappearance of rhynchosaurs (and the simultaneous disappear-

ance of many kinds of synapsids—see Chapter 19) and the radiation of dinosaurs.

An event that does appear to correspond to the waning of the rhynchosaurs is a change in the climate and flora at the end of the Triassic. During the middle of the Triassic, when rhynchosaurs were dominant, most of the southern part of the world was covered by an assemblage of plants known as the *Dicrodium* flora, which consisted of seed ferns, horsetails, cycads, ginkgoes, and conifers. The powerful and precise vertical chopping motion of the jaws of rhynchosaurs, which was guided by the groove in the tooth plates of the upper jaw, may have been suitable for cutting tough vegetation. Climatic conditions appear to have become more arid at the end of the Triassic, and the *Dicrodium* flora was replaced by a mixture of conifers and bennettitaleans (large, tree-like cycads). Rhynchosaurs probably could not rear up on their hind legs to browse on tall vegetation, and may have found their food supply greatly diminished as plants became taller and more widely spaced.

Prolacertiformes

The bar formed by the jugal and quadratojugal bones that closes the ventral side of the lower temporal opening in the diapsid skull was incomplete in the animals grouped as Prolacertiformes (Figure 13–5a). In living squamates this condition imparts a mobility to the quadrate bone that increases the efficiency of the lower jaw. Grouped among the Prolacertiformes are several medium-size tetrapods such as *Prolacerta* and *Protorosaurus* with lizard-like body proportions that appear to have been agile, terrestrial predators. Also included in the Prolacertiformes is the bizarre genus *Tanystropheus*. Several species of *Tanystropheus* are known, some of which were 6 meters in length. The body, limbs, and tail of *Tanystropheus* were of normal lizard-like proportions, but the neck, which contained 13 elongate vertebrae, was as long as the body and tail combined. A small head, its jaws armed with conical teeth, perched on the end of this remarkable neck. The two parts of the body appear so different that when the first complete skeleton of this genus was discovered, it was found that bones from the front part of the animal had previously been described as belonging to a pterosaur and bones from the trunk had been identified as being from a primitive dinosaur.

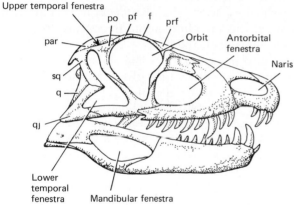

(a)

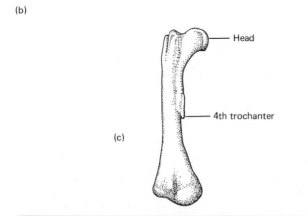

(b)

(c)

Figure 13–5. (a) Skull of *Prolacerta* showing the incomplete lower temporal arch; (b) skull of the carnosaur *Ornithosuchus* showing the characteristic features of archosaurs: two temporal arches, a keyhole-shaped orbit, and an antorbital fenestra; (c) femur of *Thescelosaurus* showing the fourth trochanter. [(a) and (b) From A. S. Romer, 1966, *Vertebrate Paleontology*, 3rd edition, University of Chicago Press, Chicago; (c) from A. S. Romer, 1956. *Osteology of the Reptiles*, University of Chicago Press, Chicago.]

Archosauria

The archosaurs are the animals most frequently associated with the great radiation of tetrapods in the Mesozoic. Dinosaurs and pterosaurs are distinctive components of many Mesozoic faunas, and other less familiar archosaurs were also abundant. The archosaurs are distinguished by the presence, in many forms, of an antorbital ("in front of the eye") fenestra. The skull was deep, the orbit of the eye was shaped like an inverted triangle rather than being circular, and the teeth were laterally compressed (Figure 13–5b). A trend toward bipedalism was widespread (but not universal) among archosaurs, and the ventral side of the shaft of the femur had a distinctive area with a rough surface, the fourth trochanter, which was the site of insertion of the powerful caudiofemoral muscle (Figure 13–5c).

Primitive archosaurs are grouped here as the Proterosuchia. This approach simplifies a complex area in which no one phylogenetic scheme is yet widely accepted. The proterosuchians in our classification include the Triassic forms known as thecodontians in earlier classifications such as that of Romer (1966). *Proterosuchus* was a quadrupedal, lizard-shaped carnivore, 2 or 3 meters long, that is known from South Africa. Related forms are known from deposits in China, Bengal, northern USSR, Australia, and Antarctica. *Erythrosuchus*, another Triassic quadruped, was twice the size of *Proterosuchus* and massively built, whereas *Eupar-*

keria was a lightly built animal about 150 centimeters long. Its hind limbs were half as long as the fore limbs, suggesting that it was capable of bipedal locomotion (Figure 13–3).

Later archosaurs are characterized by a number of features of the skull and teeth, including the loss of the postparietal bones and the absence of palatal teeth (Benton 1985a). The morphological and ecological diversity of the advanced archosaurs included semiaquatic, aquatic, and marine forms (phytosaurs and crocodilians) as well as flying animals (pterosaurs and birds) and terrestrial species (dinosaurs).

Phytosaurs and Crocodilians The archosaur stock gave rise to two lineages of aquatic fish eaters, the phytosaurs and crocodilians. The phytosaurs were the earlier radiation, and during the Triassic they were abundant and important elements of the shoreline fauna. In contrast to crocodilians, in which the nostrils are at the tip of the snout and a secondary palate separates the nasal passages from the mouth, phytosaur nostrils were located on a boss or elevation just anterior to the eyes (Box 13–1). True crocodilians appeared in the Triassic and seem to have replaced phytosaurs by the end of that period. In most respects crocodilians conform closely to the skeletal structure of archosaurs, but the skull and pelvis are specialized. Crocodilians retained the nostrils at the tip of the snout and developed a secondary palate that car-

Box 13–1. Long-Snouted Fish Eaters

Aquatic and semiaquatic tetrapods with a crocodile-like body form have evolved repeatedly among tetrapods. The distinctive feature of this specialization is an elongate snout used to capture fish with a sideward sweep of the head. Among the earliest examples of this body form were the Triassic temnospondyl trematosaurs (Chapter 10). In the Diapsida, crocodile-like animals evolved in both the lepidosauromorph and the archosauromorph lineages. The champsosaurs (Figure 13–6a) are probably lepidosauromorphs. They are known

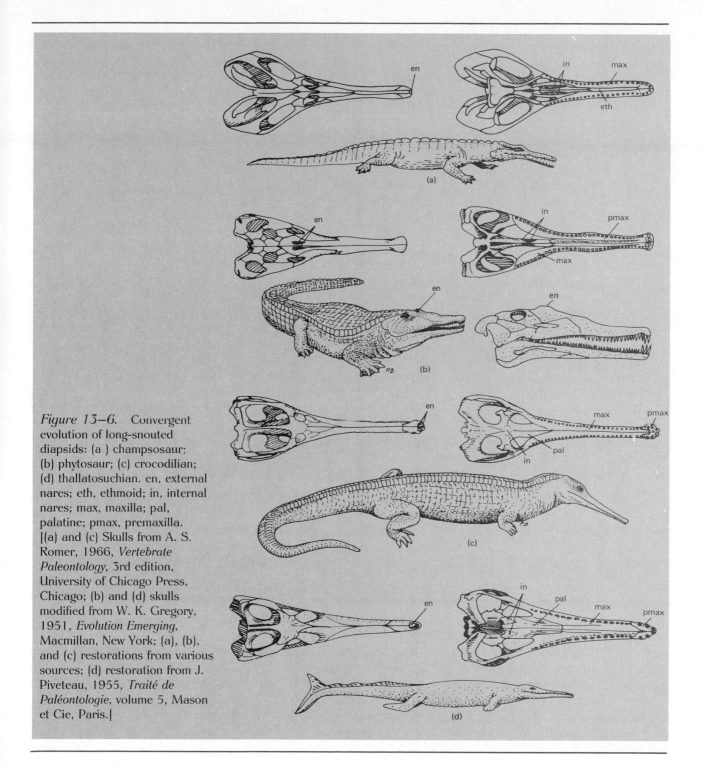

Figure 13–6. Convergent evolution of long-snouted diapsids: (a) champsosaur; (b) phytosaur; (c) crocodilian; (d) thallatosuchian. en, external nares; eth, ethmoid; in, internal nares; max, maxilla; pal, palatine; pmax, premaxilla. [(a) and (c) Skulls from A. S. Romer, 1966, *Vertebrate Paleontology,* 3rd edition, University of Chicago Press, Chicago; (b) and (d) skulls modified from W. K. Gregory, 1951, *Evolution Emerging,* Macmillan, New York; (a), (b), and (c) restorations from various sources; (d) restoration from J. Piveteau, 1955, *Traité de Paléontologie,* volume 5, Mason et Cie, Paris.]

Box 13–1. (Continued)

from Cretaceous and early Tertiary fossil beds in North American and Europe. Champsosaurs were about 2 meters long. The posterior part of the skull was broad, suggesting the presence of strong jaw muscles.

Phytosaurs (Figure 13–6b) are the sister group of crocodilians and appear to have been abundant in the Triassic. Fossils are known from North America, Europe, and India. Crocodilians (Figure 13–6c) appeared in the Triassic and radiated extensively in the Jurassic and Cretaceous. The earliest crocodilians appear to have been terrestrial. *Protosuchus* ("first crocodile"), a late Triassic form, had long legs and a short, broad skull. Later crocodilians were more aquatic and had elongate snouts. The most specialized crocodilians were the thallatosuchians ("sea crocodiles"), marine forms that lacked dermal bony armor. Thallatosuchians had paddle-like limbs and a tail fin like that of ichthyosaurs in which the lower lobe was supported by the vertebral column and the upper lobe lacked skeletal support (Figure 13–6d).

In the ancestral diapsid skull the internal nares were located in the anterior part of the mouth, close to the external nares on the snout, and air passed through the length of the mouth as it was inhaled and exhaled. That arrangement is not effective for an aquatic animal, because the mouth is often full of water. (None of these animals had lips that could exclude water from the mouth.) Thus, they faced a problem in getting air from the nostrils into the trachea without inhaling water at the same time. Two solutions to the problem emerged: Phytosaurs shifted the position of the nostrils from the tip of the snout to a location just anterior to the eyes. In some phytosaurs the nostrils were located in a volcano-shaped elevation on the front of the skull. Champsosaurs and crocodilians, in contrast, evolved a secondary palate, a shelf of bone in the roof of the mouth that separates the nasal passages from the mouth itself. In champsosaurs the maxillary and ethmoid bones formed the secondary palate, whereas in crocodilians the maxillary and palatine bones formed most of the palate. Both solutions placed the internal nares at the rear of the mouth where a fleshy valve could keep water out of the air passages.

---

ries the air passages posteriorly to the rear of the mouth. A flap of tissue arising from the base of the tongue can form a watertight seal between the mouth and throat. Thus, a crocodilian can breathe while only its nostrils are exposed without inhaling water. The increasing involvement of the premaxilla, the maxillae, and pterygoids in the secondary palate can be traced from Mesozoic crocodilians to modern forms.

Modern crocodilians are semiaquatic animals, but they have well-developed limbs and some species make extensive overland movements. Crocodilians can gallop, moving the limbs from their normal laterally extended posture to a nearly vertical position beneath the body.

The Cretaceous was the high point in crocodilian evolution. The extension of warm climates to land areas that are now in cool temperate climate zones favored both diversity and large size. *Deinosuchus* ("terrible crocodile") from the Cretaceous of Texas had a skull that was nearly 2 meters long. If this crocodilian had the same body proportions as modern forms, it would have had a total length of 12 to 15 meters and might well have preyed on dinosaurs.

The heavy, laterally flattened tail propels the body of a crocodilian in water, and the legs are held against the sides. In the late Jurassic, a lineage of specialized marine crocodiles enjoyed brief success. These thallatosuchians had long skulls with pointed snouts. They lacked the dermal body armor typical of most crocodilians and had developed a lobed tail very like that of the primitive ichthyosaurs, with the vertebral column turned downward into the lower lobe and the upper lobe supported by stiff tissue. The feet were paddle-like.

Only 21 species of crocodiles now survive. Most are found in the tropics or subtropics, but three species have ranges that extend into the temperate zone. In many respects crocodilians are the living animals most like Mesozoic forms, and they have been used as models in attempts to analyze the ecology of dinosaurs.

Systematists divide living crocodilians into three families: The Alligatoridae includes the two species of living alligators and the caimans (Figure 13–7). With the exception of the Chinese alligator, the Alligatoridae is solely a New World group. The American alligator is found in the Gulf coast states, and several species of caimans range from Mexico to South America and through the Caribbean. Alligators and caimans are freshwater forms, whereas the Crocodylidae includes species such as the saltwater crocodile that inhabits estuaries, mangrove swamps, and the lower regions of large rivers. This species occurs widely in the Indo-Pacific region and penetrates the Indo-Australian archipelago to northern Australia. In the New World, the American crocodile is quite at home in the sea and occurs in coastal regions from the southern tip of Florida through the Caribbean to northern South America.

The saltwater crocodile is probably the largest living species of crocodilian. Until recently, adults may have reached lengths of 7 meters. Crocodilians grow slowly once they reach maturity, and it takes time to attain large size. In the face of inten-

sive hunting in the last two centuries, few crocodilians now attain the sizes they are genetically capable of reaching. Not all crocodilians are giants: A number of small species live in small bodies of water. The dwarf caiman of South America and the dwarf crocodile of Africa are about a meter long as adults and live in swift-flowing streams.

The third family of crocodilians, the Gavialidae, contains only a single species—the gharial, which once lived in large rivers from northern India to Burma. This species has the narrowest snout of any crocodilian; the mandibular symphysis (the fusion between the mandibles at the anterior end of the lower jaw) extends back to the level of the 23rd or 24th tooth in the lower jaw. A very narrow snout of this sort is a specialization for feeding on fish that are caught with a sudden sideward jerk of the head. We have already called attention to the evolution of similar skull shapes in a variety of Mesozoic animals, including trematosaurs, phytosaurs, and the short-necked plesiosaurs.

Pterosaurs The archosaurs gave rise to two independent radiations of fliers. The birds are one of these lineages, and their similarity to archosaurs is so striking that had they disappeared at the end of the Mesozoic, they would be considered no more than another group of highly specialized archosaurs. The other lineage of flying archosaurs were the pterosaurs ("winged animals") of the Jurassic and Cretaceous (Figure 13–8). They ranged from the sparrow-size *Pterodactylus* to *Quetzalcoatlus,* with a wingspan of 13 meters. The wing formation of pterosaurs was entirely different from that of birds. The fourth finger of pterosaurs was elongate and supported a membrane of skin anchored to the side of the body and perhaps to the hind leg. A small splint-like bone was attached to the front edge of the carpus and probably supported a membrane that ran forward to the neck. The rhamphorhynchoid pterosaurs had a long tail with an expanded portion on the end that was presumably used for steering; the pterodactyloids lacked a tail.

Flight is a demanding means of locomotion for a vertebrate, and it is not surprising that ptero-

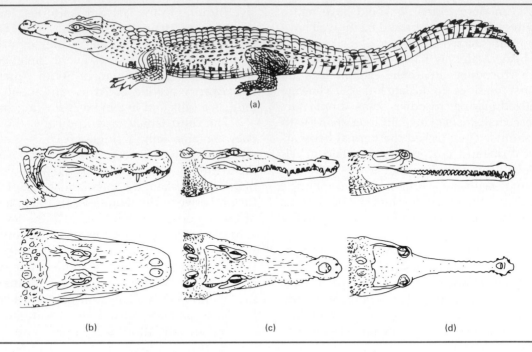

(a)

(b)   (c)   (d)

Figure 13–7. Modern crocodilians differ little from each other or from Mesozoic forms. The greatest interspecific variation in living crocodilians is seen in the shape of the head. Alligators and caimans are broad-snouted forms with varied diets. Crocodiles include a range of snout widths. The widest crocodile snouts are almost as broad as those of most alligators and caimans, and these species of crocodilians have varied diets that include turtles, fish, and terrestrial animals. Other crocodiles have very narrow snouts, and these species are primarily fish eaters. (a) Cuban crocodile; (b) Chinese alligator; (c) American crocodile; (d) gharial. (Modified from H. Wermuth and R. Mertens, 1961, *Schildkröten, Krocodile, Brückenechsen,* Gustav Fisher, Jena, East Germany.)

saurs and birds show a high degree of convergent evolution. The long bones of pterosaurs were hollow, as they are in birds and many other archosaurs, reducing weight with little loss of strength.

The sternum, to which the powerful flight muscles attach, was well developed in pterosaurs, although it lacked the keel seen in birds. The eyes were large, and casts of the brain cavities of ptero-

Figure 13–8. Pterosaurs: (a) *Rhamphorhynchus* from the Jurassic; (b) *Pteranodon* from the Cretaceous. The skulls of pterosaurs suggest dietary specializations. (c) *Anurognathus* may have been insectivorous, (d) *Eudimorphodon* may have eaten small vertebrates, (e) *Dorygnathus* may have been a fish eater, (f) *Pterodaustro* had a comb-like array of teeth that may have been used to sieve plankton, (g) *Dsungaripterus* may have pulled mollusks from rocks with a horny beak and then crushed them with its molariform teeth. (Skulls modified from D. Norman, 1985, *The Illustrated Encyclopedia of Dinosaurs,* Salamander Books, London.)

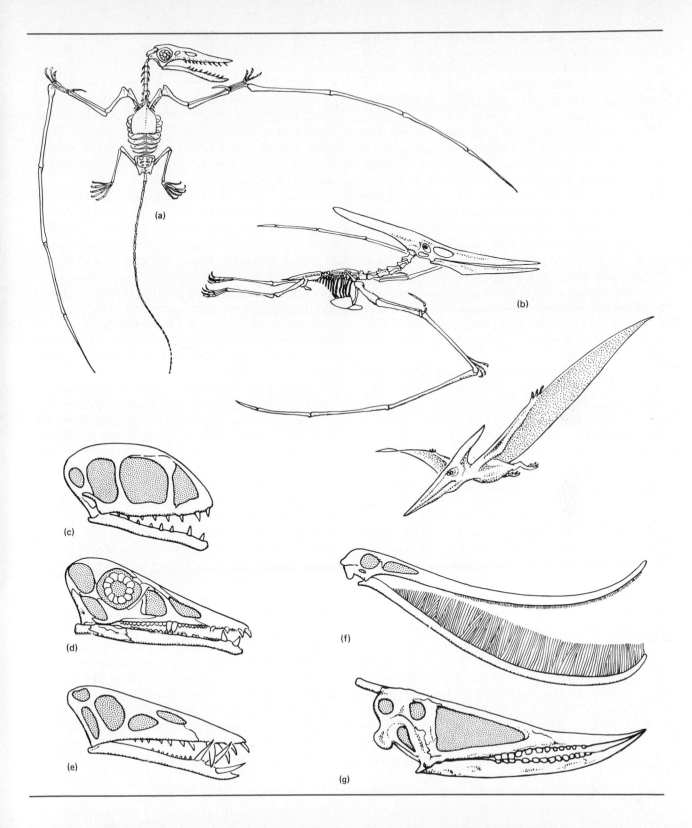

(a)

(b)

(c)

(d)

(e)

(f)

(g)

saurs show that the parts of the brain associated with vision were large and the olfactory areas small, as they are in birds. The cerebellum, which is concerned with balance and coordination of movement, was large in proportion to the other parts of the brain as it is in birds.

Some pterosaurs had lost their teeth and evolved a bird-like beak. Others had sharp, conical teeth in blunt skulls reminiscent of those of bats. Some pterosaurs with elongate skulls and stout sharp teeth may have caught fish or small tetrapods. *Pterodaustro* had an enormously long snout lined with a comb-like array of fine teeth that may have been used for sieving small aquatic organisms. *Dsungaripterus* had long jaws that meet at the tips like a pair of forceps. The tips of the jaws were probably covered with a horny beak, and blunt teeth occupied the rear of the jaw. These animals may have plucked snails from rocks with their beaks and then crushed them with their broad teeth.

Even in their physiology pterosaurs and birds may have been similar. Birds are endotherms, their feathers providing the insulation required to retain metabolically produced heat. Some fossils of pterosaurs in fine-grained sediments show impressions of structures that look like hairs (Sharov 1971). This observation suggests that pterosaurs, too, might have been endotherms. The flight capacities of pterosaurs have long been debated, and most hypotheses about their ecology have been based on the assumption that they were weak fliers. That assumption has led to suggestions of restrictions of activities and habitats of pterosaurs that seem unlikely for a group of animals that was clearly diverse and successful through much of the Mesozoic. A recent analysis suggests that the flying abilities of pterosaurs have been underestimated (Brower and Veinus 1981). This view suggests that the aerodynamic characteristics of small pterosaurs were similar to those of small birds and bats, which are fast, agile fliers. The large pterosaurs appear to have been specialized for slow, maneuverable flight, like that of frigate birds and some vultures. The structure of the pelvic girdle of pterosaurs suggests that they would have been clumsy walkers on flat surfaces but good climbers on rocks and trees (Unwin 1987).

Dinosaurs By far the most generally known of the archosaurs are the Saurischia and Ornithischia. These groups are linked in popular terminology as dinosaurs, but differ in the specializations they developed. Both groups appear to have been primitively bipedal and to have evolved some secondarily quadrupedal forms.

Many of the morphological trends that can be traced in archosaur evolution appear to be associated with increased locomotor efficiency. The two most important developments were the movement of the legs under the body and a widespread tendency toward bipedalism. Early archosauromorphs had a sprawling posture like that of many living amphibians and squamates. The humerus and femur were held out from the body and the elbow and knee were bent at a right angle. In advanced archosaurs, the legs were straight pillars held beneath the body.

In primitive tetrapods, muscles originating on the pubis and inserting on the femur protract the leg (move it forward), muscles originating on the ischium abduct it (move it toward the midline of the body), and muscles originating on the tail retract the femur (move it posteriorly). The primitive tetrapod pelvis, little changed from *Ichthyostega* through primitive archosauromorphs, was plate-like (Figure 13–9a). The ilium articulated with one or two sacral vertebrae, and the pubis and ischium did not extend far anterior or posterior to the socket for articulation with the femur (acetabulum). The pubofemoral and ischiofemoral muscles extended ventrally from the pelvis to insert on the femur. (The downward force of their contraction was countered by iliofemoral muscles that ran from the ilium to the dorsal surface of the femur.) As long as the femur projected horizontally from the body, this system was effective. The pubofemoral and caudiofemoral muscles were long enough to swing the femur through a large arc relative to the ground. As the legs were held more nearly under the body, the pubofemoral muscles became less effective. As the femur rotated closer

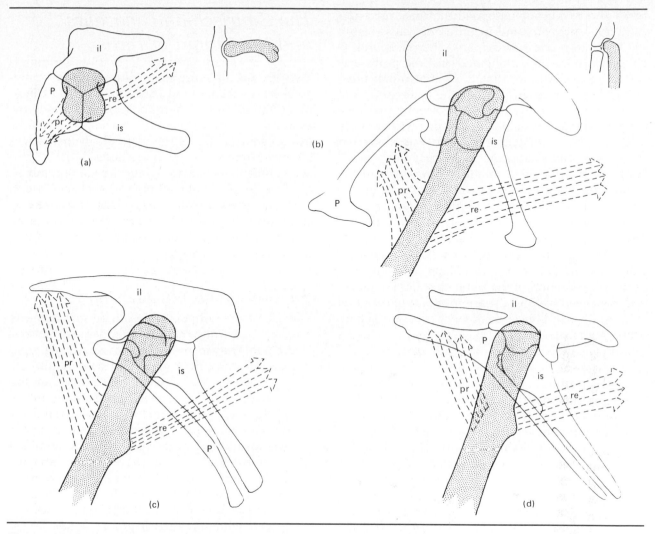

Figure 13–9. Functional aspects of the pelvises of dinosaurs. Pelvic morphology of a primitive archosaur (a, *Euparkeria*), a saurischian dinosaur (b, *Ceratosaurus*), and two ornithischian dinosaurs (c, *Scelidosaurus*; d, *Thescelosaurus*). The presumed action of femoral protractor muscles (pr) and retractors (re) is shown by arrows. Insets show an anterior view of the articulation of the femur with the pelvis. p, pubis; il, ilium; is, ischium.

to the pubis, the sites of muscle origin and insertion moved closer together and the muscles themselves became shorter. A muscle's maximum contraction is about 30 percent of the resting length; thus the shorter muscles would have been unable to swing the femur through an arc large enough

for effective locomotion had there not been changes in the pelvis associated with the evolution of bipedalism (Charig 1972).

The bipedal ornithischian and saurischian dinosaurs carried the legs completely under the body and show associated changes in pelvic structure. The two groups attained the same mechanically advantageous end in different ways (Figure 13–9). In tetrapod saurischians, the pubis and ischium both became elongated and the pubis was rotated anteriorly, so that the pubofemoral muscles ran back from the pubis to the femur and were able to protract it (Figure 13–9b). Among primitive ornithischians the pubis did not project anteriorly (Figure 13–9c). Instead, the ilium was elongated anteriorly, and it appears likely that the femoral protractors originated on the anterior part of the ilium, from which they ran posteriorly to the femur. This condition is seen in the pelvis of primitive ornithischians such as *Scelidosaurus* (Figure 13–9c), and appears to be maintained in the ankylosaurs, a group of advanced quadrupedal ornithischians. Other ornithischians developed an anterior projection of the pubis that ran parallel to and projected beyond the anterior part of the ilium (Figure 13–9d). This development occurred in both bipedal and quadrupedal lineages and provided a still more anterior origin for protractor muscles.

The trend toward bipedalism was important in terms of opening new adaptive zones to archosaurs. A fully quadrupedal animal uses its forelegs for walking, and any changes in limb morphology must be compatible with that function. As animals become increasingly bipedal, the importance of their forelegs for locomotion decreases and the scope of the specialized functions that can evolve increases. Many of the smaller carnivorous dinosaurs that were fully bipedal had forelegs that were adapted for grasping prey. The specialization of forelimbs as wings occurred twice among diapsids, once in the evolution of birds and once in pterosaurs.

Bipedal animals have hind legs that are considerably longer than their forelegs, and the degree of disproportion between hind legs and forelegs is assumed to reflect the extent of bipedalism in a given species. The quadrupedal archosaurs had longer hind legs than forelegs, and this condition is thought by most paleontologists to indicate that they were secondarily quadrupedal, having evolved from bipedal ancestors.

The Saurischian Dinosaurs and the Origin of Birds

Two groups of saurischian dinosaurs are distinguished, the **Theropoda** and the **Sauropodamorpha** (Figure 13-10). Theropods, which include the extant birds, are carnivorous bipeds, whereas the extinct sauropodamorphs were quadrupedal herbivores (Figure 13–11). Ten shared derived characters unite saurischians (Gauthier 1986); the most obvious is an elongate, mobile, S-shaped neck. This character distinguishes birds among living amniotes. Other bird-like characters of saurischians are found in modifications of the hand, skull, and post-cranial skeleton.

Sauropodamorph Dinosaurs

The earliest sauropod dinosaurs were the prosauropods, a group that was abundant and diverse in the late Triassic and early Jurassic. Three types of prosauropods are known, differing in size and tooth structure. The anchisaurids ranged in size from *Anchisaurus* (2.5 meters) to *Plateosaurus* (6 meters). The anchisaurids had long necks and small heads (Figure 13–11a), and the teeth of the best-known forms had large serrations. Modern herbivorous lizards (iguanas) have teeth with very much the same form, and anchisaurids were probably herbivorous. Supporting this view is the presence of **gastroliths** (*gast* = stomach, *lith* = stone) associated with some fossil prosauropods. These stones were probably swallowed by the dinosaurs and lodged in a muscular gizzard where they assisted in grinding plant material to a pulp that could be digested more readily—some birds use gastroliths in this manner. The body proportions of prosauropods suggest that they could stand vertically on their hind legs, but probably employed a quadrupedal posture most of the time.

The melanorosaurids were larger than the prosauropods (*Riojasaurus* from the late Triassic of Argentina was 11 meters long). No skulls of melanorosaurids have been found, so the structure of

Phylogenetic Relationships of the Saurischia

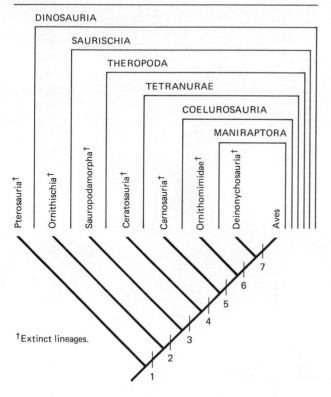

DIAPSIDA

1. Dinosauria: Characteristics of the palate, pectoral and pelvic girdles, hand, hindlimb, and foot. 2. Saurischia: Construction of the snout, extension of the temporal musculature onto the frontal bones, elongation of the neck, modifications of the articulations between vertebrae, and modifications of the hand. 3. Theropoda: Construction of the lower jaw, bones of the skull roof and palate, fenestra in the maxilla, characters of the vertebrae, neural arches and transverse processes lacking posterior to a transition point in the middle of the tail, modifications of the hand and foot, fibula and tibia closely adpressed, thin-walled (hollow) long bones. 4. Tetranurae: Large fenestra posteriorly located in the maxilla, large fanglike teeth absent from dentary, maxillary tooth row ends anterior to orbit of eye, transition point in tail is farther anterior than in theropods, expanded distal portion of pubis, characters of the foot. 5. Coelurosauria: Fenestra in roof of mouth, characters of cervical vertebrae and ribs, furcula (wishbone) formed by fused clavicles, fused bony sternal plates, elongate forelimb and hand, characters of the foot. 6. Maniraptora: Prefrontal reduced or absent, characters of the vertebrae, transition point in tail vertebrae close to base of tail, characteristics of the feet and pelvis. 7. Aves: Progressive loss of teeth on maxilla and dentary, well-developed bill, feathers, characteristics of skull, jaws, vertebrae, and axial and appendicular skeleton. (Based on J. Gauthier, 1986, pages 1–55 in *The Origin of Birds and the Evolution of Flight,* edited by K. Padian, *Memoirs of the California Academy of Sciences,* Number 8.)

Figure 13–10. Phylogenetic relationships of the Saurischia. This diagram shows the probable relationships among the major groups of saurischian dinosaurs, including birds. Extinct lineages are marked by a dagger (†). The numbers indicate derived characters that distinguish the lineages.

their teeth is unknown. The long, slender neck of *Riojasaurus* suggest that the head was small, like that of prosauropods. The yunnanosaurids were smaller than the melanorosaurids and more lightly built, and they differed from the prosauropods in having teeth shaped like flattened cylinders with a chisel-shaped tip. This is the tooth structure seen in the giant sauropod dinosaurs, and is quite distinct from that of the laterally flattened, serrated teeth of prosauropods such as *Plateosaurus*.

The long necks of all the prosauropods suggest that they were able to browse on plant material at heights up to several meters above the ground. The ability to reach tall plants might have been a significant advantage during the shift from the low-growing *Dicrodium* flora to the taller ben-

Figure 13–11. Sauropodomorph dinosaurs: (a) *Plateosaurus*; (b) *Camarasaurus*;
(c) *Diplodocus*.

nettitaleans and conifers that occurred in the late Triassic.

The advanced sauropods of the Jurassic and Cretaceous were enormous quadrupedal herbivores. The sauropods were the largest terrestrial vertebrates that have ever existed, reaching lengths of 25 meters and weighing 20,000 to 50,000 kilograms. A fossil sauropod discovered in Colorado in 1979 and not yet fully described or named may have been 30 meters long and have weighed more than 100,000 kilograms, the equivalent of 20 elephants.

Two major types of giant sauropods can be distinguished. The diplodocids include *Apatosaurus* (formerly known as *Brontosaurus*) and *Diplodocus* (Figure 13–11). These animals had long necks

(15 cervical vertebrae) and long tails (up to 80 caudal vertebrae) that ended in a thin whiplash. Their front legs were relatively short, and the trunk slanted downward from the hips to the shoulders. Their skulls were elongate, teeth were limited to the front of the mouth, and the modest development of the bones of the lower jaw suggests that the jaw muscles were not particularly powerful.

The camarasaurids and branchiosaurids had shorter necks (12 vertebrae) than the diplodocids, and their tails were shorter (about 50 vertebrae) and lacked the whiplike extension that was characteristic of the diplodocids. The forelimbs of camarasaurids were relatively long, and the vertebral column was nearly horizontal. In branchiosaurids the front legs were still longer, and the trunk sloped steeply downward from the shoulders to the hips. Camarasaurids and branchiosaurids had compact skulls with stout jaws and large chisel-shaped teeth. The teeth of *Camarasaurus* and *Branchiosaurus* show evidence of heavy wear, suggesting that they fed on abrasive material.

Both kinds of sauropods were enormously heavy, and the vertebrae show features that helped the spinal column to withstand the stresses to which it was subjected (Figure 13–12). The vertebrae themselves were massive, and the neural arches well developed. Strong ligaments transmitted forces from one arch to adjacent ones to help equalize the stress. The head and tail were cantilevered from the body, supported by a heavy spinal ligament. The sides of the neural arches and

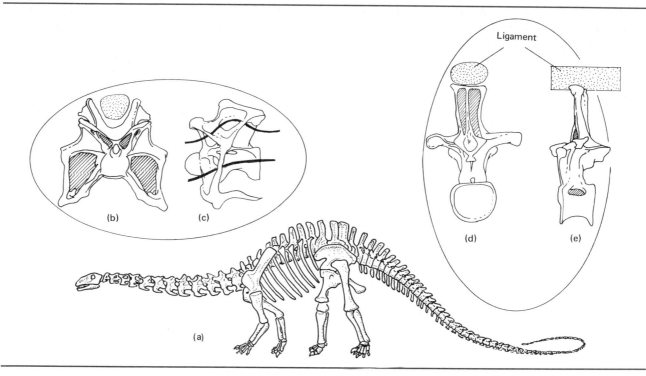

Figure 13–12. The skeletons of large diplodocid sauropods like *Apatosaurus* (a) combined lightness with strength. Vertebrae from the dorsal region (b, posterior view; c, lateral view) and neck (d, anterior view; e, lateral view) show the bony arches that acted like flying buttresses on a large building. [The black ribbons in (e) indicate the position of the arches.]

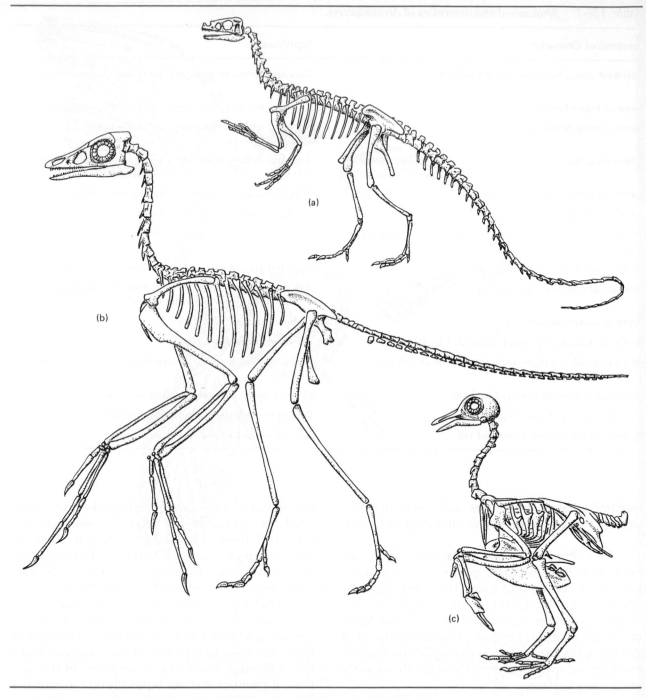

Figure 13–16. Skeleton of *Archaeopteryx* (b) compared to that of *Ornitholestes,* a maniraptor (a) and a modern bird (c). (Modified from D. Norman, 1985, *The Illustrated Encyclopedia of Dinosaurs,* Salamander Books, London.)

and a long tail that was used for balance. Most reconstructions and pictures of *Archaeopteryx* are based on this conception, which finds a parallel among living birds in the young hoatzin, which climbs about in the branches aided by functional clawed fingers on its wings.

The arboreal theory of the origin of avian flight long dominated the field. According to this view, the proavian relatives of *Archaeopteryx* were tree climbers that jumped from branch to branch and from tree to tree much as some squirrels, lizards, and monkeys do. Under selective pressures favoring increased distance and accuracy of travel between trees, structures that provided some surface area for lift would be advantageous. A functional analogy can be made to gliding lizards such as *Draco*, although the morphological structures involved in the proavian model are the forelimbs, not the ribs. By this hypothesis, the evolution of volant forms passed from gliding stages through intermediate stages, such as *Archaeopteryx*, in which gliding was aided by weak flapping flight, to fully airborne flapping fliers. However, one problem that has not been satisfactorily explained by the arboreal theory is selection for bipedalism in an arboreal habitat (Gauthier and Padian 1985). Could a two-legged creature land upright on the branch of a tree without already possessing well-coordinated, aerodynamically controlled braking ability?

If, as seems reasonable from the fossil record of the coelurosaurs and the structure of *Archaeopteryx* itself, the lineage giving rise to birds consisted of bipedal, cursorial forms, is it necessary to invoke arboreal selection pressures at all for the evolution of avian flight? Another theory postulates that flapping flight evolved directly from ground-dwelling, bipedal runners.

According to the first version of this hypothesis (the cursorial theory), proavians were fast, bipedal runners that used their primitive wings as planes to increase lift and lighten the load for running. In a later development, the wings were flapped as the animal ran to provide additional forward propulsion, much as a chicken flaps across the barnyard to escape from a dog. Finally, the pectoral muscles and flight feathers became sufficiently developed for full-powered flight through the air.

The cursorial theory in its original form failed as an explanation because in physical and mechanical terms flapping is not an effective mechanism to increase running speed. Maximum traction on the ground is required to achieve acceleration or high speed from leg movements, and this traction can be provided only by solid contact of the feet with a firm substrate. Planing with primitive wings would have reduced traction, and the push from the small surfaces of the pro-towings probably would not have compensated the loss in speed from the hind legs, much less added to acceleration.

A recently identified specimen of *Archaeopteryx*, misidentified as a coelurosaur for over 100 years, revealed some previously unknown details of the hand and led John Ostrom to modify the cursorial theory. Some elements of the manus are extremely well preserved in this specimen and show the actual horny claws on digits 1 and 3. These claws look like the talons of a predatory bird.

The similarities in morphology between the hand, metacarpus, forearm, humerus, and pectoral apparatus of *Archaeopteryx* and those of several coelurosaurs may be evidence of a similarity in biological roles for both—a grasping function for predation. Although bearing feathers, the forelimb of *Archaeopteryx* has not been structurally much modified from the skeletal condition of these small theropods, and it differs from all other known birds in lacking a number of features that are critical for powered flight (fused carpometacarpus, restricted wrist and elbow joints, modified coracoids, and a plate-like sternum with keel). In fact, the only skeletal feature suggesting flight is the well-developed **furcula** (wishbone), which was present in coelurosaurs and is present in modern birds, although reduced or absent in flightless forms. Thus, the entire pectoral appendage (skeleton and muscles) of *Archaeopteryx* appears to have been as well adapted for predation as for flight.

From these considerations, Ostrom postulated

Figure 13–18. Quadrupedal ornithischians. (a) *Stegosaurus* and (b) *Kentrosaurus* were stegosaurians, (c) *Ankylosaurus*, an ankylosaurian, and (d) *Styracosaurus*, a ceratopsoid.

gosaurus, was smaller (2.5 meters), and had a series of seven pairs of spikes that started near the middle of the trunk and extended down the tail. Anterior to these spikes were about seven pairs of plates similar to those of *Stegosaurus*, but smaller.

The function of the plates of *Stegosaurus* has been a matter of contention for decades. Originally they were assumed to have provided protection from attacks of carnosaurs, and some reconstructions have shown the plates lying flat against the sides of the body as shields. A defensive function is not very convincing, however. Whether the plates were erect or flat, they left large areas on the sides of the body and the belly unprotected. A recent suggestion (Farlow et al. 1976) proposed that the plates were used for heat exchange. Examination shows that the plates were extensively vascularized and could have carried a large flow of blood to be warmed or cooled according to the needs of the animal (Buffrénil et al. 1986). *Kentrosaurus*, the African counterpart of *Stegosaurus*, had much smaller dorsal plates than *Stegosaurus* and the plates on *Kentrosaurus* extended only from the neck to the middle of the trunk. Here the plates were replaced by a row of spikes that ran down the tail and appear to have had a primarily defensive function rather than a thermoregulatory one. It is frustrating not to be able to compare the thermoregulatory behaviors of the two kinds of stegosaurids in a controlled experiment.

The short front legs of stegosaurids kept their heads close to the ground, and their heavy bodies do not give the impression that they were able to stand upright on their hind legs to feed on trees as ornithopods and sauropods probably did. Stegosaurids may have browsed on ferns, cycads, and other low-growing plants. The skull was surprisingly small for such a large animal, and had the familiar horny beak at the front of the jaws. The teeth show none of the specializations seen in hadrosaurs or ceratopsians, and the coronoid process of the lower jaw is not well-developed. Unlike the hadrosaurs and ceratopsians, which appear to have been able to grind or cut plant material into small pieces that could be digested efficiently, stegosaurids may have eaten large quantities of food without much chewing and relied largely upon the fermentative activity of symbiotic bacteria and protozoans to aid digestion. Stegosaurids may also have used gastroliths in a muscular gizzard to pulverize plant material.

The ankylosaurs were a group of heavily armored dinosaurs that are found in Jurassic and Cretaceous deposits in North America and Eurasia. Ankylosaurs were quadrupedal ornithischians that ranged from 2 to 6 meters in length. They had short legs and broad bodies, with **osteoderms** (bones embedded in the skin) that were fused together on the neck, back, hips, and tail to produce large shield-like pieces. Bony plates also covered the skull and jaws and, in at least one form (*Euoplocephalus*), even the eyelids had bony armor. Ankylosaurs had short tails, and in some species a lump of bone at the end of the tail could apparently be swung like a club. The posteriormost caudal vertebrae of these club-tailed forms have elongated neural and hemal arches that touch or overlap the arches on adjacent vertebrae and ossified tendons running down both sides of the vertebrae. Contraction of the muscles that inserted on these tendons probably pulled the posterior caudal vertebrae together to form a stiff handle for swing-

ing the club head at the end of the tail. In all, the tail of these animals resembles nothing so much as an enormous medieval mace. Other species had spines projecting from the back and sides of the body, and ankylosaurs must have been difficult animals for tyrannosaurids to attack. The brains of ankylosaurs appear to have had large olfactory stalks leading to complex nasal passages that probably had sheets of bone that supported an epithelium with chemosensory cells. If this interpretation is correct, ankylosaurs may have had a keen sense of smell.

Euornithopods

Ornithopods from the early Jurassic, the heterodontosaurids and related groups, were mostly small (1 to 3 meters long) and bipedal. They had four toes on the hind feet and, unlike the bipedal saurischians, retained five toes on the fore feet. Their cheek teeth were specialized for grinding plant material. Some heterodontosaurids had sharp tusks that may have been better developed in males than in females. The Cretaceous hypsilophodontids had horny beaks with which they may have cropped plant material that was subsequently ground by the high-crowned cheek teeth that gave the family its name ("high-ridged tooth").

The first dinosaur fossil to be recognized as such was an ornithopod, *Iguanodon*, found in Cretaceous sediments in England (Figure 13–19). Specimens have subsequently been found in Europe and Mongolia, and related forms have been discovered in Africa and Australia. *Iguanodon* reached lengths of 10 meters, although most specimens are smaller. Iguanodontids from the early Cretaceous had large heads and elongated snouts that ended in broad, toothless beaks. Their teeth, which were in the rear of the jaws, were laterally flattened and had serrated edges, very like the lateral teeth of living herbivorous lizards like *Iguana*. The first digit on each forefoot was modified as a spine that projected upward. These spines show a striking resemblance to the spines on the forefeet of some frogs that are used as defensive weapons

(a)

(b)

(c)

(d)

and during intraspecific encounters. One or both functions may have existed in *Iguanodon*. *Ouranosaurus* is an ornithopod known from the early middle Cretaceous of Africa. It had a large sail that was supported by elongated neural spines on the vertebrae of the trunk and tail. Behavioral and thermoregulatory functions can be postulated for the sail of *Ouranosaurus* as they were for the similar structures of the spinosaurid carnosaurs.

Hadrosaurids   The advanced ornithopods include several specialized forms of hadrosaurs (duck-billed dinosaurs). Hadrosaurids were the last group of ornithopods to evolve, appearing in the middle of the Cretaceous. As their name implies, some duck-billed dinosaurs had flat snouts with a duck-like bill (Figure 13–19). These were large animals; some reached lengths in excess of 10 meters and weights greater than 10,000 kilograms. The anterior part of the jaws was toothless, but a remarkable battery of teeth occupied the rear of the jaws. On each side of the upper and lower jaws were four tooth rows, each containing about 40 teeth packed closely side by side to form a massive tooth plate. Several sets of replacement teeth lay beneath those in use, so a hadrosaur had several thousand teeth in its mouth, of which several hundred were in use simultaneously. Fossilized stomach contents of hadrosaurs consist of pine needles and twigs, seed, and fruit of terrestrial plants.

The rise of the hadrosaurs was approximately coincident with a change in the terrestrial flora during the middle Cretaceous. The bennettitaleans and seed ferns that had spread during the Triassic now were replaced by flowering plants (angiosperms). Simultaneous with the burgeoning of the angiosperms and hadrosaurian dinosaurs was a decline in the enormous sauropod dinosaurs such as *Diplodocus* and *Brachiosaurus*. Those lineages were most diverse in the late Jurassic and early Cretaceous, and only a few forms persisted after the middle of the Cretaceous.

Three kinds of hadrosaurs are distinguished—flat-headed, solid-crested, and hollow-crested (Figure 13–20). In the flat-headed forms (hadrosaurines) the nasal bones are not especially enlarged, although the nasal region may have been covered by folds of flesh. In the solid-crested forms (lambeosaurines) the nasal and frontal bones grew upward, meeting in a spike that projected over the skull roof. In the hollow-crested forms (saurolophines) a similar projection was formed by the premaxillary and nasal bones. In *Corythosaurus* those bones formed a helmet-like crest that covered the top of the skull, whereas in *Parasaurolophus* a long, curved structure extended over the shoulders. Although the crests of the lambeosaurines contained only bone, the nasal passages ran through the crests of the saurolophines. Inspired air travelled a circuitous route from the external nares through the crest to the internal nares, which were located in the palate just anterior to the eyes.

Perhaps these bizarre structures were associated with species-specific visual displays and vocalizations (Hopson 1975). The crests might have supported a frill attached to the neck, which could have been utilized in lateral displays analogous to the displays of many living lizards that have similar frills. Possibly in the noncrested forms the nasal regions were covered by extensive folds of fleshy tissue that could be inflated by closing the nasal valves. Analogous structures can be found in the inflatable proboscises of elephant seals and hooded seals. In the seals, and perhaps in the dinosaurs, the inflated structures are resonating chambers used to produce vocalizations. The size and shape of the nasal cavities of lambeosaurine hadrosaurs suggest that adults produced low-frequency sounds, but juveniles would have had higher-pitched vocalizations (Weishampel 1981).

Figure 13–19.   Bipedal ornithischian dinosaurs: (a) *Iguanodon*; (b) *Hadrosaurus*; (c) *Pachycephalosaurus*; (d) *Ouranosaurus*.

fore the results of this Cretaceous drama were revealed. Some insight into the intraspecific behavior of sauropods may be revealed by a series of tracks found in early Cretaceous sediments at Davenport Ranch in Texas. These reveal the passage of a herd of 23 brontosaur-like dinosaurs that passed by some 120 million years ago. Such a group of individuals moving in the same direction at the same time would be remarkable in most living diapsids, but the brontosaur tracks suggest that this is what happened.

Concentrations of nests and eggs ascribed to sauropods in Cretaceous deposits in southern France suggest that these animals had well-defined nesting grounds (Kerourio 1981). Eggs thought to be those of the large sauropod *Hypselosaurus priscus* are found in association with fossilized vegetation similar to that used by alligators to construct their nests. The orientation of the nests suggests that each female dinosaur probably deposited several groups of eggs in quick succession. The individual groups within a nest are spaced in one to five parallel lines, each line containing 15 to 20 eggs. The eggs had an average volume of 1.9 liters, about 40 times the volume of a chicken egg. A total of 50 of these eggs would weigh about 100 kilograms, or 1 percent of the estimated body mass of the mother. Crocodilians and large turtles have egg outputs that vary from 1 to 10 percent of the adult body mass, so an estimate of 1 percent for *Hypselosaurus* seems reasonable. The eggs might have been deposited in small groups instead of all together because 50 eggs in one clutch would have consumed oxygen faster than it could diffuse through the walls of the nest (Seymour 1979). The possibility of some degree of social behavior among juvenile sauropods is raised by the discovery of five baby prosauropod dinosaurs (*Mussaurus*) with the remains of two eggs in a nest in late Triassic deposits in Argentina (Bonaparte and Vince 1979).

Predatory Behavior of Theropods

The carnivorous saurischian dinosaurs, like the herbivorous forms, present an impression of relatively small morphological diversity. The major evolutionary lineages have already been traced—the ostrich-like ornithomimids, the giant theropods, and the deinonychosaurs that combined the small size of ornithomimids with the predatory habits of the theropods. The ornithomimids parallel flightless birds so closely in morphology that it seems reasonable to assume that they lived essentially the same life in the Cretaceous that ostriches, rheas, and related forms live now. The appearance of increasingly cursorial forms that relied on running to escape predators may indicate that habitats became increasingly open during the Mesozoic. The increase in body size seen in theropods almost certainly was a response to selective forces produced by the concomitant increase in body size of their herbivorous prey. It was advantageous for the herbivores to be large because it reduced the risk of predation by carnivores, and it was advantageous for carnivores to be large because it enabled them to prey on the large herbivores. This sort of selection can continue until the disadvantages of large body size outweigh the benefits to be gained from further increase, but there is no evidence that theropods had reached that point because the Cretaceous forms were larger than their Jurassic ancestors.

There was probably a trend to increasingly specialized methods of prey capture among carnosaurs. This hypothesis is suggested by the reduction of the size of the forelimbs in carnosaurs that presumably indicates an increased reliance on the teeth for both seizing and killing prey. The deinonychosaurs probably relied instead on fleetness of foot to capture active prey. The discovery of five *Deinonychus* skeletons in close association with the skeleton of *Tenontosaurus*, an ornithischian three times their size, might indicate that *Deinonychus* hunted in packs (Figure 13–22). Deinonychosaurs probably used their clawed forefeet to seize prey and then slashed at it with the sickle-like claws on the hind feet. This tactic appears to be illustrated by a remarkable discovery in Mongolia of a deinonychosaur called *Velociraptor*. It was preserved in combat with a *Protoceratops*, its hands grasping the head of its prey and its enormous

Figure 13–22. Dinosaurs may have had a variety of social behaviors.
(a) Dromeosaurs like *Deinonychus* may have hunted in packs to attack large prey
such as ornithopods. (b) Ceratopsians like *Chasmosaurus* may have formed a ring to
confront predators. (c) Hadrosaurs such as *Maiasaura* may have nested in colonies.
(Modified from D. Norman, 1985, *The Illustrated Encyclopedia of Dinosaurs*,
Salamander Books, London.)

for a small lizard (2 centimeters body diameter) to more than 5 days for a large dinosaur (200 centimeters body diameter). The exact values, of course, depend on the assumptions one makes about the thermal properties of dinosaur body tissue and the external conditions, but the most extreme reasonable variation in those factors changes the time constants only by a factor of 2.

These computer simulations indicate that dinosaurs could have achieved many of the benefits of homeothermy without the energy cost of endothermy as long as the daily and seasonal temperature variation was within limits that resulted in tolerable body temperatures. In an equable climate, dinosaurs had the best of both worlds—the advantages of homeothermy without the cost of endothermy (Benton 1979).

Indeed, the metabolic rates assumed by proponents of endothermy would probably have resulted in lethal overheating for animals as large as dinosaurs. Even elephants have large ears that they use to dissipate metabolic heat to avoid overheating, and some dinosaurs were many times larger than elephants. If dinosaurs did have high rates of metabolic heat production, one would expect to find that heat-dissipating devices were widespread, but they seem to have been limited to a few forms such as *Spinosaurus*, *Ouranosaurus*, and *Stegosaurus*.

The enormous ecological and phylogenetic diversity of dinosaurs has largely been overlooked in the debate about their thermoregulatory mechanisms. Not all kinds of evidence apply to all kinds of dinosaurs, and it is likely that there were substantial differences in the biology of different kinds of dinosaurs. Quite apart from any other consideration, the difference in body size of an ornithomimid and a sauropod would make them very different kinds of animals.

The high levels of activity we infer for some ornithomimids and deinonychosaurs suggest that they might have had high rates of metabolism, but those metabolic rates would have been both unnecessary and undesirable for larger dinosaurs. It is likely that dinosaurs were as diverse in their thermoregulatory physiology as they were in their morphology and ecology. Attempting to fit them into categories based on living animals may only confuse the issue.

The Lepidosauromorpha: Ichthyosaurs, Lizards, and Others

The second lineage of diapsids, the Lepidosauromorpha, is distinguished by derived characters of the skull and postcranial skeleton (Benton 1985a). These characters include entry of the postfrontal bone into the dorsal border of the upper temporal fossa, an additional articulating surface in the midline of neural arch of the vertebrae, cervical vertebrae with centra that are shorter than those of the trunk vertebrae, and single-headed ribs. The earliest lepidosauromorphs known are the Younginiformes of the late Permian and early Triassic. The terrestrial forms such as *Youngina* were about 50 centimeters long and very like *Petrolacosaurus*, but appear to have had shorter necks and limbs than that genus. Aquatic Younginiformes like *Hovasaurus* were approximately the same size as the terrestrial forms and had laterally flattened tails. The champsosaurs, *Champsosaurus* and *Simoedosaurus*, are problematic forms that may be primitive lepidosauromorphs. They were aquatic animals, about the size of modern crocodilians, and showed many of the same specializations for aquatic life, including nostrils at the tip of the snout and a secondary palate that provided a passageway for air (Box 13–1). The champsosaurs appeared in the Cretaceous and, unlike many of the large Mesozoic diapsids, persisted past the end of that period into the early Tertiary.

Marine Diapsids: Placodonts, Plesiosaurs, and Ichthyosaurs

Three groups of specialized marine tetrapods are currently grouped with the lepidosauromorphs (Sues 1987). The Triassic placodonts (''flat teeth'') were mollusk eaters specialized for crushing hard-shelled food rather than for rapid pursuit of prey (Figure 13–23). *Placodus* had large flat maxillary teeth and a heavy palate with enormous teeth on the palatine bones. The anterior teeth projected forward and might have been used to seize mussels or oysters growing on rocks and pull them loose. Another placodont, *Placochelys*, had a toothless beak instead of projecting teeth, but retained broad teeth in the rear of the mouth. In contrast, *Henodus* was almost toothless and may have crushed its food with horny plates like those of turtles. The similarity of some of these placodonts to turtles extended to their appearance as well. *Henodus* and *Placochelys* had a body armor that was as extensive as that of turtles, but it was composed of a mosaic of small polygonal dermal bones rather than of large plates like those of turtles. In *Henodus* the dermal bones were apparently covered with epidermal scutes as they are in turtles.

The plesiosaurs ("ribbon reptiles") appeared at the Triassic/Jurassic transition and persisted to the Cretaceous. Two lineages of plesiosaurs can be distinguished, and these appear to have evolved side by side, indicating a considerable ecological separation between them (Figure 13–24). One lineage comprised long-necked animals with small heads, whereas the other contained short-necked animals with long skulls. Both had heavy, rigid trunks and appear to have rowed through the water with limbs that functioned like oars and may also have acted as hydrofoils, increasing the efficiency of swimming. **Hyperphalangy**, the addition of joints to the toes, increased the size of the paddles, and some plesiosaurs had as many as 17 phalanges per digit. In both types of plesiosaurs the nostrils were located high on the head just in front of the eyes.

The long-necked plesiosaurs reached their zenith in *Elasmosaurus*, which lived in the late Cre-

Figure 13–23. Placodonts were slow-swimming animals that probably fed on mollusks. *Placodus* (a) was relatively unspecialized, but *Henodus* (b) had dermal armor that was almost as extensive as a turtle's.

(a)

(b)

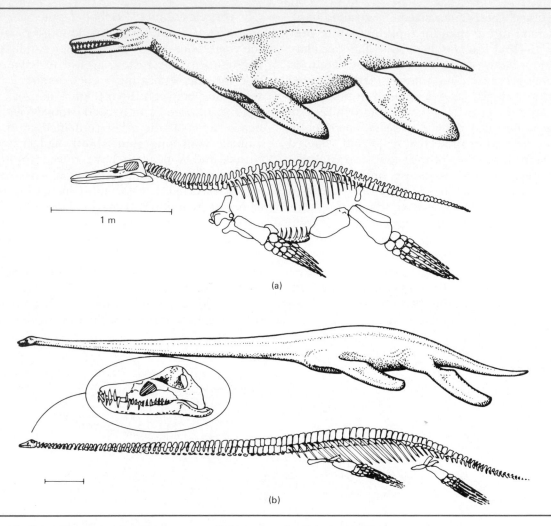

Figure 13–24.   Two radically different sorts of plesiosaurs had evolved by the Jurassic: (a) *Polycotylus*, a short-necked form; (b) *Elasmosaurus*, a long-necked form. (Body outlines modified from D. M. S. Watson, 1951, *Paleontology and Modern Biology*, Yale University Press, New Haven, Conn.; skeletons modified from J. Piveteau, 1955, *Traité de Paléontologie*, volume 5, Mason et Cie, Paris.)

taceous. The lineage can be traced back to *Plesiosaurus* in the early Jurassic. That form had 35 cervical vertebrae. The *Elasmosaurus* line is characterized by a progressive increase in the number of cervical vertebrae and a reduction in the size of the head. Not only did the number of cervical vertebrae increase, but individual vertebrae became

longer. *Microcleidus* of the middle Jurassic had 39 or 40 cervical vertebrae, and *Elasmosaurus* had 76. Even in the early forms the body was not well streamlined, and as the neck became longer, the streamlining became even poorer. The size of the paddles relative to the size of the body decreased from the Jurassic to the Cretaceous. Clearly, the

Elasmosaurus line of plesiosaurs were not rapid swimmers.

The short-necked plesiosaurs started from a form not very different from the starting point of the long-necked forms but followed a completely different evolutionary pathway, leading to increasingly streamlined forms. The neck became shorter, and the paddles larger. These animals might have captured their prey by pursuit in the manner of modern seals and sea lions. In a sense the short-necked plesiosaurs were converging on the morphological adaptations of ichthyosaurs, although they never attained the perfection of streamlining seen in advanced members of that group.

The ichthyosaurs ("fish reptiles") were the most specialized of the aquatic tetrapods of the Mesozoic (Figure 13–25). In many aspects of their body form they resemble porpoises (small toothed whales). Ichthyosaurs had a dorsal fin that was supported only by stiff tissue, not by bone, and the upper lobe of the caudal fin similarly lacked skeletal support. (The vertebral column of advanced ichthyosaurs bent sharply downward into the ventral lobe of the caudal fin.) We know of the presence of these soft tissues because many ichthyosaur fossils in fine-grained sediments near Holzmoden in southern Germany contain an outline of the entire body preserved as a carbonaceous film.

Ichthyosaurs retained both fore- and hindlimbs (unlike cetaceans, which retain only the forelimbs). The limbs of ichthyosaurs were modified into paddles by hyperphalangy and **hyperdactyly** (the addition of extra digits). Fossil ichthyosaurs with embryos in the body cavity indicate that these animals gave birth to living young. One fossil appears to be an individual that died in the process of giving birth, with a young ichthyosaur emerging tailfirst as do baby porpoises.

The stomach contents of ichthyosaurs, preserved in some specimens, include cephalopod mollusks, fish, and an occasional pterosaur (flying archosaur). Ichthyosaurs had large heads with long, pointed jaws that were armed with sharp teeth in most forms, although a few ichthyosaurs were toothless. The large eyes were supported by a ring of sclerotic bones. Triassic ichthyosaurs were poorly streamlined, and may have used angulliform locomotion. The greater streamlining of later forms may have been associated with the development of carangiform locomotion and rapid pursuit of prey like that of modern tunas. The Jurassic was the high point of ichthyosaur diversity; they were less abundant in the early Cretaceous and only a single genus remained in the late Cretaceous. Ichthyosaurs became extinct before the end of the period.

The Lepidosauria

The lepidosaurs are distinguished by a number of derived characters of both the skeleton and the soft anatomy (Benton 1985a). Among these, the long bones are capped by articulating surfaces (epiphyses) of bone or calcified cartilage, the postparietal and tabular bones have been lost from the skull, the kidney has a sexual segment, and the structures of the pituitary and adrenal glands are distinctive. Perhaps the most interesting derived character of the lepidosaurs is determinate growth. That is, growth stops when the cartilaginous plate that separates the ends of the long bones and the epiphyses becomes fully ossified. This modification of the pattern of continuous growth that is seen, for example, in crocodilians and turtles may be associated with specialization on insects as food, and the importance of not growing too large to be efficient predators of insects (Carroll 1977, 1987). Two major groups of lepidosaurs can be distinguished, the Sphenodontia (represented by the living tuatara) and the Squamata (lizards, snakes, and amphisbaenians).

Sphenodontia The correct phylogenetic allocation of the living tuatara (*Sphenodon punctatus*) and its fossil relatives (the family Sphenodontidae) has been a source of confusion for more than a century. In many books it is still listed as a rhynchocephalian. The order Rhynchocephalia was named by Albert Günther in the 1860s when he restudied the tuatara, which had previously been considered an agamid lizard. The striking anatom-

Figure 13–25. Evolutionary change in ichthyosaurs led to increasingly streamlined animals: (a) *Cymbospondylus* from the Triassic; (b) *Opthalmosaurus* from the Cretaceous. (c) The dorsal fin and the upper lobe of the caudal fin of ichthyosaurs were stiff tissue, not supported by bone.

Figure 13–27. Mosasaurs were marine form[s] lizards. (a) Skull of *Platecarpus*, about 40 centim[eters] Cretaceous Niobrara Chalk formation of wester[n] *Tylosaurus*, about 10 meters long. [(a) Courtesy Fossils, Otis, Kansas; prepared by Alan L. Deitri[ch]

Figure 13–26. Modifications of the diapsid sk[ull] diapsid forms like the Permian *Petrolacosaurus* ([a]) bone that define the temporal fenestrae. This con[dition] sphenodontid, the tuatara (b). Lizards have achie[ved] gap between the quadrate and quadratojugal and the frontal and parietal bones as shown by the m[osasaur] (e). Probable transitional stages allowing increasi[ng] nonsquamate lineage are illustrated by *Paliguana* [and] (f) skull kinesis is further increased by loss of the [amphis-] baenians (g), which use their heads for burrowing [have] akinetic skulls. f, frontal; j, jugal; p, parietal; po, p[ostorbital;] qj, quadratojugal; sq, squamosal. [(a) From R. R. [Carroll] 196:1091–1093; (b) and (c) from A. S. Romer, 1[9__,] edition, University of Chicago Press, Chicago; (f) a[fter] *Biomechanics: An Approach to Vertebrate Biology,* J[____]

ical differences between the living tuatara and living lizards seemed to Günther to warrant separating the groups at the ordinal level.

Subsequent study has changed this view: Derived characters of the tuatara and its fossil relatives link those animals to lizards. Also, differences between the sphenodontids and the other diapsids grouped with them as rhynchosaurs have become apparent and have weakened the basis for regarding rhynchocephalians as a natural group. The sphenodontids appear to be the sister group of the lizards + snakes + amphisbaenians. Sphenodontids were diverse through the early Mesozoic, and some sphenodontids are known from the Cretaceous, but there is no fossil record from the Cenozoic. The only living sphenodontid is the tuatara, which occurs on a few islands off the coast of New Zealand.

Squamata   The squamates show numerous derived characters of the cranial and postcranial skeleton and soft tissues (Benton 1985a). The most conspicuous of these is the loss of the lower temporal bar and the quadratojugal bone that formed part of that bar (Figure 13–26). This modification is part of a suite of structural changes in the skull that contribute to the development of a considerable degree of kinesis. The living tuatara shows the ancestral condition for squamates, with the quadratojugal linking the jugal and the quadrate bones to form a complete lower temporal arch. (This fully diapsid condition is not characteristic of all sphenodontids, however; some of the Mesozoic forms did not have a complete lower temporal arch.)

Early lizards are not well known. The fossil genera *Paliguana* and *Palaeagama* from the late Permian and early Triassic of South Africa are probably not squamates, but they do show changes in the structure of the skull that probably parallel the changes that occurred in early squamates. The gap between the quadrate and jugal widened, and the complexly interdigitating suture between the frontal and parietal bones on the roof of the skull became straighter and more like a hinge. Additional areas of flexion evolved at the front and rear of the

skull and in the lower jaw of some lizards. These changes were accompanied by the development of a flexible connection at the articulation of the quadrate bone with the squamosal that provided some mobility to the quadrate. This condition, known as **streptostyly**, increases the force the pterygoideus muscle can exert when the jaws are nearly closed (Smith 1980).

In snakes the flexibility of the skull was increased still further by loss of the second temporal bar, which was formed by a connection between the postorbital and squamosal bones (Figure 13–26). A further increase in the flexibility of the joints between other bones in the palate and the roof of the skull produced the extreme flexibility of modern snake skulls. The role of skull kinesis in the ecology of snakes is discussed further in Chapter 15.

The third group of squamates has a completely different sort of skull specialization. The amphisbaenians are small, legless, burrowing animals. They use their heads as rams to construct tunnels in the soil, and the skull is heavy with rigid joints between the bones (Figure 13–26). Their specialized dentition allows them to bite small pieces out of large prey.

Marine Lizards: The Mosasaurs

The mosasaurs were varanoid lizards, closely related to the living monitor lizards of the genus *Varanus*. They differ from living varanoids in their enormous size (more than 10 meters for the largest species) and their specializations for marine life. Mosasaurs are known only from the late Cretaceous, but they were a large group with nearly 20 genera. The body and tail of mosasaurs were long and the tail was laterally flattened, and the limbs were modified as flippers (Figure 13–27). The skulls of mosasaurs were very like those of varanoid lizards—elongate with sharp teeth and a flexible joint in the lower jaw between the dentary and splenial. One genus, *Globidens*, had teeth with hemispherical crowns, as its name suggests. It probably fed on mollusks. Several fossils of ammonites (free-swimming mollusks related to the

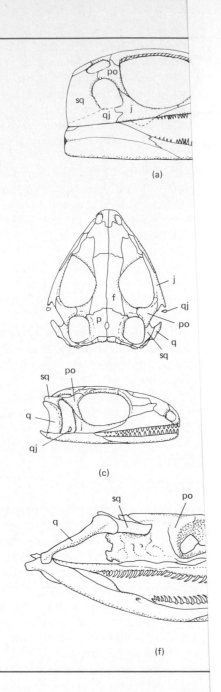

(a)

(c)

(f)

squid and nautilus) have been found bearing a pattern of tooth marks that suggests they had been bitten by another species of mosasaur, *Prognathodon overtoni*.

Mosasaurs probably lived in shallow seas. Hundreds of well-preserved specimens have been found, but none shows a trace of embryos within the body cavity. That situation contrasts with the discovery of several examples of ichthyosaurs with embryos and suggests that despite their other specializations for marine life, the mosasaurs did not develop viviparity but instead they retained the ancestral squamate characteristic of laying eggs.

Other Terrestrial Vertebrates of the Late Mesozoic

There is a tendency to look on the Jurassic and Cretaceous as the Age of Dinosaurs and to forget that there was a very considerable nondinosaur terrestrial fauna and variations in the dinosaurs in different habitats (Lucas 1981). To a certain extent the dinosaurs do form a separate unit. The large carnosaurs were the only animals capable of preying on adults of the large herbivores. Nonetheless, there must have been significant interaction between dinosaurs and nondinosaurs. As far as we know, all dinosaurs reproduced by laying eggs, and dinosaur eggs were a food source that could be exploited by small predators. The Nile monitor lizard today is a predator on eggs of crocodiles, and it is likely that Cretaceous monitor lizards had a similar taste for dinosaur eggs.

One of the distinctive features of dinosaurs is their large size. The smallest dinosaur skeletons known indicate a total adult length of about half a meter. Even 50 centimeters is large in comparison to living squamates; most lizards are smaller than 20 centimeters, and only crocodilians approach the bulk of even moderate-size dinosaurs. There were, of course, adaptive zones available for smaller vertebrates in the Jurassic and Cretaceous, and these were filled as they are now by squamates, turtles, amphibians, and mammals. Among these were the animals that might have stolen eggs from dinosaur nests and served as food for juvenile dinosaurs.

Although dinosaurs are so impressive and distinctive that there is a tendency to think of them as inhabiting a world of their own, they were in reality a part of an ecosystem that included many vertebrates that would not appear strange to us today. As Benton (1985b) has pointed out, "The assemblage of tetrapod families that includes all modern tetrapod groups—frogs, salamanders, lizards, snakes, turtles, crocodilians, birds, and mammals—arose in the late Triassic, and increased in diversity through the Jurassic and early Cretaceous. During the Cretaceous the diversity of these modern groups was approximately equal to that of . . . dinosaurs and pterosaurs. In the latest Cretaceous the diversity of the modern groups rose dramatically and became twice that of the dinosaurs and pterosaurs, long before the terminal Cretaceous extinction event."

The striking difference in the Mesozoic world was the absence of large mammals, and the occupation of that adaptive zone by large archosaurs. Figure 13–28 shows the relative abundance of different groups of vertebrates at two Cretaceous fossil localities. The Lance locality appears to have been a wooded swampy habitat with large streams and some ponds. The Bug Creek Anthills locality was probably downstream from such a swamp, in the delta of a major waterway. In both localities dinosaurs are a minor component of the community in terms of species diversity, although one dinosaur is the equivalent of a great many smaller animals in terms of biomass.

Late Cretaceous Extinctions

After thriving and dominating the terrestrial habitat for 130 million years, dinosaurs and their contemporaries, the plesiosaurs and pterosaurs, disappeared near the end of the Cretaceous. The extinctions that occurred at the Cretaceous–Tertiary transition are an example of a recurrent biological phenomenon—mass extinction. In fact,

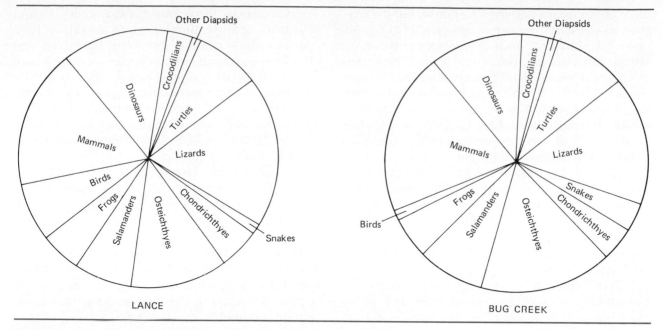

Figure 13–28. Relative abundance of genera at late Cretaceous fossil sites. Although dinosaurs were the most spectacular terrestrial vertebrates during the Mesozoic, they were surrounded by a large assortment of animals that were not very different from living species. Mammals were well represented at both sites. (Calculated from data in R. Estes and P. Berberian, 1970, *Breviora*, No. 343.)

the Cretaceous–Tertiary extinction was relatively small as mass extinctions go, but it is the most widely known because the disappearance of dinosaurs has captured the attention of both scientists and the public. In this chapter we consider the factors that might have been involved in the late Cretaceous extinctions, and in Chapter 14 we review mass extinctions more generally.

Biologists have struggled for years to explain the rapid change in faunal composition that occurred at the end of the Mesozoic, and many hypotheses, although of doubtful biological value, are classics in their own right. The catastrophism school of thought has enjoyed some vogue with the view that the giant archosaurs were wiped out by some geological or cosmic disaster. Various disasters have been proposed. One suggestion is that the moon was scooped from the Pacific Ocean basin to the accompaniment of earthquakes, tidal

waves, and volcanic eruptions. The explosion of a supernova, showering the earth with radiation, is another idea that has been advanced to explain the extinction of dinosaurs. The currently popular hypothesis in the catastrophism sweepstakes is the suggestion that the impact of a comet or an asteroid with the Earth injected enough dust into the atmosphere to blot out the sun, thereby suppressing photosynthesis and leading to the extinctions of animals.

A high concentration of iridium in some deposits at the end of the Cretaceous is the basis for the inference of an asteroid impact, but the geological evidence is not entirely consistent. High iridium levels are not always found at the Cretaceous–Tertiary boundary, and where iridium does occur, it may be of volcanic origin. The biological record also does not clearly support the asteroid hypothesis. An extraterrestrial event would produce an abrupt, worldwide response, but the Cretaceous–Tertiary transition appears to have occurred at different times in different parts of the world. For example, dinosaurs seem to have persisted in Montana 400,000 years after they became extinct in New Mexico and 800,000 years after they disappeared from Alberta. An asteroid may have struck the earth in the Late Cretaceous, but such an impact does not provide a unified explanation of biological events at that time.

Normal biological, climatic, and geological processes appear to provide the most plausible explanation for the faunal changes that occurred at the end of the Mesozoic (Hallam 1987, Kerr 1988). Because of the worldwide scale of the late Cretaceous extinctions and the wide variety of animals affected, hypotheses based on broad-scale changes in climate have been proposed repeatedly. The Mesozoic was characterized by extremely stable temperatures with little variation from day to night, from summer to winter, and from north to south. Geologic evidence indicates that climates became cooler and more variable at the end of the Cretaceous. Casper, Wyoming, is an important dinosaur fossil locality, and a reconstruction of its Mesozoic temperature regime indicates that the long-term temperature extremes probably lay between 11 and 34°C, with a mean of 22°C. Clearly, the climate has changed since then: The present temperature extremes at Casper are −27 to + 40°C, with a mean of 8°C.

In addition to these changes in climate, geological events at the end of the Cretaceous were changing the land surface and altering terrestrial habitats. A reduction in sea level of 150 to 200 meters drained the shallow inland seas that had filled the centers of the continents during the late Cretaceous. Rivers cut their way down toward the new sea level, forming valleys instead of meandering across broad floodplains. Late Cretaceous dinosaurs appear to have been concentrated in river and floodplain habitats, and these changes would have reduced the habitat available to them (Schopf 1983).

The climatic and topographic changes in the late Cretaceous were linked: Greater elevation of the land, greater differences in elevation from place to place, and the absence of the inland seas that buffered temperature change would all have increased the daily and seasonal variation in temperature. If large archosaurs depended on body temperature stability derived primarily from their enormous body size, a small increase in temperature variation would be deleterious to them. The time constant for temperature change in a dinosaur with a body diameter of 1 meter is 2 days. That is long enough to withstand a moderate day–night temperature fluctuation, but not enough to stay warm during several months of cool weather in the winter. The appearance of significant seasonal temperature variation could create a situation in which a dinosaur could not thermoregulate, and if the increased variation was accompanied by a general decline in the average temperature, the situation would be even worse.

Approximately 20 genera of dinosaurs are known from the late Cretaceous Hell Creek formation in Montana, and we assume that all of these became extinct at the end of the Cretaceous. In itself, that is not an exceptional event; from the Jurassic onward few genera of dinosaurs persisted for exceptionally long periods. Instead, there was a continuing process of extinction and replacement by new genera. A total of 19 genera of dinosaurs are known from the late Jurassic Morrison formation in Montana, and none of these survived into the Cretaceous. What seems to have happened to dinosaurs near the end of the Cretaceous was not a massive increase in extinctions so much as a reduction in the evolution of new dinosaurs to replace those that became extinct (Padian and Clemens 1985, Jablonski 1986). Furthermore, an intriguing suggestion of a gradual transition from dinosaurs to mammals is provided by the report of dinosaur fossils found in association with fossils of early ungulate mammals in strata of the Hell Creek formation that lie above the Cretaceous–Tertiary boundary (Sloan et al. 1986). These fossils may indicate that dinosaurs survived into the Tertiary, but they might also be fossils that were eroded from older sediments and redeposited in association with bones of mammals of a younger geological age.

The hypothesis of climatic change at the end of the Cretaceous that was marked primarily by increased variability in temperature and only secondarily by a reduction of the annual mean temperature can also account for the survival of squamates. These small animals have considerable ability to regulate their body temperatures (see Chapter 4). They can bask in the sun to warm up and retreat to shade to avoid getting too hot. Both of these techniques are denied to a large animal because of its size. Temperature change is slow in a large animal, so even an entire day of basking would produce relatively little change in the body temperature of a dinosaur. Furthermore, there are few places a large animal can take shelter when the sun gets too hot. A lizard can find shade under a bush or in a burrow, but a sauropod that is 25 meters long needs a small forest for shelter. The only large ectotherms that survived the Mesozoic, crocodilians and some turtles, occur only in the most equable existing climates. Crocodilians are primarily tropical, with a few subtropical forms, and are always associated with water, which buffers temperature change and provides a retreat from temperature extremes. Small turtles penetrate far north and south, but large forms occur only near the equator. Again, the presence of water as a temperature buffer is important. Sea turtles spend virtually their entire lives in water and the large terrestrial turtles occur on equatorial islands.

Although we can only speculate about the details and wonder why particular groups persisted longer than others, the hypothesis of increasing climatic variation at the end of the Cretaceous provides a context in which physical and biological stresses can be interwoven to provide a plausible explanation of why certain types of diapsids became extinct whereas others did not.

Summary

The major groups of tetrapods in the Mesozoic were members of the diapsid (two arches) lineage. This group is distinguished particularly by the presence of two fenestrae in the temporal region of the skull that are defined by arches of bone. The archosauromorph lineage of diapsids contains the most familiar of the Mesozoic tetrapods, the dinosaurs. Two major groups of dinosaurs are distinguished, the Saurischia and Ornithischia.

The saurischians included the sauropod dinosaurs—enormous herbivorous quadrupedal forms like *Apatosaurus* (formerly *Brontosaurus*), *Diplodocus*, and *Brachiosaurus*—and the theropods, which were bipedal carnivores. The carnosaurs (of which *Tyrannosaurus rex* is the most familiar example) were large theropods that probably preyed on the large sauropods. Other theropods were smaller: The ornithomimids were probably very like modern ostriches, and some had horny beaks and lacked teeth. The deinonychosaurs were fast-running predators. Ornithomimids probably seized small prey with hands that had three fingers armed with claws, whereas the deinonychosaurs probably were able to prey on dinosaurs larger than themselves. They may have hunted in packs and used the enormous claw on the second toe to slash their prey. Birds had evolved by the Jurassic: *Archaeopteryx*, the earliest bird known, is very like small coelurosaurs.

The ornithischian dinosaurs were herbivorous, and many had horny beaks on the snout and batteries of specialized teeth in the rear of the jaw. The ornithopods (duck-billed dinosaurs) and pachycephalosaurs (thick-headed dinosaurs) were bipedal, and the stegosaurians (plated dinosaurs), ceratopsoids (horned dinosaurs), and ankylosaurians (armored dinosaurs) were quadrupedal. Although the saurischians and ornithischians represent independent radiations and had different ecological specializations, they share many morphological features that appear to reflect the mechanical constraints of being very large terrestrial animals.

Dinosaurs and birds are part of the archosaur lineage of the archosauromorph diapsids. The modern crocodilians and the extinct phytosaurs are also archosaurs, as are the pterosaurs. The phytosaurs were very like crocodilians in general body form, but the external nares of phytosaurs were located immediately anterior

fidelity among ornithischian dinosaurs. *Nature* 297: 675–676.

Horner, J. R. 1984. The nesting behavior of dinosaurs. *Scientific American* 241(4):130–137.

Horner, J. R. and P. Makela. 1979. Nest of juveniles provides evidence of family structure among dinosaurs. *Nature* 282:296–298.

Jablonski, D. 1986. Mass extinctions: new answers, new questions. Pages 43–61 in *The Last Extinction*, edited by L. Kaufman and K. Mallory. M.I.T. Press, Cambridge, Mass.

Kerourio, P. 1981. Nouvelles observations sur le mode de nidification et de ponte chez les dinosauriens du Cretace terminal du Midi de la France. *Comptes Rendu Sommaire des Séances de la Societe geologique de France* No. 1, pp. 25–28.

Kerr, R. A. 1988. Was there a prelude to the dinosaurs' demise? *Science* 239:729–730.

Lang, J. W. 1986. Male parental care in mugger crocodiles. *National Geographic Research* 2:519–525.

Lucas, S. G. 1981. Dinosaur communities of the San Juan Basin: a case for lateral variations in the composition of Late Cretaceous dinosaur communities. Pages 337–393 in *Advances in San Juan Basin Paleontology*, edited by S. G. Lucas, J. K. Rigby, Jr., and B. S. Kues. University of New Mexico Press, Albuquerque, N. Mex.

Norman, D. 1985. *The Illustrated Encyclopedia of Dinosaurs*. Salamander Books, London.

Ostrom, J. H. 1969. Osteology of *Deinonychus antirrhopus*, an unusual theropod from the lower Cretaceous of Montana. *Bulletin of the Peabody Museum of Natural History* 30:1–165.

Ostrom, J. H. 1974. *Archaeopteryx* and the evolution of flight. *Quarterly Review of Biology* 49:27–47.

Ostrom, J. H. 1980. The evidence for endothermy in dinosaurs. Pages 15–54 in *A Cold Look at the Warm-Blooded Dinosaurs*, edited by R. D. K. Thomas and E. C. Olson. American Association for the Advancement of Science, Washington, D. C.

Ostrom, J. H. 1986a. The cursorial origin of avian flight. Pages 73–81 in *The Origin of Birds and the Evolution of Flight*, edited by K. Padian. *Memoirs of the California Academy of Sciences*, No. 8.

Ostrom, J. H. 1986b. Social and unsocial behavior in dinosaurs. Pages 41–61 in *Evolution of Animal Behavior, Paleontological and Field Approaches*, edited by M. H. Nitecki and J. A. Kitchell. Oxford University Press, New York.

Padian, K. (editor). 1986.*The Origin of Birds and the Evolution of Flight. Memoirs of the California Academy of Sciences*, No. 8.

Padian, K. and W. A. Clemens. 1985. Terrestrial vertebrate diversity: episodes and insights. Pages 41–96 in *Phanerozoic Diversity Patterns*, edited by J. W. Valentine. Princeton University Press, Princeton, N. J.

Peterson, A. 1985. The locomotor adaptations of *Archaeopteryx*: glider or cursor? Pages 99–103 in *The Beginning of Birds*, Proceedings of the International *Archaeopteryx Conference, Eichstätt 1984*, edited by M. K. Hecht, J. H. Ostrom, G. Viohl, and P. Wellnhofer. Jura Museum, Eichstätt, West Germany.

Pooley, A. C. and C. Gans. 1976. The Nile crocodile. *Scientific American* 234:114–124.

Regal, P. J. and C. Gans. 1980. The revolution in thermal physiology: implications for dinosaurs. Pages 167–188 in *A Cold Look at the Warm-Blooded Dinosaurs*, edited by R. D. K. Thomas and E. C. Olson. American Association for the Advancement of Science, Washington, D. C.

Reid, R. E. H. 1978. Discrepancies in claims for endothermy in therapsids. *Nature* 276:757–758.

Reid, R. E. H. 1984. The histology of dinosaur bone and its possible bearing on dinosaurian physiology. Pages 629–663 in *The Structure, Development and Evolution of Reptiles*, edited by M. W. J. Ferguson, *Symposia of the Zoological Society of London*, No. 52.

Romer, A. S. 1966. *Vertebrate Paleontology*, 3rd edition. University of Chicago Press, Chicago.

Schopf, T. J. M. 1983. Extinction of the dinosaurs: a 1982 understanding. Pages 415–422 in *Large Body Impacts and Terrestrial Evolution*, edited by L. T. Silver and P. Schultz. Geological Society of America Special Paper 190.

Seymour, R. S. 1976. Dinosaurs, endothermy and blood pressure. *Nature* 262:207–208.

Seymour, R. S. 1979. Dinosaur eggs: gas conductance through the shell, water loss during incubation and clutch size. *Paleobiology* 5:1–11.

Sharov, A. G. 1971. New flying reptiles from the Mesozoic deposits of Kasakhstan and Kirgizia. (In Russian.) *Trudy Paleontologichesky Institut Akademiya Nauk S.S.S.R. (Moscow)* 130:104–113.

Sloan, R. E., J. K. Rigby, Jr., L. M. Van Valen, and D. Gabriel. 1986. Gradual dinosaur extinction and simultaneous ungulate radiation in the Hell Creek formation. *Science* 231:1528–1533. (Criticisms of this paper and a rebuttal presenting additional information appeared in *Science* 234:1170–1175, 1986.)

Smith, K. K. 1980. Mechanical significance of streptostyly in lizards. *Nature* 283:778–779.

Spotila, J. R. 1980. Constraints of body size and environment on the temperature regulation of dinosaurs. Pages 233–252 in *A Cold Look at the Warm-Blooded Dinosaurs*, edited by R. D. K. Thomas and E. C. Olson. American Association for the Advancement of Science, Washington, D. C.

Spotila, J. R., P. W. Lommen, G. S. Bakker, and D. M. Gates. 1973. A mathematical model for body temperatures of large reptiles. *The American Naturalist* 107:391–404.

Stokes, W. L. 1964. Fossilized stomach contents of a sauropod dinosaur. *Science* 143:576–577.

Sues, H.-D. 1987. Postcranial skeleton of *Pistosaurus* and interrelationships of the Sauropterygia (Diapsida). *Zoological Journal of the Linnaean Society* 90:109–131.

Troyer, K. 1984. Microbes, herbivory and the evolution of social behavior. *Journal of Theoretical Biology* 106: 157–169.

Unwin, D. M. 1987. Pterosaur locomotion: joggers or waddlers? *Nature* 327:13–14.

Weishampel, D. B. 1981. Acoustic analyses of potential vocalization in lambeosaurine dinosaurs (Reptilia: Ornithischia). *Paleobiology* 7:252–261.

Yalden, D. W. 1985. Forelimb function in *Archaeopteryx*. Pages 91–97 in *The Beginning of Birds, Proceedings of the International* Archaeopteryx *Conference, Eichstätt 1984*, edited by M. K. Hecht, J. H. Ostrom, G. Viohl, and P. Wellnhofer. Jura Museum, Eichstätt, West Germany.

The extinction of dinosaurs discussed in the preceding chapter is one example of the phenomenon of mass extinction. Other (and larger) mass extinctions have occurred at intervals during the history of vertebrates, and the possibility that these extinctions have a regular periodicity has been suggested. In this chapter we review several of the mass extinctions that have in some cases been important events in vertebrate evolution and consider the possible causes of the extinctions. Can one, or at most a few, phenomena explain all the extinctions, or is each one unique? In either case, is the factor that triggers a mass extinction something that happens gradually, such as changing climates as continents move, or is it an abrupt event such as the impact of an extraterrestrial object with the Earth?

Geology and Ecology of Pangaea: Late Paleozoic and Mesozoic

14

Continental Geography During the Late Paleozoic and Mesozoic

The Tethys Sea closed completely during the middle to late Carboniferous. Pangaea became a single continent that persisted through the Triassic, although Siberia and China remained as discrete adjacent subcontinents until the end of the Triassic. During this long period crustal movements produced orogeny (mountain building), especially along the geosynclines sandwiched between the colliding plates. Extensive mountain chains arose, including the Appalachians and Urals. Shallow epicontinental seas so typical of the Ordovician, Silurian, and mid-Devonian periods were less extensive. Devonian, Carboniferous, and Permian geological formations contain massive deposits of evaporites, redbeds, and coal, indicative of terrestrial conditions. During the late Permian, however, the tectonic forces that created Pangaea waned, and the mountains were eroded. The resultant low profile of Pangaea permitted epicontinental seas to return, as shown by widespread deposits of limestone during portions of the Permian.

Pangaea persisted almost 200 million years, until the end of the Jurassic. During this long span

of time the major evolutionary lines of terrestrial plants, invertebrates, and vertebrates were established. Among the tetrapods, Pangaea belonged to the synapsids and diapsids, whose diversity of shapes and sizes dominated the fauna. Only in the Cretaceous did the newly forming oceans produced by the rupture of Pangaea become a barrier to dispersal. Even then, limited contact was maintained between the Northern Hemisphere and Gondwana, and plant and animal dispersals were probably still possible.

The evolution of mammals and birds is part of the radiation of synapsids and diapsids, and establishment of the various orders of birds and mammals is associated largely with the Cretaceous, and especially the Cenozoic. Here we emphasize the geology and ecology pertinent to the radiations of the amniotes that occurred in the late Paleozoic and Mesozoic.

Although temporally durable, Pangaea was not locked in position and it drifted northward from Carboniferous through Triassic time (compare Figures 14–1 and 14–2). The rate of northward drift was slow in terms of commonly used units (1.6 centimeters per year), but resulted in a migration in excess of 30° of latitude. Areas of Pangaea that were located along the equator in the Carboniferous had moved to temperate latitudes by the

Figure 14—1. Continental positions during the Carboniferous. The earliest tetrapods, the ichthyostegalians, are found in late Devonian deposits of Greenland (⊕). Tetrapods are also known from several other North American and European Carboniferous localities (temnospondyls, triangles; aïstopods and nectrideans, circles; open symbols, early Carboniferous; filled symbols, late Carboniferous). The distribution of these sites closely approximates the Carboniferous paleoequator, suggesting that the early tetrapods lived under warm, moist conditions. The distribution of Carboniferous coal deposits is shown by the letter C within the circled areas. Many of the largest coal deposits coincide with the two main Devonian floral zones. The extent of glaciation in the late Carboniferous is shown by the dashed lines, and the direction of glacial movement is shown by the small arrows.

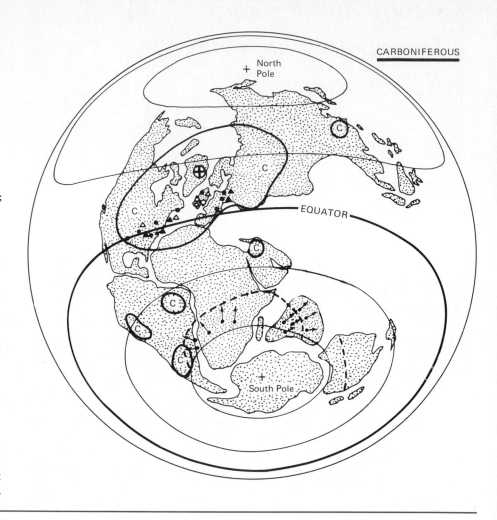

CARBONIFEROUS

Jurassic, and originally temperate locations were displaced to tropical latitudes. Thus Pangaea, a large, unified landmass through most of this time, was not static, and its drift resulted in climatic changes in particular areas.

The tectonic forces that were ultimately to rupture Pangaea began to act during the late Triassic. The first event split North America from South America and Africa (Figure 14–3) creating the North Atlantic Ocean and single North Hemispheric landmass, often called Laurasia. In comparison to the Northern Hemisphere today, Laurasia lacked only India, which historically was a part of Gondwana.

The breakup of Gondwana probably began in the late Jurassic with a continental separation that initially included South America, Antarctica, and Australia as one unit, and Africa and India as an-

508   *PART THREE Terrestrial Ectotherms: Amphibians, Turtles, Crocodilians, and Squamates*

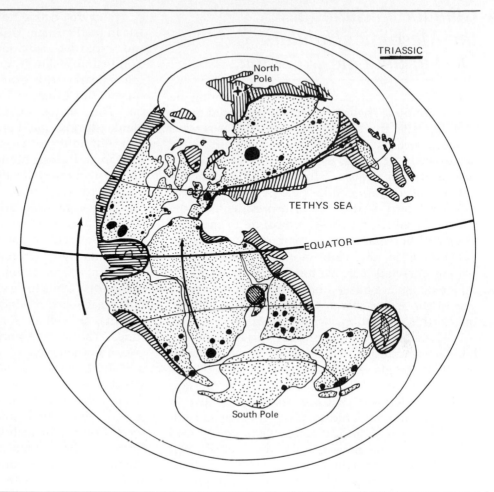

Figure 14–2. Distribution of epicontinental seas and terrestrial tetrapods during the Triassic. The topographic relief of Pangaea throughout the Triassic increased. Epicontinental seas became limited mostly to continental margins (crosshatching). Land tetrapods were widely distributed as shown by fossil sites (black dots, extensive sites shown by large dots). The large arrows indicate the general northward drift of Pangaea that took place from the late Devonian through the Triassic.

other unit (Figures 14–4 and 20–2). By the late Cretaceous, South America had separated from Antarctica, although a land bridge probably existed in the Eocene.

The splitting of Pangaea produced a series of orogenic events that continue today. Throughout Pangaea's existence a fairly continuous sequence of sediments was deposited in western North America. Near the end of the Jurassic these sediments were folded and uplifted to form the ancestral Sierra Nevadas (the Nevadan orogeny, Figure 5–4). In the Cretaceous a similar folding and uplifting of the midcontinental sediments of North America and western South America produced the Rocky Mountains and the Andes. This event, the Laramide Revolution, closed the Mesozoic Era. This Mesozoic orogeny coincides with the initial westward drift of North and South America as Pangaea broke apart.

As fascinating as the geological events are, it is the effects that these continental separations had on terrestrial organisms that most concern us. Primary effects were the isolation of continents and the influences of that isolation on climates.

Climatic Prelude to the Mesozoic: Habitats During the Late Paleozoic

Extensive forests spread across Pangaea during the Carboniferous. Their fossilization produced major coal deposits, from which the Carboniferous takes its name. Giant club mosses were dominant—large arborescent lycopods, especially the genera *Lepidodendrum* and *Sigillaria*. Horsetails 12 to 15 meters tall constituted a secondary element in these forests, and early ferns were also prevalent. By the middle Permian, however, almost all of the lycopods and horsetails had become extinct, and only *Lycopodium* in the club mosses and *Equisetum* among horsetails survive today. The extinction of this extensive nonseed-bearing flora represents one of the major changes that has taken place in the Earth's vegetation. What caused it?

There are no simple answers. In the late Carboniferous and the early Permian, however, a series of major glaciations covered much of Gondwana (Figure 14–1). In addition, evaporites, geological deposits that are usually indicative of aridity, occur over large areas of the Permian landscape between 15° and 30°N paleolatitude. A broad retreat of continental seas following general uplift of the land and the containment of water in glacial ice caps may have contributed to the demise of these forests. The occurrence of extensive cooling in the late Carboniferous and increased aridity in the Permian may have led to replacement of the water-dependent spore-bearing plants by newly evolving seed-bearing plants.

A transitional flora is fossilized in late Carboniferous rocks from most of the landmass that comprised Gondwana. Two major groups are represented: (1) elements of the early Carboniferous flora and (2) a new group of plants, the glossopterids, which are interspersed with glacial deposits, suggesting that they could endure cold conditions. By the late Permian, when glaciation had ceased, the older elements had been replaced by a pure glossopterid flora.

Four distinctive floral regions existed in Laurasia in the Permian. Unlike the sudden appearance and explosive radiation of the glossopterids in Gondwana, a slower, more orderly transition from spore bearers (for example, lycopods) to gymnosperms (for example, conifers) occurred in Laurasia. The virtual extinction of nonseed-bearing plants during the Permian, therefore, followed two different paths—the diversification of the widespread glossopterids in Gondwana, and the radiation of interrelated regional floras in Laurasia.

At the beginning of the Carboniferous the flora was composed primarily of nonvascular bryophytes and vascularized lycopods and ferns, all of which required moist conditions (Scott 1980). The extensive Carboniferous coal deposits are largely fossilized remnants of these lowland tropical plants (Dimichelle et al. 1985). The coalescence of Pangaea also produced upland areas in tropical regions, as well as in northern and southern latitudes. The late Carboniferous is noted for its increase in the diversity of seed-bearing plants (Tiffney 1982; Niklas et al., 1985). Upland regions, which are usually cool and arid, rapidly acquired dense arborescent floras that must have provided new habitats for animals, as well as sources of food. Coincident with this complex Carboniferous flora was the sudden radiation of insects, including many gigantic forms (Rolfe 1980). Prevalent were new forms with mouthparts specialized to pierce plants (Carpenter 1971).

We are left with a vista of floral and faunal complexity and diversity that suggests powerful interactions between components. New morphotypes of plants, such as arborescent forms, allowed plants to escape from herbivores and pathogens (Scott 1980). Over millions of years the evolution of flight in insects permitted them to utilize the resources of the new floral structure. We recognize such complex evolutionary interactions as coevolution. The evolutionary interactions are, of course, finely tuned by natural processes, of which predation and competition for resources are among the main selective forces that determine the survival and evolution of species.

Vertebrates exploited the food resources provided by the radiation of insects. In this generalized view, a momentous sequence of events occurred—a seed-bearing primary producer that was resistant to the dryness of terrestrial habitats provided a setting for the evolution of arid resistant herbivorous insects, which, in turn, stimulated the evolution of truly terrestrial vertebrates, the amniotes. These first fully terrestrial ecosystems were almost as complex as terrestrial ecosystems are today, for the primary consumers not only depended on their host flora for food, but the hosts themselves responded to their predation by evolving remarkable chemical defenses against their herbivorous enemies.

Mesozoic Climates and Habitats

The late Paleozoic and Mesozoic boundary in North America is marked by the Appalachian Revolution (Figure 5–4), a period of mountain building that resulted from compressional forces produced in the coalescence of Pangaea. By the end of the Permian all of North America was above sea level.

The Triassic climate reflects a continuation of the late Permian climate; overall dry, and perhaps, warmer. There is little evidence of continental glaciation. Widespread deserts and semideserts are indicated by the abundance of redbeds and dune sands, as well as by a restriction of the coal deposits so typical of the Carboniferous and early Permian periods. Nevertheless, Triassic amniotes were found throughout most of Pangaea. Geographic and climatic barriers must have acted only as a limited impediment to faunal movements.

In the Triassic the seed-bearing gymnosperms came into their own—conifers, cycads, bennettitaleans, ginkgoes, and pteridosperms (seed ferns). The unique Permian glossopterid flora of Gondwana persisted through the Triassic but disappeared as it was replaced by other gymnosperms of uncertain affinities (*Thinnfeldia, Dicroidium*). The dramatic floral changes of the Permian persisted into the Triassic with a continuing development of plants adapted to drier and cooler conditions. Surprisingly few fossils of terrestrial vertebrates have been found in tropical latitudes, probably because of poor conditions for fossilization rather than because tetrapods were absent from those areas. The existence of Pangaea as a single landmass is supported by the high percentage of taxa common to the various fossil sites (Table 14–1). This high taxonomic affinity over such a wide area can be explained only if the continents were confluent and intercontinental migration was possible.

Table 14–1. Faunal affinities of land tetrapods in the Triassic.*
(Percentages are based on the number of families that each region shares.)

	Europe	Africa	Asia	India	South America
North America	88	75	44	59	56
Europe		58	80	81	71
Africa			89	75	74
Asia				41	49
India					56

Source: C. B. Cox, 1973, in *Atlas of Palaeobiogeography*, Elsevier Scientific, Amsterdam.

*Comparison with Figure 14–3 reveals that regions with more than 50 percent affinity are geographically closer than regions with less than 50 percent affinity.

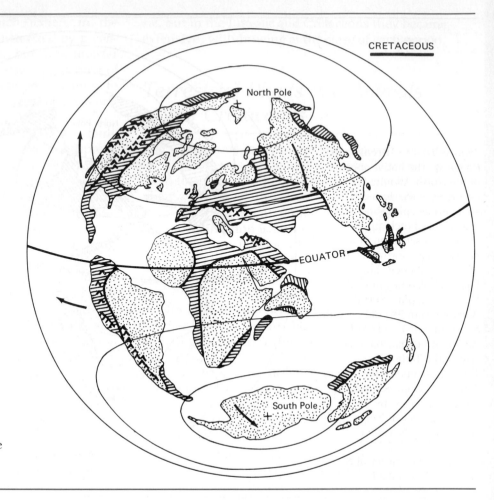

Figure 14—4. Breakup of
Laurasia and Gondwana. The
general direction of drift is
shown by the arrows.

ultraviolet light, in pigment coloration, in the production of tannins, and as protective agents
against herbivores. Chemical investigations of fossil plants suggest that the complexity of flavinoids
continued to evolve from their origins in the green
algae and the first vascular plants to culminate in
the remarkable chemical diversity seen in living
angiosperms (Niklas 1980). In fact, the lack of tannins and phenolic fungicides in the Carboniferous
lycopods may have contributed to their extinction
during the Carboniferous. These compounds,
which are present in gymnosperms and angiosperms, provide resistance to the attacks of sap-
rophytes and decomposing fungi. Again coevolu-

tionary processes are evident: In this example the
selective attacks of herbivores favored the chemical
traits that led to a temporary escape.

The earliest late Devonian insects were followed in the Carboniferous by primitive silverfish,
grasshoppers, and roaches. Flying insects, such as
dragonflies, stone flies, and beetles, evolved in the
late Carboniferous. Few new orders are recorded
in the Triassic, but a profusion of new groups appears in the Jurassic—earwigs, caddis flies, dipter-
ans (flies, mosquitos), and hymenopterans (bees,
wasps, ants). During the Cretaceous these groups
radiated into diverse forms in parallel with the diversification of angiosperms (Carpenter 1977).

Mass Extinctions

The extinction of species has occurred continuously through evolutionary time, but the rate of disappearance is usually low. This loss of individual species tends to be selective—that is, the chance that a species of snake in a tropical forest will become extinct is independent of the loss of a species of tropical bird, plant, or other member of the community. The average rate of extinction over Phanerozoic time (from the Cambrian to the present) is slightly less than one family per million years (Newell 1967). On several occasions during the Phanerozoic, however, the extinction rate has increased well above this level. Total extinction of many widespread higher taxa often has been the end result—the demise of dinosaurs at the end of the Cretaceous is the best-known example. This increased level of extinction above the typical background loss of species has been termed **mass extinction**. It is nonselective with respect to species of animals and plants, generally involves relatively abrupt disappearances, and occurs simultaneously across different groups and widespread geographic regions (Jablonski 1986).

Throughout the Phanerozoic there has been a steady increase in the diversity of life, but the increase has often been punctuated by mass extinctions (see Figure 20–4). Major episodes of extinction have occurred during the Cambrian, late Ordovician, late Devonian, late Permian, late Triassic, at the Cretaceous–Tertiary transition, and at the Eocene-Oligocene boundary. Although the mass extinction of dinosaurs has captured much attention, it by no means represents the most extensive loss of species. Prior to the origin of vertebrates, a mid-Cambrian mass extinction led to a loss of 90 percent or more of the animal species then living, and the mass extinctions at the Permian–Triassic boundary led to a loss of over 50 percent of all families of animals and as many as 80 to 90 percent of all species (Benton 1986).

Extinctions of such great magnitude imply that unusual events—perhaps even extraordinary events—have occurred. What sort of events might have such far-reaching effects on plant and animal life? Two general questions have dominated the study of mass extinctions during the past decade or more: Are the causes of these extinctions to be sought in terrestrial or in extraterrestrial events? Have mass extinctions been periodic, or have they occurred at random intervals?

The cause of the apparently sudden extinction of the dinosaurs during the late Mesozoic is an example of the controversy between proponents of terrestrial versus extraterrestrial explanations of mass extinctions. The term "sudden" must be understood in the context of geological time. Intervals shorter than 100,000 years generally cannot be defined in late Cretaceous sediments with a high degree of stratigraphic resolution (Padian and Clemens 1985). An extinction process that seems sudden from our distant perspective probably stretched over several tens of thousands or even several millions of years. This view of mass extinction implies a gradual loss of taxa that could be caused by prolonged shifts in climate rather than by extraordinary events. We already know that the Mesozoic was an era of great geographic changes, especially the rupture of Pangaea. Did the geographic isolation of the continents lead to extinctions? Did the accompanying climatic changes do so? Or did a catastrophic event temporarily disrupt Earth's climate and productivity to cause these extinctions?

The Evidence of Extraterrestrial Impacts

Studies within the last decade clearly reveal that extraterrestrial objects have impacted Earth, and these events may have caused dramatic changes in the history of life (Hallam 1979; Alvarez et al. 1980; Fisher 1981; Raup and Sepkoski 1982; Russell 1982, 1984). Here we will review the major features of the controversy with the realization that the intense scientific effort currently under way may lead to drastic revisions of currently held views (Smoluchowski et al. 1986). Several recent works have reviewed mass extinctions (Silver and Schultz 1982, Berggren and Van Couvering 1984, Nitecki

Table 14–2. Comparison of rates of extinction for different groups of organisms during those geological periods when mass extinction occurred (see Figure 14–5). Estimates of the number of families disappearing per million years and the percent extinction of families of fishes, amphibians, and "reptiles" (anapsids, synapsids, and diapsids) are from Newell (1967). The mean number of families that became extinct per million years is 0.89. The percentages of families of marine invertebrates that became extinct are from Newell (1967) and Sepkoski (1984), and for terrestrial tetrapods from Benton (1985).

| Period | Number of Families Disappearing per Million Years | Percent Extinction of Families | | | | |
		Marine Invertebrates	Fishes	Amphibians	"Reptiles"	Terrestrial Tetrapods
Cambrian	1.56	61	—	—	—	—
Ordovician	0.96	44, 27	50	—	—	—
Silurian	1.96	25	30	—	—	—
Devonian	1.50	41, 19	70	90	—	—
Permian	2.00	56, 57	55	80	80	49
Triassic	2.11	23	25	90	85	28
Cretaceous	0.74	17	35	75	80	14
Eocene–Oligocene	—	—	—	—	—	8

1984, Elliott 1985, Valentine 1985, Clube and Napier 1986, Kaufman and Mallory 1986, Patrusky 1986/87, Shoemaker and Wolfe 1986). The strengths and weaknesses of the catastrophism and gradualist explanations were reviewed by Van Valen (1984), and rates of mass extinctions have recently been reported (Table 14–2). The magnitude of extinction was far less at the end of the Cretaceous than at some of the earlier extinction boundaries, and Benton (1985) makes the point that the rate was barely above the low background extinction rate through geologic time. Nevertheless, at each of the indicated boundaries in Table 14–2, significant mass extinctions of certain groups did take place.

Extraterrestrial Impacts and Mass Extinctions
Proponents of an extraterrestrial cause of mass extinctions currently suggest that asteroids and comets strike Earth with some frequency. The impacts of such large objects are believed to inject vast quantities of dust into the air—enough dust to shade the Earth's surface and reduce photosynthesis (Urey 1973, Alvarez et al. 1980). The reduction in primary production by plants, in turn, is hypothesized to lead to mass extinctions of animals. The evidence for this hypothesis is the presence of high levels of the element iridium in sediments, and the associated presence of quartz shock crystals that result from a recrystallization of quartz after exposure to very high energies, as from the impact of an asteroid. The occurrence of iridium and shocked quartz at the Cretaceous–Tertiary boundary is widespread, and probably does indicate the impact of an extraterrestrial object (Bohor et al. 1987). The evidence for an extraterrestrial origin of the iridium is not entirely certain, however, because volcanic activity also can produce high iridium emissions (Zoller et al. 1983). Furthermore, there are many iridium-rich layers in Phanerozoic sediments, and often these thin iridium layers are multiple. For example, the Cretaceous–Tertiary transition is now believed to have multiple iridium boundaries that were de-

posited just before and just after the end of the Cretaceous. Thus, the match between the occurrence of iridium in sedimentary rocks and mass extinctions may not be as precise as proponents of the asteroid impact theory have believed.

The Eocene–Oligocene transition, like the Cretaceous–Tertiary transition, also has multiple iridium-rich bands that coincide closely with the multiple mass extinction of marine and terrestrial forms (Alvarez et al. 1984, Patrusky 1986/87). It is not easy to dismiss these correlations as mere coincidence. The fact that multiple impacts have occurred has led some paleontologists to consider the possibility that mass extinctions may find partial explanation in the effects of impacts of extraterrestrial objects. They suggest, however, that these events may create conditions in which climatic change is possible rather than being the direct cause of mass extinction (Kauffman 1984).

In summary, the evidence that asteroids and comets have impacted Earth during the Phanerozoic is not in dispute. What is debatable is whether these impacts have been a primary cause of mass extinctions. A central idea in any catastrophism hypothesis must be stressed: The arrival and impact of a large asteroid or a volcanic eruption large enough to produce mass extinction requires that its effect occurs over a very short time scale, a year or ten years at most. This period is below the usual level of stratigraphic resolution, and the fossil record should show an abrupt break. In most cases, the transitions do not appear to be so abrupt; instead, the paleontological record suggests that even mass extinctions occur over at least several thousands of years.

Furthermore, catastrophism theories share a similar weakness, whether the event postulated is the impact of an asteroid, a rain of comets, or a volcanic disaster of unusual proportions: A catastrophe should affect all taxa equally, but mass extinctions often obliterate some groups of organisms while leaving others nearly unscathed. That sort of selectivity characterizes the Cretaceous–Tertiary transition: Some groups, such as the dinosaurs and ammonites and belamites (marine mollusks), were extinguished, whereas other groups showed little or no change in diversity. What sort of catastrophe would wipe out dinosaurs and pterosaurs but leave crocodilians and birds virtually unharmed? Why would a catastrophe lead to almost complete annihilation of marsupial mammals and have almost no effect on multituberculate and placental mammals? Why were communities of marine plankton severely affected, whereas freshwater communities were unscathed? It would require a curiously selective sort of catastrophe to have such diverse consequences.

The Evidence for Terrestrial Causes of Mass Extinctions

Hypotheses of terrestrial causes of mass extinctions are more diverse than those for extraterrestrial causes, and they predict gradual and selective rather than abrupt and nonselective changes (Hallam 1987, Kerr 1988). No terrestrial factors are tightly correlated with all mass extinctions, but long-term climatic change induced by the tectonic drift of the continents does provide a general explanation that at least fits much of the paleontological evidence. For example, the regression of the epicontinental seas from the late Cretaceous continents was extensive—seas that had covered 70 percent of the land area were reduced to less than 15 percent, and existed mostly along the continental margins (Fischer 1984). Furthermore, many of the major mass extinctions that have occurred coincide with extensive regressions of shallow seas from the continents.

Withdrawal of epicontinental seas would have at least three consequences that might lead to widespread extinctions: (1) The habitat for shallow water marine organisms would be reduced; (2) rivers would cut down toward sea level, dissecting the broad coastal planes that were important terrestrial habitats; and (3) the buffering effect of the epicontinental seas on continental climates would be removed. The decrease in annual mean temperatures toward the end of the Cretaceous discussed in Chapter 13 is characteristic of several other mass extinctions (Stanley 1984).

Gradualist explanations of mass extinction also incorporate the effects of volcanism. Volcanic activity had two periods of heightened intensity during the Phanerozoic, each lasting about 150 million years (Fischer 1984, Courtillot and Besse 1987). Preceding each long volcanic episode were similar long periods of relative volcanic quiescence, when the climate was cooler.

Climatic changes produced by volcanic activity could be either gradual or catastrophic. The explosion of the volcano Krakatoa in 1883 was certainly abrupt and locally catastrophic. Although its emissions and dust cloud were observed all over earth, the eruption did not produce mass extinction of any widespread group of organisms. An eruption sufficient to produce abrupt mass extinction would have to emit a much greater volume of volcanic dust and over a more prolonged period of time than any eruption recorded in human history. Some geologists believe that such eruptions have induced significant changes in climate (Patrusky 1986/87). In fact, the massive and repetitive outflows of lava and gasses associated with the Deccan Flats in southern India possibly represent exactly this type of intense activity (McClean 1985), and they probably coincide with the Eocene–Oligocene transition (Figure 20–2).

Could volcanic activity produce the equivalent of an asteroid impact and result in mass extinctions over periods as short as 1 to 10 years? Perhaps so (Officer and Drake 1985, Carter et al. 1986), but Fischer's view of volcanic effects suggested more prolonged changes in climate—a gradualist view of mass extinction. He proposed that general climates were affected not only by continental locations, but also by volcanic emissions, especially carbon dioxide, that increased the atmospheric greenhouse effect and raised the mean surface temperature of Earth. He thus refers to icehouse and greenhouse periodic cycles which have lasted about 300 million years (Figure 5–5). (Of course, cycles of warming and cooling with much shorter periods than 300 million years have occurred within these icehouse-greenhouse cycles.)

Are Mass Extinctions Periodic?

Analysis of the Phanerozoic fossil record for periodicity in mass extinctions led Raup and Sepkoski (1984,1986) to the astonishing conclusion that they repeat every 26 to 28 million years. This conclusion is tentative, because the record of Earth's history is simply too short to provide sufficient data for a rigorous statistical analysis. Nonetheless, the possibility that major revolutions in the history of life have occurred at regular intervals has seized the fancy of both astrophysicists and paleontologists, and the evidence has been hotly debated. If a periodicity to mass extinctions could be clearly demonstrated, we might be able to draw inferences about the mechanisms of those extinctions.

Either terrestrial or extraterrestrial events could lead to periodic mass extinctions. It has been suggested that repetitive passage of our solar system through the spiral arms of the galaxy may disturb the Oort cloud (or other sources of astronomical objects) and periodically shower Earth with comets (Hsu 1980). Other astronomical phenomena, such as the orbital movements of an undiscovered planet or star, could produce a periodic rain of objects toward Earth, but the inference of periodicity in extraterrestrial impacts remains speculative, and skeptics interpret the evidence differently (Kitchell and Pena 1984, Hoffman 1985, Kerr 1985). The most recent analyses of the ages of craters on the Earth and the moon suggest that impacts have occurred in clusters, but not at regular intervals (Kerr 1987).

The important lesson gained from the gradualism–catastrophism controversy lies not so much in which hypothesis is correct, but rather in the insights that have emerged from more careful studies of the extinctions themselves. Each major extinction has resulted in the loss of particular groups of organisms. Following each loss there has been a continuing long-term increase in total diversity. The consequences of the mass extinctions, therefore, seems to have been the sudden opening of new evolutionary opportunities to the survivors—the provision of adaptive zones into which the survivors expand and speciate. This hypothesis has led paleontologists and biologists to question how evolution actually works and to suggest that the concepts of Darwinian competition between species may not be the only pattern in evolutionary history. The origin of taxa with new and innovative adaptive possibilities, in this view, is the consequence of events working outside the normal domain of competition between genetic variants of similar organisms. The great radiations of vertebrate life may be in part a response to the opportunities presented by ecological vacancies—vacancies that have resulted from mass extinctions.

Summary

The major geologic event of the Mesozoic Era was the rupture of Pangaea. Pangaea had coalesced in the Carboniferous and it was not until the Jurassic, after almost 200 million years, that it split into two huge but lesser continental masses—Laurasia to the north and Gondwana to the south. The first amniotes originated in the Carboniferous and were the dominant vertebrates during the Mesozoic Era. The early amniotes were widespread, for until the Jurassic there were few barriers to their movement across Pangaea. The ends of both the Paleozoic and the Mesozoic eras were characterized by periods of major mountain building. These orogenic events created barriers that had profound effects on continental climates. Initially warm and moist, climatic conditions in the late Carboniferous and the Permian became more arid and cooler. Major extinctions occurred among plants and invertebrates, as well as in some of the major groups of fishes and tetrapods. In the Triassic the climate was warm and arid. In the Jurassic moisture returned, and isolation of continental floras and faunas took place because of the rupture of Pangaea. Isolation of the once contiguous terrestrial tetrapods led to new adaptive radiations, typified especially in the diverse Cretaceous dinosaurs. Similar extensive radiations led to the coevolution of terrestrial flowering plants and pollinating insects. By the end of the Cretaceous the ruptures of Laurasia and Gondwana produced the continents we recognize today. The Cretaceous, warm and moist for the most part, ended with orogenies and a cooling trend. Accompanying this climatic change were the extinction of the dinosaurs and the radiations of birds and mammals. The causes of the extinction of dinosaurs remain unknown, but this mystery has led to numerous speculations. Recent geological discoveries have focused attention on the relative importance of catastrophic environmental phenomena (either extraterrestrial or terrestrial) versus longer-term climatic changes (as from continental drift) in the extinction of major groups. This often volatile scientific controversy, although far from resolved, has generated an intense interest in the causes of extinctions of major groups and is forcing a detailed reexamination of the fossil record, especially across extinction boundaries such as the Permian–Triassic and the Cretaceous–Paleocene transitions.

References

Alvarez, L. S., W. Alvarez, F. Asaro, and H. V. Michel. 1980. Extraterrestrial cause for the Cretaceous–Tertiary extinction. *Science* 208:1095–1108.

Alvarez, W., L. S. Alvarez, F. Asaro, and H. V. Michel. 1984. The end of the Cretaceous: sharp boundary or gradual transition? *Science* 223:1183–1186.

Benton, M. J. 1985. Mass extinction among non-marine tetrapods. *Nature* 316:811–814.

Benton, M. J. 1986. The evolutionary significance of mass extinctions. *Trends in Ecology and Evolutionary Biology* 1:127–130.

Berggren, W. A. and J. A. Van Couvering (editors). 1984. *Catastrophes and Earth History*. Princeton University Press, Princeton, N. J..

Bohor, B. F., P. J. Modreski, and E. F. Foord. 1987. Shocked quartz in the Cretaceous–Tertiary boundary clays: evidence for a global distribution. *Science* 236:705–709.

Carpenter, F. M. 1971. Adaptations among Paleozoic insects. Pages 1236–1251 in *Proceedings of the North American Paleontological Convention*, volume 2, edited by E. L. Yochelson, Allen Press, Lawrence, Kans.

Carpenter, F. M. 1977. Geological history and evolution of the insects. Pages 63–70 in *Proceedings of the 15th International Congress of Entomology*, Washington, D.C.

Carter, N. L., C. B. Officer, C. A. Chesner, and W. I. Rose. 1986. Dynamic deformation of volcanic ejecta from the Toba caldera. *Geology* 14:380–383. Comment by G. A. Izett and B. F. Bohor, and reply by N. L. Carter and C. B. Officer, 1987, *Geology* 5: 90–92.

Clube, S. V. M. and W. M. Napier. 1986. Giant comets and the galaxy: implications of the terrestrial record. Pages 260–285 in *The Galaxy and the Solar System*, edited by R. Smoluchowski, J. N. Bahcall, and M. S. Matthews, University of Arizona Press, Tucson, Ariz.

Courtillot, V. and J. Besse. 1987. Magnetic field reversals, polar wandering, and core-mantle coupling. *Science* 237:1140–1147.

Dimichelle, W. A., T. L. Phillips, and R. A. Peppers. 1985. The influence of climate and depositional environment on the distribution and evolution of Pennsylvanian coal-swamp plants. Pages 223–256 in *Geological Factors and the Evolution of Plants*, edited by B. F. Tiffeny, Yale University Press, New Haven, Conn.

Elliot, D. K. (editor) 1985. *The Dynamics of Extinction*. Wiley, New York.

Fischer, A. G. 1984. The two Phanerozoic supercycles. Pages 129–150 in *Catastrophies and Earth History*, edited by W. A. Berggren and J. A. Couvering. Princeton University Press, Princeton, N. J.

Fisher, A. 1981. The world's great dyings. *Mosaic* 12(2): 2–10.

Hallam, A. 1979. The end of the the Cretaceous. *Nature* 281:430–431.

Hallam, A. 1987. End-Cretaceous mass extinction event; argument for terrestrial causation. *Science* 238: 1237–1242.

Hickey, L. J. and J. A. Doyle. 1977. Early Cretaceous fossil evidence for angiosperm evolution. *Bot. Rev.* 43:3-104.

Hoffman, A. 1985. Patterns of family extinction depend on definition and geological time scale. *Nature* 315:659–662. Also a reply by J. J. Sepkoski and D. M. Raup and comments by N. L. Gilinsky, J. J. Kitchell, G. Estabrook, and A. Hoffman, 1986, Was there a 26–Myr periodicity of extinctions? *Nature* 32:533–536.

Hsu, K. J. 1980. Terrestrial catastrophe caused by cometary impact at the end of the Cretaceous. *Nature* 285:201–203.

Jablonski, D. 1986. Mass extinctions: new answers, new questions. Pages 43–61 in *The Last Extinction*, edited by L. Kaufman and K. Mallory. MIT Press, Cambridge, Mass.

Kauffman, E. G. 1984. The fabric of Cretaceous marine extinctions. Pages 151–246 in *Catastrophies and Earth History*, edited by W. A. Berggren and J. A. Van Couvering. Princeton University Press, Princeton, N. J.

Kaufman, L. and K. Mallory. 1986. *The Last Extinction*. MIT Press, Cambridge, Mass.

Kerr, R. A. 1985. Periodic extinctions and impacts challenged. *Science* 227:1451–1453.

Kerr, R. A. 1987. Impact cratering looks clustered, not periodic. *Science* 236:1426–1427.

Kerr, R. A. 1988. Was there a prelude to the dinosaurs' demise? *Science* 239:729–730.

Kitchell, J. A. and D. Pena. 1984. Periodicity of extinctions in the geologic past: deterministic versus stochastic explanations. *Science* 226:689–692.

McClean, D . M. 1985. Deccan traps mantle degassing in the terminal Cretaceous marine extinctions. *Cretaceous Research* 6:235–259.

Newell, R. E. 1967. Revolutions in the history of life. *Geological Society of America Special Report* 89:63–91.

Niklas, K. 1980. Palaeobiochemical techniques and their applications to palaeobotany. *Progress in Phytochemistry* 6:143–182.

Nitecki, M. H. (editor). 1984. *Extinctions*. University of Chicago Press, Chicago.

Officer, C. B. and C. L. Drake. 1985. Terminal Cretaceous environmental events. *Science* 227:1161–1167.

Padian, K. and W. A. Clemens. 1985. Terrestrial vertebrate diversity: episodes and insights. Pages 41–96 in *Phanerozoic Diversity Patterns*, edited by J. W. Valentine. Princeton University Press, Princeton, N. J.

Patrusky, B. 1986/87. Mass extinctions: the biological side. *Mosaic* 17:2–13.

Pearson, R. 1978. *Climate and Evolution*. Academic Press, New York.

Raup, D. M. and J. J. Sepkoski. 1982. Mass extinctions in the marine fossil record. *Science* 215:1501–1503.

Raup, D. M. and J. J. Sepkoski. 1984. Periodicities of

extinctions in the geologic past. *Proceedings National Academy Sciences USA* 81:801–805.

Raup, D. M. and J. J. Sepkoski. 1986. Periodic extinction of families and genera. *Science* 231:833–836.

Raven, J. A. 1977. The evolution of vascular land plants in relation to supracellular transport processes. *Advances in Botanical Research* 5:153–219.

Rolfe, W. D. I. 1980. Early invertebrate terrestrial faunas. Pages 117–157 in *The Terrestrial Environment and the Origin of Land Vertebrates*, edited by A. L. Panchen, Academic Press, New York.

Russell, D. A. 1982. The mass extinctions of the late Mesozoic. *Scientific American* 246(1):58–65.

Russell, D. A. 1984. Terminal Cretaceous extinctions of large reptiles. Pages 373–386 in *Catastrophies and Earth History*, edited by W. A. Berggren and J. A. Van Couvering. Princeton University Press, Princeton, N. J.

Scott, A. C. 1980. The ecology of some upper Paleozoic floras. Pages 87–115 in *The Terrestrial Environment and the Origin of Land Vertebrates*, edited by A. L. Panchen, Academic Press, New York.

Shoemaker, E. M. and R. F. Wolfe. 1986. Mass extinctions, crater ages, and comet showers. Pages 338–386 in *The Galaxy and the Solar System*, edited by R. Smoluchowski, J. N. Bahcall, and M. S. Matthews. University of Arizona Press, Tucson, Ariz.

Silver, L. T. and P. H. Schultz (editors). 1982. *Geological Implications of Large Asteroids and Comets on the Earth.* Geological Society of America Special Paper 190.

Smoluchowski, R., J. N. Bahcall, and M. S. Matthews. 1986. *The Galaxy and the Solar System*, University of Arizona Press, Tucson, Ariz.

Stanley, S. M. 1984. Marine mass extinctions: a dominant role for temperatures. Pages 69–117 in *Extinctions*, edited by M. H. Nitecki. University of Chicago Press, Chicago.

Swain, T, and G. Cooper-Driver. 1981. Biochemical evolution in early land plants. Pages 103–134 in *Paleobotany, Paleoecology, and Evolution*, volume 1, edited by K. J. Niklas, Praeger Scientific, New York.

Taylor, T. N. 1981. *Paleobotany*. McGraw-Hill, New York.

Thomas, B. A. and R. M. Spicer. 1987. *The Evolution and Paleobiology of Land Plants*. Croom Helm, Beckenham, Kent, England.

Tiffney, B. 1982. Diversity and major events in the evolution of land plants. Pages 192–230 in *Paleobotany, Paleoecology, and Evolution*, volume 2, edited by K. J. Niklas, Praeger Scientific, New York.

Urey, H. C. 1973. Cometary collisions and geological periods. *Nature* 242:32–33.

Valentine, J. W. (editor). 1985. *Phanerozoic Diversity Patterns*. Princeton University Press, Princeton, N. J.

Van Valen, L. M. 1984. Catastrophes, expectations, and the evidence. *Paleobiology* 10:121–137.

Zoller, W. H., J. R. Parrington, and J. M. Phelan Kotra. 1983. Iridium enrichment in airborne particles from Kilauea volcano: January 1983. *Science* 222:1118–1121.

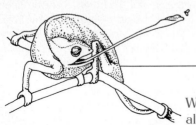

We noted in our discussion of the fauna of the Mesozoic in Chapter 13 that although terrestrial environments were dominated by dinosaurs in that era, many of the animals present would not look strange in the modern world. Mammals, birds, and squamates all originated and diversified in the Mesozoic and their overshadowing by dinosaurs is as much a matter of our perception as of biological reality. Lizards and especially snakes are elements of the diapsid lineage that have had their greatest flowering in the Cenozoic, and the species of squamate diapsids outnumber the species of mammals nearly 2 to 1.

Some aspects of the biology of squamates probably give us an impression of the ancestral way of life of amniotic tetrapods, although many of the structural characteristics of lizards and snakes are derived. One of the derived characteristics of lizards is determinate growth; that is, increase in body size stops when the growth centers of the long bones ossify. This mechanism sets an upper limit to the size of individuals of a species and may be related to the specialization of most lizards as predators of insects. The predatory behavior of lizards ranges from sitting in one place and ambushing passing insects to actively seeking food by traversing the home range in an active, purposeful way. Broad aspects of the biology of lizards are correlated with foraging mode, including morphology, exercise physiology, and social behavior. The anatomical specializations of snakes are associated with their elongate body form and include modifications of the jaws and skull that allow them to subdue and swallow large prey.

The Lepidosaurs: Tuatara, Lizards, Amphisbaenians, and Snakes

15

The Squamates

The lepidosaur lineage of diapsids includes the Squamata, which is the second largest group of living tetrapods; there are twice as many species of squamates as there are of mammals (Table 15–1). Phylogenetic distinctions within lepidosaurs are difficult because several lineages have radiated into similar adaptive zones. Squamates are defined by a variety of derived characters (see Chapter 13), of which determinate growth may have the most general significance. Crocodilians and turtles continue to grow all through their lives, although the growth rates of adults are much slower than those of juveniles (Andrews 1982). This pattern of growth can be viewed as *functionally* determinate in that the largest individuals are only half again the size of the smallest sexually mature adults, but squamates (and birds and mammals) have a *structural* type of determinate growth. The cartilaginous epiphyses of the long bones are the sites at which growth occurs as cells proliferate. Growth continues while the epiphyses are composed of cartilage, and stops completely when the epiphyses become ossified and fused to the shafts of the bones. The development of determinate growth in lepidosaurs may initially have been associated with the insectivorous diet that is presumed to have been characteristic of early lepidosaurs. Lizard-size animals can prey on insects without requiring the sorts of morphological or ecological specialization that is necessary for large insect-eating vertebrates.

Squamates have diversified greatly since their origin in the Triassic. The Sphenodontia, represented now only by the tuatara of New Zealand, is probably the sister group of the Squamata. Within the squamates, lizards can be distinguished in colloquial terms, but not phylogenetically because snakes and amphisbaenians are probably derived from lizards. Thus lizards is a paraphyletic group (not including all descendants). Nonetheless, lizards, snakes, and amphisbaenians are largely distinct in ecology and behavior, and a colloquial separation is useful in considering them.

The Radiation of Sphenodontians and the Biology of the Tuatara

The sphenodontians were a diverse group in the Mesozoic. Triassic forms were small, with body lengths of only 15 to 35 centimeters. Most of the Triassic sphenodontians had teeth that were fused to the jaws (**acrodont**) like the teeth of the modern tuatara (*Sphenodon punctatus*), but others had teeth attached to the jaw largely by their lateral surfaces (**pleurodont**), like those of some lizards. The teeth of the Triassic sphenodontians suggest that both

523

Table 15–1. Classification of the living lepidosaurs.

SPHENODONTIA

TUATARA

Sphenodontidae: one species, now restricted to islands off the coast of New Zealand.

SQUAMATA

LIZARDS (about 3300 species)

Iguania

Iguanidae: 608 species of small (10 cm) to very large (>1.5 m) terrestrial and arboreal lizards. Iguanids are primarily a New World group, but one genus occurs in Fiji and the Tonga Islands and two genera are found on Madagascar.

Agamidae: 320 species of small to large (1 m) terrestrial and arboreal lizards. Agamids are the Old World counterparts of the iguanids and are found in Africa (excluding Madagascar), Europe, Asia, and Australia.

Chamaeleonidae: 88 species, mostly specialized arboreal lizards but including a few grassland and terrestrial species. Chamaeleonids are most diverse in Africa and Madagascar, and extend into Asia. Fossil chameleons are known from the Miocene of Europe. (Note that the family Chamaeleonidae and the genus *Chamaeleo* are spelled with an *ae,* but the common name "chameleon" is spelled with an *e.*)

Gekkota

Gekkonidae: 733 species of tiny (3 cm) to medium-size (30 cm) terrestrial and arboreal lizards. Geckos are found on all continents except Antarctica.

Pygopodidae: 29 species of elongate lizards (15 to 75 cm) that lack forelimbs and have the hindlimbs reduced to flaps. Pygopodids occur in Australia and New Guinea.

Xantusiidae: 14 species of tiny to small terrestrial lizards. Xantusiids occur in North and Central America and on Cuba. Some taxonomic relationships place xantusiids with the Scincomorpha.

Scincomorpha

Scincidae: 1029 species of very small to medium-size terrestrial lizards. A few species are arboreal, and many species show some degree of specialization for fossorial life. Skinks occur on all of the continents except Antarctica.

Cordylidae: 54 species of small to medium-size terrestrial lizards with heavy osteoderms beneath the epidermal scales. Cordylids are found only in sub-Saharan Africa and on Madagascar.

Teiidae: 198 species of small to large elongate terrestrial lizards. Teiids occur primarily in South America, with some species in Central and North America.

Lacertidae: 210 species of small to medium-size elongate terrestrial lizards. The Lacertidae is the only family of scincomorph lizards that includes no legless forms. Lacertids are most abundant in Africa, and are found also in Europe and Asia. They are absent from Madagascar.

Dibamidae: legless, fossorial lizards with the pectoral girdles reduced or absent. Three species of *Dibamus* occur in the East Indies, and one species of *Anelytropsis* is found in central Mexico. Dibamids are sometimes classified with the Gekkota and sometimes placed in a group of their own, the Dibamia.

Source: Based on C. Gans, 1978, pages 16–38 in *Biology of the Reptilia,* volume 8, Academic Press, London; W. E. Duellman, 1979, *Herpetological Review* 10:83–84; D. G. Blackburn, 1982, *Amphibia-Reptilia* 3:185–205; D. G. Blackburn, 1985, *Amphibia-Reptilia* 6:259–291.

Anguinomorpha

Anguidae: 74 species of mostly small to medium-size elongate lizards with heavy osteoderms. About half the species are legless, and the European legless lizard, *Ophisaurus apodus*, may reach a length of more than a meter. Anguids are found in the Americas, Europe, and Asia. They do not occur in Australia and barely reach the northwestern tip of Africa.

Anniellidae: two species of small fossorial lizards that occur on the west coast of North America.

Xenosauridae: three species of small terrestrial lizards (*Xenosaurus*) in Mexico and one semiaquatic species (*Shinisaurus*) in China.

Helodermatidae: two species of large, heavy-bodied terrestrial lizards from the southwestern United States and western Mexico. The helodermatids are the only venomous lizards.

Lanthanotidae: one species of medium-size semiaquatic lizard known only from Borneo.

Varanidae: 30 species of medium-size to enormous lizards. Most species are terrestrial, and a few are arboreal or semiaquatic. Varanids occur in Africa, Asia, and the East Indies, but about half the species are found in Australia.

AMPHISBAENIANS (about 135 species)

Trogonophidae: six species of round-headed amphisbaenians that use an oscillating movement of the head in digging. Trogonophids are found in north Africa and western Asia.

Amphisbaenidae: 129 species of round-headed, spade-snouted, or keel-headed amphisbaenians found in southern Europe, western Asia, Africa, North and South America, and the West Indies.

SNAKES (about 2300 species)

Scolecophidia

Typhlopidae: 147 species of small (20 cm) to medium-size (75 cm) fossorial snakes with reduced eyes. Typhlopids are found on all the continents except North America and Antarctica.

Leptotyphlopidae: 64 species of tiny (10 cm) to small fossorial snakes found in Africa, southwestern Asia, South and Central America, and southwestern North America.

Anomolepidae: 20 species of small to medium-size fossorial snakes from Central and South America.

Henophidia

Acrochordidae: three species of medium-size to large (about 2 m) aquatic snakes from southern Asia, the East Indies, and northern Australia.

Aniliidae: 10 species of medium-size fossorial and semifossorial snakes. *Loxocemus* (1 species) is found in southern Mexico and Central America, *Anilius* (1 species) in northern South America, *Anomochilus* (1 species) in Sumatra, and *Cylindrophis* (7 species) in southeastern Asia.

Uropeltidae: 52 species of medium-size fossorial snakes from southern India and Sri Lanka.

Xenopeltidae: one species of medium-size terrestrial snake found in the East Indies.

Boidae: 58 species of medium-size to enormous (nearly 10 m) snakes. Boids occur on all continents except Antarctica, and include fossorial, terrestrial, arboreal, and semiaquatic species.

Boyleridae: 2 species of medium-size boa-like snakes known only from Round Island, which is near Mauritius in the Indian Ocean.

Tropidophiidae: 20 species of small terrestrial boa-like snakes from Central America, northern South America, and the West Indies.

Figure 15–1. Parallel and convergent evolution of body forms among lizards. Small, generalized insectivores: (a) spiny swift, *Sceloporus*, Iguanidae. (b) japalure, *Calotes*, Agamidae. Herbivores: (c) spiny tailed iguana, *Ctenosaura*, Iguanidae; (d) mastigure, *Uromastix*, Agamidae. Ant specialists: (e) horned lizard, *Phrynosoma*, Iguanidae; (f) spiny devil, *Moloch*, Agamidae. Nocturnal lizards: (g) gecko, *Gekko*, Gekkonidae. Arboreal lizards: (h) African chameleon, *Chamaeleo*, Chamaeleonidae. Legless lizards: (h) North American glass lizard, *Ophisaurus*, Anguidae. Large predators: (j) monitor lizard, *Varanus*, Varanidae.

can be projected forward more than a body's length to capture insects that adhere to its sticky tip. This feeding mechanism requires good eyesight, especially the ability to gauge distances accurately so that the correct trajectory can be employed. The chameleon's eyes are elevated in small cones and are independently movable. When the lizard is at rest, the eyes swivel back and forth giving the lizard a view of its surroundings. When an insect is spotted, both eyes fix on it and a cautious stalk brings the lizard within shooting range.

Most large lizards are herbivores. The iguanas (family Iguanidae) are arboreal inhabitants of the tropics of Central and South America. Large terrestrial iguanas occur on islands in the Caribbean and the Galapagos Islands, probably because the absence of predators has allowed them to spend a large part of their time on the ground. Smaller terrestrial herbivores like the spiny iguanas and the mastigures (Figure 15–1b and c) live in mainland habitats in the New World and Old World, respectively. Many species of lizards live on beaches, but very few actually enter the water. The marine iguana of the Galapagos Islands is an exception. The feeding habits of the marine iguana are unique. It feeds on seaweed, diving 10 meters or more to browse on algae growing below the tide mark.

An exception to the rule of herbivory in large lizards is found in the monitor lizards (family Var-

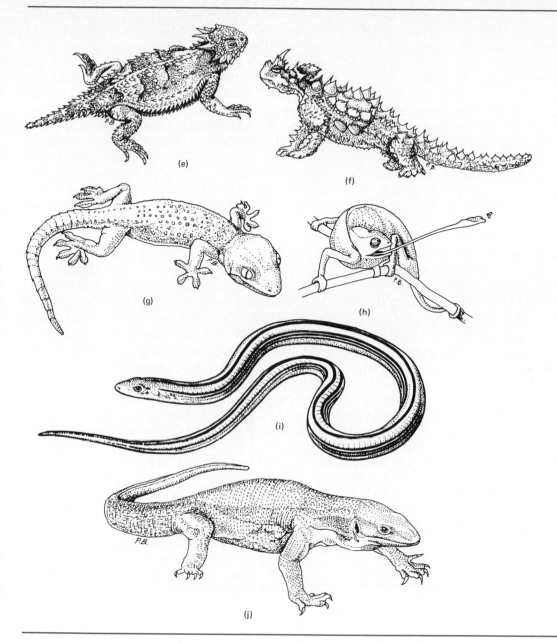

(e)

(f)

(g)

(h)

(i)

(j)

Figure 15–1. (Continued)

anidae). Varanids are active predators that feed on a variety of vertebrate and invertebrate animals including birds and mammals (Figure 15–1j). Few lizards are capable of capturing and subduing birds and mammals, but varanids have morphological and physiological characteristics that make them effective predators. The Komodo monitor lizard is capable of killing adult water buffalo, but its normal prey is deer and feral goats (Auffenberg 1981). [Large monitor lizards were widely distributed on the islands between Australia and Indonesia during the Pleistocene and may have preyed on pygmy elephants that also lived on the islands (Diamond 1987).] The hunting methods of the Komodo monitor are very similar to those employed by mammalian carnivores, showing that a primitive brain is capable of complex behavior and learning. In the late morning a Komodo monitor waits in ambush beside the trails deer use to move from the hilltops, where they rest during the morning, to the valleys, where they sleep during the afternoon. The lizards are familiar with the trails the deer use and often wait where several deer trails converge. If no deer pass the lizard's ambush, it moves into the valleys, systematically stalking the thickets where deer are likely to be found. This purposeful hunting behavior, which demonstrates familiarity with the behavior of the prey and with local geography, is in strong contrast to the opportunistic seizure of prey that characterizes the behavior of many lizards.

Leglessness has evolved repeatedly among lizards, and every continent has one or more families with legless, or nearly legless, species (Figure 15–1i). Leglessness in lizards is usually associated with life in dense grass or shrubbery in which a slim, elongate body can maneuver more easily than a short one with functional legs. Some legless lizards crawl into small openings among rocks and under logs, and a few are fossorial.

Amphisbaenians

The amphisbaenians include about 135 species of extremely fossorial squamates (*fossor* = a digger). Amphisbaenians have variously been classified as a family of lizards and as a suborder of squamates. Their morphological specializations for burrowing complicate interpretation of their phylogenetic relationships. Most amphisbaenians are legless; however, the three species in the Mexican genus *Bipes* have well-developed forelegs that they use to assist entry into the soil but not for burrowing

underground. The skulls of amphisbaenians are used for tunneling, and they are very rigidly constructed (Figure 13–26). The dental structure is also distinctive: Amphisbaenians possess a single median tooth in the upper jaw—a feature unique to this group of vertebrates (Gans 1978). The median tooth is part of a specialized dental battery that makes amphisbaenians formidable predators capable of subduing a wide variety of invertebrates and small vertebrates. The upper tooth fits into the space between two teeth in the lower jaw and forms a set of nippers capable of excising a piece of tissue from a flat surface too large for the mouth to engulf as a whole.

The skin of amphisbaenians also is distinctive (Figure 15–2). The **annuli** (rings) that pass around the circumference of the body are readily apparent from external examination, and dissection shows that the integument is nearly free of connections to the trunk. Thus, it forms a tube within which the body of the amphisbaenian can slide forward or backward. The separation of the trunk and skin is employed during the concertina locomotion that all amphisbaenians use underground. Integumentary muscles run longitudinally from annulus to annulus. Their contraction causes the skin over the area of muscular contraction to telescope and to buckle outward, anchoring that part of the amphisbaenian against the walls of its tunnel. Next, contraction of muscles that pass anteriorly from the vertebrae and ribs to the skin slide the trunk forward within the tube of integument. Amphisbaenians can move backward along their tunnels with the same mechanism by contracting muscles that pass posteriorly from the ribs to the skin. [The name "Amphisbaenia" is derived from the Greek roots *amphi* (double) and *baen* (walk) in reference to the ability of amphisbaenians to move forward and backward with equal facility.] A similar type of rectilinear (straight-line) locomotion is used by some heavy-bodied snakes, but the telescoping ability of the skin of snakes is generally restricted to the latero-ventral portions of the body, whereas the skin is loose around the entire circumference of the body of amphisbaenians.

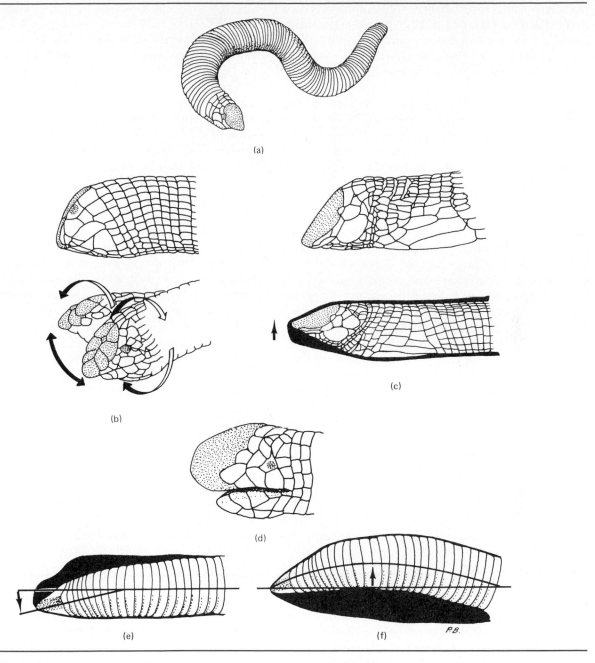

Figure 15–2. Amphisbaenians (a, *Monopeltis,* from Africa) are legless burrowing squamates; (b) blunt-snouted (*Agamodon,* from Africa); (c) shovel-snouted (*Rhineura* from Florida); (d) wedge-snouted (*Anops,* Brazil). [(b), (c), (e), and (f) modified from C. Gans, 1974, *Biomechanics,* J. B. Lippincott, Philadelphia.]

The burrowing habits of amphisbaenians make them difficult to study. Three major functional categories can be recognized: Some species have blunt heads; the rest have either vertically keeled or horizontally spade-shaped snouts. Blunt-snouted forms burrow by ramming their heads into the soil to compact it (Figure 15–2b). Sometimes an oscillatory rotation of the head with its heavily keratinized scales is used to shave material from the face of the tunnel. Shovel-snouted amphisbaenians ram the end of the tunnel, then lift the head to compact soil into the roof (Figure 15–2c). Wedge-snouted forms ram the snout into the end of the tunnel and then use the snout or the side of the neck to compress the material into the walls of the tunnel (Figure 15–2d,e,f). In many places representatives of the three types occur sympatrically and share the subsoil habitat. The unspecialized blunt-headed forms live near the surface where the soil is relatively easy to tunnel through, and the specialized forms live in deeper, more compact soil. The geographic range of the unspecialized forms is greater than that of the specialized ones, and in areas in which only a single species of amphisbaenian occurs it is a blunt-headed species.

The relationship between unspecialized and specialized burrowers is puzzling—one would expect that the specialized forms with their more elaborate methods of burrowing would replace the unspecialized ones, but this has not happened. The explanation may lie in the conflicting selective forces on the snout. On one hand it is important to have a snout that will burrow through soil, but on the other hand, it is also important to have a mouth capable of tackling a wide variety of prey. The specializations of the snout that make it an effective structure for burrowing appear to reduce its effectiveness for feeding. The blunt-headed amphisbaenians may be able to eat a wider variety of prey than the specialized forms can. Thus, in loose soil where it is easy to burrow, the blunt-headed forms may have an advantage. Only in soil too compact for a blunt-headed form to penetrate might the specialized forms find the balance of selective forces shifted in their favor.

Snakes

The 2300 species of snakes range in size from diminutive burrowing species that feed on termites and grow to only 10 centimeters to the large constrictors, which approach 10 meters in length. Three major divisions of snakes are recognized, although these may be structural grades rather than phylogenetic lineages. The Scolecophidia includes three families of small burrowing snakes with shiny scales and reduced eyes. Traces of the pelvic girdle remain in most species, but the braincase is snake-like. The Henophidia includes five families that retain a number of primitive features including vestiges of the pelvic girdle, a large left lung (the left lung is lost in advanced snakes), a coronoid process on the dentary bone, and a retina with both cones and rods. The wart snakes in the family Acrochordidae are entirely aquatic; they lack the enlarged ventral scales that characterize most terrestrial snakes and they have difficulty moving on land. Burrowing snakes in the families Aniliidae and Uropeltidae use their heads to dig through soil, and the bones of their skulls are solidly united. The sole xenopeltid, the sunbeam snake of southeastern Asia, is a ground-dwelling species that takes its common name from its highly iridescent scales. The Boidae includes the boas (New World species that give birth to living young) and the pythons (Old World species that lay eggs). The anaconda, a semiaquatic species of boa from South America, is considered the largest living species of snake. It probably approaches a length of 10 meters, and the reticulated python of southeast Asia is nearly as large. Not all boas and pythons are large, however; many secretive and fossorial species are considerably less than 1 meter long as adults.

The Caenophidia includes most of the living species of snakes, and the family Colubridae alone contains more than two-thirds of the caenophidians. The diversity of the group makes characterization difficult. Caenophidians have lost all traces of the pelvic girdle, they have only a single carotid artery (scolecophidians and henophidians have two carotid arteries), and the skull is very kinetic.

Many colubrid snakes are venomous, and snakes in the families Elapidae, Hydrophiidae, Laticaudidae, and Viperidae have hollow fangs at the front of the mouth that inject extremely toxic venom into their prey.

The body form of even a generalized snake such as the kingsnake (Figure 15–3a) is so specialized that little further morphological specialization is associated with different habits or habitats. Kingsnakes are constrictors, and they crawl slowly, poking their heads under leaf litter and into holes that might shelter prey. Olfaction is an important means of detecting prey for these snakes. Nonconstrictors such as the whipsnake (Figure 15–3b) move faster and are more visually oriented. They forage by crawling rapidly, frequently raising the head to look around. Many arboreal snakes are extremely elongate and frequently have large eyes (Figure 15–3c). Their length distributes their weight and allows them to crawl over even small twigs without breaking them. Burrowing snakes, at the opposite extreme of snake body form, are short and have blunt heads and very small eyes (Figure 15–3d). The head shape assists in penetrating soil, and a short body and tail create less friction in a burrow than would the same mass in an elongate body. Vipers, especially forms like the African puff adder (Figure 15–3e), are heavy-bodied with broad heads. The sea snakes (families Hydrophiidae and Laticaudidae, Figure 15–3f) are derived from the cobras and kraits (family Elapidae). Sea snakes, especially the hydrophiids, are characterized by extreme morphological adaptation for aquatic life: The tail is laterally flattened into an oar, the large ventral scales are reduced or absent, and the nostrils are located dorsally on the snout and have valves to exclude water. The lung extends back to the cloaca and apparently has a hydrostatic role in adjusting buoyancy during diving as well as a respiratory function. Oxygen uptake through the skin during diving has been demonstrated in sea snakes. The laticaudid sea snakes return to land to lay eggs, but the hydrophiids are so specialized for marine life that they are helpless on land, and these species are viviparous. The locomotor specializations of snakes reflect differences in their morphology associated with different predatory modes (discussed in the following section) and the properties of the substrates on which they move (Box 15–1).

Snake skeletons are delicate structures that do not fossilize readily. In most cases we have only vertebrae, and little information has been gained from the fossil record about the origin of snakes. The earliest fossils known are from Cretaceous deposits and seem to be related to boas. Colubrid snakes are first known from the Oligocene, and elapids and viperids appeared during the Miocene.

Functionally, and probably phylogenetically, snakes are extremely specialized legless lizards. They may have reached this specialization from a fossorial stage. Differences in the eyes of lizards and snakes have been interpreted as evidence of that transition. In lizards the eye is focused by distorting the lens, thus changing its radius of curvature, whereas in snakes the lens is moved in relation to the retina to bring objects into focus. The morphology of the retina of snakes also differs from that of lizards. There is no fovea centralis, the retinal cells lack colored oil droplets, and there is a unique ophidian double cone. These differences in the eyes may indicate that snake ancestors passed through a stage in which they were so specialized for burrowing that the eyes had nearly been lost. Among the most specialized living burrowing snakes and lizards, the eyes are very reduced and probably capable of little more than distinguishing between light and darkness. The eye reevolved when snakes reentered an aboveground niche, but the structural details of the original lizard eye were not exactly duplicated.

A fossorial stage is not the only plausible explanation of the origin of snakes, however. Both epigean (ground surface) and aquatic origins have been suggested. Legs are not particularly useful to a small predatory squamate in a number of habitats. Dense vegetation entangles the legs as an animal tries to draw them forward, and in small openings such as cracks in rocks there is no space to use legs. Possibly the initial radiation of snakes took advantage of these sorts of microhabitats.

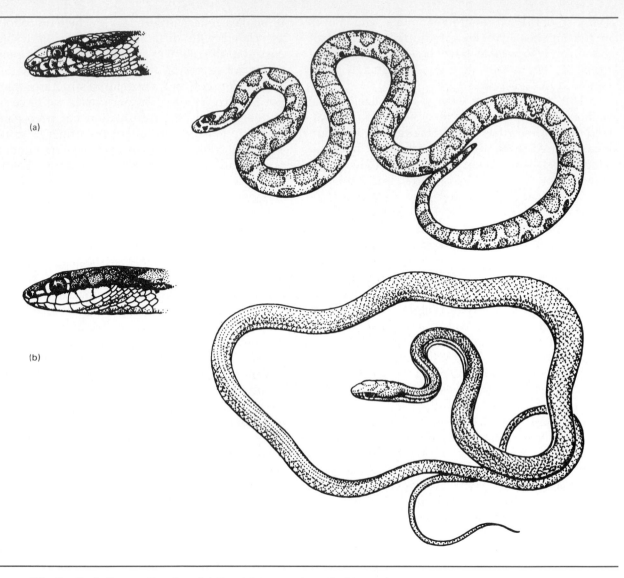

Figure 15–3. Body forms of snakes: (a) Constrictors such as the kingsnake
(Lampropeltis) crawl slowly, poking their heads into holes and under logs. They are
often nocturnal and probably rely on olfaction to detect prey that is hidden from
sight. (b) Active, visually oriented snakes like the whipsnakes *(Masticophis)* forage by
moving rapidly across the ground with their heads raised. They are diurnal and have
large eyes. Vision is probably an important sensory mode for these snakes.
(c) Arboreal snakes such as *Leptophis* are often elongate; they probably rely mainly
on vision to detect prey in their three-dimensional habitat. (d) Burrowing snakes like
Typhlops have small, rounded or pointed heads with little distinction between head
and neck, short tails, and smooth, often shiny scales. Their eyes are greatly reduced
in size. (e) Vipers, especially the African vipers like the puff adder *(Bitis arietans)*,
have large heads and stout bodies that accommodate large prey. (f) Sea snakes
such as *Laticauda* have a tail that is flattened from side to side and valves that close
the nostrils when they dive.

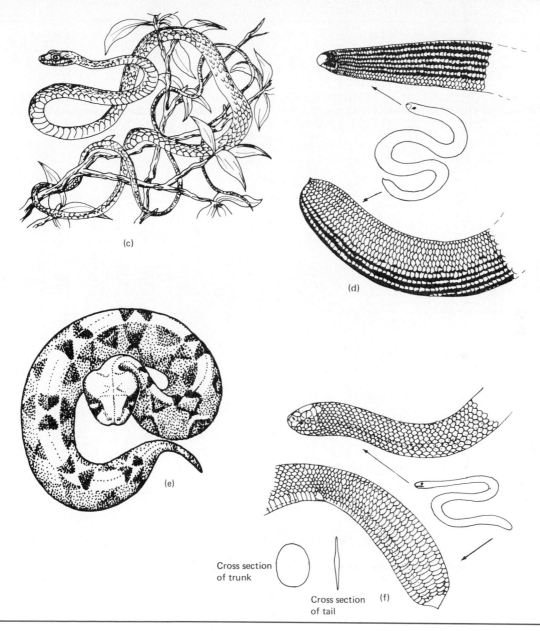

Cross section of trunk

Cross section of tail

(c)

(d)

(e)

(f)

Figure 15–3. (Continued)

Modern lizards show an array of elongate forms with reduced legs, and Mesozoic lizards were probably at least equally diverse. Alternatively, fossils of some elongate varanid-like lizards with reduced limbs from late Cretaceous marine deposits in Israel suggest that an aquatic stage is an alternative to the hypothesis of the evolution of snakes from fossorial lizards. *Ophiomorus* and

Box 15–1. The Way of a Snake

Snakes commonly employ four types of locomotion. In **curvilinear** or **serpentine** locomotion (Figure 15–4a) the body is thrown into a series of curves. The curves may be irregular, as shown in the illustration of a snake crawling across a board dotted with fixed pegs. Each curve presses backward; the pegs against which the snake is exerting force are shown in solid color. The lines numbered 1 to 7 are at 3-inch intervals, and the position of the snake at intervals of 1 second is shown.

Rectilinear locomotion (Figure 15–4b) is used primarily by heavy-bodied snakes. Alternate sections of the ventral surface are lifted

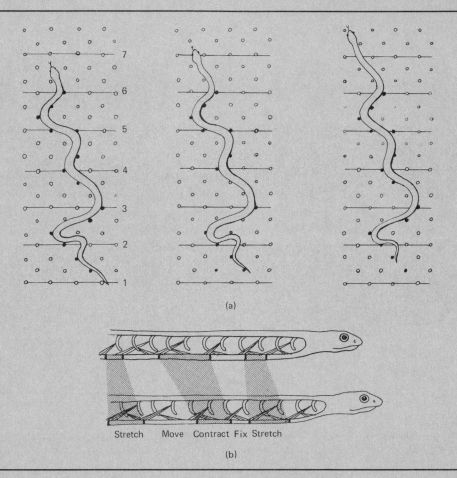

(a)

Stretch Move Contract Fix Stretch

(b)

Figure 15–4. Locomotion of snakes: (a) curvilinear; (b) rectilinear; (c) concertina; (d) sidewinding.

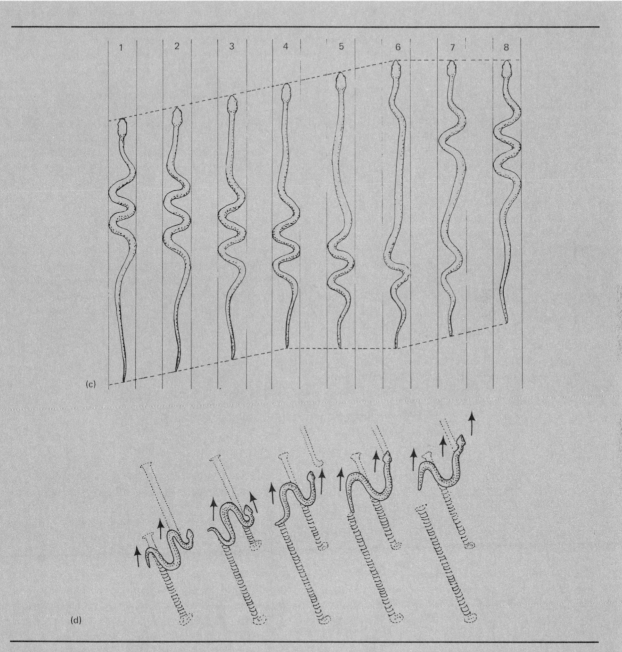

Figure 15–4. (Continued)

Box 15–1. (Continued)

clear of the ground and pulled forward by muscles that originate on the ribs and insert on the ventral scales. The intervening sections of the body rest on the ground and support the snake's body. Waves of contraction pass from anterior to posterior, and the snake moves in a straight line. Rectilinear locomotion is slow, but it is effective even when there are no surface irregularities strong enough to resist the sideward force exerted by serpentine locomotion. Because the snake moves slowly and in a straight line it is inconspicuous, and rectilinear locomotion is used by some snakes when stalking prey.

Concertina locomotion (Figure 15–4c) is used in narrow passages such as rodent burrows that do not provide space for the broad curves of serpentine locomotion. A snake anchors the posterior part of its body by pressing several loops against the walls of the burrow and extends the front part of its body. When the snake is fully extended it forms new loops anteriorly and anchors itself with these while it draws the rear end of its body forward.

Sidewinding locomotion (Figure 15–4d) is used primarily by snakes that live in deserts where windblown sand provides a substrate that slips away during serpentine locomotion. A sidewinding snake raises its body in loops, resting its weight on two or three points that are the only parts of the body in contact with the ground. The loops are swung forward through the air and placed on the ground, the points of contact moving smoothly along the body. Force is exerted downward; the lateral component of the force is so small that the snake does not slip sideward. This downward force is shown by imprint of the ventral scales in the tracks. Because the snake's body is extended nearly perpendicular to its line of travel, sidewinding is an effective means of locomotion only for small snakes that live in habitats with few plants or other obstacles.

Pachyrachis were both more than a meter long and had more than 100 presacral vertebrae. Both lack forelimbs and the pectoral girdle. *Ophiomorus* had well-developed remnants of the rear limb and pelvic girdle, but in *Pachyrachis* they were much reduced (Carroll 1987).

The specializations of snakes compared to legless lizards appear to reflect two selective pressures—locomotion and predation. Elongation of the body is characteristic of snakes. The reduction in body diameter associated with elongation has been accompanied by some rearrangement of the internal anatomy of snakes. The left lung is reduced or entirely absent, the gallbladder is poste-

rior to the liver, the right kidney is anterior to the left, and the gonads may show similar displacement.

Legless lizards face problems in capturing and swallowing prey. The primary difficulty is not the loss of limbs, because few lizards use the legs to seize or manipulate food. The difficulty stems from the elongation that is such a widespread characteristic of legless forms. As the body lengthens, the mass is redistributed into a tube with a smaller diameter. As the mouth gets smaller, the maximum size of the prey that can be eaten also decreases, and a legless animal is faced with the difficulty of feeding a large body through a small mouth. Most legless lizards are limited to eating relatively small prey, whereas snakes have evolved morphological specializations that permit them to engulf prey considerably larger than the body diameter (see the following section). This difference may be one element in the great evolutionary success of snakes in contrast to the limited success of legless lizards and amphisbaenians.

Ecology and Behavior of Squamates

The past quarter century has seen an enormous increase in the number and quality of field studies of the ecology and behavior of snakes and lizards. As a result, our understanding of how these ectothermal amniotes function has been greatly broadened. Studies of lizards have been particularly fruitful, in large measure because lizards are relatively easy to study. Field observations of animals are difficult at best, and species that have characteristics that make them relatively easy to study are overrepresented in the literature. Much less is known about species with cryptic habits. The discussions that follow rely on studies of particular species, and it is important to remember that no one species or family is representative of lizards or snakes as a group.

Foraging and Feeding

The methods that snakes and lizards use to find, capture, subdue, and swallow prey are diverse and are important in determining the interactions between species in a community. Astonishing specializations have evolved: Blunt-headed snakes with long lower jaws that can reach into a shell to winkle out a snail, nearly toothless snakes that swallow bird eggs intact and then slice them open with the sharp hemal arches on the neck vertebrae, and chameleons that project their tongues to capture insects or small vertebrates on the sticky tips are only a sample of the diversity of feeding specializations of squamates.

Studies of the trophic biology of snakes have emphasized the morphological specializations associated with different dietary habits (for examples, see Pough 1983), whereas similar studies of lizards have focused on correlations among physiological and behavioral characteristics (summarized by Huey and Bennett 1986).

Feeding Specializations of Snakes The entire skull of a snake is much more flexible than the skull of a lizard. The snake skull contains eight links, with joints between them that permit rotation (Figure 15–5). This number of links gives a staggering degree of complexity to the movements of the ophidian skull, and to make things more complicated the linkage is paired; each side of the head acts independently. Furthermore, the pterygoquadrate ligament and quadratosupratemporal ties are flexible. When they are under tension they are rigid, but when they are relaxed they permit sideward movement as well as rotation. All of this results in a considerable degree of three-dimensional movement in a snake's skull.

In lizards the mandibles are joined in a symphysis, but in snakes they are attached only by muscles and skin and can spread sideward and move forward or back independently. Loosely connected mandibles and flexible skin in the chin and throat allow the jaws to be spread to allow the widest part of the prey to pass ventral to the articulation of the jaw with the skull.

Swallowing movements take place slowly enough to be observed easily (Figure 15–6). A snake usually swallows prey head first, perhaps because that approach presses the limbs against the body out of the snake's way. Small prey may be swallowed tail first or even sideward. The teeth of one side of the jaw are anchored in the prey and the opposite jaw is advanced. The mandibles are protracted independently or together and grip the prey ventrally. Once it has reached the esophagus, the prey is forced toward the stomach by contraction of the neck musculature.

The majority of snakes seize prey and swallow it as it struggles. The risk of damage to the snake

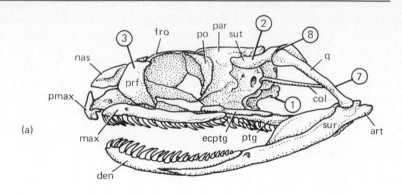

(a)

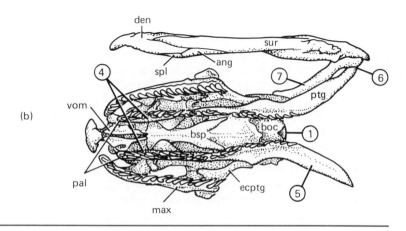

(b)

Figure 15–5. Skull of a snake in (a) lateral and (b) ventral views. A snake skull contains eight movable links: (1) braincase; (2) supratemporal; (3) prefrontal; (4) palatine; (5) pterygoid; (6) pterygoquadrate ligament; (7) quadrate; (8) quadratosupratemoral tie. ang, angular; art, articular; boc, basioccipital; bsp, basisphenoid; col, columella; den, dentary; ectpg, ectopterygoid; fro, frontal; max, maxilla; nas, nasal; pal, palatine; par, parietal; pmax, premaxilla; po, postorbital; prf, prefrontal; ptg, pterygoid; q, quadrate; spl, spenial; sur, surangular; sut, supratemporal; vom, vomer. (From C. Gans, 1961, *American Zoologist* 1:217–227.)

during this process is a real one, and various features of snake anatomy seem to give some protection from prey. The frontal and parietal bones of a snake's skull extend downward, entirely enclosing the brain and shielding it from the protesting kicks of prey being swallowed. Hypophyses project downward from the vertebrae in the neck to give additional protection to the spinal cord from bulky and struggling prey. Possibly the prey that can be attacked by snakes without a specialized feeding mechanism is limited by the snake's ability to swallow the prey without being injured in the process.

Constriction and venom are predatory specializations that permit a snake to tackle large prey with little risk of injury to itself (Greene 1983).

Constriction is characteristic of the boas and pythons as well as a number of colubrid snakes. Despite travelers' tales of animals crushed to jelly by a python's coils, the process of constriction involves very little pressure. A constrictor seizes prey with its jaws and throws one or more coils of its body about the prey. The loops of the snake's body press against adjacent loops, and friction prevents the prey from forcing the loops open. Each time the prey exhales, the snake takes up the slack by tightening the loops slightly, and in a few minutes the prey has been suffocated.

Snakes that constrict their prey must be able to throw the body into several loops of small diameter to wrap around the prey. Constrictors achieve these small loops by having short verte-

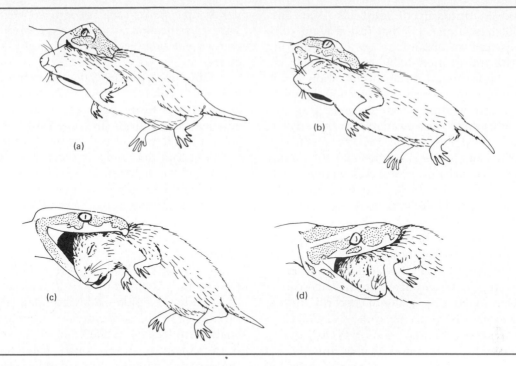

Figure 15–6. Snakes use a combination of head movements and protraction and retraction of the jaws to swallow prey. (a) Prey grasped by left and right jaws at the beginning of the swallowing process. (b) The upper and lower jaws on the right side have been protracted, disengaging the teeth from the prey. (c) The head is rotated counterclockwise, moving the right upper and lower jaws over the prey. The recurved teeth slide over the prey like the runners of a sled. (d) The upper and lower jaw on the right side are retracted, embedding the teeth in the prey and drawing it into the mouth. Notice that the entire head of the prey has been engulfed by this movement. Next the left upper and lower jaws will be advanced over the prey by clockwise rotation of the head. The swallowing process continues with alternating left and right movements until the entire body of the prey has passed through the snake's jaws. (Modified from T. H. Frazetta, 1966, *Journal of Morphology* 118:217–296.)

brae and short trunk muscles that span only a few vertebrae from the point of origin to the point of insertion. Contraction of these muscles produces sharp bends in the trunk that allow constrictors to press tightly against their prey. However, the trunk muscles of snakes are also used for locomotion, and the short muscles of constrictors limit the speed with which they can move. Rapid locomotion by snakes is accomplished by throwing the body into two or three broad loops. This is the

pattern seen in fast-moving species such as whip-snakes, racers, and mambas. The muscles that produce these broad loops are long, spanning many vertebrae and the vertebrae themselves are longer than those of constrictors.

Fast-moving snakes (colubrids) first appear in the fossil record during the Miocene, a time when grasslands were expanding. Boids (constrictors) predominated in the snake fauna of the early Miocene, but by the end of that epoch the snake fauna was composed primarily of colubrids. Alan Savitzky (1980) has suggested that fast-moving colubrid snakes had an advantage over slow-moving boids in the more open habitats that developed during the Miocene, and that the radiation of colubrids involved a complex interaction between locomotion and feeding. Rodents were probably the most abundant prey available to the snakes, and rodents are dangerous animals for a snake to try to swallow while they are alive and able to bite and scratch. Constriction provided a relatively safe way for boids to kill rodents, but the long vertebrae and long trunk muscles that allowed colubrids to move rapidly through the open habitats of the Miocene would have prevented them from using constriction to kill their prey.

Savitzky suggested that colubrids initially used venom to immobilize prey. Duvernoy's gland, found in the upper jaw of colubrid snakes, is homologous to the venom glands of vipers, elapids, and sea snakes. In many colubrids Duvernoy's gland produces a toxic secretion that immobilizes prey. Thus, the evolution of venom that could kill prey may have been a key feature that allowed Miocene colubrid snakes to dispense with constriction and become morphologically specialized for rapid locomotion in open habitats. The presence of Duvernoy's gland appears to be a primitive character for colubrid snakes, as this hypothesis predicts. Some colubrids, including the ratsnakes (*Elaphe*), gophersnakes (*Pituophis*), and kingsnakes (*Lampropeltis*), have lost the venom-producing capacity of the Duvernoy's gland, and these are the groups in which constriction has been secondarily developed as a method of killing prey.

In this context, the front-fanged venomous snakes (Elapidae, Laticaudidae, Hydrophiidae, and Viperidae) are not a new development, but instead represent alternatives for venom delivery systems. Given the ancestral nature of venom for caenophidian snakes, one would expect that different specializations for venom delivery would be widely represented in the modern snake fauna, as indeed they are.

Enlarged teeth (fangs) on the maxillae have evolved in a variety of snakes. Three categories of venomous snakes are recognized (Figure 15–7). This classification is descriptive and represents convergent evolution by different phylogenetic lineages.

Opisthoglyph (*ophistho* = behind, *glyph* = hollowed) colubrid snakes have one or more enlarged teeth near the rear of the maxilla with smaller teeth in front. In some forms the fangs are solid; in others there is a groove on the surface of the fang that may help to conduct saliva into the wound. Several African and Asian opisthoglyphs can deliver a dangerous or even lethal bite to large animals, including humans, but their primary prey is lizards or birds, which are often held in the mouth until they stop struggling and are then swallowed.

Proteroglyph snakes (*proto* = first) include the cobras, mambas, coral snakes, and sea snakes in the families Elapidae, Laticaudidae, and Hydrophiidae. The hollow fangs of the proteroglyph snakes are located at the front of the maxilla, and there are often several small, solid teeth behind the fangs. The fangs are permanently erect and relatively short.

Solenoglyph (*solen* = pipe) snakes include the pit vipers of the New World (Crotalidae) and the true vipers of the Old World (Viperidae). In these snakes the hollow fang is the only tooth on the maxilla, which rotates so that the fang is folded against the roof of the mouth when the jaws are closed. This folding mechanism permits solenoglyph snakes to have long fangs that inject venom deep into the tissues of the prey. The venom, a complex mixture of enzymes and other substances (Table 15–2), first kills the prey and then speeds its digestion after it has been swallowed.

Snakes that can inject a lethal dose of venom into their prey have evolved a very safe prey-catching method. A constrictor is in contact with its prey while it is dying and runs some risk of injury from the prey's struggles. A solenoglyph snake needs only to inject venom and allow the prey to run off to die. Later the snake can follow the scent trail of the prey to find its corpse. This is the prey-capture pattern of most of the vipers. Several features of the body form of vipers allow them to eat

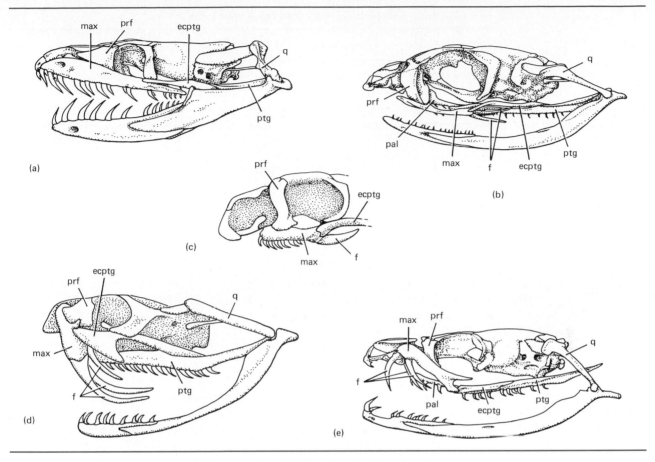

Figure 15–7. Dentition of snakes: (a) aglyphous (without fangs), African python, *Python sebae;* (b and c) opisthoglyphous (fangs in the rear of the maxilla), African boomslang, *Dispholidus typus,* and Central American false viper, *Xenodon rhadocephalus;* (d) solenoglyphous (fangs on a rotating maxilla), African puff adder, *Bitis arietans;* (e) proteroglyphous (permanently erect fangs at the front of the maxilla), African green mamba, *Dendoaspis jamesoni.* The fangs of solenoglyphs (d) are erected by an anterior movement of the pterygoid that is transmitted through the ectopterygoid and palatine to the maxilla, causing it to rotate about its articulation with the prefrontal, thereby erecting the fang. Some opisthoglyphs, especially *Xenodon* (c), have the same mechanism of fang erection. ectpg, ectopterygoid; fro, frontal; max, maxilla; pal, palatine; par, parietal; pmax, premaxilla; prf, prefrontal; ptg, pterygoid; q, quadrate; sut, supratemporal.

larger prey in relation to their own body size than nonvenomous snakes can. Many vipers, including rattlesnakes, the fer de lance, and the African puff adder and Gaboon viper, are very stout snakes. The triangular head shape that is usually associ-ated with vipers is a result of the outward exten-sion of the rear of the skull, especially the quad-rates. The wide-spreading quadrates allow pas-sage through the mouth, and even a large meal makes little bulge in the stout body and thus does

Table 15–2. Components of the venoms of squamates.

Compound	Occurrence	Effect
Proteinases	All venomous squamates, especially viperids	Tissue destruction
Hyaluronidase	All venomous squamates	Increases tissue permeability, hastens the spread of other constituents of venom through the tissues
L-Amino acid oxidase	Many snakes, but absent from vipers, elapids, and sea snakes	Attacks a wide variety of substrates, causes great tissue destruction
Cholinesterase	High in elapids, may be present in sea snakes, low in viperids	Unknown, it is not responsible for the neurotoxic effects of elapid venom
Phospholipases	All venomous squamates	Attacks cell membranes
Phosphatases	All venomous squamates	Attacks high-energy phosphate compounds such as ATP
Basic polypeptides	Elapids and sea snakes	Blocks neuromuscular transmission

not interfere with locomotion. Vipers have specialized as relatively sedentary predators that can prey even on quite large animals. The other family of terrestrial venomous snakes, the cobras, mambas, and their relatives, are primarily slim-bodied snakes.

Foraging Behavior and Energetics of Lizards
The activity patterns of lizards span a range from extremely sedentary species that spend hours in one place to species that are in nearly constant motion. Field observations of the iguanid lizard *Leiocephalus schreibersi* and the teiid *Ameiva chrysolaema* in the Dominican Republic revealed two extremes of behavior (Regal 1978, 1983). *Leiocephalus* rested on an elevated perch from sunrise to sunset and was motionless for more than 99 percent of the day. Its only movements consisted of short, rapid dashes to capture insects or to chase away other lizards. These periods of activity never lasted longer than 2 seconds, and the frequency of movements averaged 9.6 per hour. In contrast, *Ameiva* were active for only 4 or 5 hours in the middle of the day, but they were moving more than 70 percent of that time, and their velocity averaged one body length every 2 to 5 seconds.

The same difference between the two species was seen in a laboratory test of spontaneous activity: *Ameiva* was more than 20 times as active as *Leiocephalus*. In fact, the teiids were as active in exploring their surroundings as small mammals tested in the same apparatus. A xantusiid lizard tested in the laboratory apparatus had a pattern of spontaneous activity that fell approximately midway between that of the teiid and the iguanid. Thus, a spectrum of spontaneous locomotor activity is apparent in lizards that extends from species that are nearly motionless through species that move at intermediate rates to species that are as active as mammals. For convenience the extremes of the spectrum are frequently called sit-and-wait predators and widely foraging predators, respectively, and the intermediate condition has been called a cruising forager. Other field studies have shown that this spectrum of locomotor behaviors is widespread in lizard faunas. In North America, for example, fence lizards (*Sceloporus*) are sit-and-wait predators, many skinks (*Eumeces*) appear to be cruising foragers, and whiptail lizards (*Cnemidophorus*) are widely foraging predators. Regal suggested that the ancestral locomotor pattern for lizards may have been that of a cruising forager and that both sit-and-wait predation and active foraging represent derived conditions. (A spectrum of foraging modes is not unique to lizards; it probably applies to nearly all kinds of mobile animals including fishes, mammals, birds, frogs, insects, and zooplankton.)

The ecological, morphological, and behavioral characteristics that are correlated with the foraging modes of different species of lizards appear to define many aspects of the biology of these animals (Huey and Pianka 1981). For example, sit-and-wait predators and widely foraging predators consume different kinds of prey and fall victim to different kinds of predators. They have different social systems, probably emphasize different sensory modes, and differ in some aspects of their reproduction and life history.

These generalizations are summarized in Table 15–3 and are discussed in the following sections. However, a weakness of this analysis must be emphasized: Sit-and-wait species of lizards (at least, the ones that have been studied most) are primarily iguanians, whereas widely foraging species are mostly scincomorphs. That phylogenetic split raises the question of whether the differences we see between sit-and-wait and widely foraging lizards are really the consequences of the differences in foraging behavior, or if they are ancestral characteristics of iguanian versus scincomorph lizards. If that is the case, their association with different foraging modes may be misleading. In either case, however, the model presented in Table 15–3 provides a useful integration of a large quantity of information about the biology of lizards; like the cladograms in this book it represents a hypothesis that will be modified as more information becomes available.

The best test of the hypothesis that variation in ecology and behavior is truly associated with differences in foraging behavior would come from a comparison of closely related species (that is, species within one phylogenetic lineage) that differ in foraging behavior. If the differences seen in a comparison of iguanians and scincomorphs are also seen among species within just one of the lineages, those differences probably are associated with foraging behavior. Eric Pianka, Raymond Huey, Albert Bennett, Kenneth Nagy, and their associates carried out a study of this sort that employed several species of small, insectivorous lacertid lizards in the Kalahari Desert of Africa, and William Magnusson and his colleagues conducted a similar survey of three species of insectivorous teiid lizards in Brazil (Magnusson et al. 1985, Huey and Bennett 1986). These studies have shown that many of the ecological and behavioral differences seen in a comparison of iguanians and scincomorphs are also seen among species of lacertids or teiids, although the magnitudes of the differences are usually smaller for the intralineage comparisons than for comparisons between lineages.

Lizards with different foraging modes use different methods to detect prey: Sit-and-wait lizards normally remain in one spot from which they can survey a broad area. These motionless lizards detect the movement of an insect visually and capture it with a quick dash from their observation site. Sit-and-wait lizards may be most successful in detecting and capturing relatively large insects like beetles and grasshoppers. Active foragers spend most of their time on the ground surface, moving steadily and poking their snouts under fallen leaves and into crevices in the ground. They apparently rely largely on scent to detect insects, and they probably seek out local concentrations of patchily distributed prey such as termites. Widely foraging species of lizards appear to eat more small insects than do lizards that are sit-and-wait predators. Thus, the different foraging behaviors of lizards lead to differences in the diets of sit-and-wait predators compared to active foragers, even when the two kinds of lizards occur in the same habitat.

The different foraging modes also have different consequences for the exposure of lizards to their own predators. A lizard that spends 99 per-

Table 15–3. Ecological and behavioral characteristics associated with the foraging modes of lizards. Foraging modes are presented as a continuum from sit-and-wait predators to widely foraging predators. In most cases data are available only for species at the extremes of the continuum. See the text for details.

	Foraging Mode		
Character	*Sit-and-Wait*	*Cruising Forager*	*Widely Foraging*
Foraging Behavior			
Movements/hour	Few	Intermediate	Many
Speed of movement	Low	Intermediate	Fast
Sensory modes	Vision	Vision and olfaction	Vision and olfaction
Exploratory behavior	Low	Intermediate	High
Types of prey	Mobile, large	Intermediate	Sedentary, often small
Predators			
Risk of predation	Low	?	Higher
Types of predators	Widely foraging	?	Sit-and-wait and widely foraging
Body form			
Trunk	Stocky	Intermediate?	Elongate
Tail	Often short	?	Often long
Physiological characteristics			
Endurance	Limited	?	High
Sprint speed	High	?	Intermediate to low ·
Aerobic metabolic capacity	Low	?	High
Anaerobic metabolic capacity	High	?	Low
Heart mass	Small	?	Large
Hematocrit	Low	?	High
Energetics			
Daily energy expenditure	Low	?	Higher
Daily energy intake	Low	?	Higher
Social behavior			
Size of home range	Small	Intermediate	Large
Social system	Territorial	?	Not territorial
Reproduction			
Mass of clutch (eggs or embryos)			
relative to mass of adult	High	?	Low

Source: Based on data from L. J. Vitt and J. D. Congdon, 1978, *American Naturalist* 112:595–608; R. B. Huey and E. R. Pianka, 1981, *Ecology* 62:991–999; W. E. Magnusson et al., 1985, *Herpetologica* 41:324–332; R. B. Huey and A. F. Bennett, 1986, pages 82–98 in *Predator-Prey Relationships,* edited by M. E. Feder and G. V. Lauder, University of Chicago Press, Chicago.

cent of its time resting motionless is relatively inconspicuous, whereas a lizard that spends most of its time moving is easily seen. Sit-and-wait lizards are probably most likely to be discovered and captured by predators that are active searchers, whereas widely foraging lizards are likely to be caught by sit-and-wait predators as well as by actively foraging ones. As a result of this difference, foraging modes may alternate at successive levels in the food chain: Insects that move about may be captured by lizards that are sit-and-wait predators, and those lizards may be eaten by widely foraging predators. Insects that are sedentary are more likely to be discovered by a widely foraging lizard, and that lizard may be picked off by a sit-and-wait predator.

The body forms of sit-and-wait lizard predators may reflect selective pressures different from those that act on widely foraging species. Sit-and-wait lizards are often stout bodied, short tailed, and cryptically colored. Many of these species have dorsal patterns formed by blotches of different colors that probably obscure the outlines of the body as they rest motionless on rocks or trees. Widely foraging species of lizards are usually slim and elongate with long tails, and they often have patterns of stripes that may produce optical illusions as they move. However, one predator-avoidance mechanism, the ability to break off the tail when it is seized by a predator (**autotomy**), does not differ among lizards with different foraging modes (Box 15–2).

What physiological characteristics are necessary to support different foraging modes? The energy requirements of a dash that lasts for only a second or two are quite different from those of locomotion that is sustained nearly continuously for several hours. Sit-and-wait and widely foraging species of lizards differ in their relative emphasis on the two metabolic pathways that provide the ATP that is used for activity and in how long that activity can be sustained. Sit-and-wait lizards move in brief spurts, and they rely largely on anaerobic metabolism to sustain their movements. (See Chapter 4 for a discussion of aerobic and anaerobic metabolism.) Sit-and-wait lizards quickly become exhausted when they are forced to run continuously on a treadmill, whereas widely foraging species can sustain activity for long periods without exhaustion. Those differences in locomotor behavior are associated with differences in the oxygen transport systems of the lizards: Widely foraging species of lizards appear to have larger hearts and more red blood cells in their blood than

do sit-and-wait species. As a result, each beat of the heart pumps more blood and that blood carries more oxygen to the tissues in a widely foraging species of lizard than in a sit-and-wait species.

Sustained locomotion is probably not important to a lizard that makes short dashes to capture prey or to escape from predators, but sprint speed might be vitally important in both those activities. As one would predict, the sprint speed of the sit-and-wait lizards is generally greater than that of the widely foraging species.

The continuous locomotion of widely foraging species of lizards is energetically expensive. Measurements of energy expenditure of lizards in the Kalahari showed that the daily energy expenditure of a widely foraging species averaged 150 percent of that of a sit-and-wait species. However, the energy that the widely foraging species invested in foraging was more than repaid by its greater success in finding prey. The daily food intake of the widely foraging species was 200 percent that of the sit-and-wait predator. As a result, the widely foraging species had more energy available to use for growth and reproduction than did the sit-and-wait species, despite the additional cost of its mode of foraging.

The interlocking specializations of sit-and-wait and widely foraging species of lizards call into question one of the central theories of modern ecology—optimal foraging. The energy balance of the sit-and-wait and widely foraging modes of predation should be sensitive to the abundance of different sorts of prey in the habitat. If termites are abundant and beetles are scarce, wide foraging may be more effective than sit-and-wait predation. However, if the abundance of the two sorts of prey were reversed, one would predict that the relative costs and benefits of wide foraging and sit-and-wait predation would also be reversed. An area of ecological research called optimal foraging theory rests on the assumption that animals adjust their foraging behavior as the availability of different types of prey changes so that they always maximize their energy intake during the time they spend foraging. Optimal foraging theory predicts that if termites were abundant and beetles were

Territorial Behavior and Courtship Iguanian lizards employ primarily visual displays during social interactions. The iguanid genus *Anolis* includes more than 200 species of small to medium-size lizards that occur primarily in tropical America, although the geographic range of the familiar Carolina anole (*Anolis carolinensis*) extends northward to Tennessee. Male *Anolis* have gular fans, areas of skin beneath the chin that can be distended by the hyoid apparatus during visual displays. The brightly colored scales and skin of the gular fans of many species of *Anolis* make them conspicuous signaling devices, and they are used in conjunction with movements of the head and body. Figure 15–8 shows the colors of the gular fans of eight species of *Anolis* that occur in Costa Rica. No two species have the same combination of colors on their gular fans, and thus it is possible to identify a species solely by seeing the colors it displays. In addition, each species has a behavioral display that consists of raising the body by straightening the forelegs (called a pushup), bobbing the head, and extending and contracting the gular fan. The combination of these three sorts of movements allows a complex display. The three movements can be represented graphically by an upper line that shows the movements of the body and head and a lower line that shows the expansion and contraction of the gular fan. This representation is called a **display action pattern**. No two display action patterns are the same, so it would be possible to identify any of the eight species of *Anolis* by seeing its display action pattern.

The behaviors that territorial lizards use for species and sex recognition during courtship are very much like those employed in territorial defense—pushups, head bobs, and displays of the gular fan. A territorial male lizard is likely to challenge any lizard in its territory, and the progress of the interaction depends on the response it receives. An aggressive response indicates that the intruder is a male and stimulates the territorial male to defend its territory, whereas a passive response from the intruder identifies a female and stimulates the territorial male to initiate courtship. These behaviors are illustrated by the displays of a male *Anolis carolinensis* shown in Figures 15–9 and 15–10. The first response of a territorial lizard to an intruder is the assertion–challenge display shown at the top of Figure 15–9. The dewlap is extended, and the lizard bobs at the intruder. In addition the nuchal (neck) and dorsal crests are slightly raised, and a black spot appears behind the eye. The next stage depends on the sex of the intruder and its response to the challenge from the territorial male (Figure 15–10). If the intruder is a male and does not retreat from the initial challenge, both males become more aggressive. During aggressive posturing (middle panel in Figure 15–9) the males orient laterally to each other, the nuchal and dorsal crests are fully erected, the body is compressed laterally, the black spot behind the eye darkens, and the throat is swelled. All of these postural changes make the lizards appear larger and presumably more formidable to the opponent. If the intruder is a receptive female, the territorial male initiates courtship (bottom panel in Figure 15–9). Note the extension of the dewlap, the species-specific head bob, and the absence of the dorsal and nuchal crests and the eyespot.

Figure 15–8. Species-specific displays of *Anolis* lizards. Eight species of *Anolis* from Costa Rica can be separated into three groups on the basis of the size and color pattern of their gular fans. *Simple* fans are unicolored, *compound* fans are bicolored, and *complex* fans are bicolored and very large. The display action pattern of each species is shown graphically beneath the drawing of the lizard. The horizontal axis is time (the duration of these displays is about 10 seconds) and the vertical axis is vertical height. Solid line shows movements of the head; dashed line indicates extension of the gular fan. (Modified from A. A. Echelle et al. *Herpetologica* 27:221–288.)

SIMPLE   COMPOUND   COMPLEX

Orange

Red

Yellow

White

White

Yellow

Pink

Orange-amber

red

Orange-yellow

Blue

Red

Blue

Fig
sequ
terri
card
intru
anol
chal
the
dete
beha
male
Herp

tion
cour

the
whic
expe
A. cy
size,
male
adjac

sively to another male *Anolis* in its territory, and it is more aggressive toward a conspecific male than toward a male *A. cybotes*. Thus, one can manipulate the color of the gular fan of an intruder lizard and use the aggressive response by a territorial *A. marcanoi* to determine whether it has correctly identified the species identity of the intruder. Four types of encounters were staged between male lizards, and a composite score of aggressiveness was used to quantify the reaction of the lizards (Table 15–4). In the first experiment, male *A. marcanoi* with the normal red throat color were paired with territorial male *A. marcanoi*. In this situation the territorial lizards were aggressive toward the intruders (mean aggressive intensity of 17.1). Male *A. marcanoi* were less aggressive when they were paired with *A. cybotes* (mean aggressive intensity of 6.9). In other words, the male *A. marcanoi* were able to distinguish males of their own species (conspecifics) from male *A. cybotes* (heterospecifics), and they responded more aggressively to conspecifics.

Were the male *Anolis marcanoi* using the color of the gular fan to make that distinction between conspecific and heterospecific intruders? In the second set of pairings, cosmetics were used to change the color of the gular fan of one of the lizards in each pair: Lipstick was used to give *A. cybotes* a red gular fan, and clown white makeup changed the fan of *A. marcanoi* to white. When the pairings were repeated with the gular fan colors reversed, the aggressive responses of the lizards showed that they were fooled by the manipulation: The mean aggressive intensity of the conspecific encounters decreased to the level of interspecific encounters, and the intensity of the interspecific encounters increased. These experiments showed that male *A. marcanoi* were using the color of the gular fan as a species-identifying mechanism, but that was apparently not the only information they employed. Male *A. marcanoi* were not as aggressive toward *A. cybotes* with their gular fans colored red as they were toward other *A. marcanoi*. Some other signal, probably the *marcanoi* display action pattern which was not given by the altered male *A. cybotes*, is necessary before male *A. marcanoi* will react to a lizard as a conspecific.

Anolis probably use a hierarchy of species-identifying signals under different circumstances: Body size and dewlap size may be important when lizards see each other at a distance, and gular fan color may be the critical feature when they are

Table 15–4. Level of aggressive intensity of *Anolis marcanoi* during encounters.

Test	Species Pairing		Dewlap Color	Aggressive Intensity (Mean ± Standard Deviation)
1	Conspecific	*Anolis marcanoi* versus *Anolis marcanoi*	Normal	17.1 ± 8.4
2	Heterospecific	*Anolis marcanoi* versus *Anolis cybotes*	Normal	6.9 ± 3.8
3	Conspecific	*Anolis marcanoi* versus *Anolis marcanoi*	Reversed	6.7 ± 5.2
4	Heterospecific	*Anolis marcanoi* versus *Anolis cybotes*	Reversed	10.2 ± 5.5

Source: J. B. Losos, 1985, *Copeia* 1985:905–910.

closer. As an encounter progresses the display action pattern may become the most important signal for species identification. A hierarchy of that sort, combined with the difficulty lizards may experience getting clear views of each other in thick vegetation, may be the reason for the apparent redundancy in species-specific signals.

Pheromonal communication, mediated by the Jacobson's organ, probably occurs in several families of lizards, primarily scincomorphs and anguinomorphs, although recent evidence suggests that chemical cues may be more important for some iguanians than has been realized—a review of the subject can be found in Simon (1983). Experimental studies of the use of pheromones have employed several species of North American skinks (*Eumeces*). Male and female broad-headed skinks (*E. laticeps*) can detect the cloacal odors of conspecifics, and males detect skin odors. Male skinks (but not females) distinguish between the cloacal odors of male and female conspecifics, and males will follow scent trails of females but will not follow the trails of other males. Female skinks do not follow the scent trails of either sex. The discriminatory ability of male skinks is quite precise—they can distinguish the cloacal odors of conspecific females from the scents of females of other closely related species (Cooper and Vitt 1986).

Snakes probably use pheromones for sex and species recognition. Snakes in five different families have been shown to follow pheromone trails during aggregation at hibernation sites or during courtship. The discriminatory ability of the snakes can be quite precise. For example, male plains garter snakes (*Thamnophis radix*) selected the pheromone trails of females of their own species in choice experiments in preference to the pheromone trails of the sympatric eastern garter snake (*T. sirtalis*) and the partly sympatric checkered garter snake (*T. marcianus*) (Ford and Schofield 1984).

Olfactory mimicry during courtship by garter snakes has recently been described (Mason and Crews 1985). Many garter snakes mate on emergence from hibernation, and males identify females by pheromones in their skin. As female snakes emerge from the den they are courted by 10 or 20 males, and sometimes by as many as 100 males simultaneously. The males form a mating ball of snakes entwined about the female. Some males of the red-sided garter snake, *Thamnophis sirtalis parietalis*, possess a physiological feminization that results in the presence of the female pheromone in their skins and these males act as female mimics. The mimics are structurally normal and fertile, but other males respond to the mimics as they would to female snakes. Apparently, the presence of the female garter snake pheromone in the skins of the mimics confuses normal males. In simultaneous choice tests, a majority of normal males courted a female snake instead of a female mimic, but some males courted the female mimic in every test. Competitive mating trials showed that female mimics were more successful than normal males in copulating with female snakes: The female mimic mated in 69 percent of the trials compared to 21 percent for the normal males. The advantage enjoyed by female mimics may derive from confusing other males and causing them to divert some of their courtship efforts from the real female to the mimic.

Territoriality, the relative importance of vision compared to olfaction, and foraging behavior appear to be broadly correlated among lizards. Iguanids and agamids, the lizards in which territoriality is most common, are visually oriented sit-and-wait predators. The elevated perches from which they survey their home ranges allow them to see both intruders and prey, and they dash from the perch to repel an intruder or to catch an insect. In contrast, scincomorphs and anguinomorphs are almost entirely nonterritorial, and olfaction is as important as vision in their foraging behavior. These widely foraging predators spend most of their time on the ground, where their field of vision is limited and they probably have little opportunity to detect intruders.

Territoriality involves the defense of some resource that is important to the fitness of the defending individual. Two resources that are commonly involved in territorial systems are food (feeding territories) and mates (reproductive territories). Males of many species of iguanid and agamid lizards have reproductive territories, and some female iguanid and agamid lizards maintain feeding territories (Stamps 1983). The resource that male lizards defend is access to female lizards. Female lizards live in home ranges that they occupy for periods of months or years. The home ranges of adjacent females usually do not overlap, and size of a female's home range may increase or decrease as the abundance of insects changes. The territories of males are large, overlapping the home ranges of several females, and the male mates with all the females whose home ranges lie within his territory. The size of male territories does not change with changes in the abundance of insects, but may increase during the breeding season and decrease at other times of the year.

Males of nonterritorial lizards and snakes may engage in aggressive encounters during the breeding season. Lizards bite and chase each other, whereas snakes usually engage in wrestling matches in which the two individuals intertwine, each trying to pin the other to the ground.

Reproduction and Parental Care   A receptive female snake or lizard often raises its tail and bends it to the side, signaling its willingness to mate. The male responds by pressing his cloacal region against the cloacal region of the female. Male lizards often secure a grip with their jaws on the neck or shoulders of the female, but snakes rarely use their jaws during mating. Male squamates have paired copulatory organs called **hemipenes**. The name stems from the mistaken belief that the two organs were used together. In fact, only the hemipenis on the side nearest the female is used. Each hemipenis is a hollow, closed-end structure like the finger of a glove. In their retracted state, the hemipenes lie in the base of the tail just posterior to the cloaca with their closed ends pointing caudally. The walls of the hemipenes contain large sinuses, and the hemipenes are erected by filling the sinuses with blood. As the organ becomes enlarged, the propulsive muscle squeezes the hemipenis inside-out. It emerges from the posterior edge of the cloaca of the male and is inserted into the cloaca of the female. The

hemipenes of many squamates have spines on their bases that probably help to secure them within the cloaca of the female during the transfer of sperm, a process that may require several minutes or even longer. Some squamates have bifurcated hemipenes, and the structure of the hemipenis is a useful taxonomic character, especially among snakes. When copulation is finished, the hemipenis relaxes and a retractor muscle pulls it back into its resting position.

Squamates show a range of reproductive modes from oviparity (the eggs are deposited in a nest and development is supported entirely by the yolk) to viviparity (the eggs are retained in the oviducts and development is supported by transfer of nutrients from the mother to the fetuses). Intermediate conditions include retention of the eggs for a time after they have been fertilized, and the production of living young that were nourished primarily by material in the yolk (Yaron 1985). Oviparity is assumed to be the ancestral condition, and viviparity has evolved at least 45 times among lizards and 35 times among snakes (Blackburn 1982, 1985; Shine 1985). Viviparous squamates have specialized chorioallantoic placentae, and in the Brazilian skink *Mabuya heathi* more than 99 percent of the mass of the fetus results from transport of nutrients across the placenta (Blackburn et al. 1984).

Viviparity is not evenly distributed among lineages of squamates. Nearly half of the origins of viviparity have occurred in the family Scincidae, whereas viviparity is unknown in teiid lizards and occurs in only two genera of lacertids (Table 15–5). Viviparity has advantages and disadvantages as a mode of reproduction. The most commonly cited benefit of viviparity is the opportunity it provides for a female snake or lizard to use her own thermoregulatory behavior to control the temperature of the embryos during development. This hypothesis is appealing in an ecological context, because a relatively short period of retention of the eggs by the female might substantially reduce the total amount of time required for development, especially in a cold climate. Furthermore, the disadvantage of viviparity, in terms of its effect on re-

Table 15–5. Occurrence of viviparity in squamates.

Family	Probable Number of Origins of Viviparity	Viviparous Species in Family (Percent of Total)
Lizards		
Iguanidae	10	14
Agamidae	1	<1
Chamaeleonidae	1	17
Gekkonidae	2	1
Pygopodidae	0	0
Xantusiidae	1	100
Scincidae	22	43
Cordylidae	1	39
Teiidae	0	0
Lacertidae	2	1
Dibamidae	0	0
Anguidae	5	69
Anniellidae	?	100
Xenosauridae	?	100
Helodermatidae	0	0
Lanthanotidae	0	0
Varanidae	0	0
Snakes		
Typhlopidae	1	1
Leptotyphlopidae	0	0
Acrochordidae	1	100
Aniliidae	?	100
Uropeltidae	1	100
Xenopeltidae	0	0
Boidae	1	61
Colubridae	14	9
Elapidae	1	1
Hydrophiidae	3	100
Laticaudidae	0	0
Viperidae	12	83

Source: D. G. Blackburn, 1982, *Amphibia-Reptilia* 3:185–205; and 1985, *Amphibia-Reptilia* 6:259–291.

productive output, may be lower in cold regions than in warm ones. Lizards in warm habitats may produce more than one clutch of eggs in a season, but that is not possible for a viviparous species because development takes too long. However, in a cold climate lizards are not able to produce more than one clutch of eggs in a breeding season anyway, and viviparity would not reduce the annual reproductive output of a female lizard. Phylogenetic analyses of the origins of viviparity suggest that it has evolved most often in cold climates, as this hypothesis predicts, but other origins appear to have taken place in warm climates, and more than one situation favoring viviparity among squamates appears likely.

In general, large species of squamates produce more eggs or fetuses than do small species, and within one species large individuals often have more offspring in a clutch than do small individuals. Both phylogenetic and ecological constraints play a role in determining the number of young produced, however. All geckos have a clutch size of either one or two eggs, and all *Anolis* produce only one egg at a time. Lizards with stout bodies usually have clutches that are a greater percentage of the body mass of the mother than do lizards with slim bodies. The division between stout and slim bodies approximately parallels that division between sit-and-wait predators and widely foraging predators. It is tempting to infer that a lizard that moves about in search of prey finds a bulky clutch of eggs more hindrance than a lizard that spends 99 percent of its time resting motionless. However, some of the divisions among modes of predatory behavior, body form, and relative clutch mass also correspond to the phylogenetic division between iguanian and scincomorph lizards, and as a result it is not possible to decide which characteristics are ancestral and which may be derived.

All-female (**parthenogenetic**) species of squamates have been identified in six families of lizards and one snake (Darevsky et al. 1985). The phenomenon is particularly widespread in the teiids (especially *Cnemidophorus*) and lacertids (*Lacerta*) and occurs in several species of geckos. One parthenogenetic species each is known or suspected among chameleons, agamids, xantusiids, and typhlopids. However, parthenogenesis is probably more widespread among squamates than this list indicates because parthenogenetic species are not conspicuously different from bisexual species. Parthenogenetic species are usually detected when a study undertaken for an entirely different purpose reveals that a species contains no males. Confirmation of parthenogenesis can be obtained by obtaining fertile eggs from females raised in isolation, or by making reciprocal skin grafts between individuals. Individuals of bisexual species usually reject tissues transplanted from another individual because genetic differences between individuals lead to immune reactions. Parthenogenetic species, however, produce progeny that are genetically identical to the mother, so no immune reaction occurs and grafted tissue is retained.

The chromosomes of lizards have allowed the events that produced some parthenogenetic species to be deciphered (Cole 1975). Many parthenoforms appear to have had their origin as interspecific hybrids. These hybrids are diploid ($2n$) with one set of chromosomes from each parental species. For example, the diploid parthenogenetic whiptail lizard *Cnemidophorus tesselatus* is the product of hybridization between the bisexual diploid species *C. tigris* and *C. septemvittatus* (Figure 15–11). Some parthenogenetic species are triploids ($3n$). These forms are usually the result of a backcross of a diploid parthenogenetic individual to a male of one of its bisexual parental species or, less commonly, the result of hybridization of a diploid parthenogenetic species with a male of a bisexual species different from its parental species. A parthenogenetic triploid form of *C. tesselatus* is apparently the result of a cross between the partheneogenetic diploid *C. tesselatus* and the bisexual diploid species *C. sexlineatus*. It is common to find the two bisexual parental species and a parthenogenetic species living in overlapping habitats. Parthenogenetic species of *Cnemidophorus* often occur in habitats like the floodplains of rivers that are subject to frequent disruption. Disturbance of the habitat may bring together closely related bisexual species, fostering the hybridization that is the first step in establishing a parthenogenetic species. Once a parthenogenetic species has become established, its reproductive potential is twice that of a bisexual species because every individual of a parthenogenetic species is capable of producing young. Thus, when a flood or other disaster wipes out most of the lizards, a parthenogenetic species can repopulate a habitat faster than a bisexual species.

Parental care has been recorded for more than 100 species of squamates (Shine 1988). A few species of snakes and a larger number of lizards remain with the eggs or nest site. Some female skinks remove dead eggs from the clutch. Egg brooding has been reported for some species of pythons: The female coils tightly around the eggs, and in some species muscular contractions of the female's body produce sufficient heat to raise the temperature of the eggs to about 30°C, which is substantially above air temperature. One uncon-

(a) *C. tigris* (bisexual, 2n) ——→ (c) *C. tesselatus* (parthenogenetic, 2n) ——→ (e) *C. tesselatus* (parthenogenetic, 3n)

(b) *C. septemvittatus* (bisexual, 2n) ——

(d) *C. sexlineatus* (bisexual, 2n) ——

Figure 15–11. The apparent sequence of crosses leading to the formation of diploid and triploid unisexual species of *Cnemidophorus*. Hybridization of the bisexual diploid species (a) *C. tigris* and (b) *C. septemvittatus* produced a unisexual diploid form with half of its genetic complement derived from each of the parental species (an allodiploid). This parthenogenetic form is called (c) *C. tesselatus*. Hybridization between a diploid *C. tesselatus* and a male of the bisexual species (d) *C. sexlineatus* produced a unisexual triploid form with its genetic complement derived from three different parental species (an allotriploid). This parthenogenetic triploid form is also called (e) *C. tesselatus*. Thus, *C. tesselatus* consists of clones of both diploid and triploid lineages, although taxonomists will probably treat these two forms as separate species in the near future. (Photographs by C. J. Cole and C. M. Bogert, American Museum of Natural History, courtesy of C. J. Cole.)

firmed report exists of baby pythons returning at night to their empty eggshells, where their mother coiled around them and kept them warm. Little interaction between adult and juvenile squamates has been documented. In captivity female prehensile-tailed skinks (*Corucia zebrata*) have been reported to nudge their young toward the food dish, as if teaching them to eat. Prehensile-tailed skinks, which occur only on the Solomon Islands, are herbivorous and viviparous.

Free-ranging baby green iguanas have a tenuous social cohesion that persists for several months after they hatch (Burghardt 1977). The small iguanas move away from the nesting area in groups that may include individuals from several different nests. One lizard may lead the way, looking back as if to see that others are following. The same individual may return later and recruit another group of juveniles. During the first 3 weeks after they hatch, juvenile iguanas move up into the forest canopy and are seen in close association with adults. During this time the hatchlings probably ingest feces from the adults, thereby inoculating their guts with the symbiotic microbes that facilitate digestion of plant material (Troyer 1982). After their fourth week of life, the hatchlings move down from the forest canopy into low vegetation, where they continue to be found in loosely knit groups of two to six or more individuals that move, feed, and sleep together. This association might provide some protection from predators, and if the hatchlings continue to eat fecal material, it is another opportunity to ensure that each lizard has received its full complement of gut microorganisms.

Thermoregulation and the Ecology and Behavior of Squamates

The extensive repertoire of thermoregulatory mechanisms employed by ectotherms allows many species of lizards and snakes to keep body temperature within a range of a few degrees during the part of a day when they are active. Many species of lizards have body temperatures between 33 and 38°C while they are active (the **activity temperature range**), and snakes often have body temperatures between 28 and 34°C. Roger Avery (1982) has summarized the information about body temperatures of squamates in the field.

These activity temperature ranges have been the focus of much research: Field observations show that thermoregulatory activities may occupy a considerable portion of an animal's time. Less obvious, but just as important, are the constraints that the need for thermoregulation sets on other aspects of the behavior and ecology of squamates. For example, some species of lizards and snakes are excluded from certain habitats because it is impossible to thermoregulate. In temperate regions the activity season lasts only during the months when it is warm and sunny enough to permit thermoregulation; at other times of the year snakes and lizards hibernate. Even during the activity season, time spent on thermoregulation is not available for other activities: Roger Avery (1976) proposed that lizards in temperate regions show less extensive social behavior than do tropical lizards because thermoregulatory behavior in cool climates requires so much time.

The physiological advantages associated with a stable body temperature were discussed in Chapter 4. The body tissues of an organism are the site of a tremendous variety of biochemical reactions proceeding simultaneously and depending on one another to provide the proper quantity of the proper substrates at the proper time for reaction sequences. Each reaction has a different sensitivity to temperature, and regulation is greatly facilitated when temperature variation is limited. Thus, coordination of internal processes may be a major benefit of thermoregulation for squamates.

If the temperature stability that a snake or lizard achieves by thermoregulation is important to its physiology and biochemistry, one would expect to find that the internal economy of an animal functions best within its activity temperature range, and that is often the case. Examples of physiological processes that work best at temperatures within the activity range can be found at the molecular, tissue, system, and whole-animal levels of organization.

Organismal Performance and Temperature

How effectively does an organism carry out its normal activities at different temperatures? Does a difference in body temperature have a measurable effect on how fast a lizard can run, for example, or is running speed independent of temperature? These questions can be addressed with ecological analyses of organismal performance; that is, with measurements of the effectiveness of a process at different temperatures. Much information is available for squamates about the effects of temperature on processes as diverse as intracellular chemical reactions, recovery from disease, growth rates, and predatory success. [Reviews of these studies can be found in Dawson (1975) and Huey (1982).] Temperature profoundly affects the ability of squamates to carry out different activities, but not all activities are affected in parallel ways, and behavioral shifts (that is, qualitative rather than quantitative changes) are seen in some cases. Furthermore, the relationship between body temperature and physiological processes is a two-way interaction: Physiological capacities change in response to changes in body temperature, but under some conditions the sequence of cause and effect is reversed. When squamates manipulate their body temperatures in response to internal conditions such as feeding status or pregnancy.

The wandering garter snake (*Thamnophis elegans*) provides examples of the effects of body temperature on a variety of physiological and behavioral functions (Stevenson et al. 1985). Wandering garter snakes are diurnal, semiaquatic inhabitants of lakeshores and stream banks in western North America. They hunt for prey on land and in water, and feed primarily on fishes and amphibians. Olfaction is an important mode of prey detection for snakes, and is accomplished by flicking the tongue. Scent molecules are transferred from the tips of the forked tongue to the olfactory epithelium of the Jacobson's organ in the roof of the mouth. The garter snakes are diurnal; they spend the night in shelters, where their body temperatures fall to ambient levels (4 to 18°C), and emerge in the morning to bask. During activity on sunny days the snakes maintain body temperatures between 28 and 32°C.

Stevenson and his associates measured the effect of temperature on the speed of crawling and swimming, the frequency of tongue flicks, the rate of digestion, and the rate of oxygen consumption of the snakes (Figure 15–12). Crawling, swimming, and tongue flicking are elements of the foraging behavior of garter snakes, and the rates of digestion and oxygen consumption are involved in energy utilization. The ability of garter snakes to crawl and swim was severely limited at the low temperatures they experience during the night when they are inactive. At 5°C snakes often refused to crawl and at 10°C they were able to crawl only 0.1 meter per second, and could swim only 0.25 meter per second. The speed of both types of locomotion increased at higher temperatures. Swimming speed peaked near 0.6 meter per second at 25 and 30°C, and crawling speed increased to an average of 0.8 meter per second at 35°C. The rate of tongue flicking increased from less than 0.5 flick per second at 10°C to about 1.5 flicks per second at 30°C. The rate of digestion increased slowly from 10 to 20°C, and more than doubled between 20 and 25°C. It did not increase further at 30°C, and dropped slightly at higher temperatures. The rate of oxygen consumption increased steadily as temperature rose to 35°C, which was the highest temperature tested because higher body temperatures would have been injurious.

All five measures of performance by garter snakes increased with increasing temperature, but the responses to temperature were not identical. For example, swimming speed did not increase substantially above 20°C, whereas crawling speed continued to increase up to 35°C. The rate of digestion peaked at 25 to 30°C and then declined, but the rate of oxygen consumption increased steadily to 35°C. More striking than the differences among

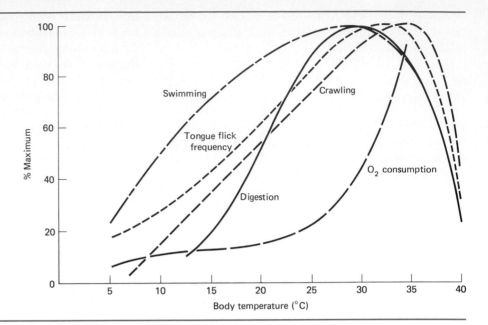

Figure 15–12. Effect of temperature on several activities by the wandering garter snake, *Thamnophis elegans vagrans.* The horizontal axis shows the body temperature of the snakes and the vertical axis shows the percent of maximum performance achieved at each temperature. (From R. D. Stevenson et al., 1985, *Physiological Zoology* 58:46–57. © 1985 by The University of Chicago. All rights reserved.)

the functions, however, is the apparent convergence of maximum performance for all of the functions on temperatures between 28.5 and 35°C. This range of temperatures is close to the body temperatures of active snakes in the field on sunny days (28 to 32°C). Anywhere within that range of body temperatures, snakes would be able to crawl, swim, and tongue flick at rates that were at least 95 percent of their maximum rates.

The relationship between the body temperatures of active garter snakes and the temperature sensitivity of various behavioral and physiological functions reported by Stevenson and his colleagues is probably common for squamates. That is, in most cases the body temperatures they maintain during activity are the temperatures that maximize organismal performance. However, at least two types of variation complicate the picture of squamate thermoregulation: changes in behavior that accompany changes in body temperature and changes in thermoregulation in response to the physiological status of an animal.

Behavioral Changes A change in the body temperature of a squamate may be accompanied

by a qualitative change in behavior instead of by the graded levels of performance shown by garter snakes. For example, *Agama savignyi* is an agamid lizard that lives in desert areas of the Middle East. It shows a pronounced temperature sensitivity of sprint speed: At a body temperature of 18°C it can run only 1 meter per second, but at 34°C it runs about 3 meters per second. *Agama savignyi* lives in open habitats where it may be some distance from shelters that could provide protection from predators. Clearly, the lizards are better able to run to a shelter when they are warm than when they are cool, and they displayed two types of defensive behavior, depending on their body temperature. At body temperatures between 18 and 26°C most *A. savignyi* did not try to run from a predator; instead, they leaped from the substrate and attempted to bite. However, at body temperatures of 30°C or above, most lizards ran away (Hertz et al. 1982). This sort of qualitative shift in behavior at different body temperatures may be a widespread response among squamates to the effects of body temperature on their ability to carry out certain activities.

Effects of Nutritional Status and Bacterial Infections Several internal states of squamates and other ectotherms have been shown to influence body temperature. A thermophilic (*thermo* = heat, *philo* = loving) response after feeding is widespread: Individuals with food in the gut maintain higher body temperatures than do individuals without food. A higher body temperature has been shown to accelerate digestion and also to increase digestive efficiency and water uptake, so a warm animal digests its food more rapidly and assimilates a higher proportion of the energy and water present in the food. Conversely, fasting animals regulate their body temperatures at low levels that reduce their metabolic rates and conserve their stored energy. Behavioral fever is another common response of ectotherms: Individuals infected by bacteria change their thermoregulatory behavior and maintain body temperatures several degrees higher than those of uninfected controls. These behavioral fevers have been demonstrated in arthropods, fishes, frogs, salamanders, turtles, and lizards. The release of prostaglandin E_1, which acts on thermoregulatory centers of the anterior hypothalamus, appears to be the immediate cause of both the behavioral fevers of ectotherms and the physiological fevers of endotherms. Survival is enhanced by fever, apparently because bacterial growth is limited by a reduction in the availability of iron at higher temperatures.

Reproductive Status Pregnancy is another condition that has been shown to affect thermoregulation of squamates. The rate of embryonic development of squamates is strongly affected by temperature, and one of the major advantages of viviparity is thought to be the opportunity it provides for the mother to control the temperature of embryos during development. The body temperatures of female squamates during pregnancy may be different from the temperatures they would normally maintain. For example, pregnant female spiny lizards (*Sceloporus jarrovi*) had an average body temperature of 32.0°C, whereas male lizards in the same habitat had an average body temperature of 34.5°C (Beuchat 1986). The female lizards changed their thermoregulatory behavior after they had given birth, and the average body temperature of postparturient female lizards was 34.5°C like that of the males. The low body temperatures of pregnant lizards were unexpected because one can easily think of reasons why giving birth as early in the year as possible would be advantageous for the lizards. That line of reasoning suggests that female lizards should maintain higher-than-normal body temperatures during pregnancy, or at least they should not reduce their body temperatures. Apparently the low body temperature of the pregnant female lizards is the result of a conflict between the thermal requirements of the mother and the best temperature for embryonic development (Beuchat and Ellner 1987).

Temperature and the Ecology of Squamates

Squamates, especially lizards, are capable of very precise thermoregulation, and microhabitats at which particular body temperatures can be maintained may be one of the dimensions that define the ecological niches of lizards (Roughgarden et al. 1983). The five most common species of *Anolis* on Cuba partition the habitat in several ways (Figure 15–13). First, they divide the habitat along the continuum, from sunny to shady: Two species (*A. lucius* and *A. allogus*) occur in deep shade in forests, one (*A. homolechis*) in partial shade in clearings and at the forest edge, and two (*A. allisoni* and *A. sagrei*) in full sun. Within habitats in the sun–shade continuum, the lizards are separated by the substrates they use as perch sites. In the forest *A. lucius* perches on large trees up to 4 meters above the ground, whereas *A. allogus* rests on small trees within 2 meters of the ground. *A. homolechis*, which does not share its habitat with another common species of *Anolis*, perches on both large and small trees. In open habitats *A. allisoni* perches more than 2 meters above the ground on tree trunks and houses, and *A. sagrei* perches below 2 meters on bushes and fenceposts.

Some species of lizards do not thermoregu-

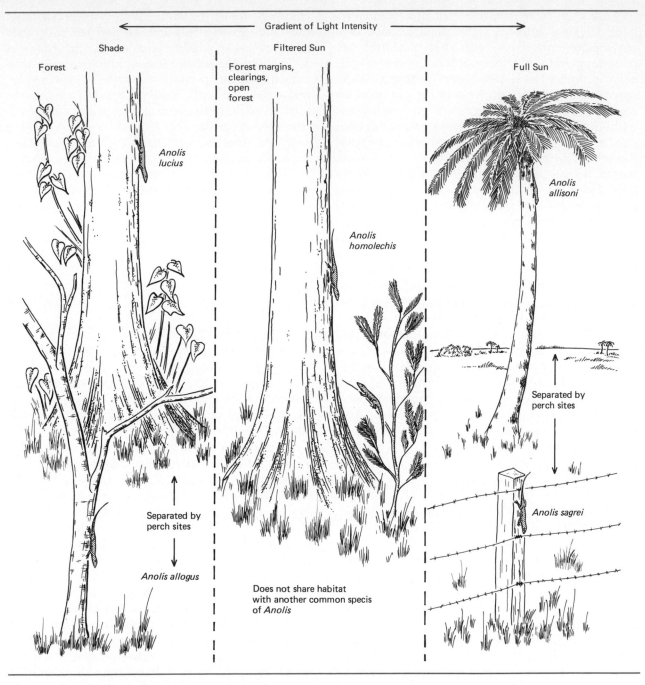

Figure 15–13. Habitat partitioning by Cuban species of *Anolis*. See text for explanation. (Based on R. Ruibal, 1961, *Evolution* 15:98–111.)

late. Lizards that live beneath the tree canopy in tropical forests often have body temperatures very close to air temperature (that is, they are thermally passive), whereas species that live in open habitats thermoregulate more precisely. The relative ease of thermoregulation in different habitats may be an important factor in determining whether a species of lizard thermoregulates or allows its temperature to vary with ambient temperature.

The distribution of sunny areas is one factor that determines the ease of thermoregulation. Sunlight penetrates the canopy of a forest in small patches that move across the forest floor as the sun moves across the sky. These patches of sun are the only sources of solar radiation for lizards that live at or near the forest floor, and the patches may be too sparsely distributed or too transient to be used for thermoregulation. In open habitats sunlight is readily available, and thermoregulation is easier. The difference in thermoregulatory behavior of lizards in open and shaded habitats can be seen even in comparisons of different populations within a species. For example, *Anolis sagrei* occurs in both open and forest habitats on Abaco Island in the Caribbean. Lizards in open habitats bask and maintain body temperatures between 32 and 35°C from 8:30 in the morning through 5 in the afternoon. Lizards in the forest do not bask, and their body temperatures vary from a low of 24°C to a high of 28°C over the same period.

The task of integrating thermoregulatory behavior with foraging is relatively simple for sit-and-wait foragers such as *Anolis*. These lizards can readily change their balance of heat gain and loss by making small movements in and out of shade or between calm and breezy perch sites while they continue to scan their surroundings for prey. Widely foraging lizards may have more difficulty integrating thermoregulation and predation. They are continuously moving between sun and shade and in and out of the wind, and their body temperatures are affected by their foraging activity. These lizards sometimes have to stop foraging to thermoregulate, resuming foraging only when they have warmed or cooled enough to return to their activity temperature range.

Body size is yet another variable that can affect thermoregulation. An example of the interaction of body size, thermoregulation, and foraging behavior is provided by three species of teiid lizards (*Ameiva*) in Costa Rica (Hillman 1969). *Ameiva* are widely foraging predators that move through the habitat, pushing their snouts beneath fallen leaves and into holes. The largest of the three species Hillman studied, *A. leptophrys*, had an average body mass of 83 grams, the middle species, *A. festiva*, weighed 32 grams, and the smallest, *A. quadrilineata*, weighed 10 grams (Figure 15–14). The study site was at the side of a road and the adjacent forest. The three species foraged in different parts of this habitat: *A. quadrilineata* spent most of its time in the short vegetation at the edge of the road, *A. festiva* was found on the bank beside the road, and *A. leptophrys* foraged primarily beneath the canopy of the forest (Figure 15–15). The different foraging sites of the three species may reflect differences in thermoregulation that result from the variation in body size.

The thermoregulatory behavior of the three species was the same: A lizard basked in the sun until its body temperature rose to 39 to 40°C, then moved through the mosaic of sun and shade as it searched for food. The body temperatures of the lizards dropped as they foraged, and lizards ceased foraging and resumed basking when their body temperatures had fallen to 35°C. Thus, the time that a lizard could forage depended on how long it took for its body temperature to cool from 39 or 40°C to 35°C. The rate of cooling for *Ameiva* in the shade was inversely proportional to the body size of the three species—*A. quadrilineata* cooled in 4 minutes, *A. festiva* in 6 minutes, and *A. leptophrys* in 11 minutes (Figure 15–16). That relationship appears to explain part of the microhabitat separation of the three species: The smallest species, *A. quadrilineata*, cools so rapidly that it may not be able to forage effectively in shady microhabitats, whereas *A. leptophrys* cools slowly and can forage in the shade beneath the forest canopy. The species of intermediate body size, *A. festiva*, uses the habitat with an intermediate amount of shade.

(a)

(b)

(c)

Figure 15–14. Three species of *Ameiva* that forage in different habitats: (a) *Ameiva leptophrys,* which has an average adult mass of 83 grams; (b) *Ameiva festiva,* average adult mass 32 grams; (c) *Ameiva quadrilineata,* average adult mass 10 grams. [(a) Courtesy of Daniel H. Janzen; (b) and (c) photographed by Michael Hopiak, courtesy of the Cornell University Herpetology Collection.]

The slow rate of cooling of *A. leptophrys* may explain why it is able to forage in the shade, but field observations indicate that its foraging is actually restricted to shade; it emerges from the forest only to bask. Does some environmental factor prevent *A. leptophrys* from foraging in the sun? The answer to that question may lie in the way the body temperatures of the three species increase when they are in open microhabitats. Body size profoundly affects the equilibrium temperature of an organism in the sun. A lizard warms by absorbing solar radiation, and as it gets warmer its heat loss by convection, evaporation, and reradiation also increases. At equilibrium, the rates of heat gain and heat loss are equal, and the body temperature does not increase further. Computer simulations by Hillman of the heating rates of the three *Ameiva* in sun showed that *A. quadrilineata* and *A. festiva* would reach equilibrium at body temperatures of 37 to 40°C, but *A. leptophrys* would continue to heat until its body temperature reached a lethal 45°C. This analysis suggests that *Ameiva leptophrys* would die of heat stress if it spent more than a few minutes in a sunny microhabitat, but the two smaller species of *Ameiva* would not have that problem.

Thus, as a result of the biophysics of heat exchange, the large body size of *A. leptophrys* apparently allows it to forage in shaded habitats (because it cools slowly), but prevents it from foraging in sunny habitats (because it would overheat). Field observations of the foraging behavior of

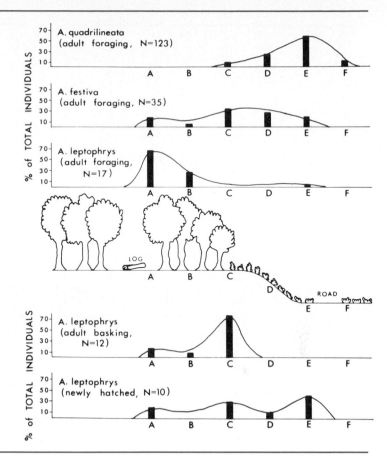

Figure 15–15. Foraging sites of three species of *Ameiva* in Costa Rica. The histograms show the number of individuals of the three species seen in each of six locations: A, a small clearing in the forest; B, immediately inside the forest edge; C, outside the edge of the forest; D, midway between the edge of the forest and open area; E, low vegetation beside a road; F, low vegetation in a large open area without trees. (From P. E. Hillman, 1969, *Ecology* 50:476–481.)

hatchling *A. leptophrys* emphasize the importance of the heat exchange in the foraging behavior of these lizards. Hatchling *A. leptophrys* forage in open habitats like *A. festiva* and *A. quadrilineata* rather than under the forest canopy like adult *A. leptophrys*. That is, the juveniles of the large species of *Ameiva* behave like adults of the smaller species, probably because of the importance of body size and heat exchange in determining the microhabitats in which lizards can thermoregulate.

The difference in the use of various microhabitats by these three species of lizards looks at first glance like an example of habitat partitioning in response to interspecific competition for food. That is, because all three species eat the same sort of prey, they could be expected to concentrate their foraging efforts in different microhabitats to reduce competition. However, Hillman's analysis of the thermal requirements of the lizards suggests that interspecific competition for food is, at most, a secondary factor.

If competition for food were important, one would not expect to find hatchling *A. leptophrys* foraging in the same microhabitat as adult *A. quadrilineata*, because the similarity in size of the two lizards would intensify competition for food. The hypothesis that competition for food determines the microhabitat distribution of the animals predicts that the forms most similar in body size should be mostly widely separated in the habitat.

In contrast, the hypothesis that energy exchange with the environment is critical in deter-

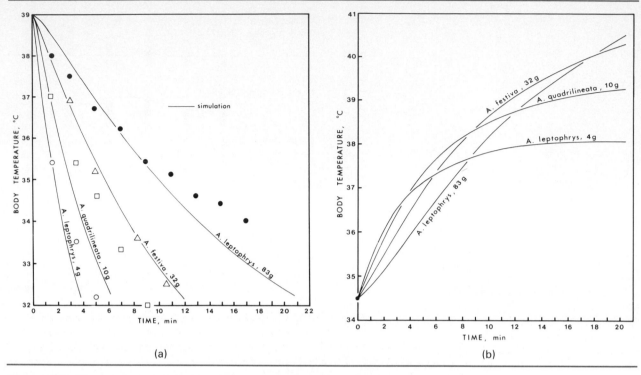

(a)

(b)

Figure 15–16. Cooling (a) and heating (b) rates of the three sympatric species of *Ameiva*. The solid lines are computer simulations of heating and cooling rates, and the symbols in (a) show the close correspondence of measured body temperatures to those predicted by the computer simulation. The largest species, *Ameiva leptophrys*, heats and cools more slowly than the smaller species. If it remained in the sun its body temperature would rise above 40°C. The smaller species heat and cool more rapidly than the large species and reach temperature equilibrium at lower body temperatures. [(a) From P. E. Hillman, 1969, *Ecology* 50:476–481; (b) courtesy of Peter E. Hillman.]

mining where a lizard can forage predicts that species of similar size will live in the same habitat, and that is approximately the pattern seen. Apparently, the physical environment (radiant energy) is more important than the biological environment (interspecific competition for food) in determining the microhabitat distributions of these lizards. That conclusion reflects the broad-scale ecological significance of the morphological and physiological differences between ectotherms and endotherms, a theme that is developed in Chapter 16.

Summary

The living lepidosaurs include the squamates (lizards, amphisbaenians, and snakes) and their sister group, the Sphenodontia. The lepidosaurs, with nearly 6000 species, form the second largest group of living tetrapods.

The tuatara of the New Zealand region is the sole living sphenodontian. It is a lizard-like animal about 60 centimeters long with a dentition and jaw mechanism that give a shearing bite. Sphenodontians were diverse in the Mesozoic and included terrestrial insectivorous and herbivorous species as well as a marine form.

Lizards range in size from tiny geckos only 3 centimeters long to the Komodo monitor lizard, which approaches a length of 3 meters. Four major lineages of lizards probably diverged in the Jurassic: The Iguania and Gekkota are composed largely of stout-bodied lizards and show considerable diversity of body form. The Scincomorpha and Anguinomorpha are elongate and show less morphological diversity than do iguanians. Differences in ecology and behavior parallel the phylogenetic divisions: Many iguanians are sit-and-wait predators that maintain territories and detect prey and intruders by vision. Iguanians often employ colors and patterns in visual displays during courtship and territorial defense. Scincomorph and anguinomorph lizards are often widely foraging predators that detect prey by olfaction and do not maintain territories. Pheromones are important in the social behaviors of many of these lizards.

Amphisbaenians are specialized burrowing squamates that are sometimes classified as a family of lizards. The skulls of amphisbaenians are solid structures that they use for tunneling through soil. Many amphisbaenians have blunt heads, but some have vertically keeled or horizontally spade-shaped snouts. The dentition of amphisbaenians appears to be specialized for nipping small pieces from prey too large to be swallowed whole. The skin of amphisbaenians is loosely attached to the trunk, and amphisbaenians slide backward or forward inside the tube of their skin as they move through tunnels with concertina locomotion.

Snakes are derived from anguinomorph lizards. Repackaging the body mass of a vertebrate into a serpentine form has been accompanied by specializations of the mechanisms of locomotion (serpentine, rectilinear, concertina, and sidewinding), prey capture (constriction and the use of venom), and swallowing (a highly kinetic skull).

Many squamates have complex social behaviors associated with territoriality and courtship, but parental care is only slightly developed. Fertilization is internal, and viviparity has evolved 80 or more times among squamates. Thermoregulation is another important behavior of squamates, and various activities are influenced by body temperature. The ecological niches of some lizards may be defined in part by the microhabitats needed to maintain particular body temperatures. Feeding status, pregnancy, and bacterial infections can change the thermoregulatory behavior of squamates, causing an affected individual to maintain a higher or lower body temperature than it otherwise would.

References

Andrews, R. M. 1982. Patterns of growth in reptiles. Pages 273–320 in *Biology of the Reptilia*, volume 13, edited by C. Gans and F. H. Pough. Academic Press, London.

Auffenberg, W. 1981. *The Behavioral Ecology of the Komodo Monitor*. University Presses of Florida, Gainesville, Fla.

Avery, R. A. 1976. Thermoregulation, metabolism, and social behaviour in Lacertidae. Pages 245–259 in *Morphology and Biology of Reptiles*, edited by A. d'A. Bellairs and C. B. Cox. Academic Press, London.

Avery, R. A. 1982. Field studies of reptilian thermoregulation. Pages 93–166 in *Biology of the Reptilia*, volume 12, edited by C. Gans and F. H. Pough. Academic Press, London.

Bellairs, A. d'A. and S. V. Bryant. 1985. Autotomy and regeneration in reptiles. Pages 301–410 in *Biology of the Reptilia*, volume 15, edited by C. Gans and F. Billett. Wiley, New York.

Benton, M. J. 1985. Classification and phylogeny of the diapsid reptiles. *Zoological Journal of the Linnean Society* 84:97–164.

Beuchat, C. A. 1986. Reproductive influence on the thermoregulatory behavior of a live-bearing lizard. *Copeia* 1986:971–979.

Beuchat, C. A. and S. Ellner. 1987. A quantitative test of life history theory: thermoregulation by a viviparous lizard. *Ecological Monographs* 57:45–60.

Blackburn, D. G. 1982. Evolutionary origins of viviparity in the Reptilia. I. Sauria. *Amphibia-Reptilia* 3:185–205.

Blackburn, D. G. 1985. Evolutionary origins of viviparity in the Reptilia. II. Serpentes, Amphisbaenia, and Ichthyosauria. *Amphibia-Reptilia* 6:259–291.

Blackburn, D. G., L. J. Vitt, and C. A. Beuchat. 1984. Eutherian-like reproductive specializations in a viviparous reptile. *Proceedings of the National Academy of Science USA* 81:4860–4863.

Burghardt, G. M. 1977. Of iguanas and dinosaurs: social behavior and communication in neonate reptiles. *American Zoologist* 17:177–190.

Carroll, R. L. 1987. *Vertebrate Paleontology and Evolution.* W. H. Freeman, New York.

Cole, C. J. 1975. Evolution of parthenogenetic species of reptiles. Pages 340–355 in *Intersexuality in the Animal Kingdom*, edited by R. Reinboth. Springer-Verlag, Berlin.

Cooper, W. E. Jr. and L. J. Vitt. 1986. Interspecific odour discriminations among syntopic congeners in scincid lizards (Genus *Eumeces*). *Behaviour* 97:1–9.

Darevsky, I. S., L. A. Kupriyanova, and T. Uzzell. 1985. Parthenogenesis in reptiles. Pages 412–526 in *Biology of the Reptilia*, vol. 15, edited by C. Gans and F. Billett. Wiley, New York.

Dawson, W. R. 1975. On the physiological significance of the preferred body temperatures of reptiles. Pages 443–473 in *Perspectives of Biophysical Ecology*, edited by D. M. Gates and R. B. Schmerl. Springer-Verlag, New York.

Diamond, J. M. 1987. Did Komodo dragons evolve to eat pygmy elephants? Nature 326:832.

Estes, R. 1983. The fossil record and early distribution of lizards. Pages 365–398 in *Advances in Herpetology and Evolutionary Biology*, edited by A. G. J. Rhodin and K. Miyata. Harvard University Press, Cambridge, Mass.

Ford, N. B. and C. W. Schofield. 1984. Species specificity of sex pheromone trails in the plains garter snake, *Thamnophis radix. Herpetologica* 40:51–55.

Gans, C. 1978. The characteristics and affinities of the Amphisbaenia. *Transactions of the Zoological Society of London* 34:347–416.

Gans, C. 1983. Is *Sphenodon punctatus* a maladapted relic? Pages 613–620 in *Advances in Herpetology and Evolutionary Biology*, edited by A. G. J. Rhodin and K. Miyata. Harvard University Press, Cambridge, Mass.

Greene, H. W. 1983. Dietary correlates of the origin and radiation of snakes. *American Zoologist* 23:431–441.

Hertz, P. E., R. B. Huey, and E. Nevo. 1982. Fight versus flight: body temperature influences defensive responses of lizards. *Animal Behaviour* 30:676–679.

Hillman, P. E. 1969. Habitat specificity in three sympatric species of *Ameiva* (Reptilia: Teiidae). *Ecology* 50:476–481.

Huey, R. B. 1982. Temperature, physiology, and the ecology of reptiles. Pages 25–91 in *Biology of the Reptilia*, volume 12, edited by C. Gans and F. H. Pough. Academic Press, London.

Huey, R. B. and A. F. Bennett. 1986. A comparative approach to field and laboratory studies in evolutionary biology. Pages 82–98 in *Predator–Prey Relationships*, edited by M. E. Feder and G. V. Lauder. University of Chicago Press, Chicago.

Huey, R. B. and E. R. Pianka. 1981. Ecological consequences of foraging mode. *Ecology* 62:991–999.

Losos, J. B. 1985. An experimental demonstration of the species-recognition role of *Anolis* dewlap color. *Copeia* 1985:905–910.

Magnusson, W. E., L. Junqueria de Paiva, R. Moreira da Rocha, C. R. Franke, L. A. Kasper, and A. P. Lima. 1985. The correlates of foraging mode in a community of Brazilian lizards. *Herpetologica* 41:324–332.

Mason, R. T. and D. Crews. 1985. Female mimicry in garter snakes. *Nature* 316:59–60.

Pough, F. H. (editor). 1983. Adaptive radiation within a highly specialized system: the diversity of feeding mechanisms of snakes. *American Zoologist* 23:339–460.

Regal, P. J. 1978. Behavioral differences between reptiles and mammals: an analysis of activity and mental capabilities. Pages 183–202 in *Behavior and Neurology of Lizards*, edited by N. Greenberg and P. D. MacLean. National Institute of Mental Health, Rockville, Md.

Regal, P. J. 1983. The adaptive zone and behavior of lizards. Pages 105–118 in *Lizard Ecology: Studies of a Model Organism*, edited by R. B. Huey, E. R. Pianka, and T. W. Schoener. Harvard University Press, Cambridge, Mass.

Rich, T. and B. Hall. 1984. Rebuilding a giant lizard: *Megalania prisca*. Pages 391–394 in *Vertebrate Zoogeography and Evolution in Australia*, edited by M. Archer and G. Clayton. Hesperian Press, Carlisle, Western Australia.

Roughgarden, J., D. Heckel, and E. R. Fuentes. 1983. Coevolutionary theory and the biogeography and community structure of *Anolis*. Pages 371–410 in *Lizard Ecology: Studies of a Model Organism*, edited by R. B. Huey, E. R. Pianka, and T. W. Schoener. Harvard University Press, Cambridge, Mass.

Savitzky, A. H. 1980. The role of venom delivery strategies in snake evolution. *Evolution* 34:1194–1204.

Shine, R. 1985. The evolution of viviparity in reptiles: an ecological analysis. Pages 606–694 in *Biology of the Reptilia*, volume 15, edited by C. Gans and F. Billett. Wiley, New York.

Shine, R. 1988. Parental care in reptiles. Pages 275–329 in *Biology of the Reptilia*, volume 16, edited by C. Cans and B. R. Huey. Alan R. Liss, New York.

Simon, C. A. 1983. A review of lizard chemoreception. Pages 119–133 in *Lizard Ecology: Studies of a Model Organism*, edited by R. B. Huey, E. R. Pianka, and T. W. Schoener. Harvard University Press, Cambridge, Mass.

Stamps, J. A. 1983. Sexual selection, sexual dimorphism, and territoriality. Pages 169–204 in *Lizard Ecology: Studies of a Model Organism*, edited by R. B. Huey, E. R. Pianka, and T. W. Schoener. Harvard University Press, Cambridge, Mass.

Stevenson, R. D., C. R. Peterson, and J. S. Tsuji. 1985. The thermal dependence of locomotion, tongue flicking, digestion, and oxygen consumption in the wandering garter snake. *Physiological Zoology* 58: 46–57.

Troyer, K. 1982. Transfer of fermentative microbes between generations in a herbivorous lizard. *Science* 216:540–542

Yaron, Z. 1985. Reptilian placentation and gestation: structure, function, and endocrine control. Pages 528–603 in *Biology of the Reptilia*, volume 15, edited by C. Gans and F. Billett. Wiley, New York.

Ectothermy is an ancestral character of vertebrates, but like many primitive characteristics it is just as effective as its derived counterpart. Furthermore, the mechanisms of ectothermal thermoregulation are as complex and specialized as those of endothermy. In Chapter 4 we examined the behavioral and physiological mechanisms that ectotherms use to control their body temperatures, and in Chapter 15 we considered the effect of these thermoregulatory mechanisms on the ecology and behavior of lizards and snakes. Here we consider the consequences of ectothermy in shaping broader aspects of the life-style of fishes, amphibians, and squamates. Ectothermal tetrapods have low energy requirements compared to endotherms, and this characteristic allows them to be successful in environments in which food is in short supply, either seasonally or chronically. Within that general characterization, the success of various species is related to specializations that facilitate finding food or mates under difficult circumstances, or allow them to endure prolonged unfavorable periods. The conclusion from this examination is that success in difficult environments is as likely to reflect the ancestral features of a group as its specializations.

Ectothermy: A Low-Cost Approach to Life

16

Vertebrates and Their Environments

Vertebrates manage to live in the most unlikely places—amphibians live in deserts where rain falls only a few times a year and several years may pass with no rainfall at all; lizards live on mountains at altitudes above 4000 meters, where the temperature falls below freezing nearly every night of the year and does not rise much above freezing during the day; birds and mammals live near the poles, where the coldest temperatures on earth are recorded in winter and the sun barely appears above the horizon for weeks on end; and fishes live in the depths of the sea, where sunlight never penetrates.

Of course, vertebrates do not seek out only inhospitable places to live—birds, lizards, mammals, and even amphibians can be found on the beaches at Malibu, sometimes running between the feet of surfers, and fishes cruise the shore. (Occasionally, a hungry shark helps itself to a surfer.) However, even this apparently benign environment is stressful to some animals, and examination of the ways that vertebrates have managed to live in stressful environments has provided much information about how they function as organisms—that is, how morphology, physiology, ecology, and behavior interact.

In some cases elegant adaptations allow specialized vertebrates to colonize stressful habitats.

More common and more impressive than these specializations, however, is the realization of how minor are the modifications of the basic vertebrate body plan that allow animals to endure environmental temperatures from -70 to $+70°C$, or water conditions ranging from complete immersion in water to complete independence of liquid water. No obvious differences distinguish animals from vastly different habitats—an arctic fox looks very much like a desert fox and a lizard from the Andes Mountains looks like one from the Atacama Desert. The adaptability of vertebrates lies in the combination of a number of minor modifications of their ecology, behavior, morphology, and physiology. A view that integrates these elements shows the startling beauty of organismal function of vertebrates.

Endotherms and ectotherms react somewhat differently to the many environmental stresses. The essence of the endothermal approach to life is the use of energy to maintain internal homeostasis. Endotherms can regulate their body temperatures and body fluid and salt concentrations with remarkable precision in the face of extreme fluctuations in their environment. Ectotherms are also capable of remarkable homeostasis, but the general characteristic of ectotherms is low rates of energy consumption (Chapter 4). In many cases ectotherms save energy by relaxing their limits of homeostasis, whereas endotherms expend energy to maintain homeostasis. This difference in the responses of endotherms and ectotherms to stressful environments has broad implications: By relaxing homeostasis ectotherms are able to do things that would not be feasible for an endotherm. On the other hand, the activity of ectotherms may be curtailed for long periods during unfavorable seasons because they are not able to maintain homeostasis. The energy requirements of endothermy are a substantial factor in many aspects of the ecology and behavior of birds and mammals, and inability to obtain enough energy may exclude endotherms from activity in certain habitats during some parts of the year: Some endotherms migrate or hibernate to avoid conditions in which they cannot get enough energy to maintain homeostasis.

In this chapter we describe some of the conditions that make certain environments stressful for vertebrates, and then consider the responses of aquatic and terrestrial ectotherms to those stressful conditions. A similar treatment of the reactions of endotherms to stressful environments (Chapter 23) follows the descriptions of the evolutionary histories and general biology of the groups.

Characteristics of Stressful Environments

The life of a vertebrate is energetically expensive; homeostasis and motility require ATP, which is derived from metabolism of food. It follows that stressful environments are those in which energy is limited or the cost of homeostasis and motility is high, and the most stressful situations are those in which both conditions apply. That qualitative generalization applies to ectotherms and endotherms, although energy is likely to become limiting for endotherms before ectotherms because endothermal life is more energetically expensive than ectothermal life.

Habitats in which food (energy) is scarce abound. In ecological terms all animals are consumers. That is, animals depend on the primary production of plants: Herbivorous animals eat plants and carnivorous animals eat herbivores. Thus, the path of energy flow through an ecological community begins with the fixation of carbon by plants, and one measure of the availability of energy in a habitat is the primary production of the plants in the habitat (Table 16–1). Tropical forests and estuarine marine environments are the most productive habitats on Earth, producing as much as 4000 or 5000 grams of carbon per square meter annually. At the opposite extreme, deserts and the open sea produce less than one-tenth as much carbon as do tropical forests and estuaries, and the annual production of arctic habitats is still lower.

The primary production of a habitat tells how much energy is potentially available to animals but does not necessarily reflect the accessibility of the energy. A tropical forest has an enormous production of new plant tissue every year, but much of that material is leaves that are in the canopy of the forest, which is 20 meters or more above the ground. Those leaves are not accessible to herbivores such as cows, which cannot climb trees. Furthermore, animals have many habitat requirements besides a source of energy: They need water, shelter from the elements, places to escape predators, microclimates that are suitable for reproduction and for rearing young, and so on. Some vertebrates use their mobility to occupy habitats that do not have all the resources they need. For example, small birds in deserts can fly to water sources that may be many kilometers away, whereas mammals of the same body size are less mobile and must subsist on the water resources they can find in a restricted area.

The availability of resources that will sustain an individual is not enough to allow a species to occupy a habitat: Reproduction is necessary for a population to be self-sustaining. This requirement means that the population density must be sufficient for males and females to locate each other during the breeding season, and that the habitat must provide conditions suitable for reproduction and the growth of young. Elaborate behavioral modifications that permit the sexes to find each other are seen in some species of vertebrates, and

	Net Primary Production (Dry Grams/Square Meter/Year)		Biomass per Unit Area (Dry Kilograms/Square Meter)	
	Range	*Mean*	*Range*	*Mean*
Terrestrial habitats				
Tropical forests	1000–5000	2000	6–80	45
Temperate forest	600–2500	1300	6–200	30
Tropical savanna	200–2000	700	0.2–15	4
Temperate grassland	150–1500	500	0.2–5	1.5
Desert	10–250	70	0.1–4	0.7
Arctic	0–10	3	0–0.2	0.02
Aquatic habitats				
Estuaries	500–4000	2000	0.04–4	1
Lakes and streams	100–1500	500	0–0.1	0.02
Open ocean	2–400	125	0–0.005	0.003

Source: R. H. Whittaker, 1970, *Communities and Ecosystems*, Macmillan, New York.

juveniles sometimes do not live in the same habitat as adults.

Stressful conditions in many habitats are seasonal; periods of plenty may alternate with periods of great scarcity. These cycles are conspicuous in arctic regions, where the long days of summer allow plants to grow rapidly and winters bring all primary production to a halt. Other habitats have annual cycles, too; even some tropical forests have pronounced wet (monsoon) and dry seasons that greatly affect plant production. The responses of animals to seasonally stressful conditions are somewhat different from their responses to chronic shortages of energy. Hibernation and estivation are ways to remain in a habitat during stressful periods, and migration allows animals to move from habitat to habitat as the availability of energy waxes and wanes.

Ectotherms in Stressful Environments

The low energy requirements of ectotherms allow them to exploit many environments in which en-

ergy is scarce seasonally or at all times. Also, the ability of ectotherms to become torpid allows them to wait out periods when the environment may be too hot or too dry to permit activity. Temporary relaxation of homeostasis—allowing physiological variables to fluctuate more widely than usual—may permit organisms to occupy habitats that would not otherwise be habitable.

The responses of ectotherms to stressful environments usually involve suites of adaptations that include behavior and physiology as well as body structure. Patterns of life history may be modified in ways that ensure that males and females can find each other to mate, or that facilitate the development of embryos and young. These phenomena are well illustrated by vertebrates in three environments: the deep sea, the desert, and freezing water.

The Deep Sea

Oceans cover 73 percent of the Earth's surface (Figure 16–1) and are classified into different depth

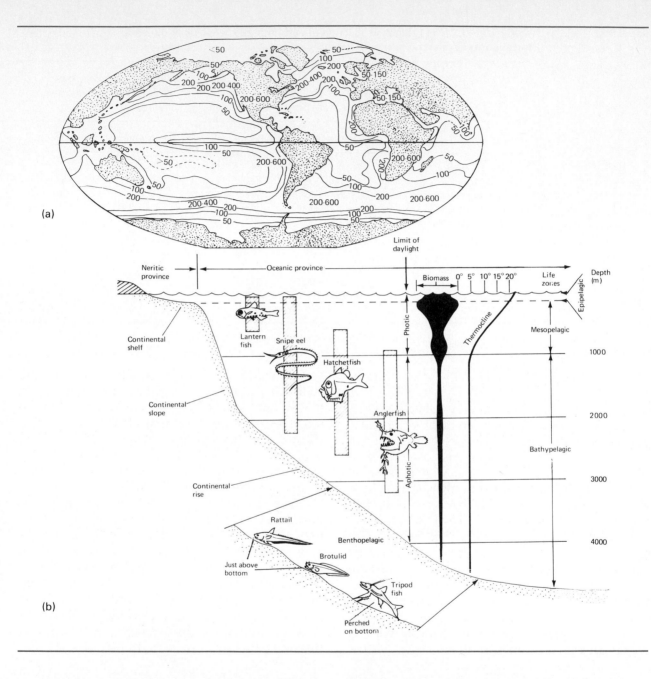

(a)

(b)

zones. Two major life zones exist: the pelagic, where organisms live a free-floating or swimming existence, and the benthic, where organisms associate with the bottom. Solar light is totally extinguished at a depth of 1000 meters in the clearest oceans and at much shallower depths in the less clear coastal seas. As a result, large portions of the oceans are aphotic, or perpetually dark (about 75 percent by volume), interrupted only by the flashes and glow of bioluminescent organisms. In those zones have evolved a distinctive and bizarre array of deep-sea fishes.

Deep-sea fishes decrease in abundance, size, and species diversity with increased depth. (This pattern also occurs among invertebrates.) These trends are not surprising, for all animals ultimately depend on plant photosynthesis, which is limited to the epipelagic regions. Below the epipelagic, animals must depend on descent of food from above—a rain of detritus from the surface into the deep sea. The decrease of fishes with depth is inevitable, for available food must diminish if it is consumed during descent. Sampling confirms this decrease in food: Surface plankton can reach biomass levels of 500 milligrams wet weight per cubic meter; at a depth of 1000 meters, where aphotic conditions commence, plankton decrease to 25 mg/m^3; at depths of 3000 to 4000 meters to 5 mg/m^3; and at 10,000 meters to 0.5 mg/m^3.

Fish diversity parallels this decrease: About 750 species of deep-sea fishes are estimated to occupy the mesopelagic zones and only 150 species the bathypelagic regions. Deep seas that lie under areas of high surface productivity contain more and larger fish species than do regions that underlie less productive surface waters. High productivity occurs in areas of upwelling, where currents recycle nutrients previously removed by the sinking of detritus. In these places deep-sea fishes tend to be most diverse and abundant.

In tropical waters, photosynthesis continues throughout the year. Away from the tropics it is more cyclic, following seasonal changes in light, temperature, and sometimes currents. Diversity and abundance of deep-sea fishes decrease away from the tropics. Over 300 species of meso- and bathypelagic fishes occur in the vicinity of tropical Bermuda. In the entire Antarctic region, only 50 species have been described. The high productivity of Antarctic waters is restricted to a few months of each year. Apparently, the sinking of detritus through the rest of the year is insufficient to nourish a diverse assemblage of deep-sea fishes.

We emphasize that availability of energy (food) is the most formidable environmental stress that deep-sea fishes encounter and, indeed, many of their specific characteristics may have been selected by food scarcity. Another major variable is hydrostatic pressure; lack of light and low temperatures are constants. Pelagic fishes also must continuously counter gravity to avoid sinking. For benthic fishes gravity presents a problem only when they venture from the seafloor.

With increased depth each species is further removed from a primary source of food. In general, mesopelagic fishes and invertebrates, which inhabit depths from 100 to 1000 meters, migrate vertically. At dusk they ascend only to descend again near dawn, apparently following light-intensity levels (Figure 16–2). Several benefits and costs probably result from this behavior. By rising at dusk, mesopelagic fishes enter a region of higher productivity where food is more concentrated. However, they also increase their exposure to predators. Furthermore, ascending mesopelagic fishes are exposed to temperature increases that

---

Figure 16–1. Life zones of the ocean depths. (a) Annual productivity at the ocean surface. Numbers are grams of carbon produced per square meter per year. Rich assemblages of deep-sea fishes occur where highly productive waters overlie deep waters. (b) Schematic cross section of the life zones within the deep sea. Various pertinent physical and biotic parameters are superimposed on the arbitrary life zones, as are the vertical ranges of several characteristic fish species, some of which migrate on a daily basis. [(a) Modified from H. Friedrich, 1973, *Marine Biology,* University of Washington Press, Seattle, Wash.; (b) modified from N. B. Marshall, 1971, *Explorations of the Life of Fishes,* Harvard University Press, Cambridge, Mass.]

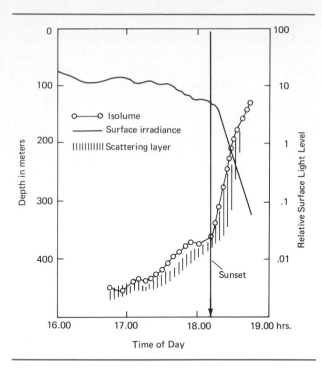

Figure 16–2. Upward migration of mesopelagic fishes as indicated by changes in depth of deep scattering layer. Sonar signals readily reflect off the swim bladders of mesopelagic fishes, and aggregations of them produce an acoustic scattering layer. At sunset the intensity of light at the surface decreases (right axis) as does the light penetrating the sea. Plotting the depth of a single light intensity (isolume, read depth on left axis) against time illustrates the progressively shallower depth at which a given light intensity occurs during dusk. The light intensity chosen for this example falls between that of starlight and full moonlight, and it corresponds closely with the upper boundary of the deep scattering layer, which rises rapidly with the upward migration of mesopelagic fishes. (Modified from E. M. Kampa and B. P. Boden, 1957, *Deep-Sea Research* 4:73–92.)

can exceed 10°C. The energy costs of maintenance at these higher temperatures can double, even triple. In contrast, daytime descent into cooler waters lowers metabolism, which conserves energy and reduces the chance of predation, for fewer predators exist at depth.

Bathypelagic Fishes

It is less certain that bathypelagic fishes undertake daily vertical migrations. There is little metabolic economy from vertical migration within this aphotic zone because temperatures are uniform (about 5°C). Furthermore, the cost and time of migration over the several thousand meters from the bathypelagic to the surface would probably outweigh the energy gained from invading the rich surface regions. As a result, bathypelagic fishes are specialized to live a less active life than those of their meso- and epipelagic counterparts.

Pelagic deep-sea fish display a series of structures related to their particular life-styles. For example, eye size and function correlate with depth. Mesopelagic fishes have large eyes, pelagic fishes have smaller eyes, and benthic fishes vary in eye size. The large pupil of mesopelagic fishes maximizes light captured by the eye, analogous to enlarging the diaphragm of a camera. In addition, the retina contains a high concentration of visual pigment, the photosensitive chemical that absorbs light in the process of vision. The visual pigments of deep-sea fishes are most efficient in absorbing blue light—the color of light most readily transmitted through clear water.

Many deep-sea fishes and invertebrates are emblazoned with startling bioluminescent designs. Photophore organs emit blue light. Bioluminescence, unlike solar illumination, does not provide a general background light against which targets are viewed. Rather it is presented as a point source against a black background. A large eye is apparently not necessary to detect a point source. Sensitivity is retained, however, in possession of a visual pigment that best absorbs the blue light of bioluminescence. Reduction of eye size provides a considerable economy for a bathypelagic fish where food is scarce, for the eye has a very high metabolic rate relative to other tissues. Retention of the eye in most bathypelagic fishes must therefore depend on the presence of bioluminescent organisms.

With increasing depth there is a reduction in bone and skeletal musculature of deep-sea fishes

(Figure 16–3). These trends relate to the scarcity of food and the problem of sinking. Cytoplasm has a specific gravity slightly greater than seawater. Because bone is very dense (specific gravity about 2.0), its reduction in deep-sea fishes decreases their total density and rate of sinking. Epipelagic fishes show little bone reduction and mesopelagic fishes show less bone reduction than bathypelagic fishes.

To act efficiently, muscle requires a stiffening framework, the vertebral column. Surface fishes such as tunas have strong ossified skeletons and large red muscles especially adapted for continuous cruising. Mesopelagic fishes, which swim mostly during vertical migration, have a more delicate skeleton and less axial red muscle. In bathypelagic fishes, the axial skeleton and the mass of muscles are greatly reduced and locomotion correspondingly limited.

A tendency toward strict carnivory occurs with depth. Jaws and teeth in deep-sea fishes are usually enlarged to ensure capture of scarce prey. If one rarely encounters potential prey, it is probably important to have a mouth large enough to engulf nearly anything one does meet and a gut that can extend to accommodate it (Figure 16–4). Increasing the chances of encountering prey is also important to survival in the deep sea. Rather than searching for prey through the blackness of the depths, the ceratioid anglerfishes dangle a bioluminescent bait in front of them. The bioluminescent lure is believed to mimic the movements of zooplankton and to lure fishes and larger crustaceans to the mouth. Prey is sucked in with a sudden opening of the gape, snared in the teeth, and then swallowed. In many anglerfishes the gape expands and the stomach stretches to accommodate prey larger than the predator. Thus deep-sea fishes, like most teleosts, show major specializations in locomotor and feeding structures. Unlike

Figure 16–3. Reduction in the amount of bone in oceanic fishes that live at increasing depths as seen in x-rays. Upper: surface-dwelling jackfish (Carangidae); middle: mesopelagic, vertically migrating lanternfish (Myctophidae); bottom: mesopelagic, deep-living bristlemouth (Gonostomatidae). Note also the change in the relative size of the eye.

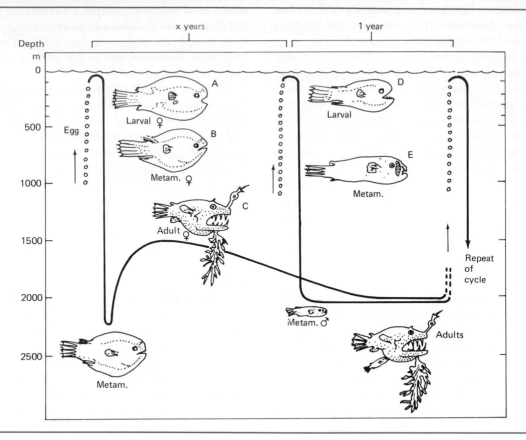

Figure 16–5. Diagrammatic representation of the life history of an anglerfish as a function of sex, age, and depth distribution: A, larva of female, 25 millimeters long; B, larva of female at 43 millimeters; C, adult female, 70 millimeters long; D, larva of male, 19 millimeters long; E, metamorphosed male, 23 millimeters long. Not drawn to scale. (Modified from E. Bertelsen, 1951, *Dana Report* 39:1–281.)

swimming period concentrated on finding a female. The search must be precarious, for males must search vast, dark regions for a single female while running a gauntlet of other deep-sea predators. In the young adults there is an unbalanced sex ratio; often more than 30 males exist for every female. Apparently, 29 of those males do not locate a female successfully. All these strategies, whether morphological (feeding), behavioral (senses), or ecological (life history), seem directed to successful reproduction.

The story does not end here, for the most bizarre specialization results in parasitism of the females by the males. Most ceratioids spawn in early summer, but the males descend months later, out of phase with this spawning, when their testes are still immature. Males, upon contacting a female, bite into her flesh and attach themselves firmly.

Preparation for this encounter begins during metamorphosis when the male's teeth degenerate and strong tooth-like bones develop at the tips of the jaws. A male remains attached to the female for life and in a parasitic state grows and the testes mature. Monogamy prevails in this parasitic marriage, for females invariably have but one attached male. As humans we consider this life-style bizarre, and, indeed, it is unknown in other vertebrates. But it has been successful—some 200 ceratioid species exist. The lesson these unique fishes provide is that vertebrate life is a plastic venture capable of adapting to extreme conditions. In the ceratioids, two features stand out: efficient energy utilization and reproduction. Actually, adaptations in all vertebrates relate to these two goals, but they are not often painted in such bold relief.

Ectotherms in the Heat: Deserts

Deserts can be produced by various combinations of topography, air movements, and ocean currents, but whatever their cause, deserts have in common a scarcity of liquid water, and that characteristic is at the root of many of the features of deserts that make them difficult places for vertebrates to live. The dry air characteristic of most deserts seldom contains enough moisture to form clouds that would block solar radiation during the day or radiative cooling at night. As a result, the daily temperature excursion in deserts is large in comparison to that of more humid areas. Scarcity of water is reflected by sparse plant life and a correspondingly low primary productivity in desert communities. Food shortages may be chronic and exacerbated by seasonal shortages and unpredictable years of low production when the usual pattern of rainfall does not develop.

Not all deserts are hot. Indeed some are distinctly cold—most of Antarctica and the region of Canada around Hudson Bay and the Arctic Ocean are deserts. The low-latitude deserts north and south of the equator are hot deserts, however, and it is in these low-latitude deserts that vertebrates encounter the most difficult problems of desert life.

The scarcity of rain contributes to the low primary production of deserts and also means that sources of liquid water for drinking are usually unavailable to small animals that cannot travel long distances. These animals obtain water from the plants or animals they eat, but plants and insects have ion balances that differ from those of vertebrates. In particular, potassium is found in higher concentrations in plants and insects than it is in vertebrate tissues, and excreting the excess potassium can be difficult if water is too scarce to waste in the production of large quantities of urine.

The low metabolic rates of ectotherms alleviate some of the stress caused by scarcity of food and water, but many desert ectotherms must temporarily extend the limits within which they regulate body temperatures or body fluid concentrations, become inactive for large portions of the year, or adopt a combination of these responses. Tortoises, lizards, and anurans from deserts illustrate these phenomena.

Balancing Water and Energy Budgets

The largest ectothermal vertebrates in the deserts of North America are tortoises. The Bolson tortoise (*Gopherus flavomarginatus*) of northern Mexico probably once reached a shell length of a meter, although predation by humans has apparently prevented any tortoise in recent times from living long enough to grow that large. The desert tortoise (*G. agassizi*) of the southwestern United States is smaller than the Bolson tortoise, but it is still an impressively large turtle (Figure 16–6). Adults reach shell lengths of 30 centimeters or more and weigh about 2 kilograms. A study of the annual water, salt, and energy budgets of desert tortoises in Nevada shows the difficulties they face in that desert habitat (Nagy and Medica 1986).

Desert tortoises construct shallow burrows that they use as daily retreat sites during the summer and deeper burrows for hibernation in winter. The tortoises in the study area emerged from hibernation in spring, and aboveground activity extended through the summer until they began hibernation again in November. Doubly labeled water was used to measure the energy expenditure of free-ranging tortoises (Box 16–1). Figure 16–8 shows the annual cycle of time spent above ground and in burrows by the tortoises, and the

(a)   (b)

Figure 16–6. (a) Desert tortoise, *Gopherus agassizi.* (b) A tortoise entering its burrow. (Photographs by R. Bruce Bury, U.S. Fish & Wildlife Service.)

Box 16–1. Doubly Labeled Water

This technique is widely employed in studies of the energy consumption of wild vertebrates because it is the only method of measuring metabolism without restraining the animal in some way. The method measures carbon dioxide production, which in turn can be used to estimate oxygen consumption. "Doubly labeled" refers to water that carries isotopic forms of oxygen and hydrogen: The oxygen atom has been replaced with its stable isotope oxygen-18 (^{18}O), and one hydrogen atom (H) has been replaced by its radioactive form, tritium (^{3}H). [Doubly labeled water is prepared by mixing appropriate quantities of water containing the hydrogen isotope (^{3}HHO) with water containing the oxygen isotope ($H_2{}^{18}O$). When this mixture is diluted by the body water of an animal the concentrations of the isotopes are so low that few individual water molecules contain both isotopes.] A measured amount of doubly labeled water is injected into an animal and allowed to equilibrate with the body water for several hours (Figure 16–7a), after which time a small blood sample is withdrawn. The concentrations of tritium and oxygen-18 in the blood are measured, and the total volume of body water can be calculated from the dilution of the doubly labeled water that was injected.

After that first blood sample has been withdrawn, the animal is released and recaptured at intervals of several days or weeks. Each time the animal is recaptured a blood sample is taken and the concentrations of tritium and oxygen-18 are measured. The calculation of the amount of carbon dioxide produced by the animal is based on the difference in the rates of loss of tritium and oxygen-18. Tritium, which behaves chemically like hydrogen, is lost as water—that is as ^{3}HHO, whereas oxygen-18 is lost both as water ($H_2{}^{18}O$) and as carbon dioxide ($C^{18}OO$). Thus, the decline in the concentration of tritium in the blood of the animal is a measure of the rate of water loss, and the decline in the concen-

annual cycles of energy, water, and salt balance. A positive balance means that the animal shows a net gain, whereas a negative balance represents a net loss. Positive energy and water balances indicate that conditions are good for the tortoises, but a positive salt flux means that ions are accumulating in the body fluids faster than they can be excreted. That situation indicates a relaxation of homeostasis and is probably stressful for the tortoises. The figure shows that the animals were often in negative balance for water or energy and accumulated salt during much of the year. Examination of the behavior and dietary habits of the tortoises through the year shows what was happening.

After they emerged from hibernation in the spring, the tortoises were active for about 3 hours every fourth day; the rest of the time they spent in their burrows. From March through May the tortoises were eating annual plants that had sprouted after the winter rains. They obtained large amounts of water and potassium from this diet, and their water and salt balances were positive. The osmotic concentration of the tortoises' body fluids increased by 20 percent during the spring, indicating that they were osmotically stressed as a result of the high concentrations of potassium they were ingesting. Furthermore, the energy content of the plants was not great enough to balance the metabolic energy expenditure of the

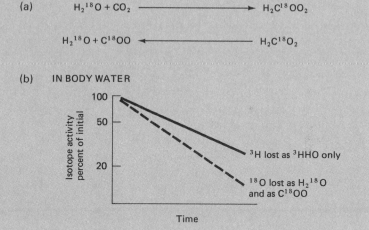

Figure 16–7. Doubly labeled water. (a) The reaction of carbon dioxide and water to produce carbonic acid, and the subsequent reconversion of carbonic acid to water and carbon dioxide produce an equilibrium of ^{18}O between H_2O and CO_2 when water labeled with the isotope is injected into a vertebrate. (b) Differential washout of hydrogen and oxygen isotopes in the body water of an animal that has been injected with doubly labeled water. (Modified from K. A. Nagy, 1988, in *Stable Isotopes in Ecological Research*, edited by J. Ehleringer, P. Rundall, and K. Nagy. Springer, New York.)

tration of oxygen-18 is a measure of the rates of loss of carbon dioxide and water (Figure 16–7b). The difference between decrease in concentration of tritium and oxygen-18, therefore, is the rate of loss of carbon dioxide, and this is proportional to the rate of oxygen consumption. A full description of the use of doubly labeled water for metabolic studies can be found in Nagy (1983).

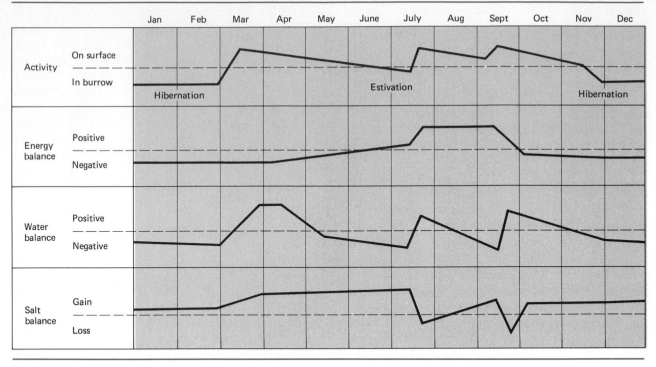

Figure 16—8. Annual cycle of desert tortoises. See text for details. (Based on
K. A. Nagy and P. A. Medica, 1986, *Herpetologica* 42:73–92.)

tortoises, and they were in negative energy balance. During this period the tortoises were using stored energy by metabolizing their body tissues.

As ambient temperatures increased from late May through early July, the tortoises shortened their daily activity periods to about 1 hour every sixth day. The rest of the time the tortoises spent estivating in shallow burrows. The annual plants died, and the tortoises shifted to eating grass and achieved positive energy balances. They stored this extra energy as new body tissue. The dry grass contained little water, however, and the tortoises were in negative water balance. The osmotic concentrations of their body fluids remained at the high levels they had reached earlier in the year.

In mid-July thunderstorms dropped rain on the study site, and most of the tortoises emerged from estivation. They drank water from natural basins, and some of the tortoises constructed bas-

ins by scratching shallow depressions in the ground that trapped rainwater. The tortoises drank large quantities of water (nearly 20 percent of their body mass) and voided the contents of their urinary bladders. The osmotic concentrations of their body fluids and urine decreased as they moved into positive water balance and excreted the excess salts they had accumulated when water was scarce. The behavior of the tortoises changed after the rain: They fed every 2 or 3 days and often spent their periods of inactivity above ground instead of in their burrows.

August was dry, and the tortoises lost body water and accumulated salts as they fed on dry grass. They were in positive energy balance, however, and their body tissue mass increased. More thunderstorms in September allowed the tortoises to drink again and to excrete the excess salts they had been accumulating. Seedlings sprouted after the rain, and in late September the tortoises started to eat them.

In October and November the tortoises continued to feed on freshly sprouted green vegetation, but low temperatures reduced their activity and they were in slightly negative energy balance. Salts accumulated and the osmotic concentrations of the body fluids increased slightly. In November the tortoises entered hibernation. Hibernating tortoises had low metabolic rates and lost water and body tissue mass slowly. When they emerged from hibernation the following spring they weighed only a little less than they had in the fall. Over the entire year, the tortoises increased their body tissues by more than 25 percent, and balanced their water and salt budgets, but they did this by tolerating severe imbalances in their energy, water, and salt relations for periods that extended for several months at a time.

The ability to tolerate physiological imbalances is an important aspect of the ability of ectothermal vertebrates to occupy habitats where seasonal shortages of food or water occur. The chuckwalla (*Sauromalus obesus*) is an herbivorous iguanid lizard that lives in the rocky foothills of desert mountain ranges (Figure 16–9). The annual cycle of the chuckwallas, like that of the desert tortoises, is molded by the availability of water. The lizards face many of the same stresses that the tortoises encounter, but their responses are different: The lizards have nasal glands that allow them to excrete salt at high concentrations, and they do not drink rainwater but instead depend on water they obtain from the plants they eat.

Two categories of water are available to an animal from the food it eats—free water and metabolic water. Free water corresponds to the water content of the food, that is, molecules of water (H_2O) that are absorbed across the wall of the intestine. Metabolic water is a by-product of the cellular reactions of metabolism. Protons are combined with oxygen during aerobic metabolism, yielding a molecule of water for every two protons. The amount of metabolic water produced can be substantial; more than a gram of water is released by metabolism of a gram of fat (Table 16–2). For animals like the chuckwalla that do not drink liquid water, free water and metabolic water are the only routes of water gain that can replace the water lost by evaporation and excretion.

Figure 16–9. Chuckwalla, *Sauromalus obesus.* (Photograph by F. Harvey Pough.)

Table 16–2. Quantity of water produced by metabolism of different substrates.

Compound	Grams of Water/Gram of Compound
Carbohydrate	0.556
Fat	1.071
Protein	0.396 when urea is the end product
	0.499 when uric acid is the end product

Chuckwallas were studied at Black Mountain in the Mojave Desert of California (Nagy 1972, 1973). They spent the winter hibernating in rock crevices and emerged from hibernation in April. Individual lizards spent about 8 hours a day on the surface in April and early May (Figure 16–10). By the middle of May air temperatures were rising above 40°C and the chuckwallas retreated into rock crevices for about 2 hours during the hottest part of the day, emerging again in the afternoon. At this time of year annual plants that sprouted after the winter rains supplied both water and nourishment, and the chuckwallas gained weight rapidly. The average increase in body mass between April and mid-May was 18 percent (Figure 16–11). The water content of the chuckwallas increased faster than the total body mass, indicating that they were storing excess water.

By early June the annual plants had withered, and the chuckwallas were feeding on perennial plants that contained less water and more ions than the annual plants. Both the body masses and the water contents of the lizards declined. The activity of the lizards decreased in June and July: Individual lizards emerged in the morning or in the afternoon, but not at both times. In late June the chuckwallas reduced their feeding activity and in July they stopped eating altogether. They spent most of the day in the rock crevices, emerging only in the late afternoon to bask for an hour or so every second or third day. From late May through autumn the chuckwallas lost water and body mass steadily, and in October they weighed an average

of 37 percent *less* than they had in April when they emerged from hibernation.

The water budget of a chuckwalla weighing 200 grams is shown in Table 16–3. In early May the annual plants it is eating contain more than 2.5 grams of free water per gram of dry plant material, and the lizard shows a positive water balance, gaining about 0.8 gram of water per day. By late May, when the plants have withered, their free water content has dropped to just under a gram of water per gram of dry plant matter, and the chuckwalla is losing about 0.8 gram of water per day. The rate of water loss falls to 0.5 gram per day when the lizard stops eating.

Evaporation from the respiratory surfaces and from the skin accounts for about 61 percent of the total water loss of a chuckwalla. When the lizards stop eating, they also become inactive and spend most of the day in rock crevices. The body temperatures of inactive chuckwallas are lower than the temperatures of lizards on the surface. As a result of their low body temperatures, the inactive chuckwallas have lower rates of metabolism: They breathe more slowly and lose less water from their respiratory passages. Also, the humidity is higher in the rock crevices than it is on the surface of the desert, and this reduction in the humidity gradient between the animal and the air further reduces evaporation. Most of the remaining water loss by a chuckwalla occurs in the feces (31 percent) and urine (8 percent). When a lizard stops eating, it also stops producing feces and reduces the amount of urine it must excrete. The combination of these effects reduces the daily water loss of a chuckwalla by almost 90 percent.

The food plants were always hyperosmotic to the body fluids of the lizards and had high concentrations of potassium. Despite this dietary salt load, the osmotic concentrations of the body fluids of the chuckwallas did not show the variation seen in tortoises because the lizards' nasal salt glands were able to excrete ions at high concentrations. The concentration of potassium ions in the salt gland secretions was nearly ten times their concentration in urine. The formation of potassium

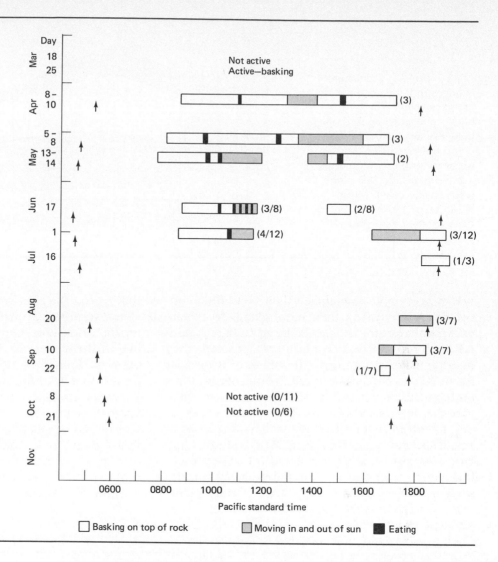

Figure 16–10. Daily behavior patterns in chuckwallas through their activity season. Arrows indicate sunrise and sunset. Numbers in parentheses for April and May indicate the number of animals whose behavior was recorded. Thereafter the fraction in parentheses indicates the number of lizards active and observed out of the number known to be present. (From K. A Nagy, 1973, *Copeia* 1973:93–102.)

salts of uric acid was the second major route of potassium excretion by the lizards, and was nearly as important in the overall salt balance as nasal secretion. The chuckwallas would not have been able to balance their salt budgets without the two extrarenal routes of ion excretion, but with them they were able to maintain stable osmotic concentrations.

Both the chuckwallas and tortoises illustrate

the interaction of behavior and physiology in responding to the stresses of their desert habitats. The tortoises lack salt-secreting glands and store the salt they ingest, tolerating increased body fluid concentrations until a rainstorm allows them to drink water and excrete the excess salt. Some of the tortoises constructed basins that collected rainwater that they drank, whereas other tortoises took advantage of natural puddles. The chuck-

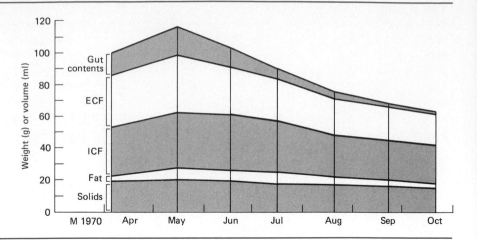

Figure 16–11. Seasonal changes in body composition of chuckwallas. The total body water is composed of the extracellular fluid (ECF: blood plasma, urine, and water in lymph sacs) and the intracellular fluid (ICF; the water inside cells). (From K. A. Nagy, 1972, *Journal of Comparative Physiology* 79:39–62.)

wallas were able to stabilize their body fluid concentrations by using their nasal glands to excrete excess salt, but they did not take advantage of rainfall to replenish their water stores. Instead, they became inactive, reducing their rates of water loss by almost 90 percent, and relying on energy stores and metabolic water production to see them through the period of drought.

Conditions for the chuckwallas were poor at the Black Mountain site during Nagy's study. Only 5 centimeters of rain had fallen during the preceding winter, and that low rainfall probably contributed to the early withering of the annual plants

that forced the chuckwallas to cease activity early in the summer. Unpredictable rainfall is a characteristic of deserts, however, and the animals that live in them must be able to adjust to the consequences. Rainfall records from the weather station closest to Black Mountain showed that in 5 of the previous 10 years the annual total rainfall was about 5 centimeters. Thus, the year of the study was not unusually harsh, and conditions are sometimes even worse—only 2 centimeters of rain fell during the winter after the study. However, conditions in the desert are sometimes good. Fifteen centimeters of rain fell in the winter of 1968, and

Table 16–3. Seasonal changes in the water balance of a 200-gram chuckwalla.

	Early May	*Late May*	*September*
Food Intake (g dry mass/day)	2.60	2.86	0.00
Water content of food (g/g dry mass)	2.53	0.96	—
Water gain (g/day)			
Free water	6.56	2.74	0.0
Metabolic water	0.68	0.68	0.20
Total water gain	7.24	3.41	0.20
Water loss (g/day)	6.41	4.26	0.52
Net water flux (g/day)	+0.81	−0.84	−0.32

Source: K. A. Nagy, 1972, *Journal of Comparative Physiology* 79:39–62.

vegetation remained green and lush all through the following summer and fall. Chuckwallas and tortoises may live for 25 or even 50 years, and their responses to the boom or bust conditions of their harsh environments must be viewed in the context of their long life spans. A temporary relaxation of the limits of homeostasis in bad years is an effective trade-off for survival that allows the animals to exploit the abundant resources of good years.

Desert Amphibians

Permeable skins and high rates of water loss are characteristics that would seem to make amphibians unlikely inhabitants of deserts, but certain species are abundant in desert habitats. Most remarkably, these animals succeed in living in the desert *because* of their permeable skins, not despite them. Anurans are the most common desert amphibians, but tiger salamanders are found in the deserts of North America and several species of plethodontid salamanders occupy seasonally dry habitats in California.

The spadefoot toads (family Pelobatidae) are the most thoroughly studied desert anurans (Figure 16-12). They inhabit the desert regions of North America, including the edges of the Algodones Sand Dunes in southern California, where the average annual precipitation is only 6 centimeters and in some years no rain falls at all. An analysis of the mechanisms that allow an amphibian to exist in a habitat like that must include consideration of both water loss and gain. The skin of desert amphibians is as permeable to water as that of species from moist regions. A desert anuran must control its water loss behaviorally by its choice of sheltered microhabitats free from solar radiation and wind movement. Different species of anurans utilize different microhabitats—a hollow in the bank of a desert wash, the burrow of a ground squirrel or kangaroo rat, or a burrow the anuran excavates for itself. All these places are cooler and wetter than exposed ground.

Desert anurans spend extended periods underground, emerging on the surface only when conditions are favorable. Spadefoot toads construct burrows about 60 centimeters deep, filling the shaft with dirt and leaving a small chamber at the bottom which they occupy. In southern Ari-

Figure 16–12. A desert spadefoot toad, *Scaphiopus multiplicatus.* (Photograph by David Dennis.)

zona the spadefoots construct these burrows in September, at the end of the summer rainy season, and remain in them until the rains resume the following July.

At the end of the rainy season when the toads first bury themselves, the soil is relatively moist. The water tension created by the normal osmotic pressure of a toad's body fluids establishes a gradient favoring movement of water from the soil into the toad. In this situation, a buried spadefoot can absorb water from the soil just as the roots of plants do. With a supply of water available, a toad can afford to release urine to dispose of its nitrogenous wastes.

As time passes, the soil moisture content decreases and the soil moisture potential becomes more negative, until it equals the water potential of the toad. At this point there is no longer a gradient allowing movement of water into the toad. When its source of new water is cut off, a toad stops excreting urine and instead retains urea in its body, increasing the osmotic pressure of its body fluids. Osmotic concentrations as high as 600 milliOsmolal have been recorded in spadefoot toads emerging from burial at the end of the dry season. The low water potential produced by the high osmotic pressure of the toad's body fluids may reduce the water gradient between the animal and the soil so that evaporative water loss is reduced. Sufficiently high internal osmotic pressures should create potentials that would allow toads to absorb water from even very dry soil.

The ability to continue to draw water from soil enables a toad to remain buried for 9 or 10 months without access to liquid water. In this situation its permeable skin is not a handicap—it is an essential feature of a spadefoot's biology. If the toad had an impermeable skin or if it formed an impermeable cocoon as some other amphibians do, water would not be able to move from the soil into the animal. Instead, the toad would have to depend on the water contained in its body when it buried. Under those circumstances spadefoot toads would probably not be able to invade the desert because their initial water content would not see them through a 9-month dry season.

A different pattern of adaptation to arid conditions is seen in a few treefrogs. The African rhacophorid *Chiromantis xerampelina* and the South American hylid *Phyllomedusa sauvagei* lose water through the skin at a rate only one-tenth that of most frogs. *Phyllomedusa* has been shown to achieve this low rate of evaporative water loss by using its legs to spread the lipid-containing secretions of dermal glands over its body surface in a complex sequence of wiping movements. These two frogs are unusual also because they excrete nitrogenous wastes as salts of uric acid rather than as urea. This uricotelism provides still more water conservation.

In view of the wide range of reproductive modes observed in anurans, it is surprising to find that species that inhabit arid regions, where the lack of water can prevent any breeding in some years, are not live-bearers, nor do they lay eggs with direct development. Instead, they adhere to the pattern of aquatic eggs that hatch into tadpoles. The modifications of the process that make them successful in desert regions involve the timing of reproduction and the rate of embryonic and larval development.

Deserts are characterized by the scantiness and unpredictability of their rainfall. Generally, there is a season when the probability of rain is higher than at other times of year, but there is no way to know within a month when rain will fall on a particular area of desert or even if any rain will fall in a given year. Because of this unpredictability, desert anurans must be ready to breed as rapidly as possible after a rain; they depend upon the temporary pools formed by rainwater to lay their eggs. The tadpoles must metamorphose and leave the pools before they dry out.

In southern Arizona, the summer rainy season extends from July through September. As July approaches, spadefoot toads have fully developed eggs and sperm. They move from the bottoms of their burrows to within a few centimeters of the surface in the evening and may rest with their eyes protruding above the ground. If it does not rain—and most days it does not—the spadefoot moves down into its burrow in the morning to escape the heat of the day. If it does rain, the spadefoots emerge almost immediately and begin to move about on the desert floor, feeding on insects. Even a very light shower initiates emergence—less than 0.5 millimeter of rain brings the animals onto the surface. Low-frequency sound—perhaps from thunder or the impact of raindrops on the ground—is apparently the stimulus for emergence.

The first spadefoots arrive at the breeding ponds the same day the rain falls, and the first clutches of eggs may be laid that night. Amphibian eggs are damaged by high temperatures. The eggs of desert species tolerate higher temperatures than those of species from moist regions, and they also develop faster. Rate of development is an important factor because the eggs become increasingly resistant to damage from high temperatures as development progresses. Because they develop rapidly, the eggs of desert anurans pass through the most temperature-sensitive stages at night, and by the time the sun rises the following morning and the water in the pool starts to get hot, the eggs are well on their way to hatching.

The rapid development of desert anuran eggs continues through hatching and metamorphosis. Spadefoot toads develop to metamorphosis in 2 or 3 weeks, compared with 10 weeks or more for species that breed in permanent ponds. Despite their fast development, groups of spadefoot tadpoles may be trapped in rapidly shrinking pools. Under crowded conditions of this sort, individual spadefoot tadpoles of some species are likely to turn into cannibals. Their horny jaws hypertrophy, and they switch from a diet of algae to killing and eating other tadpoles. The greater energy they derive from a carnivorous diet speeds their development and increases their chances of metamorphosing in time. It seems probable that the switch from herbivory to cannibalism reflects a genetic predisposition that is activated by a particular set of environmental conditions, but what those may be is not known.

None of the characteristics of desert anurans is unique. Instead, they are the maximum expression of trends seen in species from more humid regions. The toads (*Bufo*) are an example. Toads occur in terrestrial habitats ranging from wet tropical lowlands through moist temperate and subarctic regions to deserts. In all these habitats, toad eggs and larvae develop rapidly and hatch at an earlier embryonic stage than those of most other anurans.

Thus, the reproductive biology of toads in general is well suited to conditions in which rapid larval development is advantageous. It may be this characteristic of the toad lineage that allowed toads to inhabit deserts in the first place.

Ectotherms in the Cold: Supercooling, Antifreeze, and Freeze Tolerance

Temperatures drop below freezing in the habitats of many ectothermal vertebrates on a seasonal basis, and some animals at high altitudes may experience freezing temperatures on a daily basis for a substantial part of the year. Ectotherms lack the metabolic capacities to maintain body temperatures above freezing under those conditions and instead they show one of two responses—freezing avoidance (by supercooling or the use of antifreeze compounds) or tolerance of freezing and thawing.

Cold Fish

Body fluid concentrations of marine fishes are 300 to 400 milliOsmolal, whereas seawater has a concentration near 1000 milliOsmolal. The temperature at which water freezes is affected by its osmotic concentration: Pure water (0 milliOsmolal) freezes at 0°C and increasing the osmotic concentration lowers the freezing point. The osmotic concentrations of the body fluids of marine fishes correspond to freezing points of −0.6 to −0.8°C, and the freezing point of seawater is −1.86°C. The temperature of Arctic and Antarctic seas falls to −1.8°C in winter, yet the fish swim in this water without freezing.

One of the early studies of freezing avoidance

of fishes was conducted by P. F. Scholander and his colleagues in Hebron Fjord in Labrador (Scholander et al. 1957). In summer the temperature of the surface water at Hebron Fjord is above freezing, but the water at the bottom of the fjord is −1.73°C (Figure 16–13). In winter the temperature of the surface water falls to −1.73°C, like the bottom temperature. Several species of fishes live in the fjord, and some are bottom-dwellers, whereas others live near the surface. These two zones present different problems to the fishes: The temperature near the bottom of the fjord is always

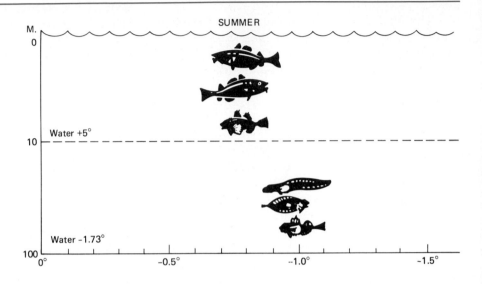

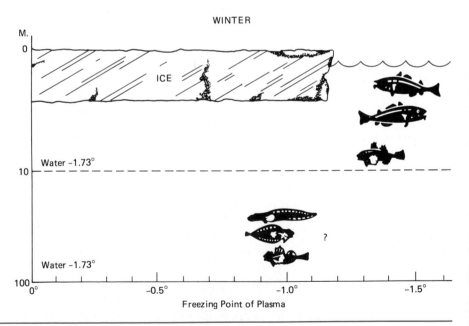

Figure 16–13. Water temperatures and distribution of fishes in Hebron Sound in summer and winter. The freezing point of the blood plasma of the fishes is indicated by the position of the symbol on the horizontal axis. Shallow-water fishes show a decrease of 0.7°C in the freezing point of their blood in winter, whereas deep-water fishes have the same freezing point all year. (From P. F. Scholander et al., 1957, *Journal of Cellular and Comparative Physiology* 79:39–62.)

below freezing, but ice is not present because ice is lighter than water and remains at the surface. Surface-dwelling fishes live in water temperatures that rise well above freezing in the summer and drop below freezing in winter, and they are also in the presence of ice.

The body fluids of bottom-dwelling fish in Hebron Fjord have freezing points of $-0.8°C$ year round. Because the body temperatures of these fishes are $-1.73°C$, the fish are supercooled. That is, the water in their bodies is in the liquid state despite the fact that it is below its freezing point. When water freezes, the water molecules become oriented in a crystal lattice. The process of crystallization is accelerated by nucleating agents that hold water molecules in the proper spatial orientation for freezing, and in the absence of nucleating agents pure water can remain liquid at $-20°C$. In the laboratory the fishes from the bottom of Hebron Fjord could be supercooled to $-1.73°C$ without freezing, but if they were touched with a piece of ice, which served as a nucleating agent, they froze immediately. At the bottom of the fjord there is no ice, and apparently the bottom-dwelling fishes exist year round in a supercooled state.

What about the fishes in the surface waters? They do encounter ice in winter when the water temperature is below the osmotically determined freezing point of their body fluids, and supercooling would not be effective in that situation. Instead, the surface-dwelling fishes synthesize antifreeze substances in winter that lower the freezing point of their body fluids to approximately the freezing point of the seawater in which they swim.

Antifreeze compounds are widely developed among vertebrates (and also among invertebrates and plants). Marine fishes have two categories of organic molecules that protect against freezing—glycoproteins with molecular weights of 2600 to 33,000 daltons, and polypeptides and small proteins, with molecular weights of 3300 to 13,000 daltons (Hew et al. 1986). These compounds are extremely effective in preventing freezing. For example, the blood plasma of the Antarctic fish *Trematomus borchgrevinki* contains a glycoprotein that is several hundred times more effective than

salt (sodium chloride) in lowering the freezing point. Apparently, the glycoprotein is absorbed on the surface of ice crystals and hinders their growth by preventing water molecules from assuming the proper orientation to join the ice crystal lattice.

Some terrestrial ectotherms also apparently rely on supercooling. Montane lizards such as Yarrow's spiny lizard (*Sceloporus jarrovi*), which lives at altitudes up to 3000 meters in western North America, are exposed to temperatures below freezing on cold nights, but sunny days permit thermoregulation and activity. These animals have osmotic concentrations of 300 milliOsmolal, which correspond to freezing points of $-0.6°C$, but they withstand substantially lower temperatures before they freeze. For example, spiny lizards supercooled to an average temperature of $-5.5°C$ before they froze. At $-3°C$ the lizards had not frozen after 30 hours. The lizards spent the nights in rock crevices that were 5 to $6°C$ warmer than the air temperature, and the combination of this protection and their ability to supercool was usually sufficient to allow the lizards to survive. However, a few individuals at the highest altitudes were found frozen in their rock crevices during most winters.

Frozen Frogs

Terrestrial amphibians, which spend the winter buried in the soil, appear to show at least two categories of responses to low temperatures. One group, which includes salamanders, toads, and aquatic frogs, buries deeply in the soil or hibernates in the mud at the bottom of ponds. These animals apparently are not exposed to temperatures below the freezing point of their body fluids, and as far as we know, they have no antifreeze substances and no capacity to tolerate freezing. However, other amphibians apparently hibernate close to the soil surface, and these animals are exposed to temperatures below their freezing points. Unlike fishes and lizards, these amphibians freeze at low temperatures, but they are not killed by freezing. Tolerance of freezing has so far been demonstrated in four species of frogs—a ranid, the wood frog (*Rana sylvatica*, Figure 16–14), and three

(a)

(b)

Figure 16–14. Wood frog (*Rana sylvatica*): (a) at normal temperature; (b) frozen. (Photographs courtesy of J. and K. Storey.)

hylids, the spring peeper (*Hyla crucifer*), the gray treefrog (*Hyla versicolor*), and the chorus frog (*Pseudacris triseriata*). These species can remain frozen at −3°C for several weeks, and they tolerate repeated bouts of freezing and thawing without damage. However, temperatures below −10°C are lethal.

Tolerance of freezing refers to the formation of ice crystals in the extracellular body fluids; freezing of the fluids inside the cells is apparently lethal. Thus, freeze tolerance involves mechanisms that control the distribution of ice, water, and solutes in the bodies of animals (Storey 1986). The ice content of frozen frogs is usually in the range 34 to 48 percent. Freezing of more than 65 percent of the body water appears to cause irreversible damage, probably because too much water has been removed from the cells.

Freeze-tolerant frogs accumulate low-molecular-weight substances in the cells that prevent intracellular ice formation. Wood frogs, spring peepers, and chorus frogs use glucose as a cryoprotectant, whereas gray treefrogs use glycerol. Glycogen in the liver appears to be the source of the glucose and glycerol. The accumulation of these substances is apparently stimulated by freezing and is initiated within minutes of the formation of ice crystals. This mechanism of triggering the synthesis of cryoprotectant substances has not been observed in any other vertebrates, or in insects.

Frozen frogs are, of course, motionless. Breathing stops, the heartbeat is exceedingly slow and irregular or may cease entirely, and blood does not circulate through frozen tissues. Nonetheless, the cells are not frozen and they have a low level of metabolic activity that is maintained by anaerobic metabolism. The glycogen content of frozen muscle and kidney cells decreased, and concentrations of two end products of anaerobic metabolism, lactic acid and alanine, increased.

The ecological significance of freeze tolerance in some species of amphibians is unclear. The four species of frogs so far identified as being freeze tolerant all breed relatively early in the spring, and shallow hibernation may be associated with early emergence in the spring. Being among the first individuals to arrive at the breeding pond may increase the chances for a male of obtaining a mate and it gives larvae as long a time as possible for development and metamorphosis, but it also entails risks. Frequently, frogs and salamanders move across snowbanks to reach the breeding ponds, and they enter ponds that are still partly covered with ice. A cold snap can lead to the entire surface of the pond freezing again, trapping some animals under the ice and others in shallow retreats under logs and rocks around the pond. Freeze tolerance may be important to these animals even after their winter hibernation is over.

The Role of Ectothermal Tetrapods in Terrestrial Ecosystems

In the previous sections we have illustrated some of the ways that an ectothermal approach to life allows animals to exploit habitats or adaptive zones that would be difficult or impossible for an endotherm to occupy. The key to this ability of ectotherms is the low energy requirement of ectothermy; an ectotherm simply requires less energy on an hourly, daily, or annual basis than does an endotherm of the same body size.

However, the amount of energy required by an ectotherm and an endotherm is not the only difference between them: Equally significant is what they do with that energy once they have it. Endotherms expend more than 90 percent of the energy they take in to produce heat to maintain their high body temperatures. Less than 10 percent, often as little as 1 percent, of the energy they assimilate is available for net conversion (that is, increasing their species' biomass by growth of an individual or production of young). Ectotherms do not rely on metabolic heat. The solar energy they use to warm their bodies is free in the sense that it is not drawn from their food. Thus, most of the energy they ingest is converted into the biomass of their species. Values of net conversion for amphibians and squamates are between 30 and 90 percent (Table 16–4). As a result of that difference in how energy is used, a given amount of chemical energy invested in an ectotherm produces a much larger biomass return than it would have from an endotherm. A study of salamanders in the Hubbard Brook Experimental Forest in New Hampshire showed that, although their energy con-

sumption was only 20 percent that of the birds or small mammals in the watershed, their conversion efficiency was so great that the annual increment of salamander biomass was equal to that of birds or small mammals. Similar comparisons can be made among lizards and rodents in deserts. [For details, see Pough (1980, 1983).]

Body size is another major difference between ectotherms and endotherms that directly affects their roles in terrestrial ecosystems. Ectotherms are smaller than endotherms, partly because the energetic cost of endothermy is very high at small body sizes. As body mass decreases the mass specific cost of living (energy per gram) for an endotherm increases rapidly, becoming nearly infinite at very small body sizes. This is a finite world, and infinite energy requirements are just not feasible. Thus, energy requirements, among other factors, apparently set a lower limit to the body size that is possible for an endotherm. The mass specific energy requirements of ectotherms also increase at small body sizes, but because the energy requirements of ectotherms are about one-tenth those of endotherms of the same body size, an ectotherm can be about an order of magnitude smaller than an endotherm. A mouse-size mammal weighs about 20 grams, and few adult birds and mammals have body masses less than 10 grams. The very smallest species of birds and mammals weigh about 3 grams, but many ectotherms are only one-tenth that size (0.3 gram). Amphibians are especially small—20 percent of the species of salamanders and 17 percent of the species of anurans have adult body masses less than 1 gram, and 65 percent of salamanders and 50 percent of anurans are smaller than 5 grams. Squamates are generally larger than amphibians, but 8 percent of the species of lizards and 2 percent of snakes weigh less than 1 gram.

Small amphibians and squamates occupy a key position in terms of energy flow through an ecosystem: Because they are so small, they can capture tiny insects and arachnids that are too small to be eaten by birds and mammals. Because they are ectotherms, they are efficient at converting the energy in the food they eat into their own

Table 16–4. Efficiency of biomass conversion by ectotherms and endotherms. These are net conversion efficiencies calculated as (energy converted/energy assimilated) × 100.

Ectotherms		Endotherms	
Species	*Efficiency*	*Species*	*Efficiency*
Red-backed salamander *Plethodon cinereus*	48	Kangaroo rat *Dipodomys merriami*	0.8
Mountain salamander *Desmognathus ochrophaeus*	76–98	Field mouse *Peromyscus polionotus*	1.8
Panamanian anole *Anolis limifrons*	23–28	Meadow vole *Microtus pennsylvanicus*	3.0
Side-blotched lizard *Uta stansburiana*	18–25	Red squirrel *Tamiasciurius hudsonicus*	1.3
Hognose snake *Heterodon contortrix*	81	Least weasel *Mustela rixosa*	2.3
Python *Python curtus*	6–33	Savanna sparrow *Passericulus sandwichensis*	1.1
Adder *Vipera berus*	49	Marsh wren *Telmatodytes palustris*	0.5
Average of 12 species	50	Average of 19 species	1.4

Source: F. H. Pough, 1980, *The American Naturalist* 115:92–112.

tissues. As a result, the small ectothermal vertebrates in terrestrial ecosystems can be viewed as repackaging energy into a form that avian and mammalian predators can exploit. In other words, if you put yourself in the position of a shrew or a bird searching for a meal in the Hubbard Brook Forest, you would be well advised to eat salamanders. In this context, frogs, salamanders, lizards, and snakes occupy a position in terrestrial ecosystems that is important both quantitatively (in the sense that they constitute a substantial energy resource) and qualitatively (in that endotherms are not able to exploit the food resources used by the ectotherms).

In a very real sense small ectotherms can be thought of as living in a different world from that of endotherms. As we saw in the case of the three species of *Ameiva* lizards in Costa Rica (Chapter 15), interactions with the physical world may be more important in shaping the ecology and behavior of small ectotherms than biological interactions such as competition. In some cases these small vertebrates may have their primary predatory and competitive interactions with insects and arachnids rather than with other vertebrates. For example, orb-web spiders and *Anolis* lizards on some Caribbean islands are linked by both predation (adult lizards eat spiders and spiders may eat hatchling lizards) and competition (lizards and spiders eat many of the same kinds of insects). When lizards were removed from experimental plots the abundance of insect prey increased and the spiders consumed more prey and survived longer (Schoener and Spiller 1987).

Thus, ectothermy and endothermy represent fundamentally different approaches to the life of a terrestrial vertebrate, each with its own advantages and disadvantages. An appreciation of ectotherms and endotherms requires understanding the functional consequences of the differences between them. Ectothermy is primitive in the evolutionary sense of being an ancestral character of vertebrates, but it is also a very effective way of life in modern ecosystems.

Summary

Ectotherms do not use chemical energy from the food they eat to maintain high body temperatures. The results of that primitive vertebrate characteristic are far-reaching for modern ectothermal vertebrates, and ectotherms and endotherms represent quite different approaches to vertebrate life.

Because of their low energy requirements, ectotherms can colonize habitats in which energy is in short supply, either seasonally (deserts) or chronically (the deep sea). Ectotherms are able to extend some of their limits of homeostasis to tolerate high or low body temperatures and high or low body water contents when doing so allows them to survive in difficult conditions.

When food is available, ectotherms are efficient at converting the energy it contains into their own body tissues for growth or reproduction. Net conversion efficiencies of ectotherms average 50 percent of the energy assimilated compared to an average of 1.4 percent for endotherms.

Ectotherms can be smaller than endotherms because their mass-specific energy requirements are low, and many ectotherms weigh less than a gram, whereas most endotherms weigh more than 10 grams. As a result of this difference in body size many small ectotherms, such as salamanders, frogs, and lizards, eat prey that is too small to be consumed by endotherms. The efficiency of energy conversion by ectotherms and their small body sizes lead to a distinctive role in modern ecosystems, one that is in many respects quite different from that of terrestrial ectotherms. Understanding these differences is an important part of understanding the organismal biology of terrestrial ectothermal vertebrates.

References

Hew, C. L., G. K. Scott, and P. L. Davies. 1986. Molecular biology of antifreeze. Pages 117–123 in *Living in the Cold: Physiological and Biochemical Adaptations*, edited by H. C. Heller, X. J. Musacchia, and L. C. H. Wang. Elsevier, New York.

Nagy, K. A. 1972. Water and electrolyte budgets of a free-living desert lizard, *Sauromalus obesus*. *Journal of Comparative Physiology* 79:39–62.

Nagy, K. A. 1973. Behavior, diet and reproduction in a desert lizard, *Sauromalus obesus*. *Copeia* 1973:93–102.

Nagy, K. A. 1983. The doubly labeled water ($^3HH^{18}O$) method: a guide to its use. *UCLA Publication 12-1417*. University of California, Los Angeles.

Nagy, K. A. and P. A. Medica. 1986. Physiological ecology of desert tortoises in southern Nevada. *Herpetologica* 42:73–92.

Pough, F. H. 1980. The advantages of ectothermy for tetrapods. *The American Naturalist* 115:92–112.

Pough, F. H. 1983. Amphibians and reptiles as low-energy systems. Pages 141–188 in *Behavioral Energetics: Vertebrate Costs of Survival*, edited by W. P. Aspey and S. I. Lustick. Ohio State University Press, Columbus, Ohio.

Schoener, T. W. and D. A. Spiller. 1987. Effect of lizards on spider populations: manipulative reconstruction of a natural experiment. *Science* 236:949–952.

Scholander, P. F., L. Van Dam, J. W. Kanwisher, H. T. Hammel, and M. S. Gordon. 1957. Supercooling and osmoregulation in Arctic fish. *Journal of Cellular and Comparative Physiology* 49:5–24.

Storey, K. B. 1986. Freeze tolerance in vertebrates: Biochemical adaptation of terrestrially hibernating frogs. Pages 131–138 in *Living in the Cold: Physiological and Biochemical Adaptations*, edited by H. C. Heller, X. J. Musacchia, and L. C. H. Wang. Elsevier, New York.

PART FOUR

Terrestrial Endotherms: Birds and Mammals

Birds and mammals are the vertebrates with which people are most familiar, partly because many species are large and diurnal, and partly because birds and mammals have colonized nearly every terrestrial habitat on Earth. The success of birds and mammals in many habitats is related to their endothermy, which allows them to be active at night and in cold weather. Those are conditions in which terrestrial ectotherms find it difficult or impossible to thermoregulate and, consequently, are inactive.

Flight dominates the biology of birds: Most of the morphological features of birds are directly or indirectly related to the requirements of flight, and many of the distinctive aspects of their behavior and ecology stem from the mobility that flight provides. Migration, for example, is a particularly avian characteristic because of all the terrestrial vertebrates, birds are best able to move long distances.

No one feature of mammals characterizes the group and dominates its biology as flight does for birds, but sociality comes close. Many features of the biology of mammals are related to their interactions with other individuals of their species, ranging from the period of dependence of young on their mother to life-long alliances between individuals that affect their social status and reproductive success in a group.

Humans differ from other vertebrates in the extent to which they have come to dominate all of the habitats of Earth, and in their effect, directly and indirectly, on other vertebrates. Unfortunately, that effect is almost always deleterious. Human population growth and its consequent demands for more space and more resources is wiping other species from the face of the Earth. The current crisis in vertebrate evolution is directly attributable to us, and only we can change it. Unfortunately, remedial efforts have generally been too little and too late. In this portion of the book we explore the evolution of birds and mammals, the adaptive zones opened to them by their distinctive characteristics, and the origin of humans and the consequences of human life for the lives of other vertebrates.

The linear progression of a book does not lend itself to following the simultaneous evolution of several different phylogenetic lineages. At this point we must turn our attention back to the conditions in the middle of the Mesozoic that were described in Chapter 14 and consider the diversification of another group of diapsids, the birds, from their origin in or before the Jurassic. *Archaeopteryx* is the oldest fossil of a bird that is known, but by the time it lived in the Jurassic it was probably an archaic relic that existed contemporaneously with more derived birds. The debate about the role of flight in the origin of birds was described in Chapter 13. Despite our uncertainty about the route by which birds took to the air, the demands of flight have clearly shaped many aspects of the morphology of birds. In this chapter we consider the relationships between the body forms of birds and the physical and biological requirements of flight.

Characteristics of Birds: Specializations for Flight

17

Birds as Flying Machines

In some respects birds are variable: Beaks and feet are specialized for different modes of feeding and locomotion, the morphology of the intestinal tract is related to dietary habits, and wing shape reflects flight characteristics. Despite that variation, however, the morphology of birds is more uniform than that of mammals, and much of this uniformity is a result of the specialization of birds for flight (King and King 1979).

Consider body size as an example: Flight imposes a maximum body size on birds. The muscle power required to take off increases by a factor of 2.25 for each doubling of body mass. That is, if species B weighs twice as much as species A, it will require 2.25 times as much power to fly at its minimum speed. If the proportion of the total body mass allocated to flight muscles is constant, the muscles of a large bird must work harder than the muscles of a small bird. In fact, the situation is still more complicated because the power output is a function of both muscular force and wingbeat frequency. Large birds have lower wingbeat frequencies than small birds for mechanical and aerodynamic reasons. As a result, if species B weighs twice as much as species A, it will develop only 1.59 times as much power from its flight muscles, although it needs 2.25 times as much power to fly. Therefore, large birds require longer takeoff runs than small birds, and ultimately a body mass is reached at which any further increase in size would move a bird into a realm in which its flight muscles were not able to provide enough power to take off.

Calculations of this maximum size from aerodynamic principles suggest that it lies near 12 kilograms and that estimate corresponds reasonably well with the observed body masses of birds. The mute swan weighs about 12 kilograms, and the trumpeter swan is about 17 kilograms. The largest volant bird known was a giant condor that had a wingspan estimated to be 7 meters and a possible body mass of 20 kilograms. Large pterosaurs also had body masses estimated to be in the region of 20 kilograms.

Flightless birds are spared the mechanical constraints associated with producing power for flight, but they still do not approach the body sizes of mammals. The largest living bird is the flightless ostrich, which weighs about 150 kilograms, and the largest bird known, one of the extinct elephantbirds, weighed an estimated 450 kilograms. In contrast, the largest mammal, the blue whale, weighs more than 135,000 kilograms. If one restricts the comparison to quadrupedal, terrestrial mammals, the elephant weighs some 5000 kilograms.

The structural uniformity of birds is seen even more clearly if their body shapes are compared to those of other diapsids such as the dinosaurs. Even those species of birds that have become secondarily flightless retain many ancestral characters and the constraints that are associated with them. In this chapter we consider the body form and function of birds, especially in relation to the requirements of flight.

Feathers and Flight

Feathers develop from pits or follicles in the skin, generally arranged in tracts or **pterylae**, which are separated by patches of unfeathered skin, the **apteria** (Figure 17–1). In some species, such as ratites, penguins, and mousebirds, pterylae are absent and the feathers are uniformly distributed over the skin.

For all their structural complexity, feathers are remarkably simple and uniform in chemical composition. More than 90 percent of the substance of a feather consists of beta-keratin, a protein related to the keratin of scales and to the hair and horn of mammals. About 1 percent of a feather consists of lipids, about 8 percent is water, and the remaining fraction consists of small amounts of other proteins and pigments, such as melanin.

Basic Types of Feathers

Ornithologists usually distinguish five types of feathers: (1) **contour feathers**, including typical body feathers and the flight feathers (**remiges** and **rectrices**); (2) **semiplumes**; (3) **down feathers** of several sorts; (4) **bristles**; and (5) **filoplumes**. Contour feathers (Figure 17–2) include a short, tubular base, the **calamus**, which remains firmly implanted within the follicle until molt occurs. Distal to the calamus there is a long, tapered **rachis**, which bears closely spaced side branches called **barbs**, the lowermost of which externally mark the division between calamus and rachis. The barbs on either side of the rachis constitute a surface called a **vane**. Vanes may be symmetrical or asymmetrical. The proximal portions of the vanes of a feather have a downy or plumulaceous texture, being soft, loose, and fluffy. This gives the plumage of a bird its excellent properties of thermal insulation. The more distal portions of the vanes have a pennaceous or sheet-like texture, firm, compact, and closely knit. This exposed part provides an airfoil, protects the downy undercoat, sheds water, reflects or absorbs solar radiation, and may have a role in visual or auditory communication. The barbules are structures that maintain the pennaceous character of the feather vanes. They are arranged in such a way that any physical disruption to the vane is easily corrected by the bird's preening behavior, in which the bird realigns the barbules by drawing its slightly separated bill over them.

The **remiges** (wing feathers, singular **remex**)

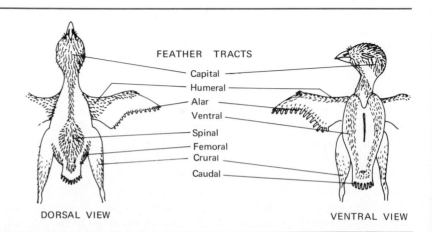

FEATHER TRACTS

Capital
Humeral
Alar
Ventral
Spinal
Femoral
Crural
Caudal

DORSAL VIEW

VENTRAL VIEW

Figure 17–1. Feather tracts of a typical songbird.

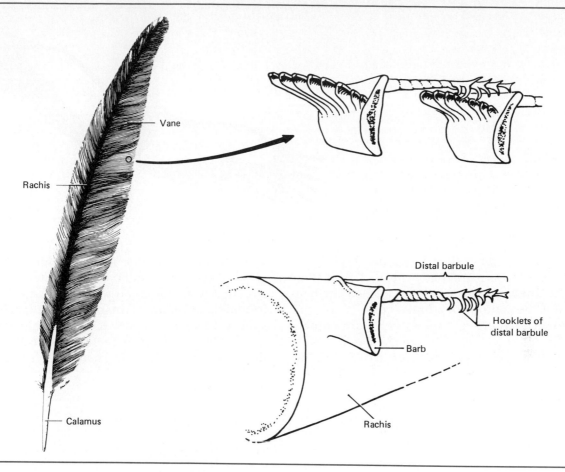

Figure 17–2. Typical contour feather (a wing quill) showing its main structural features. The inset shows details of the interlocking mechanism of the proximal and distal barbules. (From A. M. Lucas and P. R. Stettenheim, 1972, *Avian Anatomy and Integument*, Agriculture Handbook 362, United States Department of Agriculture, Washington, D.C.; and H. E. Evans, 1982, in *Diseases of Cage and Aviary Birds*, 2nd edition, edited by M. L. Petrak, Lea & Febiger, Philadelphia.)

and **rectrices** (tail feathers, singular **rectrix**) are large, stiff, mostly pennaceous contour feathers that are modified for flight. For example, the distal portions of the outer primaries of many species of birds are abruptly tapered or notched, so that when the wings are spread the tips of these primaries are separated by conspicuous gaps or slots (Figure 17–3). This condition reduces the drag on the wing and, in association with the marked asymmetry of the outer and inner vanes, allows the feather tips to twist as the wings are flapped and to act somewhat as individual propeller blades (see Figure 17–6).

Semiplumes are feathers intermediate in structure between contour feathers and down feathers. They combine a large rachis with entirely plumulaceous vanes and can be distinguished from down feathers by the fact that the rachis is

Figure 17–3. Flight silhouette of the California condor, showing the extreme development of slotting in the outer primaries.

longer than the longest barb (Figure 17–4). Semiplumes are mostly hidden beneath the contour feathers. They provide thermal insulation and help to fill out the contour of a bird's body.

Down feathers of various types are entirely plumulaceous feathers in which the rachis is shorter than the longest barb or entirely absent.

Down feathers provide insulation for adult birds of all species. In addition, natal down, which is structurally simpler than adult down, provides an insulative covering on many birds at hatching or shortly thereafter. Natal downs usually precede the development of the first contour feathers, and down feathers are associated with apteria (the

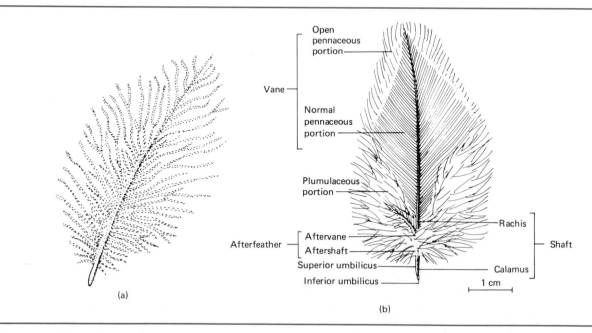

Figure 17–4. Comparison of a semiplume (a) and body contour feather (b).

spaces between the contour feather tracts). Definitive downs are those that develop with the full body plumage. Uropygial gland downs are associated with the large sebaceous gland found at the base of the tail in most birds. The papilla of the gland usually bears a tuft of modified, brush-like down feathers that aid in transferring the oily secretion from the gland to the bill to provide waterproof dressing to the plumage.

Powder down feathers, which are difficult to classify by structural type, produce an extremely fine, white powder composed of granules of keratin. The powder, which is shed into the general plumage, is nonwettable and is therefore assumed to provide another kind of waterproof dressing for the contour feathers. All birds have powder down, but it is best developed in herons.

Bristles are specialized feathers with a stiff rachis and barbs only on the proximal portion or none at all (Figure 17–5). Bristles occur most commonly around the base of the bill, around the eyes, as eyelashes, and on the head or even on the toes of some birds. The distal rachis of most bristles is colored dark brown or black by melanin granules. The melanin not only colors the bristles but also adds to their strength, resistance to wear, and resistance to photochemical damage. Bristles and structurally intermediate feathers called semibristles screen out foreign particles from the nostrils and eyes of many birds and act as tactile sense organs and possibly as aids in the aerial capture of flying insects, as for example, the long bristles at the edges of the jaws in nightjars and flycatchers.

Filoplumes are fine, hair-like feathers with a

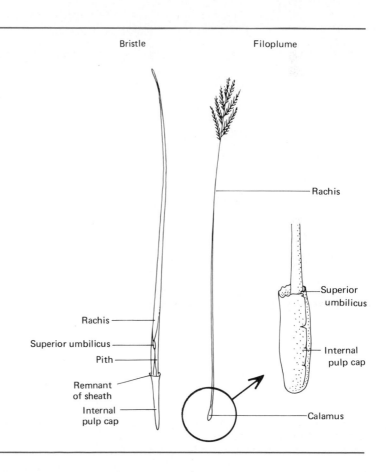

Figure 17–5. Comparison of a bristle (left) and a filoplume (right).

few short barbs or barbules at the tip (Figure 17–5). In some birds, such as cormorants and bulbuls, the filoplumes grow out over the contour feathers and contribute to the external appearance of the plumage, but usually they are not exposed. Filoplumes are sensory structures that aid in the operation of other feathers. Filoplumes have numerous free nerve endings in their follicle walls, and these nerves connect to pressure and vibration receptors around the follicles. Apparently, the filoplumes transmit information about the position and movement of the contour feathers via these receptors. This sensory system probably plays a role in keeping the contour feathers in place and adjusting them properly for flight, insulation, bathing, or display.

Aerodynamics of the Avian Wing Compared to Fixed Airfoils

Unlike the fixed wings of an airplane, the wings of a bird function both as an airfoil (lifting surface) and as a propeller for forward motion. The avian wing is admirably suited for these functions, consisting of a light, flexible airfoil (Figure 17–6). The primaries, inserted on the hand bones, do most of the propelling when a bird flaps its wings, and the secondaries along the arm provide lift. Removal of flight feathers from the wings of doves and pigeons shows that when only a few of the primaries

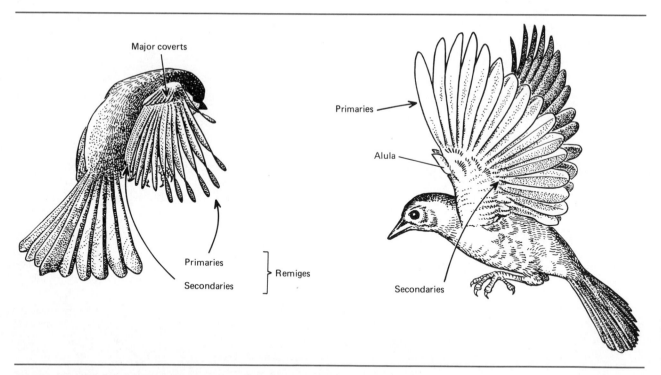

Figure 17–6. Drawings from high-speed photographs show the twisting and opening of the primaries during flapping flight. (From A. C. Thompson, 1964, *A New Dictionary of Birds*, McGraw-Hill, New York.)

are pulled out the bird's ability to fly is greatly altered, but a bird can still fly when as much as 55 percent of the total area of the secondaries has been removed.

A bird also has the ability to alter the area and shape of its wings and their positions with respect to the body. These changes in area and shape cause corresponding changes in velocity and lift that allow a bird to maneuver, change direction, land, and take off. Moreover, a bird's wing is not a solid structure like a conventional airfoil (for example, an airplane wing) but allows some air to flow through and between the feathers.

Obviously, the functioning of a bird's wing in flight—even in nonflapping flight—is vastly more complex than the functions of a fixed wing on an airplane or glider. Nevertheless, it is instructive to consider a bird's wing in terms of the basic performance of a fixed airfoil. [See Pennycuick (1975), Norberg (1985), and Rayner (1988) for reviews.] Although a bird's wing actually moves forward through the air, it is easier to think of the wing as stationary with the air flowing past. The flow of air produces a force, which is usually called the **reaction**. It can be resolved into two components: the **lift**, which is a vertical force equal to or greater than the weight of the bird, and the **drag**, which is a backward force opposed to the bird's forward motion and to the movement of its wings through the air.

When the leading end of a symmetrically streamlined body cleaves the air, it thrusts the air equally upward and downward, reducing the air pressure equally on the dorsal and ventral surfaces. No lift results from such a condition. There are two ways to modify this system to generate lift. One is to increase the **angle of attack** of the airfoil, and the other is to modify its surface configuration.

When the contour of the dorsal surface of the wing is convex and the ventral surface is concave (a **cambered airfoil**) the air pressure against the two surfaces is unequal because the air has to move farther and faster over the dorsal convex surface relative to the ventral concave surface (Figure 17–7). The result is reduced pressure over the wing, or lift. When the lift equals or exceeds the bird's body weight, the bird becomes airborne. The camber of the wing varies in birds with different flight characteristics and also changes along the length of the wing. Camber is greatest close to the body and decreases toward the wingtip. This change in camber is one of the reasons why the proximal part of the wing generates greater lift than the distal part.

If the leading edge of the wing is tilted up so that the angle of attack is increased, the result is increased lift up to an angle of about 15°, the **stalling angle**. This lift results more from a decrease in pressure over the dorsal surface than from an in-

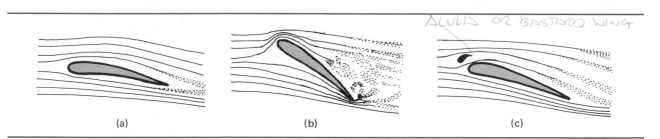

(a)   (b)   (c)

Figure 17–7. Diagram of airflow around a cambered airfoil. At a low angle of attack (a) the air streams smoothly over the upper surface of the wing and creates lift. When the angle of attack becomes steep (b), air passing over the wing becomes turbulent, decreasing lift enough to produce a stall. A wing slot (c) helps to prevent turbulence by directing a flow of rapidly moving air close to the upper surface of the wing.

crease in pressure below the airfoil. If the smooth flow of air over the wing becomes disrupted, the airflow begins to separate from the wing because of the increased air turbulence over the wing. The wing is then stalled. Stalling can be prevented or delayed by the use of slots or auxiliary airfoils on the leading edge of the main wing. The slots help to restore a smooth flow of air over the wing at high angles of attack and at slow speeds. The bird's **alula** has this effect, particularly during landing or takeoff (Figure 17–7). Also, the primaries act as a series of independent, overlapping airfoils, each tending to smooth out the flow of air over the one behind.

Another characteristic of an airfoil has to do with wingtip vortexes—eddies of air resulting from outward flow of air from under the wing and inward flow from over it. This is **induced drag**. One way to reduce the effect of these wingtip eddies and their drag is to lengthen the wing so that the tip vortex disturbances are widely separated and there is proportionately more wing area where the air can flow smoothly. Another solution is to taper the wing, reducing its area at the wingtip where induced drag is greatest. The ratio of length to width is called the **aspect ratio**. Long, narrow wings have high aspect ratios and high lift-to-drag (L/D) ratios. High-performance sailplanes and albatrosses, for example, have aspect ratios of 18:1 and L/D ratios in the range of about 40:1.

Wing loading is another important consideration. This is the weight of the bird divided by the wing area. The lighter the loading, the less is the power needed to sustain flight. There is a general relationship between wing loading and body size: Small birds usually have lighter wing loading than large ones, but wing loading is also related to specializations for powered versus soaring flight. The comparisons in Table 17–1 illustrate both of these trends. Small species such as hummingbirds, barn swallows, and mourning doves have lighter wing loading than large species such as the peregrine, golden eagle, and mute swan; yet the 3-gram hummingbird, a powerful flier, has a heavier wing loading than the more buoyant, sometimes soaring barn swallow, which is more than five times heavier. Similarly, the rapid-stroking peregrine has a heavier wing loading than the larger, often soaring golden eagle.

Gliding Flight

All birds can stretch their wings and glide, and some species depend primarily on this mode of flight to soar on rising air currents. The conditions of a stable glide, one in which airspeed remains constant in still air, are like those for level flight except that the flight path is inclined downward at some angle that is determined by the lift-to-drag ratio. Obviously, a glide can continue without loss of airspeed only by loss of altitude, in which case the drag multiplied by distance traveled along the glide path will equal the loss of potential energy, which is the weight multiplied by the loss in

Table 17–1. Wing loading of some North American birds.

Species	Body Mass (g)	Wing Area (cm²)	Wing loading (g/cm²)
Ruby-throated hummingbird	3.0	12.4	0.24
Barn swallow	17.0	118.5	0.14
Mourning dove	130.0	357.0	0.36
Peregrine falcon	1,222.5	1,342.0	0.91
Golden eagle	4,664.0	6,520.0	0.71
Mute swan	11,602.0	6,808.0	1.70

Source: E. L. Poole, 1938, *Auk* 55:511–517.

height:

$$\frac{\text{weight}}{\text{drag}} = \frac{\text{distance moved}}{\text{height lost}}$$

A high L/D ratio means a flat glide angle for a given airspeed, and a low **sinking speed** means that a bird can move a long distance horizontally relative to the distance dropped. Since the L/D ratio is importantly influenced by the aspect ratio of the wing and by size and weight of the flying body, it can be expected that different species of birds will have vastly different gliding performances depending on their sizes and the shapes of their wings.

Studies of gliding by different kinds of birds have been carried out in wind tunnels and by observing the actions of wild birds in free flight in comparison with a sailplane flying under the same conditions. The **glide polar** is a curve showing the relationship between sinking speed and forward speed through the air and is a convenient way of summarizing the gliding performance of a bird or aircraft. Figure 17–8 shows the main characteristics of the glide polar. The forward speed at which loss of height is minimal is the speed for minimum sink (V_{ms}). This speed is somewhat higher than the stalling speed (V_{min}). Still higher is the speed for best glide ratio (V_{bg}), which can be determined by

drawing a tangent to the curve from the origin of the graph. By flying at V_{bg} a bird or aircraft covers the greatest horizontal distance for a given loss of height. Figure 17–9 presents generalized glide polars for several species of birds. In general, soaring birds such as vultures and albatrosses have gliding performances superior to birds such as pigeons and gulls, which rely mainly on flapping flight.

Flapping Flight

Flapping flight is remarkable for its automatic, unlearned performance. A young bird on its maiden flight uses a form of locomotion so complex that it defies precise analysis in physical and aerodynamic terms. The nestlings of some species of birds develop in confined spaces such as burrows in the ground or cavities in tree trunks in which it is impossible for them to spread their wings and practice flapping before they leave the nest. Despite this seeming handicap, many of them are capable of flying considerable distances on their first flights. Young whale birds and diving petrels may fly as far as 10 kilometers the first time out of their burrows. On the other hand, young birds reared in open nests frequently flap their wings vigorously in the wind for several days before flying—especially large birds such as albatrosses, storks, vultures, and eagles. Such flapping may help to

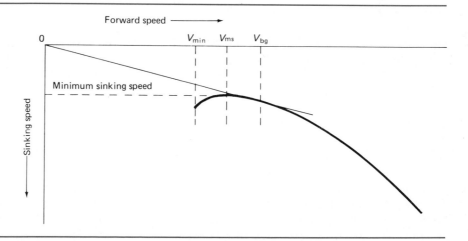

Figure 17–8. Main parameters of the glide polar; V_{min} = minimum gliding speed necessary to avoid stalling, V_{ms} = speed at which the rate of sinking is minimum, V_{bg} = speed at which bird travels farthest.

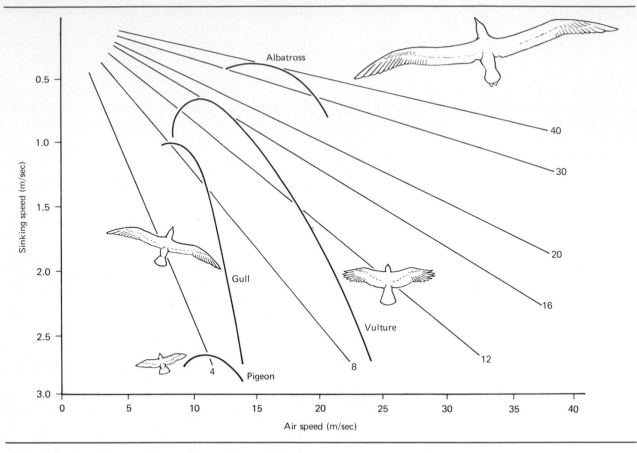

Figure 17–9. Generalized glide polars for several species of birds. The diagonal straight lines represent points with the L/D values indicated. (From V. A. Tucker and G. C. Parrot, 1970, *Journal of Experimental Biology* 52:345–367.)

develop musculature, but it is doubtful whether these birds are learning to fly; however, a bird's flying abilities do improve with practice for a period after it leaves the nest.

There are so many variables involved in flapping flight that it becomes difficult to understand exactly how it works. A beating wing is flexible and porous and yields to air pressure, unlike the fixed wing of an airplane. Its shape, wing loading, camber, sweepback, and the position of the individual feathers all change remarkably as a wing moves through its cycle of locomotion. This is a formidable list of variables, and it is no wonder

that flapping flight has not yet fully yielded to explanation in aerodynamic terms; however, the general properties of a flapping wing can be described (see Figure 17–10).

We can begin by considering the flapping cycle of a small bird in flight. A bird cannot continue to fly straight and level unless it can develop a force or thrust to balance the drag operating against forward momentum. The flapping of the wings, especially the wingtips (primaries), produces this thrust, whereas the inner wings (secondaries) are held more nearly stationary with respect to the body and generate lift. It is easiest to

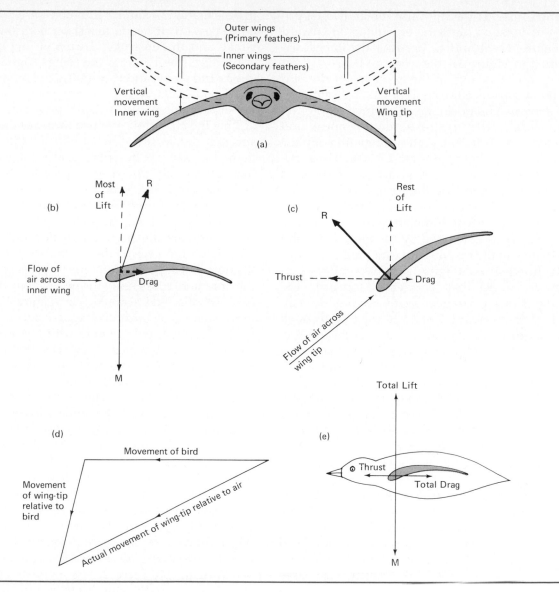

Figure 17–10. Generalized diagrams of forces acting on the inner and outer wings and the body of a bird in flapping flight. M, gravitational force; R, resultant force. See text for explanation.

consider the forces operating on the inner and outer wing separately. Most of the lift, but also most of the drag, comes from the forces acting on the inner wing and body of a flying bird. The forces on the wingtips derive from two motions that have to be added together. The tips are moving forward with the bird, but at the same time they are also moving downward relative to the bird. The wingtip would have a very large angle of attack and would stall if it were not flexible. As it is, the forces on the tip cause the individual primaries to twist as the wing is flapped downward (see Figure 17–6) and to produce the forces diagrammed in Figure 17–10c.

The forces acting on the two parts of the bird combine to produce the conditions for equilibrium flight shown. The positions of the inner and outer wings during the upstroke (dotted lines) and downstroke reveal that vertical motion is applied mostly at the wingtips (Figure 17–10a). Thus, the inner wing (the secondaries) acts as if the bird were gliding to generate the forces shown in Figure 17–10b, and the outer wing generates the force shown in part c. Most of the lift to counter the pull of gravity (M) is generated by the inner wing and body. Canting of the wing tip (the primaries) during the downstroke (Figure 17–10c) produces a resultant force (R) that is directed forward. The movement of the wingtip relative to the air is affected by the forward motion of the bird through the air (Figure 17–10d). As a result of this motion and the canting of the wing tips during the downstroke, the flow of air across the primaries is different from the flow across the secondaries and the body (Figure 17–10e). When flight speed through the air is constant, the forces acting on the inner wing and the body and on the outer wing combine to produce a set of summed vectors in which thrust exceeds total drag and lift at least equals the body mass.

As the wings move downward and forward on the downstroke, which is the power stroke, the trailing edges of the primaries bend upward under air pressure, and each feather acts as an individual propeller biting into the air and generating thrust. Contraction of the **pectoralis major**, the large breast muscle, produces the downstroke during level flapping flight. During this downbeat, the thrust is greater than the total drag, and the bird accelerates. In small birds, the return stroke, which is upward and backward, provides little or no thrust and is mainly a passive recovery stroke. The bird decelerates during the recovery stroke.

In larger birds with slower wing actions, the time of the upstroke is too long to spend in a state of deceleration. A similar situation exists when any bird takes off: It needs thrust on both the downstroke and the upstroke. Thrust on the upstroke is produced by bending the wings slightly at the wrists and elbow and by rotating the humerus upward and backward. This movement causes the upper surfaces of the twisted primaries to push against the air and to produce thrust as their lower surfaces did in the downstroke. In this type of flight the wingtip describes a rough figure eight through the air. As speed increases the figure-eight pattern is restricted to the wingtips.

A powered upstroke results mainly from the contraction of the **supracoracoideus**, a deep muscle underlying the pectoralis major and attached directly to the keel of the sternum. It inserts on the dorsal head of the humerus by passing through the foramen triosseum, formed where the coracoid, furcula, and scapula join (Figure 17–11). In most species of birds the supracoracoideus is a relatively small, pale muscle with low myoglobin content, easily fatigued. In species that rely on a powered upstroke for fast, steep takeoffs, for hovering, or for fast aerial pursuit, the supracoracoideus is relatively larger. The ratio of weights of the pectoralis major and the supracoracoideus is a good indication of a bird's reliance on a powered upstroke; such ratios vary from 3:1 to 20:1. The total weight of the flight muscles is also indicative of the extent to which a bird depends upon powered flight. Strong fliers such as pigeons and falcons have breast muscles comprising more than 20 percent of body weight, whereas in some owls, which have very light wing loading, the flight muscles make up only 10 percent of total weight.

A flying bird increases its speed by increasing the amplitude of its wing beats, but the frequency of wing beats remains nearly constant at all speeds during level flight. Large birds have slower wing beat frequencies than those of small birds, and strong fliers usually have slower beat frequencies than those of weak fliers. The frequency of respiration of many birds during flight appears to have a constant relationship to the wing beat frequency. Some birds, especially those that have low wing

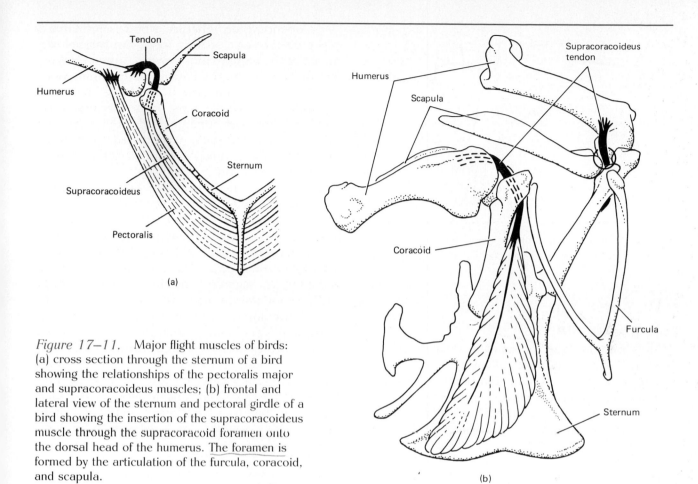

Figure 17–11. Major flight muscles of birds: (a) cross section through the sternum of a bird showing the relationships of the pectoralis major and supracoracoideus muscles; (b) frontal and lateral view of the sternum and pectoral girdle of a bird showing the insertion of the supracoracoideus muscle through the supracoracoid foramen onto the dorsal head of the humerus. The foramen is formed by the articulation of the furcula, coracoid, and scapula.

beat frequencies, breathe once per wing beat cycle, whereas birds with high wing beat frequencies have breathing cycles that span several wing beats. In general, inspiration appears to occur during or at the end of a wing upstroke, with expiration at the end of a downstroke.

A brief description of the flight of hummingbirds serves to indicate an extreme of specialization of flapping flight. Hummingbirds can hover and also fly backward. The hand bones form most of the wing skeleton of hummingbirds, the forearm and upper arm are very short. In addition, there is little articulation in the elbow and wrist joints, so that the entire wing is essentially an inflexible framework that can be moved very freely and in almost any direction at the shoulder joint. Thus, the entire wing functions essentially as a variable-pitch propeller. The wings beat in a horizontal plane as the bird hovers, and both the upstroke and the downstroke are powered (Figure 17–12). The supracoracoideus muscle is large in relation to the pectoralis, and the entire flight muscle mass is large in relation to body size, comprising 30 percent of total body weight. Hovering requires very high energy utilization because no lift is generated by forward movement through the air.

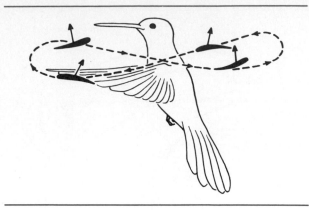

Figure 17–12. Diagrammatic representation of the wing movements of a hummingbird in hovering flight.

Wing Structure and Flight Characteristics

Wings may be large or small in relation to body size, resulting in light wing loading or heavy wing loading. They may be long and pointed, short and rounded, highly cambered or relatively flat, and the width and degree of slotting are additional important characteristics.

Depending on whether a bird is primarily a powered flier or a soaring form, the various segments of the wing (hand, forearm, upper arm) are lengthened to different degrees. Hummingbirds have very fast, powerful wing beats, requiring maximum propulsive force from the primaries. The hand bones of hummingbirds are longer than the forearm and upper arm combined. Most of the flight surface is formed by the primaries, and hummingbirds have only six or seven secondaries. Frigate birds are marine species with long, narrow wings specialized for powered flight as well as for gliding and soaring. All three segments of the forelimb are about equal in length. The soaring albatrosses have carried lengthening of the wing to the extreme found in birds: The humerus or upper arm is the longest segment, and there may be as many as 32 secondaries in the inner wing (Figure 17–13).

Ornithologists recognize four structural and functional types of wings (Figure 17–14). Birds that live in forests and woodlands where they must maneuver around obstructions have **elliptical wings**. These wings have a low aspect ratio, tend to be highly cambered, and usually have a high degree of slotting in the outer primaries. These features are generally associated with slow flight

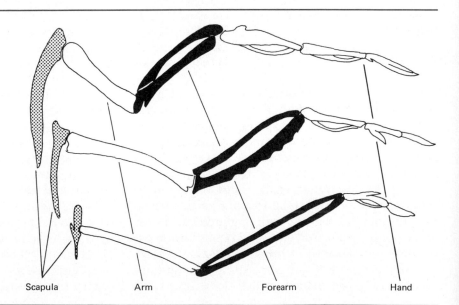

Figure 17–13. Comparison of the relative lengths of the proximal, middle, and distal elements of the wing bones of a hummingbird (top), frigate bird (middle), and albatross (bottom) drawn to the same size.

Scapula   Arm   Forearm   Hand

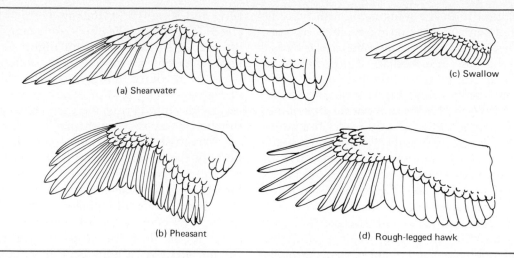

Figure 17–14. Comparison of four basic types of bird wings: (a) high aspect ratio; (b) elliptical; (c) high speed; (d) slotted.

and a high degree of maneuverability. Although some species with elliptical wings, notably upland gamebirds such as pheasants and grouse, have fast takeoff speeds, they maintain rapid flight only for short distances.

Many birds that are aerial foragers, make long migrations, or have a heavy wing loading that is related to some other aspect of their lives, such as diving, have **high-speed wings.** These wings have a moderately high aspect ratio, taper to a pointed tip, have a flat profile (little camber), and often lack slots in the outer primaries. In flight they show the swept-back attitude of jet fighter plane wings. All fast-flying birds have converged on this form.

Seabirds, particularly those such as albatrosses and shearwaters that rely on **dynamic soaring**, have long, narrow, flat wings lacking slots in the outer primaries. Some albatrosses have aspect ratios of 18:1 and lift-to-drag ratios similar to those of high-performance sailplanes. Dynamic soaring is possible only where there is a pronounced vertical wind gradient, with the lower 15 or so meters of air being slowed by friction against the ocean surface. Furthermore, dynamic soaring is feasible only in regions where winds are strong and persistent, such as in the latitudes of the Roaring Forties. This is where most albatrosses and shear-

waters are found. Starting from the top of the wind gradient, an albatross glides downwind with great increase in ground speed (kinetic energy). Then, as it nears the surface, it turns and gains altitude while gliding into the wind. Because the bird flies into wind of increasing speed as it rises, its loss of airspeed is not as great as its loss of ground speed, and consequently it does not stall until it has mounted back to the top of the wind gradient, where the air velocity becomes stable. At that point, the bird has converted much of its kinetic energy to potential energy and it turns downwind to repeat the cycle.

Slope soaring is a second type of soaring that is used by albatrosses and large petrels. In slope soaring a bird glides parallel to the windward face of a wave using slope lift that is caused by upward deflection of the wind by the wave. If the vertical component of the rising air exceeds the sinking speed of the bird, it is carried upward.

Large species of albatrosses weigh 10 to 12 kilograms and have wing spans greater than 3 meters, making them among the largest flying birds. Because their pectoral muscles are relatively small and weak, they can take off only by running or paddling and flapping into a strong wind, or by launching forth from the brink of a precipice.

The slotted **high-lift wing** is a fourth type. It is associated with static soaring typified by vultures, eagles, storks, and some other large birds. This wing has an intermediate aspect ratio between the elliptical wing and the high aspect ratio wing, a deep camber, and marked slotting in the primaries. When the bird is in flight the tips of the primaries turn markedly upward under the influence of air pressure and body weight. Static soarers remain airborne mainly by seeking out and gliding in air masses that are rising at a rate faster than the bird's sinking speed. Hence, a light wing loading and maneuverability (slow forward speed and small turning radius) are advantageous. Broad wings provide the light wing loading, and the highly developed slotting enhances maneuverability by responding to changes in wind currents with changes in the positions of individual feathers instead of movements of the entire wing. A bird cannot soar in tight spirals at high speed, and flying slowly with enough lift to prevent stalling requires a high angle of attack. The deeply slotted primaries apparently make the combination of low speed and high lift possible. The distal, emarginated portion of each primary produces lift by acting as a separate high aspect ratio airfoil set at a high angle of attack. This design reduces induced drag.

In regions where topographic features and meteorological factors provide currents of rising air, static soaring is an energetically cheap mode of flight. By soaring rather than flapping, a large bird the size of a stork can decrease by a factor of 20 or more the energy required for flight per unit of time, whereas the saving is only one-tenth as much for a small bird such as a warbler. It is little wonder, then, that most large land birds perform their annual migrations by soaring and gliding as much of the time as possible, and some condors and vultures cover hundreds of kilometers each day soaring in search of food.

Body Form and Flight

Many aspects of the morphology of birds appear to have been molded by aerodynamic forces. [See King and King (1979), Raikow (1985), and Rayner (1988) for reviews.] Feathers, for example, provide lift and streamlining during flight. Feathers are light, yet they are strong and resilient for their weight. Of course, flying is not the only function of feathers—they also provide the insulation necessary for endothermy and their colors and shapes function in crypsis and display.

Weight reduction can be seen in several aspects of the structure of birds. The avian skeleton is lighter in relation to the total body mass of the bird than is the skeleton of a mammal of similar size. The skeleton of a pigeon is about 4.4 percent of the total body mass, whereas the skeleton of a white rat accounts for 5.6 percent of the mass. A frigate bird with a wingspan of 1.5 meters has a skeleton that weighs only 114 grams, which is less than the weight of the bird's feathers. The toothlessness of birds is another contribution to weight reduction: The skull and teeth of a fox comprise about 1.25 percent of its total body mass, whereas the skull and horny beak of a golden eagle are only 0.2 percent of its total mass.

Characteristics of some of the organs of birds contribute to their relative lightness. For example, birds lack urinary bladders and most species have only one ovary (the left). The gonads of both male and female birds are usually small; they hypertrophy during the breeding season and regress when breeding has finished.

Power-producing features are equally important components of the ability of birds to fly. The pectoral muscles of a strong flier may account for 20 percent of the total body mass. The power output per unit mass of the pectoralis major of a turtle dove during level flight has been estimated to be 10 to 20 times that of most mammalian muscles. Birds have large hearts and high rates of blood flow and complex lungs that use cross-current flows of air and blood to maximize gas exchange and to dissipate the heat produced by high levels of muscular activity during flight. The brains of birds are similar in size to the brains of rodents, and the forebrain and cerebellum are well developed. Birds rely heavily on visual information and the optic lobes are especially large. The sense of smell is not well developed in most birds and the olfactory lobes are correspondingly small.

Streamlining

Birds are the only vertebrates that move fast enough in air for wind resistance and streamlining to be important factors in their lives. Many passerine birds are probably able to fly 50 kilometers per hour or even faster when they must, although their normal cruising speeds are lower. Ducks and geese can fly at 80 or 90 kilometers per hour, and peregrine falcons reach speeds as high as 200 kilometers per hour when they dive on prey. Fast-flying birds have many of the same structural characters as those seen in fast-flying aircraft. Contour feathers make smooth junctions between the wings and the body and often between the head and body as well, eliminating sources of turbulence that would increase wind resistance. The feet are tucked close to the body during flight, further improving streamlining.

At the opposite extreme, some birds are slow fliers. Many of the long-legged, long-necked wading birds such as spoonbills and flamingos fall in this category. Their long legs trail behind them as they fly and their necks are extended. They are far from streamlined, although they may be strong fliers.

Skeleton and Muscles

The hollow, air-filled (pneumatic) bones of birds are probably an ancestral character of the archosaur lineage, not a derived character of birds. Nonetheless, the combination of lightness and strength provided by this bone structure is probably useful in flight (Figure 17–15). However, not all birds have pneumatic bones. In general, pneumatization of bones is better developed in large birds than in small ones. Diving birds (penguins, grebes, loons) have little pneumaticity, and the bones of diving ducks are less pneumatic than those of nondivers.

The distribution of pneumaticity among the bones of the skeleton also varies. The skull is

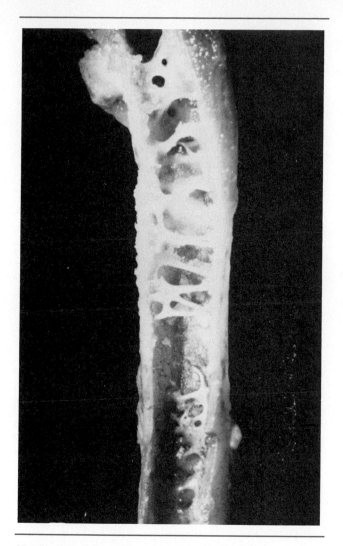

Figure 17–15. The hollow bones of birds are reinforced by struts.

pneumatic in nearly all birds, although the kiwi, a flightless bird, lacks air spaces in the skull. The sternum, pectoral girdle, and humerus are pneumatic and are part of a system of interconnected air sacs that allow a one-way flow of air through the lungs (discussed in the following section). Pneumaticity extends through the rest of the appendicular skeleton of some birds, even into the phalanges.

Except for the specializations associated with flight, the skeleton of a bird is very much like that of a small, bipedal archosaur (Figure 17–16). The pelvic girdle of birds is elongated, and the ischium and ilium have broadened into thin sheets that are firmly united with a **synsacrum** that is formed by the fusion of 10 to 23 vertebrae. The long tail of ancestral diapsids has been shortened in birds to about five free caudal vertebrae and a **pygostyle** formed by the fusion of the remaining vertebrae. The pygostyle supports the tail feathers (retrices). The thoracic vertebrae are joined by strong ligaments that are often ossified. The relatively immobile thoracic vertebrae, the synsacrum, and the pygostyle in combination with the elongated, roof-like pelvis produce a nearly rigid vertebral column. Flexion is possible only in the neck, at the joint between the thoracic vertebrae and the synsacrum, and at the base of the tail. The rigid trunk is balanced on the legs. The femur projects anteriorly and its articulation with the tibiotarsus and fibula is close to the center of gravity of the bird.

The wings are positioned above the center of gravity. The sternum is greatly enlarged compared to other vertebrates, and (except in flightless birds) it bears a keel from which the pectoralis and supracoracoideus muscles originate (Figure 17–11) Strong-flying birds have well developed keels and their flight muscles are large. The scapula extends posteriorly above the ribs and is supported by the coracoid, which is fused ventrally to the sternum. Additional bracing is provided by the clavicles, which, in most birds, are fused at their distal ends to form the **furcula** (wishbone).

The relative size of the leg and flight muscles of birds is related to their primary mode of locomotion. Flight muscles comprise 25 to 35 percent of the total body mass of strong fliers such as hummingbirds and swallows. These species have small legs and the leg muscles account for as little as 2 percent of the body mass. Predatory birds such as hawks and owls use their legs to capture prey. In these species the flight muscles make up about 20 percent of the body mass and the limb muscles are 10 percent. Swimming birds, ducks and grebes, for example, have an even division between limb and

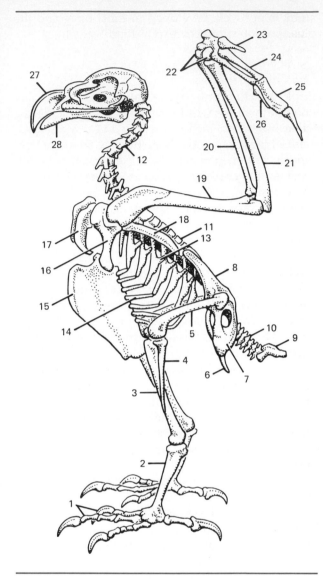

Figure 17–16. Skeleton of a raptor. 1, Toes; 2, tarsus; 3, tibia; 4, fibula; 5, femur; 6, pubis; 7, ischia; 8, ilium; 9, pygostyle; 10, caudal vertebrae; 11, thoracic vertebrae; 12, cervical vertebrae; 13, uncinate process of rib; 14, sternal rib; 15, sternum; 16, coracoid; 17, furcula; 18, scapula; 19, humerus; 20, radius; 21, ulna; 22, carpal bones; 23, first digit; 24, metacarpal; 25, second digit; 26, third digit; 27, upper mandible; 28, lower mandible. (From J. Dorst, 1974, *The Life of Birds*, Columbia University Press, New York.)

flight muscles, and the combined mass of these muscles may be 30 to 60 percent of the total body mass. Birds such as rails, which are primarily terrestrial and run to escape from predators, have limb muscles that are larger than their flight muscles.

Muscle fiber types and metabolic pathways also distinguish running birds from fliers. The familiar distinction between the light meat and dark meat of a chicken reflects those differences. Fowl, especially domestic breeds, rarely fly, but they are capable of walking and running for long periods. The dark color of the leg muscles reveals the presence of myoglobin in the tissues and indicates a high capacity for aerobic metabolism in the limb muscles of these birds. The white muscles of the breast lack myoglobin and have little capacity for aerobic metabolism. The flights of fowl (including wild species such as pheasants, grouse, and quail) are of brief duration and are used primarily to evade predators. The bird uses an explosive take-off, fueled by anaerobic metabolic pathways, followed by a long glide back to the ground. Birds that are capable of strong sustained flight have dark breast muscles with high aerobic metabolic capacities.

The skulls of most birds consist of four bony units that can move in relation to each other. This skull kinesis is important in some aspects of feeding. The upper jaw flexes upward as the mouth is opened, and the lower jaw expands laterally at its articulation with the skull (Figure 17–17). The flexion of the upper and lower jaws increases the bird's gape in both the vertical and horizontal planes and probably assists in swallowing large items.

Heart, Lungs, and Gas Exchange

The heart of birds is composed of morphologically distinct left and right atria and ventricles. The separation of the ventricle into left and right halves is an archosaurian character that is seen in crocodilians as well as in birds. However, birds lack the capacity possessed by crocodilians to shunt blood between the pulmonary and systemic circulation.

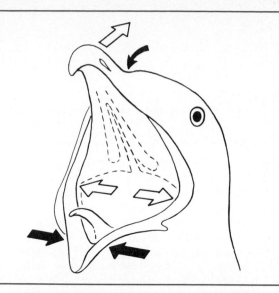

Figure 17–17. Skull and jaw kinesis of birds. A yawning herring gull (*Larus argentatus*) shows the kinetic movements of the skull and jaws that occur during swallowing. White arrows show the positions of outward flexion and black arrows show inward flexion. (From P. Böhler, 1981, in *Form and Function in Birds*, volume 2, edited by A. S. King and J. McLelland, Academic Press, New York.)

In birds, as in mammals, the blood must flow in series through the two circuits.

The respiratory system of birds is unique among living vertebrates (Figure 17–18). The air sacs that occupy much of the dorsal part of the body and extend into the pneumatic spaces in many of the bones provide a system in which airflow through the lungs is unidirectional instead of tidal (in and out) as it is in mammalian lungs. The air sacs are poorly vascularized, and gas exchange occurs only in the parabronchial lung. The combined volume of the air sacs is about nine times the volume of the parabronchial lung itself. The flow of air through this extensive respiratory system may help to dissipate the heat generated by high levels of muscular activity during flight.

Air flows through the parabronchial lung in the same direction during both inspiration and exhalation (Figure 17–19). During inspiration the volume of the thorax increases, drawing air through

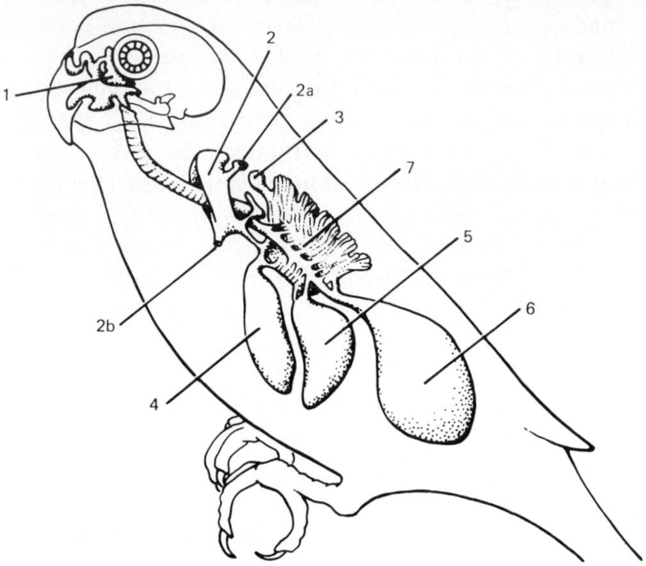

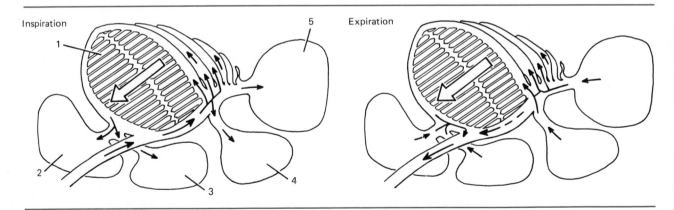

Figure 17–18. The lung and air sac system of the budgerigar; only the left side is shown. 1, Infraorbital sinus; 2, clavicular air sac; 2a, axillary diverticulum to the humerus; 2b, sternal diverticulum; 3, cervical air sac; 4, cranial thoracic air sac; 5, caudal thoracic air sac; 6, abdominal air sacs; 7, parabronchial lung. (From H. E. Evans, 1982, in *Diseases of Cage and Aviary Birds*, 2nd edition, edited by M. L. Petrak, Lee & Febiger, Philadelphia.)

the bronchus and into the posterior thoracic and abdominal air sacs and the parabronchial lung. Simultaneously, air from the parabronchial lung is drawn into the clavicular and anterior thoracic sacs. On expiration the volume of the thorax de-creases, air from the posterior thoracic and abdominal sacs is forced into the parabronchial lung, and air from the clavicular and thoracic sacs is forced out through the bronchus.

Gas exchange takes place in a network of

Figure 17–19. Pattern of airflow during inspiration and expiration. Note that air flows through the parabronchial lung during both phases of the respiratory cycle. 1, Parabronchial lung; 2, clavicular air sac; 3, cranial thoracic air sac; 4, caudal thoracic air sac; 5, abdominal air sacs. (From P. Scheid, 1982, in *Avian Biology*, volume 6, edited by J. R. King and K. C. Parkes, Academic Press, New York.)

tiny, blind-end air capillaries that intertwine with equally fine blood capillaries. Air flow and blood flow pass in opposite directions, and the capillary pairs compose a cross-current exchange system (Figure 17–20). This arrangement is more effective at exchanging lung gases than is the mammalian lung, and the efficiency of gas exchange in the parabronchial lung may allow birds to breathe at higher altitudes than mammals (Box 17–1). Birds do sometimes penetrate to very high altitudes, either as residents or in flight. For example, radar tracking of migrating birds shows that they sometimes fly as high as 6500 meters, the alpine chough lives at altitudes around 8200 meters on Mount

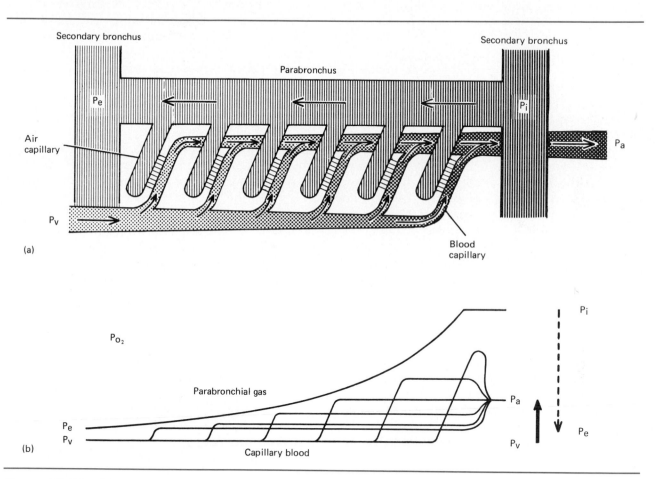

Figure 17–20. Diagram of gas exchange in a cross-current lung. Air flows from right to left in this diagram and blood flows from left to right. P_e = oxygen pressure in the air exiting the parabronchus, P_v = oxygen pressure in the mixed venous blood entering the blood capillaries, P_a = oxygen pressure in the blood leaving the blood capillaries P_i = oxygen pressure in the air entering the parabronchus. Top: General pattern of air and blood flow through the parabronchial lung. Bottom: Diagrammatic representation of cross-current gas exchange. (From P. Scheid, 1982, in *Avian Biology*, volume 6, edited by J. R. King and K. C. Parkes, Academic Press, New York.)

Box 17-1. High-Flying Birds

Birds regularly fly at altitudes higher than human mountain climbers can ascend without using auxiliary breathing apparatus, and the ability of birds to sustain activity at high altitudes is a result of the morphological characteristics of their pulmonary systems. Mammals have difficulty breathing at high altitudes because of the low oxygen pressure in the air. The atmosphere is most dense at the surface of the earth (where the entire weight of the atmosphere is pressing down on it), and it becomes increasingly less dense at higher altitudes. At sea level atmospheric pressure is 760 millimeters of mercury (760 torr in the units of the International System). The composition of dry air by volume is 79.02 percent nitrogen and other inert gases, 20.94 percent oxygen, and 0.04 percent carbon dioxide. These gases contribute to the total atmospheric pressure in proportion to their abundance, so the contribution of oxygen is 20.94 percent of 760 torr, or 159.16 torr. The pressure exerted by an individual gas is called the partial pressure of that gas. The rate and direction of diffusion of gas between the air in the lungs and the blood in the pulmonary capillaries is determined by the difference in the partial pressures of the gas in the blood and in the lungs. Oxygen diffuses from air in the lungs into blood in the pulmonary capillaries because oxygen has a higher partial pressure in the air than in the blood, whereas carbon dioxide diffuses in the opposite direction because its partial pressure is higher in blood than in air.

At higher altitudes the atmospheric pressure is lower. At 7700 meters, the atmospheric pressure is only 282 torr, and the partial pressure of oxygen in dry air is about 59 torr. As a result of the low atmospheric pressure at this altitude the driving force for diffusion of oxygen into the blood is small. [The actual pressure differential is reduced even below this figure because air in the lungs is saturated with water, and water vapor contributes to the total

Everest, and bar-headed geese pass directly over the summit of the Himalayas at altitudes of 9200 meters during their migrations.

Vocalization is a second function of the respiratory systems of many vertebrates, including birds. However, the vocal organ of birds, the **syrinx** (plural syringes), is unique. The syrinx lies at the base of the trachea where the two bronchi diverge. Several of the cartilaginous rings that support the trachea and bronchi are modified and surrounded by muscles. The anatomy of the syrinx varies among birds—some are purely tracheal, some involve both the trachea and the bronchi, and others are purely bronchial. Two membranes are associated with the syrinx, one in the base of each bronchus. Song birds (oscines) have five to nine (usually seven) pairs of syringeal muscles, parrots and lyrebirds have three pairs, falcons have two, and most other birds have only a single pair of syringeal muscles. Storks, vultures, ratites, and most pelicans lack syringeal muscles.

The mechanism of avian song production is an area of controversy. One view holds that sound is produced by vibration of the syringeal membranes and that modulations of that sound are the result of changes in the vibration of the membranes (Greenewalt 1968, 1969). An alternative hypothesis maintains that pulsatile activity of the abdominal muscles and resonance in the trachea are important in modulating the amplitude and frequencies produced by the syringeal membranes (Gaunt et al. 1982, Nowicki 1987).

pressure in the lungs. The vapor pressure of water at 37°C is 47 torr. Thus, the partial pressure of oxygen in the lungs is 20.94 percent × (282 − 47 torr) = 49 torr.]

The tidal ventilation pattern of the lungs of mammals means that the partial pressure of oxygen in the pulmonary capillaries can never be higher than the partial pressure of oxygen in the expired air. The best that a tidal ventilation system can accomplish is to equilibrate the partial pressures of oxygen in the pulmonary air and in the pulmonary circulation. In fact, failure to achieve complete mixing of the gas within the pulmonary system means that oxygen exchange falls short even of this equilibration, and blood leaves the lungs with a partial pressure of oxygen slightly lower than the partial pressure of oxygen in the exhaled air. The cross-current blood flow system in the parabronchial lung of birds ensures that the gases in the air capillaries repeatedly encoun-

ter a new supply of deoxygenated blood (Figure 17–20, left side of diagram).

When blood enters the system (on the left side of the diagram) it has the low oxygen pressure of mixed venous blood (P_v). The blood entering the leftmost capillary is exposed to air that has already had much of its oxygen removed farther upstream. Nonetheless, the low oxygen pressure of the mixed venous blood ensures that even in this part of the parabronchus, oxygen uptake can occur. Blood flowing through capillaries farther to the right in the diagram is exposed to higher partial pressures of oxygen in the parabronchial gas and takes up correspondingly more oxygen. The oxygen pressure of the blood that flows out of the lungs (P_a) is the result of mixing of blood from all the capillaries. The oxygen pressure of the mixed arterial blood is higher than the partial pressure of oxygen in the exhaled air.

Feeding, Digestion, and Excretion

The digestive systems of birds show some differences from those of other vertebrates, although the distinctions are less pronounced than the differences in the respiratory systems. With the specialization of the forelimbs as wings, which largely precludes their having any role in prey capture, birds have concentrated their predatory mechanisms in their beaks and feet. The absence of teeth prevents birds from doing much processing of food in the mouth, and the gastric apparatus takes over some of that role.

Beaks and Tongues The presence of a horny beak in place of teeth is not unique to birds—we

have noted the same phenomenon in turtles, rhynchosaurs, dinosaurs, pterosaurs, and in the dicynodonts—but the diversity of beaks among birds is remarkable. The range of morphological specializations of beaks defies complete description, but some categories can be recognized (Figure 17–21).

Insectivorous birds such as warblers, which find their food on leaf surfaces, usually have short, thin, pointed bills that are adept at seizing insects, whereas aerial sweepers such as swifts, swallows, and nighthawks that catch their prey on the wing have short, weak beaks and a wide gape. Kinesis of the lower jaw substantially increases the gape of the nightjar; the distance between the two rami of the lower jaw increases from 12.5 millimeters

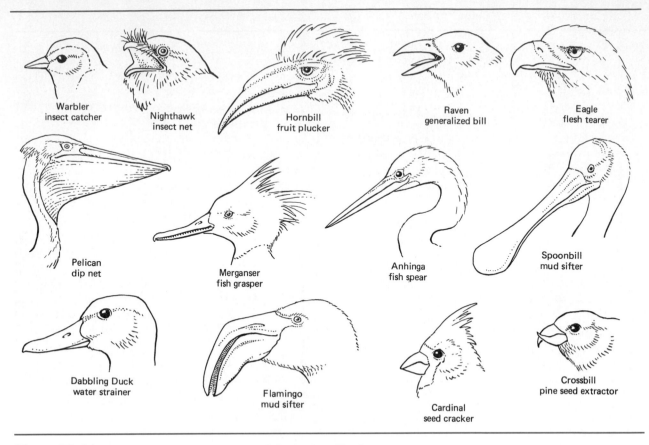

Figure 17–21. Examples of specializations of the beaks of birds.

when the mouth is closed to 40 millimeters when it is opened. Stiff feathers at the corners of the mouth further increase the insect-trapping area. It is always the case that one can find exceptions to generalizations about the correspondence between morphology and behavior, and the bills of insectivorous birds reveal many of these. For example, some hornbills use their heavy, simitar-shaped beaks to pick up termites and small insects from the ground, and flycatchers, which snap up insects in flight, have hooked beaks.

Many carnivorous birds, such as gulls, ravens, crows, and roadrunners, use their heavy pointed beaks to kill their prey, but most hawks, owls, and eagles kill prey with their talons and employ their beaks to tear off pieces small enough to swallow.

Falcons stun prey with the impact of their dive, and then bite the neck of the prey to disarticulate the cervical vertebrae. The true falcons (the genus *Falco*) have a structure called a tomial tooth that aids in the process. This tooth actually is a sharp projection from the upper mandible that matches with a corresponding notch on the bottom mandible. Shrikes, a group of predatory passerine birds that employ neck biting to kill prey, also have a tomial tooth. Fish-eating birds such as cormorants and pelicans have beaks with a sharply hooked tip that is used to seize fish, and mergansers have long, narrow bills with a series of serrations along the sides of the beak in addition to a hook at the tip. Darters and anhingas have harpoon-like bills that they use to impale fish.

Many aquatic birds strain small crustaceans or plankton from water or mud with bills that incorporate some sort of filtering apparatus. Spoonbills have flattened bills with broad tips that they use to skim the water surface. Dabbling ducks have bills with horny lamellae that form crosswise ridges and their tongues also have horny projections. The tongue and bill are densely invested with sensory corpuscles and form a filter system that allows ducks to scoop up a billfull of water and mud, filter out the prey, and allow the debris to escape. Flamingos have a similar system. The bill of a flamingo is sharply bent and the rostral part is held in a horizontal position when the flamingo lowers its head to feed. The lower jaw of a flamingo is smaller than the upper jaw, and it is the upper jaw that moves during feeding while the lower jaw remains motionless. (This reversal of the usual vertebrate pattern is possible because of the kinetic skull of birds.)

Seeds contain the energy and nutrients that plants have invested in reproduction, and seeds are usually protected by hard coverings (husks) that must be removed before the nutritious contents can be eaten. Specialized seed-eating birds use one of two methods to husk seeds before swallowing them. One group holds the seed in its beak and slices it by making fore-and-aft movements of the lower mandible, whereas birds in the second group hold the seed against ridges on the palate and crack the husk by exerting an upward pressure with their robust lower mandible. After the husk has been opened, both kinds of birds use their tongues to remove the contents. Other birds have different specializations for eating seeds: Crossbills extract the seeds of conifers from between the scales of the cones, using the diverging tips of their bills to pry the scales apart. Woodpeckers, nuthatches, and chickadees may wedge a nut or acorn into a hole in the bark of a tree and then hammer at it with their sharp bill until it cracks.

The production of fruits is a strategy of seed dispersal for plants, and birds are important dispersal agents. The bright colors of many fruits advertise their presence or ripeness and attract birds that eat them. The pulp surrounding the seed is digested during its passage through the bird's intestine, but the seeds are not damaged. The birds may discard the seed before eating the pulp, or the seeds may pass through the gut to be voided with the feces. In some cases the chemical or mechanical stress of its passage through the digestive system of a bird facilitates subsequent sprouting by the seed. Fruit-eating birds do not have to penetrate a hard covering, but the very size of the fruit may make it hard to swallow. Skull kinesis can be important for fruit eaters; the widening of the gape as the mouth opens may allow them to swallow large items.

Many birds use their beaks to probe for hidden food. Long-billed shorebirds probe in mud and sand to locate worms and crustaceans. These birds display a form of skull kinesis in which the flexible zone in the upper jaw has moved toward the tip of the beak, allowing the tip of the upper jaw to be lifted without opening the mouth (Figure 17–22). This mechanism enables long-billed waders to grasp prey under the mud.

Tongues are an important part of the food-gathering apparatus of many birds. Woodpeckers drill holes into dead trees and then use their long

Figure 17–22. Long-billed wading birds that probe for worms and crustaceans in soft substrates can raise the tips of the upper bill without opening their mouth. (From P. Böhler, 1981, in *Form and Function in Birds*, volume 2, edited by A. S. King and J. McLelland, Academic Press, New York.)

tongues to investigate passageways made by wood-boring insects. The tongue of the green woodpecker, which extracts ants from their tunnels in the ground, extends four times the length of its beak. The hyoid bones that support the tongue are elongated and housed in a sheath of muscles that passes around the outside of the skull and rests in the nasal cavity (Figure 17–23). When the muscles of the sheath contract, the hyoid bones are squeezed around behind the skull, and the tongue is projected from the bird's mouth. The tip of a woodpecker tongue has barbs that impale insects and allow them to be pulled from their tunnels. Nectar-eating birds such as hummingbirds and sunbirds also have long tongues and a hyoid apparatus that wraps around the back of the skull. In nectar-eating birds the tip of the tongue is divided into a spray of hair-thin projections, and capillary attraction causes nectar to adhere to the tongue.

The Buccal Cavity Birds swallow their prey intact or in large pieces; there is no chewing of the food of the sort seen in mammals. Salivary glands in the mouth produce secretions that lubricate food and assist swallowing. Woodpeckers have a large, distinctive gland that produces a sticky secretion that coats the tongue and probably helps to trap prey. Some jays have specialized mandibular glands that produce a sticky secretion used to form compact balls of food that are attached to twigs as a food-storage mechanism. The sublingual glands of swifts produce a sticky glycoprotein that is used to bind sticks together during construction of the nest. These cement-producing glands enlarge during the breeding season. The edible swiftlet of the East Indies builds its nest entirely from dried cement, and a culinary delicacy, bird's-nest soup, is produced by cooking the nests.

The Esophagus and Crop Birds often gather more food than they can process in a short period, and the excess is held in the esophagus. Many birds have a **crop**, an enlarged portion of the esophagus that is specialized for temporary storage of food. The crop of some birds is a simple

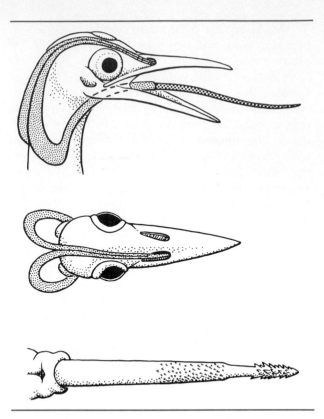

Figure 17–23. Hyoid apparatus of a woodpecker, showing the mechanism for protrusion of the tongue beyond the tip of the mandibles. The tongue itself is about the length of the bill, and it can be extended well beyond the tip of the beak by muscles that move the elongated hyoid apparatus. The detail of the tongue shows the barbs on the tip that impale prey.

expansion of the esophagus, whereas in others it is a unilobed or bilobed structure (Figure 17–24). An additional function of the crop is transportation of food for nestlings. When the adult returns to the nest it regurgitates the material from the crop and feeds it to the young. In doves and pigeons the crop of both sexes produces a nutritive fluid (crop milk) that is fed to the young. The milk is produced by fat-laden cells that detach from the squamous epithelium of the crop and are suspended in an aqueous fluid. Crop milk is rich in lipids and proteins, but contains no sugar. Its

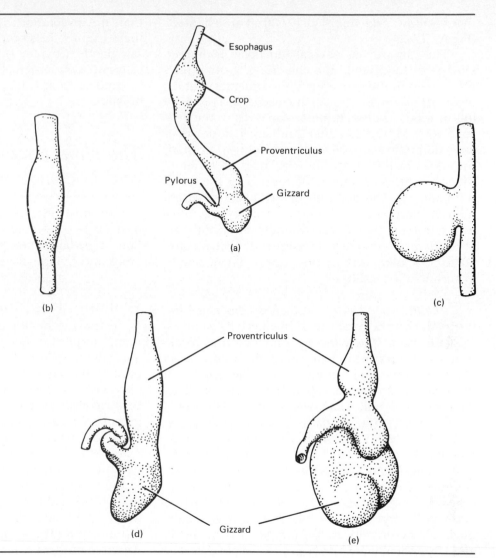

Figure 17–24. Anterior digestive system of birds. (a) The relationship among the parts. The relative sizes of the proventriculus and gizzard vary in relation to diet. Carnivorous and fish-eating birds like the great cormorant have a relatively small crop (b) and gizzard (c), whereas seed-eaters and omnivores like the peafowl have a large crop (d) and muscular gizzard (e). (From J. McLelland, 1979, *Form and Function in Birds*, edited by A. S. King and J. McLelland, Academic Press, London.)

chemical composition is similar to that of mammalian milk, although it differs in containing intact cells. The proliferation of the crop epithelium and the formation of crop milk is stimulated by prolactin, as is the lactation of mammals. A nutritive fluid produced in the esophagus is fed to hatchlings by greater flamingos and by the male emperor penguin.

The Gastric Apparatus The form of the stomachs of birds is related to their dietary habits. Carnivorous and piscivorous (fish-eating) birds need expansible storage areas to accommodate large volumes of soft food, whereas birds that eat insects or seeds require a muscular organ that can contribute to the mechanical breakdown of food. Usually, the gastric apparatus of birds consists of two relatively distinct chambers, an anterior glandular stomach (**proventriculus**) and a posterior muscular stomach (**gizzard**). The proventriculus contains glands that secrete acid and digestive enzymes. The proventriculus is especially large in species that swallow large items such as fruit or fish intact (Figure 17–24).

The gizzard has several functions, including storage of food while the chemical digestion that was begun in the proventriculus continues, but its most important function is the mechanical processing of food. The thick, muscular walls of the gizzard squeeze the contents, and small stones that are held in the gizzards of many birds help to grind the food. In this sense the gizzard is performing the same function that is performed by the teeth of mammals. The pressure that can be exerted on food in the gizzard is intense. A turkey's gizzard can grind up two dozen walnuts in as little as 4 hours, and can crack hickory nuts that require 50 to 150 kilograms of pressure to break under experimental conditions.

The Intestine, Ceca, and Cloaca   The small intestine is the principal site of chemical digestion as enzymes from the pancreas and intestine break down the food into small molecules that can be absorbed across the intestinal wall. The mucosa of the small intestine is modified into a series of folds, lamellae, and villi that increase its surface area. The large intestine is relatively short, usually less than 10 percent of the length of the small intestine. Passage of food through the intestines of birds is quite rapid: Transit times for carnivorous and fruit-eating species are in the range of a few minutes to a few hours. Passage of food is slower in herbivores and may require a full day. Birds generally have a pair of ceca at the junction of the small and large intestines. The ceca are small in carnivorous, insectivorous, and graminivorous species, but they are large in herbivorous and omnivorous species such as cranes, fowl, ducks, geese, and the ostrich. Symbiotic microorganisms in the ceca apparently ferment plant material.

The cloaca temporarily stores waste products while water is being reabsorbed. The precipitation of uric acid in the form of urate salts frees water from the urine, and this water is returned to the bloodstream. Species of birds that have salt-secreting glands can accomplish further conservation of water by reabsorbing some of the ions that are in solution in the cloaca and excreting them in more concentrated solutions through the salt glands (Chapter 4). The mixture of white urate salts and dark fecal material that is voided by birds is familiar to anyone who has washed an auto-. mobile.

The Hind Limbs and Locomotion

Unlike most tetrapods, birds usually are specialized for two or more different modes of locomotion—bipedal walking or swimming with the hind limbs and flying with the forelimbs.

Walking, Hopping, and Perching

Terrestrial locomotion may involve walking or running, supporting heavy bodies, hopping, perching, climbing, wading in shallow water, or supporting the body on insubstantial surfaces. It is instructive to consider cursorial adaptations first because birds evolved from bipedal dinosaurs and because the principles of cursorial adaptation have been well worked out in the quadrupedal mammals. Modifications usually associated with running in quadrupeds are (1) a progressive increase in the lengths of the distal limb elements relative to the proximal ones, (2) a decrease in the area of the foot surface that makes contact with the ground, and (3) a reduction in the number of toes. All three of these cursorial trends are expressed to varying degrees in running birds; however, problems of balance are more critical for bipeds than for quadrupeds, and these problems may have restricted the evolution among bipeds of some of the cursorial adaptations found in quadrupeds.

Because the center of gravity must lie over the feet of a biped to maintain balance, a reduction in the surface area in contact with the ground can be achieved only by some sacrifice in stability. No bird has reduced the length and number of toes in contact with the ground to the extent that the hooved mammals have, but the large, fast-running ostrich has only two toes on each foot and many other cursorial species have only the three forward-directed toes in contact with the ground (Figure 17–25).

Graviportal characters such as large, heavy leg bones arranged in vertical columns as supports of great body weight are well demonstrated in some large mammals, but no surviving birds show specializations of this kind. Some of the large, flightless rails show modest graviportal development of the legs, but these features were most fully expressed by some of the large, flightless terrestrial birds of earlier times (Feduccia 1980). The extinct elephantbirds (Aepyornithiformes) of Madagascar and moas (Dinornithiformes) of New Zealand were graviportal herbivores that evolved on oceanic islands in the absence of large carnivorous mammals and survived there until contact with primitive humans in the post-Pleistocene period (Figure 17–26). The giant, carnivorous *Diatryma* (Gruiformes) and related species were successful continental forms in the Americas until the appearance of large mammalian carnivores. Among the large, flightless terrestrial birds, only some of the fleet-footed cursors such as the rheas and os-

trich have managed to survive in the presence of large placental carnivores.

Hopping, a succession of jumps with the feet moving together, is a special form of pedal locomotion found mostly in perching, arboreal birds. It is most highly developed in the passerines, and only a few nonpasserine birds regularly hop. Many passerines cannot walk and hopping is their only mode of terrestrial locomotion. Some groups of passerines have a more terrestrial mode of existence and these birds possess a walking gait as well as the ability to hop. Larks, pipits, starlings, and grackles are examples. The separation between walking and hopping passerines cuts across families. For example, in the family Corvidae, the ravens, crows, and rooks are walkers, whereas the jays and magpies are hoppers.

The most specialized avian foot for perching on branches is one in which all four toes are free and mobile, of moderate length, and with the hind toe well developed, lying in the same plane as the forward three, and opposable to them (**ansiodactyl**). Such a foot produces a firm grip and is highly developed in the passerine birds. The **zygodactylous** condition, with two toes forward and oppos-

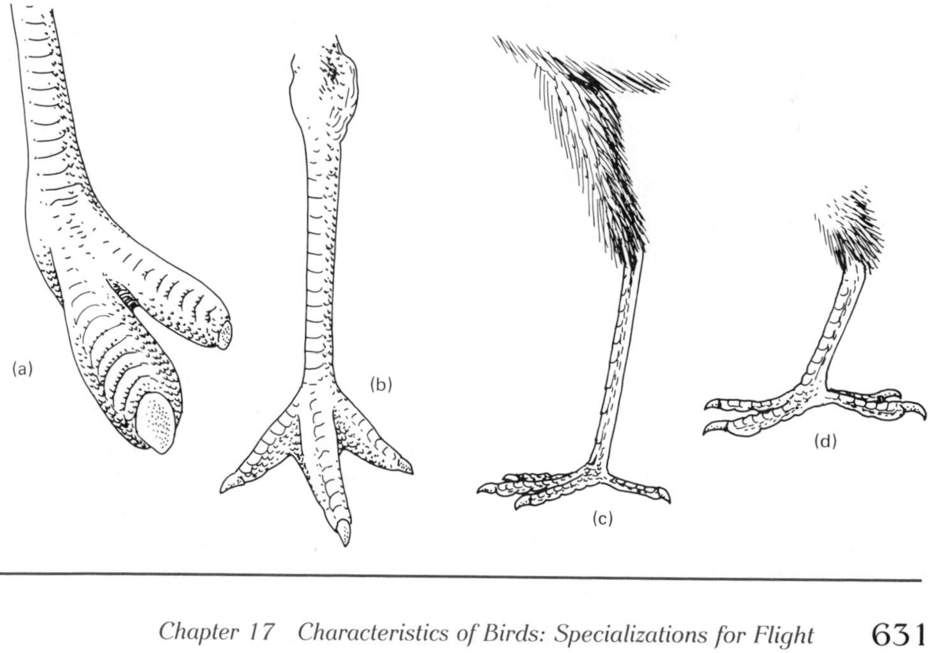

Figure 17–25. Avian feet with various specializations for terrestrial locomotion: (a) ostrich, with only two toes; (b) rhea, with three toes; (c) secretary bird, with a typical avian foot; (d) roadrunner, with zygodactyl foot. (Not drawn to scale.)

Figure 17–26. Graviportal birds: (a) elephantbird of Madagascar and
(b) *Diatryma*, compared with (c) the cursorial ostrich.

able to two extending backward, provides an even firmer grip, but seems to be better adapted for birds such as parrots and woodpeckers, which climb or perch on vertical surfaces.

The tendons involved in flexing the toes of a perching bird show modifications that lock the foot in a tight grip so that the bird does not fall off its perch when it relaxes or goes to sleep. The plantar tendons, which insert on the individual phalanges of the toes, slip in grooves and sheaths that are positioned anterior to the knee joint and posterior to the ankle joint in such a way that the tendons tighten and flex the toes around the perch just by the weight of the resting bird; thus, muscular contraction is not required to hold the toes closed. Furthermore, the tendons lying underneath the toe bones have hundreds of minute, rigid, hobnail-like projections that mesh with ridges on the inside surface of the surrounding tendon sheath. The projections and ridges lock the tendons in place in the sheaths and help to hold the toes in their grip around the branch.

Birds that wade are specialized for walking at various depths and over various types of bottoms.

Every shoreline community has its complement of short-legged, intermediate, and long-legged waders, each exploiting a somewhat different wading and feeding niche. A similar stratification of species can often be seen in mixed species aggregations of herons in a southern swamp such as the Florida Everglades (Figure 17–27). If the bottom substrate is soft mud, wading birds require feet with longer toes, as seen in herons, or with partial webbing between short toes (flamingos, avocets, and others).

Birds that walk on insubstantial surfaces usually have long toes (Figure 17–28). The jacanas or lily-trotters, which walk on floating vegetation, have the longest toes and claws relative to their body size of any birds. Certain sandgrouse have especially short and broad claws and densely feathered toes, the whole foot forming a broad base for the bird when walking in loose sand. The winter plumage of northern grouse has similar modifications of the feet for walking on snow.

Some birds have poorly developed pedal locomotion. Highly aerial species such as swifts, swallows, and some goatsuckers use their feet for little

Figure 17–27. The water depths at which wading birds forage are determined in part by the length of their legs. From left to right this figure shows species that feed in water of progressively greater depth: green heron, little blue heron, reddish egret, common egret, and great blue heron.

Figure 17–29. Webbed and [...] of some aquatic birds: (a) duck [...] partial webbing; (b) cormorant, [...] totipalmate webbing. The lobed [...] grebe showing how it is rotated [...] stroke: (c) position of toes durin[...] backward power stroke in side v[...] front view; (e) side view of the r[...] foot during the forward recovery [...] (From R. T. Peterson, 1978, *The* [...] 2nd edition, Time-Life Books, N[...]

Diving and Swimming Under Water

The transition from a surface [...] subsurface swimmer has occ[...] mentally different ways: eith[...] ization of a hindlimb already[...] ming or by modification of t[...] flipper under water. Highl[...] propelled divers have evolve[...] the grebes, cormorants, loon[...] Hesperornithidae. All these [...] loons include some flightless f[...] Wing-propelled divers have e[...] lariiformes (the diving petrels[...] formes (penguins), and in t[...] (auks and related forms). Onl[...] fowl are there both foot-prope[...] pelled diving ducks, but none [...]

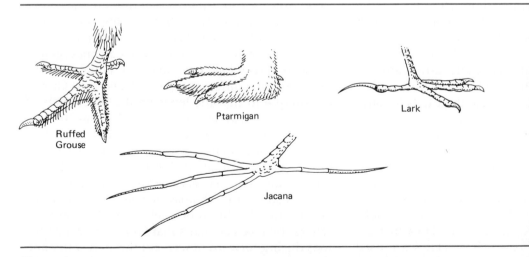

Ruffed Grouse

Ptarmigan

Lark

Jacana

Figure 17–28. Specializations of the feet of birds that walk on insubstantial surfaces like snow (ruffed grouse, ptarmigan), sand (lark), or lily pads (jacaña).

more than clinging
cialized aquatic forr
have their legs posit
propulsion through
walk upright on lar
upright, but their leg
move very fast. The
tobogganing over sr
modified wings as we
one of the few case
used by birds for no:

Climbing

Birds climb on tree tru
by using their feet, t.
forelimbs. Several dis
have independently
climbing and foraging
cies such as woodp
which use their toes
near the base of a tre
vertically upward, he
with strong feet on si
used as a prop to brac
erful pecking exertion
gostyle and free caud.
are much enlarged at
feathers. A similar mc
same function is found
on cave walls and insi

Nuthatches and si
on trunks and rock wa.
head-downward direct
these species, which d
port, the claw on the
on the forward-direc
curved.

Although a few n
with their wings, only
America has evolved a
forelimbs for climbing.
on the wing bear large
moved by special mus
for grasping branches
leaves the nest. Later i

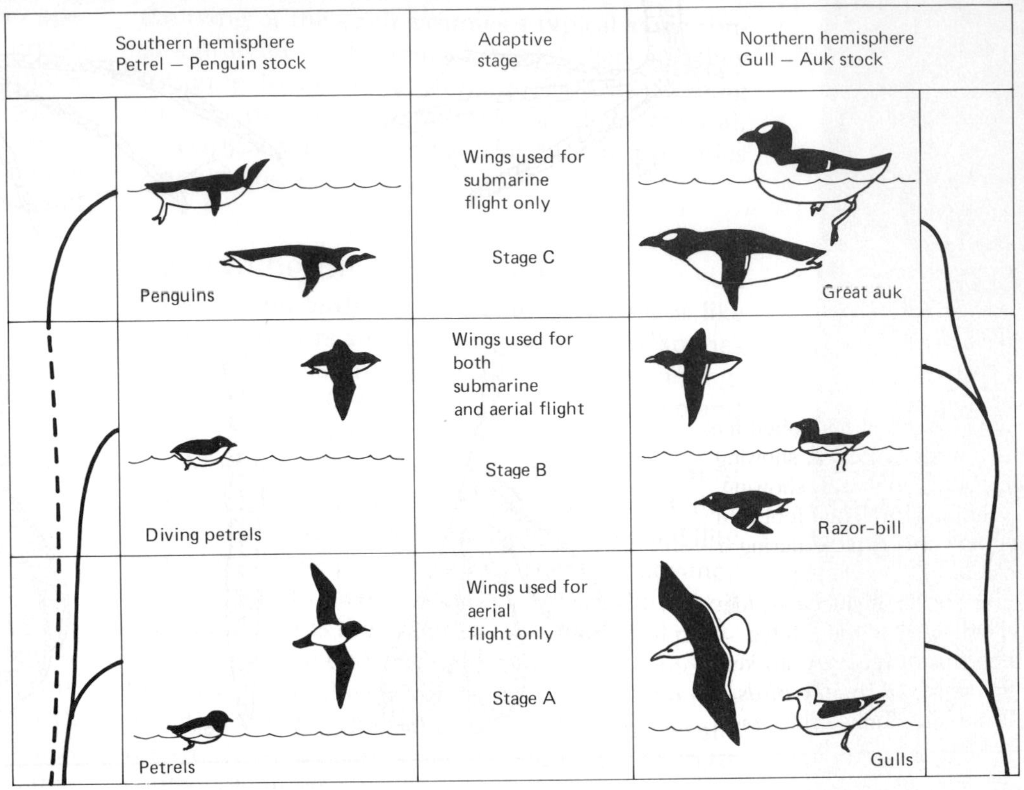

Figure 17–30.   Parallel evolution of swimming and diving birds in the Southern Hemisphere (Procellariiformes and Sphenisciformes) and the Northern Hemisphere (Charadriiformes). (From R. W. Storer, 1971, in *Avian Biology*, volume 1, edited by D. S. Farner and J. R. King, Academic Press, New York.)

a high oxygen-carrying capacity and muscles that are especially rich in myoglobin, they are able to tolerate high carbon dioxide levels in blood, and they can obtain considerable energy from anaerobic metabolism. These features allow diving times from 1 to 3 minutes, with a maximum recorded survival time of 15 minutes (Kooyman 1975, Butler and Jones 1982).

The Sensory Systems

A bird moves rapidly through three-dimensional space and requires a continuous flow of sensory information about its position and the presence of obstacles in its path. Vision and hearing are the senses best suited to provide this sort of information on a rapidly renewed basis, and birds have well-developed visual and auditory systems. Their predominance is reflected in the brain: The optic lobes are large and the tectum is an important area for processing visual and auditory information. Olfaction is relatively unimportant for most birds and the olfactory lobes are small. The cerebellum, which coordinates body movements, is large. The cerebrum is less developed in birds than it is in mammals and is dominated by the corpus striatum. Many aspects of the behavior of birds are relatively stereotyped in comparison to the more plastic behavioral responses of mammals, and that difference may reflect the greater development of the neopallium of mammals.

Vision

The eyes of birds are large—so large that the brain is displaced dorsally and caudally and in many species the eyeballs meet in the midline of the skull. The eyes of some hawks, eagles, and owls are as large as the eyes of humans. In its basic structure the eye of a bird is like that of any other vertebrate, but the shape varies from a flattened sphere to something approaching a tube (Figure 17–31). The geometry of the eyes of birds may be the result of fitting large eyes into small skulls, but other features of the eyes have functional significance. The flat eyes that are characteristic of most birds provide a wide field of view; many birds have a binocular view of objects in front of them and a monocular field of view to the sides. Some birds have a full 360° visual field—that is, they can see in all directions. The asymmetry of the retina of many birds may be related to maximizing the width of the visual field.

Birds with high visual acuity, such as hawks and eagles, have eyes in which the distance from the lens to the retina is approximately equal to the width of the eye. This shape increases visual acuity because the image formed by the lens is focused on a wider area of the retina and is thus spread over more retinal photoreceptors. Birds that see at low light levels, owls for example, have eyes that are still more tube-shaped than those of most raptors. This geometry apparently increases the range of intensities of illumination over which the owl's eye can function. Horizontally oriented cells within the retina form a network that may allow groups of adjacent photoreceptor cells to interact. By connecting a group of individual photoreceptors, these cells may be able to change the diameter of the photoreceptive area. At low light intensities the output of many photoreceptors can be summed to give an effective photoreceptor that receives stimuli from a wide angle and provides high sensitivity. The tube-shaped eye of an owl spreads the visual image over a large number of retinal cells, perhaps allowing effective photoreceptors of various sizes to be formed to match different light conditions (Martin 1983).

The pecten is a conspicuous structure in the eye of birds. It is formed exclusively of blood capillaries surrounded by pigmented tissue and cov-

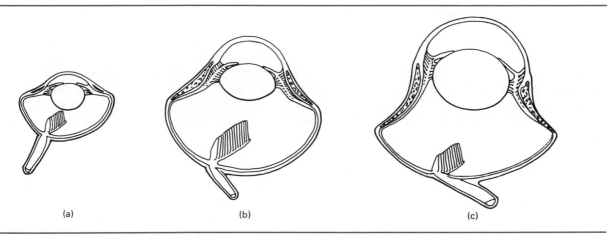

(a)   (b)   (c)

Figure 17–31. Variation in the shape of the eye of birds: (a) flat, typical of most birds; (b) globular, found in most falcons; (c) tubular, characteristic of owls and some eagles.

ered by a membrane; it lacks muscles and nerves. The pecten arises from the retina at the point where the nerve fibers from the ganglion cells of the retina join to form the optic nerve. In some species of birds the pecten is small, but in other species the pecten extends so far into the vitreous humor of the eye that it almost touches the lens. The function of the pecten remains uncertain after 200 years of debate. Possible functions for the pecten in vision have been proposed (reduction of glare, a mirror to reflect objects above the bird, production of a stroboscopic effect, a visual reference point), but none seems very likely. The pecten is formed largely by blood capillaries, and its highly vascularized structure suggests that it may provide nutrition for the retinal cells and help to remove metabolic waste products that accumulate in the vitreous humor.

Oil droplets are found primarily in the cone cells of the avian retina, as they are in other vertebrates. The droplets range in color from red through orange and yellow to green; red droplets occur only in birds and turtles. Different color oil droplets are associated with various types of cone cells, and these cells are localized in different areas of the retina. For example, the central dorsal part of the retina of the pigeon has a predominance of photoreceptors with red droplets and the ventral and lateral parts of the retina have cells with yellow droplets. The oil droplets act as filters, absorbing some wavelengths of light and transmitting others. The function of the oil droplets is unclear and is certainly complex because the various colors of droplets are combined with different kinds of photoreceptor cells and different visual pigments. Birds like gulls, terns, gannets, and kingfishers that must see through the surface of water have a preponderance of red droplets. Aerial hawkers of insects (swifts, swallows) have predominantly yellow droplets.

Hearing

In birds, as in other diapsids, the columella and its cartilaginous extension, the extracolumella, transmit vibrations of the tympanum to the oval window of the inner ear. The cochlea of birds has the same basic structure as that of other vertebrates (Chapter 3), but it appears to be specialized for fine distinctions of the frequency and temporal pattern of sound. The cochlea of a bird is about one-tenth the length of the cochlea of a mammal, but it has about ten times as many hair cells per unit of length. The space above the basilar membrane (the scala vestibuli) is nearly filled in birds by a folded, glandular tegmentum. This structure may damp sound waves, allowing the ear of a bird to respond very rapidly to changes in sounds.

The openings of the external auditory meatus are small, only a few millimeters in diameter in most birds, and are covered with feathers that may ensure laminar airflow across the opening during flight. The columellar muscle inserts on the columella. On contraction this muscle draws the columella away from the oval window, decreasing the sound energy transmitted to the inner ear. This mechanism may protect the ear against the noise of wind during flight and it may also help to tune the auditory system. In starlings contraction of the columellar muscle increases the effect of the middle ear as a filter for sounds of different frequencies. The sounds that most readily pass the filter are those in the range of greatest auditory sensitivity, which, in most birds, corresponds to the frequency range of their own vocalizations.

Localization of sounds in space can be difficult for small animals such as birds. Large animals localize the source of sounds by comparing the time of arrival, intensity, or phase of a sound in their left and right ears, but none of these methods is very effective when the distance between the ears is small. The pneumatic construction of the skulls of birds may allow them to use sound that is transmitted through the air-filled passages between the middle ears on the two sides of the head to increase their directional sensitivity. If this is true, internally transmitted sound would pass from the middle ear on one side to the middle ear on the other side and reach the *inner* surface of the contralateral tympanum. Here it would interact with the sound arriving on the external surface of the tympanum via the external auditory meatus. The vibration of each tympanum would be the product of the combination of pressure and phase of the internal and external sources of sound energy, and the magnitude of the cochlear response would be proportional to the difference in pressure across the tympanic membrane. Localization of sound by owls is apparently further aided by asymmetry of the external auditory openings (Box 17–2).

The sensitivity of the auditory system of birds is approximately the same as that of humans, despite the small size of birds' ears. Most birds have tympanic membranes that are large in relation to the size of the head. A large tympanic membrane enhances auditory sensitivity, and owls (which have especially sensitive hearing) have the largest tympani relative to their head size among birds. Sound pressures are amplified during transmission from the tympanum to the oval window of the cochlea because the area of the oval window is smaller than the area of the tympanum. The reduction ratio for birds ranges from 11 to 40. High ratios suggest sensitive hearing, and the highest values are found in owls; songbirds have intermediate ratios (20 to 30). (The ratio is 36 for cats and 21 for humans.) The inward movement of the tympanum as sound waves strike it is opposed by air pressure within the middle ear and birds show a variety of features that reduce the resistance of the middle ear. The middle ear is continuous with the dorsal, rostral, and caudal air cavities in the pneumatic skulls of birds. In addition to potentially allowing sound waves to be transmitted to the contralateral ear, these interconnections increase the volume of the middle ear and reduce its stiffness, thereby allowing the tympanum to respond to faint sounds. Owls have more spacious middle ear cavities and more extensive communication of the middle ear with the air spaces of the skull than do other birds.

Olfaction

The sense of smell is well developed in some birds and poorly developed in others; a review of the olfactory systems of birds can be found in Bang and Wenzel (1985). The size of the olfactory bulbs is a rough indication of the sensitivity of the olfactory system. Relatively large bulbs are found in ground nesting and colonial nesting species, species that are associated with water, and carnivorous and piscivorous species of birds. Some birds use scent to locate prey. The kiwi, for example, has nostrils at the tip of its long bill and finds earthworms underground by smelling them. Turkey vultures follow airborne odors of carrion to the vicinity of a carcass, which they then locate by sight. Sponges soaked in fish oil and placed on floating buoys attracted shearwaters, fulmars, albatrosses, and petrels even in the dark (Wenzel 1986).

Olfaction probably plays a role in the orientation and navigation abilities of some birds. The tube-nosed seabirds (albatrosses, shearwaters, fulmars, and petrels) nest on islands, and when they return from foraging at sea they approach the islands from downwind. The well-developed nasal bulbs of colonial nesting species of birds suggest the possibility that olfaction is used for social functions such as recognition of individuals, but this has never been demonstrated.

Other Senses

Birds use a variety of cues for navigation, and many of these are extensions of the senses we have discussed into regions beyond the sensitivity of mammals. For example, birds can detect polarized light when it falls on the area of the retina that normally receives skylight, but other parts of the retina cannot detect polarization. Ultraviolet light (light with a wavelength of less than 400 nanometers) is not detected by most mammals, but pigeons are more sensitive to ultraviolet light in the region of 350 nanometers than they are to light at any wavelength in the visible part of the spectrum.

Birds have been shown to be sensitive to very small differences in air pressure. Experiments by Melvin Kreithen (1983) showed that pigeons were able to detect the difference in air pressure between the ceiling and floor of a room, and he calculated that a pigeon flying upward at a rate of 4 centimeters per second would be able to detect a

Box 17–2. Not Hearing Straight: Ear Asymmetry of Owls

Owls are acoustically the most sensitive of birds. At frequencies up to 10 kilohertz (10,000 cycles per second), the auditory sensitivity of an owl is as great as that of a cat. Owls have large tympanic membranes, large cochleae, and well-developed auditory centers in the brain. Some owls are diurnal, others crepuscular (active at dawn and dusk), and some are entirely nocturnal. In an experimental test of their capacities for acoustic orientation, barn owls were able to seize mice in total darkness (Konishi 1973). If the mice towed a piece of paper across the floor behind them, the owls struck the rustling paper instead of the mouse, showing that sound was the cue they were using.

A distinctive feature of many owls is the facial ruff that is formed by stiff feathers. The ruff acts as a parabolic sound reflector, focusing sounds with frequencies above 5 kilohertz on the external auditory meatus and amplifying them by 10 decibels. The ruffs of some owls are asymmetric, and that asymmetry appears to enhance the ability of owls to locate prey. When the ruffs were removed from the barn owls in Konishi's experiments, the owls made large errors in finding targets.

Asymmetry of the aural system of owls goes beyond the feathered ruff. The skull itself is markedly asymmetric in many owls (Figure 17–32), and these are the species with the greatest auditory sensitivity (Norberg 1977). The asymmetry ends at the external auditory meatus; the middle and inner ears of owls are bilaterally symmetric. R. Å. Norberg (1978) has shown that the asymmetry of the external ear openings of owls assists with localization of prey in the horizontal and vertical axes. The time at which sounds are received by the two ears can tell the horizontal direction of the source, and owls are capable of detecting differences of a few hundredths of a millisecond

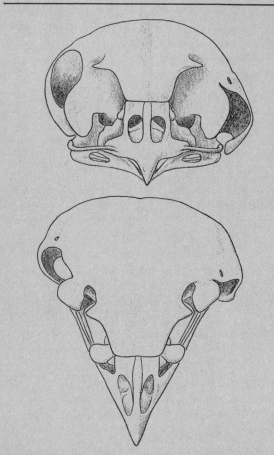

Figure 17–32. The skulls of many owls show pronounced asymmetry in the position of the external auditory meatus that assists in localization of sound. (From R. Å. Norberg, 1978, *Philosophical Transactions of the Royal Society of London* B 282:325–410.)

in the arrival times of a sound at the left and right ears. The vertical direction of a sound source can be determined by the differential sensitivity of the two ears to sounds coming from above and below the level of the owl. The soft tissue and feathers surrounding the face of an owl produce a situation of reversed asymmetry in which the actual directional sensitivity of the ears is the reverse of what would be expected from examination of the skull. The left ear, which opens low on the side of the head, is most sensitive to sounds that originate *above* the level of the owl's head, and the right ear is sensitive to sounds coming from *below* the level of the head. This auditory asymmetry applies only to sounds with frequencies above 6000 hertz; sensitivity to lower frequencies is bilaterally symmetrical.

change in its altitude of 4 millimeters. This sensitivity may be useful during flight, and may also play a role in allowing birds to anticipate changes in weather patterns that are important in the timing of migration.

Another kind of sensory information that may tell birds about the weather is infrasound—very low frequency sounds (less than 10 cycles per second) that are produced by large-scale movements of air. Thunderstorms, winds blowing across valleys, and other geophysical events produce infrasound that is propagated over thousands of miles. The lower-frequency limit of human hearing is around 300 hertz, but birds can hear well into the region of infrasound. Pigeons, for example, can detect frequencies as low as 0.05 hertz (3 cycles per *minute*).

The magnetic field of the earth provides a cue that could be used for orientation if an animal were able to perceive it, and considerable evidence suggests that birds are able to detect magnetic fields. The orientation of several species of birds during their migratory periods can be adjusted in predictable ways by placing them in an artificial magnetic field. The mechanism by which the magnetic field is detected is not known, but deposits of magnetite recently described in the heads of pigeons may be involved (Walcott et al. 1979, Presti 1985, Wiltschko and Wiltschko 1988).

Summary

Feathers are the distinguishing character of birds, and flight is the distinctive mode of avian locomotion. In many respects the morphology of birds is shaped by the demands of flight. Flapping flight is a more complicated process than flight with fixed wings like those of aircraft, but it can be understood in aerodynamic terms. Feathers compose the aerodynamic surfaces responsible for lift and propulsion during flight and they also provide streamlining. There are many variations and specializations within the four basic types of wings, but in general high speed and elliptical wings are used in flapping flight, whereas high aspects and high-lift wings are used for soaring and gliding. The hollow birds of bones, probably an ancestral character of the archosaur lineage, combine lightness and strength. The air sacs, some of which fill several of the pneumatic bones, create a one-way flow of air through the lung that enhances oxygen uptake, and the air sacs probably also help to dissipate the heat produced by muscles during flight.

The skeletons of birds are light in relation to the

sizes of their bodies, and some of this lightness has been achieved by modification of the skull. Several skull bones that are separate in other diapsids are fused in birds, air sacs are extensively developed, and teeth are absent. The function of teeth in processing food has been taken over by the muscular gizzard, which is part of the stomach of birds. The gizzard is well developed in birds that eat hard items such as seeds and it may contain stones that probably assist in grinding food.

Vision is the primary sense for most birds, and visual capacities extend to detection of ultraviolet and polarized light. Hearing is also important and the auditory system of birds is capable of very precise discriminations of the frequencies and temporal patterns of sound. Sensitivity extends downward to include sounds with frequencies lower than 10 cycles per second. Sensitivity to magnetism has been demonstrated for several species of birds, but the mechanisms involved are unknown. Olfactory sensitivity is variably developed among birds: Some species use scent to locate food and for navigation, but other species appear not to use olfaction at all.

References

Bang, B. G. and B. M. Wenzel. 1985. Nasal cavity and olfactory system. Pages 195–225 in *Form and Function in Birds*, volume 3, edited by A. S. King and J. McLelland. Academic Press, London.

Butler, P. J. and D. R. Jones. 1982. The comparative physiology of diving in vertebrates. Pages 179–264 in *Advances in Comparative Physiology and Biochemistry*, volume 8, edited by O. Lowenstein. Academic Press, New York.

Feduccia, A. 1980. *The Age of Birds*. Harvard University Press, Cambridge, Mass.

Gaunt, A. S., S. L. L. Gaunt, and R. M. Casey. 1982. Syringeal mechanics reassessed: evidence from *Streptopelia*. *Auk* 99:474–494.

Greenewalt, C. H. 1968. *Bird Song: Acoustics and Physiology*. Smithsonian Institution Press, Washington, D.C.

Greenewalt, C. H. 1969. How birds sing. *Scientific American* 221:126–139.

King, A. S. and D. Z. King. 1979. Avian morphology: general principles. Pages 1–38 in *Form and Function in Birds*, volume 1, edited by A. S. King and J. McLelland. Academic Press, London.

Konishi, M. 1973. How the owl tracks its prey. *American Scientist* 61:414–424.

Kooyman, G. L. 1975. Behaviour and physiology of diving. Pages 115–137 in *The Biology of Penguins*, edited by B. Stonehouse. University Park Press, Baltimore.

Kreithen, M. L. 1983. Orientational strategies in birds: a tribute to W. T. Keeton. Pages 3–28 in *Behavioral Energetics: The Cost of Survival in Vertebrates*, edited by W. P. Aspey and S. I. Lustick. Ohio State University Press, Columbus, Ohio.

Martin, G. R. 1983. Eye. Pages 311–373 in *Form and Function in Birds*, volume 3, edited by A. S. King and J. McLelland. Academic Press, London.

Norberg, R. Å. 1977. Occurrence and independent evolution of bilateral ear asymmetry in owls and implications for owl taxonomy. *Philosophical Transactions of the Royal Society of London* B280:375–408.

Norberg, R. Å. 1978. Skull asymmetry, ear structure and function, and auditory localization in Tengmalm's owl, *Aegolius funereus* (Linné). *Philosophical Transactions of the Royal Society of London* B282:325–410.

Norberg, U. M. 1985. Flying, gliding, and soaring. Pages 129–158 in *Functional Vertebrate Morphology*, edited by M. Hildebrand, D. M. Bramble, K. F. Liem, and D. B. Wake. Harvard University Press, Cambridge, Mass.

Nowicki, S. 1987. Vocal tract resonances in oscine bird sound production: evidence from birdsongs in a helium atmosphere. *Nature* 325:53–55.

Pennycuick, C. J. 1975. Mechanics of flight. Pages 1–75 in *Avian Biology*, volume 5, edited by D. S. Farner, J. R. King, and K. C. Parkes. Academic Press, New York.

Presti, D. E. 1985. Avian navigation, geomagnetic field sensitivity, and biogenic magnetite. Pages 455–482 in *Magnetite Biomineralization and Magnetoreception in Organisms*, edited by J. L. Kirschvink, D. S. Jones, and B. J. McFadden. Plenum Press, New York.

Raikow, R. J. 1985. Locomotor system. Pages 57–147 in *Form and Function in Birds*, volume 3, edited by A. S. King and J. McLelland. Academic Press, London.

Rayner, J. M. V. 1988. Form and function in avian flight. Pages 1–66 in *Current Ornithology*, volume 5, edited by R. J. Johnston. Plenum Press, New York.

Scheid, P. 1982. Respiration and control of breathing. Pages 406–453 in *Avian Biology*, volume 6, edited by D. S. Farner, J. R. King, and K. C. Parkes. Academic Press, New York.

Walcott, C., J. L. Gould, and J. L. Kirschvink. 1979. Pigeons have magnets. *Science* 205:1027–1029.

Wenzel, B. M. 1986. The ecological and evolutionary challenges of procellariiform olfaction. Pages 357–368 in *Chemical Signals in Vertebrates,* Volume 4, *Ecology, Evolution, and Comparative Biology*, edited by D. Duvall, D. Möller-Schwarze, and R. M. Silverstein. Plenum Press, New York.

Wiltschko, W. and R. Wiltschko. 1988. Magnetic orientation in birds. Pages 67–121 in *Current Ornithology*, volume 5, edited by R. F. Johnston. Plenum Press, New York.

urine lineage included *Archaeopteryx* and a group of primitive birds known as Enantiornithes (opposite birds). The anterior thoracic vertebrae were fused in all sauriurines, and an anteromedial process of the scapula abutted these vertebrae. The Enantiornithians were volant forms with well-developed wings, but the keel on the sternum was not large and the coracoids may have been important sites of muscle origins. The enantiornithians had a bird-like wing skeleton that included a fused carpometacarpus, but the metatarsal morphology was so dinosaur-like that their identification as birds was initially received with skepticism. Enantiornithians are known from fossil deposits in both the Northern and Southern Hemispheres, and the material from Argentina is particularly diverse. Enantiornithians were apparently predators and one form had raptorial claws like those of modern hawks. Most enantiornithians were crow-size to heron-size, but *Alexornis* was the size of a sparrow. *Gobipteryx* is an enantiornithian known from skulls found in late Cretaceous deposits in Mongolia. *Gobipteryx* is unique among Cretaceous birds in lacking teeth. In the same deposits that yielded fossils of adult *Gobipteryx*, the Polish Mongolian Expedition found fossilized eggs, some of which contained well-preserved skeletons with skulls very like those of adult *Gobipteryx*. The skeletons of these embryos were well developed, and it is likely that the chicks were precocial at hatching.

The other half of the Mesozoic avian dichotomy proposed by Martin is the Ornithurae, which includes all the living birds. The earliest ornithurines known are the Hesperornithiformes (western birds). *Enaliornis* establishes the presence of hesperornithiforms in the early Cretaceous, and they were well established by the late Cretaceous. The hesperornithiforms were medium-size to large flightless birds that were specialized for foot-propelled diving (Figure 18–1). The body and neck were elongate, the sternum lacked a keel, and the bones were not pneumatic. Teeth remained in the maxilla and dentary. Feathers preserved with two specimens of *Parahesperornis* were plumulaceous and the birds may have had a furry appearance somewhat like that of the living kiwis. The feet were placed far posteriorly on the body (a position that is characteristic of many foot-propelled diving birds) and the toes had lobes like those seen in modern grebes. The femur and tibiotarsus were locked in place and could not be rotated under the body. As a result, hesperornithiforms would not have been able to walk on land and must have pushed themselves along by sliding on their stomachs. The lateral placement of the feet would have made it possible for hesperornithiforms to exert force directly backward during swimming and diving without an upward component that would have tended to drive them toward the surface. Pachyostosis (an increased density of bone) gave hesperornithiforms a high specific gravity that would have facilitated diving. Coprolites (fossilized feces) found in association with *Hesperornis* and *Baptornis* contain the remains of small fishes.

Another toothed bird, *Ichthyornis*, is known from the same late Cretaceous deposits in Kansas that contain *Hesperornis*, but the ichthyornithiforms were flying birds with a well-developed keel on the sternum. Several species of *Ichthyornis* have been named, mostly on the basis of differences in size. In general, ichthyornithiforms were the size of gulls and terns, and they may have had similar habits. By the late Cretaceous these toothed birds were probably oceanic relicts like modern frigate birds.

The Evolution of Modern Orders and Families of Birds

Two modern orders of birds have been identified in the late Cretaceous, the Charadriiformes (wading birds) and perhaps the Procellariiformes (tube-nosed seabirds). The charadriiforms are abundant and diverse in deposits from New Jersey and Wyoming, whereas the presence of Procellariiformes is deduced from the presence of only two bones in the New Jersey deposits.

Most orders of birds had evolved by the end

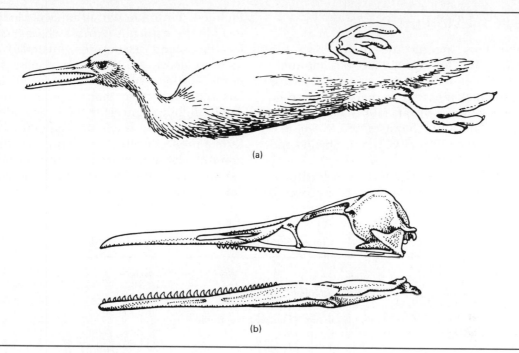

(a)

(b)

Figure 18–1. The hesperornithiforms were flightless, toothed birds.
(a) Restoration of *Hesperornis;* (b) skull of *Parahesperonsis.* Note the teeth in the
maxilla and dentary bones. [(a) From A. Feduccia, 1980, *The Age of Birds*, Harvard
University Press, Cambridge, Mass.; (b) from L. D. Martin, 1983, in *Perspectives in
Ornithology*, edited by A. H. Brush and G. A. Clark, Jr., Cambridge University
Press, Cambridge.]

of the Eocene some 40 to 50 million years ago.
Groups associated with deposits of Eocene age include the ostriches (Struthioniformes), the rheas
(Rheiformes), the recently extinct elephant birds of
Madagascar (Aepyornithiformes), the penguins
(Sphenisciformes), the tube-nosed marine birds
(Procellariiformes), the kingfisher-like birds (Coraciiformes), the woodpeckers (Piciformes), and the
perching birds (Passeriformes). The first diurnal
birds of prey (Falconiformes) had already appeared in the Paleocene. Five other orders first appear in the fossil record around the end of the
Eocene or the beginning of the Oligocene: the
doves and pigeons (Columbiformes), cuckoos
(Cuculiformes), goatsuckers (Caprimulgiformes),
swifts (Apodiformes), and trogons (Trogoni-

formes), so that by the beginning of the Oligocene
at least 26 orders of birds had become differentiated.

Two major radiations of avian families occurred during the Tertiary. The Eocene was the
epoch of greatest diversification of birds. More surviving families arose in that period than at any
other time. Most of these families consist of additional water birds and nonpasserine forest dwellers. A second radiation occurred in the Miocene
and included a few additional families of water
birds, but mostly land-dwelling passerines that
were adapted to drier, less forested environments.
Most families of birds had evolved by the end of
the Miocene, and many still-existing genera and
some species were present by the Pliocene.

Phylogeny of Modern Birds

The relationships among the living birds are poorly known and are the subject of continuing controversy. Several views of these relationships and the difficulties inherent in their study can be found in reviews by Cracraft (1981, 1986), Olson (1982, 1985), Sibley and Ahlquist (1983, 1986), Raikow (1985), Houde (1986, 1987), and Shields and Helm-Bychowski (1988).

One of the major controversies in bird phylogeny centers on the relationships of a group of flightless birds known as ratites. Modern ratites include the ostriches (Africa), rheas (South America), emus and cassowaries (Australia), and kiwis (New Zealand). All of these land masses were part of the southern supercontinent Gondwana, and it has been suggested that the ratites arose from a single flightless ancestor that was widely distributed on Gondwana. By this hypothesis the ratites form a monophyletic group and their current geographic distribution reflects the fractionation of Gondwana in the late Mesozoic and early Cenozoic.

Table 18–1. Classification of living birds to the level of orders.

Aves (about 9000 species)

Superorder Palaeognathae (ratites)
  Order Tinamiformes (tinamous)
    (1 family, 9 genera, 42 species; Neotropics)
  Order Rheiformes (rheas)
    (2 families, 1 fossil, 2 living genera, 2 species; South America)
  Order Struthioniformes (ostriches)
    (1 family, 1 species, Africa)
  Order Casuariiformes (emus and cassowaries)
    (3 families, 1 fossil, 2 living genera, 5 species; Australia, New Guinea)
  Order Dinornithiformes (moas; kiwis)
    (3 families, 2 fossil, 1 living genus, 3 species; New Zealand)
Superorder Neognathae
  Order Podicipediformes (grebes)
    (1 family, 5 genera, 18 species; worldwide)
  Order Sphenisciformes (penguins)
    (1 family, 6 genera, 17 species; southern oceans)
  Order Procellariiformes (albatrosses, shearwaters, petrels)
    (4 families, 23 genera, about 95 species; all oceans)
  Order Pelecaniformes (tropic birds, boobies, gannets, cormorants, pelicans, frigate birds)
    (9 families, 3 fossil, 10 living genera, 58 species; worldwide, primarily marine)
  Order Anseriformes (screamers and waterfowl)
    (2 families, 47 genera, 153 species; worldwide)
  Order Phoenicopteriformes (flamingos)
    (6 families, 5 fossil, 3 living genera, 6 species; all continents but Australia and Antarctica)
  Order Ciconiiformes (herons, bitterns, whale-head stork, storks, ibises, spoonbills)
    (5 families, 65 genera, 117 species; worldwide, especially tropical and north temperate)
  Order Falconiformes (condors, hawks, eagles, kites, falcons, caracaras)
    (6 families, 1 fossil, 81 living genera, 287 species; worldwide)
  Order Galliformes (upland gamebirds and fowl)
    (5 families, 95 genera, 275 species; worldwide)

An alternative view proposes that the similarities, of modern ratites are primitive characters that were found in many lineages of birds. The ratites and another group of birds, the tinamous of Central and South America, share a paleognathous palatal structure. This palate is characterized by long prevomers that extend posteriorly to articulate with the palatines and pterygoids and by large basipterygoid processes that articulate with the pterygoids. The paleognathus palate is distinguished from the neognathus palate that is characteristic of other birds. Paleognathous palatal characters are seen in toothed birds such as *Hesperornis* and in the primitive toothless bird *Gobipteryx*, and they are present in early stages of the embryonic development of neognathus birds. Thus, the paleognathus condition of ratites and tinamous could have evolved by neoteny from a neognathus form. This possibility indicates that the paleognathus birds are not necessarily a monophyletic lineage. Furthermore, the discovery of birds related to ostriches in Paleocene and Eocene deposits in North America and Europe casts doubt on the assumption that the current distribution of

---

Order Gruiformes (rails, sungrebes, kagus, sunbitterns, roatelos, buttonquail, cranes, limpkins, trumpeters, seriemas, phororhacids, bustards)
   (22 families, 10 fossil, 81 living genera, 205 species; worldwide, but many families endemic to restricted regions)
Order Charadriiformes (shorebirds, plovers, sandpipers, gulls, terns, auk, murres, puffins, sand grouse)
   (19 families, 2 fossil, about 84 living genera, 314 species; worldwide, especially northern oceans)
Order Gaviiformes (loons)
   (1 family, 1 genus, 4 species; Holarctic)
Order Columbiformes (doves, pigeons, dodos)
   (3 families, 2 fossil, 59 living genera, 306 species; worldwide)
Order Psittaciformes (parrots)
   (1 family, 80 genera, 339 species)
Order Cuculiformes (turacos, cuckoos)
   (2 families, 44 genera, 150 species; 1 family restricted to Africa, the other worldwide)
Order Strigiformes (owls)
   (3 families, 1 fossil, 26 genera, 146 species; worldwide)
Order Caprimulgiformes (oilbirds, goatsuckers)
   (5 families, 24 genera, 94 species; 1 family restricted to South America, the other worldwide)
Order Apodiformes (swifts, hummingbirds)
   (4 families, 1 fossil, 152 living genera, 411 species; hummingbirds restricted to New World, swifts worldwide)
Order Trogoniformes (trogons)
   (1 family, 8 genera, 34 species; pantropical except Australia)
Order Coliiformes (mousebirds)
   (1 family, 1 genus, 6 species; Africa)
Order Coraciiformes (kingfishers, todies, motmots, bee-eaters, rollers, hoopoe, hornbills)
   (10 families, 49 genera, 195 species; worldwide, especially Old World tropics and subtropics)
Order Piciformes (jacamars, barbets, honeyguides, toucans, woodpeckers)
   (6 families, 72 genera, 391 species; worldwide, especially tropics)
Order Passeriformes (perching birds, including the songbirds)
   (60 families, 1 fossil and 46 in the diverse songbird suborder, 328 nonsongbird genera, more than 570 songbird genera, 1117 nonsongbird species, more than 4050 songbird species; worldwide)

ratites is the result of events associated with the fractionation of Gondwana (Houde 1986).

The phylogeny and zoogeography of ratites are of considerable significance in current biochemical studies of the phylogeny of birds, because Charles Sibley and Jon Ahlquist based the calibration of the DNA molecular clock on the assumption that the separation of the lineages of ratites was caused by the breakup of Gondwana. If the ratites do represent two or more independent origins, or if the origins of some ratites were in the Northern Hemisphere rather than in Gondwana, the calibration of the DNA clock is erroneous and conclusions about relationships among other groups of birds that were based on that calibration are weakened. Further discussion of this controversy can be found in Olson (1985) and Houde (1987).

Disagreement over the phylogenetic relationships of birds has led to a proliferation of new classifications and a failure of biologists to adopt any one system. The most commonly used classification is based on work carried out in the late nineteenth century by Hans Gadow. The lack of correspondence between this classification and a true phylogeny of birds is widely recognized. Nonetheless, the Gadow–Wetmore classification presented in Table 18–1 provides the organizational basis for most of the current literature of ornithology. As such, it is functional for day-to-day use, despite its shortcomings.

Divergence and Convergence in Feeding

Birds show a great variety of morphological, physiological, and behavioral specializations associated with feeding on diverse sources of food. Modifications of the beak and digestive tract are associated with behaviors that also involve the locomotor appendages and sense organs.

Major types of avian feeders are (1) fish and aquatic invertebrate feeders, (2) aquatic filter feeders, (3) carnivorous or predatory birds (which eat other tetrapods, especially birds and mammals),

(4) carrion eaters, (5) insectivores and terrestrial invertebrate eaters, (6) pollen and nectar feeders, (7) fruit eaters, (8) seed eaters, and (9) grazers and browsers. Many species employ more than one type of feeding, especially at different times of year.

Fish and Aquatic Animal Feeders

Feeding in aquatic environments has led to a variety of specializations. Frigate birds capture flying fish in the air, and a number of species, such as fulmars, skuas, jaegers, some large gulls, and the fishing eagles, as well as frigate birds, frequently obtain food by forcing other birds to disgorge or drop their prey.

More frequently, birds obtain food in flight by capturing prey at or just under the surface of the water. Dipping involves picking up food in the beak from the surface or just below the water while the bird is in flight; it is the principal method used by many terns and small petrels. The hydrobatid petrels feed mainly by pattering, using their feet in addition to their wings to maintain a precise height above uneven water surfaces, a method that allows the birds to pick many small prey from the surface in rapid succession. Only the specialized tern-like birds called skimmers regularly use the technique of flying with the lower mandible immersed in water to make contact with and then seize individual prey. This mode of feeding is effective only in calm water, but it has the advantage of permitting location of prey by touch and making capture possible in turbid water or at night.

Birds feed below the surface by diving from the air into water (plunging), by actively swimming under water, either by foot-propelled or wing-propelled diving, by wading into shallow water, by extending the head and neck under water while floating on the surface, and in the case of the water ouzels or dippers, by diving under water and walking on the bottom. Plunging birds such as terns, gulls, the brown pelican, kingfishers, and some others penetrate the surface by a body length or less, whereas others, such as boobies and gannets, which have dense bodies, penetrate several meters from the momentum of their dives to capture prey (Figure 18–2). Most of the birds that use foot-propelled or wing-propelled diving depart from the water surface to pursue prey under water. Such birds actively chase fishes, cephalopods, crustaceans, and other actively moving animals at various depths in the water or they dive to the bottom to feed on shelled mollusks and other bottom-dwelling organisms in relatively shallow water. The latter group includes some of the diving ducks (eiders, old-squaw, scoters), some cormorants, and auks. Those that feed on shelled mollusks have especially enlarged and muscular gizzards that crush and grind up the shells.

Many kinds of aquatic birds capture and eat fish, and the shapes and sizes of their beaks and tongues are specialized for particular methods of seizing, holding, and ingesting their prey. Most fish eaters seize their prey in a forceps-like action; many are specialized to hold onto slippery bodies. The bills of mergansers, fish-eating ducks, have hook-like structures along their edges derived from the rostral lamellae that characterize the beaks of other ducks and waterfowl. Serrations on the edge of the bill occur in many other fish eaters such as the tropic birds, boobies, anhingas, and some auks. The pelicans have distensible throat pouches that aid in capturing fish. A few birds, anhingas and the western grebe, actually spear fish on their long, dagger-like beaks.

Some birds of prey use specialized feet to catch fish. The widely distributed osprey or fish hawk dives from the air and catches fish by plunging feet-first into the water and seizing a fish in its talons. The scales on the plantar surfaces of its toes

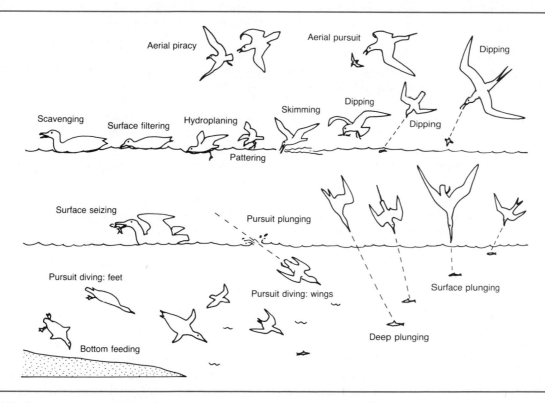

Figure 18–2. Modes of feeding of aquatic birds in water too deep for wading.

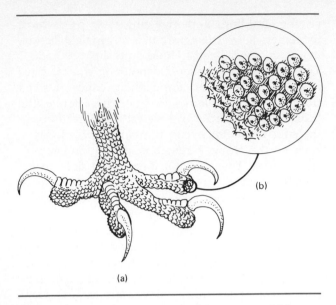

Figure 18–3. Foot structure of the osprey, which catches fish with its feet. (a) Reversible zygodactyl foot. (b) Detail of fish-holding spicules on foot pads.

bear small spines that aid in holding slippery fishes, and the outer toe is reversible so that prey can be held in a zygodactylous grip (Figure 18–3).

Aquatic Filter Feeders

Filter feeding involves the intake of numerous small food particles along with water or other materials such as sand or mud that are separated from food and expelled. Visual or other sensory detection of each bit of food does not occur. Small animals and plants such as diatoms and algae and organic detritus are the types of food usually consumed by filter-feeders. Some of the marine fulmar-like birds have wide-spreading lower mandibles, distensible gular pouches, and an associated muscular system for sucking zooplankton from the sea surface. Some storm petrels may also use suction to obtain animal oils from slicks on the surface. The specialized petrels called prions rest on their breasts on the surface, holding their wings outstretched and their bills under water, and propel themselves with their feet. Flamingos and many ducks strain out small items by forcing water through a series of lamellae in the bill with the broad, fleshy tongue.

Carnivorous or Predatory Birds

The diurnal birds of prey (hawks, eagles, and falcons) and the nocturnal birds of prey (owls) are the most specialized predators among birds, but many other avian orders and families contain species that are specialized to some extent for feeding on terrestrial vertebrates. Frigate birds or man-of-war birds often snatch young of other seabirds in their beaks as they fly by nesting colonies; the skuas and jaegers and some of the large gulls feed on rodents (lemmings) and small birds, including penguin chicks.

Carnivorous species are also found scattered through other such orders as the Gruiformes (South American seriemas, Cariamidae), Cuculiformes (roadrunners and coucals), and the Coraciiformes (some kingfishers, rollers, and motmots). In these cases, snakes, lizards, and occasional amphibians are the animals most often preyed upon. The large and diverse order of perching birds contains several families with species that prey to a degree on terrestrial vertebrates, including birds and mammals. The most specialized passerines in this respect are the true shrikes or butcherbirds of the genus *Lanius*. Less specialized but nevertheless very capable predators are also found among the corvids, especially some of the jays, magpies, and ravens.

The predatory mode of feeding involves hunting, capturing, killing, and eating prey which usually is too large to be swallowed whole. The various carnivorous birds have become specialized to varying degrees and adapted in different ways to perform these functions (Figure 18–4).

Most avian predators are aerial, diurnal, and visual hunters; but some are mainly terrestrial cursors, and some, especially the owls, are nocturnal and use auditory cues as well as vision for locating prey in darkness. A bird's habitat also strongly influences its hunting behavior. Most species that hunt in open habitats such as savannas, prairies,

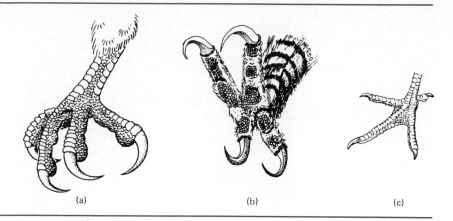

Figure 18–4. Feet of predatory birds (a, goshawk; b, great horned owl) compared to the foot of a fowl of the same body size (c).

(a)   (b)   (c)

or marshes are active foragers that search for prey by flying and begin their attacks in the air. Large raptors such as eagles and the broad-winged hawks typically soar on thermals at heights up to several hundred meters looking for suitable quarry. The bird-hunting falcons attack flying prey from the air. When prey is spotted the falcon folds its wings close to its sides, dives at a steep angle, and closes on its intended victim with great speed, either striking the prey as they dive past or more often pulling up behind the fleeing bird and overtaking it with the speed gained from the dive. Some birds of prey are able to attack flying birds, bats, and large insects by direct, flapping pursuit through the air. This method is particularly characteristic of some of the strong, fast-flying falcons (gyrfalcon, peregrine, merlin, bat falcon).

Several unrelated predatory birds have the ability to maintain a more or less stationary position in the air by facing into the wind and hovering on beating wings to search the ground or water below before dropping down. Well-known examples include the osprey and the small falcons called kestrels or wind-hovers. Other airborne predators such as the harriers and some open-country owls quarter back and forth only a few meters above the ground and drop suddenly down onto prey caught unaware of the predator's approach.

In wooded or forested habitats, predatory birds usually hunt by perching and waiting for an unsuspecting animal to reveal itself or move into a vulnerable position (sit-and-wait foraging). Most of the bird-eating hawks hunt in this fashion, as do many other diurnal birds of prey ranging in size from pygmy falcons (weighing 50 grams) to forest eagles (for example, the harpy eagle, crowned eagle, and the monkey-eating eagle), which weigh 6 kilograms or more. Most owls and shrikes and nearly all of the unspecialized woodland predators—kingfishers, rollers, coucals, jays—use some variant of this method, which is highly successful in habitats with dense cover.

A number of predatory birds in grasslands and semideserts have become essentially terrestrial and catch most of their prey by running animals down or chasing them into holes, crevices, or crannies where they can be caught. The crane-like secretary bird of Africa is a prime example among the Falconiformes; it runs after lizards and snakes, small mammals, occasional birds, and large insects on the ground. The extinct diatrymas and phororhachids were the most spectacular examples of terrestrial predatory birds (Box 18–1).

Carrion Eaters

The bodies of dead animals provide food for certain birds, such as vultures, condors, some eagles, caracaras, and even some storks, the large marabous, and adjutants (Figure 18–6). The basic sim-

Box 18–1. Giant Predatory Birds

One does not think of predatory birds as being enormous, although some eagles have wingspans of 2 meters or more. However, the extinction of dinosaurs at the end of the Mesozoic left empty the adaptive zone that had been filled by bipedal carnivores, and giant flightless birds appear to have filled this role from the Paleocene until the Pleistocene.

The earliest of these dinosaur analogs were the diatrymas (terror cranes), which are known from Paleocene and Eocene deposits in North America and Europe. The diatrymas were 2 meters tall and had massive legs and toes with enormous claws. The head was huge—nearly as large as that of a modern horse—and had an enormous, hooked beak (Figure 18–5).

South America was isolated from the northern continents in the early Cenozoic, and another lineage of giant predatory birds, the phororhachids, evolved on that continent. The phororhachids, which were 1.5 to 2.5 meters tall, were more lightly built than the diatrymas and were probably faster runners. Like the diatrymas, the phororhachids had huge hooked beaks and powerful claws. Phororhachids are known from South America from the Oligocene to the end of the Pliocene; they disappeared about the time of the great faunal interchange between North and South America. However, a phororhachid is known from Pleistocene deposits in Florida, apparently representing a genus that moved north across the Central American land bridge.

Figure 18–5. Terror crane, *Diatryma*, attacking a condylarth.

Figure 18–6. White-backed vultures *(Gyps bengalensis)* and Ruppell's vultures *(Gyps ruppelli)* feeding at a carcass. (Photograph by Sara Cairns.)

ilarities, independently evolved, in these carrion feeders are (1) naked heads and necks or much reduced feathering in these areas, especially immediately behind the beak and around the face, presumably an adaptation that reduces the chances of befouling the plumage with juices, blood, and grease from the carcass, and (2) in the case of falconiform scavengers, feet more adapted for walking on the ground than for holding prey.

Most carrion eaters are large, soaring birds that spot carcasses from a great height by keen vision, although some of the vultures also use smell at closer range. In the East African savannas where many large ungulates die, carrion-eating niches are highly diversified and as many as nine species of birds can be found on the same carcass. Variations in body size and in beak size and shape are specializations for feeding on different-size carcasses or on different parts of the same carcass. The large, heavy-beaked species (lappet-faced vulture, white-headed vulture) are the ones that first open up a carcass of a large animal like an ungulate because they are capable of tearing through the skin, whereas the long-necked griffons reach deep into the body cavity once the animal has been opened up.

Feeders on Insects and Other Terrestrial Invertebrates

Terrestrial arthropods, especially insects, and annelid worms constitute a vast array of prey for birds, and it has been estimated that as many as 60 percent of all birds are insectivorous to some degree. An informative way to approach the diversity of foraging specializations among insectivorous birds is to group them by functional similarities. Groups of birds that exploit the same types of food species in a similar way have been called **guilds**. Although some species overlap two guilds, the following major groups find representatives in various families and orders of birds.

1. **Gleaners** pick visually located insects from foliage, trunks, rocks, and ground by a quick peck from a perched position. This method of capture is the most widespread among insectivorous birds, and it is probably the ancestral feeding mode of insectivorous birds. The American wood warblers are typical foliage gleaners, as are their ecologically close counterparts, the Old World warblers and many other passerine birds in wooded or forested environments throughout the world. Typically,

they have short, thin, pointed bills that are adept at picking up stationary insects.

2. **Probers** are probably derived from gleaning ancestors. These are birds that poke their beaks into existing holes, cracks, or other enclosed cavities in plants, rocks, or ground or probe directly into the soil, for hidden insects, their eggs, larvae, and pupae. **Gapers** poke their closed beaks into the ground and then open the mandibles against the resistance of the soil, thereby creating an opening that can be explored by eye and tongue for edible objects. Flower probers are principally specialized for nectar feeding and secondarily for insect eating. Some of the Old World starlings and the New World meadowlarks, grackles, and relatives are gapers; they have their eyes set forward in the skull, enabling them to see into the opening made by the bill, and have very powerful muscles for opening the mandibles.

3. **Wood drillers** are birds with powerful neck muscles and chisel-shaped beaks that are used to excavate holes, usually into dead wood, to expose hidden insects; the **bark scalers** are a variant of this general guild. The main group of wood-drilling and bark-scaling birds are the woodpeckers, with long, protrusible, barbed tongues (Figure 17–23).

4. **Diggers** are birds that excavate the soil or forest litter with beaks or feet to expose subterranean items of food. Their origins are multiple, and they probably have been derived from both gleaners and probers. Diggers range in size from gallinaceous birds such as turkeys to small passerines.

5. Various hawks, falcons, and owls catch insects in their feet, either on the ground or in the air; they can be termed **grabbers**. It must be remembered that none of these species is exclusively insectivorous.

6. Certain cursorial birds run after insects disturbed in flight and catch them as they alight or sometimes in the air; these birds can be grouped together as **ground chasers.**

7. **Flycatchers** are birds that sally forth from a perched position to capture visually located insects at some distance away in the air and then return to a perch before launching the next attack. The flycatchers are as varied as the gleaners and show as many examples of parallel evolution of similar behavior and structure in different taxa. The tyrant flycatchers of the New World and the muscicapid flycatchers of the Old World are a prime example of independent evolution of the same specializations for the flycatching guild, but many other groups of birds contain flycatchers, too. The beaks of these birds tend to be long, slightly hooked, and often with long bristles at the gape, all specializations associated with catching insects in the air (Figure 18–7).

A few groups of flycatching birds have evolved specializations for eating bees, wasps, and other stinging hymenopterans, whereas others have developed an ability to distinguish nonvenomous castes (drones) from venomous forms (workers), apparently by differences in the sounds they produce in flight. The bee-eaters, jacamars, and shrikes all feed regularly and extensively on a variety of bees, wasps, and hornets. After capturing and killing a stinging insect in their beaks, bee-eaters and shrikes move the prey laterally through the bill until the beak comes into contact with the tip of the insect's abdomen; several bites are delivered to this region, causing any venom in the poison gland to be expelled as a droplet on the sting itself. Then the bird vigorously wipes the sting against a branch several times and swallows the insect.

8. **Aerial sweepers** are birds that fly high and fast through swarms of flying insects and make repeated captures in their beaks while on the wing, perhaps in part by random contact with the prey but certainly also in many cases by visual detection of individual prey. There are various intermediate modes of hunting between flycatchers and sweepers, and it is likely that sweepers have evolved from flycatching ancestors. The diurnal sweepers include the swifts, swallows, and the Australian wood swallows, fast-flying birds with long, pointed wings. Their nocturnal counterparts are the somewhat owl-like nightjars or goatsuckers, with soft plumage and a silent, slower flight. Typically, sweepers have a short, weak beak with a very wide gape and expanded palatine bones

Figure 18–7. Eastern phoebe shows the long, narrow bill typical of flycatchers. (Photograph by Lang Elliott, courtesy of the Cornell Laboratory of Ornithology.)

that act as a brace against the impact of a flying insect when it strikes the mouth.

Pollen and Nectar Feeders

Most, but not all, of the pollen and nectar feeders are probers with long bills and tongues (Figure 18–8). The tongues of the most specialized pollen and nectar feeders are tipped with brush-like structures, which collect the food, or the tongues are tubular (the tubes being formed in different ways in the various groups), or both conditions are combined. Even within hummingbirds beaks are highly variable from species to species in relative length, shape, and curvature, and adapted to particular kinds of flowers, which are quite often pollinated by the birds. Many short-billed nectar feeders pierce the base of the corolla of a tubular flower, making a hole for the tongue to obtain nectar.

Fruit Eaters

Fruit-eating species of birds are widely distributed among both nonpasserine and passerine families. Most of these species feed insects to their young, thereby providing the protein necessary for growth because most fruits are deficient in amino acids. Although bill shapes and sizes vary enormously among the different fruit eaters, these may be molded by selective forces other than those associated with eating fruit. Specialization for this diet occurs mainly in the digestive tract, which tends to be short and tubular and without a well-developed stomach. The food is passed through the digestive tract rapidly; only the pulp of the fruit is digested and even this is usually only partially digested.

Granivorous Birds

Seeds are a rich source of nutrients for birds and are often available at seasons of the year when

Figure 18–8. Ruby-throated hummingbird feeding from a flower. (Photograph by Michael Hopiak, courtesy of the Cornell Laboratory of Ornithology.)

insects and other foods are scarce. With food ranging in size from tiny grass and forb seeds less than a millimeter in greatest dimension to some very large nuts and pits with dimensions of 50 millimeters or more, seed-eating birds show a variety of adaptations for exploiting seeds as food.

Seed-eating birds can be divided into three functionally different groups: (1) those that eat the seeds and nuts entire, including the outer husk or shell, and depend upon powerful gizzards for grinding up and mechanically processing the food; (2) those that hammer seeds and nuts open with their beaks and then pick out the meat; and (3) those that have evolved special beaks, palates, and jaw muscles for husking and cracking open seeds before they are swallowed. Bill size correlates generally with the size of seeds fed upon, so that in regions where several kinds of seed-eating birds occur sympatrically there is usually a gradation of body and bill sizes that is associated with average differences in the kinds of seeds eaten by the different species.

Grazers and Browsers

Some of the cranes, geese, and ducks eat the roots, rhizomes, and bulbous parts of aquatic plants, but only a few kinds of birds feed extensively on grass, leaves, buds, and twigs. Some of the geese and goose-like waterfowl crop grass on land. The peculiar South American hoatzin feeds mainly on the tough rubbery leaves, flowers, and fruits of arum plants and the white mangrove. Most mechanical processing occurs in the hoatzin's crop, which is large and has thick muscular walls and a horny internal lining. The New Zealand kakapo or owl parrot extracts the juices from leaves, twigs, and young shoots by chewing on them without detaching them from the plant, or fills its large crop with browse and retires to a roost where it grinds the plant material, swallowing the juices and ejecting the fiber as dry balls.

Grouse browse on the buds, leaves, and twigs of willows and other plants of low nutritive value. They have extremely large ceca in which large volumes of vegetable materials are digested by symbiotic bacteria, a system analogous to that found in many herbivorous mammals.

Foraging Behavior of Birds: The Concept of Optimization

Refined morphological specializations, such as the diversity of beak shapes of birds, demonstrate such appropriate design that we are intuitively led to hypothesize the existence of similar perfection in physiological and behavioral adaptations to the environment. There is a general hypothesis implicit in much of modern biology that over evolutionary time natural selection has favored organisms with genotypes providing them with characteristics that solve environmental problems in an optimal way for survival and reproduction. Over the past quarter century this general hypothesis has seen vigorous testing and modification, in large part by ecologists working with birds and **optimal foraging theory** (OFT) (Krebs et al. 1983, Stephens and Krebs 1987). Numerous other possible optimalities have been investigated, primarily those involving behavioral characteristics of animals, all borrowing from economics the procedures of cost/benefit analysis. However, the concept of optimalization in biology remains controversial, and OFT is criticized by some ecologists as strongly as it is defended by others (Pierce and Ollason 1987, Stearns and Schmid-Hempel 1987).

The technique of OFT allows us to compare the predictions of a theoretical model to actual observations in the natural world. If the model and nature resemble one another, we have probably correctly considered the essential elements of the organism's adaptations in constructing the model; if they do not match well, the nature of the mismatch is a guide to other factors that must be considered key to the particular adaptation under study. Lack of correspondence between model and nature leads us to understand better the currencies and constraints influencing behavior. OFT has been used to investigate what animals feed on, where they go to feed, and how they search for food. It is possible to describe the rules animals use to make decisions to eat one food item and ignore another, or to hunt here and not there.

Most food resources used by birds occur in patches, and the food supply within a patch generally diminishes with time (seeds become depleted or insects take evasive action after becoming aware of a foraging predator). A foraging bird must decide how long to feed in one patch before moving to the next, taking into account that no food intake at all occurs during the move from one patch to another.

Theoretically a forager should behave as though it estimates the average rate of its food intake in the environment in which it is feeding and compares its current food intake with the average, including the cost of moving from one patch to another. When the rate of food capture in a patch falls to the environmental average, it is time to switch. This hypothesis is known as the **marginal value theorem** (Figure 18–9). The graph shows the rate of prey capture for a predator foraging in

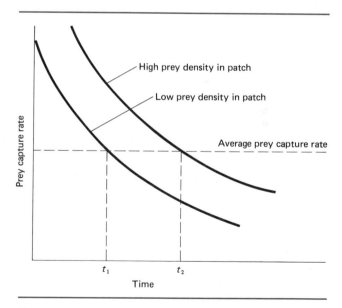

Figure 18–9. Marginal value theorem of optimal foraging.

patches where the prey density is high or low. As the bird continues to forage in a patch, the rate of prey capture decreases. After a period of feeding (t_1 or t_2) the rate of prey capture in the patch will have decreased until it equals the average capture rate for the entire habitat, shown by the dotted horizontal line. Optimal foraging theory predicts that a predator will move to a new patch when its capture rate has dropped to the average for the habitat. The graph shows that a predator behaving in that manner will forage longer in a patch where prey is abundant than in a patch where prey is scarce, thereby maximizing its energy intake during the time it spends foraging.

Most field studies of this hypothesis have been unsuccessful because the researcher has little way of knowing the food abundance in a patch. On the other hand, laboratory tests of this aspect of foraging have been rewarding. By creating patches with known prey numbers and varying not only those numbers but the time required to change patches, researchers have shown that there is a remarkable correspondence between the predictions of the model and the behavior of foraging birds.

What food to eat depends on the net energy content of the food after the energy costs of search and handling (the energy required to husk or kill) have been subtracted. OFT assumes that it is advantageous for a predator to maximize its energy intake per unit of time (Figure 18–10). The rate of energy intake for prey of different sizes depends on both the energy content of the prey and the time it takes to handle (subdue and swallow) the prey item. Large prey individuals contain more energy than small individuals, but they also require longer handling times. The graph shows this relationship for a wagtail eating dungfly larvae. Larvae that are 7 millimeters long provide the greatest energy return per second of handling time. Because food patches in nature are composed of several types of edible material, a foraging organism must decide whether to eat an item or ignore it. Optimal foraging theory predicts that a low net energy item should be rejected if the predator could expect to find an energetically superior tidbit

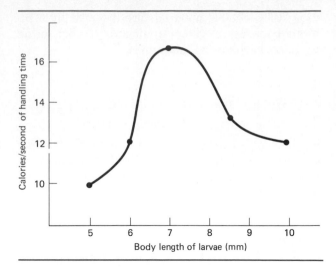

Figure 18–10. Rate of energy intake for prey of different sizes. (Data from N. B. Davies, 1977, *Journal of Animal Ecology* 46:37-57.)

soon after rejecting the lower-quality item. Even if a feeding patch is rich in inferior food items, theoretically they should be completely ignored if there is a greater net energy gain from rejecting them in favor of some higher-quality item also in the patch. However, when high-quality items are scarce, the wise forager eats what is at its bill.

These predictions of what prey should be eaten and what should be ignored are excellent beginnings for laboratory experiments, and they have also been tested successfully in the field. J. D. Goss-Custard (1977) studied the European redshank (*Tringa totanus*); these sandpipers feed on polychaete worms by probing in estuarine intertidal mud with their long bills. Such habitats often contain large numbers of several species of these worms but little else that appeals to a redshank. Goss-Custard was able to study them where only two species of worm occurred in significant numbers, a large species with a patchy distribution and ubiquitous small species. As food items for the redshank these two species seem to differ only in the greater quantity of nutritious flesh in the larger species. The rate at which large worms were captured increased as the density of large worms in a patch increases (Figure 18–11a). When the rate of capture of large worms is low the birds eat many small worms, but they ignore small worms when they are able to capture large worms (Figure 18–11b). That is, they are ignoring less profitable prey items (small worms) when more profitable prey items (large worms) are readily available. As the rate of capture of large worms increased the redshanks became more and more selective no matter what the density of small worms was, as the theoretical predictions indicate they should.

Not all studies of the prey optimization of foragers have matched theoretical predictions. Several of these inconsistencies can be resolved by recognizing that energy content is not the only value of food. Nutrient constraints also affect the value of a given food. Trace nutrients or the relative proportions of carbohydrate, fat, and protein may add to the value of a food item, or toxins may detract from its value.

One aspect of the biology of many birds is transport of food to their young, which are confined to a nest for some period of time. In terms of optimal foraging this means the parent must take into account the cost in energy and time of a round trip between its nest and the feeding site.

Logically, it would seem that the farther from the nest a parent must forage, the larger the load it should bring to its young to offset the costs of travel. On the other hand, the efficiency of capturing or holding prey probably decreases as the load size increases and the cost of transport over long distances may become significant. Thus, the ability of parent birds to maximize the rate at which they can bring food to the nest by selecting increasingly larger prey will ultimately be constrained by prey size.

Tests of the **central place foraging** problem of parent birds have so far proven difficult because of the problems of accurately measuring the necessary parameters in the field. Nevertheless, birds that catch only a single insect before returning to the nest have been shown to prefer larger insects the farther they must fly from the nest. At least one species that loads itself with many prey items (the European starling, *Sternus vulgaris*) increases its load size with its distance from the nest when feeding at artificial food sources. Birds also appear to invest more search effort close to the nest and less at a greater distance. Despite the paucity of data, the challenge of central place foraging is undoubtedly a significant one for many birds and deserves further attention.

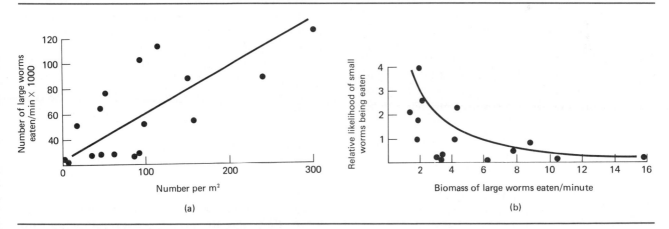

Figure 18–11. Redshanks feeding on polychaete worms illustrate the behavior expected from an optimally foraging predator. (Data from J. D. Goss-Custard, 1977, *Animal Behaviour* 25:10-29.)

Perhaps the greatest value of OFT has been in sharpening our perspective of the trade-offs necessary in adaptation to the complex environments in which animals live. However, the skepticism about OFT expressed by Pierce and Ollason (1987) is shared by many ecologists. Few behaviors match our simplistic models; animals must make compromises among simultaneous and conflicting demands. Adaptations also may be greatly affected by phylogenetic constraints. These complications must be remembered when the results of OFT studies appear to demonstrate that animals are behaving optimally.

Social Behavior and Reproduction

Vision and hearing are the major sensory modes of birds as they are of humans, and one result of this correspondence has been the important role played by birds in behavioral studies. Most birds are active during the day, and they are relatively easy to observe. A tremendous amount of information has been accumulated about the behavior of birds under natural conditions, and this background has contributed to the design of experimental studies in the field and in the laboratory.

The activities associated with reproduction are among the most complex and conspicuous behaviors of birds, and much of our understanding of the evolution and function of the mating systems of vertebrates is derived from studies of birds. Classic work in avian ethology, such as Konrad Lorenz's studies of imprinting and Niko Tinbergen's demonstration of innate responses of birds to specific visual stimuli, has formed a basis for current studies of behavioral ecology.

Vocalization and Visual Displays in Pair Formation

Birds use colors, postures, and vocalizations for species, sex, and individual identification. Studies of birdsong have contributed greatly to our un-derstanding of communication by vertebrates, and important general concepts such as species specificity in signals and innate predisposition to learning were first developed in studies of birdsong. Recent studies of the neural basis of song are leading to a close integration of behavior and neurobiology. [See Konishi (1985) for a review.]

Birdsong has a specific meaning that is distinct from a birdcall. The song is usually the longest and most complex vocalization produced by a bird. In many species songs are produced only by mature males, and only during the breeding season. Song is a learned behavior that is controlled by a series of song control regions (SCRs) in the brain. The SCRs are under hormonal control, and in many species of birds the SCRs of males are larger than those of females and have more and larger neurons and longer dendritic processes (Nottebohm and Arnold 1976). The vocal behavior of female birds varies greatly across taxonomic groups: In some species females produce only simple calls, whereas in other species the females engage with males in complex song duets. The SCRs of females of the latter species are very similar in size to those of males (Table 18–2).

A birdsong consists of a series of notes with intervals of silence between them (Figure 18–12). Changes in frequency (frequency modulation) are conspicuous components of the songs of many birds, and we have noted that the avian ear may be very good at detecting rapid changes in frequency (Chapter 17). Birds often have more than one song type, and some species may have repertoires of several hundred songs. Males and females of some species sing duets (Figure 18–13). Birdsongs are species specific, and they often show regional dialects. These dialects are transmitted from generation to generation as young birds learn the songs of their parents and neighbors. In the indigo bunting, one of the best studied species, song dialects that were characteristic of small areas persisted up to 15 years, which is substantially longer than the life of an individual bird (Payne et al. 1981). Birdsongs also show individual variation that allows birds to recognize the songs of residents of adjacent territories and to distin-

Table 18–2. Sexual dimorphism in the song control regions of the brains of birds. The average ratio of the volumes of five SCRs in males compared to females parallels the difference in the sizes of the song repertoires of males and females.

	Zebra Finch	Canary	Chat	Bay Wren	Buff-Breasted Wren
SCR volume ratio	4.0	3.1	2.3	1.3	1.3
Song repertoire	Males only	Males ⟩⟩⟩ females	Males ⟩⟩ females	Males = females	Males = females

Source: Modified from E. A. Brenowitz, A. P. Arnold, and R. N. Levin, 1985, *Brain Research* 343:104–112.

guish the songs of these neighbors from those of intruders.

The songs of male birds identify their species, sex, and occupancy of a territory. Territorial males respond to playbacks of the songs of other males with vocalizations, aggressive displays, and even attacks on the speaker. These behaviors repel intruders, and broadcasting recorded songs in a territory from which the male has been removed delays the occupation of the vacant territory by a new male. In addition to its roles in male–male interactions, the song of a male bird may convey information about his fitness that a female can use in selecting a mate. Some studies have shown that the breeding success of males with large song repertoires is greater than that of individuals with smaller repertoires. For example, male great tits (*Parus major*) with intermediate or large song repertoires produced heavier fledglings and were more likely to father offspring that survived to breed than were males with fewer song types (McGregor et al. 1981). However, female great tits did not mate preferentially with males with large repertoires.

Partial evidence for positive gains in fitness associated with song complexity has been provided by studies of the European sedge warbler (*Acrocephalus schoenbaenus*). Sedge warblers sing songs made up of about 50 syllables and compose new themes as they sing. By recording songs from males in the field and playing them back to females in the laboratory, it was shown that females responded most strongly to the songs of males that

had been successful in pairing earliest in the season. The females showed less interest in the songs of males that acquired mates only late in the season (Catchpole 1980, Catchpole et al. 1984). The songs of the earliest paired males were more complex than those of males that paired later in the year. That observation suggests a mechanism that could allow a female to increase her fitness by choosing a male with a complex song: Birds that pair early are more likely to be successful in fledging a brood because they have time to nest again if the first brood is lost. Even so, these experiments are not a clear demonstration of increased female fitness based on song choice, but only show that female choice could increase fitness. They provide no information about *why* males with complex songs pair early in the year and males with less complex songs pair later.

Visual displays are often associated with songs—a particular body posture that displays colored feathers may accompany singing. Male birds are often more brightly colored than females and have feathers that have become modified as the result of sexual selection. In this process, females mate preferentially with males that have certain physical characteristics. As a result of that response by females, those physical characteristics contribute to the reproductive fitness of males, although they may have no useful function in any other aspect of the ecology or behavior of the animal. The colorful speculum on the secondaries of male ducks, the red epaulets on the wings of male red-winged blackbirds, the red crowns on kinglets,

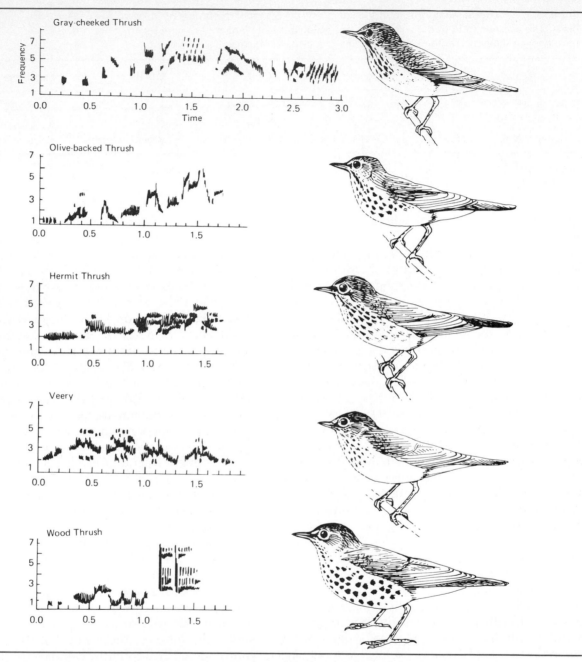

Figure 18–12. Difference in the songs of some species of thrushes from eastern North America. The sonographs are plots of sound frequency in kiloHertz versus time in seconds. They show differences in loudness (amplitude modulation) by varying degrees of darkness and change in frequency (frequency modulation) by upward or downward slopes within a syllable. (From W. C. Dilger, 1956, *Auk*, 73:313–353 and R. C. Stein, 1956, *Auk* 73:503–512.)

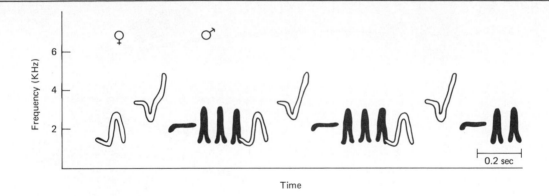

Figure 18–13. A duet by male and female bay wrens. The female (shown in white) initiates the duet and the male (in black) responds. The female and male parts are so closely interdigitated as to sound like a single bird. (Courtesy of Rachel N. Levin.)

and the elaborate tails of male peacocks are familiar examples of specialized areas of plumage that are involved in sexual behavior and display. It has been suggested that the bright colors of male birds may be an indication of good nutritional status, and thus could provide a basis for females to evaluate the merits of several potential mates, although this hypothesis is still controversial.

Conspicuous or aerodynamically cumbersome feathers can make a male bird vulnerable to capture by visually guided predators, and the bright colors and special adornments of the breeding season are often discarded for a more sober, even cryptic, appearance during the rest of the year. Thus, the male African standard-wing nightjar has specially elongated and flagged second primaries that are used in flight displays during courtship, but these feathers probably slow the male's flight and make it easier for an aerial predator to capture him (Figure 18–14). As soon as courtship is over, the male bites off the projecting parts of the feathers, leaving the stubs in the wings. The pattern of molting is so arranged that the primaries are not replaced until just before the next breeding season. This pattern of molting differs from that of all other caprimulgids and from that of female standard-wing nightjars. The usual pattern for caprimulgids

is to begin molt in the spring with the outermost primary and to move sequentially through the primaries to the tenth, and this is the pattern followed by the female standard-wing nightjar. In the case of the males, however, molt begins in the center of the wing with the fifth and sixth primaries and ends with the tenth and the stump of the second. That sequence leaves just enough time for the second primary to grow to its full length at the beginning of the next courtship season.

Other species of birds, especially ground-dwelling and ground-nesting species that are vulnerable to visual predators, have feathers with cryptic colors, patterns, and textures. Ground-dwelling birds often match the general color of their habitat and hide effectively by remaining motionless and blending with the background. The ptarmigan is white in winter and matches its snowy background and molts to a mottled brown plumage during the spring and summer that matches the tundra vegetation. The females turn brown first and begin sitting on their nests, situated on snow-free tundra, while the males are still white and very conspicuous in their courtship and territorial displays. When a hunting gyrfalcon appears, the males seek out the last small patches of snow on which to hide from the predator.

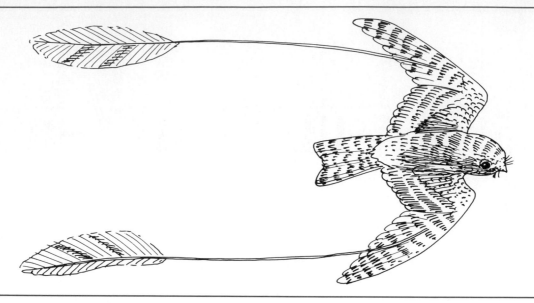

Figure 18–14. Male standard-wing nightjar, showing the elongated second primaries that are used in an aerial courtship display.

Vocalizations are not the only sounds that birds use in courtship; nonvocal sounds are produced by the feathers of some species—the drumming of male grouse in the spring is a familiar example of a nonvocal sound that plays a role in courtship. Sounds are often produced incidentally to the beating of a bird's wings in flight, and only slight modification in the shapes of primaries or tail feathers is needed to produce the characteristic whistling sounds made by certain kinds of ducks, bustards, and hummingbirds when they fly. Such sounds may be used in territorial advertisement or as individual location signals among birds flying at night or in heavy fog. Other species of birds have undergone more specific modification of their flight feathers to produce sounds used in displays. Among the tropical American manakins one finds not only narrowed and stiffened primaries involved in the production of sounds during displays, but also secondaries with thickened, clublike shafts that apparently act like castanets to produce sounds when the wings are moving (Figure 18–15). Other species, including goatsuckers, owls, doves, and larks, clap their wings together in flight, producing characteristic sounds associated with courtship or territorial defense.

Mating Systems and Parental Investment

The mating systems of vertebrates are believed to reflect the distribution of food, breeding sites, and potential mates. These resources affect individuals of the two sexes differently, and some types of resource distributions give one sex the opportunity to achieve multiple matings by controlling access to the resources. Studies of birds have contributed very largely to the development of theories of sexual strategies (Emlen and Oring 1977, Oring 1982).

The energy cost of reproduction for males of most species of vertebrates is probably lower than the cost to females. Sperm are small and cheap to produce compared to eggs that must supply the nutrients required for embryonic development. Furthermore, a male does not necessarily have any

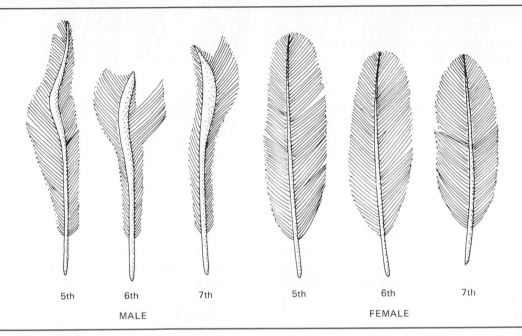

Figure 18–15. Secondaries of the male (left) and female (right) South American manakin. The shafts of the male's feathers are thickened and produce sounds when the wings are moved in display.

commitment beyond insemination, whereas a female must at least carry the eggs until they are deposited, and often is involved in brooding eggs and caring for the young as well. Courtship and territorial displays may be energetically expensive (see Chapter 11), but even if that is generally true, most male vertebrates can potentially mate more often than females. A male is ready to mate again very quickly after inseminating a female, but a female must ovulate and yolk a new clutch of eggs before she can attempt a second mating. Because of this disparity in the costs of reproduction for males and females, the routes to maximizing reproductive success may be different for the two sexes. For males the most productive strategy may be to mate with as many females as possible, whereas a female may maximize her success by devoting time to careful choice of the best male and to care of her young.

The extent to which males can achieve multiple matings depends on ecological conditions and especially on the availability of food and nest sites, which are the resources most needed by females. Theoretically, a male could increase his opportunities to mate by defending these resources—excluding other males and mating with all the females in the area. His ability to do that will depend on the spatial distribution of resources. If food and nest sites are more or less evenly distributed through the habitat, it is unlikely that a male could control a large enough area to monopolize a large number of females. Under those conditions, all males will have access to the resources and to the females. On the other hand, if resources are clumped in space with barren areas between the patches, the females will be forced to aggregate in the resource patches and it will be possible for a male to monopolize several females by defending a patch. Males that are able to defend good patches should attract more females than males defending patches of lower quality.

The temporal distribution of breeding females

is also important in determining the potential for a male to achieve multiple matings. A male can mate with only one female at a time, so an important consideration is the number of receptive females relative to the number of breeding males at any one time. This ratio is called the **operational sex ratio** (OSR). If the actual ratio of males to females in a population is 1:1 and if all the females become receptive at the same time, there would be one receptive female for each breeding male and the OSR would be 1:1. In that situation, males have little opportunity to monopolize a large number of females and the variance in mating success among males will be low. That is, most males will mate with one female and relatively few males will have no matings or more than one mating. On the other hand, if females become receptive over a period of weeks, the number of breeding males at any given time will be larger than the number of females and the OSR will be greater than 1:1. In that situation it is possible for one male to mate with many females over the course of the breeding season. As the OSR rises above 1:1, competition between males for mates increases, the variation in reproductive success of individual males increases, and sexual selection is likely to become more intense. These are the elements that contribute to determining the mating strategies of males and females.

Social vertebrates exhibit one of two broad categories of mating systems—monogamy or polygamy. Monogamy (*mono* = one, *gamy* = marriage) refers to a pair bond between a single male and a single female. The pairing may last for part of a breeding season, an entire season, or for a lifetime. Polygamy (*poly* = many) refers to a situation in which an individual has more than one mate in a breeding season.

Polygamy can be exhibited by males, females, or both sexes. In polygyny (*gyn* = female) a male mates with more than one female, whereas in polyandry (*andr* = male) a female mates with two or more males. Promiscuity is a mixture of polygamy and polyandry in which both males and females mate with several different individuals.

Monogamy is the dominant mating system of birds. David Lack (1968) found that 92 percent of all bird species are monogamous. In monogamous mating systems both parents usually participate in caring for the young, and 93 percent of the species of birds that produce altricial young (which require extensive parental care) are monogamous compared to 83 percent monogamy among species with precocial young.

Promiscuity is the second most common mating system for birds, accounting for 6 percent of the living species of birds. Two percent of the species of birds are polygynous and only 0.4 percent are polyandrous. Despite their relative rarity, promiscuous, polygynous, and polyandrous species of birds have been extensively studied because these unusual mating systems can reveal much about the mechanisms of sexual selection and evolution.

Monogamy Monogamy occurs in so many species of birds in so many different ecological conditions that no one mechanism is likely to explain its prevalence. Resource distribution and the degree of parental care appear to be two factors that are frequently important in determining the importance of monogamy. When nest sites and food are evenly distributed through a habitat, a male or female cannot control access to these resources. If neither sex has the opportunity to monopolize additional members of the opposite sex by controlling resources, monogamy is the reproductive strategy that maximizes the fitness of individuals. When the territory quality of one male is much like that of all other males, a female can probably maximize her reproductive success by pairing with an unmated male. Perhaps a more important incentive for monogamy for many species of birds is the need for attendance by both parents to raise a brood to fledging. Dramatic examples include situations in which continuous nest attendance by one parent is necessary to protect the eggs or chicks from predators while the other parent forages for food. This situation is commonly observed in seabirds that nest in dense colonies that sometimes include mixtures of two or more species. In the absence of an attending parent, neighbors raid the nest and kill the eggs or chicks. The male and female alternate periods of nest attendance and foraging, and some species engage in elaborate displays when the parents switch duties (Figure 18–16). A third situation that could make monogamy advantageous would be a sex ratio that deviates widely from 1:1. When one sex is in short supply, individuals of that sex may be the resource that individuals of the other sex defend. For example, female ducks suffer higher mortality than males and the sex ratios of ducks are biased toward males. As a result of the shortage of female ducks, competition between males for mates is intense. A male duck pairs with a female several months before the breeding season begins and defends her against other males.

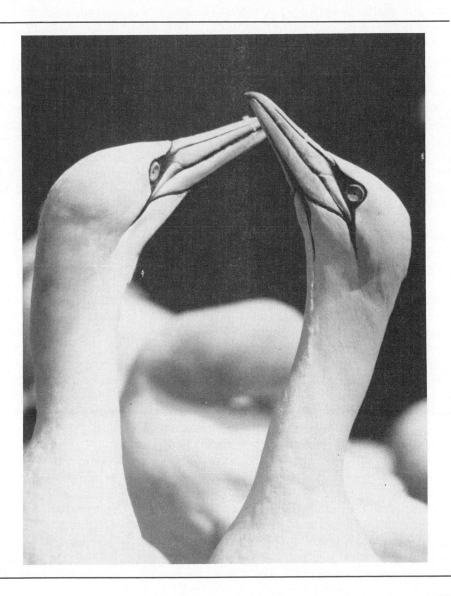

Figure 18–16. Nest exchange display of northern gannets. (Photograph by Mary Tremaine, courtesy of the Cornell Laboratory of Ornithology.)

Polygyny When an individual male can control or gain access to several females, the male can increase his reproductive success by mating with more than one female. In **resource defense polygyny** males control access to females by monopolizing critical resources such as nest sites or food that have patchy distributions. A male that stakes out its territory in a high-quality patch can attract many females. For this system to work, a female must benefit from mating with a male that already has one mate. That is, the reproductive fitness of a female must be greater as a secondary mate on a high-quality territory than it would be as a primary mate on a territory of lower quality. Redwinged blackbirds are a familiar example of resource defense polygyny (Orians 1980). Male blackbirds arrive at their marshy breeding areas before females and compete for territories. When the females arrive, they have a choice among a variety of territories of different quality, each defended by a male blackbird. A female should choose to mate polygynously if the difference in quality of the territories was large enough so that she would raise more young than she would by mating monogamously with a male on a poorer territory. This hypothesis is known as the **polygyny threshold model** (Orians 1969).

In **male dominance polygyny** males are not defending females, nor are they defending a resource that females require. Instead, males compete for females by establishing patterns of dominance or by demonstrating the quality through displays. This type of reproductive system is typical of situations in which the male is not involved in parental care and no potential exists for controlling resources or mates. Birds with precocial young in rich habitats often show male dominance polygyny, although this mating system is not limited to such species. The sizes of male territories in male dominance polygyny and the degree of aggregation of males are not set directly by the resources in the habitat, as is the case with resource defense polygyny. Instead, the distribution of males is determined by the sizes of the home ranges of females. Aggregations of many males in a small area are called leks. The prairie chicken of western North America is a well-studied lekking species (Wiley 1973). During the breeding season male prairie chickens congregate in traditional lek sites. Each male occupies a small territory (from 13 to 100 square meters in area) in the lek, and within this territory performs a courtship display that includes elaborate postures (Figure 18–17). Two sacs that are outgrowths from the esophagus are filled with air and project through the breast feathers. Air is expelled from these sacs with a popping sound. Females visit the leks and copulate with a single male. The central sites on the lek appear to be the most favored and the 10 percent of the males in the most central sites obtain 75 percent of the matings.

In **harem defense polygyny** males control access to females that are gregarious and form groups. This mating system is rare among birds, but well known among mammals—harems of females guarded by a male are typical of some bats, many species of pinnipeds (seals, sea lions, and the walrus), and some ungulates (deer and elk). The first well-documented example of harem defense polygyny among birds has been described for the Montezuma oropendola (*Psarocolius montezuma*) by Michael Webster. These oropendolas are large tropical passerine birds; females weigh about 250 grams and males 520 grams. The females breed colonially and one tree may contain 30 or more nests. The aggressive activities of the males are not directed to defending any resource in the environment that the females require; instead, the males defend groups of females. High-ranking males exclude low-ranking males from the area of the colony, and mating success of males is correlated with the time they spend in the colony.

Male Incubation and Polyandry Incubation and care of the young by both parents is considered to be the primitive condition for birds, and it occurs in the vast majority of living species of birds. However, biparental care is not always required, and situations in which it is possible for one parent to care for the eggs and young allow the development of the polygynous mating strategies we have discussed. Males of polygynous species of birds

Figure 18–17. Male greater prairie chicken displaying on a lek. (Photograph by Mary Tremaine, courtesy of the Cornell Laboratory of Ornithology.)

do not participate in parental care, and all the tasks of incubation and rearing the young are performed by the female. Less commonly it is the female that is emancipated and the male that assumes parental responsibilities. The rarity of this situation probably reflects the relative parental investments of the two sexes. A male bird can leave a newly laid clutch of eggs and breed again as soon as he locates a receptive female, whereas a female who leaves her nest must ovulate and yolk a new clutch of eggs before she can obtain a second breeding. Thus, a female has more to lose by abandoning her eggs that does a male.

Despite the imbalance of energy investment in reproduction by males and females, there are a few situations in which the ability of a female bird to increase her fitness by multiple breedings equals or exceeds that of males, and these are the cases in which polygamy is balanced between males and females (promiscuity) or favors females (polyandry). Several species of charadriiforms show this pattern of breeding, including jacanas, stints, sandpipers, and phalaropes.

In **rapid multiple clutch polygamy** both sexes have an opportunity to increase their fitness by multiple breedings in rapid succession, and both males and females incubate eggs. Temminck's stint (*Calidris temminckii*), a charadriiform that occurs in northern Europe, provides one of the best-studied examples (Hildén 1975). Male stints establish territories and display to attract females. Every female breeds in rapid succession with two males on different territories, and every male breeds with two females. The first clutch of eggs is incubated by the male, and the second clutch is incubated by the female. The evolution of this reproductive system appears to depend on the ready availability of resources that allow single-parent care to lead to successful fledging.

In polyandrous mating systems females control or gain access to multiple males. **Resource defense polyandry**, like its counterpart resource defense polygyny, is based on the ability of one sex to control access to a resource that is critical for the other sex. This pattern of breeding among birds seems to be typical of situations in which the cost of each reproductive effort for the female is low (because food is abundant and a clutch contains only a few small eggs) and the probability of successful fledging is small. Spotted sandpipers (*Ac-*

titis macularia) provide an example of this mating strategy (Oring 1982). Predation on sandpiper nests is high, and the resource that female sandpipers control is replacement clutches for males that have lost their clutches to predators. Male spotted sandpipers form territories and incubate the eggs. A female spotted sandpiper mates with a male and remains in his territory at least until she has laid three eggs. After that, she may move away to breed with other males, leaving parental care to the male, or she may remain in the territory. If she remains she may or may not participate in parental care. A female that has moved away and bred with other males may return and breed again with the original male if their first clutch is destroyed.

Female access polyandry has been described for the gray phalarope (*Phalaropus fulicarius*). This shorebird occurs in flocks with large home ranges that include feeding areas separated by hundreds of meters. Defense of a resource is not possible, but females initiate courtship and may limit access to males by interactions among themselves (Kistchinski 1975). The male phalarope incubates the eggs, driving the female away after she has finished laying.

In **cooperative polyandry** a single female and a group of males form a communal breeding unit in which all males have an opportunity to mate. Females are able to maintain a male group because of the advantages to males in cooperating as a breeding unit. In the Galapagos hawk (*Buteo galapagoensis*), for example, female mortality is higher than male mortality and the breeding ratio on Santiago Island is 2.3 males per female. The habitat is saturated with territories, and territories remain stable for periods exceeding the life spans of individual birds. Some females do not breed, apparently because they cannot obtain territories. Breeding groups persist from year to year, changing only with the death of a member. Monogamous males have higher reproductive success than males that are part of a breeding group (an average of 1.0 versus 0.67 fledged young per year on Santiago Island), but joining a group may help males to establish territories initially, and some polyan-

drous males appear to become monogamous after several years. Consequently, the best chance for a male Galapagos hawk to increase its lifetime fitness may be initially to become a member of a breeding group (Faaborg et al. 1980). The Tasmanian native hen (*Tribonyx mortierii*) shows a variation on this pattern: The sex ratio is about 1.5 males per female, and the breeding population consists of pairs and trios. A trio is usually composed of two brothers and an unrelated female. The trios form when the birds are about a year old, and they remain together for life. Trios have larger territories than pairs, and trios have larger clutches and fledge more young per year than do pairs (Ridpath 1972).

The conspicuousness of birds and the relative ease with which they can be studied has made them a mainstay of sociobiological research. The diversity of avian mating systems and the correlations between ecological conditions and certain types of mating systems have contributed largely to our current understanding of vertebrate behavior. Recent work has begun to emphasize the roles of individual experience and of lability of breeding systems. If environmental conditions determine the relative advantages of different mating systems, how should organisms respond to variation in these conditions? One possible response is a flexible mating system that responds in ecological time to changes in ecological conditions. Investigation of the short-term causes and consequences of variation in avian mating systems is emerging as an area of increasing importance for both ornithologists and behaviorists (Oring 1982).

Oviparity, Nesting, and Brooding Eggs

Elaborate and diverse behaviors are associated with egg laying and parental care. Nest preparation by birds runs the gamut from nothing more than the fairy tern's selection of a branch on which it balances its egg, to the multiroom communal nests of weaver birds, which are used by generation after generation. Incubation provides heat for the development of eggs and the presence of a parent is a deterent to many predators. However, some birds leave their eggs for periods of days while they forage, and brood parasites deposit their eggs in the nests of other species of birds and play no role in brooding or rearing their young.

Oviparity In contrast to the diversity of mating strategies of birds, their mode of reproduction is limited to laying eggs. No other group of vertebrates that contains such a large number of species is exclusively oviparous—why is this true of birds?

Constraints imposed on birds by their specializations for flight are often invoked to explain the failure of birds to evolve viviparity, but those arguments are not particularly convincing when one remembers that bats have successfully combined flight and viviparity. Furthermore, flightlessness has evolved in at least 15 families of birds, but none of these flightless species has evolved viviparity.

Oviparity is presumed to be the ancestral reproductive mode for diapsids, and it is retained by both living groups of archosaurs, the crocodilians and the birds. However, viviparity has evolved nearly 100 times in the other major lineage of living diapsids, the lepidosaurs (Chapter 15), so the capacity for viviparity is clearly present in diapsids. A key element in the evolution of viviparity among lizards and snakes appears to be the retention of eggs in the oviducts of the female for some period before they are deposited. This situation occurs when the benefits of egg retention outweigh its costs. For example, the high incidence of viviparity among snakes and lizards in cold climates may be related to the ability of a female ectotherm to speed embryonic development by thermoregulation. A lizard that basks in the sun can raise the temperature of eggs retained in her body, but after the eggs are deposited in a nest the mother no longer has any control over their temperature and rate of development. Birds are endotherms and brood their eggs, thereby controlling their temperature after the eggs are laid. Thus egg retention provides no thermoregulatory advantage for a bird.

Broad aspects of the biology of birds may create an unfavorable balance of costs and benefits of egg retention, thereby making it unlikely that any lineage of birds would take the first step in an evolutionary process that has repeatedly led to viviparity among snakes and lizards. If birds are viewed as being specialized for the production of one relatively large egg at a time and for complex egg incubation and parental care, the potential advantages of egg retention are greatly diminished and the costs of decreased fecundity and increased risk of maternal mortality are increased (Blackburn and Evans 1986). Perhaps it is this balance of costs and benefits rather than any single factor that is responsible for the retention of the ancestral reproductive mode by all living birds. The same line of reasoning probably can be applied to crocodilians, which construct nests and care for their young, and it can be extended with caution to speculations about the reproductive mode of dinosaurs.

Nesting Construction of nests is an important aspect of avian reproduction because nests provide protection for the eggs from such physical stresses as heat, cold, or rain and from predators. Bird nests range from shallow holes in the ground to enormous structures that represent the combined efforts of hundreds of individuals over many generations (Figure 18–18). The nests of passerines are usually cup-shaped structures composed of plant materials that are woven together. Swifts use sticky secretions from buccal glands to cement material together to form nests, and grebes, which are marsh-dwelling birds, build floating nests from the buoyant stems of aquatic plants. A recent review of bird nests can be found in Collias and Collias (1984).

Most birds nest individually: Only 16 percent of passerines nest in colonies, but 98 percent of seabirds are colonial nesters (Wittenberger and Hunt 1985, Kharitonov and Siegel-Causey 1988). Nesting colonies of some species of penguins, petrels, gannets, gulls, terns, and auks contain hundreds of thousands of individuals. Colonies are smaller in most other groups of birds; colonies

Figure 18–18. The nests of birds show great diversity. Some nests are no more than shallow depressions, whereas other birds build elaborate structures: (a) colony of northern gannets. The nests, which are only shallow scrapes, are located two beak-lengths apart; (b) nest of the killdeer plover; (c) house swallow nest, built of mud; (d) northern oriole nest, woven from vegetation. [(a) photograph by L. B. Chapmann; (b,d) photographs by Michael Hopiak, all courtesy of the Cornell Laboratory of Ornithology; (c) photograph by Michael A. Recht, Ph.D.]

of herons, storks, doves, swifts, and passerines contain a few tens of nests. Colonial nesting offers advantages and disadvantages. A colony is a concentration of potential prey that may attract predators, but the density of nesting birds may provide a degree of protection. In many colonies the nests are located two neck lengths apart, and an intruder is menaced from all sides by snapping beaks. Centrally placed nests may be better protected against predators than nests on the periphery of the colony.

Mixed colonies of two or more species of seabirds occur, but at least some of these may be transitional situations in which one species is in the process of displacing the other. For example, on the eastern coast of North America the greater black-backed gull (*Larus marinus*) is extending its range southward, apparently in response to the abundance of food available in garbage dumps. As it moves south, it is invading the breeding colonies of herring gulls (*Larus argentatus*). Great black-backed gulls, which are larger than herring gulls and breed earlier in the year, appear to be displacing herring gulls from some of their traditional breeding sites.

Incubation   The megapodes, known as mound birds, bury their eggs in sand or soil and rely on heat from the sun or rotting vegetation for incubation, and the Egyptian plover buries its eggs in sand, but all other birds are believed to brood their eggs using metabolic heat. Some species of birds begin incubation as soon as the first egg is laid and others wait until the clutch is complete. Starting incubation immediately may protect the eggs, but it means that the first eggs in the clutch hatch while the eggs that were deposited later are still developing, forcing the parents to divide their time between incubation and gathering food for the hatchlings. Furthermore, the eggs that hatch last produce young that are smaller than their older nestmates and these young probably have less chance of surviving to fledge. Most passerines, as well as ducks, geese, and fowl, do not begin incubation until the next-to-last or last egg has been laid.

Prolactin, secreted by the pituitary gland, suppresses ovulation and induces brooding behavior, at least in those species of birds that wait until a clutch is complete to begin incubation. The insulative properties of feathers that are so important a feature of the thermoregulation of birds become a handicap during brooding when it is necessary for the parent to transfer metabolic heat from its own body to the eggs. Prolactin plus estrogen or androgen stimulates the formation of brood patches in female and male birds, respectively. These brood patches are areas of bare skin on the ventral surface of a bird. The feathers are lost from the brood patch and blood vessels proliferate in the dermis, which may double in thickness and give the skin a spongy texture. Not all birds develop brood patches, and in some species only the female has a brood patch, although the male may share in incubating the eggs. Ducks and geese create brood patches by plucking the down feathers from their breasts; they use the feathers to line their nests. Some penguins lay a single egg that they hold on top of the feet and cover with a fold of skin on the belly that envelopes the egg as it rests on the bird's feet.

The temperature of eggs during brooding is usually maintained within the range 33 to 37°C, and some eggs can withstand periods of cooling when the parent is off the nest. [See Drent (1975) for details.] Tube-nosed seabirds (Procellariiformes) are known for the long periods that adults spend away from the nest during foraging. Fork-tailed storm petrels (*Oceanodroma furcata*) lay a single egg in a burrow or rock crevice. Both parents participate in incubation, but the adults forage over vast distances and both parents may be absent from the nest for periods of hours or even for several days at a time. The mean period of parental absence was 11 days (during an incubation period that averaged 50 days) for storm petrels studied in Alaska, and eggs were exposed to ambient temperatures of 10°C while the parents were away. Experimental studies showed that storm petrel eggs were not damaged by being cooled to 10°C every 4 days (Vleck and Kenagy 1980). The pattern of development of chilled eggs was like that of eggs incubated continuously at 34°C, except that each day of chilling added about one day to the total time required for the eggs to hatch.

Parent birds turn the eggs as often as several times an hour during incubation, and individual eggs are moved back and forth between the center and the edge of the clutch. Temperature variation exists within a nest and shifting the eggs about may ensure that they all experience approximately the same average temperature. In addition, turning the eggs may help to prevent premature fusion of the chorioallantoic membrane with the inner shell membrane. Eggs attain a stable orientation during incubation—that is, the same side is usually uppermost. This position is dictated by the asymmetric distribution of embryonic mass within the egg, and the process of turning the egg allows the embryo to assume its equilibrium position. Apparently this mechanism assures that when the chorioallantoic and inner shell membranes fuse (approximately midway through incubation), the embryo is in a position that will facilitate hatching.

Incubation periods are as short as 10 to 12 days for some species and as long as 60 to 80 days for others. In general, large species of birds have longer incubation periods than small species, but ecological factors also contribute to determining the length of the incubation period. The effect of parental absence in slowing development has already been mentioned, and the amount of time a parent is absent may depend on its foraging success. A high risk of predation may favor rapid development of the eggs. Among tropical tanagers, species that build open-topped nests near the ground are probably more vulnerable to predators than are species that build similar nests farther off the ground. The incubation periods of species that nest near the ground are short (11 to 13 days) compared to those of species that build nests at greater heights (14 to 20 days). Species of tropical tanagers with roofed-over nests have still longer incubation periods—17 to 24 days.

The inorganic part of eggshells contains about 98 percent crystalline calcite, $CaCO_3$, and the embryo obtains about 80 percent of its calcium from the eggshell. An organic matrix of protein and mucopolysaccharides is distributed through the shell and may serve as a support structure for the growth of calcite crystals. Eggshell formation begins in the isthmus of the oviduct. Two shell membranes are secreted to enclose the yolk and albumen, and carbohydrate and water are added to the albumen by a process that involves active transport of sodium across the wall of the oviduct followed by osmotic flow of water. The increased volume of the egg contents at this stage appears to stretch the egg membranes taut. Organic granules are attached to the egg membrane, and these **mammillary bodies** appear to be the sites of the first formation of calcite crystals (Figure 18–19). Some crystals grow downward from the mammillary bodies and fuse to the egg membranes, and other crystals grow away from the membrane to

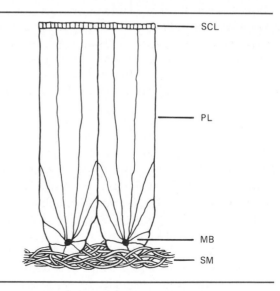

Figure 18–19. Diagram of the crystal structure of an avian eggshell. Crystallization begins at the mammillary bodies (MB) and crystals grow into the outer shell membrane (SM) and upward to form the palisade layer (PL). Changes in the chemical composition of the fluid surrounding the growing eggshell are probably responsible for the change in crystal form in the surface crystalline layer (SCL). (From C. Carey, 1983, in *Current Ornithology*, volume 1, edited by R. F. Johnston, Plenum Press, New York.)

form cones. The cones grow vertically and expand horizontally, fusing with crystals from adjacent cones to form the pallisade layer. Changes in the ionic composition of the fluid surrounding the egg during shell formation lead to an increase in the concentrations of magnesium and phosphorus and a change in the pattern of crystallization in the surface layers of the shell.

The eggshell is penetrated by an array of pores that allow oxygen to diffuse into the egg and carbon dioxide and water to diffuse out. Pores occur at the junction of three calcite cones, but only 1 percent or less of those junctions form pores; the rest are fused shut. Pores occupy about 0.02 percent of the surface of an eggshell. The morphology of the pores varies in different species of birds: Some pores are straight tubes, whereas others are branched. The openings of the pores on the surface of the eggshell may be occluded to varying degrees with organic or crystalline material. Additional information about the structure and function of avian eggs can be found in Carey (1980, 1983).

Water evaporates from an egg during development and the loss of water creates an air cell at the blunt end of the egg. The embryo penetrates the membranes of this air cell with its beak 1 or 2 days before hatching begins, and ventilation of the lungs begins to replace the chorioallantoic membrane in gas exchange. Pipping, the formation of the first cracks on the surface of the eggshell, follows about half a day after penetration of the air cell, and actual emergence begins half a day later. Shortly before hatching the chick develops a horny projection on its upper mandible. This structure is called the egg tooth and it is used in conjunction with a hypertrophied muscle on the back of the neck (the hatching muscle) to thrust vigorously against the shell. The egg tooth and hatching muscle disappear soon after the chick has hatched. In those species of birds that delay the start of incubation until all the eggs have been laid, an entire clutch nears hatching simultaneously. Hatching may be synchronized by clicking sounds that accompany breathing within the egg, and both acceleration and retardation of individual eggs may

be involved. A low-frequency sound produced early in respiration, before the clicking phase is reached, appears to retard the start of clicking by advanced embryos. That is, the advanced embryos do not begin clicking while other embryos are still producing low-frequency sounds. Subsequently, clicking sounds or vocalizations from advanced embryos appear to accelerate late embryos. Both effects were demonstrated by Vince (1969) in experiments with bobwhite quail eggs. She found that she could accelerate the hatching of a late egg by 14 hours when she paired it with an early egg that had started incubation 24 hours sooner, and the presence of the late egg delayed hatching of the early egg by 7 hours.

Parental Care

The ancestral form of reproduction in the archosaur lineage appears to consist of the deposition of eggs in a well-defined nest site, attendance at the nest by one or both parents, hatching of precocial young, and a period of association between the young and one or both parents. All of the crocodilians that have been studied conform to this pattern, and evidence is increasing that at least some dinosaurs remained with their nests and young.

Modern birds follow these ancestral patterns, but not all species produce precocial young. Instead, hatchling birds show a spectrum of maturity that extends from precocial young that are feathered and self-sufficient from the moment of hatching to altricial forms that are naked and entirely dependent on their parents for food and thermoregulation (Figure 18–20 and Table 18–3). The most precocial birds at hatching are the megapodes, and these show the most ancestral form of nesting, burying their eggs in nests made from mounds of soil and vegetation very much like those of crocodilians. Newly hatched megapodes scramble to the surface already feathered and capable of flight. Most precocial birds are covered with down at hatching and can walk, but are not able to fly.

The distinction between precocial and altricial birds extends back to differences in the amount of

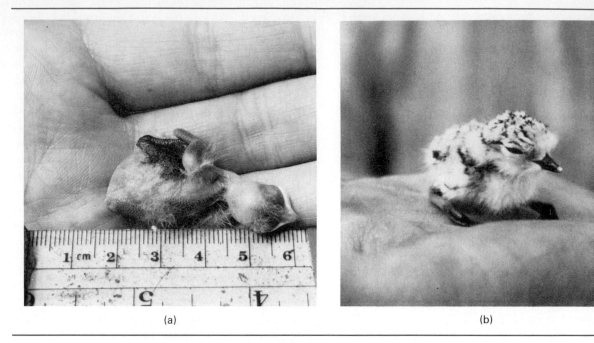

(a)   (b)

Figure 18–20. Altricial chicks (a) such as that of the tree swallow *(Tachycinta bicolor)* are entirely naked when they hatch and unable even to stand up. Precocial species (b) such as the snowy plover *(Charadius alexandrinus)* are covered with down when they hatch and can stand erect and even walk. The plover in this photograph has just hatched; it retains the egg tooth on the tip of its bill, whereas the tree swallow chick is 5 days old. The dark color on the leg of the swallow is ink, used to identify individual hatchlings for a study of parental care. (Photographs by David Winkler.)

Table 18–3. Maturity of birds at hatching.

Precocial: eyes open, covered with feathers or down, leave nest after 1 or 2 days
1. Independent of parents: megapodes
2. Follow parents, but find their own food: ducks, shorebirds
3. Follow parents and are shown food: quail, chickens
4. Follow parents and are fed by them: grebes, rails

Semiprecocial: eyes open, covered with down, able to walk but remain at nest and are fed by parents: gulls, terns

Semialtricial: covered with down, unable to leave nest, fed by parents
1. Eyes open: herons, hawks
2. Eyes closed: owls

Altricial: eyes closed, little or no down, unable to leave nest, fed by parents: passerines

Source: Modified from M. M. Nice, 1962. *Transactions of the Linnaean Society of New York* 8:1–211.

yolk originally in the eggs and includes differences in the relative development of organs and muscles at hatching, and the rates of growth after hatching (Table 18–4). Robert Ricklefs (1979) has proposed that the physical maturity of tissues at hatching, especially skeletal muscles, can be used to subdivide the growth patterns of birds. Mature tissues may not be capable of rapid growth, and these mature tissues may set the limits to the growth of other tissues. Ricklefs emphasized the relationship between mode of development and food supply, and David Winkler and Jeffrey Walters (1983) have pointed to a strong phylogenetic influence on development modes. Most differences in developmental mode occur between orders, not within them, even when species within an order differ substantially in their ecology.

After altricial young have hatched they are guarded and fed by one or both parents. In some species the parents are assisted by nest helpers (Box 18–2). Adults of some species of birds carry food to nestlings in their beaks, but many species swallow food and later regurgitate it to feed the young. Hatchling altricial birds respond to any disturbance that might signal the arrival of a parent at the nest by gaping their mouths widely. The sight of an open mouth appears to stimulate a par-

Table 18–4. Comparison of altricial and precocial birds.

Amount of yolk in eggs	precocial > altricial
Amount of yolk remaining at hatching	precocial > altricial
Size of eyes and brain	precocial > altricial
Development of muscles	precocial > altricial
Size of gut	altricial > precocial
Rate of growth after hatching	altricial > precocial

ent bird to feed it, and the young of many altricial birds have brightly colored mouth linings. Ploceid finches have covered nests, and the mouths of the nestlings of some species are said to have luminous spots that have been likened to beacons showing the parents where to deposit food in the gloom of the nest.

The duration of parental care is quite variable: The young of small passerines fledge (leave the nest) about 2 weeks after hatching and are cared for by their parents for an additional week. Larger species of birds such as the tawny owl spend a month in the nest and receive parental care for an additional 3 months after they have fledged, and the young of the wandering albatross requires a year to become independent of its parents.

Box 18–2. Built-In Babysitters: Nest Helpers

A peculiar feature of the reproductive biology of more than 200 species of birds is the existence of nest helpers that provide care to offspring that are not their own. Most examples of nest helpers occur in Australia or the tropics, and few are known from Europe or North America. The mating systems of species with helpers vary from monogamy (the most common situation) through species with multiple breeders of either sex. Most species that have nest helpers are territorial, but some are colonial. Helper systems are characterized by regular involvement of the helpers in feeding and care of the young. Helpers defer their own breeding for one or more years while they assist in raising the offspring of other birds.

The peculiarity of helper systems lies in the expenditure of time and energy by helpers in caring for young that are not genetically their own. This altruistic behavior would appear to reduce the fitness of helpers, and that paradox has stimulated many studies. The

Box 18–2. (Continued)

concept of **kin selection** has contributed substantially to understanding helper systems. Stated in simplified form, this hypothesis proposes that an individual can increase its fitness by providing assistance to a related individual because relatives share alleles of common descent, and these alleles (not individuals) are the units of inheritance. Thus, the **inclusive fitness** of an individual consists of (1) its own reproductive success, plus (2) the additional reproductive success of relatives that results from the altruistic behavior of the individual multiplied by the fraction of alleles shared with each relative, minus (3) any decrease in the reproductive success of the individual that results from its altruistic behavior.

The hypothesis of kin selection predicts that nest helpers will be related to the individuals they help, and this is often the case. For example, almost 50 percent of the cases of nest helpers among Florida scrub jays involved birds helping their own parents, and 25 percent involved birds helping a parent and a stepparent. Birds helping their own siblings accounted for 20 percent of the nest helpers, and less than 5 percent of the examples involved helping entirely unrelated individuals (Woolfenden 1975, 1981). However, this pattern is not universal, and other examples of nest helpers involve more complicated genetic relationships among the participants, including cases where the helpers are unrelated to the individuals they help.

Nest helpers do help; nests with helpers almost always fledge more young than nests without helpers. Thus, kin selection could produce some benefit to the helpers, but why would the helpers not increase their fitness still more by breeding themselves? In other words, why *do* helpers help? Also, why do helpers *ever* help nonrelatives?

Studies of these questions have stimulated much discussion and various points of views; a summary can be found in Oring (1982). Several general hypotheses have been proposed:

1. A shortage of breeding territories, nest sites, or potential mates may make it difficult for young birds to breed. Helping to raise younger siblings may be the best way to mark time until an opportunity to breed presents itself.
2. Becoming a nest helper may be a way to gain access to a territory and, eventually, to a mate.
3. Some components of parental care are learned by experience, and birds that act as helpers for one or more breeding seasons may fledge more young when they do reproduce as a result of the experience they have gained.

These hypotheses are not mutually exclusive—they all may apply to some species—and hypotheses 2 and 3 suggest that some advantage could be gained from helping even unrelated individuals.

Migration and Navigation

The mobility that characterizes vertebrates is perhaps most clearly demonstrated in their movements over enormous distances. These displacements, which may cover half the globe, require both endurance and the ability to navigate. Other vertebrates migrate, even over enormous distances, but migration is best known among birds.

Long-Distance Movements of Vertebrates

Four major kinds of long-distance movements by vertebrates can be distinguished, although the categories blend into each other. **Dispersal movements** are universal among animals, even among species in which individuals typically occupy limited home ranges or territories. These dispersal movements of individuals, often by juveniles or nonreproductive adults, may be related to intraspecific competition for resources and can result in expansion of the species' geographic range. Thus, overcrowding in the center of the range may lead some individuals to move to new areas where they may establish new populations.

Nomadism is a more or less irregular or random movement by a population of individuals (usually organized in herds, flocks, or schools) to favorable feeding or breeding areas that are of unpredictable occurrence in time or space. Nomadism allows animals to exploit local and temporary sources of food. Such movements are especially characteristic of ungulates living in environments subject to harsh and variable conditions such as Arctic tundras, arid grasslands, and deserts. Most nomadic mammals have favored calving grounds to which females return on a regular basis. Some pelagic fishes and a few species of birds in boreal forests and in austral deserts are also nomadic.

Emigration, also referred to as **invasion** or **irruption**, is another kind of movement characteristic of species in unpredictable environments. These are irregular movements of large numbers of individuals into areas where the species is not usually found. Irruptions result from periods of favorable feeding and breeding conditions that allow populations to increase in numbers until food gives out. The mass exodus that follows is usually accompanied by high mortality. Norwegian lemmings normally dwell in alpine tundra, but occasionally make irruptive movements down fjords to the sea. In the boreal forest, crossbills and nutcrackers, which respond to superabundant but irregular production of spruce nuts, and also goshawks and snowy owls, which depend on snowshoe hares and lemmings for food, are additional examples of species that show emigration.

Migration, on the other hand, refers to a regular, predictable movement between two home ranges or territories, typically between a summer breeding area and a winter, nonbreeding area. It is important to emphasize, however, that true migrations are separated only by definition from emigration and nomadism. All sorts of gradations and combinations occur, even within a single species. Arctic caribou are at times nomadic, but they also make regular migratory movements to and from winter and summer feeding grounds. Some segments of the snowy owl and goshawk populations are regularly migratory, but occasional large-scale invasions bring them far south of their usual wintering ranges.

Migrations often involve movements over thousands of kilometers, especially in the case of birds nesting in northern latitudes, some marine mammals, sea turtles, and fishes. Short-tailed shearwaters, for example, make an annual migration between their breeding range in southern Australia and the North Pacific that requires a round trip of more than 30,000 kilometers (Figure 18–21). Gray whales make round trips of 9000 kilometers or more between their feeding grounds in the Bering Sea and calving areas in bays on the coast of Baja California. The 600-kilometer migration by North American caribou probably is the longest distance regularly traveled by a land mammal. Less mobile vertebrates such as amphibians may migrate a kilometer or more from nonbreeding habitats to breeding areas.

Migratory Movements of Birds

Few migratory movements are as dramatic as those of birds, and in no other group has migration been so well studied. Migration is a widespread phenomenon among birds—about 40 percent of the bird species in the Palearctic are migratory, and an estimated total of some 5 billion birds migrate from the Palearctic every year. Data accumulated from recaptures of banded birds have established the origins, destinations, and migratory pathways

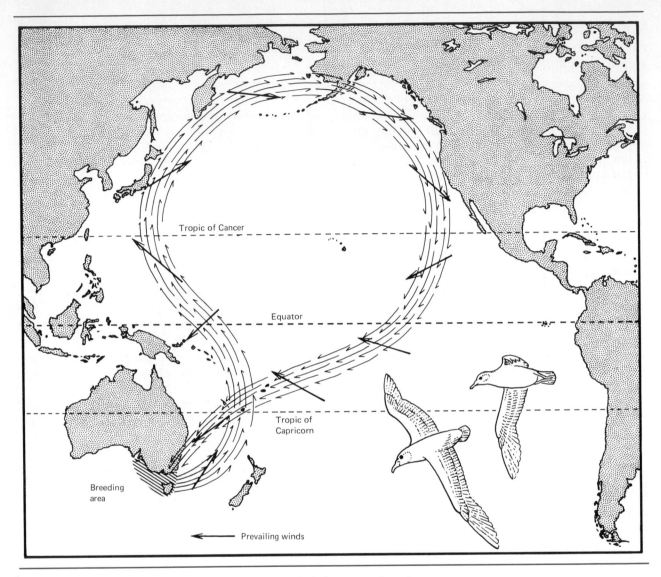

Figure 18–21. The migratory path of the short-tailed shearwater from its
Australian breeding area to its northern range takes advantage of the prevailing
winds in the Pacific region to reduce the energy cost of migration. (From A. J.
Marshall and D. L. Serventy, 1956, *Proceedings of the Zoological Society of London*
127:489–510.)

for many species. A remarkable feature of most of
these migrations is their relatively recent origin:
Migrations are responses to seasonal changes in
the availability of resources. These variations in
the resource base are, in turn, the result of seasonal cycles in the climate, and worldwide patterns of climate have changed frequently during
the past 2 million years. Current avian migratory
patterns are probably no more than 15,000 years
old (Moreau 1972).

The Advantages of Migration The high energy costs of migration must be offset by energy gained as a result of moving to a different habitat. The normal food sources for some species of birds are unavailable in the winter, and the benefits of migration for those species are starkly clear. Other species may save energy mainly by avoiding the temperature stress of northern winters. In other cases the main advantage of migration may come from breeding in high latitudes in the summer when the long days provide more time to forage than the birds would have if they remained closer to the equator.

Competition may also play a role in migration. Direct evidence of competition for food resources is lacking, but indirect evidence was provided by a study of birds that migrate between the Neotropics and North America (Cox 1968). The extent of differentiation of the beak is considered to be a measure of the difference in feeding niches of birds. Birds that have bills of different sizes and shapes are usually assumed to be exploiting different sources of food even when they forage in the same place, whereas birds with bills of the same size and shape are assumed to eat the same kinds of food and must forage in different habitats to avoid competition. During the breeding season, when food requirements are high, competition should be more intense between species with bills of similar shape than between species with bills of different shapes. Cox compared the bill morphology of resident and migrant tropical birds: If competition is irrelevant to migration the similarity of bill shape should be the same for resident and migrant species. If competition is important in determining migration, one would predict that species with similar bill shapes would migrate, because they are less able to separate their feeding niches ecologically.

Cox found that variation in bill form was greatest in resident tropical birds, and decreased as the proportion of migratory species in a group increased (Figure 18–22). This result is consistent with the predictions of the hypothesis that competition is an important force in migration. Cox proposed that species of birds that are not ecolog-

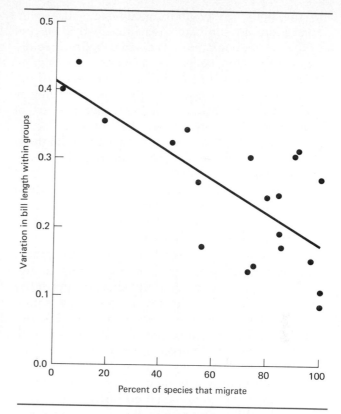

Figure 18–22. Birds that winter in Central America show an inverse relationship between the amount of variation in beak length within a group and the percent of species in that group that migrate to North America during the breeding season. Variation in beak length is expressed as coefficient of variation for each group. (Data from G. W. Cox, 1968, *Evolution* 22:180–192.)

ically separated by bill morphology avoid competition for food during the breeding season by migrating northward.

Physiological Preparation for Migration Migration is the result of a complex sequence of events that integrate the physiology and behavior of birds. Fat is the principal energy store for migratory birds, and birds undergo a period of heavy feeding and premigratory fattening (*Zugdisposition*, migratory preparation) in which fat deposits in the body cavity and subcutaneous tissue increase tenfold, ultimately reaching 20 to 50 percent of the nonfat body mass. Fat is metabolized rapidly when migration begins, and many birds migrate at night and eat during the day. Even diurnal migrants divide the day into periods of migratory flight (usually early in the day) and periods of feeding. In addition, pauses of several days to replenish fat stores are a normal part of migration. *Zugdisposition* is followed by **Zugstimmung** (migratory mood), in which the bird undertakes and maintains migratory flight. In caged birds, which are prevented from migrating, this condition results in the well-known phenomenon of **Zugunruhe** (migratory restlessness).

Preparation for migration must be integrated with environmental conditions, and this coordination appears to be accomplished by the interaction of internal rhythms with an external stimulus. Daylength is the most important cue for *Zugdisposition* and *Zugstimmung* for birds in north temperate regions. Northward migration in spring is induced by increasing daylength (Figure 18–23). The direction in which migratory birds orient during *Zugunruhe* depends on their physiological condition. In this experiment, photoperiod was manipulated to bring one group of indigo buntings into their autumn migratory condition at the same time that a second group of birds was in its spring migratory condition. When the birds were tested under an artificial planetarium sky, the birds in the spring migratory condition oriented primarily in a northeasterly direction (Figure 18–23a), whereas birds in the fall migratory condition oriented in a southerly direction (Figure 18–23b). Dark circles show the mean nightly headings pooled for several observations for each of six birds in the spring migratory condition and five birds in the fall condition.

After breeding, many species of birds enter a refractory period in which they are unresponsive to long daylengths, their gonads regress to the nonbreeding condition, and they molt. Decreasing photoperiods in autumn may accelerate the southward migration. Many birds are again refractory to photoperiod stimulation for several weeks after the autumnal migration and require an interval of several weeks of short photoperiods before they can again be stimulated by long photoperiods.

Underlying the responses of birds to changes in daylength is an endogenous (internal) rhythm. This circannial (about a year) cycle can be demonstrated by keeping birds under constant conditions. Fat deposition and migratory restlessness coincide in most species and alternate with gonadal development and molt as they do in wild birds. When the rhythms are free-running (that is, when they are not cued by external stimuli), they vary between 7 and 15 months. In other words, the birds' internal clocks continue to run, but in the absence of the cue normally provided by changing daylength the internal rhythms drift away from precise correspondence with the seasons. Further discussion of this topic can be found in Meir and Fivizzani (1980), Farner (1986), and Bluhm (1988).

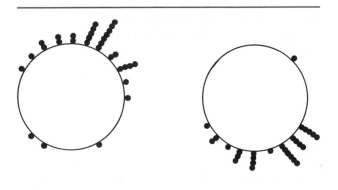

SPRING CONDITION   AUTUMN CONDITION

Figure 18–23. Direction in which migratory birds orient during *Zugunruhe*. (Data from S. T. Emlen, 1969, *Science* 165:716–718.)

Orientation and Navigation

The seasonal migrations of vertebrates that cover thousands of kilometers and, especially, their ability to return regularly to the same locations, year after year, pose an additional question—how do they find their way? Various hypotheses propose mechanisms to explain how animals navigate on these journeys and the explanations fall into two general categories: (1) Long-distance migration is an extension of the tendency to explore territory beyond the local home range, learning to recognize landmarks as one goes along or, (2) the ability to home through unfamiliar territory results from an internal navigational system (Able 1980, Baker 1982). There are no clear answers, but experiments reveal that many vertebrates can find their way home when they are displaced and that a number of different mechanisms and sensory modalities are involved. A general review of animal migration and navigation can be found in Gauthreaux (1980).

The homing pigeon has become a favorite experimental animal for studies of navigation. As long as people have raised and raced pigeons, it has been known that birds released in unfamiliar territory vanish from sight flying in a straight line, usually in the direction of home. How do pigeons accomplish this feat? There is no complete answer yet, but experiments have shown that navigation by homing pigeons (and presumably by other vertebrates as well) is complex and can be based on a variety of sensory cues. On sunny days pigeons vanish toward home and return rapidly to their lofts. On overcast days vanishing bearings are less precise and birds more often get lost. These observations led to the idea that pigeons use the sun as a compass.

Of course, the position of the sun in the sky changes from dawn to dusk. That means that a bird must know what time of day it is to use the sun to tell direction, and this time-keeping ability requires some sort of internal clock. If that hypothesis is correct, it should be possible to fool a bird by shifting its clock forward or backward. For example, if one turns on the lights in the pigeon loft 6 hours before sunrise every morning for about 5 days, the birds will become accustomed to that artificial sunrise and at any time during the day they will assume that the time is 6 hours later than it actually is. When those birds are released they will judge direction by the sun, but their internal clocks will be wrong by 6 hours. As a consequence

of that error, the birds should fly off on courses that are displaced by 90° from the correct course for home. Clock-shifted pigeons react differently under sunny and cloudy skies, indicating that pigeons have at least two mechanisms for navigation (Figure 18–24). Each dot in these plots shows the direction in which a pigeon vanished from sight when it was released in the center of the large circle. The home loft is straight up in each diagram. The solid bar extending outward from the center of each circle is the average vector sum of all the individual vanishing points. When they are able to see the sun, control birds that have been kept on the normal photoperiod orient predominantly in the direction of home (Figure 18–24a). Birds that have had their photoperiods shift 6 hours fast disappear on bearings westward of the true direction of home (Figure 18–24b). However, when the sun is obscured by clouds the birds cannot use it for navigation, and must rely on other mechanisms. Under those conditions both control and clockshifted birds orient correctly toward home (Figure 18–24c and d).

Polarized light is another cue vertebrates use to determine directions. In addition, some vertebrates have been shown to detect ultraviolet light and to sense extremely low frequency sounds, well below the frequencies humans can hear (Kreithen 1983). Those sounds are generated by ocean waves and air masses moving over mountains and can signal a general direction over thousands of kilometers, but their use as cues for navigation remains obscure. The senses that have been shown to be involved in navigation by pigeons do not end here. Apparently pigeons can also navigate by recognizing airborne odors as they pass over the terrain (Papi et al. 1972). Even magnetism is implicated: On cloudy days pigeons wearing small magnets on their heads have their ability to navigate disrupted, but on sunny days magnets have no effect. When it is clear, pigeons apparently rely on their sun compass and magnetic cues are ignored (Keeton 1969, 1971).

Results of this sort are being obtained with other vertebrates as well, and lead to the general conclusion that a great deal of redundancy is built

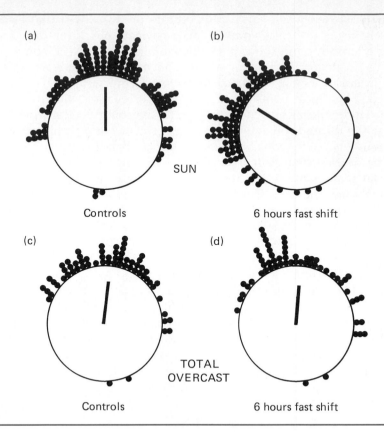

(a)

(b)

SUN

Controls

6 hours fast shift

(c)

(d)

TOTAL
OVERCAST

Controls

6 hours fast shift

Figure 18–24. Orientation of clock-shifted pigeons under sunny and cloudy skies. The line shows the average direction chosen by the birds. (From W. T. Keeton, 1969, *Science* 165:922–928.)

into navigational systems. Apparently, there is a hierarchy of useful cues. For example, a bird that relies on the sun and polarized light to navigate on clear days could switch to magnetic direction sensing on heavily overcast days. For both conditions, it might use local odors and recognition of landmarks as it approaches home.

Many birds migrate only at night. Under these conditions a magnetic sense of direction might be important, but it has not yet been demonstrated. It has been shown that several species of nocturnally migrating birds use star patterns for navigation. Apparently, each bird fixes on the pattern of particular stars and uses their motion in the night sky to determine a compass direction. As is the case for sun compass navigation, an internal clock is required for this sort of celestial navigation, and

artificially changing the time setting of the internal clock produces predictable changes in the direction in which a bird orients.

Despite numerous studies the complexities of navigational mechanisms of vertebrates are far from being fully understood and much controversy surrounds some hypotheses. [See Gould (1982), Griffin (1982), and Wallraff et al. (1982) for additional details.] The built-in redundancy of the systems makes it difficult to devise experiments that isolate one mechanism. When it is deprived of the use of one sensory modality, an animal is likely to have one or several others it can use instead. This redundancy itself and the remarkable sophistication with which many vertebrates navigate show the importance of migration in their lives.

Summary

The fossil record reveals that birds radiated in the Jurassic. *Archaeopteryx*, the earliest fossil bird known, was probably a late-surviving relict that was contemporaneous with more advanced birds. It retained a large number of primitive characters and, except for the presence of feathers, was very like a small dinosaur. True flying birds were widespread and diverse by the Cretaceous. Modern orders of birds had evolved by the end of the Eocene, but the phylogenetic relationships among the groups of modern birds are poorly understood.

The ecology and behavior of birds are directly influenced by their ability to fly. The mobility of birds allows them to exploit food supplies that have patchy distributions in time and space and to feed and reproduce in different areas. Migration is the most dramatic manifestation of the mobility of birds, and some species travel tens of thousands of kilometers in a year. Migrating birds use a variety of cues for navigation, including the position of the sun, polarized light, the Earth's magnetic field, and infrasound.

Many of the complex social behaviors of birds are associated with reproduction, and birds have contributed greatly to our understanding of the relationship between ecological factors and the mating systems of vertebrates. All birds are oviparous, perhaps because egg retention is the first step in the evolution of viviparity and the specializations of the avian way of life do not make egg retention advantageous for birds. Monogamy, with both parents caring for the young, is the most common mating system for birds, but polygyny (a male mating with several females) and polyandry (a female mating with several males) also occur.

References

Able, K. P. 1980. Mechanisms of orientation, navigation, and homing. Pages 282–373 in *Animal Migration, Orientation, and Navigation*, edited by Sidney A. Gauthreaux, Jr. Academic Press, New York.

Baker, R. R. 1982. *Migration Paths Through Time and Space*. Hodder and Stoughton, Sevenoaks, Kent, England.

Blackburn, D. G. and H. E. Evans. 1986. Why are there no viviparous birds? *American Naturalist* 128:165–190.

Bluhm, C. K. 1988. Temporal patterns of pair formation and reproduction in annual cycles and associated endocrinology in waterfowl. Pages 123–185 in *Current Ornithology*, volume 5, edited by R. F. Johnston. Plenum Press, New York.

Carey, C. (editor). 1980. Physiology of the avian egg. *American Zoologist* 20:325–484.

Carey, C. 1983. Structure and function of avian eggs. Pages 69–103 in *Current Ornithology*, volume 1, edited by Richard F. Johnston. Plenum Press, New York.

Catchpole, C. K. 1980. Sexual selection and the evolution of song in European warblers of the genus *Acrocephalus*. *Behaviour* 74:149–166.

Catchpole, C. K., J. Dittami, and B. Leisler. 1984. Differential responses to male song repertoires in female songbirds implanted with oestradiol. *Nature* 312:563–564.

Collias, N. E. and E. C. Collias. 1984. *Nest Building and Bird Behavior*. Princeton University Press, Princeton, N. J.

Cox, G. W. 1968. The role of competition in the evolution of migration. *Evolution* 22:180–192.

Cracraft, J. A. 1981. Toward a phylogenetic classification of the recent birds of the world (Class Aves). *Auk* 98:681–714.

Cracraft, J. A. 1986. The origin and early diversification of birds. *Paleobiology* 12:383–399.

Drent, R. 1975. Incubation. Pages 333–420 in *Avian Biology*, volume 5, edited by D. S. Farner, J. R. King, and K. C. Parkes. Academic Press, Orlando, Fla.

Emlen, S. T. and L. W. Oring. 1977. Ecology, sexual selection, and the evolution of mating systems. *Science* 197:215–223.

Faaborg, J., Tj. deVries, C. B. Patterson, and C. R. Griffin. 1980. Preliminary observations on the occurrence and evolution of polyandry in the Galapagos hawk (*Buteo galapagoensis*). *Auk* 97:581–590.

Farner, D. S. 1986. Generation and regulation of annual cycles in migratory passerine birds. *American Zoologist* 26:493–501.

Gauthreaux, S. A., Jr. (editor). 1980. *Animal Migration, Orientation, and Navigation*. Academic Press, New York.

Goss-Custard, J. D. 1977. Optimal foraging and size selection of worms by redshank, *Tringa totanus*. *Animal Behaviour* 25:10–29.

Gould, J. L. 1982. The map sense of pigeons. *Nature* 296:205–211.

Griffin, D. R. 1982. Ecology of migrations: is magnetic orientation a reality? *Quarterly Review of Biology* 57:293–295.

Hildén, O. 1975. Breeding system of Temminck's stint, *Calidris temminckii*. *Ornis Fennica* 52:117–146.

Houde, P. 1986. Ostrich ancestors found in the Northern Hemisphere suggest new hypothesis of ratite origins. *Nature* 324:563–565.

Houde, P. 1987. Critical evaluation of DNA hybridization studies in avian systematics. *Auk* 104:17–32.

Keeton, W. T. 1969. Orientation by pigeons: is the sun necessary? *Science* 165:922–928.

Keeton, W. T. 1971. Magnets interfere with pigeon homing. *Proceedings of the National Academy of Science, USA* 68:102–106.

Kharitonov, S. P. and D. Siegel-Causey. 1988. Colony formation in seabirds. Pages 223–272 in *Current Ornithology*, volume 5, edited by R. F. Johnston. Plenum Press, New York.

Kistchinski, A. A. 1975. Breeding biology and behaviour of the grey phalarope, *Phalaropus fulicarius*, in east Siberia. *Ibis* 117:285–301.

Konishi, M. 1985. Birdsong: from behavior to neuron. *Annual Review of Neurosciences* 8:125–170.

Krebs, J. R., D. W. Stephens, and W. J. Southerland. 1983. Perspectives in optimal foraging. Pages 165–216 in *Perspectives in Ornithology*, edited by A. H. Brush and G. A. Clark, Jr. Cambridge University Press, Cambridge.

Kreithen, M. L. 1983. Orientational strategies in birds: a tribute to W. T. Keeton. Pages 3–28 in *Behavioral Energetics: Vertebrate Costs of Survival*, edited by W. P. Aspey and S. I. Lustick. Ohio State University Press, Columbus, Ohio.

Lack, D. 1968. *Ecological Adaptations for Breeding in Birds*. Methuen, London.

Martin, L. D. 1983a. The origin and early radiation of birds. Pages 291–338 in *Perspectives in Ornithology*, edited by A. H. Brush and G. A. Clark, Jr. Cambridge University Press, Cambridge.

Martin, L. D. 1983b. The origin of birds and of avian flight. Pages 105–129 in *Current Ornithology*, volume 1, edited by Richard F. Johnston. Plenum Press, New York.

McGregor, P. K., J. R. Krebs, and C. M. Perrins. 1981.

Song repertoires and lifetime reproductive success in the great tit (*Parus major*). *American Naturalist* 118:149–159.

Meir, A. H. and A. J. Fivizzani. 1980. Physiology of migration. Pages 225–282 in *Animal Migration, Orientation, and Navigation*, edited by Sidney A. Gauthreaux, Jr. Academic Press, New York.

Moreau, R. E. 1972. *The Palearctic–African Bird Migration Systems*. Academic Press, New York.

Nottebohm, F. and A. P. Arnold. 1976. Sexual dimorphism in vocal control areas of the songbird brain. *Science* 194:211–213.

Olson, S. L. 1982. A critique of Cracraft's classification of birds. *Auk* 99:733–739.

Olson, S. L. 1985. The fossil record of birds. Pages 79–238 in *Avian Biology*, volume 8, edited by D. S. Farner, J. R. King, and K. C. Parkes. Academic Press, Orlando, Fla.

Orians, G. H. 1969. On the evolution of mating systems in birds and mammals. *American Naturalist* 103:589–603.

Orians, G. H. 1980. *Some Adaptations of Marsh-Nesting Blackbirds*. Princeton University Press, Princeton, N. J.

Oring, L. W. 1982. Avian mating systems. Pages 1–92 in *Avian Biology*, volume 6, edited by D. S. Farner, J. R. King, and K. C. Parkes. Academic Press, New York.

Ostrom, J. H. 1974. *Archaeopteryx* and the origin of flight. *Quarterly Review of Biology* 49:27–47.

Ostrom, J. H. 1979. Bird flight: how did it begin? *American Scientist* 67:46–56.

Papi, F., L. Fiore, V. Fiaschi, and S. Benvenuti. 1972. Olfaction and homing in pigeons. *Monitore Zoologia Italiana* (N.S.) 6:85–95.

Payne, R. B., W. L. Thompson, K. L. Fiala, and L. L. Sweany. 1981. Local song traditions in indigo buntings: cultural transmission of behavior patterns across generations. *Behaviour* 77:199–221.

Pierce, G. J. and J. G. Ollason. 1987. Eight reasons why optimal foraging theory is a complete waste of time. *Oikos* 49:111–118.

Raikow, R. J. 1985. Problems in avian classification. Pages 187–212 in *Current Ornithology*, volume 2, edited by Richard F. Johnston. Plenum Press, New York.

Ricklefs, R. E. 1979. Adaptation, constraint, and compromise in avian postnatal development. *Biological Review* 54:269–290.

Ridpath, M. G. 1972. The Tasmanian native hen, *Tribonyx mortierii*. I. Patterns of behaviour. *CSIRO Wildlife Research* 17:1–51.

Shields, G. F. and K. M. Helm-Bychowski. 1988. Mitochondrial DNA of birds. Pages 273–295 in *Current Ornithology*, volume 5, edited by Richard F. Johnston. Plenum Press, New York.

Sibley, C. G. and J. E. Ahlquist. 1983. Phylogeny and classification of birds based on the data of DNA–DNA hybridization. Pages 245–292 in *Current Ornithology*, volume 1, edited by Richard F. Johnston. Plenum Press, New York.

Sibley, C. G. and J. E. Ahlquist. 1986. Reconstructing bird phylogeny by comparing DNAs. *Scientific American* 254(2):82–92.

Stearns, S. C. and P. Schmid-Hempel. 1987. Evolutionary insights should not be wasted. *Oikos* 49:118–125.

Stephens, D. W. and J. R. Krebs. 1987. *Foraging Theory*. Princeton University Press, Princeton, N. J.

Vince, M. A. 1969. Embryonic communication, respiration, and the synchronization of hatching. Pages 233–260 in *Bird Vocalizations*, edited by R. A. Hinde. Cambridge University Press, Cambridge.

Vleck, C. M. and G. J. Kenagy. 1980. Embryonic metabolism of the fork-tailed storm petrel: physiological patterns during prolonged and interrupted incubation. *Physiological Zoology* 53:32–42.

Wallraff, H. G., S. Benvenuti, F. Papi, and J. Gould. 1982. The homing mechanism of pigeons. *Nature* 300:293–294.

Wiley, R. H. 1973. Territoriality and non-random mating in the sage grouse, *Centrocercus urophasianus*. *Animal Behaviour Monographs* 6:87–169.

Winkler, D. W. and J. R. Walters. 1983. The determination of clutch size in precocial birds. Pages 33–68 in *Current Ornithology*, volume 1, edited by Richard F. Johnston. Plenum Press, New York.

Wittenberger, J. F. and G. L. Hunt, Jr. 1985. The adaptive significance of coloniality in birds. Pages 1–78 in *Avian Biology*, volume 8, edited by D. S. Farner, J. R. King, and K. C. Parkes. Academic Press, Orlando, Fla.

Woolfenden, G. E. 1975. Florida scrub jay helpers at the nest. *Auk* 92:1–15.

Woolfenden, G. E. 1981. Selfish behavior by Florida scrub jay helpers. Pages 257–260 in *Natural Selection and Social Behavior*, edited by R. D. Alexander and D. W. Tinkle. Chiron Press, New York.

Again we must backtrack, this time to the end of the Paleozoic (Chapters 9 and 10) to find the origins of the final lineage of vertebrates, the synapsids. The synapsids actually had their first major radiation in the Paleozoic, before the radiations of the diapsids we have already discussed. However, the second radiation of the synapsid lineage consisted of the mammals and this radiation reached its peak in the Cenozoic. Nonetheless, through the late Paleozoic and early Mesozoic the synapsid lineage was becoming increasingly mammal-like and mammals and dinosaurs both appeared on the scene in the Triassic.

In Chapter 4 we discussed the paradox of the origin of endothermy: The two components of endothermal thermoregulation are a high metabolic rate that produces heat and insulation that retains that heat in the body, and neither characteristic is advantageous without the previous existence of the other. We suggested that the solution of the paradox might lie in the evolution of high rates of metabolism for some purpose other than thermoregulation. One possible benefit of a high rate of metabolism would be to enhance the locomotor endurance of predatory synapsids, and it is gratifying to see in the fossil record just the sorts of structural changes in the limbs and vertebral column of synapsids that might be expected to accompany an increasingly active form of locomotion and foraging. The three groups of therian mammals—monotremes, marsupials, and placentals—had evolved by the late Mesozoic, and they were accompanied by several groups of nontherian mammals that are now extinct.

The Synapsida and the Evolution of Mammals

Terrestrial Vertebrates of the Late Paleozoic

The synapsid lineage was the first group of amniotes to radiate widely in terrestrial habitats. During the late Carboniferous and the first half of the Permian the pelycosaurs were the most abundant terrestrial vertebrates, and from the mid-Permian into the Triassic the therapsids were the top carnivores in the food web. Both pelycosaurs and the early therapsids were large animals. By the late Triassic many of the synapsid lineages had disappeared, and the surviving forms were small and perhaps nocturnal.

The appearance of many mammalian skeletal characters among synapsids can be linked to increasing capacity for locomotor activity. The legs became longer and were progressively held more nearly under the body with accompanying changes in the pelvic girdle. The teeth and the jaw muscles became increasingly differentiated and probably permitted more effective chewing. These structural characters suggest that advanced synapsids were active predators, and their foraging behavior may have required a high aerobic metabolic capacity (see Chapter 4). The metabolic rates of advanced synapsids were probably sufficient to allow endothermal thermoregulation, and the early mammals appear to have been nocturnal. The sensory requirements of nocturnal activity may have led to an increase in the sizes of the brains of early mammals compared to their diurnal contemporaries, the lizards and sphenodontians.

The Synapsid Skull

In primitive amniotes the skull was solid and muscles ran inside the dermal bones. This is the **anapsid** (without arches) skull condition. Among the earliest amniotes was a late Carboniferous group known as limnoscelids. These anapsids had an open suture between the skull roof and the cheek region. It seems likely that the edges of the bones bordering this slit—the supratemporal and postorbital above and the squamosal below—were mechanically advantageous sites for attachment of the jaw muscles. The edges of a temporal fenestra may provide a more secure attachment for jaw muscles than does a flat surface. If some of the adductor muscles of the lower jaw found a particularly favorable origin on this region, selection could have favored the enlargement of the gap, giving a progressively larger area for muscle attachment and ultimately forming the temporal fenestra characteristic of the synapsids (Figure 19–1).

Pelycosaurs

The most generalized synapsids were the ophiacodontid pelycosaurs. Some of these were large

691

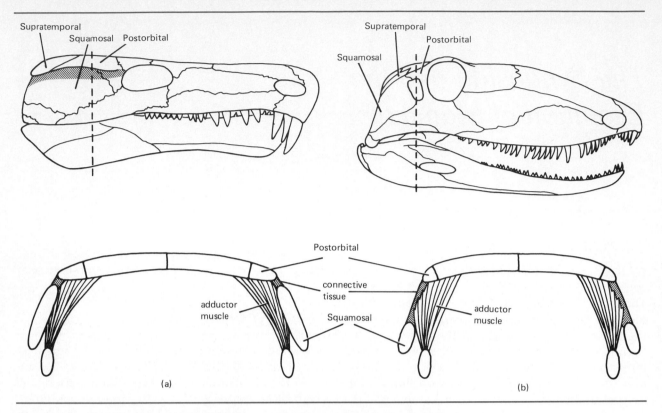

Figure 19–1. Hypothetical origin of the synapsid temporal fenestra from a suture between the cheek and roof of an anapsid skull: (a) *Limnoscelis*, anapsid; (b) *Ophiacodon*, synapsid. (Modified from T. S. Kemp, 1982, *Mammal-like Reptiles and the Origin of Mammals*, Academic Press, London; and A. S. Romer, 1966, *Vertebrate Paleontology*, 3rd edition, University of Chicago Press, Chicago.)

animals; *Ophiacodon major* was 3 meters long and may have weighed 200 kilograms, and *Dimetrodon grandis* may have been still larger. The ophiacodontids appear to have been semiaquatic fish eaters. Their heads were long and slender and the upper jaws contained 40 or more small, sharp teeth. Three groups of pelycosaurs can be distinguished (Figure 19–2), but the phylogenetic relationship among the groups is not certain (Reisz 1980, Kemp 1982). The sphenacodonts, and to a lesser extent the ophiacodontids, became increasingly specialized for predation, whereas the edaphosaurids and caseids were herbivores. Carnivorous pelycosaurs had long, narrow skulls. Their

dentition was heterodont, with large canine-like teeth. Herbivorous pelycosaurs had short, broad skulls. *Edaphosaurus* had homodont dentition with peg-like teeth, whereas *Cotylorhynchus* was heterodont. Both herbivores had well-developed teeth in the palate.

Carnivorous Pelycosaurs: The Sphenacodonts

Sphenacodonts probably preyed on a variety of small and large animals, including other sphenacodonts as well as the herbivorous caseids and

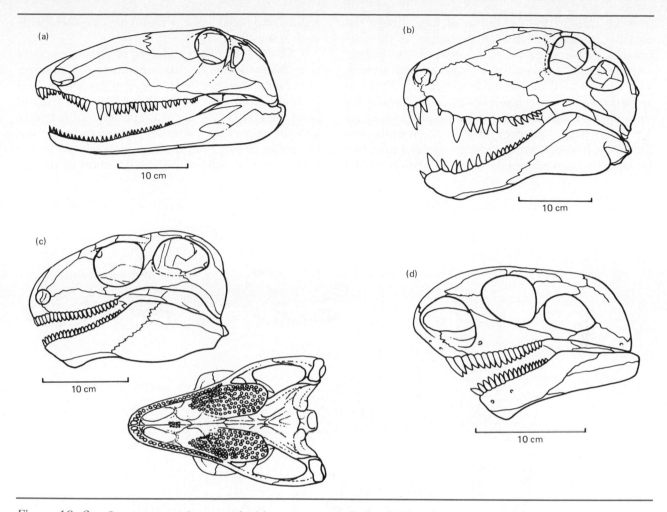

Figure 19–2. Carnivorous pelycosaurs had long, narrow skulls (a, *Ophiacodon*, an ophiacodontid; b, *Dimetrodon*, a sphenacodont). Their dentition was heterodont, with well developed canine-like teeth. Herbivorous pelycosaurs had short, broad skulls (c, *Edaphosaurus*, an edaphosaurid; d. *Cotylorhynchus*, a caseid). *Edaphosaurus* had homodont dentition with peg-like teeth; *Cotylorhynchus* was heterodont. Both herbivores had teeth in the palate. (Modified from A. S. Romer, 1966, *Vertebrate Paleontology*, 3rd edition, University of Chicago Press, Chicago.)

edaphosaurids. The progressive changes that can be seen in the structure of the jaws and teeth apparently reflect selection for effective predation (Figure 19–2). The anterior teeth became larger, the posterior premaxillary teeth formed a graded series and their number was reduced until, in *Dimetrodon*, the large anterior premaxillary teeth were separated from the maxillary teeth by a gap. The enlarged teeth on the lower jaw fitted into this space when the mouth was closed.

Other changes in the form of the sphenacodont skull reflect the mechanical requirements of the specialized dentition. The enlarged maxillary teeth seen in *Dimetrodon* have deep roots in the maxilla, and in the evolution of this lineage the maxilla gradually increased in depth. As the pre-maxilla and maxilla grew downward, the palate, which was flat in primitive genera, became arched. The significance of this arch is twofold: First, an arched palate is mechanically stronger than a flat one. Second, the arch of the palate provides space for air to pass over prey held in the mouth. The arched palate of sphenacodonts was the first step toward the development of a separation of the mouth and nasal passages seen in some therapsids and in mammals.

The postcranial skeleton of sphenacodonts also changed in ways that appear to reflect selection for increased effectiveness of predation. The legs, although they were still held out horizontally from the body in the ancestral pattern, were longer and slimmer in the advanced sphenacodonts than in earlier forms, suggesting increasingly active search and pursuit of prey. Reduction of the in-

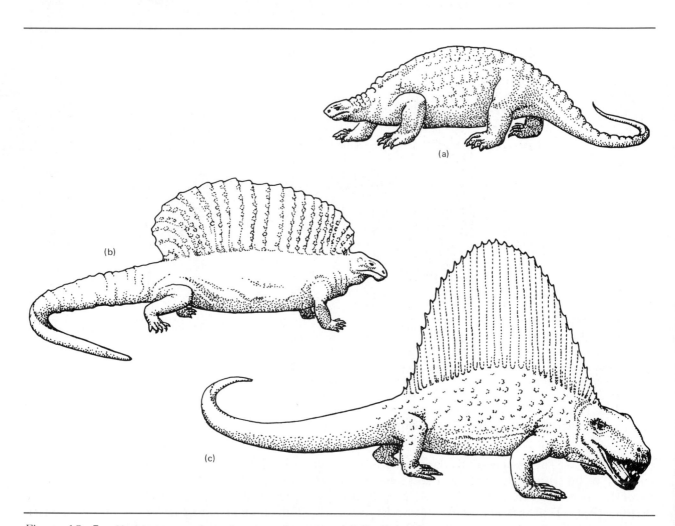

Figure 19–3. Herbivorous and carnivorous pelycosaurs: (a) *Cotylorhynchus*, a caseid; (b) *Edaphosaurus*, an edaphosaurid; (c) *Dimetrodon*, a sphenacodont.

tercentra and development of slanting (instead of horizontal) articulations between adjacent vertebrae indicates that lateral undulation of the vertebral column during locomotion was decreased in these predators.

A remarkable feature of some sphenacodonts, including *Dimetrodon*, was the elongation of the neural spines of the trunk region (Figure 19–3). In a *Dimetrodon* that was 3 meters long, the spines rose as much as 1.5 meters above the vertebrae and supported a crest that may have been a temperature-regulating device. Pelycosaurs probably relied upon absorbing solar energy to raise their body temperatures to activity levels, just as living reptiles do. Marks of blood vessels on the spines indicate that the tissue they supported was heavily vascularized.

The potential blood flow through the crest, indicated by the extent of vascularization, so greatly exceeds any reasonable metabolic requirement for such a tissue that it seems likely that the animals shunted blood through the crest in response to their thermoregulatory requirements. In the morning a *Dimetrodon* could orient its body perpendicular to the sun's rays and allow a large volume of blood to flow through the crest where the blood would be warmed by the sun and the heat carried back into the animal's body. When a *Dimetrodon* was warm enough, blood flow through the crest could be restricted, and the heat would be retained within the body.

One mathematical model of thermoregulation by crested pelycosaurs suggests that the crest could have increased rates of heating in the morning and the maximum body temperatures obtainable (Table 19–1). The effect of the crest might have been particularly significant for large species such as *Dimetrodon limbatus* and *D. grandis*. The rates of heating calculated for these animals are two or three times faster with the crest than without it, and the maximum body temperatures they could achieve with a crest (36 to 37°C) are 5°C higher than those they would have reached without a crest (Haack 1986). However, mathematical models of energy exchange are sensitive to the assumptions on which they are based, and a different analysis concluded that the presence or absence of a crest would have had little effect on the rate of heating (Tracy et al. 1986).

Herbivorous Pelycosaurs: The Edaphosaurids and Caseids

While the carnivores evolved a body form that made them more effective predators, another lineage of pelycosaurs was evolving in a very different direction to become specialized herbivores.

Table 19–1. Effect of the crest on the heating rates of pelycosaurs (*Dimetrodon*).

| | | Estimated Time Required to Raise Body Temperature (hr) | | | |
| | | 5°C | | 10°C | |
Species	Estimated Body Mass (kg)	With Crest	Without Crest	With Crest	Without Crest
D. milleri	50	2	4	4	6.5
D. limbatus	140	4	8	8	—
D. grandis	250	3	10.5	11	—

Source: S. R. Haack, 1986, *Paleobiology* 12:450–458.

In general body form the caseids and edaphosaurids were similar to each other and to the herbivorous cotylosaurs and *Diadectes* (Figure 19–3). All were heavy-bodied animals with short, sturdy legs sprawled out from the trunk. The skull was small in proportion to the body, and the dentition was specialized for crushing and grinding vegetation rather than for killing and tearing apart prey.

In caseids and edaphosaurids the facial region was short (Figure 19–2). This morphology increases the biting force at the front of the jaws, although it does so at the cost of a reduction in the speed of jaw closure. For an herbivore, however, force is more important than speed. Plant material is difficult to chew, and the jaw articulation of herbivorous vertebrates is frequently displaced from the horizontal plane of the tooth row. This displacement may increase the leverage of the jaw muscles, and it also produces a horizontal displacement between the upper and lower rows of teeth when the jaws close. The grinding action that occurs as the teeth in the lower jaw move past those in the upper jaw may be helpful in chewing plants.

Edaphosaurids had teeth of uniform size along the margins of the jaws (**homodonty**) and large tooth-bearing areas in the palate. The dentition of caseids was **heterodont,** with the largest teeth in the front of the jaws. Caseids had palatal teeth like those of edaphosaurids. The neural spines of some edaphosaurids were elongate, giving them a dorsal crest like that of *Dimetrodon*; in *Edaphosaurus* the spines had lateral projections.

Therapsids and Therosaurs

From the mid-Permian to the end of the Triassic there was a flourishing fauna of mammal-like animals that are grouped under the general name of therapsids (Kemp 1982, Hotton et al. 1986). Taxonomic arrangements and phylogenetic inferences are in a state of flux. Therapsids are related to sphenacodont pelycosaurs and, like the pelycosaurs, radiated into herbivorous and carnivorous forms. Therapsids, however, showed much greater diversity of cranial and postcranial morphology than had pelycosaurs. Some of the herbivorous therapsids were large, heavy-bodied, slow-moving animals that probably congregated in herds as modern antelopes do. Other herbivorous therapsids were small and superficially very like rodents. The carnivorous therapsids were more agile animals than the large herbivores. The pattern of increasing effectiveness as terrestrial predators, which we have traced from the earliest anthracosaurs through sphenacodont pelycosaurs, continued in the carnivorous therapsids.

The success of therapsids was noteworthy. Late Permian fossil deposits all over the world indicate that the ecosystems of the time were based on these animals. The remarkable similarity of faunas in different parts of the world at the end of the Paleozoic testifies not only to the continuity of landmasses and climates, but also to the very high degree to which therapsids were successful in exploiting the resources of their environments.

The transition from the Permian to the Triassic was a period of great extinction, actually greater in magnitude than the better-known extinction of many diapsids at the end of the Mesozoic. The known genera of tetrapods represented by fossils dropped from 200 in the late Permian to 50 in the early Triassic. The pattern of extinction was not regular—some old groups persisted while seemingly more progressive forms disappeared. Among the herbivorous therapsids, only the tusked dicynodonts survived into the Triassic in abundance. The most progressive groups of carnivorous therapsids survived the Permo-Triassic transition and diversified in the early Triassic (Figure 19–4).

Early Therapsids: Dinocephalians, Anomodonts, and Eotheriodonts

In the late Permian a group of large therapsids, the dinocephalians, dominated land faunas worldwide. The earliest dinocephalians were the brithopodids, large carnivorous animals up to 3 meters long that showed clear affinities to carnivorous

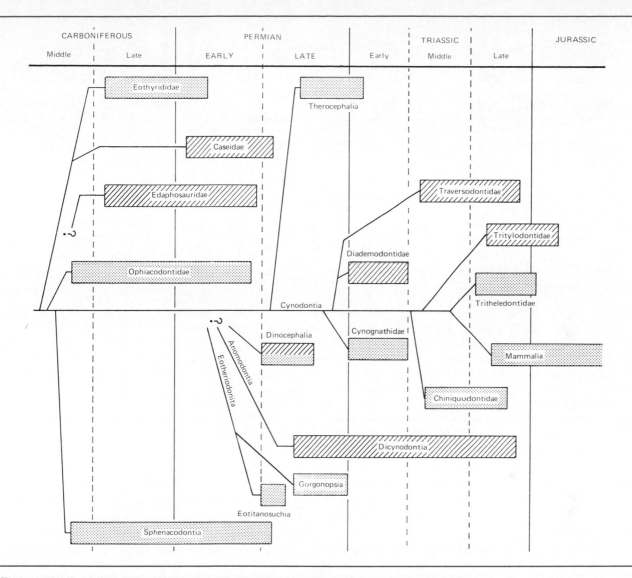

Figure 19—4. Phylogeny of the synapsids. Herbivorous groups are indicated by crosshatching, carnivorous ones by stippling. (From T. S. Kemp, 1982, *Mammal-like Reptiles and the Origin of Mammals*, Academic Press, London.)

pelycosaurs (Figure 19–5). Later in the Permian the brithopodids were replaced by the still-larger anteosaurids. These carnivorous dinocephalians had heavy bodies, short limbs, and long tails. Lateral undulation during locomotion appears to have been reduced still further from the condition in

sphenacodont pelycosaurs. Dentition was concentrated at the front of the mouth: The canines and incisors were enlarged and the postcanine teeth reduced in size and number. The long jaws of brithopodids and anteosaurids would have produced a large kinetic force as they snapped closed,

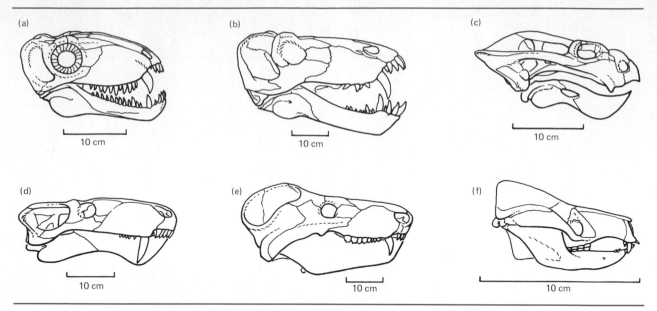

Figure 19–5. Skulls of therapsids. Early carnivorous therapsids like the
(a) eotitanosuchian *Phthinosuchus* and the (b) brithopodid dinocephalian
Titanophoneus had skulls that closely resemble those of sphenacodont pelycosaurs.
The herbivorous dicynodonts (c, *Dicynodon*) had lost most or all incisors and cheek
teeth, retaining only the tusk-like canines. Gorgonopsids (d, *Scymognathus*) and
cynodonts (e, *Cynognathus*) were carnivores specialized for attacking large prey. The
tritylodonts (f, *Oligokyphus*) had very mammal-like dentition with a diastema
between the canines and cheek teeth. (Modified from A. S. Romer, 1966, *Vertebrate
Paleontology*, 3rd edition, University of Chicago Press, Chicago.)

embedding the canine teeth deep in the flesh of
their prey. Then the interdigitating upper and
lower incisors would have permitted the animal to
tear off bite-sized pieces.

The herbivorous dinocephalians, the tapinocephalids, were huge, clumsy-looking animals
(Figure 19–6). Some, like *Moschops*, were nearly
3 meters long and more than 1.5 meters at the
shoulder. These herbivores had anterior teeth that
decreased gradually in size posteriorly. Teeth in
the upper and lower jaws interdigitated and vegetation was torn into pieces between adjacent
teeth. All of the dinocephalians, and particularly
the tapinocephalids, show a degree of thickening
of the roof of the skull that may indicate head-
butting behavior associated with intraspecific combat.

The dinocephalians flourished in the early
part of the late Permian, but had disappeared before the end of that period. Other therapsids that
appeared in the middle of the Permian, however,
persisted until the late Triassic. These were the
anomodonts, a collection of herbivores that were
in many respects the most successful of all therapsids. In the late Permian they dominated terrestrial faunas in diversity and in the number of specimens found as fossils. The most diverse and
lasting of the anomodonts were the animals
known as dicynodonts.

A distinctive feature of dicynodonts was the
extreme specialization of the skull for an herbivorous diet. In most forms all of the teeth were lost,
except for the upper canines, which were retained
as a pair of tusks (Figure 19–5). The jaws are pre-

Figure 19–6. Herbivorous and carnivorous therapsids. Herbivores: (a) *Moschops*, a dinocephalian; (b) *Lystrosaurus*, a dicynodont. Carnivores: (c) *Lycaenops*, a gorgonopsid; (d) *Cynognathus*, a cynodontid.

sumed to have been covered with a horny beak like that of modern-day turtles. A beak has some advantages over teeth as a structure for grinding plant material. Unlike a row of teeth, a horny beak can provide a continuous cutting surface and it can be replaced continuously as it is worn away. The structure of the jaw articulation of dicynodonts permitted extensive fore-and-aft movement of the lower jaw, shredding the food between the two cutting plates.

A third group of therapsids was the eotitan-osuchids and their descendants, the gorgonopsids. These were carnivorous forms, apparently specialized to feed on prey at least as large as themselves. The canine teeth were long and the incisors were well developed (Figure 19–5). The postcanine teeth, however, were small and the gorgonopsids probably relied upon the anterior teeth to tear chunks of flesh from their prey.

The gorgonopsids were lightly built animals the size of coyotes (*Lycaenops*) or wolves (*Cynognathus*) (Figure 19–6). The long tail characteristic of pelycosaurs was greatly shortened in the gorgon-opsids (Figure 19–7). The lateral undulation of the trunk that was characteristic of the locomotion of pelycosaurs had probably been abandoned by gorgonopsids. The trunk vertebrae had no intercentra, and the articulating surfaces of the zygapo-physes connecting adjacent vertebrae were nearly vertical. The forelimbs were probably still held in a sprawled posture, but the hindlimbs could apparently be brought beneath the body. This arrangement would allow the hindlimbs to provide greater thrust for rapid locomotion.

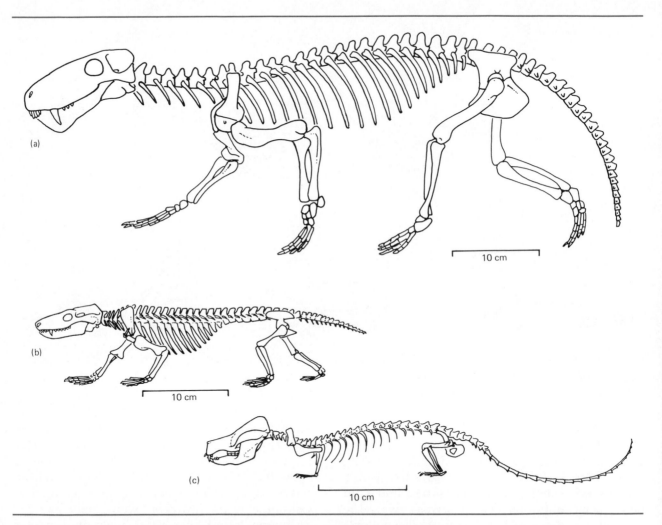

Figure 19–7.   Post-cranial skeletons of therapsids. *Lycaenops* (a), a gorgonopsid, had ribs that extended the full length of the trunk like those of modern lizards. The cynodonts *Thrinaxodon* (b) and *Oligokyphus* (c) showed a reduction of the size in the ribs in the lumbar region like that seen in mammals. (Modified from A. S. Romer, 1966, *Vertebrate Paleontology,* 3rd edition, University of Chicago Press, Chicago.)

Triassic Therosaurs: The Cynodonts

Cynodonts were the last of the major groups of therapsids to evolve. By the end of the Triassic they included some very mammal-like groups and had given rise to the mammals themselves. A marked reduction in body size characterized the cynodont lineages. Some early cynodonts were as big as large dogs, but by the middle Triassic, cynodonts had skulls only 70 millimeters long, and in the latest Triassic the total body length of the earliest mammals was less than 100 millimeters.

Both carnivorous and herbivorous forms are represented among the cynodonts. The carnivorous cynodonts include the early Triassic cynognathids, the middle Triassic chiniquodontids, and the late Triassic tritheledontids. These animals had well-developed canine teeth and incisors, like the earlier groups of carnivorous therapsids, but in advanced cynodonts the post-canine teeth also were well developed (Figure 19–5). Traces of wear on the surface of the post-canine teeth of *Cynognathus* indicate that the upper and lower teeth had a tearing action aided by a cutting function of serrations on the cusps of the teeth. This arrangement would have allowed cynognathids to use the cheek teeth to cut off bite-size pieces of meat. In contrast, the carnivorous dinocephalians and gorgonopsids had poorly developed cheek teeth and apparently used the incisors to nip off bits of flesh.

Surprisingly, these changes in dentition may have reflected selective forces that acted initially on the ability of cynodonts to hear. The jaw muscles of cynodonts showed progressive differentiation of the superficial masseter muscle in association with the development of a pronounced coronoid process of the dentary bone (Figure 19–8). These changes probably improved the bite force of the jaws, and also allowed a reduction in the size of the articular and quadrate bones that lay posterior to the dentary bone in the lower jaw. The articular and quadrate of cynodonts are homologous with the malleus and incus of the middle ear of modern mammals (Box 19–1), and it is likely that at the cynodont stage they already

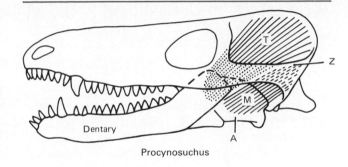

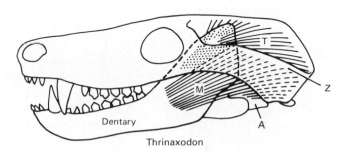

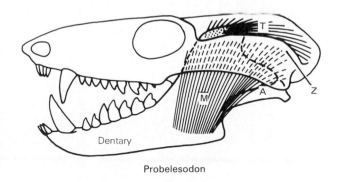

Figure 19–8. In cynodonts the dentary bone increased in size relative to the other bones of the lower jaw. As this happened, the masseter muscle, one of the major jaw-closing muscles, progressively changed its insertion from the angular bone to the dentary. A, Angular bone; Z, zygomatic arch; M, masseter muscle; T, temporalis muscle. (From T. S. Kemp, 1982, *Mammal-like Reptiles and the Origin of Mammals.* Academic Press, London.)

Box 19–1. The Evolution of the Mammalian Middle Ear

More than a century ago embryological studies demonstrated that the malleus and incus of the middle ear of mammals were homologous with the articular and quadrate bones that formed the primitive jaw joint of amniotes. More recently the transition has been traced in fossils from its beginning in pelycosaurs through therapsids (Allin 1975).

The jaw joint of pelycosaurs was formed by the articular and quadrate bones, which were sturdy structures that resisted the compressive forces applied to them during biting. A key to the change in the jaw joint, and the reduction in the size of the articular and quadrate, was a change in the mechanics of the jaw in the synapsid lineage that reduced the stress on the articulation (Bramble 1978).

The transition from the synapsid to the mammalian condition can be visualized by comparing the posterior half of the skull of *Thrinaxodon*, a cynodontid (Figure 19–9a), with that of *Didelphis*, the Virginia opossum (Figure 19–9b). The jaw joint between the articular and quadrate of *Thrinaxodon* was also part of the auditory apparatus. The reflected lamina of the angular bone is a distinctive feature of the jaw of synapsids. In therapsids the reflected lamina was probably the principal support of the tympanic membrane, and vibrations were transmitted to the stapes via the articular and quadrate. The mammalian jaw joint is formed by the dentary and squamosal bones. The tympanum and middle ear ossicles lie behind the jaw joint and are much reduced in size, but

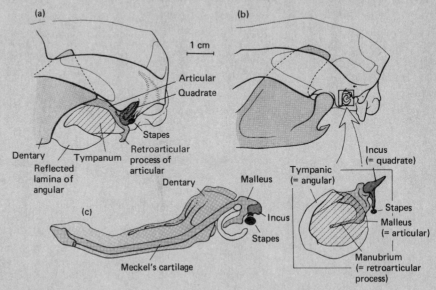

Figure 19–9. Origin of the middle ear bones of mammals: (a) *Thrinaxodon*, a cynodont; (b) *Didelphis*, the Virginia opossum; (c) embryonic mammal. (From A. W. Crompton and F. A. Jenkins, Jr. 1979, Origin of mammals, pages 59–73 in *Mesozoic Mammals: The First Two-Thirds of Mammalian History*, edited by J. A. Lillegraven, Z. Kielan-Jaworowska, and W. A. Clemens, University of California Press, Berkeley, Calif.)

they retain the same relation to each other as they had in cynodonts. The lower jaw of a fetal mammal viewed from the medial side shows that the angular (tympanic), articular (malleus), and quadrate (incus) develop in the same positions they had in the cynodontid skull (Figure 19–9c). The homolog of the angular, the tympanic bone, supports the tympanum of modern mammals. The retroarticular process of the articular bone is the manubrium of the malleus, and the primitive jaw joint persists as the articulation between the malleus and incus.

played a role in sound transmission (Crompton 1985). If this was the case, lighter bones would be more effective than heavy bones at transmitting vibrations picked up by the tympanic membrane, but a reduction in the size of the bones would have reduced the strength and load-bearing capacity of the jaw joint. Thus, the reorganization of the jaw muscles and the development of the coronoid process of the dentary bone seen in cynodonts could have been advantageous because they reduced the forces applied to the jaw joint during biting, allowing the articular and quadrate bones to be small. In this analysis, selective pressures to improve hearing are seen as being the important factors in the reorganization of the mammalian jaw.

However, the improvement in hearing would have some cost associated with it. A weak jaw would have limited the ability of cynodonts to chew their food: The canines and incisors could puncture and tear prey, but the cheek teeth could not process food effectively because those chewing motions (especially unilateral chewing) would exert too much stress on the jaw joint. Ultimately the cynodont lineage evolved a new load-bearing jaw joint between the dentary and squamosal bones that was strong enough to resist the forces exerted during strong biting. This is the mammalian type of jaw suspension. A classic example of an evolutionary intermediate is provided by the trithel-odontid *Diarthrognathus*. This animal gets its name from its double jaw joint (*di* = two, *arthro* = joint, *gnath* = jaw). It has both the ancestral articular–quadrate joint and, next to it, the mammalian dentary–squamosal articulation.

The herbivorous cynodonts include the early Triassic diademodontids, the middle Triassic traversodontids, and the late Triassic tritylodontids. As the names suggest, features of their teeth are particularly significant in these groups. The surfaces of the cheek teeth developed patterns of cusps associated with crushing, cutting, and grinding plant material.

The arrangement of teeth in the jaws of herbivorous cynodonts was also quite mammalian. For example, the tritylodont *Oligokyphus* had a large diastema separating the anterior teeth from the cheek teeth (Figure 19–5). The postcanine teeth were multirooted like the cheek teeth of mammals, and had crescentic cusps on their surfaces forming opposing pairs of cutting edges suitable for shredding plant material. *Oligokyphus* was about 50 centimeters in total length, and nearly half of that was tail.

The postcranial skeleton of the cynodonts was also verging on the mammalian condition (Figure 19–7). The structure of the pelvis and femur indicates that the hindlimb was held almost directly beneath the body and moved in a parasagittal plane (parallel to the long axis of the body). The ability to shift between a sprawling and an upright posture that was present in gorgonopsids had been lost in cynodonts. The forelimbs of cynognathids and diademodontids appear to have retained a widespread position, but changes in the shape of the scapula suggest that the speed of the stride was increased. In the tritylodont *Oligokyphus* the shoulder girdle was very like that of mammals, and the forelimb probably moved in a parasagittal plane. Taken together, these changes indicate increased power and agility in the cynodonts.

Some features of the soft anatomy of therapsids can be inferred from traces left on the bones. The smooth surfaces of the premaxilla, maxilla, and dentary adjacent to the teeth suggest that cynodonts had muscular lips like those of primitive mammals. The lips of therapsids could have functioned in conjunction with the cusped cheek teeth and, probably, a muscular tongue to manipulate food as it was chewed.

An abrupt change in the size and shape of ribs in the lumbar region suggests that cynodonts had a muscular diaphragm like that of mammals (Figure 19–7). Furthermore, ridges in the nasal passages suggest that turbinal bones may have been present. These features are likely to have been associated with ventilation of the lungs by large volumes of air, and that inference in turn suggests that cynodonts were capable of high rates of metabolism. In mammals the turbinal bones support tissues that, in addition to having olfactory cells, warm and humidify inhaled air before it reaches the lungs and then recover some of that heat and water when the air is exhaled. This countercurrent recycling system can conserve enough heat to be an important part of the energy budget of a mammal. The presence of turbinal bones in cynodont therapsids would suggest that their thermoregulation was basically like that of mammals.

The First Mammals

The origin of mammals is often treated as if it were a discrete event that happened suddenly sometime in the late Triassic or early Jurassic, but this is an oversimplified view. The animals we know as mammals are the products of an evolutionary lineage that stretches back to the division between diapsids and synapsids in the Carboniferous. The anatomical structures that characterize mammals evolved in mosaic fashion through the Permian and Triassic, and their original functions were not necessarily the same as those they have today. The biology of animals in the synapsid lineage was also changing during the late Paleozoic and early Mesozoic. We have traced the evidence for increasing locomotor agility and increased specialization of the jaws and teeth in cynodonts, and the most advanced cynodonts are little different from the earliest mammals. A list of the derived characters of mammals (Table 19–2) is useful as a summary, but it emphasizes the differences between mammalian and nonmammalian synapsids, whereas a review of their evolution stresses the continuous process of transition.

Tritheledontids: The Sister Group of Mammals?

The Tritheledontidae (also known as the Diarthrognathidae) is a poorly known group of small cynodonts from the middle and late Triassic of South Africa. Tritheledontids may be the cynodonts most closely related to mammals. *Diarthrognathus*, the best known of the tritheledontids, possessed the mammalian dentary-squamosal joint and also the nonmammalian articular-quadrate joint. The postcranial skeleton of tritheledontids is not known, but the skull and teeth showed derived mammalian characters: The postorbital and prefrontal bones were absent and the zygomatic arch was slender. Furthermore, tritheledontids are the only cynodonts known in which the teeth were covered with prismatic enamel like that of mammals. Thus, the tritheledontids may be the sister group of mammals, but that relationship is by no means certain and two other groups might be more closely related to mammals than are the tritheledontids. The tritylodontids have derived mammalian characters in the skull and postcranial

Table 19–2. Derived characters of mammals.

Character	Ancestral Condition	Derived Condition
Jaw articulation	Articular-quadrate	Dentary-squamosal
Lower jaw	Contains several bones	Contains only the dentary
Teeth	Tooth row undifferentiated (homodonty), single cusps, single roots	Tooth row differentiated (heterodonty), multiple cusps, multiple roots
Tooth replacement	Continuous (polyphyodont)	Replaced once (diphyodont)
Tooth occlusion	No occlusion of teeth	Cusps of upper and lower teeth occlude
Occipital condyle	Single	Double
Pectoral girdle	Large coracoids	Reduced coracoids
Vertebrae	Little regional differentiation	Cervical, thoracic, and lumbar regions distinct
Ribs	Thoracic and lumbar	Thoracic only
Diaphragm	Absent	Present
Pelvic girdle	Ilium oriented posteriorly	Ilium oriented anteriorly
Feet	Talus and calcaneus adjacent, both in sole of foot	Talus superimposed on calcaneus

skeleton, and *Probainognathus*, a cynodont from the middle Triassic of South America, has cheek teeth very like those of the earliest mammals (the Triconodonta, represented by the family Morganucodontidae). We have placed the tritheledontids as the sister group of mammals (Figure 19–10), largely on the basis of the prismatic enamel on their teeth, but a strong case can be made for giving that position to the tritylodontids (Kemp 1983).

Nontherian and Therian Mammals

The difficulty in identifying the sister group of mammals is a reflection of the fragility and small size of the skeletons of these small animals. Most fossil material consists only of scraps, and the postcranial skeletons of several groups are unknown. The fossils are often so small that they must be examined under a microscope, and they are found by sifting and washing tons of dirt and rock. In fact, searching for fossils of early mammals is very much like searching for gold.

Teeth are hard structures that are more likely to be fossilized than other bones, and much of the phylogeny of mammals is based on the patterns of tooth cusps. Differentiation of the teeth into incisors, canines, and molariform teeth was evident in sphenacodont pelycosaurs and continued in the cynodonts. The molar teeth of the advanced cynodonts were not simple conical pegs as they were in pelycosaurs and early therapsids. They had various complex patterns of cusps and double or triple roots. The interlocking cusps guided the closing jaws into alignment and sheared food, rather than tearing it as had the teeth of early cynodonts. Tooth morphology is complex and difficult to visualize, but it is basic to understanding the classification and appreciating the life-styles of early mammals.

Phylogenetic Relationships of the Synapsida

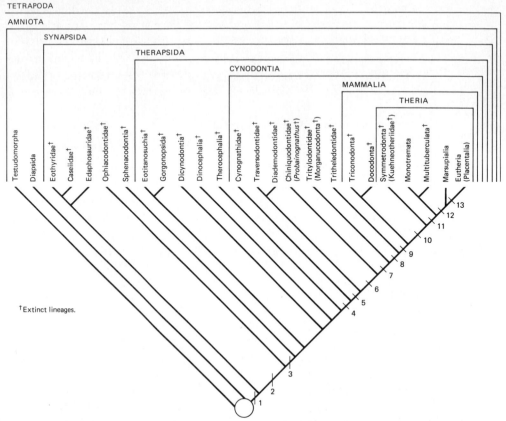

1. **Synapsida:** Lateral skull fenestra located ventrally between skull roof and cheek. **2.** Enlarged canine-like teeth. **3.** Canine-like teeth larger and reduced in number, reflected lamina on the angular bone. **4. Cynodontia:** Enlarged dentary bone, reduced post-dentary bones, jaw joint is primarily between quadrate and articular plus a secondary joint between surangular and squamosal, true occlusion of upper and lower post-canine teeth. **5.** Enlarged dentary with angular and articular processes, very reduced post-dentary bones, further development of a secondary jaw articulation between surangular and squamosal. **6.** Characters of palate, lower jaw, pectoral girdle, and ribs. **7.** Characters of skull, wall of braincase, palate, and postcranial skeleton. **8.** Prismatic enamel.

9. **Mammalia:** Dentary replaces surangular in jaw articulation, action of molars involves lateral movement of the lower jaw. **10.** Triangular pattern of main molar cusps. **11. Theria:** Ear ossicles, characters of the braincase, expanded neopallium in brain, characteristics of vertebrae and long bones, probably tribosphenic molars. **12.** Characteristics of braincase, vertical tympanic membrane, tribosphenic molars here if not in (11). **13.** Chorioallantoic placenta, long gestation. (Based on T. S. Kemp, 1982, *Mammal-like Reptiles and the Origin of Mammals,* Academic Press, London; T. S. Kemp, 1983, *Zoological Journal of the Linnaean Society* 77:353–384; and Z. Kielan-Jaworowska, A. W. Crompton, and F. A. Jenkins, Jr., 1987, *Nature* 326:871–873.)

Figure 19–10. Phylogenetic relationships of the Synapsida. This diagram shows the probable relationships among the major groups of synapsids. Extinct lineages are marked by a dagger (†). The numbers indicate derived characters that distinguish the lineages. The circle indicates that the relationships of the lineages at that node cannot yet be defined.

The study of teeth has played a large role in understanding mammalian evolution, in large part because little material except for teeth has been available. Mammals with teeth that have three cusps aligned anteroposteriorly (the **triconodont** pattern) have been distinguished from forms that have three cusps in a triangular (**tribosphenic**) pattern (Figure 19–11). The tribosphenic molar differs from the ancestral mammalian molar in having a much-enlarged talanid process on the posterior margin of the lower molars. The talonid process contains a basin that receives the inner cusp (the protocone) of the corresponding upper molar. Together they mash and grind food caught between them.

The cynodont-mammal transition appeared to incorporate a dichotomous radiation with the morganucodontids (triconodont molars) closely related to nontherian mammals, and the kuehneotheriids (tribosphenic molars) related to therian mammals. New discoveries of fossil mammals are suggesting that this scheme may be oversimplified and a new classification will be needed when more information is available. Material from early Jurassic deposits in Arizona (*Dinnetherium*), Switzerland (*Hallautherium*), and China (*Sinoconodon*) has revealed unsuspected diversity. None of these fossils appears to have been closely related to the morganucodontids or kuehneotheriids. Apparently, mammals had undergone an extensive radiation before the earliest mammals we have yet found (Jenkins 1984). For the present, however, a distinction between therian and nontherian mammals persists in the paleontological literature.

Nontherian Mammals Morganucodontid mammals (triconodonts) and the poorly known docodonts had three cusps on the molar teeth arranged in a line from anterior to posterior; the middle cusp was the largest. When the jaws closed the inner faces of the cusps of the upper molars slipped down over the external faces of the cusps of the lower molars and produced a shearing action like that of the blades of a pair of scissors. A piece of food could be processed by shearing it repeatedly, cutting it into smaller and smaller bits. The post-

cranial skeleton of morganucodontids had many mammalian features (Figure 19–12). The joint between the atlas and axis was apparently capable of as much rotation as that of modern mammals, and the joint between the atlas and the double occipital condyles allowed flexion and extension. The cervical, thoracic, and lumbar vertebrae were well differentiated, and the cervicothoracic flexure indicates that the head and neck were held in an upright posture. The lumbar vertebrae lacked ribs, as they do in modern mammals. The anterior neural spines were angled posteriorly, and the posterior spines were angled anteriorly. Three transitional vertebrae are evident in the posterior thoracic region. This pattern of vertebrae in modern mammals is associated with flexion and extension of the trunk in a vertical plane in contrast to horizontal flexion. The pelvic girdle of *Morganucodon* was fully mammalian: It had a narrow, rod-like ilium that was directed anterodorsally, a large obturator foramen, and a reduced pubis. The form of the limbs suggests that morganucodontids were active foragers with climbing abilities similar to those of many small mammals today.

The kuehneotheriids are probably included among the Symmetrodonta of the middle Jurassic and early Cretaceous. These fossils are known largely from teeth, which have three cusps arranged in the triangular tribosphenic pattern. The symmetrodonts appear to be the primitive sister group of the therian mammals.

Therian Mammals Multituberculates were a diverse and successful group of mammals. They included nearly 50 genera and survived for 100 million years, from the late Jurassic into the Oligocene. Some multituberculates were the size of small mice and others grew as large as a woodchuck. The skulls of multituberculates have a rodent-like appearance that is produced by a pair of long incisors separated by a diastema from the cheek teeth (Figure 19–13). The lower premolars of most multituberculates were specialized as large shearing blades (**plagiaulacoid** premolars); similar premolars are found among some living marsupials. Multituberculates were probably omnivorous,

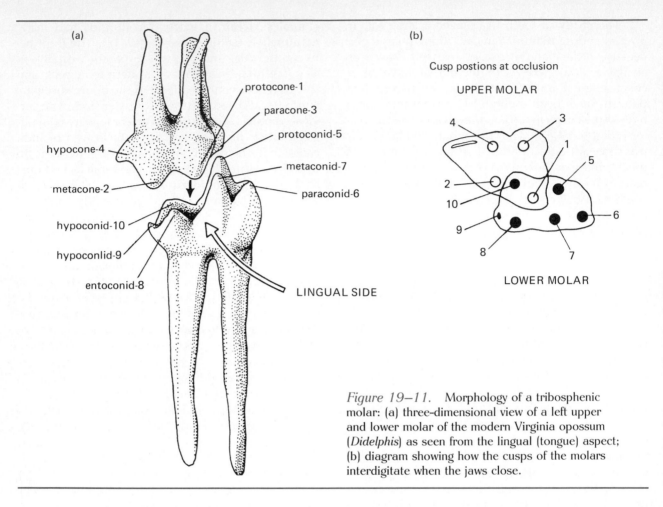

(a)

protocone-1
paracone-3
protoconid-5
metaconid-7
paraconid-6
hypocone-4
metacone-2
hypoconid-10
hypoconlid-9
entoconid-8
LINGUAL SIDE

(b)

Cusp postions at occlusion

UPPER MOLAR

LOWER MOLAR

Figure 19–11. Morphology of a tribosphenic molar: (a) three-dimensional view of a left upper and lower molar of the modern Virginia opossum (*Didelphis*) as seen from the lingual (tongue) aspect; (b) diagram showing how the cusps of the molars interdigitate when the jaws close.

much like modern rodents that include both plant and animal material in their diet.

The living mammals Monotremata (the duck-billed platypus and echidnas), Marsupialia (the marsupials), and Eutheria (placental mammals) have long been considered to represent surviving lineages of nontherians (the monotremes) and therians (the marsupials and eutherians). However, echidnas have extremely reduced teeth and the duck-billed platypus lacks teeth altogether, so it has not been possible to link the tooth structure of modern monotremes with the triconodont pattern of nontherian fossil mammals. The recent dis-covery of a fossil monotreme from Cretaceous deposits in Australia indicates that early monotremes had a triangular arrangement of tooth cusps (Kielan-Jaworowska et al. 1987). The teeth are not true tribosphenic molars, but they seem most likely to be derived from the pattern of tooth cusps seen in therian mammals. Furthermore, a second character used to distinguish nontherian and therian mammals, the construction of the sidewall of the braincase, now appears less convincing than it was once considered (Kemp 1982, 1983). In this context, monotremes seem likely to be therian mammals.

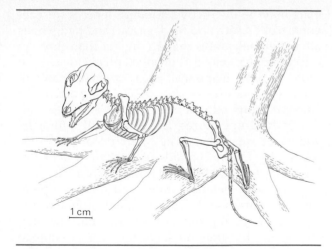

Figure 19–12. Reconstruction of *Megazostrodon*, an early Jurassic morganucodontid. (From F. A. Jenkins, Jr. and F. R. Parrington, 1976, *Philosophical Transactions of the Royal Society of London* B273:387–431.)

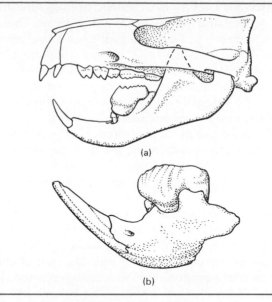

Figure 19–13. Multituberculates: (a) skull of ptilodontid multituberculate; (b) fragment of the lower jaw of the multituberculate *Mesodma formosa* showing the plagaulacoid premolar. (From W. A. Clemens, 1979, in *Mesozoic Mammals: The First Two-Thirds of Mammalian History*, edited by J. A. Lillegraven, Z. Kielan-Jarowoska, and W. A. Clemens, University of California Press, Berkeley, Calif.)

The Adaptive Zone of Early Mammals

The structure of the teeth of Mesozoic mammals suggests that they occupied at least three general feeding niches: The shearing dentition of triconodonts probably indicates that they preyed on other vertebrates, whereas the interdigitating cusps of the teeth of symmetrodonts may have been suited to puncturing and crushing the exoskeletons of insects. Multituberculates were the most diverse of the Mesozoic mammals, and appear to have included forms that may have been primarily insectivorous as well as species that were probably omnivorous or herbivorous.

Mammals were far from being the only small tetrapods with these dietary habits in the Triassic and Jurassic. Dinosaurs and rhynchosaurs were substantially larger than any of the Mesozoic mammals, but the early mammals shared their habitats with other synapsids—cynodonts such as the chiniquodontids (probably carnivorous) and traversodontids (probably herbivorous)—and with the diapsids that were continuing their radiation—lizards (mostly insectivorous) and sphenodontians (insectivorous and herbivorous). Some ecological separation among these groups may have been related to body size: The chiniquodontids and traversodontids were larger than most of the other forms, but mammals, lizards, and sphenodontians were all about the same size. The ecological distinctions among these groups may have rested in the times of day during which they were active—the diapsids during the day and the mammals at night.

Changes in the auditory apparatus and in the size of the brains of mammals compared to cynodonts and diapsids suggest that the mammals were active at dusk or at night. Vision has the advantage of revealing a large amount of information about an animal's surroundings. For example, it is readily apparent visually that some objects are closer to the observer than others, or that some are approaching and others are moving away. These spatial relationships are more difficult to perceive in the dark, when an animal must depend primarily on olfaction or hearing. Those senses require that spatial relationships be inferred by comparing the intensity of stimuli over intervals of time: A sound that becomes louder may mean that the source of the noise is approaching, whereas a sound that grows fainter means that the source is moving away. Similarly, an odor that changes in intensity may mean that its source is approaching or receding.

Making sequential comparisons of the intensity of olfactory and auditory stimuli over time may require more neural integration in the brain than do visual comparisons, and the size of the brain in early mammals has been interpreted as evidence of nocturnal activity. The volume of the braincase of the earliest mammals appears to have been four or five times the volume of the braincases of cynodonts such as *Probainognathus*. The olfactory bulbs and cerebral hemispheres were large (Kielan-Jaworowska 1986). Jerison (1973) suggested that this increase in the size of the brain may be correlated with the ability of early mammals to process more sensory information than could cynodonts, especially information from the nasal capsules and the inner ear (see Chapter 21). That interpretation is speculative, but it is consistent with what is known of the structure of the earliest mammals.

Endothermy would have been an important component of the nocturnality of early mammals, allowing them to maintain body temperatures above ambient levels when the sun had set and ectotherms had been forced to cease activity (Crompton 1980). In this context the thermoregulatory capacities of some living nocturnal mammals may be instructive. Modern tenrecs (insectivores from Africa and Madagascar) and hedgehogs have metabolic rates that are higher than those of ectothermal amniotes, but substantially lower than those of most mammals. These animals maintain body temperatures of 30 to 35°C during activity in ambient temperatures of 20 to 25°C. Despite their low body temperatures, tenrecs are active foragers and cover areas as large as 2 hectares in the course of a night. During periods of inactivity they enter torpor and allow their body temperatures to fall.

The low metabolic rates and body temperatures of tenrecs and hedgehogs are probably specializations of the modern species, not ancestral characters, but they do show that it is possible to be an endothermal homeotherm with a low metabolic rate. A similar way of life has been suggested for Mesozoic mammals: If the modifications for sustained, rapid locomotion suggested by the skeletons of cynodonts were reflected by increased aerobic metabolic capacity, early mammals might well have had metabolic rates like those of tenrecs, and probably would have had similar thermoregulatory capacities. The capacity for endothermal heat production that was provided by the high (relative to lizards) metabolic rates of Mesozoic mammals could have allowed them to adopt nocturnal habits. Nocturnal activity was probably not possible for the ectothermal lizards and sphenodontians, and a difference in the time of day when they were active may have been the major ecological separation of early mammals and their diapsid contemporaries. If the large size of the brains of early mammals was a consequence of the need to integrate olfactory and auditory information about their environment, as Jerison believes, the nocturnal habits of early mammals may have been a key factor in the subsequent radiation of mammals as large-brained, social vertebrates (see Chapter 22).

Summary

The synapsid lineage is characterized by a fenestra high on the side of the skull bordered dorsally by the supratemporal and postorbital bones and ventrally by the squamosal bone. The first synapsids were the pelycosaurs of the late Carboniferous and early Permian—the ophiacodontids and sphenacodonts (both carnivores), and the edaphosaurids and caseids (herbivores). Pelycosaurs were large animals; some may have weighed as much as 250 kilograms. Progressive changes in the jaws and teeth of sphenacodonts indicate increasingly effective predatory capacities. The limbs became longer and slimmer in later forms, suggesting the development of a more active mode of foraging. From the mid-Permian through the Triassic these trends continued in the therapsids, particularly the cynodonts. By the end of the Triassic advanced cynodonts had a mosaic of mammalian and nonmammalian characters. A substantial reduction in body size accompanied the development of mammalian characters, and early mammals were only 100 millimeters long and probably weighed less than 50 grams.

Most information about these mammals comes from fossilized teeth, which are the most durable parts of the skeleton. Early mammals have traditionally been divided into two major lineages on the basis of tooth structure: The morganucodonts had triconodont molars in which three cusps were arranged in a straight anteroposterior line, whereas the kuehneotheriids had tribosphenic molars with three cusps forming a triangle. These two groups have been designated nontherian and therian mammals, respectively. However, recently discovered fossils of Jurassic mammals show that mammalian diversity in the early Mesozoic was greater than has been appreciated, and this newly recognized diversity weakens the hypothesis of a simple dichotomous division among early mammals. Views of mammalian phylogeny are likely to be revised in coming years as these and other fossils are studied. Currently, it seems most likely that the three groups of living mammals—the monotremes, marsupials, and eutherians—are all derived from the same lineage.

The tooth structures of Mesozoic mammals suggest that they included insectivorous, carnivorous, omnivorous, and herbivorous species. Contemporaries of the mammals included diapsids (lizards and sphenodontians) that had similar body sizes and dietary habits. Mesozoic mammals were probably nocturnal, using heat produced by metabolism and insulation provided by hair to sustain body temperatures several degrees above ambient temperature. The large brains of early mammals (compared to those of their diapsid contemporaries) may have been associated with the ability to process olfactory and auditory information.

References

Allin, E. F. 1975. Evolution of the mammalian middle ear. *Journal of Morphology* 147:403–437.

Bramble, D. M. 1978. Origin of the mammalian feeding complex: models and mechanisms. *Paleobiology* 4: 271–301.

Crompton, A. W. 1980. Biology of the earliest mammals. Pages 1–11 in *Comparative Physiology: Primitive Mammals*, edited by K. Schmidt-Nielsen, L. Bolis, and C. R. Taylor. Cambridge University Press, Cambridge.

Crompton, A. W. 1985. Origin of the mammalian temporomandibular joint. Pages 1–18 in *Development of Temporomandibular Joint Disorders*, edited by D. S. Carlson, J. McNamara, and K. A. Ribbens. Monograph 16, Craniofacial Growth Series, University of Michigan Press, Ann Arbor.

Haack, S. C. 1986. A thermal model of the sailback pelycosaur. *Paleobiology* 12:450–458.

Hotton, N., III, P. D. MacLean, J. J. Roth, and E. C. Roth. 1986. *The Ecology and Biology of Mammal-Like Reptiles*. Smithsonian Institution Press, Washington, D.C.

Jenkins, F. A., Jr. 1984. A survey of mammalian origins. Pages 32-47 in *Mammals*, edited by T. W. Broadhead. University of Tennessee Department of Geological Science, Studies in Geology, volume 8. University of Tennessee Press, Knoxville, Tenn.

Jerison, H. J. 1973. *Evolution of Brain and Intelligence*. Academic Press, New York.

Kemp, T. S. 1982. *Mammal-Like Reptiles and the Origin of Mammals*. Academic Press, London.

Kemp, T. S. 1983. The relationships of mammals. *Zoological Journal of the Linnaean Society* 77:353–384.

Kielan-Jaworowska, Z. 1986. Brain evolution in Mesozoic mammals. Pages 21–34 in *Vertebrates, Phylogeny, and Philosophy*, edited by K. M. Flanagan and J. A. Lillegraven. Contributions to Geology, Special Paper 3, University of Wyoming, Laramie, Wyo.

Kielan-Jaworowska, Z., A. W. Crompton, and F. A. Jenkins, Jr. 1987. The origin of egg-laying mammals. *Nature* 326:871–873.

Reisz, R. R. 1980. The Pelycosauria: a review of phylogenetic relationships. Pages 553–596 in *The Terrestrial Environment and the Origin of Land Vertebrates*, edited by A. L. Panchen. Academic Press, London.

Tracy, C. R., J. S. Turner, and R. B. Huey. 1986. A biophysical analysis of possible thermoregulatory adaptations in sailed pelycosaurs. Pages 195–206 in *The Ecology and Biology of Mammal-Like Reptiles*, edited by N. Hotton III, P. D. MacLean, J. J. Roth, and E. C. Roth. Smithsonian Institution Press, Washington, D.C.

We have stressed the role of Earth history in shaping the evolution of vertebrates. Heretofore our emphasis has been primarily on the positions of continents as they affected climates, although we also noted the importance of Pangaea in allowing the spread of synapsids to nearly all parts of the world from their origin in the late Paleozoic through the middle of the Mesozoic. However, by the late Mesozoic Pangaea no longer existed as a single entity. Epicontinental seas cut across the centers of North America and Eurasia and the southern continents were separating from the northern continents and from each other. Continental drift in the late Mesozoic and early Cenozoic moved most of the landmasses of the Earth from the tropics into temperate regions. This movement, plus changes in patterns of ocean currents, resulted in a climatic deterioration culminating in a series of ice ages that began in the Pleistocene and continue to the present. (We are currently in a relatively ice-free interglacial period.) Both the fragmentation of the landmasses and the changes in climate during the Cenozoic have been important factors in the evolution of mammals.

Geology and Ecology of the Cenozoic

20

Continental Geography During the Cenozoic

You will recall that the breakup of Pangaea in the Jurassic was signaled by the movement of North America to open the ancestral Atlantic Ocean (Figure 14–3). Later in the Jurassic, rifts began to open in Gondwana, and India moved northward on a separate oceanic plate (Figures 20–1 and 20–2), eventually to coalesce with Eurasia. The inertial consequences of this event produced the Himalayas, the highest mountain range in the world today.

It was not until the middle to late Cretaceous that South America–Antarctica–Australia separated from Africa as a single continental unit (Figures 14–3 and 14–4). Land connections persisted across this platform during the Paleocene and, perhaps, the early Eocene (Figure 20–1). In the middle to late Eocene, Australia separated from Antarctica and, like India, drifted northward (Figure 20–3). Intermittent land connections between South America and Antarctica apparently were retained via the Scotia Island arc until recently.

In the Northern Hemisphere a land bridge between Alaska and Siberia—the trans-Bering Bridge—connected North America to Asia (Figures 20–1 and 20–3). Intermittent land connections also persisted between eastern North America and Europe via Greenland from the Cretaceous to the early Eocene. In Laurasia, then, overland migration was still possible until the early Eocene. Apparently in the late Eocene, and definitely in the Oligocene, the land bridges were submerged (Figure 20–3), but they emerged again in the Miocene.

These massive geographic rearrangements also modified the oceans and placed continental interiors closer to the air-moistening seas. This geography would favor more equable continental climates than had existed in central Pangaea, because water evaporated over the sea would have returned as rain on nearby landmasses.

Whatever the effects of the changing seas and continental land areas, the warm and humid conditions typical of the Jurassic and early to middle Cretaceous changed toward the end of the Cretaceous, when widespread moderate cooling took place. This trend continued in the Paleocene but was followed by warming in the Eocene. From the Oligocene to Pliocene, general cooling returned, and was capped in the Pleistocene by a period of major glaciation. Overall, the Cenozoic climate became drier and cooler and habitats changed.

Periods of Major Glaciation

We have emphasized in earlier chapters that continental movements have affected evolution by causing major changes in climate, in local geography, in habitat, and in isolation of vertebrates on landmasses and in seas of highly variable extent.

713

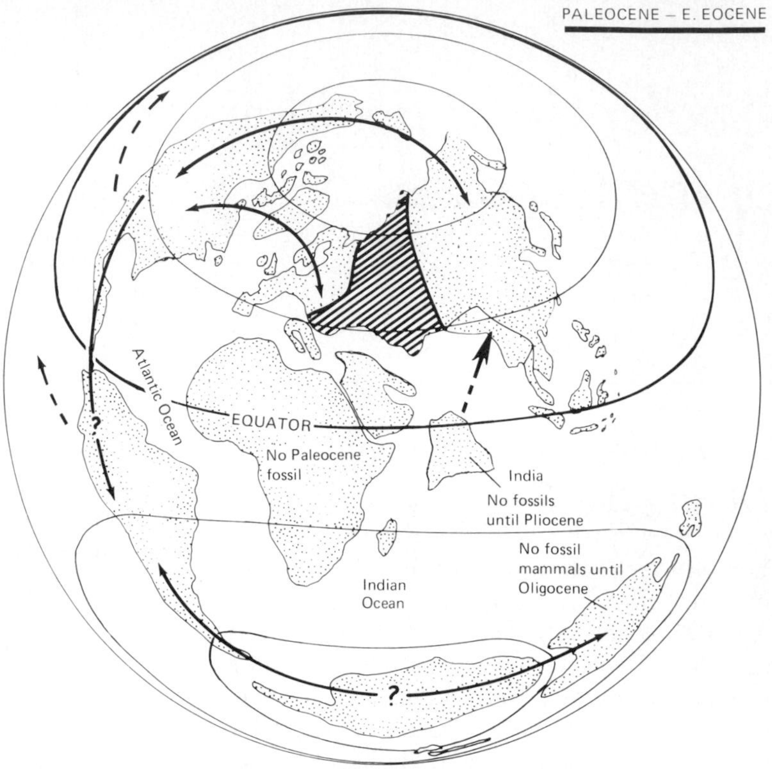

Figure 20–1. Continental positions in the late Paleocene and the early Eocene. Shaded areas are land, hatched areas epicontinental seas. Cenozoic fossil mammals are known in Laurasia and South America from the beginning of the Paleocene, in Africa from the Eocene, in Australia from the Oligocene, and in India from the Pliocene. Solid arrows are major land bridges. Dashed arrows show direction of drift.

Some of the major features of change (world climates, the extent of epicontinental seas, periods of major mountain building, and major glaciation) coincide with reductions in worldwide temperatures and considerable declines in sea level. These changes occupy relatively short spans of geological time, unlike the much slower and longer-lasting ecological trends associated with drift. How might these abrupt changes have influenced vertebrate evolution? This is an especially important point to consider because human culture emerged during the Pleistocene, a short period in which there was little continental displacement but precipitous changes in worldwide climate (Table 20–1).

The fossil evidence gathered about *Homo* over the last decade continues to push our generic origin further into the past (Susman and Stern 1982).

Some anthropologists believe that humans originated 3 to 4 million years ago, others only a million or fewer years ago. However ancient our generic origin, as a species we evolved within the Pleistocene and the Holocene epochs, periods that cover only 1.6 million years of the 500 million years of vertebrate history. That is far less than 1 percent of the total period of vertebrate evolution. Yet the extensive continental glaciers that characterize the Pleistocene and so closely parallel human evolution are rare geological events. Similar extensive glacial periods occurred only at three other times during the evolution of vertebrates (Figure 20–4) and those glaciations were confined mostly to the southern continents. Only in the Pleistocene has extensive glaciation occurred in the Northern Hemisphere.

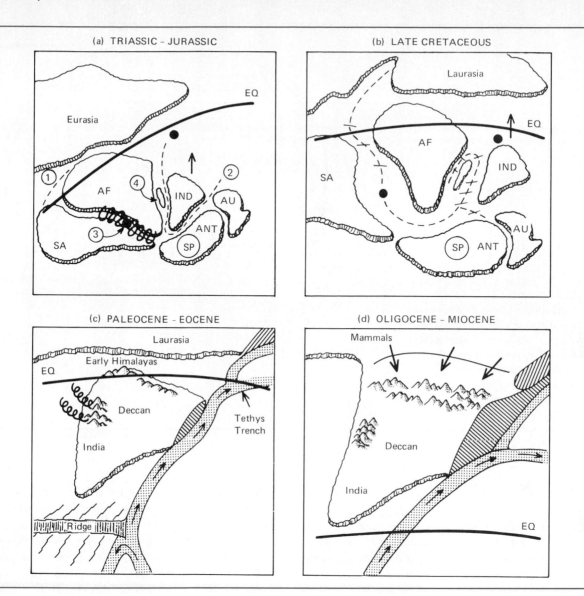

Figure 20–2. Positions of India during the Mesozoic and Cenozoic. (a) The rifts that ruptured Gondwana separated India by middle Jurassic. One through four refer to the position and chronology of these rifts. (b) A suggested chronology of the breakup of Gondwana was (1) India, (2) South America + Antarctica + Australia, (3) Madagascar from Africa. (c) India contacted the Tethys Trench in its northward movements. The Deccan plateau formed in passing over an Indian Ocean hot spot. (d) Formation of the Himalayas from the Indian plate overriding the Tethys Trench and later collision with Laurasia. The collision commenced approximately 45 million years ago during the late Eocene. The Himalayas apparently acted as a barrier to the infiltration of Laurasian mammals until the Pliocene. EQ, Equator; SP, South Pole; black circles, volcanic hot spots.

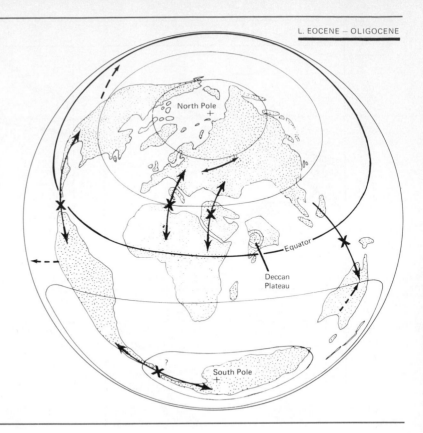

Figure 20–3. Continental positions in the late Eocene and the Oligocene. Shaded areas are land. Solid arrows stress migratory routes for mammals; cross marks on arrows represent occasional or chance dispersal routes. Dashed arrows indicate direction of continental movement.

Ice Ages

What causes an ice age? A long-standing theory suggests that the amount of solar radiation impinging on Earth varies sufficiently to affect Earth's climate. As early as the late eighteenth and early nineteenth centuries famous scientists such as Lagrange, Laplace, and Lyell suggested that eccentricities in Earth's orbit could lead to climatic changes and, perhaps, even periodically precipitate an ice age. In 1930 and 1938 the Yugoslavian astronomer Milutin Milankovitch proposed that ice ages are caused by small changes in (1) Earth's elliptical orbit around the sun, (2) the tilt of Earth's rotational axis, and (3) the precession (wobble) of Earth's rotational axis (Covey 1984, Imbrie and Berger 1984). This view of astronomical effects correlates well with paleoclimatological evidence from various sources. Earth's orbital changes result in three dominant periodicities for the amount of insolation impinging on Earth's surface. Eccentricity of Earth's orbit varies with a cycle of 100,000 years, tilt varies with a cycle of 40,000 years, and precession has a cycle of 26,000 years. However, each of these orbital properties produces different effects. Change in tilt and precession modify the distribution of sunlight (**insolation**) with respect to season and latitude, but not total global insolation, whereas changes in Earth's orbit result in minute changes in global insolation. These changes in insolation seem insignificant as a cause of the advance and retreat of the massive glaciers of the Pleistocene ice age. To explain this paradox, Milankovitch suggested that the critical factor that leads to glaciation is not a change in total global

Table 20–1. Glacial and interglacial stages from two northern hemisphere regions.*

Age† (millions of years ago)	Midcontinent of North America	European Alps
0.03	Wisconsin—late *Interstadial II*	Würm—late
0.06	Wisconsin—middle *Interstadial I*	Würm—early *Riss-Wurm*
0.12	Wisconsin—early *Sagamonian*	Riss *Mindel-Riss*
0.50	Illinoian *Yarmouthian*	Mindel *Günz-Mindel*
1.30	Kansan *Aftonian*	Günz *Donau-Günz*
1.60	Nebraskan	Donau

*Interglacials in italics.

†Age refers to beginning of each glacial episode in millions of years before the present.

insolation, but a change in the amount of summer insolation at high latitudes.

Initially, this hypothesis was received with skepticism, partly because the historical record of glacial periodicities was imprecise. Furthermore, why were there only four major periods of past glaciation, when the cycles predict an ice age every 100,000 years or so? Modern geochemical evidence and improved knowledge of past continental locations have provided an answer to this question. Using isotope and radioactive dating methods, it has been possible to date the history of Pleistocene glaciers accurately (Hays et al. 1976). A glacier forms from compaction of snow when yearly precipitation exceeds melting. The source of water to form the massive glaciers of the ice ages was the sea. When evaporation takes place, more water vapor containing light oxygen molecules (atomic weight 16) than heavy oxygen molecules (atomic weight 18) escapes from the sea surface. Seawater becomes enriched in heavy oxygen. Today, for example, mollusk shells contain relatively fewer heavy oxygen molecules than mollusks did 20,000 years ago at the height of the Wisconsin glaciation. The isotope method used to determine the amounts of heavy and light oxygen in a fossil shell is very accurate and, when coupled with radioactive dating of the sediments in which the shells were embedded, yields a precise time record of sea volume and, indirectly, of glacial ice volume.

What accounts for the absence of significant ice ages prior to the Carboniferous–Permian boundary, when the predicted glaciation periodicity is every 100,000 years? The most plausible explanation lies in the positioning of the drifting continents. The approach of North America and Eurasia toward the North Pole 30 million years ago may have induced sufficient changes in atmospheric and ocean circulations (planetary heat transport) to explain the general cooling of climates that began in the Oligocene and strongly influenced mammalian evolution (Chapter 21). Large landmasses must be positioned near the polar regions, as Antarctica and Greenland are at present, for ice sheets to form. In the Carboniferous, South Africa was located at the South Pole and was scoured by glaciers (Figure 14–1). Sophisticated models of climates that incorporate the positions of the continents and their surface areas further substantiate the Milankovitch theory (Covey 1984). These models predict that more extensive glaciation should have occurred in North America than in Eurasia, as the record confirms. This distribution of glaciers resulted because summers were warmer in Eurasia than in North America.

An additional question remains: How well do the glacial fluctuations of the Pleistocene correlate with the periods for insolation of 100,000, 40,000, and 26,000 years? When the global ice volume curve is subjected to Fourier analysis (a mathematical technique that dissects a complex waveform into its major periodicities), three dominant cycles are revealed. The main cycle has a period of 100,000 years, the second cycle has a period of 43,000 years, and the third cycle has a period of 24,000 years. This is an astonishing correlation between independently derived data sets, and that

Box 20—1. Are Disasters Periodic?

Since the late Cambrian, the evolution of vertebrates has been punctuated by extinctions, often of entire taxa, and replacement by new groups. The gradualist view of vertebrate evolution holds that most of these changes can be attributed to the action of selective forces such as predation, competition, and environmental factors. Changes in climate, in the degree of isolation of populations, and in the diversity of communities which could lead to more or less competition, certainly have influenced the course of vertebrate evolution. In recent years, however, scientists have tended to focus on the mode and the tempo of evolution, and to develop new images of how evolution works. This has been especially true with respect to the hypothesis of punctuated equilibrium (Chapter 1), which proposes that new taxa often originate rapidly when an opportunity suddenly opens. An extension of this portrayal of rapid surges in evolution is the speculation that sudden events, such as extraterrestrial bodies hitting Earth, might lead to the extinction of entire groups of organisms and provide new opportunities to the survivors (Chapter 14).

Although such concepts are currently in vogue, the idea that events outside of our everyday experiences modulate the course of evolution is not new. Various scientists have proposed that repetitive extraterrestrial phenomena may induce major changes in Earth's environment. A major problem with all such theories is the multitude of astronomical periodicities that can be considered to have an effect, some intense, others weaker, some within our solar system, others within our galaxy, and even some beyond the galaxy. As a result of the multiplicity of natural cycles, almost any sequence of events on Earth could be fitted to some extraterrestrial periodicity.

Some of these major periodicities and relationships are summarized in Figure 20–4. A periodic increase and decrease in sea level and the intensity of volcanism seems to correlate with changes in continental drift. Periods of fragmentation or dispersion of the ancient landmasses are associated with increases in sea level and volcanism, and periods of formation or coalescence of a supercontinent correlate with decreases in sea level and volcanism. This cycle seems to recur approximately every 300 million years and corresponds to the length of the cosmic year, the time required for our solar system and Earth to complete one galactic revolution. Along this galactic trajec-

Figure 20—4.   Summary of periodic astronomical, geological, and climatic phenomena that may have influenced the evolution of vertebrates and other organisms. Note that glacial episodes occurred at slightly different times in different regions. Estimated periodicities vary from 300 million years to 26 or 28 million years. (Compiled from various sources, especially N. D. Newell, 1967, *Geological Society of America Special Report* 89:63–91; R. Pearson, 1978, *Climate and Evolution*, Academic Press, London; D. M. Raup and J. J. Sepkoski, 1984, *Proceedings of the National Academy of Sciences USA* 81:801–805; and A. G. Fischer, 1984, in *Catastrophes and Earth History*, edited by W. A. Berggren and J. A. Couvering, Princeton University Press, Princeton, N.J.)

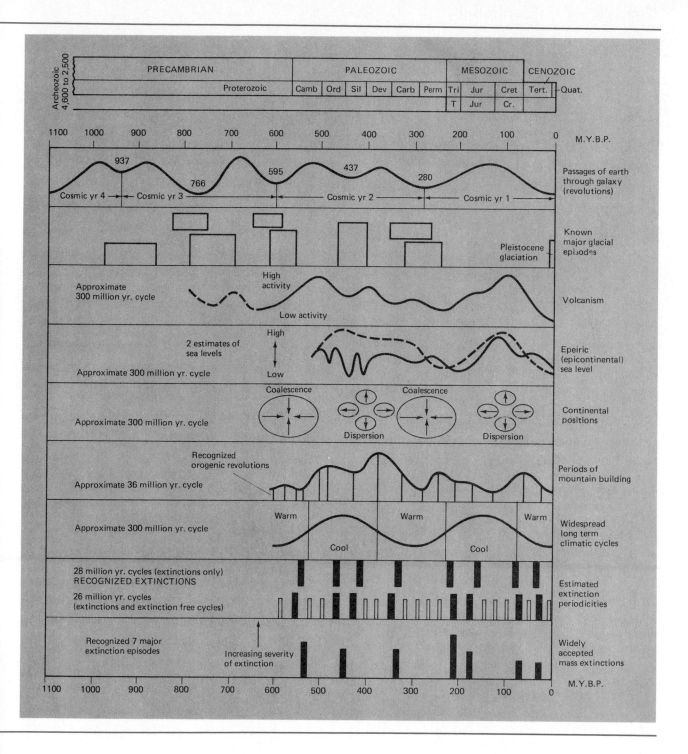

Box 20–1. (Continued)

tory Earth is subjected to changes in the galactic force, which might induce changes in the flow of Earth's core and mantle (Fischer 1984). The correlation indicates increased plate tectonic activity when the force increases and reduced activity when the force diminishes. Changes in sea level and volcanism are predicted by this hypothesis, as are long-term changes in climate (Figure 20–4). Many of the glacial episodes recorded in Earth's rocks occurred when the galactic force is calculated to have been at a minimum. The seven major extinctions of life that are recognized in the Phanerozoic do not, however, show any obvious correlation with the cosmic year and its possible consequences. The 26 to 28 million year periodicity of mass extinctions postulated by Raup and Sepkoski (1984; see Chapter 14), was believed to result from the impact of massive objects from space. Explanations for the proposed periodicity for mass extinctions rely on events that occur within the reaches of our solar system, such as the periodic passage of Earth through the spiral arms of the galaxy.

correspondence has convinced many climatologists that periodic and predictable perturbations in Earth's orbit explain the Pleistocene ice ages. But why, then, should the 100,000 year eccentricity cycle dominate the periodic appearance of ice sheets when it has the least overall effect on insolation of the three orbital perturbations involved? There is no final solution, but several factors are persuasive. First, the 100,000-year periodicity has not always been dominant; during the earlier periods of the Pleistocene a 41,000-year cycle was most evident. Second, the Earth's atmosphere and ocean may possess an inherent resonance near 100,000 years. An astronomical push in the amount of insolation every 100,000 years would reinforce the inherent resonant cycle and be instrumental in perturbing the transport of heat over Earth's surface. Third, as glaciers accumulate, their weight causes the underlying land to settle and altitudinal relief changes. Climatic models that include this interaction actually predict that the dominant climate cycles shift from 20,000 and 40,000 years to 100,000 years. These models are examples of the view that major events in the history of the Earth occur periodically as a result of the interaction of terrestrial and extraterrestrial cycles (Box 20–1).

The Pleistocene: A Series of Ice Ages

Today glaciers cover 10 percent of the land surface of the earth, mostly within polar regions. In the Pleistocene glaciers covered not 10 but 30 percent of the land and, in North America, extended southward as a solid sheet of ice from the North Pole to 38°N latitude, that is, into southern Illinois (Figure 20–5). A similar ice sheet covered all of northern Europe. Judging from modern remnants, this Pleistocene ice mass was very thick; modern glaciers in Antarctica and Greenland are between 3 and 4 kilometers thick.

Glaciers and Sea Levels At present the volume of ice on earth constitutes about 26 million cubic kilometers; in the Pleistocene there were 77 million cubic kilometers of ice. An enormous volume of water was, and still is, locked up in glaciers. Melting of the Pleistocene glaciers to their

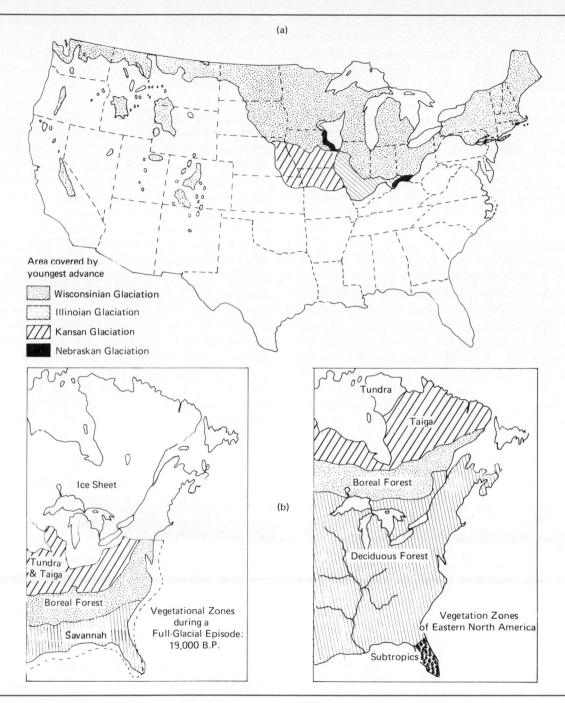

(a)

Area covered by youngest advance

Wisconsinian Glaciation

Illinoian Glaciation

Kansan Glaciation

Nebraskan Glaciation

Ice Sheet

Tundra & Taiga

Boreal Forest

Savannah

Vegetational Zones during a Full-Glacial Episode: 19,000 B.P.

(b)

Tundra

Taiga

Boreal Forest

Deciduous Forest

Vegetation Zones of Eastern North America

Subtropics

Figure 20–5. (a) Extent of Pleistocene glaciations in the Northern Hemisphere and (b) effects of glaciers on the position of biomes in the United States. During a representative full glacial episode (b, left) the deciduous forests native to most of eastern North America were reduced to small isolated pockets (not indicated) in moist regions of savannah.

721

present size has caused the sea level to rise relative to a coastline by about 140 meters, almost half the height of the Empire State Building. Since the end of the last glacial maximum some 18,000 years ago, the sea surface has been rising at an average rate of 1 centimeter per year.

If the modern glaciers were to melt, sea level would rise at least another 50 meters. Most of the coastal cities of the world would be submerged and close to 25 percent of the current land surface would be beneath the sea. Such an event—catastrophic to modern civilization—has been a consequence of every previous glacial period.

Glaciers and World Climates Continental glaciation has a pronounced influence on world climates. For example, in the last episode of the Wisconsin glaciation the cyclonic storm belts were farther south than today and so was the temperate zone. Large biomes occurred much father south in North America, suggesting a generally cooler climate (Figure 20–5). Surface temperatures of tropical seas in the Caribbean were 4 to 5°C lower than today, as were average air temperatures. The climatic effects of glaciation therefore were widespread—generally, climates were cooler.

A vivid example emphasizes this point. Many of the equatorial regions that today are occupied by lowland rain forests were then arid. The forests in the Amazon and Congo Basins contracted into several isolated refugia with each glacial episode (Figure 1–10). During interglacial periods, the forests again spread over the basins. As a result, the plant and animal populations were repeatedly isolated and remixed. In genetic theory, isolation and remixing of populations bring about rapid speciation. In the refugia all populations, including vertebrates, suffered a reduction in total numbers and therefore also in their total genetic diversity. Random isolation, genetic drift, and natural selection of each species within the several isolated refugia should have led to new species. Theoretically, the number of species should have increased and, apparently, it did. Species diversity of birds in these tropical forest basins is very high. The same applies to mammals, but less dramatically. The effect of glaciation on equatorial aridity and the resultant contraction of moist habitats is the most logical mechanism for high species diversity in these areas. Glaciation, therefore, can affect the lives of vertebrates everywhere, not just those in close contact with the spreading ice.

Summary

The evolution of birds and mammals was as much a consequence of the geological events of the late Mesozoic and early Cretaceous as of competition with the Mesozoic diapsids. Fossil birds and mammals present us with a reasonable picture of how geologic events probably influenced their radiation. The central event was the Jurassic rupture of Pangaea initially isolating the huge northern landmass of Laurasia from its southern counterpart, Gondwana. By the early Cretaceous the North Atlantic Ocean and the Caribbean Sea were formed as North America separated from South America and Europe. By the middle Cretaceous, India and Africa had split away from the remainder of Gondwana. South America, Antarctica, and Australia, however, retained land connections until the early Cenozoic.

In general, drift and separation of the continents led to gradual changes in climates. Dramatic changes in worldwide climates, as exemplified by ice ages, have occurred infrequently in Earth's history. Yet ice ages have had enormous effects on the evolution of vertebrates.

The Pleistocene, which began about 1.6 million years ago, is typified by a series of climatic oscillations in which glaciers advanced southward across the northern continents during cooler periods, only to retreat during warmer interglacial periods. Associated with these glaciations were precipitous changes in sea level and worldwide reduction in temperature. Both floristic and faunistic assemblages were repeatedly forced to contract their geographic ranges, a circumstance that led to the genetic isolation of populations and, therefore, an increased chance for speciation.

In the last decade interest in the causes of cyclic events such as the Pleistocene glaciation has widened the scope of inquiry to include studies of astronomical events that, at first glance, seem to have little to do with biology. This renewed awareness of periodic phenomena has led to more exacting studies of paleoclimates and their possible influence on evolution. The rapid accumulation of new evidence is forcing reconsideration of how environmental conditions might affect the mode and tempo of vertebrate evolution.

References

Covey, C. 1984. The Earth's orbit and the ice ages. *Scientific American* 250:58–66.

Fischer, A. G. 1984. The two Phanerozoic supercycles. Pages 129–150 in *Catastrophes and Earth History*, edited by W. A. Berggren and J. A. Couvering. Princeton University Press, Princeton, N. J.

Hays, J. D., J. Imbrie, and N. J. Shackelton. 1976. Variations in the Earth's orbit: pacemaker of the ice ages. *Science* 194:1121–1132.

Imbrie, J. and A. Berger (editors). 1984. *Milankovitch and Climate Change*. Elsevier, Amsterdam.

Raup, D. M. and J. J. Sepkoski. 1984. Periodicities of extinctions in the geologic past. *Proceedings of the National Academy of Sciences USA* 81:801–805.

Susman, R. L. and J. T. Stern. 1982. Functional morphology of *Homo habilis*. *Science* 217:931–933.

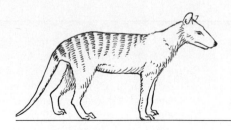

The diversification of therian mammals in the Cenozoic has continued some
trends that were pointed out among Paleozoic and Mesozoic synapsids in
Chapter 19. In particular, the specialization of the jaws and dentition
that allowed more effective processing of food has continued in mammals,
and a useful perspective on mammalian diversity can be gained by examining
their feeding specializations. This survey reveals that morphology of the
skulls, jaws, and teeth of mammals is clearly related to dietary habits.
Furthermore, very similar specializations have evolved convergently or in
parallel in different evolutionary lineages in different parts of the world. This
situation reflects the fragmentation of landmasses during the Cenozoic
discussed in Chapter 20, which left different combinations of mammalian
stocks on different continents. Thus, to take just one example, saber-tooth
predators evolved among placental mammals in North America and among
marsupials in South America. In this chapter we provide an overview of
the diversity of living mammals and compare them to the even more diverse
mammal faunas of the middle and late Cenozoic.

Characteristics of Mammals

21

The Major Lineages of Mammals

Mammals alone widely exploit the resources of earth, from pole to pole, mountaintop to deep sea, and even the night sky. In this chapter we survey the major groups of mammals and emphasize some characteristics of living mammals, especially the structure and function of the integument, the feeding system, the nervous system, and the reproductive system. Two other salient characteristics of mammals, their sociality and endothermy, are treated in detail in Chapters 22 and 23.

Although it has often been suggested that the mammals may have had a polyphyletic origin, current opinion favors a monophyletic origin, as depicted by Figure 21-1 and Table 21-1. The most likely sister group of mammals is the trithelodontid cynodonts (Chapter 19). Some comments on the cladogram of Figure 21–1 are in order. First, mammalian systematics is in a state of flux (Benton 1988). A recent and generally accepted phylogeny of mammals (McKenna 1975) has been modified by Eisenberg (1981) and quite recently by Novacek and Wyss (1987). Figure 21–1 presents an amalgamation of these phylogenies. You will notice that the cladogram in Figure 21–1 differs slightly from the one presented in Figure 19–10 in its treatment of monotremes, multituberculates, and some other primitive mammals. Those differences reflect divergence of opinions among various authorities who do not attach the same importance to characters. Remember that cladograms are hypotheses of relationships; they are always subject to revision in the light of new information or new interpretations of existing material.

We recognize 20 orders of extant mammals. There are, however, unresolved relationships between many of the orders and what we have depicted will certainly change over the next few years. Perhaps the most extensive modifications in the phylogeny will involve the relations among the Lagomorpha + Rodentia + Macrocelidea, Carnivora, Tubulidentata, Insectivora, and Archonta (which includes the Primates, Scandentia, Dermoptera, and Chiroptera). Although these groups have certain internal affinities, their broad interrelationships have not been elucidated. The lagomorph–rodent–macroscelid lineage has been separated from the other orders for clarity of presentation in Figure 21–1, but no obvious shared derived characters support that separation. Although we cannot resolve the exact affinities of several of the mammalian groups, we can examine characteristics of some of the major living mammal groups and orders.

Modern mammals are divisible into three major groups, separated on the basis of reproductive biology. The Prototheria, or monotremes, survive as about six species isolated in Australia and New Guinea (Griffiths 1979). They are grouped in two suborders—the spiny anteaters and the duck-

Phylogenetic Relationships of the Mammalia

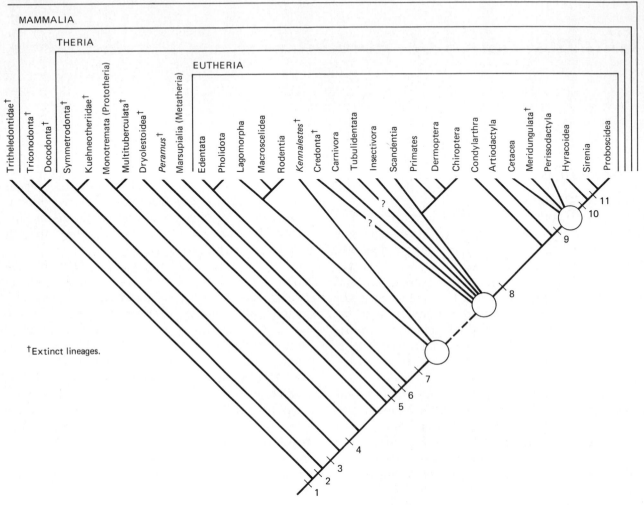

†Extinct lineages.

billed platypuses. Prototherians lay eggs that are incubated and hatched outside the reproductive tract of the females. Nevertheless, echidnas and platypuses are hairy, endothermic, milk-producing vertebrates with only the dentary in the lower jaw, and thus qualify as mammals.

The remaining two groups of living mammals are closely related but have had separate evolu-

tionary histories since the earliest Cretaceous. The 250 or so species of Metatheria (marsupials) are distinguished by their short gestation periods, tiny, feebly developed offspring, and (in many) a protective pouch, the **marsupium**. This pouch lies over the mammary glands of the female and the young crawl into it immediately after birth to feed and complete development. The marsupials are

1. Prismatic enamel on teeth.   2. Dentary replaces sur-angular in jaw articulation, action of molars involves lateral movement of the lower jaw.   3. Triangular pattern of main molar cusps.   4. Theria: ear ossicles, characters of braincase, expanded neopallium in brain, characters of vertebrae and long bones, probably tribosphenic molars. 5. Loss of eggshell, characters of braincase, vertical tympanic membrane, tribosphenic molars here if not at node 4.   6. Chorioallantoic placenta, long gestation.   7. Stapes stirrup-shaped (i.e., penetrated by a fenestra).   8. Clavicle reduced or absent, bunodont crowns on molars.   9. Expanded auditory capsule roof.   10. Mastoid bone not visi-ble on external surface of skull.   11. Orbits anteriorly placed, bilophodont teeth. (Based on M. C. McKenna, 1975, pages 21–46 in *Phylogeny of the Primates,* edited by W. P. Luckett and F. S. Szalay, Plenum Press, New York; J. F. Eisenberg, 1981, *The Mammalian Radiations,* The University of Chicago Press; T. S. Kemp, 1982, *Mammal-like Reptiles and the Origin of Mammals,* Academic Press, London; T. S. Kemp, 1983, *Zoological Journal of the Linnaean Society* 77:353–384; M. S. Novacek and A. R. Wyss, 1987, *Cladistics* 2:257–287; and Z. Kielan–Jaworowska, A. W. Crompton, and F. A. Jenkins, Jr., 1987, *Nature* 326:871–873.)

Figure 21–1. The phylogenetic relationships of the Mammalia. The diagram shows the probable relationships among the major groups of mammals. Extinct lineages are marked by a dagger (†). The numbers indicate derived characters that distinguish the lineages. The circles indicate that the relationships of the lineages at that node cannot yet be defined, and the dashed line connecting the circles that lie between characters 7 and 8 indicates that all of the lineages originating from both circles are included in that uncertainty.

restricted to the Australian region and the New World.

The eutherians, or placental mammals, which include about 3800 species, are born at a more advanced state of development than marsupials. Many are able to run or swim by their mother's side within minutes of birth, and most of the species with long periods of postnatal development are weaned sooner after birth than marsupial young.

These differences between the reproductive patterns of Metatheria and Eutheria reflect different evolutionary responses to the stresses of the environment (Lillegraven 1974). Eutherians produce advanced young with superior survival potential, but at high and prolonged cost to the

Table 21–1.   Classification of mammals with orders and approximate numbers of species of living forms.

Classification	Major Examples

Class Mammalia
"Subclass Prototheria"
  Infraclass †Eotheria
    Order †Triconodonta
    †Docodonta	

Infraclass †Trituberculata
    Order †Symmetrodonta
    †Pantotheria	

Subclass Theria
  Infraclass †Allotheria
    Order †Multituberculata	

Infraclass Ornithodelpia Order Monotremata (3 species)	Spiny anteaters, duck-billed platypus; 2 to 10 kg; Australian region.
Infraclass Metatheria Order Marsupialia (260 species)	Opossums, bandicoots, koala, wombats, kangaroos, wallabies; primarily 5 gm to 5 kg, some Australian forms are large (90 kg); Nearctic, Neotropical, and Australian regions.
Infraclass Eutheria Order Edentata (29 species)	Anteaters, sloths, armadillos; 20 gm to 33 kg; Nearctic and Neotropical regions.
Pholidota (7 species)	Pangolins; 2 to 33 kg; Ethiopian and Oriental regions.
Lagomorpha (65 species)	Pikas, rabbits, hares; 180 gm to 7 kg; worldwide except Antarctica, introduced in Australia by humans.
Rodentia (1752 species)	Rats, mice, squirrels, agouti, capybara; 7 gm to over 50 kg; worldwide except Antarctica.

†Extinct.

Source: J. A. Hopson, 1970, *Journal of Mammalogy* 51:1–9; M. C. McKenna, 1975, in *Phylogeny of the Primates,* edited by W. P. Luckett and F. S. Szalay, Plenum Press, New York; T. A. Vaughn, 1978, *Mammalogy,* 2nd edition, W. B. Saunders, Philadelphia; J. F. Eisenberg, 1981, *The Mammalian Radiations,* University of Chicago Press, Chicago; R. M. Nowak and J. L. Paradiso, 1983, *Walker's Mammals of the World,* 4th edition, Johns Hopkins University Press, Baltimore; S. Anderson and J. K. Jones, Jr., 1984, *Orders and Families of Recent Mammals of the World,* Wiley-Interscience, New York; M. J. Novacek and A. R. Wyss, 1987, *Cladistics* 2:257–287.

mother. The metatherian female invests much less before the birth of her young and can quickly recycle her reproductive apparatus to produce subsequent broods if appropriately stimulated by environmental conditions (Parker 1977).

Nevertheless, eutherians are the dominant mammals of recent times. Most eutherian species are contained in about 18 orders. Over one-third of the living species of eutherian mammals are insectivores and bats. Because of their small size and nocturnal habitats, these mammals are not well known to the general public. The Insectivora,

Table 21–1. Classification of mammals with orders and approximate numbers of species of living forms.

Classification	Major Examples
Macroscelidea (19 species)	Elephant shrews; 25 to 500 gm; Ethiopian region with one species in Morocco and Algeria.
Insectivora (391 species)	Hedgehogs, moles, shrews; 2 gm to 1 kg; worldwide except Antarctica.
Scandentia (16 species)	Tree shrews; 400 gm; Oriental region.
Primates (180 species)	Lemurs, monkeys, great apes, humans; 85 gm to over 275 kg; primarily Oriental, Ethiopian, and Neotropical regions, humans are now worldwide.
Dermoptera (2 species)	Flying lemurs; 1 to 2 kg; Oriental region.
Chiroptera (917 species)	Bats; 4 gm to 1.4 kg; worldwide except Antarctica.
Carnivora (269 species)	Dogs, bears, racoons, weasels, hyaenas, cats, sea lions, walrus, seals (these last three are often assigned to the order Pinnipedia); 70 gm to 760 kg, some marine forms over 1000 kg; worldwide.
Tubulidentata (1 species)	Aardvark; 64 kg; Ethiopian region.
Hyracoidea (7 species)	Hyraxes; 4 kg; Ethiopian and southern Saharan regions.
Sirenia (4 species)	Dugongs, manatees; 140 to over 1000 kg; coastal waters and estuaries of all tropical and subtropical oceans except the eastern Pacific (in the Atlantic drainage they enter rivers).
Proboscidea (2 species)	Elephants; 4500 to 7000 kg; Ethiopian and Oriental regions.
Perissodactyla (18 species)	Horses, tapirs, rhinoceroses; 200 to 3600 kg; Africa, southern and central Asia, Central and northern South America.
Artiodactyla (185 species)	Swine, hippopotamuses, deer, giraffe, antelope, sheep, goats, cattle; 2.5 to 3600 kg; worldwide except Antarctica (introduced into Australia and New Zealand by humans).
Cetacea (78 species)	Porpoises, toothed whales, baleen whales; 20 to 200,000 kg; worldwide in oceans and in some rivers and lakes in Asia, South America, northern North American and Eurasia.

which feed on the flesh of soft-bodied invertebrates or pierce the exoskeletons of insects to eat their soft parts, may provide us with models of the life of the earliest mammals. The winged Chiroptera generally feed on flying insects or on fruit and nectar. They have superb specializations that challenge our understanding and have led researchers to significant advances in the areas of biophysics, acoustics, and neurobiology.

By far the most familiar wild mammals are species of Rodentia. These small, gnawing, plant-eating eutherians are among the most successful of mammals. They are also humanity's greatest mammalian competitors and constant, if uninvited, companions.

In contrast, domesticated mammals are found among the species of plant-eating Artiodactyla, upon whom we depend for much of our animal food, and the flesh-eating Carnivora, from which we gain cooperative assistance, protection, and endless hours of enjoyment.

Finally, two orders of mammals, although unknown in the wild to most people, are of great interest to almost every person. The fish-, squid-, and plankton-eating Cetacea include the largest vertebrates ever known to have lived on earth. The primates, with their generally catholic tastes in food, are humankind's closest relatives. As such they are a perennial attraction and an important guide to understanding our own aptitudes and deficiencies.

Although we will draw examples from other orders in our discussion of mammals, these six orders—Insectivora, Chiroptera, Primates, Rodentia, Carnivora, and Artiodactyla—receive most of our attention.

The Mammalian Integument

We have emphasized that endothermy is an energetically expensive process; a bird or mammal may use nearly all the energy in the food it eats merely to keep itself warm. Clearly, reducing the heat lost to the environment can be an important factor in the energy balance of an endothermal homeotherm. Part of this reduction is accomplished behaviorally (for example, by seeking a sheltered microclimate) and part by physiological mechanisms such as directing blood flow away from the body surface. Nonetheless, the characteristics of an animal's integument are major factors that affect its heat exchange with the environment. Much of the ability of mammals to live in harsh climates is attributable to properties of their integument. The skin, its hair, glands, and sense cells interact with every other aspect of mammal-

ian life. Some small rodents have exceedingly delicate epidermis only a few cells thick. Human epidermis varies from a few dozen cells thick over much of the body to over a hundred cells thick on the palms and soles. Elephants, rhinoceroses, hippopotamuses, tapirs, and pigs were once classified as pachyderms because their epidermis is several hundreds of cells thick. The texture of the external surface of the epidermis varies from smooth (in fur-covered skins and the hairless skin of cetaceans) to rough, dry, and crinkled (many hairless terrestrial mammals). The tail of many rodents is textured with epidermal scales very like those of lizards.

Integumentary Derivatives

The uniqueness of the mammalian integument lies in its derivatives: growing, replaceable hair; lubricant- and oil-producing holocrine (sebaceous) glands; apocrine and exocrine glands that secrete volatile substances, water, and ions; the unique mammary glands; scales, nails, claws, hoofs, and horns. The integument is also a contributing factor in the growth of antlers. Not every species has each class of integumentary appendage, but the various types often occur in unrelated mammals, attesting to their functional significance. These appendages of the mammalian skin are derived from epidermis but supported, nourished, and innervated through the dermis. Figure 21–2 shows the superficial layer of the skin, the **epidermis**, which is formed by dead cells that are derived from the underlying **dermis**. Beneath that, the **hypodermis** is rich in subcutaneous fat cells. Specialized structures of the skin include guard hairs, undercoat or wool hairs, sebaceous glands, apocrine glands, and sweat glands. Sensory nerve endings include free nerve endings (probably pain receptors), beaded nerve nets around blood vessels, Meissner's corpuscles (touch receptors), Pacinian corpuscles (pressure receptors), nerve terminals around hair follicles, and warmth and cold receptors. Vascular plexuses (intertwined blood vessels) of the skin are involved in thermoregulation.

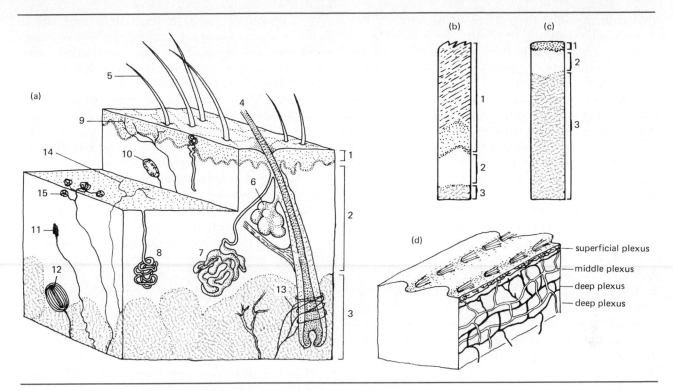

Figure 21–2. Structure of mammalian skin. (a) Composite of skin and appendages showing: (1) layers and surface configuration of the epidermis, (2) the relatively cell-poor dermis, (3) the hypodermis rich in subcutaneous fat cells, (4) guard hairs, (5) undercoat or wool hairs, (6) sebaceous gland, (7) apocrine gland, (8) sweat gland, (9) free nerve endings, (10) beaded nerve nets around blood vessels, (11) Meissner's corpuscles, (12) Pacinian corpuscles, (13) nerve terminals around a hair follicle, (14) heat, and (15) cold receptors. (b) Thick epidermis of the skin on the sole of the human foot. (c) Thin, hairy skin and subcutaneous tissue from the human thigh. (d) Canine skin showing three of the vascular plexuses involved in thermoregulation. (Modified from various sources, especially A. W. Ham and D. W. Cormack, 1973, *Histology*, 8th edition, J. B. Lippincott, Philadelphia; and R. J. Harrison and E. W. Montagna, 1973, *Man*, 2nd edition, Appleton-Century-Crofts, New York.)

Hair Hair has a variety of functions in a modern mammal, including concealing an animal from predators and signaling to other members of its species, but the basic function of hair in nearly all mammals is insulation. How did hair evolve?

One possibility is that hair functioned initially as part of the sensory system rather than as insulation. Hairs are richly innervated around their base even when insulation seems to be their prime function. Highly specialized tactile hairs are localized on the legs, nose, and around the mouth and eyes of many mammals. These **vibrissae**, which have bulbous complex roots, vascular beds, and a multitude of nerve endings, are extremely sensitive to deflection of the tips of their stiff rod-like shafts. P. F. A. Maderson (1972) has suggested that thin, rod-like projections from germinal epidermal cells were the first hairs. Basal innervation of these rods, which were distributed over the body, provided tactile sensory input from bending and distortion of the rod. Therefore, like the lateral-line receptors, hairs evolved as epidermal mechanoreceptors. At some point these sensory hairs were altered to result in a localized multiplication of the units, which perhaps improved the precision of the information provided by these tactile receptors. Whatever the selective forces, Maderson argues that dense sense organs produced a pelage with insulative properties that presumably became the focus of subsequent selection.

Prominent features of the pelage of living mammals are its growth, replacement, color, and mobility. A hair grows from a deep invagination of the germinal layer of the epidermis called the hair follicle (Figure 21–3) and is composed of keratin, pigments, and in some cases, tiny encapsulated air bubbles. These substances are common to all tetrapods, but their organization and chemistry make hair uniquely mammalian. The keratin of hair is of two types: soft and hard. **Soft keratin**, found in the core or medulla of a hair shaft, is similar to that of the epidermis but is often interrupted by masses of trapped air bubbles. **Hard keratin** in the cortex covering the medulla and the sheath that encapsulates the hair surface results from a gradual fusion of living epidermal cells into a hard, chemically inert homogeneous mass that forms stiff hair, bristles, and quills. (Bird feathers and leg scales, as well as the scales of snakes and lizards, are hard keratin.) When soft keratin is dominant, a supple pelage such as wool is produced.

The color of hair depends on the quality and quantity of melanin injected into the forming hair by melanocytes at the base of the hair follicle. Black hair has dense melanin deposits in both cortex and medulla; the white hair of humans has an unpigmented cortex but pigmented medulla. When the medullary portion of the hair shaft is absent, a fine, often blond or reddish pelage (depending on the melanin) results. The color patterns of mammals are built up by the colors of individual hairs.

Most mammals are countershaded (dark above and lighter beneath), a combination that is inconspicuous under most conditions, whereas other species (skunks, for example) that rely on advertising their noxious properties to deter predators have strongly contrasting patterns that are visible even at low levels of illumination.

Because exposed hair is nonliving, it wears and bleaches. Replacement occurs by growth of an individual hair or by **molting**, in which old hairs are replaced. Human hair lengthens about one-third of a millimeter per day during active growth. When it becomes quiescent, the follicle produces a bulbous anchor that firmly locks the hair in place. Although in some mammals, like humans, quiescent and active hair follicles occur in a mosaic over the entire body, most mammals have pelage that grows and rests in seasonal phases. Molting usually occurs once or twice a year. Only by this mechanism can mammals like the snowshoe hare and ermine change from a summer camouflage of brown to a winter coat of white. Because the hairs are tightly bound to the dermis, only skins with quiescent (generally, winter) pelage are of value in the fur trade.

The insulative effectiveness of hair depends upon its ability to trap air, and this is proportional to the length of the hairs. Mammals from the polar and temperate regions have longer coats in winter than in summer. The closeness of hairs is also important because it creates smaller entrapped air spaces; winter coats are thicker than summer pelage. A duplex pelage often occurs: Guard hairs are long, cylindrical, straight shafts, whereas curly underfur or wool hairs are shorter, more numerous, and flattened.

Most hairs are not perpendicular to the surface epithelium but angle in the same direction over a given region of the body. Attached about midway along the length of each hair follicle is a bundle of smooth muscle, the **erector pili** (Figure 21–3). This muscle pulls the hair into a nearly perpendicular orientation to the skin. The result is to thicken the pelage and thus entrap a larger volume of air. A curious side effect noticeable in near-naked mammals such as humans are the dimples

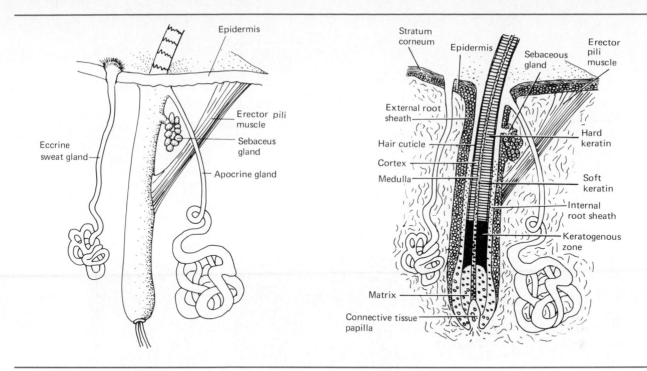

Figure 21–3. The structure of mammalian hair and associated glands. Composite view of hair follicle (left) and surrounding structures with dermis dissected away and a longitudinal section (right) to show the internal structure of a typical hair. (Modified from various sources, especially A. W. Ham and D. W. Cormack, 1973, *Histology*, 8th edition, J. B. Lippincott, Philadelphia, and R. J. Harrison and E. W. Montagna, 1973, *Man*, 2nd edition, Appleton-Century-Crofts, New York.)

(goose pimples) on the skin's surface over the insertion of contracted erector pili muscles. Cold stimulates a general contraction of the erector pili via the sympathetic nerves, as do other stressful conditions such as fear and anger. The significance of these behaviorally induced erections of the pelage appears to lie in increasing the apparent size and, therefore, the apparent strength, of the excited mammal. Often such displays are limited to specific regions, such as the hackle and tail hair erections of dogs. Displays such as those of a porcupine are impressive and are clearly understood by a wide range of other animals.

Glandular Structures Secretory structures of the skin develop from the epidermis. **Sebaceous**

glands open into the neck of a hair follicle and often lie in the space bounded by the hair shaft, the superficial epidermis, and the erector pili muscle. When the muscle contracts, the sebaceous gland is squeezed and its oily contents flow into the follicle and thence onto the skin surface (Figure 21–3). Sebaceous glands may also be found in hairless areas of the mouth and lips, the end of the penis, and around the vulva, the nipples of the mammary glands, and the edges of the eyelids. The oils lubricate and waterproof the hair and skin. Specialized sebaceous glands produce lanolin in sheep.

A second type of gland common to many mammals is the **apocrine gland**. These glands frequently open near but not into the hair follicles

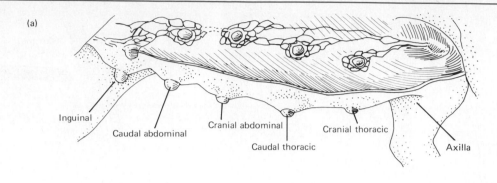

(a)

Inguinal

Caudal abdominal

Cranial abdominal

Caudal thoracic

Cranial thoracic

Axilla

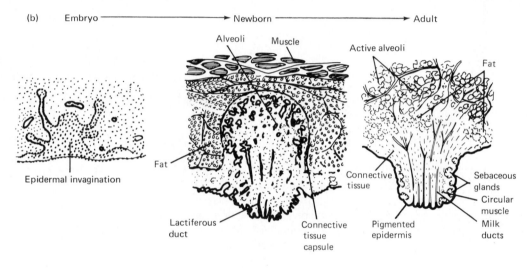

(b)   Embryo ──────────► Newborn ──────────► Adult

Alveoli   Muscle

Active alveoli

Fat

Epidermal invagination

Fat

Lactiferous duct

Connective tissue capsule

Connective tissue

Pigmented epidermis

Sebaceous glands

Circular muscle

Milk ducts

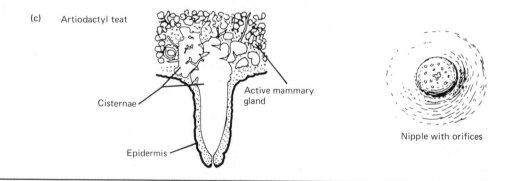

(c)   Artiodactyl teat

Cisternae

Active mammary gland

Epidermis

Nipple with orifices

and also occur in skin that lacks hair. Apocrine glands are tubular epidermal invaginations that extend deep into the dermis. The exact functions of apocrine secretions are not well understood. Although the secretion from human apocrine glands is odorless, bacteria on the body surface convert it to an odorous product. Various musk and scent glands so characteristic of mammals are thought to be enlarged, modified apocrine glands. In their development, their structure, and their mode of secretion, **mammary glands** closely resemble apocrine glands.

Mammary glands (Figure 21–4) develop as paired ridges of hypertrophied ventral epithelium of the fetus that run from the axillary to the inguinal regions. This mammary ridge differentiates into discrete localized thickenings in different positions. Manatees (Order Sirenia) have one pair of axillary mammae, primates have one pair of thoracic mammae, some artiodactyls have abdominal mammae, and perissodactyls have inguinal mammae. The number of mammae roughly corresponds to the number of young born in a litter. Certain opossum-like marsupials are reported to have nearly 20 mammae. In male mammals further development of the mammae does not occur after birth. In females the thickened epithelium produces numerous elongate cords that branch and proliferate into the dermis. Hollow sacs develop at the terminal ends of these cords and become continuous with channels that develop through the cords of epithelium to produce a highly branched

duct system. Under proper hormonal stimulation the terminal sacs become more numerous, enlarge greatly, and the cells of the sac walls begin to secrete milk. Milk is a water-based solution of proteins (primarily the phosphoprotein calcium caseinate), the sugar lactose, and lipids in the form of several different fatty acids combined in suspended droplets. The fat droplets often are capped by or contain cell fragments, indicating a mode of secretion similar to that of apocrine glands, which also lose portions of cells during secretion. The exact proportions of these primary constituents vary greatly from species to species (Table 21–2). The concentrations of the major components are far above those of the maternal blood and, in general, growth of a newborn mammal is positively correlated with the protein content of the milk.

Pregnancy hormones stimulate the mammary glands into a physiological state capable of lactation, but milk does not flow until some time after suckling has begun. Monotremes lack highly specialized openings for the mammae, but other mammals have nipples that are complexly innervated epidermal organs. Certain artiodactyls have analogous but much larger structures called teats.

Similar in structure to the apocrine and mammary glands and often grouped with them are the **sweat** or **eccrine glands**. The coiled tubules of sweat glands are never associated with hair follicles but are prevalent on the hairless surfaces that come in contact with the substrate: soles of feet and prehensile tails. Secretions maintain skin pli-

---

Figure 21–4. Mammary glands. (a) The positions of nipples in eutherians as illustrated by a dissection of the dog. Only the axillary mammae fail to develop in these large-litter bearing canids. (b) Stages in the development of the human mammary gland resemble those of apocrine and eccrine glands. An embryo 15 centimeters in length has a simple proliferation and invagination of epidermal cells (left). A newborn infant (middle) has the basic elements of a nipple, multiple lactiferous ducts, and nonsecretory alveoli. The tip of the mature, lactating primate breast shows the addition of circular constrictor muscles, sebaceous glands, and swollen active alveoli which vent their product via multiple openings on the nipple. (c) The artiodactyl teat delivers a much larger volume of milk in a shorter time via the development of large milk-holding cisternae adjacent to the elongate teat, which has a single wide terminal opening.

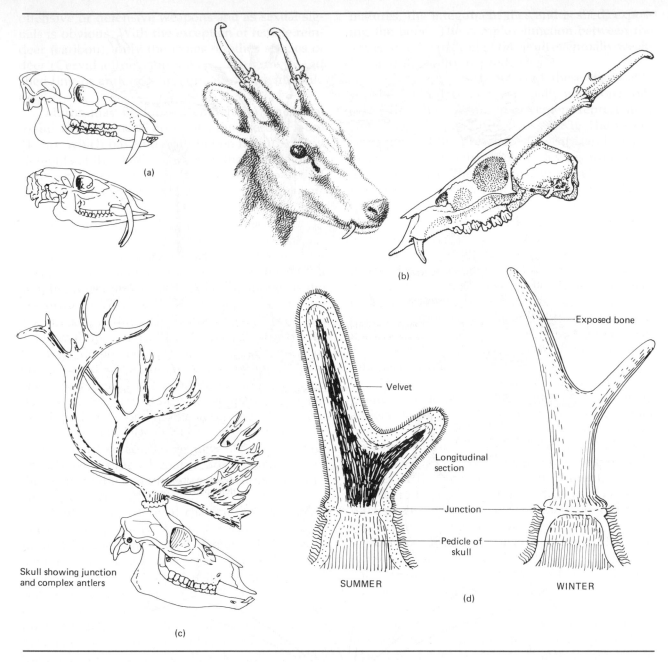

Figure 21–7. Cephalic appendages: antlers. See text for details.

Labels within figure:
- (a)
- (b)
- Exposed bone
- Velvet
- Longitudinal section
- Junction
- Pedicle of skull
- SUMMER
- WINTER
- (d)
- Skull showing junction and complex antlers
- (c)

lers, and sexual selection seems likely to be among them. This is another example of the maxim that the evolutionary origin of a character may be quite different from its current use.

Mammalian Food and Feeding Specializations

An efficient means of obtaining and processing food to sustain high energy needs is absolutely essential to an endotherm. Mammals meet these needs in various ways, as seen in their trophic, or nutritional, adaptations. The trophic apparatus includes the teeth and jaws, the muscles involved in chewing, the alimentary canal, and even locomotion and social behavior.

Mammalian Teeth

Each tooth is attached in its jaw socket by a type of bone known as **cement**, hard and rigid like other bone, but wearing rapidly when exposed. The core of a tooth is hollow and filled with nerves, blood vessels, and cells that produce a substance called **dentine** that forms the inner layer of the tooth. Dentine contains more mineralized material and less organic matter than bone and is harder, heavier, and more resistant to wear than cement. The exposed part of the tooth is encased in **enamel**, a unique substance that is the hardest, heaviest, and most friction-resistant tissue evolved by vertebrates. Unlike both cement and dentine, enamel is ectodermal, not mesodermal, in origin. Enamel is totally acellular and is nearly devoid of organic matter, being composed of large, uniformly oriented calcium phosphate crystals.

The teeth of a generalized mammal may be designated from anterior to posterior on each side of each jaw as: **incisors** (usually two to five), **canines** (never more than one), **premolars** (generally two to four), and **molars** (variable but often three). Most mammals have two sets of teeth during an animal's lifetime, a condition known as **diphyodonty** (*di* = two, *phyo* = grow, produce, *dont* =

tooth). The lacteal (milk or deciduous) dentition appears first in a generally regular order from anterior to posterior. The molars are adult teeth that function throughout life, but ontogenetically they are posterior, unreplaced members of the lacteal dentition. In many species the premolars have taken a form very similar to that of the molars, a process known as molarization of the premolars. It is often difficult to distinguish premolars from molars in an intact skull, especially in herbivores.

Because digestion is basically a chemical process, increasing the area for contact and penetration of digestive enzymes allows rapid chemical breakdown of food. Most vertebrates use their teeth to catch, grasp, or crop food but not to process it. Mammals are different: The anterior teeth (**incisors** and **canines**) of mammals retain this basic function, but the posterior teeth (**premolars** and **molars**) are modified to **masticate** foodstuffs. During mastication, the food is repeatedly worked between interlocking **occlusion surfaces** of the upper and lower dentition by the action of the mobile, sensitive tongue and a characteristic mammalian facial structure, the cheeks, which assist in retaining food in the mouth when chewing. The food thus becomes a loose pulp that readily absorbs gastric juice in the stomach.

Dentition

The most primitive living eutherians, the insectivores, feed upon invertebrates and small vertebrates as their late Cretaceous counterparts probably did. From these stocks have evolved carnivores capable of bringing down prey many times their own size, strainers of minute floating organisms, nectar drinkers, and grazers upon dry and brittle grasses. Generalized carnivores have formed the basis of the trophic radiations that lead to various forms of herbivory. The specialized dentition and digestive modifications needed to consume plants directly have apparently locked mammalian herbivores out of further trophic diversification.

The skulls of living mammals (Figure 21–8) can be arranged to illustrate a functional classifi-

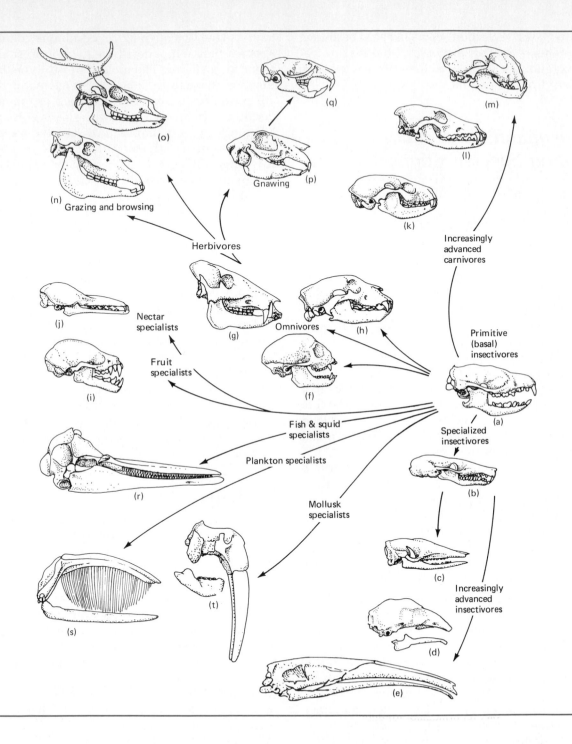

(o)

(q)

(m)

(l)

(n) Grazing and browsing

(p)
Gnawing

(k)

Herbivores

Increasingly
advanced
carnivores

(j)

Nectar
specialists

(g)

Omnivores

(h)

Primitive
(basal)
insectivores

(i)

Fruit
specialists

(f)

(a)

Fish & squid
specialists

Specialized
insectivores

(r)

Plankton specialists

(b)

Mollusk
specialists

(c)

Increasingly
advanced
insectivores

(s)

(t)

(d)

(e)

cation of recent eutherians based on their trophic specializations, especially their teeth. Such an arrangement is not phylogenetic; instead, it provides an ecological perspective of mammalian trophic diversity. The earliest therian mammals had a dentition similar to that found in relatively unspecialized insectivores living today. The hedgehog, *Erinaceus*, has a dentition with piercing cusps on most of the teeth (Figure 21–8a). More specialized is the battery of sharp teeth of the mole, *Scalopus* (Figure 21–8b), which is used for piercing and holding worms and insects.

Myrmecophagous forms (*myrme* = ant, *phago* = to eat) like the armadillo (*Dasypus*), the anteater (*Cyclopes*), and the giant anteater (*Myrmecophaga*) feed on ants and termites in their nests (Figure 21–8c to e). Hence, in numerous unrelated mammalian groups, flat crushing teeth or no teeth at all are to be found, a peculiarity coupled with long, mobile, and worm-like tongues with exceptional protrusibility. Enlarged salivary glands produce a viscous, sticky secretion that coats the tongue. Myrmecophagous mammals also have elongate snouts and digging specializations.

Omnivorous mammals from many orders retain some piercing and ripping cusps in the anterior teeth but have flat broad crushing cusps posteriorly. Diverse mammals such as humans and many monkeys [e.g., marmoset (*Saguinus*)], peccary (*Tayassu*), and bears (*Ursus*) have an omnivorous diet, including plant as well as animal tissue (Figure 21–8f to h). Because of their tough cellulose cell walls, many plant tissues are resistant to efficient digestion. Grinding plant matter between rough surfaces is the usual method of disrupting cell walls. The cheek teeth of these omnivores are **brachyodont** (*brachy* = short, *dont* = tooth) and **bunodont** (*buno* = a hill or mound), meaning that the cusps are rounded. These teeth

process soft animal material that can be ground (not sliced) into digestible pieces and plant material like soft roots, tubers, and berries.

A similar dentition with much enlarged anterior biting teeth occurs in the fruit-eating bat, *Artebius* (Figure 21–8i), which bites chunks from fruit and crushes the pulp for its juices with the broad flat posterior teeth. Another bat, *Choeronycterus* (Figure 21–8j), uses a long tongue to extract nectar from flower and has greatly reduced dentition.

Herbivores of two very different types show similarities in the great relative size of the flat grinding cheek teeth and the development of a gap (a **diastema**) between these cheek teeth and the anterior food-procuring teeth. A diastema moves the cutting apparatus away from the face and allows the snout to penetrate narrow openings or reach close to the ground during feeding. The rabbits (order Lagomorpha; Figure 21–8p) and the rodents (order Rodentia; Figure 21–9q) both have a greatly enlarged pair of incisors in both the upper and the lower jaw. These incisors are used to gnaw through hard plant coverings to reach tender material inside, as well as for nibbling grasses and shrubs, and they continue to grow throughout life. Behind the incisors, a long diastema separates the gnawing apparatus from the plant-crushing cheek teeth. A soft fold of cheek skin is puckered inward across the diastema while gnawing and closes off the mouth from flying particles of bark and wood. The incisors of rodents have enamel only on their anterior surfaces. The enamel is the hardest part of the tooth and wears more slowly than the dentine behind it, producing a self-sharpening chisel edge. Lagomorph incisors are completely encased in enamel (except where it wears off at the tips). In addition, lagomorphs have a second set of small incisors immediately behind the first.

Some ungulates, such as the horse, *Equus* (Fig-

Figure 21–8. Feeding specializations of teeth and skulls of mammals:
(a) hedgehog, (b) mole, (c) armadillo, (d) anteater, (e) giant anteater, (f) marmoset, (g) peccary, (h) bear, (i) fruit-eating bat, (j) nectar-eating bat, (k) raccoon, (l) coyote, (m) mountain lion, (n) the grazing horse, (o) deer, (p) jackrabbit, (q) woodrat, (r) porpoise, (s) right whale, (t) walrus.

ure 21–8n), have canines that are very similar in shape and function to incisors and have migrated forward to join the incisors in a single functional cutting edge. In artiodactyls, such as cows and deer, (Figure 21–8o), the upper incisors are usually absent, as are the upper canines. The lower incisors and incisiform canines bite against a strongly cornified (calloused) palatal plate. This apparatus, in which the very protrusible tongue plays a significant role, enables the rapid plucking of a large quantity of grass, with the tongue and lips pulling blades toward the clipping jaw tips. The food thus rapidly procured is later regurgitated, thoroughly masticated, and then returned to the stomach for the second time and final digestion (Box 21–1).

Carnivores that feed on flesh cut or sheared from the carcasses of prey have developed the fourth upper premolar and first lower molar into a scissors-like pair of shearing blades known as the **carnassial apparatus**. The crushing posterior molars are less important, and the carnassials enlarge as the proportion of meat in the diet increases, as shown by the carnivore series raccoon (*Procyon*), coyote (*Canis*), and mountain lion (*Felis*) (Figure 21–8k to m).

Marine habits have produced some highly specialized forms, such as the fish-trap dentition of the porpoise, *Delphinus*, the toothless plankton-straining right whale, *Eubalaena*, and the mollusk-digging and crushing dentition of the walrus,

Box 21–1. Herbivores, Microbes, and the Ecology of Digestion

Plant food is much more abundant than animal food, and it does not run away, but the energy content of plant material is lower than that of animal tissues. The protein content of leaves and stems is usually low, and the protein is enclosed by a tough carbohydrate cell wall. Although specialized teeth can rupture the leaves and expose the cells, only enzymes can break through the cellulose protecting the cytoplasm. However, no multicellular animal has the ability to synthesize cellulases. Thus efficient use of plants as food requires indirect attack on the cell walls by enzymes produced by symbiotic microorganisms that live in the guts of herbivorous animals. The complex gut morphologies characteristic of herbivorous vertebrates provide fermentation chambers that promote the action of these microorganisms.

The many separate evolutions of fermentative digestion among vertebrates have resulted in distinctly different solutions to the problems posed by plants as food (Janis 1976). Two different fermentative systems have evolved among ungulates. Horses and other perissodactyls are examples of **monogastric**, **caecalid**, or **hindgut** fermenters. These have a simple stomach and an enormous caecum—a closed-end sac at the junction of the small and large intestines (Figure 21–9). Other ungulate hindgut fermenters include rhinoceroses, tapirs, and elephants. Hindgut fermentation is also known among birds, lizards, turtles, and fishes. Even omnivorous and carnivorous mammals such as humans and dogs obtain measurable amounts of nutrients from hindgut fermentation.

Cows are examples of **ruminant** or **foregut** fermenters, in which the nonabsorptive forestomach is divided into three chambers that store and process food, followed by a fourth chamber in which digestion occurs (Figure 21–9). Some other ruminants have only three-chambered stomachs.

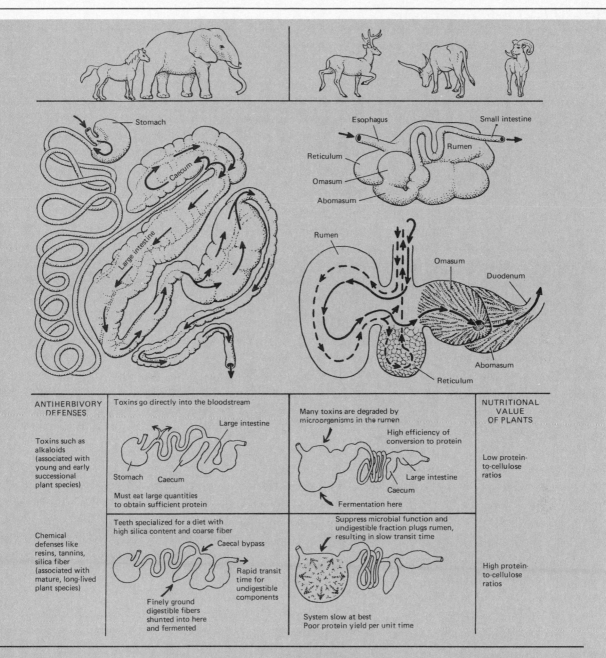

ANTIHERBIVORY DEFENSES			NUTRITIONAL VALUE OF PLANTS
Toxins such as alkaloids (associated with young and early successional plant species)	Toxins go directly into the bloodstream Large intestine Stomach Caecum Must eat large quantities to obtain sufficient protein	Many toxins are degraded by microorganisms in the rumen High efficiency of conversion to protein Large intestine Caecum Fermentation here	Low protein-to-cellulose ratios
Chemical defenses like resins, tannins, silica fiber (associated with mature, long-lived plant species)	Teeth specialized for a diet with high silica content and coarse fiber Caecal bypass Rapid transit time for undigestible components Finely ground digestible fibers shunted into here and fermented	Suppress microbial function and undigestible fraction plugs rumen, resulting in slow transit time System slow at best Poor protein yield per unit time	High protein-to-cellulose ratios

Figure 21–9. Monogastric and digastric digestive systems. The monogastric system (left): Fermentation occurs in the enlarged caecum and large intestine. The digastric (ruminant) system (right): Food passes slowly through the four chambers of the stomach of a ruminant.

Box 21-1. (Continued)

Hindgut fermenters chew their food as they eat and digestion is initiated by enzymes in the saliva and continued when the food reaches the stomach, which is acid. The stomach is not disproportionately large, and the partly digested mass of food moves into the small intestine rather rapidly as new food is eaten. Absorption of nutrients occurs in the small intestine. At the end of the small intestine, finely ground particles of food pass into the caecum and larger food particles move through the large intestine and are passed as feces. In the caecum microorganisms attack the cell walls, releasing nutrients and forming volatile fatty acids. The products of this fermentation and the symbiotic microorganisms pass into the large intestine and some of the material is absorbed across the intestinal wall.

Foregut fermenters (ruminants) swallow partly chewed food into the first chamber of the forestomach, the **rumen**, which is enormously enlarged. Here the food is moistened and churned, mixing thoroughly with the symbiotic microorganisms that live in the rumen. Large particles of food float on top of the rumen fluid and are passed to the **reticulum**, a blind-end sac with honeycomb partitions in its walls. Here small masses (**cuds**) of moist plant material are formed. These cuds are regurgitated when the animal is at rest, chewed again, and swallowed into the rumen. Only after material is finely ground does it pass from the rumen through the **omasum** and into the acid stomach (**abomasum**), where it is processed by the usual digestive enzymes of vertebrates. These enzymes act on the plant material, on the products of fermentation from the rumen, and on the microorganisms themselves, rupturing their cell walls and releasing the protein and carbohydrate they contain.

From the acid stomach the digesta passes into the small intestine, where the products of microbial digestion and acid digestion are both absorbed.

Distinct advantages and disadvantages attach to the two kinds of fermentative digestion. Foregut fermentation can be extremely efficient because the microorganisms attack the plant material before it reaches the small intestine, where most absorption takes place. In a hindgut fermenter, the food has already passed the small intestine before it is mixed with microorganisms in the caecum. Furthermore, the microorganisms from the rumen are digested in the acid stomach, and the material that moves into the intestine contains the protein and carbohydrates the microorganisms have synthesized as well as the material released from the plants by fermentation. A hindgut fermenter does not digest the microorganisms from the caecum, so this potential source of energy and nutrients is not exploited. In addition, the microorganisms in the rumen are able to detoxify many chemical compounds that would be harmful to a vertebrate; a hindgut fermenter receives no such benefit and must absorb allelochemicals from plants into its bloodstream and transport them to its liver for detoxification.

On the other hand, a hindgut fermentation system processes material rapidly. Food moves through the gut of a horse in 30 to 45 hours, compared to 70 to 100 hours for a cow. Hindgut fermentation works well with food that has relatively high concentrations of protein, because a large volume of food can be processed rapidly. The system is not efficient at extracting energy from the food, but by processing a large volume of food rapidly a horse can obtain a large quantity of energy in a short time. In addition, a hindgut system is effective when the food contains a large proportion of indigestible material such as silica, resins, or tannins, because these compounds pass quickly through the gut without entering the caecum. In contrast, a foregut system is slow because food cannot pass out of the rumen until it has been ground into very fine particles. Ruminants do not do well on diets that contain high levels of tannins or resins because these compound suppress microbial function in the rumen, and plants with high silica contents break down so slowly that they impede the movement of food out of the rumen.

These differences in digestive physiology are reflected in the ecology of foregut and hindgut fermenters. Hindgut fermenters can survive on very low quality food such as straw as long as it is available in large quantities.

Consequently, perissodactyls can survive on dry pastures in regions of seasonal drought, whereas ruminants cannot process low-quality food fast enough to subsist.

Some rodents and lagomorphs with well-developed caecal fermentation produce feces from which relatively little of the products of fermentation has been absorbed. These animals regularly eat some of the feces they deposit in their burrows, thereby retrieving the products of microbial fermentation as well as the protein and carbohydrates synthesized by the microbes (**coprophagy**). Lagomorphs produce special moist fecal pellets at night which they subsequently eat. Young koalas feed for an extended period on the feces of their mother. The end result of all gut fermentation is the gleaning of more protein and energy from plant material, especially from grasses, than simple direct digestion would produce.

Odobenus (Figure 21–8r to t). Porpoises are toothed whales (order Cetacea, suborder Odontoceti). Most toothed whales feed on fishes and their teeth consist of a long series of nearly identical sharp cones (Figure 21–8r). This dental pattern is convergent with the fish-trap type of dentition already described for ichthyosaurs, crocodilians, and gars. A second type of specialized dentition of cetaceans is represented by the baleen or whalebone whales (suborder Mysticeti) of which the right whale is an example (Figure 21–8s). The teeth of mysticetes have been replaced by sheets of a fibrous, stiff, horn-like epidermal derivative known as baleen which extend downward from the upper jaw. Baleen whales are filter feeders, straining small organisms known collectively as plankton from the water by use of their baleen sieves. The ten species of baleen whales include the largest mammal (and the largest vertebrate), the blue whale, which may reach a mass of 160,000 kilograms. Walruses use their canines to dislodge the mollusks on which they feed.

Types of Mammalian Teeth

Mammalian teeth are structures with well-defined growth characteristics. Because the texture and quality of mammalian foods differ enormously, cheek teeth are variously specialized. Generalized mammalian dentitions have brachyodont cheek teeth with rectangular crowns that do not protrude much above the gums. The occlusal (grinding) surface is not smooth, but is marked by one to several sharp cusps that interlock (occlude) when the jaws close. This cusp pattern is referred to as **tubercular** (*tuber* = a knob or hump) and permits both piercing and crushing of the food.

Most teeth have a discrete period of growth during ontogeny, and growth is not resumed if the tooth is subsequently worn or broken. An immature incisor tooth has an open pulp cavity, indicating that it is growing (Figure 21–10a), whereas a mature incisor has a narrow, canal-like pulp cavity that restricts nutrient supply and growth (Figure 21–10b). The incisor teeth of rodents have a persistently open pulp cavity, grow throughout life, and have enamel only on the outer face (Figure 21–10c). A typical molar tooth has multiple roots (Figure 21–10d). The low, rounded cusps identify the tooth as **bunodont**. If the cusps of the cheek teeth fuse into ridges, useful in grinding plant material, a **lophodont** tooth is formed (Figure 21–10e). Two further specializations for herbivory are evident in the dentition of the horse: The crowns stand high above the gums (the **hypsodont** condition, *hyps* = high), and the occlusal surfaces of the cheek teeth show complex ridges composed of dentine adjacent to ridges of enamel and areas of exposed cement. As a hypsodont tooth wears, the various hardnesses of tooth material wear differentially to maintain an uneven grinding surface

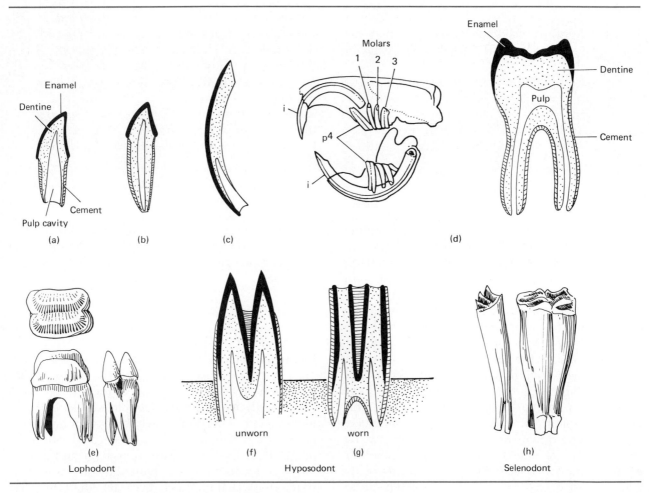

Figure 21–10. Mammalian teeth. See text for details.

(Figure 21–10f and g). Similar specializations can be seen in the dentition of cattle (order Artiodactyla). In artiodactyls, however, the ridges are all derived from longitudinal cusps and are termed **selenodont** (*selen* = a crescent moon; Figure 21–10h). As these complex teeth wear down, the difference in hardness of the cement, enamel, and dentine produces a rough, self-sharpening surface.

Carnivores must be able to shear or cut away chunks of flesh from a large carcass. Once obtained, such pieces are readily digested without much mastication. Hence, the occlusal surfaces of carnivore dentition are sharp and knife-like and present little grinding surface. The teeth of advanced carnivores are called **sectorial** (*sect* = a knife). In living carnivores sectorial teeth reach their height of development in the shearing **carnassial** apparatus of the fourth upper premolar and the first lower molar, which slide past one another like scissor blades to cut flesh.

Figure 21–11 shows the canine, premolar, and molar teeth of a series of species of Carnivora that have very different diets: (a) bear, *Ursus*; (b) dog,

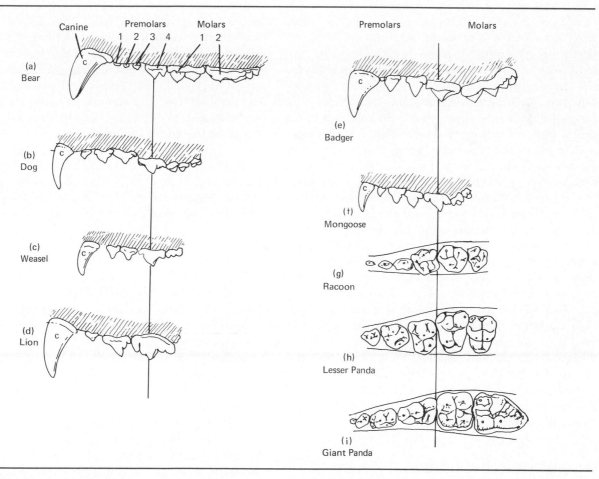

Figure 21–11. Partial dentition of the upper left jaw of various members of the order Carnivora. (After B. Peyer, 1968, *Comparative Odontology.* University of Chicago Press, Chicago.)

Canis; (c) weasel, *Mustela*; (d) lion, *Panthera*; (e) badger, *Meles*; (f) mongoose, *Herpestes*; (g) raccoon, *Procyon*; (h) lesser panda, *Ailurus*; (i) giant panda, *Ailuropoda*. The vertical lines locate the main cusp of the premolar 4, which forms the carnassial apparatus when it is present. A series grading from omnivore to advanced carnivore (a to d) shows how the crushing, bunodont molars are reduced and eliminated and the cusps of the fourth premolar aligned and fused to form a blade. Crushing mastication is of so little importance to cats, which are exclusively flesh-eaters, that only a vestigial first molar occurs in the upper jaw. The carnassial apparatus is the major part of the cheek dentition of cats. Canids have more diverse diets than cats and a correspondingly less specialized dentition. The anterior cheek teeth of dogs are sectorial, and the carnassial apparatus occurs between the upper fourth premolar and the lower first molar, but well-developed molars with tubercular cusps follow these slicing teeth and allow dogs to crush bones for their marrow. This explains why a dog often holds a bone between its paws and gnaws with the bone far back in the corner of its mouth. When meat is on the bone, the dog does not place the bone as far back in its mouth and uses the carnassials to slice off the meat.

Similar dentitions may be developed independently using different elements. The posterior crushing elements of the badger are the result of enlargement of the first molar, whereas in the mongoose both molars 1 and 2 are retained.

Not all carnivores eat meat, and some are specialized herbivores (Box 22–1). This change in diet has been accompanied by changes in dentition: Cheek teeth seen in occlusal view show that the premolars have been molarized and the molars have increased in size in the omnivorous raccoon, the herbivorous lesser panda, and the bamboo-eating giant panda.

Nontrophic Functions of Teeth

Many wild swine and some domestic breeds have greatly enlarged and often recurved upper and lower canines. They are used for rooting and digging in the soil for the nutritious storage roots and tubers of plants. In addition, the tusks are larger in the males of many species than in the females. In some species these tusks have lost their digging function and must be considered sexually dimorphic characters, probably connected with aggressive and/or sexual display.

Tusks in other mammals are also modified teeth. The canines of the walrus (order Pinnipedia) are massive (Figure 21–8t). Walruses are marine and crawl out of the sea onto drifting ice or onto the shore using their tusks as levers to lift their massive bodies from the water. Major functions of the tusks may be in social communication as well as gathering mollusks from the seabed. Other well-known tusks are those of elephants. Here the incisors of the upper jaw form the tusk, which is pure dentine (ivory). Both sexes of the African elephant have tusks, but female Indian elephants are usually tuskless. The elephant's tusks are used in defense and to hold down branches pulled into reach by the trunk for browsing.

In many primates, the canines are enlarged and are larger in males than in females. By rolling back the very flexible lips, these primates present a fierce aggressive display. It has been suggested that the large roots of human canines are a remnant of teeth that once functioned in these displays.

Mammalian Mastication

Teeth, of course, are only part of a mammal's adaptation to a specialized diet. Other skeleto-muscular specializations allow them to utilize teeth effectively in mastication. Many variations in mammalian mastication can be resolved by the study of five pairs of muscles, their sites of origin and insertion, and the articulation between the mandible and the skull (Figure 21–12a). Depression of the lower jaw requires little muscular force because gravity does most of the work. The paired **digastric** muscles (having two fleshy parts; *di* = two, *gaster* = belly) originate in the ear region and insert on the inner ventral borders of each side of the mandible. Each digastric passes posterior to

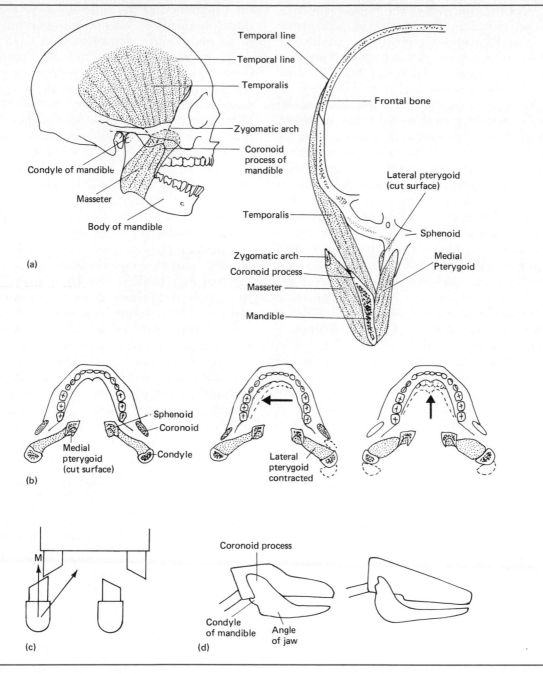

Figure 21–12. Masticatory muscles and jaw movements. See text for details.

the angle of the jaw, where the muscle constricts to a ligamentous neck. Contraction of the digastric appears to aid forceful opening of the mouth.

Mastication is a complex activity that involves several pairs of muscles (Figure 21–12a). The **temporalis** originates on the skull roof and inserts on the coronoid process of the mandible. In addition, another bilaterally paired muscle, the **masseter**, originates on the zygomatic arch and inserts on the posterior half of the lateral surface of the mandible. The fibers run obliquely from the arch posteriorly and down to the mandible. A final set of muscles of mastication is hidden from superficial view but is extremely important in adding new dimensions to the possibilities for mastication. These **pterygoideus** muscles originate on the base of the skull posterior to the palate; the fibers run laterally and obliquely to insert on the medial surface of the angle of the mandible (Figure 21–12b). Contraction of the muscles of one side pulls the mandible toward the opposite side of the head. Acting together, they protrude the lower jaw. The pterygoid complex allows the lower jaw to be moved in a grinding rotary path relative to the upper jaw or to apply force on one side of the jaw only.

The shape of the lower jaw (mandible) and of the mandibular condyle and its fossa, as well as the orientation of the fibers in these masticatory muscles, clearly reflect specialization for different functions. Some mammalian jaws have the condylar processes in or very near the occlusal plane (Figure 21–13). Such jaws close like scissors when the strongly developed temporalis and masseter muscles contract. The posterior teeth come into contact before the anterior ones. Other mammals have condylar processes high above the occlusal plane and the jaws close in a nearly parallel fashion, all the occlusal surfaces approaching each other at about the same time. The former orientation is ideal for slicing and cutting, the latter for crushing, and with the addition of lateral motion, permits grinding along the whole of the tooth row.

Because carnivores generally attack, hold, and kill with well-developed canines, it is not surprising that they usually have large temporalis muscles (Table 21–3) attached to a prominent coronoid process. The mandibular condyles of carnivores are cylindrical bars set deeply in fossae that allow no anterior–posterior movement and little lateral shifting of the lower jaws. These features help prevent dislocation of the jaw by struggling prey. In contrast, both the temporalis muscles and the coronoid process are reduced in highly specialized herbivores, but the ramus of the jaw is very deep,

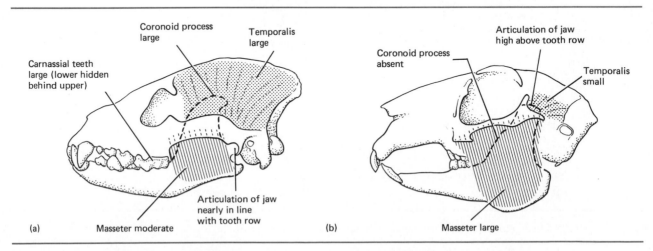

Figure 21–13.   The shape of the jaws is influential in producing the differing jaw actions of carnivore (a) and herbivore (b).

Table 21–3. Weights of the jaw-closing muscles of some mammals (weight of each muscle as a percentage of the total).

	Temporalis	Masseter	Pterygoideus
Carnivores			
Tiger	48	45	7
Bear	64	30	7
Dog	67	23	10
Herbivores			
Zebra	10	50	40
European bison	10	60	30
Horse	11	57.5	31.5

Source: R. McN. Alexander, 1968, *Animal Mechanics*, University of Washington Press, Seattle, Wash.

to accommodate insertion of the large masseter muscles needed to apply force to the cheek teeth during mastication. The mandibular condyles of herbivores are of various shapes and orientations, permitting anterior–posterior and lateral movement. The human jaw, condyles, and musculature are intermediate between these extremes.

Coevolution of Ungulates and Grasses

The impact that grazing herbivores can have on vegetation is well known. Plants have reacted to those attacks in a multitude of ways, including the evolution of chemical and mechanical defenses. Allelochemicals are compounds that are distasteful or toxic to herbivores that are deposited by plants in otherwise edible tissues. Mechanical defenses include thorns, spines, and structures in the cell walls.

The mechanical defenses of the grasses (family Graminae) provide an excellent illustration of plant–animal coevolution. The 700 genera and perhaps 9000 species of grasses are worldwide in occurrence and are considered the zenith of flowering plant evolution. The first undoubted fossil grasses are from mid-Tertiary beds. It is clear that a number of features of grasses evolved simulta-neously with the evolution and radiation of herbivorous mammals, and many descendents of this herbivore radiation today depend on grasses.

Reconstruction of primitive characters of living Graminae indicate that the first grasses were low-growing tufts. Instead of growing from the tip of the stem as other plants do, grasses grow from the base of the leaf blade. A grass, therefore, elevates the oldest, nongrowing part of its foliage, and this can be removed without stopping growth. Most grazing mammals cannot graze close enough to the ground to remove all the leaf tissue, yet from the earliest radiations of herbivores the muzzle has shown elongation permitting deeper and deeper cropping of plant tufts (Figure 21–14). Grasses have repeatedly evolved nearly stemless growth forms with growing portions very close to the ground, out of reach of even the most specialized grazers.

Two chemical defenses of grasses have been effective against herbivores and have shaped the evolution of herbivorous mammals since the Miocene. Plant cells are encased in a rigid cell wall. To obtain the cell protoplasm, it is necessary for an herbivore to rupture this cell wall and expose the cell contents. Grasses have made this as unrewarding as possible by incorporating crystalline silica into the fibrous cell wall. Silica is a hard mineral and grinds teeth to useless, flat nubbins.

Cellulose and lignin are common constituents of all plant cell walls. Cellulose is a complex carbohydrate closely related to starch but distinctive in its indigestibility by vertebrates. Grasses often have greatly elongated cells and thickened cell walls, making the amount of directly digestible foodstuffs a small proportion of the total leaf. It seems improbable that vertebrates should evolve complex specializations to enable them to feed on plants with such tooth-destroying texture and low nutritional value, yet the orders Artiodactyla, Perissodactyla, Sirenia, Hyracoidea, Proboscidea, Rodentia, Lagomorpha, Primates, and Marsupialia contain many species that are partially or wholly adapted to grasses for their food. Why should this be so?

The whole of the Mesozoic was warm by

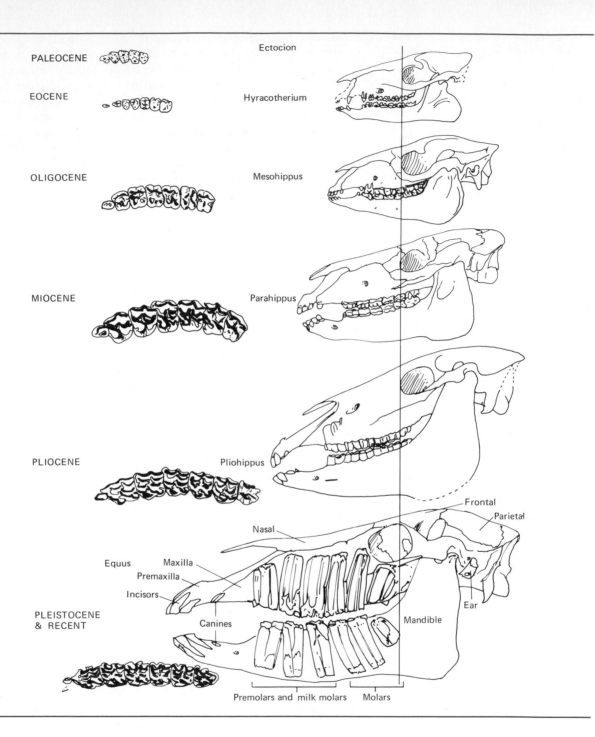

PALEOCENE

Ectocion

EOCENE

Hyracotherium

OLIGOCENE

Mesohippus

MIOCENE

Parahippus

PLIOCENE

Pliohippus

Frontal

Parietal

Nasal

Equus

Maxilla

Premaxilla

Incisors

Canines

Ear

Mandible

PLEISTOCENE
& RECENT

Premolars and milk molars   Molars

present climatic standards, and the climate became increasingly moist as the period progressed. During the Cretaceous ferns and cycads similar to those now living in subtropical South America were distributed from the paleo-equator to nearly 90° N and S latitude. As described in Chapters 14 and 20, the late Cretaceous was a period of gradual withdrawal of the shallow seas from much of the land and a decided cooling and drying of the Earth's climate. Except for a brief period of warming between the Paleocene and Eocene, this deterioration of climate and increase in seasonality has continued for 70 million years. These changes restricted the geographic ranges of the nonflowering plants of the Cretaceous, leaving many as relict populations in protected habitats and forcing others to extinction. The apparently more adaptable angiosperms, present since the Jurassic, radiated and were followed in short order by the radiation of mammals. By the Miocene, even angiosperm forests had been somewhat diminished, and grasslands dominated the worldwide terrestrial flora. Survival of some phylogenetic lineages could have depended on grasses as the predominant food source despite the relative difficulty in eating and absorbing them.

Specializations of the Digestive System of Herbivorous Mammals

Grazing mammals have teeth with large, rough occlusal surfaces that can rupture the thick cell walls of leaves and grasses. Additional specializations include an overall increase in the size of the teeth, increased molarization of the premolars, and development of complex folds of enamel, dentine, and cement that result in all three substances being exposed on the occlusal surface simultaneously. To deal with the exceedingly abrasive silica, the cheek teeth became elongate, permitting them much longer service before they were worn away. Ultimately, the cheek teeth of various lineages of herbivorous mammals evolved either of two specializations: In some herbivores, tooth eruption patterns allow a basically diphyodont mammal to be functionally polyphyodont (having several successive replacement teeth). Examples are the serially replaced teeth of proboscideans and a similar arrangement in some sirenians. In these mammals only a few cheek teeth are exposed above the gums at a time. As these teeth wear down they migrate forward along the jaws and finally are shed. New teeth erupt posteriorly to replace them. Other herbivores (many perissodactyls, artiodactyls, lagomorphs, and rodents) have continuously growing cheek teeth that renew themselves from below as they are worn away from above.

Several clades of mammals have independently evolved a complex forestomach that allows symbiotic microorganisms to convert the cellulose and lignin of plant cell walls into digestible nutrients. This is the **ruminant** digestive system, and the animals that possess it (camels, giraffes, antelope, cattle, sheep, goats, and deer) are called rum-

---

Figure 21–14. Development of hypsodont dentition in horse-like perissodactyls. Note that this is not a phylogenetic lineage, but it illustrates changes that took place in a number of lineages that became specialized to feed on abrasive vegetation during the Miocene and Pliocene. The molars became broader and flatter, increasing their total surface area, and the premolars became identical in form to the molars. The pattern of enamel, dentine, and cement on the occlusal surfaces of the teeth became increasingly complex, producing a self-sharpening grinding surface. The teeth became increasingly hyposodont with long-persisting open roots that allowed continued growth to replace wear. The skull of the modern horse *Equus* is dissected around the base of the teeth to show the extreme hyposodonty. (Modified from B. Peyer, 1968, *Comparative Odontology,* University of Chicago Press, Chicago.)

inants or cud-chewing mammals. Similar specializations are found in a wide variety of other animals, including kangaroos, koalas, colobine monkeys, rodents, lagomorphs, perissodactyls, and many nonruminant artiodactyls. In these forms large sac-like portions of the gut, either the forestomach or the cecum (at the junction of the small intestine and colon), provide storage space for large quantities of masticated herbage where gut microbial symbionts can ferment plant compounds to digestible end products (Box 21–1).

Cursorial Specializations of Ungulates and Their Predators

Grasses flourish in open habitats that place a premium on efficient, rapid locomotion over generally flat terrain. With nowhere to hide, most large grazing animals run to escape predators, and many species migrate long distances to find grass at the proper stage of growth. Such locomotion is termed **cursorial** and is characteristic of most large herbivores, represented today primarily by the ungulates (orders Perissodactyla and Artiodactyla).

The primitive foot posture of mammals is the **plantigrade** type (Figure 21–15), in which the sole of the foot rests flat on the ground and the entire foot skeleton supports the weight of the body. In the case of the hind foot illustrated, support extends from the calcaneus to the terminal phalanges. Several mammals, especially stealthy or moderately cursorial (running) carnivores, support their weight only on the phalanges. This **digitigrade** posture reduces friction and increases the length or reach of the stride by an amount equal to the length of the vertical metapodial and the mesopodial elements.

Efficiency in cursorial locomotion is promoted by a long stride length, and speed is further enhanced by increasing the number of strides per unit of time. The ultimate in cursorial adaptations is the **unguligrade** posture achieved independently by numerous primarily herbivorous clades of mammals. Here the weight of the body is supported entirely by the hoof-clad terminal phalanx or phalanges, and the effective stride length includes the contribution of the second and third phalanges. Increased efficiency is fostered by reducing the distal mass, and therefore the inertia of the limbs, by concentrating muscles in the upper limbs and using tendons to transmit muscular forces to the slim lower limbs. The muscles of the upper limb insert very close to the joint over which they act, increasing their effective speed at the cost of reduced force. Elastic ligaments running across the joints of the limbs are stretched when those joints are bent by the weight of the body and they snap back when the weight is released, straightening the joint and returning much of the energy stored in them. Behavioral characteristics also aid long-distance travel. The stride length of the gait used in steady travel coincides with the natural oscillation period of a pendulum with dimensions

Figure 21–15. Contrasting specializations of the limbs of mammals. (a, left to right) **Plantigrade**, **digitigrade**, and **unguligrade** postures. (b) The skeletal elements of the left forelimb of a deer (cursorial) and the armadillo (fossorial) compared. In many ways, specializations of fossorial (digging) limbs are the opposite of those for efficient cursorial limbs. The length of reach when digging (equivalent to length of stride) is usually unimportant so that the limbs are generally short. Power, however, is very important. Thus the muscles and their bony attachments are enlarged and generally insert far from the joint over which they act, enhancing their leverage. The distal mass of the limb is often increased by broadening and flattening the bony elements. Code: C = calcaneus, MS = mesopodial elements, MT = metapodial elements. (From various sources, including M. Hildebrand, 1982, *Analysis of Vertebrate Structure*, 2nd Edition, Wiley, New York.)

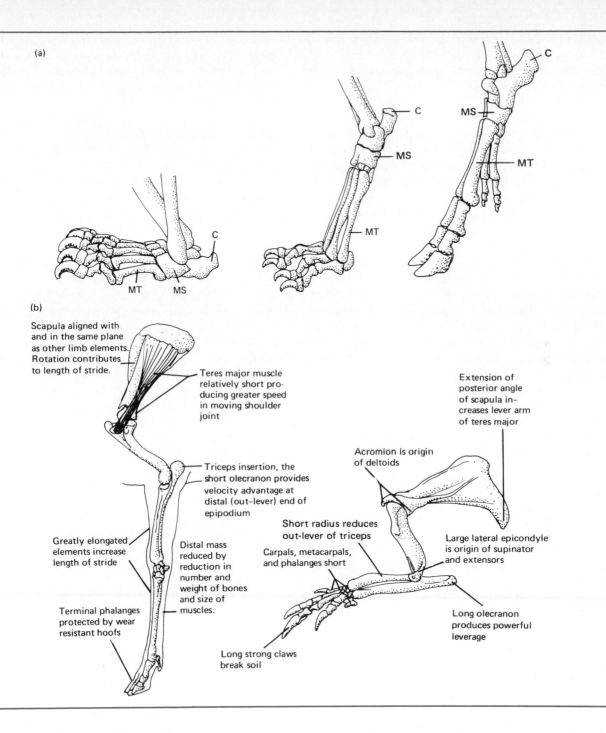

(a)

C

MS

MT

MT   MS

C

MS

MT

C

(b)

Scapula aligned with
and in the same plane
as other limb elements.
Rotation contributes
to length of stride.

Teres major muscle
relatively short pro-
ducing greater speed
in moving shoulder
joint

Extension of
posterior angle
of scapula in-
creases lever arm
of teres major

Acromion is origin
of deltoids

Triceps insertion, the
short olecranon provides
velocity advantage at
distal (out-lever) end of
epipodium

Short radius reduces
out-lever of triceps

Greatly elongated
elements increase
length of stride

Distal mass
reduced by
reduction in
number and
weight of bones
and size of
muscles.

Carpals, metacarpals,
and phalanges short

Large lateral epicondyle
is origin of supinator
and extensors

Terminal phalanges
protected by wear
resistant hoofs

Long strong claws
break soil

Long olecranon
produces powerful
leverage

similar to those of the limb. Thus, little energy is spent in reversing the motion of the limb at the extremes of its excursion—gravity and elastic recoil do much of the work.

Such extensive locomotor specializations have obviously had an effect on the carnivores that prey on grassland herbivores. Three families of the order Carnivora have members that are specialized to prey on large grazing animals: Hyaenidae (hyaenas), Canidae (dogs and wolves), and Felidae (cats). Both hyaenas and dogs hunt in packs and use stamina and cooperation to exhaust and kill large prey. Many of the specializations for speed and efficiency seen in ungulates have analogs in cursorial carnivores, especially the canids. Few felids run down their prey. Cats stalk by vision and, when close to their prey (usually less than 100 meters), they attack with a brief burst of speed followed by a spring. Only the lion hunts in groups, and its social system is accordingly complex. The cheetah is very different from other cats and is distinct from all other mammals in being a solitary, cursorial predator. Although it rarely runs for distances greater than 600 meters, it is the fastest terrestrial animal, having been clocked at 112 kilometers per hour.

Evolution of the Mammalian Nervous and Sensory Systems

Throughout mammalian evolution the central nervous system and its sensory apparatus have shown progressive enlargement and increased complexity. Unquestionably, this expansion has been independent in each of several orders of mammals involved. Nevertheless, the detailed similarities in the morphological and behavioral end products are intriguing. To unravel these parallel changes, we must compare which parts of the brain have enlarged during the last 200 million years and which sensory modalities are most characteristic of mammals.

The basic structure of vertebrate nervous systems and the regions of the generalized vertebrate

brain, all of which are retained in mammals, have been reviewed. Two characteristics of the mammalian brain that set it apart from other vertebrates are (1) the evolution of a superficial mat of gray matter from the primitive condition in which the nuclei were buried within the white matter, and (2) the specialization of that coat of gray matter in the telencephalon of mammals to form the **neopallium**.

Thus the mammalian brain is enlarged primarily by the evolution of extensive layers of neuron cell bodies on the surface of a more primitive grain stem. These new layers are especially thick over the telencephalon of the anteriormost part of the brain. This trend is evident from endocasts of Mesozoic mammals. These small nocturnal creatures evolved from lineages whose nervous systems have left convincing evidence of diurnal activity with a dependence on bright light vision (Jerison 1973). Which of the five classic senses to which the early mammals were heir—sight, smell, touch, taste, and hearing—might best have replaced the bright light vision so important as a distance sense to the synapsid and therapsid ancestors of mammals?

The Distance Senses of Early Mammals

One might first suppose that evolution would have probably reworked the visual system given the extensive correlates of vision, brain, and behavior already inherent as distance sense in the mammalian genotype. However, there is one major limitation to the information-rich visual system that mammals inherited. Cones have relatively low photosensitivity; they are at least two orders of magnitude less sensitive than rods. An eye adapted for acute vision required the brightness of daylight to function effectively, but early mammals are believed to have been nocturnal (Chapter 19). Thus nocturnal mammal-like tetrapods would have required different visual systems, or even a different sensory modality than their synapsid ancestors to obtain and process information from a distance.

Vision   The most obvious adaptations of a primitive mammal to the new sensory demands of a life at night would be those of a nocturnal eye, that is, one with a pure rod retina. Living members of the order Insectivora seem to fit this hypothesis, and many otherwise specialized mammals also have a predominantly nocturnal eye. However, the quality of image formed by the eyes of living nocturnal vertebrates is poor. If these eyes are representative of those of early mammals, the usefulness of vision as a distance sense would not have been great.

Acute vision does occur in several mammalian orders, and it is clear that many independent alterations of the primitive mammalian eye have occurred. The overwhelming majority of mammalian orders appears to have retinas that are dominated by rods. Primates are the only mammals with well-developed trichromatic color discrimination comparable to that found in other vertebrate classes. This kind of color vision is produced by three visual pigments that have different absorption spectra and, consequently, are sensitive to different wavelengths of light. A trichromatic retina is not the only way to see color, however. Certain diurnal mammals like squirrels appear to have good dichromatic color vision, which is based on cones with two visual pigments. The cones of mammals show such distinct differences from the cones of other vertebrates that they probably are not homologous structures. Mammalian cones probably evolved from pure rod retinas during the first major post-archosaurian radiation of mammalian types. Only recently, during the Cenozoic radiations of diurnal mammals, have retinal systems similar to those of diapsids evolved in some mammals. The vast majority of mammals have rather poor visual acuity. Elephants cannot distinguish two objects separated by less than 10 minutes and 15 seconds of arc on the retina. Chimpanzees, however, can distinguish points separated by only 26 seconds of arc. The evidence indicates that early mammals probably had a visual system with low acuity that was not appropriate for a primary distance sense.

Touch   Mechanoreception of the immediate environment surrounding it may well have been very important to early mammals. Pits in the rostral region of fossil skulls suggest that some synapsids might have had vibrissae. In living mammals vibrissae detect obstacles near the face. However, the range of sensitivity of mammalian touch is restricted, and touch also seems a poor candidate for a distance sense.

Taste and Smell   Taste can be discounted as contributing significantly to the expansion of the mammalian brain. It probably was, however, the major sensory modality involved in the coevolution of chemical defenses of plants and vertebrate herbivores. Nevertheless it requires direct application of the taste buds to the environment and this greatly curtails its value as a distance sense.

Olfaction, on the other hand, detects substances borne by air currents. Thus, unlike touch and taste stimuli, olfactory stimuli can originate at a distance. Olfactory stimuli are not blocked by most natural obstacles, and olfaction functions either in daylight or at night. A very small amount of stimulant is often detectable. Endocasts of early mammal brains indicate the presence of enlarged olfactory regions. However, olfactory information is not transmitted rapidly, and the direction that scent particles travel depends on the vagaries of air currents. An additional problem with olfaction as a primary distance sense is its poor resolution of spatial information about the source. Although olfaction has great significance to vertebrates, especially to mammals, it is a rather poor candidate for a rapid nocturnal distance sense to replace diurnal vision.

Hearing   Audition has several advantages as a distance sense compared to vision and olfaction. Sound is not readily blocked by obstacles in the environment as is light and it is transmitted more directionally and much faster than odors. The mammalian ear shows the importance of audition. The phylogenetically oldest portion of the auditory system (the inner ear) is derived and elaborate as are the newer structures (the tympanic cavity and middle ear). The outer portion (the pinna and auditory tube) is unique to mammals. The combined effect of these specialized structures is exceptional frequency discrimination, broad sensitivity to various intensities of sound, and precision of directionality. Cranial casts of early mammals show that auditory areas of the brain were large. The advantages of hearing as a distance sense are emphasized by those mammals that have specialized in this mode of probing the world about them.

The Evolution of Mammalian Brain Size

Why did the earliest mammals have brain volumes four or five times those of other tetrapods and why do living mammals have brain volumes four to five times those of the earliest mammals (Figure 21–16)? Examination of the specializations of mam-

mals and their sensory structures in particular suggests an explanation. The resulting hypothesis, although entirely speculative and admittedly oversimplified, seems plausible.

Active organisms require a three-dimensional method of perception of the world around them. In diapsids, which are predominantly diurnal, vision provides this information. We have pointed out, however, that vision would not have been an adequate sense for early mammals, which were nocturnal. Smell and hearing appear to have been more likely senses for obtaining information about the world, and casts made of the inside of brain cases of early mammals indicate that the regions associated with those sensory modalities were indeed enlarged. Three-dimensional information about the world must be extracted from scent or sound largely by comparing the strength of a stimulus from one moment to the next, or as the head is turned from one side to the other. This means having the ability to monitor incoming information, store it, and then to compare it with subsequent information. Evaluations about positions in three-dimensional space are based on the

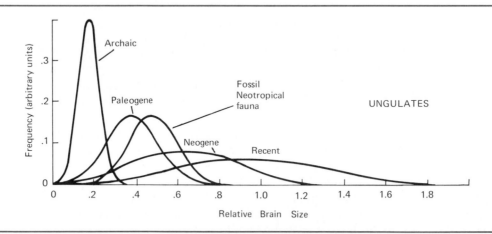

Figure 21–16. Changes in the relative brain size of ungulates of the Northern Hemisphere and South America during the Cenozoic. The Paleogene includes species from the Paleocene through the Oligocene epochs, the Neogene species from the Miocene through the Pleistocene epochs. (After H. J. Jerison, 1973, *Evolution of the Brain and Intelligence*, Academic Press, New York. Also see L. B. Radinsky, 1978, *American Naturalist*, 112:815-831.)

differences in signals received through time. Thus, processing the information about its world would have required greatly enhanced integrative ability in the central nervous system (CNS) of an early mammal.

This integrative ability probably had strong selective value in its own right. It should permit more complex and particularly more flexible behavior patterns. The ability to make associations between past events and a present situation is the basis of learning. Selection for this sort of integrative ability may be the force that produced a second four- to fivefold increase in mammalian brain size that occurred during the Cenozoic. All archaic mammals tended to have brains smaller than more recent mammals. As mammals evolved, the average brain size increased as did the range of relative brain sizes. Increase in brain size may reflect the coevolution of prey and predator and the increasing complexity of defensive and offensive behaviors. The increase in the breadth of variation in brain sizes probably reflects differentiation into some niches which placed less selective value on enlarged neural capacity. Such niches might include evolution of formidable size in ungulates or a scavenging habit in carnivores. Interestingly the unique mammals that evolved on the isolated South American landmass did not develop progressively larger brains. They have all become extinct since the late Cenozoic and the immigration of the northern fauna.

Integrative and associative capacity appears to be carried to the extreme in humans, although it would be rash to underestimate the complexity of the neural capacities of other mammals. Recent successes in communicating with chimpanzees and gorillas (Patterson and Linden 1981, Premack and Premack 1983) and studies of the many lines of evidence for cognitive behavior by toothed cetaceans (Schusterman et al. 1986) clearly demonstrate that many of our integrative and associative capabilities are not uniquely human. The features that we proudly consider attributes of human intelligence may be approached more closely by other vertebrates than we realize because their features, like ours, evolved through natural selective

processes that probably affected all mammals. Nobel laureate Konrad Lorenz (1977) said: "I . . . consider human understanding in the same way as any other phylogenetically evolved function which serves the purposes of survival, that is, as a function of a natural physical system interacting with a physical external world." In Chapter 24 we examine more closely the precise physical external world of the ancestors of modern humans to gain insight into the evolution of our species.

Specialization of the Auditory System: Echolocation

Many mammals are more sensitive than humans to one or another sensory modality: The olfactory sensitivity of dogs is probably the most familiar example of the sensory capacity of a mammal that exceeds our own. Equally impressive is the use that bats and cetaceans have made of hearing as a distance sense used for navigation. An examination of echolocation illustrates the way in which a sensory capacity that is ancestral for mammals can be elaborated, and how environment and phylogeny interact.

Several modern mammals emit sounds above 20 kilohertz (20,000 cycles per second) in frequency and thus above the range of normal human sensitivity (the 10 octaves between 0.002 and 20 kilohertz). Infant rodents emit such sounds, which seem to elicit retrieval behavior by adults. Some adult rodents, a few specialized marsupials, dermopterans, pinnipeds, many insectivores, microchiropteran bats, and odontocete cetaceans emit such sounds. The most thoroughly studied echolocating mammals are the microchiropteran bats and the toothed cetaceans.

Bat Echolocation Because of their nocturnal and secretive habits, bats have until recently been little understood. How do bats avoid obstacles under conditions where vision can be of little or no use? Between 1793 and 1799, Lazzaro Spallanzani, an Italian naturalist and physiologist, elegantly tested the auditory contribution to obstacle avoidance of bats. "I had two slender conical tubes soldered from very thin brass plate and introduced them with their thinner end in front into the ears and auditory openings of a bat. The tubes were externally covered with pitch, which served to fill up the space between tube and the deep concavity of the external ear and to attach the tube to the ear. In this way the air had no passage to the internal ear other than through the tubes. The animal (which could see) showed no influence of this impediment during its flight. But when I closed the tubes with pitch so that the air could no longer enter the auditory duct, the animal did not fly at all, or its flights were short and uncertain, and it frequently fell. This experiment which is so decisively in favor of hearing has been repeated with equal results both in blinded bats and in seeing ones.''

Spallanzani also captured bats from the belfry of the cathedral at Pavia, removed their eyes, released them, and 4 days later captured some of these bats at the same roost. These blinded bats were as full of insects as any sighted bat, indicating that blinded bats could use some sense other than vision to capture insects on the wing and return to a home roost. Spallanzani thought that they heard the buzz of the insects' wings and that obstacle avoidance was based on a bat's ability to hear the sounds of its wings and body as they were reflected from objects. At the time of Spallanzani's studies there was a general ignorance of the properties of sound; that sound existed below and above the range of human audition was alien to scientific thought. Unfortunately this surprisingly modern, experimentally tested hypothesis was eclipsed by the scientific politics of the day. The eminent and influential French anatomist Baron Georges Cuvier derided both the experiments and conclusions of the Italian without providing any new experimental evidence.

At the beginning of the twentieth century, several researchers, often unaware of Spallanzani's work, independently concluded from experiments that ears were important for bat navigation. In 1938, G. W. Pierce and D. R. Griffin, with newly developed high-frequency acoustic detectors, reported that bats emit intense ultrasonic sound syn-chronously with a spitting motion. In the following years Griffin and R. Galambos at Harvard and S. Dijkgraff at Utrecht independently discovered that flying bats produce ultrasonic cries that echo back from obstacles and that bats have ears that are sensitive to these sound frequencies. They also rediscovered Spallanzani and, much to their credit, brought his 150-year-old work to the attention of the twentieth century (Dijkgraff 1960). Subsequently, Griffin and Galambos showed that basically the same echolocation behavior is used by bats for navigation and for capturing food (Griffin et al. 1960).

The weight gained by bats that had been allowed to feed for a known period of time in a swarm of tiny insects showed that they must average as many as 10 mosquito or 14 fruit fly captures per minute. High-speed photography has shown two separate catches in 0.5 second. By use of electronic transducers capable of detection of high-frequency sound, a three-phase hunting pattern has been defined (Figure 21–17).

The initial search phase of the little brown bat (*Myotis lucifugus*) is characterized by fairly straight flight and the emission of ten or so pulsed sounds separated by silent periods of more than 50 milliseconds. Each of the 10 pulses in a call is about 2 milliseconds in duration, and each pulse constitutes a downward sweep of frequencies starting at about 85 kilohertz and ending near 35 kilohertz. These bat calls are therefore frequency modulated (FM). Other bats use different pulse lengths and frequencies that vary from family to family of bats; some produce constant frequency (CF) calls that either terminate in a short downward FM sweep or, like the calls of some FM species, include several simultaneous harmonics. Despite our inability to hear their high-frequency cries, many bats produce sounds of extraordinary loudness. Sound intensities higher than 200 dynes per square centimeter (as loud as a jet airplane) have been measured 5 centimeters from the mouths of some bats.

The second phase begins as a bat detects an insect. Fruit flies and mosquitoes can be detected from a distance of about a meter. The interval be-

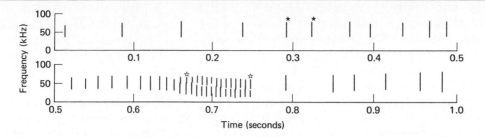

Figure 21–17. Bat echolocation sounds. Sound spectrograph analysis of the frequency-modulated (FM) pulses emitted by the little brown bat, *Myotis lucifugus*, during an interception maneuver. Frequency in kilohertz is plotted against time during the continuous one second record. Filled stars indicate typical loud pulses near the time of detection of the target; open stars indicate the terminal part of the buzz. (After M. S. Gordon et al. 1982, *Animal Physiology, Principles and Adaptations*, 4th edition, Macmillan, New York.)

tween pulses shortens, the silent intervals falling to less than 10 milliseconds in *M. lucifugus*; 100 cries per second, each lasting only 0.5 to 1 millisecond, are typical as the bat alters its flight path to intercept its prey.

Finally, the terminal phase of the hunt is characterized by a buzz-like emission of ultrasound. The intervals between pulses are less than 10 milliseconds, the pulse duration is about 0.5 millisecond, and the frequencies drop to 25 to 30 kilohertz. The exact details vary from species to species, but the general behaviors are consistent. When the bat is within a few millimeters of the prey, it often scoops with a wing or with the membrane between its legs and pulls the insect toward the mouth.

To accomplish these amazing feats, bats must hear and recognize high-frequency echoes bounced off the bodies of insects, determine target direction and distance with great accuracy, and be able to orient, approach, and capture the insect even though it may be moving. How do they do it?

The bat larynx is typically mammalian. Only in mammals is the larynx an organ capable of producing sounds of complex frequency and temporal modulation and variation. Situated just posterior to the hyoid, this box of cartilages and muscles encircles the esophageal end of the trachea. All inspired and expired air passes through its confines. The lumen of the larynx can be occluded by one or two pairs of folds of tissue that oppose each other across the tracheal lumen. The deepest of these folds form the vocal cords; they have thickened lumenal edges and considerable associated musculature. Sounds are produced by forcing air through the slit between these tensed sheets of tissue. Modulation of frequency and loudness is accomplished by alteration of the tension on the vocal cords and entire larynx, the amount of air expelled per unit time through the structure, and sometimes by extralaryngeal resonating chambers. Bats produce echolocation sounds without major modification of this basic mammalian larynx, although the whole structure is enlarged and the fleshy folds of the vocal cords are exceptionally thin. The production of such brief, rapidly repeated and precisely patterned sounds seems a gargantuan task for the tiny muscles controlling the vocal cords, but large muscles would contract and relax too slowly to produce rapid pulses.

The sounds produced by the larynx are emitted either through the open mouth or the nose, depending on the family of bat. Mouth calls have a wide angle of dispersion (180° or more). The noses of those bats that use it to broadcast are complex structures with epidermal flaps and a nostril spacing that concentrates and focuses the sound in a narrow cone (less than 90°) in front of the bat, much like a megaphone. The calls travel through the air in radially expanding waves at 34 centimeters per millisecond. Because of the dispersion pattern, the amount of sound energy striking a target decreases as the square of the distance traveled. A small object intercepts very little sound energy and thus can reflect very little. As the echo is reflected back toward the bat, its energy continues to diminish as the square of the distance.

Thus, the returned sound—despite its initial loudness—is exceedingly faint. In addition, only those wavelengths in the emitted call that are equal to or shorter than the diameter of the reflecting object will be returned. Despite these problems, bats can detect and locate remarkably small objects. Little brown bats can detect wires 1 millimeter in diameter from a distance of 2 meters and wires only 0.08 millimeter in diameter from shorter distances. Even irregularities on surfaces can be located. Fish-eating bats apparently locate fish by detecting ripples on the water surface, and bats use echolocation to find the cracks in rocks where they cling while they sleep.

Many features of an echo convey information. An object's size is indicated by the frequencies in the echo; large objects reflect longer wavelengths (lower frequencies) than small ones. The extremely high frequencies of the emitted calls are necessary to detect very small objects. The character of the reflecting surface is indicated by the character of the echo. A smooth, hard surface such as the exoskeleton of a beetle returns a sharp echo, whereas a blurred echo indicates a rough surface like the body of a moth. The time required for an echo to return is directly proportional to the distance from the bat to the target, and the change in return time between successive calls can indicate the relative movement of the bat and its target. As a bat approaches a target, the call repetition rate increases, giving the bat more and more precise information about the target's location.

Several features of the morphology and neurology of the auditory system of echolocating bats contribute to the sensitivity of their hearing and their ability to process the information contained in echoes. The tympanic membranes and ear ossicles are small and light and are easily set into motion. Contraction of the middle ear muscles briefly damps the sensitivity of the ear as each cry is emitted; thus, the bat does not deafen itself. A padding of blood sinuses, fat, and connective tissue isolates the bony labyrinth of the inner ear from the rest of the skull and reduces the direct conduction of sound into the inner ear.

Perception of the direction of a returning echo is aided by the large, complex pinnae and by a neural mechanism known as contralateral inhibition. Stimulation of cells sensitive to a particular wavelength in the inner ear on one side of the head produces a transient desensitization of the cells that respond to the same wavelength in the ear on the other side of the head. The effect of that desensitization is to increase the contrast between the intensity of sound perceived by the two ears and thus to permit more precise determination of the direction of an echo.

One question that has long puzzled scientists is how an individual bat can discriminate between the echoes of its own call and those of other bats. When thousands of bats fly out of a roost in the evening the din must be enormous. One mechanism that probably contributes to a bat's ability to recognize the echoes of its own calls has recently been described. There is a brief neural sensitization following emission of a call to sounds of the same wavelengths that were emitted. The sensitive period begins about 2 milliseconds after the end of the call and lasts about 20 milliseconds. Because sound travels 34 centimeters per millisecond in air, this timing means that a bat is especially sensitive to echoes of its own call from objects at distances between 30 centimeters and 4 meters. This same mechanism probably helps a bat on its final approach to prey.

The behavior of some insects that are potential prey for bats indicates that a degree of coevolution has occurred. Some noctuid moths are sensitive to the ultrasonic sounds produced by hunting bats. Low-intensity ultrasonic sounds cause the moths to fly away from the source, whereas loud sounds cause them to stop flying and drop to the ground. Some arctiid moths produce ultrasonic sounds themselves, and these sounds should be audible to bats. This moth family includes species that contain chemical compounds that make them unpalatable, and it has been suggested that their ultrasonic emissions advertise their distastefulness to bats, just as the bright colors and conspicuous displays of other insects advertise their unpalatability to predators that hunt by vision.

Cetacean Echolocation　A large number of modern mammals live in close association with water. Some are aquatic, and a few are truly pelagic. The most specialized aquatic mammals are the members of the order Cetacea—the whales and porpoises. The terrestrial ancestors of cetaceans are little known. Anatomical and embryological evidence suggests that they may have evolved from Paleocene subungulates. The demands of aquatic life have so entirely reshaped cetaceans that there is little external indication of this relationship.

The two suborders of living cetaceans are the Mysticeti (the baleen whales) and the Odontoceti (the toothed whales). They appear to have radiated independently in the Eocene from a third suborder of primitive toothed whales, all of which are now extinct. Baleen whales, which filter small organisms from the water, are not known to use echolocation, although they produce intricate vocalizations that probably are related to their complex social behavior. These are the whales whose songs are available on records. Calculations of the sound energy emitted suggests that their songs travel at least hundreds of miles through the sea. It is not inconceivable that all the mysticete whales of a particular species in a particular ocean of the world could be in indirect vocal communication with each other.

The toothed whales share the same general modifications of body form for aquatic life that are seen in baleen whales, but are otherwise different animals. The toothed whales are much smaller than baleen whales. Although the largest species, the sperm whale, reaches a length of 18 meters and a mass of 53,000 kilograms, most odontocetes are considerably smaller. Many toothed whales inhabit the littoral zone, and there are even some purely freshwater species—the family Platanistidae has species in the Amazon and Orinoco Rivers of South America, the Ganges River in India, and Lake Tung'ing in China. Most toothed whales eat fishes or squids and their distributions are limited to areas of high productivity where food is abundant. Frequently, productive waters are also murky; thus many odontocetes must hunt in water in which vision is limited. Other species, such as the sperm whale which feeds on squid, hunt at extreme depths where sunlight does not penetrate. Dives of 1000 meters are probably routine for sperm whales. Odontocete evolution has solved the sensory problems in these habitats by an echolocation system as sophisticated as that of bats.

The difficulties inherent in any echolocation system are complicated for cetaceans by the acoustical properties of water. Sound travels about five times as fast in water as in air, and therefore the wavelength of any sound frequency is five times longer in water than in air. Because objects reflect only those wavelengths equal to or shorter than their diameters, cetaceans must use exceedingly high-frequency sounds to produce wavelengths short enough for the detection of small objects. Not surprisingly, the echolocation clicks of the bottlenose porpoise include frequencies from 20 to 220 kilohertz.

A second problem faced by mammals when in water is the difficulty of matching the acoustic impedance of the ear to that of water. Sound energy is not transmitted well across an air–water interface. About 99.9 percent of the energy that reaches such an interface is reflected and only 0.1 percent crosses the boundary. Terrestrial vertebrates face the same problem because the inner ear is water filled. The middle ear converts sound energy transmitted through air to mechanical movement of bones, and thence to displacement of the fluid in the inner ear. The system does not work underwater, however, because a new water–air interface is created at the tympanic membrane and most of the sound energy is reflected from the tympanum back into the water.

A related problem is the efficiency of transmission of sound energy from water into body tissues, especially fat and muscle. A very high proportion of the acoustic energy that impinges on the body surface is absorbed and propagates through the tissues, echoing back and forth to produce a diffuse buzz that conveys no directional information. Cetaceans have apparently solved these twin problems by abandoning the middle ear as a sound-receiving organ, isolating the inner ear from sound propagated through the body tissues, and conducting sound to the inner ear via a special fat body that extends from the lower jaw to the auditory bulla. The bulla is composed of extremely dense bone, which does not transmit sound readily and is separated from the rest of the skull by soft sound-absorbing tissues. Thus the inner ear is isolated from sound approaching from other parts of the body. Kenneth Norris has proposed that the intramandibular fat body receives sound energy through an acoustic window formed by an area of very thin bone on the side of the mandible (Figure 21–18). Sound penetrates this thin bone readily and enters the fat body, which serves as a waveguide conducting the sound energy back to the bulla. The fat body terminates on the bulla itself where the bone of the bulla is thin. Norris's hypothesis is that the sound a cetacean perceives is not received via the middle ear but instead comes from the lower jaw. In a test of the hypothesis, Theodore Bullock and Alan Grinnell used a speaker to beam sound energy at restricted por-

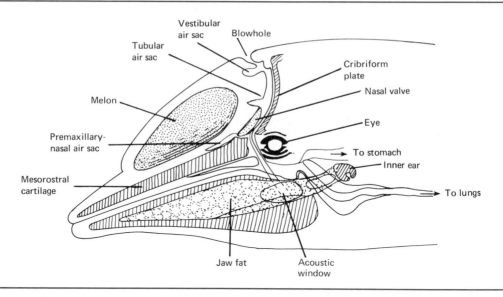

Figure 21–18. Proposed odontocete echolocation sound production and reception apparatus shown for the porpoise (*Tursiops*). Sound produced by air shuttled between air sacs through the nasal valve is focused by the oil of the melon into a forwardly directed beam. Some sound may also be guided by the mesorostral cartilage. Returned sound is channeled through the mandibular (jaw) fat bodies and especially the acoustic window of the lower jaw to the otherwise isolated and fused middle and inner ear. (Modified after K. S. Norris, 1966, in *Evolution and Environment*, edited by E. T. Drake. Peabody Museum Centenary Volume, Yale University, New Haven, Conn.)

tions of a porpoise's head while they recorded nerve cell activity in the auditory region of the brain. Sound beamed at the acoustic window of the lower jaw caused a large increase in nerve cell activity in the brain, but sound impinging on the side of the head did not.

Sound production underwater also poses problems not experienced by animals in air. Diving animals submerge with a limited supply of air, and it is probably undesirable to waste any of the oxygen supply by bubbling air out of the mouth in the process of generating sounds. Norris has suggested that the site of sound production in cetaceans has shifted from the larynx to the complex system of nasal passages and diverticulae that extend upward from the pharynx to the blowhole, the single external naris of odontocetes, which is located on the top of the head. Air is forced back and forth in these passages to produce sound energy which is reflected forward from the broad anterior surface of the skull and probably focused by an acoustic lens formed by the oil-filled **melon**. This fat body is a characteristic feature of odontocete cetaceans, and it is responsible for the external shape of the head. The skull itself occupies a relatively small portion of the head. The melon of the sperm whale is the source of sperm oil that was especially valued for the clear smokeless flame it produced when burned in oil lamps. As much as 30 barrels of oil can be obtained from the melon of a large sperm whale.

Sound passes from one medium to another as it leaves a cetacean's head. Because the acoustic coupling between oil and water is good, little of the sound energy is lost by reflection as it would be in an air-to-water transmission, but the sound waves may be bent. It seems likely that the melon serves as a flexible lens, changing its shape from moment to moment to beam the sound energy in different directions. The melon of some cetaceans changes shape very conspicuously during echolocation. The echolocation sounds of the bottlenose porpoise are broadband clicks (20 to 220 kilohertz). The highest frequencies in the click are emitted in a narrow beam directly forward and on a level with the animal's head. The lower-frequency sounds are dispersed in wider vertical and horizontal arcs. In addition, an echolocating porpoise moves its head around as if scanning with its sound beam.

The ability of porpoises to detect objects and to distinguish between similar objects is remarkable. Many of Norris's early studies were conducted with a bottlenose porpoise named Alice. Although she was frequently described affectionately by Norris and his associates as the world's dumbest porpoise, when she was blindfolded, Alice was able to distinguish between 2 steel balls, one 57 millimeters in diameter and the other 64 millimeters, from a distance of 2 meters in 1.5 seconds (Norris 1974). Blindfolded she could tell the difference between different species of fish or between fresh fish and day-old fish, and she was able to pick up her vitamin pill (5 millimeters in diameter) from the concrete bottom of her tank.

In their natural surroundings porpoises use their echolocation abilities to navigate and to find food. The low-frequency sounds that are beamed in a broad arc would reflect from large objects and are probably the important components of navigation, whereas the high-frequency sounds that are concentrated in a narrow beam are probably used to locate prey. The best sound-reflecting surface in fishes is the air bladder, and it is probably this structure that porpoises detect. It has been suggested that the reduced air bladders of some fast-swimming open-water fishes such as tunas make them harder for cetaceans to detect by echolocation.

Mammalian Reproduction

Juvenile mammals depend on one or both parents for periods of weeks or months after birth. This enforced association with parents and siblings may be an important part of the sociality that is such a distinctive feature of mammals (Chapter 22). The dependence of young mammals on their parents is a result of the reproductive specializations of mammals. Understanding mammalian reproductive anatomy and physiology is not easy, but it is pivotal to understanding mammals.

Modes of Reproduction

A major achievement of mammals has been the refinement of viviparity into a process in which food for a developing embryo is supplied continuously by the mother via a chorioallantoic placenta. Unique to mammals is the evolution of lactation, whereby the newborn mammal is supplied with food (as milk) by the mother (Pond 1975). As might be expected, mammals show a gradation from oviparity to viviparity.

Monotremes The platypus and echidna retain their heritage as egg layers (Figure 21–19a and b). The ovaries are bigger than in other mammals for they produce eggs with large amounts of yolk. As in lizards there are two oviducts, each opening into the urogenital sinus. The ovulated eggs are fertilized prior to entrance into the uterus. There they begin to differentiate and are coated with a leathery, mineralized shell. Although both oviducts are present in monotremes, only the left duct functions, as in birds the right oviduct is reduced in size. Unlike birds, however, a recognizable shell gland secretion section is absent. There is no distinct separation between tube, isthmus, and uterus (Figure 21–19b).

The platypus generally lays two eggs, which are incubated in a nest. Echidnas lay a single egg and carry it about in a metatherian-like pouch. A shell tooth, like that of birds, allows the embryo to break out of its shell.

Marsupials The reproductive tract of female marsupials combines structural features found in both monotremes and eutherians (Figure 21–19c and d). Following copulation, sperm pass from the vagina (the usual site of ejaculation of sperm in mammals) through the uterus and into the fallopian tubes, where fertilization takes place. Secretions of the oviduct assist sperm in reaching the egg. Although the penis of mammals is a single organ, in many marsupials its distal end is forked to fit into each lateral vagina.

The amount of yolk present in a marsupial egg, although greater than the yolk in a placental mammal's egg, nevertheless is limited. A yolk sac placenta is often formed, but actual implantation, if it occurs at all, is brief. Gestation takes 13 days in the opossum and only several weeks in kangaroos. The degree of prematureness at birth is probably best visualized by considering that the newborn of a large kangaroo, with adult size equaling that of a human, is only 2 to 3 centimeters long. Usually, the medial vagina serves as the birth canal. Following birth the embryos crawl from the vaginal region into the marsupial pouch (marsupium), where they grasp a mammary gland teat with their mouth. The teat swells so that the baby cannot drop off. Usually, more zygotes develop into embryos than there are teats. In the opossum, for example, there are about 13 nipples, but between 20 and 40 eggs are ovulated, fertilized, and gestated. This apparent waste may reflect the difficulty that a newborn faces in finding and entering the marsupium and securing a teat.

Placentals In placentals the paired oviducts are fused to form a common vagina and sometimes a common large uterus, as in the human female (Figure 21–19d). In many placentals (some bats, rodents, and carnivores) the uteri remain distinct as right and left halves, or are partly fused to form a T-shaped uterus. In all placentals the **fallopian tubes** are small, short, and separate. As a rule fertilization of the egg takes place in the fallopian tube after the sperm travels through the cervix and uterus and up the tubes. The fertilized egg is propelled into the uterus by contractions of the fallopian tubes and implants on the uterine wall, where it develops the embryonic–maternal connections called the **placenta**. As embryonic–maternal dependency evolved in mammals the importance of yolk diminished and eutherian eggs are small.

The placenta is a special vascular extraembryonic structure derived from the extraembryonic membranes of the amniotic egg (Figure 21–20). Placentae are either nondecidual or decidual. In the nondecidual placenta all the fetal and uterine membrane linings remain intact throughout gestation. As a result, six distinct layers separate the fetal blood from the maternal blood. In a decidual

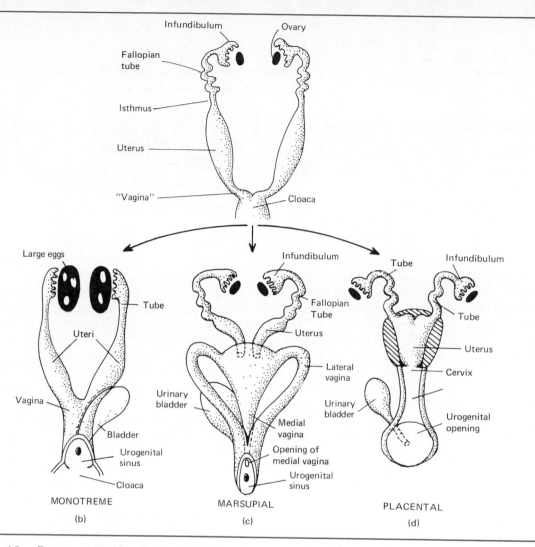

Figure 21–19. Representative female reproductive tracts of ancestral and major living mammals: (a) hypothetical ancestral mammal similar to lizard; (b) egg-laying monotreme; (c) marsupial showing complicated vaginal structures; (d) placental mammal as exemplified by an advanced primate.

placenta varied degrees of erosion of the membranes occur and the number of layers decreases. In humans, only two layers remain; and in rabbits and some rodents, such as the guinea pig, the fetal and maternal bloodstreams are separated only by the fetal endothelial lining of the capillaries. The concept that reducing the number of membranes

across the fetal-maternal barrier improves the efficiency of exchange sounds plausible, but has little evidence to support it (Mossman 1987). Perhaps of greater importance is the total area and the amount of vascular penetration associated with the fetal–maternal or placental junctions. Variation of placental morphology among eutherians defies

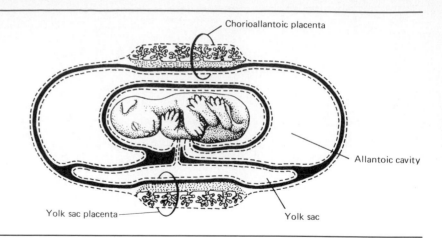

Figure 21–20. Two basic types of placental structures as seen in the cat. Both a yolk sac and chorioallantoic placenta are shown. The chorioallantoic placenta grows outward and supersedes the function of the earlier, primitive yolk sac placenta. (Modified after W. W. Ballard, 1964, *Comparative Anatomy and Embryology*, Ronald Press, New York.)

Chorioallantoic placenta

Allantoic cavity

Yolk sac placenta

Yolk sac

simple description (see Mossman 1987, for details), but in most species the placental exchange structures between fetus and mother appear to function as a countercurrent exchange mechanism (Figure 21–21). This anatomical arrangement is functionally similar to the countercurrent exchange arrangement in the gills of teleost fishes. Thus, the fetal capillary blood flow is opposite to the maternal capillary blood flow, and in this manner oxygen and carbon dioxide, nutrients, and wastes are supplied to and removed from the fetus in an efficient manner.

In contrast to birds and monotremes, the placentals do not invest energy in laying down chemically specialized yolk for embryonic development because they supply energy via the placenta. However, the placenta itself represents a large amount of energy invested by the mother in the growth of a specialized tissue. Eutherian embryos are afforded the protection of the womb and develop at the high body temperatures typical of a mammal. Brooding eggs to speed embryonic development, as in birds, becomes unnecessary. The fetus can achieve a much greater size at birth and therefore a more advanced degree of development. Nevertheless, even in eutherians, adulthood comes slowly and parental postnatal nurture and protection are crucial to survival. The true value of mammalian viviparity is hard to ascribe, therefore, to any single feature (Lillegraven et al. 1987). It correlates, however, with the facts that parents can direct more attention to fewer, larger offspring and that the offspring do benefit by learning complex behaviors during the period of parental care. The

Figure 21–21. Generalized placental blood flow showing the counter current structure. Exchange occurs across the capillaries of fetus and mother in the *zona intima*. Attachment of the placenta to the uterus occurs at the *decidua basalis*. The gray area in (c) labeled the trophospongium contains the maternal venous return vessels. The simpler anatomy of (a) is typical of a labyrinthine type arrangement as found in carnivores, rodents, chiropterans, and lagomorphs. The highly lobulated placenta of (b) is typical of most ungulates, cetaceans, and primates and increases the total exchange area by folding the *zona intima*. The cross section in (c) reveals the architecture of the maternal circulation and the fetal circulation. (Modified after H. W. Mossman, 1987, *Vertebrate Fetal Membranes*, Rutgers University Press, New Brunswick, N.J.)

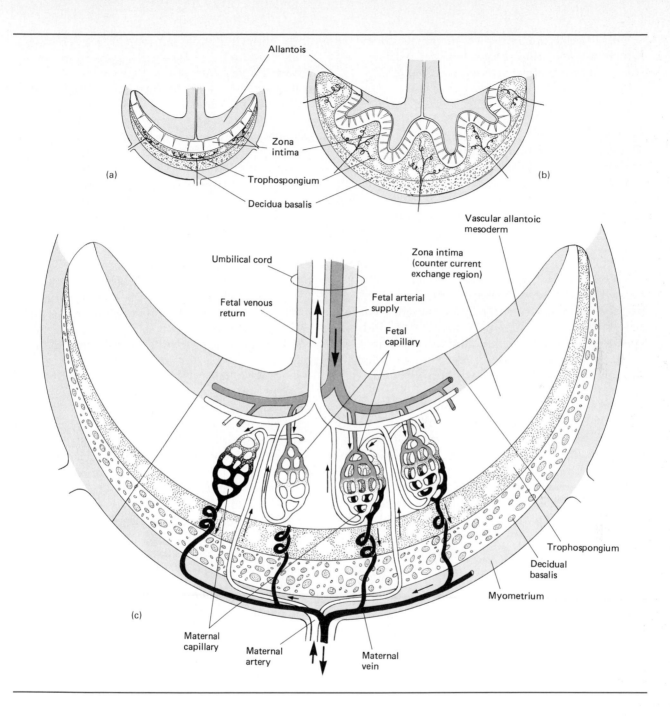

Allantois

Zona
intima

Trophospongium

Decidua basalis

(a)

(b)

Vascular allantoic
mesoderm

Umbilical cord

Zona intima
(counter current
exchange region)

Fetal venous
return

Fetal arterial
supply

Fetal
capillary

Trophospongium

Decidual
basalis

Myometrium

(c)

Maternal
capillary

Maternal
artery

Maternal
vein

diverse social behaviors of mammals, many of which involve interactions between parents and their offspring, are discussed in Chapter 22.

Hormonal Control of Reproduction

The sex of a vertebrate normally is a matter of inheritance. Whether marsupial or placental, the long-term behavioral and physiological roles of mammalian parents with respect to their offspring are precisely defined and absolutely essential for reproductive success. Only the birds have a similarly demanding parental role. A female mammal must be finely tuned physiologically and behaviorally if her investment in offspring is to survive. In species where the males participate in reproduction after insemination, they too must coordinate specific behaviors with their mates and offspring. Reproductive cycles and their synchronism with environmental conditions are controlled by the release of sex hormones (Blum 1986).

Reproductive Cycles A single reproductive cycle of the female mammal is called an **estrous cycle** and is divided into four periods—proestrus, estrus, metestrus, and diestrus, each defined by events in the mature ovary. In addition, specific reproductive events are associated with each period of the cycle, the durations of which vary greatly in different mammals. **Proestrus** starts each reproductive cycle, and is typified by initial growth of an ovarian follicle and its secretion of estrogen. **Estrus** begins with **ovulation** and modification of the follicle into the corpus luteum, which secretes progesterone. In many mammals estrus is the period of maximum female receptivity to the sexual overtures of males. **Metestrus** is a period of regression of the corpus luteum, with a declining secretion of progesterone. **Diestrus** is a variable period of quiescence before the next proestrus. In many primates metestrus is terminated by a menses, a discharge of blood and cellular debris from the endometrial lining of the uterus. The reproductive cycle is therefore known as the **menstrual cycle**.

In many mammals diestrus is prolonged throughout the year and females are receptive to males only once a year. Such species are **monestrous**. **Polyestrous** species commence proestrus following a short diestrus and are receptive several times during a breeding season or cycle continuously throughout the year. Except for the higher primates, most female mammals respond to copulation attempts by males only during estrus, when ovulation takes place. The chance for fertilization of a ripe egg is thus improved.

Four hormones regulate the various periods of estrus—the anterior pituitary hormones, follicular stimulating hormone (FSH), and luteinizing hormone (LH); and the ovarian follicle hormones, estrogen and progesterone (Figure 21–22). The four hormones interact in a negative feedback loop to synchronize the various periods of estrus. Following the sequence of events from proestrus through diestrus in a polyestrous mammal such as the human female provides insight into the complexity of vertebrate reproductive cycles.

Menstruation in the human coincides with diestrus and is associated with low levels of all four hormones in the bloodstream (Figure 21–22). During this period the uterine lining is thinnest. The hypothalamus and the pituitary gland, collectively termed the **pituitary axis**, are involved in initiation of estrus (Figure 21–23). A low concentration of estrogen in the blood is sensed by neurons in the hypothalamus, which respond by secreting a neurohormone (FSH releasing hormone) into the pituitary blood portal system. Upon transport to the anterior pituitary, this neurohormone influences the release of FSH into the blood. As menstruation (diestrus) proceeds, circulating FSH levels therefore increase. FSH has as its specific target the follicles of the ovary. Their growth does not proceed in the absence of FSH (Figure 21–22).

Proestrus commences as FSH induces follicular development and secretion of estrogen. Estrogen has three primary effects in the estrus cycle. First, it causes the uterine lining to grow and to enrich its vascularization. This process prepares for the possible implantation of a fertilized egg. Second, the increased levels of estrogen feed back on the hypothalamus to shut off secretion of FSH

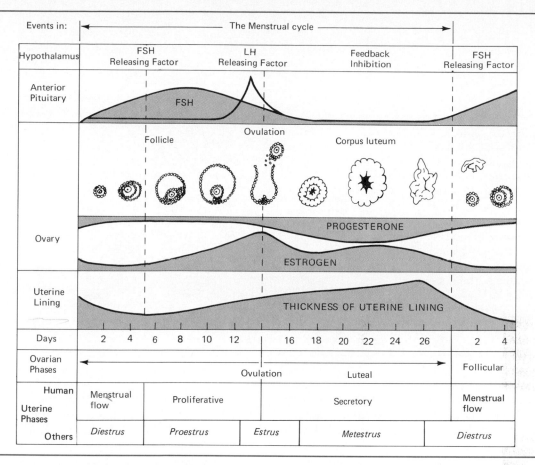

Figure 21-22. Relationships among reproductive structures and hormones during the menstrual cycle in the human female. Correspondence between the human menstrual cycle and the estrus cycle typical of most mammals is indicated at the bottom (uterine phases). The height of shading is proportional to the circulating blood levels of the indicated hormones. (Modified after R. P. Shearman, 1972, *Human Reproductive Physiology*, Blackwell Scientific, Oxford.)

releasing hormone. Third, the higher levels of estrogen also stimulate the hypothalamic neurons to secrete a second releasing factor, LH releasing hormone. In a manner analogous to FSH releasing hormone, this neurohormone arrives in the anterior pituitary and causes release of the gonadotropic (gonad regulating) hormone LH.

At approximately this stage the reproductive cycle can be considered to pass into estrus. Follicular growth reaches the stage where ovulation oc-

curs and the ripe ovum is shed into the coelom. LH directly stimulates the remaining follicle cells to enlarge and form the **corpus luteum**. This ovarian structure manufactures not only estrogen but also progesterone, which prepares the uterus for implantation by inducing uterine secretions and inhibiting contractions of its smooth muscle. The high levels of estrogen and progesterone feed back on the pituitary axis, inhibiting the release of FSH and LH. If pregnancy does not occur, the corpus

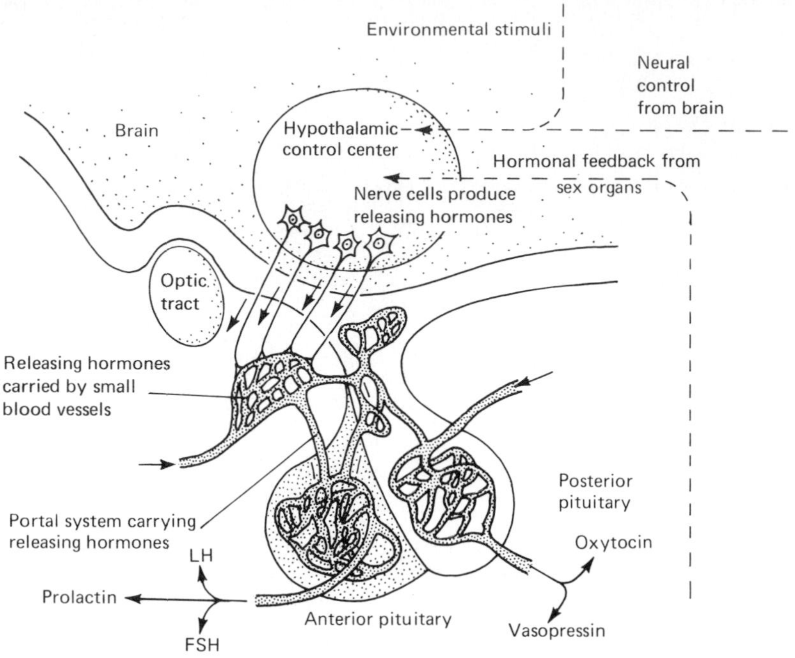

Figure 21–23. Relationships of the pituitary axis, which is composed of the hypothalamus and pituitary gland, in the neurohumoral control of reproductive cycles. Note that blood flowing to the anterior pituitary must pass through the hypophysial portal blood system. Blood flow to the posterior pituitary lobe may or may not pass through the portal system. (Modified after R. P. Shearman, 1972, *Human Reproductive Physiology*, Blackwell Scientific, Oxford.)

luteum begins to degenerate and reduced secretion of both estrogen and progesterone results in collapse of the thick uterine linings, which sloughs off, forming the menstrual flow. This process is assisted by the release of vasodilators such as histamine and prostaglandins, which promote endometrial hemorrhage (Heap and Flint 1984). Metestrus can therefore be considered to end and pass into diestrus as the uterine lining begins to collapse.

In the human male both gonadotropins, FSH and LH, are secreted by the pituitary axis. Their target organs, as in the human female, are the gonads. FSH stimulates the production of sperm in the seminiferous tubules of the testes. LH causes the interstitial cells of the testes to secrete the steroid hormone testosterone, which (with FSH) is essential to the maturation of sperm. Testosterone, like estrogen, feeds back on the pituitary axis and can reduce gonadotropic output. Despite this negative feedback, testosterone is relatively inactive in depressing the function of the pituitary, and

a delicate balance is achieved: Sperm and testosterone production proceed continuously. In several mammals, such as humans, a male reproductive cycle is therefore usually not described. In many mammal species, however, breeding activity and spermatogenesis are seasonal and correlate roughly with the timing of estrus (Short 1984a).

If implantation occurs following copulation, the female estrus cycle is interrupted and metestrus is prolonged. The ovarian corpus luteum does not degenerate, but continues to produce high levels of estrogen and progesterone. Maintenance of the corpus luteum past its usual nonpregnant functional period is achieved by the secretion of HCG (human chorionic gonadotropin) from the placenta. HCG, although chemically different from LH, has a similar effect on the corpus luteum and induces continued secretion of estrogen and progesterone. Unlike LH, however, HCG secretion is not inhibited by high estrogenic levels in the blood, and a continued increase in blood levels of estrogen and progesterone follows. As the placenta grows, it becomes a new source of estrogen and progesterone (Figure 21–24). These placental secretions of estrogenic substances assure that the uterus retains its vascularization and nutrient supply to the developing embryo. As the placental secretion of estrogen and progesterone increases, the levels of HCG secretion diminish and the importance of the ovary's corpus luteum in maintaining pregnancy vanishes.

Just prior to birth of the embryo (parturition) the hypothalamic peptide hormone oxytocin is secreted from the posterior pituitary gland of the mother (Figure 21–23). This hormone stimulates contractions of the uterus (onset of labor), which lead to expulsion of the fetus and its placenta.

Recent studies of sheep and cattle have demonstrated that the fetus may play a role in the initiation of delivery. Normal function of the fetal pituitary axis and adrenal glands is required for normal birth to occur. Cortisol secreted by the fetal adrenal glands causes increased activity of en-

zymes that convert progesterone to estrogen in the placenta. The result is a decrease in progesterone production and an increase in estrogen. The higher levels of estrogen probably stimulate production of prostaglandin in the fetal membranes and in the lining of the uterus. In turn, prostaglandin increases the excitability of the uterus and promotes the uterine contractions that expel the fetus. This mechanism by which the fetus gives the signal for delivery has been demonstrated only in ruminants so far, but it may be more widespread among mammals. For example, there is evidence for increased fetal adrenal growth just prior to delivery in primates.

Lactation   Postnatal growth of the young mammal is supported by the production of maternal milk in the mammary glands and by parental care. Lactation, like the estrus cycle and pregnancy, is largely regulated by hormones and usually requires the presence of suckling young.

The high levels of circulating estrogen and progesterone during pregnancy induce enlargement of the mammary tissues and the accumulation of fats and connective tissue. Milk production does not take place until parturition has occurred. A rapid decline in estrogen and progesterone levels occurs when their source, the placenta, is expelled in the afterbirth. The hypothalamus, thus released from feedback inhibition, secretes pituitary releasing hormones. In a manner analogous to FSH and LH releasing hormones, another releasing neurohormone, prolactin releasing factor, is secreted into the pituitary portal system (Figures 21–23 and 21–25). This factor instigates secretion of prolactin by the anterior pituitary gland. Prolactin stimulates milk production, but does not itself cause the delivery of milk. Suckling by the newborn infant stimulates nerve receptors in the nipple, and this information is transmitted to the hypothalamus and on to the posterior pituitary. Oxytocin, the hormone involved in the induction of labor, is released and causes milk ejection by inducing contractions of the smooth muscles of the breast. Suckling by young maintains lactation. In many mammals, estrus is not initiated during the

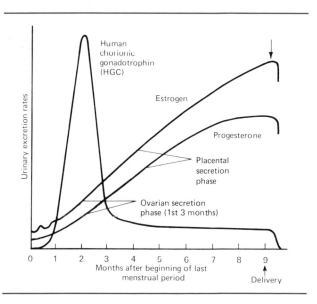

Figure 21–24.   Changes in circulating hormone levels during pregnancy as indicated by their presence in urine. (Redrawn from R. P. Shearman, 1972, *Human Reproductive Physiology*, Blackwell Scientific, Oxford.)

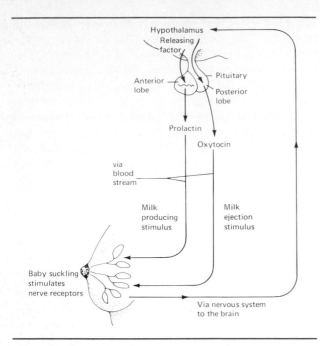

Figure 21–25. Role of the pituitary axis in lactation. Note that milk production and secretion are promoted by the anterior pituitary lobe hormone prolactin, and milk ejection by the posterior lobe hormone oxytocin. Both of these processes are initiated via neural connection from the nipple to the hypothalamus, a complex reflex which is stimulated by suckling. (Modified after R. P. Shearman, 1972, *Human Reproductive Physiology*, Blackwell Scientific, Oxford.)

period of lactation, for FSH secretion is low. In others, however, estrus quickly begins again. In the human female individuals vary. In some mothers, menstruation is delayed for 3 to 4 months during lactation; in others, the menstrual cycle commences soon after birth.

Metatherians and Eutherians

Current theories suggest that the marsupials and placentals evolved in isolation. Gestation in marsupials seldom lasts more than 30 days. The fetus does not usually implant in the uterus until the last few days of gestation. Instead, a shell-like membrane surrounds the embryo and most of prenatal development depends on consumption of yolk (**lecithotrophy**). The embryo is born at an early stage of development, and unless it finds its way into the marsupium and a mammary teat, its survival is impossible. In most placentals the embryo is born at a comparatively advanced stage of development, and although dependence upon parental care is often prolonged, the young may be quickly weaned and begin to cope directly with the world. Humans are one of the notable exceptions to the general mammalian pattern of early weaning and independence. The studies of Geoffrey Sharman (1976) and his colleagues in Australia and subsequent work by other researchers have shown that although marsupials and placentals have invested approximately the same amount of energy in an offspring by the time it is weaned, the patterns of energy investment are different for the two groups. Marsupial development is largely a postnatal cost (lactation), whereas placental development is largely a prenatal cost (placentation). This difference may have important ecological and evolutionary implications in times of energy stress.

The placental reproductive mode involves a large maternal investment in intrauterine structures to assure development. When food is scarce, it is usually the mother who suffers because the embryo's demands continue in spite of external environmental conditions. In placental mammals natural abortion during food shortage is rare. In many rodent populations, for example, limited food resources may restrict the initial number of zygotes implanted but not the number of fetuses aborted. However, in marsupials unfavorable food resources result in what might be termed marsupium abortion, for lactation declines. The young, therefore, suffer directly and can be entirely abandoned. After parturition, placentals, of course, can also abandon young and do, but the energy already invested in reproduction is high. The marsupial reproductive plan may be more successful than that of placentals when environmental conditions at the time of birth are poor, for less energy has been invested in prenatal development (Parker 1977).

Timing of Reproduction Although the investment in a fetus at birth is greater for a placental than for a marsupial, both placentals and marsupials have ways of coping with varying, harsh, and unpredictable condi ons. One such response takes the form of timing copulations so they result in births during the season of the year when food is most plentiful. To accomplish this, copulation should occur during an estrus cycle when the gestation period typical of the species will produce newborn in spring or summer in temperate regions or during the rainy season in the tropics. The length of gestation in placental mammals varies greatly and generally correlates with size. Most large placentals, for example, carry embryos from 6 months to a year or even longer. As a result, copulations that take place in late summer or autumn result in births the following spring or summer when food presumably is adequate to sustain lactation and the weaning of young. With gestation lengths of 100 days or less, smaller species may have a problem in timing parturition to the more productive spring and summer season. Also, most small mammals are seasonally monoestrous or polyestrous and ovulate only at a specific time each year, so females have few chances to become pregnant. Copulation in summer or even fall would result in a litter just as food is becoming scarce and winter approaches.

Embryonic Diapause Several species of marsupials and placentals use **delayed implantation** or **embryonic diapause** to assure that young are born at favorable periods of the year or to avoid the birth of a second litter before the first litter has been weaned. Although details of the mechanisms for inducing embryonic diapause are not fully understood, two modes of control appear to be involved, one biological and the other environmental. A few examples will clarify how the reproductive cycle can be adjusted to accommodate ecological conditions.

Several species in the eutherian order Carnivora halt development of the new, unattached blastocyst immediately following fertilization. Growth is dependent on uterine secretions, and at this stage of pregnancy uterine secretions are controlled by hypothalamic and pituitary outputs. Thus, the growth of the blastocyst can be arrested via central nervous system control. This mechanism does not work after implantation of the blastocyst on the uterine wall when fetal growth becomes solely dependent on the passage of nutrients across the placental barrier. Carnivorous placentals such as martens, skunks, and badgers breed in the summer and employ embryonic diapause to prolong gestation. The diapause may last 10 months, assuring that birth occurs during the spring of the following year. Diapause is apparently triggered by changes in day length, and results in a cessation of uterine thickening and secretions associated with estrus. The diapause may result from melatonin release and ultimate suppression of the secretion of progesterone by the corpus luteum following ovulation.

Many species of bats in temperate zones employ embryonic diapause, but as a consequence of their winter hibernation. Lowering the body temperature slows all physiological processes, including the rate of fetal growth. However, the California leaf nose bat slows the rate of embryonic development and spreads it over a 9-month gestation period without diapause. The mechanism in this case is unknown (Short 1984b). Bats feed on nocturnal flying insects, which are unavailable in cold weather. Thus, the delayed implantation shown by bats clearly fits the pattern of adjusting the time of birth to correspond with a season of abundant food, whatever its mechanism.

Among rodents and insectivores diapause serves a different purpose—the temporal separation of litters to avoid competition for resources. Small polyestrous forms can enter estrus every 4 to 5 days (Table 21–4), have gestation lengths of around 20 days, and become pregnant even when feeding the previous litter. Weaning a litter usually requires a month. Arrival of a second litter before the first litter has been weaned would overstrain the mother's ability to provide enough milk. To avoid this energetic dilemma diapause by lactation delays the development of the second litter until the first litter has finished suckling. The hormonal

Table 21–4. Length of the gestation period and estrous cycle in some mammals (days).

Nonprimate Species	Gestation	Estrous Cycle	Primates	Gestation	Estrous Cycle
Mouse	20	4-5	Lemur	120-140	
Rat	22	4-5	Tarsier	180	24
Rabbit	63	*	Wooly monkey	139	
Cat	63	18	Rhesus monkey[†]	150-180	27
Dog	63	235	Chacma baboon[†]	180-190	20-36
Lion	110	21	Langur[†]	170-190	
Goat	150	20	Gibbon[†]	210	
Domestic cattle	278	21	Orangutan[†]	220-270	
Horse	340	20	Chimpanzee[†]	216-260	36
Dolphin	360		Gorilla[†]	250-290	
African elephant	660	42	Human[†]	267 avg.	28

*Continuous; induced ovulation.

[†]Old World monkeys and the great apes menstruate externally. Rhesus monkeys and humans show maximal behavioral changes during menstruation; other species have maximal change during estrus.

Source: Modified from R. J. Harrison and W. Montagna, 1973, *Man,* 2nd edition. Appleton-Century Crofts, New York; and S. A. Asdell, 1964, *Patterns of Mammalian Reproduction,* 2nd edition, Cornell University Press, Ithaca, N.Y.

mechanisms are not totally clear, but suckling apparently inhibits LH secretion from the pituitary and the secretion of estrogen from the ovary.

Embryonic diapause reaches its zenith in the large macropods of Australia. The red kangaroo, which inhabits the climatically unpredictable arid desert regions of Australia, can have an arrested blastocyst in the uterus, an immature suckling joey tucked away in its marsupium, and a nearly weaned juvenile that still needs occasional sips of milk to supplement its newly acquired diet of vegetation. This suite of reproductive interactions is induced by the joey, whose suckling produces nervous stimulation of the hypothalamus. The pituitary release of FSH and LH diminishes and reduces follicular development and subsequent ovulation. The output of estrogen and progesterone by the corpus luteum is suppressed also by the increased prolactin release. The red kangaroo represents the utmost in opportunistic reproduction: As long as environmental conditions are suitable, the litters can develop for long periods in the pouch before they are weaned. If conditions are poor, which can happen quickly and unpredictably in the Australian desert, a female can abandon her joeys with no delay in producing the next litter. The diapausing blastocysts immediately begin to grow and are born within about 30 days (Table 21–4). The red kangaroo is a prime example of biological control of diapause, where the proximal cause is the presence of a suckling joey. But even in this example, the ultimate control remains the environment and its vagaries.

Although there are differences in reproductive modes of marsupials and placental mammals, it should be clear that there are also many common features. The array of reproductive variations evolved by the viviparous mammals, be they marsupials or placentals, emphasizes the need of each species to adjust its biology to enhance reproductive fitness. Certainly, there are constraints imposed by body size, by mode of locomotion, and by phylogenetic history, but it is clear that the environment plays a major role in regulating many aspects of mammalian reproduction.

Cenozoic Mammals and Vicariance Biogeography

During the Cenozoic era the geography and faunas of the world changed to assume their present characters. Following their origin in the Triassic and humble persistence through the Mesozoic, in the Cenozoic mammals evolved the basic higher taxa we recognize today. Extinction of the dinosaurs in the late Cretaceous may have opened niches to mammals. More than 150 genera of fossil mammals are recognized in Paleocene rocks, documenting the expansion of this group during the Tertiary (Figure 21–26). The radiation of mammals occurred in two waves: an early Tertiary radiation of rather primitive mammals and a later one when the primitive forms were replaced by modern mammals. There were also two peaks in total numbers of genera—the first in the Eocene and the second in the Miocene–Pliocene.

During the Cenozoic the Earth's landmasses were more widely separated than ever before or since. Eurasia was divided by a great north-south epicontinental sea into a European and an Asian portion. In addition, North America, South America, Africa, Antarctica, and Australia were all separate and each had its own complement of early mammals and its own flora, and each region showed responses of these elements to the cooler and drier Oligocene climate. The radiation of mammals occurred at the same time as the fractionation of the Earth's landmasses. As a result, various mixtures of the various basal stocks of mammals were isolated on different continents. As the continents drifted and connections be-

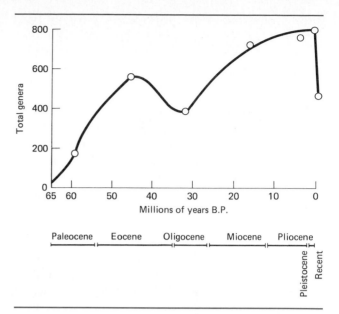

Figure 21–26. Increase in mammalian genera during the Cenozoic. Note the bimodal peak in total genera.

tween them were made and broken a series of isolations and mergers of faunas took place. This phenomenon is called **vicariance biogeography**—the movement of faunas as a result of geological processes. Vicariance can have an enormous influence on evolution, as illustrated by the parallelisms and convergences among Cenozoic mammals.

Today, the mammals of the northern hemisphere, Africa, South America, and Australia, are each strikingly different. They were also distinct in the early Cenozoic. These four geographic groupings fall into three faunal lineages: (1) a Laurasian fauna, including Africa (often considered a separate Ethiopian fauna); (2) a South American fauna; and (3) an Australian fauna. In Laurasia and Africa the first Cenozoic mammals were predominantly eutherians. In South America a mixed fauna of eutherians and metatherians existed. Australia began with a metatherian lineage that was altered only slightly by eutherian migrants

Convergent Evolution of Mammalian Feeding Types

We have emphasized that the trophic specializations of mammals have ramifications in many areas of their biology. The interaction of feeding specializations with other aspects of morphology and with behavior and ecology is strikingly illustrated by the convergent and parallel evolution of mammals that occurred during the Cenozoic as different groups of mammals evolved on continents newly isolated by the late Mesozoic fractionation of Pangaea. Such similarities of widely divergent phylogenetic lines on isolated landmasses reflect the adaptive potential of mammals and the way in which the course of evolution is influenced by the morphological and genetic characteristics of the basal stock. Figure 21–27 shows some of these examples of convergence. Grazing herbivores include the extinct litopterns (Figure 21–27a) of South America, the Holarctic artiodactyls (Figure 21–27b), the horse-like perissodactyls, and the Australian diprotodonts such as the kangaroo (Figure 21–27c). Although the modes of locomotion of these groups are different, their jaws, teeth, and feeding mechanisms are very similar.

The carnivorous mammals gave rise to similar convergences. For example, the true wolf (Figure 21–27d) in the Holarctic, the marsupial Tasmanian wolf (Figure 21–27e) in Australia and Tasmania, and the extinct marsupial borhyaenids (Figure 21–27f) of South America have similar body morphology and tooth form (see also Figure 21–28a). The Pliocene South American marsupial saber-tooth was very similar to the Holarctic placental saber-tooth (Figure 21–28b).

Mammals specialized for feeding on ants and termites include the giant anteater in South America (Figure 21–27g), the aardvark in Africa (Figure 21–27h), the pangolin in Asia, Africa, and Europe (Figure 21–27i), and the spiny anteater in Australia (Figure 21–27j). Ants and termites are social insects and employ group defense. They often build impressive earthen nests containing thousands of individuals, some of which are strong-jawed soldiers. Specializations of the teeth, jaws, and skull to myrmecophagy have already been discussed. Convergent specializations of the forelimbs for digging (Figure 21–14) include a reduced number of enlarged digits with powerfully developed curved claws.

Sometimes similar forms evolved under what appears to be much less isolated conditions. Thus hares and rabbits (Figure 21–27k) evolved in the Holarctic where other gnawing herbivores, the rodents, also occurred. Other forms evolved in one area but migrated and survived only in another. Examples are horses and rhinoceroses (Figure 21–27l, m), which originated in North America but now are found as native animals only in Asia and Africa. Isolated land masses also saw the evolution of unique forms that seem to have had few close ecological counterparts. These include the sirenians (Figure 21–27n) of African origin but now found in North and South America and Asia as well, and the extinct desmostylians (Figure 21–27o), which were large freshwater or shallow-water marine forms that are thought to have evolved on the coast of Africa and dispersed to the Pacific coasts of North America and Asia.

The convergences and parallelisms between some of the mammals that evolved on these isolated landmasses were so great that systematists formerly assumed that they must be closely related. For example, South American litopterns and North American perissodactyls converged on the same general body form, tooth morphology, and foot morphology (Figure 21–29) despite the fact that they belong to very different ungulate clades.

Early Tertiary Mammals in Laurasia

The first Tertiary mammals, from early Paleocene rocks in North America, were of three major types. Primitive marsupials appear to have been omnivorous and arboreal and not unlike modern opos-

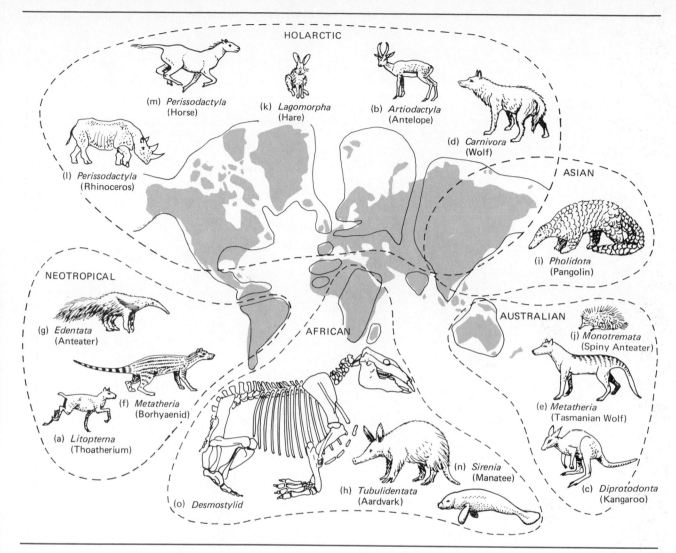

Figure 21–27. Adaptive radiation and convergence of mammals evolving in isolation during the Cenozoic. Superimposed on a map of the *recent* relative positions of the continents (shading) are the land areas (outlines) thought to have existed during the early Eocene. Note that these areas do not occupy positions they may have had during the Eocene. Isolating distances may be underestimated by this projection. Mammals that evolved during various periods of the Cenozoic (not just the Eocene) are illustrated with the landmasses on which they probably originated. Mammals of Northern Hemisphere origin are grouped as Holarctic.

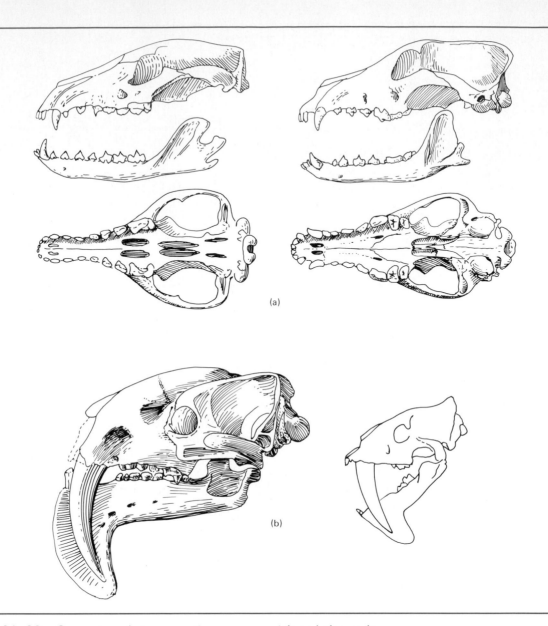

Figure 21–28. Convergence between carnivorous marsupials and placentals evolved on isolated continents (not to scale): (a) Skull of the Tasmanian wolf, *Thylacinus cynocephalus* (left), in lateral and palatal view compared with that of a placental dog, *Canis familiaris*; (b) Pliocene South American marsupial saber-tooth, *Thylacosmilus atrox* (left), compared with the Eocene-to-Oligocene Holarctic placental saber-tooth cat, *Eusmilus* (right).

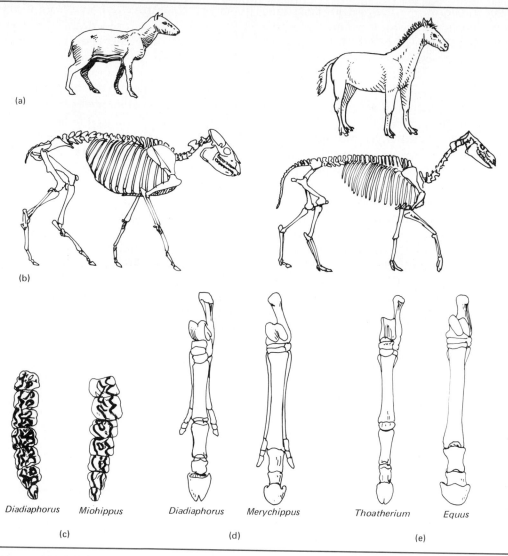

Diadiaphorus   Miohippus

(c)

Diadiaphorus   Merychippus

(d)

Thoatherium   Equus

(e)

Figure 21–29. Convergence between unrelated mammals from isolated continents to grazing and sustained running over hard ground. South American litopterns (left) and North American (Holarctic) perissodactyls (right). (Not to scale.) (a) Reconstructions of the Miocene South American litoptern *Diadiaphorus* and a late Miocene and Pliocene North American and Asian three-toed horse, *Hypohippus*. (b) Skeletons of *Diadiaphorus* and *Hypohippus*. (c) Occlusal surface of the cheek teeth of *Diadiaphorus* and comparable developments in perissodactyls from the mid-Oligocene to Miocene *Miohippus*. (d) Frontal view of the left hind feet of the three-toed forms *Diadiaphorus* and *Merychippus* (= *Protohippus*) from the mid-Miocene to the Pliocene. (e) Frontal views of the hind foot of the Miocene one-toed litoptern *Thoatherium* and the Pleistocene one-toed horse *Equus.* The litopterns reached a degree of specialization that the horses did not attain until millions of years later.

sums. A second group were small placentals, mostly shrew-like insectivorous mammals (leptictids and deltatheridiids), from which the greatest variety of living mammals have evolved. A third group, the Multituberculata with rodent-like dental arrangements, probably filled the ecological niche the rodents occupied after multituberculate extinction later in the Cenozoic.

In North America eutherian and metatherian fossils are recognizable as separate mammalian lineages. By the early Tertiary the eutherians included primates, condylarths, carnivores, artiodactyls, and perissodactyls. The multituberculates can be traced back to the early Jurassic, where they occurred with the pantotheres. The fossil locations of these general groupings of mammals reveal a broad biogeographic distribution (Savage and Russell 1983). In the early Cretaceous rocks of England, Mongolia, and Texas, unspecialized therians occur in association with several nontherian mammals characteristic of the Jurassic: triconodonts, symmetrodonts, and multituberculates. The earliest known marsupials are found in Texas. Did the Metatheria and Eutheria originate in North America, or at least in the Western Hemisphere? The paleontological record is not yet complete enough to provide an answer, but it appears likely that is the case.

A distinct change in the mammal fauna is observed in the continuous depositional sequence of North America that extends from the late Cretaceous into the Tertiary. The common and diverse late Cretaceous marsupial faunas became impoverished and the eutherian faunas became very rich. These changes in faunal balance suggest that the marsupials were replaced by eutherians in North America in the late Cretaceous and the Paleocene. If that is so, why were marsupials so successful in North America in the middle Cretaceous and then suddenly much less successful in the Paleocene? Their decline in North America in the early Cenozoic is especially puzzling in the light of recent discoveries of fossils that show marsupials spreading across the northern hemisphere at the same time they were apparently becoming less abundant in North America. The earliest fossils of marsupials currently know from Europe are from the late Paleocene. African marsupial fossils are known from the Paleocene–Eocene boundary, and the first Asian fossil to be discovered comes from the early Eocene (Benton 1985).

One explanation of the decline of marsupials in North America suggests that the earliest eutherians originated and first radiated in Eurasia, while the marsupials first radiated in North America, in isolation from eutherians (Clemens 1970, Keast 1972, Lillegraven et al. 1979). G. G. Simpson (1965) believed that invasion of early eutherians into North America via the Bering land bridge from Asia was necessary to explain the increase in eutherians in North America because their diversification occurred too rapidly to be explained by a radiation originating in North America. He suggested that the arrival of placental mammals upset the community balance between marsupials and the other Cretaceous mammals. This argument implies, however, that marsupials cannot compete with placentals for the same or closely related adaptive zones. As we shall see, this alleged inferiority of marsupials is hard to substantiate and probably does not exist.

By the late Paleocene and during the Eocene a wide variety of different eutherians inhabited Laurasia. Most can be traced from the leptictid insectivores from which bats and primates clearly arose. Rodents, having evolved in Asia, were common in the early Eocene and appear to have quickly displaced the North American multituberculates after their arrival in the Western Hemisphere (Flanagan and Lillegraven 1986). From the deltatheridiids arose the carnivorous creodonts and the herbivorous condylarths (Figure 21–30). Both carnivorous and herbivorous forms had large canines, but the cheek teeth of the herbivorous forms were modified for crushing plant materials. Although most early forms were clawed, some condylarths developed hoofs on their toes suggesting they filled an early ungulate-like niche. The creodonts and condylarths were modest in size, but some of the late Paleocene and Eocene forms, the uintatheres and pantodonts, were huge—many as large as a modern rhinoceros.

Figure 21–30. Some extinct mammals of the Cenozoic: (a) *Phenacodus*, a Paleocene and Eocene herbivorous condylarth of Europe and North America. South American mammals evolved on that continent during its long isolation; (b) one of the armored glyptodonts, mid-Eocene to Pleistocene edentates; (c) tapir-like *Pyrotherium*, a late Eocene amblypod; (d) rhinoceros-like *Toxodon*, a Pliocene and Pleistocene notoungulate; (e) *Andrewsarchus*, a late Eocene condylarth from Mongolia that appears to have been carnivorous. Not to the same scale.

These early eutherians occupied much of Laurasia, but most were extinct by the end of the Eocene.

Tertiary Mammals of Australia and South America

Most of us are aware that the mammals of Australia differ from those of Asia, Europe, Africa, and North America. The mammals of South America also differ from those of the Northern Hemisphere, but less so than do the Australian mammals (Keast 1972). These two continents provide particularly clear examples of the effects of vicariance biogeography. Both continents appear to have been invaded by animals that moved southward, and the timing of those movements in relation to the changing connections between continents illustrates the role that chance events play in biogeography and evolution.

Tertiary Mammals of South America The paleontological record of South America is extensive and shows a history of connection, separation, and reconnection to North America. The effects of these geological events can be traced in the changing fauna of South American mammals. South America was isolated from North America by a seaway between Panama and the northwestern corner of South America from the late Jurassic until the end of the Tertiary. In the Pliocene, about 3 million years ago, the Central American land bridge was established between North and South America and faunal elements of the two continents were free to mix for the first time in more than 100 million years (Marshall et al. 1982). The result was a massive exchange of elements between two radically different groups of animals that enriched the faunas of both continents (Mares and Genoways 1982).

Three major groups of Tertiary mammals can be distinguished in South America (Patterson and Pascual 1972). The earliest inhabitants—marsupials, edentates, and condylarths—appear to have been derived from North American forms that entered South America in the late Cretaceous before the connection between North and South America was broken, or moved from island to island while the seaway was still narrow. (This method of crossing a water barrier is known as island hopping.) Equally important were the groups of North American mammals that failed to colonize South America at this time, notably the multituberculates, insectivores, carnivores, and rodents. We have no explanation of why these groups did not succeed in populating South America, but their absence is important because it left adaptive zones open and marsupials radiated into them during the isolation of South America in the Tertiary.

The result of these radiations was the evolution in South America of a varied and complex fauna of marsupials, edentates, and ungulates.

These animals showed extensive convergence with mammals in North America and the Old World in their general body form and ways of life (see Savage, 1986, for restorations of the South American mammals). The resemblances between the mammals of North and South America in the Tertiary were only general, however, not exact. Enough differences existed so that most North and South American forms that came into contact in the Pliocene (after the Central American land bridge arose) did so with little evidence of direct competition or displacement.

The South American marsupials present in the Cretaceous radiated into adaptive zones that were occupied in North America by insectivores, carnivores, multituberculates, and rodents. The caenolestid lineage included forms with teeth like multituberculates and others that were rodent-like animals. The argyrolagids were saltatory animals, with long hind legs and tails reminiscent of such placental mammals as kangaroo rats and jerboas. Most spectacular of the marsupials of South America were the borhyaenids. Some borhyaenids were generalized predators with head-plus-body lengths up to 75 centimeters, whereas others, represented by *Thylacosmilus*, converged on the skull features of the saber-tooth cats of the Northern Hemisphere (Figure 21–28b). All the borhyaenids were relatively short-legged animals that probably stalked their prey rather than pursuing it. The adaptive zone for large cursorial predators in Tertiary South America was apparently occupied by the psilopterids and phororhachids, flightless birds standing 1.5 meters or taller (Figure 18–5). These were swift animals with long legs, and they appear to have been so successful that cursorial mammalian predators with body forms like wolves or cheetahs did not evolve in South America during the Tertiary.

The edentates are one of the success stories of South American mammals. Armadillos were the first group of South American edentates to appear in the fossil record and they became very diverse in the Miocene. In the late Pliocene and Pleistocene the armadillo *Pampatherium* was as large as a rhinoceros. A second group of armored edentates, the glyptodonts (Figure 21–30), appeared in the Eocene and radiated in the Miocene and Pliocene. Glyptodonts were encased in bone; a carapace covered the body, a separate bony casque protected the head, and bony rings encircled the tail. In one species the tail terminated in a spiked knob like a medieval mace. In the Pliocene, some glyptodonts achieved lengths in excess of 4 meters. Large flightless birds, the brontornithids, with massive beaks and heavy legs might have been specialized predators of glyptodonts.

A third group of successful edentates, the ground sloths, are well represented by fossils. Early members appeared in the Oligocene, and by the Pleistocene some forms were the size of elephants. *Nothrotherium*, one of the megatheriid sloths that spread into North America, was more than 3 meters long, and a South American ground sloth, *Megatherium*, reached 6 meters. These animals were tall enough to rest on their hind legs and browse on tree branches. Dung and even mummified body tissues of ground sloths have been found in caves in arid regions of North and South America, sometimes in association with human artifacts. Two other groups of edentates, the tree sloths and the anteaters, spread from South America to North America, but have only sparse fossil records on either continent.

The third group of early colonizers of South America was the condylarths and ungulates. These animals radiated into several orders, many of them containing extremely large animals. The litopterns were cursorial herbivores ranging in size from the small, delicately built adianthids through the middle-size proterotheriids to the large macrauchenids. The proterotheriids were horse-like in their general body form and showed the same pattern of reduction of the digits (Figure 21–29). They lacked the high-crowned (hypsodont) teeth of equids, however. The macraucheniids, some of which stood 1.5 meters at the shoulder, were somewhat camel-like, but the nostrils were located on the top of the head. Some paleontologists interpret this character as indicating that macrauche-

niids had an elongated proboscis, whereas others suspect that the dorsal position of the nostrils permitted macraucheniids to breathe while they waded in shallow water eating floating vegetation.

The astrapotheres and pyrotheres were two orders of large, heavy-bodied herbivores with small proboscises and tusks (Figure 21–30). Pyrotheres appeared in the Eocene and increased progressively in size until the middle Oligocene when they became extinct. Astrapotheres, also of Eocene origin, persisted through the Miocene.

The notoungulates showed an early dichotomy; the orders Typoptheria and Hegetotheria were rodent- and rabbit-like herbivores. Some of these were quite large: *Hegetotherium* of the Oligocene and Miocene was the size of a small dog, and *Typotherium* was the size of a bear. The order Toxodontia contained large hippopotamus-like animals. *Toxodon*, an abundant animal in the Pleistocene, was nearly 3 meters in length (Figure 21–26).

A second category of South American mammals made their way to South America between the early Oligocene and the Pliocene by island hopping. These were placental mammals and included the caviomorph rodents (guinea pigs and Patagonian hares) and platyrrhine primates (New World monkeys), which arrived in the Oligocene and were followed by the procyonid carnivores in the middle Pliocene. The caviomorph rodents diversified in the Miocene and Pliocene to include the largest rodents that have ever lived. *Telicomys*, a dinomyid (terrible mouse), was the size of a small rhinoceros.

A striking feature of the South American fauna in the middle Tertiary was the variety of large herbivorous mammals (Mares and Genoways 1982). In South America seven orders of ungulates fell into this category, plus five families of edentates and two families of rodents. In North America, in contrast, just three orders of ungulates (artiodactyls, perissodactyls, and proboscideans) plus the ground sloths (which had invaded from South America by island hopping during the Miocene) filled the large herbivore adaptive zone.

What happened when the establishment of a land bridge in the Pliocene allowed these two very different faunas to mix? The results were less dramatic than one might have expected. Animals from North America moved southward, and South American forms moved northward. On each continent the newcomers and the native fauna appeared to coexist; for the most part the immigrants enriched the existing fauna rather than displacing it.

Several groups of large South American mammals, including the astrapotheres and some notoungulates, disappeared in the Miocene and early Pliocene before the land bridge formed. These extinctions were probably associated with the Miocene orogeny that lifted the Andean Cordillera to its present height. The rain shadow on the eastern side of the mountains created a new habitat consisting of grassy plains, the pampas. The loss of some native South American forms at this time was balanced by the evolution of new ones, particularly ground sloths and glyptodonts. The native South American mammals that were present at the end of the Pliocene largely persisted through the Pleistocene. The dramatic extinctions of South American mammals occurred at the end of the Pleistocene. Large species, both natives and immigrants, were most affected. These faunal changes in South America seem to be part of a worldwide pattern because similar extinctions occurred in North America and Africa at the same time and may be related to human activities.

The sole exception to coexistence of native forms and immigrants in South America appears to be the displacement of the marsupial borhyaenid carnivores by North American placental carnivores. Replacement of marsupials by placentals was not a general phenomenon, however; it was confined to the borhyaenids. Other marsupials have fared well in South America. In the late Pliocene, when the land connection to North America was established, marsupials comprised 19 percent of the genera of mammals in South America. Despite the loss of the borhyaenids, marsupials now include 17 percent of the genera of Recent mammals in South America, an insignificant decline. One familiar South American marsupial invader, the Virginia opossum, has extended its range northward to Canada (Figure 21–31). Thus, the history of South America provides no evidence to support the simplistic notion of a general inferiority of marsupials in comparison to placental mammals (Kirsch 1977).

Cenozoic Isolation of Australia The mammalian fauna of Australia from its beginnings has been composed almost entirely of marsupials rather than placentals (Rich and van Tets 1985). The earliest marsupial fossils currently known from Australia are of late Oligocene age, and there are abundant and diverse remains from the Miocene onward. In contrast, placentals did not enter Australia until late in the Cenozoic.

Two routes of movement for marsupials from North America to Australia appear plausible in the light of our understanding of continental positions in the late Mesozoic and early Cenozoic: Marsupials might have moved across the Northern Hemisphere from North America through Europe to Asia and ultimately southward to Australia. Alternatively, they could have moved from North America through South America and across Antarctica [which was warm and moist until about 35 million years ago (Kerr 1987)] to Australia. The position of continents in the late Mesozoic and early Cenozoic appears to make the southern route more likely because South America, Antarctica, and Australia were still close together, whereas the route from Asia would apparently have required movement across water barriers. Marsupial fossils from the Cretaceous are known from Peru and Bolivia in South America, and the discovery in 1982 of a fossil South American marsupial in Eocene deposits in Antarctica strengthened the case for the southern route. Furthermore, Australian marsupials appear to have their closest phylogenetic affinities with South American forms. Recently, the discovery of a marsupial in early Oligocene sediments in Soviet central Asia demonstrates for the first time the presence of marsupials on the Asian continent. However, the discovery does not greatly strengthen the case for the northern route. The Asian marsupial was a didelphid, a family that

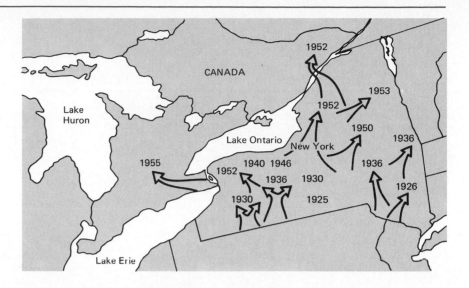

Figure 21–31. Northward invasion of the opossum (*Didelphis virginiana*) in eastern North America. Capture dates are placed over localities from which specimens were first obtained. Direct introductions into northern areas made during the early 1900s were always unsuccessful. Subsequently, naturally occurring invasions into these same northern areas by expanding, locally adapted populations have been very successful. (Data from W. J. Hamilton, 1958, *Cornell University Agricultural Experiment Station Memoirs* 354:1–48; and R. L. Peterson, 1966, *The Mammals of Eastern Canada*, Oxford University Press, Toronto.)

is characteristic of the Northern Hemisphere and has no particular affinities to Australian marsupials. Furthermore, the Asian fossil is approximately the same age as the earliest Australian marsupials. Older Asian fossils would be needed to strengthen the evidence for the northern route. At present it seems most likely that Australia was populated by marsupials that moved from South America sometime during the Eocene (Benton 1985).

However marsupials reached Australia, once they arrived they enjoyed the advantages of long-term isolation and a radiation of diverse types followed. Kangaroos and wallabies, the wombats and koala, marsupial mice and marsupial moles, bandicoots and the nearly extinct Tasmanian wolf evolved to fill a variety of niches, with food habits ranging from complete herbivory to carnivory (Archer and Clayton 1984).

In the Miocene and on several occasions since, placental rodents have founded colonies in Australia by island hopping from Southeast Asia. The only other major group of placentals to reach Australia was the bats. Endemism has evolved in Aus-

tralian bats, but as one might expect, it is less prevalent in these flying and constantly intermingling forms than in the less mobile Australian rodents. These later Cenozoic incursions have probably had little overall effect on the Australian marsupials. A far greater threat has been the prehistoric invasion by humans and dogs, which already has led to the near extinction of the Tasmanian wolf and endangers numerous other forms.

Chance in Evolution

The remarkable convergences of the mammalian faunas of North America, South America, and Australia clearly show the role that chance events play in evolution. For example, an adaptive zone exists for predatory mammals and the group that radiates to fill that zone depends in large measure on what phylogenetic lineage is in the right place at the right time. Thus, in North America placental canids gave rise to a variety of dog-like carnivores ranging in size from foxes to wolves, whereas in South America and Australia it was borhyaenid

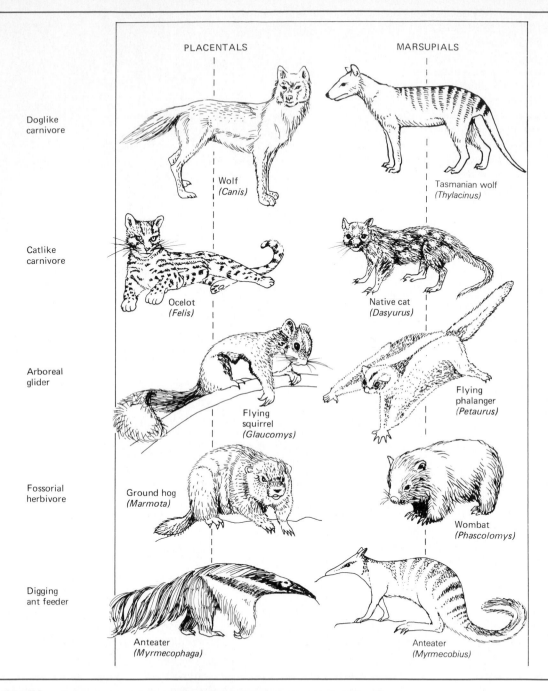

PLACENTALS   MARSUPIALS

Doglike
carnivore
                    Wolf   Tasmanian wolf
                   (Canis)   (Thylacinus)

Catlike
carnivore
                   Ocelot   Native cat
                   (Felis)   (Dasyurus)

Arboreal
glider
                                                Flying
                                                phalanger
                   Flying   (Petaurus)
                   squirrel
                  (Glaucomys)

Fossorial
herbivore
            Ground hog
            (Marmota)   Wombat
                                               (Phascolomys)

Digging
ant feeder
            Anteater   Anteater
          (Myrmecophaga)   (Myrmecobius)

Figure 21–32.   Convergences in body form and habits between placental and
marsupial mammals. (From G. G. Simpson, G. S. Pittendrigh, and L. H. Tiffany,
1957, *Life*, Harcourt, Brace, Jovanovich, New York.)

and dasyurid marsupials that produced dog-like predators (Figure 21–32). The same comparisons can be made between other mammalian adaptive zones: North American artiodactyls, South American litopterns, edentates, and ground sloths, and Australian kangaroos and giant wombats were all medium-size or large herbivores. The North American lagomorphs, the South American typotheres, and the Australian quokkas were small, rabbit-like herbivores. Not all the parallels are complete: The North American saber-tooth cats and the South American saber-tooth borhyaenids were convergent, but the Australian marsupial lion (*Thylacoleo*), although it may have been a cat-like carnivore, lacked the enormous canines of the saber-tooth forms.

The picture that emerges is of a mammalian body plan that can evolve into a wide variety of specialized forms when given the opportunity. The events that create an opportunity are largely the results of chance: The coincidence in time between the formation or breaking of connections between landmasses and the presence or absence of particular phylogenetic lineages at those times determines which lineages will dominate a particular adaptive zone on a particular landmass.

The importance of chance and vicariance is not limited to mammals; it is merely particularly clearly demonstrated by a comparison of the mammalian faunas of the northern and southern hemispheres. Exactly the same phenomena are revealed by the distributions of hadrosaurian dinosaurs (eastern North America and Europe) and ceratopsians (western North American and Asia), and for exactly the same reason: When these groups evolved in the Cretaceous, both North America and Eurasia were divided by epicontinental seas. Movement and colonization was possible between eastern North America and Europe, or between western North American and Asia, but it was not possible to move by land between eastern and western North America or between Europe and Asia.

Summary

The first mammals were small and probably insectivorous. Because of their endothermic ability to maintain a relatively constant internal body temperature, they were able to be active at night and to remain active under cool conditions.

A cladistic classification of mammals retains the three widely recognized mammal groupings of the Prototheria (the egg-laying monotremes), Metatheria (the marsupials), and the Eutheria (the placental mammals). Eighteen orders of living mammals and slightly more than 4000 species are recognized. Mammalian diversity is astonishing, and mammals have radiated into almost every habitat on earth.

A few selected aspects of the adaptive radiation of modern mammals give an appreciation for the versatility of mammalian structure and function. The ways mammals interact with their environments, satisfy their nutritional demands, sense the world around them, and reproduce are areas of particular importance.

The mammalian integument is a unique structure that plays a major role in mammalian homeostasis, especially thermoregulation, and in mammalian social behavior. Its derivatives include hair, scales, claws, hoofs, horns, and glands that secrete substances ranging from milk to pheromones and sensory transducers that provide information about the environment.

The specializations of mammalian teeth are basic to mammalian evolution. Jaw, muscle, and locomotor anatomy also reflect feeding adaptations. A case can be made for plant evolution having stimulated changes in herbivore evolution that, in turn, affected carnivore evolution.

Evolutionary advances in mammals also occurred in the evolution of the brain, especially the increased size of the cerebral hemispheres. Mammals evolved from earlier synapsids that could process three-dimensional visual reconstructions of the world. The original sensory demands made upon a nocturnal mammal-like therapsid may have acted as an evolutionary stimulus to brain enlargement. Auditory specialists that echolocate (bats and cetaceans) demonstrate the analytical power of the mammalian nervous system.

Reproduction stands at the base of the complex adaptations of mammals; a prolonged, plentiful energy supply to the developing mammal is essential to accomplish its complicated construction. The placenta and lactation provide energy under the control of complex neurohormonal patterns, but the reproductive strategies of the Metatheria (marsupials) and Eutheria (placentals) are quite different. Marsupial reproduction appears to reduce the energy investment in the embryo and concentrate the maternal energy contribution in lactation, whereas placental reproduction emphasizes energy investment *in utero*.

Mammals provide one of the clearest examples of vicariance biogeography—the effect of events such as continental movements on the geographic distribution of animals or plants. Much of the diversification of mammals occurred during the early Cenozoic, a period when the continents were more fragmented than ever before or since. Different lineages of mammals were isolated on various continents and evolved into similar adaptive zones, showing remarkable examples of convergence and parallelism.

References

Archer, M. and G. Clayton (editors). 1984. *Vertebrate Zoogeography and Evolution in Australasia*. Hesperian Press, Carlisle, Western Australia.

Barrette, C. 1977. Fighting behavior of muntjac and the evolution of antlers. *Evolution* 31:169–176.

Benton, M. J. 1985. First marsupial fossil from Asia. *Nature* 318:313.

Benton, M. J. 1988. The relationships of the major group of mammals: new approaches. *Trends in Ecology and Evolution* 3:40–45.

Blum, V. 1986. *Vertebrate Reproduction*. Springer-Verlag, Berlin.

Clemens, W. A. 1970. Mesozoic mammalian evolution. *Annual Review of Ecology and Systematics* 1:357–390.

Dijkgraff, S. 1960. Spallanzani's unpublished experiments on the sensory basis of object perception in bats. *Isis* 51:9–20.

Eisenberg, J. F. 1981. *The Mammalian Radiations: An Analysis of Trends in Evolution, Adaptation, and Behavior*. University of Chicago Press, Chicago.

Flanagan, K. M. and J. A. Lillegraven (editors). 1986. *Vertebrates, Phylogeny and Philosophy*. Contributions to Geology, Special Paper 3. University of Wyoming, Laramie, Wyo.

Goss, R. J. 1983. *Deer Antlers*. Academic Press, New York.

Griffin, D. R., F. A. Webster, and C. R. Michael. 1960. The echolocation of flying insects by bats. *Evolution* 8:141–154.

Griffiths, M. 1979. *Biology of the Monotremes*. Academic Press, New York.

Heap, R. B. and A. P. F. Flint. 1984. Pregnancy. Pages 153–194 in *Reproduction in Mammals,* volume 3, *Hormonal Control of Reproduction*, edited by C. R. Austin and R. V. Short. Cambridge University Press, London.

Janis, C. 1976. The evolutionary strategy of the Equidae and the origins of rumen and cecal digestion. *Evolution* 30:757–774.

Jerison, H. J. 1973. *Evolution of the Brain and Intelligence*. Academic Press, New York.

Keast, A. L. 1972. Comparisons of contemporary mammal faunas of southern continents. Pages 433–501, in *Evolution, Mammals, and Southern Continents*, edited by A. Keast, F. C. Erk, and B. Glass. State University of New York Press, Albany, N.Y.

Kerr, R. A. 1987. Ocean drilling details steps to an icy world. *Science* 236:912–913.

Kirsch, J. A. W. 1977. The six-percent solution: second thoughts on the adaptedness of the Marsupialia. *American Scientist* 65:276–288.

Lillegraven, J. A. 1974. Biological considerations of the marsupial-placental dichotomy. *Evolution* 29:707–722.

Lillegraven, J. A., Z. Kielan-Jaworowska, and W. A. Clemens (editors). 1979. *Mesozoic Mammals: The First Two-Thirds of Mammalian History*. University of California Press, Berkeley, Calif.

Lillegraven, J. A., S. D. Thompson, B. K. McNab, and J. L. Patton. 1987. The origin of eutherian mammals. *Biological Journal of the Linnaean Society* 32:281–336.

Lorenz, K. 1977. *Behind the Mirror: A Search for a Natural History of Human Knowledge*, first American edition, translated by R. Taylor. Harcourt Brace Jovanovich, New York.

Maderson, P. F. A. 1972. When? Why? and How? Some speculations on the evolution of the vertebrate integument. *American Zoologist* 12:159–171.

Mares, M. A. and H. H. Genoways (editors). 1982. *Mammalian Biology in South America*, volume 6, Special

Publication Series. Pymatuning Laboratory of Ecology, University of Pittsburgh, Pittsburgh, Pa..

Marshall, L. G., S. D. Webb, J. J. Sepkoski, and D. M. Raup. 1982. Mammalian evolution and the great American interchange. *Science* 215:1351–1357.

McKenna, M. C. 1975. Toward a phylogenetic classification of the Mammalia. Pages 21–46 in *Phylogeny of the Primates*, edited by W. P. Luckett and F. S. Szalay. Plenum Press, New York.

Mossman, H. W. 1987. *Vertebrate Fetal Membranes*. Rutgers University Press, New Brunswick, N.J.

Norris, K. S. 1974. *The Porpoise Watcher*. George J. McLeod, Toronto.

Novacek, M. J. and A. R. Wyss. 1987. Higher-level relationships of the recent eutherian orders: morphological evidence. *Cladistics* 2:257–287.

Parker, P. 1977. An ecological comparison of marsupial and placental patterns of reproduction. Pages 273–286, in *Biology of the Marsupials*, edited by B. Stonehouse and D. Gillmore. University Park Press, Baltimore.

Patterson, B. and R. Pascual. 1972. The fossil mammal fauna of South America. Pages 274–309, in *Evolution, Mammals and Southern Continents*, edited by A. Keast, F. C. Erk, and B. Glass. State University of New York Press, Albany.

Patterson, F. and E. Linden. 1981. *The Education of Koko*. Holt, Rinehart and Winston, New York. Also see articles "Conversations with a gorilla," *National Geographic Magazine* 154(4) 1978, and "Koko's kitten," *National Geographic Magazine* 167(1) 1985.

Pond, C. M. 1975. The significance of lactation in the evolution of mammals. *Evolution* 31:177–199.

Premack, D. and A. J. Premack. 1983. *The Mind of an Ape*. W. W. Norton, New York.

Rich, P. V. and G. F. van Tets. 1985. *Kadimakara, Extinct Vertebrates of Australia*. Pioneer Design Studio, Lilydale, Victoria, Australia.

Savage, D. E. and D. E. Russell. 1983. *Mammalian Paleofaunas of the World*. Addison-Wesley, Reading, Mass.

Savage, R. J. G. 1986. *Mammalian Evolution: An Illustrated Guide*. Facts on File Publications, New York. Excellently and profusely illustrated survey of fossil mammals.

Schusterman, R. J., J. A. Thomas, and F. G. Wood (editors). 1986. *Dolphin Cognition and Behavior: A Comparative Approach*. Lawrence Erlbaum, Hillsdale, N.J.

Sharman, G. B. 1976. Evolution of viviparity in mammals. Pages 32–70 in *Reproduction in Mammals*, volume 6, *The Evolution of Reproduction*, edited by C. R. Austin and R. V. Short. Cambridge University Press, London.

Short, R. V. 1984a. Oestrous and menstrual cycles. Pages 115–152 in *Reproduction in Mammals*, volume 6, *The Evolution of Reproduction*, edited by C. R. Austin and R. V. Short. Cambridge University Press, London.

Short, R. V. 1984b. Species differences in reproductive mechanisms. Pages 24–61 in *Reproduction in Mammals*, volume 4, *Reproductive Fitness*, edited by C. R. Austin and R. V. Short. Cambridge University Press, London.

Simpson, G. G. 1965. *The Geography of Evolution: Collected Essays*. Chilton Books, Philadelphia.

In the preceding chapter we noted the increase in brain size that has occurred during the evolution of mammals and suggested that part of the origin of the derived features of the mammalian brain might be sought in the nocturnal habits that are postulated for Mesozoic mammals. Relying on scent or hearing instead of vision to interpret their surroundings, ancestral mammals may have experienced selection for an increased ability to associate and compare stimuli received over intervals of time. The associative capacity resulting from changes in the brain during the evolution of mammals might also contribute to more complex behavior, and the association of mother and young during nursing could provide an opportunity to modify behavior by learning.

Indeed, social behaviors and interactions between individuals play a large role in the biology of mammals. These behaviors are modified by the environment, and clear-cut relationships between energy requirements, resource distribution, and social systems can often be demonstrated. In this chapter we consider some examples of those interactions that illustrate the complexity of the evolution of mammalian social behavior. In addition, we consider the social behavior of several species of primates. The social behavior of many primates is elaborate, but not necessarily more complex than that of some other kinds of mammals, including canids. However, primates have been the subjects of more field studies than have other mammals, with the result that we know a great deal about their social behavior, its consequences for the fitness of individuals, and even a little about the way some species of primates view their own social systems.

Body Size, Ecology, and Sociality of Mammals

22

Social Behavior

Sociality means living in structured groups and some form of group living is found in nearly all kinds of vertebrates. However, the greatest development of sociality is found among mammals, and much of the biology of mammals can best be understood in the context of what sorts of groups form, the advantages of group living for the individuals involved, and the behaviors that stabilize groups.

Mammals may be particularly social animals as a result of the interaction of several mammalian characteristics, no one of which is directly related to sociality, that in combination create conditions in which sociality is likely to evolve. Thus, the relatively large brains of mammals (which presumably facilitate complex behavior and learning), long gestation periods and (in some species) continued growth of the central nervous system after birth, prolonged association of parents and young, and high metabolic rates and endothermy (with the resulting high resource requirements) may be viewed as conditions that are conducive to the development of interdependent social units (Eisenberg 1981).

Of course, not all mammals are social. Mammalian characteristics do not necessarily lead to sociality: Solitary and social species are known among marsupials and placentals. Monotremes appear to be solitary, but the three living monotreme species provide too small a sample to form a basis for speculations about the phylogenetic origins of sociality among mammals (Grant 1983). Of course, the social behavior of mammals does not operate in a vacuum—it is only one part of the biology of a species. Social behavior interacts with other kinds of behavior such as food gathering and predator avoidance, with the morphological and physiological characteristics of a species, and with the distribution of the resources in the habitat. Our emphasis in this chapter is on that interaction, and we illustrate the interrelationships of behavior and ecology with examples drawn from both predators and their prey.

The social behavior of mammals is an area of active research, and our treatment is necessarily limited. Additional information can be found in reviews by Alexander (1974), Morse (1980), Eisenberg (1981), and Eisenberg and Kleiman (1983). Some interesting examples of the costs and benefits of social behavior were briefly reviewed by Lewin (1987).

Population Structure and the Distribution of Resources

From an ecological perspective, the distribution of resources needed by a species is usually a major factor in determining its social structure. If resources are too limited to allow more than one individual of a species to inhabit an area, there is little chance of developing social groupings. Thus, the distribution of resources in the habitat and the amount of space needed by an individual to fill its resource requirements are important factors influencing the sociality of mammals.

Most animals have a **home range**, an area within which they spend most of their time and find the food and shelter they need. Home ranges are not defended against the incursions of other individuals (an area that is defended is called a **territory**), and the value of a home range probably lies in the familiarity of an individual animal with the locations of food and shelter. Many species of mammals and birds employ a type of foraging known as trap lining, in which they move over a regular route and visit specific places where food is available. For example, a mountain lion may carefully approach a burrow where a marmot lives, beginning its stalk long before it can actually see whether the marmot is outside its burrow. This kind of behavior demonstrates a familiarity with the home range and with the resources that are likely to be available in particular places.

The **resource dispersion hypothesis** predicts that the size of the home range of an individual animal will depend primarily on two factors: (1) the resource needs of the individual, and (2) the distribution of resources in the environment. That is, individuals of species that require large quantities of a resource such as food should have larger home ranges than individuals of species that require less food. Similarly, the home ranges of individuals should be smaller in a rich environment than in one where resources are scarce. The resource dispersion hypothesis is a very general statement of an ecological principle. It applies equally well to any kind of animal and to any kind of resource. The resources usually considered are food, shelter, and access to mates. In Chapter 18 we considered the role of monopolization of resources by individuals in relation to the mating systems of birds, and here we discuss the role of resource dispersion in relation to home range size and sociality of mammals.

Body Size and Resource Needs

Studies of mammals have concentrated on food as the resource of paramount importance in determining the sizes of home ranges. Recent reviews by Gittleman and Harvey (1982), Clutton-Brock and Harvey (1983), Macdonald (1983), and McNab (1983) will provide additional details. Food requirements are assumed to be equivalent to energy requirements, and we discussed the relationship between energy requirements and body size in Chapter 4.

You will recall that the energy consumption of vertebrates increases in proportion to body mass raised to a power that is usually between 0.75 and 1.0. If one assumes that energy requirements determine home range size, one can predict that home range size will also increase in proportion to the 0.75 or 1.0 power of body mass. That prediction appears to be correct in general, but perhaps wrong in detail (Figure 22–1). That is, the sizes of the home ranges of mammals do increase with increasing body size, but the rates of increase (the slopes in Figure 22–1) are somewhat greater than expected. Home ranges appear to be proportional to body mass raised to powers between 0.75 and 2.0. This relationship between energy requirements and home range sizes suggests that energy needs are important in determining the size of the home range, but that additional factors are involved. One possibility is that the efficiency with which an animal can find and utilize resources decreases as the size of a home range increases. If that is true, the sizes of home ranges would be expected to increase with the body sizes of animals more rapidly than energy requirements increase with body size.

The failure of the resource dispersion hypothesis to predict the exact relationship between body size and the size of home ranges indicates that we have more to learn about how animals use the resources of their home ranges. So far we have been assuming that resources are distributed evenly throughout the home range, but that assumption overlooks the structural complexity of most habitats. What insights can be obtained from a more realistic consideration of how mammals gather food?

The Availability of Resources

Three factors seem likely to be important in determining the availability of food to mammals—what they eat, whether their food is evenly dispersed through the habitat or is found in patches, and how they gather their food. We will consider examples of each of these factors.

Dietary Habits   Figure 22–1 shows that home range size increases with body size, and that dietary habits also affect the size of the home range. Herbivores have smaller home ranges than do omnivores of the same body size, and carnivores have larger home ranges than do herbivores or omnivores. For example, the home range of an elk (an herbivore) that weighs 100 kilograms is approximately 100 hectares. A bear (an omnivore) of the same body size has a home range larger than 1000 hectares, and a tiger (a carnivore) has a home range of more than 10,000 hectares.

The relationship between home range size and dietary habits of mammals probably reflects the abundance of different kinds of food. The grasses and leaves eaten by some herbivores are nearly ubiquitous, and a small home range provides all the food an individual requires. The plant materials (seeds and fruit) eaten by omnivores are less abundant than leaves and grasses, and different species of plants produce seeds and fruit at different seasons. Thus, a large home range is probably necessary to provide the food resources needed by an individual omnivore. The vertebrates that are eaten by carnivores are still less abundant, and a correspondingly larger home

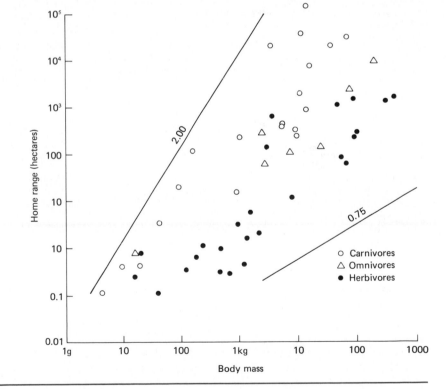

Figure 22–1.   Home-range size of mammals as a function of body mass. The lines have slopes of 0.75 and 2.0 which illustrate allometric increases in home-range size on this double logarithmic scale. (From B. K. McNab, 1983, in *Advances in the Study of Mammalian Behavior,* edited by J. F. Eisenberg and D. G. Kleiman, Special Publication 7, The American Society of Mammalogists.)

range is apparently needed to ensure an adequate food supply. Furthermore, not all mammals classified as members of the order Carnivora are flesh eaters (Box 22–1). For species within the Carnivora, the size of the home range increases in proportion to the importance of flesh in the diet.

Distribution of Resources   We have continued to assume that resources are evenly distributed through the habitat and that one part of a home range is equivalent to another part in terms of the availability of food. This assumption may be valid for some grazing and browsing herbivorous mam-

Box 22–1. The Herbivorous Carnivores

The mammals grouped as the Carnivora are not by any means all flesh eaters. The proportion of flesh in the diets of carnivores ranges from 100 percent to 0 percent (Table 22–1). Most felids feed almost entirely on flesh, but other families of carnivores include species that have little or no flesh in their diets. The giant panda (a bear) is entirely herbivorous, and most other species of bears eat little flesh.

Among the Carnivora, species that include a large proportion of flesh in their diets have larger home ranges than species that are omnivorous, insectivorous, or herbivorous. This relationship within the Carnivora repeats the correlation between diet and home range size previously noted in a comparison of different orders of mammals (Figure 22–1).

Table 22–1.   Diets of representative Carnivora.

Species	Flesh as Percent of the Diet	Species	Flesh as Percent of the Diet
Felidae		Hyaenidae	
African lion (*Panthera leo*)	100	Spotted hyaena (*Crocuta crocuta*)	100
Tiger (*Panthera tigris*)	98	Striped hyaena (*Hyaena hyaena*)	53
Bobcat (*Felis silvestris*)	94	Aardwolf (*Proteles cristatus*)	6
Serval (*Leptailurus serval*)	84	Viverridae	
Canidae		White-tailed mongoose (*Ichneumia albicauda*)	61
Timber wolf (*Canis lupus*)	100	African palm civet (*Nandinia binotata*)	15
Fox (*Vulpes vulpes*)	64	Banded mongoose (*Mungos mungos*)	9
Indian jackal (*Canis aureus*)	58	Procyonidae	
African big-eared fox (*Otocyon megalotis*)	14	Cacomistle (*Bassariscus astutus*)	31
Mustelidae		Raccoon (*Procyon lotor*)	27
Ermine (*Mustela erminea*)	95	Ursidae	
American marten (*Martes americana*)	73	Polar bear (*Thalarctos maritimus*)	80–95
American badger (*Taxidea taxus*)	80	Alaska brown bear (*Ursus arctos*)	5
Old World badger (*Meles meles*)	22	Giant panda (*Ailuropoda melanoleuca*)	0
Striped skunk (*Mephitis mephitis*)	14		

Source: Modified from J. L. Gittleman and P. H. Harvey, 1982, *Behavioral Ecology and Sociology* 10:57–63.

mals, but it is clearly not true for mammals that seek out fruiting trees (which represent patches of food) or for any carnivorous mammal that preys on animals that occur in groups. The sizes of the home ranges of animals that utilize food that occurs in patches should reflect the quality of the habitat—home ranges should be small if concentrations of food are abundant and large if concentrations of food are widely dispersed.

That relationship was well illustrated by a study of the home ranges of arctic foxes (*Alopex lagopus*) in Iceland (Hersteinsson and Macdonald 1982). The foxes lived in social groups that consisted of one male and two females plus the cubs from the current year. The home ranges of the individuals of a group overlapped widely with each other, and there was very little overlap between the home ranges of members of different groups. The home ranges of the foxes were located along the coast and did not extend far into the uplands (Figure 22–2). Between 60 and 80 percent of the diet was composed of items the foxes found on the shore—the carcasses of seabirds, seals, and fishes, and invertebrates from clumps of seaweed that had been washed up on the beach. Little food was available for foxes in the uplands. The foxes concentrated their foraging on the beach during the 3 hours before low tide, which is the best time for beachcombing. The foxes approached the shore carefully, stalking along gullies. They crept out on the beach carefully, apparently looking for birds that were resting or feeding. If birds were present, the foxes would stalk them. If no birds were present, the foxes began to search the beach for carrion.

Three groups of foxes were studied in detail, using radiotelemetry to follow the movements of individuals. The areas of the home ranges varied more than twofold, from 8.6 to 18.5 square kilometers (Table 22–2). The sizes of the home ranges were more similar when only the coastline was considered: Each territory included between 5.4 and 10.5 kilometers of coastline.

The coastline was, of course, the source of most of the food the foxes were eating, but not all areas of the coastline accumulated floating objects.

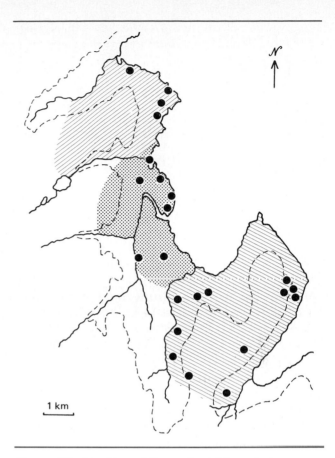

Figure 22–2. Map of the territorial boundaries of three groups of arctic foxes in Iceland. The 200-meter contour line is shown. Black dots mark the sites of dens used by the foxes. (From P. Hersteinsson and D. W. Macdonald, 1982, in *Telemetric Studies of Vertebrates*, edited by C. L. Cheeseman and R. B. Mitson, *Symposia of the Zoological Society of London, No. 49*, Academic Press, London.)

The distribution of food on the beaches was patchy and depended on the directions of currents. As a result, some parts of the shore were more productive than others. The length of productive coastline occupied by each group of foxes was quite similar—from 5.4 to 6.0 kilometers. Farmers in Iceland use driftwood to make fenceposts, and the amount of driftwood that was harvested by the farmers from the coasts in the home ranges of the three

Table 22–2.   Home ranges of three groups of arctic foxes in Iceland.

	Group 1	Group 2	Group 3	Average
Total area (km²)	10.3	8.6	18.5	12.5
Length of coastline (km)	5.6	5.4	10.5	7.2
Length of productive coastline (km)	5.6	5.4	6.0	5.7
Driftwood productivity logs/year)	1800	1800	2100	1900

Source: P. Hersteinsson and D. W. Macdonald, 1982, pages 259–289 in *Telemetric Studies of Vertebrates,* edited by C. L. Cheeseman and R. B. Mitson, Symposia of the Zoological Society of London, No. 49, Academic Press, London.

groups of foxes varied only from 1800 to 2100 logs per year. Because both driftwood and carrion are moved by currents and deposited on the beaches, the harvest of driftwood by farmers probably reflects the harvest of carrion by the foxes. Thus, the sizes of the home ranges of the three groups appear to match the distribution of their most important food resource, and the productive areas of the home ranges of the three groups are very similar despite the twofold difference in total areas of their home ranges.

Group Size and Hunting Success   It is readily apparent that the average size of the home range of a species of mammal can influence the social system of that species. Individuals of a species probably encounter each other frequently when home ranges are small, whereas individuals of species that roam over thousands of hectares may rarely meet. Thus, the distribution of resources in relation to the resource needs of a species is one of the factors that can set limits to the degree to which social groupings can occur. However, sociality may influence resource distribution if groups of animals are able to exploit resources that are not available to single individuals.

The influence of sociality on resource distribution is most readily apparent among predatory animals that can hunt individually or in groups. Some species of prey are too large for an individu-

al predator to attack, but are vulnerable to attack by a group of predators. For example, spotted hyaenas (*Crocuta crocuta*) weigh about 50 kilograms. When hyaenas hunt individually they feed on Thomson's gazelles (*Gazella thomsoni*, 15 kilograms) and juvenile wildebeest (*Connochaetes taurinus*, about 30 kilograms) (Figure 22–3). However, when hyaenas hunt in packs they feed on adult wildebeest (about 200 kilograms) and zebras (*Equus burchelli*, about 220 kilograms). Some species of prey have defenses that are effective against individual predators but less effective with groups of predators. For example, the success rate for solitary lions (*Panthera leo*) hunting zebras and wildebeest is only 15 percent, whereas lions hunting in groups of six to eight individuals are successful in up to 43 percent of their attacks. Groups of lions make multiple kills of wildebeest more than 30 percent of the time, but individual lions kill only a single wildebeest.

The relationship of sociality and body size of a predator to the size of its prey is shown in Figure 22–4: Social predators (defined in this study as those that hunt in groups of eight to ten individuals) attack larger prey than less social predators (average group sizes of 1.6 to 3.1 individuals), and these weakly social predators attack larger prey than do solitary predators (average group sizes of 1.0 to 1.3 individuals). Thus, one consequence of sociality for predatory mammals appears to be an

Figure 22–3. Spotted hyaenas (a) prey on small animals like the Thomson's gazelle (b) when they hunt individually. When they hunt in packs, they attack larger prey such as the wildebeest (c) and zebras (d). (Photographs by Sara Cairns.)

increase in the potential food resources of an environment: Individual predators may be able to extend the range of prey species they can attack by hunting in groups. Of course, the food requirement of a group of predators is the sum of the individual requirements of the members of the group, and the home range required to sustain a group is greater than the range needed by an individual. When species of carnivorous mammals that hunt in groups are included in the analysis, home range size appears to be proportional to the energy requirements of an individual multiplied by the number of individuals in the group.

Advantages of Sociality

The advantages of cooperative hunting may provide a basis for sociality among some carnivorous mammals, but sociality is not limited to species of mammals that hunt in groups and the potential advantages of sociality are not confined to preda-

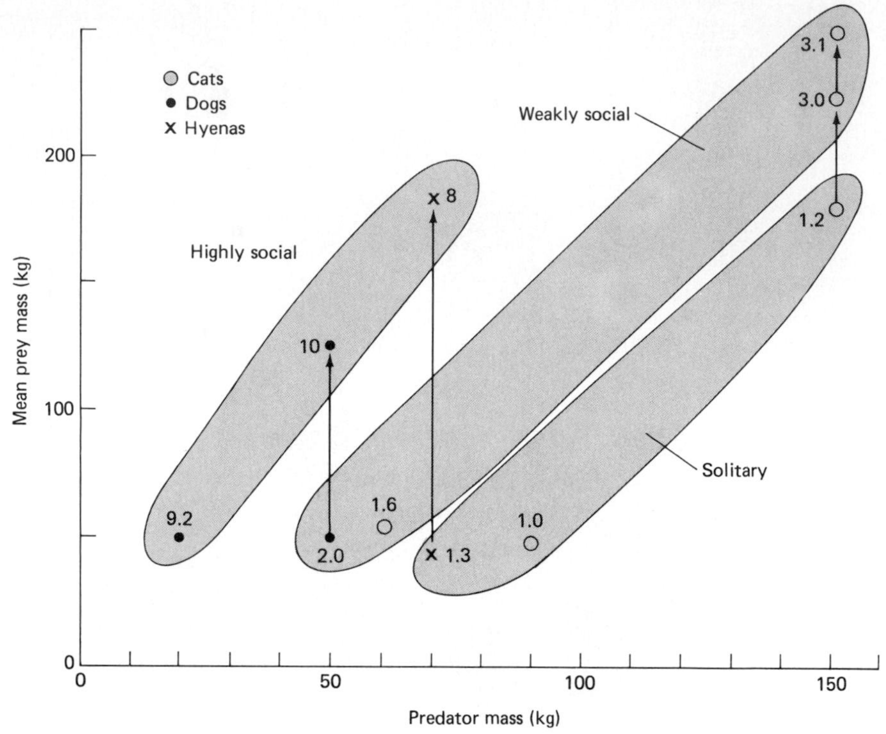

Figure 22–4. Size of prey in relation to predator mass for solitary predators and for predators that hunt in small or large groups. Vertical lines connect points for species that hunt in groups of variable size. (From B. K. McNab, 1983, in *Advances in the Study of Mammalian Behavior,* edited by J. F. Eisenberg and D. G. Kleiman, Special Publication 7, The American Society of Mammalogists.)

tory behavior. Mammals may derive benefits from sociality in terms of avoiding predation and in reproduction and care of young.

Defenses Against Predators

One probable advantage of sociality is a reduction in the risk of predation for an individual that is part of a group compared to the risk for a solitary individual. The benefits of sociality in avoiding predation take many forms. A group of animals may be more likely to detect the approach of a predator than an individual would be, simply because a group has more eyes, ears, and noses to keep watch. Alternatively, an individual in a group may be able to devote a larger proportion of its time to feeding and less to watching for predators than a solitary individual can. Mammals that live in groups generally occur in open habitats, whereas solitary species are usually found in for-

ests, and that relationship may reflect in part the antipredator aspects of group living. It is important to note that the benefits of sociality in predator avoidance result from the reduction in the risk of predation for an *individual* that is part of a group, not for the species as a whole. This view was articulated most clearly by William Hamilton (1971), who coined the term **selfish herd** to describe the protection received by an individual that stays close to others.

Sociality and Reproduction

Groupings of animals are important factors in mating systems and in care of young. In Chapter 18 we described mating systems in the context of avian biology, and most aspects of that discussion apply equally well to the mating systems of mammals. In the next section we extend the analysis by considering the specific relationships among body size, habitat, diet, antipredator behavior, and mating systems of several species of African ungulates.

The extensive period of dependence of many young mammals on their parents provides a setting in which many benefits of sociality can be manifested. Maternal care of the offspring is universal among mammals, and males of many species also play a role in parental care. Group living provides opportunities for more complex interactions among adults and juveniles that involve various sorts of **alloparental behavior** (care provided by an individual that is not a parent of the young receiving the care). Collaborative rearing of young of several mothers is characteristic of lions and of many canids. Frequently, nonbreeding individuals join the mothers in caring for the young. Among dwarf mongooses (*Helogale undulata*) this kind of behavior extends to the care of sick adults, and reports of similar behaviors exist for mammals as diverse as elephants and cetaceans. Many social groups of mammals consist of related individuals, and these helpers may increase their inclusive fitness by assisting in rearing the offspring of their kin. (See Chapter 18 for a discussion of helpers.)

Body Size, Diet, and the Structure of Social Systems

The complex relationships among body size, sociality, and other aspects of the ecology and behavior of herbivorous mammals are illustrated by the variation in social systems of African antelopes (family Bovidae) (Estes 1974, Jarman 1974). The smallest species of these bovids have adult weights of 3 to 4 kilograms (the dik-diks, *Madoqua*, and some duikers, *Cephalophus*), and the largest (the African buffalo, *Syncerus caffer*) weighs 900 kilograms (Figure 22–5). The smallest species are forest animals that browse on the most nutritious parts of shrubs, live individually or in pairs, defend a territory, and hide from predators (Table 22–3). The largest species (including the 300-kilogram eland, *Taurotragus oryx*, and the African buffalo)

are grassland animals that graze unselectively, live in large herds, are migratory, and use group defense to deter predators. Species with intermediate body sizes are also intermediate in these ecological and behavioral characteristics. It seems likely that the correlated variation in body size, ecology, and behavior among these bovids reveals functional relationships among these aspects of their biology. How might such diverse features of mammalian biology interact?

The feeding habits of these antelope appear to provide a key that can be used to understand other aspects of their ecology and behavior. The food habits of the different species are closely correlated with body size and the habitats in which the species live. In turn, those relationships are important in setting group size. The size of a group determines the distribution of females in time and space, and this is a major factor in establishing the mating system used by males of a species. Group size also plays an important role in determining the appropriate antipredator tactics for a species. Mating systems and antipredator mechanisms are central factors in the social organization of a species.

Body Size and Food Habits

Antelope are ruminants, relying on symbiotic microorganisms in the rumen to convert cellulose from plants into compounds that can be absorbed by the vertebrate digestive system (Box 21–1). The effectiveness of ruminant digestion is proportional to body size. This relationship exists because the volume of the rumen in species with different body sizes is proportional to body mass, whereas metabolic rates are proportional to the 0.75 power of body mass. The ecological consequence of this difference in allometric slopes is illustrated in Figure 22–6: A large ruminant has proportionately more capacity to process food than a small ruminant. As one moves downward to animals of very small body size, the metabolic requirements become high in relation to the volume of rumen available to ferment plant material.

Because of this relationship, small ruminants

Figure 22–5. The family Bovidae includes species with a wide range of adult body sizes. The dik-dik (a) is among the smallest species, the impala (b) is medium-size, and the African buffalo (c) is one of the largest. [(a) Photographed by Jack Cranford; (b,c) by Sara Cairns.]

must be more selective feeders than large ruminants. That is, a large ruminant has so much volume in its rumen that it can afford to eat large quantities of food of low nutritional value. It does not extract much energy from a unit volume of this food, but it is able to obtain its daily energy requirements by processing a large volume of food.

Small ruminants, in contrast, must eat higher-quality food and rely upon obtaining more energy per unit volume from the smaller volume of food that they can fit into their rumen in a day. In fact, 40 kilograms is the approximate lower limit of body size at which an unselective ruminant can balance its energy budget—species larger than 40

Table 22–3. Correlations of the ecology and social systems of African ungulates.

Diet Type	Examples	Body Mass (kg)	Food Habits	Group Size	Mating System	Predator Avoidance
I	Dikdik, some duikers	3–20	Highly selective browser	1 or 2	Stable pair, territorial	Hide
II	Thomson's gazelle, impala	20–100	Moderately selective browser and grazer	2 to 100	Male territorial in breeding season, temporary harems	Flee
III	Wildebeest, hartebeest	100–200	Grazers, selective for growth stage	Large herds	Nomadic, temporary harems	Flee, hide in herd, threaten predator
IV	Eland, buffalo	300–900	Grazers, unselective	Large herds	Male hierarchy	Group defense

Source: Modified from P. J. Jarman, 1974, *Behaviour* 58:215–267.

kilograms can be unselective grazers, whereas smaller species must eat only the most nutritious parts of plants (Van Soest 1982).

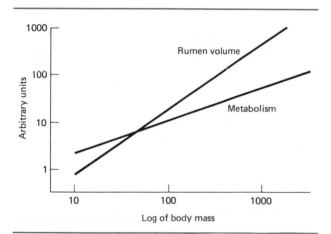

Figure 22–6. Illustration of the consequences of the different allometric relationships between rumen volume, which increases in proportion to body size (an allometric slope of 1), and energy requirement, which is proportional to metabolism (an allometric slope of 0.75). Both axes are drawn with logarithmic scales, and the scale of the vertical axis is in arbitrary units.

The species of antelopes in this example can be divided into four feeding categories:

Type I species are very selective grazers and browsers. They feed preferentially on certain species of plants, and they choose the parts of those plants that provide the highest-quality diet—new leaves (which have a higher nitrogen content than mature leaves) and fruit. Dik-diks and duikers fall in this category, and they have adult body masses between 3 and 20 kilograms.

Type II species are moderately selective grazers and browsers. They eat more parts of a plant than the type I species, and they may have seasonal changes in diet as they exploit the availability of fresh shoots or fruits on particular species of plants. Thomson's gazelle (*Gazella thomsoni*) and the impala (*Aepyceros melampus*) weigh 20 to 100 kilograms and have type II diets.

Type III species are primarily grazers that are unselective for species of grass, but selective for the parts of the plant. That is, they eat the leaves and avoid the stems. Hence, they are selecting for a growth stage: They avoid grass that is too short, because that limits food intake, and also avoid grass that is too long (because it has too many stems that are low-quality food). Wildebeest (*Connochaetes taurinus*) and hartebeest (*Alcelaphus buselaphus*), which weigh about 200 kilograms, are type III feeders,

Type IV species are unselective grazers and browsers. They eat all species of plants and all parts of the plant. Eland and buffalo (300 and 900 kilograms) are type IV species.

Food Habits and Habitat

The food habits of the different species of antelope are important in determining what sorts of habitats provided the resources they need. Selective feeding operates at three levels: the vegetation type, the species and individual groups of plants, and the parts of plants eaten. The type of vegetation present largely depends on the habitat—forests contain shrubs and bushes, whereas the plains are covered with grass. The resources needed by species with type I diets are found in forests where the presence of a diversity of species with different growth seasons ensures that fresh leaves and fruits will be available at all times of the year. Species with type II diets are found in habitats that are a mosaic of woodland and grassland, and type III species (which are primarily grazers) are found in savanna and grassland. Both kinds of species may move from place to place in response to patterns of rainfall. For example, wildebeest require grass that has put out fresh new growth, but that has not had time to mature. To find grass at this growth stage, wildebeest have extensive nomadic movements that follow the seasonal pattern of rain on the African plains.

Type IV feeders eat almost any kind of plant material, and they can find something edible in almost any habitat. They occupy a range of habitats, including grassland and brush, and do not have nomadic movements.

Habitat and Group Size

The habitats in which antelope feed and the types of food they eat set certain constraints on the sorts of social groupings that are possible. For example, species with type I diets live in forests and feed on scattered, distinct items. They eat an entire leaf or fruit at a bite and they must move between bites. A type I feeder completely removes the items it eats, so it changes the distribution of resources in its habitat (Figure 22–7). A second individual cannot feed close behind the first, because the food resources of an individual bush or shrub are entirely consumed by the first individual to feed there. As a result, the feeding behavior of species with type I diets makes it impossible for a group of animals to feed together. If one individual attempts to follow behind another to feed, the second animal must search to find food items over-

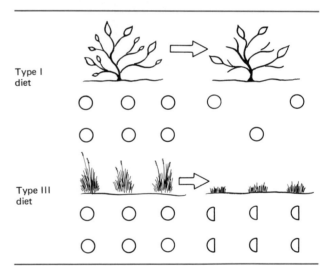

Figure 22–7. Diagrammatic illustration of the effect of feeding by a selective browser with a type I diet (a) and a grazer with a type III diet (b). The browser removes entire food items, thereby changing the distribution of food in the habitat, as well as the abundance of food. The grazer removes part of a grass clump, changing the abundance of food in the habitat but not its distribution. (From P. J. Jarman, 1974, *Behaviour* 58:215–267.)

looked by the individual ahead, and consequently it falls behind. Alternatively, it can move aside from the path of the first animal to find an area that has not already been searched. In either case, small animals in dense habitats rapidly lose track of each other and no cohesive group structure is maintained.

Instead, type I species are solitary or occur in pairs, and the individuals of a pair are only loosely associated as they feed. A type I diet places a premium on familiarity with a home range because a tree or bush is a patch of food that must be visited repeatedly to harvest fruit or new leaves as they appear. Also, the resources represented by the patches of food can be defended against intruders, and species of antelope with type I diets are territorial. The resources of a territory do not change very much from one season to the next, and the only seasonal variation in the group size of species with type I diets results from the presence of a juvenile in its parents' territory for part of the year.

Species with type II and type III diets are less selective than species with type I diets, and their feeding has less impact on the distribution of resources. These species do not remove all of the food resource in an area, and other individuals can feed nearby. Type III feeders, in particular, graze as they walk—taking a bite of grass, moving on a few steps, and taking another bite. This mode of feeding changes the abundance of food, but not its distribution in space, and herds of wildebeest graze together, all moving in the same direction at the same speed and maintaining a cohesive group. Rainfall is a major determinant of the distribution of suitable food in the habitat of these species. The rainstorms that stimulate the growth of grass are erratic, and the patch sizes in which food resources occur are enormous—hundreds of square kilometers of new grass where rain fell are separated by hundreds of kilometers of old, dry grass that did not receive rain. Instead of having home ranges or territories, species with type II and type III diets move nomadically with the rains. Group sizes change as the distribution of resources changes, from half a dozen to 60 individuals for species with type II diets, and from herds of 300 or 400 to

superherds of many thousands of individuals during the nomadic movements of wildebeests.

Species of bovids with type IV diets are so unselective in their choice of food that they can readily maintain large groups, and herds of buffalo number in the hundreds. Because these species can eat almost any kind of vegetation, the distribution of resources does not change seasonally and the size of the herds is stable.

Group Size and Mating Systems

The mating systems used by African antelope are closely related to the size of their social groups and the distribution of food because those are the major factors that determine the distribution of females and the potential for males to obtain opportunities to mate by monopolizing resources that females need. The females of species with type I diets are dispersed because the distribution of resources in the habitat does not permit groups of individuals to form. A male of a type I species can defend food resources, but individuals must disperse through the territory to feed and it is not feasible for a male to maintain a harem of females. Males of type I species pair with one female; the male defends its territory year round, the pair bond with an individual female appears to be stable, and offspring are driven out of the territory as they mature.

A group of individuals of type II species contains several males and females. The evenly distributed nature of food of these species makes it difficult for a male to monopolize resources. Only some of the males are territorial, and even this territoriality is manifested for only part of the year. A territorial male tries to exclude other males from its territory and to gather harems of females. Exclusive mating rights are achieved by holding a patch of ground and containing females within it, driving them back if they try to leave. These species have no long-term association between a male and a particular female.

Type III species are nomadic, and males establish territories only when the herd is stationary. During these periods male wildebeest hold harems of females within their territories, but the association between a male and females is broken when the herd moves on. However, mothers and their daughters maintain associations for 2 or 3 years. Unmated male wildebeest form bachelor herds with hierarchies, and individuals at the top of a bachelor hierarchy try to displace territorial males. If a territorial male is displaced, it joins a bachelor herd at the bottom of the hierarchy and must work its way up to the top before it can challenge another territorial male.

The social structure of buffalo herds differs in two respects from that of wildebeests: (1) Each herd includes many mature males that form a dominance hierarchy. The individuals near the top of the hierarchy have access to receptive females, but no territoriality or harem formation is seen. (2) The female membership of the herd is fixed, and this situation results in a degree of genetic relatedness among all the members of a herd of buffalo. That genetic relationship among individuals creates situations in which kin selection may be a factor in the social behavior of the African buffalo.

Mating Systems and Predator Avoidance

Prey species have a variety of ways to avoid predators, but only some will work in a given situation. In general, a prey species can (1) avoid detection by a predator, (2) flee after it has been detected but before the predator attacks, (3) flee after the predator attacks, or (4) threaten or attack the predator. Body size, habitat, group size, and the mating system all contribute to determining the risk of predation faced by a species and which predator avoidance methods are most effective.

Predators usually attack prey that are the same size as the predator or smaller. Small species of animals potentially have more species of predators than do large species. Species of antelopes with type I diets are small and, consequently, they are at risk from many species of predators. Furthermore, small antelopes may not be able to run fast enough to escape a predator after it has at-

tacked. On the other hand, these small antelopes live in dense habitats, where they are hard to see. They are cryptically colored and secretive, and they rely on being inconspicuous to avoid detection by predators. If they are pursued, they may be able to use their familiarity with the geography of their home range to avoid capture.

Groups of animals are more conspicuous to predators than are individuals, but groups also have more eyes to watch for the approach of a predator. Species of antelopes with type II diets live in small groups in open habitats, where they can detect predators at a distance. These antelopes avoid predators by fleeing either before or after the predator attacks (Box 22–2). Small predators may be attacked by the antelope, but usually only when a member of the group has been captured, and this sort of defense is normally limited to a mother protecting her young; the rest of the group does not participate.

Species of antelope in the type III diet category are large enough to have relatively few predators, and in a group they may be formidable enough to scare off a predator. Wildebeest sometimes form a solid front and walk toward a predator: This behavior is effective in deterring even lions from attacking. Many predators of wildebeest focus their attacks on calves, and defense of a calf is usually undertaken only by the mother. Much of the antipredator behavior of wildebeest depends on the similarity of appearance of individuals in the herd to each other. Field observations have shown that the individuals in a group of animals that are distinctive in their markings or behavior are most likely to be singled out and captured by predators. One of the unavoidable events that makes a female wildebeest distinctive is giving birth to a calf, and the reproductive biology of the species has specialized features that appear to minimize the risks associated with giving birth. The breeding season and birth of wildebeests are highly synchronized. Mating occurs in a short interval, and as a consequence 80 percent of the births occur within a period of 2 or 3 weeks. Furthermore, nearly all of the births that will take place on a day occur in the morning in large aggregations of females, all giv-

Box 22–2. Unprofitable Prey?

A distinctive behavior—stotting, which consists of leaping vertically into the air—is used by some species of antelope when they are being pursued by a predator. The function of stotting is unclear: It may be an alarm signal that alerts other individuals of the species to the presence of a predator, but the advantage to the individual that gives the warning is not clear. Altruistic behavior of this type is usually associated with kin selection, but the individuals in groups of antelope with type II diets are not closely related and do not show other types of altruistic behavior such as group defense of young. It has recently been suggested that some behaviors of prey species that had been considered to be altruistic alarm signals are really signals directed to the predator by fleet-footed prey.

Alarm signals are given by many other kinds of vertebrates. A familiar example is the white underside of the tail of deer (Figure 22–8). A deer that sees a predator at a distance does not immediately flee but stands watching the predator. It may flick its tail up and down, exposing the white ventral surface in a series of flashes. This kind of behavior is not limited to mammals. For example, several related species of fleet-footed lizards that live in open desert habitats have dorsal colors that blend with the substrate on which they live, and a pattern of white with black bars on the underside of the tail. These lizards stand poised for flight as a predator approaches, looking back over their shoulder at the predator. The tail is curled upward, exposing the contrasting black and white pattern on its ventral surface, and

Figure 22–8. Alarm signals to conspecifics or signals to a predator? The white-tail deer displays the white underside of its tail when it detects a predator.

waved from side to side. Is it possible that in behaviors of this sort the prey animal is signaling to the predator that it has been detected, and to attack it will be unprofitable because the prey is ready to flee? This intriguing hypothesis requires experimental test of the prediction that predators are less likely to attack prey individuals that engage in these behaviors than individuals that do not signal their awareness of the predator's approach. However, if the hypothesis can be supported by experiments, some puzzling examples of apparently altruistic signals can be reinterpreted as behavior that benefits the individual giving the signal.

ing birth at once. A female wildebeest who is slightly out of synchrony with other members of her group can interrupt delivery at any stage up to emergence of the calf's head so as to join the mass parturition. Presumably, this remarkable synchronization and control of parturition reflects the advantage of presenting predators with a homogeneous group of cows and calves rather than a group with only a few calves that could readily be singled out for attack.

Buffalo are formidable prey even for a pride of lions and they escape much potential predation simply as a result of their size. When buffalo are attacked, they engage in group defense, and if a calf is captured its distress cries bring many members of the group to its defense. This altruistic behavior probably represents kin selection, because the stability of the female membership of buffalo herds results in genetic relationships among the individuals.

Phylogeny of Social Behavior of Antelope

It is tempting to view the continuum of increasing sociality from antelopes with type I diets through species with type IV diets as an evolutionary progression from simple social systems to complex ones. However, this view would almost certainly be wrong. The relationships among body size, diet, habitat selection, group size, mating system, and antipredator behavior form a web, not a ladder. If any one aspect of the biology of antelopes can be seen as limiting the range of possibilities for other features of their biology, body size is probably the key factor. The relationships among body size, rumen size, and metabolic requirements appear to define the four dietary types we have discussed and to determine what sorts of habitats supply the resources they require. In turn, the habitats in which different species of antelope live constrain the sorts of social systems they can exhibit.

If that is so, a progression from simple social systems to more complex ones is unlikely. Rumi-

nant digestion is most advantageous for medium-size animals, and the first ruminants probably weighed about 100 kilograms and had diets like those of impalas and gazelles (Van Soest 1982). If dietary habits and habitat selection were important influences on the ecology and behavior of the early ruminants, it is likely that the ancestral social system was some form of temporary territoriality and harem formation. The monogamous mating system seen in dik-diks and the huge herds of buffalo sharing some degree of kinship probably both represent derived behaviors.

Social Systems Among Primates

The phylogenetic relationship of humans to other primates has led some biologists to assume that these are the animals that should have the most elaborate social systems, and that the study of the social systems of primates will provide information about the evolution of human behavior. Both assumptions are controversial: Increasing information indicates that complex social systems exist among many kinds of vertebrates other than primates. For example, nothing like the cooperative hunting and food sharing that is known for lions and some canids has been reported for primates. Interpretation of primate behavior in the context of human evolution is fraught with difficulty and must be approached cautiously. Nonetheless, primates do have elaborate and complex social systems, and more long-term research has focused on the social systems of primates than on any other vertebrates.

The approximately 200 species of primates (Table 22–4) are ecologically diverse. They live in habitats ranging from lowland tropical rain forests, to semideserts, to northern areas that have cold, snowy winters. Some species are entirely arboreal, whereas others spend most of their time on the ground. Many are generalist omnivores that eat fruit, flowers, seeds, leaves, bulbs, insects, bird eggs, and small vertebrates, but many of the colobus (*Colobus*) and howler monkeys (*Alouatta*) are specialized folivores (leaf eaters) with sacculated stomachs in which bacteria and protozoans ferment cellulose, and some of the small prosimians and callithricid monkeys are insectivores.

R. W. Wrangham (1982) proposed that the social systems of primates can best be classified on the basis of the amount of movement of females that occurs between groups (Table 22–5). Four categories can be defined on this basis:

1. **Female transfer systems.** In species with this type of social organization most females move away from the group in which they were born to join another social group. Because of this migration of females among groups, the females in a group are not closely related to each other. In contrast, males often remain with their natal groups and associations of male kin may be important elements of the social behaviors of these species of primates. Male chimpanzees, for example, cooperate in defending their territories from invasion by neighboring males. The majority of species of primates with female transfer systems live in relatively small social groups.

2. **Nonfemale transfer systems**. Most females of these species spend their entire lives in the group in which they were born. Social relations among the females in a group are complex and are based on kinship. Males of these species emigrate from their natal group as adolescents and may continue to move between groups as adults. In some of these species a single male lives with a group of females until he is displaced by a new male, whereas in other species several males may be part of the group and maintain an unstable dominance hierarchy among themselves. Group size is usually larger for nonfemale transfer species than for species with female transfer.

3. **Monogamous species.** These are found in pairs, sometimes accompanied by juvenile offspring. These species of primates show little sexual dimorphism, the sexes share parental care and territorial defense, and the offspring are expelled from the parents' territory during adolescence.

4. **Solitary species.** These live singly or as females with their infants and juvenile offspring.

Male prosimians maintain territories that include the home ranges of several females and exclude other males from their territories, whereas male orangutans do not defend territories, but instead repulse other males when a female within the male's home range comes into estrus.

Three ecological factors appear to be particularly important in shaping the social systems of primates, as they are for other vertebrates.

1. The defensibility of food resources appears to determine whether individuals will benefit from (a) not attempting to defend resources, (b) defending individual territories, or (c) forming long-term relationships with other individuals and jointly defending resources.
2. The distribution of food in time and in space may determine how large a group can be, and whether the group can remain stable or must break into smaller groups when food is scarce.
3. The risk of predation may determine whether individuals can travel alone or require the protection of a group, whether the benefit of the additional protection that is provided by a large versus a small group outweighs the added competition among individuals in a large group, and whether the presence of males is needed to protect young.

Life within a group of primates is a balance between competition and cooperation (Figure 22–9). Competition is manifested by aggression. Some aggression—for example, the defense of food, rest, sleeping sites, or mates—is closely linked to resources. Other types of aggression involve establishing and maintaining dominance hierarchies, which may be an indirect form of resource competition if high-ranking individuals have preferential access to resources.

Cooperation, too, is diverse. Grooming behavior in which one individual picks through the hair of another, removing ectoparasites and cleaning wounds, is the most common form of cooperation. Other types of cooperation include sharing food or feeding sites, collective defense against predators, collective defense of a territory or a re-

Table 22–4. Taxonomy and social organization of living primates.

Taxon	Social Organization
Lemuroidea 　Aye-aye (*Daubentonia*, 1 species) 　Lemurs (*Lemur* and 9 other genera, 18 species) 　Indri (*Indri*, 1 species) 　Sifaka (*Propithecus*, 2 species)	Largely solitary or monogamous
Lorisoidea 　Bushbabies (*Galago*, 8 species) 　Lorises (*Loris*, 1 species; *Nycticebus*, 2 species) 　Potto (*Perodicticus*, 1 species) 　Angwantibo (*Arctocebus*, 1 species)	Largely solitary
Tarsoidea 　Tarsiers (*Tarsius*, 3 species)	Solitary or monogamous
Ceboidea = Platyrrhines (New World monkeys) 　Callimiconidae 　　Goeldi's marmoset (*Callimico*, 1 species)	Small groups with one resident male
Callithrichidae 　　Marmosets and tamarins (4 genera, 　　　15 species)	Largely monogamous pairs
Cebidae 　　Howler monkeys (*Alouatta*, 6 species) 　　Spider monkeys (*Ateles, Brachyteles, Lagothrix,* 　　　7 species) 　　Capuchin monkeys (*Cebus*, 4 species) 　　Squirrel monkeys (*Samiri*, 2 species) 　　Owl monkey (*Aotus*, 1 species) 　　Uakaris (*Cacajao*, 2 species) 　　Titis (*Callicebus*, 3 species) 　　Sakis (*Chiropotes* and *Pithecia*, 6 species)	Monogamous pairs, or small to large 　groups

Source: Modified from B. B. Smuts et al., editors, 1987, *Primate Societies,* University of Chicago Press, Chicago.

source within a home range, and the formation of alliances between individuals. Two-way, three-way, and even more complex alliances that function during competition within a group are common among primates.

Kinship and the concept of inclusive fitness play important roles in the interpretation of primate social behavior. A behavior must not decrease the fitness of the individual exhibiting the behavior if it is to persist in the repertoire of a species. Fitness is nearly impossible to demonstrate in wild populations, and behaviorists normally search for effects that are likely to be correlated with fitness, such as access to females (for males), interbirth interval (for females), or the probability that offspring will survive to reproductive age. Behaviors that increase these measures are assumed to increase fitness. The behaviors may directly benefit the individual displaying the behavior (personal fitness) or they may be

Table 22–4. (Continued)

Taxon	Social Organization
Catarrhini (Old World monkeys and apes)	
Cercopithecoidea	
Cercopithecidae	Mostly small to large groups
Vervet monkey, guenons, and others (*Cercopithecus*, 17 species)	
Mangabeys (*Cercocebus*, 5 species)	
Macaques (*Macaca*, 12 species)	
Baboons (*Papio*, 4 species; *Theropithecus*, 1 species)	
Yellow or savanna baboon (*Papio cynocephalus*)	
Hamadryas baboon (*Papio hamadryas*)	
Drill (*Papio leucophaeus*)	
Mandrill (*Papio sphinx*)	
Gelada baboon (*Theropithecus gelada*)	
Colobus monkeys (*Colobus*, 7 species)	
Langurs (*Nasalis, Presbytis, Pygathrix, Rhinopithecus*, 20 species)	
Hominoidea (apes and humans)	
Hylobatidae	Monogamous
Gibbons (*Hylobates*, 9 species)	
Pongidae	
Orangutan (*Pongo*, 1 species)	Solitary
Panidae	
Gorilla (*Gorilla*, 1 species)	Small groups with a variable number of resident males
Chimpanzee (*Pan*, 2 species)	Closed social network containing several breeding males and females
Hominidae	Closed social network containing several breeding males and females
Human (*Homo*, 1 species)	

costly to the personal fitness of the individual but sufficiently beneficial to its close relatives to offset the cost to the individual (inclusive fitness).

Social Relationships Among Primates

Four general types of relationships among individuals have been described in the social behavior of primates. [For more details see Altmann (1980), Hinde (1983), Watts (1985), Cheney et al. (1986), Goodall (1986), and Smuts et al. (1987).] Most of our knowledge of primate social behavior comes from studies of cercopithecids that live in multimale groups (female transfer and nonfemale transfer systems).

Adult–Juvenile Associations Primates are born in a relatively helpless state compared to many mammals, and they are dependent on adults for

Table 22–5. Characteristics of the social systems of primates.

System	Group Size	Number of Males in Group	Male Behavior	Examples
Female transfer	Small	One or many	Territorial, harems, sometimes male kinship groups	Chimpanzee, gorilla, howler monkeys, hamadryas baboons, colobus monkeys, some langurs
Nonfemale transfer	Large	One or several	Male hierarchy, whole group (males and females) may exclude conspecifics from food sources	Most cercopithecines: yellow baboons, mangabeys, macaques, guenon monkeys
Monogamous	Male and female, plus juvenile offspring	One	Both sexes participate in territorial defense and parental care	Gibbons, marmosets, tamarins, indri, titis
Solitary	Individual, or female plus juvenile offspring	—	Range of male overlaps ranges of several females	Bushbabies, tarsiers, lorises, orangutans

Source: Based on R. W. Wrangham, 1982, pages 269–289 in *Current Problems in Sociobiology,* edited by King's College Sociobiology Group, Cambridge University Press, Cambridge.

(a)   (b)

Figure 22–9. Social behaviors of yellow baboons: (a) male friend grooming a female baboon in estrus; (b) aggression among male baboons. (Photographs by Carol Saunders.)

unusually long periods. The relationship of a mother to her infant is variable within a species— some mothers are protective, whereas others are permissive. Permissive mothers often wean their offspring earlier than protective mothers and may have shorter intervals between the birth of successive offspring, although this relationship has not been observed in all species. The offspring of permissive mothers may suffer higher rates of mortality than the offspring of protective mothers, and the incompetence of some first-time mothers appears to lead to high mortality among firstborn offspring. Older siblings often participate in grooming and carrying an infant, but they may also assault, pinch, and bite the infant while it is being fed or groomed by the mother. Allomaternal behavior provided by an adult female who is not the mother includes cuddling, grooming, carrying, and protecting an infant. Several factors seem to influence allomaternal behavior: Young infants are preferred to older ones, infants of high-ranking mothers receive more attention and less abuse than infants of low-ranking mothers, and siblings may participate more than unrelated females in allomaternal behavior (Nicolson 1987). Males of the monogamous New World primates participate extensively in caring for infants, carrying them for much of the day and sharing food with them, whereas the relationships of males of Old World primates with infants are more often characterized by proximity and friendly contact than by care.

Female Kinship Bonds   The females of some species of semiterrestrial Old World primates live in groups that include several males and females: This social organization is typical of yellow baboons (*Papio cynocephalus*), several species of macaques (*Macaca*), and vervet monkeys (*Cercopithecus aethiops*). Females of these species remain for their entire lives in the troops in which they were born, whereas males migrate to other troops when they mature. The role of female kinship bonds is much smaller in female transfer systems because the females in a group are not closely related.

The females within a group form a dominance hierarchy and compete for positions in the hierarchy. Related females within a group are referred to as **genealogies**. Females consistently support their female relatives during encounters with members of other genealogies. The supportive relationship among females within a genealogy is an important element of the social structure of a group. For example, when their female kin are nearby, young animals can dominate older and larger opponents from subordinate genealogies. Furthermore, high-ranking females retain their position in the hierarchy even when age or injury reduces their fighting ability. An adolescent female yellow baboon normally attains a rank in the group just below that of her mother, and this inheritance of status provides stability in the dominance relationships among the females of a group. However, the social rank of genealogies is not fixed: Low-ranking female yellow baboons, with their female kin, may challenge higher-ranking individuals, and if they are successful their entire genealogy may rise in rank within the group.

The female kinship bonds are clearly important elements of the social structure of nonfemale transfer systems, but the exact contribution of the long-term relationships among females to the fitness of individual females is not clear. In some troops high-ranking females are young when they first give birth and have short interbirth intervals and high infant survival, but those correlations are not present in all the troops that have been studied. Furthermore, female kinship bonds are manifested weakly if at all in female transfer systems, which include most species of apes and many species of monkeys.

Male–Male Alliances   Male primates in nonfemale transfer species often form dominance hierarchies, but male rank depends mainly on individual attributes and is therefore less stable than female dominance systems based on genealogies. Young adult males, which are usually recent immigrants from another group, have the greatest fighting ability and usually achieve the highest rank. Some older males achieve stable alliances with each other that enable them to overpower younger and stronger rivals in competition for opportunities to court receptive females. These males probably achieve greater mating success by engaging in these reciprocal alliances than they would achieve on the basis of their individual ranks in the hierarchy.

Cooperative relationships among males are most common in female transfer systems, because the males of these species remain in their natal group. As a result, kin relationships exist among the males in a group. In red colobus monkeys (*Colobus badius*), for example, only males born in the group appear to be accepted by the adult male subgroup, and the membership of this subgroup can remain stable for years. Adult males spend much of their time in close proximity to each other and cooperate in aggression against males of a neighboring group. Male chimpanzees (*Pan troglodytes*) spend more time together than do females and engage in a variety of cooperative behaviors, including greeting, grooming, and sharing meat.

Male–Female Friendships Among Baboons Barbara Smuts's (1985) observations of a group of yellow baboons revealed that interactions between individual male and female baboons were not randomly distributed among members of the group. Instead, each female had one or two particular males called friends. Friends spent much time near each other and groomed each other often. These friendships lasted for months or years, including periods when the female was not sexually receptive because she was pregnant or nursing a baby. Male friends were solicitous of the welfare of their female friends and of their infants. Similar male–female friendships have been described in mountain gorillas (*Gorilla gorilla beringei*), gelada baboons (*Theropithecus gelada*), hamadryas baboons (*Papio hamadryas*), rhesus macaques (*Macaca mulatta*), and Japanese macaques (*Macaca fuscata*).

The advantage for a female of these friendships appears to lie in the protection that males provide both the females and their offspring from predators and from other members of the group. The advantage for a male of friendship with a female is less apparent. If the female's offspring had been sired by the male, protecting it would contribute to the male's fitness. However, in Smuts's study of yellow baboons, only half the friendships between males and infants involved relationships in which the male was the likely sire of the infant. The other friendships involved males that had never been seen mating with the mother of the infant. The advantage of friendship for males may depend on long-term associations with females. Barbara Smuts noted that males who participated in a friendship with a female had a significantly increased chance of mating with that female many months later when she was again receptive.

How Do Primates Perceive Their Social Structure?

The summary of primate social structures presented above represents the results of tens of thousands of hours of observations of individual animals over periods of many years. Statistical analyses of interactions between individuals— grooming sessions, aggression, defense—reveal correlations associated with factors including age, personality, kinship, and rank. Do the animals themselves recognize those relationships?

That is a fascinating but difficult question, especially with studies of free-ranging animals, but observations are accumulating that suggest that primates probably do recognize different kinds of relationships among individuals. For example, when juvenile rhesus macaques are threatened by another monkey, they scream to solicit assistance from other individuals who are out of sight. The kind of scream they give varies depending on the intensity of the interaction (threat or actual attack) and the dominance rank and kinship of their opponent. Furthermore, a mother baboon appears to be able to interpret the screams of her juvenile and to respond more or less vigorously depending on the nature of the threat her offspring faces. When tape-recorded screams were played back to the mothers, the mothers responded most strongly to screams that were given during an attack by a higher-ranking opponent, less strongly to screams that were given in response to interactions with lower-ranking opponents, and least strongly to screams that were given in interactions with relatives.

In a similar experiment with free-ranging vervet monkeys, the screams of a juvenile were played back to three females, one of whom was the mother of the juvenile. The mother responded more strongly to the screams than did the other two monkeys, as one would expect if females can recognize the voices of their own offspring. However, the other two monkeys responded to the screams by looking toward the mother, suggesting that they were able not only to associate the screams with a particular juvenile, but also to associate that juvenile with its mother.

Observations of redirected aggression also suggest that some primates classify other members of a group in terms of genealogies and friendships. When a baboon or macaque has been attacked and routed by a higher-ranking opponent, the victim frequently attacks a bystander who took no part in the original interaction. This behavior is known as redirected aggression, and the targets of redirected aggression are relatives or friends of the original opponent more frequently than would be expected by chance. Vervet monkeys show still more complex forms of redirected aggression: They are more likely to behave aggressively toward an individual when they have recently fought with one of that individual's close kin. Furthermore, an adult vervet is more likely to threaten a particular animal if that animal's kin and one of its own kin fought earlier that same day. This sort of feud is seen only among adult vervets, suggesting that it takes time for young animals to learn the complexities of the social relationships of a group.

These sorts of observations suggest that adult primates have a complex and detailed recognition of the genetic and social relationships of other individuals in their group. Furthermore, they may be able to recognize more abstract categories, such as *relative* versus *nonrelative*, *close relative* versus *distant relative*, or *strong friendship bond* versus *weak friendship bond*, that share similar characteristics independent of the particular individuals involved.

Summary

Sociality, the formation of structured groups, is a prominent characteristic of the behavior of many species of mammals. However, social behavior is only one aspect of the biology of a species, and social behaviors coexist with other aspects of behavior and ecology, including finding food and escaping from predators.

The size and geography of an animal's home range is related to the distribution and abundance of resources, the body size of the animal, and its feeding habits. Large species have larger home ranges than small species, and for any given body size the sizes of home ranges are in the order carnivores > omnivores > herbivores.

Social systems are related to the distribution of food resources and to the opportunities for an individual (usually, a male) to increase access to mates by controlling access to resources. Dietary habits, the structural habitat in which a species lives, and its means of avoiding predators are closely linked to body size and mating systems. These aspects of the biology form a web of interactions, each influencing the others in complex ways.

The social systems of primates, especially cercopithecoid monkeys, have been the subjects of field studies and more information about social behavior under field conditions is available for primates than for other mammals. The social systems of primates are complex, but not unique among mammals. Some primates are solitary or monogamous, but others live in groups and display behaviors that suggest not just recognition of other individuals, but also recognition of the genetic and social relationships among other individuals. Studies of other kinds of mammals will probably reveal similar phenomena, and understanding the behavior of mammals requires a broad understanding of their ecology and evolutionary histories.

References

Alexander, R. D. 1974. The evolution of social behavior. *Annual Review of Ecology and Systematics* 5:325–383.

Altmann, J. 1980. *Baboon Mothers and Infants*. Harvard University Press, Cambridge, Mass.

Cheney, D., R. Seyfarth, and B. Smuts. 1986. Social relationships and social cognition in nonhuman primates. *Science* 234:1361–1366.

Clutton-Brock, T. H. and P. H. Harvey. 1983. The functional significance of variation in body size among mammals. Pages 632–663 in *Advances in the Study of Mammalian Behavior*, edited by J. F. Eisenberg and D. G. Kleiman. Special Publication 7, The American Society of Mammalogists.

Eisenberg, J. F. 1981. *The Mammalian Radiations*. University of Chicago Press, Chicago.

Eisenberg, J. F. and D. G. Kleiman. 1983. *Advances in the Study of Mammalian Behavior*. Special Publication No. 7, The American Society of Mammalogists.

Estes, R. D. 1974. Social organization of the African Bovidae. Pages 166–205 in *The Behavior of Ungulates and Its Relation to Management*, edited by V. Geist and F. Walther. International Union for the Conservation of Nature, Morges, Switzerland.

Gittleman, J. L. and P. H. Harvey. 1982. Carnivore home-range size, metabolic needs, and ecology. *Behavioral Ecology and Sociobiology* 10:57–63.

Goodall, J. 1986. *The Chimpanzees of Gombe*. Harvard University Press, Cambridge, Mass.

Grant, T. R. 1983. Behavioral ecology of monotremes. Pages 360–394 in *Advances in the Study of Mammalian Behavior*, edited by J. F. Eisenberg and D. G. Kleiman. Special Publication 7, The American Society of Mammalogists.

Hamilton, W. D. 1971. Geometry for the selfish herd. *Journal of Theoretical Biology* 31:295–311.

Hersteinsson, P. and D. W. Macdonald. 1982. Some comparisons between red and arctic foxes, *Vulpes vulpes* and *Alopex lagopus*, as revealed by radio tracking. Pages 259–289 in *Symposia of the Zoological Society of London*, No. 49, edited by C. L. Cheesman and R. B. Mitson. Academic Press, London.

Hinde, R. A. (editor). 1983. *Primate Social Relationships*. Sinauer Associates, Sunderland, Mass.

Jarman, P. J. 1974. The social organization of antelope in relation to their ecology. *Behaviour* 58:215–267.

Lewin, R. 1987. Social life: a question of costs and benefits. *Science* 236:775–777.

Macdonald, D. W. 1983. The ecology of carnivore social behavior. *Nature* 301:379–381.

McNab, B. K. 1983. Ecological and behavioral consequences of adaptation to various food resources. Pages 664–697 in *Advances in the Study of Mammalian Behavior*, edited by J. F. Eisenberg and D. G. Kleiman, Special Publication 7, The American Society of Mammalogists.

Morse, D. H. 1980. *Behavioral Mechanisms in Ecology*. Harvard University Press, Cambridge, Mass..

Nicolson, N. A. 1987. Infants, mothers, and other females. Pages 330–342 in *Primate Societies*, edited by B. Smuts, D. Cheney, R. Seyfarth, R. Wrangham, and T. Struhsaker. University of Chicago Press, Chicago.

Smuts, B. 1985. *Sex and Friendship in Baboons*. Aldine, Hawthorne, N.Y.

Smuts, B., D. Cheney, R. Seyfarth, R. Wrangham, T. Struhsaker (editors). 1987. *Primate Societies*. University of Chicago Press, Chicago.

Van Soest, P. J. 1982. *Nutritional Ecology of the Ruminant.* O & B Books, Corvallis, Ore.

Watts, E. S. (editor). 1985. *Nonhuman Primate Models for Human Growth and Development.* Alan R. Liss, New York.

Wrangham, R. W. 1982. Mutualism, kinship, and social evolution. Pages 269–289 in *Current Problems in Sociobiology,* edited by King's College Sociobiology Group. Cambridge University Press, Cambridge.

Endothermy is a second salient characteristic of mammals, and of birds as well. The two lineages have evolved endothermy independently, but the costs and benefits are the same for both. Endothermy is a superb way to become independent of many of the stresses of the physical environment, especially cold. Birds and mammals can live in the coldest habitats on Earth—provided that they can find enough food. That qualification expresses the major problem of endothermy; it is energetically expensive. Endotherms need a reliable supply of food. As a result, the conspicuous interactions of endotherms are often with their biological environment—predators, competitors, and prey—rather than with the physical environment as was often the case for ectotherms. Because energy intake and expenditure are often important factors in the daily lives of endotherms, calculations of energy budgets can help us to understand the consequences of some kinds of behavior.

When all efforts at homeostasis are inadequate, endotherms have two more methods of dealing with harsh conditions: Birds and large species of mammals can migrate, and many species of small mammals and some birds can become torpid. This response, a temporary drop in body temperature, conserves energy and prolongs survival at the cost of abandoning the benefits of homeothermy.

820

Endothermy: A High-Energy Approach to Life

23

Costs and Benefits

Endothermy has both benefits and costs compared to ectothermy. On the positive side, endothermy allows birds and mammals to maintain high body temperatures when solar radiation is not available or is insufficient to warm them—at night, for example, or in the winter. The thermoregulatory capacities of birds and mammals are astonishing; they literally can live in the coldest places on Earth. On the negative side, endothermy is energetically expensive. We have pointed out that the metabolic rates of birds and mammals are nearly an order of magnitude greater than those of amphibians and reptiles. The energy to sustain those high metabolic rates comes from food, and endotherms need more food than ectotherms.

Of course, a host of other differences distinguish the ecology and behavior of endotherms and ectotherms, and these also can be considered costs or benefits of the different methods of thermoregulation. In this chapter we concentrate on how endotherms use energy, the ways in which endotherms thermoregulate in cold and in hot environments, and how endotherms control their water gains and losses. These topics are intimately related, because the high metabolic rates of endotherms are associated with more precise homeostatic control of their internal environment than is necessary for many ectotherms. For example, rising body temperature can be countered by evaporative cooling (sweating or panting), but too much evaporative cooling depletes water stores and leads to other problems. Body size plays a large role in determining the stresses to which endotherms are subjected and the responses that are possible for them.

The same sorts of habitats that are stressful for ectotherms (Chapter 16) are also difficult for endotherms, although not always for exactly the same reasons. Cold temperatures, for example, make ectothermal thermoregulation difficult or impossible and may present a risk of freezing. Endotherms have sufficient insulation and thermogenic (heat-producing) capacity to survive low temperatures, but they need a plentiful and regular supply of food to do that. Indeed, their high energy requirements appear to shape several aspects of the biology of endotherms, such as the relationship among body size, diet, and home range discussed in the preceding chapter. The role that energy gain and use play in the day-to-day lives of endotherms can be illustrated by an energy budget.

Energy Budgets of Vertebrates

An understanding of the costs of living for an animal can be obtained by constructing an energy budget for a species. An energy budget, like a financial budget, shows income and expenditure but uses units of energy as currency. The energy costs of different activities and alternative behaviors can be evaluated. This quantitative approach to the study of animal behavior offers the promise of understanding some of the reasons why animals behave as they do.

A Daily Energy Budget: The Vampire Bat

Studies of vampire bats (*Desmodus rotundus*) by Brian McNab (1973) have revealed a clear-cut relationship between energy intake, energy expenditure, and the species' geographic range. These bats of the suborder Microchiroptera inhabit the Neotropics and are specialized to feed exclusively on blood (Figure 23–1). Their daily pattern of activity is simple: They spend about 22 hours in their caves, fly out at night to a feeding site, and return after they have fed. Typically, a vampire flies about 10 kilometers round trip at 20 kilometers per hour to find a meal. Thus, a bat spends half an hour per day in flight. The remaining time outside the cave may be spent in feeding.

The vampire's food is convenient for energetic calculations because of the relatively constant caloric content of blood. The information needed to construct an energy budget is the following:

I = ingested energy (blood). The blood a bat drinks must provide the energy needed for all of its life processes: maintenance, activity, growth, and reproduction.

E = excreted energy. As in any animal, not all of the food ingested is digested and taken up by the bat. The energy contained in the feces and urine is lost.

$I - E$ = assimilated energy. This is the energy actually taken into the bat's body.

M = metabolism. This can be subdivided into M_i (metabolism while the bat is inside the cave) and M_o (nonflight metabolism while the bat is outside the cave).

A = cost of activity, a half hour of flight per day.

B = biomass increase. This term is the profit a bat shows in its energy budget. It may be stored as fat or used for growth or for reproduction (production of gametes, growth of a fetus, or nursing a baby).

In its simplest form, the energy budget is

energy in = energy out ± biomass change

The biomass term appears as ± because an animal metabolizes some body tissues when its energy expenditure exceeds its energy intake. This is what every dieter hopes to do in order to lose weight. Translating this general equation into the terms defined gives

$$I - E = M_i + M_o + A \pm B$$

All these terms can be measured and expressed as kilojoules per bat per day (kJ/bat · day). These calculations are based on McNab's studies and apply to a Brazilian vampire bat weighing 42 grams.

Ingested energy: In a single feeding a vampire can consume 57 percent of its body mass in blood, which contains 4.6 kJ/g. Thus the ingested energy is

42 g × 57% × 4.6 kJ/g blood = 110.2 kJ

Excreted energy: A vampire excretes 0.24 g urea in the urine plus 0.95 g of feces daily. Urea contains 10.5 kJ/g and the feces contain 23.8 kJ/g. Thus the excreted energy is

0.24 g urea × 10.5 kJ/g + 0.95 g feces × 23.8 kJ/g = 2.5 kJ + 22.6 kJ = 25.1 kJ

Assimilated energy: The energy the bat actually assimilates from blood equals the energy ingested (110.2 kJ) minus that excreted (25.1 kJ). Thus a vampire's energy intake is 85.1 kJ/day:

110.2 kJ − 25.1 kJ = 85.1 kJ

Figure 23–1. The geographic range of the vampire bat, *Desmodus rotundus* (shaded area), closely approximates the 10°C isotherm for the minimum average temperature during the coldest month of the year (dashed line) at the northern limit of its range (in Mexico) and the southern limit (in Uruguay, Argentina, and Chile). The positions of the 10°C isotherm and the altitudinal range of the bats in the Andes Mountains are not known and are indicated by question marks. (Based on B. K. McNab, 1973, *Journal of Mammalogy* 54:131–144.)

Metabolism: In a tropical habitat, 20°C is a reasonable approximation of the temperature a bat experiences both inside and outside the cave. While at rest in the laboratory at 20°C a vampire's metabolic rate is 3.8 ml O_2/g · hr. The terms for metabolism can be calculated and converted to joules using the energy equivalent of oxygen (20.1 J/ml O_2).

$$M_i = 42 \text{ g} \times 3.8 \text{ ml } O_2/\text{g} \cdot \text{hr} \times 20.1 \text{ J/ml } O_2$$
$$\times 22 \text{ hr/day} = 70.6 \text{ kJ}$$
$$M_o = 42 \text{ g} \times 3.8 \text{ ml } O_2/\text{g} \cdot \text{hr} \times 20.1 \text{ J/ml } O_2$$
$$\times 1.5 \text{ hr/day} = 4.8 \text{ kJ}$$

Activity: The metabolism of a bat flying at 20 km/hr is three times its resting metabolic rate (3 x

3.8 ml $O_2/g \cdot hr = 11.4$ ml $O_2/g \cdot hr$). The cost of the round trip from the cave to the feeding site is

$$
\begin{aligned}
A &= 42 \text{ g} \times 11.4 \text{ ml } O_2/g \cdot hr \times 20.1 \text{ J/ml } O_2 \\
&\quad \times 0.5 \text{ hr} \\
&= 4.8 \text{ kJ}
\end{aligned}
$$

Biomass change: The quantities calculated so far are fixed values that the bat cannot avoid. The biomass change is a variable value. If the assimilated energy is greater than the fixed costs, this energy profit can go to biomass increase. Fixed costs that exceed the assimilated energy are reflected as a loss of biomass. For the situation described there is an energy profit:

$$
\begin{aligned}
I - E &= M_i + M_o + A \pm B \\
110.2 \text{ kJ} - 25.1 \text{ kJ} &= 70.6 \text{ kJ} + 4.8 \text{ kJ} + 4.8 \text{ kJ} \pm B \\
B &= 4.9 \text{ kJ/bat} \cdot day
\end{aligned}
$$

These calculations show that vampires can live and grow under the conditions assumed. What happens if we change some of the assumptions? McNab points out that the northern and southern limits of the geographic range of vampires conform closely to the winter isotherms of 10°C. That is, the minimum temperature outside the cave during the coldest month of the year is 10°C; the bats do not occur in regions where the minimum temperature is lower. Is this coincidence, or is 10°C the lowest the bats can withstand? Calculating an energy budget for a vampire under these colder conditions provides an answer.

Caves have very stable temperatures that usually do not vary from summer to winter. We will assume that temperature remains constant at 20°C in the cave our imaginary vampires inhabit. In that case only the conditions a bat encounters outside the cave are altered. Because of limitations of stomach capacity ingestion cannot increase beyond 57 percent of body mass, the value assumed in the previous calculation. Therefore, we need recalculate only M_o, A, and B.

Metabolism outside: At 10°C a bat must increase its metabolic rate to maintain its body temperature, and laboratory measurements indicate the resting metabolic rate increases to 6.3 ml $O_2/g \cdot hr$.

$$
\begin{aligned}
M_o &= 42 \text{ g} \times 6.3 \text{ ml } O_2/g \cdot hr \\
&\quad \times 20.1 \text{ J/ml } O_2 \times 1.5 \text{ hr} \\
&= 7.9 \text{ kJ}
\end{aligned}
$$

Activity: The cost of activity will not change because the metabolic rate of the bat during flight (11.4 ml $O_2/g \cdot hr$) is higher than the resting metabolic rate needed to keep it warm (6.3 ml $O_2/g \cdot hr$). Only the term M_o changes, increasing from 4.8 to 7.9 kJ, and the sum of the energy costs becomes 83.3 kJ/bat $\cdot$ day.

Because the assimilated energy remains at 85.1 kJ/bat day, only 1.8k J is available for biomass increase. The assumptions in these calculations introduce a degree of uncertainty, and probably 1.8 kJ is not different from 0 kJ. At 10°C a bat uses all its energy staying alive. Thus a vampire bat could live under those conditions, but it would have no energy surplus for growth or reproduction. If the temperature outside the cave were lower than 10°C, the bat would have a negative energy balance and would lose weight with each meal. The agreement between our calculations and the actual geographic distribution of the bats suggests that energy may be a biologically significant factor in limiting their northward and southward spread.

Additional calculations reveal more about the selective forces that shape the life of vampire bats. A bat's stomach can hold a volume of blood equal to 57 percent of its body mass, but a bat cannot fly with that load. The maximum flight load is 43 percent of the body mass. Before it can take off to start the flight back to its cave, therefore, a bat must reduce the weight gained from its meal. Vampires do this by rapidly excreting water. Within 2 minutes after it begins to eat, a vampire starts to emit a stream of dilute urine. Experiments reveal that a vampire produces urine at a maximum rate of 0.24 milliliter per gram body weight $\cdot$ hour. Thus in the hour and a half the bat may spend in feeding, it could excrete as much as 15 grams of water—more than enough to allow it to fly (McFarland and Wimsatt 1969, Busch 1988).

Although rapid excretion of water solves the bat's immediate problem, it introduces another. The bat is left with a stomach full of protein-rich food that will yield a large amount of urea. To excrete this urea, the bat needs water to form urine. By the time a vampire gets back to its cave it is facing a water shortage instead of a water excess. Unlike many mammals, vampires seldom drink water but depend instead on blood for their water requirements. Like other mammals adapted to conditions of water scarcity, vampire bats have kidneys capable of producing very concentrated urine to conserve water (see Table 4–4). As a result of its unusual ecology and behavior, a vampire bat can be considered to live in a desert of its own making in the midst of a tropical forest.

Endotherms in the Cold: The Arctic

As the energy budget for the vampire bat revealed, endotherms expend most of the energy they consume just keeping themselves warm even in the moderate conditions of tropical and subtropical climates. Nonetheless, endotherms have proven themselves very adaptable in extending their thermoregulatory responses to allow them to inhabit even arctic and antarctic regions. Not even small body size is an insuperable handicap to life in these areas: Redpolls and chickadees that weigh only 10 grams overwinter in central Alaska.

Aquatic life in cold regions places still more stress on an endotherm. Because of the high heat capacity and conductivity of water, an aquatic animal may lose heat at 50 to 100 times the rate it would if it were moving at the same speed through air. Even a small body of water is an infinite heat sink for an endotherm; all of the matter in its body could be converted to heat without appreciably raising the temperature of the water. How, then, do endotherms manage to exist in such stressful environments?

Increased Heat Production Versus Decreased Heat Loss

There are potentially two solutions to the problems of endothermal life in cold environments and the special problems of aquatic endotherms in particular. A stable body temperature could be achieved by increasing heat production or by decreasing heat loss. On closer examination the option of increasing heat production does not seem particularly attractive. Any significant increase in heat production would require an increase in food intake. This scheme poses obvious ecological difficulties in terrestrial arctic and antarctic habitats where primary production is extremely low, especially during the coldest parts of the year. For most polar animals the quantities of food necessary would probably not be available, and a number of studies have shown that metabolic rates of most polar endotherms are similar to those of related species from temperate regions.

Because they lack the option of increasing heat production significantly, conservation of heat within the body is the primary thermoregulatory mechanism of polar endotherms. Insulative values of pelts from arctic mammals are two to four times as great as those from tropical mammals. In arctic species insulative value is closely related to the length of the fur (Figure 23–2). Small species such as the least weasel and the lemming have fur only 10 or 15 millimeters long. Presumably, the thickness of their fur is limited because longer hair would interfere with the movement of their legs. Large mammals (caribou, polar and grizzly bears, dall sheep, and arctic fox) have hair 3 to 7 centimeters long. There is no obvious reason why their hair could not be longer; apparently, further insulation is not needed. The insulative values of pelts of short-haired tropical mammals are similar to those measured for the same hair lengths in arctic species. Long-haired tropical mammals, like the sloths, have less insulation than arctic mammals with hair of similar length.

A comparison of the lower critical temperatures of tropical and arctic mammals illustrates the effectiveness of the insulation provided (Figure

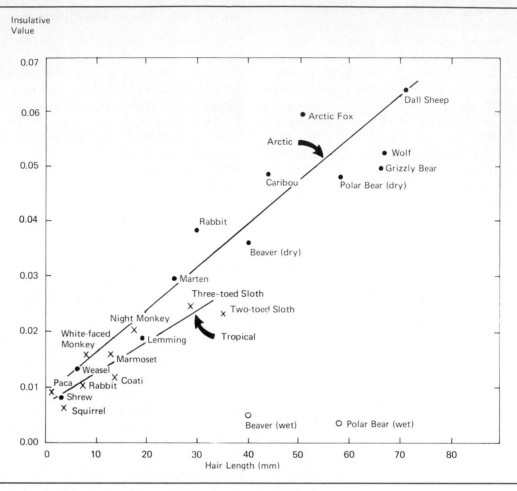

Figure 23–2. The insulative values of the pelts of arctic mammals (•) are proportional to the length of the hair. Pelts from tropical mammals (×) have approximately the same insulative value as those of tropical mammals at short hair lengths, but long-haired tropical mammals like sloths have less insulation than arctic mammals with hair of the same length. Immersion in water greatly reduces the insulative value of hair, even for such semiaquatic mammals as the beaver and polar bear (○). (Modified from P. F. Scholander et al., 1950, *Biological Bulletin* 99:237–258.)

23–3). **Lower critical temperature** is the environmental temperature at which energy utilization must rise above its basal level to maintain a stable body temperature. A number of tropical mammals have lower critical temperatures between 20 and 30°C. As air temperatures fall below their lower critical temperatures, the animals are no longer in their thermoneutral zones and must increase their metabolic rates to maintain normal body temperatures. For example, a tropical raccoon has in-

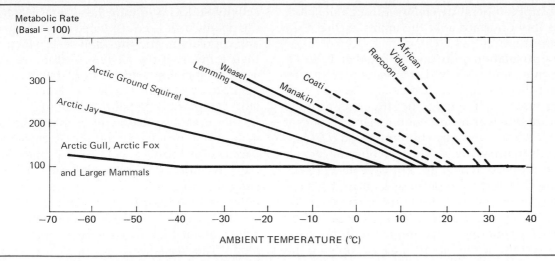

Figure 23–3. Lower critical temperatures for birds and mammals. Solid lines, arctic birds and mammals; dashed lines, tropical birds and mammals. The basal metabolic rate for each species is considered to be 100 units to facilitate comparisons among species. The metabolic rate of a species rises above its basal rate at the lower critical temperature for the species. Because of their effective insulation arctic birds and mammals can maintain resting metabolic rates at lower environmental temperatures than tropical species, and show smaller increases in metabolism (i.e., flatter slopes) below the lower critical temperature. (Modified from P. F. Scholander et al., 1950, *Biological Bulletin* 99:237–258.)

creased its metabolic rate approximately 50 percent above its standard level at an environmental temperature of 25°C.

Arctic mammals are much better insulated; even small species like the least weasel and the lemming have lower critical temperatures in still air that are between 10 and 20°C, and larger mammals have thermoneutral zones that extend well below freezing (Chappell 1980). The arctic fox, for example, has a lower critical temperature of −40°C, and at −70°C (approximately the lowest air temperature it ever encounters) has elevated its metabolic rate only 50 percent above its standard level. Under those conditions the fox is maintaining a body temperature approximately 110°C above air temperature. Arctic birds are equally impressive. An arctic jay has a lower critical temperature below 0°C in still air and can withstand

−70°C with a 150 percent increase in its metabolic rate, whereas an arctic gull, like the arctic fox, has a lower critical temperature near −40°C and can withstand −70°C with a modest increase in metabolism.

Clearly, hair or feathers can provide superb insulation for a terrestrial endotherm. These external insulative coverings are of limited value to aquatic animals, however, because when air trapped between hairs is displaced by water they lose most of their insulative value. The insulation of polar bear hair falls almost to zero when it is wet, and even seal hair loses much of its insulative value (Figure 23–2). In water, fat is a far more effective insulator than hair, and aquatic mammals have thick layers of blubber. This blubber forms the primary layer of insulation; skin temperature is nearly identical to water temperature and there is an abrupt temperature change through the blubber. Its inner surface is at the animal's core body temperature.

The insulation provided by blubber is so effective that pinnipeds and cetaceans require special heat-dissipating mechanisms to avoid overheating when they engage in strenuous activity, or venture into warm water or onto land. This heat dissipation is achieved by shunting blood into capillary beds in the skin outside the blubber layer and into the flippers, which are not covered by blubber. Selective perfusion of these capillary beds enables a seal or porpoise to adjust its heat loss to balance its heat production. When it is necessary to conserve energy, a countercurrent heat exchange system in the blood vessels supplying the flippers is brought into operation; when excess heat is to be dumped, blood is shunted away from the countercurrent system into superficial veins.

The effectiveness of the insulation of marine mammals is graphically illustrated by the problems experienced by the northern fur seal (*Callorhinus ursinus*) during its breeding season. Northern fur seals are large animals; males attain weights in excess of 250 kilograms. Unlike most pinnipeds, fur seals have a dense covering of fur, which is probably never wet through. They are inhabitants of the North Pacific. For most of the year they are pelagic, but during summer they breed on the Pribilof Islands in the Bering Sea north of the Aleutian Peninsula. Male fur seals gather harems of females on the shore. Here they must try to prevent the females from straying, chase away other males, and copulate with willing females.

George Bartholomew and his colleagues have studied both the behavior of the fur seals and their thermoregulation (Bartholomew and Wilke 1956). Summers in the Pribilof Islands (which are near 57° N latitude) are characterized by nearly constant overcast and air temperatures that rise only to 10°C during the day. These conditions are apparently close to the upper limits the seals can tolerate. Almost any activity on land causes the seals to pant and to raise their hind flippers (which are abundantly supplied with sweat glands) and wave them about. If the sun breaks through the clouds

activity suddenly diminishes—females stop moving about, males reduce harem guarding activities and copulation, and the adult seals pant and wave their flippers. If the air temperature rises as high as 12°C, females, which never defend territories, begin to move into the water. Forced activity on land can produce lethal overheating.

Professional seal hunters herd the bachelor males from the area behind the harems inland preparatory to killing and skinning them. Bartholomew recorded one drive that took place in the early morning of a sunny day while the air temperature rose from 8.6°C at the start of the drive to 10.4°C by the end. In 90 minutes the seals were driven about 1 kilometer with frequent pauses for rest. "The seals were panting heavily and frequently paused to wave their hind flippers in the air before they had been driven 150 yards from the rookery. By the time the drive was half finished most of the seals appeared badly tired and occasional animals were dropping out of the pods [groups of seals]. In the last 200 yards of the drive and on the killing grounds there were found 16 'roadskins' (animals that had died of heat prostration) and in addition a number of others prostrated by overheating." The average body temperature of the roadskins of this drive was 42.2°C, which is 4.5°C above the 37.7°C mean body temperature of adults not under thermal stress.

Fur seals can withstand somewhat higher environmental temperatures in water than they can in air because of the greater heat conduction of water, but they are not able to penetrate warm seas. Adult male fur seals apparently remain in the Bering Sea during their pelagic season. Young males and females migrate into the North Pacific, but they are not found in waters warmer than 14 to 15°C, and they are most abundant in water of 11°C. Examples of physiological characteristics setting clear-cut geographic limits to animal distributions are relatively rare. Bartholomew concluded that northern fur seals represent such a situation. Their inability to regulate body temperature during sustained activity and their sensitivity to even low levels of solar radiation and moderate air temperatures probably restrict the location of potential breeding sites and their movements during their pelagic periods. Summers in the Pribilofs are barely cool enough to allow the seals to breed there.

Migration to Avoid Stressful Conditions

Every environment has unfavorable aspects for some species, and these unfavorable conditions are often seasonal, especially in latitudes far from the Equator. The primary cause of migrations is usually related to seasonal changes in climatic factors such as temperature or rainfall. In turn, these conditions influence food supply and the occurrence of suitable breeding conditions. The catalog of vertebrates that migrate is lengthy and includes representatives from all the classes and many of the orders (Gauthreaux 1982). Aquatic and flying animals are best represented, especially when one considers long-distance migrations, probably because these animals can use the seas or the air as highways and are not limited to the narrower choice of routes that terrestrial habitats provide.

We have seen that locomotion is energetically costly. We can consider the costs and benefits of investing energy in migration by considering two kinds of animals that represent extremes of body size. The baleen whales are the largest animals that have ever lived, and hummingbirds are among the smallest endotherms, yet both whales and hummingbirds migrate.

Whales

The annual cycle of events in the lives of the great baleen whales is particularly instructive in showing how migration relates to the use of energy and how it correlates with reproduction in the largest of all animals. Most baleen whales summer in polar or subpolar waters of either the Northern or the Southern Hemisphere, where they feed on krill or other crustaceans that are abundant in those cold, productive waters. For 3 or 4 months each year a whale consumes a vast quantity of food that is converted into stored energy in the form of blubber and other kinds of fat. During this same time pregnant female whales nurture their unborn young, which may grow to one-third the length of their mothers before birth.

Near the end of summer the whales begin migration toward tropical or subtropical waters where the females bear their young and nurse them for a period of time before making the reverse migration. During this winter sojourn in warm waters, some of the whales also mate. The young grow rapidly on the rich milk provided by their mothers, and by spring the calves are mature enough to travel with their mothers back to Arctic or Antarctic waters. The calves are weaned about the time they arrive in their summer quarters. From a bioenergetic and trophic point of view the remarkable feature of this annual cycle is that virtually all of the energy required to fuel it comes from ravenous feeding and fattening during the 3 or 4 months spent in polar seas. Little or no feeding occurs during migration or during the winter period of calving and nursing. Energy for all these activities comes from the abundant stores of blubber and fat.

The gray whale (*Eschrichtius robustus*) of the Pacific Ocean has one of the longest and best known migrations (Figure 23–4). The summer feeding waters are in the Bering Sea and the Chukchi Sea north of the Bering Strait in the Arctic Ocean. A small segment of the population moves down the coast of Asia to Korean waters at the end of the Arctic summer, but most gray whales follow the Pacific Coast of North America, moving south to Baja California and adjacent parts of western Mexico. They arrive in December or January, bear their young in shallow, warm waters, and then depart northward again in March. Some gray whales make an annual round trip of at least 9000 kilometers and, as far as scientists have been able to determine, the adults do not eat for the 8 months they are away from their northern feeding grounds.

The amount of energy expended by a whale in this annual cycle is phenomenal. The basal met-

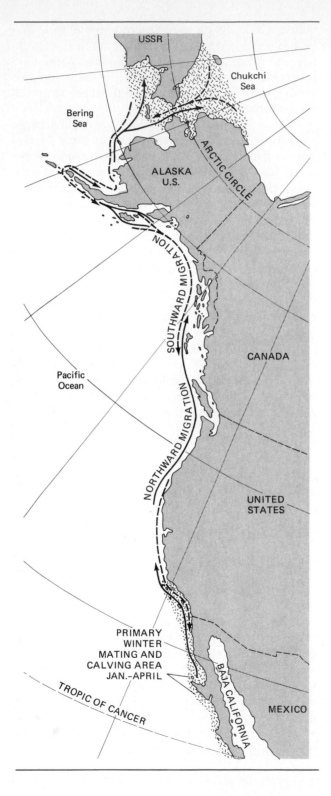

abolic rate of a gray whale with a fat-free body mass of 50,000 kilograms is approximately 979,000 kilojoules per day. If the metabolic rate of a free-ranging whale, including the locomotion involved in feeding and migrating, is about three times the basal rate (a typical level of energy use for mammals), the whale's average daily energy expenditure is around 2,937,000 kilojoules. Body fat contains 38,500 kilojoules per kilogram, so the whale's daily energy expenditure is equivalent to metabolizing over 76 kilograms of blubber or fat per day. Assuming an energy content of 20,000 kilojoules per kilogram for krill and a 50 percent efficiency in converting the gross energy intake of food into biologically usable energy, the energy requirement for existence is equivalent to a daily intake of 294 kilograms of food.

To exist for 245 days during the migration and calving in midlatitudinal waters without eating, the whale must metabolize a minimum of 18,375 kilograms of fat. Accumulating that amount of fat in 120 days of active feeding in Arctic waters at a conversion efficiency of 50 percent requires the consumption of 70,438 kilograms of krill, or 586 kilograms per day. The total food intake per day on the feeding grounds to accommodate the whale's daily metabolic needs plus energy storage for the migratory period would be not less than 294 + 586 = 880 kilograms of krill per day.

This is a minimum estimate for females because the calculations do not include the energetic costs of the developing fetus or the cost of milk production. Nor does it include the cost of transporting 20,000 kilograms of fat through the water. But a large whale can do all this work and more without exhausting its insulative blanket of blubber because nearly half the total body mass of a large whale consists of blubber and other fats.

Why does a gray whale expend all this energy to migrate? The adult is too large and too well insulated ever to be stressed by the cold arctic and subarctic waters that do not vary much from 0°C

Figure 23–4. Migratory route of the gray whale between the Arctic Circle and Baja California.

from summer to winter anyway. It seems strange for an adult whale to abandon an abundant source of food and go off on a forced starvation trek into warm waters that may cause stressful overheating. The advantage probably accrues to the newborn young, which, though relatively large, lacks an insulative layer of blubber. If the young whale were born in cold northern waters it would probably have to utilize a large fraction of its energy intake (milk produced from its mother's stored fat) to generate metabolic heat to regulate its body temperature. That energy could otherwise be used for rapid growth. Apparently, it is more effective, and perhaps energetically more efficient, for the mother whale to migrate thousands of kilometers into warm waters to give birth and nurse in an environment where the young whale can invest most of its energy intake in rapid growth.

Hummingbirds

At the opposite end of the size range, hummingbirds are the smallest endotherms that migrate. Ornithologists have long been intrigued by the ability of the ruby-throated hummingbird (*Archilochus colubris*), which weighs only 3.5 to 4.5 grams, to make a nonstop flight of 800 kilometers during migration across the Gulf of Mexico from Florida to the Yucatan Peninsula. For many years it seemed impossible that such a small bird with such a high resting metabolic rate could store enough energy for such a long flight.

Measurements of body fat, oxygen consumption during flight, and speed of flight have finally solved the riddle. Like most migratory birds, ruby-throated hummingbirds store subcutaneous and body fat by feeding heavily prior to migration. A hummingbird with a lean mass of 2.5 grams can accumulate 2 grams of fat. Measurements of the energy consumption of a hummingbird hovering in the air in a respirometer chamber in the lab indicate an energy consumption of 2.89 to 3.10 kilojoules per hour. Hovering is energetically more expensive than forward flight, so these values represent the maximum energy used in migratory flight. Even so, 2 grams of fat produces enough

energy to last for 24 to 26 hours of sustained flight. Hummingbirds fly about 40 kilometers per hour, so crossing the Gulf of Mexico requires about 20 hours. Thus, by starting with a full store of fat, they have enough energy for the crossing with a reserve for unexpected contingencies such as a headwind that slows their progress. In fact, most migratory birds wait for weather conditions that will generate tailwinds before they begin their migratory flights, thereby further reducing the energy cost of migration.

Torpor as a Response to Low Temperatures and Limited Food

We have stressed the high energy cost of endothermy because the need to collect and process enough food to supply that energy is a central factor in the lives of many endotherms. In extreme situations environmental conditions may combine to overpower a small endotherm's ability to process and transform enough chemical energy to sustain a high body temperature through certain critical phases of its life. For diurnally active birds, long cold nights during which there is no access to food can be lethal, especially if the bird has not been able to feed fully during the daytime. Cold winter seasons usually present a dual problem for resident endotherms: (1) the necessity to maintain high body temperature against a greatly increased temperature difference between its internal temperature and the ambient temperature, and (2) a relative scarcity of food. In response to such problems, some birds and mammals have mechanisms that permit them to avoid the energetic costs of maintaining a high body temperature under unfavorable circumstances by entering a state of torpor (adaptive hypothermia). By entering torpor an endotherm is giving up many of the advantages of endothermy, but in exchange it realizes an enormous saving of energy.

Physiological Adjustments During Torpor

When an endotherm becomes torpid profound changes occur in a variety of physiological functions (Heller 1987). Although body temperatures may fall very low during torpor, temperature regulation does not entirely cease. In **deep torpor** an animal's body temperature drops to within 1°C or less of the ambient temperature, and in some cases (bats, for example) extended survival is possible at body temperatures just above the freezing point of the tissues. Oxidative metabolism and energy use are reduced to as little as one-twentieth of the rate at normal body temperatures. Respiration is low, and the overall breathing rate can be less than one inspiration per minute. Heart rates are drastically reduced and blood flow to peripheral tissues is virtually shut down, as is blood flow posterior to the diaphragm. Most of the blood is retained in the core of the body. In this respect, the cardiovascular adjustments that occur during torpor are like those seen in diving animals (Jones et al. 1988).

Deep torpor is a comatose condition much more profound than the deepest sleep. Voluntary motor responses are reduced to sluggish postural changes, but some sensory perception of powerful auditory and tactile stimuli and ambient temperature changes is retained. Perhaps most dramatically, a torpid animal can arouse spontaneously from this state by endogenous heat production that restores the high body temperature characteristic of a normally active endotherm. Some endotherms can rewarm under their own power from the lowest levels of torpor, whereas others must warm passively with an increase in ambient temperature until some threshold is reached at which arousal starts.

There are all sorts of torpor, from the deepest states of hypothermia to the lower range of body temperatures reached by normally active endotherms during their daily cycles of activity and sleep. Many birds and mammals, especially those with body weights under 1 kilogram, undergo **circadian temperature cycles**. These cycles vary from 1 to 5°C or more between the average high temperature characteristic of the active phase of the daily cycle and the average low temperature characteristic of rest or sleep (Aschoff 1982). Small birds (sunbirds, hummingbirds, chickadees) and small mammals (especially bats and rodents) may drop their body temperatures during quiescent periods from 8 to 15°C below their regulated temperature during activity. Even a bird as large as the turkey vulture (about 2.2 kilograms) regularly drops its body temperature at night. When all these different endothermic patterns are considered together, no really sharp distinction can be drawn between torpidity and the basic daily cycle in body temperature that characterizes most small to medium-size endotherms.

Body Size and the Occurrence of Torpor

Species of endotherms capable of deep torpor are found in a number of groups of mammals and birds. All three subclasses of mammals include species capable of torpor. The echidna, the platypus, and several species of small marsupials possess various patterns of hypothermia, but the phenomenon is most diverse among placentals, particularly among bats and rodents. Certain kinds of hypothermia have also been described for some insectivores, particularly the hedgehog, some primates, and some edentates. Deep torpor, contrary to popular notion, is not known for any of the carnivores despite the fact that some of them den in the winter and remain inactive for long periods. Among birds, deep torpor occurs in some of the goatsuckers or nightjars and in hummingbirds, swifts, mousebirds, and some passerines (sunbirds, swallows, chickadees, and others). Other species, including larger ones like turkey vultures, show varying depths of hypothermia at rest or in sleep but are not in a semicomatose state of deep torpor.

Torpor and body size are closely related, especially among mammals (French 1986). The largest mammals that undergo deep torpor are marmots, which weigh 5 kilograms. The limitation on body size reflects a balance between the energy expenditure at normal body temperatures and during torpor and the time and energy spent in entry into torpor and in arousal. Torpor is not as energetically advantageous for a large animal as for a small one. In the first place, the energetic cost of maintaining high body temperature is relatively greater for a small animal than for a large one and, as a consequence, a small animal has more to gain from becoming torpid. Second, a large animal cools off more slowly than a small animal and does not lower its metabolic rate as rapidly.

Furthermore, large animals have more body tissue to rewarm on arousal, and their costs of arousal are correspondingly larger than those of small animals (Figure 23–5). An endotherm weighing a few grams, such as a little brown bat or a hummingbird, can warm up from torpidity at the rate of about 1°C per minute, and be fully active within 30 minutes or less depending on the depth of hypothermia. A 100-gram hamster requires more than 2 hours to arouse, and a marmot takes many hours. Entrance into torpor is slower than arousal. Consequently, daily torpor is feasible only for very small endotherms; there would not be enough time for a large animal to enter and arouse from torpor during a 24-hour period. Moreover, the energy required to warm up a large mass is very great. Oliver Pearson calculated that it costs a 4-gram hummingbird only 0.48 kilojoule to raise its body temperature from 10°C to 40°C. That is 1/85 of the total daily energy expenditure of an active hummingbird in the wild. By contrast, a 200-kilogram bear would require 18,000 kilojoules to warm from 10°C to 37°C, the equivalent of a full day's energy expenditure. The smaller potential savings and the greater costs of arousal make daily torpor impractical for any but small endotherms.

Medium-size endotherms are not entirely excluded from the energetic savings of torpor, but the torpor must persist for a longer period to realize a saving. For example, ground squirrels and woodchucks enter prolonged torpor during the winter **hibernation** when food is scarce. They spend several days at very low body temperatures (in the region of 5°C), then arouse for a period

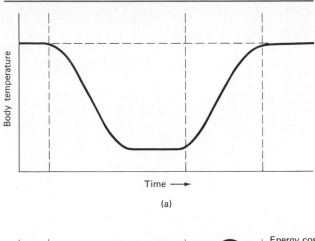

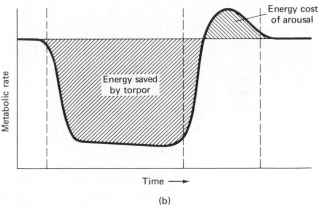

Figure 23–5. Changes in body temperature and metabolic rate (energy consumption) during torpor. A decrease in metabolic rate precedes a fall in body temperature to a new set point. An increase in metabolism produces the heat needed to return to normal body temperatures; the metabolic rate during arousal briefly overshoots the resting rate.

before becoming torpid again (Box 23–1). Still larger endotherms would have such large costs of arousal (and would take so long to warm up) that torpor is not feasible for them even on a seasonal basis. Bears in winter dormancy, for example, lower their body temperatures only about 5°C from normal levels, and metabolic rate decreases about 50 percent. Even this small drop, however,

Box 23–1. Waking Up is Hard Work:
The Cost of Arousal

Hibernation is an effective method of conserving energy during long winters, but hibernating animals do not remain at low body temperatures for the whole winter. Periodic arousals are normal, and these arousals consume a large portion of the total amount of energy used by hibernating mammals (French 1986). An example of the magnitude of the energy cost of arousal is provided by Lawrence Wang's study of the Richardson's ground squirrel (*Spermophilus richardsonii*, Figure 23–6) in Alberta, Canada (Wang 1978).

The activity season for ground squirrels in Alberta is short: They emerge from hibernation in mid-March and adult squirrels reenter hibernation 4 months later, in mid-July. Juvenile squirrels begin hibernation in September. When the squirrels are active they have body temperatures of 37 to 38°C, and their temperatures fall as low as 3 to 4°C when they are torpid. Figure 23–7 shows the body temperature of a juvenile male ground squirrel from

Figure 23–6. Richardson's ground squirrel, *Spermophilus richardsonii.* (Photograph by Gail R. Michener.)

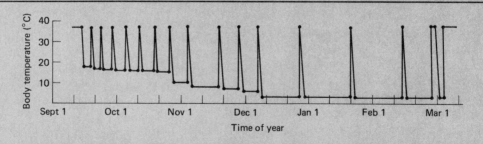

Figure 23–7. Record of body temperature during a complete torpor season for a Richardson's ground squirrel. Torpor cycles are initially short and become longer as winter progresses, then shorten again as spring approaches. (From L. C. H. Wang, 1978, *Strategies in Cold: Natural Torpidity and Thermogenesis,* edited by L. C. H. Wang and J. W. Hudson, Academic Press, New York.)

September through March; periods of torpor alternate with arousals all through the winter. Hibernation began in mid-September with short bouts of torpor followed by rewarming. At this time the temperature in the burrow was about 13°C. As the winter progressed and the temperature in the burrow fell, the intervals between arousals lengthened and the body temperature of the torpid animal declined. By December, January, and February the burrow temperature had dropped to 0°C and the periods between arousals averaged 14 to 19 days. In late February the periods of torpor became shorter, and in early March the squirrel emerged from hibernation.

A torpor cycle consists of entry into torpor, a period of torpidity, and an arousal (Figure 23–8). Entry into torpor began shortly after noon on February 16, and 24 hours later the body temperature had stabilized at 3°C. This period of torpor lasted until late afternoon on March 7, when the squirrel started to arouse. In 3 hours the squirrel warmed from 3°C to 37°C. It maintained that body temperature for 14 hours, and then began entry into torpor again.

These periods of arousal account for most of the energy used during hibernation (Table 23–1). The energy costs associated with arousal include the cost of warming from the hibernation temperature to 37°C, the cost of sustaining a body temperature of 37°C for several hours, and the metabolism above torpid levels as the body temperature slowly declines during reentry into torpor. For the entire hibernation season the combined metabolic ex-

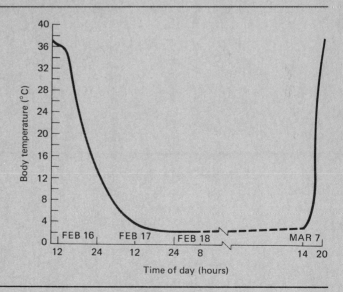

Figure 23–8. Record of body temperature during a single torpor cycle for a Richardson's ground squirrel. (From L. C. H. Wang, 1978, *Strategies in Cold: Natural Torpidity and Thermogenesis*, edited by L. C. H. Wang and J. W. Hudson, Academic Press, New York.)

Box 23–1. (Continued)

Table 23–1. Use of energy during different phases of the torpor cycle by Richardson's ground squirrel.

| | *Percent of Total Energy per Month* | | | |
Month	Torpor	Warming	Intertorpor Homeothermy	Reentry
July	8.5	17.2	56.5	17.8
September	19.2	15.2	49.9	15.7
November	20.8	23.1	43.1	13.0
January	24.8	24.1	40.0	11.1
March	3.3	14.0	76.4	6.3
Average for season	16.6	19.0	51.6	12.8

Source: L. C. H. Wang, 1978, pages 109–145 in *Strategies in Cold: Natural Torpidity and Thermogenesis*, edited by L. C. H. Wang and J. W. Hudson, Academic Press, New York.

penditures for those three phases of the torpor cycle account for an average of 83 percent of the total energy the squirrel uses.

Surprisingly we have no clear idea of why a hibernating ground squirrel undergoes these arousals that increase its total winter energy expenditure nearly fivefold. Ground squirrels do not store food in their burrows, so they are not using the periods of arousal to eat. They do urinate during arousal, and it is possible that some time at a high body temperature is necessary to carry out physiological or biochemical activities such as resynthesizing glycogen, redistributing ions, or synthesizing serotonin. Arousal may also allow a hibernating animal to determine when environmental conditions are suitable for emergence. Whatever their function, the arousals must be important because the squirrel pays a high energy price for them during a period of extreme energy conservation.

amounts to a large energy saving through the course of a winter for an animal as large as a bear.

Energy Relations of Daily Torpor

Recent studies of daily torpor in birds have emphasized the flexibility of the response in relation to the energetic stress faced by individual birds. Susan Chaplin's work with chickadees provides an example (Chaplin 1974). These small (10- to 12-gram) passerine birds are winter residents in northern latitudes, where they regularly experience ambient temperatures that do not rise above freezing for days or weeks (Figure 23–9).

Chaplin found that in winter chickadees around Ithaca, New York, allow their body temperatures to drop from the normal level of 40 to 42°C that is maintained during the day to 29 to 30°C at night. This reduction in body temperature permits a 30 percent reduction in energy con-

Figure 23–9. The black-capped chickadee, *Parus atricapillus*, is a small bird that winters in cold climates. (Photograph by Michael Hopiak, courtesy of the Cornell Laboratory of Ornithology.)

sumption. The chickadees rely primarily on fat stores they accumulate as they feed during the day to supply the energy needed to carry them through the following night. Thus the energy available to them and the energy they utilize at night can be estimated by measuring the fat content of birds as they go to roost in the evening and as they begin activity in the morning. Chaplin found that in the evening chickadees had an average of 0.80 gram of fat per bird. By morning the fat store had decreased to 0.24 gram. The fat metabolized during the night (0.56 gram per bird) corresponds to the metabolic rate expected for a bird that had allowed its body temperature to fall to 30°C.

Chaplin's calculations show that this torpor is necessary if the birds are to survive the night. It would require 0.92 gram of fat per bird to maintain a body temperature of 40°C through the night. That is more fat than the birds have when they go to roost in the evening. If they did not become torpid, they would starve before morning. Even with torpor, they use 70 percent of their fat reserve

in one night. They do not have an energy supply to carry them far past sunrise and chickadees are among the first birds to start to forage in the morning. They also forage in weather so foul that other birds, which are not in such precarious energy balance, remain on their roosts. The chickadees must reestablish their fat stores each day if they are to survive the next night.

Hummingbirds, too, may depend on the energy they gather from nectar during the day to carry them through the following night. These very small (4- to 10-gram) birds have extremely high energy expenditures and yet are found during the summer in northern latitudes and at high altitudes. An example of the lability of torpor in hummingbirds was provided by studies of nesting broad-tailed hummingbirds at an altitude of 2900 meters near Gothic, Colorado (Calder and Booser 1973). Ambient temperatures drop nearly to freezing at night, and under these conditions hummingbirds normally become torpid. Calder and Booser were able to monitor the body temperatures of nesting birds by placing an imitation egg containing a temperature-measuring device in the nest. These temperature records showed that hummingbirds incubating eggs normally did not become torpid at night. The reduction of egg temperature that results from the parent bird's becoming torpid does not damage the eggs, but it slows development and delays hatching. Presumably, there are advantages to hatching the eggs as quickly as possible, and as a result brooding hummingbirds expend energy to keep themselves and their eggs warm through the night, provided that they have the energy stores necessary to maintain the high metabolic rates needed.

On some days bad weather interfered with foraging by the parent birds, and as a result they apparently went into the night with insufficient energy supplies to maintain normal body temperatures. In this situation the brooding hummingbirds did become torpid for part of the night. One bird that had experienced a 12 percent reduction in foraging time during the day became torpid for 2 hours, and a second that had lost 21 percent of its foraging was torpid for 3.5 hours. Thus, torpor can be a flexible response that integrates the energy stores of a bird with environmental conditions and biological requirements such as brooding eggs.

Endotherms in the Heat: Deserts

Hot, dry areas place a more severe physiological stress on endotherms than do the polar conditions we have already discussed. The difficulties endotherms encounter in deserts result from a reversal of their normal relationship to the environment. Endothermal thermoregulation is most effective when an animal's body temperature is higher than the temperature of its environment. In this situation heat flow is from the animal to its environment, and thermoregulatory mechanisms achieve a stable body temperature by balancing heat production and heat loss. Very cold environments merely increase the gradient between an animal's body temperature and the environment.

The example of arctic foxes with lower critical temperatures of $-40°C$ illustrates the success that endotherms have had in providing sufficient insulation to cope with enormous gradients between high core body temperatures and low environmental temperatures.

In a desert the gradient is not increased; it is reversed. Desert air temperatures can climb to 40 or 50°C during summer, and the ground temperature may exceed 60 or 70°C. Instead of losing heat to the environment an animal is continually absorbing heat, and that heat must somehow be dissipated to maintain the animal's body temperature in the normal range. It can be a greater challenge for an endotherm to maintain its body temperature 10°C below the ambient temperature than to maintain it 100°C above ambient.

Evaporative cooling is the major mechanism an endotherm uses to reduce its body temperature. The evaporation of water requires approximately 2427 kilojoules/kilogram. (The exact value varies with temperature.) Thus, evaporation of a liter of water dissipates 2427 kilojoules, and evaporative cooling is a very effective mechanism as long as an animal has an unlimited supply of water. In a hot desert, however, where the thermal stress is greatest, water is a scarce commodity and its use must be carefully rationed. Calculations show, for example, that if a kangaroo rat were to venture out into the desert sun, it would have to evaporate 13 percent of its body water per hour to maintain a normal body temperature. Most mammals die when they have lost 10 to 20 percent of their body water, and it is obvious that, under desert conditions, evaporative cooling is of limited utility except as a short-term response to a critical situation.

Unable to rely on evaporative cooling, endotherms have evolved a number of other responses that have allowed a diverse assemblage of birds and mammals to inhabit deserts. The mechanisms they use are complex and involve combinations of ecological, behavioral, morphological, and physiological mechanisms that act together to enhance the effectiveness of the entire system. As a start toward unravelling some of these complexities, we can categorize three major classes of responses of endotherms to desert conditions as follows:

1. Some endotherms manage to avoid desert conditions by behavioral means. They live in deserts but are rarely exposed to the full stress of desert life.
2. Other endotherms have relaxed the limits of homeostasis. They manage to survive in deserts by tolerating greater ranges of variation in characters such as body temperature or body water content than normal.
3. Specializations such as torpidity in response to shortages of food or water and a reduced standard metabolic rate (and consequently, a reduction in the amount of metabolic heat an animal must dissipate) are combined with some of the characters mentioned in 1 and 2 in some desert endotherms.

Large Mammals in Hot Deserts

Large animals, including humans, have specific advantages and disadvantages in desert life that are directly related to body size. A large animal has nowhere to hide from desert conditions. It is too big to burrow underground, and few deserts have vegetation large enough to provide useful shade to an animal much larger than a jackrabbit. Thus a large animal cannot avoid desert condi-

tions. On the other hand, large body size offers some options not available to smaller animals. Large animals are mobile and can travel long distances to find food or water, whereas small animals may be limited to home ranges only a few meters or tens of meters in diameter. Large animals have small surface/mass ratios and consequently absorb heat from the environment slowly. The specific heat of animal tissue is approximately 3.4 kilojoules/kilogram, and a large body mass gives an animal a large thermal inertia; it can absorb a large amount of heat before its body temperature rises.

The Camel The dromedary camel (*Camelus dromedarius*) is the classic large desert animal (Figure 23–10). The biology of camels and their domestication have been reviewed by Gauthier-Pilters and Dagg (1981) and by Yagil (1985). The reputation of the camel as the ship of the desert has lost nothing in the telling, but there are authentic records of journeys in excess of 500 kilometers, lasting 2 or 3 weeks, during which the camels did not have an opportunity to drink. The longest trips are made in winter and spring when air temperatures are relatively low and scattered rainstorms may have produced some fresh vegetation that provides a little food and water for the camels. Gauthier-Pilters studied camels in the northern Sahara, traveling with the nomadic

Figure 23–10. Dromedary camels, *Camelus dromedarius.* In the heat of the day, these camels are facing into the sun, reducing the amount of solar radiation they receive, and pressed against each other to reduce the heat they gain by convection and reradiation from the hot ground. (Photograph by David A. Robertshaw.)

herdsmen. She found that in winter the camels were independent of drinking water and had to be herded continuously to keep them from straying. In summer the camels returned voluntarily to camp to be watered every 2 to 5 days.

The pioneering studies of camels' adaptations to desert life were conducted by Knut Schmidt-Nielsen and a number of associates in North Africa in the 1950s and in Australia in 1962. Subsequent work by other biologists has filled in additional details. Camels are large animals; adult weights of dromedary camels are 400 to 450 kilograms for females and up to 500 kilograms for males. The North African desert is cold in the winter, and the camels grow heavy coats; in summer they shed the winter coat but retain hair 50 to 60 millimeters long on the back and up to 110 millimeters long over the hump. On the ventral surface and legs the hair is only 15 to 20 millimeters long. The nature of the camel's adaptation to desert life is revealed by comparing the daily cycle of body temperature in a camel that receives water daily and one that has been deprived of water (Figure 23–11). The watered camel shows a small daily cycle of body temperature with a minimum of 36°C in the early morning and a maximum of 38°C in mid afternoon. When a camel is deprived of water, the daily temperature variation triples. Body temperature is allowed to fall to 34.5°C at night and climbs to 40.5°C during the day.

The significance of this increased daily fluctuation in body temperature can be assessed in terms of the water that the camel would have had to expend to prevent the 6°C rise by evaporative cooling. With a specific heat of 3.4 kilojoules/kilogram · °C, a 6°C increase in body temperature for a 500-kilogram camel represents storage of 10,000 kilojoules of heat in the body. Evaporation of a

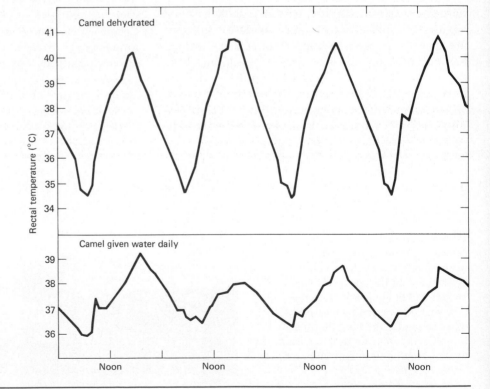

Figure 23–11. Daily cycles of body temperature for a dehydrated camel (top) and a camel with daily access to water (bottom). (From K. Schmidt-Nielsen et al., 1957, *American Journal of Physiology* 188:103–112.)

kilogram of water dissipates approximately 2427 kilojoules. Thus a camel would have to evaporate slightly more than 4 liters of water to maintain a stable body temperature at the nighttime level; by tolerating hyperthermia during the day it can conserve that water.

In addition to the direct saving of water not used for evaporative cooling, the camel receives an indirect benefit from tolerating hyperthermia in a reduction of energy flow from the air to the camel's body. As long as the camel's body temperature is below air temperature a gradient exists that causes the camel to absorb heat from the air. At a body temperature of 40.5°C the camel's temperature is equal to that of the air for much of the day, and no net heat exchange takes place. Thus, the camel saves an additional quantity of water by eliminating the temperature gradient between its body and the air. The combined effect of these measures on water loss is illustrated by data from a young camel (Table 23–2). When deprived of water the camel reduced its evaporative water loss by 64 percent and reduced its total daily water loss by half. This water economy was achieved primarily by relaxation of the limits of thermoregulation. By permitting itself to become hypothermal at night and hyperthermal during the day, the camel stored a significant quantity of heat each day that could be dissipated by conduction, convection, and radiation the following night rather than by evaporation of water. At the same time its elevated body temperature during the hottest part of the day reduced the amount of heat it gained from the hot environment.

Several morphological, behavioral, and phys-iological features of camel biology are integrated with this basic response and enhance the benefits a camel can gain. The thick fur on the body, for example, covers the dorsal regions and shields them from the sun. Fur surface temperatures reach 70 to 80°C, but skin temperatures are only 40°C. As a result of the insulation provided by the fur, much of the solar energy that strikes a camel's fur never reaches its skin. Furthermore, sweat glands release their secretions onto the skin surface beneath the hair. The energy that evaporates the sweat comes from the camel's metabolism. For a bare-skinned animal in the sun, most of the heat that is dissipated by evaporation of sweat comes from the absorption of solar radiation by the skin. Thus, a camel's long hair allows it to make effective use of the evaporative cooling that it does employ.

Behavioral mechanisms aid dehydrated camels in reducing their heat load. Early in the morning camels lie down on surfaces that have cooled overnight by radiation of heat to the night sky. The legs are tucked beneath the body and the ventral surface, with its short covering of hair, is placed in contact with the cool ground. In this position a camel exposes only its well-protected back and sides to the sun and places its lightly furred legs and ventral surface in contact with cool sand, which may be able to conduct away some body heat. Gauthier-Pilters has reported that camels may assemble in small groups and lie pressed closely together through the day. Spending a day in the desert sun squashed against a sweaty camel may not sound pleasant, but in this posture a camel reduces its heat gain because it keeps its side

Table 23–2. Daily water loss of a 250 kilogram camel.

| Condition | Water Loss (liters/day) by Different Routes | | | |
	Feces	Urine	Evaporation	Total
Drinking daily (8 days)	1.0	0.9	10.4	12.3
Not drinking (17 days)	0.8	1.4	3.7	5.9

in contact with another camel (both at about 40°C) instead of allowing solar radiation to raise the fur surface temperature to 70°C or above.

Despite the mechanisms that reduce water loss, camels do continue to lose water. The conservation mechanisms of the camel Schmidt-Nielsen studied reduced its rate of water loss from nearly 5 percent of body weight per day when it was hydrated to just over 2 percent per day during dehydration. The reduced rate of water loss prolongs the period before the water loss becomes critical. In addition, camels appear to tolerate a greater amount of dehydration than most mammals. In full sun during the summer Schmidt-Nielsen's camels lost water equivalent to more than 25 percent of their hydrated body weight without ill effect; other mammals die when they have lost half that much water.

The camel's tolerance of dehydration appears to be related to its ability to maintain its blood plasma volume near normal levels during dehydration. In most mammals dehydration affects the plasma volume disproportionately, with the result that the blood becomes increasingly viscous. Eventually, the increasing viscosity of the blood prevents the circulatory system from transporting metabolic heat from the core of the body to the surface. The body temperature rises abruptly and death follows rapidly. In camels, water is lost primarily from interstitial and intercellular fluids during dehydration, and plasma volume is maintained. The basis of this physiological difference between camels and other mammals is not clear. Schmidt-Nielsen points out that a theoretical analysis of the osmotic and water relationships within an animal's body suggests that all animals should respond as a camel does. The difficulty lies not in explaining why a camel maintains its plasma volume but rather in explaining why other mammals do not. Whatever its physiological basis, however, the significance of a camel's tolerance of dehydration is readily apparent.

The decline in urinary water loss of a dehydrated camel is the result of changes in kidney function (Table 23–3). Blood flow to the kidney decreases and the glomerular filtration rate is reduced, so less urine is formed initially, and the concentration of antidiuretic hormone (ADH) rises. The ADH increases the permeability of the collecting duct to water (see Chapter 4), causing more water to be resorbed from the urine. In addition to its effect of reabsorption of water, ADH increases the permeability of the collecting duct to urea, and the reabsorption of urea also increases in dehydrated camels. At first glance that appears paradoxical, but camels are ruminants and the urea reabsorbed from the urine is recycled into the rumen where it is used by the symbiotic microorganisms to synthesize protein. As the contents of the rumen pass into the stomach and then into the small intestine the microorganisms are digested, the protein and water are again absorbed from the intestines, and are again recycled into the rumen. As a result of this process, both the camel and its

Table 23–3. Kidney function of camels during dehydration and rehydration.

Condition	Blood Flow to Kidney (ml/min)	Glomerular Filtration Rate (ml/min)	Urine Flow Rate (ml/min)
Control	1856 ± 174	235 ± 30	3.9 ± 1.9
Dehydrated	420 ± 88	73 ± 13	0.7 ± 0.07
0.5 hr rehydration	1374 ± 339	181 ± 15	2.1 ± 0.09
2 hr rehydration	2678 ± 735	213 ± 42	1.7 ± 0.12

Source: Z. Etzion and R. Yagil, 1986, *Physiological Zoology* 59:558–562.

rumen microorganisms continue to receive some nutrients even when the camel is not eating.

Despite their ability to reduce water loss and to tolerate dehydration, the time eventually comes when even camels must drink. These large, mobile animals can roam across the desert seeking patches of vegetation produced by local showers and move from one oasis to another, but when they come to drink they face a problem they share with other grazing animals: Water holes can be dangerous places. Predators frequently center their activities around water holes, where they are assured of water as well as a continuous supply of prey animals. Reducing the time spent drinking is one method of reducing the risk of predation, and camels can drink remarkable quantities of water in very short periods. A dehydrated camel can drink as much as 30 percent of its body weight in 10 minutes. (A very thirsty human can drink about 3 percent of body weight in the same time.) Schmidt-Nielsen's measurements of dehydrated camels' body weights just before they were allowed to drink and again immediately after they had finished drinking showed that if they had been only moderately dehydrated (moderate dehydration was defined as a loss of up to 20 percent of initial body weight), they drank enough water to restore the loss in one session. If they had been subjected to greater dehydration, they required two drinking sessions several hours apart to restore the deficit fully.

The water a camel drinks is rapidly absorbed into its blood. The renal blood flow and glomerular filtration rate increase and urine flow returns to normal within a half hour of drinking. The urine changes from dark brown and syrupy to colorless and watery. Aldosterone stimulates sodium reabsorption, which helps to counteract the dilution of the blood by the water the camel has drunk. Nonetheless dilution of the blood causes the red blood cells to swell as they absorb water by osmosis. Camel erythrocytes are resistant to this osmotic stress, but other desert ruminants have erythrocytes that can burst under hypotonic conditions. Bedouin goats, for example, have fragile erythrocytes, and the water a goat drinks is absorbed

slowly from the rumen. Goats require 2 days to return to normal kidney function after dehydration.

Other Large Mammals Richard Taylor has investigated the temperature and water relations of several species of African antelope that live in arid grasslands or desert regions (Taylor 1972). These animals, which range in size from the 20-kilogram Thomson's gazelle and 50-kilogram Grant's gazelle (*Gazella thomsoni* and *G. granti*) to the 100-kilogram oryx (*Oryx beisa*) and 200-kilogram eland (*Taurotragus oryx*) utilize heat storage like the dromedary but allow their body temperatures to rise considerably above the 40.5°C level recorded in the camel. Taylor recorded rectal temperatures of 45°C in the oryx and 46.5°C in the Grant's gazelle. (Thomson's gazelles, which do not penetrate into desert areas, began to pant at air temperatures above 42°C and used evaporative cooling to maintain body temperature below air temperature.)

Body temperatures above 43°C rapidly produce brain damage in most mammals, but Grant's gazelles maintained rectal temperatures of 46.5°C for as long as 6 hours with no apparent ill effects. These antelope keep brain temperature below body temperature by using a countercurrent heat exchange to cool blood before it reaches the brain. In ungulates the blood supply to the brain passes via the external carotid arteries (Figure 23–12). At the base of the brain these arteries break into a *rete mirabile* that lies in a venous sinus. The blood in the sinus is venous blood, returning from the walls of the nasal passages where it has been cooled by the evaporation of water. This cool venous blood thus cools the warmer arterial blood before it reaches the brain. A mechanism of this sort has been demonstrated in sheep and goats and is probably widespread (Kamau et al. 1984).

Unlike the dromedary camel, the antelopes are apparently independent of drinking water even during summer. Taylor suggests that one of the mechanisms that permits this independence is behavioral. The leaves of a desert shrub, *Diasperma*, are an important part of the diet of the antelopes. During the day, when air temperature

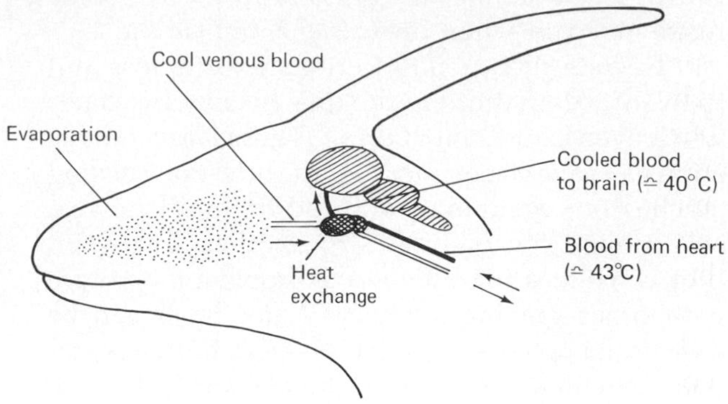

Figure 23–12. Countercurrent heat-exchange mechanism that may cool blood going to a gazelle's brain. (Modified from C. R. Taylor, 1972, in *Comparative Physiology of Desert Animals*, edited by G. M. O. Maloiy, Academic Press, London.)

is high and humidity low, these leaves contain about 1 percent water by weight. They are so dry that they disintegrate into powder when they are touched. At night, however, as air temperatures fall and relative humidity increases, the leaves take up water and after 8 hours have a water content of 40 percent. If they ate the leaves at night rather than in the daytime, the antelope might obtain enough water from their food to be independent of water holes.

Large animals such as those we have discussed illustrate one approach to desert life. Too large to escape the stresses of the environment, they have adapted to them by tolerating a temporary relaxation of homeostasis. Their success under the harsh conditions in which they live is the result of complex interactions between diverse aspects of their ecology, behavior, morphology, and physiology. Only when all of these features are viewed together does an accurate picture of an animal emerge.

Birds in Desert Regions

Although birds are relatively small vertebrates, the problems they face in deserts are more like those experienced by camels and antelope than like those of small mammals. Birds are predominantly diurnal and few seek shelter in burrows or crevices. Thus, like large mammals, they meet the stresses of deserts head on and face the antago-

nistic demands of thermoregulation in a hot environment and the need to conserve water.

Also like large mammals, birds are mobile. It is quite possible for a desert bird to fly to a mountain range on a daily basis to reach water. For example, mourning doves in the deserts of North America congregate at dawn at water holes, some individuals flying 60 kilometers or more to reach them. The normally high and labile body temperatures of birds give them an advantage in deserts that is not shared by mammals. With body temperatures normally around 40°C, birds face the problem of a reversed temperature gradient between their bodies and the environment for a shorter portion of each day than would a mammal. Furthermore, birds' body temperatures are normally variable, and birds tolerate moderate hyperthermia without apparent distress. These are all preadaptations to desert life that are present in virtually all birds. Neither the body temperatures nor the lethal temperatures of desert birds are higher than those of related species from nondesert regions.

The mobility provided by flight does not extend to fledgling birds, and the most conspicuous adaptations of birds to desert conditions are those that ensure a supply of water for the young. Altricial fledglings, those that need to be fed by their parents after hatching, receive the water they need from their food. One pattern of adaptation in desert birds ensures that reproduction will occur at a time when succulent food is available for fledglings. In the arid central region of Australia, bird reproduction is precisely keyed to rainfall. The sight of rain is apparently sufficient to stimulate courtship, and mating and nest building commence within a few hours of the start of rain. This rapid response ensures that the baby birds will hatch in the flush of new vegetation and insect abundance stimulated by the rain.

A different approach, very like that of mammals, has been evolved in columbiform birds (pigeons and doves), which are widespread in arid regions. Fledglings are fed on pigeon's milk, a liquid substance produced by the crop under the stimulus of prolactin. The chemical composition of pigeon's milk is very similar to that of mammalian milk; it is primarily water plus protein and fat, and it simultaneously satisfies both the nutritional requirements and the water needs of the fledgling. This approach places the water stress on the adult, which must find enough water to produce milk as well as meeting its own water requirements.

Seed-eating desert birds with precocial young, like the sandgrouse found in the deserts of Africa and the Near East, face particular problems in providing water for their young. Baby sandgrouse begin to find seeds for themselves within hours of hatching, but they are unable to fly to water holes as their parents do and seeds do not provide the water they need. Instead, adult male sandgrouse transport water to their broods. The belly feathers, especially in males, have a unique structure in which the proximal portions of the barbules are coiled into helices. When the feather is wetted, the barbules uncoil and trap water. The feathers of male sandgrouse hold 15 to 20 times their weight of water, and the feathers of females hold 11 to 13 times their weight (Cade and Maclean 1967).

Male sandgrouse in the Kalahari Desert of southern Africa fly to water holes just after dawn and soak their belly feathers, absorbing 25 to 40 milliliters of water. Some of this water evaporates on the flight back to their nests, but calculations indicate that a male sandgrouse could fly 30 kilometers and arrive with 10 to 28 milliliters of water still adhering to its feathers. As the male sandgrouse lands, the juveniles rush to him, and seizing the wet belly feathers in their beaks, strip the water from them with downward jerks of their heads. In a few minutes, the young birds have satisfied their thirst and the male rubs itself dry on the sand.

Small Mammals in Hot Deserts

Rodents are the preeminent small mammals of arid regions (Figure 23–13). It is a commonplace observation that population densities of rodents may be higher in deserts than in moist situations. A number of features of rodent biology can be viewed as preadaptive for extending their geographic ranges into hot, arid regions. Among the most important of these preadaptations are the normally nocturnal habits of many rodents and their practice of living in burrows. A burrow provides ready escape from the heat of a desert, giving an animal access to a microenvironment within its thermoneutral zone while soil temperatures on the surface climb above 60°C. Rodents that live in burrows during the day and emerge to forage at night escape the desert heat so successfully that their greatest temperature stress may be cold. Because of the normal absence of clouds, deserts cool rapidly after sundown and many deserts are distinctly chilly at night during much of the year.

Although retreat to a burrow during the day provides direct escape from heat, it is not, by itself, a solution to the other major challenges of desert life, the chronic shortages of food and water. What the burrow does provide is the shelter and microclimate an animal needs in order to solve the other problems. We pointed out earlier that a kangaroo rat would reach its lethal limit of dehydration in less than 2 hours if it had to rely on evaporative cooling to maintain its body temperature at normal levels during the day. By retreating into its burrow, a kangaroo rat avoids that use of water. Indeed, the water savings of a burrow probably go beyond that. As a rodent in a burrow loses water by evaporation, the air in the burrow becomes hu-

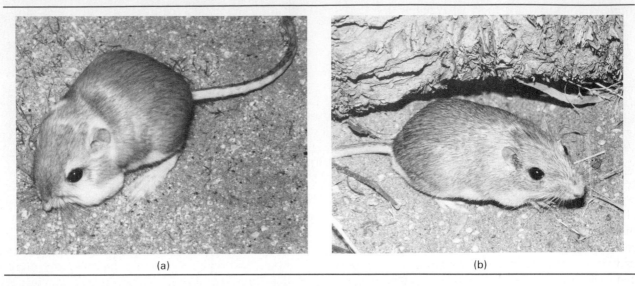

(a) (b)

Figure 23–13. Nocturnal desert rodents: (a) Merriam's kangaroo rat, *Dipodomys merriami*; (b) pocket mouse, *Perognathus longimembris*. (Photographs by Michael A. Recht, Ph.D.)

mid and may approach saturation. At the same time of day, the relative humidity of air outside the burrow may be only 20 to 30 percent. The higher humidity of burrow air reduces an animal's evaporative water loss.

A further saving is achieved in some animals (including birds and lizards in addition to mammals) by a countercurrent water recycler in the nasal passages. The air an animal exhales leaves the nares at a temperature lower than that at which it left the lungs. The phenomenon has important implications in terms of energy and water balance.

A brief consideration of the respiratory cycle illustrates the mechanism involved. As air is inhaled, it passes over moist tissues in the nasal passages. The nasal passages themselves are narrow and the wall surface area is large. As the air passes over these moist surfaces, it is warmed and humidified so that when it enters the lungs it is saturated with water at the animal's core body temperature. Temperature equilibration and saturation with water vapor are essential to protect the lungs, which are delicate structures that would be damaged if they were exposed to dry air. As the relatively dry inhaled air passes over the moist tis-

sues of the nasal passages, evaporation cools the walls of the nasal passages. When the warm, saturated air from the lungs is exhaled, water from the air condenses on the cool walls. This process of evaporation on inhalation and condensation on exhalation saves water and, because a large quantity of energy goes into evaporating the water, it also saves energy. It should be emphasized that this countercurrent exchange of heat and energy is not an adaptation to desert life. It is an inevitable consequence of the anatomy and physiology of the nasal passages. However, a preliminary comparison of the nasal heat and water exchange of five species of rodents suggests that interspecific differences may be related to the aridity of the habitat (Welch 1984). Two desert rodents, the kangaroo rat (a heteromyid) and the Australian hopping mouse (a murid), were especially good at recovering water when they were breathing air of low humidity. In contrast, two murid rodents from moist habitats, the deer mouse and the house mouse, were good at recovering heat when they were breathing cold air.

The importance of the nasal countercurrent in water recovery can be illustrated by calculations

Table 23–4. Water recycled by nasal countercurrent exchange in a kangaroo rat.

Conditions	Inhaled Air				Exhaled Air			
	Temperature (°C)	RH (%)	Water Content (mg/liter)	Water Added (mg/liter)	Temperature (°C)	RH (%)	Water Content (mg/liter)	Water Recovered (mg/liter)
Daytime, in burrow	30	80	24	22	31	100	31	15
Night, on surface	15	20	2.5	43.5	14.5	100	12	34

(Table 23-4). During the day a kangaroo rat in its burrow inhales air that is at 30°C and 80 percent relative humidity. This air contains 24 milligrams of water per liter of air. (Saturated air, at 100 percent relative humidity, contains 30 milligrams of water per liter at 30°C.) The kangaroo rat must warm this air to core temperature (38°C) and add enough water to raise the relative humidity at that temperature to 100 percent. At 38°C saturated air contains 46 milligrams water per liter, so the amount of water that must be evaporated in the nasal passages is 22 milligrams per liter of inhaled air.

Under the conditions we have assumed, a kangaroo rat exhales air at 31°C. That is, as the air travels out through the nasal passages it is cooled 7°C by contact with the walls. The exhaled air is still saturated with water, but at 31°C saturated air contains only 31 milligrams of water per liter instead of the 46 milligrams/liter it contained at 38°C. Thus, a kangaroo rat evaporates 22 milligrams of water per liter of air to saturate the air before it enters the lungs, and recovers 15 milligrams of water per liter of air on exhalation. The nasal countercurrent system reduces its respiratory water loss by nearly 70 percent from that which the kangaroo rat would experience if the air were exhaled at core body temperature. The savings amounts to 15 milligrams of water per liter of air under the burrow conditions we have assumed.

The water saving achieved at night when the rat is outside its burrow is still greater. The outside air is cool (15°C) and dry (20 percent relative hu- midity). Consequently the kangaroo rat must evaporate 43.5 milligrams of water per liter of air to bring it to saturation at core temperature. The increased evaporation cools the kangaroo rat's nose to 14.5°C, and air exhaled at this temperature contains only 12 milligrams of water per liter. Thus, under nighttime conditions, a kangaroo rat recovers 74 percent of the water evaporated in the nasal passages, a saving of 34 milligrams of water per liter of inhaled air.

Evaporation of water from the respiratory passages is one major avenue of water loss, and water excreted with the urine and feces is another. Rodents in general have the ability to produce relatively dry feces and concentrated urine. The laboratory white rat, for example, can produce urine with twice the osmotic concentration humans can achieve. The dromedary camel has a urine-concentrating ability approximately equivalent to that of a rat, and so do dogs and cats. In desert rodents, such as kangaroo rats, sand rats, and jerboas, urine concentrations of 3000 to 6000 milliosmoles/liter are commonly observed, and the current world champion urine concentrator appears to be the Australian hopping mouse, which can produce urine concentrations in excess of 9000 milliosmoles/liter. As we pointed out in Chapter 4, high urine concentrations in mammals are associated with long loops of Henle in the kidney which enhance the countercurrent multiplier function.

As a result of their low evaporative water losses and ability to concentrate urine and produce relatively dry feces, many desert rodents are com-

Box 23–2. (Continued)

pends on windspeed as well as air temperature. Thus, measuring air temperature provides only part of the information needed to assess just one of the three important routes of heat exchange. Consequently, air temperature is not a very useful measure of heat stress.

If air temperature is unsatisfactory as a measure of environmental heat load because it makes only a small contribution to the overall energy exchange, perhaps a measurement of the major source of heat is what is needed. **Solar insolation** in this arid habitat is the major source of heat stress, and the magnitude of the insolation (Q_r) can be measured with a device called a pyranometer. Solar insolation rises from 0 at dawn to about 900 watts per square meter in midday, and falls to 0 at sunset. This measurement provides information about how much solar energy is available to heat an animal, but that is still only one component of the energy exchange that determines the heat stress.

The **effective environmental temperature** (T_e) combines the effects of air temperature, ground temperature, solar insolation, and wind velocity. The effective environmental temperature is measured by making an exact copy of the animal (a mannikin), equipping it with a temperature sensor such as a thermocouple, and putting the mannikin in the same place in the habitat that the real animal occupies. Frequently, taxidermic mounts are used as mannikins: The pelt of an animal is stretched over a framework of wire or a hollow copper mold of the animal's body. Because the mannikin has the same size, shape, color, and surface texture as the animal, it responds the same way as the animal to solar insolation, infrared radiation, and convection. The equilibrium temperature of the mannikin is the temperature that a metabolically inert animal would have as a result of the combination of radiative and convective heat exchange. At Deep Canyon the temperatures of mannikins of antelope ground squirrels increased more rapidly than air temperatures and stabilized near 65°C from midmorning through late afternoon. The temperatures of the mannikins were 15°C higher than air temperature, showing that the heat load experienced by the ground squirrels was much greater than that estimated from air temperature alone.

Because the mannikin is a hollow shell—a pelt stretched over a supporting structure—it does not duplicate the thermoregulatory processes that have important influences on the thermoregulation of a real animal. Metabolic heat production increases the body temperature of a ground squirrel, evaporative water loss lowers body temperature, and changes in peripheral circulation and raising or lowering the hair change the insulation. The effects of these factors can be incorporated mathematically if the appropriate values for metabolism and insulation are known. The result of this calculation is the **standard operative temperature** (T_{es}); an explanation of how to calculate T_{es} can be found in Bakken (1980). For the ground squirrels in our example the standard operative temperature was nearly 10°C higher than the effective temperature and about 25°C higher than the air temperature. For most of the day the T_{es} of a ground squirrel in the sun at Deep Canyon was 30°C or more above the squirrel's upper critical temperature of 43°C. Similar calculations can provide values for T_{es} in other microenvironments the squirrels might occupy—in the shade of a bush, for example, or in a burrow. This is the information needed to evaluate the behavior of the squirrels to determine if their activities are limited by the need to avoid overheating.

squirrels are heat stressed for most of the day. Standard operative temperature rises above the thermoneutral zone of ground squirrels within 2 hours after sunrise, and the squirrels follow a bimodal pattern of activity that peaks in midmorning and again in the late afternoon. Relatively few squirrels were active in the middle of the day. The body temperatures of antelope ground squirrels are labile, and body temperatures of individual squirrels varied as much as 7.5°C (from 36.1 to 43.6°C) during a day. The squirrels use this lability of body temperature to store heat during their periods of activity.

The high operative temperatures limit the time that squirrels can be active in the open to no more than 9 to 13 minutes. The squirrels sprint furiously from one patch of shade to the next, pausing only to seize food or to look for predators. They minimize their exposure to the highest temperatures by running across open areas, and seek shade or their burrows to cool off. On a hot summer day a squirrel can maintain a body temperature below 43°C (the maximum temperature it can

tolerate) only by retreating every few minutes to a burrow deeper than 60 centimeters where the soil temperature is 30 to 32°C. The body temperature of an antelope ground squirrel shows a pattern of rapid oscillations, rising while the squirrel is in the sun and falling when it retreats to its burrow (Figure 23–16). Ground squirrels do not sweat or pant; instead, they use this combination of transient heat storage and passive cooling in a burrow to permit diurnal activity. The strategy the antelope ground squirrel uses is basically the same as that employed by a camel—saving water by allowing the body temperature to rise until the heat can be dissipated passively. The difference between the two animals is a consequence of their difference in body size: A camel weighs 500 kilograms and can store heat for an entire day and cool off at night, whereas an antelope ground squirrel weighs about 50 grams and heats and cools many times in the course of a day.

The tails of many desert ground squirrels are wide and flat, and the ventral surfaces of the tails are usually white. The tail is held over the squir-

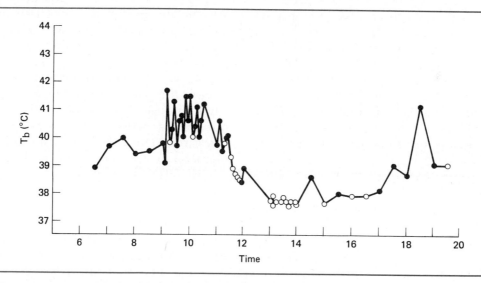

Figure 23–16. Short-term cycles of activity and body temperature of an antelope ground squirrel: (●), Active; (○), inactive. (From M. A. Chappell and G. A. Bartholomew, 1981, *Physiological Zoology*, 54:81–93. © 1981 by The University of Chicago. All rights reserved.)

(a)   (b)

Figure 23–17. Cape ground squirrel (*Xenurus inauris*) using its tail as a parasol: (a) The erected tail shades the dorsal surface of the animal; (b) the tail is held over the back of a horizontal squirrel, shading its head and body. (Photographs courtesy of Albert F. Bennett; from A. F. Bennett, R. B. Huey, H. John-Alder, and K. A. Nagy, 1984, *Physiological Zoology,* 57:57–62. © 1984 by The University of Chicago. All rights reserved.)

rel's back with its white ventral surface upward. In this position it acts as a parasol, shading the squirrel's body and reducing the standard operative temperature. The tail of the antelope ground squirrel is relatively short, extending only halfway up the back, but the shade it gives can reduce the standard operative temperature by as much as 6 to 8°C. The Cape ground squirrel (*Xenurus inauris*) of the Kalahari Desert has an especially long tail that can be extended forward nearly to the squirrel's head. Cape ground squirrels use their tails as parasols (Figure 23–17), and observations of the squirrels indicated that the shade may significantly extend their activity on warm days (Bennett et al. 1984).

The Use of Torpor by Desert Rodents The significance of daily torpor as an energy conservation mechanism in small birds was illustrated earlier. Many desert rodents have the ability to become torpid. In most cases the torpidity can be induced by limiting the food available to an animal. When the food ration of the California pocket mouse (*Perognathus californicus*) is reduced slightly below its daily requirements, it enters torpor for a part of the day. In this species even a minimum period of torpor results in an energy saving. If a pocket mouse were to enter torpor and then immediately arouse, the process would take 2.9 hours. Calculations indicate that the overall energy expenditure during that period would be reduced 45 percent compared to the cost of maintaining a normal body temperature for the same period. In this animal, the briefest possible period of torpor gives the animal an energetic saving, and the saving increases as the time spent in torpor is lengthened.

The duration of torpor is proportional to the severity of food deprivation in the pocket mouse. As its food ration is reduced, it spends more time each day in torpor and conserves more energy. Adjustment of the time spent in torpor to match the availability of food may be a general phenomenon among seed-eating desert rodents. These animals appear to assess the rate at which they accumulate food supplies during foraging rather than their actual energy balance. Species that accumulate caches of food will enter torpor even with large quantities of stored food on hand if they are unable to add to their stores by continuing to forage. When seeds were deeply buried in the sand, and thus hard to find, pocket mice spent more time in torpor than they did when the same quantity of seed was close to the surface (Reichman and Brown 1979). This general response probably represents a response to the chronic food shortage that may face desert rodents because of the low primary productivity of desert communities and the effects of unpredictable variations from normal rainfall patterns, which may almost completely eliminate seed production by desert plants in dry years.

Conclusion

The energetically expensive lives of endothermal vertebrates provide enormous freedom from environmental constraints. Endotherms are most impressive in their capacity to live in the coldest environments on earth; hot environments are more troublesome for endotherms, but many species can survive even in deserts.

Because endotherms achieve their independence of environmental conditions largely by using energy to maintain internal homeostasis they are more clearly constrained by the availability of food than are ectotherms. A comparison of the physiological ecology of endotherms and ectotherms emphasizes different relationships between the organisms and their environments: We saw in Chapter 16 that interactions with the physical world, especially energy exchange, can be seen to affect the day-to-day lives of ectotherms. Similar cases can be found among endotherms, for example the antelope ground squirrels in Deep Canyon, but in general the biological environment appears to be reflected more clearly than the physical environment in the biology of endotherms.

The high energy requirements of endotherms—as illustrated, for example, by the relationship between home range size, body size, and diet—set habitat requirements for some species of endotherms. Large endotherms, especially carnivores, require enormous home ranges and they must be able to move freely about those ranges. Nomadic herbivores also require huge areas to seek out grass at the proper stage of growth. These specializations of endotherms evolved during the Cenozoic when the only constraints on movement were set by geography and climate. Since the late Pleistocene the spread of human populations has drastically altered the kinds and sizes of habitats available for other species of vertebrates. These alterations have had particularly serious consequences for some large endotherms because the space they need is no longer unrestrictedly available. Human evolution and the consequences of human activities for other vertebrates are considered in Chapter 24.

Summary

Endothermy is an energetically expensive way of life. It allows organisms considerable freedom from the physical environment, especially low temperatures, but it requires a large base of food resources to sustain high rates of metabolism. Energy budgets—calculations that quantify the energy expenditures and energy returns of specific activities—can provide insights about the energy implications of many behaviors of endotherms.

Endothermy is remarkably effective in cold environments; some species of birds and mammals can live in the coldest temperatures on Earth. The insulation provided by hair, feathers, or blubber is so good that little increase in metabolic heat production is needed to maintain body temperatures 100°C above ambient temperatures. In fact, some aquatic mammals, such as northern fur seals, are so well insulated that overheating is a problem when they are on land or in water warmer than 10 or 15°C.

Endothermy is most effective in situations in which an animal is warmer than its environment. In this situation metabolism and insulation are adjusted to balance heat production and heat loss. In hot environments the temperature gradient can be reversed—the environment is hotter than the organism—and this condition creates problems for endotherms. Evaporative cooling is effective as a short-term response to overheating, but it depletes the body's store of water and creates new problems. Small animals, rodents for example, can often avoid much of the stress of hot environments by spending the day underground in burrows and emerging only at night when it is cool. Larger animals have nowhere to hide and must meet the stress of hot environments head-on. Camels and other large mammals of desert regions relax their limits of homeostasis when they are confronted by the twin problems of high temperatures and water shortage: They allow their body temperatures to rise during the day and fall at night. This physiological tolerance is combined with behavioral and morphological characteristics that reduce the amount of heat that actually reaches their bodies from the environment.

Mobility is an important part of the response of large endotherms to both hot and cold environments: Seasonal movements away from unfavorable conditions (migration) or regular movements between scattered oases that provide water and shade are options available to medium-size or large mammals. The great mobility of birds makes these sorts of movements feasible even for relatively small species.

When the stresses of the environment overwhelm the regulatory capacities of an endotherm and resources to sustain high rates of metabolism are unavailable, many small mammals (especially rodents) and some birds enter torpor, a state of adaptive hypothermia. During torpor the body temperature is greatly reduced and the animal becomes inert. Periods of torpor can be as brief as a few hours (nocturnal hypothermia is widespread), or can last for many weeks. Mammals that hibernate (enter torpor during winter) arouse at intervals of days or weeks, warming to their normal temperature for a few hours and then returning to a torpid condition. Torpor conserves energy at the cost of forfeiting the benefits of endothermy.

The most remarkable feature of the ability of birds and mammals to live in diverse climates is not the specializations of arctic or desert animals, remarkable as they are, but the realization that only minor changes in the basic endothermal pattern are needed to permit existence over nearly the full range of environmental conditions on Earth.

References

Aschoff, J. 1982. The circadian rhythm of body temperature as a function of body size. Pages 173–188 in *A Companion to Animal Physiology*, edited by C. R. Taylor, K. Johansen, and L. Bolis. Cambridge University Press, Cambridge.

Bakken, G. S. 1980. The use of standard operative temperature in the study of the thermal energetics of birds. *Physiological Zoology* 53:108–119.

Bartholomew, G. A. and F. Wilke. 1956. Body temperature in the northern fur seal, *Callorhinus ursinus*. *Journal of Mammalogy* 37:327–337.

Bennett, A. F., R. B. Huey, H. John-Alder, and K. A. Nagy. 1984. The parasol tail and thermoregulatory behavior of the Cape ground squirrel *Xerus inauris*. *Physiological Zoology* 57:57–62.

Busch, C. 1988. Consumption of blood, renal function, and utilization of free water by the vampire bat, *Desmodus rotundus*. *Comparative Biochemistry and Physiology* 90A:141–146.

Cade, T. J. and G. L. Maclean. 1967. Transport of water by adult sandgrouse to their young. *Condor* 69: 323–343.

Calder, W. A. and J. Booser. 1973. Hypothermia of broad-tailed hummingbirds during incubation in nature with ecological correlations. *Science* 180: 751–753.

Chaplin, S. B. 1974. Daily energetics of the black-capped chickadee, *Parus atricapillus*, in winter. *Journal of Comparative Physiology* 89:321–330.

Chappell, M. A. 1980. Thermal energetics and thermoregulatory costs of small Arctic mammals. *Journal of Mammalogy* 1:278–291.

Chappell, M. A. and G. A. Bartholomew. 1981a. Standard operative temperatures and thermal energetics of the antelope ground squirrel *Ammospermophilus leucurus*. *Physiological Zoology* 54:81–93.

Chappell, M. A. and G. A. Bartholomew. 1981b. Activity and thermoregulation of the antelope ground squirrel *Ammosphermophilus leucurus* in winter and summer. *Physiological Zoology* 54:215–223.

French, A. R. 1986. Patterns of thermoregulation during hibernation. Pages 393–402 in *Living in the Cold: Physiological and Biochemical Adaptations*, edited by H. C. Heller, X. J. Musacchia, and L. C. H. Wang. Elsevier, New York.

Gauthier-Pilters, H. and A. I. Dagg. 1981. *The Camel*. University of Chicago Press, Chicago.

Gauthreaux, S. A., Jr. 1982. The ecology and evolution of avian migration systems. Pages 93–168 in *Avian Biology*, volume 6, edited by D. S. Farner, J. R. King, and K. C. Parkes. Academic Press, New York.

Heller, H. C. (editor). 1987. Living in the cold. *Journal of Thermal Biology* 12, no. 2. (The entire issue is devoted to this topic.)

Jones, D. R., W. K. Milsom, and N. H. West (editors). 1988. The comparative physiology and biochemistry of cardiovascular, respiratory, and metabolic responses to hypoxia, diving, and hibernation. *Canadian Journal of Zoology* 66:3–200.

Kamau, J. M. Z., J. N. Maina, and G. M. O. Maloiy. 1984. The design and the role of the nasal passages in temperature regulation in the dik-dik antelope (*Rhynchotragus kirkii*) with observations on the carotid rete. *Respiration Physiology* 56:183–194.

McFarland, W. N. and W. A. Wimsatt. 1969. Renal function and its relation to the ecology of the vampire bat, *Desmodus rotundus*. *Comparative Biochemistry and Physiology* 28:985–1006.

McNab, B. K. 1973. Energetics and distribution of vampires. *Journal of Mammalogy* 54:131–144.

Reichman, O. J. and J. H. Brown. 1979. The use of torpor by *Perognathus amplus* in relation to resource distribution. *Journal of Mammalogy* 60:550–555.

Taylor, C. R. 1972. The desert gazelle: a paradox resolved. Pages 215–217 in *Comparative Physiology of Desert Animals*, edited by G. M. O. Maloiy. *Symposia of the Zoological Society of London, No. 31.*

Wang, L. C. H. 1978. Energetic and field aspects of mammalian torpor: the Richardson's ground squirrel. Pages 109–145 in *Strategies in Cold: Natural Torpidity and Thermogenesis*, edited by L. C. H. Wang and J. W. Hudson. Academic Press, New York.

Welch, W. R. 1984. Temperature and humidity of expired air: interspecific comparisons and significance for loss of respiratory heat and water from endotherms. *Physiological Zoology* 57:366–375.

Yagil, R. 1985. *The Desert Camel, Comparative Physiological Adaptation*. S. Karger, Basel.

Humans are primates and have complex social systems like those discussed in Chapter 22 as an ancestral character. The fossil record indicates that the first stone tools are about 2.5 million years old and are probably associated with the first appearance of the genus *Homo*. Molecular techniques of studying genetic relationships among organisms have recently been applied to human evolution and have challenged some prevailing views: The molecular studies suggest that the separation of humans from the African great apes was more recent than had been inferred from anatomical studies and that the closest living relatives of humans are chimpanzees, not gorillas. Furthermore, application of cladistic methods to human fossils reveals that humans retain more ancestral morphological character states than the apes.

The growth of human populations, especially in the past 50,000 years, is a unique phenomenon in the history of vertebrates. Never before has a single species so dominated the Earth. The consequences of this domination, and the environmental changes that have resulted, have been catastrophic for many other species of vertebrates. The first extinctions of other species of vertebrates that can be traced to humans may have occurred during the Pleistocene when humans first entered North America, although this hypothesis is controversial. Documented extinctions can be traced back to the Age of Exploration, and the situation continues to worsen. In this chapter we trace the origin of humans and the spread of modern humans, and discuss the impact of humans on other vertebrates and the prospects and vital importance of reversing the increasing rate of extinction of species.

856

Homo sapiens *and the Vertebrates*

24

The Origin of Humans

Human beings are primates. Our close living relatives, the chimpanzees and gorillas of Africa, and we have descended from arboreal ancestors that lived in the early Tertiary forests 65 million years ago. What is the basis for these statements, which Darwin's Victorian contemporaries found so startling and upsetting? In the following sections we summarize the evidence and the inferences about human phylogeny and evolution.

Characteristics of Primates

Humans share many biological traits with animals variously called apes, monkeys, and prosimians. Using the principle of homology, which we discussed in Chapter 1, comparative anatomists recognized that all of these mammals can be grouped together in the order Primates. Obvious traits that characterize primates are (1) retention of the clavicle (which is reduced or lost in many mammalian orders) as a prominent element of the pectoral girdle; (2) a shoulder joint allowing a high degree of limb movement in all directions and an elbow joint permitting rotation of the forearm; (3) general retention of five functional digits on the fore and hind limbs; (4) enhanced mobility of the digits, especially the thumb and big toes, which are usually opposable to the other digits; (5) claws modified into flattened nails; (6) sensitive tactile pads

developed on the distal ends of the digits; (7) reduced snout and olfactory apparatus with most of the skull posterior to the orbits; (8) reduction in number of teeth compared to stem mammals but retention of simple molar cusp patterns; (9) complex visual apparatus with increased acuity and color perception and varying degrees of development of forward-directed binocular eyes; (10) enlarged brain relative to body size, particularly enlarged is the cerebral cortex; (11) trend toward advanced fetal nourishment mechanisms; (12) only two mammary glands (some exceptions); (13) typically, only one young per pregnancy associated with prolonged infancy and preadulthood; and (14) trend toward holding the trunk of the body upright leading to facultative bipedalism (Szalay and Delson 1979). Note that several of the above characteristics can be identified as trends within the primates and thus may not be apparent in primitive members of the order.

These traits have long been attributed to an arboreal life. All of the basic modifications of the appendages can be seen as functional for arboreal locomotion, as can the stereoscopic depth perception that results from binocular vision and the enlarged brain for neuromuscular coordination between visual perception and locomotory response. Most primates are arboreal, but some have become secondarily terrestrial, humans most of all. Even so, many of the traits that are most distinctively human have been said to derive from earlier ar-

Phylogenetic Relationships of the Primates

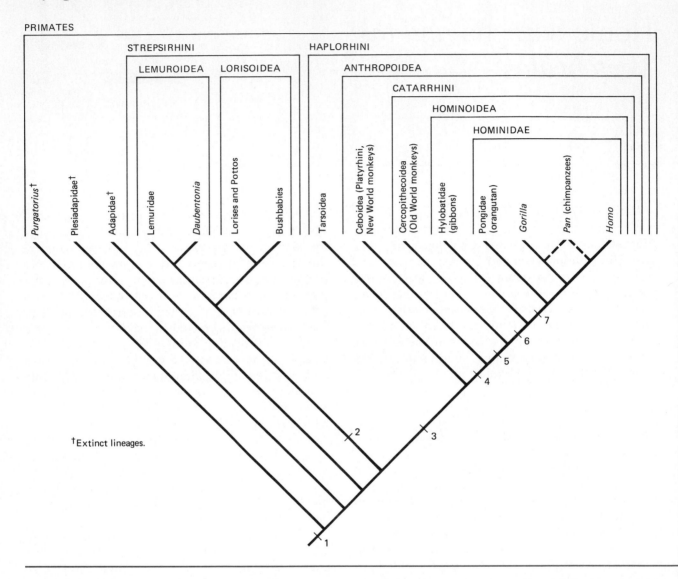

boreal specializations. However, a moment's reflection indicates that arboreality cannot be the entire basis for these primate characteristics. Squirrels provide a telling counter example; they are arboreal but show few of the specializations seen in primates (Cartmill 1974). Although squirrels do have a clavicle and good mobility of the shoulder and elbow joint, the range of movement is not as great as that of primates. The thumbs of squirrels are greatly reduced, there is no opposability, and the remaining digits have long, sharp, recurved claws in addition to lacking tactile pads at their

1. Primates: Cheek teeth bunodont or brachydont; limbs plantigrade; a nail (instead of a claw) always present at least on the pollex (thumb).   2. Strepsirhini (lemurs and lorises): Cranium elongate; orbit and temporal fossa broadly confluent.   3. Haplorhini [tarsiers plus anthropoids]: Cranium short; orbit and temporal fossa separated ventrally by a postorbital plate.   4. Anthropoidea: Tubular external auditory meatus; broad interorbital bony pillar and prominent thickening [the glabella] medially above orbits distinct from brow ridges [the tori]; dental formula 2-1-2-3; third premolar single-cusped, bilaterally compressed; fourth premolar bicuspid, slightly longer than broad; lower molars increase in size posteriorly, the third only slightly larger than the second, all with five cusps, the hypoconulid small; second upper molar larger than first or third; distinctive muscle, nerve and articular modifications of the humerus; characteristic ankle and thumb bones.   5. Catarrhini [Old World monkeys, apes, and humans]: Nasal aperture high and oval-shaped; premaxilla short and broad; upper incisors of similar size, the first spatulate and the second caniform; canine slender, high crowned relative to length and sexually dimorphic; third and fourth upper premolars broad in buccolingual direction; upper molars similarly broad with four cusps, the hypocone small; molar enamel thin; tooth rows form a V-shape; palate shallow and longer than broad; distinctive characters of humerus, radius, and ulna.   6. Hominoidea [apes and humans]. Frontal bone wide posteriorly; cusps on upper premolars nearly uniform; third lower premolar low-crowned; modifications of the vertebral column foreshadowing those seen in Recent hominoids [see below]; scapula with elongated vertebral border and robust acromion; humeral head rounded, medially oriented, and longer than femoral head. Additional characters of hominoids absent in the problematic fossil *Proconsul* include enlarged sinuses; first upper incisor as broad as high; lower molars broad; palate deep; clavicle elongated; shaft of ulna bowed; radial head rounded; blade of ilium broadened; thoracic vertebrae protrude into thoracic cavity; five lumbar, four or five sacral, and six caudal vertebrae, no tail; various characteristics of muscle insertions; several derived features of bones of the appendages.   7. Hominidae [apes and humans]: Canines robust and long relative to their height; lower third molar robust, not laterally compressed; tooth rows form a U-shape. Additional characters of Recent hominids include mastoid process distinct; medial and lateral pterygoids of equal size; maxillary sinus enlarged; orbits higher than broad; lengthened premaxilla; second upper incisor spatulate; third and fourth lower premolars elongated; enamel on molars thick; ischial tuberosities absent; four or five lumbar vertebrae; several derived characters of trunk and arm muscles and arm bones. (Based on R. L. Cichon, 1983, pages 783–837 in R. L. Cichon and R. S. Corruccini, *New Interpretations of Ape and Human Ancestry,* Alan R. Liss, Inc., New York; and on data from various sources in E. Delson, editor, 1985, *Ancestors: The Hard Evidence,* Alan R. Liss, Inc., New York.)

Figure 24–1.   Phylogenetic relationships of the Primates. This diagram shows the probable relationships among the major groups of primates. Extinct lineages are marked with a dagger (†). The dashed lines indicate two possible relationships for chimpanzees.

tips. Squirrels have large olfactory organs and snouts, laterally directed eyes, and no notable brain enlargement compared with fully terrestrial diurnal rodents. Their life histories are typical of small mammals: They produce large litters of fast-growing young. Nevertheless, squirrels are fully arboreal and can match or exceed the climbing skills of similar-size primates. If primate characters are not required by an arboreal mammal, what do they indicate about our early evolution?

Small marsupials and true (African) chameleons may give a clue, because they are also arboreal and share more convergences with primitive primates than do squirrels. These three convergent taxa (stem primates, small marsupials, and chameleons) are all visually directed predators on arboreal insects. They all use grasping digits during cautious well-controlled pursuit of insects on slender branches. Thus, we may attribute many of the basic characters of primates not just to arboreal habits but to very specialized ones: visual location and manual capture of arboreal insect prey.

Evolutionary Trends and Diversity in Primates

Table 22–4 presents a traditional classification of modern primates, and Figure 24–1 presents a simplified cladogram. It is generally agreed that the first primates evolved from a line of arboreal proto-insectivores not unlike the present-day tree shrews (Scandentia: Tupaiidae). These small, agile southeast Asian mammals prefer low tropical forest growth. They are slender with long bushy tails and rather short limbs, particularly the forelegs, and are more like squirrels than monkeys (Figure 24–2). In fact, the name Tupaiidae comes from the Malay word for squirrel. The skull and teeth show most of the unspecialized characteristics of primitive mammals, including nostrils at the end of a pronounced moist snout, a large gape, lateral eyes, and occipital condyles located at the extreme posterior of the skull with the head carried fully out in front of the vertebral column. Further, tree shrews have lost only one incisor and one premolar from the primitive mammalian dental formula of three incisors, one canine, four premolars, and three molars. The cheek teeth also retain the primitive tricuspid condition (Luckett 1980).

The first primates appear at the Cretaceous–Tertiary boundary and show considerable diversity by the Eocene. These fossil primates are most closely related to modern lemurs, pottos, aye-ayes, lorises, and galagos. The first primates have long been grouped with these living forms and the tarsiers as "prosimians." The earliest primates were rather like tree shrews with, perhaps,

some development toward prehensile fingers and toes, opposability of the first digits, and enlarged brains.

In a short period of geological time, prosimians spread over large portions of the Old and New Worlds and differentiated into ecological forms ranging from insectivorous to vegetarian in their diets and from nocturnal to diurnal in their daily activity cycles. Most were primarily arboreal and retained certain primitive features such as long snouts and long tails (Figure 24–3). They became more clearly identifiable as primates by changes leading to increased range of movement in their appendicular skeleton, the tendency for the eyes to move forward, and—most notably—larger brain capacity. The late Eocene prosimians possessed brains that were larger in relation to their body sizes than any other contemporary mammals (Romer 1966). This allometric enlargement was probably associated with the sophisticated neuromuscular coordination required for complicated arboreal insect-eating activities, complex visual functions, and control of a grasping, manipulative hand. Today, the prosimians are considered paraphyletic. The species generally considered prosimian retain a great many primitive primate features. They are restricted to parts of Africa and Southeast Asia and to the island of Madagascar, where the lemurs and their relatives continued to maintain a diversity of forms in the absence of predators and other competing primates. Today many of the lemurs are threatened with extinction, owing to destruction of their forest habitat (Jolly 1980). By the Oligocene the first simians had evolved in most parts of the world, and the prosimians began to disappear.

There is much disagreement about the evolutionary transition from the prosimians to the higher primates (Ciochon and Fleagle 1987). Fossil primates from the critical Oligocene and Miocene periods are rare. Most of the fossils from the Miocene are, in fact, already well advanced New World (platyrrhine) or Old World (catarrhine) monkeys or apes (hominoids). The morphological features of these groups and their fossil histories suggest that they evolved from a tarsier-like ances-

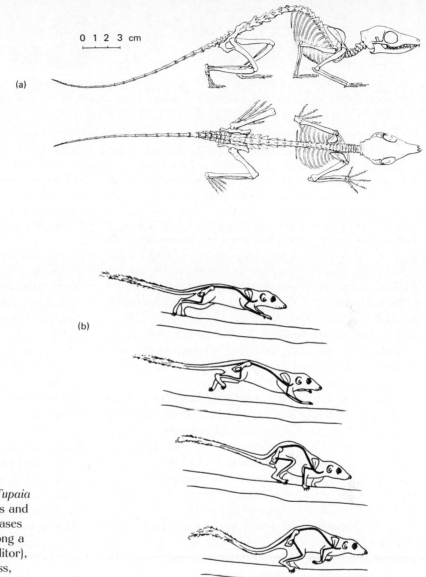

Figure 24–2.   Skeleton of tree shrew (*Tupaia glis*) (a) in typical posture with flexed limbs and quadrupedal stance and (b) sequential phases of the bounding run used at top speed along a limb. [Modified from F. A. Jenkins, Jr. (editor), 1974, *Primate Locomotion*, Academic Press, New York.]

tor in Asia in the late Eocene. Today's tarsiers are tiny nocturnal arboreal insect eaters highly specialized for jumping locomotion. Although not the only derived feature uniting tarsiers and advanced species, one curious shared unique feature for these primates is the condition of the nose. The nostrils of tarsiers and all monkeys and apes are surrounded by smooth dry skin or hair rather than the moist, glandular skin of the nose in more primitive primates.

A rapid invasion of Africa followed the evolution of the advanced primates. Dispersal of some

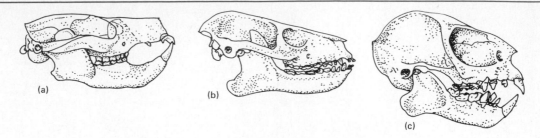

Figure 24–3. Early primate fossils, showing variation in dentition with feeding habits: (a) *Plesiadapis*, a late Paleocene form, was rodent-like; (b) *Notharctus*, an Eocene lemur, was more omnivorous and carnivorous, while (c) *Tetonius*, a more advanced Eocene tarsier-like form, had highly specialized teeth suggesting a basically predatory mode of life, and its large eye orbits clearly indicate nocturnal habits. Not to same scale. (Modified from A. S. Romer, 1966, *Vertebrate Paleontology*, 3rd edition, University of Chicago Press, Chicago.)

of the early undifferentiated advanced primates across the South Atlantic ocean (which was much narrower then) had occurred by the early Oligocene. Rodents ancestral to the South American guinea pigs and their relatives (caviomorphs) also arose in Africa and crossed the Atlantic at about the same time.

All known platyrrhine anthropoids have been restricted to the American tropics since the first record in the late Oligocene. All catarrhines with the exception of *Homo sapiens* have been similarly restricted to the Old World. But Oligocene platyrrhines from Argentina resemble Egyptian catarrhines of the same period in a number of dental features, indicating close relationship.

The New World monkeys do not share the derived features of Old World monkeys, apes, and humans but have unique features of their own. As the group name indicates, these monkeys have flat noses, with nostrils that are wide apart and face outward. They are on average smaller than the catarrhines; there is little tendency toward reduction in their primitively long tail, which is specialized as a prehensile organ in a number of the medium and large platyrrhines. The pollex (thumb) is not fully opposable to the other fingers and is reduced or absent in some species. However, the hallux (big toe) is fully opposable. There are three pre-

molars present in each jaw. In addition to the typical South American monkeys, the platyrrhines also include the squirrel-like marmosets (callithricids). The marmosets are specialized dwarf primates, some with a special taste for the edible gums exuded from the bark of certain trees. Marmoset toes, except the hallux, have secondarily claw-like nails that dig into bark and aid in climbing.

The catarrhines include the Old World monkeys, the apes, and humans. Their nostrils are close together, open forward and downward, and have a relatively smaller bony opening from the skull than is the case for platyrrhines. There is a trend toward large body size, culminating in the great apes and humans. Although catarrhines are mostly arboreal, there is a marked tendency toward upright carriage of the trunk and secondary specializations for terrestrial locomotion. The tail is often short or absent, and prehensility has never evolved. The second premolar is always absent, giving a dental formula of

$$\frac{2\text{-}1\text{-}2\text{-}3}{2\text{-}1\text{-}2\text{-}3}$$

a condition traceable to fossils in the Oligocene. The group consists of two clades: the Old World monkeys (Cercopithecoidea); and the apes and humans (Hominoidea), including the gibbons (Hylobatidae), the orangutan (Pongidae) and the African great apes, *Australopithecus*, and humans (Hominidae). The Cercopithecidae constitute a spectrum of nearly 80 species of often highly colorful, sometimes impressively fierce, and always behaviorally enthralling vertebrates (Chapter 22). They are almost universally social animals living in habitats that range from open grassy plains to the tops of mature rain forests and from the highlands of the Himalayas to intertidal mangrove swamps.

Evolution and Phylogeny of the Hominoidea

Apes and humans are placed in the superfamily Hominoidea. The fossil record for this clade is better than for other lineages of primates despite a gap of 4 million years beginning about 8 million years ago.

The hominoids are more diversified in morphology than their nearest relatives, the cercopithecoid monkeys. There is a greater range in body size, the small Asian gibbons being dwarfed by the huge African gorillas, and the dense hair covering of gibbons stands in contrast to the sparse remnants of hair in humans. The gibbons move through the trees most frequently by brachiation and are among the most versatile arboreal acrobats alive. Gibbons become bipedal when moving on the ground, holding their arms outstretched like a tightrope walker. The larger and more sluggish orangutans rarely swing by their arms and prefer slow quadrupedal climbing among the branches of trees. The African gorillas and chimpanzees are more at home on the ground, where they most often progress quadrupedally, supported on their forelimbs by their knuckles. Although all modern hominoids can stand erect and walk to some degree on their hind legs, only humans have evolved an erect bipedal mode of striding locomotion involving a specialized structure of the hindlimbs, thereby freeing the forelimbs from obligatory functions of support and locomotion.

Hominoids are morphologically distinguished from other recent anthropoids, including their sister taxon the cercopithecoids, by a pronounced widening and dorsoventral flattening of the trunk relative to body length so that the shoulders, thorax, and hips have become proportionately broader than in monkeys. As a consequence the clavicles are elongated, the iliac blades of the pelvis are broad, and the sternum is a broad structure, the bony elements of which fuse soon after birth to form a single flat bone. The shoulder blades of hominoids lie over a broad, flattened back, in contrast to their lateral position next to a narrow chest in monkeys (Figure 24–4). The pelvic and pectoral girdles are relatively closer together in hominoids, owing to the shortening of the lumbar region of the vertebral column. The caudal vertebrae have become reduced to vestiges in all recent hominoids, and normally no free tail appears postnatally. After birth, there is a tendency for the spinal column to arch and especially in the lumbar region to approach the center of gravity of the body when it is held upright. This is augmented by the abovementioned flattening of the thorax, which moves the center of gravity back toward the vertebral column. These and other anatomical specializations of the trunk are common to all hominoids and help to maintain the erect postures that these primates assume during sitting, climbing, and walking bipedally (Campbell 1985). Secondary curves of the spine are the consequence of bipedal locomotion, forming as soon as the human infant learns to walk.

Brain enlargement has been a dominant evolutionary force molding the shape of the hominoid skull, especially in the last 2 million years, and has tended to increase the neurocranial vault more than its base. The skulls of hominoids also differ from those of other catarrhine monkeys by their extensive formation of sinuses, hollow air-filled spaces lined with mucous membranes that develop between the outer and inner surfaces of skull bones. Chimpanzees, gorillas, and humans share the derived character of large frontal sinuses.

In the identification of fossils, differences in the cheek teeth between hominoids and Old

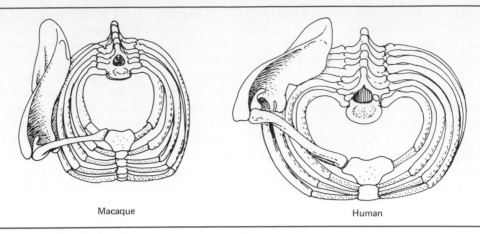

Figure 24—4. Cephalic view of chest and pectoral girdle (right half) of macaque (cercopithecoid monkey) and human (hominoid).

World monkeys are of paramount significance because many primate fossils are represented by individual teeth, or teeth in fragments of jawbone. The Old World monkeys have lower molars (M_1 and sometimes M_2) with four cusps, one at each corner of a rectangular shaped crown; both mesial and distal pairs of cusps are connected by ridges. Since the Miocene hominoid lower molars have had crowns with five cusps, but in modern humans (not fossil ones) this pattern is frequently obscured by crenulations or variations in the number of cusps. When five cusps do appear the grooves between the cusps usually resemble the letter Y, with the open part of the Y embracing the hypoconid cusp (Figure 24–5). This molar pattern, called Y-5, has persisted among hominoids for more than 20 million years (Day 1986).

A phylogenetic tree of the Hominoidea (Figure 24–6) indicates that the human, chimpanzee, and gorilla are more closely related to each other, in terms of recency of their common ancestor, than they are to orangutans or gibbons. These relationships among living apes and humans were first determined on the basis of comparative anatomy by T. H. Huxley, who published his ideas in 1863. They have gained additional support and clarity from recent biochemical, serological, and cytolog-

ical comparisons as well as from the fossil record of the earlier hominoids. The majority of molecular studies support a close relationship between chimpanzees and humans, whereas anatomical studies place humans and gorillas closer to each other (Lewin 1987).

The earliest indications of both the cercopithecoid and the hominoid lineages are fossil teeth and jaws found in the Egyptian Fayum deposits of the Oligocene, approximately 35 million years old. Fossils assigned to the genus *Propliopithecus* and others of similar structure, especially *Aegyptopithecus*, are sufficiently generalized to be considered ancestral for both clades of catarrhines. *Oligopithecus* occurs in the same deposit. If its molars do indicate that it is near the stock that gave rise to the cercopithecoid monkeys as has been claimed, we may assume that the division between the cercopithecoids and hominoids occurred near the Eocene–Oligocene boundary.

By the Miocene the ancestral hominoids had diversified into a number of ecological types and had spread widely over the Old World, including Africa, Europe, and Asia. One genus, *Pliopithecus* (Figure 24–7), has a skull like that of the gibbons but its appendicular and trunk skeletons are generalized with the fore- and hindlimbs about equal

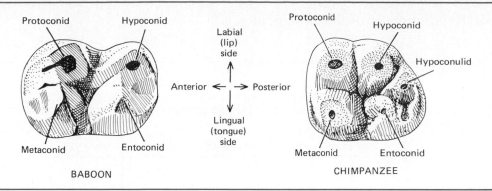

Figure 24–5. Right lower molar patterns of Old World monkey (baboon) and hominoid (chimpanzee) compared. Shaded bars indicate low crown areas or grooves between major cusps and ridges illustrating the Y-5 configuration of hominoids.

in length. It was not as specialized for brachiation and arboreal life as the modern gibbons or the somewhat later *Oreopithecus* (Figure 24–8). *Oreopithecus*, an Italian fossil primate, resembles the gibbons in its postcranial skeleton but has a less specialized skull. Thus *Oreopithecus* is considered to be a cercopithecoid monkey that may have lived a life like modern gibbons.

Of great interest are Miocene fossils variously referred to as dryopithecids and ramapithecids. These primates occurred across Eurasia and Africa and have been assigned various generic names: *Dryopithecus, Proconsul, Sivapithecus, Kenyapithecus,* and *Ramapithecus* (*Proconsul* is usually synonymized with *Dryopithecus,* and *Ramapithecus* often included in *Sivapithecus,* illustrating the lumping typical of recent primate paleontology). They are all closely related hominoids, but paleoclimatic and paleofaunistic evidence suggests they occupied a wide range of ecological niches (Lovejoy 1981) and some genera span an enormous range of geologic time.

Some early Miocene dryopithecids lived in heavily forested areas of Africa and later Eurasia. The later ramapithecines first appear in the fossil record 17 million years ago during the Miocene when the heavily forested areas of the Old World

were giving way to mixed environments consisting of patches of forests, savanna woodlands, and open grasslands. The kinds of mammalian fossils found with ramapithecines suggest that these primates lived on the edge of the forests and open areas and derived their livelihood from both. If true, then some of them may have been evolving toward a terrestrial mode of locomotion, and it is among the ramapithecines that we can expect to find the ancestors of the great apes and humans.

Origin and Evolution of Humans

Humans differ enough in their anatomy from the apes so that they have usually been placed in a separate family, called the Hominidae, although biochemical evidence suggests a much closer relationship between apes and humans. The jaw in the human clade is shorter in association with the shortening of the muzzle, and certain teeth are smaller and the entire dentition is more uniform in size and shape. In particular the longer canines of apes and monkeys came to lie on the same occlusal plane with the incisors and cheek teeth. The prominent diastema between the canines and in-

Phylogenetic Relationships of the Hominoidea

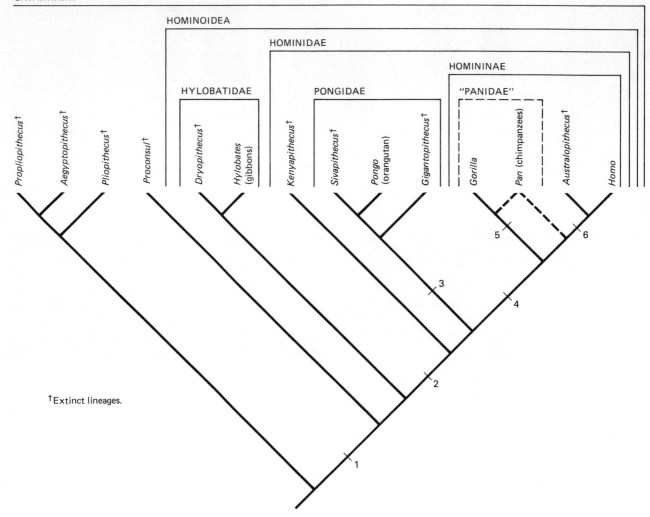

†Extinct lineages.

1. Hominoidea: See node 6 in Figure 24–1. **2.** Hominidae: See node 7 in Figure 24-1. **3.** Ponginae (orangutan and fossil relatives): Zygomatic arch flattened, facing anteriorly or even slightly superiorly; loss of glabellar thickening; narrow interorbital distance; small incisive foramen, first upper incisor much larger than second; distinctive junction of premaxilla with maxilla and 18 additional derived characters. **4.** Homininae (African apes, australopithecines, and humans): Large sphenopalatine fossae; several characteristics of the nasopalatine region; presence of a fronto-ethmoid sinus; lower molar trigonids markedly reduced; several derived characters of limbs indicative of vertical climbing and incipient bipedal locomotion. **5.** African apes: Thin enamel caps on molars; extensive modification of the hands and less extensive modifications of the arms reflecting the occurrence of knuckle-walking. **6.** Australopithecines and humans: Progressive increase in cranial capacity; anterior shift in position of foramen magnum and occipital condyles; reduction in size and sexual dimorphism of canines; reduction in diastema posterior to canines; decreased molar length relative to width; increased molar crown height; distinctive enamel prism patterns; progressive reduction in length of forelimb relative to length of trunk; shortened ischium; broad ilium; pronounced lumbosacral curve in vertebral column; modifications of arm and leg muscles. (Based on R. L. Cichon, 1983, pages 783–837 in R. L. Cichon and R. S. Corruccini, *New Interpretations of Ape and Human Ancestry,* Plenum Press, New York; and on data from various sources in E. Delson, editor, 1985, *Ancestors: The Hard Evidence,* Alan R. Liss, Inc., New York.)

Figure 24–6. Phylogenetic relationships of the Hominoidea. This diagram shows the probable relationships among the major groups of hominoids. Extinct lineages are marked by a dagger (†). The quotation marks indicate that the family "Panidae" is paraphyletic if chimpanzees are the sister group of hominids.

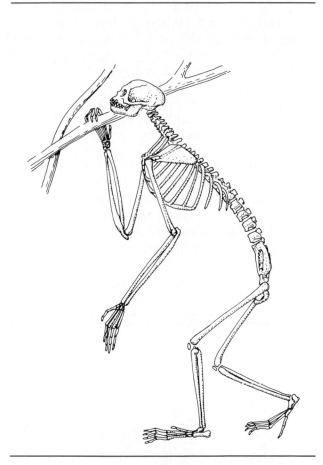

Figure 24–7. Reconstruction of *Pliopithecus,* a Miocene ape.

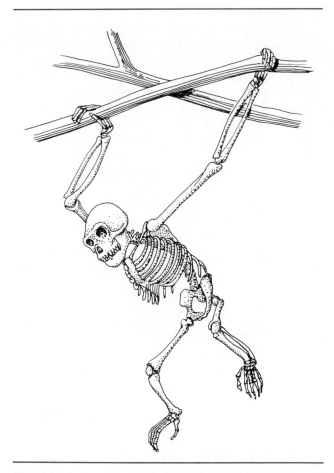

Figure 24–8. Reconstruction of *Oreopithecus,* a gibbon-like primate of the Miocene.

cisors to accommodate occlusion of the jaws in apes has disappeared in humans; normally, all the teeth touch their adjacent members (Figure 24–9). The jaws of an ape are rectangular or U-shaped, with the four incisors lying at right angles to the canines and cheek teeth, which form nearly parallel lines on the two sides. The jaw of members of the genus *Homo* is bow-shaped, with the teeth running in a curve that is widest at the back of the mouth. The ancestors of the hominids, the hominoids such as *Sivapithecus*, had a V-shaped jaw. However, some of the earliest hominids had a U-shaped jaw. The human palate is prominently arched, whereas the ape palate is flatter with parallel columns of cheek teeth.

Various evolutionary trends within the hominids can also be identified. The articulations of the skull with the vertebral column (**occipital condyles**) and the **foramen magnum** (for the passage of the spinal cord) shifted from the rear of the braincase to a position under the braincase. This change balances the skull on top of the vertebral column, in association with an upright, vertical posture. The braincase itself became greatly enlarged in association with an increase in forebrain size. By the end of the middle Pleistocene a prominent vertical forehead developed, in contrast to the sloping foreheads of the apes. The brow ridges and crests for muscular attachments on the skull became reduced in size in association with the reduction in size of the muscles that once attached to them. The human nose became a more prominent feature of the face, with a distinct bridge and tip.

The most radical changes in the human postcranial skeleton are associated with the assumption of a fully erect, bipedal stance: the S-shaped curvature of the vertebral column; the modification of the pelvis and position of the **acetabulum** (socket in the hip for the ball joint with the femur) in connection with upright bipedal locomotion; the lengthening of the leg bones and their positioning as vertical columns directly under the head and trunk. The feet of humans show drastic modification for bipedal striding locomotion, having become flattened except for a tarso-metatarsal arch with corresponding changes in the shapes and positions of the tarsals (ankle bones) and with close, parallel alignment of all five metatarsals and digits; the big toe is no longer opposable as in apes and monkeys (Figure 24–10), although it may still have had the capacity to diverge from the rest of the toes in early hominids.

It is important to emphasize that among mammals the bipedal, striding mode of terrestrial locomotion is unique to the human line of evolution and that it may have been a key change that made possible the evolution of other distinctively human traits, such as tool-using hands, and thus indirectly stimulated the evolution of large forebrains. In addition to changes in the foot, evolution of a bipedal, striding gait from the anthropoid quadrupedal mode required major reorganization in the structure and function of the pelvis and hind leg

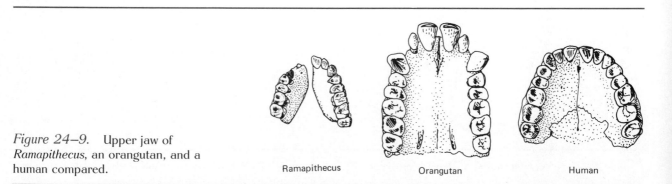

Figure 24–9. Upper jaw of *Ramapithecus*, an orangutan, and a human compared.

Ramapithecus

Orangutan

Human

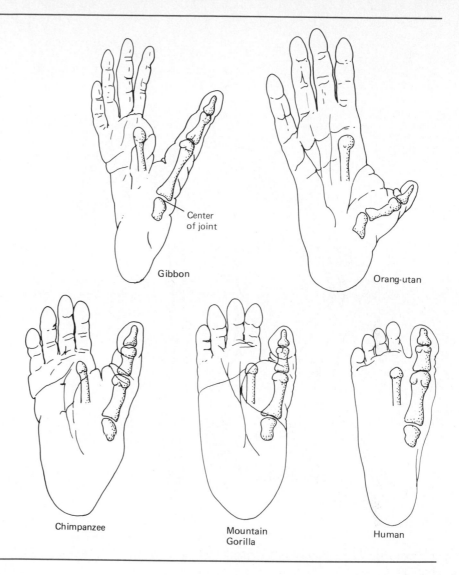

Figure 24–10. Feet of modern hominoids compared, showing some skeletal parts (metatarsals for digits in I and II and phalanges of digit I) in relation to foot form. Note difference in arboreal specializations of gibbon and orangutan, and the degrees of specialization for terrestrialism in chimpanzee, gorilla, and human.

bones, and their associated muscles. Figure 24–11 contrasts the shape and orientation of the gorilla and human pelvis to show the postural differences between quadrupedal and bipedal locomotion. In the gorilla the ischium protrudes backward prominently and the long ilium extends forward and lateral to the vertebral column; the whole pelvis is tilted toward the horizontal. In humans the ischium is much shorter, the broadened ilium extends forward, and the pelvis is vertical.

What has been the evolutionary history and origin of our distinctively human traits? As animals with a great curiosity about our ancestry, we want to know what evolutionary events turned the human into an obligate terrestrial biped from ancestors that clearly were arboreal, and how our tool-using hands and our big brain with its associated implications for intelligence, advanced social organization, and culture evolved.

The Miocene, like the Pleistocene (Chapter 20)

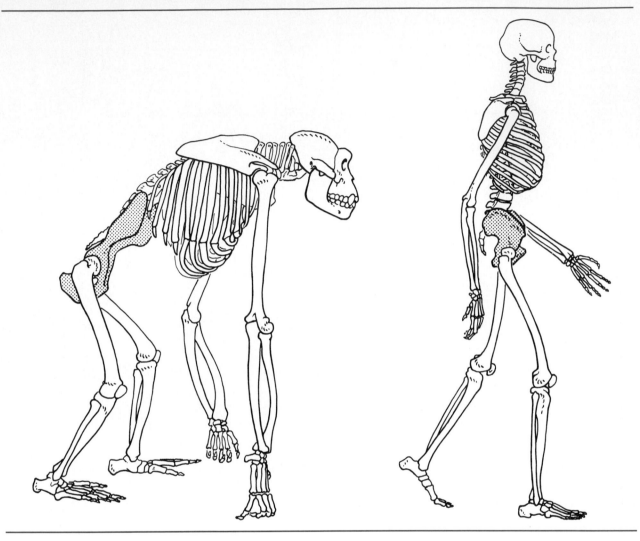

Figure 24—11.   Shape and orientation of the pelvis (shaded) in gorilla and human.

after it, was a period of increasing seasonality of climate. Seasonality favors herbs and annual plants over woody perennials. As tropical forests yielded to more open woodland savannas and grasslands in the Miocene, terrestrial vertebrates, including various clades of hominoids, came under selective forces favoring terrestrialism and in the hominids this evolved as upright bipedalism. Some hominids of this time had dentition suited

for grinding coarse substances and may have been omnivorous, foraging on seeds, nuts, and other small food items gleaned from the ground. They probably also scavenged small animal carcasses and preyed on vulnerable animals. This much of the story is inferred from the combination of ape-like and human-like traits found in various Miocene and early Pliocene hominoid fossils.

We can speculate further that once freed of

locomotor requirements, the forelimbs and hands could respond to selective forces that would shape them into increasingly effective manipulative feeding organs. One advantage may have been increased survival of the young, because the foraging members of the family groups could gather and carry back quantities of food for nursing mothers and their dependent offspring (Lovejoy 1981). The prolongation of gestation, increase in intervals between births, decrease in number of young per pregnancy, lengthening of time to reach sexual maturity, and postreproductive survival of old individuals constitute unique changes in primate evolution culminating in the human line (Figure 24–12). They are traits that favor the evolution of a well-developed social organization. Increase in brain capacity—permitting functions such as communication, eye–limb coordination, manual dexterity, memory, and other functions associated with the neocortex—would be favored by social cooperation in food gathering, hunting, or defense against powerful predators. With increase in body size, tool use, and intelligence in the advanced lines of hominids known from the late Pliocene and Pleistocene, this large brain also permitted the cooperative hunting of large mammals.

Before Darwin's time it was believed that the modern human species was only a few thousand years old, and by counting the genealogy of the Old Testament, it was possible to fix the date of Adam's creation in the Garden of Eden rather precisely. Even after the discovery of the first human fossil bones in Europe and later in Asia and Africa, it was long thought that true human beings (genus *Homo*) were no more than about 500,000 years old, roughly mid-Pleistocene in origin. Better methods of dating fossils have revealed that the original estimates of the age of some fossils were erroneously short. More important, older and older fossils with hominid characteristics have been uncovered, so that it becomes more difficult to say when the first true humans appeared.

The oldest fossils that show great ape/human characteristics are some teeth and jaw fragments from the Siwalik Hills of Pakistan and India; Fort Ternan, Kenya; and localities in Europe and

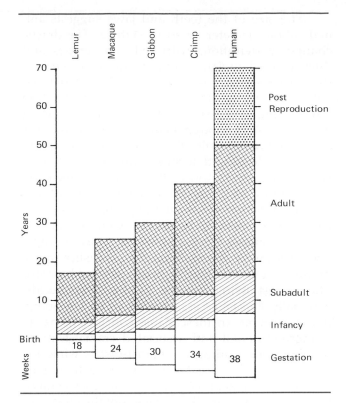

Figure 24–12. Progressive changes in the temporal length of life phases and gestation in primates. Note the proportional increase in the length of the post-juvenile stages as gestation increases. Only humans have a significant postreproductive stage.

China. They span an enormous time period for primate remains, from about 17 to 9 million years ago. They have been assigned to several genera known as the god apes because several of the generic names are derived from the names of Hindu gods. *Sivapithecus* and the possibly congeneric *Ramapithecus* are the best known, and were contemporary with other Miocene forms of sivapithecids. It is widely accepted that these are the probable ancestral stock from which the great apes and human primates evolved. The shape and relative size of the teeth are more like those of humans than like those of apes, but the lower tooth row is V-shaped unlike humans *or* apes.

Phylogenetic Relationships Within the Hominidae

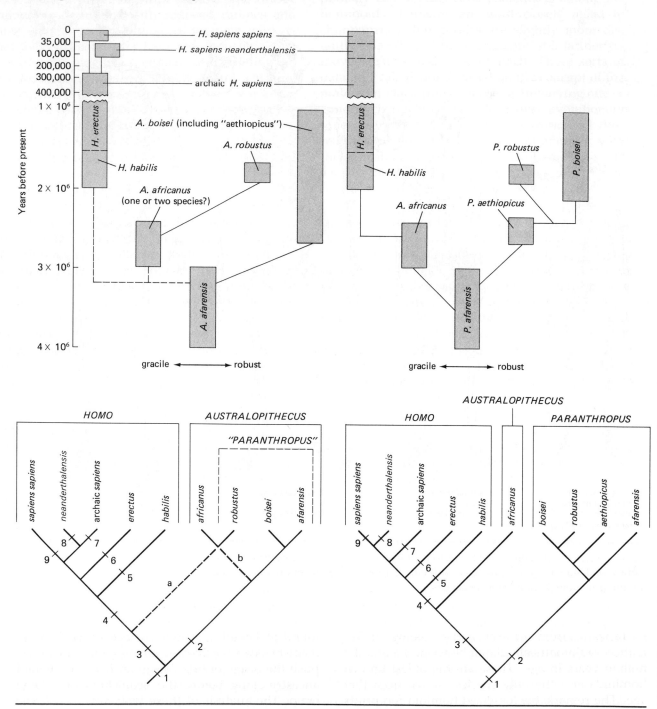

1. Australopithecines and humans: See node 6 in Figure 24-6.   2. *"Paranthropus"*: An extensively pneumatized and projecting mastoid process incorporating on its expanded medial wall the occipitomastoid crest that provides a broad origin for the digastric muscle; an enlarged occipital-marginal venous sinus system; keystone-shaped nasal bones that extend beyond the fronto-maxillary suture to the glabella; more than 20 characters of the dentition.   3. *Australopithecus* and *Homo*: Tooth rows convex laterally; lower third premolar nearly bicuspid; hypoconulids of first and second lower molars small; mastoid not greatly enlarged; origin of digastric muscle restricted to a strong tendon with its base ossified as a juxtamastoid ("beside the mastoid") eminence; nasal bones triangular with the apex at the fronto-maxillary suture and distinctly beneath the glabella.   4. *Homo.* Increased mobility of the elbow joint and solidity of the knee joint (the reverse of the condition in early australopithecines, especially *A. afarensis*) indicating a shift from suspension or climbing to terrestrial bipedal locomotion; associated features of the femur and pelvis; reduced face size and projection associated with reduced robustness of the jaw and dentition; increase in brain size.   5. *Homo habilis:* Cheek teeth narrow in buccolingual direction, palate retracted under the nose, brain size increased by 45% over *Australopithecus africanus*.   6. *Homo erectus:* Distinguished on the basis of frontal proportions; parietal measurements; occipital curvature; the anatomy of the gleonoid cavity and the tympanic plate; straightness of the basicranium; the range of brain sizes; thickness of the cranial bones; a keeled, tent-shaped cranial vault; and a robust pelvis.   7. Archaic *Homo sapiens:* Basicranial flexion essentially modern with corresponding changes in facial prognathism and the angulation of the tympanic and petrous bones.   8. *Homo sapiens neanderthalensis:* Nose large relative to face; midface projection and large sinuses; long pubic bone; relatively short, stout limbs.   9. *Homo sapiens sapiens:* Long, curved parietal bones contribute to a spherical cranial vault and a short, high skull with its maximum dimension high on the head; a weak brow ridge but prominent, strongly developed chin; gracile (lightly built) long-boned postcranial skeleton. (Based on E. Delson, editor, 1985, *Ancestors: The Hard Evidence,* Alan R. Liss, Inc., New York.)

Figure 24–14.   Various hypotheses of the phylogenetic relationships within the Hominidae. They differ in two respects: (1) The relationship of modern humans (*Homo sapiens sapiens*), Neandertals (*H. sapiens neanderthalensis*), and archaic *H. sapiens*. The diagram on the left shows modern humans as the sister group of archaic *sapiens* plus Neandertals. In the hypothesis presented in the diagram on the right, archaic *sapiens* are the sister group of modern humans plus Neandertals. (2) An independent question arises in the relationships of early humans. The dashed lines marked (a) and (b) in the diagram on the left show two possible relationships for *A. africanus* and *A. robustus*. If hypothesis (a) is correct, the genus *Australopithecus* would be paraphyletic and the species *bosei* and *afarensis* should be referred to the genus *Paranthropus*. The genus *Australopithecus* is restricted to the form *africanus* and the remaining fossil species are referred to the genus *Paranthropus*. (Based in part on E. Delson, 1987, *Nature* 327:654–655.)

ash beds indicates that they do not differ substantially from modern human trails made on a similar substrate. Indeed, the upright stance associated with bipedal striding and the consequent physical requirement to support the head squarely on top of the vertebral column may have been prerequisites that permitted the later evolution of a large forebrain.

Lucy's species represents a line of near-*Homo* primates that stretches back into the Pliocene toward the separation of the pongid and hominid clades. Lucy and her kin were either contemporary with, or gave rise to, other hominids called *Australopithecus africanus*, another Pliocene hominid perhaps ancestral to *Homo*. It now seems possible, however, that Lucy's species may have given rise to both the later *Australopithecus* and directly or indirectly to *Homo* (Figure 24–14). Some of the fossil teeth and jaws of 1.8 million years ago are sufficiently like those of modern humans that an ancestral species—*Homo habilis*—may have existed as early as 2 million years ago and even been responsible for the nearly simultaneous demise of some species of *Australopithecus*. The structure of the hand, wrist, and long and powerful arms in both *Australopithecus* and *H. habilis* indicates the retention of some arboreal capabilities (Susman and Stern 1982, Lewin 1987b). The lower limbs are also much like Lucy's, suggesting the possibility that these small species of hominids (*Australopithecus afarensis* and *Homo habilis*) were differentiated by cranial capacity and not body size or locomotor morphology. Cranial capacities of *Australopithecus afarensis* are the smallest in the hominid fossil record and fall between 350 and 450 cubic centimeters in the three specimens measured; cranial capacities of *Homo habilis* are between 500 and 750 cubic centimeters (Campbell 1985, Falk 1985). Some paleoanthropologists believe that the brain of *A. afarensis* was still ape-like in appearance and had not undergone the cortex reorganization characteristic of *Homo*.

The later australopithecines were distributed in East and South Africa from about 2.5 to 1 million years ago. Neither they nor *Homo habilis* are known from elsewhere. The earliest stone tools date to perhaps 2.5 million years ago, but most paleoanthropologists believe they were made by *Homo*, not *Australopithecus* (Lewin 1988, but see Susman 1988). The later australopithecines are represented by two different types: In South Africa and in East Africa there are large, *robust* forms with sagittal crests on the skulls and enormous grinding postcanine teeth; and smaller, more delicate, *gracile* forms, very *Homo*-like in build with fully upright posture and lacking sagittal crests and massive eye ridges (Edey 1976, Johanson and Edey 1981, Lewin 1986a, Delson 1987).

Current interpretations recognize one gracile species and three robust species. *Australopithecus africanus* was the gracile form that was probably very close to the *Homo* lineage. *Australopithecus africanus* may have been a wide-ranging hunter and gatherer with an omnivorous diet. Three robust forms, *Australopithecus robustus* of South Africa and *A. aethiopicus* and *A. boisei* of East Africa, were terrestrial savanna-dwelling vegetarians, somewhat analogous to the forest-inhabiting gorillas. The gracile species was either replaced by *Homo* or according to some theories evolved into *Homo*. The actual taxonomic status of the highly variable remains that have been called *H. habilis* is much debated (Wood 1987). Whatever occurred, *A. africanus* disappeared from the fossil record around 2 million years ago. The two robust types persisted to about 1 million years ago and were contemporary for at least 1 million years with *Homo erectus*, an advanced form in the direct line of modern humans (Harris et al. 1988).

An intriguing question about human evolution is that of how much phyletic gradualism (change through time among continually interbreeding populations) rather than allopatric speciation (the spatial and temporal separation of populations, Chapter 1) accounts for the changes revealed by hominid fossils from the Pliocene to the Recent. According to one view, changes in the sequence from *Australopithecus africanus* to *Homo habilis* to *Homo erectus* to *Homo sapiens* resulted from phyletic gradualism. Others think that the suddenness of these changes in the fossil record (punctuations) indicate frequent geographic isolation and allopatric speciation in which adaptive advantage evolved in some isolates among these early hominid populations and led to consequent extinction of the less-fit groups when they came into geographic contact with the better-adapted forms (Gould 1979).

Homo erectus, originally described as *Pithecanthropus erectus*, was the first intercontinentally distributed human originating in East Africa. About 1 million years ago *H. erectus* appears to have spread to Asia and subsequently into Europe. The species disappeared between 200,000 and 300,000 years ago. This hominid had a considerably larger brain than any of the australopithecines, larger than that estimated for *H. habilis* and approaching the lower limit of brain capacity in modern humans. Chimpanzees have a cranial capacity ranging from 280 to 500 cubic centimeters; gorillas, from 350 to 750 cm³; australopithecines, from 350 to 700 cm³; *Homo erectus*, from 775 to 1100 cm³; and *Homo sapiens* from 950 to 2200 cm³. The facial features of *H. erectus* remained primitive—a massive prognathous (projecting) jaw, almost no chin, large teeth, sloping forehead, prominent bony eyebrow ridges, and broad, flat nose (Figure 24–15). This hominid made stone tools and may have known the use of fire. A cave near Zhoukoudian, China, about 50 kilometers southwest of Beijing was occupied by *H. erectus* almost continuously for more than 200,000 years beginning at least 460,000 years ago (Wu and Lin 1982). From these and other sites it can be said that *H. erectus* appears to have changed gradually (unless fragments of unrecognized species have been lumped with *H. erectus*; Lewin 1986b). Later individuals tend to have larger cranial capacities but the change is not great. The tools of *H. erectus*, known as the Acheulean tool industry, changed very little in style for 1.5 million years, although the materials chosen changed and later samples have a much higher proportion of small tools than do early examples. The lack of dramatic advance in tool manufacture over this immense period is surprising.

Homo erectus is probably the sister group of *Homo sapiens*, which came into existence around 200,000 to 300,000 years ago, perhaps as a result of increases in overall body size and relative cranial capacity in some geographically isolated population of *Homo erectus*. Unfortunately for ease of understanding, there are no clear, derived characters yet defined that separate the latest fossils assigned to *Homo erectus* from fossils believed to be early *Homo sapiens*. The latter have a further modest brain increase, a greater thickening and robustness of the skull and teeth, and a reduction in the degree of prognathism (projection of the jaw beyond the plane of the upper face). The characters used to separate the two taxa are all measurements—

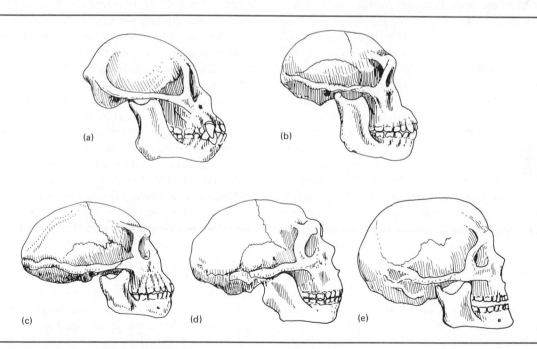

Figure 24–15. Reconstructed skulls of hominids and ancestral hominoid compared: (a) *Dryopithecus*; (b) *Australopithecus africanus*; (c) *Homo erectus*; (d) *Homo sapiens neanderthalensis*; (e) *Homo sapiens sapiens*.

not presence or absence of a definitive structure. These metric characters show overlap between the two groups; the clearest distinction is the reduction in prognathism.

Whatever is eventually understood about the transition to modern humans, it will undoubtedly be accompanied by shifts in the assignment of names to these fossils. This has already occurred with the most famous of fossil humans, the Neandertals. The first fossils of *Homo sapiens* were from the Neander Valley in western Germany, and used to be called Neandertal Man (for a time considered a separate species, *Homo neanderthalensis*). The Neandertal features first appear in fossils about 100,000 years ago. Neandertals were the same height as *Homo erectus*, 1.5 to 1.7 meters and still had receding foreheads, prominent brow ridges, and weak chins, but their brains were as large or larger than those of modern-day *Homo sapiens*. They were advanced stone tool makers, producing tools known as the Mousterian tool industry, with a well-organized society and rapidly advancing culture. The Neandertal people present a curious puzzle, because since their initial discovery in 1856 more modern-appearing forms of *Homo sapiens* have been unearthed dating to about 200,000 years before the appearance of the Neandertals.

It is now widely (but by no means universally) accepted that *H. sapiens* arose from *H. erectus* possibly in Africa about 300,000 years ago (Stringer and Andrews 1988). The early forms are known as **archaic** *H. sapiens* and share many primitive features with *H. erectus*. About 200,000 years later a distinct population evolved numerous derived characters that may be interpreted in part as progressively specializing them for resisting a cold climate. Some similar traits are seen in certain modern *H. sapiens sapiens* whose populations are cold adapted. But perhaps due to severe natural selection and restricted gene flow, these humans, known as Neandertals, became much more distinctive and advanced in their peculiar characters than any living population. They are thus considered an isolated subspecies of humans *H. sapiens neanderthalensis* by most anthropologists.

On the average Neandertals appear to have been much stronger than modern humans. Another striking difference between Neandertals and modern humans is the difference in the size of facial features: Neandertals had thick brow ridges, large nose, and broad midface regions (Lewin 1986c). The postcranial skeleton shows relatively short limbs with attachments for massive muscles. Neandertals had well-developed incisor and canine teeth, and the facial features of Neandertals may have been associated with structural modifications that supported these large teeth. The front teeth of Neandertals typically show very heavy wear, sometimes down to the roots. Were Neandertals processing tough food between their front teeth, or perhaps chewing hides to soften them as do some modern aboriginal peoples? Fossilized feces attributable to both archaic *H. sapiens* and *H. sapiens neanderthalensis* ranging from 300,000 to 50,000 years ago appear to indicate an exclusively meat diet (Kurtén 1986). Of course, the sample size is small and we have no way of knowing if such habits were merely seasonal. Neandertals hunted wild horse, mammoth, bison, giant deer, and woolly rhinoceros from close range: Their Mousterian-style hunting tools were of the punching, stabbing, and hacking type, requiring close encounters. Although almost all skeletal remains of Neandertals show evidence of serious injury during life, one in five persons was over 50 years old at the time of death. It was not until after the Middle Ages that historic human populations again achieved this longevity. The Neandertals were the first humans known to bury their dead, often after considerable ritual. Of special importance are burials at Shanidar cave in Iraq that include a variety of plants recognized in modern times for their medicinal properties. The emerging picture of *H. sapiens neanderthalensis* is quite different from that proposed for much of the time since their discovery in 1856; they seem now to have been adept hunters who probably lived in a complex society. The Neandertals vanished between 40,000 and 34,000 years ago, long after modern *Homo sapiens sapiens* appeared. Whether the observed rapid shift from

Neandertal to fully modern humans was a result of competitive exclusion, inbreeding of subspecies leading to the disappearance of one phenotype, or behavioral shifts leading to transformation of phenotypic characters to a modern morph is debated (Delson 1985, Gould 1988). Anatomically modern humans, *Homo sapiens sapiens,* known in southern Europe as Cro-Magnon people, are distinguished by suites of subtle skeletal features equivalent to those distinguishing modern races. It is with the Cro-Magnons that an unbroken trail of evidence and history leads to the present.

Origins of Human Technology and Culture

Humans are tool users and tool makers *par excellence*. Some other animals also use tools to a limited extent, usually in rather stereotyped and instinctive ways. Egyptian vultures open ostrich eggs by picking up stones in their beaks and dropping them on the shells; a Galapagos finch holds twigs and cactus spines in its beak to probe for insects in holes or under bark; and both baboons and chimpanzees use sticks and stones as weapons very much in the way that primitive humans must have done. But only humans, especially *Homo sapiens*, through the use of tools, have achieved an entirely new order of relationships with the world.

One of the remarkable conclusions from the recent paleoanthropological findings is that the use of tools, chipped stones, and possibly modified bone and antler, antedates the origin of the big-brained *Homo sapiens* by at least a million and a half years. There is evidence of the occurrence of modified stone tools 2.5 million years old. Later in the fossil record such tools are found in association with remains of *Homo habilis*. The use of tools by primitive hominids may have been a major factor in the evolution of the modern *Homo* type of cerebral cortex. Once the use and making of tools began to increase fitness, it would seem likely that there would be a high selection pressure for neural mechanisms promoting improved crafting and use of tools. In fact, the elaborate brain of *Homo sapiens* may be the consequence of culture as much as its cause (but see Susman 1988).

By at least 750,000 years ago *Homo erectus* had perfected stone tools and had also learned to control and use fire, but perhaps not to make fire. Actual hearths in caves are not widely recognized before 500,000 years ago. With fire humans could cook their food, increasing its digestibility and nutritional value and preserving meat for longer periods than it would remain usable in a fresh state, they could keep themselves warm in cold weather, they could ward off predators, and they could light up the dark to see, work, and talk.

Neandertal people practiced ritual burial in Europe and the Near East at least 60,000 years ago, suggesting that religious beliefs had developed by that time. Not long afterward the cave bear became the focus of a cult in Europe. By 40,000 years ago, Cro-Magnon people began constructing their own dwellings and living in communities—things Neandertal peoples also probably did. The domestication of animals and plants, development of agriculture, and the dawn of civilization were soon to follow (Leakey 1981).

The Human Race and the Future of Vertebrates

The preceding chapters have surveyed the great diversity of body forms and ways of living that have been evolved by vertebrates. Some 50,000 species of living vertebrates have been described by scientists, and new species are still being discovered, especially among oceanic and tropical fishes. Even in such a well-studied group as the birds an occasional new species is still found.

Living species, of course, represent only a small fraction of the diversity and number of vertebrate species that have existed on earth. Since the evolution of the first ostracoderms, hundreds of thousands of vertebrate species have evolved and then become extinct. Just as death is the natural end of the life of an individual, so extinction of a species is a natural result of the evolutionary process. What is new is the extent to which one species, *Homo sapiens*, has been responsible for recent extinctions and for endangering many forms of life. Past extinctions probably resulted mainly from climatic and geological changes that altered living conditions to the disadvantage of some species while favoring others. The emergence of human beings as a single dominant species on earth adds a new dimension to the problems of survival for other species (Kaufman and Mallory 1986). The impact of humans on other vertebrates may have begun as early as the Pleistocene and has increased in historical and modern times as the worldwide population of humans has mushroomed (Figure 24–16). Several million years of hominoid evolution in Africa led to a population of *Homo* estimated at 125,000 during the middle of the Pleistocene. As the human population spread into Europe and Asia, it slowly increased to 1 million individuals during the next 700,000 years. By the end of the last glacial advance, about 10,000 years ago, hunting and food-gathering techniques had allowed humans to spread over most of the earth and to reach a level of more than 5 million persons. By 2000 years ago, developing civilizations in the Old World had further increased the world population to more than 100 million, but in the New World human populations were still counted in the few millions. Not until after 1800 did the human population reach 1 billion individuals; by 1950 it had more than doubled again to nearly 2.5 billion; by 1987 it was 5 billion. By 2000 it could be more than 6 billion, unless catastrophe or self-control intervenes.

Humans and the Pleistocene Extinctions of Mammals

The number of genera of Cenozoic mammals reached a peak in the Pleistocene (Figure 21–26). Only 60 percent of the known fossilized Pleistocene genera are living now. The dramatic and re-

YEARS AGO	CULTURAL STAGE	ASSUMED DENSITY PER SQUARE KILOMETER	TOTAL POPULATION (MILLIONS)
1,000,000	Lower Paleolithic	.00425	
35,000–40,000	Lower Mesolithic	.02	
10,000	Mesolithic	.04	
2,000–9,000	Village farming and Urban, Religious, and Government centers	1.0	133
160	Farming and Industrial	6.2	906
A.D. 2000	Technological	46.0	6,270

Figure 24–16. Human population growth and dispersion from lower Paleolithic time through different cultural stages projected to 2000. Each grid unit represents an assumed population density of one human per square kilometer. (Based on E. S. Deevey, 1960, *Scientific American* 203(3):194–204.)

cent decline in certain mammals is emphasized by calculating extinction rates of different mammalian orders (Figure 24–17). Extinction seems to have occurred mostly in terrestrial mammals. For example, the fissipeds, which are the closest terrestrial relatives of the marine pinnipeds, show a large increase in the rate of extinction, whereas the pinnipeds do not. For all mammals extinction increased from 25 genera per million years in the Miocene to 40 in the Pliocene and to over 200 in the Pleistocene.

What caused the extinctions? Some feel that the ice ages and climatic change were the root cause, others that the extinctions were largely caused by another phenomenon of the Pleistocene—the evolution of hunting humans with their newly manufactured tools and social skills. The latter view has been pursued intensively by P. S. Martin at the University of Arizona (1973, Martin

and Wright 1967, Mosimann and Martin 1975, Diamond 1983, Martin and Klein 1984). What types of information lead to these conclusions? The most important points in favor of human causation are that the highest rates of extinction (1) were largely a postglacial event, and (2) occurred more frequently in large mammals than in small mammals. The climatic changes in postglacial time have left a more subtle record than the human spear points found in association with mammal remains and are consequently more difficult to evaluate.

The Human Factor   Clearly, the high rates of mammalian extinction occurred only during the Wisconsin and, especially, in postglacial time (Figure 24–18). Examples of extinction are mastodons, mammoths, and ground sloths. Various camels and antelopes also became extinct, but their remains are rarely found in association with human artifacts. It is Martin's contention that large mammals were preyed upon by invading nomadic humans and archaeological data support this view. Presumably, large mammals would be more susceptible to extinction by humans than small mammals. The changing numbers of small- versus large-bodied genera do show differences (Figure 24–19), but should the high extinction rate of large mammals be attributed entirely to humans?

D. H. Janzen (1983) proposed that humans had help from large contemporary carnivores. He agreed with Martin that humans migrated into an area that had abundant large herbivorous mammals and began extensive hunting. The presence of humans and their kills would have provided more carrion to the carnivores, whose population densities, Janzen pointed out, would have increased in response to the food supply. The combined predation pressure of increasing human and carnivore populations could have led to further decimation of the large herbivores. When the prey populations became critically low the nomadic hunters would have moved on and the populations of carnivores declined because their resource base was gone.

Norman Owen-Smith (1987) suggested that the elimination of large herbivorous mammals

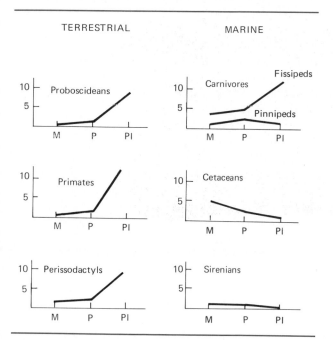

Figure 24–17.   Generic extinction rates for various terrestrial and marine mammalian groups. Numbers are the estimated number of genera that became extinct per million years. M = Miocene; P = Pliocene; Pl = Pleistocene.

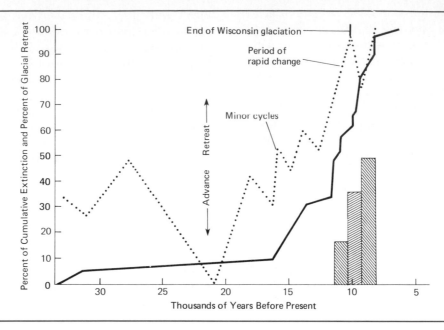

Figure 24–18. Coincidence between the extinction of North American mammals, glacial retreat, and the population of humans. Solid line represents the cumulative percentage extinction of 40 mammalian species; dashed line is the percentage of withdrawal of the late Wisconsin glaciation; histograms are arbitrary units of relative abundance of evidence for the occurrence of humans in North America. [From J. J. Hester, 1967, pages 169–192 in P. S. Martin and H. E. Wright, Jr. (editors), *Pleistocene Extinctions: The Search for a Cause,* Yale University Press, New Haven, Conn.]

(megaherbivores) by humans could have had a cascading effect on smaller species because grazing by the megaherbivores was a major factor in creating and maintaining spatial diversity of habitats. Once the megaherbivores were gone, habitats would have become more uniform and this change would have been detrimental for the small herbivores that were dependent on nutrient-rich and spatially diverse habitats.

A relation can be shown between the arrival of prehistoric humans in different regions and the extinction of large herbivorous mammals. It would appear that gradually a hunting technology developed in Africa during the middle Wisconsin glaciation and then spread northward and burst into the Western Hemisphere as the late Wisconsin gla-

cier retreated (Table 20–1). In the New World humans were present at least 25,000 years ago (Patrusky 1980, Adovasio and Carlisle 1986) and possibly more than 40,000 years ago. But evidence of extensive hunting of mammals began in the colder north, perhaps with a new wave of humans immigrating from Asia some 13,000 years ago (Mosimann and Martin 1975). No extinction of similar proportion has occurred in Africa, where apparently human hunting evolved side by side with large mammals. A late extinction of the marsupial megafauna of Australia also occurred (Gillespie et al. 1978, Martin and Klein 1984). As in Africa, however, these large Australian mammals coexisted with aboriginal humans for well over 7000 years. Researchers conclude therefore that

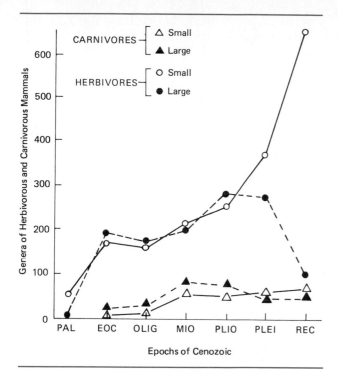

Figure 24–19. Changes in the number of genera of carnivorous and herbivorous mammals known from the epochs of the Cenozoic. Mammals of 20 kilograms adult body weight or less are considered small.

the model of overkill of large mammals by humans does not explain megafaunal extinction in Australia.

The Climatic Factor   Could human hunters have been the primary cause of extinction of large mammals in the Pleistocene? The evidence shows that prehistoric humans did hunt mammals, and probably often with devastating results. Fire was used to stampede large herds of ungulates over cliffs to their deaths—a wasteful technique, for only a small portion of the carcasses could be utilized. Nevertheless, other scientists do not attribute the demise of the large mammals solely to human influence, but suggest that changing climates and their influence on habitat availability were as important as humans in producing extinctions. In

their view, large mammals are more susceptible to environmental stress. Severe restrictions of forage, available water, and elimination of migratory corridors to more suitable habitat would affect large mammals with their correspondingly large requirements more than smaller mammals. Thus, large North American mammals may well have been on the decline when intensive human hunting began.

R. D. Guthrie (in Martin and Klein 1984) has pointed out that most climatic theories of Pleistocene extinction are too ambiguous and diffuse to account for the final extinction event, however effective they are at depicting environmental stress. In the general trend of replacement of closed-canopy forests by more open, grass-dominated steppes and savannas that began in the early Tertiary, Guthrie sees increasing seasonality as being of primary significance. The climatic trend throughout the Cenozoic was toward a shorter growing season. At first this trend increased the mosaic patchwork of forest, scrub, and grassland over much of the world. This increasingly complex pattern of vegetation supported a large, diverse fauna.

Indeed, the spread of grasses promoted faunal diversification. Responding to the spread of grasslands, the generally large-bodied ungulates reached maximum diversity in the late Miocene and Pliocene, contributing significantly to the total number of mammalian genera known from those epochs. But as the trend continued toward seasonal extremes of temperature and aridity the diversity of vegetation declined, especially locally. The flora over broad expanses of continents changed from a mosaic of plant associations to a latitudinal and altitudinal zonation of stripes of plant communities with low diversity. The lower overall diversity of floras and greater regional homogeneity reached its maximum after the Wisconsin glacial advance. The reduction of plant diversity may have eliminated many herbivores, especially larger forms, and restricted all but a few species to limited ranges.

The postglacial period beginning 15,000 to 20,000 years ago appears to have been more stressful to large mammal faunas than the previous 2.5 million years of the Pleistocene. Shorter growing seasons meant less plant productivity to support herbivores, even in the restricted ranges where they found suitable floras. Guthrie argues that the rapid advent of short growing seasons and heightened seasonality critically decreased the resources available to many species of large herbivorous mammals with long periods of growth to maturity and long gestation periods. These changes constituted environmental challenges to which they could not adapt physiologically or genetically. The fossil record attests to the resulting decrease in local faunal diversity, distributional ranges, and frequently to the resulting extinctions without the influence of human hunting.

Many large mammals persisted in some regions despite prehistoric human hunting pressure. In the New World camels and horses became extinct, but they survived in Europe. Others, like bison and elk, survived in the New World in spite of humans. There is currently no final answer; a combination of unique new climatic stress *and* human predation seems most plausible.

Human Settlement of Oceanic Islands

Recent analyses of fossil deposits from islands in the Pacific, Indian, Caribbean, and Mediterranean Oceans indicate that the arrival of humans on these islands was accompanied by large-scale extinctions of native animals (James et al. 1987). Excavation of fossils preserved in a tube-like cave in a lava flow of Haleakala Volcano on the Hawaiian island of Maui has provided a record of faunal changes over the past 7750 years. This record shows that before humans reached the islands the bird fauna included three times as many native species as are present now. At least two-thirds of the native species of birds became extinct following settlement of the island by Polynesians between 1000 and 400 years ago. Similar results from other islands in the Pacific suggest that the total number of species of birds living in the recent past

may have been much larger than the approximately 9000 species that now exist.

The Age of Exploration and Discovery

Since 1600, more than 200 extinctions have been documented among vertebrates, mostly birds and mammals; all can be attributed directly or indirectly to the activities of humans. The history of extinctions among the endemic birds of the Indian Ocean Mascarene Islands off the African coast is a particularly dismal but revealing example. The island of Mauritius was uninhabited by humans until discovered in 1505 by the Portuguese. The endemic dodo, a large flightless bird related to pigeons, had become extinct by 1681, the victim of slaughter by sailors for food and of predation on its eggs and young by introduced pigs and monkeys. A similar bird, the solitaire, had vanished on Reunion by 1750, and another, on Rodriguez, before 1800. In all, as many as 24 endemic birds have become extinct in the Mascarene Islands since 1600, more than 50 percent of the original avifauna. Today, many of the remaining endemic species of parrots, pigeons, and falcons number only a handful of individuals, so severe has been the human destruction of the Mascarene habitat.

The pattern of destruction of the vertebrate faunas of the Mascarene islands is characteristic of the effects of European exploration and discovery in the fifteenth through nineteenth centuries. Island endemics were brought to extinction either directly, through slaughter for food, or indirectly as the result of introductions of feral populations of domestic animals that altered island ecosystems. Subsequently, not only island but continental faunas and floras began to be affected.

Commercial Exploitation of Vertebrates

Explorers were soon followed by fur trappers, commercial hunters, fishermen, whalers, plume hunters, and a host of others who made their living by exploiting wild animal populations for food, leather, fur, oil, ivory, and other commercial products. Hunting was often done by companies with small armies of organized hunters. These enterprises have caused great destruction among vertebrate populations but have actually brought few species to extinction because it usually becomes uneconomical to hunt out the last few hard-to-reach individuals of a species.

When Vitus Bering voyaged among the islands of the Bering Sea and North Pacific Ocean he found millions of fur seals breeding on the Pribilof Islands, and great numbers of sea otters spread over an even wider range (Figure 24–20). The furs of these marine mammals are among the most luxuriant known and were highly valued in China and Europe. Soon after the reports of the Bering expedition circulated the sealers and otter hunters began their work.

Between 1745 and 1867 Russian explorers and later the Russian-American Company shipped 260,790 sea otter pelts from Alaska. In 1757 alone the Company killed more than 7700 sea otters. From 1823 to 1830 fewer than 100 otters per year were taken, and the Russian-American Company expanded its activity to other resources.

For a time when so few pelts were received in trade people assumed the species had become extinct, but fortunately a few pairs managed to survive in isolated parts of the range. After complete protection in North America in 1911 and careful transplantation of individuals since 1965, the otter population has slowly increased. The wild population may now total 150,000. Sea otters are important predators in subtidal kelp beds and have an influence on the distribution and abundance of other species of animals and even plants in the community. Indeed, around Amchitka Island the sea otter shows signs of overpopulation and in southern parts of the range sea otters are in conflict with fishermen over a favorite food—abalone.

Between 1908 and 1910 Japanese seal hunters slaughtered 4 million fur seals in the Pribilofs. Fortunately, through diplomatic negotiations undertaken by the United States, the Fur Seal Treaty of 1911 was signed by Japan, Russia, Canada, and the United States. The first international agreement to protect a marine resource, it provides protection, management, and harvest by the United States with quotas of the harvested pelts going to the signatories. As a consequence, the northern fur seal population rebounded in numbers and has

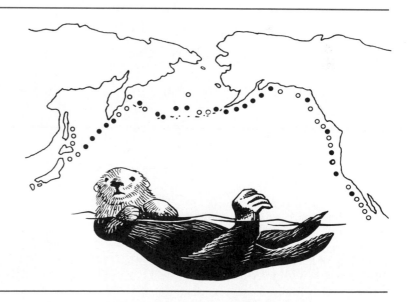

Figure 24–20. Original distribution of the sea otter (open circles) compared with its present relict occurrence (closed circles). (From V. Ziswiler, 1967, *Extinct and Vanishing Animals*, Springer-Verlag, New York.)

been harvested annually on a planned and scientifically managed basis for more than 50 years. Unfortunately, southern fur seal species were not so lucky, and they exist today only as remnant populations on a few islands scattered from southern California to Chile. The impact of fur seal populations on their ecosystem is also appreciable. Today's large populations are in serious political trouble with the Bering Sea fishermen because of their alleged efficiency in competing for the same seafood resources as humans.

A similar history can be related about the northern elephant seals, which come ashore to calve and breed on islands adjacent to southern California and western Mexico (Figure 24–21). They were slaughtered in the tens of thousands for oil and dog food, until there were only a few, perhaps as few as 20 individuals, left in the early 1930s. After protection by the governments of the United States and Mexico, the populations have increased, slowly at first and then more rapidly, until now there are more than 100,000 of these great, lumbering beasts. They are still increasing, having recently begun breeding on the California mainland south of San Francisco. However, biochemical studies reveal virtually no genetic variation in the population, apparently as a result of its having been reduced to so few individuals. The evolutionary consequences of this lack of genetic variation remain to be seen.

These three marine mammals are good examples of how well vertebrate populations can recover from greatly reduced numbers after heavy killing stops. A different story has to be told for the great whales. All of the large baleen whales, except the gray whale, which has responded to protection, are critically endangered as a consequence of the relentless use by the whaling nations of highly efficient modern techniques for capture and processing. The blue whale, perhaps the largest animal ever to have existed, escaped persecution during the early days of whaling. When whalers traveled in sailing ships and harpooned from open boats, blue whales simply were too fast and too heavy to be handled. The modern catcher ship, armed with cannon-launched harpoons that explode inside the animal, proved to be the nemesis of the blue whale. International treaties have so far

Figure 24–21. A breeding colony of elephant seals on Isla Año Nuevo off the coast of California. A male (foreground) displays in front of a group of females and pups. (Photograph by F. Harvey Pough.)

not been very effective in controlling whaling; some of the signatories of the treaties have recently announced that despite an international ban on whaling to which they agreed, they will continue to hunt whales for what they call research purposes. It is unclear whether the blue whale, fin whale, bowhead whale, and the right whales can ever recover from their pathetically reduced numbers.

Many avian species have been overharvested for commercial purposes. The passenger pigeon was probably brought to extinction as a result of commercial hunting, although the destruction of the nut-producing hardwood forests in the eastern United States was indirectly contributory to its demise. Roosting and nesting in dense flocks of many millions of birds made the passenger pigeon especially vulnerable to shot and nets. In 1813, John James Audubon made a 55-mile trip along the Ohio from his home to Louisville, Kentucky. Wild passenger pigeons were migrating through the region in such numbers that "the light of noonday was obscured as by an eclipse." Audubon estimated the flock to be 1 mile wide with an average density of two birds per square yard. During a 3-hour period he calculated that more than 1,100,000,000 passenger pigeons must have flown by, and the flight lasted for 3 days.

Even if we reduce Audubon's estimate by half, the number of pigeons must have been truly astounding. Passenger pigeons were social nesters. The last great nesting occurred in 1878 near Petoskey, Michigan, and covered more than 40,000 hectares of forest; many millions of birds were present. Unfortunately, passenger pigeons were so tasty that the demand for them in urban markets seemed insatiable. The adult birds were shot, netted, and removed from nests by the millions. The New York City market alone would receive 100 barrels a day, week after week, without a drop in price. Chicago, St. Louis, Boston, and all the eastern cities, great and small, joined in the demand for pigeon flesh. Some people still speculate about the disappearance of the passenger pigeon—that they all drowned in a great storm over Lake Michigan—but the true story is revealed by the statistics of the marketplace. The last bird died in the Cincinnati Zoo in 1914.

There is nothing biologically wrong about harvesting wild animal populations except when we exercise no wisdom or restraint. It is our responsibility to prevent overexploitation and depletion of our prey species. Animal populations produce an excess of individuals above the number that can be sustained by the ecosystem. This surplus of animals above the number required to replace losses in the breeding population provides a legitimate harvest for humans, and our responsibility is to determine what share of that surplus we can take for ourselves without diminishing animal populations over the long term, including those of other animals that depend on the species we are harvesting.

Extinction of a species or population results when the mortality of individuals is continually greater than the production of new individuals. Generally, when excessive mortality caused by humans is eliminated, even drastically reduced populations rebound to the limits imposed by factors other than human predation. However, this resilient capacity of vertebrate populations has limits that vary for different species and seem to be determined by a *critical minimum population size*, below which the species cannot recover because reproduction has been affected independently of mortality. In such cases, even if mortality can be greatly reduced, reproduction is still insufficient to result in a net increase in numbers and the species becomes extinct (Frankel and Soulé 1981).

Highly social and gregarious species seem to have large critical minimum population sizes, whereas widely dispersed, territorial species and island endemics may have very small critical minimum numbers. The passenger pigeon is an example of the first group. After the last great nesting in 1878 in Michigan, many millions of pigeons still existed despite the slaughter that occurred there, but even though significant human depredations on the passenger pigeons stopped after that time, the remaining flocks were never again able to reproduce in sufficient numbers to replace the natural losses in the adult population, and the species had become extinct in the wild by the turn of the century. The small amount of recorded information on the bird's breeding habits indicates that mating and nesting may have been initiated by social stimulation involving communal displays of large numbers of birds closely packed together. Other species seem to have such a low critical population size that the concept has no application to them. Certainly, this would seem to be the case of the Laysan duck, which appears to have been reduced in 1930 to a single adult female and her clutch of fertile eggs. By the late 1950s there were again an estimated 580 to 740 ducks on Laysan Island, although the wild population has since dropped back to a lesser number of birds (Zimmerman 1975).

Several factors probably play a role in determining the critical minimum population size for a species. The population density may become too low for males to have much chance to find females, a possible problem now for some of the great whales, especially because their sound communication channels are becoming more and more jammed by shipping noise. Population density could become too low for adequate stimulation of courtship and mating, as indicated for the passenger pigeon. The population might become too small to counterbalance mortality from large numbers of surviving predators, parasites, or competitors. (For example, the rare Kirtland's warbler has had a problem with nest parasitism by the brownheaded cowbird in recent years.)

The total population size might become too small to survive a natural disaster such as a flood, fire, or cyclone, particularly if the population is localized. Remnant populations of island endemics are especially vulnerable to this factor. The last few surviving Laysan honeycreepers were blown off the island by a storm in 1923, and the same fate could be in store for the relict populations of the kestrel, pink pigeon, and parakeet on the island of Mauritius, which has a history of periodic cyclones.

Reduction of fecundity or increased susceptibility to diseases and parasites owing to genetic deterioration from inbreeding are other factors to consider when a species population becomes reduced to a very few individuals; the Laysan duck may be suffering from such problems at the present time. Reduced fecundity is sometimes characteristic of highly inbred captive populations of vertebrates. [See Schonewald-Cox et al. (1983).] Exactly which problems may be affecting an endangered species can be determined only by intensive study of the species and its environment, a costly and time-consuming process—and one in which we are already far behind.

The Impact of Agriculture and Land Development

Human use and abuse of land for agriculture, timbering, mining, and suburban living are the most detrimental of our activities to wild animal populations, primarily because of the inexorable destruction of natural habitat. This would be evolutionarily comparable to the spread of grasslands in the Miocene or any of the great climatic changes that have precipitated biotic revolutions throughout the history of vertebrate life except for two exceedingly critical factors: No natural habitat is unaffected or favored by twentieth-century practices and the rate at which habitats of every kind are being profoundly altered and destroyed is unprecedented. Some species have been directly destroyed in a process called "reclaiming the land" for human uses. Although as yet only a few species of vertebrates have been brought to total extinction as a result of human land uses, many have suffered major and perhaps irreversible reductions in distribution and density, and no doubt many others will join the ranks of threatened or endangered wildlife in the next few decades as a result of the diminution and deterioration of natural habitats. The actual magnitude of species extinctions due to twentieth-century habitat alteration will not become clear until well into the twenty-first century.

Perhaps the earliest and most persistent conflict between humans and wild animals has been with other predators, particularly the large carnivores. Early humans were efficient and destructive predators in their own right, and with some justification they regarded the carnivores as competitors for the great ungulates that humans hunted for food. Humans surely had to protect their kills from scavenging predators, and they no doubt robbed other predators of kills. Direct interactions between individual humans and carnivores must have been common in the Pleistocene. From these remote times human attitudes toward other large and powerful predators have been molded by fear and by the idea that other predators compete directly for available food. Except for isolated instances, neither belief is justified by objective analysis, but they still influence many contemporary minds.

With domestic livestock and pastoral nomadism as a way of life, humans began between 40,000 and 10,000 years ago a systematic destruction of large carnivores that has continued and accelerated to the present. In North America, the timber wolf, mountain lion, black bear, and grizzly bear have been steadily retreating to more remote areas as cattlemen and sheep herders arrived in the Midwest and West. The same can be said of the lion in Africa and Asia, of the tiger and wildcat in Asia, and of the jaguar in the Americas. Tillage and development of intensive farming beginning 10,000 years ago provided the final insults for large predators, which simply are not compatible with intensive, disruptive human uses of the land.

Unconscionably, the carnivores continue to be pursued by humans, and governments are still persuaded to spend millions of dollars a year on misguided and unjustified predator control programs. Predator control usually results in a need for rodent control, but rodents are much more difficult and costly to control than predators. It is ecologically and economically wise to accept the loss of a few pheasants and lambs killed by foxes and coyotes in exchange for the greater benefits of their impact on rodent and rabbit populations.

Other vertebrates have frequently been considered to be competitors of humans. The Boer farmer who settled in South Africa regarded the existing wild ungulate herds as competitors for their cattle, sheep, and goats, and systematically destroyed ungulates, as they did the Cape lion. The blue buck had been exterminated by 1800; the last of the quagga, a once common zebra, disappeared in 1878; Burchell's zebra had been eliminated from all areas other than game parks by 1920; and three other ungulates, the bontebok, the white-tailed gnu, and the Cape mountain zebra, exist today only as remnant populations in wildlife reservations.

The loss of natural environments from their conversion to intensive agriculture and other disruptive uses of land and water is by far the most important factor in the overall reduction of wild vertebrate populations. This will continue to be the overriding problem for the survival of many species in the closing decade of the twentieth century and the beginning of the twenty-first. The cutting of the original eastern hardwood forests in North America for timber, firewood, and agricultural clearing had an effect on the passenger pigeon at least equal to the killing of these birds for food. The pigeons fed on the seeds of oaks, hickory, and other nut-producing trees and also depended on the trees for nesting sites. As the eastern forests were demolished during the eighteenth and nineteenth centuries, the pigeons were forced westward and northward into marginal habitats, which probably were not favorable for the mass reproductive efforts that characterized the species. Today, much of the eastern United States is again covered by hardwood forest, but it is nearly all second-growth stands, different in species composition and habit from the late pre-Columbian forest.

The prairie grasslands have been the most disrupted natural communities in North America, especially the tall grass prairie, which was ploughed under and no longer exists. Even the short grass prairie is no longer truly natural. It has been invaded by many exotic species of grasses and weeds introduced through human agency and has been drastically altered by overgrazing of domestic livestock. Many of these western ranges are receiving additional insults from extensive mining processes, especially strip mining for coal.

Continued growth of the human population

means that more and more land must be devoted to agriculture and other disruptive uses required to sustain unprecedented numbers of humans. Thus, more natural habitat is lost, and fewer places are left for wild animals, except for the very limited number of species that can adjust to the altered conditions created by humans. Most of the agriculturally productive areas of the world have been developed, but good agricultural lands are being reduced by suburban developments, businesses, roads, airports, and artificial lakes.

Farmers worldwide must increasingly turn their attention to marginal lands. Deserts, for example, can be highly productive under irrigation. Much of the water in such areas is fossil water trapped in isolated aquifers from prehistoric times when local climates were wetter. This water is no longer being replenished by regional rainfall. Thus where fossil water has been used it is becoming an increasingly scarce commodity. In some places irrigation followed by rapid evaporation into the dry atmosphere has increased the salinity of the soil, greatly reducing the harvest. The most critical agricultural devastation in the late 1900s is taking place in the tropical rain forests of South America, Africa, and Asia. When the forest is cut, the soils do not sustain intensive agriculture for more than a few years. Once abandoned these soils are frequently so changed in texture and composition that the original forest trees cannot come back; instead, there is a new climax of grasses and shrubs, which are marginal pastures for livestock. Gone

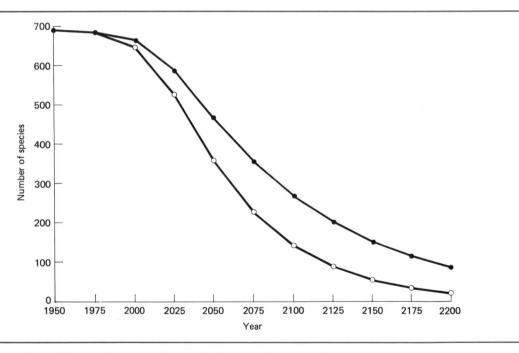

Figure 24—22. Predicted loss of mammalian species from tropical moist forests in the New World due to the reduction in extent of their habitat. The projected rate of deforestation is based on a continuation of the current rate. The upper curve represents the immediate loss of species expected. The lower curve is the final number of species expected at equilibrium if deforestation were to cease at that date. No date for reaching this equilibrium is indicated by these data. (Based on data from the Society for Conservation Biology.)

forever is the rich diversity of tree species and associated flora and the animals of these tropical forests (Figure 24–22). Genetic resources of potential value to agriculture and medicine are being lost along with complex ecosystems as tropical forest disappear (Myers 1984).

In addition to a loss of species diversity, failure to keep large portions of some natural ecosystems intact may have serious consequences on climate, composition of the atmosphere, and on the distribution and quality of water, perhaps the most limiting resource for life. This has been dramatically demonstrated in recent years in the sub-Saharan regions of Africa (Soulé 1986). There is growing evidence, too, that cutting the Amazon tropical forest is having an impact on rainfall patterns in that region (Myers 1984). Deforestation can have consequences well beyond the actual area of forest that is cut (Figure 24–23). Clearcutting can create desert-like regions, and the run-off of heavy tropical rains from these barren areas can cause flash floods that destroy habitats many miles downstream. On the positive side, restoration projects can reestablish some tropical habitats when they are undertaken with the necessary biological and political skill (Janzen 1988).

Global consequences of human activity are also becoming apparent. The exchange of materials between terrestrial ecosystems and the atmosphere is increasingly dominated by human activities, and rising concentrations of many compounds have resulted from human biological or industrial processes (Mooney et al. 1987). Acid precipitation, which results when oxides of sulfur and nitrogen produced by combustion of fossil fuels combine with water vapor in the atmosphere, has acidified soils and water systems in North America and Europe, and acid precipitation has recently been discovered in China (Galloway et al. 1987, Schindler 1988).

Chlorofluorocarbons of the sort used to produce plastic foams, for dry cleaning, and in refrigeration systems drift into the stratosphere where ultraviolet radiation breaks down the chlorofluorocarbon molecules, releasing chlorine atoms as one of the products of the breakdown. In turn, the chlorine atoms react with ozone, which is the principal component of the atmosphere that blocks potentially damaging ultraviolet radiation from reaching the Earth's surface. A single chlorine atom can catalyze the destruction of 100,000 molecules of ozone. Over the last decade the concentration of ozone in the stratosphere has dropped by 3 percent, and the decline is continuing. A major reduction in the concentration of ozone in the stratosphere could affect plant and animal life worldwide, and the growing hole in the ozone layer that appears annually over Antarctica may be associated with the release of these fluorocarbons into the atmosphere (Kerr 1987, Bowman 1988).

Many other human practices threaten wildlife: the pet trade and the traffic of wild animals, the introduction of exotic plants and animals, the modification of waterways and water distribution by canals and dams, and chemical pollution (Kaufman and Mallory 1986). Humans have had, and continue to have, disastrous effects on other vertebrate species. The *Global 2000 Report* predicted that more species will become extinct between 1980 and the year 2000 than in previous recorded history if current trends continue.

Human Concern for Vanishing Wildlife

Human impact on other forms of life has led to a new classification of species. The categories include "rare species," "unique species," "threatened species," and "endangered species." The International Union for the Conservation of Nature and Natural Resources, in cooperation with organizations such as the World Wildlife Fund and the International Council for Bird Preservation, attempts to keep current a catalog of specific details on all rare, severely threatened, or endangered species of plants and animals. The information is published in an ongoing series of *Red Data Books*, arranged taxonomically. Over 1000 species of vertebrates are now included. Increasingly in recent years governments have become concerned with such species too. Public Law 93-205, the United

Figure 24–23. Neotropical forests: (a and b) undisturbed forest have a high diversity of plant and animal species; (c and d) after destruction the forests may be unable to regenerate their original diversity. (Photographs by John B. Heiser.)

Table 24–1. Number of endangered species of vertebrates officially listed by the United States Department of the Interior in 1987 and (1982).*

Taxonomic Category	USA Only	U.S. and Foreign	Foreign Only	Total
Mammals	26(15)	20(18)	242(223)	288(256)
Birds	61(52)	16(14)	141(144)	218(210)
Reptiles	8(8)	6(6)	60(55)	74(69)
Amphibians	5(5)	0(0)	8(8)	13(13)
Fishes	39(28)	4(4)	11(11)	54(43)
Totals	139(108)	46(42)	462(441)	647(591)

*Data from *Endangered Species Technical Bulletin*, VII:11, November 1982, and XII:4, April 1987. U.S. Fish and Wildlife Service, Washington, D.C.

States Endangered Species Act of 1973, gives authority to the Secretary of the Interior to designate endangered and threatened species (Table 24-1), to promote ways to preserve them from further diminution and, if possible, to increase their numbers. There are also important international agreements, such as the Convention on International Trade in Endangered Species of Wild Fauna and Flora. Many states and provinces also now have their own endangered species laws. Although the executive branch of the U.S. government has been slow to put the provisions of the Endangered Species Act into operation, public interest in the survival of endangered species has increased greatly.

Characteristics of Endangered Species

Are there predisposing characteristics of a species that are likely to lead to its becoming an endangered species? Species of birds and mammals that are restricted to islands are highly vulnerable, but there are many other variables that influence the survival potential of a species. By comparing the characteristics of threatened and endangered species with related and apparently thriving species, we can begin to identify the main factors involved (Table 24–2). Theoretically, one could check each

species against this list of variables and come up with a statistical estimate of its chances for continued survival in the modern world. At present our knowledge of the vast majority of living species is so minimal that the best we can do is to make a qualitative appraisal.

David Ehrenfeld (1970) has constructed an interesting description of the hypothetical *most endangered animal*: "It turns out to be a large predator with a narrow habitat tolerance, long gestation period, and few young per litter. It is hunted for a natural product and/or for sport, but is not subject to efficient game management. It has a restricted distribution, but travels across international boundaries. It is intolerant of humans, reproduces in aggregates, and has nonadaptive behavioral idiosyncrasies." Although there really is no such animal, the polar bear comes close to approximating this model. Conversely, if one takes the opposite characteristics (those listed under "safe" in Table 24–2), a composite picture of a typical wild animal of the twenty-first century develops. Familiar existing approximations are the herring gull, house sparrow, gray squirrel, Virginia opossum, Norway rat, and carp.

In short, the result of human influence on environment and our wanton destruction of animal life is a greatly reduced diversity in faunas and floras. Increasingly, there is an emergence of a relatively few, superabundant, highly successful species that dominate ecosystems. Unless sufficiently large tracts of natural ecosystems are left intact to maintain some possibility of ecosystem balance, these organisms probably are going to interact with one another and with human crops in boom and bust oscillations in numbers.

Actions to Save Threatened Fauna and Flora

What can we do to prevent other species from becoming rare, endangered, or extinct? Two cardinal requirements must be met if natural biotas and ecosystems are to remain intact for future generations: (1) preservation of natural habitats and eco-

Table 24–2. Characteristics of species that decrease or increase survival potential.

Endangered	Safe
Individuals of large size (cougar)	Individuals of small size (coyote)
Predator (hawk)	Grazer, scavenger, insectivore (turkey vulture)
Narrow habitat tolerance (orangutan)	Wide habitat tolerance (macaques)
Valuable fur, hide, oil (chinchilla)	Not a source of natural products and not exploited for research or pet purposes (gray squirrel)
Hunted for the market or hunted for sport where there is no effective game management (passenger pigeon)	Commonly hunted for sport in game management areas (mourning dove)
Has a restricted distribution: island, desert watercourse, bog (Bahamas parrot)	Has broad distribution (herring gull)
Lives largely in international waters, or migrates across international boundaries (green sea turtle)	Has populations that remain largely within the territory(ies) of a specific country(ies) (snapping turtle)
Intolerant of the presence of humans (grizzly bear)	Tolerant of humans (black bear)
Species reproduction in one or two vast aggregates (West Indian flamingo)	Reproduction by solitary pairs or in many small- or medium-size aggregates (house sparrow)
Long gestation period; one or two young per litter, and/or extended maternal care (giant panda)	Short gestation period; more than two young per litter, and/or young become independent early and mature quickly (raccoon)
Has behavioral idiosyncrasies that are nonadaptive today (redheaded woodpecker—flies in front of cars)	Has behavior patterns that are particularly adaptive today (burrowing owl: highly tolerant of noise and low-flying aircraft; lives near the runways of airports)
Top-of-the-food-chain predator that is subject to the effects of biological magnification of chemical pollutants (peregrine)	Lower trophic level predator that feeds on less contaminated prey (kestrel)

Source: D. W. Ehrenfeld, 1970, *Biological Conservation*, Holt, Rinehart, and Winston, New York.

systems on a large and representative scale, and (2) control of human population growth.

The most immediately critical action, for the short term, is habitat preservation. Preservation must be of large blocks of natural ecosystems—10,000 to 20,000 square kilometers or more. These blocks must be large enough for the natural communities of plants and animals to remain self-per-petuating. Unfortunately, there is nothing approaching scientific understanding of "how large is large enough?," the immediate socioeconomic question that arises (Soulé 1986). The larger national parks and game preserves of the world show what can be done when natural areas are protected against major disruptive land use by humans.

The Kruger National Park was the first game

park established in Africa; portions of it have been in preserve status since 1895; its present boundaries, encompassing more than 12,000 square kilometers of lowland bush-veld along the border with Mozambique, have been intact since 1926. Although situated squarely in one of the most critical sociopolitical trouble centers of the world and surrounded by lands that are heavily used by humans for agriculture and stock raising, the park remains relatively inviolate and is fenced around its entire perimeter. Kruger has lost only one species of plant or animal from its natural ecosystem, and it has been in existence long enough to give assurance that as long as it is properly cared for and preserved it can continue to hold its living treasures indefinitely. One can see and study in Kruger more than 44 species of fish, 29 species of amphibians, nearly 100 species of squamates, more than 440 species of birds, and 114 species of mammals, including 16 species of large ungulates, lions, leopards, cheetahs, hyenas, and wild dogs.

Any conservation issue ultimately rests on a solution to the problems that have been created by the human population explosion. The human population has been increasing at an exponential rate for thousands of years, a pattern of growth that is unprecedented in other animal populations and that *cannot* continue indefinitely. An increase in the numbers of human beings leads to exponential rates of increase in the use of resources—food, fiber, fuel—and the production of waste products that pollute the global ecosystem: wastes from human food consumption, from use of fossil fuels, from fertilizing agricultural lands, and from synthetic chemical products and other sources. It is important to understand that these rates of use and waste frequently increase by exponents that are larger than the exponent of population increase. This is especially true as global communication (television most especially) exposes the overwhelming majority of humans to a Western middle-class life-style, raising worldwide expectations and aspirations. Typically, in a modern technological society, use of materials increases four to five times faster than population growth, a rate far faster than in pretechnological societies.

Many people, especially in North America and Western Europe, still are not especially worried about human population growth or exponential rates of increase in the uses of land and resources. They believe that technology and production can keep pace with population growth. New resources of energy can be tapped, more green revolutions can be brought about, and more efficient occupancy of space for human dwellings can be designed. We are still wedded to the progressivist philosophy of the eighteenth and nineteenth centuries, a philosophy that followed in the wake of the scientific and industrial revolutions. It is the philosophy of the ever-expanding economy; it is the philosophy of the TV commercials; "Progress is our most important product," where progress equates with growth and expansion.

Many biologists and conservationists are not convinced that more technology is needed as much as a new frame of reference built on restraint—restraint on the use of resources and, above all, restraint on human procreation. A respect and reverence for all life, not just human life, must be realized before acceptance of such a world view can be expected to have an effect.

There is nothing bad about technology per se. The same technology that can be used to destroy the world can also very probably be used to recreate it except in one instance: When a species is lost to extinction, it is lost forever.

Conclusion

In July 1987 the world human population passed the 5,000,000,000 mark. One billion *more* people will be added by 1999. Nine-tenths of these humans live in what is known as the Third World. The situation is grave. Allison Jolly (1985) has calculated that the total world population of all wild primates combined is less than that of any of the Earth's major cities. The entire living populations of many species are no larger that that of a small town. Conditions are equally critical for many other vertebrates. Jolly (1980) has expressed the problems we face as biologists.

"This realization has been painful. It began for me in Madagascar, where the tragedy of forest felling, erosion, and desertification is a tragedy without villains. Malagasy peasant farmers are only trying to change wild environments to feed their own families, as mankind has done everywhere since the Neolithic Revolution. The realization grew in Mauritius, where I watched the world's last five echo parakeets land on one tree and knew they will soon be no more. It has come through an equally painful intellectual change. I became a biologist through wonder at the diversity of nature. I became a field biologist because I preferred watching nature go its own way to messing it about with experiments. At last I understood that biology, as the study of nature apart from man, is a historical exercise. From the Neolithic Revolution to its logical sequels of twentieth-century population growth, biochemical engineering of life forms, and nuclear mutual assured destruction, human mind has become the chief factor in biology. . . . the urgent need in [vertebrate] studies is conservation. It is sheer self-indulgence to write books to increase understanding if there will soon be nothing left to understand."

Summary

Evidence for the origin of *Homo sapiens* comes from the Cenozoic fossil record of primates and from comparative study of existing monkeys, apes, and human beings. The traits that characterize the order Primates have evolved in relation to an arboreal, insect-grasping mode of existence; secondarily, some primates are terrestrial, humans most of all.

The first primates were probably much like the tree shrews, but the earliest known primates (from the late Paleocene) were similar to the modern lemurs, pottos, and galagos. These prosimians quickly spread over the Old and New Worlds and differentiated into a variety of ecological forms. All were arboreal and some possessed somewhat larger brains in relation to body size.

By the Oligocene two different groups of higher primates had evolved—the so-called platyrrhine monkeys of the New World tropics and the Old World catarrhine monkeys and apes. The anthropoid apes and human-like species, including *Homo sapiens*, are grouped in the superfamily Hominoidea. The fossil record for this clade is richer than for other lineages of primates. Many morphological features distinguish the hominoids from other catarrhines, and enlargement of the brain has been a major evolutionary force molding the shape of the hominoid skull, especially in the later part of human evolution.

The first known hominoids occur in the Oligocene, 35 million years ago. By the Miocene hominoids had diversified into a number of environments and had spread widely over Africa, Europe, and Asia. Fossils showing the derived characters of the great apes and humans have been found in China, Pakistan, India, Kenya, Turkey, and Hungary and date from about 17 to 9 million years ago. Humans retain primitive characters of these early forms that the apes have lost in the evolution of more derived characters.

A variety of hominid (human) fossils occur in late Pliocene and early Pleistocene deposits of Kenya and Ethiopia, some nearly 4 million years old. The earliest, with small cranial volumes and often heavy features of skull, tooth, and jaw, are assigned to the genus *Australopithecus*. The first stone tools are found in East Africa dating from about 2 million years ago. As more of these early human-like fossils are unearthed, it becomes increasingly difficult to separate them from the genus *Homo* because some very *Homo*-like fossils, currently called *Homo habilis,* are older than some forms of *Australopithecus*.

Homo erectus ranged across Africa and Eurasia from about 1.6 million years to 300,000 years ago. This hominid had a brain capacity approaching the lower range of modern *Homo sapiens*, made stone tools, and used fire. *Homo sapiens*, the only surviving species of the family Hominidae, came into existence around 200,000 to 300,000 years ago. By 40,000 to 60,000 years ago some populations of *Homo* were barely distinguishable in structure from modern humans. They had a well-organized society with a rapidly developing culture, especially obvious in their use of stone tools.

The biological success of human beings is measured

by their universal distribution over the earth and by the fact that humans derive some portion of their livelihood from every ecosystem and habitat in the biosphere. This all-pervasive status of *Homo sapiens* has had a dramatic impact on environments and on other animals, particularly on other vertebrates. Although extinction of species is a natural result of evolutionary processes through time, there is no precedent for the extent to which one species—*Homo sapiens*—has been responsible for extinctions of many other forms of life.

The first significant human impacts on other vertebrates may have occurred in prehistoric time. Extinction of many species of large vertebrates in the late Pleistocene may be the result of hunting by humans, although other predators and changes in climate may also have been significant factors. Modern industrialized people have exerted drastic influences on populations of other animals. The influence of humans on vertebrate species within historical time has occurred in three overlapping phases: First, the period of exploration and discovery in the fifteenth through nineteenth centuries when many island endemics were brought to extinction.

Second, and following close on the heels of the explorers, were the fur trappers, commercial hunters, fishermen, and other harvesters who exploited wild vertebrate populations for commerce. Third and most serious are the long-term effects of human agriculture and other land uses that destroy natural habitats. If current rates continue, more species will join the ranks of endangered wildlife and will become extinct in the next few decades than in all the centuries since 1600, owing to the diminution and deterioration of natural habitats. Such losses will be particularly severe in the biotically rich tropical forests of South America, Africa, and Asia.

Two cardinal requirements must be met if we are to save a significant vestige of natural biotas and ecosystems intact for future generations of humans to know: preservation of natural habitats and ecosystems on a large and representative scale, and control of human population growth. Ultimately, and soon, human beings must learn to accept a morality of restraints—restraint on the use of resources and, above all, restraint on human procreation.

References

Adovasio, J. M. and R. C. Carlisle. 1986. The first Americans, Pennsylvania pioneers. *Natural History* 95:(12)20–27.

Bowman, K. P. 1988. Global trends in total ozone. *Science* 239:48–50.

Bruce, E. J. and F. J. Ayala. 1979. Phylogenetic relationships between man and the apes: electrophoretic evidence. *Evolution* 33:1040–1056.

Campbell, B. 1985. *Human Evolution, an Introduction to Man's Adaptations*, 3rd edition. Aldine, Hawthorne, N.Y.

Cartmill, M. 1974. Rethinking primate origins. *Science* 1984:436–443.

Ciochon, R. L. and R. S. Corruccini (editors). 1983. *New Interpretations of Ape and Human Ancestry*. Plenum Press, New York.

Ciochon, R. L. and J. G. Fleagle. 1987. *Primate Evolution and Human Origins*. Aldine, Hawthorne, N.Y.

Day, M. H. 1986. *Guide to Fossil Man*, 4th Edition. University of Chicago Press, Chicago.

Delson, E. 1985. *Ancestors: The Hard Evidence*. Alan R. Liss, New York.

Delson, E. 1987. Evolution and palaeobiology of robust *Australopithecus*. *Nature* 327:654–655.

Diamond, J. M. 1983. Extinctions, catastrophic and gradual. *Nature* 304:396–397. Review of symposium presentations [see Martin and Klein (1984) below] on extinctions in the geological record, including those thought to have been brought about by humans.

Edey, M. A. 1976. *Three-Million-Year-Old Lucy*. Pages 19–31, in *Nature/Science Annual*, edited by J. D. Alexander. Time-Life Books, New York.

Ehrenfeld, D. W. 1970. *Biological Conservation*. Modern Biology Series. Holt, Rinehart and Winston, New York.

Falk, D. 1985. Hadar AL162-28 endocast as evidence that brain enlargement preceded cortical reorganization in hominid evolution. *Nature* 313:45–47. See also *Nature* 321:536–537, 1986.

Frankel, O. H. and M. E. Soulé. 1981. *Conservation and Evolution*. Cambridge University Press, Cambridge.

Galloway, J. N., Z. Dianwu, X. Jiling, and G. E. Likens.

1987. Acid rain: China, United States, and a remote area. *Science* 236:1559–1562.

Gillespie, R., D. R. Horton, P. Ladd, P. G. Macumber, T. H. Rich, R. Thorne, and R. V. S. Wright. 1978. Lancefield swamp and the extinction of the Australian megafauna. *Science* 200:1044–1048.

Gould, S. J. 1979. Our greatest evolutionary step. *Natural History* 88(6):40–44.

Gould, S. J. 1988. A novel notion of Neanderthal. *Natural History* 97(6):16–21.

Harris, J. M., F. H. Brown, M. G. Leakey, A. C. Walker, and R. E. Leakey. 1988. Pliocene and Pleistocene hominid-bearing sites from west of Lake Turkana, Kenya. *Science* 239:27–33.

James, H. F., T. W. Stafford, Jr., D. W. Steadman, S. L. Olsen, P. S. Martin, A. J. T. Jull, and P. C. McCoy. 1987. Radiocarbon dates on bones of extinct birds from Hawaii. *Proceedings of the National Academy of Sciences USA* 84:2350–2354.

Janzen, D. H. 1983. The Pleistocene hunters had help. *American Naturalist* 121:598–599.

Janzen, D. H. 1988. Tropical ecology and biocultural restoration. *Science* 239:243–244.

Johanson, D. C. and M. A. Edey. 1981. *Lucy, The Beginnings of Humankind*. Simon and Schuster, New York.

Jolly, A. 1980. *A World Like Our Own: Man and Nature in Madagascar*. Yale University Press, New Haven, Conn.

Jolly, A. 1985. *The Evolution of Primate Behavior*, 2nd Edition. Macmillan, New York.

Kaufman, L. and K. Mallory. 1986. *The Last Extinction*. MIT Press, Cambridge, Mass.

Kerr, R. A. 1987. Has stratospheric ozone started to disappear? *Science* 237:131–132.

Kurtén, B. 1986. *How to Deep-Freeze a Mammoth*. Columbia University Press, New York.

Leakey, R. E. 1981. *The Making of Mankind*. Michael Joseph, London.

Lewin, R. 1984. DNA reveals surprises in human family tree. *Science* 226:1179–1182.

Lewin, R. 1986a. New fossil upsets human family. *Science* 233:720–721.

Lewin, R. 1986b. Recognizing ancestors is a species problem. *Science* 234:1500.

Lewin, R. 1986c. A new look at an old fossil face. *Science* 234:1326.

Lewin, R. 1987a. My close cousin the chimpanzee. *Science* 238:273–275.

Lewin, R. 1987b. The earliest "humans" were more like apes. *Science* 238:1061–1063.

Lewin, R. 1988. A new tool-maker in the hominid record? *Science* 240:724–725.

Lovejoy, C. O. 1981. The origin of man. *Science* 211:341–350.

Luckett, W. P. 1980. *Comparative Biology and Evolutionary Relationships of Tree Shrews*. Plenum Press, New York.

Martin, P. S. 1973. The discovery of America. *Science* 179:969–974.

Martin, P. S. and R. G. Klein (editors). 1984. *Quaternary Extinctions, a Prehistoric Revolution*. University of Arizona Press, Tucson, Ariz. Symposium covering organisms, worldwide sites, geologic, climatic, and cultural models of Quaternary, primarily Pleistocene, mammal and bird extinctions.

Martin, P. S. and H. E. Wright, Jr. (editors). 1967. *Pleistocene Extinctions: The Search for a Cause*. Yale University Press, New Haven, Conn. A collection of symposium papers from the first major meeting where varied views on the extinctions of the larger mammals during the Pleistocene were presented.

Mosimann, J. E. and P. S. Martin. 1975. Simulating overkill by paleoindians. *American Scientist* 63:304–313.

Mooney, H. A., P. M. Vitousek, and P. A. Matson. 1987. Exchange of materials between terrestrial ecosystems and the atmosphere. *Science* 238:926–932.

Myers, N. 1984. *The Primary Source, Tropical Forests and Our Future*. W. W. Norton, New York.

Owen-Smith, N. 1987. Pleistocene extinctions: the pivotal role of megaherbivores. *Paleobiology* 13:352–362.

Patrusky, B. 1980. Pre-Clovis man: sampling the evidence. *Mosaic* 11(5):2–10.

Romer, A. S. 1966. *Vertebrate Paleontology*. 3rd Edition. The University of Chicago Press, Chicago.

Schindler, D. W. 1988. Effects of acid rain on freshwater ecosystems. *Science* 239:149–157.

Schonewald-Cox, C. M., S. M. Chambers, B. MacBride, and L. Thomas (editors). 1983. *Genetics and Conservation*. Benjamin-Cummings, Menlo Park, Calif.

Soulé, M. E. (editor). 1986. *Conservation Biology, the Science of Scarcity and Diversity*. Sinauer Associates, Sunderland, Mass.

Stringer, C. B. and P. Andrews. Genetic and fossil evidence for the origin of modern humans. *Science* 239:1263–1268.

Susman, R. L. 1988. Hand of *Paranthropus robustus* from

member 1, Swartkrans: fossil evidence for tool behavior. *Science* 240:781–784.

Susman, R. L. and J. T. Stern. 1982. Functional morphology of *Homo habilis*. *Science* 217:931–934.

Szalay, F. S. and E. Delson. 1979. *Evolutionary History of the Primates*. Academic Press, New York.

White, T. D. 1980. Evolutionary implications of Pliocene hominid footprints. *Science* 208:175–176.

Wood, B. 1987. Who is the "real" *Homo habilis*? *Nature* 327(6119):187–188.

Wu, R. and S. Lin. 1983. Peking man. *Scientific American* 248(6):86–94.

Yunis, J. J. and O. Prakash. 1982. The origin of man: a chromosomal legacy. *Science* 215:1525–1530.

Zimmerman, D. R. 1975. *To Save a Bird in Peril*. Coward, McGann & Geoghegan, New York.

Glossary

Technical terms, and some words with special biological meanings, are included in this glossary. In general, if a word is defined in the text and appears only on the page where it is defined and immediately adjacent pages, we have not attempted to define it again in the glossary. Similarly, if the definition in a standard dictionary is adequate, we have not included it here. However, certain words have slightly different meanings in different contexts. In those cases we have included the word in this glossary and limited the definition to the use of the word in this book.

abduction   Movement away from the midventral axis of the body (*see also* adduction).

acetabulum   Depression on the pelvic girdle that accommodates the head of the femur.

adduction   Movement toward the midventral axis of the body (*see also* abduction).

aestivation (estivation)   Form of torpor, usually a response to high temperatures or scarcity of water.

allochthonous   With an origin somewhere other than the region where found.

allopatry   Situation in which two or more populations or species occupy mutually exclusive, but often adjacent, geographic ranges.

ammonotelic   Excreting nitrogenous wastes primarily as ammonia.

amniotes   Those vertebrates whose embryos possess an amnion chorion, and allantois (i.e., turtles, lepidosaurs, crocodilians, birds, and mammals).

amphicoelus   Vertebral centrum with both the anterior and posterior surfaces concave.

anadromus   Migrating up a stream or river from a lake or ocean to spawn (of fishes).

angiosperm   Most advanced and recently evolved of the vascular plants characterized by production of seeds enclosed in tissues derived from the ovary, the combination of ovary and/or seed being of major importance to many vertebrates as food. The success of the angiosperms has had important consequences for the evolution of terrestrial vertebrates.

aphotic   "Without light"—for example, in deep-sea habitats or caves.

apocrine gland   Type of gland in which the apical part of the cell from which the secretion is released breaks down in the process of secretion (*see also* holocrine gland).

apomorphic   A character that is changed from its preexisting condition (*see also* autapomorphy).

aposematic   Device (color, sound, behavior) used to advertise the noxious qualities of an animal.

arcade   Curve or arch in a structure such as the tooth row of humans.

archaic   Form typical of an earlier evolutionary time. When capitalized, referring to the Archaeozoic period, previous to 2.5 billion years before the present.

archipterygium   Fin skeleton, as in a lungfish, consisting of symmetrically arranged rays that extend from a central skeletal axis.

aural   Of the external or internal ear or sense of hearing.

autapomorphy   An attribute unique to one group of organisms.

autochthonous   With an origin in the region where found.

benthic   Living at the soil/water interface at the bottom of a body of water.

bilateral symmetry (bisymmetry)   Characteristic of a body which can be divided into mirror-image halves by a single plane of space.

biomass   Living organic material in a habitat (available as food for other species).

biome   Biogeographic region defined by a series of spatially interrelated and characteristic life forms (e.g., tundra, mesopelagic zone, tropical rain forest, coral reef).

brachial   Pertaining to the forelimb.

branchial   Pertaining to the gills.

branchiomeric   Segmentation of structures associated with, or derived from, the ancestral pharyngeal arches (*see also* metameric).

carapace   Dorsal shell, as in a turtle.

catadromous   Migrating down a river or stream to a lake or ocean to spawn (of fishes).

catastrophism   Hypothesis of major evolutionary change as a result of unique catastrophic events of broad geographic and thus ecologic effect.

centrum   Vertebral element formed in or around the notochord (plural, centra).

cephalic   Pertaining to the head.

ceratotrichia   Keratin fibers that support the web of the fins of chondrichthyes.

choana   Internal nares (plural, choanae).

chondrification   Formation of cartilage.

clade   Phylogenetic lineage originating from a common ancestral taxon and including all descendants (*see also* grade).

cladistic   Pertaining to the branching sequences of phylogenesis.

cladogram   Branching diagram representing the hypothesized relationships of taxa.

cleidoic egg   One independent of environment except for heat and gas (carbon dioxide, oxygen, water vapor) exchange. Characteristic of amniotes.

cline   Change in a biological character along a geographic gradient.

coelom (celom)   Body cavity, lined with tissue of mesodermal origin.

coevolution   Complex biotic interaction through evolutionary time resulting in the adaptation of the interacting species to unique features of the life histories of the other species in the system.

cones   Photoreceptor cells in the vertebrate retina that are differentially sensitive to light of different wavelengths.

conspecific   Belonging to the same species as that under discussion (*see also* heterospecific).

convergent evolution   Evolution of similar characters in phyletically distantly related or unrelated forms.

cosmine   Form of dentine containing branching canals characteristic of the cosmoid scales of crossopterygian fishes and early dipnoans.

countershaded   Color pattern in which the aspect of the body that is more brightly lighted (normally, the dorsal surface) is darker colored than the less brightly illuminated surface. The effect of countershading is to make an animal harder to distinguish from its background.

cranial   Pertaining to the cranium or skull, a unique and unifying characteristic of all vertebrates.

demersal   More dense than water and therefore sinking, as in the eggs of many fishes and amphibians.

detritus   Particulate organic matter that sinks to the bottom of a body of water.

deuterostomy   Condition in which the embryonic blastopore forms the anus of the adult animal; characteristic of chordates (*see also* protostomy).

double cone   Type of retinal photoreceptor in which two cones share a single axon (*see also* cones).

durophagous   Feeding upon hard material.

ecosystem   Community of organisms and their entire physical environment.

ectoderm   One of the embryonic germ layers, the outer layer of the embryo.

edentulous   Lacking teeth.

endemism   Property of being endemic (i.e., found only in a particular region).

endoderm   Innermost of the germ cell layers of late embryos.

epicontinental sea (epeiric sea)   Sea extending within the margin of a continent.

epigenetic   Pertaining to an interaction of tissues during embryonic development that results in the formation of specific structures.

epiphysis   (1) Pineal organ, an outgrowth of the roof of the diencephalon. (2) (plural, epiphyses) Accessory center of ossification at the ends of the long bones of mammals, birds, and some squamates. When the ossifications of the shaft (diaphysis) and epiphysis meet, further lengthwise growth of the shaft ceases. This process produces a determinate growth pattern.

epiphyte   Plant that grows nonparasitically upon another plant.

estuarine   Pertaining to, or formed in, a region where the freshwater of rivers mixes with the seawater of a coast.

euryhaline   Capable of living in a wide range of salinities (see also stenohaline).

euryphagous   Eating a wide range of food items; a food generalist (see also stenophagous).

eurythermal   Capable of tolerating a wide range of temperatures (see also stenothermal).

eurytopy   Capable of living in a broad range of habitats.

extraperitoneal   Positioned in the body wall beneath the lining of the coelom (the peritoneum) in contrast to being suspended in the coelom by mesentaries.

fossorial   Burrowing through the soil.

fovea centralis   Area of the vertebrate retina containing only cone cells, where the most acute vision is achieved at high light intensities.

furcula   Avian wishbone formed by the fusion of the two clavicles at their central ends.

geosyncline   Portion of the earth's crust that has been subjected to downward warping. Sediments frequently accumulate in geosynclines.

gestation   Period during which an embryo is developing in the reproductive tract of the mother.

gill arch   Assemblage of tissues associated with a gill; the term may refer to the skeletal structure only or to the entire epithelial muscular and connective tissue complex.

Gondwana   Supercontinent that existed either independently or in close contact with all other major continental land masses throughout vertebrate evolution until the middle of the Mesozoic and was composed of all the modern Southern Hemisphere continents plus the subcontinent of India.

grade   A level of morphological organization achieved independently by different evolutionary lineages (see also clade).

gymnosperms   Group of plants in which the seed is not contained in an ovary—conifers, cycads, and ginkos.

hemal arch   Structure formed by paired projections ventral to the vertebral centrum and enclosing caudal blood vessels.

hermaphroditic   Having both male and female gonads.

heterocoelus   Having the articular surfaces of the vertebral centra saddle-shaped, as in modern birds.

heterospecific   Belonging to a different species from that under discussion (see also conspecific).

heterosporous plants   Plants with large and small spores; the smaller give rise to male gametophytes and the larger to female gametophytes (equivalent to protogymnosperms).

heterotrophic   Capable of using only organic materials as a source of energy.

holocrine gland   Type of gland in which the entire cell is destroyed with the discharge of its contents (see also apocrine gland).

homologous   Characters having a common origin.

homology   Pertaining to the fundamental similarity of individual structures that belong to different species within a monophyletic group.

hydrosphere   Free liquid water of the earth—oceans, lakes, rivers, etc.

hyperdactyly   Increase in the number of digits.

hyperphalangy   Increase in the number of bones in the digits.

hypertrophy   Increase in the size of a structure.

hypotremate   Having the main gill openings on the ventral surface and beneath the pectoral fins as in skates and rays (see also pleurotremate).

inguinal   Pertaining to the groin.

interspecific   Pertaining to phenomena occurring between members of different species.

intraspecific   Pertaining to phenomena occurring between members of the same species.

isohaline   Of the same salt concentration.

isostasy   Condition of gravitational balance between segments of the earth's crust or of return to balance after a disturbance.

isostatic movement   Vertical displacement of the lithosphere due to changes in the mass over a point or region of the earth.

isotherm   Line on a map that connects points of equal temperature.

leptocephalus larva   Specialized, transparent, ribbon-shaped larva of tarpons, true eels, and their relatives.

lithosphere   Crust of the earth.

littoral   Pertaining to the shallow portion of a lake, sea, or ocean where rooted plants are capable of growing.

lophophorate   Pertaining to several kinds of marine animals that possess ciliated tentacles (lophophores) used to collect food (e.g., pterobranchs).

meninges   Sheets of tissue enclosing the central nervous system. In mammals these are the dura mater, arachnoid, and pia mater.

mesentaries   Membranous sheets derived from the mesoderm that envelope and suspend the viscera from the body wall within the coelom.

mesoblast   Mesodermal cell.

mesoderm   Central of three germ layers of late embryos.

metameric   Pertaining to ancestral segmentation, used in reference to serially repeated units along the body axis.

monophyletic lineage   A taxon composed of a common ancestor and all its descendants.

monophyly   Relationship of two or more taxa having a common ancestor.

morph   Genetically determined variant in a population.

morphotypic   Type of classification based entirely on physical form.

neoteny   Retention of larval or embryonic characteristics past the time of reproductive maturity (*see also* paedomorphosis and progenesis).

neural arch   Dorsal projection from the vertebral centrum that, at its base, encloses the spinal cord.

neural crest   Embryonic cells unique to vertebrate animals, associated with the neurectoderm but subsequently widely migrating to participate in the formation of many tissues and structures which are characteristic of the subphylum.

neurocranium   Portion of the head skeleton encasing the brain.

niche   Pertaining to the functional role of a species or other taxon in its environment—the ways in which it interacts with both the living and nonliving elements.

occipital   Pertaining to the posterior part of the skull.

ontogenetic   Pertaining to the development of an individual organism.

operculum   Flap or plate of tissue covering the gills.

orogeny   Process of crustal uplift or mountain building.

osseous   Bony.

out-group   Group of organisms that is related to but removed from the group under study. One or more outgroups are examined to determine which character states are evolutionary novelties (apomorphies).

paedomorphosis   Condition in which a larva becomes sexually mature without attaining the adult body form. Paedomorphosis may be achieved by neoteny or by progenesis.

palatoquadrate   Upper jaw element of primitive fishes and chondrichthyes, portions of which contribute to the palate, jaw articulation, and middle ear of other vertebrates.

Pangaea (Pangea)   Single supercontinent that existed during the mid-Paleozoic and consisted of all modern continents apparently in direct physical contact with a minimum of isolating physical barriers.

paraphyletic   Taxon that includes the common ancestor and some but not all of its descendants.

Phanerozoic   Period since the Cambrian.

pharyngotremy   Condition in which the pharyngeal walls are perforated by slit-like openings; found in chordates and hemichordates.

photophore   Light-emitting organ.

phylogenetic   Pertaining to the development of an evolutionary lineage (*see also* ontogenetic).

physoclistic   Lacking a connection from the gut to the swim bladder as adults (of fishes).

physostomous   Having a connection between the swimbladder and gut in adults (of fishes).

piloerection   Contraction of muscles attached to hair follicles resulting in the erection of the hair shafts.

piscivorous   Fish-eating.

placoid scales   Primitive type of scale found in elasmobranchs and homologous with vertebrate teeth.

plastron   Ventral shell, as of a turtle.

plate tectonics   Theory of earth history in which the lithosphere is continually being generated from the underlying core at specific areas and reabsorbed into the core at others resulting in a series of conveyor-like plates which carry the continents across the face of the earth.

plesiomorphic   The ancestral character from which an apomorphy is derived.

pleurotremate   Having the main gill openings on sides of the body anterior to the pectoral fins as in sharks (*see also* hypotremate).

polymorphism   Simultaneous occurrence of two or more distinct phenotypes in a population.

polyphyletic   A taxon that does not contain the most recent common ancestor of all the subordinate taxa of the taxon.

portal system   Portion of the venous system specialized for the transport of substances from the site of production to the site of action. A portal system begins and ends in capillary beds.

postzygapophysis   Articulating surface on the posterior face of a neural arch.

prezygapophysis   Articulating surface on the anterior face of a neural arch.

progenesis   Accelerated development of reproductive organs relative to somatic tissue, leading to paedomorphosis.

Proterozoic   Later part of the Precambrian, from about 1.5 billion years ago until the beginning of the Cambrian 500 million years ago (*see also* Phanerozoic).

protostomy   Condition in which the embryonic blastopore forms the mouth of the adult animal (*see also* deuterostomy).

protraction   Movement away from the center of the body (*see also* retraction).

protrusible   Capable of being moved away (protruded) from the body.

refugium   Isolated area of habitat fragmented from a formerly more extensive biome.

rete mirabile   "Marvelous net," a complex mass of intertwined capillaries specialized for exchange of heat and/or dissolved substances between countercurrent flowing blood.

retraction   Movement toward the center of the body (*see also* protraction).

rod   Photoreceptor cell in the vertebrate retina specialized to function effectively under conditions of dim light.

rostrum   Snout; especially an extension anterior to the mouth.

scapulocoracoid cartilage   In elasmobranchs and certain primitive gnathostomes, the single solid element of the pectoral girdle.

scutes   Scales, especially broad or inflexible ones.

serial   Repeated, as are the body segments of vertebrates.

sinus   Open space in a duct or tubular system.

sister group   Group of organisms most closely related to the study taxa, excluding their direct descendants.

somite   Member of a series of paired segments of the embryonic dorsal mesoderm of vertebrates.

speciose   Refers to a taxon that contains a large number of species.

squamation   Scaly covering of the body.

stenohaline   Capable of living only within a narrow range of salinity of surrounding water; not capable of surviving a great change in salinity (*see also* euryhaline).

stenophagous   Eating a narrow range of food items; a food specialist.

stenothermal   Capable of living or of being active in only a narrow range of temperatures (*see also* eurythermal).

stratigraphy   Classification, correlation, and interpretation of stratified rocks.

stratum   Layer of material (plural, strata).

sympatry   Occurrence of two or more species in the same area.

symplesiomorphy   Character shared by a group of organisms that is found in their common ancestor.

symphysis   A joint between bones formed by a pad or disc of fibrocartilage that allows a small degree of movement.

talonid   Basinlike heel on a lower molar tooth, found in certain mammals.

tarsometatarsus   Bone formed by fusion of the distal tarsal elements with the metatarsals in birds and some dinosaurs (*see also* tibiotarsus).

taxon   Any scientifically recognized group of organisms.

thecodont teeth   Teeth set in bony sockets in the jaw.

tibiotarsus   Bone formed by fusion of the tibia and proximal tarsal elements in birds and some dinosaurs (*see also* tarsometatarsus).

troglodyte   Organism that lives in caves.

trophic   Pertaining to feeding and nutrition.

ureotelic   Excreting nitrogenous wastes primarily as urea.

uricotelic   Excreting nitrogenous wastes primarily as uric acid and its salts.

urogenital   Pertaining to the organs, ducts, and structures of the excretory and reproductive systems.

vacuoles   Membrane-bound spaces within cells containing secretions, storage products, etc.

viscera   Internal organs of the coelom.

visceral skeleton   Skeleton primitively associated with the pharyngeal arches, uniquely derived from the neural crest cells and forming in mesoderm immediately adjacent to the endoderm lining the gut.

zygapophysis   Articular process of the neural arch of a vertebra (*see also* postzygapophysis and prezygapophysis).

zygodactylus   Type of foot, in which the toes are arranged in two opposable groups.

Author Index

Subject Index

Ansiodactyl foot, 631
Antarctic ice fish, 116
Anteater, 742, 743, 781, 790
Antelope ground squirrels, 848, 849, 850, 851–852
Antelope, 738, 739, 755–756, 803–810, 843–844, 881
Anterior pituitary hormones, 772, 773, 774
Antesaurids, 697
Anthracosaurs, 341, 349, 351, 352, 360–363, 367, 370
Antiarchs, 239
Antibody immunity, 117
Antidiuretic hormone, 121, 140, 144, 163
Antifreeze compounds, in vertebrates, 595
Antigen, 117
Antlers, 737–738, 740–741
Ants, 780, 781
Anurans, 6, 7, 341, 349, 351, 382–385, 386, 392–404
  body size of, 597
  cardiovascular system in, 112
  in desert, 583, 591–593
  origin, 373, 374–375
  parental care by, 400–401
  reproduction by, 398–400
  saltatory locomotion by, 382–383
  tadpoles, 401–404
  taxonomic arrangement of, 378–379
  vocalizations by, 392–398
  see also Frogs; Toads; Treefrogs
Anurognathus, 456, 457
Anus, 106
Aorta, in frog heart, 114
Aortic arches, evolution of, 112
Aotus, 812
Apatosaurus, 462–463, 484
Apes, 813, 861, 862, 863, 865, 867, 868
  see also Hominoids; Primates
Aphaneramma, 356, 357
Aphotic zone, 576–577
Aplocheilus lineatus, 283, 284
Apnea, 114–115
  in turtles, 430–431
Apocrine glands, 733–734
Apodiforms, 647, 649
Apophyses, vertebrae and, 100
Aposematic colors, 410, 412
Appalachian Revolution, 511
Appalachians, 507
Appendicular skeleton, 97, 101, 102
  see also Skeletal system
Apteria, 604, 606

Aquatic endotherms, 825–829
Aqueous humor, 116, 133
Arandaspis, 208–209
Arapaima, 309
Arawana, 309, 310
Arboreal snakes, 533, 534, 535
Arboreal theory, avian flight and, 473
Archaeopteris, 336
Archaeopteryx, 12, 468–474, 645, 646
  lithographica, 470
Archaic Homo sapiens, 878
Archegosaurus, 356, 357
Archeria, 363
Archeriids, 363
Archilochus colubris, 831
Archinephric ducts, 118, 148–149
Archipallium, 126, 127, 128
Archipterygium, 344, 345
Archosauromorphs, 7, 46, 449, 452
  prolacertiformes, 451
  rhynchosaurs, 449–451
  see also Archosaurs
Archosaurs, 451, 452, 458, 459
  birds and, 455, 458
  crocodilians, 452–455, 456
  heart in, 111
  phytosaurs, 452–455
  pterosaurs, 455–458, 468, 474
  reproduction in, 677
  see also Dinosaurs
Arctic, endotherms in, 825–829
Arctic foxes, 798–800, 825
Arctocebus, 812
Area opaca, 81
Area pellucida, 81
Argyrolagids, 786
Aristostomias grimaldi, 580–581
Armadillos, 742, 743, 756, 757, 786–787
Artebius, 743
Arteries, 108, 110
Arterioles, 93, 108
Arteriovenous anastomoses, 93, 108, 115
Arthrodires, 237, 239
Arthropods, 68
Artiodactyls, 729, 734–735, 738, 744, 749, 753, 755, 756, 780, 781, 784, 787, 791
Aspect ratio, 610
  drag, 278
Aspidorhynchus, 298
Assimilated energy, in energy budget, 822
Astrapotheres, 787, 788
Astraspis, 208–209
Astroscopus, 287

Ateles, 812
Atherinomorphs, 284, 294, 312, 316
Atlantic Ocean, 713
Atlantic silverside, 320–321
Atlas, as cervical vertebra, 100
ATP, 185–186
Atriopore, 64
Atrioventricular valve, in frog heart, 114
Atrium, 64
  of fish heart, 111
  of frog heart, 114
Audition. See Hearing
Auditory nerve, 129
Auditory placode, embryonic, 84
Auks, 635, 636, 651, 673
Australia
  fauna of, 779
  Tertiary mammals of, 788–789
Australian hopping mice, 162, 846, 847
Australian lungfish, 325, 326
Australian snake-necked turtles, 421
Australian wood swallows, 656
Australopithecus, 863, 875, 876, 877
  aethiopicus, 876
  afarensis (Lucy), 872, 873, 874, 875, 876
  africanus, 128, 873, 874, 875, 876, 877
  boisei, 128, 873, 876
  robustus, 874, 875, 876
Autapomorphy, 49
Autonomic nervous system, 124
Autonomy, of appendages, 547, 548–549
Axial skeleton, 97
  see also Skeletal system
Axillary amplexus, 398
Axis, as cervical vertebra, 100
Axons, 124
Aye-aye, 812, 860

B lymphocytes, 117
Baboons, 813, 865, 879
Badgers, 749, 750, 777
Balaenoptera musculus, 128
Baleen, 747
Baleen whales, 747, 765, 829–831, 886
Balistiform, 274, 276
Balistiform swimming, 278
Baltica, 197, 198
Banded mongoose, 798
Bandicoots, 789
Bannertail kangaroo rats, 848
Baptornis, 646
Barasaurus besairiei, 368, 369
Barbs, 604, 605
Bar-headed geese, 624

Bark scalers, 656
Barn swallows, 610
Barracudas, 312
Basal elements, 101, 102
Basal nuclei, 126, 127
Basement membrane, 121
Basilar membrane, 138
Basking shark, 251, 259
Bassariscus astutus, 798
Bat falcon, 653
Bathypelagic fishes, 577, 578–583
Bathypelagic zone, 577
Batoids, 240
Batrachotoxin, 411
Bats, 43, 670, 742, 743, 777, 784, 789, 832
  echolocation in, 761–765
Bay wrens, 663, 665
Beaks, of birds, 625–627
Bears, 742, 743, 749, 753, 797, 798, 833, 836
Bedouin goats, 843
Bee-eaters, 656
Benthic zone, 576, 577
Benthosuchus, 346
Bettas, 268
Bichirs, 306, 307, 308
Big-headed turtle, 431
Bile, 106
Bile duct, 106
Billfish, 312
Binomial nomenclature, of Linnaean
    system, 41
Biogeography, vicariance in mammals,
    779–791
Biological species, 22
Biological variation, 16–17
Bioluminescence, of deep-sea fishes,
    578, 581
Biomass change, in energy budget, 822,
    824
Bipedalism, evolution of, 459–460
Bipes, 530
Bipolar cell, 134
Birds, 6, 7, 12, 447
  arctic, 827
  beak, 625–627
  brain, 126–128, 618
  buccal cavity, 628
  carnivorous, 168
  ceca, 630
  cloaca, 630
  coelurosaurs and, 466
  dermis of, 93
  digestion, 628–630
  as dinosaur model, 482–484, *see also*
    Dinosaurs

diurnality of, 644
eggs of, 79, 81, 148, 363
environments of, 573
  in hot desert, 844–845
esophagus and crop, 628–629
evolution of, 645–650, *see also*
    Dinosaurs
  modern orders and families, 646–
    647
  phylogeny of modern, 648–650
  primitive, 645–646
feeding by, 650–662
  aerial sweepers, 656–657
  aquatic filter feeders, 652
  carnivorous or predatory birds,
    652–653, 654
  carrion eaters, 653, 655
  diggers, 656
  fish and aquatic animal feeders,
    650–652
  flycatchers, 656, 657
  food gathering, 625–628
  fruit eaters, 657
  gleaners, 655–656
  grazers and browsers, 658–659
  insectivores and terrestrial inverte-
    brate eaters, 655–657
  optimal foraging theory and, 644,
    659–662
  pollen and nectar feeders, 657, 658
  probers, 656
  seed eaters, 657–658
  wood drillers, 656
flightless, 603
gastric apparatus, 622–625
heart in, 111
intestines of, 105
ion concentration in, 155
Jacobson's organ in, 449
migration and navigation, 644, 680–
    686
nitrogen excretion in, 167–170
parental care, 4, 482, 484
Pleistocene and, 35
pterosaurs and, 455, 458
reproduction. *See* social behavior and
    reproduction, *below*
rete mirabile in, 108
sex determination in, 147
social behavior and reproduction, 4,
    662–680
  incubation, 668, 670–672, 675–677
  male incubation and polyandry,
    668, 670–672
  mating systems, 666–672

monogamy, 668–669
nesting, 673–675
oviparity, 673
parental care, 677–680
polygyny, 668, 670
visual displays, 663, 665–666, 667
vocalization, 662–663, 664, 665, 666
stomachs of, 105
thermoregulation by, 174, 180–184
  torpor and, 831–838
tongues of, 627–628
vascular systems of, 114–115
ventilation in, 107
vision in, 132
vocalization, 107–108, 624
Birds, flight in, 7, 12, 271, 603–643
  body form and, 618–630
    gas exchange, 622–625
    heart, 618, 621
    lungs, 3, 618, 621–622
    muscles, 618, 620–621
    olfaction, 618–619
    power and, 618–619
    size and, 603
    skeleton, 619–620, 621
    streamlining, 619
    weight, 618
  feathers and, 604–608, 618
  high-flying birds, 624–625
  locomotion, 630–636
    climbing, 634
    diving, 635–636
    hopping, 631
    perching, 631–632
    swimming, 634–636
    walking, 632–634
  sensory systems, 636–641
    air pressure, 639–641
    hearing, 622, 636, 638–639, 640–641
    infrasound, 641
    magnetic fields, 641
    olfaction, 618–619, 639
    vision, 618, 622, 636–638
  wing, 608–618
    fixed airfoil and, 608–610
    flapping flight, 611–616
    gliding flight, 610–611, 612
    structure of and flight characteris-
      tics, 616–618
Birds of prey, 652–653, 654
Birdsong, 662–663, 664, 665, 666
Birkenia, 210
Bison, 884
Bitis arietans, 534, 543
Black bear, 889

Black-backed gull, 675
Blackbirds, 663, 670
Black-capped chickadee, 836–837
Blackfooted albatross, 169
Black-throated green warbler, 34
Blastocoel, 80, 81, 82
Blastoderm, 81
Blastodisc, 80, 82
Blastopore, 82
Blastula, 80, 81, 82
Blood, 115–118
    see also Cardiovascular system
Blood plasma. *See* Plasma
Blood shunts, 429–431
Blood vessels, 93, 108–111
Blubber, 95, 827–828
Blue buck, 889
Blue chromis, 322, 323
Blue shark, 259
Blue whales, 128, 277, 603, 747, 886–887
Bluejay, 410
Boas, 540
Bobcat, 798
Bobwhite quail eggs, 677
Body fluids. *See* Water and ion exchange
Body size
    desert environment and, 838–853
    energy use and, 796
    food habits and, 803–810
    home range and, 796, 797
    resource needs and, 796–797
    sociality and prey size and, 800–801, 802
    standard metabolic rate and, 188–189
    torpor and, 832–836
    see also Energy use
Body wall, embryonic, 84
Boids, 525, 532, 541, 542, 557
Bolitoglossine plethodontid salamanders, 378, 380
Bolson tortoise, 419, 583
Bombina, 375
Bone
    early vertebrates and, 229–230
    mineralized tissue in, 90, 95
    origin of, 69, 73–74
    ossification, 95, 96–97
    see also Skeletal system
Bonefish, 309
Bontebok, 889
Bony fishes. *See* Osteichthyes
Bony labyrinth, 138, 139
Bony tongues, 271
Boobies, 650–651
Boomslang, 543

Boreaspis, 220
Borhyaenid marsupials, 780, 781, 786, 788, 789, 791
Boron analysis of early vertebrate habitat, 207
Bothriolepis, 238, 239
Bothrocara, 298
Bottlenecks, speciation and, 24
Bottlenose porpoise, 162, 767
Bovids, 738, 803–806
Bowfin, 5, 276, 297, 308, 309
Box turtles, 421, 425, 432
Boxfish, 270
Boylerids, 525
Brachiation, 60
Brachiopods, 65
Brachycephalids, 379
Brachyodont, 743
Brachyopids, 356
Brachyteles, 812
Brains
    of birds, 618
    in early mammals, 710
    embryonic, 83
    hominoids and, 863, 871
    human, 879
    mammalian, 758, 760–761
Branchiosaurs, 354, 463
Branchiosaurus, 356, 463, 479
Branchiostoma lanceolatum, 61
Bristlemouth, 579
Bristles, 604, 607
Bristletails, in Devonian, 336
Brithopodids, 696, 697, 698
Broad adaptations, 241, 243
Broad-tailed hummingbirds, 837–838
Brontornithids, 787
Brontosaurs. See Apatosaurus
Brooding, by birds, 675–677
Brook salamanders, 376
Brown pelican, 650
Brown-headed cowbird, 888
Browsers, birds as, 658–659
Bryophytes, in Devonian, 334–335
Bryozoans, 65
Buccal cavity, of birds, 628
Buccal cirri, 64
Budgerigar, 622
Buffalos, 805, 806
Buff-breasted wren, 663
Buffer systems, 115
Bufo, 384, 392, 398, 593
    americanus, 604
    marinus, 392
    microscaphus, 392

quercicus, 392
Bufonids, 379, 383
Bulbar conjunctiva, 133
Bulbul, 608
Bulbus cordis, 111, 113
Bull shark, 157, 158
Bullfrog, 392
Bunodont teeth, 743, 748, 750
Burchell's zebra, 889
Bushbabies, 812, 814
Bustards, 666
Butcherbirds, 652
Buteo galapagoensis, 672
Butterflyfish, 324

Cacajao, 812
Cacomistle, 798
Cacops, 354, 355, 367
Caecalid fermenters, 744–747
Caecilians, 6, 7, 341, 349, 351, 378, 385–386, 387
    origin, 373, 374–375
    reproduction in, 387–388
        testes, 148
    taxonomic arrangements, 378
Caecum (caeca), 744–747. *See also* Cecum
Caenolestids, 786
Caenophidians, 526, 532–533, 542
Caimans, 168, 455, 456
Calamus, 604, 605
Calcichordates, 70, 72–73
Calcification, 96
Calcitonin, 140
Caledonian orogeny, 200
Calidris temminckii, 671
California condors, 606
California pocket mice, 852–853
Callicebus, 812
Callimico, 812
Callithricid monkeys, 811, 812, 862
Callorhinus ursinus, 828
Calotes, 528
Camarasaurids, 463
Camarasaurus, 462–463
Cambered airfoil, 609
Cambrian, 36, 49, 74
    climate in, 200
    fossil vertebrates in, 205–207
    land masses during, 195, 198
    see also Early Cambrian; Late Cambrian
Camels, 162, 755–756, 839–843, 881, 884
Camelus dromedarius, 162, 839–843
Canaliculi, 95, 96
Canary, 663

Cancellous bone, 95
Canids, 11, 750, 758, 798, 803, 810
Canines, 741, 749, 750, 752
Canis, 744, 790
  aureus, 798
  familiaris, 782
  lupus, 798
Cape ground squirrel, 852
Cape jumping hare, 45
Cape lion, 889
Cape mountain zebra, 889
Capillaries, 93, 108, 110, 111
Caprimulgids, 665
Caprimulgiforms, 647, 649
Capsule parietal epithelium, renal, 121
Captorhinids, 367, 368
Capuchin monkeys, 812
Caracaras, 653
Carangids, 579
Carangiform fish, 269, 270, 273, 276
Carangiform swimming, 275, 276, 278
Carapace, 419, 423, 425
Carbonic anhydrase, 116
Carboniferous, 336
  continental drift during, 507–508
  flora and fauna in, 510
  glaciations in, 510
  insects in, 514
  Pangaea and, 195
  tetrapods of, 367–370
  see also Late Carboniferous
Carcharhinid sharks, 260
Carcharhinus, 249
Carcharodon carcharias, 257
Cardiac muscle, 111
Cardinal, 626
Cardinalfish, 322, 323, 324
Cardiovascular system, 62, 108–115
  embryonic origins, 90
  evolution of, 113–115
  in turtles, 429–431
  vascular circuits, 111, 112
  vessels, 108–111
  see also Heart
Caretta caretta, 169, 421
Carettochelyids, 420, 436
Cariamids, 652
Caribbean islands, human settlement of, 884
Caribou, 682, 738, 740, 825
Carnassial apparatus, 744, 749, 750
Carnivora, 781
Carnivores, 3, 729, 730, 741, 744, 758, 768, 784, 786
  convergences in, 780, 781, 782

dentition in, 749–750
diapause in, 777
herbivorous, 798
home range of, 797–798
humans and, 888–889
jaws of, 752–753
placenta in, 770, 771
Carnosaurs, 464, 465–466, 486, 498
Carolina anole, 550, 552, 553
Carotid arch, in frog heart, 114
Carp, 187, 304, 305, 317, 318, 320
Carpet shark, 260
Carretochelys, 425
Carrion, 639
Carrion eaters, birds as, 653, 655
Cartilaginous fishes. *See* Chondrichthyes
Cascade frog, 404
Caseids, 692, 694, 695, 696, 697
Cassowaries, 648
Casuariiforms, 648
Catadromous eels, 309
Catarrhines. *See* Apes; Humans; Old World monkeys
Catastrophism theories, for mass extinctions, 515–517
Catecholamines, 143–144
Catfish, 314–315
  electric, 285, 286
Catharus guttatus, 412
Cats, 3, 162, 736, 750, 758, 770, 778
Cattle, 755–756, 775, 778
  dentition of, 749
  freemartins in, 146–147
Caudal autonomy, of squamates, 547, 548–549
Caudal vertebrae, 101
Caudata. *See* Salamanders
Caviomorphs, 787, 862
Cavum arteriosum, 426
Cavum pulmonale, 426
Cavum venosum, 426
Cebids, 812
Cebus, 812
Cecum (ceca), 106, 630. *See also* Caecum
Cement, bone as, 741
Cenozoic, 37, 38
  climate of, 713
  continental geography during, 713, 714, 715, 716
  convergences of mammals in, 38
  see also Glaciation; Mammals
Central nervous system, 124
Central place foraging, 661
Centrolenella, 399
Centrolenids, 379

Centrum, 100
Cephalaspids, 210, 216, 219–221, 222, 230
Cephalochordates, 55, 61, 64, 65, 80
Cephalodiscus, 69
Cephalophus, 803
Ceratioids, 581–583
Ceratophrys, 373, 374, 384
Ceratopsians, 474, 476, 477, 481–482, 487, 488, 491
Ceratosaurs, 464, 465–466
Ceratosaurus, 459
Ceratotrichia, 247
Cerberus rhynchops, 169
Cercocebus, 813
Cercopithecoids, 813, 814, 863, 864, 865
Cercopithecus, 813
  aethiops, 815
Cerebellum, 126
Cerebral hemispheres, of primates, 59
Cerebral vesicle, 65
Cerebrospinal fluid, 116
Cervical vertebrae, 100
Cervids, 738
Cetaceans, 634, 729, 730, 747, 828
  brain of, 761
  echolocation in, 765–767
  placenta in, 770, 771
Cetorhinus maximus, 251
Chamaeleo, 528–529
Chameleons, 524, 527–529, 557, 558, 859–860
Champsosaurs, 452, 453, 454, 490
Champsosaurus, 490
Channichthys, 116
Chara, 200
Characins, 314–315
Charadius alexandrinus, 678
Charadriiforms, 635, 636, 645, 646, 649, 671
Chasmosaurus, 481, 487
Chat, 663
Cheetah, 758
Cheirolepis, 295
Chelids, 436
Chelodina, 421
Chelonia mydas, 433, 434–436
Chelonians, 420, 422, 436
Chelydrids, 419, 420, 436
Chelys, 421, 422
Chemoreceptors, 130
  see also Sense organs
Chianocarpus, 70
Chiasmodus niger, 580–581
Chickadees, 627, 825, 832, 836–837

El Niño Years, 24, 26
Elands, 803, 805, 806, 843
Elaphe, 542
Elapids, 526, 533, 542, 557
Elasmobranchs, 240
    earliest, 243–245
    electroreception by, 254–256
    evolutionary specializations of, 240–243
    ion concentrations in, 155
    osmotic regulation in, 157–158
    second radiation, 245–248
    see also Rays; Sharks; Skates
Elasmosaurus, 491–493
Elastin, in arteries, 110
Electric catfish, 285, 286, 287
Electric eels, 266, 268, 285, 286, 287
Electric fish, 260, 266, 285–289
Electric organ discharge, 288–289
Electric rays, 260
Electrocytes, in electric fishes, 286
Electrophorus, 287
Electroreception, by fishes, 285–289
Elephant seals, 479, 886
Elephantbirds, 603, 631, 632, 647
Elephant-nose fish, 309, 310
Elephants, 603, 730, 736, 744, 750, 759
Elephant-trunk snakes, 168
Eleutherodactylus
    coquí, 393
    jasperi, 401
Elk, 797, 884
Elliptical wings, 616–617
Elopidae, 309
Elopomorphs, 105, 294, 309, 311
Embolomeres, 363
Embryogenesis, 79–84
Embryonic diapause, 777–779
Emigration, 681
Emperor penguins, 629
Emus, 648
Emydids, 168, 419, 420, 436
Emys orbicularis, 437
Enaliornis, 645, 646
Enamel, 74, 93, 94–95, 96, 741
Enantiorniths, 646
Endangered species, 888, 891, 893–895
Endochondral ossification, 97
Endocrine system, 63, 126, 139–144
    liver and, 106
    see also Hormones; *specific glands and
        hormones*
Endoderm, 81–84
Endolymph, 139
Endoskeleton, 97
    see also Skeletal system

Endostyle, 65
Endothelial cells, 110
Endothermy, 173, 180–182, 573–574,
        821–855
    cold and, 825–836
        hibernation, 833–836
        migration and, 829–831
        torpor and, 831–838
    dinosaurs and, 488–490
    ectothermy compared to, 597–599
    energy budget and, 821–825
    first mammals and, 710
    in hot deserts, 838–853
        birds, 844–845
        large mammals, 839–844
        small mammals, 845–848, 851–853
    mammalian integument and, 730
Energy budget, 821–825
Energy use, 184–191
    aerobic metabolism, 185–187
    anaerobic metabolism, 185–187
    ectotherms and, 597–599
    energy cost of activity, 189–191
    metabolic levels, 187–189
    see also Endothermy
Environments, stressful, 574–575
    see also Endothermy; Habitat
Eocaptorhinus, 367, 368
Eocene, mammals in, 779
Eogyrinids, 363
Eogyrinus, 363, 364
Eoherpetontids, 363, 367
Eotherians, 728
Eotitanosuchids, 698, 699
Epiblast, 81
Epicontinental seas, 36, 517–518, 714,
        719, 791
Epidermis, 91, 92–93, 730, 731
Epinephrine, 141, 142, 143
Epipelagic eggs, 581
Epipelagic zone, 577
Epipharyngeal groove, 65
Epochs, 36
Eptatretus, 225
Equus, 125, 742, 743–744, 745–755, 783
    burchelli, 800
Erector pili, 733–734
Erinaceus, 743
Ermines, 732, 798
Erpetoichthys, 307
Eryops, 352, 353, 354, 355
Erythrocytes, 115, 116
Erythropoietin, 141
Erythrosuchus, 452
Eschrichtius robustus, 829–831

Esocids. *See* "Protacanthopterygii"
Esophagus, 105
    of birds, 628–629
Estivation, 575
Estradiol, 141
Estrogen, 146, 147, 772–773, 775
Estrus, 772, 773–777
Estuaries, 575
Etmopterus vierens, 250, 251
Eubalaena, 744
Eumeces, 545, 555
    laticeps, 555
Euoplocephalus, 477
Euornithopods, 474, 477–480
Euparkeria, 447, 448, 452, 459
Eurasia, 779
European bison, 753
European salamander, 391
European starling, 661
Euryapsids, 449
Eurycea, 376
    bislineata, 389
Euryhaline, fishes, 154
Euryphagous vertebrates, 247
Eurypharynx pelecanoides, 580–581
Eusmilus, 782
Eustachian tube, 138
Eusthenopteron, 346, 352, 353
Euteleosts, 294, 310
    acanthopterygians, 294, 311–312, 316
        on coral reefs, 322–325
    ostariophysans, 294, 310, 314–315
    paracanthopterygians, 294, 311, 316
    phylogenetic relationships of, 312–313
    protacanthopterygians, 294, 310–311,
        312–313, 315
    scopelomorphs, 294, 311, 315
Eutheria (placental mammals). *See*
        Mammals, placental
Evaporation, 176
Evaporative cooling, 838, 841
Evolution, 1, 12–38
    biological variation, 16–17
    of cardiovascular systems, 113–115
    chance in, 789–791
    Christian theology and, 12, 13
    climate and, 36–38
    Darwin, 13–14
    definition, 16
    Earth history and, 36–38
    of endothermy, 183–184
    evolving views of, 13–15
    genetic drift and, 23
    gradualist view of, 718
    Lamarck, 14

Evolution (*Continued*)
  macroevolution, 14–15
  microevolution, 14–15
  molecular clock hypothesis and, 31–33
  natural selection, 1, 2, 13, 14, 17
    differential reproduction and, 17
    directional, 17–18
    disruptive, 18
    fitness and, 17, 19
    genic, 18
    glaciation and, 722
    intensity of, 23
    interdemic, 19
    levels of, 18–19
    species level, 19
    stabilizing, 18
    variation and, 20–22
  neo-Darwinism, 14
  periodic phenomena and, 718–720
  Platonic views, 12, 16
  population size and, 23–24
  punctuated equilibrium and, 718
  stochastic element of, 38
  see also Birds; Classification; Variation
Evolution: The Modern Synthesis (Huxley),
    14
Evolutionary systematics. *See* Traditional
    systematics
Excreted energy, in energy budget, 822
Excretory system, 63
  see also Kidneys
Exocrine secretions, from liver, 106
Exploration, extinctions and age of, 884
Explosive radiation, 27, 30–31
External auditory meatus, 138
External ear. *See* Pinna
Extinction
  climate and, 883–884
  endangered species, 888
  of South American mammals, 788
  see also Humans; Mass extinctions
Extra embryonic membrane, 80
Extraterrestrial objects, mass extinction
    and, 515–517
Eyes. *See* Vision

Facial nerve, 129, 130
Facultative air breathers, 268
Falco, 626
  peregrinus, 22
  punctatus, 22
Falconiforms, 647, 648, 653
Falcons, 22, 624, 626, 637, 652, 653, 656
Fallopian tubes, 768, 769
False viper, 543

Families, names of, 10
Fat bodies, in amphibians, 375
Feathers, 618
  flight and, 604–608
  origin of, 184
  see also Birds, flight in
Feces, 106–107
Feeding, 104
  see also Digestive system; Food habits
Feeding guild, 29
Felids, 758, 798
Felis, 744, 790
  domesticus, 162
  silvestris, 798
Female access polyandry, 672
Female kinship bonds, in primates, 815
Female transfer systems, in primates,
    811, 814
Fence lizards, 545
Fer de lance, 543
Fibroblasts, of dermis, 93
Field mice, 598
Filoplumes, 604, 607–608
Filter feeding
  by birds, 652
  ciliary, 209
Finches, 679
  Galapagos, 24–26, 31, 879
Fin-fold theory, of paired appendages,
    236
Fingerfish, 298
Fingernails. *See* Nails
Fins
  of hybodonts, 247
  origin of, 235–237
Fire salamander, 377
Fire-bellied toad, 412
First mammals, 691, 704–710
  adaptive zone of, 709–710
  braincase of, 710
  derived characters of, 704, 705
  endothermy of, 710
  nontherian, 707, 708, 709
  teeth and, 705, 707, 708
  therian, 707–708, 709
  tritheledontids, 704–705, 706
  see also Synapsids
Fish hawk, 22, 651–652
Fishes, 6, 744
  advanced bony. *See* Teleosts
  appendicular skeleton, 101, 102
  archinephric ducts of, 148–149
  bathypelagic, 577, 578–583
  birds that eat, 650–652
  bony. *See* Osteichthyes

cartilaginous. *See* Chondrichthyes
circulation in, 109, 111, 113, 187
in cold, 593–595
deep-sea, 575–583
earliest, 207
epidermis of, 92–93
fins of, 101
girdles of, 101
intestines of, 105
ion and body fluid regulation in, 154,
    155, 156
jawed. *See* Gnathostomes
jawless (agnathans). *See* Earliest
    vertebrates
locomotion (swimming) by, 269–278
  drag and, 273–278
  gravity and, 270–273
  thrust and, 273–278
neopterygian, 30, 31
nephrotome region in, 120
nitrogen excretion by, 159, 161
osmoregulation and evolution of,
    158–159
ostracoderms. *See* Earliest vertebrates
parental care, 4
respiration in, 107
  gills, 91, 266–268, 269
rete mirabile in, 108
senses in, 278, 281–289
  displacement, 282–289
  electroreception, 285–289
  light, 278
  sight, 278, 281
  smell, 132–133
  taste, 281–282
skeletomuscular function in, 102–103,
    104
  vertebrae and ribs, 100
sound conduction transmission in,
    44
standard metabolic rate in, 170–172
stomach of, 105
temperature regulation in, 173, 174
testes in, 148–149
venoms in, 105
Fissipeds, 881
Fitness, 17
  inclusive, 19
  natural selection and, 17
  primates and, 812–813
Flamingos, 168, 619, 626, 627, 652
Flapping flight, 611–616
Flatfish, 266
Fleet-footed lizards, 809
Flehmen response, 134–135

Horses, 125, 744–747, 748–749, 753, 755, 778, 781, 884
Horsetails, 335, 510
House swallows, 674
Hovasaurus, 490
Howler monkeys, 810–811, 812, 814
Human blood plasma, 115, 736
Human chorionic gonadotropin (HCG), 774
Humans, 736, 743, 813, 862, 863
  brain of, 761
  extinction of vertebrates, 879–895
    actions to save, 893–895
    agriculture, land development, 888–890
    commercial exploitation of vertebrates, 884–888
    endangered species, 888, 891, 893–895
    exploration and discovery, 884
    human concern for, 891–892
    human settlement of oceanic islands, 884
    Pleistocene mammals, 880–884
  origin, 865, 868–879
    see also Hominoids; Primates
  population, 880, 895
  reproduction in, 768, 769, 776
    male, 774
    menstruation, 772, 773, 774, 776
  technology and culture origins, 879
  as typical vertebrate, 58–60
Hummingbirds, 610, 615–616, 620, 628, 657–658, 666, 831, 832, 833, 837–838
Hyaena hyaena, 798
Hyaenas, 758, 798
Hybodus, 245–248
Hydrogen ion, in body fluids, 115
Hydrolagus colliei, 261
Hydromantes, 380
Hydrophiids, 168, 526, 533, 542, 557
Hyla
  crucifer, 596
  versicolor, 396–398, 596
Hylids, 379, 383, 384, 385, 402, 407, 592
Hylobates, 813
Hylobatids, 813, 863, 872
Hylonomus lyelli, 368
Hymenochirus, 378
Hynobiids, 378, 388
Hyohippus, 783
Hyoid arch, jaw hinge and, 99
Hyomandibular, 99
Hyostylic jaw suspension, 248

Hyperdactyly, 493
Hyperodapedon, 450
Hyperoliids, 379
Hyperosmotic, 154
Hyperphalangy, 491
Hypoblast, 81
Hypocercal, 212
Hypodermis, 92, 95–96, 730, 731
Hypoglossal nerve, 129, 130
Hyposmotic, 154
Hypothalamus, 126, 140, 142, 143, 180
  lactation and, 775, 776
  menstruation and, 772, 773
  pituitary axis and, 774
Hypotremate elasmobranchs, 240, 260
  see also Rays; Skates
Hypselosaurus piscus, 486
Hypsilophodontids, 477
Hypsodont dentition, 754–755
Hypsodont tooth, 748–749
Hyracoids, 729, 753

Icarosaurus, 469
Ice ages, 716–722
Icehouse-greenhouse cycles, mass extinction and, 518
Ichneumia albicauda, 798
Ichthyophiids, 378
Ichthyornis, 168, 646
Ichthyornithiforms, 646
Ichthyosaurs, 493, 494, 634
Ichthyostega, 344–345, 346, 366, 367, 458
Ichthyostegids, 334, 344–345, 347, 349, 351–354
Iguana, 366, 477
Iguanas, 187, 488, 528
  Galapagos marine, 178, 179, 481, 524, 527, 528
Iguanids, 524, 527, 528, 544, 545, 548, 549, 550–551, 555, 557
Iguanodon, 477, 478–479
Illinoian glaciation, 717, 721
Immune systems, 111, 117–118
Immunoglobulins, 117
Impalas, 804, 805, 810
Incisors, 741, 748
Inclusive fitness, 19
  nest helpers and, 680
Incubation, by birds, 668, 670–672, 675–677
Incus, 43, 45, 137, 138
Indeterminate cleavage, 80
Index of refraction, 278
India, 713, 715
Indian jackals, 798

Indian Ocean, settlement of islands of, 884
Indigo buntings, 662–663, 684
Individual variation, 20
Indri, 812, 814
Induced drag, 610
Inertial drag, 275
Infrared, energy exchange in the, 176
Infrasound, birds detecting, 641
Ingested energy, in energy budget, 822
Inguinal amplexus, 398
Iniopterygia, 262
Inner ear, 137, 139
Insectivores, 728, 729, 741, 743, 759, 786
  birds as, 655–657
  diapause in, 777–778
  torpor in, 832
Insects
  amniotes and, 365, 367
  in Carboniferous, 510, 514
  in Devonian, 336
  radiation of, 510–511
Insolation, glaciation and, 716–717, 720
Insulin, 141
Integration, 124
  see also Endocrine system; Nervous system; Sense organs
Integumentary system, 62, 91–97
  coloration, 94–95
  dermis, 91–92, 93–95
  embryonic origins of, 90
  epidermis, 91, 92–93
  hypodermis, 92, 95–96
  mineralization, 93, 94–95, 96
  see also Mammals
Interatrial system, in frog heart, 114
Intercalated disks, 111
Interdemic selection, 19
Intermediate mesoderm, 83
Intermuscular ribs, 100
Internal endothelium, 110
Interstitial cells, 148
Interstitial fluid, 115–116
Intervertebral disks, 100
Intestines, 105
  of birds, 630
  see also Digestive system
Intramembranous ossification, 96, 97
Intromittent organs, 148–149
Invasion. See Emigration
Involuntary muscles, 101
Ion exchange. See Water and ion exchange
Iris of eye, 132, 133
Irruption. See Emigration

Mousebirds, 604, 832
Mouth brooders, 29
Movement, 91
    see also Integumentary system; Muscular system; Skeletal system
Moythomasia, 295, 297, 301
Mucus, feeding mechanisms and, 65
Mud turtles, 421, 425
Mudpuppy, 376, 377
Mudskippers, 349
Mugger crocodile, 483
Mullerian ducts, 149
Mullets, 284, 312
Multituberculates, 707–708, 709, 728, 784, 786
Mungos mungos, 798
Muntjac, 738, 740
Murid rodents, 45
Muscicapid flycatchers, 656
Muscular system, 62, 101–104
    aerobic and anaerobic metabolism and, 185
    of birds, 618, 620–621
    red, 185
    voluntary muscles, 91
    white, 185
Musk turtles, 419, 431
Muskellunges, 311
Mussaurus, 486
Mussels, 317
Mustela, 749
    erminea, 798
    rixosa, 598
Mustelids, 798
Mutation, 17
Mute swans, 603, 610
Myctophids, 579
Myctophiform fishes. *See* Lanternfish
Myelencephalon, 126
Myelin, 124
Myelin sheath, 124
Myobatrachids, 379
Myocommas, 61
Myoglobin, aerobic and anaerobic metabolism and, 185
Myosepta, 100
Myotis lucifugus, 762–764
Myotome, 61, 63, 84
Myriapods, 335
Myrmecobius, 790
Myrmecophaga, 743, 790
Mysticetes, 747, 765
Myxinoids, 6, 73, 75, 158, 223

Nails, 736–737

Nandinia binotata, 798
Narcine, 286
Nasal countercurrent exchange, water recovery and, 846–847
Nasal placode, embryonic, 84
Nasalis, 813
Natal down, 606
National parks, importance in conservation, 894–895
Native cat (*Dasyrus*), 790
Natural selection. *See* Evolution
Natural system, of classification, 46
Navigation
    by birds, 684–686
    by turtles, 434–436
Neandertals, 874, 875, 878–879
Nebraskan glaciation, 717, 721
Nectar, birds feeding on, 657, 658
    bats feeding on, 742, 743
Nectophrynoides, 379
Nectrideans, 349, 351, 358–359
Necturus, 376, 377
Negaprion brevirostris, 250
Negative pressure mechanism, ventilation as, 107
Neobatrachus wilsmorei, 409
Neoceratodus forsteri, 325, 326
Neo-Darwinism, 1, 14
Neognathous birds, 648
Neopallium, 58, 60, 126–128, 758
Neopterygians, 5, 155, 158, 288, 294, 297, 299, 300, 301, 302, 303, 305, 308, 309
    jaw, 297
Neoselachii, 248
Neoteny, 68
Neotoma albigula, 162
Nephric ridge, embryonic, 84
Nephrons, 119, 120, 121, 122, 123, 161, 162–163, 166, 167
    see also Kidneys
Nephrotome, 84, 120
Nerves. *See* Neurons
Nervous system, 62, 124–130
    autonomic, 124
    brain, 124–128
    central, 124
    cranial nerves, 128–130
    embryonic origins, 90
    mammalian, 758, 760–761
    neural crest and, 87, 88, 90
    peripheral, 124
    spinal cord, 124–125
    spinal nerves, 128–130
Nervus terminalis, 129, 130

Nest helpers, birds having, 679–680
Nesting
    by birds, 673–675
    by crocodilians, 482–484
    by turtles, 436–438
Net primary production, for ecosystems, 574–575
Neural arch, 100
Neural crest, 71
Neural crest cells
    in dermis, 93–94
    in vertebrate development, 86–89, 90
Neural ectoderm, 81
Neural folds, 82
Neural spine, 100
Neural tube, 82, 83, 89
Neurobiology, 139, 142
Neurocranium, 98, 343
Neurohumors, 142
Neuromast organs, 256, 282–284
Neurons, 93, 124
    see also Nervous system
Neurotransmitters, 142
Neurulation, 82
Nevadan orogeny, 509
New Guinea river turtle, 425
New Synthesis of evolution, 1
New World meadowlarks, 656
New World monkeys, 59, 787, 812, 815, 862
Niche, 56
Nictitating membrane, 257
Nidimental glands, 258
Nighthawks, 625
Nightjars, 625–626, 656, 832
Nitrogen excretion. *See* Water and ion exchange
Nocturnal, fishes, 324
Nomadism, 681
Nomenclature, 10–11
Nonamniotes, epidermis of, 92
Nonamniotic tetrapods. *See* Tetrapods, origin of
Nonfemale transfer systems, in primates, 811, 814
Nontherian mammals, 707, 708
Norepinephrine, 141, 143
Northern fur seals, 828–829
Northern gannets, 669, 674
Northern grouse, 632
Northern orioles, 674
Nose. *See* Olfaction
Notaden nicholsi, 409
Notharctus, 862
Nothrotherium, 787

Notochord, 55–56, 82, 97, 99–100
  of amphioxus, 61, 64
  embryonic, 84, 89
Notomys alexis, 162
Notophthalmus viridescens, 389, 390
Notoungulates, 785, 787, 788
Nurse shark, 260
Nutcrackers, 681
Nuthatches, 627, 634
Nycticebus, 812
Nyctimystes, 402, 403

Oak toads, 392
Obligate air breathers, 268
Occipital condyles, 868
Occlusion surfaces, 741
Oceanodroma furcata, 675–676
Oceans
  climate and, 36–37
  ectotherms in, 575–583
Ocelot, 790
Odobenus, 747
Odontoblasts, 94
Odontocetes, 747, 765
Old Red Continent. See Laurussia
Old-squaw, 651
Old World badgers, 798
Old World minnows, 298
Old World monkeys, 59, 813, 815, 862–863, 864, 865
Old World starlings, 656
Old World warblers, 655
Olfaction, 127, 132–136
  in birds, 618–619, 639
  in mammals, 759
Olfactory nerves, 129, 136
Oligocene, continental positions in, 713, 716
Oligokyphus, 698, 700, 703, 704
Oligolecithal, eggs and zygotes as, 80
Oligopithecus, 864
Olm, 376, 377
Omasum, 746
Omnivores, home range of, 797
Open circulatory system, 108
Operational sex ratio, 668
Operculum-plectrum complex, 374
Ophiacodon, 692, 693
  major, 692
Ophiacodontids, 692, 693
Ophiderpeton, 359
Ophiomorus, 535, 538
Ophisaurus, 528–529
  apodus, 525
Opisthoglyph snakes, 542, 543

Opisthonephros, 118, 119
Opisthotics, 351
Opposability, by human hand, 60
Opossums, 60, 128, 395, 768, 788–789
Opsanus, 269
Opthalmosaurus, 494
Optic cup, embryonic, 84
Optic nerve, 132, 134
Optimal foraging theory, 547, 548
  birds and, 644, 659–662
  lizards and, 547–548
Orangutans, 811, 813, 814, 863, 864, 869, 872
Orb-web spiders, 598
Orders, names of, 10
Ordovician, 74
  land masses during, 195, 198
  plants in, 199–200
  vertebrates of, 199, 205–209
  see also Late Ordovician
Oreopithecus, 865, 867
Orestias, 31
Organ of Corti, 138
Organ systems, 57, 62–63, 79–150
  embryogenesis, 79–84
  organogenesis, 79, 85
  pharyngula, 87–88, 91
  wandering cell populations, 85–87, 88–89, 90
  see also Cardiovascular system; Digestive system; Endocrine system; Integumentary system; Kidneys; Muscular system; Nervous system; Reproductive system; Respiratory system; Sense organs; Skeletal system
Organogenesis, 79, 85
Organs, 85
Orientation, navigation by birds and, 684–686
Origin of Species by Means of Natural Selection, The (Darwin), 13
Origin of vertebrates, 55–77
  bone, 69, 73–74
  characteristics, 55, 56
  chordates and, 55, 61
  environment, 74–76
  freshwater, 74–75
  invertebrates and, 68–69, 70–73
    calcichordates, 70, 72–73
    chordates, 69, 70, 71–72
    deuterostomes, 68–69, 70, 71, 72
    lophophorates, 69
    paedomorphosis, 70, 71
  marine, 75

ostracoderm armor and, 74, 75–76
  relatives, 57, 61, 63–68
    amphioxus, 61, 63–65
    tunicate larvae, 65–68
  similarity and differences and, 55–57
  vertebrate body plan, 57, 61
  see also Organ systems
Ornithischian dinosaurs. See Dinosaurs
Ornithodelpians, 728
Ornitholestes, 472
Ornithomimids, 464, 465, 486
Ornithominosaurs, 466
Ornithomimus, 465, 466
Ornithopods. See Euornithopods
Ornithosuchus, 451
Ornithurines, 645, 646
Orogeny, 507
  see also Mountain building
Oropendolas, 670
Oryx, 843
  beisa, 843
Os cornu, 738, 739
Oscine birds, 624
Osmosis, 75, 154, 155
Ospreys, 22, 651–652, 653
Ossification, 95, 96–97
Ostariophysans, 134, 294, 310, 314–315
Osteichthyes, 265, 293–296
  acanthodians as, 290–291
  advanced. *See* Teleosts
  buoyancy of, 271
  ceca in, 106
  ears of, 319
  eggs of, 79–80
  fins of, 101
  gills of, 266, 268
  intestines of, 105
  musculature in, 103
  primitive, 5–7, 295, *see also* Actinopterygians; Sarcopterygians
  respiration in, 107
  testes of, 148
Osteocyte, 95
Osteoderms, 7, 477
Osteoglossomorphs, 294, 309, 310
Osteoglossum, 309, 310
Osteolepiforms, 7
  early tetrapods and, 341–345, 346, 347, 348–349, 352, 353, 354
  jaws of, 366
Osteolepis, 295, 346
Osteostracans, 216, 220–221
Ostium, 149
Ostraciiform swimming, 269, 270, 276
Ostracoderms. *See* Earliest vertebrates

Ostrich, 168, 603, 631, 632, 647, 648, 649
Otocyon megalotis, 798
Otoliths, 319
Ouranosaurus, 478–479, 490
Outgroup, 40
Ouzels, 635
Ova. *See* Eggs
Ovale, 273
Ovaries, 141, 144, 145, 147–148
Oviducts, 149
Oviparous species, 149, 258, 768
   actinopterygians as, 312
   birds as, 673
Ovoviviparous species, 258
Ovulation, 148, 772
Owl monkey, 812
Owl parrots, 658
Owls, 620, 626, 637, 638, 640–641, 653, 656, 666
Oxytocin, 140, 144, 775, 776
Ozone layer, destruction of, 891

Pachycephalosaurs, 474, 478–479, 480–481, 488
Pachycephalosaurus, 478–479
Pachyrachis, 535, 538
Pacific islands, human settlement of, 884
Pack rats, 162
Paddlefish, 306, 308–309
Paedogenesis, 68
Paedomorphosis, 66–67, 68, 70, 71
   in salamanders, 375, 391–392
Painted turtles, 432, 433, 438
Paired appendages, origin of, 235–237
Palaeagama, 495, 496–497
Palaeognathious birds, 648
Palaeogyrinus, 366
Palatoquadrate, 99
Palearctic, bird species in, 681
Paleomagnetism, Pangaea and, 197–198
Paleoniscoids, 158, 294, 296–299
Paleopallium, 126, 127–128
Paleostegalians, 351, 359–360
Paleothyris, 366
Paleozoic
   fishes in. *See* Earliest vertebrates
   ostracoderms in, 211
   Pangaea and, 197
   vertebrates of, 221, 222
   see also Early Paleozoic; Late Paleozoic
Paliguana, 495, 496–497
Pampas, 788
Pampatherium, 786–787
Pan, 813
Pan troglodytes, 816

Panamanian anoles, 598
Panamanian frogs, 400
Pancake tortoises, 419, 421, 425
Pancreas, 106, 142
Pancreatic islets, 141
Pandion haliaetus, 22
Pangaea, 36, 195, 197, 200, 247, 507–509, 510, 511, 512, 713, 780
Pangolin, 780, 781
Panids, 813
Panthera
   leo, 798, 800
   tigris, 798
Pantodonts, 784
Pantotherians, 728, 784
Papilla amphibiorum, 374–375
Papio, 813
   cynocephalus, 813, 814, 815
   hamadryas, 813, 816
   leucophaeus, 813
   sphinx, 813
Paracanthopterygians, 294, 311, 316
Parachordals, 98
Parahesperornis, 646, 647
Parakeets, 888
Parallelisms, 45
   in Cenozoic mammals, 779–780
Paranthropus, 875
Paraphyletic groups, 46, 50
Parasaurolophus, 479, 480
Parasympathetic division, 124
Parathormone (PTH), 140
Parathyroid glands, 140
Paratype, 12
Paraxial mesoderm, 83
   embryonic, 89
Pareiasaurs, 368, 369, 370
Pareiasaurus karpinskyi, 368, 369
Parental care, 4
   by anurans, 400–401
   by birds, 482, 484, 677–680
   by crocodilians, 482–484
   by mammals, 802–803
   by ornithischians, 488
   by primates, 813, 815
   by squamates, 549, 558–560
Parietal peritoneum, embryonic, 84
Parrotfish, 274, 323, 324
Parrots, 624, 632
Parthenogenetic species, of squamates, 558, 559
Parturition, 775
Parulid warblers, 34
Parus
   atricapillus, 836–837

major, 663
Passenger pigeons, 887–888, 889
Passerculus sandwichensis, 162, 598
Passeriforms, 649
Patagonian hares, 787
Peacocks, 665
Peccary, 742, 743
Pecten, 637–638
Pectoralis major, 614, 615
   of birds, 618
Pedetes, 45
Pedicel, 121
Pedicellate teeth, 374
Pelagic spawning, 312–313, 315–317
Pelagic zone, 576
   fishes in, 577, 578
Pelamis platurus, 169, 526
Pelecaniforms, 648
Pelecanus erythrorhynchos, 162
Pelican eels, 580–581
Pelicans, 162, 168, 624, 626
Pelobatids, 379, 591–593
Pelodytids, 379
Pelomedusids, 420, 436
Peltocystis, 70
Pelusios, 421, 425
Pelvic patch, 409
Pelycosaurs, 368, 369, 691–696, 702
   carnivorous (sphenacodonts), 692–695, 697
   herbivorous, 692, 693, 694, 695–696
      caseids, 692, 693, 694, 695, 696, 697
      edaphosaurids, 692, 693, 694, 695, 696, 697
Penguins, 168, 604, 619, 634, 635, 636, 647, 673, 675
Pentaceratops, 481
Peptides, 142
Perch, 103, 278, 312
Perching, by birds, 631–632
Perciformes, 294, 312
Peregrine falcons, 22, 610, 619, 653
Perilymph, 139
Periods, geologic, 36
Periosteum, 95
Peripheral nervous system, 124
Perissodactyls, 729, 734, 744–747, 753, 754–755, 756, 780, 781, 783, 784, 787
Peristalsis, 105
Permian
   floral regions in, 510
   neopterygians in, 299, 300
   tetrapods in, 367–370
Permitfish, 298

Perodicticus, 812
Perognathus
  californicus, 852–853
  longimembris, 846
Peromyscus polionotus, 598
Petaurus, 790
Petrels, 617, 636, 639, 650, 652, 673
Petrolacosaurus, 448, 449, 490, 496–497
  kansensis, 368
Petromyzon marinus, 228
Petromyzontids, 6, 223
Phagocytosis, 117
Phalaropes, 671, 672
Phalaropus fulicarius, 672
Phanerozoic, 36
  extinctions in, 515, 518
Pharyngeal arches, 88, 90
Pharyngeal (branchiomeric) muscles,
  91
Pharyngeal clefts, embryonic, 55, 84
Pharyngeal grooves, 88
Pharyngeal jaws, 303
Pharyngeal pouches, 87–89
Pharyngeal skeleton, 97
Pharyngeal slits, 55
Pharyngolepis, 222
Pharyngotremy, 69
Pharyngula, 87–88, 91
Pharynx, embryonic, 84, 89
Phascolomys, 790
Pheasant, 617
Phenacodus, 785
Phenetic methods, of classification, 45
Phenotypes, 16, 17
Phenotypic variation, 16–17
Pheromones, 134–135
  in lizards, 555
  in red-back salamanders, 381
  in snakes, 555
  in squamates, 548, 549
  in turtles, 432
Philander opossum, 395
Phoenicopteriforms, 648
Pholidotans, 728, 781
Phomeryele, 261
Phoronid worms, 65
Phororhachids, 653, 654, 786
Photophores, 311
Phrynosoma, 528–529
Phthinosuchus, 698
Phyletic gradualism, human evolution
  and, 876–877
Phyllobates, 374, 411
  terribilis, 411
Phyllomedusa sauvagei, 170, 407, 592

Phylogenetic interpretation, of classifica-
  tion, 42–45
Phylogenetic relationships, 38–39
Phylogenetic systematics. *See* Cladistics
Phylogeny of Vertebrates, The (Lovtrup),
  72
Physalaemus, 399
  pustulosus, 394–395, 399
Physoclistic fishes, 272
Physostomus fishes, 271–272
Phytoplankton, in Ordovician, 199
Phytosaurs, 452–455
Piciforms, 647, 649
Pickerel, 311
Pigeons, 618, 628, 638, 641, 685, 686,
  845, 887–888, 889
Pigs, 730, 736
Pike, 311, 315
Pineal organ, 126, 141, 212
Pink pigeon, 888
Pinna, 137
Pinnipeds, 670, 750, 828, 881
Pipa, 378, 384, 399, 400–401
Pipids, 378, 383, 384, 402
Pipits, 631
Pipping, 677
Pit vipers, 526, 542
Pithecanthropus erectus. See Homo erectus
Pithecia, 812
Pituitary axis, 142, 143, 772, 773, 774,
  775, 776
Pituitary gland, 126, 140, 142, 143
Pituophis, 542
Placenta, 141, 768–771, 774–775
Placental mammals. *See* Mammals
Placentotrophic viviparity, 259
Placochelys, 491
Placoderms, 235–236, 237–239
  cladistics and, 240–241
  intestines of, 105
  ptyctodontids, 261–262
Placodes, in vertebrate development, 87,
  90
Placodonts, 490–491
Placodus, 491
Placoid scales, 253
Plagiaulacoid premolars, 707
Plagiosaurids, 356
Plains garter snakes, 555
Plankton, 577
Plantigrade posture, 756–757
Plants
  in Devonian, 199–200, 334–336
  in Ordovician, 199–200
Plasma, 115, 736

Plastron, 419
Platanistids, 765
Platelets, 116
Plateosaurus, 460, 461, 462–463
Platocarpus, 497
Platypus. *See* Duck-billed platypus
Platyrrhines. *See* New World monkeys
Platysternids, 420
Platysternon megacephalum, 431
Pleistocene, 4
  extinctions in, 880–884
  glaciation in, 26, 29, 34, 35, 714–722
  Homo sapiens origination in, 714
Plesiadapis, 862
Plesiosaurs, 491–493
Plesiosaurus, 492
Plethodon, 377
  cinereus, 380–382, 391, 410, 412–414,
  598
Plethodontid salamanders, 376–382,
  389–392, 591
Pleural cavity, 107
Pleurodire turtles, 420, 422, 423
Pleurodont teeth, 523
Pleurotremates, 240, 259
  see also Sharks
Pliopithecus, 864–865, 867
Plovers, 168
Pneumatic bones, of birds, 619
Pneumatic duct, 271, 272
Pocket mice, 846
Podicipediforms, 648
Podocyte, 121
Poecilia, 316
Poikilotherm, 170, 173
Poison-dart frogs, 400, 404, 410
Polar bears, 798, 825, 827
Polar (cold) conditions. *See* Endothermy
Polarity of characters, 40
Pollen, birds feeding on, 657, 658
Polyandry, 668
  in birds, 668, 670–672
Polycotylus, 492
Polyestrous species, 772
Polygamy, 668
Polygyny, 668
  in birds, 668, 670
Polygyny threshold model, 670
Polymorphism, 20–22
Polyodon spathula, 308
Polyodontids, 308
Polypeptides, freezing depression and,
  595
Polyplocodus, 346
Polypterids, 155, 294, 296, 305–307, 308

Swimming *(Continued)*
energy costs of, 190–191
see also Fishes
Swine, 750
Swordfish, 266, 270, 277
Swordtails, 312
Symmetrodonts, 707, 709, 718, 784
Sympathetic motor division, 124
Symplesiomorphies, 49
Synapomorphies, 46, 49
Synapse, 142
Synapsids, 450, 691, 709
phylogeny of, 697, 706–707
skull of, 691, 692
sound transmission in, 44
see also Pelycosaurs; Therapsids
Syncerus caffer, 803, 804
Synsacrum, 620
Syrinx, 107–108, 624
Systema Naturae (Linnaeus), 41
Systematic biology, 40
Systematics, 39, 40
evolutionary. *See* Traditional systematics
phylogenetic. *See* Cladistics
Systemic arch, in frog circulation, 114
Systemic circuit, 111
Systole, 110

T lymphocytes, 117–118
Tachycinta bicolor, 678
Tadpoles, 7, 383, 400, 401–404
Tailwalk, 391
Talonid process, 707
Tamarins, 812, 814
Tamiasciurius hudsonicus, 598
Tanagers, 676
Tanystropheus, 451
Tapetum lucidum, 256
Tapinocephalids, 698
Tapirs, 730, 744
Tarbosaurus, 465
Tarpon, 297, 309, 311
Tarrasius, 298
Tarsiers, 59, 812, 814, 860–861
Tarsius, 812
Tarsoids, 812
Tasmanian native hen, 672
Tasmanian wolf, 780, 782, 789, 790
Taste, 130, 131
in mammals, 759
Taste buds, 130, 131
Taurotragus oryx, 803, 843
Tawny owls, 679
Taxidea taxus, 798

Taxon (taxa), 10, 41
Taxonomic levels, 41
Taxonomy, 40
see also Classification
Tayassu, 743
Teats, 734, 768
Tectum, 126
Teeth
in hominoids, 863–864, 865, 870
in humans, 865, 868, 871, 872
mammalian, 741–744, 747–750
evolution and, 705, 707, 708
mineralized tissue in, 95
pedicellate, 374
in primates, 862–863, 865, 868, 871–872
Teiids, 524, 527, 544, 545, 556, 557, 558, 565
Telencephalon, 126–127, 128
Teleology, 57
Teleonomy, 57
Teleostomes, 289, 291
gills of, 266, 267
phylogenetic relationships of, 290–291
swim bladder of, 271
see also Acanthodians; Osteichthyes
Teleosts, 6–7, 289, 294, 298–301, 309–312
body fluid concentrations in, 154
brain of, 125, 128
cichlids, 7, 27–29
cladistics and, 240–241
cleavage and gastrulation in, 81
clupeomorphs, 294, 310
eggs of, 80
elopomorphs, 294, 309, 311
ion concentration in, 155
jaws of, 302, 303–304
musculature in, 103
nephron structure in, 123
osmotic regulation in, 154–158
osteoglossomorphs, 294, 309, 310
relationships among major groups of, 312–313
testes in, 148
see also Euteleosts
Telicomys, 787
Telmatodytes palustris, 598
Telolecithal eggs, 80
Temminck's stint, 671
Temnospondyls, 341, 349, 351, 352, 354–357, 367, 370
Temperate forest, 575
Temperature measures, 849, 850
see also Climate
Temperature, responses to, 170–184
in fishes, 170–172

high temperature advantages, 182–183
standard metabolic rate, 170–171
see also Ectothermy; Endothermy; Squamates
Temperature-dependent sex determination, 320–321, 436–438
Temporalis, 752
Tendons, 104
Tenontosaurus, 486
Tenrecs, 710
Terminalis nerve, 129, 130
Termites, 780, 781
Terns, 638, 650, 673
Terrapene, 421, 425
Territoriality, in lizards, 550–556
see also Competition
Territory, 796
Terror cranes, 653, 654
Tersomius, 354
Tertiary, mammals in, 779, 780, 784–789
Testes, 141, 148
formation of, 144, 145
Testosterone, 141, 146, 147, 774
Testudinids, 419, 420, 436
Testudo, 421, 425
Tethys Sea, 333, 334, 507, 512
Tethys Trench, 715
Tetonius, 862
Tetrapods, origin of, 333–336, 341–371
amniotic, 341
Carboniferous, 367–370
egg of, 363–365
insects and, 365, 367
jaws and, 366, 367
Permian, 367–370
evolution of, 345, 347–349
on land, 348–349
in water, 347–348
nonamniotic, 34, 349–363, *see also* Amphibians; Anurans; Caecilians; Salamanders
aïstopods, 349, 351, 357–358, 359
anthracosaurs, 341, 349, 351, 352, 360–363, 367, 370
diadectomorphs, 341, 350, 351, 360–363, 364
ichthyostegids, 347, 349, 351–354, 366, 367
loxommatoids, 351
microsaurs, 349, 351, 357, 358
nectridia, 349, 351, 358–359
paleostegalians, 351, 359–360
temnospondyls, 341, 349, 351, 352, 354–357, 367, 370

osteolepiform fishes and, 341–345, 346, 347, 348–349, 351, 352, 353, 354

  phylogenetic relationships of, 350–351

  sarcopterygians and, 341, 344, 345

Texas blind salamanders, 375, 376

Thalarctos maritimus, 798

Thallatosuchians, 453, 454, 455

Thamnophis

  elegans vagrans, 561–562

  marcianus, 555

  radix, 555

  sirtalis, 555

    parietalis, 555

Thecodontians, 452

Thelodonti, 215

Therapsids, 691, 696–704

  carnivorous, 696, 697, 699

  cynodonts, 701–704

  dinocephalians, 696–699

  herbivorous, 696, 697, 698, 699

  skulls of, 698

Therian mammals, 707–708, 709, 728, 743, 784

Thermoneutral zone, 181

Thermoregulation. *See* Temperature, responses to

Theropithecus, 813

  gelada, 813, 816

Theropods, 464–467, 485, 486–488

  see also Birds

Therosaurs, cynodont, 701–704

Thescelosaurus, 451, 459

Thinnfeldia, 511

Third law of motion, fish locomotion, 273

Thoatherium, 781, 783

Thomson's gazelles, 800, 801, 805, 843

Thoracic vertebrae, 100

Thrinaxodon, 700, 702

Thrombocytes, 116

Thrushes, 664

Thrust, fish locomotion, 273–278

Thylacinus cynocephalus, 782, 790

Thylacoleo, 791

Thylacosmilus, 786

  atrox, 782

Thymus, 116, 118, 141

Thyreophora, 474, 476–477

Thyroid gland, 140, 142

Thyroid stimulating hormone (TSH), anuran metamorphosis and, 405

Thyrotropin, 140, 143

Thyrotropin releasing hormone, 140

Thyroxine, anuran metamorphosis and, 405

Thysanurans, in Devonian, 336

Tiger salamanders, 377, 591

Tiger shark, 253

Tigers, 753, 797, 798, 889

Timber wolves, 798, 889

Tinamiforms, 648

Tinamous, 649

Tissues, formation of, 85

Titanophoneus, 698

Titanosaurus, 512

Titis, 812, 814

Toadfish, 266, 269

Toads, 382, 386, 404, 593

  see also Anurans

Tomial tooth, 626

Tongues, of birds, 627–628

Tools, humans and, 879

Toothed whales, 765

Tornaria, 69

Torpedinids, 260

Torpedo, 286, 287

Torpedo rays, 260, 285, 286, 287

Torpor

  in cold, 831–838

  in heat, 852–853

Tortoises, 421, 425

  conservation of, 439–441

  courtship in, 432–433

  desert, 583–587, 589

  temperature regulation in, 431

  see also Turtles

Totipotent cells, 80

Touch, 130

  in mammals, 759

Toxodon, 785, 787

Trabeculae (embryonic cartilages), 88, 98

Trachinotus, 298

Trachops cirrhosus, 395

Trachydosaurus rugosus, 162

Traditional classification, cladistics compared with, 8–9

Traditional systematics, 4–5, 8–9, 45–47, 49

Trans-Bering Bridge, 713, 714, 716

Trap lining behavior, foraging, 796

Traversodontids, 703, 709

Tree shrews, 59, 860, 861

Tree swallows, 678

Treefrogs, 383, 400, 407, 592

  see also Anurans

Tremataspis, 220

Trematomus borchgrevinki, 595

Trematosaurs, 356, 452

Triadobatrachus, 373, 376

Triassic

  climates in, 198, 200, 511

  continental drift in, 507–509

  cynodonts of, 701–704

  faunal affinities of land tetrapods in, 511

  turtles in, 419

  see also Late Triassic

Tribonyx mortierii, 672

Tribosphenic pattern, 707, 708

Triconodon, 128

Triconodonts, 705, 707, 709, 728, 784

Trigeminal nerve, 129, 130

Tringa totanus, 660–661

Trionychids, 419–420, 436

Trionyx, 421

Tritheledontids, 701, 703, 704–705, 706

Trituberculates, 728

Triturus, 375

Tritylodonts, 698, 703, 704

Trochlear process, 423

Troglodyty, 376

Trogoniforms, 647, 649

Trogonophids, 525

Trogons, 647

Tropical forests, 574, 575, 890–891, 892

  glaciation and, 722

Tropidophiids, 525

Tropins, 142

Trout, 270, 276, 278

True shrikes, 652

True toads, 384

True vipers, 542

Trumpeter swan, 603

Trunkfish, 276

Tseajaia, 362

Tseajaiids, 367

Tuatara, 7, 167–170, 447, 449, 493, 495, 496–497, 523, 524, 526–527. *See also Sphenodon punctatus*

Tube-nosed seabirds, 645, 675

Tubercular cusp pattern, 747–748

Tuberous organs, of electric fish, 288

Tubulidentates, 729, 781

Tuditanomorphs, 357

Tuna, 108, 173, 174, 183, 185, 266, 270–271, 273, 276, 277, 278, 312, 579

Túngara frogs, 394–395, 404

Tunicate larvae, 65–68

Tunicates, 65, 317

Tupaia glis, 860, 861

Tupaiids, 860

Turkey vultures, 639, 832

Viperids, 526, 533, 542–544, 557
Vipers, 533, 534, 535, 543
Virginia opossum, 702, 708, 788, 789
Visceral arches, 88, 90
Visceral nerves, 128–129
Visceral peritoneum, embryonic, 84
Visceral skeleton, 97
Viscous drag, 275
Vision, 130–132, 133, 134, 135
  of bathypelagic fish, 578, 581
  in birds, 618, 636–638, 662
  embryonic eyes, 88
  in mammals, 709–710, 758–759
Visual displays, by birds, 663, 665–666, 667
Viverrids, 798
Viviparity, 149, 259, 673
  mammalian reproduction and. *See*
    Mammals
Vocalizations, 107–108
  by anurans, 392–398
  by birds, 662–663, 664, 665, 666
  by crocodilians, 484
  by seals, 479
Volcanism
  mass extinction and, 518
  periodicity of, 719, 720
Voluntary muscles, 91, 101–102
Vomeronasal organ, 135
Vulpes vulpes, 798
Vultures, 168, 611, 624, 653

Walking, by birds, 632–634
Walking catfish, 349
Wallabies, 789
Walrus, 670, 742, 743, 744, 747, 750
Wandering cells, 85–87, 88–89, 90
Wandering garter snake, 561–562
Warblers, 34, 625, 626
Wart snakes, 532
Water
  doubly labeled, 583, 584–585
  evaporation of, 176
  free, 587
  land animal evolution in, 347–348
  life in, 265–266, *see also* Amphibians;
    Fishes
  metabolic, 587

Water and ion exchange, 154–170
  nitrogen excretion and, 159–170
    by diapsids and turtles, 167–170
    by fishes and amphibians, 159, 161
    by mammals, 161–167
    by uricotelic frogs, 169–170
  regulation of, 154–159
    by amphibians, 158
    by elasmobranchs, 157–158
    evolution of fish and, 158–159
    by freshwater organisms, 154–157
    by marine organisms, 157
    by teleosts, 157
Weasel, 749–750
Webbing, in birds, 634, 635
Weberian apparatus, 310, 314–315
Western grebes, 651
Whale birds, 611
Whale shark, 5, 251–252, 260
Whalebone whales, 747
Whales, 736, 742, 743, 744, 747, 765
  hunting for, 886–887
  migration by, 829–831
Wheel organ, 65
Whipsnakes, 533, 534, 541
Whiptail lizards, 545
White blood cells. *See* Leucocytes
White muscle, 185
White rats, 162
White-backed vultures, 655
White-headed vulture, 655
White-tail deer, 809
White-tailed gnu, 889
White-tail mongoose, 798
Wildcats, 889
Wildebeest, 800, 801, 805, 806, 807–808,
    810
Wing loading, 610
Wings. *See* Birds, flight in
Wingtip vortex, 610
Wisconsinian glaciation, 717, 721, 722,
    882
Wolffian ducts, 148–149
Wolves, 758, 780, 781, 789, 790
Wombats, 12, 38, 789, 790
Wood drillers, bird foraging, 656
Wood frogs, 595, 596

Wood warblers, 34
Woodchucks, 833
Woodcreepers, 634
Woodpeckers, 627–628, 632, 634, 647,
    656
Woodrats, 742, 743
Wrasses, 274, 304, 305, 322, 323, 324
Würm glaciation, 717

X chromosome, 146–147
Xantusiids, 524, 527, 544, 557, 558
Xenacanthus, 242–243, 244–245
Xenodon, 543
  rhadocephalus, 543
Xenopeltids, 525, 532, 557
Xenopus, 378, 384
  laevis, 282
Xenosaurids, 557
Xenosaurus, 525
Xenurus inauris, 852
X-ray diffraction, early vertebrates
    habitats and, 207

Y chromosome, 146–147
Yarrow's spiny lizard, 595
Yaw, fish locomotion, 273
Yellow baboon, 813, 814, 815, 816
Yellow perch, 81
Yolk, 79–80, 81, 82, 148
Yolk sac, 80
Youngina, 490
Younginiformes, 490
Yunnanosaurids, 461

Zapodids, 45
Zebra finch, 663
Zebras, 753, 800, 889
Zone of chemical thermogenesis, 181
Zone of evaporative cooling, 182
Zone of physical thermoregulation. *See*
    Thermoneutral zone
Zone of tolerance, 181
Zugdisposition, 683–684
Zugstimmung, 684
Zugunruhe, 684
Zygodactylous foot, 527, 631–632
Zygote, 17, 79